The Organic Chemistry of Drug Design and Drug Action

The Organic Chemistry of Drug Design and Drug Action

Third Edition

Richard B. Silverman
Northwestern University
Department of Chemistry
Department of Molecular Biosciences
Chemistry of Life Processes Institute
Center for Molecular Innovation and Drug Discovery
Evanston, Illinois, USA

Mark W. Holladay
Ambit Biosciences Corporation
Departments of Drug Discovery and Medicinal Chemistry
San Diego, California, USA

AMSTERDAM • BOSTON • HEIDELBERG • LONDON
NEW YORK • OXFORD • PARIS • SAN DIEGO
SAN FRANCISCO • SINGAPORE • SYDNEY • TOKYO
Academic Press is an imprint of Elsevier

Academic Press is an imprint of Elsevier
525 B Street, Suite 1900, San Diego, CA 92101-4495, USA
225 Wyman Street, Waltham, MA 02451, USA

Library of Congress Cataloging-in-Publication Data
Silverman, Richard B., author.
The organic chemistry of drug design and drug action. -- Third edition / Richard B. Silverman, Mark W. Holladay.
pages cm
Includes bibliographical references and index.
ISBN 978-0-12-382030-3 (alk. paper)
1. Pharmaceutical chemistry. 2. Bioorganic chemistry. 3. Molecular pharmacology. 4. Drugs--Design. I. Holladay, Mark W., author. II. Title.
RS403.S55 2014
615.1'9--dc23
2013043146

British Library Cataloguing in Publication Data
A catalogue record for this book is available from the British Library

For information on all Academic Press publications visit our web site at store.elsevier.com

ISBN: 978-0-12-382030-3

Dedications

RBS

To the memory of Mom and Dad, for their love,
their humor, their ethics, and their inspiration.
To Barbara, Matt, Mar, Phil, Andy, Brooke, Alexander,
Owen, Dylan, and, hopefully, more to come,
for making life a complete joy.

MWH

To my wonderful wife, Carol, and our awesome kids,
Tommy and Ruth.

Contents

4. Enzymes 165

5. Enzyme Inhibition and Inactivation 207

Please look for the Student/companion site at http://booksite.elsevier.com/9780123820303
Please look for the Instructor site (coming soon) at http://textbooks.elsevier.com/web/manuals.aspx?isbn=9780123820303

Preface to the First Edition

From 1985 to 1989, I taught a one-semester course in medicinal chemistry to senior undergraduates and first-year graduate students majoring in chemistry or biochemistry. Unlike standard medicinal chemistry courses that are generally organized by classes of drugs, giving descriptions of their biological and pharmacological effects, I thought there was a need to teach a course based on the organic chemical aspects of medicinal chemistry. It was apparent then, and still is the case now, that there is no text that concentrates exclusively on the organic chemistry of drug design, drug development, and drug action. This book has evolved to fill that important gap. Consequently, if the reader is interested in learning about a specific class of drugs, its biochemistry, pharmacology, and physiology, he or she is advised to look elsewhere for that information. Organic chemical principles and reactions vital to drug design and drug action are the emphasis of this text with the use of clinically important drugs as examples. Usually only one or just a few representative examples of drugs that exemplify the particular principle are given; no attempt has been made to be comprehensive in any area. When more than one example is given, generally it is to demonstrate different chemistry. It is assumed that the reader has taken a one-year course in organic chemistry that included amino acids, proteins, and carbohydrates and is familiar with organic structures and basic organic reaction mechanisms. Only the chemistry and biochemistry background information pertinent to the understanding of the material in this text is discussed. Related, but irrelevant, background topics are briefly discussed or are referenced in the general readings section at the end of each chapter. Depending on the degree of in-depthness that is desired, this text could be used for a one-semester or a full-year course. The references cited can be ignored in a shorter course or can be assigned for more detailed discussion in an intense or full-year course. Also, not all sections need to be covered, particularly when multiple examples of a particular principle are described. The instructor can select those examples that may be of most interest to the class. It was the intent in writing this book that the reader, whether a student or a scientist interested in entering the field of medicinal chemistry, would learn to take a rational physical organic chemical approach to drug design and drug development and to appreciate the chemistry of drug action. This knowledge is of utmost importance for the understanding of how drugs function at the molecular level. The principles are the same regardless of the particular receptor or enzyme involved. Once the fundamentals of drug design and drug action are understood, these concepts can be applied to the understanding of the many classes of drugs that are described in classical medicinal chemistry texts. This basic understanding can be the foundation for the future elucidation of drug action or the rational discovery of new drugs that utilize organic chemical phenomena.

Richard B. Silverman
Evanston, Illinois
April 1991

Preface to the Second Edition

In the 12 years since the first edition was written, certain new approaches in medicinal chemistry have appeared or have become commonly utilized. The basic philosophy of this textbook has not changed, that is, to emphasize general principles of drug design and drug action from an organic chemical perspective rather than from the perspective of specific classes of drugs. Several new sections were added (in addition to numerous new approaches, methodologies, and updates of examples and references), especially in the areas of lead discovery and modification (Chapter 2). New screening approaches, including high-throughput screening, are discussed, as are the concepts of privileged structures and drug-likeness. Combinatorial chemistry, which was in its infancy during the writing of the first edition, evolved, became a separate branch of medicinal chemistry and then started to wane in importance during the twenty-first century. Combinatorial chemistry groups, prevalent in almost all pharmaceutical industries at the end of the twentieth century, began to be dissolved, and a gradual return to traditional medicinal chemistry has been seen. Nonetheless, combinatorial chemistry journals have sprung up to serve as the conduit for dissemination of new approaches in this area, and this along with parallel synthesis are important approaches that have been added to this edition. New sections on SAR by NMR and SAR by MS have also been added. Peptidomimetic approaches are discussed in detail. The principles of structure modification to increase oral bioavailability and effects on pharmacokinetics are presented, including log *P* software and "rule of five" and related ideas in drug discovery. The fundamentals of molecular modeling and 3D-QSAR are also expanded. The concepts of inverse agonism, inverse antagonism, racemic switches, and the two-state model of receptor activation are introduced in Chapter 3. In Chapter 5 efflux pumps, COX-2 inhibitors, and dual-acting drugs are discussed; a case history of the discovery of the AIDS drug ritonavir is used to exemplify the concepts of drug discovery of reversible enzyme inhibitors. Discussions of DNA structure and function, topoisomerases, and additional examples of DNA-interactive agents, including metabolically activated agents, are new or revised sections in Chapter 6. The newer emphasis on the use of HPLC/MS/MS in drug metabolism is discussed in Chapter 7 along with the concepts of fatty acid and cholesterol conjugation and antedrugs. In Chapter 8 a section on enzyme prodrug therapies (ADEPT, GDEPT, VDEPT) has been added as well as a case history of the discovery of omeprazole. Other changes include the use of both generic names and trade names, with generic names given with their chemical structure, and the inclusion of problem sets and solutions for each chapter.

The first edition of this text was written primarily for upper class undergraduate and first-year graduate students interested in the general field of drug design and drug action. During the last decade it has become quite evident that there is a large population, particularly of synthetic organic chemists, who enter the pharmaceutical industry with little or no knowledge of medicinal chemistry and who want to learn the application of their skills to the process of drug discovery. The first edition of this text provided an introduction to the field for both students and practitioners, but the latter group has more specific interests in how to accelerate the drug discovery process. For the student readers, the basic principles described in the second edition are sufficient for the purpose of teaching the general process of how drugs are discovered and how they function. Among the basic principles, however, I have now interspersed many more specifics that go beyond the basics and may be more directly related to procedures and applications useful to those in the pharmaceutical industry. For example, in Chapter 2 it is stated that "Ajay and coworkers proposed that *drug-likeness* is a possible inherent property of some molecules,[a] and this property could determine which molecules should be selected for screening." The basic principle is that some molecules seem to have scaffolds found in many drugs and should be initially selected for testing. But following that initial statement is added more specifics: "They used a set of one- and two-dimensional parameters in their computation and were able to predict correctly over 90% of the compounds in the Comprehensive Medicinal Chemistry (CMC) database.[b] Another computational approach to differentiate druglike and nondruglike molecules using a scoring scheme was developed,[c] which was able to classify correctly 83% of the compounds in the Available Chemicals Directory (ACD)[d] and 77% of the compounds in the World

---

[a]Ajay; Walters, W P.; Murcko, M. A. *J. Med. Chem.* 1998, *41*, 3314.

[b]This is an electronic database of Volume 6 of *Comprehensive Medicinal Chemistry* (Pergamon Press) available from MDL Information systems, Inc., San Leandro, CA 94577.

[c]Sadowski, J.; Kubinyi, H. *J. Med. Chem.* 1998, *41*, 3325.

[d]The ACD is available from MDL Information systems, Inc., San Leandro, CA, and contains specialty and bulk commercially available chemicals.

Drug Index (WDI).[e] A variety of other approaches have been taken to identify druglike molecules."[f] I believe that the student readership does not need to clutter its collective brain with these latter specifics, but should understand the basic principles and approaches; however, for those who aspire to become part of the pharmaceutical research field, they might want to be aware of these specifics and possibly look up the references that are cited (the instructor, for a course who believes certain specifics are important may assign the references as readings).

For concepts peripheral to drug design and drug action, I will give only a reference to a review of that topic in case the reader wants to learn more about it. If the instructor believes that a particular concept that is not discussed in detail should have more exposure to the class, further reading can be assigned.

To minimize errors in reference numbers, several references are cited more than once under different endnote numbers. Also, although multiple ideas may come from a single reference, the reference is only cited once; if you want to know the origin of discussions in the text, look in the closest reference, either the one preceding the discussion or just following it. Because my expertise extends only in the areas related to enzymes and the design of enzyme inhibitors.

I want to thank numerous experts who read parts or whole chapters and gave me feedback for modification. These include (in alphabetical order) Shuet-Hing Lee Chiu, Young-Tae Chang, William A. Denny, Perry A. Frey, Richard Friary, Kent S. Gates, Laurence H. Hurley, Haitao Ji, Theodore R. Johnson, Yvonne C. Martin, Ashim K. Mitra, Shahriar Mobashery, Sidney D. Nelson, Daniel H. Rich, Philippa Solomon, Richard Wolfenden, and Jian Yu. Your input is greatly appreciated. I also greatly appreciate the assistance of my two stellar program assistants, Andrea Massari and Clark Carruth, over the course of writing this book, as well as the editorial staff (headed by Jeremy Hayhurst) of Elsevier/Academic Press.

Richard B. Silverman
Still in Evanston, Illinois
May 2003

---

[e]The WDI is from Derwent Information.

[f](a) Walters, W. P.; Stahl, M. T.; Murcko, M. A. *Drug Discovery Today* 1998, *3*, 160. (b) Walters, W. P.; Ajay; Murcko, M. A. *Curr. Opin. Chem. Biol.* 1999, *3*, 384. (c) Teague, S. J.; Davis, A. M.; Leeson, P. D.; Oprea, T. *Angew.Chem. Int. Ed. Engl.* 1999, *38*, 3743. (d) Oprea, T. I. J. *Comput.-Aided Mol. Des.* 2000, *14*, 251. (e) Gillet, V. J.; Willett, P. L.; Bradshaw, J. *J. Chem. Inf. Comput. Sei.* 1998, *38*, 165. (f) Wagener, M.; vanGeerestein, V. J. *J. Chem. Inf. Comput. Sei.* 2000, *40*, 280. (g) Ghose, A. K.; Viswanadhan, V.N.; Wendoloski, J. J. *J. Comb. Chem.* 1999, *1*, 55. (h) Xu, J.; Stevenson, J. *J. Chem. Inf. Comput. Sei.* 2000, *40*, *Uli*. (i) Muegge, I.; Heald, S. L.; Brittelli, D. *J. Med. Chem.* 2001, *44*, 1841. (j) Anzali, S.; Barnickel, G.; Cezanne, B.; Krug, M.; Filimonov, D.; Poroikiv, V. *J. Med. Chem.* 2001, *44*, 2432. (k) Brstle, M.; Beck, B.; Schindler, T.; King, W; Mitchell, T.; Clark, T. *J. Med. Chem.* 2002, *45*, 3345.

Preface to the Third Edition

Ten years have rolled by since the publication of the second edition, and the field of medicinal chemistry has undergone a number of changes. To aid in trying to capture the essence of new directions in medicinal chemistry, I decided to add a coauthor for this book. Mark W. Holladay was my second graduate student (well, that year I took four graduate students into my group, so he's actually from my second class of graduate students), and I knew from when he came to talk to me, he was going to be a great addition to the group (and to help me get tenure!). In my naivete as a new assistant professor, I assigned Mark a thesis project to devise a synthesis of the newly-discovered antitumor natural product, acivicin, which was believed to inhibit enzymes catalyzing amido transfer reactions from L-glutamine that are important for tumor cell growth. That would be a sensible thesis project, but I told him that the second part of his thesis would be to study its mechanism of action, as Mark had indicated a desire to do both organic synthesis and enzymology. Of course, this would be a 10-year doctoral project if he really had to do that, but what did I know then? Mark did a remarkable job, independently working out the total synthesis of the natural product (my proposed synthetic route at the beginning failed after the second step) and its C-5 epimer, and he was awarded his Ph.D. for the syntheses. He moved on to do a postdoc with Dan Rich, the extraordinary peptide chemist now retired from the University of Wisconsin, and joined Abbott Laboratories as a senior scientist. After 15 years at Abbott, and having been elected to the Volwiler Society, an elite honor society at Abbott Labs for their most valuable scientists, he decided to move to a smaller pharmaceutical environment, first at SIDDCO, then Discovery Partners International, and now at Ambit Biosciences. Because of his career-long association with the pharmaceutical industry (and my knowledge that he was an excellent writer), I invited him to coauthor the third edition to give an industrial pharmaceutical perspective. It has been a rewarding and effective collaboration. Although both of us worked equally on all of the chapters, I got the final say, so any inconsistencies or errors are the result of my oversight.

Richard B. Silverman

As was the case for the second edition, the basic philosophy and approach in the third edition has not changed, namely, an emphasis on general principles of drug design and drug action from an organic chemistry perspective rather than a discussion of specific classes of drugs. For didactic purposes, directed at the industrial medicinal chemist, more depth was added to many of the discussions; however, for the student readers, the basic principles are sufficient for understanding the general process of drug discovery and drug action. For a full-year course, the more in-depth discussions may be appropriate; the professor teaching the course should indicate to the class the depth of material that the student is expected to digest. In addition to an update of all of the chapters from those in the second edition with new examples incorporated, several new sections were added, some sections were deemphasized or deleted, and other sections were reorganized. As a result of some of the comments by reviewers of our proposal for the third edition, two significant changes were made: we expanded Chapter 1 to make it an overview of topics that are discussed in detail throughout the book, and the topics of resistance and synergism were pulled out of their former chapters and combined, together with several new examples, into a new chapter, Drug Resistance and Drug Synergism (now Chapter 7). Sections on sources of compounds for screening, including library collections, virtual screening, and computational methods, as well as hit-to-lead and scaffold hopping, were added; the sections on sources of lead compounds, fragment-based lead discovery, and molecular graphics were expanded; and solid-phase synthesis and combinatorial chemistry were deemphasized (all in Chapter 2). In Chapter 3, other drug-receptor interactions, cation-π and halogen bonding, were added, as was a section on atropisomers and a case history of the insomnia drug suvorexant as an example of a pharmacokinetically-driven drug project. A section on enzyme catalysis in drug discovery, including enzyme synthesis, was added to Chapter 4. Several new case histories were added to Chapter 5: for competitive inhibition, the epidermal growth factor receptor tyrosine kinase inhibitor erlotinib and Abelson kinase inhibitor imatinib, both anticancer drugs, were added; for transition state analogue inhibition, the purine nucleoside phosphorylase inhibitors, forodesine and DADMe-ImmH, both antitumor agents, were added, as well as the mechanism of the multisubstrate analog inhibitor isoniazid; the antidiabetes drug saxagliptin was added as a case history for slow, tight-binding inhibition. A section on toxicophores and reactive metabolites was added to Chapter 8, and the topic of antibody-drug conjugates was incorporated into Chapter 9.

As in the case of the second edition, many peripheral topics are noted but only a general reference is cited. If an instructor wants to pursue that topic in more depth, additional readings can be assigned. To minimize errors in reference numbers, some references are cited more than once with different reference numbers. Also, when multiple ideas are taken from the same reference, the reference is cited only once; if a statement appears not to have been referenced, try looking at a reference just prior to or following the discussion of that topic.

We want to thank several experts for their input on topics that needed some strengthening: Haitao (Mark) Ji, now in the Department of Chemistry at the University of Utah, for assistance in 3D-QSAR and for assembling the references for computer-based drug design methodologies at the end of Chapter 2; Eric Martin, Director of Novartis Institutes of BioMedical Research, for assistance in the 2D-QSAR section of Chapter 2; and Yaoqiu Zhu, President, MetabQuest Research and Consulting, for input on the metabolism methodology section of Chapter 8. The unknown outside reviewers of Chapters 1, 2, and 5 made some insightful comments, which helped in strengthening those respective sections. Finally, this project would have been much more onerous if it were not for Rick Silverman's remarkable program assistant, Pam Beck, who spent countless hours organizing and formatting text, renumbering structures, figures, and schemes when some were added or deleted, getting permissions, coordinating between the two authors, and figuring out how to fix problems that neither author wanted to deal with. We also thank the Acquisitions Editor, Katey Birtcher, the Editorial Project Manager, Jill Cetel, and, especially, the Production Manager, Sharmila Vadivelan, for their agility and attention to detail in getting the third edition in such a beautiful form.

Richard B. Silverman
Evanston, Illinois (for over 37 years!)

Mark W. Holladay
San Diego, California, February, 2014

Chapter 1

Introduction

Chapter Outline

1.1. OVERVIEW

Medicinal chemistry is the science that deals with the discovery and design of new therapeutic chemicals or biochemicals and their development into useful medicines. *Medicines* are the substances used to treat diseases. *Drugs* are the molecules used as medicines or as components in medicines to diagnose, cure, mitigate, treat, or prevent disease.[1] Medicinal chemistry may involve isolation of compounds from nature or the synthesis of new molecules; investigations of the relationships between the structure of natural and/or synthetic compounds and their biological activities; elucidations of their interactions with receptors of various kinds, including enzymes and DNA; the determination of their absorption, transport, and distribution properties; studies of the metabolic transformations of these chemicals into other chemicals, their excretion and toxicity. Modern methods for the discovery of new drugs have evolved immensely since the 1960s, in parallel with phenomenal advances in organic chemistry, analytical chemistry, physical chemistry, biochemistry, pharmacology, molecular biology, and medicine. For example, *genomics*,[2] the investigations of an organism's *genome* (all of the organism's genes) to identify important target genes and *gene products* (proteins expressed by the genes) and *proteomics*, the characterization of new proteins, or the abundance of proteins, in the organism's *proteome* (all of the proteins expressed by the genome)[3] to determine their structure and/or function, often by comparison with known proteins, have become increasingly important approaches to identify new drug targets.

Today, harnessing modern tools to conduct *rational drug design* is pursued intensely in the laboratories of pharmaceutical and biotech industries as well as in academic institutions and research institutes. Chemistry, especially organic chemistry, is at the heart of these endeavors, from the application of physical principles to influence where a drug will go in the body and how long it will remain there, to the understanding of what the body does to the drug to eliminate it from the system, to the synthetic organic processes used to prepare a new compound for testing, first in small quantities (milligrams) and ultimately, if successful, on multikilogram scale.

First, however, it needs to be noted that drugs are not generally discovered. What is more likely discovered is known as a *lead compound* (or *lead*). The lead is a prototype compound that has a number of attractive characteristics, including the desired biological or pharmacological activity, but may have other undesirable characteristics, for example, high toxicity, other biological activities, absorption difficulties, insolubility, or metabolism problems. The structure of the lead compound is, then, modified by synthesis to amplify the desired activity and to minimize or eliminate the unwanted properties to a point where a *drug candidate*, a compound worthy of extensive biological and pharmacological studies, is identified, and then a *clinical drug*, a compound ready for clinical trials, is developed.

The Organic Chemistry of Drug Design and Drug Action. http://dx.doi.org/10.1016/B978-0-12-382030-3.00001-5

The chapters of this book describe many key facets of modern rational drug discovery, together with the organic chemistry that forms the basis for understanding them. To provide a preview of the later chapters and to help put the material in context, this chapter provides a broad overview of modern rational drug discovery with references to later chapters where more detailed discussions can be found. Prior to launching into an overview of modern rational drug discovery approaches, let us first briefly take a look at some examples of drugs whose discoveries relied on circumstances other than rational design, that is, by happenstance or insightful observations.

1.2. DRUGS DISCOVERED WITHOUT RATIONAL DESIGN

1.2.1. Medicinal Chemistry Folklore

Medicinal chemistry, in its crudest sense, has been practiced for several thousand years. Man has searched for cures of illnesses by chewing herbs, berries, roots, and barks. Some of these early clinical trials were quite successful; however, not until the last 100–150 years has knowledge of the active constituents of these natural sources been known. The earliest written records of the Chinese, Indian, South American, and Mediterranean cultures described the therapeutic effects of various plant concoctions.[4–6] A Chinese health science anthology called *Nei Ching* is thought to have been written by the Yellow Emperor in the thirteenth century B.C., although some believe that it was backdated by the third century compilers.[7] The Assyrians described on 660 clay tablets 1000 medicinal plants used from 1900 to 400 B.C.

Two of the earliest medicines were described about 5100 years ago by the Chinese Emperor Shen Nung in his book of herbs called *Pen Ts'ao*.[8] One of these is *Ch'ang Shan*, the root *Dichroa febrifuga*, which was prescribed for fevers. This plant contains alkaloids that are used in the treatment of malaria today. Another plant called *Ma Huang* (now known as *Ephedra sinica*) was used as a heart stimulant, a diaphoretic agent (perspiration producer), and recommended for treatment of asthma, hay fever, and nasal and chest congestion. It is now known to contain two active constituents: ephedrine, a drug that is used as a stimulant, appetite suppressant, decongestant, and hypertensive agent, and pseudoephedrine, used as a nasal/sinus decongestant and stimulant (pseudoephedrine hydrochloride (**1.1**) is found in many over-the-counter nasal decongestants, such as Sudafed). Ephedra, the extract from *E. sinica*, also is used today (inadvisably) by some body builders and endurance athletes because it promotes thermogenesis (the burning of fat) by release of fatty acids from stored fat cells, leading to quicker conversion of the fat into energy. It also tends to increase the contractile strength of muscle fibers, which allows body builders to work harder with heavier weights.

Theophrastus in the third century B.C. mentioned opium poppy juice as an analgesic agent, and in the tenth century A.D., Rhazes (Persia) introduced opium pills for coughs, mental disorders, aches, and pains. The opium poppy, *Papaver somniferum*, contains morphine (**1.2**), a potent analgesic agent, and codeine (**1.3**), prescribed today as a cough suppressant. The East Asians and the Greeks used henbane, which contains scopolamine (**1.4**, truth serum) as a sleep inducer. Inca mail runners and silver miners in the high Andean mountains chewed coca leaves (cocaine, **1.5**) as a stimulant and euphoric. The antihypertensive drug reserpine (**1.6**) was extracted by ancient Hindus from the snake-like root of the *Rauwolfia serpentina* plant and was used to treat hypertension, insomnia, and insanity. Alexander of Tralles in the sixth century A.D. recommended the autumn crocus (*Colchicum autumnale*) for relief of pain of the joints, and it was used by Avrienna (eleventh century Persia) and by Baron Anton von Störck (1763) for the treatment of gout. Benjamin Franklin heard about this medicine and brought it to America. The active principle in this plant is the alkaloid colchicine (**1.7**), which is used today to treat gout.

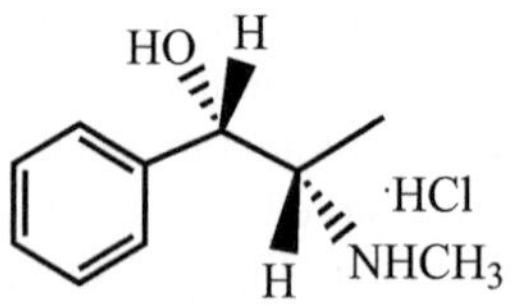

Pseudophedrine hydrochloride
1.1

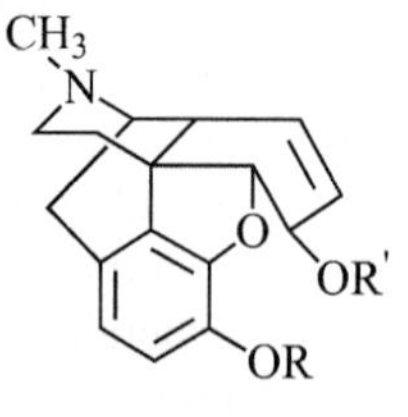

1.2, Morphine (R = R' = H)
1.3, Codeine (R = CH_3, R' = H)

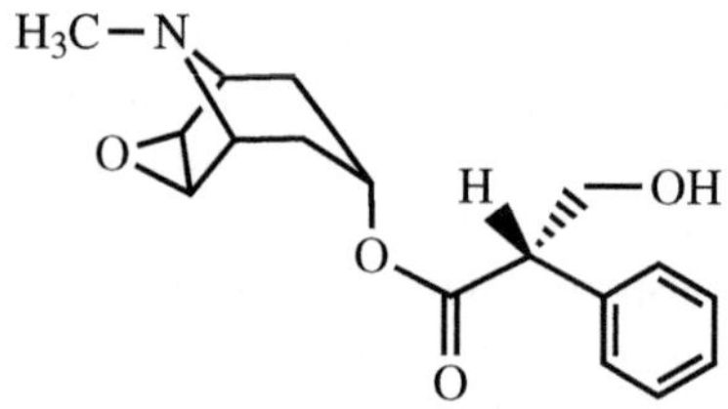

Scopolamine
1.4

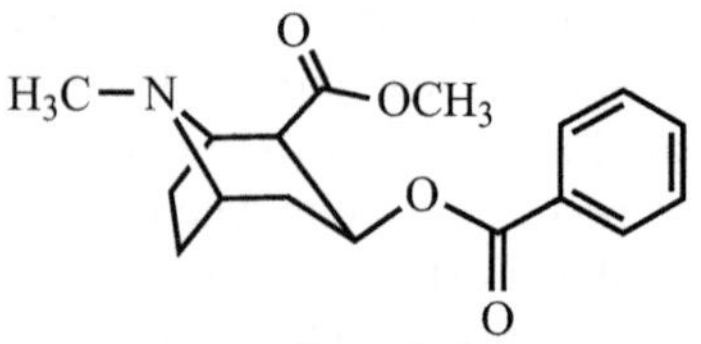

Cocaine
1.5

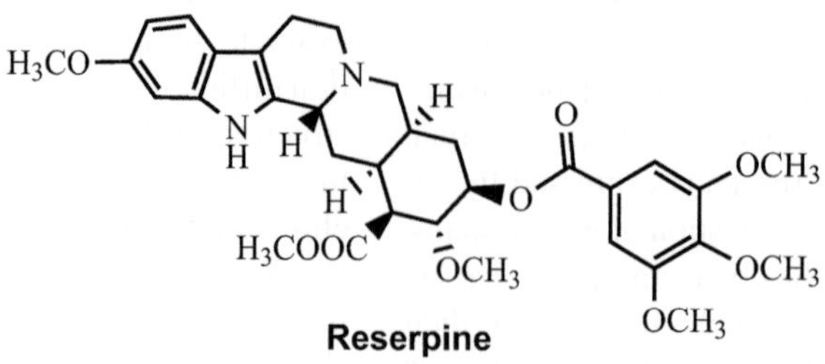

Reserpine
1.6

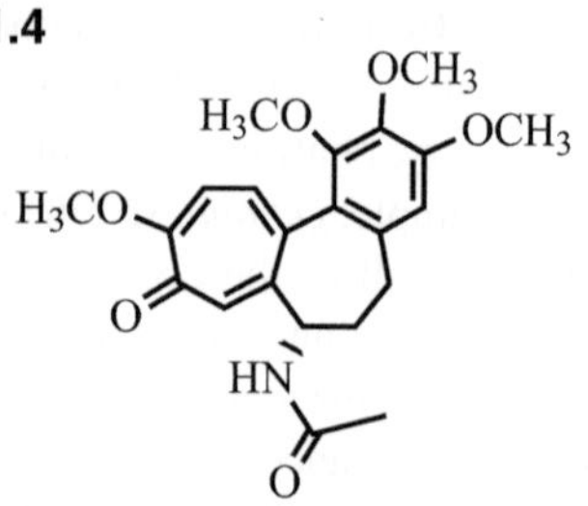

Colchicine
1.7

'That's Dr Arnold Moore. He's conducting an experiment to test the theory that most great scientific discoveries were hit on by accident.'
Drawing by Hoff; © 1957
The New Yorker Magazine, Inc.

FIGURE 1.1 Parody of drugs discovered without rational design.

In 1633, a monk named Calancha, who accompanied the Spanish conquistadors to Central and South America, introduced one of the greatest herbal medicines to Europe upon his return. The South American Indians would extract the cinchona bark and use it for chills and fevers; the Europeans used it for the same and for malaria. In 1820, the active constituent was isolated and later determined to be quinine (**1.8**), an antimalarial drug, which also has antipyretic (fever-reducing) and analgesic properties.

Quinine
1.8

Modern therapeutics is considered to have begun with an extract of the foxglove plant, which was cited by Welsh physicians in 1250, named by Fuchsius in 1542, and introduced for the treatment of dropsy (now called edema) in 1785 by Withering.[5,9] The active constituents are secondary glycosides from *Digitalis purpurea* (the foxglove plant) and *Digitalis lanata*, namely, digitoxin (**1.9**) and digoxin (**1.10**), respectively; both are important drugs for the treatment of congestive heart failure. Today, digitalis, which refers to all of the cardiac glycosides, is still manufactured by extraction of foxglove and related plants.

Digitoxin (R = H) **Digoxin (R = OH)**
1.9 **1.10**

1.2.2. Discovery of Penicillins

In 1928, Alexander Fleming noticed a green mold growing in a culture of *Staphylococcus aureus*, and where the two had converged, the bacteria were lysed.[10] This led to the discovery of penicillin, which was produced by the mold. Actually, Fleming was not the first to make this observation; John Burdon-Sanderson had done so in 1870, ironically also at St. Mary's Hospital in London, the same institution where Fleming made the rediscovery![11] Joseph Lister had treated a wounded patient with *Penicillium*, the organism later found to be the producer of penicillin (although the strains discovered earlier than Fleming did not produce penicillin, but, rather, another antibiotic, mycophenolic acid). After Fleming observed this phenomenon, he tried many times to repeat it without success; it was his colleague, Dr Ronald Hare,[12,13] who was able to reproduce the observation. It only occurred the first time because a combination of unlikely events all took place simultaneously. Hare found that very special conditions were required to produce the phenomenon initially observed by Fleming. The culture dish inoculated by Fleming must have become accidentally and simultaneously contaminated with the mold spore. Instead of placing the dish in the refrigerator or incubator when he went on vacation, as is normally done, Fleming inadvertently left it on his lab bench. When he returned the following month, he noticed the lysed bacteria. Ordinarily, penicillin does not lyse these bacteria; it prevents them from developing, but it has no effect if added after the bacteria have developed. However, while Fleming was on vacation (July–August), the weather was unseasonably cold, and this provided the particular temperature required for the mold and the staphylococci to grow slowly and produce the lysis. Another extraordinary circumstance was that the particular strain of the mold on Fleming's culture was a relatively good penicillin producer, although most strains of that mold (*Penicillium*) produce no penicillin at all. The mold presumably came from the laboratory just below Fleming's where research on molds was going on at that time.

Although Fleming suggested that penicillin could be useful as a topical antiseptic, he was not successful in producing penicillin in a form suitable to treat infections. Nothing more was done until Sir Howard Florey at Oxford University reinvestigated the possibility of producing penicillin in a useful form. In 1940, he succeeded in producing penicillin that could be administered topically and systemically,[14] but the full extent of the value of penicillin was not revealed until the late 1940s.[15] Two reasons for the delay in the universal utilization of penicillin were the emergence of the sulfonamide antibacterials (sulfa drugs, **1.11**; see Chapter 5, Section 5.2.2.3) in 1935 and the outbreak of World War II. The pharmacology, production, and clinical application of penicillin were not revealed until after the war to prevent the Germans from having access to this wonder drug. Allied scientists, who were interrogating German scientists involved in chemotherapeutic research, were told that the Germans thought the initial report of penicillin was made just for commercial reasons to compete with the sulfa drugs. They did not take the report seriously.

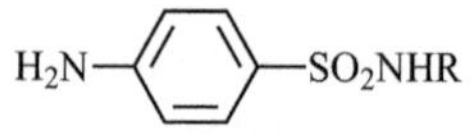

$H_2N-C_6H_4-SO_2NHR$

Sulfa drugs
1.11

The original mold was *Penicillium notatum*, a strain that gave a relatively low yield of penicillin. It was replaced by *Penicillium chrysogenum*,[16] which had been cultured from a mold growing on a grapefruit in a market in Peoria, Illinois!

For many years, there was a raging debate regarding the actual structure of penicillin (**1.12**),[17] but the correct structure was elucidated in 1944 with an X-ray crystal structure by Dorothy Crowfoot Hodgkin (Oxford); the crystal structure was not published until after the war in 1949.[18] Several different penicillin analogs (R group varied) were isolated early on; only two of these early analogs (**1.12**, R=$PhOCH_2$, penicillin V and **1.12**, R=$PhCH_2$, penicillin G) are still in use today.

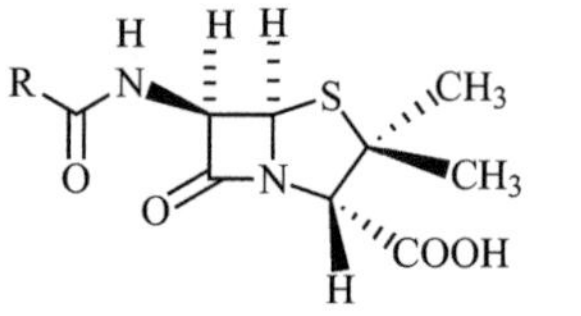

Penicillin V (R = $PhOCH_2$)
Penicillin G (R = CH_2Ph)
1.12

1.2.3. Discovery of Librium

The first benzodiazepine tranquilizer drug, Librium (7-chloro-2-(methylamino)-5-phenyl-3H-1,4-benzodiazepine 4-oxide; chlordiazepoxide HCl; **1.13**), was discovered serendipitously.[19]

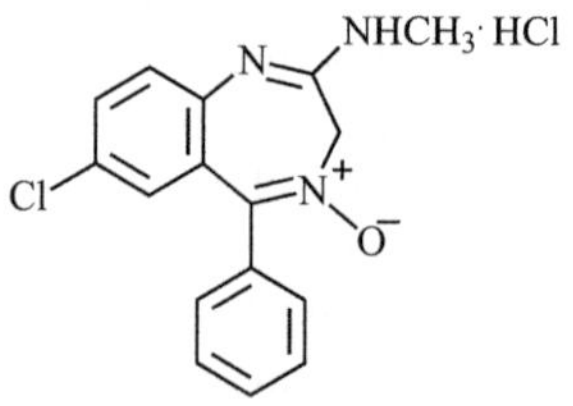

Chlordiazepoxide HCl
1.13

Dr. Leo Sternbach at Roche was involved in a program to synthesize a new class of tranquilizer drugs. He originally set out to prepare a series of benzheptoxdiazines (**1.14**), but when R^1 was CH_2NR_2 and R^2 was C_6H_5, it was found that the actual structure was that of a quinazoline 3-oxide (**1.15**). However, none of these compounds gave any interesting pharmacological results.

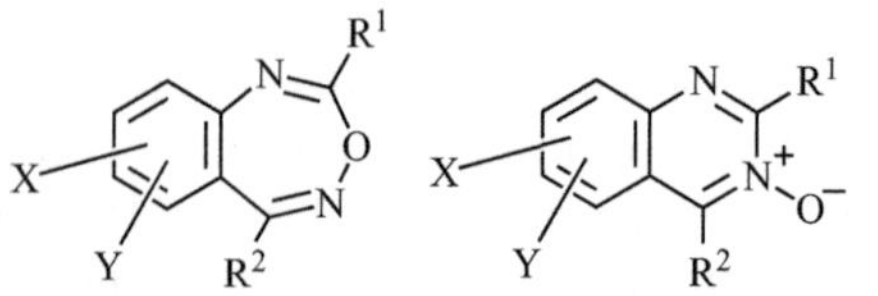

1.14 **1.15**

The program was abandoned in 1955 in order for Sternbach to work on a different project. In 1957, during a general laboratory cleanup, a vial containing what was thought to

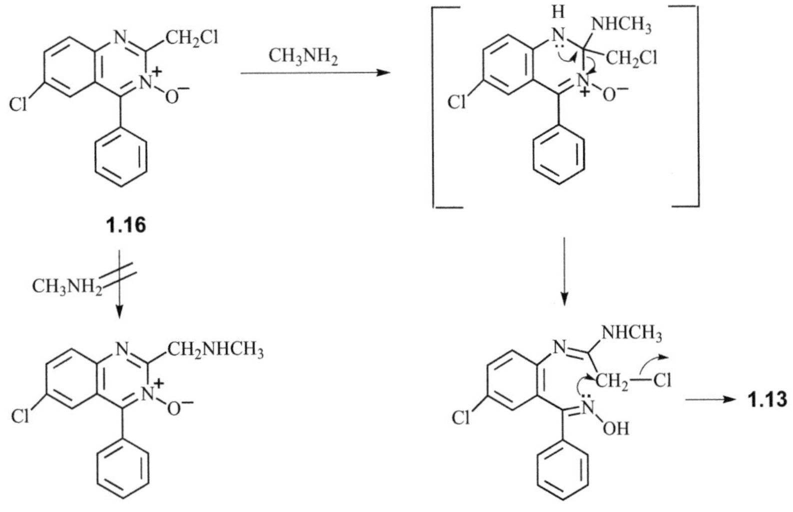

SCHEME 1.1 Mechanism of formation of Librium

be **1.15** (X=7-Cl, $R^1=CH_2NHCH_3$, $R^2=C_6H_5$) was found and, as a last effort, was submitted for pharmacological testing. Unlike all of the other compounds submitted, this one gave very promising results in six different tests used for preliminary screening of tranquilizers. Further investigation revealed that this compound was not a quinazoline 3-oxide, but, rather, was the benzodiazepine 4-oxide (**1.13**), presumably produced in an unexpected reaction of the corresponding chloromethyl quinazoline 3-oxide (**1.16**) with methylamine (Scheme 1.1). If this compound had not been found in the laboratory cleanup, all of the negative pharmacological results would have been reported for the quinazoline 3-oxide class of compounds, and benzodiazepine 4-oxides may not have been discovered for many years to come.

Penicillin V and Librium are two important drugs that were discovered without a lead. However, once they were identified, they then became lead compounds for second generation analogs. There are now a myriad of penicillin-derived antibacterials that have been synthesized as the result of the structure elucidation of the earliest penicillins. Valium (diazepam, **1.17**) was synthesized at Roche even before Librium was introduced onto the market; this drug was derived from the lead compound, Librium, and is almost 10 times more potent than the lead.

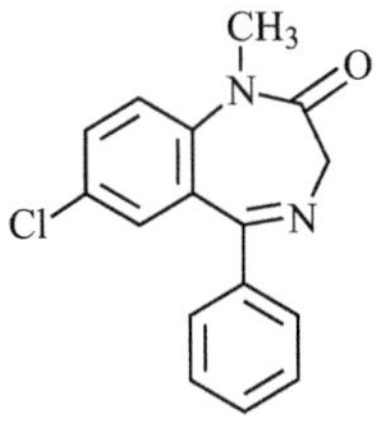

Diazepam
1.17

1.2.4. Discovery of Drugs through Metabolism Studies

During drug metabolism studies (Chapter 7), *metabolites* (drug degradation products generated in vivo) that are isolated are screened to determine if the activity observed is derived from the drug candidate or from a metabolite. For example, the anti-inflammatory drug sulindac (**1.18**; Clinoril) is not the active agent; the metabolic reduction product (**1.19**) is responsible for the activity.[20] The non-sedating antihistamine terfenadine (**1.20**; Seldane) was found to cause an abnormal heart rhythm in some users who also were taking certain antifungal agents, which were found to block the enzyme that metabolizes terfenadine. This caused a build-up of terfenadine, which led to the abnormal heart rhythms (Chapter 7). Consequently, Seldane was withdrawn from the market. However, a metabolite of terfenadine, fexofenadine (**1.21**; Allegra), was also found to be a nonsedating antihistamine, but it can be metabolized even in the presence of antifungal agents. This, then, is a safer drug and was approved by the Food and Drug Administration (FDA) to replace Seldane.

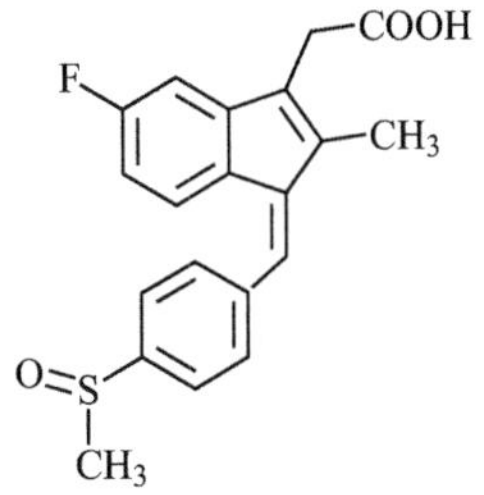

Sulindac
1.18

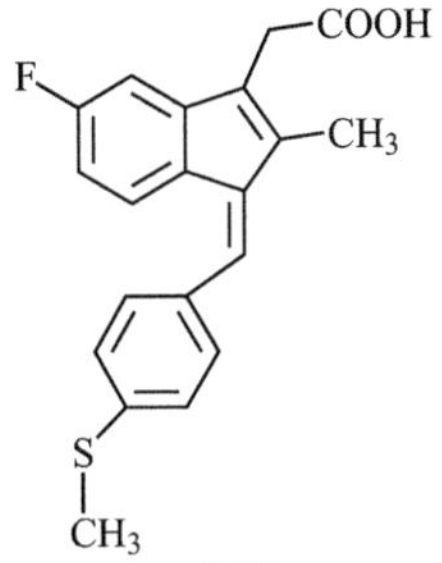

1.19

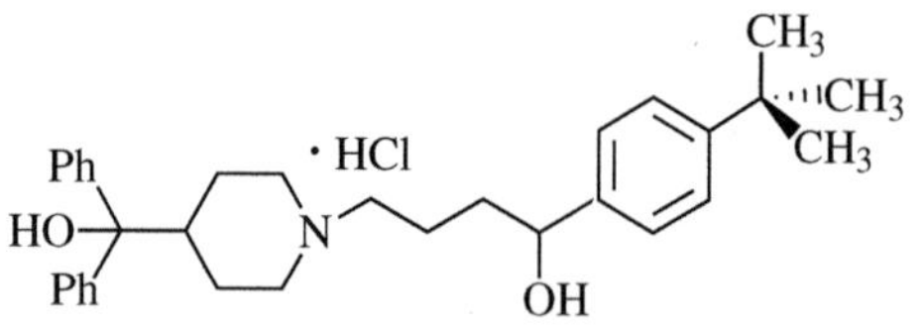

Terfenadine HCl
1.20

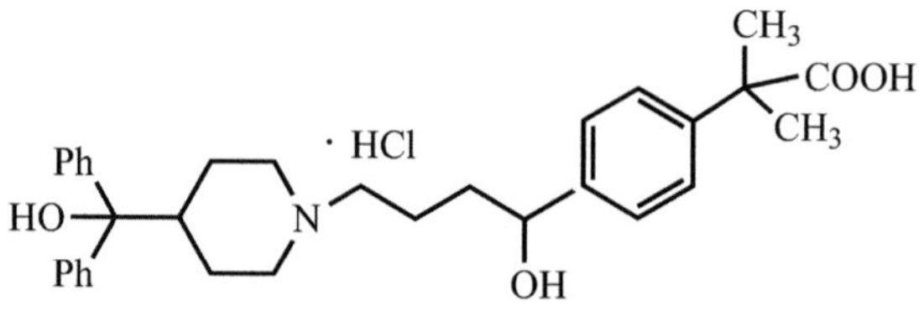

Fexofenadine HCl
1.21

1.2.5. Discovery of Drugs through Clinical Observations

Sometimes a drug candidate during clinical trials will exhibit more than one pharmacological activity, that is, it may produce a side effect. This compound, then, can be used as a lead (or, with luck, as a drug) for the secondary activity. In 1947, an antihistamine, dimenhydrinate (**1.22**; Dramamine) was tested at the allergy clinic at Johns Hopkins University and was found also to be effective in relieving a patient who suffered from car sickness; a further study proved its effectiveness in the treatment of seasickness[21] and airsickness.[22] It then became the most widely used drug for the treatment of all forms of motion sickness.

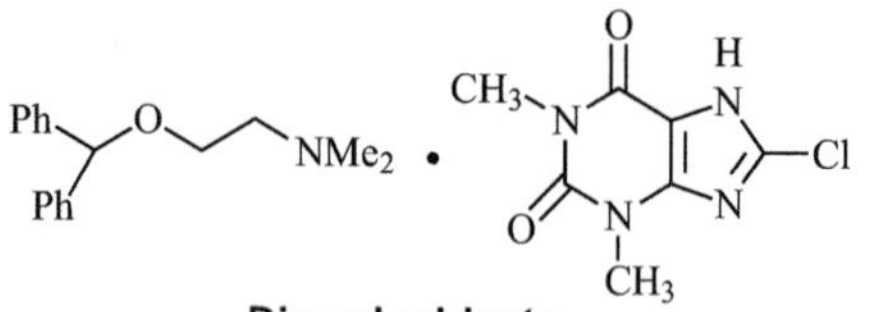

Dimenhydrinate
1.22

There are other popular examples of drugs derived from clinical observations. Bupropion hydrochloride (**1.23**), an antidepressant drug (Wellbutrin), was found to help patients stop smoking and became the first drug marketed as a smoking cessation aid (Zyban). The impotence drug sildenafil citrate (**1.24**; Viagra) was designed for the treatment of angina and hypertension by blocking the enzyme phosphodiesterase-5, which hydrolyzes cyclic guanosine monophosphate (cGMP), a vasodilator that allows increased blood flow.[23] In 1991, sildenafil went into Phase I clinical trials for angina. In Phase II clinical trials, it was not as effective against angina as Pfizer had hoped, so it went back to Phase I clinical trials to see how high of a dose could be tolerated. It was during that clinical trial that the volunteers reported increased erectile function. Given the weak activity against angina, it was an easy decision to try to determine its effectiveness as the first treatment for erectile dysfunction. Sildenafil works by the mechanism for which it was designed as an antianginal drug, except it inhibits the phosphodiesterase in the penis (phosphodiesterase-5) as well as the one in the heart (Figure 1.2).

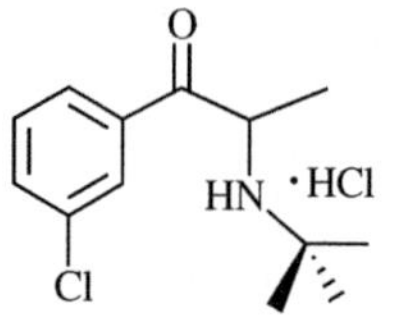

Bupropion HCl
1.23

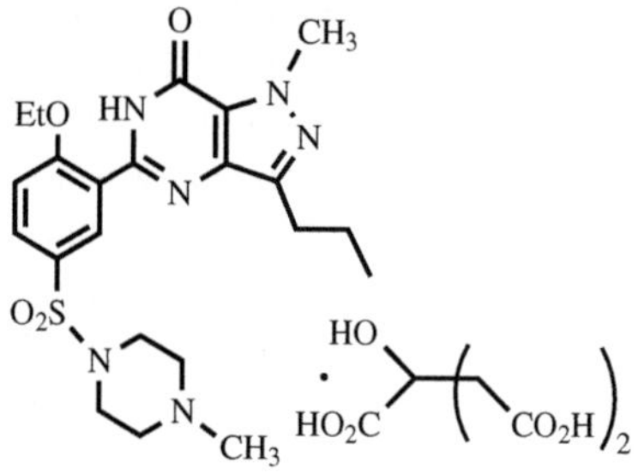

Sildenafil citrate
1.24

Sexual stimulation causes release of nitric oxide in the penis. Nitric oxide is a second messenger molecule that turns on (pun intended) the enzyme guanylate cyclase, which converts guanosine triphosphate to cGMP. The vasodilator cGMP

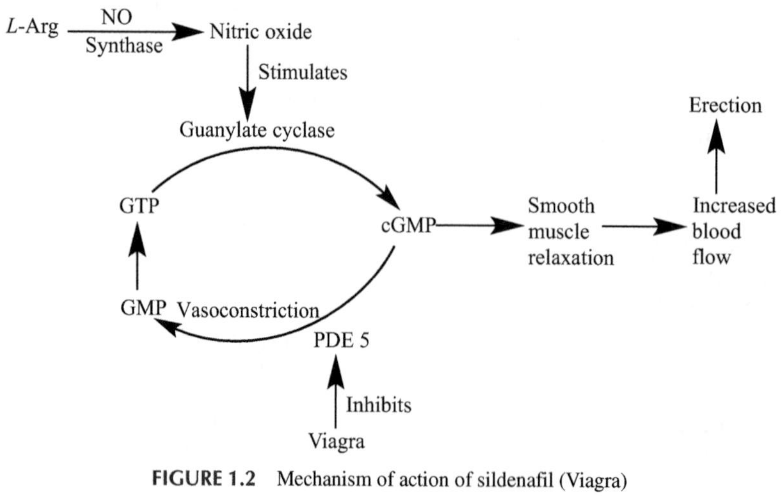

FIGURE 1.2 Mechanism of action of sildenafil (Viagra)

relaxes the smooth muscle in the *corpus cavernosum*, allowing blood to flow into the penis, thereby producing an erection. However, phosphodiesterase-5 (PDE-5) hydrolyzes the cGMP, which causes vasoconstriction and the outflow of blood from the penis. Sildenafil inhibits this phosphodiesterase, preventing the hydrolysis of cGMP and prolonging the vasodilation effect.

1.3. OVERVIEW OF MODERN RATIONAL DRUG DESIGN

The two principal origins of modern pharmaceutical industries are apothecaries, which initiated wholesale production of drugs in the mid-nineteenth century, and dye and chemical companies that were searching for medical applications for their products in the late nineteenth century.[24] Merck started as a small apothecary shop in Germany in 1668 and started wholesale production of drugs in the 1840s. Other drug companies, such as Schering, Hoffmann-La Roche, Burroughs Wellcome, Abbott, Smith Kline, Eli Lilly, and Squibb, also started as apothecaries in the nineteenth century. Bayer, Hoechst, Ciba, Geigy, Sandoz, and Pfizer began as dye and chemical manufacturers.

During the middle third of the twentieth century, antibiotics, such as sulfa drugs and penicillins (Section 1.2.2), antihistamines, hormones, psychotropics, and vaccines were invented or discovered. Death in infancy was cut by 50% and maternal death from infection during childbirth decreased by 90%. Tuberculosis, diphtheria, and pneumonia could be cured for the first time in history. These advances mark the beginning of the remarkable discoveries made today, not only in the pharmaceutical industry but also in academic and government laboratories.

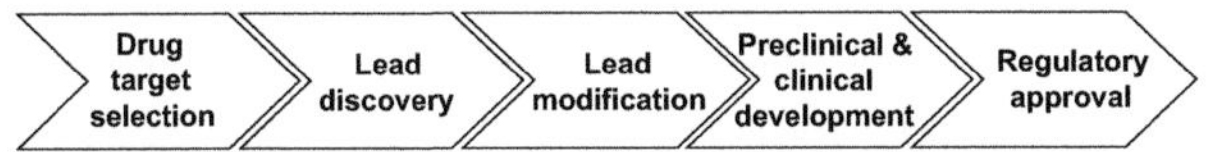

FIGURE 1.3 Typical stages of modern rational drug discovery and development

Figure 1.3 shows the typical stages of modern rational drug discovery and development. Below we present an overview of each of these steps to provide context for the concepts discussed in subsequent chapters. Among these topics, the interactions of drugs with their targets, the rationale and approaches to lead discovery, and the strategies underlying lead modification have a strong basis in physical and mechanistic organic chemistry and, hence, will be the central themes of subsequent chapters.

1.3.1. Overview of Drug Targets

The majority of drugs exert their effects through interactions with specific macromolecules in the body. Many of these macromolecular drug targets are proteins. You may recall that proteins are long polymer chains of amino acid residues that can loop and fold to produce grooves, cavities, and clefts that are ideal sites for interactions with other large or small molecules (Figure 1.4). Other drugs exert their effects by interacting with a different class of macromolecules called nucleic acids, which consist of long chains of nucleotide residues. Figure 1.5 shows the model

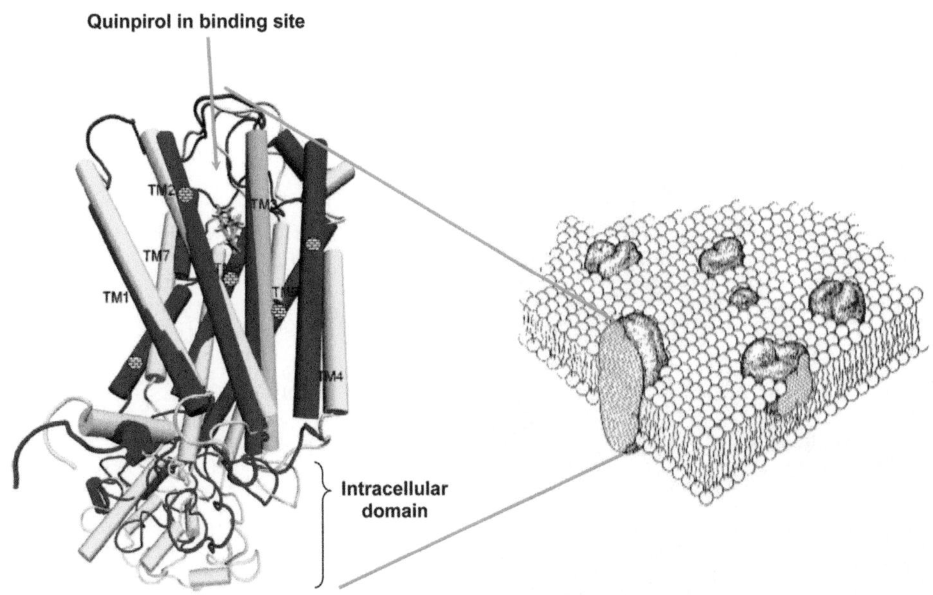

FIGURE 1.4 Small molecule drug (quinpirol) bound to its protein target (dopamine D_3 receptor). The cartoon on the right shows how a protein, such as the D_3 receptor, spans the membrane of a cell. The D_3 receptor in red depicts its conformation when the drug is bound. The D_3 receptor in yellow depicts its conformation when no drug is bound. "TM" designates a transmembrane domain of the protein. Note the significant differences between the red and yellow regions on the *intracellular* side of the membrane, prompted by the binding of quinpirol from the *extracellular* side (Ligia Westrich, et al. *Biochem. Pharmacol.* **2010**, *79*, 897–907.) On the right is a molecular representation of the fluid mosaic model of a biomembrane structure. *From Singer, S. J.; Nicolson, G L.* Science. ***1972***, 175, *720. Reprinted with permission from AAAS.*

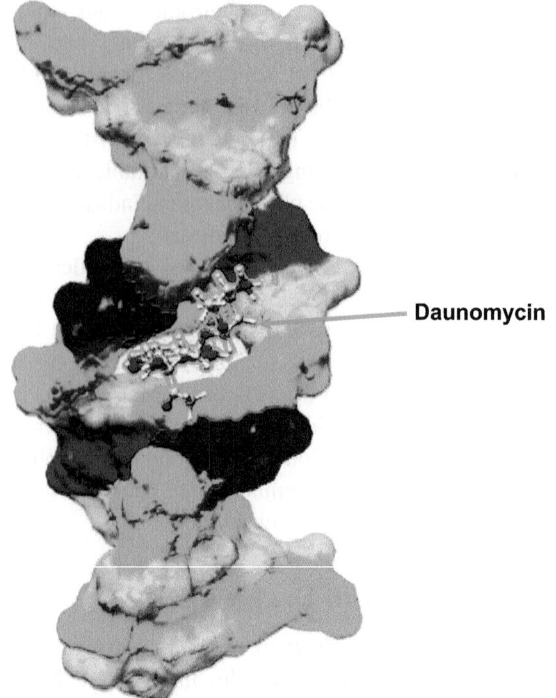

FIGURE 1.5 Small molecule drug (daunomycin) bound to its nucleic acid target (DNA). The different colors represent C (yellow), G (green), A (red), and T (blue). Mukherjee, A.; Lavery, R.; Bagchi, B.; Hynes, J. T. On the molecular mechanism of drug intercalation into DNA: A computer simulation study of the intercalation pathway, free energy, and DNA structural changes. J. Am. Chem Soc. ***2008***, 130, *9747. Reprinted with permission from Dr. Biman Bagchi, Indian Institute of Science, Bangalore, India.* Journal of the American Chemical Society *by American Chemical Society. Reproduced with permission of American Chemical Society in the format republish in a book via Copyright Clearance Center.*

of a small molecule drug (daunomycin) interacting with a nucleic acid target.

While some drugs form covalent bonds with their targets, in the majority of cases, including those in Figures 1.4 and 1.5, noncovalent interactions are responsible for the affinity between the drug and the target. The main classifications of such noncovalent attractive forces are ionic interactions, ion–dipole interactions, dipole–dipole interactions, hydrogen bonding, charge–transfer complexes, hydrophobic interactions, cation–π interactions, halogen bonding, and van der Waals forces. For example, a negatively charged moiety on the drug will be attracted to a positively charged residue on the target, or a phenyl ring on the drug will be attracted to the hydrophobic side chains of amino acids such as phenylalanine, leucine, valine, and others. Figure 1.6 shows schematically the multiple noncovalent interactions of the drug zanamivir (Relenza) with its target, neuraminidase, an enzyme that is critical in the reproductive cycle of the influenza virus. Figure 1.6 illustrates how multiple noncovalent interactions can combine to result in a high affinity of the drug for the target. Noncovalent interactions that are important for drug–target interactions are discussed in more detail in Chapter 3, Section 3.2.2.

Certain proteins are attractive as drug targets because of the critical roles they play in the body (Table 1.1). Receptors are proteins whose function is to interact with ("receive") another molecule (the receptor ligand), thereby inducing the receptor to perform some further action. Many receptors serve the role of translating signals from outside the cell to actions inside the cell. Figure 1.4 depicts a receptor protein that spans the membrane of a cell. The receptor ligand binds

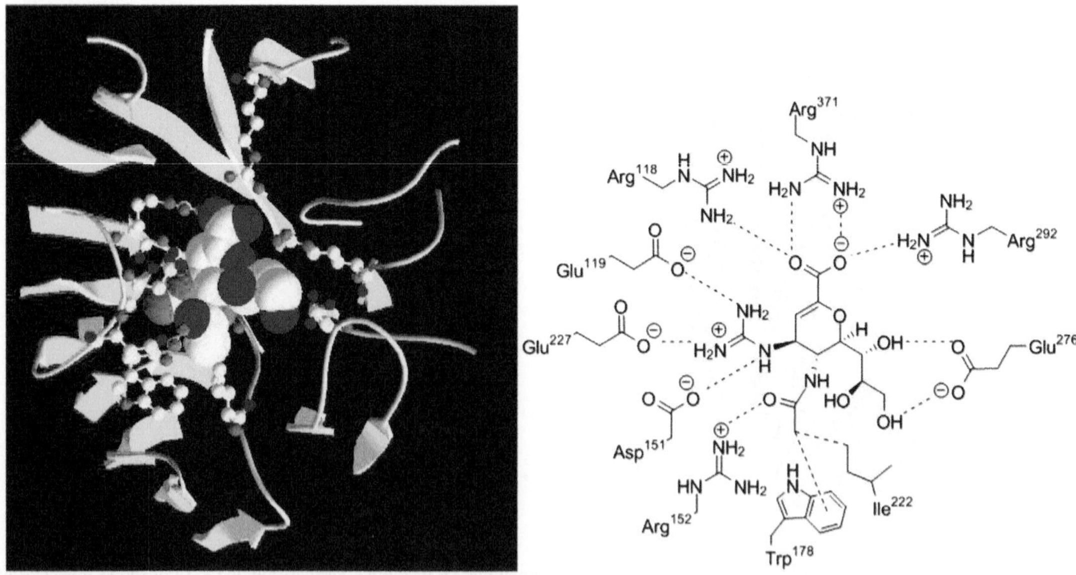

FIGURE 1.6 Interaction of the drug zanamivir with its enzyme target neuraminidase. (a) Model derived from an X-ray crystal structure; zanamivir is depicted as a space-filling model at center: carbon (white), oxygen (red), nitrogen (blue), and hydrogen (not shown). Only the regions of the enzyme that are close to the inhibitor are shown: small ball and stick models show key enzyme side chains (b) Schematic two-dimensional representation showing noncovalent interactions (dotted-lines) between zanamivir and the enzyme.

TABLE 1.1 Important Classes of Protein Drug Targets

Important Classes of Protein Drug Targets	Role or Function
Receptors	Transmit biological signals. Binding of certain ligands stimulates receptors to conduct a further action
Transporters	Facilitate transport of substances across cell membranes
Enzymes	Catalyze the transformation of substrate(s) to product(s)

to the region of the protein that is outside the cell, causing changes to the region of the protein that is inside the cell, thereby triggering further intracellular events (events inside the cell). Depending on the disease, it may be desirable to design drugs that either promote this trigger (receptor agonists) or block it (receptor antagonists). The organic chemical basis for the design and action of drugs that promote or inhibit the actions of receptors is discussed in more detail in Chapter 3.

Other proteins act as transporters. These proteins also span cell membranes, where their role is to carry or transport molecules or ions from one side of the cell to the other. Examples of drugs that modulate transporter action are discussed in Chapter 2.

Enzymes are another class of proteins that serve as very important drug targets. The formal name of an enzyme usually ends in the suffix "-ase". Enzymes are biological catalysts that facilitate the conversion of one or more reactants ("substrates") to one or more new products. For example, the enzyme acetylcholinesterase catalyzes the breakdown of the excitatory neurotransmitter acetylcholine (Scheme 1.2), which is important for learning and memory (among other actions). This breakdown of acetylcholine by acetylcholinesterase is the mechanism by which the effect of acetylcholine is turned off by the body. A drug that inhibits this enzyme should prolong the action of acetylcholine. Thus, for example, acetylcholinesterase inhibitors such as rivastigmine (Exelon) have been used for treatment of the symptoms of Alzheimer's disease (Chapter 2, Section 2.1.2.1). Another important drug target is HMG-CoA reductase, an enzyme in the pathway of cholesterol biosynthesis (Scheme 1.3). Inhibitors of this enzyme serve to reduce the production of cholesterol and are, therefore,

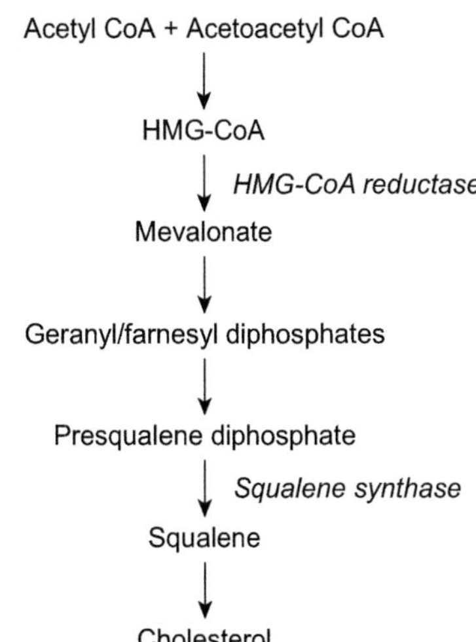

SCHEME 1.3 Pathway for cholesterol biosynthesis showing the role of the enzyme HMG-CoA reductase. *Adapted from http://www.jneuroinflammation.com/content/figures/1742-2094-4-5-1.jpg.*

important drugs for patients with excessive cholesterol in their bloodstreams (Chapter 5, Section 5.2.4.3). Note that in the foregoing examples, enzyme inhibition was a strategy to *promote* the action of acetylcholine (by preventing its breakdown), but to *impede* the action of cholesterol (by impeding its biosynthesis). Further examples of the organic chemistry of enzyme inhibitor design and action are discussed throughout Chapters 4 and 5.

Nucleic acids, for example, DNA, have an important role in cell replication, and drugs that bind to DNA can disrupt this function. This mechanism is responsible for the action of some anticancer and anti-infective drugs that disrupt the replication of, respectively, cancer cells and infectious organisms. The organic chemical basis for the design and action of drugs that disrupt nucleic acid function is discussed in Chapter 6.

1.3.2. Identification and Validation of Targets for Drug Discovery

In modern rational drug design, there are a number of key tools useful for uncovering, or at least hypothesizing, the role of potential drug targets in disease.[25] This exercise is sometimes referred to as *target validation* although many investigators do not consider a target truly validated until its role in human disease has been convincingly demonstrated in clinical trials. It has been estimated that there are only 324 drug targets for all classes of approved drugs (266

Acetylcholine —Acetylcholinesterase→ +

Acetylcholine

SCHEME 1.2 Reaction catalyzed by the enzyme acetylcholinesterase are human-genome derived proteins; the rest are pathogen targets) and only 1357 unique drugs, of which 1204 are small molecules and 166 are biologics.[26] Of the small molecule drugs, only 803 can be administered orally. One approach to identify targets related to a disease is to compare the genetic make-up of a large number of patients with the disease with that of a large number of normal patients, and identify which genes, and therefore the corresponding proteins, are consistently different in the two sets. Given that there are about 20,500 genes in the human genome,[27] there are many potential sites for mutations, leading to a disease. However, only about 7–8% of human genes have been explicitly associated with a disease. Another approach is to apply one of the several methods of selectively eliminating the function of a particular protein and observing the consequence in an isolated biochemical pathway or a whole animal.[28] Among prominent methods to achieve this, gene knockout[29] or knockdown using small interfering RNA (siRNA) technology[30] are important ones (RNA interference has an important role in directing the development of gene expression). Alternatively, antibodies to a specific protein can be developed that block the function of the protein.[31] The direct use of siRNA as a therapeutic agent is under intense investigation; similarly, a number of antibodies to proteins are already in active use as therapeutic agents.[32] But, at least to date, rarely do these modes of therapy entail simply swallowing a pill once or twice a day, so these therapies have significant limitations. Sometimes, a small molecule that very specifically modifies the function of a target may serve to establish the role of that target, even if it is not itself suitable as a drug.

The more simple approach to target identification, rather than attempting to uncover a new one, is to use a target that has already been validated in the clinic. It has been estimated that the probability of getting a compound for a novel target into preclinical (animal) development is only 3%, but it is 17% for an established target.[33] However, the use of a well-established target can result in *me-too drugs* (drugs that are structurally very similar to already known drugs and act by the same mechanism of action), producing more drugs of the same class. With appropriate marketing, a company is able to benefit economically from the "me-too" approach although society may not realize a significant benefit. On the other hand, a novel target can lead to drugs that have novel properties that can treat diseases or subpopulations of diseases not previously treated. While this latter approach is more expensive and usually has a lower probability of success, it is also potentially more rewarding both for society and also for the finances of the company that established the new mechanism of treatment.

The target-based approach sometimes gives surprises when it turns out that, after a drug is in clinical trials or on the market, its mechanism of action is found to be completely different from what the drug was designed for. For example, the cholesterol-lowering drug ezetimibe (**1.25**, Zetia) was designed as an inhibitor of acyl-coenzyme A cholesterol acyltransferase (ACAT), the enzyme that esterifies cholesterol, which is required for its intestinal absorption; inhibition of ACAT should lower the absorption of cholesterol.[34] It was found that its in vivo activity did not correlate with its in vitro ACAT inhibition; ezetimibe was later found to inhibit the transport of cholesterol through the intestinal wall rather than inhibit ACAT.[35] Pregabalin (**1.26**, Lyrica), a drug for the treatment of epilepsy, neuropathic pain, fibromyalgia, and generalized anxiety disorder, was found to be an activator of the enzyme glutamate decarboxylase in vitro, and that was thought to be responsible for its anticonvulsant activity; the mechanism of action was later found to be antagonism of the $\alpha_2\delta$-subunit of a calcium channel.[36]

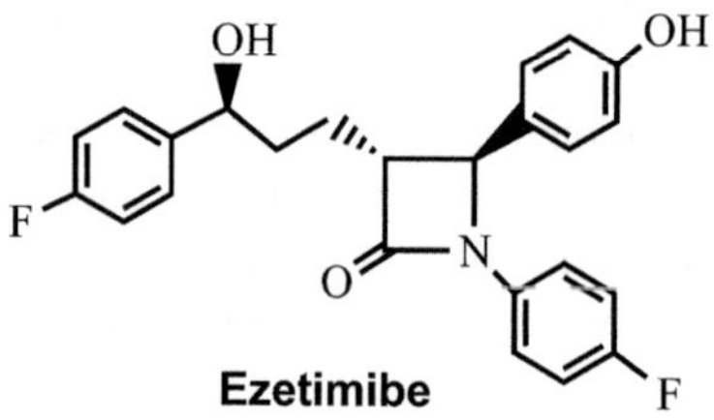

Ezetimibe 1.25

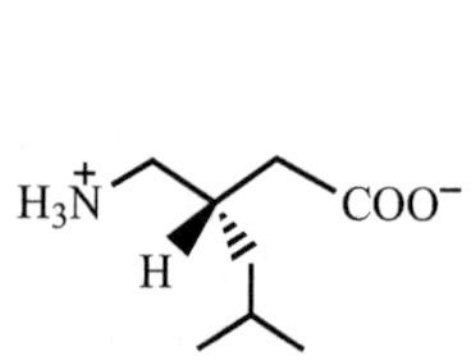

Pregabalin 1.26

Modern rational drug discovery usually begins with identification of a suitable biological target whose actions may be amenable to enhancement or inhibition by a drug, thereby leading to a beneficial therapeutic response. But how does one start in the search for the molecule that has the desired effect on the target? And what properties, other than exerting the desired action on its target, must the drug have? The typical approach is to first identify one or more lead compounds (defined in Section 1.1), i.e. molecular starting points, the structures of which can be modified ("optimized") to afford a suitable drug. In Section 1.3.4 there is a brief overview of methods of lead discovery, followed by a short overview of considerations underlying lead modification (Section 1.3.5). Chapter 2 will discuss the organic chemistry behind these topics in more detail.

1.3.3. Alternatives to Target-Based Drug Discovery

As discussed above (Sections 1.3.1 and 1.3.2), the most common approach to drug discovery involves initial identification of an appropriate biological target. Sams-Dodd[37] notes that diseases can be thought of as abnormalities at the mechanistic level, for example, abnormalities in a gene, a receptor, or an enzyme. This mechanistic abnormality can then result in a functional problem, for example, an abnormal function of the mitochondria, which causes a functional problem with an organ. These abnormalities produce physiological symptoms of diseases. Therefore, drug discovery approaches can be based on mechanism of action (screening compounds for their effect on a particular biological target, as discussed above), on function (screening compounds for their ability to induce or normalize functions, such as growth processes, hormone secretion, or apoptosis (cell death)), or on physiology (screening compounds in isolated organ systems or in animal models of disease to reduce symptoms of the disease). The latter approach, using animal models, was actually the first drug discovery approach, but it is now generally used as a last resort because of the low screening capacity, its expense, and the difficulty to identify the mechanism of action.

1.3.4. Lead Discovery

As noted in Section 1.1, drugs are generally not discovered; lead compounds are discovered. In the modern drug discovery paradigm that we are discussing, a lead compound typically has most or all of the following characteristics:

- It interacts with the target in a manner consistent with that needed to achieve the desired effect.
- It is amenable to synthetic modifications needed to improve properties.
- It possesses, or can be modified to possess, physical properties consistent with its ability to reach the target after administration by a suitable route. For example, evidence suggests that compounds with a high molecular weight (>~500), many freely rotatable bonds, high lipophilicity, and too many hydrogen bond-forming atoms have a reduced probability of being well absorbed from the gastrointestinal tract after oral administration. Therefore, it is desirable for a lead compound of a drug that is to be administered orally to either already possess the necessary properties or be amenable to modification to incorporate them.

Common sources of lead compounds are the following:

- *The natural ligand or substrate for the target of interest.* For example, dopamine (**1.27**) is the natural ligand for the family of dopamine receptors. Increasing dopamine concentrations is an important aim for the treatment of Parkinson's disease. Therefore, dopamine was the lead compound for the discovery of rotigotine (**1.28**), a drug used for the treatment of Parkinson's disease and restless leg syndrome.[38]

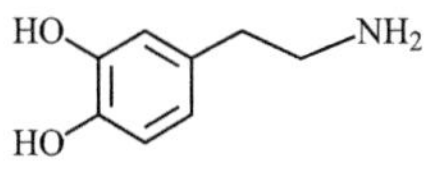

HO HO NH_2 OH N CH_3 S

Dopamine
1.27

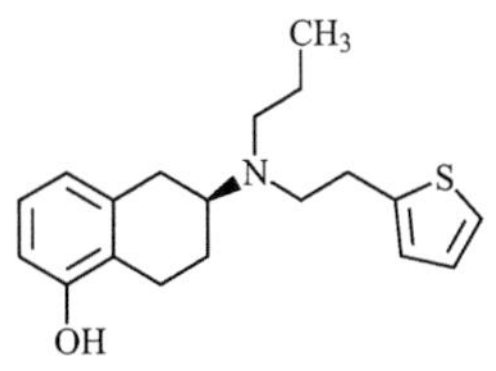

Rotigotine
1.28

- *Another substance already known to interact with the target of interest.* For example, the plant alkaloid cytisine (**1.29**) was known to interact with nicotinic acetylcholine receptors. Another well-known plant alkaloid, nicotine (**1.30**), also interacts with these receptors. Cytisine was the lead compound used for the Pfizer's development of varenicline (**1.31**, Chantix), a drug that helps patients quit smoking.[39] Comparing the three structures, one can also imagine that the structure of nicotine inspired some of the ideas for the modifications of cytisine on the way to the discovery of varenicline.

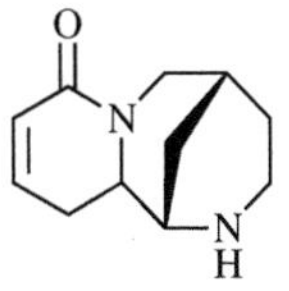

O N N H   N N CH_3 N N NH

Cytisine
1.29

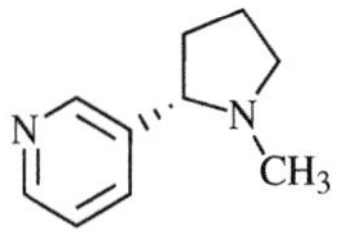

Nicotine
1.30

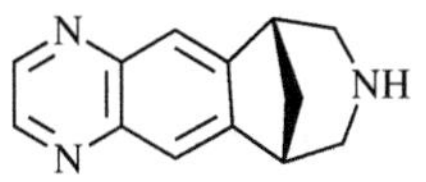

Varenicline
1.31

- *Random or targeted screening.* Screening refers to the exercise of conducting a biological assay on a large collection of compounds to identify those compounds that have the desired activity. Initially, these compounds may bind weakly to the target and are known as *hits*. Hits can be considered as predecessors to leads (the hit to lead process is discussed in Chapter 2, Section 2.1.2.3.5). Assays that rapidly measure binding affinities to targets of interest, called *high-throughput screens*, have been commonly used for this purpose since the early 1990s. Alternatively, cellular responses that are influenced by the target of interest may be measured. For example, activation of some receptors, such as dopamine receptors, is known to result in an increase in the concentration of Ca^{2+} ions inside the cell. Therefore, measurement of changes in the intracellular Ca^{2+} concentration in cells (with Ca^{2+}-sensitive dyes) that express dopamine receptors (either naturally or by transfection) can be used to identify ligands for these receptors. Such biochemical and cellular methods have largely supplanted the earlier practice of screening compounds in whole animals or in sections of tissue. *Random screening* implies that there is no effort to bias the set of screened compounds based on prior knowledge of the target or its known ligands; therefore, random compounds are screened. *Targeted screening* implies application of some prior knowledge to intelligently select compounds that are judged most likely to interact with the target.
- *Fragment-based screening.* Several screening methods using, for example, X-ray crystallography or NMR spectrometry have been developed to identify simple molecules (fragments) possessing typically modest affinity for

a target, with the intent of connecting two or more of these fragments to create a useful lead compound (Chapter 2, Section 2.1.2.3.6).

- *Computational approaches.* Given knowledge of the binding site on the target (for example through X-ray crystallography) or of the structure of several known ligands, computational approaches may be used to design potential lead compounds (Chapter 2, Section 2.2.6).

With respect to random screening, a major consideration is the source of the large number of compounds usually required to identify good leads, and it is an important role of organic chemists to address this question. For the targeted approach, the intelligent selection of compounds to be screened is an additional consideration requiring the attention of organic and computational chemists. Further aspects of these topics will be discussed in detail in Chapter 2.

1.3.5. Lead Modification (Lead Optimization)

Once one or more lead compounds have been identified, what more needs to be done before you have a viable drug candidate? Typically it is necessary, or at least advantageous, to optimize at least one, but more often several, of a number of key parameters to have the highest probability of identifying a successful drug. As discussed in more detail below, the most notable parameters that may need to be optimized include: potency; selectivity; absorption, distribution, metabolism, and excretion (ADME); and intellectual property position. This process normally involves synthesizing modified versions (*analogs*) of the lead compound and assessing the new substances against a battery of relevant tests. It is not uncommon to synthesize and test hundreds of analogs in the lead optimization process before a drug candidate (a compound worthy of extensive animal testing) is identified.

1.3.5.1. Potency

Potency refers to the strength of the biological effect, or put another way, how much (what concentration) of the compound is required to achieve a defined level of effectiveness. Thus, all other things being equal, the more potent a drug, the less will need to be administered to achieve the desired effect. Administering less drug is desirable from a number of viewpoints, including minimizing the cost per dose of the drug and maximizing the convenience of administration, that is, avoiding overly large pills, a need to take a large number of pills at the same time, or the necessity to take the drug more than twice a day. Perhaps more importantly, if lower doses of the drug can be administered to achieve a desired effect, then the probability should be lower that other unintended sites of action ("off-targets"), especially those unrelated to the desired target, will be affected, which can lead to unwanted side effects. Sometimes interactions with unrelated targets are not detected until they are revealed in advanced studies involving, for example, chronic administration in animals or studies in humans. Such late-stage discoveries can be costly indeed!

1.3.5.2. Selectivity

Unintended sites of action, noted above, refer to interactions with unidentified or unexpected targets. In addition, there may be off-targets that are related to the intended target, with which it would be disadvantageous for the drug to interact. For example, the dopamine D_3 receptor discussed above has related family members, namely, the dopamine D_1, D_2, D_4, and D_5 receptors, all of which utilize dopamine as the endogenous ligand but can mediate different responses.[40]

There are other well-known off-targets that should be avoided. One example is the cytochrome P450 (CYP) family of enzymes, which are responsible for the metabolism of many drugs (Chapter 7). Inhibiting a CYP enzyme can inhibit the metabolism of other drugs that someone may be taking at the same time, leading to dramatic changes in the levels of the other drugs. The result, referred to as drug–drug interactions, can severely limit the drugs that you can take at the same time or can cause, sometimes, fatal accumulation of other drugs.

Table 1.2 summarizes common targets against which selectivity would be desirable during lead optimization. If a lead compound interacts potently with any of these targets, then assessment of the newly synthesized compounds against the affected target(s) often occurs early

TABLE 1.2 Common "Off-Targets" that should be Avoided During Lead Modification

Off-Target	Role or Reason for Avoiding as Off-Target
Related family members	Although targets may be related, their actions may be quite different from, or even opposed to, those of the primary target, leading to undesired effects
Cytochrome P450 enzymes	Assist in eliminating drugs from the system. Inhibiting these off-targets can result in drug–drug interactions
Transporters	Transporters may be involved in regulating the extent to which drugs are concentrated inside vs outside of cells or the extent to which drugs are absorbed from the intestine. Inhibiting these off-targets can result in drug–drug interactions
hERG channel	Has a role in maintaining proper heart rhythm; inhibition can lead to fatal arrhythmias

in the testing process, with the objective of identifying which structural features are responsible for the undesired interactions.

1.3.5.3. Absorption, Distribution, Metabolism, and Excretion (ADME)

Absorption refers to the process by which a drug reaches the bloodstream from its site of administration. Frequently, the term is presumed to refer to absorption from the gastrointestinal tract after oral administration because this is often the preferred route of administration. However, it can also apply to absorption after other routes of administration, for example, nasal, oral inhalation, vaginal, rectal, subcutaneous, or intramuscular administration. In essentially every case, other than intravenous administration, a drug must pass through cell membranes on its way to the bloodstream. In the case of oral administration, a drug entering the bloodstream is funneled immediately through the liver, where it may be subject to extensive metabolism (see below) before passing into the systemic circulation.

Distribution refers to what "compartments" in the body the drug goes. For example, some drugs stay primarily in the bloodstream, while others distribute extensively into tissues. Physical properties of the compounds, such as aqueous solubility and partition coefficient (a measure of affinity for organic vs aqueous environment), can have a significant effect on drug distribution, and therefore are key parameters that are frequently monitored and modified during lead optimization.

Metabolism refers to the action of specific enzymes on a drug to convert it to one or more new molecules (called *metabolites*). Together with excretion of the intact drug (see below), metabolism is a major means by which the body clears a drug from the system. A common overall objective in drug discovery is to identify a compound for which therapeutic (but not toxic) levels in the system can be maintained following a convenient dosing schedule (for example, once or twice a day). This may entail identifying a drug that lasts long enough, but not too long. Therefore, understanding and controlling the metabolism of a drug are frequently major objectives of a lead optimization campaign. Moreover, metabolites may themselves be biologically active, leading in favorable cases to an increase or prolongation of the desired activity, or in unfavorable cases to undesired side effects. Chapter 8 discusses the organic chemistry of metabolic processes, and thereby provides key concepts for rational approaches to address metabolism issues during lead optimization.

Excretion refers to means by which the body eliminates an unchanged drug or its metabolites. The major routes of excretion are in the urine or feces. Exhalation can be a minor route of excretion when volatile metabolites are produced.

1.3.5.4. Intellectual Property Position

Discovering a new drug and bringing it to market is an exceptionally expensive endeavor, with some cost estimates ranging from $1.2–1.8 billion for each successful drug.[41] To recover the costs and also be able to appropriately compensate investors who are financing the research (and incentivize potential new investors), it is imperative to obtain a patent on a drug that is progressing toward drug development. The patent gives the patent holder the legal means to prevent others from making, selling, or importing the drug, effectively granting the holder a monopoly, for a limited period of time, on selling the drug. To obtain the most useful form of a patent, the chemical structure must be novel and nonobvious compared to publicly available information. It is within the scope of responsibilities of the medicinal chemist to conceive and synthesize the substances that meet the potency, selectivity, and ADME criteria discussed above while being novel and nonobvious. The successful accomplishment of all of those stringent criteria requires innovation, highly creative thinking, and superior synthetic skills.

1.3.6. Drug Development

Drug development normally refers to the process of taking a compound that has been identified from the drug discovery process described above through the subsequent steps necessary to bring it to market. Typically, these additional major steps include preclinical development, clinical development, and regulatory approval.

1.3.6.1. Preclinical Development

Preclinical development is the stage of research between drug discovery and clinical development, which typically entails:

- Development of synthetic processes that will enable the compound to be manufactured in reproducible purity on large (multikilogram) scale.
- Development of a formulation, in most cases a solution or suspension of the drug that can be administered to animals in toxicity tests and a solution or suspension or pill that can be administered to humans in clinical trials.
- Toxicity testing in animals under conditions prescribed by the regulatory authorities in the region where the clinical trials will occur (the FDA in the US; the European Medicines Agency in Europe; the Japanese Ministry of Health and Welfare in Japan).
- Following toxicity studies, gaining permission from the regulatory authorities to administer the drug to humans. In the US, such permission is obtained through the submission to the FDA of an Investigational New Drug (IND) application, which summarizes the discovery and preclinical development research done to date.

1.3.6.2. Clinical Development (Human Clinical Trials)

Clinical development is normally conducted in three phases (Phases I–III) prior to applying for regulatory approval to market the drug:

- *Phase 0* trials, also known as human microdosing studies, were established in 2006 by the FDA for exploratory, first-in-human trials.[42] They are designed to speed up the development of promising drugs or imaging agents from preclinical (animal) studies. A single subtherapeutic dose of the drug is administered to about 10–15 healthy subjects to gather preliminary human ADME data on the drug and to rank order drug candidates that have similar potential in preclinical studies with almost no risk of side effects to the subjects.
- *Phase I* evaluates the safety, tolerability (dosage levels and side effects), pharmacokinetic properties, and pharmacological effects of the drug in about 20–100 individuals. These individuals are usually healthy volunteers although actual patients may be used when the disease is life threatening. A key objective of these studies is to attempt to correlate the results of the animal toxicity studies (including levels of the drug in blood and various tissues) with findings in humans to help establish the relevance of the animal studies. Phase I generally lasts a few months to about a year and a half.
- *Phase II* assesses the effectiveness of the drug, determines side effects and other safety aspects, and clarifies the dosing regimen in a few hundred diseased patients. These studies typically provide an initial sense of effectiveness of the drug against the disease, but, because of the limited size and other factors, are not generally regarded as definitive to establish drug efficacy. Phase II typically lasts from 1 to 3 years.
- *Phase III* is a larger trial typically with several thousand patients that establishes the efficacy of the drug, monitors adverse reactions from long-term use, and may compare the drug to similar drugs already on the market. Appropriate scientific controls are included to allow statistically meaningful conclusions to be made on the effectiveness of the drug. Phase III typically requires about 2–6 years to be completed.

1.3.6.3. Regulatory Approval to Market the Drug

In the US, regulatory approval requires submission to the FDA of a New Drug Application (NDA), summarizing the results from the clinical trials. This can now be done electronically; previously, it would require, literally, a truckload of paper describing all of the preclinical and clinical studies. On the basis of these data, the FDA decides whether to grant approval for the drug to be prescribed by doctors and sold to patients. Once the drug is on the market, then it is possible to assess the real safety and tolerability of a drug because it is taken by hundreds of thousands, if not millions, of people. Such postmarketing surveillance activities are often referred to as *Phase IV* studies because this is when statistically insignificant effects in clinical trials can become significant with a large and varied patient population, leading to side effects not observed with relatively small numbers of patients in Phase III trials. On the other hand, Phase IV studies may reveal new indications for a drug with patients having symptoms from other diseases.

1.4. EPILOGUE

It should be appreciated from the foregoing discussions that the drug discovery and development process is a long and arduous one, taking on average from 12 to 15 years, a time that has been constant for over 30 years. For approximately every 20,000 compounds that are evaluated in vitro, 250 will be evaluated in animals, 10 will make it to human clinical trials, to get one compound on the market at a cost estimate of $1.2–1.8 billion (in 1962 it was only $4 million!). Drug candidates (or *new chemical entities* or *new molecular entities* as they are often called) that fail late in this process result in huge, unrecovered financial losses for the company. Furthermore, getting a drug on the market may not be so rewarding; it has been estimated that only 30% of the drugs on the market actually make a profit.[43] This is why the cost to purchase a drug is so high. It is not that it costs that much to manufacture that one drug, but the profits are needed to pay for all of the drugs that fail to make it onto the market after large sums of research funds have already been expended or that do not make a profit once on the market. In addition, funds are needed for future research efforts. As a result, to minimize expenses, outsourcing has become an important economic tool.[44] Not only are labor rates significantly lower in Eastern Europe and Asia than in the United States and Western Europe, but outsourcing also allows a company to have more flexibility to manage its staffing needs compared to hiring full time staff. Interestingly, the rise in drug discovery costs has *not* been accompanied by a corresponding increase in the number of new drugs being approved for the market. In 1996, 53 drugs were approved by the FDA, and in 2002, only 16 drugs were approved; 2002 was the first time in the US that generic drug sales were greater than nongeneric drug sales. From 2004 to 2010, 20–28 drugs per year were approved by the FDA,[45] and many of these are just new formulations or minor modifications of existing drugs; in general, only five or six of the new drugs approved each year are first-in-class. Possible contributors to this lower-drug-approval-at-higher-cost trend (other than inflation) include increasingly higher regulatory hurdles, for example, greater safety regulations for drug approval, as well as recent efforts to tackle increasingly difficult therapeutic objectives, such as curing cancers or halting the progression of Alzheimer's disease.[46] In 2011 and 2012 new drug approvals rose to 30 and 39, respectively, suggesting a possible effect of some of the more modern approaches discussed in this book. Unfortunately, in 2013 that number dropped to 27, indicating we still have a lot of work to do.

Mechanistic and synthetic organic chemistry play a central role in numerous critical aspects of the drug discovery process, most prominently in generating sufficient numbers of compounds for lead discovery, in effectively optimizing compounds for potency, selectivity, and intellectual property position, and in understanding factors governing ADME. The ensuing chapters will delve in detail into the organic chemistry of these critical aspects of drug design and drug action.

1.5. GENERAL REFERENCES

Journals and Annual Series

ACS Chemical Biology
ACS Medicinal Chemistry Letters
Advances in Medicinal Chemistry
Annual Reports in Medicinal Chemistry
Annual Review of Biochemistry
Annual Review of Medicinal Chemistry
Annual Review of Pharmacology and Toxicology
Biochemical Pharmacology
Biochemistry
Bioorganic and Medicinal Chemistry
Bioorganic and Medicinal Chemistry Letters
Chemical Biology and Drug Design
Chemical Reviews
Chemistry and Biology
ChemMedChem
Current Drug Metabolism
Current Drug Targets
Current Genomics
Current Medicinal Chemistry
Current Opinion in Chemical Biology
Current Opinion in Drug Discovery and Development
Current Opinion in Investigational Drugs
Current Opinion in Therapeutic Patents
Current Pharmaceutical Biotechnology
Current Pharmaceutical Design
Current Protein and Peptide Science
Drug Design and Discovery
Drug Development Research
Drug Discovery and Development
Drug Discovery Today
Drug News and Perspectives
Drugs
Drugs of the Future
Drugs of Today
Drugs under Experimental and Clinical Research
Emerging Drugs
Emerging Therapeutic Targets
European Journal of Medicinal Chemistry
Expert Opinion on Drug Discovery
Expert Opinion on Investigational Drugs
Expert Opinion on Pharmacotherapy
Expert Opinion on Therapeutic Patents
Expert Opinion on Therapeutic Targets
Future Medicinal Chemistry
Journal of Biological Chemistry
Journal of Chemical Information and Modeling
Journal of Medicinal Chemistry
Journal of Pharmacology and Experimental Therapeutics
MedChemComm
Medicinal Research Reviews
Methods and Principles in Medicinal Chemistry
Mini Reviews in Medicinal Chemistry
Modern Drug Discovery
Modern Pharmaceutical Design
Molecular Pharmacology
Nature
Nature Chemical Biology
Nature Reviews Drug Discovery
Nature Medicine
Perspectives in Drug Discovery and Design
Proceedings of the National Academy of Sciences
Progress in Drug Research
Progress in Medicinal Chemistry
Science
Science Translational Medicine
Trends in Pharmacological Sciences
Trends in Biochemical Sciences

Books

Abraham, D. J.; Rotella, D. P. (Eds.) *Burger's Medicinal Chemistry and Drug Discovery*, 7th ed., Wiley & Sons, New York, 2010, Vols. 1–8.

Albert, A. *Selective Toxicity*, 7th ed., Chapman and Hall, London, 1985.

Ariëns, E. J. (Ed.) *Drug Design*, Academic, New York, 1971–1980, Vols. 1–10.

Borchardt, R. T.; Freidinger, R. M.; Sawyer, T. K. *Integration of Pharmaceutical Discovery and Development: Case Histories*, Plenum Press, 1998.

Bruton, L.; Chabner, B.; Knollman, B. (Eds.) *Goodman and Gilman's The Pharmacological Basis of Therapeutics*, 12th ed., McGraw-Hill, New York, 2010.

Kerns, E. H.; Di, L. *Drug-like Properties: Concepts, Structure, Design, and Methods*, Elsevier: Amsterdam, 2008.

Lednicer, D. *Strategies for Organic Drug Synthesis and Design*, 2nd ed., Wiley, New York, 2009.

Lednicer, D. Mitscher, L. A. *The Organic Chemistry of Drug Synthesis*, seven-volume set, Wiley, New York, 2008, Vol. 7.
Lemke, T. L.; Williams, D. A.; Roche, V. F.; Zito, S. W. (Eds.) *Foye's Principles of Medicinal Chemistry*, 7th ed., Lippincott Williams & Wilkins, Philadelphia, 2012.
O'Neil, M. J. (Ed.) *The Merck Index*, 14th ed., Merck & Co., Inc., Whitehouse Station, NJ, 2006.
Taylor, J. B.; Triggle, D. J. (Eds.) *Comprehensive Medicinal Chemistry II*, Elsevier, Amsterdam; 2007.
Wermuth, C. G. (Ed.) *The Practice of Medicinal Chemistry*, 3rd ed., Academic Press, San Diego, 2009.

Blogs

www.biospace.com
www.genengnews.com
pipeline.corante.com/
www.sciencedaily.com/new/health_medicine

Links

American Chemical Society Division of Medicinal Chemistry. www.acsmedchem.org
American Chemical Society Division of Organic Chemistry. www.organicdivision.org
BioChemWeb—Chemical Biology subpage. www.biochemweb.org/chemical.shtml
Broad Institutes ChemBank (small molecule guide for drug discovery). www.broadinstitute.org/chembank
ChemSpider. www.chemspider.com
ClinicalTrials.gov. www.clinicaltrials.gov
Emolecules. www.emolecules.com
EMBL-EBI European Bioinformatics Institute. www.ebi.ac.uk
US Food and Drug Administration. www.fda.gov
KinasePro (kinase chemistry). www.Kinasepro.wordpress.com
NCBI (National Center for Biotechnology Information). www.ncbi.nlm.nih.gov/
NCBI PubMed (literature search). www.ncbi.nlm.nih.gov/pubmed
European Patent Office. www.epo.org
US Patent Office. www.uspto.gov/
Bordwell pKa Table. www.chem.wisc.edu/areas/reich/pkatable/inde.htm
Protein Data Bank (protein crystal structure database). www.pdb.org

1.6. PROBLEMS (ANSWERS CAN BE FOUND IN THE APPENDIX AT THE END OF THE BOOK)

1. Define the following terms.
   a. Lead compound
   b. High-throughput screen
   c. Fragment-based screening
   d. Metabolite
   e. Transporter
   f. Enzyme
   g. Absorption
2. Describe ways in which lead compounds are obtained.
3. List some ways that drugs can be discovered without rational design.
4. Name the noncovalent interactions.
5. What problems are associated with compounds that have low potency for their target?
6. What problems arise from poor selectivity of compounds for a target?
7. Why is it important to patent your drug?

REFERENCES

1. Merriam-Webster Online Dictionary, http://www.merriam-webster.com/dictionary/drug.
2. Choudhuri, S.; Carlson, D. B. (Eds.) *Genomics: Fundamentals and Applications*, Informa Healthcare, New York, 2009.
3. Wilkins, M. R.; Appel, R. D.; Williams, K. L.; Hochstrasser, D. F. (Eds.) *Proteome Research: Concepts, Technology and Application (Principles and Practice)*, 2nd ed., Springer-Verlag, Berlin, 2007.
4. Bauer, W. W. *Potions, Remedies and Old Wives' Tales*, Doubleday, New York, 1969.
5. Withering, W. *An Account of the Foxglove and Some of its Medicinal Uses: With Practical Remarks on Dropsy and Other Diseases*, Robinson, C. G. J., Robinson, J., London, 1785; reprinted in *Med. Class.* **1937**, *2*, 305.
6. Sneader, W. *Drug Discovery: The Evolution of Modern Medicines*, Wiley, Chichester, 1985.
7. Nakanishi, K. In *Comprehensive Natural Products Chemistry*, Barton, D.; Nakanishi, K. (Eds.), Elsevier, Amsterdam and New York, 1999, Vol. 1, pp. xxiii–xl.
8. Chen, K. K. A pharmacognostic and chemical study of ma huang (*Ephedra vulgaris* var. Helvetica). *J. Am. Pharm. Assoc.* **1925**, *14*, 189.
9. Burger, A. In *Burger's Medicinal Chemistry*, Wolff, M. E. (Ed.), 4th ed., Wiley, New York, 1980, Part I, Chap. 1.
10. Fleming, A. The antibacterial action of cultures of a *Penicillium*, with special reference to their use in the isolation of *B. influenzae*. *Br. J. Exp. Pathol.* **1929**, *10*, 226–236.
11. Stone, T.; Darlington, G. Pills, *Potions and Poisons. How Drugs Work*, Oxford University Press, Oxford, 2000, p. 255.
12. Hare, R. *The Birth of Penicillin*, Allen & Unwin, London, 1970.
13. Beveridge, W. I. B. *Seeds of Discovery*, W. W. Norton, New York, 1981.
14. Abraham, E. P.; Chain, E.; Fletcher, C. M.; Gardner, A. D.; Heatley, N. G.; Jennings, M. A.; Florey, H. W. Further observations on *Penicillin*. *Lancet.* **1941**, *2*, 177–188.
15. Florey, H. W.; Chain, E.; Heatley, N. G.; Jennings, M. A.; Sanders, A. G.; Abraham, E. P.; Florey, M. E. *Antibiotics*, Oxford University Press, London, 1949, Vol. 2.
16. Moyer, A. J.; Coghill, R. D. Penicillin: IX. The laboratory scale production of penicillin in submerged cultures by *Penicillium notatum* westling (NRRL 832). *J. Bacteriol.* **1946**, 51, 79–93.

17. (a) Sheehan, J. C. The Enchanted Ring: the Untold Story of Penicillin, MIT Press, Cambridge, MA, 1982. (b) Williams, T. I. *Robert Robinson: Chemist Extraordinary*, Clarendon Press, Oxford, 1990. (c) Todd, A. R.; Cornforth, J. S. In *Robert Robinson*. 13 September 1886–8 February 1975. *Biogr. Mem. R. Soc.* **1976**, *22*, 490.
18. Hodgkin, D. C.; Bunn, C.; Rogers-Low, B.; Turner-Jones, A. In *Chemistry of Penicillin*, Clarke, H. T.; Johnson, J. R.; Robinson, R. (Eds.), Princeton University Press, Princeton, NJ, 1949.
19. Sternbach, L. H. The benzodiazepine story. *J. Med. Chem.* **1979**, *22*, 1–7.
20. Shen, T. Y. In Clinoril in the Treatment of Rheumatic Disorders, Huskisson, E. C.; Franchimont, P. (Eds.), Raven Press, New York, 1976.
21. Gay, L. N.; Carliner, P. E. The prevention and treatment of motion sickness I. Seasickness. *Science*, **1949**, *109*, 359.
22. Strickland, B. A., Jr.; Hahn, G. L. The effectiveness of dramamine in the prevention of airsickness. *Science*, **1949**, *109*, 359–360.
23. (a) Corbin, J. D.; Francis, S. H. Cyclic GMP phosphodiesterase-5: Target of sildenafil. *J. Biol. Chem.* **1999**, *274*, 13729–13732. (b) Palmer, E. Making the love drug. *Chem. Br.* **1999**, *35*, 24–26.
24. Daemmrich, A.; Bowden, M. E. A rising drug industry. *Chem. Eng. News*. **2005** (June 20), *83*, 3.
25. (a) Lindsay, M. A. Innovation: target discovery. *Nat. Rev. Drug Discov.* **2003**, *2*, 831–838. (b) Peet, N. P. What constitutes target validation? *Targets*. **2003**, *2*, 125–127. (c) Roses, A. D.; Burns, D. K.; Chissoe, S.; Middleton, L.; St Jean, P. Disease-specific target selection: a critical first step down the right road. *Drug Discov. Today*. **2005**, *10*, 177–189.
26. Overington, J. P.; Al-Lazikani, B.; Hopkins, A. L. How many drug targets are there? *Nat. Rev. Drug Discov*. **2006**, *5*, 993–996.
27. Clamp, M.; Fry, B.; Kamal, M.; Xie, X.; Cuff, J.; Lin, M. F.; Kellis, M.; Lindblad-Toh, K.; Lander, E. S. Distinguishing protein-coding and noncoding genes in the human genome. *Proc. Natl. Acad. Sci. U.S.A.* **2007**, *104*(49), 19428–19433.
28. Sioud, M. Main approaches to target discovery and validation. In *Methods Mol. Biol. (Totowa, NJ, United States)* **2007**, *360* (Target Discovery and Validation, Vol. 1), 1–12.
29. (a) Zambrowicz, B. P.; Sands, A.T. A path to innovation: gene knockouts model new drug action. *Ann. Rep. Med. Chem.* **2009**, *44*, 475–497, Chap. 23. (b) Kuehn, R.; Wurst, W. (Eds.), *Gene Knockout Protocols (Methods in Molecular Biology)*, 2nd ed., Humana Press, 2009.
30. (a) Khurana, B.; Goyal, A. K.; Budhiraja, A.; Arora, D.; Vyas, S. P. siRNA delivery using nanocarriers – an efficient tool for gene silencing. *Curr. Gene Ther.* **2010**, *10*(2), 139–155. (b) Whitehead, K. A.; Langer, R.; Anderson, D. G. Knocking down barriers: advances in siRNA delivery. *Nat. Rev. Drug Discov*. **2009**, *8*(2), 129–138.
31. Lichtlen, P.; Auf der Maur, A.; Barberis, A. Target validation through protein-domain knockout – applications of intracellularly stable single-chain antibodies. *Targets*, **2002**, *1*(1), 37–44.
32. Leavy, O. *Nat. Rev. Immunol.* **2010**, *10*(5), entire issue.
33. Accenture and CMR International, Rethinking innovation in pharmaceutical R&D, **2005**.
34. (a) Clader, J. W. The discovery of ezetimibe: a view from outside the receptor. *J. Med. Chem.* **2004**, *47*, 1–9. (b) Sliskovic, D. R.; Picard, J. A.; Krause, B. R. ACAT inhibitors: the search for a novel and effective treatment of hypercholesterolemia and atherosclerosis. *Prog. Med. Chem.* **2002**, *3*, 121–171.
35. (a) Van Heek, M.; France, C. F.; Compton, D. S.; McLeod, R. L.; Yumibe, N. P.; Alton, K.B.; Sybertz, E. J.; Davis, H. R. In vivo metabolism-based discovery of a potent cholesterol absorption inhibitor, SCH58235, in the rat and rhesus monkey through the identification of the active metabolites of SCH48461. *J. Pharmacol. Exp. Ther.* **1997**, *283*, 157–163. (b) Van Heek, M.; Farley, C.; Compton, D. S.; Hoos, L.; Alton, K. B.; Sybertz, E. J.; Davis, H. R. Comparison of the activity and disposition of the novel cholesterol absorption inhibitor, SCH58235, and its glucuronide, SCH60663. *Br. J. Pharmacol.* **2000**, *129*, 1748–1754.
36. Silverman, R. B. From basic science to blockbuster drug: the discovery of Lyrica. *Angew. Chem. Int. Ed.* **2008**, *47*, 3500–3504.
37. Sams-Dodd, F. Drug discovery: selecting the optimal approach. *Drug Discov. Today*. **2006**, *11*, 465–472.
38. Chen, J. J.; Swope, D. M.; Dashtipour, K.; Lyons, K. E. Transdermal rotigotine: a clinically innovative dopamine-receptor agonist for the management of Parkinson's disease. *Pharmacotherapy*. **2009**, *29*, 1452–1467.
39. Dwoskin, L. P.; Smith, A. M.; Wooters, T. E.; Zhang, Z.; Crooks, P. A.; Bardo, M. T. Nicotinic receptor-based therapeutics and candidates for smoking cessation. *Biochem. Pharmacol*. **2009**, *78*, 732–743.
40. Missale, C.; Nash, S. R.; Robinson, S. W.; Jaber, M.; Caron, M. G. Dopamine receptors: from structure to function. *Physiol. Rev*. **1998**, *78*, 189–225.
41. (a) Adams, C. P.; Brantner, V. V. Spending on new drug development. *Health Econ.* **2010**, *19*, 130–141. (b) Paul, S. M.; Mytelka, D. S.; Dunwiddie, C. T.; Persinger, C. C.; Munos, B. H.; Lindborg, S. R.; Schacht, A. L. How to improve R&D productivity: the pharmaceutical industry's greatest challenge. *Nat. Rev. Drug Discov*. **2010**, *9*, 203–214.
42. Garner R. C. Practical experience of using human microdosing with AMS analysis to obtain early human drug metabolism and PK data. *Bioanalysis*. **2010**, *2*, 429–440.
43. www.phrma.org/files/attachments/Cost_of_Prescription_Drugs.pdf.
44. Kalorama Information's Outsourcing in Drug Discovery, 3rd ed., 2008.
45. *Ann. Rep. Med. Chem.* Academic Press: San Diego, CA, **2005–2010**.
46. Scanelli, J. W.; Blanckley, A.; Boldon, H.; Warrington, B. Diagnosing the decline in pharmaceutical R&D efficiency. *Nat. Rev. Drug Discov*. **2012**, *11*, 191–200.

Chapter 2

Lead Discovery and Lead Modification

Chapter Outline

The Organic Chemistry of Drug Design and Drug Action. http://dx.doi.org/10.1016/B978-0-12-382030-3.00002-7

2.1. LEAD DISCOVERY

2.1.1. General Considerations

As discussed in the drug discovery overview in Chapter 1, identification of suitable lead compounds provides starting points for lead optimization, during which leads are modified to achieve requisite potency and selectivity, as well as absorption, distribution, metabolism, and excretion (ADME), and intellectual property (patent) position. Given the hurdles often presented by these multiple and diverse objectives, identification of the best lead compounds can be a critical factor to the overall success of a drug discovery program. The approach to lead identification taken in a given drug discovery program will usually take into account any known *ligand* (a smaller molecule that binds to a receptor) for the target. At one extreme, if there are already marketed drugs for a particular target, these may serve as lead compounds; however, in this case, establishing a suitable intellectual property position may be the greatest challenge. On the other hand, whereas the *endogenous ligand* (the molecule that binds to a biological target in an organism and is believed to be responsible for the native activity of the target) has provided good lead structures for many programs, the endogenous ligand for a new biological target may not be well characterized, or the only known ligand may not be attractive as a lead compound. For example, if an endogenous ligand is a complex molecule that is not readily amenable to synthetic modification or has some other undesirable properties that are not reasonably addressable, it may not be attractive as a lead, and other approaches to lead discovery must be considered. In the next few sections, we will first provide additional examples of endogenous or other known ligands as lead compounds to complement the examples given in Chapter 1, and then we will turn to a more detailed discussion of alternative approaches to lead discovery.

2.1.2. Sources of Lead Compounds

Lead compounds can be acquired from a variety of sources: endogenous ligands, e.g., substrates for enzymes and transporters or agonists for receptors; other known ligands, including marketed drugs, compounds isolated in drug metabolism studies, and compounds used in clinical trials; and through screening of compounds, including natural products and other chemical libraries, either at random or in a targeted approach.

2.1.2.1. Endogenous Ligands

Rational approaches are important routes to lead discovery. The first step is to identify the cause for the disease state. Many diseases, or at least the symptoms of diseases, arise from an imbalance (either excess or deficiency) of a particular chemical in the body, from the invasion of a foreign organism, or from aberrant cell growth. As will be discussed in later chapters, the effects of the imbalance can be corrected by antagonism or agonism of a receptor (see Chapter 3) or by inhibition of a particular enzyme (see Chapter 5); interference with deoxyribonucleic acid (DNA) biosynthesis or function (see Chapter 6) is another important approach to treating diseases arising from microorganisms or aberrant cell growth. Once the relevant biochemical system is identified, initial lead compounds become the endogenous receptor ligands or enzyme substrates. In Chapter 1, the example of dopamine as a lead compound for the discovery of rotigotine (**1.28**) was presented. Dopamine is the endogenous ligand for dopamine receptors, including the D_3 receptor, which is the target of rotigotine. Dopamine is one of a number of important *neurotransmitters*, substances released by nerve cells (*neurons*) that interact with receptors on the surface of nearby neurons to propagate a nerve signal (Figure 2.1). Endogenous neurotransmitters have served as lead compounds for many important drugs. Table 2.1 shows

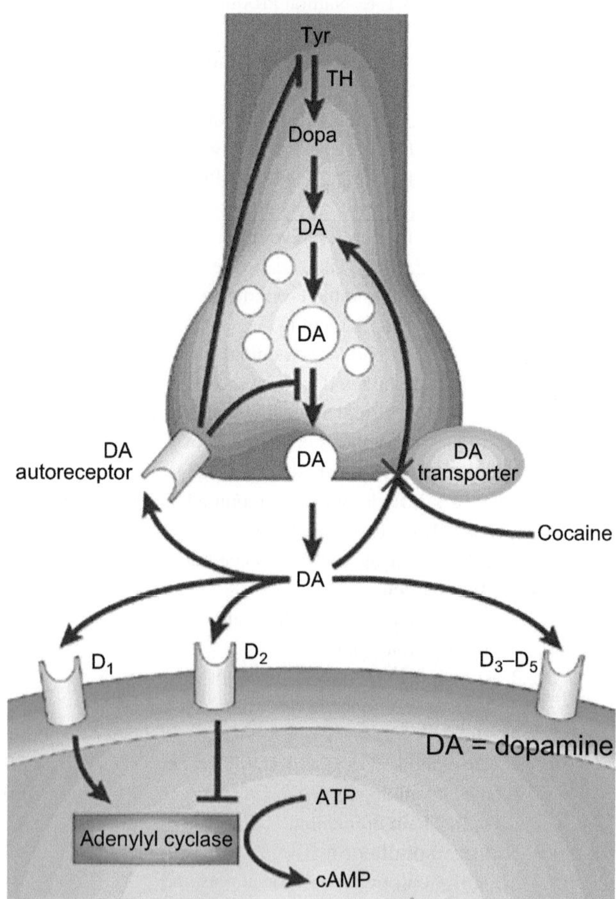

FIGURE 2.1 Depiction of dopamine (DA) in its role as a neurotransmitter. DA is released by a neuron prior to interacting with dopamine receptors (D1–D5) on the surface of another nearby neuron. Also shown is the dopamine transporter, which terminates the action of dopamine by transporting the released neurotransmitter from the synaptic cleft back into the presynaptic neuron. *Reprinted by permission from Macmillan Publishers Ltd: Nature Reviews Drug Discovery (Kreek, M. J.; LaForge, K. S.; Butelman, E. Pharmacotherapy of addictions. Nat. Rev. Drug Discov.* ***2002****, 1, 710–726) Copyright 2002.*

TABLE 2.1 Examples of Endogenous Neurotransmitter Ligands That Have Served as Lead Compounds for Drug Discovery

Endogenous Ligand	Marketed Drug
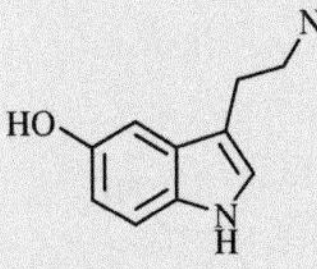 Serotonin	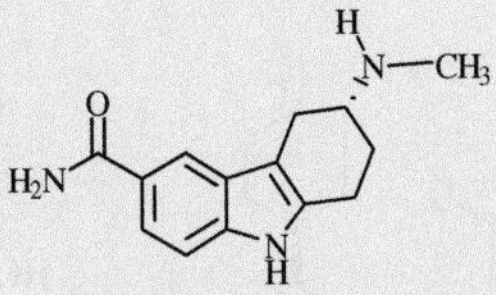 Frovatriptan (antimigraine)
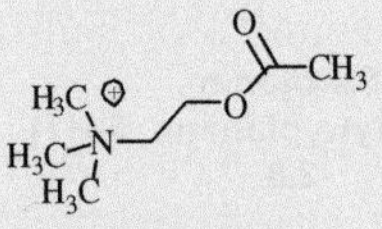 Acetylcholine	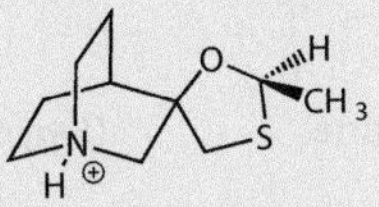 Cevemaline (dry mouth treatment)
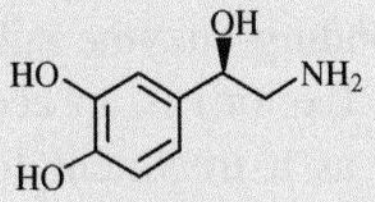 Norepinephrine	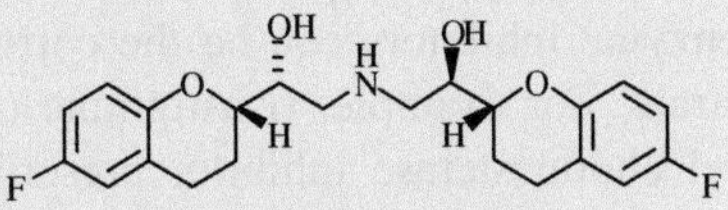 Nebivolol (antihypertensive)

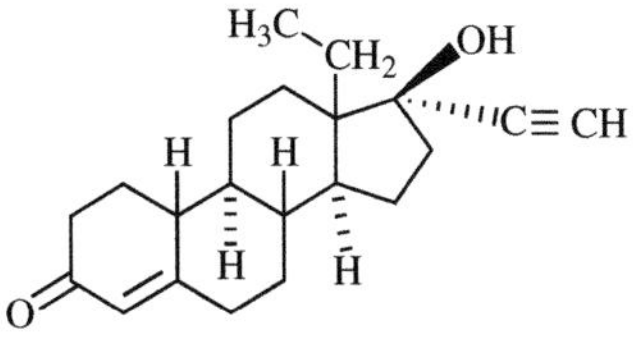

Norgestrel
2.1

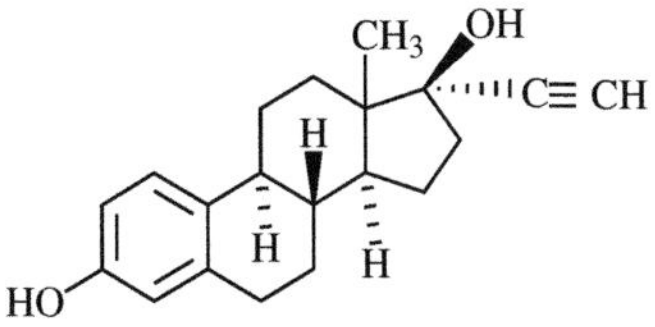

17α-Ethynyl estradiol
2.2

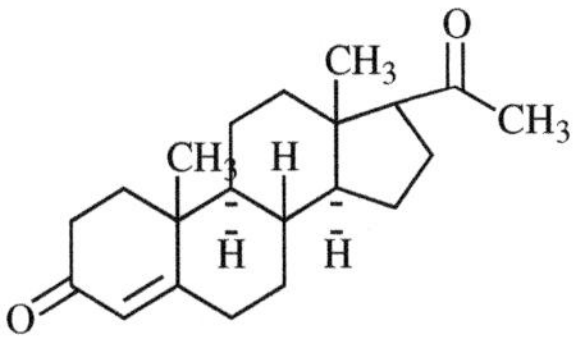

2.3a

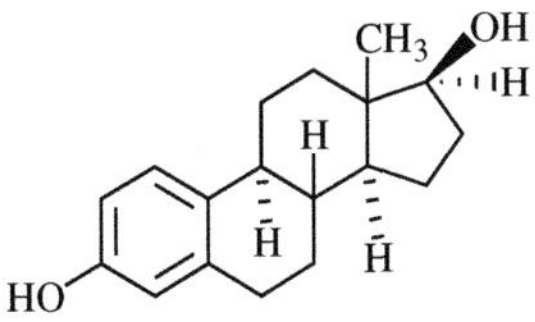

2.3b

examples of the drugs that evolved from the structures of the endogeous neurotransmitters serotonin, acetylcholine, and norepinephrine.

Hormones are another important class of endogenous substances that have served as lead compounds for drug discovery. Like neurotransmitters, hormones are released from cells and interact with receptors on the surface of other cells. However, whereas receptors for neurotransmitters are close to the site of neurotransmitter release, hormone receptors can be at quite some distance from the site of hormone release, so hormones have to travel to their site of action through the bloodstream. Steroids are one important class of hormones; lead compounds for the contraceptives (+)-norgestrel (**2.1**, Ovral) and 17α-ethynyl estradiol (**2.2**, Activella) were the steroidal hormones progesterone (**2.3a**) and 17β-estradiol (**2.3b**), respectively. The endogenous steroid hormones (**2.3a** and **2.3b**) show weak and short-lasting effects, whereas oral contraceptives (**2.1** and **2.2**) exert strong progestational activity of long duration.

Peptides constitute another broad class of hormones. Peptides, like proteins, consist of a sequence of amino acid residues, but are smaller than proteins (in the range of two to approximately 100 amino acids). Most peptides have low stability in plasma as a result of the ubiquitous presence of *peptidases* (enzymes that catalyze hydrolysis of peptides into smaller peptides or constituent amino acids). Moreover, peptides usually cannot be delivered orally because of low permeability across gut membranes (as a result of their charge and polarity) and because of instability to gut peptidases. However, incorporation of disulfide bonds to cross-link a peptide can confer enzymatic stability, e.g., linaclotide (**2.4**, Linzess) used to treat bowel diseases. Considerable effort has been devoted to the goal of using natural peptides as lead compounds for the discovery of derivatives with improved properties. One successful drug that resulted from these endeavors is lanreotide (**2.5**, Somatuline),[1] a long-acting analog of the peptide hormone somatostatin (**2.6**), which is administered by injection to treat *acromegaly* (thickening of skin and enlargement of hands and feet from overproduction of growth hormone).

The discussion of endogenous ligands so far has focused on leads for drugs designed to interact with receptor targets. Endogenous ligands for other types of drug targets, including transporters and enzymes, have also served as valuable starting points for drugs. As mentioned in Chapter 1, transporters are proteins that help transport substances across cell membranes. One important class of transporters is responsible for neurotransmitter reuptake.[2] As illustrated in Figure 2.1 for the neurotransmitter dopamine, after dopamine is released into the synaptic cleft, excess neurotransmitter is transported back into the neuron that released it (the presynaptic neuron) by specific transporter proteins, which serves to deactivate the signal

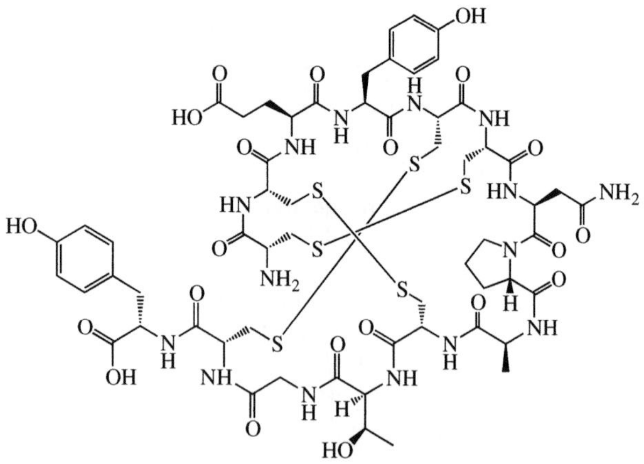

Linaclotide
2.4

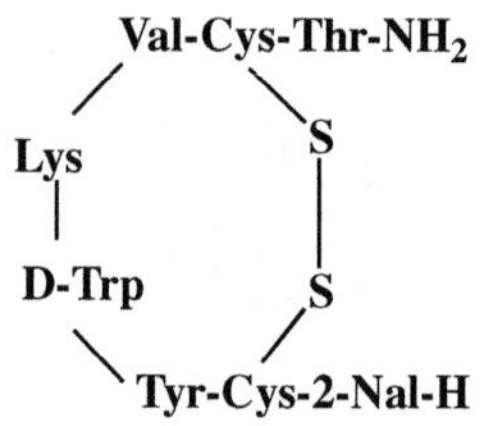

Val-Cys-Thr-NH_2
Lys
D-Trp
Tyr-Cys-2-Nal-H

Lanreotide
(Nal = 2-naphthylalanine)
2.5

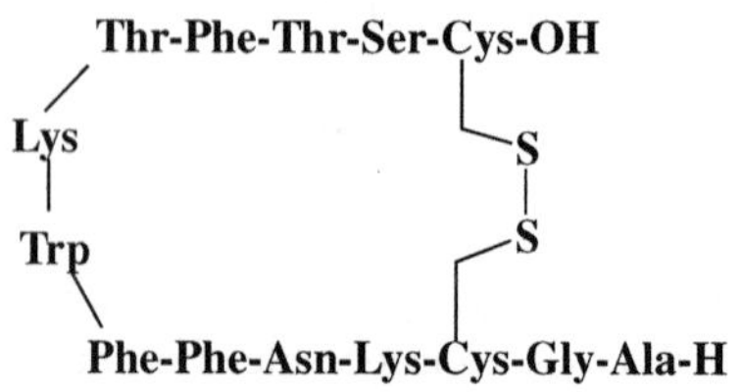

Thr-Phe-Thr-Ser-Cys-OH
Lys
Trp
Phe-Phe-Asn-Lys-Cys-Gly-Ala-H

Somatostatin
2.6

carried by the neurotransmitter. Therefore, an inhibitor of a neurotransmitter reuptake transporter would have the effect of prolonging the action of the neurotransmitter. Cocaine exerts its effects by inhibiting the dopamine reuptake transporter. Inhibitors of the reuptake transporters for other important neurotransmitters, such as norepinephrine and serotonin, comprise important classes of antidepressant drugs. The leads for many of these reuptake inhibitors were the transporter ligands, that is, norpinephrine or serotonin. Paroxetine (**2.7**, Paxil) is an example of a selective serotonin reuptake inhibitor marketed as an antidepressant drug with considerable structural resemblance to serotonin (**2.8**). Transporters of glucose have recently been targeted for the treatment of type 2 diabetes.[3]

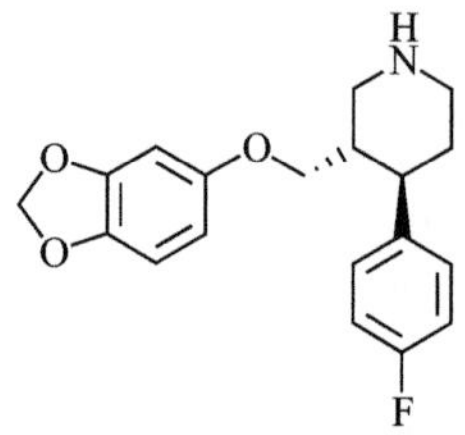

Paroxetine
2.7

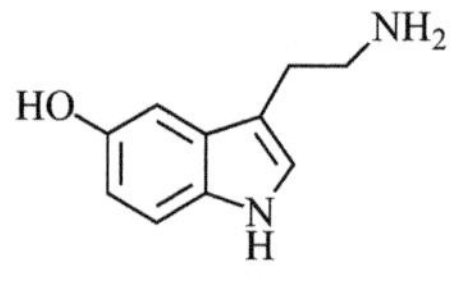

Serotonin
(5-hydroxytryptamine, 5-HT)
2.8

Similarly, an important source of leads for the design of enzyme inhibitors can be the corresponding enzyme substrate. For example, rivastigmine (**2.9**, Exelon) is an acetyl cholinesterase inhibitor prescribed as a treatment for dementia, for which the ultimate starting point was acetylcholine (Table 2.1), although in actuality, the evolution of rivastigmine occurred across several generations of drugs (you are probably thinking it is hard to see how this structure could come from acetylcholine, but that is how lead optimization evolves new structures).

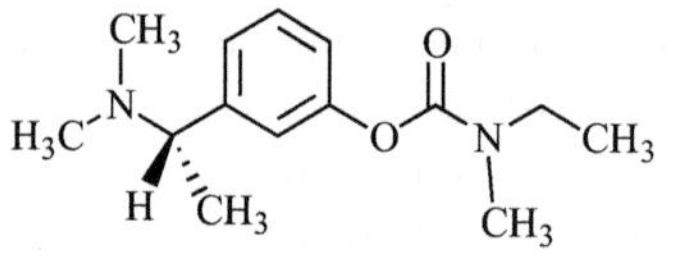

Rivastigmine
2.9

Another example of using an enzyme substrate as a lead for drug discovery is in the design of kinase inhibitors. Kinases catalyze the transfer of the terminal phosphate group of adenosine triphosphate (ATP) and related molecules usually to the hydroxyl group of another molecule (Scheme 2.1), for example, to the hydroxyl group on the tyrosine residue of a substrate protein (protein tyrosine kinase). Thus, kinases have two substrates, ATP (the phosphate donor) and the phosphate acceptor. Many kinase inhibitors were ultimately designed based on the structure of ATP, for example, gefitinib (**2.10**, Iressa), which is used for the treatment of lung cancer.

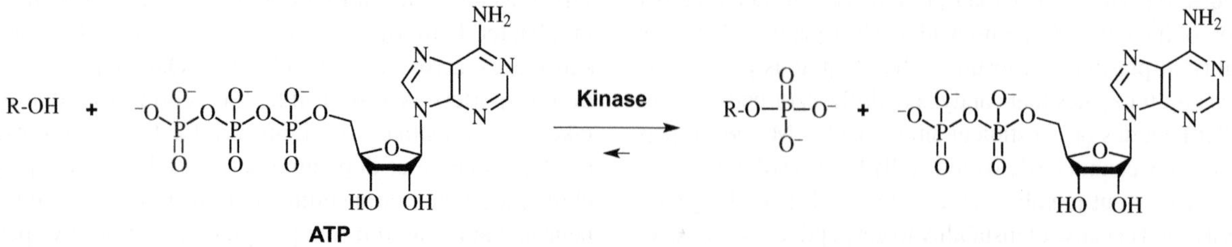

R-OH + **ATP** —**Kinase**→ R-O-P(O)(O⁻)-O⁻ + ADP

SCHEME 2.1 Reaction catalyzed by the kinase class of enzymes. Kinases catalyze the transfer of the terminal phosphate group of ATP or related molecules acceptor to the group of a substrate, in this case, an alcohol (ROH).

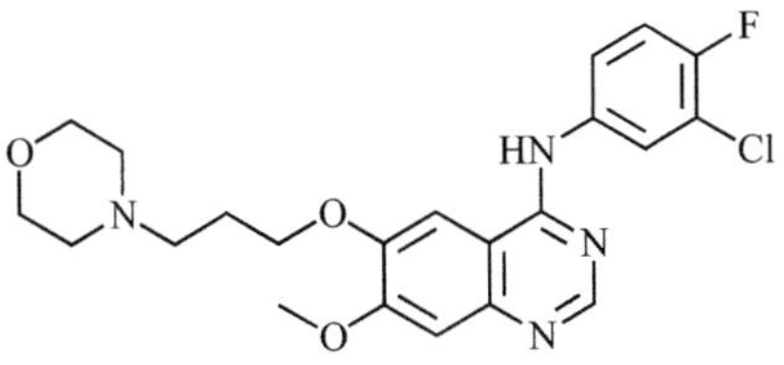

Gefitinib
2.10

Currently, rational approaches to drug discovery are most relevant to the earlier stages of the process, most notably including target identification, lead discovery, and optimization of molecular interactions with the target during lead optimization. Later stages of drug discovery presently remain much more empirical owing to the difficulties in accurately predicting toxicities, anticipating transport properties, accurately predicting the full range of ADME properties of a drug, and numerous other factors. However, active ongoing research is attempting to increase the degree of rationality even for these complex facets of drug behavior. In addition to rational approaches, particularly when no target protein is known or little structural information is available for rational design, other less rational approaches can be taken to get a starting point for lead discovery using other known ligands or screening approaches.

2.1.2.2. Other Known Ligands

In Chapter 1, the example of using the plant alkaloid cytisine (**1.29**) as the starting point for discovery of the smoking cessation agent varenicline (**1.31**, Chantix) was described. Another variant of using a known ligand as a starting point is the use of an established drug as a lead toward development of the next generation of compounds.[4] One example is diazepam (**1.17**, Valium), as described in Chapter 1, Section 1.2.3, which was derived from the marketed drug Librium (**1.13**) and is almost 10 times more potent than the lead. Another example is zoledronic acid (**2.11**, Zometa), which is used to treat *osteoporosis* (loss of bone density) and *hypercalcemia*, a condition resulting in high blood calcium levels due to cancer, and to delay bone complications resulting from multiple myeloma and bone metastases. This is a second-generation drug derived from pamidronate disodium (**2.12**, Aredia), also used for treating hypercalcemia from malignancy.

Known drugs can also be *repurposed* (the identification and development of new uses for existing or abandoned drugs; also called *repositioned*) for a completely different indication.[5] The advantage of a repurposed drug is that the cost to bring it to market is diminished because the safety and pharmacokinetic profiles have already been established for its original indication. A *library* (a collection of compounds) of 3665 Food and Drug Administration (FDA)-approved and investigational drugs was tested for activity against hundreds of targets, from which 23 new drug–target relationships were confirmed.[6] For example, the reverse transcriptase inhibitor and acquired immune deficiency syndrome (AIDS) drug delavirdine (**2.13**, Rescriptor) was found to antagonize the histamine H_4 receptor, which is a target for the potential treatment of asthma and allergies. Isradipine (**2.14**, Dynacirc), an antihypertensive drug, is in clinical trials as a treatment for Parkinson's disease.[7] The antidepressant drug duloxetine (**2.15,** Cymbalta) has been approved to treat chronic lower back pain.[8] A common dilemma to the repurposing of marketed drugs is that if the repurposed drug is used directly for a new indication, then only a new *method of use patent* (a patent that covers the new use for the molecule) application can be filed; however, it is best to own the rights to a molecule for *any* purpose (*composition of matter patent*), which an altered structure would allow. Therefore, using a known drug as a lead to discover a novel compound could warrant independent patent protection for the new structure. An important advantage to repurposed drugs is that whereas only 10% of new drugs in Phase I clinical trials and 50% of Phase III drugs make it to the market, the rates for repurposed drugs are 25 and 65%, respectively.

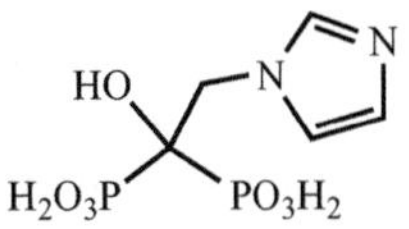

Zoledronic acid
2.11

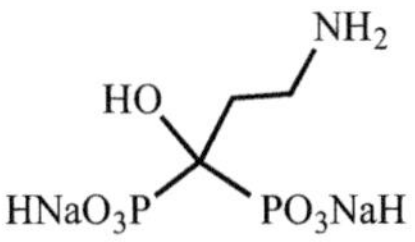

Pamidronate disodium
2.12

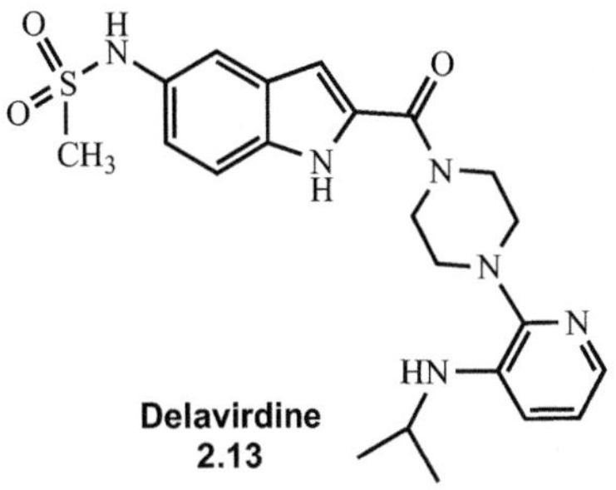

Delavirdine
2.13

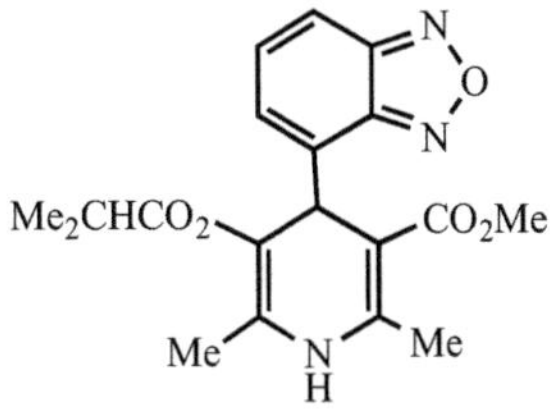

Isradipine
2.14

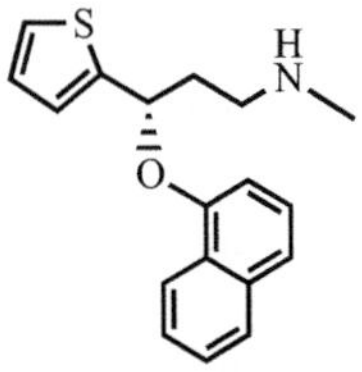

Duloxetine
2.15

Other sources of lead compounds, as described in Chapter 1, Sections 1.2.4 and 1.2.5, are metabolism studies and clinical trials. The cases cited in those sections involved the identification of new drugs from metabolism or from the clinic, some with novel indications; however, it is possible that the metabolite from a drug metabolism study or a compound in a clinical trial might act as a lead compound for a new indication requiring modification to enhance its potency or diminish undesirable properties.

2.1.2.3. Screening of Compounds

Endogenous or other ligands may not be known for a target of interest. Alternatively, known ligands for a target may not be well suited as starting points for discovery of drugs that will ultimately possess the desired properties. For example, many endogenous ligands are large proteins, which are not usually good leads when the goal is to discover an orally administered drug. For these reasons, screening for leads has played a central role in drug discovery for decades, although technological advances in the past 20 years have markedly changed how these screens are conducted, as discussed below.

The first requirement for a screening approach is to have a means to assay compounds for a particular biological activity, so that researchers will know when a compound is active. *Bioassay* (or *screen*) is a means of determining in a biological system, relative to a control compound, if a compound has the desired activity, and if so, what the relative potency of the compound is. Note the distinction between the terms activity and potency. *Activity* is the particular biological or pharmacological effect (for example, antibacterial activity or anticonvulsant activity); *potency* is the strength of that effect.

Until the late 1980s many screening efforts were conducted using whole animals or whole organisms, for example, screening for antiepileptic activity by assessing the ability of a compound to prevent an induced seizure in a mouse or rat, or for antibacterial activity by measuring the effect of test compounds on the growth of bacterial cultures in glass dishes. Especially when screening in whole animals, efforts have often been hampered by the comparatively large quantities of test compound required and by the fact that the results depended on other factors apart from the inherent potency of the compound at its intended target (*pharmacodynamics*), for example, the ability of the compound to be absorbed, distributed, metabolized, and excreted (*pharmacokinetics*). Thus, in general, in vitro tests have fewer confounding factors and are also quicker and less expensive to perform. The downside to this approach, however, is that you may identify a very potent compound for a target, but it may not have the ability to be absorbed or is rapidly metabolized. This more rapid screening method then requires additional studies of pharmacokinetics once the appropriate pharmacodynamics has been established. Pharmacokinetic aspects are discussed further throughout the chapter.

An exciting approach for screening compounds that might interact with an enzyme in a metabolic pathway was demonstrated by Wong, Pompliano, and coworkers for the discovery of lead compounds that block bacterial cell wall biosynthesis (as potential antibacterial agents).[9] Conditions were found to reconstitute all six enzymes in the cell wall biosynthetic pathway so that incubation with the substrate for the first enzyme led to the formation of the product of the last enzyme in the pathway. Then by screening compounds and looking for the buildup of an intermediate it was possible to identify compounds that blocked the pathway (and prevented the formation of the bacterial cell wall) and also which enzyme was blocked (the buildup of an intermediate meant that the enzyme that acted on that intermediate was blocked).

Compound screening also can be carried out by electrospray ionization mass spectrometry (MS)[10] (the technique for which John Fenn received the Nobel Prize in 2002) and by nuclear magnetic resonance (NMR) spectrometry (the technique for which Richard Ernst and Kurt Wüthrich received Nobel Prizes in 1991 and 2002, respectively).[11] Tightly bound noncovalent complexes of compounds with a macromolecule (such as a receptor or enzyme) can be observed in the mass spectrum. The affinity of the ligand can be measured by varying the collision energy and determining at what energy the complex dissociates. This method also can be used to screen mixtures of compounds, provided they have different molecular masses and/or charges, so that m/z for each complex with the biomolecule can be separated in the mass spectrometer. By varying the collision energy, it is possible to determine which test molecules bind to the biomolecule best. The ^1H NMR method exploits changes in either relaxation rates or diffusion rates of small molecules when they bind to a macromolecule. This method can also be used to screen mixtures of compounds to determine the ones that bind best.

High-throughput screening (HTS),[12] from which greater than two-thirds of drug discovery projects now originate,[13] was initially developed in the late 1980s employing very rapid and sensitive in vitro screens, which could be carried out robotically. According to Drews,[14] the number of compounds assayed in a large pharmaceutical company in the early 1990s was about 200,000 a year, which rose to 5–6 million during the mid-1990s, and by the end of the 1990s it was >50 million! HTS can be carried out robotically in 1536- or 3456-well titer plates on small (submicrogram) amounts of compound (dissolved in submicroliter volumes). With these ultrahigh throughput screening approaches of the early part of the twenty-first century,[15] it is possible to screen 100,000 compounds in a day! In 2010, an HTS method using *drop-based microfluidics* (the ability to manipulate tiny volumes of liquid) was reported that allowed a 1000 times faster screening (10 million reactions per hour) with 10^{-7} times the reagent volume and at one-millionth the cost of conventional techniques.[16] In this technique, drops of aqueous fluid dispersed in fluorocarbon oil replace the microtiter plates, which allows analysis and compound sorting in picoliter volume reactions while reagents flow through channels. A silicone sheet of lenses can be used to cover the microfluidic arrays, allowing fluorescence measurements of 62 different output channels simultaneously and analysis of 200,000 drops per second.[17] Therefore, screening compounds is no longer the slow step in the lead discovery process!

Because of the ease of screening vast numbers of compounds, early in the application of HTS, every compound in the company library, regardless of its properties, was screened. By the early part of the first decade of the twenty-first century, because an increase in the number of useful lead compounds was not forthcoming despite the huge rise in the application of screening, it was realized that the physicochemical properties of molecules were key for screening compounds.[18] Therefore, additional considerations for HTS became the sources and selection of compounds to be screened and the development of effective methods for processing and utilizing the screening data that were generated. Medicinal chemists have an important role in these activities, which we discuss in more detail in the next several sections. A keyword search for "high-throughput screening" in the *Journal of Medicinal Chemistry* website (http://pubs.acs.org/journal/jmcmar) readily retrieves a multitude of examples in which HTS played a central role in lead discovery. Representative examples are shown in Table 2.2, together with

TABLE 2.2 Examples of Hits from HTS and Analogs Resulting from Subsequent Optimization Efforts

Biological Target	HTS Hit	Representative Structure after Initial or Full Optimization
Peroxisome proliferator-activated receptor (PPAR) δ (target class: nuclear hormone receptor)	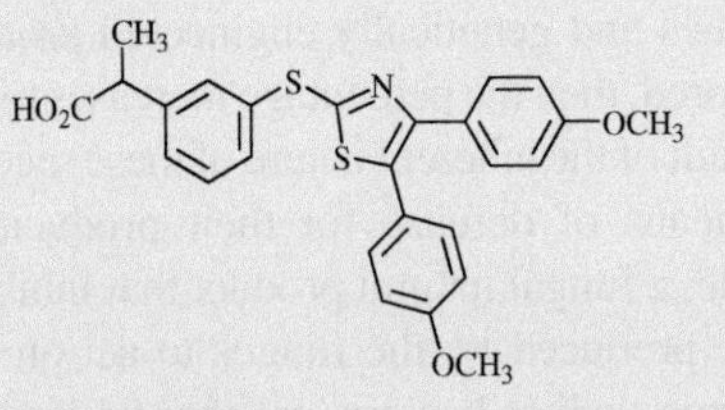 EC_{50} = 3200 nM	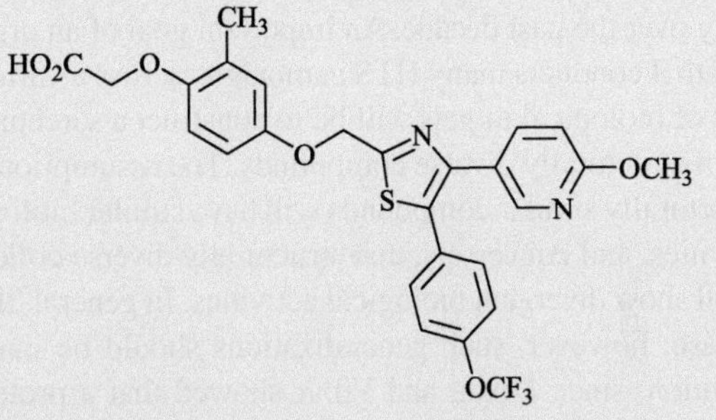 EC_{50} = 17 nM
Rho kinase (target class: enzyme)	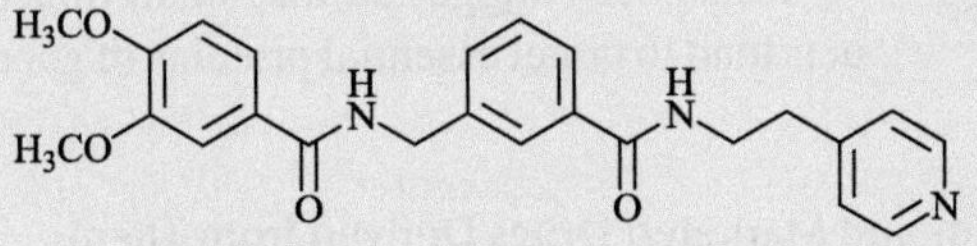 IC_{50} = 2300 nM	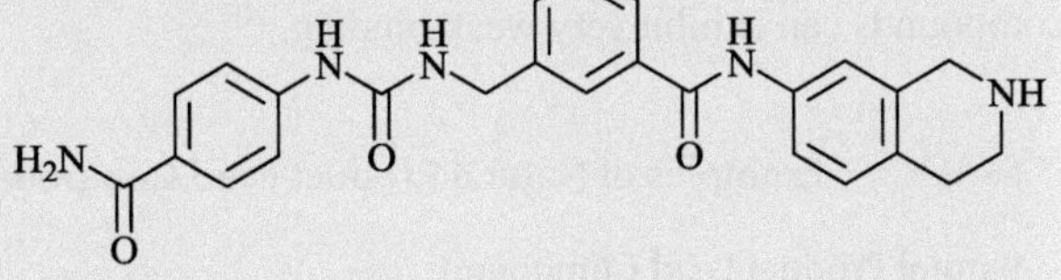 IC_{50} = 4 nM
KCNQ2/Q3 potassium channels (target class: ion channel)	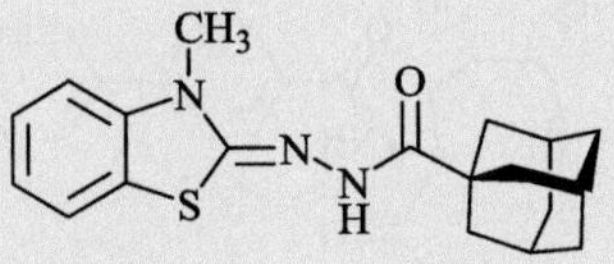 EC_{50} = 27 nM	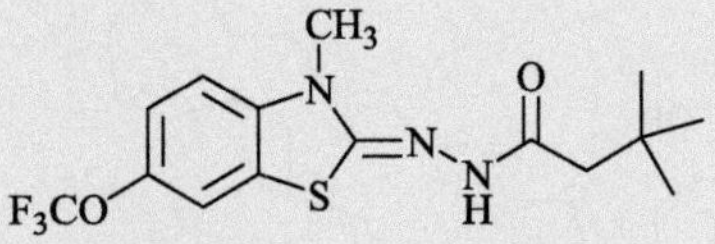 EC_{50} = 49 nM Significantly increased oral efficacy
Influenza A (H1N1) virus	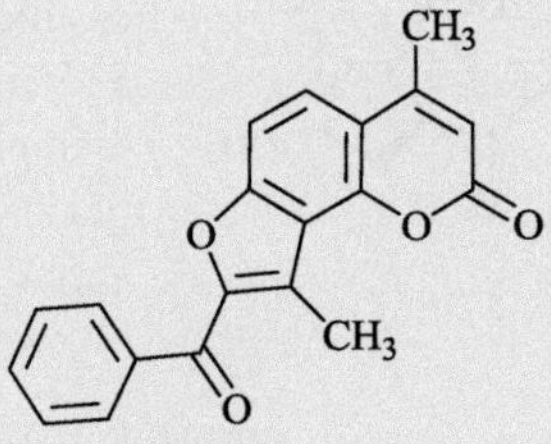 IC_{50} = 4500 nM	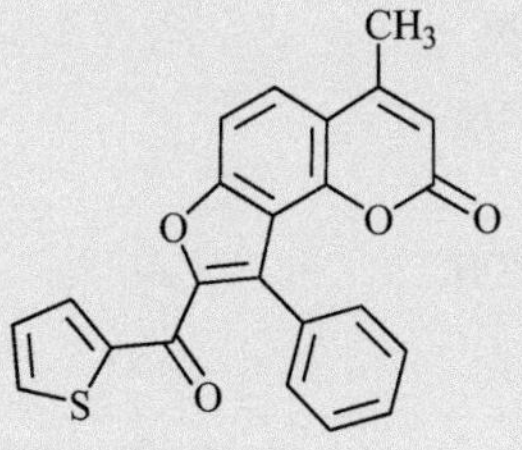 IC_{50} = 70 nM
Vascular endothelial growth factor receptor 2 kinase domain (target class: enzyme)	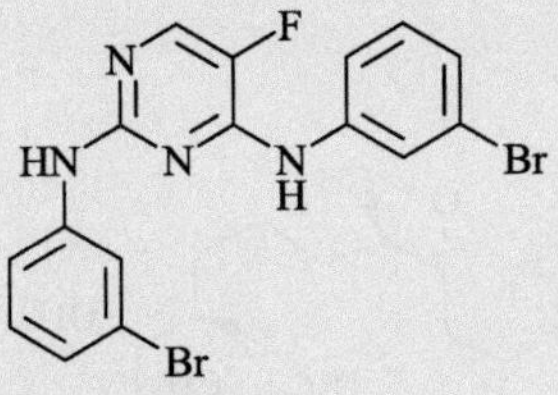 IC_{50} ~ 400 nM	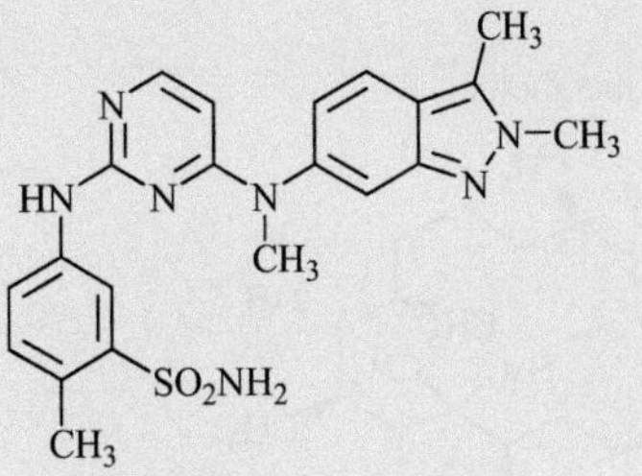 IC_{50} = 30 nM Marketed drug (pazopanib)

structures of products from subsequent lead optimization activities.[19] See Section 2.2 for what properties need to be considered prior to and during the lead optimization process.

2.1.2.3.1. Sources of Compounds for Screening

As stated above, besides a high-throughput assay, an essential second requirement for HTS is a large number of suitable compounds for screening. In the following several subsections, we discuss the most common sources of compounds for HTS. The criteria for selecting compounds to be added to a general screening collection and for improving the selection of specific compounds for a given screen have evolved considerably over the past decade. An important goal of an organization that conducts many HTS campaigns across a variety of types of biological targets will be to construct a screening library of structurally diverse compounds. The assumption is that structurally similar compounds will have similar biological activities, and conversely, that structurally diverse collections will show divergent biological activities. In general, this is the case; however, such generalizations should be made with caution, since Dixon and Villar showed that a protein can bind a set of structurally diverse molecules with similar potent binding affinities, and analogs closely related to these compounds can exhibit very weak binding.[20]

2.1.2.3.1.1. Natural Products Nature is still an excellent source of drug precursors, or in some cases, of actual drugs. Although endogenous ligands discussed earlier are technically also natural products, the present category is intended to encompass products from nonmammalian natural sources, for example, plants, marine organisms, bacteria, and fungi. Nearly half of the new drugs approved between 1994 and 2007 are based on natural products, including 13 natural product-related drugs approved from 2005 to 2007.[21] More than 60% of the anticancer and antiinfective agents that went on the market between 1981 and 2006 were of natural product origin or derived from natural products; if biologicals, for example, antibodies and genetically engineered proteins, and vaccines are ignored, then the percentage increases to 73%.[22] This may be a result of the inherent nature of these secondary metabolites as a means of defense for their producing organisms; for example, a fungal natural product that inhibits cell replication may be produced by the fungus to act on potential invading organisms such as bacteria or other fungi.[23] Table 2.3 shows two examples of recently approved drugs that were derived from natural product lead compounds[24]; many others are currently in various stages of clinical development.

It has been suggested that small molecule natural products tend to target essential proteins of genes from organisms

TABLE 2.3 Examples of Natural Product Lead Compounds and Marketed Drugs Derived from Them

Natural Product Lead Compound	Marketed Drug
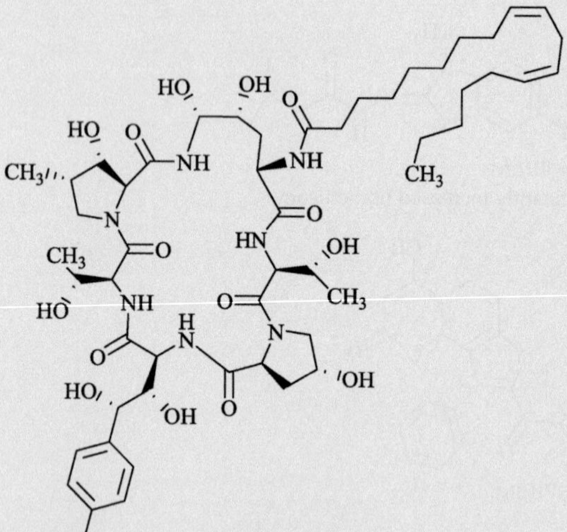 Echinocandin B (a fungal metabolite)	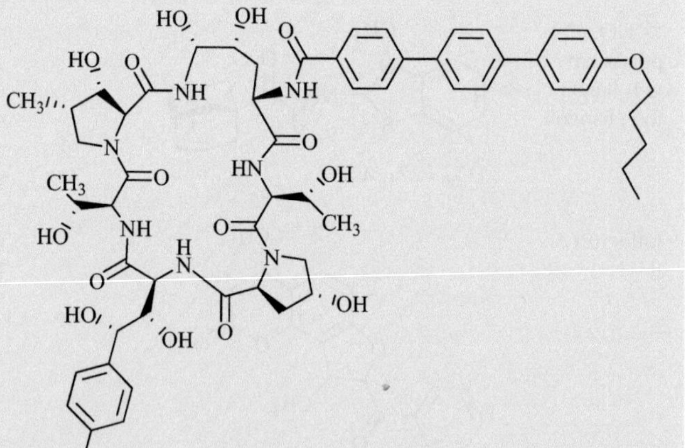 Anidulafungin (antifungal)
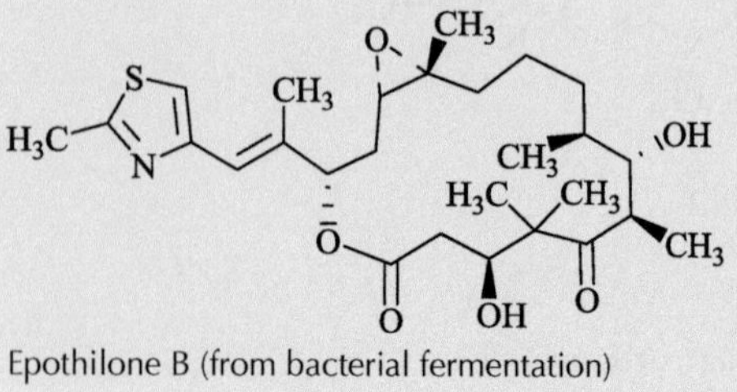 Epothilone B (from bacterial fermentation)	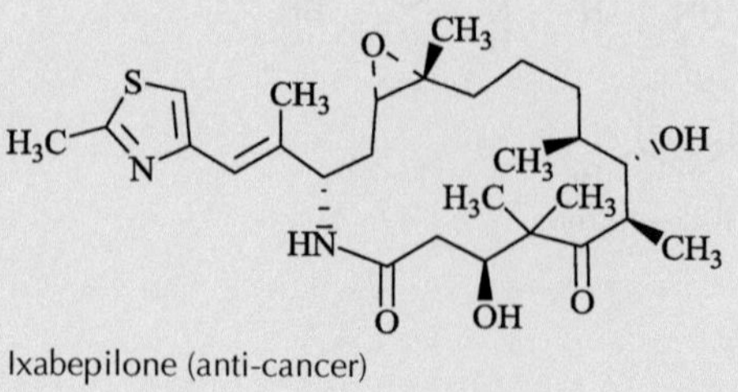 Ixabepilone (anti-cancer)

with which they coevolved, rather than those involved in human disease, and the reverse is true of synthetic drugs.[25] According to this hypothesis, natural products should be important molecules to combat microorganisms or aberrant (tumor) cell growth, but they should not be expected to be effective for other diseases, such as central nervous system (CNS) or cardiovascular diseases. However, genomes and biological pathways can be conserved across a variety of organisms. Furthermore, evolution over billions of years has produced these natural products to bind to specific regions in targets, and these binding regions can be very similar in targets for human disease as well as in microorganisms.

Because natural products often have the ability to cross biological barriers and penetrate cells, they often have desirable pharmacokinetic properties, which makes them good starting points for lead discovery. In fact, several structural neighbors of active natural products were shown to retain the same activity as the natural product.[26] One measure of the potential oral bioavailability of a compound is a set of guidelines called the *Rule of 5* (see Section 2.1.2.3.2). About 60% of the 126,140 natural products in the *Dictionary of Natural Products* had no violations of these guidelines, and many natural products remain bioavailable despite violating these rules.[27] This supports natural products as being an important source of lead compounds.

Frequently, screening of natural products has been done on semipurified extracts of sources such as plant materials, marine organisms, or fermentation broths. A significant challenge in screening of natural products in this way is that when activity is found, there is still considerable work to be done to isolate the active component and determine its structure. When HTS of chemical libraries started, such slower, more tedious screening methods were often put aside. However, because of the earlier success with natural product screening, the natural product approach has begun to return to the drug discovery process.

2.1.2.3.1.2. Medicinal Chemistry Collections and Other "Handcrafted" Compounds Many large, established pharmaceutical companies have been synthesizing compounds in one-at-a-time fashion for decades as part of their overall drug discovery efforts. In most cases, these institutions have had long-standing compound inventory management systems, such that samples of compounds prepared many years ago are still available for screening. One advantage of using these compounds for screening is that they are frequently close analogs of compounds that progressed substantially through the drug discovery process and thus have a reasonable probability of possessing biological activity and drug-like properties. One disadvantage, though, is that these compounds may be structurally biased toward the limited proteins that these companies have targeted over the years. Large companies may possess up to several million compounds in their corporate compound collections; however, most companies have substantially trimmed their collections used for screening, leaving only compounds that have good drug-like properties for lead discovery (see Section 2.1.2.3.2).

Another source of handcrafted compounds is samples from academic or nonpharmaceutical synthetic laboratories. Some businesses have been established to purchase such samples and market them to drug discovery organizations.

2.1.2.3.1.3. High-Throughput Organic Synthesis To provide the large number of compounds needed to feed ultrahigh throughput screening operations, enormous efforts during the 1990s turned toward developing methods for high-throughput organic synthesis (HTOS). HTOS had its origins in the techniques of *solid-phase synthesis* (synthesis carried out on a polymer support, which makes removal of excess reagents and by-products from the desired product easier), and many drug discovery organizations established internal HTOS groups to supply compounds for screening using solid-phase chemistry. Millions of compounds were synthesized for HTS campaigns using these HTOS methods. The synthesis of large numbers of related compounds has now declined substantially in favor of smaller sets,[28] and this evolution has been accompanied by a dramatic shift of emphasis from solid-phase methods back to solution-phase chemistry. One approach taken to create more diversity in chemical libraries called *diversity-oriented synthesis*, the synthesis of numerous diverse scaffolds from a common intermediate, has had limited success.[29] Below we briefly review key aspects of the HTOS approach of the 1990s and early 2000s and its relationship to HTS during these years, because some of the lessons learned during this period serve as key concepts in the present practices of lead discovery.

2.1.2.3.1.3.1. Solid-Phase Library Synthesis The most widely practiced methods in the early application of HTOS centered on the simultaneous synthesis of large collections (*libraries*) of compounds using solid-phase synthesis techniques. The synthesis of large numbers of compounds generally relied on a *combinatorial* strategy, that is, the practice of combining each member of one set of building blocks (i.e., reactants) with each member of one or more additional sets of building blocks (see examples below).[30] The beginnings of combinatorial chemistry are attributed to Furka,[31] with applications in peptide synthesis by Geysen and coworkers[32] and by Houghten.[33] These initial efforts in peptide library synthesis were followed by the synthesis of peptoids by Zuckermann and coworkers[34] and of small molecule nonpeptide libraries by Ellman and coworkers[35] and Terrett and coworkers.[36]

The efficiency of HTOS in producing large numbers of compounds relies, among other factors, on the ability to conduct reactions on multiple different (albeit often related) reactants in parallel. Solid-phase synthesis[37] is carried out by covalently attaching the starting material to a polymeric solid support and conducting a sequence of reactions while the corresponding intermediates and product remain attached to the solid phase, ultimately followed by a cleavage step to release the product into solution. Classically, functionalized polystyrene beads (polystyrene resin) were used as the solid support, although many other polymeric materials have since been developed expressly for the purpose of increasing the versatility of the solid-phase methodology. To minimize unreacted starting material, excess reagents are usually used, which are then easily removed along with any solution-phase by-products by filtration and repeated washing of the solid-phase material. This type of reaction workup is well suited to parallel processing and automation, accounting for its initial broad implementation for synthesis of large libraries. Somewhat less well advertised during the early hype of solid-phase combinatorial chemistry was the fact that side reactions can and do occur during solid-phase synthesis just as they do in solution, and the resulting polymer-bound side products are retained as impurities throughout the solid-phase process. Monitoring reactions on solid phase is not as straightforward as it is for solution-phase reactions; it requires either specialized methods such as Fourier transform infrared spectroscopy or separate cleavage of an aliquot of a polymer-bound intermediate to release it into solution so it can be analyzed by conventional methods such as thin-layer chromatography or high-performance liquid chromatography (HPLC). Nevertheless, since the early days of solid-phase peptide synthesis (the Merrifield synthesis[38]) carried out through sequential amide couplings and amine deprotections, a remarkably wide variety of reactions have been adapted to solid-phase methods.[39]

An early example of using solid-phase methodology to synthesize a nonpeptide library was the preparation of benzodiazepines as shown in Scheme 2.2.[40] Key reactions on solid phase include a Stille coupling to form ketone **2.18**, an amide coupling followed by an *N*-deprotection to form aminoketone **2.20** (note that by-products from Fmoc cleavage are soluble and thus readily removed), acid-promoted intramolecular imine formation to give polymer-bound benzodiazepine **2.21**, and an *N*-alkylation to form the polymer-bound version (**2.23**) of the final product. The *p*-alkoxybenzyl linker **2.16** serves two purposes: (1) the *p*-alkoxyl substituent promotes the release of the final product from the polymer under acid conditions and (2) it acts as a spacer, moving the sites of the reactions in the synthetic sequence away from the surface of the resin to avoid steric hindrance to reaction and to facilitate access to the reaction sites by reactants in solution. In this solid-phase synthesis, there are three *diversity elements* (R^1, R^2, and R^3), which are correspondingly introduced by three sets of building blocks (also known as *monomers*), namely, a set of acid chlorides **2.17**, a set of Fmoc-protected amino acids **2.19**, and a set of alkylating agents **2.22**. The theoretical

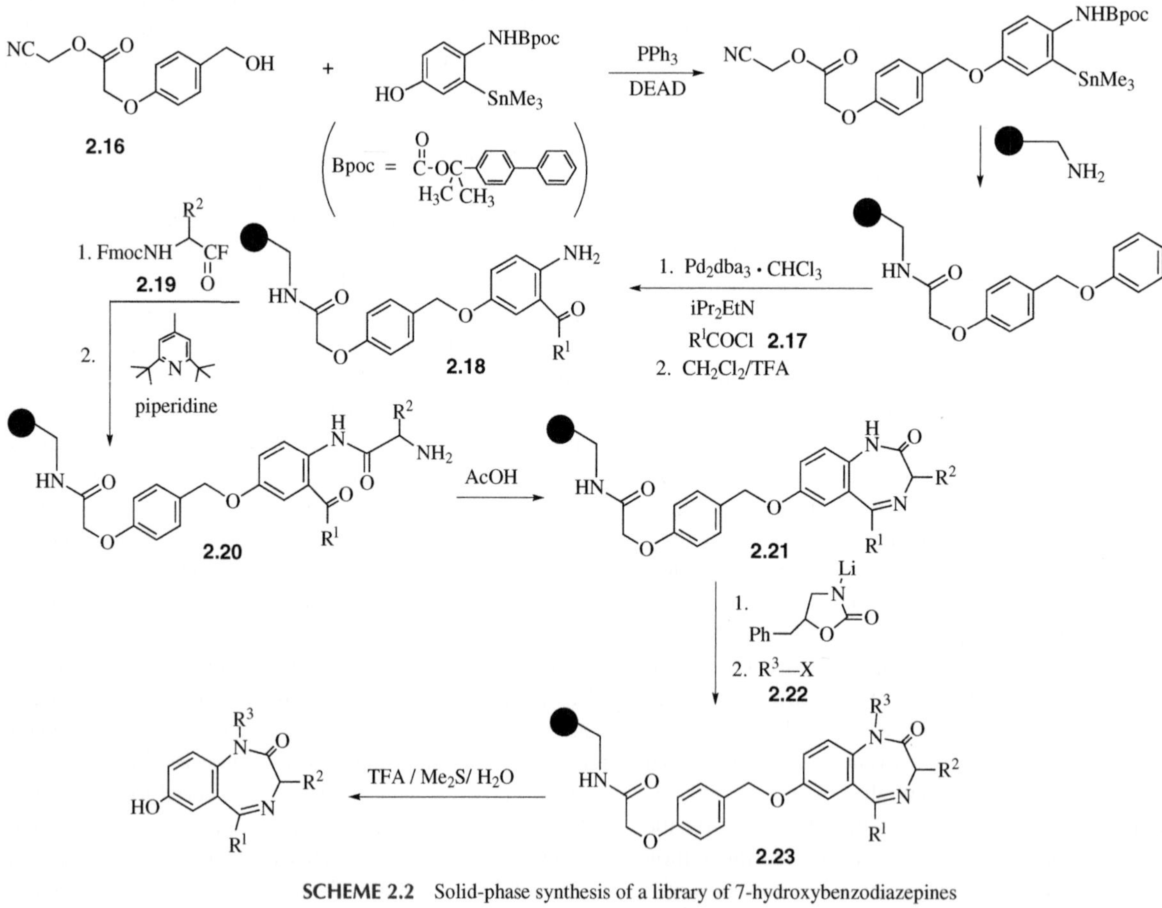

NC O O OH + HO NHBpoc SnMe$_3$ —PPh$_3$, DEAD→ NC O O O NHBpoc SnMe$_3$

2.16

(Bpoc = C(=O)-OC(H$_3$C)(CH$_3$)-C$_6$H$_4$-C$_6$H$_5$)

↓ (polymer)–NH$_2$

1. Pd$_2$dba$_3$ · CHCl$_3$, iPr$_2$EtN, R^{1}COCl **2.17**
2. CH$_2$Cl$_2$/TFA

2.18

1. FmocNH–CH(R^2)–C(=O)F **2.19**
2. (2,6-di-tert-butyl-4-methylpyridine), piperidine

2.20 —AcOH→ **2.21**

1. Li oxazolidinone (Ph)
2. R^3–X **2.22**

2.23 —TFA / Me$_2$S/ H$_2$O→ HO (7-hydroxybenzodiazepine, R^1, R^2, R^3)

SCHEME 2.2 Solid-phase synthesis of a library of 7-hydroxybenzodiazepines number of products equals the *product* of the number of each type of building block used; for example, 10 of each type of building block in Scheme 2.2 would theoretically afford 1000 (10×10×10) final products. Alternatively, 10 R^1 building blocks, 20 R^2 building blocks, and 50 R^3 building blocks would theoretically afford 10,000 products (10×20×50). This comparison underscores the combinatorial power of combinatorial chemistry (in the above examples, a total of 30 monomers (10+10+10) leads to 1000 different products, whereas adding only 50 monomers leads to an additional 9000 products!). It should be noted that all final products from Scheme 2.2 have a hydroxyl substituent on the benzo portion of the benzodiazepine; this is an artifact that was required for linkage to the solid phase via spacer **2.16**. Accordingly, the products of this work are technically a library of 7-hydroxybenzodiazepines.

The efficiencies inherent in conducting many reactions simultaneously in separate reaction vessels (termed *in parallel*[41]) on solid phase include efficient use of time, simplified workups (filtration and washing), and no need to perform chromatography, recrystallization, or distillation of intermediates (not because the intermediates are necessarily highly pure, but because these techniques are not applicable to polymer-bound intermediates). Since it is generally not practical to obtain and critically assess NMR spectra or elemental analysis data on so many final compounds, these steps are usually bypassed in favor of HPLC and MS as the sole methods for final compound analysis.

As an example, the chemistry in Scheme 2.3 was used to synthesize over 17,000 discrete compounds in parallel.[42] First, multiple Boc-4-alkoxyproline derivatives **2.24** were prepared in solution using a modified Williamson reaction at the 4-hydroxyl group, and the products were then coupled to polymer-bound phenolic hydroxyl groups to give polymer-bound activated esters **2.25**. A test for free phenolic hydroxyl groups on the polymer using $FeCl_3$/pyridine qualitatively showed that most of the free sites had been acylated, and the gain in resin weight was consistent with this conclusion. Acid-mediated cleavage of the Boc protecting group of **2.25** followed by functionalization of the resulting secondary amine with diverse reagents gave diverse resin-bound products **2.26**. In this library synthesis, the primary and secondary amines (**2.27**) that provide the final diversity element also cleave the products from the solid phase via reaction with the activated ester linkage to result in product amides **2.28** in solution. The final products need to be separated from the excess amine reactants. This can be accomplished by filtering the reaction mixtures through diatomaceous earth (Celite®) impregnated with aqueous acid, effectively sequestering the excess basic amines (**2.27**) onto the diatomaceous earth while the neutral library products (**2.28**) pass through with the filtrate. This procedure demonstrates the feasibility of performing solution phase workups in a parallel fashion, foreshadowing the ultimate emergence of *solution-phase parallel synthesis* as the dominant HTOS method (next section).

The foregoing library synthesis is an example of *parallel synthesis*. In contrast, a special variant of solid-phase combinatorial synthesis called *mix and split synthesis* (also known as *split and pool synthesis*) should be mentioned.[43] This technique is applicable to making very large libraries (10^4–10^6 compounds) as a collection of polymer beads, each containing, in principle, one library member, i.e., one bead, one compound. An important consideration is that for the one bead, one compound result to hold, each synthetic step must proceed reproducibly with very high conversion, even higher than in the synthesis of discrete compounds, to a single product.[44] Each bead carries only about 100–500 pmol of product, and special methods must be employed to determine which product is on a given bead. For simple compounds, mass spectrometric methods can be used,[45] but this is not applicable if the library

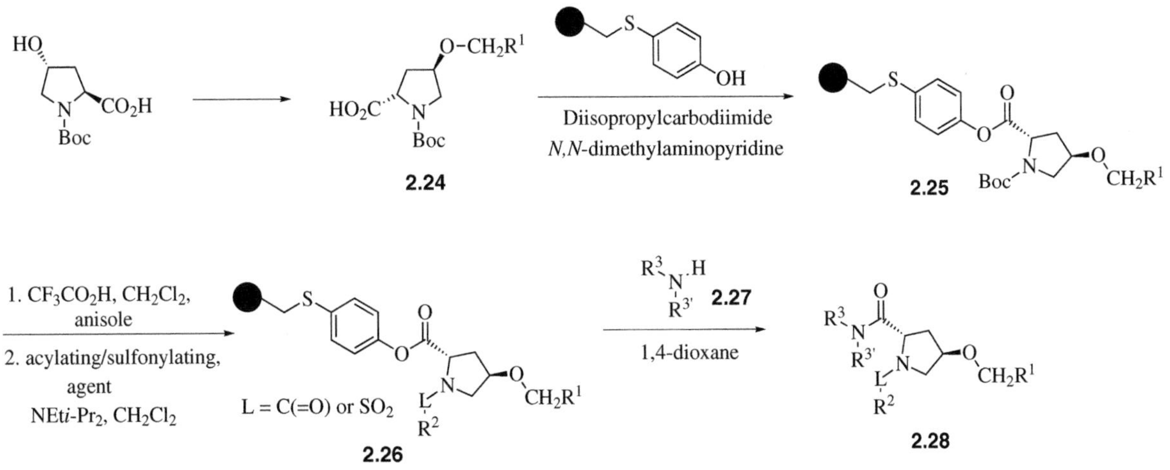

HO, N, Boc, CO_2H → **2.24** ($O-CH_2R^1$, HO_2C, N, Boc) → (Diisopropylcarbodiimide, *N,N*-dimethylaminopyridine) **2.25**

1. CF_3CO_2H, CH_2Cl_2, anisole; 2. acylating/sulfonylating, agent, NEt*i*-Pr_2, CH_2Cl_2 → **2.26** (L = C(=O) or SO_2) → ($R^3R^{3'}NH$ **2.27**, 1,4-dioxane) **2.28**

SCHEME 2.3 Solid-phase synthesis of a library of 4-alkoxyproline derivatives contains many thousands or millions of members that may not be pure or are isomeric with other library members. In that case, encoding methods need to be utilized. Although the structure of the actual compound might not be directly elucidated, the structure of certain tag molecules attached to the polymer that encode the structure can be determined.[46] One important approach that involves the attachment of unique arrays of readily analyzable, chemically inert, small molecule tags to each bead in a split synthesis was reported by Still and coworkers.[47] In this method, groups of tags are attached to a bead at each combinatorial step in a split synthesis, which create a record of the building blocks used in that step. At the end of the synthesis, the tags are removed and analyzed, which decodes the structure of the compound attached to that bead. Ideal encoding tags must survive organic synthesis conditions, not interfere with screening assays, be readily decoded without ambiguity, and encode large numbers of compounds; the test compound and the encoding tag must be able to be packed into a very small volume.

Although combinatorial chemistry was a common approach for about 15 years (from the late 1980s to the early 2000s), only one new de novo drug is believed to have resulted from this massive effort, namely, the antitumor drug sorafenib (**2.29**, Nexavar).[48] As will be discussed in more detail in Section 2.1.2.3.1.3.3, since about 2003–2005, solid-phase methods have been much less frequently used for HTOS than the solution-phase methods described in the next section.

Sorafenib
2.29

2.1.2.3.1.3.2. Solution-Phase Library Synthesis Parallel library synthesis of up to a few thousand compounds at a time can frequently be carried out entirely by solution-phase parallel methods[49]; Scheme 2.4 summarizes the methods used to prepare a several thousand-member library in solution phase.[50] This library is derived from D-glucose, so it could be characterized as being derived from a natural product. In the first step, the free hydroxyl group of diacetone D-glucose is alkylated with different alkyl halides to form a series of ethers varied at R^1. These intermediates are then selectively hydrolyzed (aq. HOAc) to the corresponding 1,2-diols, which are oxidatively cleaved with periodate to form aldehydes **2.30**. In this solution-phase library example, the subsequent reactions are run in parallel in microtiter plates (Figure 2.2), which facilitates convenient tracking of the individual reactions using plate positions in place of physical labels on reaction flasks. Thus, each aldehyde (**2.30**) is added to multiple wells of a microtiter plate and treated with different secondary amines under reductive amination conditions ($NaBH(OAc)_3$) to give aminomethyl derivatives **2.31**. Workup can be accomplished sequentially using two different *solid-phase scavenger resins* (a polymer-supported molecule that can react with excess reagents in solution, thereby removing them from solution), followed by filtration. Thus, after completion of the reductive amination reactions, the mixtures are first treated with Amberlite IRA743 resin to scavenge borate anion (derived from $NaBH(OAc)_3$). This scavenging agent contains polymer-bound *N*-methylglucosamine, which chelates with borate anion and is highly effective for removing borate from solution (Figure 2.3).[51] The Amberlite scavenger resin is removed by filtration using a 96-well filter plate (Figure 2.4; you can use an eight-channel pipettor to transfer contents of the microtiter plate eight wells at a time to the filter plate, which has various sorbents or filters, collecting the filtrate in another microtiter plate). The filtrates are treated with a polystyrene-bound isocyanate, which reacts with the excess secondary amine used in each

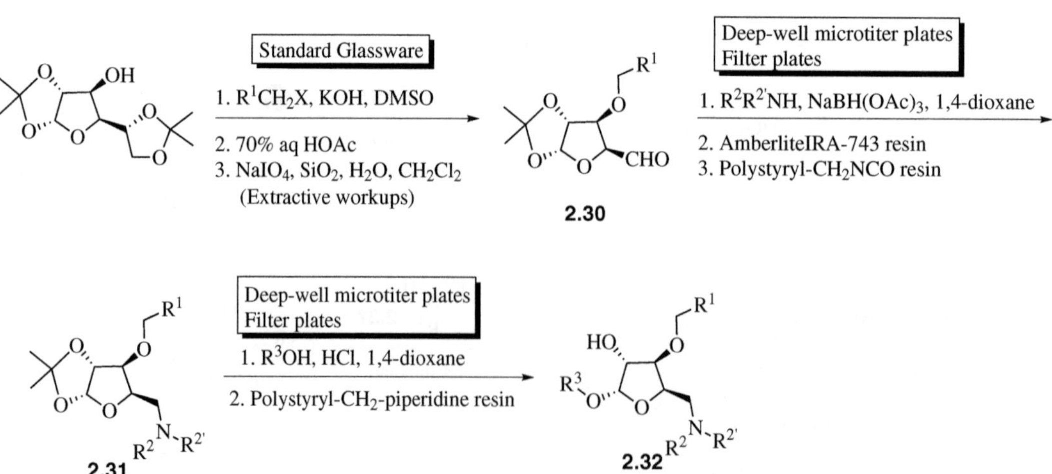

SCHEME 2.4 Solution-phase synthesis of a library of furanose derivatives

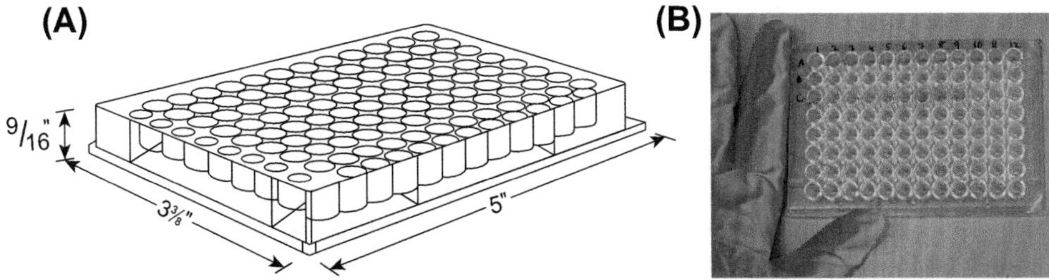

FIGURE 2.2 (A) Schematic of a typical 96-well microtiter plate. *(Reprinted with permission from Custom Biogenic Systems (http://www.biomedical-marketing.com/CBS/MicrotiterCRacks.html).)* (B) Picture of a 96-well microtiter plate taken by Jeffrey M. Vinocur, 4/21/06, published on Wikipedia Commons (http://commons.wikimedia.org/wiki/File:Microtiter_plate.JPG)

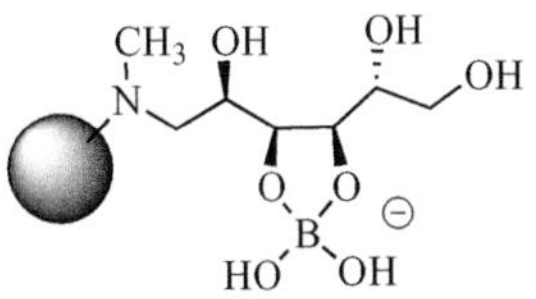

FIGURE 2.3 Product of polymer-bound *N*-methylglucosamine with borate anion

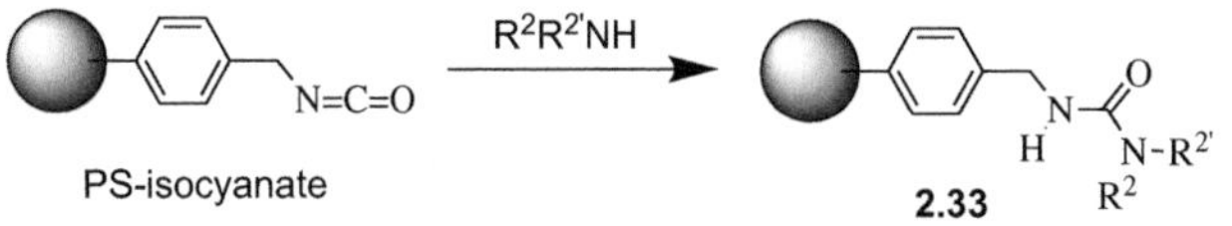

FIGURE 2.4 Image of 96-well filter plates. *Reprinted with permission from Norgen Biotek Corp.*

reaction, to form polymer-bound urea **2.33** (Scheme 2.5), effectively removing the amine from solution. The mixtures are again filtered (filter plate) to remove the polymeric scavenger. In the preceding step, 1,4-dioxane (freezing point 12 °C) is used as the reaction and rinse solvent. After the second filtration, the filtrates are frozen on Dry Ice, and the solvents are removed by sublimation under vacuum (called *lyophilization*). Introduction of a third point of diversity is effected by treatment of products **2.31** with an alcohol in the presence of hydrogen chloride to form hydroxyl ethers **2.32**, followed by evaporation of volatile components under vacuum. The resulting residues are dissolved in 1,4-dioxane/THF and treated with polystyrene-bound piperidine to remove residual HCl; omitting removal of residual HCl leads to poor stability of the products to storage and moisture. Finally, the products are frozen and lyophilized to afford library products as residues in the wells of the 96-well plates. These compounds often are then purified by reverse-phase liquid

PS-isocyanate $\xrightarrow{R^2R^{2'}NH}$ **2.33**

SCHEME 2.5 Use of a solid-phase scavenger in solution-phase synthesis. In this example, a polymer-bound isocyanate is used to scavenge excess primary or secondary amine from a solution by forming the corresponding polymer-bound urea.

chromatography. It is important to point out that for each step in the sequence, it is necessary to first evaluate a number of conditions to identify those conditions that give the highest purity of products across a number of representative building blocks. Therefore, although library production is rapid once the conditions are worked out, the myriad of process development trials must be factored in when assessing the overall efficiency gained by parallel synthesis.

Many of the techniques illustrated in the above example have gained considerable use in the parallel synthesis of smaller libraries as well, many of which may have only one or two points of diversity. Use of two points of diversity can reasonably support the synthesis of a library containing more than a 1000 compounds, for example, a 20 × 96 array (1920 compounds). When large libraries of analogs are needed, developmental work is often done in-house; then the library production can be outsourced to lower the cost of generating the library and to free up the time of the in-house chemist for new design and developmental studies.

2.1.2.3.1.3.3. Evolution of HTOS The use of solid-phase methods to synthesize large combinatorial libraries was in widespread practice during the 1990s and the early 2000s, but is currently not favored. Although obtaining large numbers of compounds for HTS was the initial driver for the technology, some investigators began to question whether the effort to collect and analyze HTS data on thousands, much less tens of thousands or millions, of compounds that are necessarily related by virtue of their common method of synthesis was truly an efficient use of resources. The structural diversity is limited in many cases not only by the fundamental chemistry used to prepare a library but also by the fact that diversity in commercially available building blocks did not always translate to a high level of diversity in the corresponding substituents of the final products. This is because the building blocks that were *successfully* incorporated into final products were more frequently those with simpler, less reactive functionality (like substituted phenyl compared to a heterocycle). Furthermore, the large numbers of compounds generated usually precluded individual purification and weighing of final products; therefore, the screening samples were usually of only approximate purity and concentration. Moreover, although the incorporation of three or more diversity elements in a library contributed greatly to combinatorial power and the number of compounds in the library, this also tended to yield compounds of molecular weight (MW) higher than that of most orally active drugs (see Section 2.1.2.3.2). Because of this observation, several groups began to define what properties a compound should possess to make it drug-like or lead-like. Among the several properties considered, MW less than about 500 Da and CLog*P* (a term related to lipophilicity of the compound; see Sections 2.2.5.2.2 and 2.2.5.2.3) less than 5 emerged as central criteria. Many of the libraries most amenable to large-scale synthesis by solid-phase combinatorial methods did not meet either of these criteria for a significant proportion of library members. For example, consider a library with a scaffold having a MW of 149 (see Scheme 2.4, **2.32**, where R^1CH_2, R^2, $R^{2'}$, R^3 all = H) and incorporating three diversity elements; the average contribution of the diversity elements to the MW of a given product must be <117 to keep the MW of the product molecule under 500.

Consequently, several significant changes to the common practice of HTOS began to evolve, including the synthesis of fewer compounds per library and the decision to purify final products, for example, by preparative reverse-phase HPLC. Once a final purification step was incorporated into the process, there developed a tendency to work on a larger scale to make up for mechanical purification losses. The prospect of obtaining a larger quantity of each purified product inspired a desire to store some of the material as dry solid, enabling more extensive follow-up studies in case interesting biological activity could be identified. It then became difficult for solid-phase synthesis to be applicable to these new objectives because the reaction scale is limited by the amount of solid support that could fit into reaction vessels of manageable size.

Although solid-phase methodology offers a strong advantage when the objective is to synthesize very large numbers of unpurified compounds in limited quantities and with a distinct tendency toward high MWs, the disadvantages of each of these characteristics led to the decline of its use in lead discovery. The synthesis of smaller libraries of compounds in larger quantities is usually well accommodated by parallel solution-phase chemistry, and its inherently greater flexibility with respect to scale, variety of reaction conditions accommodated, ability to analyze reaction mixtures, and option to purify intermediates made it the method of choice for high-throughput synthesis of lead discovery libraries. Moreover, solution-phase parallel synthesis using scavenger resins, disposable reaction vessels, specialized liquid transfer methods, automated purification, and other tools is applicable not only to the preparation of libraries for lead discovery but also to the downstream medicinal chemistry objectives, for example, during hit-to-lead (see Section 2.1.2.3.5) or lead modification activities (Section 2.2.).[52] In these latter contexts, it is most common to prepare libraries of only about 10–200 compounds.

2.1.2.3.2. Drug-Like, Lead-Like, and Other Desirable Properties of Compounds for Screening

As discussed in Chapter 1, lead compounds often require optimization with respect to not only their activity against a biological target but also a number of pharmacokinetic parameters, including ADME characteristics. If these properties could be predicted from the structure of a compound, then they could be taken into account at an early stage, even including the design and selection of compounds for a screening collection. Lipinski[53] proposed *the Rule of 5* as a guide to predict oral bioavailability. On the basis of a large database of known drugs, the Rule of 5 states that it is highly likely (>90% probability) that compounds with two or more of the following characteristics will have *poor* oral absorption and/or distribution properties:

- The MW is >500
- The log*P* is >5 (log*P* is a measure of the lipophilicity, discussed in Section 2.2.5.2.2); conveniently, the value can be predicted computationally, as described in Section 2.2.5.2.3.
- There are more than 5 H-bond donors (expressed as the sum of OH and NH groups)
- There are more than 10 H-bond acceptors (expressed as the sum of N and O atoms)

In 2006, it was determined that 885 (74%) of all small molecule drugs pass the Rule of 5; 159 of the orally administered small molecules fail at least one of the Rule of 5 parameters.[54]

Gleeson compared results of about 10 ADME assays with many compounds from GlaxoSmithKline and found that MW (<400), log *P* (<4), and ionization state are the most important molecular properties that affect ADME parameters.[55] To get a drug across the blood–brain barrier, the upper limits really should be 3 H-bond donors and 6 H-bond acceptors.[56] Some drugs, for example, certain antibiotics, antifungal drugs, vitamins, and cardiac glycosides, have active transporters to carry them across membranes, so lipophilicity is less relevant in those cases. Because active transporters allow molecules with poor physicochemical parameters to cross membranes readily, it is possible to design compounds with groups that are recognized by one of these transporters to aid in their bioavailability.[57] In the absence of a transporter, it is useful to understand what properties of a molecule promote good oral bioavailability (oral bioavailability is usually expressed as a percent; 100% bioavailable means that all the administered drug reached the systemic blood circulation).

In contrast to the Rule of 5, Veber and coworkers[58] measured the oral bioavailability of 1100 drug candidates and found that reduced molecular flexibility, as determined by the number of rotatable bonds (10 or fewer), and low *polar surface area* (PSA, the sum of surfaces of polar atoms, usually oxygens, nitrogens, and attached hydrogens, in a molecule) favored good oral bioavailability. The three-dimensional (3D)-PSA can be readily calculated and is referred to as the *topological polar surface area* (TPSA).[59] Veber and coworkers determined that a PSA ≤ 140 Å^2 (for intestinal absorption; ≤70 Å^2 to cross the blood–brain barrier[60]) or a total hydrogen bond count (≤ a total of 12 donors and acceptors) are important predictors of good oral bioavailability *independent of MW*. Both the number of rotatable bonds and hydrogen bond count tend to increase with MW, which may explain Lipinski's first rule. Lower PSA was found to correlate better with increased membrane permeation than did higher lipophilicity. The charge on molecules at physiological pH affects the PSA range that is important.[61] The fraction of anions with >10% F (F is the symbol for oral bioavailability) falls from 85% when the PSA is ≤75 Å^2 to 56% when 75 Å^2 < PSA < 150 Å^2. For neutral, zwitterionic, and cationic compounds that pass the Rule of 5, 55% have >10% *F*, but for those that fail the Rule of 5, only 17% have >10% *F*. A group at AstraZeneca found that two physicochemical properties unrelated to molecular size or lipophilicity, but related to molecular topology, namely, the fraction of the molecular framework (f_{MF}) and the fraction of sp^3-hybridized carbon atoms (Fsp^3) are important to ADME and toxicity.[62] The f_{MF} refers to the size of the molecule without side chains (the core ring structure) relative to its overall size (or the number of heavy atoms in the molecular framework divided by the total number of heavy atoms in the molecule)[63]; Fsp^3 is the number of sp^3-hybridized carbon atoms divided by the total number of carbon atoms.[64] Aqueous solubility, Caco-2 permeability, plasma protein binding, human ether à go-go-related gene (hERG; see Section 2.1.2.3.5) potassium channel inhibition, and cytochrome P450 (CYP3A4) inhibition are all influenced by molecular topology, some favorably and others unfavorably by increased f_{MF} and Fsp^3. Important considerations for assessing potential oral bioavailability of compounds were assembled in the form of a road map for oral bioavailability with emphasis on absorption (permeability and solubility) and metabolism properties.[65] Analogously, a group at Pfizer used six physicochemical parameters to construct a drug likeness algorithm for CNS drugs and applied it to marketed CNS drugs, CNS candidate compounds, and a diverse set of compounds.[66] This CNS multiparameter optimization algorithm showed that 74% of the marketed CNS drugs received a high score (≥4 out of 6). Of the compounds with a score >5, 91–96% displayed high passive permeability into the CNS, low efflux liability (ejection from the CNS), favorable metabolic stability, and high cellular viability.

Compounds that meet the Lipinski or Veber criteria are frequently referred to as *drug-like molecules*. However, the physicochemical properties of marketed orally administered drugs are generally more conservative than these rules allow compared to nonorally administered or nonmarketed drugs, e.g., lower MW, fewer H-bond donors and acceptors, and rotatable bonds.[67] Over the years, certain physicochemical properties of oral drugs change and others do not. Up through 2003 (the time frame of the Veber study), mean values of lipophilicity, PSA, and H-bond donor count were the same, which implies that they are the most important properties of oral drugs; however, MW, numbers of O and N atoms, H-bond acceptors, rotatable bonds, and number of rings increased between 1983 and 2002 (13–29%).[68] Fewer than 5% of marketed oral drugs have more than 4 H-bond donors; only 2% have a combination of MW > 500 and >3 H-bond donors. The balance between polar and nonpolar properties seems to be quite important for oral drugs.

Ajay and coworkers proposed that *drug-likeness* is a possible inherent property of some molecules,[69] and this property could determine which molecules should be selected for screening. They used a set of one-dimensional and two-dimensional (2D) parameters in their computation and were able to predict correctly over 90% of the compounds in the Comprehensive Medicinal Chemistry (CMC) database.[70] Another computational approach to differentiate drug-like and nondrug-like molecules using a scoring scheme was developed,[71] which was able to classify correctly 83% of the compounds in the Available Chemicals Directory (ACD)[72] and 77% of the compounds in the World Drug Index.[73] A variety of other approaches have been taken to identify drug-like molecules.[74]

It is now a common practice to bias screening collections in favor of drug-like molecules, particularly when the ultimate objective is development of orally bioavailable drugs.[75] Teague and coworkers[76] have taken the concept a step further to describe *lead-like molecules*. These authors note that during lead optimization, an increase in MW by up to 200 Da and increase of CLog*P* by up to 4 units frequently occur. Therefore, in order for an optimized compound to stay within, or close to, drug-like parameters, a lead compound should have a MW of 100–350 Da and a CLog*P* value of 1–3, and the authors propose that screening collections should be more heavily populated with compounds possessing these lead-like properties. As already noted, in the parallel synthesis of compounds for screening libraries, the more points of diversity, the greater the MW; therefore, there is always a balance between increasing diversity and MW.

Another approach to bias screening collections in favor of molecules likely to show biological activity is to consider *privileged structures*.[77] Evans and coworkers at Merck first introduced this term for certain molecular scaffolds that appear to be capable of binding to multiple receptor targets, and, consequently, with appropriate structure modifications, could exhibit multiple pharmacological activities.[78] This phenomenon was earlier mentioned by Ariëns and coworkers without referring to them as privileged structures.[79] The Merck group used benzodiazepines as a primary example of this phenomenon, because the benzodiazepine scaffold is found not only in antianxiety and anticonvulsant drugs that act through the γ-aminobutyric acid-activated ion channel but also in compounds that interact with opioid and cholecystokinin receptors. The latter two receptors are members of another major class of drug targets, the G-protein-coupled receptors (GPCRs; see Chapter 3, Section 3.1), which are quite distinct in their macromolecular structure from ion channels. Note that library synthesis around the benzodiazepine scaffold was the focus of Scheme 2.2; the privileged structure concept formed the basis for this scaffold. The commonality of molecular features in a variety of drugs is apparent by the revelation that only 32 scaffolds describe half of all known drugs.[80] In recognizing a molecule containing a privileged structure, it is important to note that the privileged components frequently consist of two or three rings linked by single bonds or by ring fusion, which constitute a substantial part of the overall size of the compound; otherwise, the contribution of the privileged structure to the activity of the compound would be questionable.[81] Additional examples of privileged structures include indoles, purines, dihydropyridines, spiropiperidines, benzimidazoles, benzofurans, and benzopyrans. Examples of indoles, dihydropyridines, and benzimidazoles that interact with diverse biological targets are shown in Figure 2.5.

Analogous to the small number of scaffolds found in a large number of drugs, there are a small number of moieties that account for a large majority of the side chains found in drugs.[82] The average number of side chains per molecule is four. If the carbonyl side chain is ignored, then 73% of the side chains in drugs are from the top 20 most common side chains. Accordingly, efforts to incorporate privileged scaffolds and privileged side chains are common considerations when identifying compounds to add to a screening collection.

An additional filter for many screening collections is to remove (or at least flag) compounds containing functional groups viewed as undesirable in a drug, usually because these groups have been found, or can be hypothesized, to have undesirable effects in vivo. These so-called *toxicophoric* groups can generally be classified into one of two different types: (1) those functional groups that may have undesirable effects by their own right and (2) those functional groups that can be converted by metabolic processes to moieties that may have undesirable effects[83]; representative examples of each type of toxicophore are shown in Tables 2.4 and 2.5, respectively. One approach to identify toxicophoric groups is illustrated in a study by Kazius et al.[84] The investigators took a *chemoinformatics* approach by computationally comparing the structures of over 4000 compounds, about half of which were mutagenic and half of which were nonmutagenic, and ascertaining which substructures were prominent in the mutagenic set. It should be noted, however, that the presence of a so-called toxicophoric group in a molecule does not imply that the substance is necessarily unsafe for human consumption. For example, an alkyl halide is frequently considered to be a toxicophoric group (because it is an electrophile), yet this has not prevented the FDA-approved human consumption of the popular artificial sweetener sucralose (**2.34**, Splenda).

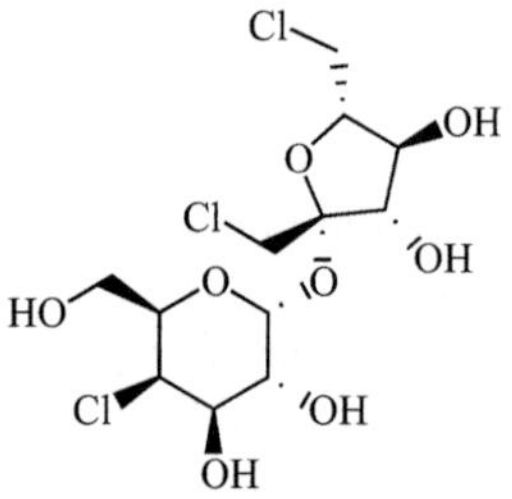

Cl O OH Cl O OH HO O Cl OH OH

Sucralose
2.34

A further approach to optimize a screening collection is to minimize the number of compounds that will ultimately prove to be *false positives* across many different high-throughput screens. False positives are compounds that appear to be *hits* (compounds that have a level of activity that the researcher believes is sufficient to pursue further), but upon additional investigation are found to be inactive against the target. Shoichet and coworkers[85] and others[86] have shown that a frequent source of false positives is the formation of colloidal aggregates of compounds in the screening mixture. Such aggregates frequently interact with targets in a nonspecific manner, and hence the component compounds have been characterized as *promiscuous binders*. The activity observed for such compounds may be counteracted by the addition of a detergent in the screening solution, which provides the basis for a straightforward method to identify this source of false positives at an early stage. Aggregate formation in the screening medium can be detected using an NMR assay.[87] Of course, another source of false positives is impurities in the samples, supporting the necessity to screen pure samples whenever practical.

Etodolac
Cyclooxygenase 2 inhbitor
Target class: enzyme

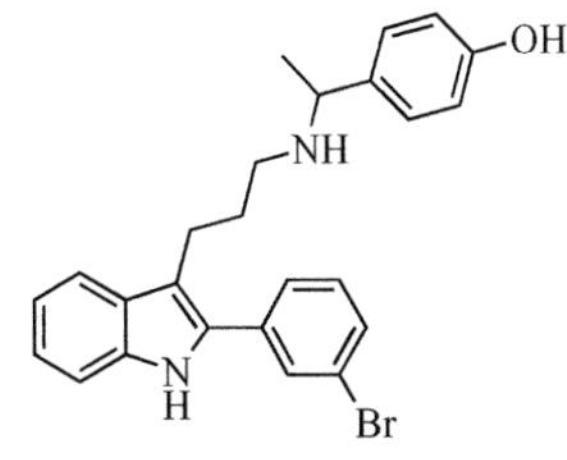

Lead compound (0.8 nM)
Neuropeptide Y5 ligand
Target class: GPCR

Felodipine
Calcium channel antagonist
Target class: ion channel

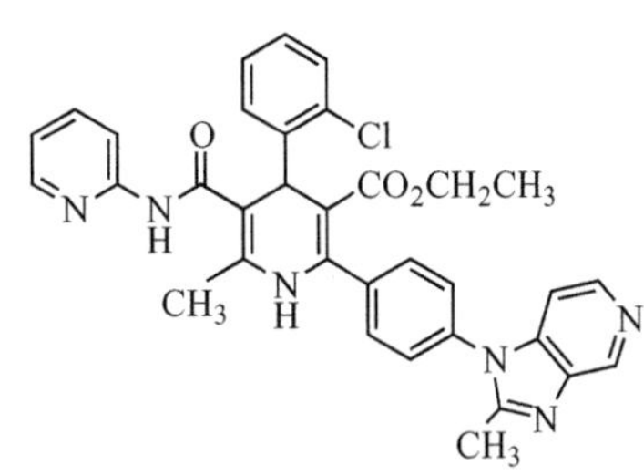

Optimized lead (4 nM)
Platelet-activating factor receptor antagonist
Target class: GPCR

BIBR 953 (4.5 nM)
Thrombin inhibitor
Target class: enzyme

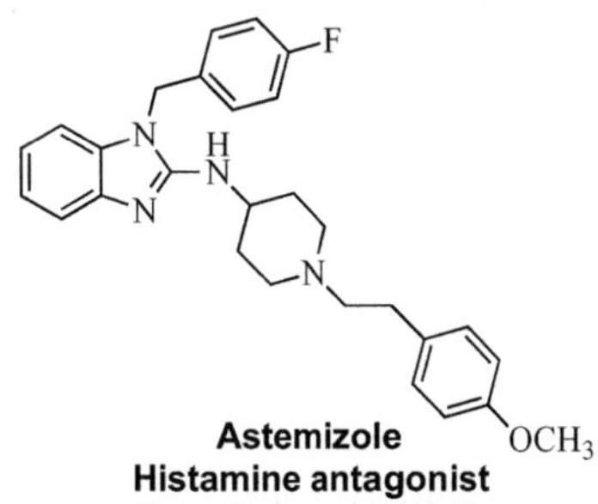

Astemizole
Histamine antagonist
Target class: GPCR

FIGURE 2.5 Pairs of compounds containing a privileged structure (indole, dihydropyridine, or benzimidazole) and binding to diverse target classes

TABLE 2.4 Representative Groups Viewed as Toxicophoric Because of the Reactivity

Toxicophoric Group	Rationale
EWG **EWG = electron withdrawing group, e.g., carbonyl, cyano, etc.**	Michael acceptor; electrophilic group that can alkylate biological nucleophiles, for example, cysteine -SH
R–epoxide (O)	Epoxide; electrophilic group that can alkylate biological nucleophiles
R–N imidazole (N)	Imidazole; can chelate metals, for example, iron in heme proteins such as cytochrome P450 enzymes

TABLE 2.5 Representative Groups Viewed as Toxicophoric Because They May be Metabolized to Undesirable Moieties

Toxicophoric Group	Rationale
 X = O or S; Metabolic activation	Furans and thiophenes; tend to be metabolized to electrophilic epoxides
 Metabolic activation	Thioamides and thioureas; tend to be metabolized to electrophilic imines
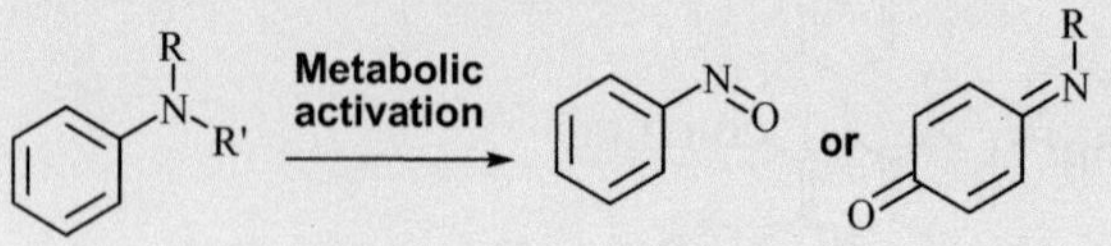 Metabolic activation; or	Anilines; tend to be metabolized to electrophilic nitroso or quinone derivatives

2.1.2.3.3. Random Screening

Given a high-throughput assay and access to an appropriate collection of compounds, how do you select which compounds to screen? In the absence of known drugs and other compounds with desired activity or structural information about the target, a random screen is the most common approach. *Random screening* in its simplest form involves no intellectualization; compounds are tested in the bioassay without regard to their structures. However, as discussed above, it is desirable to maximize lead-like and drug-like molecules in your random screening library.

Prior to 1935 (the discovery of sulfa drugs), this was essentially the only approach; today this method is still a very important approach to discover leads, particularly because it is now possible to screen such large numbers of compounds rapidly with (ultra) high-throughput screens.

An example of a random screen of synthetic and natural compounds is the "war on cancer" declared by Congress and the National Cancer Institute (NCI) in the early 1970s. Any new compound submitted was screened in a mouse tumor bioassay. Few new anticancer drugs resulted from that screen, but many known anticancer drugs also did not show activity in the screen used, so a new set of screens was devised, which gave more consistent results. In the 1940s and 1950s, a random screen of soil samples by various pharmaceutical companies in search of new antibiotics was undertaken. However, in this case, not only were numerous leads uncovered, but two important antibiotics, streptomycin and the tetracyclines, were found. Screening of microbial broths, particular strains of *Streptomyces*, was a common random screen methodology prior to 1980; it is now regaining importance in the search for new leads.

In recent years, attempts have been made to increase the efficiency of random screening by using computational methods to select a representative subset of compounds from a compound collection. This usually entails grouping (*clustering*) compounds that are structurally similar, and then choosing a few members from each cluster for screening. Methods for quantifying the similarity between molecules are discussed in the next section. If hits are identified in the initial screen, then further screening of other compounds that are structurally similar to the initial hits, a technique known as *hit-directed nearest neighbor screening,* is often productive for identification of additional hits.[88] This subsequent round of screening is a special case of targeted (or focused) screening, which is also discussed in the next section.

Another technique proposed to increase efficiency has been to screen mixtures of compounds, generated either as a result of the synthetic method[89] or by intentionally mixing pure compounds. However, as noted above (Section 2.1.2.3.2), many screening collections contain a significant number of compounds that tend to aggregate, leading to false positives. When mixtures of compounds are used, these aggregates can also mask the identification of compounds that are active when screened alone.[90] Therefore, the likelihood of a high rate of *false negatives* (an active compound that does not show activity) is also considerable when screening mixtures.

2.1.2.3.4. Targeted (or Focused) Screening, Virtual Screening, and Computational Methods in Lead Discovery

Information about one or more known ligands for a target or about the structure of the target itself may be used to narrow a large screening collection to a smaller set of compounds that may be more likely to hit the target, thereby saving screening resources. The screen is then regarded as *targeted or focused*, in contrast to the random approach discussed in the previous section. The most common computational method for selection of the compounds is called *virtual*

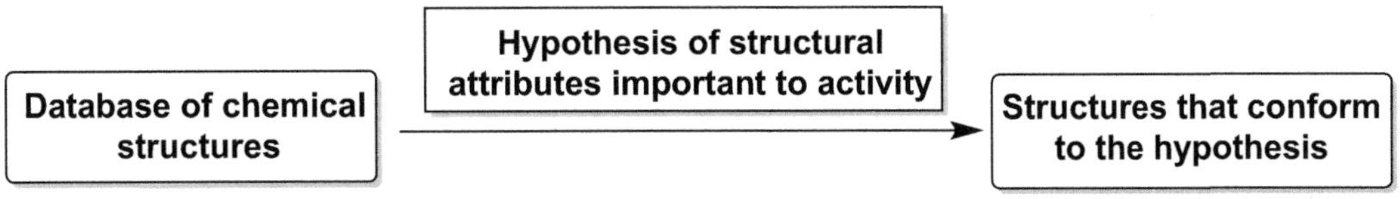

SCHEME 2.6 The process of virtual screening to identify compounds that conform to a hypothesis specifying properties (that are discernible from a compound's structure) that are required for activity

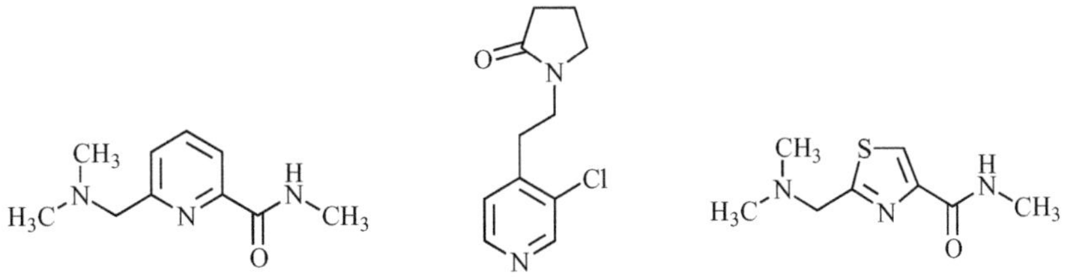

FIGURE 2.6 Hypothetical example illustrating that substructure search (e.g., using pyridine as the search query) may not retrieve the most structurally similar compounds in a compound collection

screening, which involves the rapid in silico (by computer) assessment of large libraries of chemical structures to identify those structures that most likely bind to a drug target, such as a protein receptor or enzyme.[91] The goal is to identify new scaffolds, especially ones that may be in the existing collection. In its most general form, virtual screening can be described by the process shown in Scheme 2.6.

Two components are needed: (1) a database of structures in a form that can be computationally analyzed for structural attributes and (2) a hypothesis or model of the structural attributes that are important for activity, for example, the hypothesis that structural similarities to a known active ligand should yield similarly active compounds or a hypothesis of the shape and charge density of a binding pocket that defines what features a complementary ligand structure should have (see discussions below).

2.1.2.3.4.1. Virtual Screening Database A key criterion for the structures that will be virtually screened is that physical samples of the compounds will be available if they are identified as compounds of interest by the virtual screen. This criterion would generally argue for the inclusion of compounds in an organization's corporate collection as well as compounds that are offered for sale commercially. Saario et al.[92] used two databases of structures in a virtual screen for fatty acid amide hydrolase inhibitors, one representing compounds in the LeadQuest collection offered commercially by Tripos (St Louis, MO, USA) and another screening collection offered commercially by Maybridge (Cornwall, England, UK). Databases that compile compounds from the catalogs of many vendors include the commercial Accelrys ACD with almost 4 million chemicals or the free "ZINC" database with almost 19 million commercially available compounds, 4 million lead-like compounds, and over 13 million drug-like compounds.[93] The virtual screening database might also contain other compounds that could be considered reasonably accessible. For example, compounds believed to be easily synthesizable might be included in a virtual screening database. Such compounds may range from those that have been previously synthesized in the organization, and for which detailed procedures are available, to members of combinatorial libraries for which general synthetic procedures have been reported in the literature.[94] Toward the latter set of compounds, the reviews by Dolle et al.[95] provide detailed lists of published combinatorial library syntheses and a rich source for generation of such virtual compounds. Among published library syntheses are many that target privileged structures, which should be of particular interest.

2.1.2.3.4.2. Virtual Screening Hypothesis Many methods have been developed to describe properties against which a compound might be assessed to estimate the likelihood that it will interact with a given target. For example, if a known ligand for the target exists, then searching a database of compounds for structures that are similar to the known ligand is a reasonable approach. Although this, in principle, could be accomplished by a seasoned medicinal chemist by visual inspection, to do so for many thousands of compounds is clearly impractical; moreover, computers can sometimes discern similarity features that the naked eye would miss. One simple and easily understood method is searching for other molecules that contain a *substructure* (part of the total structure) in common with the active molecule. To understand the shortcomings of this approach, assume that the structures in Figure 2.6 are three among thousands of compounds in a screening collection. A substructure search of the corresponding database using the structure of pyridine as the query would retrieve compounds **2.35** and **2.36**, but not **2.37**. Yet inspection of the structures might reasonably suggest that

2.35 and **2.37** would be more likely to share similar biological activity than **2.35** and **2.36**. Therefore, computational methods more sophisticated than the substructure approach have been developed. These methods can be generally categorized according to the following models:

1. 2D similarity models
2. 3D-QSAR models
3. Structure-based pharmacophore models and computational docking

Each of the above methods is discussed in this section. Companies such as Tripos (St Louis, MO), Accelrys (San Diego, CA), and Chemical Computing Group (Montreal, Quebec, Canada) have specialized in developing sophisticated computational chemistry software to assist in the use of such models.

Two-dimensional similarity models (2D because they mirror the similarity between flat structures, as drawn on paper) for assessing similarities between two molecules typically rely on defining a set of so-called 2D *descriptors* and then assessing how a given molecule conforms to each descriptor. Many types of descriptors have been developed and applied,[96] but simple examples include properties such as "contains an NC(O)O fragment", "contains a sulfur-containing heterocycle", or "contains a group IIIA element", as part of a set containing, say, 80–150 descriptors. Frequently, the descriptors are formulated in such a way that, for a given molecule, the assessment results are an answer of either "yes" or "no" or, in computer language, "1" or "0". Then, for a set of descriptors listed in a given order, a corresponding sequence of 1's and 0's can be generated that defines a *fingerprint* for that molecule. The concept is illustrated in Table 2.6, where the fingerprints are shown for two compounds (**1** and **2**) as defined by a set of 18 descriptors A through R (again, in most real-life cases, the number of descriptors used is considerably larger). The extent to which the two compounds share the same property is noted according to how often both molecules have a value of 1 for a given descriptor (gray areas). The *Tanimoto coefficient T* is a frequently used index to quantify similarity[97] and is defined as:

$$T = \frac{N_{11}}{n - N_{00}}$$

where N_{11} is the number of descriptors for which both values are 1, N_{00} is the number of descriptors for which both values are 0, and n is the total number of descriptors used. For the example in Table 2.6, $T = 7/(18-4) = 0.50$ (50% structurally similar). A computer can quickly determine 2D fingerprints for each structure in a database, and from these, quickly determine the level of similarity to a query molecule, for example, a known active ligand, to help select a set of compounds to be assayed in a real screen. This is a widely used similarity search method in the early stages of lead discovery when there are limited SAR and target structure data available. It allows the identification of a few actives that can be used in more sophisticated 3D virtual screening approaches, such as pharmacophore mapping and docking.[98] *Extended-connectivity fingerprints*, topological fingerprints designed to capture molecular features relevant to drug activity, were developed for substructure and similarity searching and are available in the commercial software called Pipeline Pilot (Accelrys, San Diego, CA, USA).[99]

It bears repeating that the assumption that compounds with similar structures are likely to have similar biological activity must be exercised with some caution. It has been shown that only 30% of compounds considered to be at least 85% structurally similar ($T \geq 0.85$) to an active compound will themselves have the same activity.[100] Adding just one methylene group to a 4-hydroxypiperidine analog changed it from a poor binder of the chemokine receptor CCR1 into a potent binder.[101] Nevertheless, given an active compound, the use of these methods to select additional active compounds from a data set is still far superior to random selection.

Three-dimensional quantitative structure–activity relationships (3D-QSARs) quantitative structure–activity relationship (QSAR) analysis is a method that permits correlations between different series of molecular structures and their biological function at a particular target. Various QSAR methods, which have served as valuable predictive tools for the design of drug candidates, have been developed over more than a 100 years. Classical 2D QSAR methods considered only 2D structures and are discussed later in this chapter (Section 2.2.4.2) as part of a historical overview of computational methods in lead modification.

TABLE 2.6 Illustration of Data Used to Calculate Tanimoto Similarity

Descriptor	A	B	C	D	E	F	G	H	I	J	K	L	M	N	O	P	Q	R
Compound 1	0	1	1	0	0	1	1	0	0	0	1	1	0	1	0	1	1	0
Compound 2	1	1	1	1	0	1	1	0	1	1	1	0	0	1	0	1	0	1

A set of descriptors is assigned a value of either 1 or 0, depending on whether that descriptor applies or does not apply, respectively, to the molecule. The string of 0s and 1s found for each molecule defines its descriptor-based fingerprint.

Three-dimensional QSAR was a natural extension of 2D-QSAR and was first proposed in the 1980s. The general approach of 3D-QSAR is to select a group of molecules, each of which has been assayed for a particular activity; align the 3D conformations of the molecules according to some predetermined orientation rules; calculate a set of spatially dependent parameters for each molecule determined in the receptor space surrounding the aligned series; derive a function that relates each molecule's spatial parameters to its respective biological property; and establish self-consistency and predictability of the derived function. There are a variety of computer-based methods that have been used to correlate molecular structure with receptor binding, and, therefore, activity. Some are mentioned here; others are cited in the General References at the end of the chapter.

Crippen and coworkers[102,103] devised a linear free energy model, termed the *distance geometry* approach, for calculating QSAR from receptor binding data. The distances between various atoms in the molecule, compiled into a table called the distance matrix, define the conformation of the molecule. Rotations about single bonds change the molecular conformation and, therefore, these distances; consequently, an upper and lower distance limit is set on each distance. Experimentally determined free energies of binding of a series of compounds to the receptor are used with the distance matrix of each molecule in a computerized method to deduce possible binding sites in terms of geometry and chemical character of the site, thereby defining a 3D pharmacophore. This approach requires considerably more computational effort and adjustable parameters, but it is thought to give good results on more difficult data sets.

The distance geometry approach was extended by Sheridan et al.[104] to treat two or more molecules as a single ensemble. The ensemble approach to distance geometry can be used to find a common pharmacophore for a receptor with unknown structure from a small set of biologically active molecules. A virtual screen of this type of model was used to identify inhibitors of human immunodeficiency virus type 1 integrase (HIV-1 IN) as potential anti-AIDS drugs.[105] HIV-1 IN mediates the integration of HIV-1 DNA into host chromosomal targets and is essential for effective viral replication. From a known inhibitor of HIV-1 IN, a pharmacophore hypothesis was proposed. On the basis of this hypothesis, a 3D search of the NCI database of compounds was performed, which produced 267 structures that matched the pharmacophore; 60 of these were tested against HIV-1 IN, and 19 were found to be active. The relevance of the proposed pharmacophore was tested using a small 3D validation database of known HIV-1 IN inhibitors, which had no overlap with the group of compounds found in the initial search. This new 3D search supported the existence of the postulated pharmacophore and also suggested a possible second pharmacophore. Using the second pharmacophore in another 3D search of the NCI database, 10 novel, structurally diverse, HIV-1 IN inhibitors were found.

Hopfinger[106] developed a set of computational procedures termed molecular shape analysis for the determination of the active conformations and, thereby, molecular shapes during receptor binding. Common pairwise overlap steric volumes calculated from low-energy conformations of molecules are used to obtain 3D molecular shape descriptors, which can be treated quantitatively and used with other physicochemical parameter descriptors.

Two other descriptors for substructure representation, the atom pair[107] and the topological torsion,[108] have been described by Venkataraghavan and coworkers. These descriptors characterize molecules in fundamental ways that are useful for the selection of potentially active compounds from hundreds of thousands of structures in a database. The atom pair method can select compounds from diverse structural classes that have atoms within the entire molecule similar to those of a particular active structure. The topological torsion descriptor is complementary to the atom pair descriptor, and focuses on a local environment of a molecule for comparison with active structures.

One of the most widely used computer-based 3D-QSAR methodologies, developed by Cramer and coworkers,[109] is termed *Comparative Molecular Field Analysis* (CoMFA).[110] In this method, the molecule–receptor interaction is represented by the steric and electrostatic fields exerted by each molecule. A series of active compounds are identified, and 3D structural models are constructed. These structures are superimposed on one another and placed within a regular 3D grid. A probe atom, with its own energetic values, is placed at lattice points on the grid, where it is used to calculate the steric and electrostatic potentials between itself and each of the superimposed structures. At each lattice point, one steric value and one electrostatic value are saved for each inhibitor in the series. The results are represented as a 3D contour map in which contours of various colors represent locations on the structure where lower or higher steric or electrostatic interactions would increase binding. However, because simple steric and electrostatic fields are unlikely to represent a complete description of a drug–receptor interaction, alternative and modified forms have been proposed.[111] Because it is assumed that the molecules bind with similar orientations in the receptor, which may not necessarily be the case, correct alignments are almost impossible, particularly for compounds with a large number of rotatable bonds, which limits the applicability of CoMFA. *Comparative Molecular Similarity Indices Analysis* (CoMSIA) is similar to CoMFA in the aspect of atom probing.[112] However, CoMSIA uses a different potential function; therefore, not only steric and electrostatic, but also hydrophobic, fields can be calculated. Different from CoMFA and CoMSIA, which are ligand-based approaches, *Comparative Binding Energy Analysis*

(COMBINE) takes advantage of structural data of ligand–receptor complexes and applies them to a 3D-QSAR paradigm.[113] This technique is based on the hypothesis that the free energy of binding can be correlated with a subset of energy components calculated from the structures of receptors and ligands in bound and unbound forms.

CoMFA, CoMSIA, and COMBINE require molecular alignment prior to the calculation of descriptors. If the structures of the macromolecules are known, the alignment can be guided by the binding conformations of receptor–ligand complexes (COMBINE is only useful when the protein structure is known). Otherwise, when CoMFA and CoMSIA are employed for 3D-QSAR analyses, purely computational alignment has to be postulated to superimpose all ligand structures in space. The 3D descriptors and their corresponding 3D-QSAR models, therefore, are related to molecular rotation and translation. In the past two decades, much effort has been made to develop 3D-QSAR models that are independent of subjective alignment rules. Several methods have been proposed, including *Comparative Molecular Moment Analysis* (CoMMA),[114] EVA,[115] Weighted Holistic Invariant Molecular (WHIM) descriptors,[116] and Grid-independent descriptors (GRIND).[117] CoMMA, EVA, and WHIM do not give an intuitively 3D display of the resulting models. In contrast to CoMMA, EVA, and WHIM, GRIND was devised to overcome the problem of interpretability that is common to alignment-independent descriptors.

Another popular 3D-QSAR method is an approach known as *topomer similarity searching*.[118] A *topomer* is a *molecular descriptor* (any property, measured or calculated, of a molecule, such as melting point or PSA) that focuses on the shape of a molecule, as represented by a combination of the shapes of different fragments of the molecule. This is a method to search 3D molecular structures in conventional structural databases and compare them as sets of fragments (or topomers) by superimposition of their fragmentation bonds, which allows comparison of the molecules by their pharmacophoric features. This method is an improvement over the 2D-QSAR similarity metric, Tanimoto coefficient[119] (see Section 2.1.2.3.4.2). CoMFA and topomer similarity technologies were merged by Cramer[120] into a 3D-QSAR methodology called *Topomer CoMFA*. In this approach, structures in a series are each broken into two or more fragments at central acyclic single bonds while removing core fragments that are common to the series. The method requires a common scaffold among the molecules in the series, but the commonality can be as simple as a key sp^3 carbon. Topomer 3D models are constructed for each fragment, and a set of steric and electrostatic fields is generated for each topomer set. The Topomer CoMFA results can be used to query virtual libraries already composed of topomer structures to identify fragment structures having increased potency. The advantages of this method are that it minimizes the preparation needed for 3D-QSAR analysis, automates the creation of models for predicting biological activity, which are created much quicker than traditional CoMFA, and is more user friendly than traditional CoMFA analysis. Other popular 3D methods that focus on shape or volume similarity between molecules include *Surflex-Sim and Flex-S*.[121] *Hologram QSAR* uses molecular holograms and partial least squares (PLS) analysis to generate a fragment-based SAR but does not require the alignment of molecules, which allows for automated analysis of very large data sets.[122]

A *pharmacophore* model is a 3D representation of the regions of ligands that are believed to be responsible for interactions with the biological target. An example[123] is shown in Figure 2.7. When such a model is derived from known ligands for the target, it is called a *ligand-based pharmacophore model* (in contrast to a *structure-based pharmacophore model,* which is based on knowledge of the receptor structure, see below). Computer software, such as Catalyst (Accelrys, Inc., San Diego, CA, USA), DISCO (Tripos, Inc., St Louis, MO, USA), LigandScout (Inteligand, Wien, Austria), Phase (Schrodinger, Portland, OR, USA), or MOE (Chemical Computing Group, Montreal, Canada), is used to generate one or more models, given the structures of a collection of known ligands, often including ligands with diverse structures. This technique is called *receptor mapping*.[124] It is founded on the premise that receptor topography is complementary to that of drugs, but in this case the structure of the lock is deduced from the shape of the keys that fit it. A variety of receptor mapping techniques have been described. An approach termed *steric mapping*[125] uses molecular graphics to combine the volumes of compounds known to bind to the target receptor. This composite volume generates a receptor-excluded volume map, which defines that region of the binding site available for binding by drug analogs and, therefore, not occupied by the receptor itself. The same procedure is, then, carried out for similar molecules that are inactive. The composite volume is inspected for regions of volume overlap common to all the inactive analogs. These are the receptor-essential regions, sites required by the receptor itself and unavailable for occupancy by ligands. Any other molecule that overlaps with these regions should be inactive. This approach has been termed an Active Analog Approach.[126] A pharmacophore-based virtual screen, then, would involve the identification of compounds possessing the appropriate pharmacophore that filled the receptor-excluded regions and that avoided the receptor-essential regions.

Some types of targets, specifically soluble enzymes such as kinases and proteases, are amenable to crystallization and hence to structure determination by X-ray crystallography. The structure of a protein that is similar to one for which the crystal (or NMR) structure has been determined can sometimes be deduced with a reasonable degree of

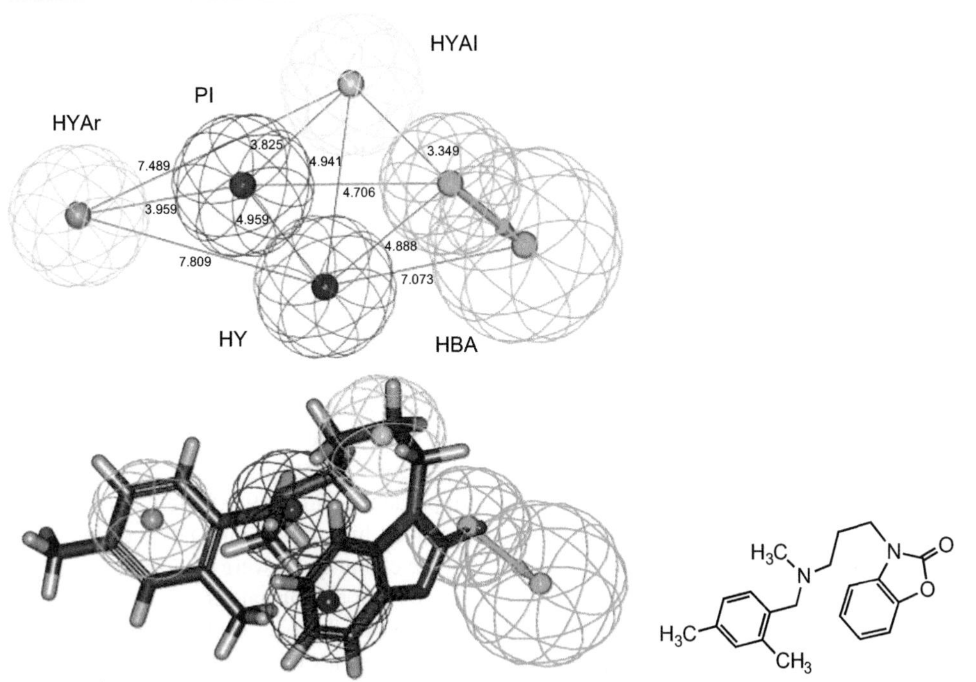

FIGURE 2.7 Example of a computer-generated pharmacophore model. *From Laurini et al Bioorg. Med. Chem. Lett. 2010 (Ref. 111)*

accuracy using a process known as *homology modeling*.[127] This technique involves the alignment of the amino acid sequence of the protein of unknown structure onto the corresponding positions in the experimentally determined structure (the template structure), followed by energy minimization. Naturally occurring homologous proteins have similar protein structure, and 3D protein structures are evolutionarily more conserved than expected because of sequence conservation.[128] The sequence alignment onto the template structure can be used to produce a structural model of the target. Membrane-associated proteins, such as GPCRs and ion channels, are much less amenable to crystallization, but steady progress has been made to experimentally determine the structures of these targets as well.[129] When the structure of a biological target, or preferably of the target complexed with a ligand, is available, then the information can be used to develop models for use in virtual screening. A model that has been derived based on an experimentally determined *target structure* is referred to as a *structure-based* model (see below).

When the X-ray crystallographic structure or the NMR solution structure of a target receptor is known, an analysis of the active site can be performed to facilitate drug design. Two popular approaches, GRID[130] and multiple copy simultaneous search (MCSS),[131] have been employed to identify the energetically favored sites in the active site for ligand binding. Goodford's program *GRID* uses a grid force field that includes a very good description of hydrogen bonding.[132] Because the energetics and shape complementarity of a drug–receptor complex are vital to its stability, this method simultaneously displays the energy contour surfaces and the macromolecular structure on the computer graphics system. This allows both the energy and shape to be considered together when considering the design of molecules that have an optimal fit in the receptor, and it determines probable interaction sites between various functional groups on the ligand and the enzyme surface. MCSS uses numerous small chemical group copies simultaneously, each transparent to the others (i.e., noninteracting) but each subjected to full force minimization in the receptor. This approach provides exhaustive information of the possible binding sites and orientations for small chemical groups in a known protein structure.

After an analysis of the active site, sophisticated computational programs such as DOCK,[133] AutoDock,[134] FlexX,[135] Glide,[136] GOLD,[137] Surflex,[138] and MolDock,[139] are capable of *docking* (inserting, on the computer, an unbound ligand into the binding site of the target) ligands into the biological target.[140] The ability for the software to independently dock a ligand in a way that corresponds closely to an experimentally determined structure for the same complex serves as validation of the docking method used. It is important to recognize, however, that the lowest energy structure of the ligand does not have to be the one that binds to the receptor; that is, the *bioactive conformation* can be a higher energy conformation of the molecule.[141] Currently available software programs are capable of carrying out virtual docking experiments across large numbers of compounds, including multiple conformations of each molecule, in a short period of time.

The algorithm DOCK, which was originally restricted to rigid ligands and receptors, was modified[142] for flexible ligands by representing the ligand as a small set of rigid fragments. This approach focuses on molecular shapes, and like most docking methods, DOCK ranks molecules based on polar, steric, hydrophobic, and solvation terms. Starting with a high-resolution structure (X-ray crystal structure or NMR spectral structure) of the receptor *with a bound ligand*, the ligand is removed from the binding site on the graphic display; then DOCK fills the binding site with sets of overlapping spheres, where a set of sphere centers serve as the negative image of the binding site. When a crystal structure of a receptor is available, but without a ligand bound, DOCK characterizes the entire surface of the receptor with regard to grooves that could potentially form target-binding sites, which are filled with the overlapping spheres. Next, DOCK matches structures of putative ligands to the image of the receptor on the basis of a comparison of internal distances and searches 3D databases of small molecules and ranks each candidate on the basis of the best orientations that can be found for a particular molecular conformation.[143] The drawbacks of this approach are the assumptions that binding is determined primarily by shape complementarity and that only small changes in the shape of the receptor occur upon ligand binding. An important advantage, though, is that this method is not limited to docking of known ligands. A library of molecular shapes can be scanned to determine which shapes best fit a particular receptor-binding site. In fact, DOCK was used to identify the antipsychotic drug haloperidol (**2.38**, Haldol)[144] and fullerenes[145] as potential inhibitors of HIV-1 protease.

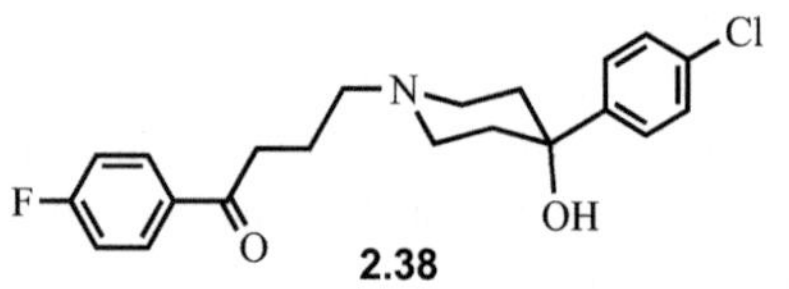

F O N Cl OH

2.38

An example of the application of DOCK to the identification of new leads for the ubiquitous GPCRs, which have been an important focus of the pharmaceutical industry for many years, came from Shoichet and coworkers.[146] Because few crystal structures of GPCRs are available, lead discovery efforts have largely been ligand based. Crystal structures of the β_2-adrenergic receptor with two partial inverse agonists (see Chapter 3, Section 3.2.3) bound[147] allowed a structure-based approach. About 1 million commercially available lead-like molecules were docked into this structure; the 25 top hits were tested, and 25% of them were active inverse agonists of this receptor. Impressively, one of them had a K_i of 9 nM, the most efficacious inverse agonist for the β_2-adrenergic receptor to that date. A crystal structure of this high-potency molecule bound to the β_2-adrenergic receptor[148] revealed the same overall fold observed for the previous crystal structures and exhibited the same binding conformation predicted by DOCK.

Given the wide variety of models and methods that are available for virtual screening, it is of both theoretical and practical interest to understand which ones are most effective. Somewhat surprisingly, systematic comparisons have frequently led to the conclusions that 2D similarity methods have similar effectiveness to 3D similarity methods, and that ligand-based pharmacophore models are frequently as effective as structure-based models.[149]

A comparison of hits obtained by HTS and by virtual screening of the same compound library against the protein cruzain, a target for Chagas disease, revealed the strengths and weaknesses of the two approaches and demonstrated the power of integrating the two.[150] Experiments by both approaches with a 198,000-member library led to 146 well-behaved hits, representing five different chemotypes. Two of the chemotypes were discovered through HTS alone, two came from the virtual screen, and one resulted from a combination of the two methods. Testing of these compounds gave potencies ranging from 65 nM to 6 μM. Integration of these two approaches can be very beneficial to identify and prioritize hits.

Another dramatically different computational approach for lead identification is *structure-based* de novo *design.* This approach is used to design, from scratch (i.e., de novo), a bioactive compound that does not exist in your known compound libraries. It is often applied when the 3D structure of the target protein or a specific set of pharmacophores is known. It therefore provides an opportunity to explore and utilize other areas of chemical space that have not been explored by your known compound libraries. *De novo* design approaches were first proposed in 1980s and can primarily be divided into structure-based approaches and ligand-based approaches. In the former case, the 3D structure of the receptor is known or can be modeled by homology modeling (vide supra), and the de novo design is based on the structural information of the target. In the latter case, the structure of the target is unknown, and the pharmacophore information of ligands is used to guide the design of new structures. Five different approaches have been developed for receptor-based de novo design depending on the method of structure sampling[151]: (1) planar structure fitting, (2) atom or fragment growing, (3) fragment linking, (4) target protein lattice-based sampling, (5) and molecular dynamics simulation-based sampling. Sophisticated computational programs for this approach include LUDI, LEGEND, and BOMB (Biochemical and Organic Model Builder).

The program *LUDI*[152] uses statistical analyses of nonbonded contacts in crystal packings of organic molecules to establish a set of rules that define the possible nonbonded contacts between proteins and ligands. Using these rules it also can search databases to find structures that fit a particular binding site in a protein based not on shape, as in DOCK, but on physicochemical properties, such as hydrogen bonding, ionic interactions, and hydrophobic interactions.

Some software programs grow molecules from atoms added into receptor structures. LEGEND[153] grows molecules by adding atoms one by one up to the specified molecular size using random numbers and force field energy calculations. BOMB,[154] another de novo ligand-growing lead discovery program, grows molecules by adding substituents to a core that is isolated or that has been placed in a binding site. BOMB has a library of about 700 possible substituents, including the most common heterocycles and substituted phenyl groups. The core may be as simple as ammonia or benzene or it may represent a polycyclic framework of a lead series. The user specifies a template, which includes the core, the topology, and the substituents, and all molecules corresponding to the template are grown. The template is generally selected because it conforms to the geometry of the target-binding site and because of synthetic ease. A thorough conformational search is performed for each molecule that is grown, and the dihedral angles for the conformers are optimized along with their position and orientation in the binding site. The resultant lowest energy conformer is evaluated with a docking-like scoring function to predict activity.

New developments in the field of de novo design have led to the generation of scaffold hopping (see Section 2.2.6.3) and fragment hopping. Different from structure-based de novo design, which aims to generate entire ligands, scaffold hopping is an attempt to replace only the core motif of a known ligand, while conserving key substituents.[155] This approach can lead to the identification of compounds that have similar biological activities, but totally different scaffolds. A pharmacophore-driven de novo design strategy for fragment-based drug discovery (see Section 2.1.2.3.6) is fragment hopping.[156] The core of this approach is the derivation of the minimal pharmacophoric elements for each pharmacophore. The minimal pharmacophoric element can be an atom, a cluster of atoms, a virtual graph, or vectors. The new fragments that match the requirements of the minimal pharmacophoric elements are generated and hopped onto the corresponding position in the active site. After linking the fragments, new inhibitors with novel scaffolds can be generated. Key features for both ligand-binding affinity and isozyme selectivity (when there are multiple isozymes of the target protein) can be included in the definition of minimal pharmacophoric elements, which leads to the generation of new inhibitors with diverse scaffolds and greater isozyme selectivity.

Although the interaction of a drug with multiple protein targets generally leads to side effects, many diseases, such as CNS diseases, infectious diseases, and cancer, involve multiple proteins. In these cases, it would be desirable to have a drug that can interfere with more than a single target. A computational method for the design of small molecules that bind to multiple desirable targets, in favor of proteins that could cause side effects, was developed; 800 ligand–target predictions were tested experimentally of which 75% were confirmed.[157]

Structure-based drug design has broader applications than just virtual screening for lead discovery and is discussed further in the context of lead modification in Section 2.2.6.

2.1.2.3.5. Hit-To-Lead Process

The *hit-to-lead* phase of the drug discovery process is the follow-up to HTS, where a *hit* is any compound that exhibits a level of activity that the researcher believes is worth pursuing further. Large-scale HTS campaigns generate enormous amounts of data that must be processed, analyzed, and ultimately acted upon if the program is to move forward. Because of this, certain activities need to be carried out to help avoid potential pitfalls and improve the chances that downstream efforts will ultimately result in a successful drug candidate, ideally within a reasonable time frame.[158] The main focus of such hit-to-lead efforts is not to identify the best compound, which normally takes place during the later lead optimization stage, but rather to provide data from the hits and related compounds that will support a decision to advance one or more series into the lead optimization stage. A central tenet of the hit-to-lead process is that identification of liabilities that are significant enough to disqualify a series of compounds for further work is of greatest value prior to lead optimization efforts. The precise activities undertaken during a hit-to-lead process may vary according to the organization carrying out the work. The following activities of a hit-to-lead phase are typical:

- *Confirmation of the structure, purity, and activity of the compound (hit confirmation).* Does the screening sample still contain the expected compound? Are there other compounds in the sample that might be responsible for the observed activity? It is useful to repeat the assay with a range of doses using freshly prepared solutions made from pure material that has been stored as a powder and for which purity and identity can be verified by NMR spectroscopy, MS, and HPLC. If dry pure sample is not available, it is often prudent to resynthesize or reisolate the compound.
- *Computational assessments.* Computational support can be applied to several aspects of hit-to-lead evaluations.[159] It is common to organize the hits into groups of similar compounds (*clusters*) in order to organize the information around structure classes. The finding that a number of structurally similar compounds possess similar biological activity lends credence to the data arising from any given hit within the cluster. By contrast, data for *singletons* (a compound that has no other similar structures among the hits) have no such substantiation, which increases the

need for gathering independent verifications of structure, purity, and activity at an early stage. Computation is also applied to calculating properties such as CLog *P* and PSA that are believed to correlate with drug-like properties and oral bioavailability (see Section 2.1.2.3.2). More sophisticated calculations that can be used to assess hits include predictions of solubility[160] and membrane permeability. Predicting these properties computationally can save time compared to determining them experimentally. Poor aqueous solubility affects results in biological assays and in absorption and distribution; the two most important descriptors to predict aqueous solubility are the aromatic proportion[161] of the molecule and the MW.[162]

- *Early ADME-tox assessments.* Measurement of the stability of hit compounds and close analogs by incubation with liver microsome preparations gives an early indication of the degree of metabolic stability in vivo. *In vitro* systems for measuring membrane permeability (e.g., with human epithelial (Caco-2) cells[163]) are also available and can provide an early indication of the likelihood that compounds will be absorbed from the gastrointestinal (GI) tract. Assessment of hits for inhibition of cytochrome P450 enzymes, important enzymes that metabolize drugs,[164] gives an early indication of potential drug–drug interactions.[165] Ways to reduce cytochrome P450 inhibition include lowering the lipophilicity of the molecule, adding steric hindrance, and adding an electron-withdrawing substituent (e.g., a halogen) to reduce the pK_a.[166] Assessment of hits for interactions with hERG[167] potassium ion channels gives an early indication of potential adverse cardiac toxicity; inhibition of the hERG channel is a major cause for compound attrition and withdrawal from the market.[168] While improving ADME-tox properties is frequently a major objective of the lead optimization phase, such early assessments can help in the prioritization of different series and further give an early indication of what parameters should be of concern during lead optimization.
- *Intellectual property assessments.* Patent searches are time consuming and thus difficult to conduct thoroughly on a large number of hits. Nevertheless, an early evaluation of whether the chemical space around a hit is very crowded or less so can be obtained by carrying out substructure searches across the Chemical Abstracts Registry File, noting how many of the retrieved publications are patent documents.
- *Early structure–activity relationship* (SAR) *assessments, synthetic accessibility, and ligand efficiency* (LE). Once the activity and identity of a hit have been verified, it is common practice to synthesize a number of close analogs of the hit for biological assessment. Such a set of analogs is frequently termed a *focused library* around the hit; the most efficient and desirable way to accomplish this is by parallel synthesis (see Sections 2.1.2.3.1.3.1 and 2.1.2.3.1.3.2). It is helpful to observe a range of biological activities among the analogs, which lends confidence to the prospect of eventually increasing potency through structure modifications. This process also helps to prioritize a series on the basis of synthetic accessibility since, other factors being equal, those series that are more easily synthesized can generally proceed more rapidly through the lead optimization process.

While the natural inclination is to place the highest value on the most potent compounds, the concept of LE offers an interesting alternative perspective, one that takes into account not only potency but also MW, which we found (Section 2.1.2.3.2) might be related to oral bioavailability.[169] LE is defined (Eqn (2.1)) as the binding energy per ligand atom:

$$\text{Ligand efficiency} = \Delta G/N \qquad (2.1)$$

where $\Delta G = -RT(\ln K_d)$, and N = the number of nonhydrogen atoms in the ligand. Thus, a small ligand with moderate potency could have a higher LE than a more potent, but significantly larger, molecule. Accordingly, LE is a way of normalizing potency at a target (pharmacodynamics) and molecular size (a contributor to pharmacokinetics), and is therefore useful for comparing compounds with a range of potencies and MWs. In refinements of the concept, the term ΔG in the above equation can be substituted by the pK_i or pIC_{50} (where IC_{50} is the concentration that gives 50% inhibition) and N may be replaced by terms for CLog *P*, MW, or PSA to normalize potency against these other parameters that are critical to drug-likeness.[170] The LE should remain relatively constant during optimization if the scaffold is preserved and optimal substitutions are incorporated into the lead. In comparing the properties of a set of drugs with the leads from which they were derived, in general, pK_i (drug) ≫ pK_i (lead), but the CLog *P* (drug) = CLog *P* (lead), resulting in LLE (drug) ≫ LLE (lead),[171] where the LLE[172] is the *ligand lipophilicity efficiency* = $pK_i - \text{CLog}\,P$. One of the keys to success in a lead optimization program is the maintenance of low levels of lipophilicity as the MW inevitably increases. The LLE links the potency and lipophilicity to estimate drug-likeness of compounds. However, the LLE does not include the LE term; therefore, an alternative term can be used that stresses the importance of lipophilicity and LE, called LELP (*ligand efficiency and log P*), which is $\log P/\text{LE}$.[173] The higher the LELP, the less drug-like is the lead. The accepted lower limit of LE for a lead is 0.3, and lead-like compounds have $-3 < \log P < 3$; therefore $-10 < \text{LELP} < 10$ is an acceptable LELP range for leads. In general, the closer the LELP is to zero in the positive range, the better. If a good hit or lead has an LE > 0.4 and $0 < \log P < 3$, then an LELP between 0 and 7.5 is an excellent range. With regard to their impact on ADME, safety properties, and binding thermodynamics, both LLE and LELP are helpful in identifying higher quality compounds; however, LLE is not as useful as LELP with fragment-based hits (see section below).[174]

2.1.2.3.6. Fragment-based Lead Discovery

Despite several successes,[175] HTS has not yet completely fulfilled the original expectations of bringing medicines to the market rapidly,[176] because HTS has some inherent fundamental limitations. First, a typical HTS campaign utilizes approximately 10^5–10^6 compounds, which is much less than the potential chemical diversity space, estimated to be about 10^{60} molecules containing ≤ 30 nonhydrogen atoms.[177] Second, corporate libraries are filled with compounds that have drug-like rather than lead-like properties, i.e., having relatively high MWs (on average, 400 Da) and high lipophilicity,[178] which limit lead optimization efforts. Finally, for many targets, suitable lead molecules will be absent from the compound collections or the HTS hit rate will be very low.

The awareness of concepts such as lead-likeness[179] and drug-likeness[180] and their importance in the construction of compound collections should yield improved success rates in HTS-based lead discovery efforts.[181] Hann et al.[182] showed that poor ligand–receptor interactions increase exponentially with the size and complexity of the ligand, suggesting that the probability of small, simple molecules binding to the receptor, although with low affinity, is much higher than HTS-sized compounds. Indeed, LE (see Section 2.1.2.3.5) calculations[183] of HTS hit compounds show that the average contribution to binding per atom can be rather modest, suggesting that small molecules might have greater potential as starting points for lead optimization.

Fragment-based lead discovery[184] involves the screening of low-MW building blocks (*fragments*), followed by the application of various methods to increase potency. Typically, the focus is on ligand efficiencies of fragments, rather than potency, when prioritizing hits for follow-up. The interactions with the individual fragments are often rather weak since the small molecular structure usually offers only a small number of points of contact with the target. A significant rationale for the fragment-based discovery approach is that interactions with a biological target might be identified in an isolated fragment, whereas such interactions might have been obscured if the same fragment were part of a larger molecule containing structural elements that interfere with binding to the target. The molecular mass of these fragments is typically in the range of 150–300, having less functionality; fragments are expected to be much less potent (millimolar to 30 μM potency range) than hits from HTS campaigns (30 μM to nanomolar potency range). Because of the poor binding affinities of fragments, standard assay methods generally cannot be used, as they are not sufficiently sensitive. Attempts made to utilize standard screening approaches with the fragments at high concentrations (millimolar rather than the typical micromolar concentrations commonly used for normal HTS) are usually unsuccessful. Therefore, one of the serious limitations of fragment-based methods is the requirement to implement sensitive biophysical techniques, such as NMR spectroscopy,[185] X-ray crystallography,[186] MS,[187] and surface plasmon resonance[188] to screen fragments because of their weak binding to the target.

Nonetheless, fragment-based screening offers a number of attractive features compared to HTS. First, the larger compounds typically found in HTS libraries are less able to adapt to a variety of binding sites; however, a high proportion of the atoms of a fragment directly interact with the receptor, which allows for optimal positioning in the binding site. Therefore, a hit fragment generally has a higher LE.[189] Second, the number of potential fragments with ≤12 nonhydrogen atoms (<160 Da) has been estimated to be about 14 million.[190] Therefore, the number of fragments that need to be screened is only in the range of hundreds to a few thousands, which still explores a much larger percentage of fragment chemical space (14 million) relative to the percentage of drug-like space (10^{60} compounds) that a million compounds screened in an HTS campaign explores. This also leads to much higher hit rates for fragment screens than HTS screens (in one report, 10–1000 times higher hit rates).[191] Furthermore, developing and maintaining a small set of fragments is easier than maintaining a large HTS library. Third, the subsequent structural optimization of a hit fragment has many more options and can result in a higher success rate for generating novel chemical structures. Finally, starting with a low molecular mass and low lipophilic fragment is likely to produce leads with small, simple structures, allowing for the typical molecular mass and, if necessary, lipophilicity (CLog *P*) increases during the lead optimization process.[192]

Intelligent construction of fragment screening collections is beneficial for more rapid lead discovery. One approach is to focus on fragments containing moieties that are frequently found in known drugs or other compounds that interact with proteins, since they have already passed toxicity and ADME studies.[193] In analogy to the Rule of 5 for drug-like molecules (Section 2.1.2.3.2), a *Rule of 3* (MW < 300 Da, CLog *P* ≤ 3, number of hydrogen bond donors and acceptors each ≤3, number of rotatable bonds ≤3, and PSA ≤ 60 Å^2) has been proposed as a guideline for the selection of fragments.[194] Such constraints should, in principle, enhance the probability that drug-like molecules will result after the fragments are linked. Virtual screening methods discussed earlier (Section

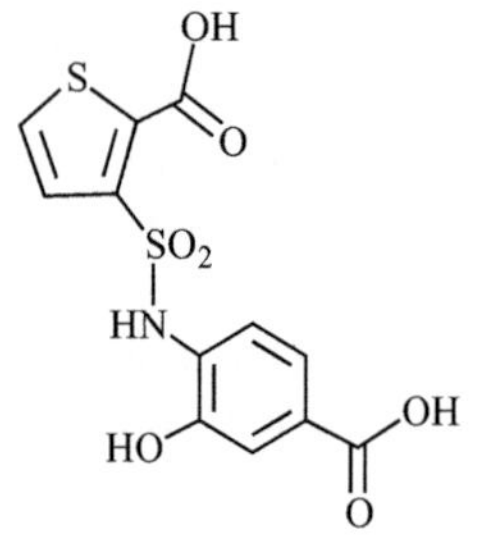

2.39 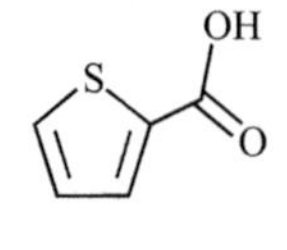 **2.40** 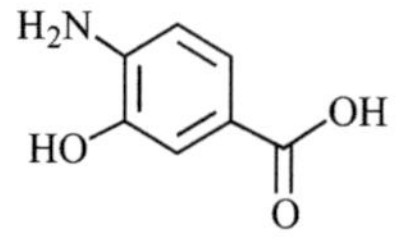 **2.41** 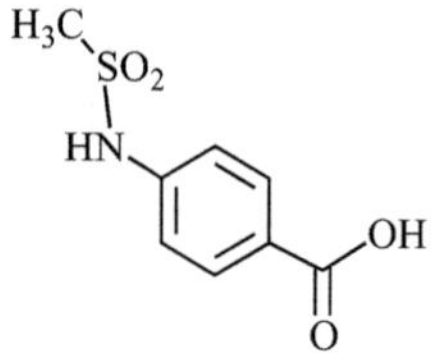 **2.42**

2.1.2.3.4) may be productively applied to computationally predicting fragments that are likely to interact with the target; indeed, because small fragments are likely to be less conformationally mobile than larger molecules, virtual screening has at least one less confounding factor when applied to fragments as opposed to larger, more flexible molecules.

A retrospective analysis of 18 different drug leads confirmed that fragments should not be larger than 20 nonhydrogen atoms or about 300 Da (for some targets, the upper limit was set to 250 Da). However, a lower limit to the MW also should be taken into account in a fragment library[195] because smaller, less complex fragments that only contain single rings with small substituents have a greater likelihood of binding in multiple orientations; therefore elaborated fragments may have different binding geometries than unelaborated fragments.[196] For example, the crystal structure of **2.39** bound to AmpC β-lactamase (K_i 1 μM) was compared to the crystal structures of fragments (**2.40** (K_i 40 mM), **2.41** (K_i 19 mM), and **2.42** K_i 10 mM)) derived from **2.39**. None of the fragments bound in the corresponding positions when they were part of **2.39**. In fact, they were in different orientations, and the fragments bound in two entirely different binding sites. It is normally assumed that the geometries of the parts of larger, more potent molecules from elaboration of fragments are the same as the fragments from which they were derived. However, by this converse experiment, deconstruction of molecules into fragments, it is apparent that small fragments can bind differently than those fragments bind when part of a more complex molecule, implying that there will be some potentially good inhibitors missed in molecules constructed from different fragments in a fragment-based approach. Because of the potential for different orientations of small fragments, a lower limit for fragment sizes of approximately 150 Da minimizes the chance that a fragment might bind in a different orientation in the target upon elaboration.[197] On the other hand, similar larger molecules seem to have a high degree of structural conservation to a binding site. A survey of the *Protein Data Bank* (PDB), which stores experimentally determined protein structures, showed that the binding orientation of a majority of structurally similar ligands in a protein is conserved, especially when the MWs are greater than 370 Da; however, binding site side-chain movements occur in half of the ligand pairs.[198] This supports the tenet in drug design that making small modifications to lead molecules will retain activity. For simple fragments, effective molecular recognition elements are important. Hydrophobic and electrostatic interactions are two important forces between ligands and proteins (discussed further in Chapter 3). Most structures in a generic fragment library, therefore, should include a hydrophobic group[199] and a strong hydrogen bonding or charged group.[200]

A comparison of fragment-based drug design (FBDD) approaches and HTS is given in Table 2.7.[201] The theory upon which FBDD is based can be tracked to Jencks in 1981,[202] who showed that binding efficiencies can be thought of as a combination of two or more moieties of the molecule; experimental evidence was provided by Nakamura and Abeles when they rationalized the potency of the first statin, mevastatin, as a combination of two "fragments" binding into separate, but adjacent, binding pockets.[203]

The actual exploitation of the method came in 1996 with a report by Fesik and coworkers at Abbott Laboratories of a new technique called *SAR by NMR*, an approach for screening fragments and elaborating them into a potent lead using NMR spectrometry.[204] The first step of the process (Figure 2.8) involves screening a library of small compounds, 10 at a time, by observing a ^{15}N-chemical shift in the heteronuclear single quantum coherence NMR spectrum for a specific amide nitrogen of the protein. Once a fragment is identified that causes a notable change in this chemical shift, a library of similar analogs is screened to identify compounds with optimal binding at that site. Then, with a saturating (excess) concentration of the first optimized ligand, a second library of compounds is screened to find a compound that binds at a nearby site, and then the second compound is optimized by screening a library of related compounds. On the basis of the NMR spectrum of the ternary complex of the protein and the two bound ligands, the location and orientation of each ligand is determined, and compounds are synthesized

TABLE 2.7 Comparison of Fragment-Based Approaches and High-throughput Screening

Fragment-Based Approaches	HTS
Emphasis on efficiency	Emphasis on potency
Screen a few hundred to a few 1000 compounds	Screen hundreds of thousands of compounds
MW range 120–250	MW range 250–600
Hit activity millimolar–30 μM	Hit activity 30 μM–nanomolar
High proportion of atoms in pharmacophore, i.e., high ligand efficiency	Hits contain groups that contribute poorly to binding or act as scaffold; low ligand efficiency
Biophysical screening techniques (NMR, X-ray, surface plasmon resonance) required because of weak binding	In vitro screening; often generates false positives and high attrition during validation
Protein structure-based information key to validation and prioritization of hits	Chemical (re)synthesis required for validation and prioritization of hits
Hit to lead usually requires synthesis of only a few compounds	Usually requires several iterations of high-throughput chemistry; protein structure can lower this
Design intensive	Resource intensive
Requires expertise and knowledge in protein-structure and protein–ligand interactions	HTS requires extensive infrastructure for storing and handling compound collections, screening, automation, data processing, and chemistry

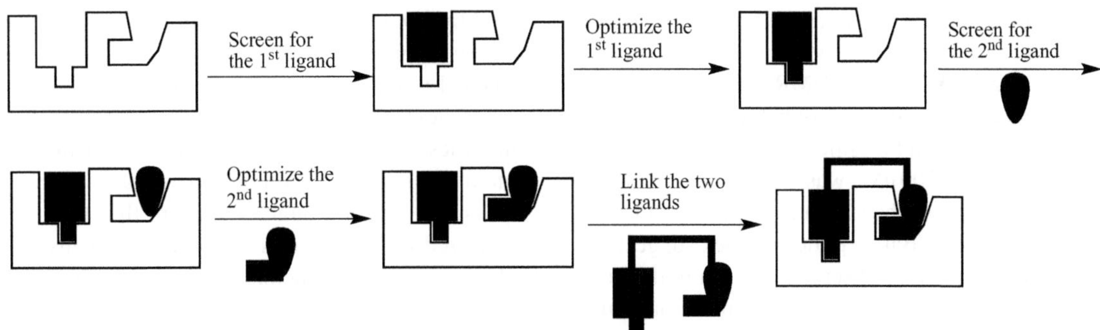

FIGURE 2.8 SAR by NMR methodology

in which the two ligands are covalently attached. When two low-affinity fragments are linked into a single molecule that effectively delivers each fragment to its respective site of interaction with the target, then a compound with *much* higher affinity results. This is because the free energy of binding becomes the sum of three free energies: those of the two ligands plus a free energy to reflect the effect of linking (note that the *sum* of the free energies of the two ligands translates to the *product* of their binding affinities!). The free energy from linking likely has numerous components that might individually result in either a positive or negative effect, but a positive entropic effect (reduced "randomness" of the individual fragments) is likely a major contributor. An example of this is the identification of the first potent inhibitor of the enzyme stromelysin, a *matrix metalloprotease* (a family of zinc-containing hydrolytic enzymes responsible for degradation of extracellular matrix components, such as collagen and proteoglycans, in normal tissue remodeling and in many disease states such as arthritis, osteoporosis, and cancer),[205] as a potential antitumor agent.[206] Matrix metalloproteases are generally inhibited by compounds that contain a hydroxamate moiety to bind to the zinc ion. A library of hydroxamates was screened, and acetohydroxamic acid (**2.43**) was identified with a K_d of 17 mM (generally regarded as exceedingly weak binding affinity). A focused screen of hydrophobic compounds was carried out in the presence of saturating amounts of acetohydroxamic acid, and biphenyl analogs were identified; optimization led to **2.44** with a K_d of 20 μM. From the NMR spectrum, the best site for a linker was expected to be between the methyl of acetohydroxamic acid and the hydroxyl group of **2.44**. Consequently, alkyl linkers of varying chain length were

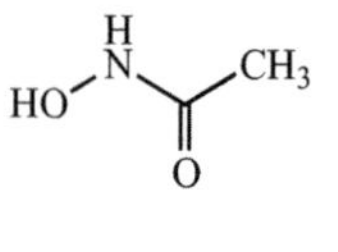

2.43 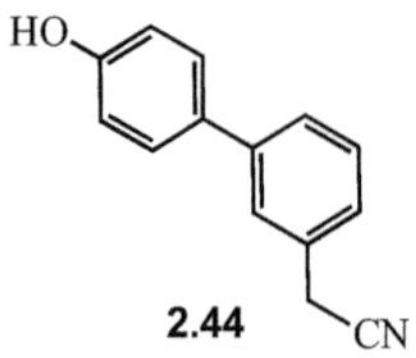 **2.44** 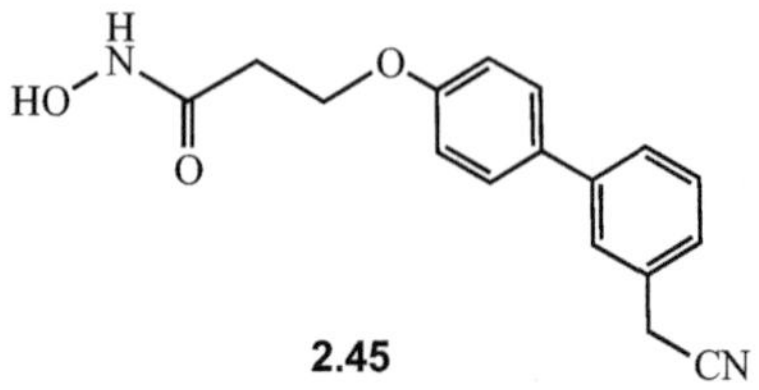 **2.45**

tried, and the best was a one-carbon linker, giving **2.45** having a K_d of 15 nM! The ΔG for **2.43** is −2.4 kcal/mol, for **2.44** is −4.8 kcal/mol, and for the linker is −2.6 kcal/mol; the total, therefore, is −9.8 kcal/mol. It took about six months to identify this inhibitor; prior to this study, 115,000 compounds had been screened with no leads.

The first compound in clinical trials derived from the SAR by NMR method is navitoclax (**2.46**), an anticancer drug that inhibits the protein Bcl-x_L, an antiapoptotic *B*-cell lymphoma protein.[207] Normal cellular homeostasis is regulated by expression of antiapoptotic proteins, such as Bcl-x_L, Bcl-2, and Bcl-w and proapoptotic proteins, such as Bak, Bax, and Bad.[208] For some cancers, *apoptosis* (programmed cell death) is circumvented by overexpression of the antiapoptotic proteins Bcl-2 or Bcl-x_L, which makes them targets for the development of new anticancer drugs.[209] There is a hydrophobic groove on the surface of these proteins to which the proapoptotic proteins bind, and the 3D structure of this binding region was determined by NMR spectrometry.[210] SAR by NMR was used to identify inhibitors that bind in the hydrophobic groove of Bcl-x_L.[211] A 10,000-compound fragment library was screened, and **2.47**, with a K_i of 300 μM, was identified as a ligand for Bcl-x_L. Comparison of the structure of this ligand complex to that of the Bcl-x_L/Bak peptide complex suggested a proximal second site. A second screen was run in the presence of an excess of **2.47** using a 3500-fragment compound library, which identified **2.48**, K_i 6 mM. A variety of possible linkers were assessed, and a *trans*-olefin was deemed best, giving **2.48a** with a K_i of 1.4 μM. Further synthetic manipulation resulted in **2.49**, K_i 36 nM. It was found that **2.49** was too hydrophobic, making it poorly aqueous soluble and tightly bound to serum albumin. Consequently, the polarity of **2.49** was increased for improved pharmacokinetics, leading to the antilymphoma drug **2.46**. Note that, despite the relatively good pharmacokinetic properties of **2.46**, its molecular mass (974 Da) far exceeds the Rule of 5 maximum of 500 Da for good oral bioavailability. In Phase II clinical trials it was found that **2.46** caused thrombocytopenia (low platelet count), and it was terminated. The cause for the platelet loss was found to be inhibition of Bcl-x_L; this led the Abbott group to modify **2.46** in search of a selective Bcl-2 inhibitor. With the aid of cocrystal structures of small molecules in Bcl-2, the first-in-class Bcl-2-selective inhibitor, ABT-199 (**2.50**) was developed, which showed potent antitumor activity (chronic lymphocytic leukemia) without platelet loss.[212] Several other drugs discovered from fragment-based approaches, rather than high-throughput screens, are reaching clinical trials.[213]

Sounds simple, doesn't it? But let's think about what is involved in carrying out SAR by NMR. The method requires screening compounds and observing a specific ^{15}N-amide chemical shift for binding. Where did the ^{15}N come from? This had to be incorporated into the protein because natural abundance ^{15}N is not sufficiently high to detect. To incorporate ^{15}N, it is necessary to be able to express the protein in a microorganism, and then grow the microorganism on $^{15}NH_4Cl$ as its sole nitrogen source. This gives the protein with all ^{15}N-containing amino acids. To perform the NMR experiments, large amounts of soluble (>100 μM) protein (>200 mg per spectrum) are needed; therefore, an efficient overexpression system for the protein is needed. Then the protein has to be purified, and its complete structure determined by 3D and 4D NMR techniques, so that the position of every amino acid residue in the protein is known (which is needed to determine when the two ligands are bound in nearby sites). This means that the protein target should have a mass less than about 40 kDa (the current limit for rapid protein NMR spectra, although spectra of larger proteins is possible[214]). Although it appears that this is a highly specialized technique, it is used widely because molecular biology and protein chemistry techniques have been well developed, making overexpression of proteins in microorganisms and their purification routine.[215] NMR instrumentation and methods also have made structure determination plausible. If the structure can be determined, SAR by NMR provides a technique to screen, by automation, about 1000 compounds a day and identify, relatively rapidly, potent protein binders.[216] Integration of a medicinal chemist's input into computational methods can accelerate fragment-based lead discovery.[217]

Ellman and coworkers have developed a combinatorial lead optimization approach using the basic principles described above for SAR by NMR, except without the use of NMR spectrometry and without the need for any structural or mechanistic information about the target protein![218] First, a diverse library of compounds is synthesized in which each molecule incorporates a common chemical linkage group (Figure 2.9). Next, the library is screened to identify any member that shows even weak binding to the target. Third, a new library is constructed containing all combinations of any two of the active compounds linked to

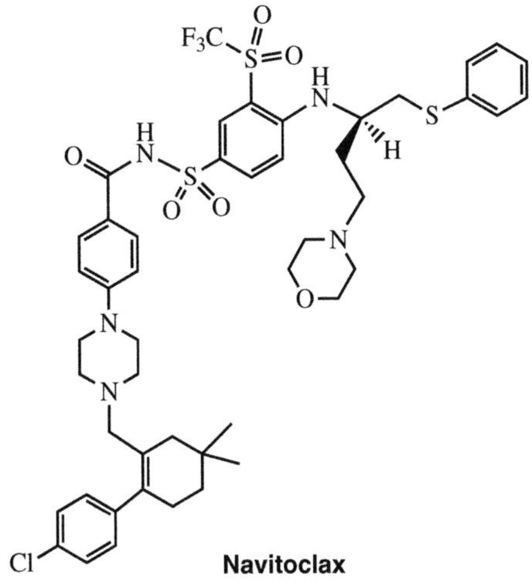

Navitoclax
2.46

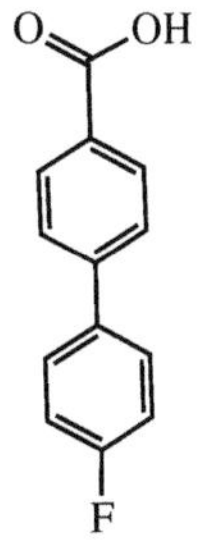

2.47 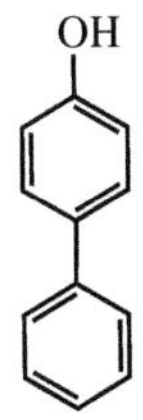 **2.48** 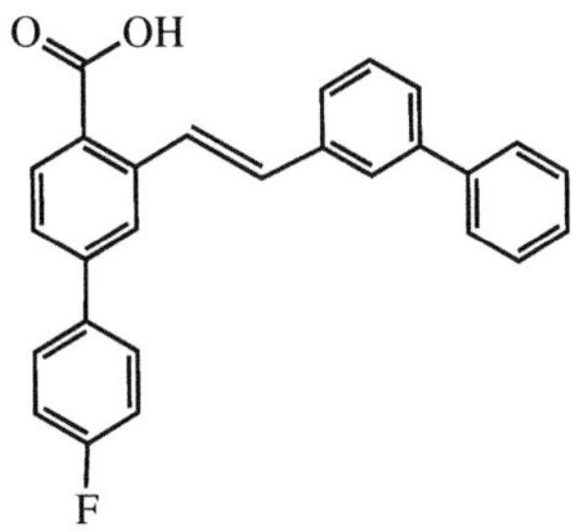 **2.48a** 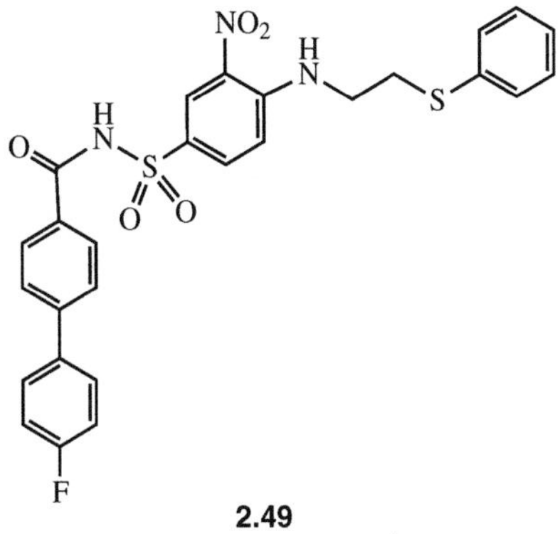 **2.49**

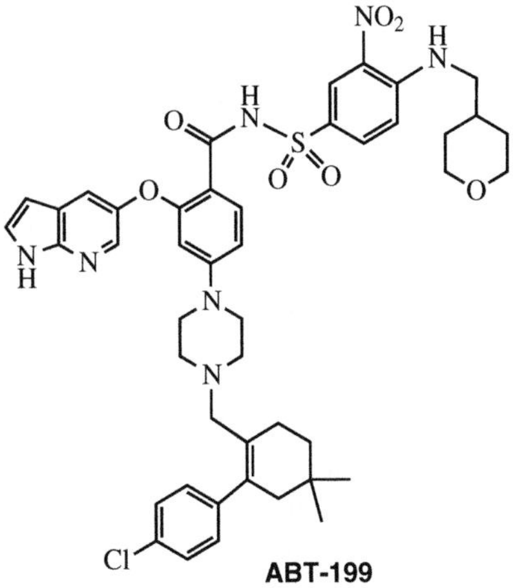

ABT-199
2.50

each other by the common chemical linkage group through a set of flexible linkers. Then this combinatorial library is screened to identify the most potent analog. The method depends on two analogs binding in nearby sites (although it is not known which two will bind or where the sites are) and finding the appropriate linker size combinatorially so the linked active compounds take advantage of the additive free energy gain of the three elements, the two compounds and the linker. This approach was used to identify a potent (IC_{50} 64 nM) and selective inhibitor of one type of tyrosine kinase.

A complementary method to SAR by NMR is *SAR* by *MS*.[219] This is a high-throughput MS-based screen that quantifies the binding affinity, stoichiometry, and specificity over a wide range of ligand-binding energies. A set of diverse compounds is screened by MS to identify those that bind to the receptor. Competition experiments are used to identify the ones that bind to the same site and those that do

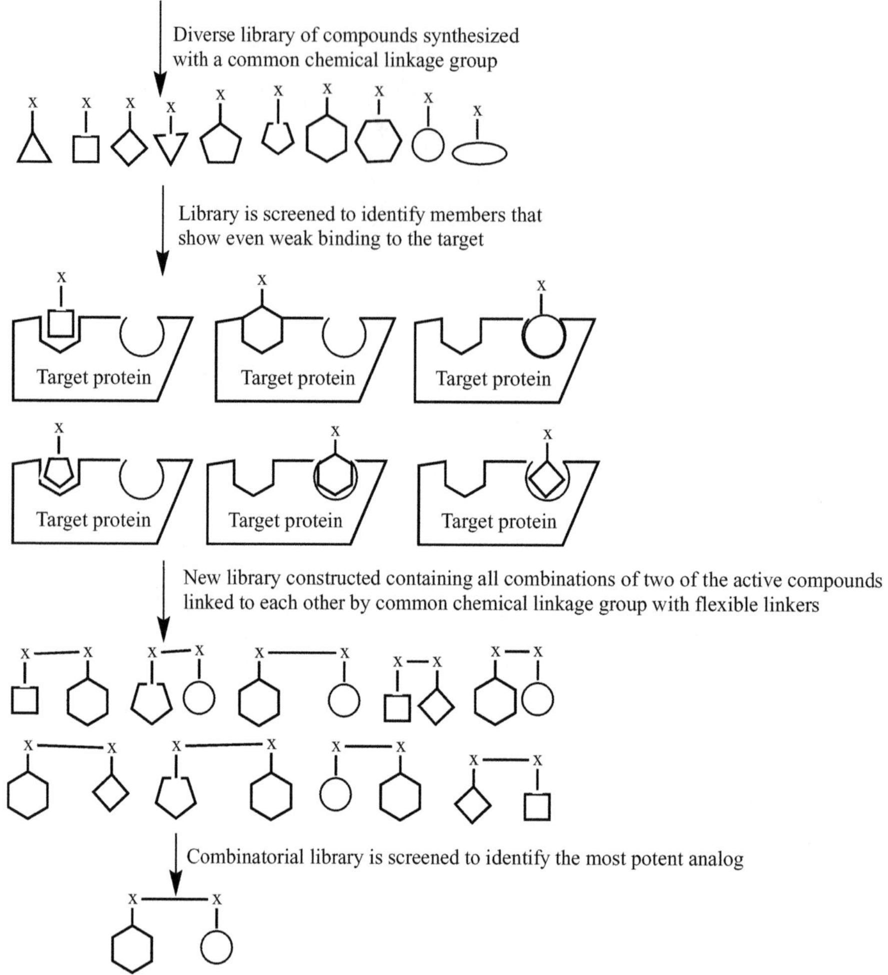

FIGURE 2.9 Ellman combinatorial methodology for lead generation with an unknown or impure protein

not. If two compounds bind at different binding sites, then a ternary complex of the two molecules plus the receptor is detected in the mass spectrum. If the two compounds bind at the same site, the tighter binding molecule displaces the other from the binding site, and only a binary complex is detected. By varying the substituent size on various classes of compounds and rescreening, it is possible to identify those molecules that bind at nearby sites as the ones that become competitive once a larger substituent is appended to one of the molecules. Once adjacent binding sites are realized, then the same methodology as for SAR by NMR, namely, attaching the two or more molecules to each other with linkers, can be employed. This approach was applied to the development of a new class of small molecules with high affinity for the hepatitis C virus-internal ribosome entry site IIA subdomain, which mediates initiation of viral-ribonucleic acid (RNA) translation.[220] MS of the company's compound collection (180,000 compounds) led to the identification of a benzimidazole analog with activity, which was optimized to submicromolar binding affinity for the IIA RNA construct using SAR by MS. The optimized benzimidazoles reduced viral RNA in a cellular replicon assay at concentrations comparable to the binding constants observed in the MS assay.

Ellman and coworkers developed a substrate-based fragment identification method for protease inhibitors, called *substrate activity screening* (SAS).[221] This method addresses two key challenges in fragment-based screening: (1) the efficient identification of weak binding fragments and (2) the rapid optimization of the initial weak binding fragments into high-affinity compounds. SAS has three steps (Figure 2.10): (1) a library of substrates consisting of the substrate-catalytic functionalities, in this case, the amide of the acylaminocoumarin and diverse, low-MW fragments, in this case, the R groups, is screened using a single-step, high-throughput fluorescence-based substrate assay; (2) the activity of the substrate is optimized by rapid analog synthesis and evaluation; (3) the optimized substrates are converted to inhibitors by replacement of the substrate-catalytic functionality with inhibitor pharmacophores, which match the catalytic residues in the active site (in this case, the aminocoumarin

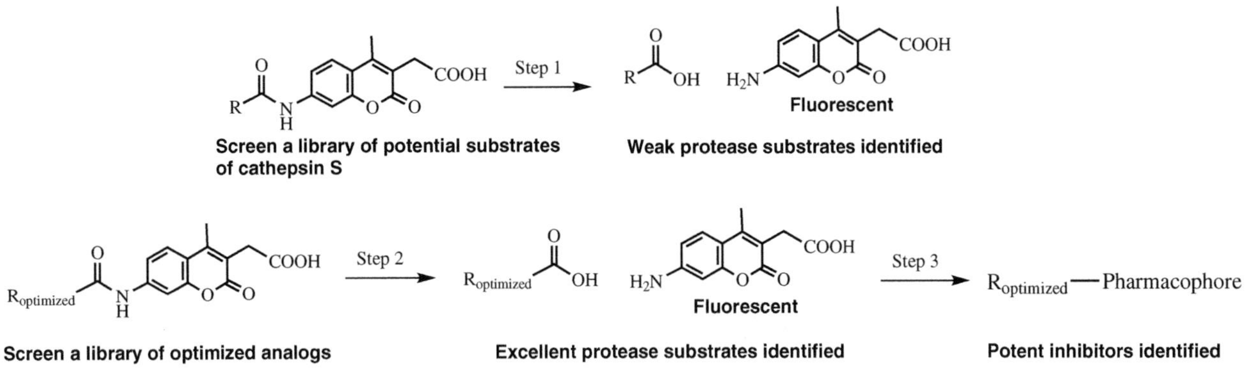

Step 1

Fluorescent

Screen a library of potential substrates of cathepsin S

Weak protease substrates identified

Step 2

Step 3

$R_{optimized}$—Pharmacophore

Fluorescent

Screen a library of optimized analogs

Excellent protease substrates identified

Potent inhibitors identified

FIGURE 2.10 Steps of SAS for identification of protease inhibitors

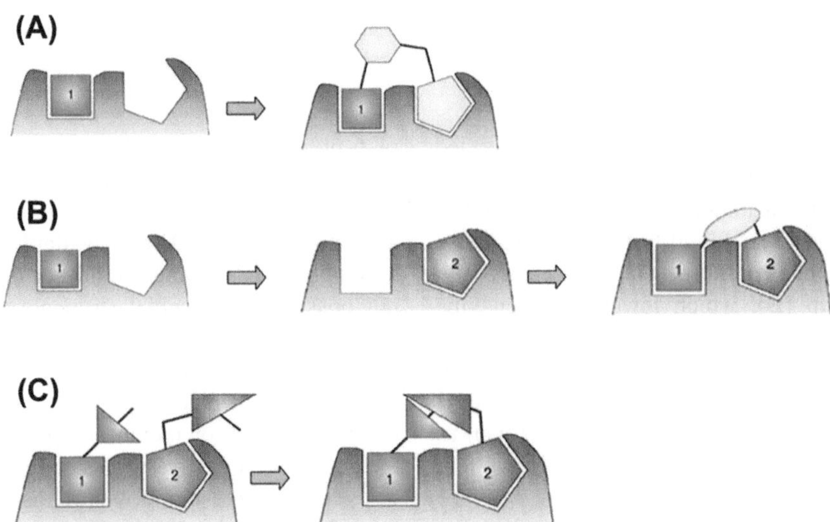

FIGURE 2.11 Three approaches to linking fragments: (A) fragment evolution, (B) fragment linking, and (C) fragment self-assembly. *Reprinted with permission from Macmillan Publishers Ltd: Nature Reviews Drug Discovery (Reese, D. C.; Congreve, M.; Murray, C. W.; Carr, R. Fragment-based lead discovery. Nat. Rev. Drug Discov.* ***2004****, 3, 660–672) Copyright 2004.*

was replaced by H to give an aldehyde, a known functionality for cathepsin inhibitors). In SAS, both an active enzyme and productive active site binding are required for catalytic function. However, SAS has some prominent advantages, such as being able to detect weak binding fragments because catalytic substrate turnover results in signal amplification (from release of a fluorescent molecule), and therefore, even very weak substrates can be identified at concentrations where only minimal binding to the enzyme occurs. Also, it is a high-throughput and straightforward technique to perform. Using this method, a 9 nM inhibitor of cathepsin S, which has been implicated in autoimmune diseases such as rheumatoid arthritis and multiple sclerosis,[222] was identified.

As illustrated in the foregoing examples, after the fragment hits are identified, the next step is to transform them into a lead structure while maintaining drug-like properties in the generated molecule. There are three general strategies for converting fragments into a drug-like lead compound: (1) *fragment evolution*, (2) *fragment linking*, and (3) *fragment self-assembly* (Figure 2.11). If the target structure is available, then elaboration of the fragment can be guided by the X-ray crystallographic or NMR spectral data of the fragment bound to the target. The fragment must also be optimized for pharmacokinetic properties.

Fragment evolution (Figure 2.11(A)) involves the addition of functionality to the fragment to allow for binding to additional pockets in the target. An example of fragment-based lead discovery incorporating fragment evolution (followed by lead modification to an optimized compound) is shown in Figure 2.12. Note that LE was used to guide the overall process. Ultimately, a balance had to be reached between potency for cyclin-dependent kinase (CDK2) inhibition, pharmacokinetics, and tumor cell activity. Changes in structure **A** in Figure 2.12 did not lead to large increases in potency, so a different strategy was taken, i.e., removal of the benzene ring from the benzpyrazole to give **B**, which had much lower potency

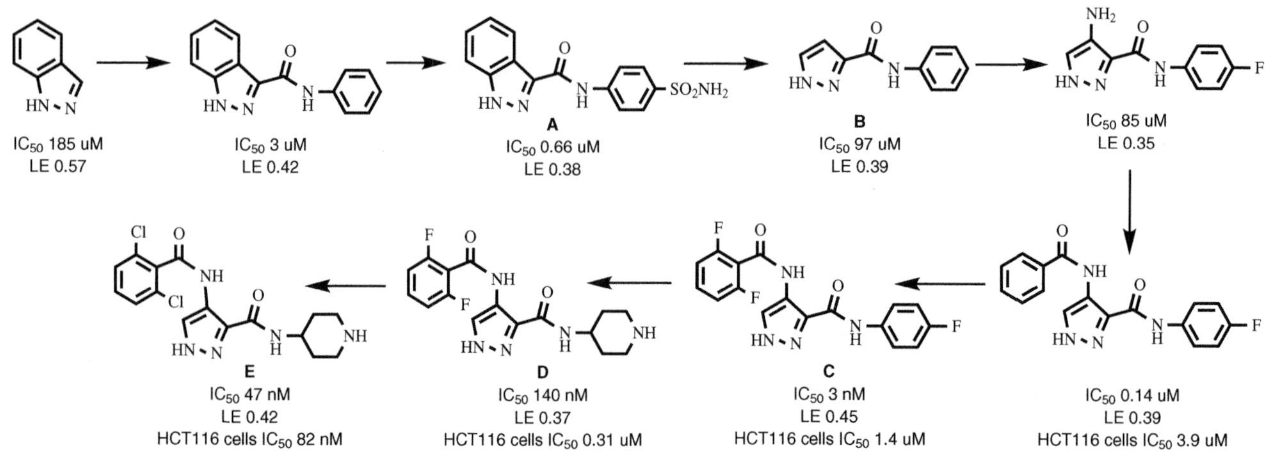

IC_{50} 185 uM LE 0.57 → IC_{50} 3 uM LE 0.42 → **A** IC_{50} 0.66 uM LE 0.38 → **B** IC_{50} 97 uM LE 0.39 → IC_{50} 85 uM LE 0.35 → IC_{50} 0.14 uM LE 0.39 HCT116 cells IC_{50} 3.9 uM → **C** IC_{50} 3 nM LE 0.45 HCT116 cells IC_{50} 1.4 uM → **D** IC_{50} 140 nM LE 0.37 HCT116 cells IC_{50} 0.31 uM → **E** IC_{50} 47 nM LE 0.42 HCT116 cells IC_{50} 82 nM

FIGURE 2.12 Example of fragment-based lead discovery incorporating the fragment evolution approach followed by lead modification to an optimized compound. Ligand efficiencies help guide the overall process.

(but similar LE). Growing from the pyrazole ring led to **C**, which was potent and had good pharmacokinetic properties, but activity in tumor cells was only moderate. The measured log *P* was found to be >4. To increase polarity, the *p*-fluorophenyl substituent was replaced by a 4-piperidinyl group (**D**), which lowered the enzyme inhibitory potency, but increased tumor cell activity. Further increases in lipophilicity and size by conversion of the 2,6-difluorophenyl ring of **D** to a 2,6-dichlorophenyl ring in **E** increased enzyme inhibitory potency as well as antitumor cell activity. This compound showed excellent in vivo activity and entered clinical trials.

As the name implies, *fragment linking* (Figure 2.11(B)) involves the linkage of two or more fragments that bind in proximal pockets, leading to higher affinity. The stromelysin example given in the discussion of SAR by NMR above was fragment linking: fragments in two adjacent binding pockets were linked to produce a 10^6 increase in potency relative to the hydroxamate fragment.

Fragment self-assembly (Figure 2.11(C)) is when fragments with complementary functional groups are allowed to react within the binding sites of the target. An example of this is the origins of "click chemistry", where a series of alkyne analogs and azide analogs were incubated with acetylcholinesterase; the alkyne analog and azide analog that bound in adjacent binding pockets were held in the optimal position to react and give the corresponding triazoles, one of which had femtomolar inhibitory potency (Figure 2.13).[223]

A lead discovery example that illustrates the fragment-based approach as well as several other concepts discussed throughout this chapter follows. *Trypanosoma brucei* is the causative parasite of African sleeping sickness, one of the most widespread and lethal diseases in Africa. Ruda, et al.[224] set out to discover a new inhibitor of 6-phosphogluconate dehydrogenase (6PGDH), a key enzyme for the function and survival of *T. brucei*. An X-ray crystal structure of 6PGDH in complex with a known inhibitor (**2.51**) was available. It is noteworthy that although **2.51** is a potent inhibitor of the enzyme, it does not possess trypanocidal (trypanosome-killing) activity. This deficit is attributed to the inability of the inhibitor to pass through membranes of the organism, thereby preventing it from reaching the enzyme target (pharmacokinetics). The poor membrane permeability is attributed to the double-negatively charged phosphate moiety (the basis for such reasoning will be discussed further in Sections 2.2.5.3 and 2.2.5.4). The crystal structure revealed that the enzyme contains a cluster of positively charged moieties that interact with the negatively charged phosphate. Therefore, the objective was to identify a new class of inhibitors that still contained a negatively charged moiety (to retain binding properties), but that was less likely to preclude permeability through membranes. The researchers started with an electronic database of commercially available chemicals and filtered it to retain only molecules that had MW < 320 and one of the following negatively charged groups: phosphonate, sulfonate, sulfonamide, carboxylic acid, or tetrazole.[225] This operation resulted in a set of 64,000 compounds. The compounds were computationally docked into the active site of the enzyme, with the requirement that the negatively charged group docked into the same region as the phosphate group of **2.51**. To validate the docking method, it was demonstrated that when **2.51** was docked computationally, a model resulted that closely resembled the crystallographically determined enzyme–inhibitor complex. About 6000 compounds gave reasonable docking poses with the enzyme. Using a computationally determined similarity approach, the 6000 compounds were divided into similar

FIGURE 2.13 Example of fragment-based lead discovery incorporating the fragment self-assembly approach: click chemistry

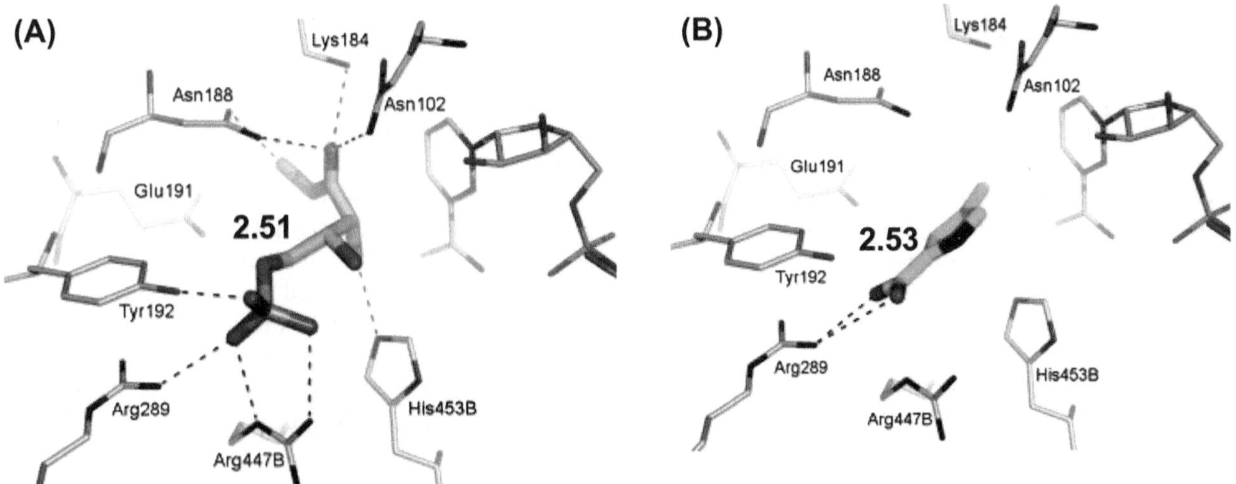

FIGURE 2.14 (A) Structure of **2.51** complexed with 6-phosphogluconate dehydrogenase (6PGDH) determined by X-ray crystallography. (B) Structure of **2.53** complexed with 6PGDH predicted by computational docking. *From Ruda, et al. Bioorg. Med. Chem.* ***2010****, 18, 5056–5062.*

groups (clusters), and 71 molecules were selected for purchase. The 71 compounds were tested as enzyme inhibitors, first at a very high concentration (200 μM), and then promising compounds were tested at a range of doses to determine binding potency. In this way, compounds **2.52**, **2.53**, and **2.54** were identified as fragments with moderate affinities but high ligand efficiencies, and thus as reasonable starting points for further modification. A comparison of the computationally derived docking pose of **2.53** with the complex of the protein and **2.51** (Figure 2.14, B vs A)) suggests where potential substituents could be added to **2.53** for additional productive interactions with the enzyme.

Another fragment-based HTS approach, rather than starting with a diverse random fragment library, is to start

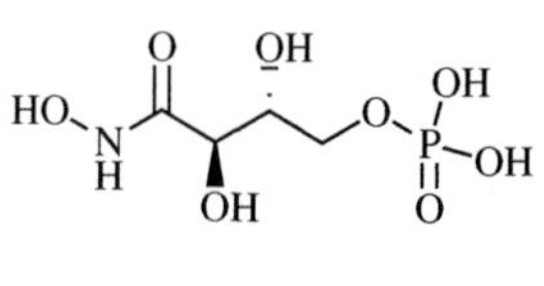

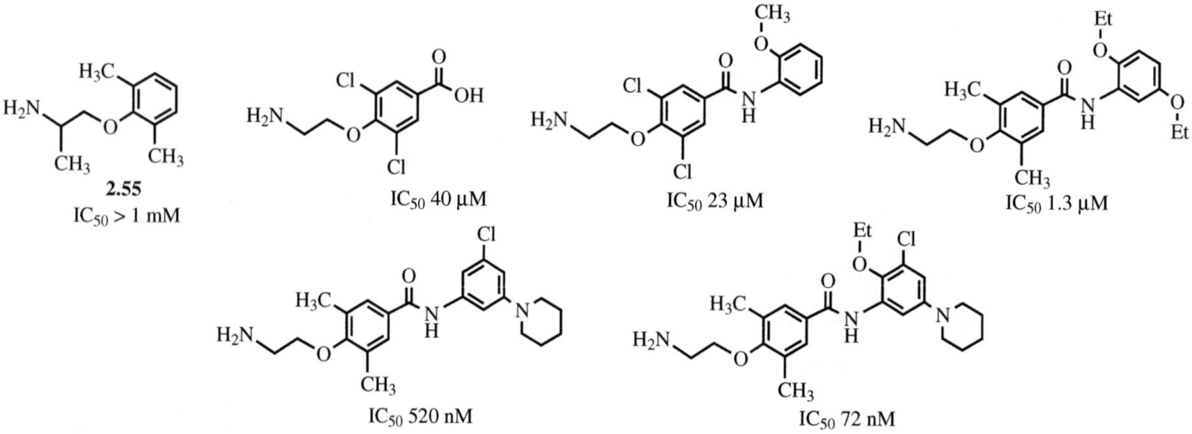

FIGURE 2.15 Progression from mexiletin (**2.55**, identified by fragment-based screening) to a potent orally bioavailable uPA

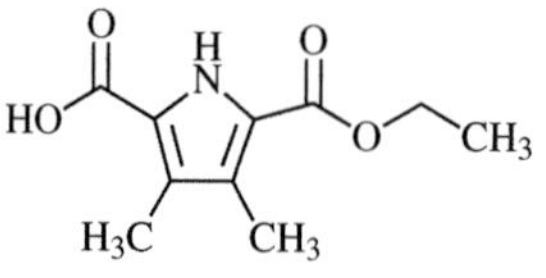

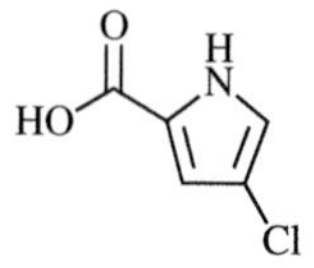

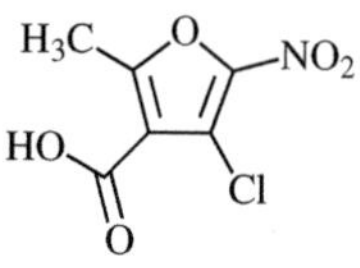

with a library of small known drugs, which have already been shown to have drug-like pharmacokinetic and safety characteristics (because they already are drugs) and see if they have other activities, a method Wermuth has called *selective optimization of side activities* (SOSA).[226] Because essentially all drugs can bind to more than one target, this approach searches for the minor off-target hits, and then optimizes the side activity into the main activity, diminishing (or eliminating) the original target activity. A library of small drug molecules can be purchased from Prestwick Chemical (Washington, DC, USA).[227] For example, the antiarrhythmic drug mexiletin (**2.55**, Fig. 2.15, Mexitil, MW 179) was found to be a weak inhibitor (IC_{50} > 1 mM) of urokinase-type plasminogen activator (uPA),[228] a serine protease that, when bound to its receptor, catalyzes the conversion of plasminogen to plasmin, which is responsible for a variety of proteolytic processes in the extracellular matrix.[229] Therefore, uPA is implicated in the progression of disease states associated with abnormal tissue destruction, such as multiple sclerosis[230] and cancer.[231] Figure 2.15 shows the structural progression from mexiletin to a potent orally bioavailable uPA inhibitor, using X-ray crystallography to guide the optimization.

The earlier discussion on the hit-to-lead process (Section 2.1.2.3.5) also applies in part to fragment-based lead discovery. For example, confirmation of activity, many computational methods, and early SAR assessments are already an inherent part of fragment-based approaches. On the other hand, taking the opportunity to perform basic calculations of physical properties and to conduct early ADME-tox and intellectual property assessments on leads discovered by fragment-based methods is still a good idea before proceeding into full-scale lead modification.

The SOSA example is a segue to the next section (Section 2.2), where we take a detailed look at how leads described in Section 2.1 are modified, resulting in a drug candidate ready for advanced preclinical studies. During this next phase of the drug discovery process, there is enhanced concern with pharmacokinetic (ADME) and toxicological properties as the potency at the intended target (pharmacodynamics) is being increased.

2.2. LEAD MODIFICATION

Once your lead compound is in hand, how do you know what to modify in order to improve the desired pharmacological, toxicological, and pharmacokinetic properties? The lead modification process, often referred to as *lead optimization*, can be context dependent, that is, the approach may vary depending on what property or properties most require improvement. In the discussion below, as well as in subsequent chapters, general principles and case examples are presented that provide a flavor for how specific challenges might be approached by the medicinal chemist.

2.2.1. Identification of the Active Part: The Pharmacophore

Interactions of drugs with receptors, known as *pharmacodynamics*, are very specific (see Chapter 3). Therefore, depending on how the lead was discovered, only a small part of the lead compound may be involved in the appropriate receptor interactions. The relevant groups on a molecule that interact with a receptor and are responsible for the activity are collectively known as the *pharmacophore*. The other atoms in the lead molecule, sometimes referred to as the *auxophore*, may be extraneous. Some of the atoms, of course, are essential to maintain the integrity of the molecule and hold the pharmacophoric groups in their appropriate positions. Some of these extraneous atoms, however, may be interfering with the binding of the pharmacophore, and those atoms need to be excised from the lead compound. Other atoms in the auxophore may be dangling in space within the receptor and neither binding to the receptor nor preventing the pharmacophoric atoms from binding. Although these atoms appear to be innocuous, it is important to know which atoms these are, because these are the ones that can be modified without loss of potency. As previously noted (Sections 1.3.5 and 2.1.2.3.2), there are other aspects to lead modification that are as important as increasing binding to the target receptor, such as *pharmacokinetics* (ADME). Modification of the atoms that are not directly involved in interactions with the biological target could be very important for solving pharmacokinetic problems.

By determining which are the pharmacophoric groups and which are the auxophoric groups on your lead compound, and of the auxophoric groups, and which are interfering with lead compound binding and which are not detrimental to binding, you will know which groups must be excised and which you can retain or modify as needed. One approach in lead modification to help make this determination is to cut away sections of the lead molecule and measure the effects of those modifications on potency. As an example of how this might be done, consider this artificial example involving in vivo effects; pharmacokinetic differences, therefore, are ignored. Assume that the addictive analgesics morphine (**2.56**, R=R′=H), codeine (**2.56**, R=CH_3, R′=H), and heroin (**2.56**, R=R′=$COCH_3$) are the lead compounds, and we want to know which groups are pharmacophoric and which are auxophoric. The morphine family of analgesics bind to the μ opioid receptors. The pharmacophore is known and is shown as the darkened part in **2.56**. A decrease in potency on removal of a group will suggest that it may have been pharmacophoric, an increase in potency means it was auxophoric and interfering with proper binding, and essentially no change in potency will mean that it is auxophoric but not interfering with binding.

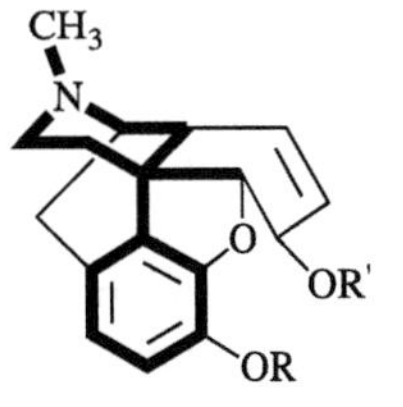

Morphine (R = R' = H)
Codeine (R = CH_3, R' = H)
Heroin (R = R' = $COCH_3$)
2.56

Let's start by excising the dihydrofuran oxygen atom, which is not in the pharmacophore. This may not seem to be sensible because that atom connects the cyclohexene ring to the benzene ring; its removal will result in a change in the conformation of the cyclohexene ring and an increase in the degrees of freedom of the molecule. Excision of the dihydrofuran oxygen gives morphinan (**2.57**, R=H)[232]; the hydroxyl analog, levorphanol[233] (**2.57**, R=OH, Levo-Dromoran) is three to four times *more* potent than morphine as an analgesic, but it retains the addictive properties (note that in **2.57** the cyclohexene ring conformation has not been changed for ease of comparison with **2.56**; surely, a lower energy conformer will be favored). Possibly, the additional conformational mobility allowed the molecule to better approximate its bioactive conformation. Removal of half of the cyclohexene ring (also not in the pharmacophore), leaving only methyl substituents, gives benzomorphan (**2.58**, R=CH_3).[234] This compound shows some separation of analgesic and addictive effects; pentazocine (**2.58**, R=$CH_2CH=C(CH_3)_2$; component of Talwin) is less potent than morphine (about as potent as codeine), but has a much lower addiction liability. Remember, your goal is both to increase potency and decrease adverse effects, such as addictive properties. Although this analog is not more potent than morphine, it is less addicting. Cutting away the methylene group of the cyclohexane fused ring (**2.59**) also, surprisingly, has little effect on the analgesic activity in animal tests. Again, this excision removes the rigidity of the parent structure. Removal of all fused rings, for example, in the case of meperidine (**2.60**, Demerol), gives an analgesic still possessing 10–12% of the overall potency of morphine.[235] Although the potency is lower, it certainly will be much easier to synthesize analogs of meperidine than those of morphine. Even acyclic analogs are active. Dextropropoxyphene (**2.61**, Darvon; again note the side chain is left in a conformation to resemble the structure of morphine) is one-half to two-thirds as potent as codeine. Both morphine and dextropropoxyphene bind to the μ opioid receptors, so the activity of dextropropoxyphene can be ascribed to the fact that it can assume a conformation related to that of the morphine pharmacophore. It is unlikely that anyone seeing dextropropoxyphene written in a more energetically favorable conformation would ever make the connection between this structure and that of morphine. By cutting pieces off of the lead compound, one gains new perspectives on possible active structures, which should open up completely new scaffolds to consider. In a sense, this is the medicinal chemistry analogy to a retrosynthesis; you may not know where to begin a synthesis based on the structure of the target molecule, but by working backward from the target molecule, you uncover new structures that were not obvious initially. This, of course, could be beneficial in terms of new intellectual property considerations (new scaffolds are uncovered) as well as in ease of synthesis. Another acyclic analog is methadone (**2.62**; Methadose), which is as potent an analgesic as morphine; the (−)-isomer is used in the treatment of opioid abstinence syndromes in heroin abusers because it is eliminated from the body more slowly than morphine, allowing the body to adapt to the falling levels of drug gradually.

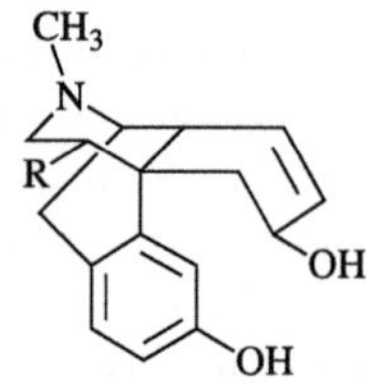

Levorphanol (R = OH)
2.57

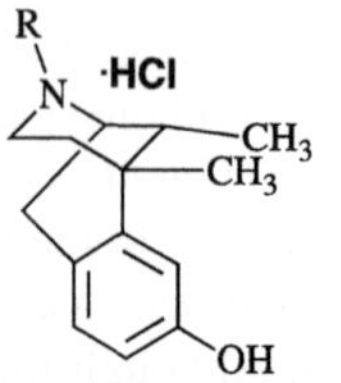

Pentazocine HCl (R = $CH_2CH{=}C(CH_3)_2$)
2.58

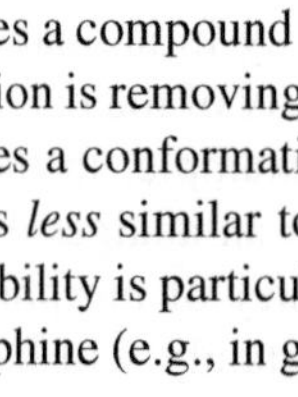

2.59

What if every cut in the lead produces a compound with lower potency? Then, either every excision is removing part of the pharmacophore or each cut causes a conformational change that results in a structure that is *less* similar to the bioactive conformation. The latter possibility is particularly relevant to rigid structures, such as morphine (e.g., in going from **2.56** to **2.57**). In the case of morphine, groups can be added to the lead structure to *increase* the pharmacophore. For example, oripavine derivatives such as etorphine (**2.63**, R=CH_3, R′=C_3H_7; Immobilon), which has a two-carbon bridge and substituent not in morphine, is 3200 times more potent than morphine[236] and is used in veterinary medicine to immobilize large animals. The related analog, buprenorphine (**2.63**, R=CH_2-cyclopropyl,; R′=*t*-Bu; double bond reduced; Buprenex) is 10–20 times more potent than morphine and has a very low level of dependence liability. Apparently, the rigidity of the oripavine derivatives increases the appropriate receptor interactions.

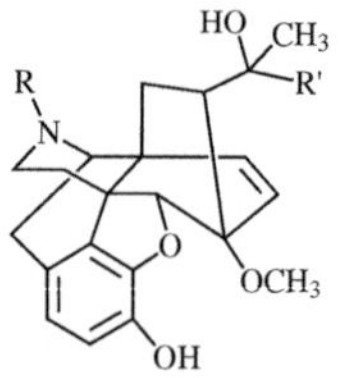

Etorphine (R = CH_3; R' = C_3H_7)
2.63

The activity and potency of a molecule is related to the interactions of the pharmacophoric groups with groups on the biological target (Chapter 3, Section 3.2.2). The binding constants of 200 drugs and potent enzyme inhibitors were used by Andrews and coworkers[237] to calculate the average binding energies of common functional groups; these energies can be used to determine how well a new molecule binds to its receptor. If the test molecule has a measured binding energy that is lower than the calculated average value, it suggests that the molecule contains groups that do not interact with the receptor (are not in the pharmacophore). These groups, then, could be excised without loss of potency, giving a simplified lead for further structural modification. This *Andrews analysis* was carried out on a highly substituted lead compound, leading to a more simple analog structure, which was modified to give molecules with enhanced potency.[238] If the test compound has a binding energy greater than the calculated average value, then the molecule may bind differently than suspected, leading to enhanced binding interactions. This indicates that manipulation of functional groups is an important lead modification approach.

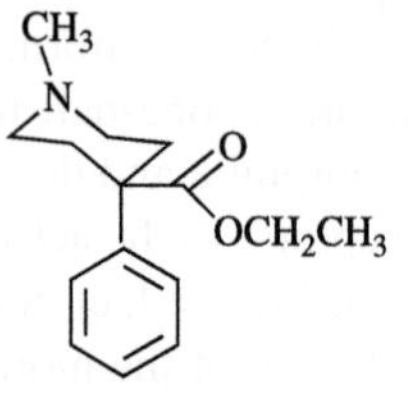

Meperidine
2.60

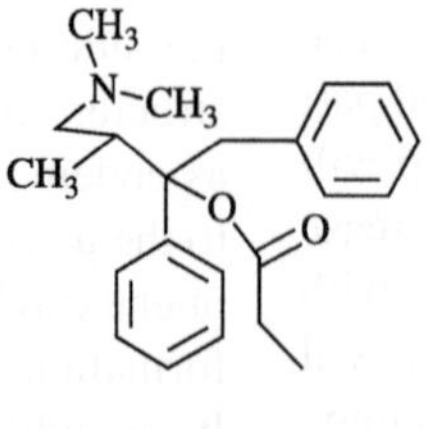

Dextropropoxyphene
2.61

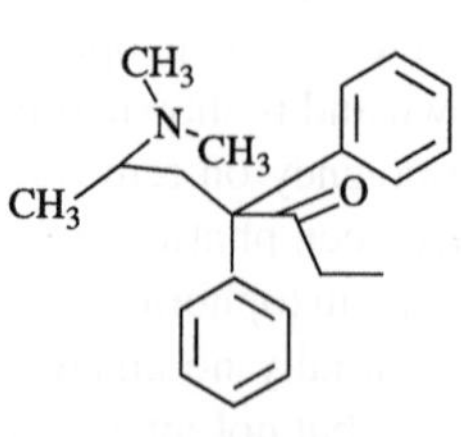

Methadone
2.62

2.2.2. Functional Group Modification

The importance of functional group modification is demonstrated by **2.64**. The antibacterial agent, carbutamide (**2.64**, R = NH_2), was found to have an antidiabetic side effect; however, it could not be used as an antidiabetic drug because of its antibacterial activity, which could lead to bacterial resistance (see Chapter 7). The amino group of carbutamide was replaced by a methyl group to give tolbutamide (**2.64**, R = CH_3; Orinase), and in so doing the antibacterial activity was separated away from the antidiabetic activity. In some cases, an experienced medicinal chemist knows what functional group will elicit a particular effect. Chlorothiazide (**2.65**; Aldocor) is an antihypertensive agent that has a strong diuretic (increased urine excretion) effect as well. It was known from sulfanilamide work in the 1930s and 1940s that the primary aminosulfonyl side chain group can give diuretic activity (see Section 2.2.3). Consequently, diazoxide (**2.66**; Hyperstat) was prepared as an antihypertensive drug without diuretic activity.

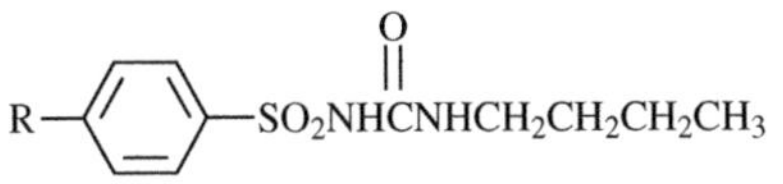

R—C_6H_4—$SO_2NHCONHCH_2CH_2CH_2CH_3$

Tolbutamide (R = CH_3)
2.64

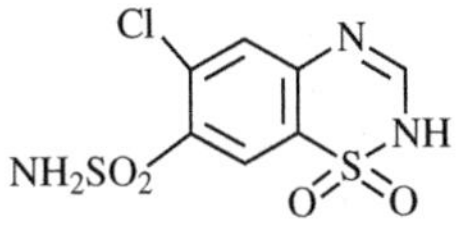

Chlorothiazide
2.65

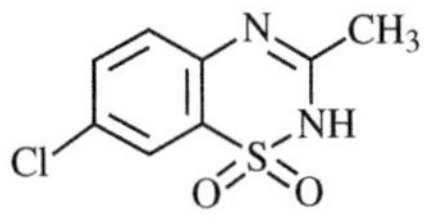

Diazoxide
2.66

Obviously, there is a relationship between the molecular structure of a compound and its activity. This phenomenon was first realized about 145 years ago.

2.2.3. Structure–Activity Relationships

In 1868, Crum-Brown and Fraser[239] suspected that the quaternary ammonium character of curare (Figure 2.16), the name for a variety of South American quaternary alkaloid poisons that cause muscle paralysis known since the sixteenth century, when they were used on arrowheads, may be responsible for its muscular paralytic properties (it blocks the action of the excitatory neurotransmitter acetylcholine at muscle receptors). Consequently, they examined the neuromuscular blocking effects of a variety of simple quaternary ammonium salts and quaternized alkaloids in animals. From these studies they concluded that the physiological action of a molecule was a function of its chemical constitution. Shortly thereafter, Richardson[240] noted that the hypnotic activity of aliphatic alcohols was a function of their MW. These observations were the basis for the future focus by medicinal chemists on SARs.

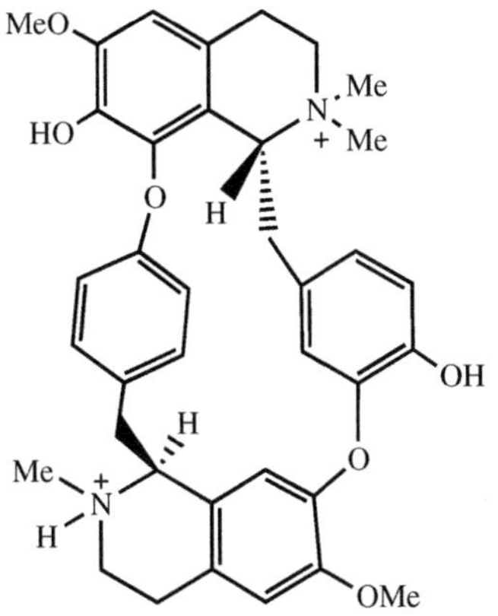

FIGURE 2.16 Structure of D-tubocurarine, a constituent of curare

Drugs can be classified as being structurally specific or structurally nonspecific. *Structurally specific drugs*, which most drugs are, act at specific sites, such as a receptor or enzyme. Their activity and potency are very susceptible to small changes in chemical structure; molecules with similar biological activities tend to have common structural features. *Structurally nonspecific drugs* have no specific site of action and usually have lower potency. Similar biological activities may occur with a variety of structures. Examples of these drugs are gaseous anesthetics, sedatives and hypnotics, and many antiseptics and disinfectants.

Even though only a part of the molecule may be associated with its activity, there is a multitude of molecular modifications that could be made. The hallmark of SAR studies is the synthesis of numerous analogs of the lead compound and their testing to determine the effect of structure on potency for a particular activity. Once enough analogs are prepared and sufficient data accumulated, conclusions can be made regarding SARs. In practice, ease of synthesis, rather than a cogent rationale, is sometimes the main guiding force behind the choice of analogs to synthesize. While this approach might yield an optimized molecule with the desired properties, it can often be disadvantageous and nonproductive to solely pursue this approach at the expense of tackling more challenging syntheses that are directed at addressing key questions. A common way to track structural changes and develop an SAR is by employing the *activity landscape concept*, any graphical representation that correlates structural similarity and potency for a particular activity.[241] Generally, these are computer-based representations, giving specific structural and potency information, which then can be visualized for a comparison of structure and potency. Using an activity landscape representation, regions of SAR continuity, where gradual structural changes lead to moderate potency changes, can easily be discerned, which can be used for potency predictions. However, sometimes a discontinuity in the SAR can occur, in which two compounds having a similar structure can have markedly different potencies. This discontinuity is known as an *activity cliff*.[242] These activity cliffs, such as the addition of a single methyl group to an active molecule that destroys its activity,[243] can provide valuable information regarding binding phenomena.

The development of the sulfonamide antibacterial agents (sulfa drugs) illustrates the analysis and description of SARs. After a number of analogs of the lead compound sulfanilamide (**2.67**, R = H) were prepared, clinical trials determined that compounds of this general structure exhibited diuretic and antidiabetic activities as well as antimicrobial activity. Compounds with each type of activity eventually were shown to possess certain structural features in common. On the basis of the biological results of greater than 10,000 compounds, several SAR generalizations have been made.[244] For example, antimicrobial agents have structure **2.68** (R = SO_2NHR' or SO_3H) where (1) the amino and sulfonyl groups on the benzene ring should be para; (2) the anilino amino group may be unsubstituted (as shown) or may have a substituent that is removed in vivo; (3) replacement of the benzene ring by other ring systems, or the introduction of additional substituents on it, decreases the potency or abolishes the activity; (4) R in **2.68** may be any of the alternatives shown below **2.68**, but the potency is reduced in most cases; (5) *N'*-monosubstitution (R = SO_2NHR') results in more potent compounds, and the potency increases when R′ is a heteroaromatic ring; (6) *N'*-disubstitution (R = $SO_2NR'_2$), in general, leads to inactive compounds.

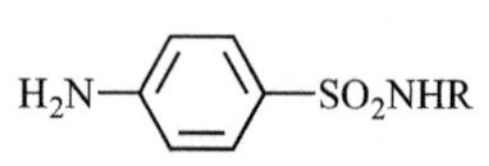

H_2N–C6H4–SO_2NHR

Sulfa drugs
2.67

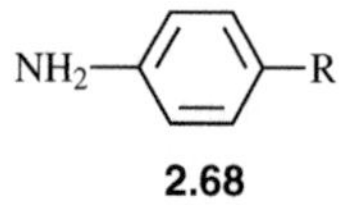

NH_2–C6H4–R

2.68

SO_2–C6H4–NH_2 SO–C6H4–NH_2

C(=O)NH_2 C(=O)NHR'   C(=O)–C6H4–R''

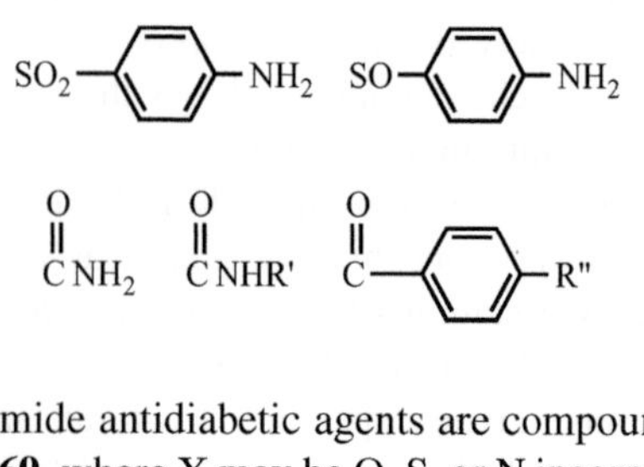

Sulfonamide antidiabetic agents are compounds having structure **2.69**, where X may be O, S, or N incorporated into a heteroaromatic structure, such as a thiadiazole or a pyrimidine or in an acyclic structure, such as a urea or thiourea. In the case of ureas, the N-2 should carry as a substituent (R') a chain of at least two carbon atoms.[245]

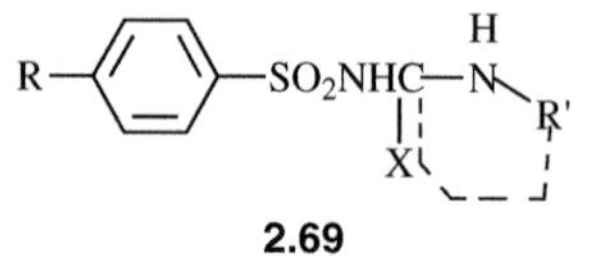

R–C6H4–SO_2NHC(=X)–NH–R'

2.69

Sulfonamide diuretics are of two general structural types, hydrochlorothiazides (**2.70**) and the high ceiling type[246] (**2.71**). The former compounds have 1,3-disulfamyl substitution on the benzene ring and R^2 is an electronegative group such as Cl, CF_3, or NHR. The high ceiling compounds contain 1-sulfamyl-3-carboxy substitution. Substituent R^2 is Cl, Ph, or PhZ, where Z may be O, S, CO, or NH and X can be at position 4 or 5 and is normally NHR, OR, or SR.[247]

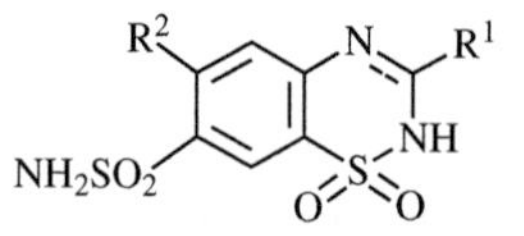

2.70 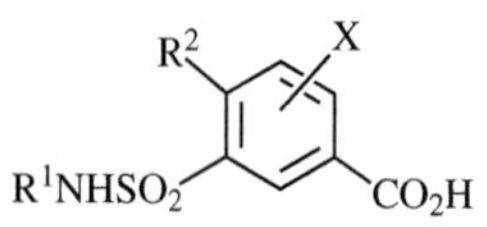   **2.71**

A more recent example of SAR is that of the natural product anticancer drug paclitaxel (**2.72**; Taxol), which was the first anticancer compound found to act by promoting the assembly of tubulin into microtubules, thereby blocking mitosis.[248] After a large number of modifications were introduced,[249] many SAR conclusions could be made. A simple manual SAR representation to track the effect of structural changes on potency is a *molecular activity map* (Figure 2.17), a structural drawing of a lead compound annotated to show where in the molecule specific structural changes affect potency measured in a particular bioassay. These maps may also document changes in other properties, for example, onset of unexpected toxicity. These maps may depict the results of a long-lasting drug discovery effort involving numerous chemists (and the biologists who do the screens). Their main virtues are that they can concisely summarize a huge number of facts relating structures with their activities and potencies, and they may direct the chemists' creativity to unexplored regions of the lead compound yielding novel structural changes. Another way to visualize your SAR data is with an *SAR map*,[250] which allows analysis of large data sets with multiple R groups, correlation of substituent structure and biological activity, determination of additivity of substituent effects, identification of missing analogs and screening data, and creation of graphical representations. Once you have obtained an SAR, an *SAR index* can be calculated,[251] which quantitatively describes the nature of the SAR and allows

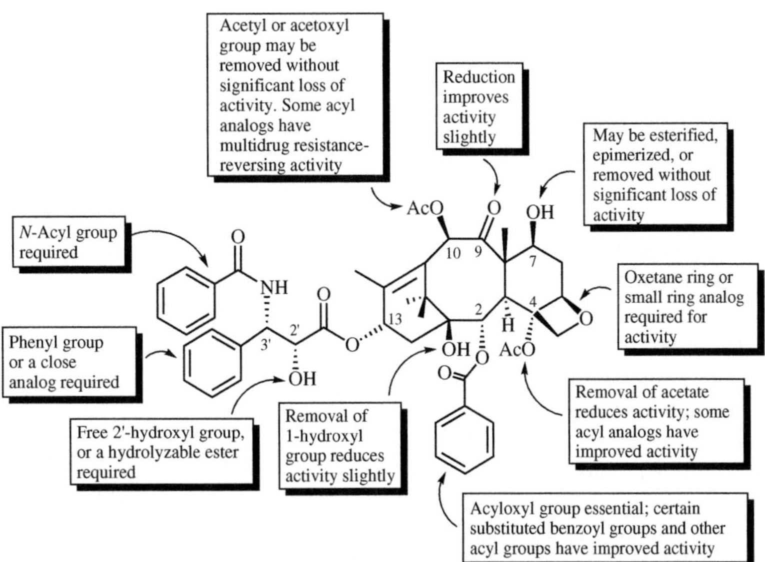

FIGURE 2.17 SAR for paclitaxel (Taxol)

you to estimate how likely it is to identify structurally distinct molecules with similar activity.

Paclitaxel
2.72

The above examples provide strong evidence to support the notion that a correlation does exist between the structure of a compound and its activity at a given biological target; furthermore, when a compound interacts with more than one biological target, the SAR at each target may be different, corresponding to differences in the topography of the targets at the sites of interaction.

2.2.4. Structure Modifications to Increase Potency, Therapeutic Index, and ADME Properties

How do you know what molecular modifications to make to fine-tune the lead compound? In the preceding section, it was made clear that structure modifications were the keys to activity and potency manipulations. After years of SAR studies, various standard molecular modification approaches have been developed for the systematic determination of SARs. Frequently, an important goal during lead optimization is improvement of the *therapeutic index* (also called the *therapeutic ratio*), which is a measure of the safety of a drug as determined from the ratio of the concentration of a drug that gives undesirable effects to that which gives desirable effects. The therapeutic index can be determined by any method that measures undesirable and desirable drug effects, but often it is taken as the dose-limiting toxicity versus the desirable pharmacological effect, initially in appropriate animal models, and then later in humans once clinical data are available. For example, the therapeutic index could be the ratio of the LD_{50} (the lethal dose for 50% of the test animals) to the therapeutic ED_{50} (the effective dose that produces the maximum therapeutic effect in 50% of the test animals; see Chapter 3, Section 3.2.3); a toxic ED_{50} (the dose that produces toxicity in 50% of the test animals) may substitute for the LD_{50}. The larger the therapeutic index, the greater the margin of safety of the compound; ideally one would like to have to administer gram quantities of the drug before any undesirable effects are observed, but milligrams of the drug to attain the desirable effects. There is no specific minimum value for a therapeutic index that must be attained before a drug can be approved; it depends on the disease that is being treated and whether there are already other therapies available. A low therapeutic index is tolerable for lethal diseases, such as cancer or AIDS (maybe even as low as 1–5), especially if there is no other treatment available, or if the side effect is minor compared with the treatment benefit. For less threatening diseases therapeutic indices on the order of 10–100 may be reasonable. As an example, the therapeutic index for the antitumor agent chlorambucil (**2.73**, Leukeran) is 23[252]; for ethanol, it is 10 (sorry to put a damper on your next party).[253] The Merck Index is a good source to obtain LD_{50} data for drugs in animals.

Chlorambucil
2.73

One source of toxicity (mutagenicity, hepatotoxicity, cardiotoxicity) derives from particular substructures of molecules, known as *toxicophores or structural alerts* (see Section 2.1.2.3.2). These are groups that are either reactive themselves or can be metabolized to reactive species (see Chapter 8, Section 8.4.4), leading to covalent attachment to macromolecules or to radicals that cause toxicity.[254] Typical toxicophores include aromatic nitro, aromatic amine (anilines), epoxides, azidirines, nitroso, azo, aliphatic halides, heteroatom–heteroatom bonds, and polycyclic aromatic systems.[255] For many years, it has been known that in lead modification approaches, these toxicophoric groups should generally be avoided to minimize the toxicological potential, especially given that almost one-third of drug attrition during drug development derives from toxicity issues. However, it was found on comparing drugs that were recalled from the market because of idiosyncratic (unpredictable) toxicity with the 200 top-selling drugs on the market in 2009 that there was no obvious link between the idiosyncratic toxicity and specific physicochemical properties of the molecules, such as MW or lipophilicity.[256] About half of the 200 top drugs also contained structural alerts. The difference between the drugs that were withdrawn and those that did not cause idiosyncratic toxicity appears to be the daily dose required for the drug and the clearance rate; low-dose drugs (about 10 mg), despite containing toxicophores, do not have the toxicity issues of high-dose drugs (several 100 mg/day). That makes the decision to retain a toxicophore or not during lead modification even more difficult.

A concept related to therapeutic index is the *NOAEL*, no observed adverse effect level. This is the maximum amount of compound administered for which no statistically significant adverse effects are observed. NOAEL of a new drug is generally assessed in laboratory animals prior to initiation of clinical trials to establish a safe starting dose to administer to human subjects.

As noted above, while establishing an SAR for the desired activity, you may make structure modifications that reveal an SAR for an undesired activity or toxicity. If these two SARs are different for different activities, then rational design of compounds with the optimal activity–safety profile becomes a possibility.

In addition to making structural changes to influence the activity and safety profiles of a compound, it is often also important to make changes to influence the ADME profile. An important (and in many cases rationally adjustable) component of improving the overall ADME profile of a molecule is adjusting the membrane permeability, since this can significantly affect the absorption and distribution properties of the molecule. Some of the modifications routinely made during lead optimization discussed below can affect membrane permeability, and these will be pointed out when applicable. Subsequently (Section 2.2.5), some well-developed principles regarding the effects of electronic properties, lipophilicity, and charge on membrane permeability are discussed.

2.2.4.1. Homologation

A *homologous series* is a group of compounds that differ by a constant unit, generally a CH_2 group. As will become more apparent in Section 2.2.5.2, biological properties of homologous compounds often show regularities of increase and decrease. For many series of compounds, lengthening of a saturated carbon side chain from one (methyl) to five to nine atoms (pentyl to nonyl) produces an increase in pharmacological effects; further lengthening results in a sudden decrease in potency (Figure 2.18). In Section 2.2.5.2.2 it will be shown that this phenomenon corresponds to increased lipophilicity of the molecule, which permits penetration into cell membranes until its lowered water solubility becomes problematic in its transport through aqueous media or its high solubility in membranes. In the case of aliphatic amines, another problem is micelle formation, which begins at about C_{12}. A typical micelle in aqueous solution consists of a collection of molecules with their hydrophobic regions aggregated toward the center of spherelike structure with hydrophilic regions on the outside in contact with the aqueous environment. Micelle formation effectively removes the compound from potential interaction with the appropriate receptors. An early example of this potency versus chain length phenomenon was reported by Richardson,[257] who was investigating the hypnotic activity

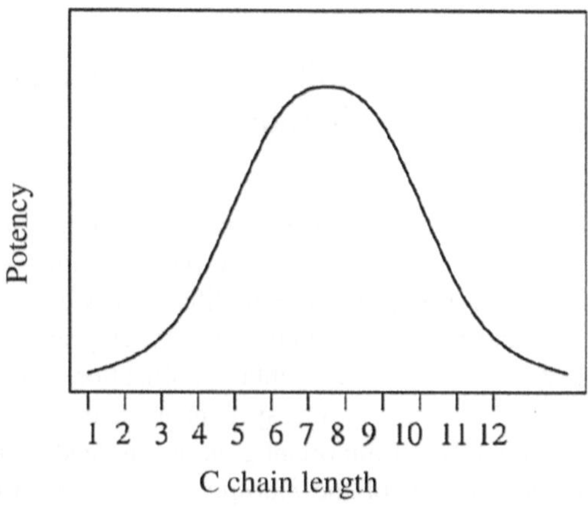

FIGURE 2.18 General effect of carbon chain length on drug potency

of alcohols. The maximum effect occurred for 1-hexanol to 1-octanol; then the potency declined on chain lengthening until no activity was observed for hexadecanol. Figure 2.18, therefore, can be rationalized on the basis of either pharmacodynamics or pharmacokinetics (or both). On the basis of *pharmacodynamics,* as the chain length increases, it may fit into a hydrophobic binding pocket with increasing interactions (see Chapter 3, Section 3.2.2), thereby increasing potency, until the size of the substituent exceeds the size of the binding pocket, and the potency decreases. On the basis of *pharmacokinetics,* as the chain length increases, the lipophilicity increases, allowing better absorption through membranes, increasing the potency, until the chain length and lipophilicity increase to the point where the molecule becomes insoluble in the aqueous medium or becomes stuck in the membrane, and the apparent potency decreases.

A study by Dohme et al.[258] on 4-alkyl-substituted resorcinol derivatives showed that the peak antibacterial activity occurred with 4-*n*-hexylresorcinol (see Table 2.8), a compound now used as a topical anesthetic in a variety of throat lozenges. Funcke et al.[259] found that the peak spasmolytic activity of a series of mandelate esters occurred with the *n*-nonyl ester (see Table 2.8).

TABLE 2.8 Effect of Chain Length on Potency

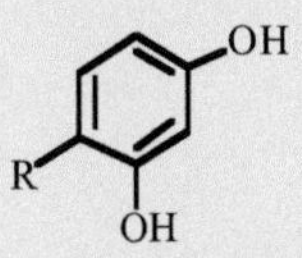

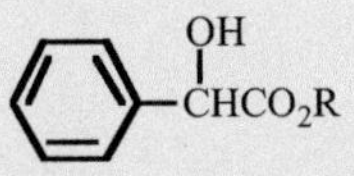

R	Phenol coefficient[1]	% Spasmolytic activity[2]
Methyl	–	0.3
Ethyl	–	0.7
n-propyl	5	2.4
n-butyl	22	9.8
n-pentyl	33	28
n-hexyl	51	35
n-heptyl	30	51
n-octyl	0	130
n-nonyl	0	190
n-decyl	0	37
n-undecyl	0	22
i-propyl	–	0.9
i-butyl	15.2	8.3
i-amyl	23.8	28
i-hexyl	27	–

Antibacterial activity of 4-alkylresorcinols and spasmolytic activity of mandelate esters.
[1]*The ratio of the bactericidal potency of the compound relative to that of phenol.*
[2]*Relative to 3,3,5-trimethylcyclohexanol, set at 100%.*

2.2.4.2. Chain Branching

When a simple lipophilic relationship is important, as described above, then chain branching lowers the potency of a compound because a branched alkyl chain is less lipophilic than the corresponding straight alkyl chain as a result of larger molar volumes and shapes of branched compounds. This phenomenon is exemplified by the lower potency of the compounds in Table 2.8 having isoalkyl chains, relative to the corresponding *n*-alkyl chains containing the same number of carbon atoms. Effects on lipophilicity, and hence membrane permeability, may be a factor (a pharmacokinetic argument). Another possible explanation for lower potency with branching could be that chain branching may interfere with receptor binding (a pharmacodynamics argument). For example, phenethylamine ($PhCH_2CH_2NH_2$) is an excellent substrate for monoamine oxidase, but α-methylphenethylamine (amphetamine) is a poor substrate. Primary amines often are more potent than secondary amines, which are more potent than tertiary amines. For example, the antimalarial drug primaquine phosphate (**2.74**; Primaquine) is much more potent than its secondary or tertiary amine homologs.

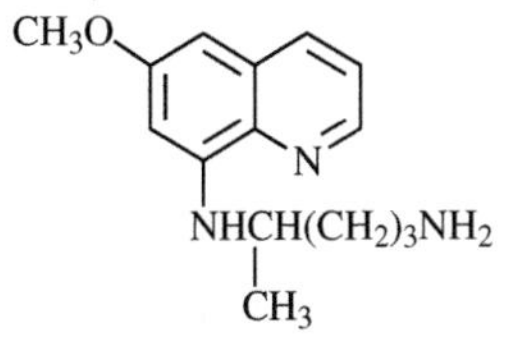

CH_3O … $NHCH(CH_2)_3NH_2$ / CH_3

Primaquine
2.74

Major pharmacological changes can occur with chain branching or homologation. Consider the 10-aminoalkylphenothiazines (**2.75**, X=H). When R is $-CH_2CH(CH_3)N(CH_3)_2$ (promethazine HCl; Phenergan), antispasmodic and antihistaminic activities predominate. However, the straight-chain analog **2.75** with R being $-CH_2CH_2CH_2N(CH_3)_2$ (promazine) has greatly reduced antispasmodic and antihistaminic activities, but sedative and tranquilizing activities are greatly enhanced. In the case of the branched-chain analog **2.75** with R equal to $-CH_2CH(CH_3)CH_2N(CH_3)_2$ (trimeprazine) (next larger branched-chain homolog), the tranquilizing activity is reduced and antipruritic (anti-itch) activity increases. This indicates that there are multiple receptors involved, and branching or homologation can cause the molecule to bind more or less well to the receptors responsible for antispasmodic activity, antihistamine activity, tranquilizing activity, or antipruritic activity.

Promethazine (R = $CH_2CH(CH_3)N(CH_3)_2$)
Promazine (R = $CH_2CH_2CH_2N(CH_3)_2$)
Trimeprazine (R = $CH_2CH(CH_3)CH_2N(CH_3)_2$)
2.75

2.2.4.3. Bioisosterism

Bioisosteres are substituents or groups that have chemical or physical similarities and related molecular shapes, and which produce roughly similar biological properties.[260] Bioisosterism, a term coined by Harris L. Friedman[261] and extended by Alfred Burger,[262] is an important lead modification approach that has been shown to be useful to attenuate toxicity or to modify the activity of a lead, and may have a significant role in the alteration of the pharmacokinetics of a lead. There are classical isosteres[263] and nonclassical isosteres.[264] Langmuir conceptualized *isosterism* in 1919[265]; in 1925, Grimm[266] formulated the *hydride displacement law* to describe similarities between groups that have the same number of valence electrons, but may have a different number of atoms. Erlenmeyer[267] later broadened the concept of isosteres as atoms, ions, or molecules in which the peripheral layers of electrons can be considered to be identical. These two definitions describe *classical isosteres*; examples are shown in Table 2.9.

Nonclassical bioisosteres do not have the same number of atoms and do not fit the steric and electronic rules of the classical isosteres, but do often produce similar biological activities. Examples of these are shown in Table 2.10. There are hundreds of examples of compounds that differ by a bioisosteric interchange[268]; some examples are shown in Table 2.11.

Bioisosterism sometimes can lead to significant changes in activity or potency. For example, if the sulfur atom of the phenothiazine neuroleptic drugs (**2.75**) is replaced by –CH=CH– or –CH_2CH_2– bioisosteres, then dibenzazepine antidepressant drugs (**2.76**) result. As another example, when the thiazolone ring in a series of antiinflammatory

2.76

TABLE 2.9 Classical Isosteres

1. Univalent atoms and groups

a. CH_3 NH_2 OH F Cl
b. Cl PH_2 SH
c. Br *i*-Pr
d. I *t*-Bu

2. Bivalent atoms and groups

a.	—CH_2—	—NH—	—O—	—S—	—Se—
b.	—$COCH_2R$	—CONHR	—CO_2R	—COSR	

3. Trivalent atoms and groups

a.	—CH=	—N=
b.	—P=	—As=

4. Tetravalent atoms

a.	—C— (tetravalent)	—Si— (tetravalent)	
b.	=C=	$=\overset{+}{N}=$	$=\overset{+}{P}=$

5. Ring equivalents

a.	—CH=CH—	—S—	(e.g., benzene, thiophene)		
b.	—CH=	—N=	(e.g., benzene, pyridine)		
c.	—O—	—S—	—CH_2—	—NH—	(e.g., tetrahydrofuran, tetrahydrothiophene, cyclopentane, pyrrolidine)

TABLE 2.10 Nonclassical Isosteres

1. Carbonyl Group

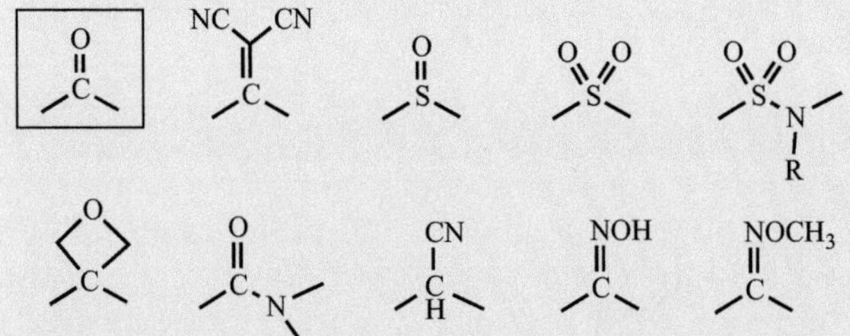

2. Carboxylic acid group

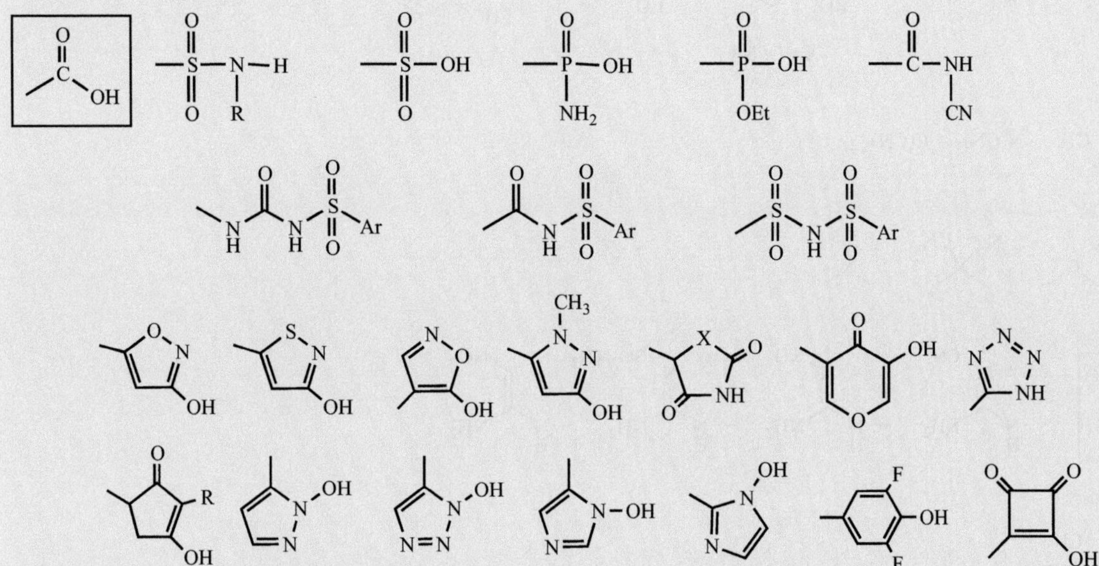

3. Amide group

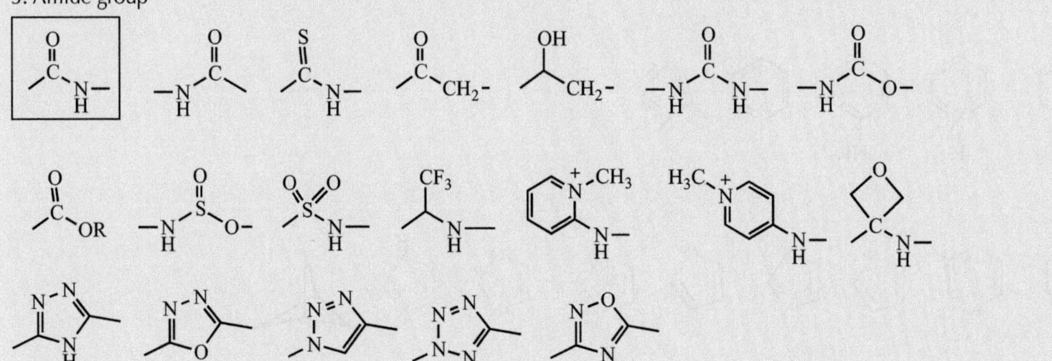

4. Ester group

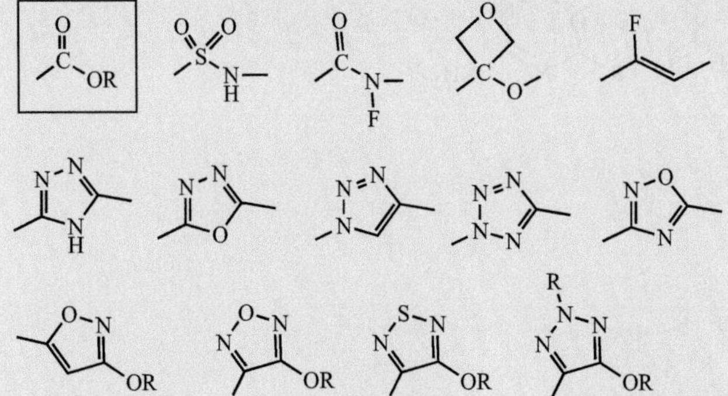

Continued

TABLE 2.10 Nonclassical Isosteres—Cont'd

5. Hydroxyl group

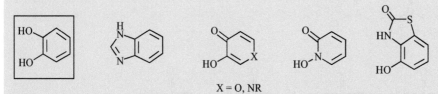

—OH (boxed) —NHC(O)R —$NHSO_2R$ —CH_2OH —$NHC(O)NH_2$

—NHCN —$CH(CN)_2$

6. Catechol

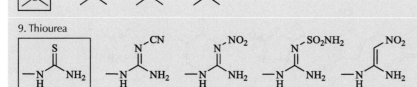

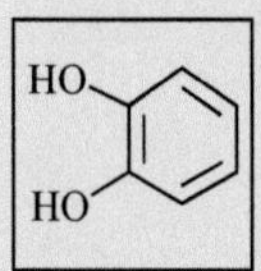

X = O, NR

7. Halogen

X (boxed) CF_3 CN $N(CN)_2$ $C(CN)_3$

8. Thioether

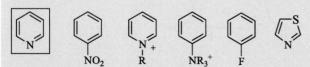

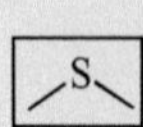

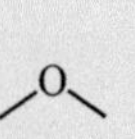

9. Thiourea

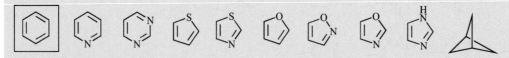

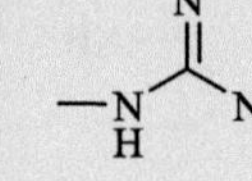

10. Azomethine

—N= (boxed) —C(CN)=

11. Pyridine

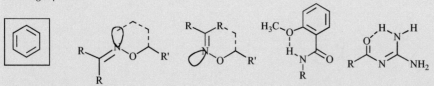

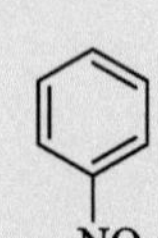

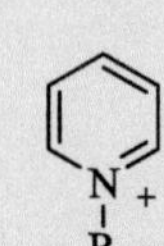

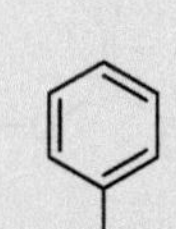

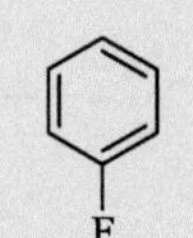

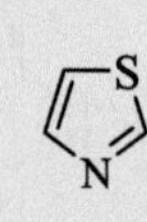

12. Benzene

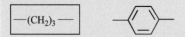

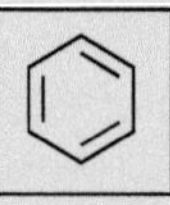

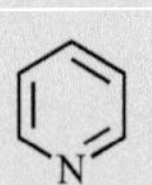

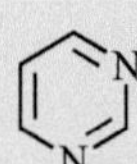

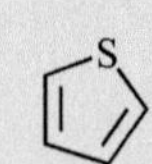

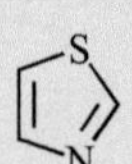

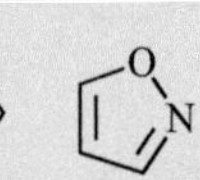

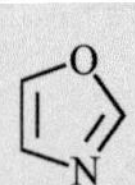

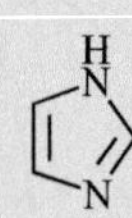

13. Ring equivalents

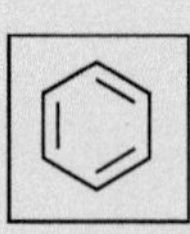

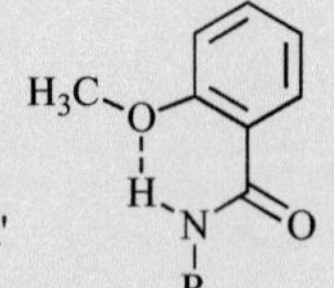

14. Spacer group

—$(CH_2)_3$— (boxed)

15. Hydrogen

H (boxed) F D

TABLE 2.11 Examples of Bioisosteric Analogs

1. Neuroleptics (antipsychotics)

$(CH_2)_3$—X— R, N, NH

X = C=O or CHCN

2. Anti-inflammatory agents

CH_3O, X, O, CH_3, N, O, Cl

X = OH (indomethacin)

= NHOH

= (tetrazolyl: N–N, N, N, H)

OH, Y, O, CH_3, Z

Y = CH_3O Z = Cl

Y = F Z = SCH_3 (sulindac)

3. Antihistamines

R—X—$(CH_2)_n$—Y

X = NH, O, CH_2

Y = N$(CH_3)_2$ (n = 2)

(imidazolyl: N, N, H) (n = 1)

(imidazolyl: H, N, N) (n = 1, 2)

compounds that are selective for the cyclooxygenase-2 isozyme over the cyclooxygenase-1 isozyme (see Chapter 5, Section 5.3.2.2.2) was substituted by an oxazolone ring (i.e., the S was replaced by O), the selectivity for the two isozymes was reversed.[269]

Perusal of Table 2.9, and especially of Table 2.10, makes it clear that in making a bioisosteric replacement, one or more of the following parameters will change: size, shape, electronic distribution, lipid solubility, water solubility, pK_a, polarizability, chemical reactivity, and hydrogen bonding. It is because of these subtle changes that bioisosterism is effective. This approach allows the medicinal chemist to tinker with only some of the parameters to augment the potency, selectivity, and duration of action and to reduce toxicity. Thus, bioisosteric modifications made to a molecule may have one or more of the following effects:

1. *Structural.* If the moiety that is replaced by a bioisostere has a structural role in holding other functionalities in a particular geometry, then size, shape, polarizability, and hydrogen bonding may be affected.
2. *Receptor interactions.* If the moiety replaced is involved in a specific interaction with a receptor or enzyme, then all of the parameters except lipid and water solubility may be affected.

3. *Pharmacokinetics.* If the moiety replaced is necessary for absorption, transport, or excretion of the compound, then lipophilicity, hydrophilicity, pK_a, and hydrogen bonding may be affected.
4. *Metabolism.* If the moiety replaced is involved in blocking or aiding metabolism, then the metabolic reactivity may be affected.

Multiple alterations may be necessary to counterbalance effects. For example, if modification of a functionality involved in binding also decreases the lipophilicity of the molecule, thereby reducing its ability to penetrate cell walls and cross other membranes, the molecule can be modified at a different site by substitution with a more lipophilic group to increase absorption. But where can these bioisosteric replacements be made? A pharmacophore study (see Section 2.2.1) could have identified the auxophoric groups that could be modified without an effect on receptor binding (the scissions that led to little change in potency); those are sites that often can be bioisosterically replaced with retention of activity. During a lead optimization effort, it is typical to try a few bioisosteric replacements wherever they are reasonably accessible synthetically. When the SAR indicates that such changes are tolerated, or especially if one or more properties are thereby improved, additional bioisosteric replacements are warranted, even if they are more synthetically challenging. Fluorine is an important bioisostere of hydrogen.[270] Because of its high electronegativity, fluorine can cause a change in geometry of the molecule and influence a preferred conformation stabilized by dipole–dipole interactions[271] or hydrogen bonding.[272] Fluorine also can reduce basicity of neighboring amino groups[273] or increase acidity of alcohols and acids as a result of its electron-withdrawing ability, and it has an important pharmacokinetic effect by blocking metabolism and enhancing absorption.[274] An unusual bioisostere for a phenyl group is bicyclo[1.1.1]pentane, which has comparable dihedral angles and similar distances between substituents as phenyl, but it is more water soluble and metabolically stable.[275] Oxetanes (four-membered cyclic ethers) are very versatile bioisosteres for carbonyl groups (ketones, esters, amides).[276] They have high polarity, can accept hydrogen bonds, and act as electron-withdrawing groups,[277] just like carbonyl groups. They also have lower lipophilicity than gem-dimethyl groups but more lipophilicity than carbonyl groups, greater aqueous solubility, and better metabolic and chemical stability.[278] Bioisosteric replacements might also be useful to improve an intellectual property position on a lead series.

2.2.4.4. Conformational Constraints and Ring-Chain Transformations

Another type of modification that is commonly employed for lead optimization is the incorporation of conformational constraints into the lead molecule. Molecules with rotatable bonds can adopt multiple conformations, and the ideal conformation of the ligand when bound to the target (the *bioactive conformation*) is often unclear. One approach for elucidating the bioactive conformation of a ligand is to determine the structure of the ligand in complex with its target by X-ray crystallography or by use of sophisticated NMR techniques. However, when such experimentally determined structural information is unavailable, another approach is to synthesize analogs in which conformational mobility has been reduced. If such a conformationally constrained analog is biologically active at the target of interest, you can feel confident that the conformation around the constrained bonds closely approximates the bioactive conformation around those bonds. The converse, however, is not necessarily true; that is, if a conformationally constrained molecule is synthesized and found to be poorly active, it is not clear whether this is because the constrained conformation is incorrect for proper binding or whether the extra atoms added to impose the constraint are not sterically accommodated by the target. For example, consider structures **2.77a** and **2.77b** (Figure 2.19), which represent two different conformations of **2.77**. Compound **2.77** was reported as a lead in the search for antimigraine compounds that act through a target known as the calcitonin gene-related peptide receptor.[279] Each of the bonds labeled (**t–z**) is rotatable, so that many reasonably stable conformations of **2.77** are possible. Structure **2.77a** differs from **2.77b** only by rotation about bond **t–u** by 180°. Compound **2.78** constrains the rotation of bonds **t–u** and **u–v** by replacing the amide bond with a pyridine ring; this compound approximates the angles shown in **2.77a**, and binds to the receptor with affinity similar to that of **2.77**. Compound **2.79** constrains bonds **t–u** and **u–v** by replacing the amide bond with an imidazole ring such that the angles shown in **2.77b** are approximated; **2.79** binds to the receptor with much lower affinity than does **2.77**. The favorable biological activity of **2.78** compared to **2.77** thus suggests that the bioactive conformation of **2.77**, at least with respect to bonds **t–u** and **u–v**, is similar to that shown in **2.77a**. Compounds **2.78** and **2.79** are referred to as *conformationally rigid, conformationally restricted, or conformationally constrained analogs* (see Chapter 3, Section 3.2.5.4).

Understanding the bioactive conformation of a molecule can help in the design of additional analogs. In some cases, the constrained conformation interacts with the receptor *more* favorably than does the unconstrained ligand, resulting in a boost in potency. Because the bioactive conformation does not have to be the lowest energy conformation, an increase in binding would occur if the bioactive conformer were a high-energy conformer, and the conformation of the constrained analog mimicked that conformation. Therefore, the constrained analog would be in the bioactive conformation 100% of the time, but the unconstrained compound would rarely be in the bioactive conformation. Of course, the opposite would have been true in this case

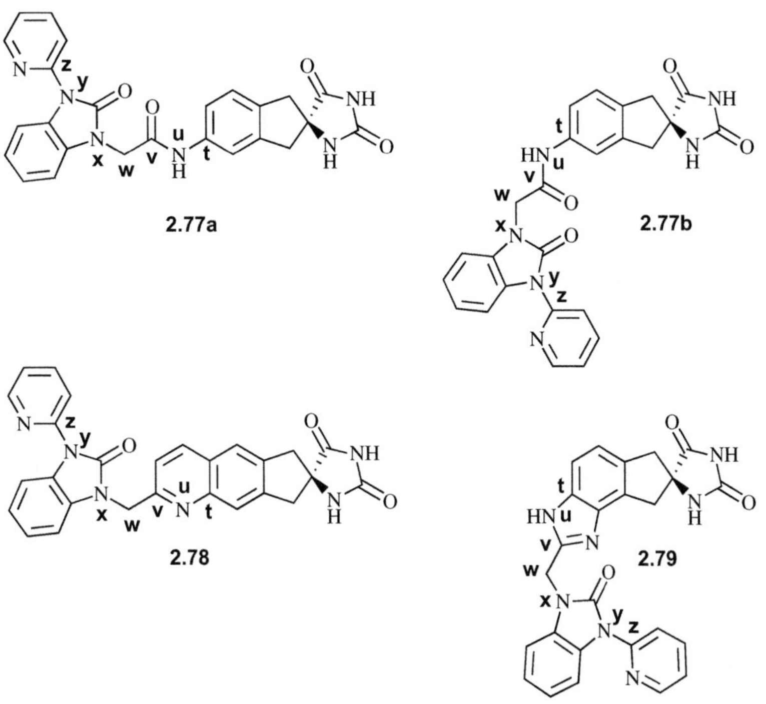

FIGURE 2.19 Conformationally rigid analogs to determine bioactive conformation

if a low-energy conformation were important to binding. In addition, conformational constraints can lead to differentiation in the affinity of a ligand to different targets, resulting in improved selectivity for one target over the other. Moreover, if introduction of a conformational constraint reduces affinity for an enzyme responsible for metabolizing the drug, an increase in metabolic stability may result. As discussed in Section 2.1.2.3.2, it has been asserted that too many rotatable bonds can reduce the likelihood that a compound will be absorbed from the GI tract after oral administration. Thus, reducing the number of rotatable bonds through incorporation of one or more conformational constraints is also a strategy for improving the oral activity of a drug. The technique of introducing conformational constraints has been applied heavily in efforts to improve the properties of peptides as drugs, as discussed further in the next section (Peptidomimetics, Section 2.2.4.5). In Chapter 3 (Section 3.2.5.4), the effect of conformational constraints on interactions with target macromolecules is discussed further.

A straightforward approach to attain conformational constraint is to connect alkyl substituents to give the corresponding cyclic analogs, an approach known as a *ring-chain transformation* (in this case, it probably should be called a chain-ring transformation, but that is not the nomenclature). Chlorpromazine (**2.80**; Thorazine) and **2.81** are equivalent as tranquilizers in animal tests. The branched methyl group and one of the dimethylamino methyl groups of trimeprazine (**2.82**, Vallergan or Temaril) could be connected to give methdilazine (**2.83**; Dilosyn), which has similar antipruritic activity to **2.82** in man. Although large changes in potency were not observed, this ring-chain transformation would have the effect of increasing lipophilicity, which could make the drug more effective in vivo, for reasons discussed in more detail in Section 2.2.5.2.

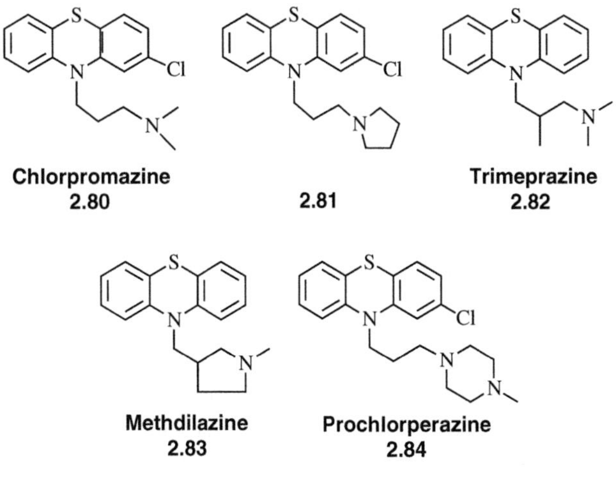

Chlorpromazine **2.80** | **2.81** | Trimeprazine **2.82** | Methdilazine **2.83** | Prochlorperazine **2.84**

When the dimethylamino group of the tranquilizer chlorpromazine (**2.80**) is substituted by a methylpiperazine ring to give prochlorperazine (**2.84**), antiemetic (prevents nausea and vomiting) activity is greatly enhanced. In this case, however, an additional methylamino group also is added, which may have contributed to changing the activity. A ring-chain transformation also can have a significant effect on toxicity.[280] Animals dosed with **2.85** showed severe weight loss, bone marrow toxicity, and death. However, conformationally constrained, ring-chain transformation analog **2.86** exhibited only mild adverse effects with no negative clinical signs.

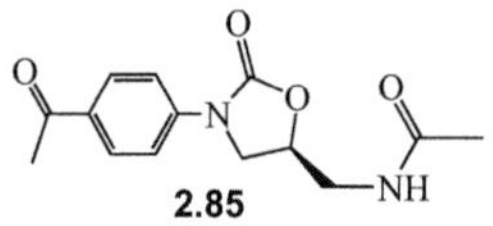

2.85

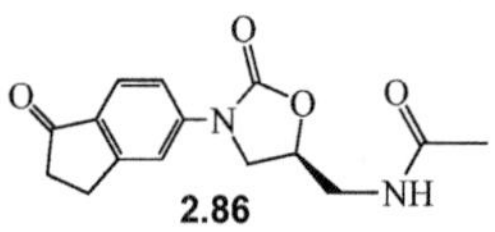

2.86

2.2.4.5. *Peptidomimetics*

Peptides are a very important class of endogenous molecules that bind to a variety of receptors in their action as neurotransmitters, hormones, and neuromodulators,[281] and there are numerous enzymes that are involved in the biosynthesis and catabolism of these peptides. Plants and animals,[282] including human skin,[283] contain a variety of antibiotic peptides. Endogenous peptides such as endorphins, enkephalins, substance P, cholecystokinin, oxytocin, vasopressin, and somatostatin, to name a few, serve as neuropeptides or peptide hormones, with diverse activities including, for example, analgesia,[284] effects on blood pressure,[285] and antitumor activity.[286] However, peptides generally do not make good drug candidates, especially as orally administered drugs, because they are rapidly proteolyzed in the GI tract and serum, and they are poorly bioavailable and rapidly excreted. In addition, many peptides can bind to multiple receptors, especially, to multiple members of the same receptor family. What is needed is a compound that mimics or blocks the biological effect of a peptide by interacting with its receptor or enzyme, but does not have the undesirable characteristics of peptides. This is a *peptidomimetic.*

One way to inhibit the action of an endogenous substance, including an endogenous peptide, is to inhibit an enzyme on the pathway to its biosynthesis. This is the basis for the action of inhibitors of angiotensin-converting enzyme (ACE), which catalyzes the formation of the blood pressure-elevating peptide angiotensin II from its inactive peptide precursor, angiotensin I. Therefore, inhibiting the formation of angiotensin II has the effect of lowering blood pressure (antihypertensive effect). In Chapter 5 (Section 5.2.4.2), captopril (**2.87**) will be discussed in detail as the first orally active ACE inhibitor to reach the market and the first billion-dollar drug for Squibb and Sons (now Bristol-Myers Squibb). Captopril was designed in part from angiotensin I, the peptide substrate of the ACE enzyme, and in part from peptide leads identified from snake venom, taking into account the presence of a key zinc atom in the active site.[287] The successful design of this important orally active drug starting from peptide leads catalyzed major investments during the 1980s and the early 1990s to capitalize on the potential therapeutic properties of endogenous neuropeptides and peptide hormones. The design of peptidomimetics offered an approach toward improving the oral activity, plasma half-life, and receptor selectivity of these peptide leads.

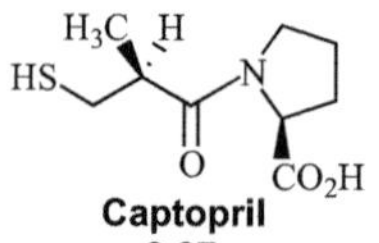

Captopril
2.87

Earlier in this chapter (Section 2.2.1), morphine and morphine analogs (**2.56**) were discussed as potent binders to the μ opioid receptor. In the mid-1970s Hughes and coworkers[288] showed that the endogenous peptides, the enkephalins and β-endorphin, also bound to the same site on the opioid receptor as did morphine. A remarkable resemblance was demonstrated between the *N*-terminal tyrosine structure of these opioid peptides and the morphine phenol ring system, which suggested why they all interacted with these receptors in a similar way.[289] Farmer then proposed that this may be a general phenomenon and that other nonpeptide structures may mimic natural peptide effectors.[290] His postulate was that peptide mimetics (which later became "peptidomimetics") could be designed that would replace peptide backbones while retaining the appropriate topography for binding to a receptor; this initiated the field of peptidomimetics research.

The impetus was in place to attempt designing of peptidomimetics using a native peptide as the lead compound and applying the general principles and approaches for lead modification described in this chapter. For example, lead refinement often includes the formation of conformationally restricted analogs (see Section 2.2.4.4), which hold appropriate pharmacophoric groups in the bioactive conformation for binding to the target receptor.[291] Conformational constraint of single amino acids that are incorporated into the peptidomimetic offer one approach. Examples of conformationally restricted analogs of phenylalanine are shown in Figure 2.20.[292] Alternatively, modifications incorporating conformational restriction across two adjacent amino acid residues may be tried, as exemplified in Figure 2.21.[293]

Another approach involves the design of conformationally restricted analogs that mimic the tertiary structure of the endogenous peptide,[294] such as β-turns (**2.88**,[295] **2.89**,[296] Figure 2.22), α-helices (**2.90**),[297] Ω-loops (**2.91**),[298] and β-strands (**2.92**).[299] This idea can be extended to *scaffold peptidomimetics* in which important pharmacophoric residues are held in the appropriate orientation by a rigid template. Compounds that block the binding of fibrinogen to its receptor (glycoprotein IIb/IIIa) can prevent platelet aggregation and are of potential value in the treatment of strokes and heart attacks.[300] A common β-turn motif that has been found to bind to GPIIbIIIa is arginine–glycine–aspartic acid (or in the one-letter amino acid code, RGD, **2.93**, Figure 2.23). Consequently, a variety of scaffold peptidomimetics for RGD have been designed based on the hypothesis

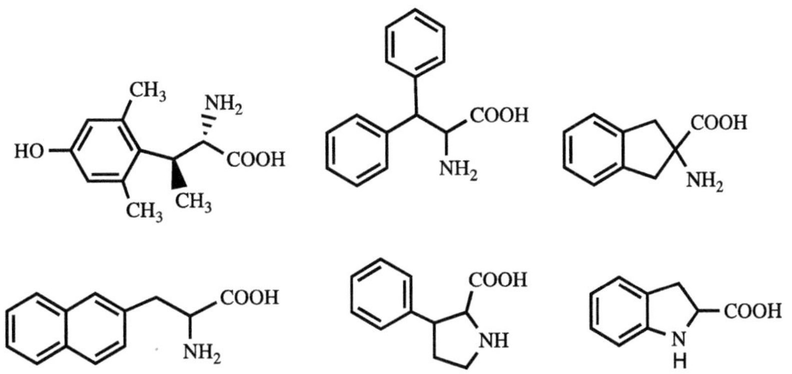

FIGURE 2.20 Conformationally restricted phenylalanine analogs

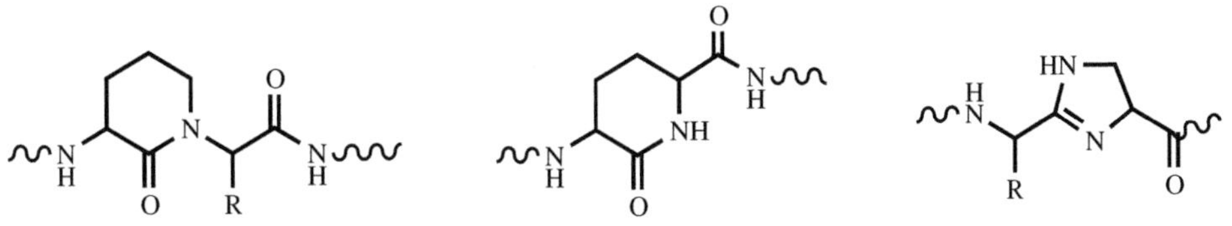

FIGURE 2.21 Conformationally restricted dipeptide analogs

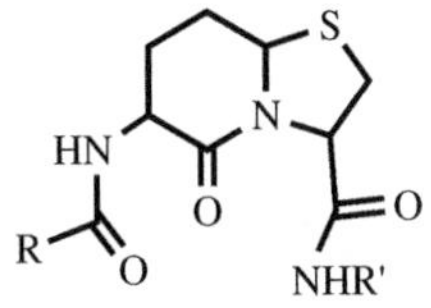

2.88

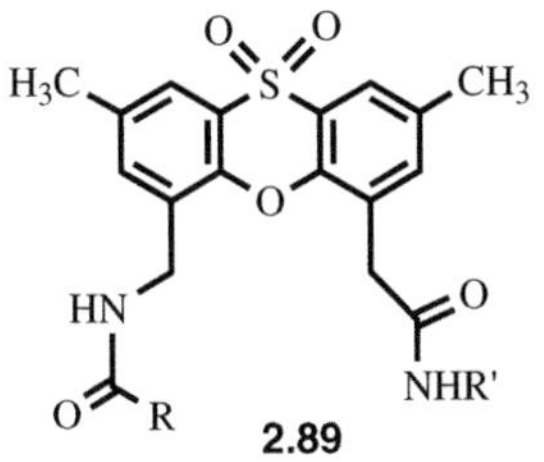

2.89

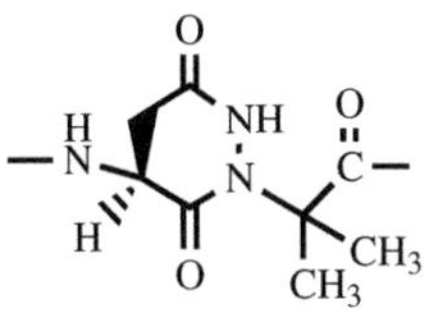

2.90

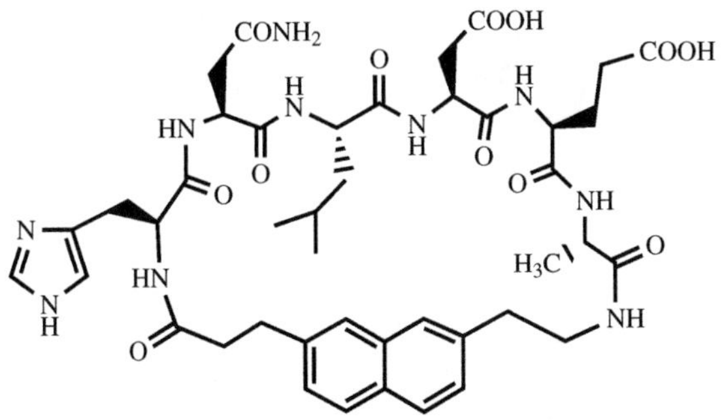

2.91

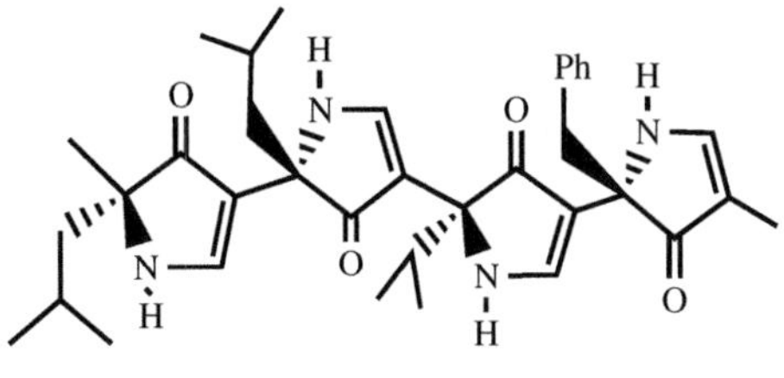

2.92

FIGURE 2.22 Conformationally restricted secondary structure peptidomimetics

that the glycine residue only represents a spacer between the two important recognition residues, arginine and aspartate. Several potent binders to this receptor have been found by replacement of the glycine with more rigid mimics, such as steroid (**2.94**),[301] tetrahydroisoquinolone (**2.95**),[302] and benzodiazepinedione (**2.96**)[303] spacers. A β-D-glucose-based nonpeptide scaffold (**2.98**) was designed as a mimic (note the darkened groups) of the potent somatostatin agonist (see Chapter 3, Section 3.2.3 for a discussion of agonism) **2.97**.[304] A target peptide for the treatment of cognitive disorders, such as Alzheimer disease, is thyrotropin-releasing hormone (pyroGlu-His-ProNH_2, **2.99**)[305]; a scaffold peptidomimetic for this hormone is **2.100**.[306]

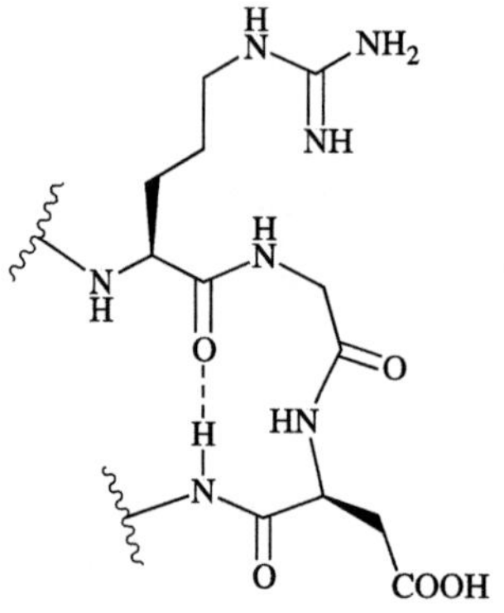

2.93

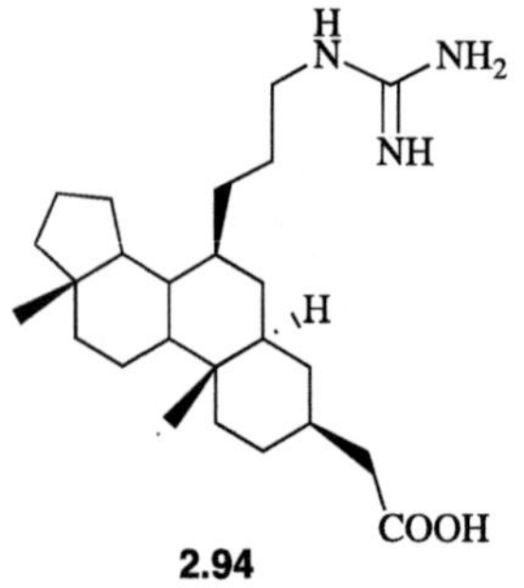

2.94

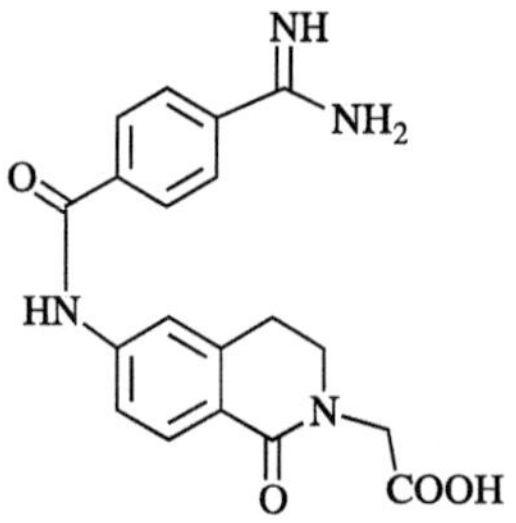

2.95

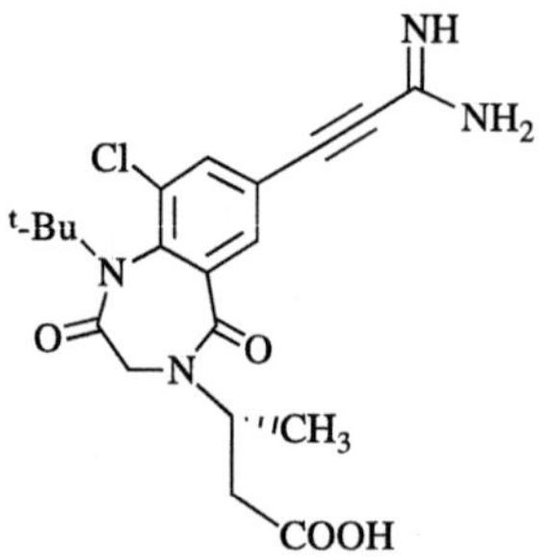

2.96

FIGURE 2.23 RGD scaffold peptidomimetics

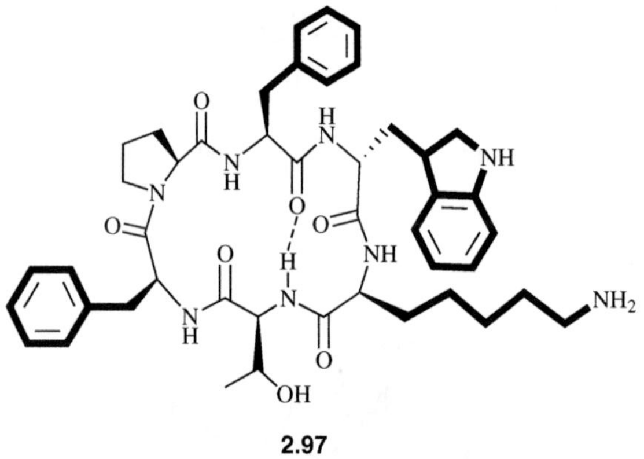

2.97

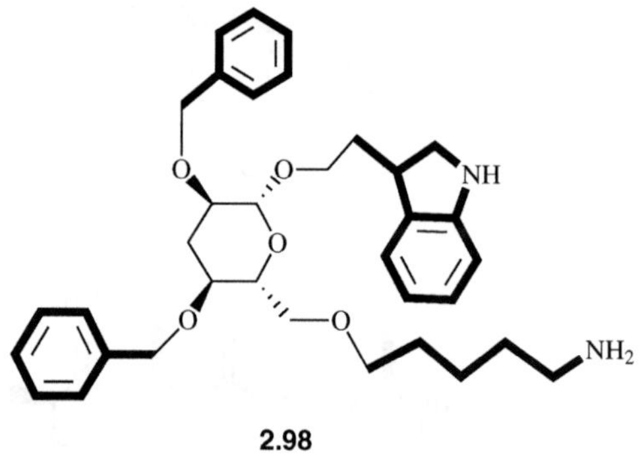

2.98

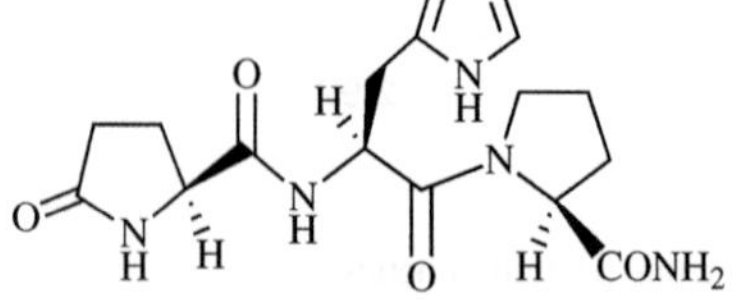

2.99

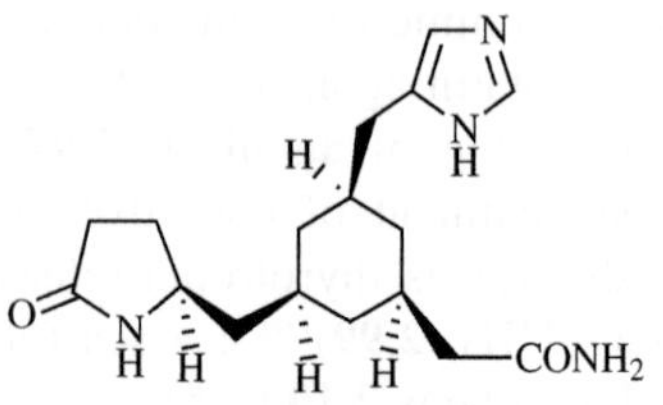

2.100

A goal for peptidomimetics is to replace as much of the peptide backbone as possible with nonpeptide fragments while still maintaining the pharmacophoric groups (usually the amino acid side chains) of the peptide. This also frequently makes the compound more lipophilic, which may increase its bioavailability. Replacement of the amide bond with alternative groups, furthermore, prevents proteolysis and promotes metabolic stability. A common and important approach for the conversion of a peptide lead into a peptidomimetic is the use of peptide backbone isosteres (Table 2.12). Peptides in which the amide bonds are replaced with alternative groups are known as *pseudopeptides*.[307] Several variants of azapeptides (**2.101**, in which one or more of the α-carbons are replaced by *N*)[308] include azatides (**2.102**; azapeptides in which *all*

TABLE 2.12 Peptide Backbone Isosteres for Peptidomimetics

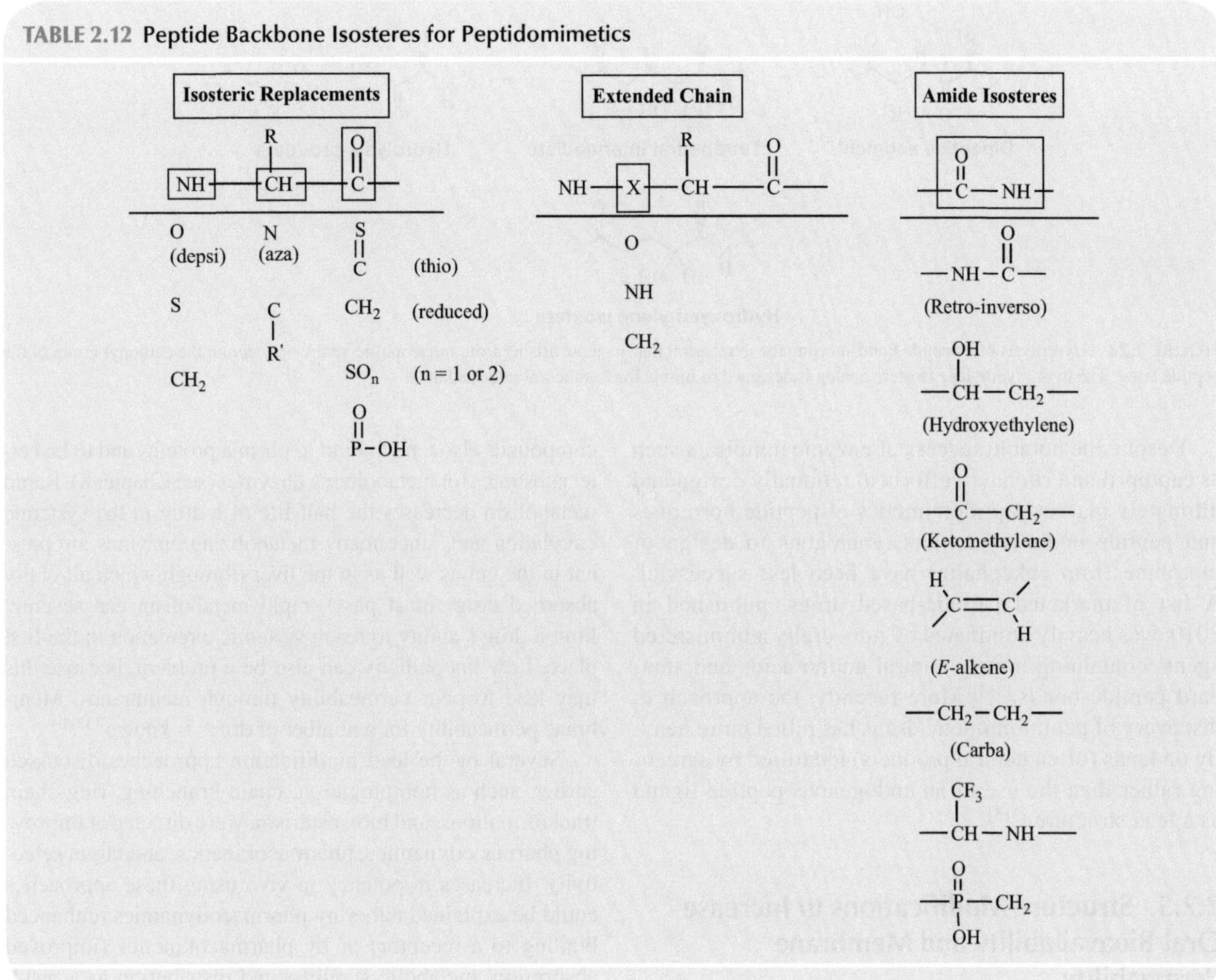

the α-carbons are replaced by *N*)[309] and peptoids (**2.103**, in which the α-CHR groups are replaced by NR units and the NH groups are replaced by CH_2 units).[310] Ultimately, the successful generation of peptidomimetics is aided by an understanding of the conformational, topochemical, and electronic properties of the lead peptide when bound to its target receptor or enzyme.[311]

A notable success story in the design of peptidomimetic drugs includes the discovery and successful commercial launch of human immunodeficiency virus (HIV) protease inhibitors. HIV protease is an enzyme (an aspartyl protease, see Chapter 5, Section 5.2.5) that is critical to the life cycle of HIV, the causative agent of AIDS. The design principles[312] for a number of HIV protease inhibitors had as a major component the replacement of the scissile amide bond (the site of the enzyme-catalyzed hydrolysis) of a peptide substrate with an isosteric group, for example, the hydroxyethylene isostere[313] (Table 2.12); this isostere is representative of a number of variants, all intended to mimic the tetrahedral intermediate formed during enzyme-catalyzed amide bond hydrolysis (Figure 2.24). A more detailed description of the discovery of ritonavir, one of the first HIV protease inhibitors, together with the concept of designing enzyme inhibitors by mimicking an enzyme-bound intermediate, is presented in Chapter 5 (Section 5.2.5).

2.101 **2.102** **2.103**

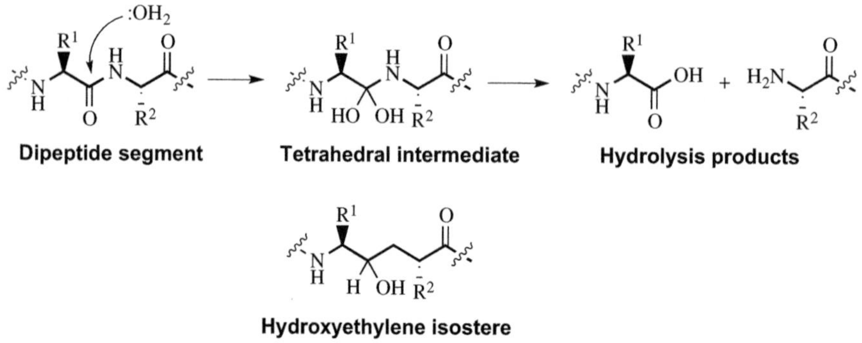

FIGURE 2.24 Hydrolysis of a peptide bond showing the tetrahedral intermediate arising from nucleophilic attack of water on the carbonyl group of the peptide bond. The hydroxyethylene isostere analog is designed to mimic the tetrahedral intermediate.

Despite the notable success of enzyme inhibitors such as captopril and ritonavir, efforts to rationally design and ultimately market peptidomimetics of peptide hormones and peptide neurotransmitters (analogous to design of morphine from enkephalin) have been less successful. A list of marketed peptide-based drugs, published in 2010, was heavily dominated by non–orally administered agents containing mostly natural amino acids and standard peptide bonds.[314] More recently, the approach to discovery of peptidomimetic drugs has relied more heavily on leads (often natural products) identified by screening rather than the use of an endogenous peptide ligand as a lead structure.[315]

2.2.5. Structure Modifications to Increase Oral Bioavailability and Membrane Permeability

Less than 10% of drug candidates entering clinical trials become marketed products. Because of the huge waste of time and resources by having a drug candidate fail late in the drug discovery process because of pharmacokinetic issues, it is prudent to examine pharmacokinetic and metabolic aspects of molecules as early as possible.[316] The use of MS[317] for this purpose is discussed in Chapter 8 (Section 8.3.3). Numerous computational methods have been devised to calculate various properties purported to help predict pharmacokinetic behavior,[318] but reliable predictive capabilities across diverse compound sets are often still lacking. On the other hand, use of in vitro assays to predict in vivo activity has met with some success. Thus, a graphical model for estimating high, medium, or low oral bioavailability of drugs in humans, rats, dogs, and guinea pigs, based on their in vitro permeability through human intestinal epithelial (Caco-2) cells together with their in vitro liver enzyme metabolic stability rates, gave excellent results.[319]

Low water solubility of a compound, often associated with high lipophilicity, can be a limiting factor in absorption, for example, from the GI tract,[320] and highly lipophilic compounds also tend to bind to plasma proteins and to be better substrates for metabolizing enzymes (see Chapter 8). Rapid metabolism decreases the half-life of a drug in the systemic circulation and, since many metabolizing enzymes are present in the gut as well as in the liver (through which all orally absorbed drugs must pass), rapid metabolism can severely limit a drug's ability to reach systemic circulation in the first place. Low lipophilicity can also be a problem, because this may lead to poor permeability through membranes. Membrane permeability for a number of drugs is known.[321]

Several of the lead modification approaches discussed earlier, such as homologation, chain branching, ring-chain transformations, and bioisosterism, were directed at improving pharmacodynamics, pharmacokinetics, and target selectivity. Increases in potency in vivo using these approaches could be explained either by pharmacodynamics (enhanced binding to a receptor) or by pharmacokinetics (improved absorption, metabolic stability, and distribution as a result of, for example, optimized lipophilicity).

Because of the importance of lipophilicity in drug design,[322] it is essential to understand how to determine lipophilicities of compounds. However, because SAR studies often involve modifying, adding, or removing substituents around the central core of a lead compound, it is also important to determine lipophilicities of substituents in order to understand the effect of individual substituents on lipophilicity. The basis for the determination of the lipophilicities of substituents, as presented by Corwin Hansch and coworkers,[323] is derived from the earlier postulate by L. P. Hammett on how the electronic effects of substituents affect the reactivity of organic molecules, known as the Hammett equation. Those of you who know how to derive this equation can skip ahead to Section 2.2.5.2.

2.2.5.1. Electronic Effects: The Hammett Equation

Hammett's postulate was that the electronic effects (both the inductive and resonance effects) of a set of substituents should be similar on different organic reactions.

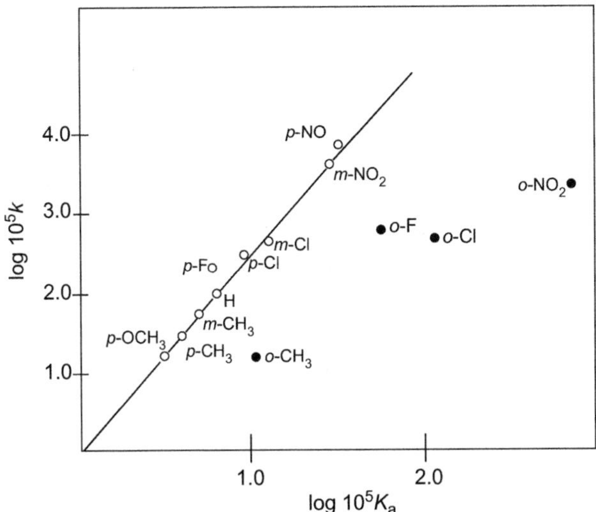

FIGURE 2.25 A typical linear free energy relationship for the dissociation of substituted benzoic acids (K_a) vs the rates of alkaline hydrolysis of substituted ethyl benzoates (k)

Therefore, if values could be assigned to substituents in a standard organic reaction, these same values could be used to estimate rates in a new organic reaction. This was the first approach that allowed the prediction of reaction rates. Hammett chose benzoic acids as the standard system. Consider the reactions shown in Scheme 2.7. Intuitively, it seems reasonable that as X becomes electron withdrawing (relative to H), the equilibrium constant (K_a) should increase (the reaction should be favored to the right) because X is inductively pulling electron density from the carboxylic acid group, making it more acidic (a reactant argument); it is also stabilizing the negative charge on the carboxylate group in the transition state (a product argument). Conversely, when X is electron donating, the equilibrium constant should decrease. A similar relationship should exist for a rate constant (k) where charge develops in the transition state (consider ground-state and transition-state stabilizations). Hammett chose the reactions shown in Schemes 2.7 and 2.8 as the standard system; note that the same benzoate products are obtained in both. If K_a is measured from Scheme 2.7 and k from Scheme 2.8

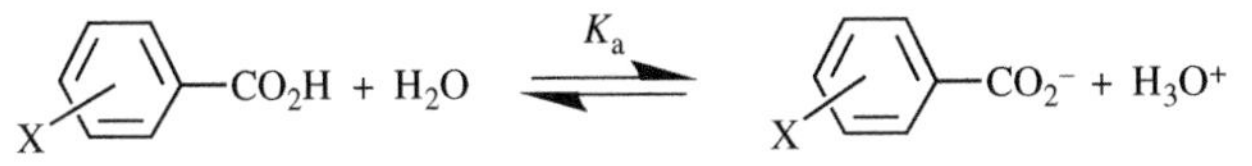

$$X\text{-}C_6H_4\text{-}CO_2H + H_2O \underset{}{\overset{K_a}{\rightleftharpoons}} X\text{-}C_6H_4\text{-}CO_2^- + H_3O^+$$

SCHEME 2.7 Equilibrium constants for ionization of substituted benzoates

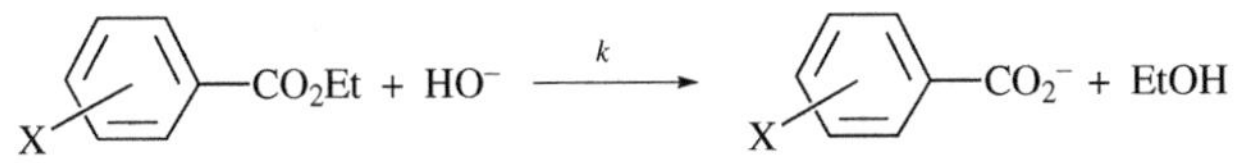

$$X\text{-}C_6H_4\text{-}CO_2Et + HO^- \xrightarrow{k} X\text{-}C_6H_4\text{-}CO_2^- + EtOH$$

SCHEME 2.8 Rate constants for saponification of substituted ethyl benzoates

for a series of substituents X, and the data are expressed in a double logarithmic plot (Figure 2.25), then a straight line can be drawn through most of the data points. This is known as a *linear free energy relationship*. When X is a meta- or para-substituent, then virtually all the points fall on the straight line; the ortho-substituent points are badly scattered. The inital Hammett relationship does not hold for ortho-substituents because of steric interactions and polar effects. The linear correlation for the meta- and para-substituents is observed for rate and equilibrium constants for a wide variety of organic reactions. The straight line can be expressed by Eqn (2.2), where the two variables are

$$\log k = \rho \log K + C \qquad (2.2)$$

log k and log K. The slope of the line is ρ, and the intercept is C. When there is no substituent, i.e., when X=H, then Eqn (2.3) holds. Subtraction of Eqn (2.3) from equation

$$\log k_o = \rho \log K_o + C \qquad (2.3)$$

2.2 gives Eqn (2.4), where k and K are the rate and equilibrium constants, respectively, for compounds with a

$$\log k/k_o = \rho \log K/K_o \qquad (2.4)$$

substituent X, and k_o and K_o are the rate and equilibrium constants, respectively, for the parent compound (X=H). If $\log K/K_o$ is defined as σ, then Eqn (2.4) reduces to Eqn (2.5), the *Hammett equation*. The *electronic parameter*, σ, depends on the

$$\log k/k_o = \rho\sigma \qquad (2.5)$$

electronic properties and position of the substituent on the ring and therefore, is also called the *substituent constant*. The more electron withdrawing a substituent, the more positive is its σ value (relative to H, which is set at 0.0); conversely, the more electron donating, the more negative is its σ value. The meta σ constants result from inductive effects, but the para σ constants correspond to the net inductive and resonance effects. Therefore, σ_{meta} and σ_{para} for the same substituent, generally, are not the same.

The ρ values (the slope) depend on the particular type of reaction and the reaction conditions (e.g., temperature and solvent) and, therefore, are called *reaction constants*. The importance of ρ is that it is a measure of the sensitivity of the reaction to the electronic effects of the meta- and para-substituents. A large ρ, either positive or negative, indicates great sensitivity to substituent effects. Reactions that are favored by high electron density in the transition state (such as reactions that proceed via carbocation intermediates) have negative ρ values (i.e., the linear free energy relationship has a negative slope); reactions that are aided by electron withdrawal (such as reactions that proceed via carbanion intermediates) have positive ρ values.

2.2.5.2. *Lipophilicity Effects*

2.2.5.2.1. Importance of Lipophilicity

Hansch believed that, just as the Hammett equation relates the electronic effects of substituents to reaction rates, there should be a linear free energy relationship between lipophilicity and biological activity. Hansch proposed that the first step in the overall process was a diffusion process, in which the drug made its way from a dilute solution outside of the cell to a particular site in the cell. This was visualized as being a relatively slow process, the rate of which is highly dependent on the molecular structure of the drug. For the drug to reach the site of action, it must be able to interact with two different environments, namely, a lipophilic environment (e.g., membranes) and an aqueous environment (e.g., the cytoplasm inside the cell and the extracellular fluid outside the cell). The cytoplasm of a cell is essentially a dilute solution of salts in water; all living cells are surrounded by a nonaqueous phase, the membrane. The functions of membranes are (1) to protect the cell from water-soluble substances, (2) to form a surface to which enzymes and other proteins can attach to produce a localization and structural organization, and (3) to separate solutions of different electrochemical potentials (e.g., in nerve conduction). One of the most important membranes is known as the *blood–brain barrier*,[324] a membrane that surrounds the capillaries of the circulatory system in the brain and protects the CNS from passive diffusion of undesirable polar chemicals from the bloodstream. This is an important prophylactic boundary, but it can also block the delivery of CNS drugs to their site of action.

A simple model of a cell membrane is the fluid mosaic model (Figure 2.26).[325] In this depiction, integral proteins are embedded in a lipid bilayer; peripheral proteins are associated with only one membrane surface. The structure of the membrane is primarily determined by the structure of the lipids of which it is composed. The principal classes of lipids found in membranes are neutral cholesterol (**2.104**) and the ionic phospholipids, e.g., phosphatidylcholine (**2.105**, R=$CH_2CH_2N(CH_3)_3^+$), phosphatidylethanolamine (**2.105**, R=$CH_2CH_2NH_3^+$), phosphatidylserine (**2.105**, R=$CH_2CH(COO^-)NH_3^+$), phosphatidylinositol (**2.105**, R=inositol), and sphingomyelin (**2.106**, R=P(=O)(O^-)$OCH_2CH_2N(CH_3)_3^+$); R′CO and R″CO in **2.105** and **2.106** are derived from fatty acids. Glycolipids (**2.106**, R=sugar) are also important membrane constituents. All of these lipids are amphipathic, which means that one end of the molecule is hydrophilic (water soluble) and the other is hydrophobic or, if you prefer, lipophilic (water insoluble; soluble in organic solvents). Thus, the hydroxyl group in cholesterol, the ammonium groups in the phospholipids, and the sugar residue in the glycolipids are the polar, hydrophilic ends, and the steroid and hydrocarbon moieties are the lipophilic ends. The hydrocarbon part (R′ and R″) actually can be a mixture of chains from 14 to 24 carbon atoms long; approximately 50% of the chains contain a double bond. The polar groups of the lipid bilayer are in contact with the aqueous phase; the hydrocarbon chains project toward each other in the interior with a space between the layers. The stability of the membrane arises from the stabilization of the ionic charges by ion–dipole interactions (see Chapter 3, Section 3.2.2.3) with the water and from association of the nonpolar groups. The hydrocarbon chains are relatively free to move; therefore, the core is similar to a liquid hydrocarbon.

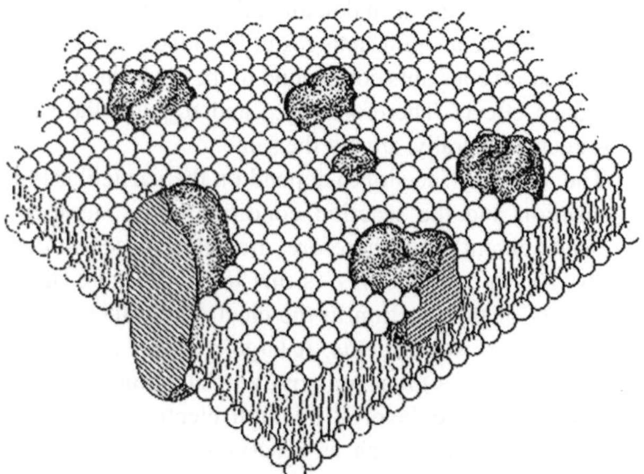

FIGURE 2.26 Fluid mosaic model of a membrane. The balls represent polar end groups, and the wavy lines are the hydrocarbon chains of the lipids. The masses embedded in the lipid bilayer are proteins. *From Singer, S. J. and Nicolson, G. L. (1972).* Science ***175****, 720. Reprinted with permission from AAAS.*

2.104

2.105

2.106

2.2.5.2.2. Measurement of Lipophilicities

It occurred to Hansch that the fluidity of the hydrocarbon region of the membrane may help explain the correlation noted by Richet,[326] Overton,[327] and Meyer[328] between lipid solubility and biological activity. He first set out to measure the lipophilicities of various compounds then to determine lipophilicities of substituents. But how should the lipophilicities be measured? The most relevant method would be to determine their solubility in membranes or vesicles. However, as an organic chemist, Hansch probably realized that if he set a scale of lipophilicities based on membrane solubility, which required the researcher to prepare membranes or vesicles, there was no way organic chemists, especially in the 1960s, would ever bother to use this method, and it would become very limited. So he decided to propose a model for a membrane and determine lipophilicities by a simple methodology that organic chemists would not hesitate to use. The model for the first step in drug action (transport to the site of action) would be the solubility of the compound in 1-octanol, which would simulate a lipid membrane, relative to water (or aqueous buffer, the model for the cytoplasm). 1-Octanol has a long saturated alkyl chain and a hydroxyl group for hydrogen bonding, and it dissolves water to the extent of 1.7 M (saturation). This combination of lipophilic chains, hydrophilic groups, and water molecules gives 1-octanol properties that reasonably simulate those of natural membranes.

As a measure of lipophilicity, Hansch proposed the *partition coefficient, P*, between 1-octanol and water,[329,330] and P was determined by Eqn (2.6), where α is the degree of dissociation of the compound in water calculated from ionization constants; ionization makes the compound more

$$P = \frac{[\text{compound}]_{\text{oct}}}{[\text{compound}]_{\text{aq}}\,(1-\alpha)} \tag{2.6}$$

soluble in water than the structure appears, so this has to be taken into account. The concept of partition coefficient was discussed earlier in this chapter (Section 2.1.2.3.2) in the context of assessing compounds for drug-likeness, supporting the importance of this parameter in drug discovery. The partition coefficient is derived experimentally by placing a compound in a shaking device (like a separatory funnel) with varying volumes of 1-octanol and water, and after good mixing, determining the concentration of the compound in each layer (by gas chromatography or HPLC), and then employing Eqn (2.6) to calculate P.[331] The value of P varies slightly with temperature and concentration of the solute, but with neutral molecules in dilute solutions (<0.01 M) and small temperature changes (±5 °C), variations in P are minor.

Collander[332] had shown previously that the rate of movement of a variety of organic compounds through cellular material was approximately proportional to the logarithm of their partition coefficients between an organic solvent and water. Therefore, as a model for a drug traversing through the body to its site of action, the relative potency of the drug, expressed as log $1/C$, where C is the concentration of the drug that produces some standard biological effect, was related by Hansch et al.[333] to its lipophilicity by the parabolic expression shown in Eqn (2.7).

$$\log 1/C = -\mathrm{k}(\log P)^2 + k'(\log P) + k'' \tag{2.7}$$

On the basis of Eqn (2.6), it is apparent that if a compound is more soluble in water than in 1-octanol, $P<1$, and, therefore, $\log P$ is negative. Conversely, a molecule more soluble in 1-octanol than in water has a $P>1$, and $\log P$ is positive. Therefore, the more positive the $\log P$, the more lipophilic it is. The larger the value of P, the more there will be an interaction of the drug with the lipid phase (i.e., membranes). As P approaches infinity, micelles will form and/or the drug interaction will become so great that the drug will not be able to cross the aqueous phase, and it will localize in the first lipophilic phase with which it comes into contact. As P approaches zero, the drug will be so water soluble that it will not be capable of crossing the lipid phase and will localize in the aqueous phase. Somewhere between $P=0$ and $P=\infty$, there will be a value of P such that drugs having this value will be least hindered in their journey through macromolecules to their site of action. This value is called $\log P_0$, the logarithm of the optimum partition coefficient for biological activity. This random walk analysis supports the parabolic relationship (Eqn (2.7)) between potency ($\log 1/C$) and $\log P$ (Figure 2.27). Note the correlation of Figure 2.27 with the generalization regarding homologous series of compounds (Section 2.2.4.1; Figure 2.18). An increase in the alkyl chain length increases the lipophilicity of the molecule; apparently, the $\log P_0$ often occurs in the range of five to nine carbon atoms. Hansch et al.[334] found that a number of series of nonspecific hypnotics had similar $\log P_0$ values, approximately 2, and they suggested that this is the value of $\log P_0$ needed for penetration into the CNS, i.e., for crossing the blood–brain barrier. If a hypnotic agent has a $\log P$ considerably different from 2, then its activity probably is derived from mechanisms other than just lipid transport. If a lead compound has modest CNS

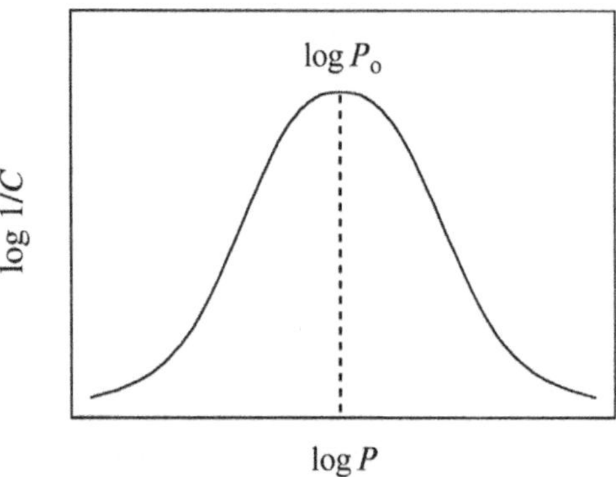

FIGURE 2.27 Effect of $\log P$ on biological response. P is the partition coefficient, and C is the oncentration of the compound required to produce a standard biological effect. Log P_0 is the optimal log P for biological activity.

TABLE 2.13 Change in log *D* as a Function of pH for Metoprolol (2.107)[1]

(Structure of metoprolol: H_3CO–CH₂CH₂–C₆H₄–O–CH₂–CH(OH)–CH₂–$NHCH(CH_3)_2$)

2.107

pH	log *D*
2.0	−1.31
3.0	−1.31
4.0	−1.31
5.0	−1.28
5.5	−1.21
6.0	−1.05
6.5	−0.75
7.0	−0.34
7.5	0.12
8.0	0.59
8.5	1.03
9.0	1.39
10.0	1.73

[1]*The authors are grateful to Karolina Nilsson and Ola Fjellström (AstraZeneca) for providing the log D values as a function of pH using ACD software.*

activity and has a log *P* value of 0, it would be reasonable to synthesize an analog with a higher log *P*.

The term *α* in Eqn (2.6) attempts to correct for the degree of ionization of ionizable substances such as carboxylic acids and amines. However, the degree of ionization depends not only on the pK_a of the substance but also on the pH of the aqueous phase, and it is useful to consider that the pH inside the GI tract varies widely from the stomach (pH 1–2) through to the end of the small intestine (pH ~8). The term log *D* (the log of the distribution coefficient, generally between 1-octanol and aqueous buffer) describes the log *P* of an ionizable compound at a particular pH.[335] For example, log $D_{4.5}$ is the log *P* of an ionizable compound at pH 4.5. The log *D* value will change for ionizable compounds as a function of pH, whereas log *P* of nonionizable compounds will be independent of pH. Table 2.13 shows how log *D* for the antihypertensive drug metoprolol (**2.107**, Toprol-XL) changes as a function of pH. Note that at low pH values, the amine is protonated, lowering the log *D* value. As the pH is increased, the equilibrium starts to favor the neutral free base form, which is more lipophilic.

Although it is valuable to be able to determine the lipophilicity of a molecule, for lead modification purposes you need to be able to predict, prior to synthesis of the compound, what the lipophilicity of an unknown molecule will be. To do that, it is necessary to know the lipophilicities of substituents and atoms. In the same way that substituent constants were derived by Hammett for the electronic effects of atoms and groups (*σ* constants), Hansch and coworkers[336] derived substituent constants for the contribution of individual atoms and groups to the partition coefficient. The lipophilicity substituent constant, *π*, is defined by Eqn (2.8),

$$\pi = \log P_X - \log P_H = \log \frac{P_X}{P_H} \tag{2.8}$$

which has the same derivation as the Hammett equation. The term P_X is the partition coefficient for the compound with substituent X and P_H is the partition coefficient for the parent molecule (X=H). As in the case of the Hammett substituent constant *σ*, *π* is additive and constitutive. *Additive* means that multiple substituents exert an influence equal to the sum of the individual substituents. *Constitutive* indicates that the effect of a substituent may differ depending on the molecule to which it is attached or on its environment. Alkyl groups are some of the least constitutive groups. For example, methyl groups attached at the meta- or para-positions of 15 different benzene derivatives had π_{CH_3} values with a mean and standard deviation of 0.50±0.04. Because of the additive nature of *π* values, π_{CH_2} can be determined as shown in Eqn (2.9), where the log *P* values are obtained from standard tables.[337]

$$\begin{aligned}\pi_{CH_2} &= \log P_{\text{nitroethane}} - \log P_{\text{nitromethane}} \\ &= 0.18 - (-0.33) = 0.51\end{aligned} \tag{2.9}$$

Because, by definition, $\pi_H = 0$, then $\pi_{CH_2} = \pi_{CH_3}$. However, beware that π_{CH_2OH} does not equal π_{CH_3O} because there is a difference in hydrogen bonding for substituents with and without hydroxyl groups, and, therefore, a difference in water solubility between these two substituents. Note that *π* represents the lipophilicity of a substituent and log *P* is the lipophilicity of a compound.

As was alluded to in the section on chain branching (Section 2.2.4.2), branching in an alkyl chain lowers the log *P* or *π* as a result of larger molar volumes and shapes of branched compounds. As a rule of thumb, the log *P* or *π* is lowered by 0.2 unit per branch. For example, the $\pi_{i\text{-Pr}}$ in 3-isopropylphenoxyacetic acid is 1.30; $\pi_{n\text{-Pr}}$ is 3(0.5) = 1.50, or 0.2 greater than $\pi_{i\text{-Pr}}$. Another case where *π* values are fairly constant is conjugated systems, as exemplified by $\pi_{CH=CHCH=CH}$ in Table 2.14.

Inductive effects are quite important to lipophilicity.[338] In general, electron-withdrawing groups increase *π* when a hydrogen-bonding group is involved. For example, π_{CH_2OH} varies as a function of the proximity of an electron-withdrawing phenyl group (Eqn (2.10)), and π_{NO_2} varies as a function of the inductive effect of the nitro group on the

TABLE 2.14 Constancy of π for –CH=CH–CH=CH–[1]

			$\pi_{CH=CHCH=CH}$		
log *P* (N, H)	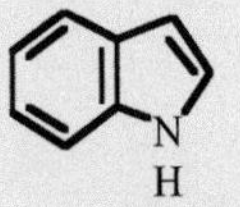 –log *P* (N, H)	 =	2.14 – 0.75	=	1.39
log *P* (N)	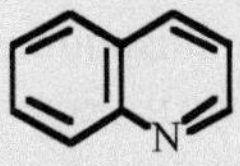 –log *P* (N)	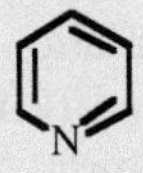 =	2.03 – 0.65	=	1.38
log *P* (N)	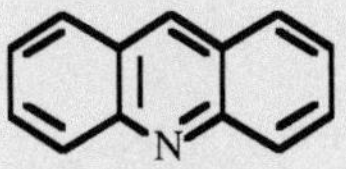 –log *P* (N)	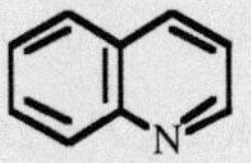 =	3.40 – 2.03	=	1.37
log *P* (O)	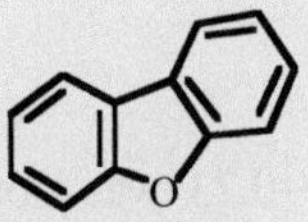 –log *P* (O)	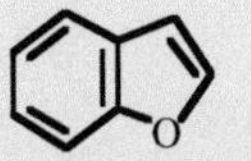 =	4.12 – 2.67	=	1.45
log *P* (S)	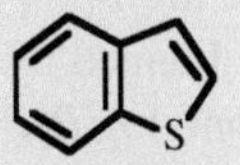 –log *P* (S)	 =	3.12 – 1.81	=	1.31
log *P*	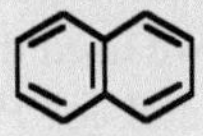 –log *P*	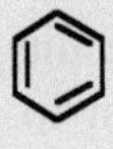 =	3.45 – 2.13	=	1.32
2/3 log *P*		=	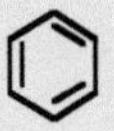 2/3(2.13)	=	1.42
log *P* (OH)	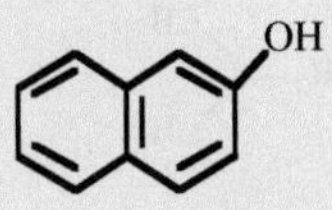 –log *P* (OH)	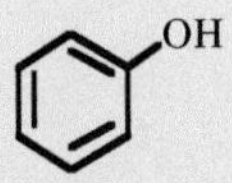 =	2.84 – 1.46	=	1.38
Average					1.38 ± 0.046

[1]*Reprinted with permission from Hansch, C.; Steward, A. R.; Anderson, S. M.; Bentley, D. J. Med. Chem.* **1968**, *11, 1. Copyright 1968 American Chemical Society.*

hydroxyl group (Eqn (2.11)). The electron-withdrawing inductive effects of the phenyl group (Eqn (2.10))[339] and the nitro group (Eqn (2.11))[340] make the nonbonded electrons on the OH group less available for hydrogen bonding,

$$\begin{aligned}\pi_{CH_2OH} &= \log P_{Ph(CH_2)_2OH} - \log P_{PhCH_3} = -1.33 \\ \pi_{CH_2OH} &= \log P_{PhCH_2OH} - \log P_{PhH} = -1.03\end{aligned} \quad (2.10)$$

$$\begin{aligned}\pi_{NO_2} &= \log P_{PhNO_2} - \log P_{PhH} = -0.28 \\ \pi_{NO_2} &= \log P_{4-NO_2PhCH_2OH} - \log P_{PhCH_2OH} = 0.11\end{aligned} \quad (2.11)$$

thereby reducing the affinity of this functional group for the aqueous phase. This, then, increases the log *P* or π. Also note in Eqns 2.10 and 2.11 that, because $\pi_H = 0$ by definition, $\log P_{benzene} = \pi_{Ph}$. These examples reinforce the notion of the constitutiveness of π values.

Resonance effects are also important to the lipophilicity much the same way as are inductive effects. Delocalization of nonbonded electrons into aromatic systems decreases their availability for hydrogen bonding with the aqueous phase, and, therefore, increases the π. This is supported by the general trend that aromatic π_X values are greater than aliphatic π_X values, again emphasizing the constitutive nature of π and log *P*.

Steric effects are variable. If a group sterically shields nonbonded electrons, then aqueous interactions will decrease, and the π value will increase. However, crowding of functional groups involved in hydrophobic interactions (Chapter 3, Section 3.2.2.6) will have the opposite effect.

Conformational effects can also affect the π value. The π_X values for $Ph(CH_2)_3X$ are consistently lower (more water soluble) than the π_X values for $CH_3(CH_2)_3X$ (Table 2.15).

TABLE 2.15 Effect of Folding of Alkyl Chains on π

X	π_x (Aromatic)[1]	π_x (Aliphatic)[2]	$\Delta\pi_x$
OH	−1.80	−1.16	0.64
F	−0.73	−0.17	0.56
Cl	−0.13	0.39	0.52
Br	0.04	0.60	0.56
I	0.22	1.00	0.78
COOH	−1.26	−0.67	0.59
CO_2CH_3	−0.91	−0.27	0.64
$COCH_3$	−1.26	−0.71	0.55
NH_2	−1.85	−1.19	0.66
CN	−1.47	−0.84	0.63
OCH_3	−0.98	−0.47	0.51
$CONH_2$	−2.28	−1.71	0.57
Average			0.60±0.05

[1] $\log P_{Ph(CH_2)_3X} - \log P_{Ph(CH_2)_3H}$
[2] $\log P_{CH_3(CH_2)_3X} - \log P_{CH_3(CH_2)_3H}$

This phenomenon is believed to be the result of folding of the side chain onto the phenyl ring (**2.108**), which means a smaller apolar surface for organic solvation. The folding may be caused by the interaction of the CH_2-X dipole with the phenyl π-electrons and by intramolecular hydrophobic interactions.

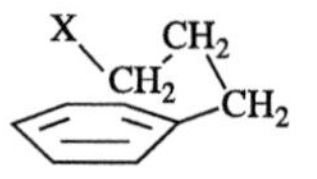

2.108

Two examples follow to show the additivity of π constants in predicting $\log P$ values. A calculation of the $\log P$ for the anticancer drug, diethylstilbestrol (**2.109**) is shown in Eqn (2.12). In Eqn (2.12), $\pi_{CH=CH} = 1/2(\pi_{CH=CHCH=CH})$, which, upon substituting the average value from Table 2.14,

$$\begin{aligned} \text{Calc.log } P &= 2\pi_{CH_3} + 2\pi_{CH_2} + \pi_{CH=CH} + 2 \log P_{PhOH} - 0.40 \\ &= 2(0.50) + 2(0.50) + 0.69 + 2(1.46) - 0.40 \\ &= 5.21 \end{aligned} \qquad (2.12)$$

gives 1/2(1.38) = 0.69; −0.40 is added into the equation to account for two branching points (each end of the alkene). The calculated $\log P$ value of 5.21 is quite remarkable considering that the experimental $\log P$ value is 5.07.

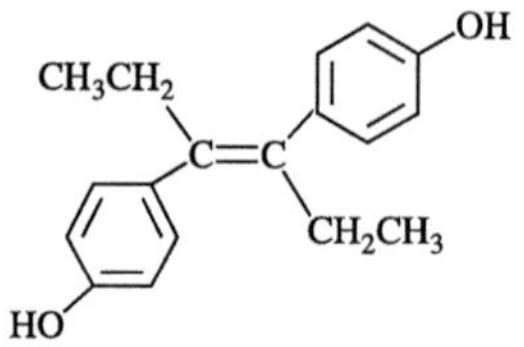

Diethylstilbestrol
2.109

Calculation of the $\log P$ for the antihistamine, diphenhydramine (**2.110**; Benadryl), is shown in Eqn (2.13). In this equation, 2.13 is $\log P$ for benzene, which is the same as π_{Ph},

$$\begin{aligned} \text{Calc.log } P &= 2\pi_{Ph} + \pi_{CH} + \pi_{OCH_2} + \pi_{CH_2} + \pi_{NMe_2} - 0.2 \\ &= 2(2.13) + 0.50 - 0.73 + 0.50 - 0.95 - 0.2 \\ &= 3.38 \end{aligned} \qquad (2.13)$$

0.50 is π_{CH} (same as π_{CH_3}), −0.73 was obtained by subtracting 1.50 ($2\pi_{CH_3}+\pi_{CH_2}$) from $\log P_{CH_3CH_2OCH_2CH_3}$ (= 0.77), −0.95 is the value for π_{NMe_2} obtained by subtracting $\pi_{Ph(CH_2)_3}$ (2.13 + 3(0.5) = 3.63) from $\log P_{Ph(CH_2)_3NMe_2}$ (= 2.68), and −0.2 is for branching at the CH (note that there is no branching at the $N(CH_3)_2$ because we used that whole substituent to obtain π). The experimental $\log P$ value is 3.27.

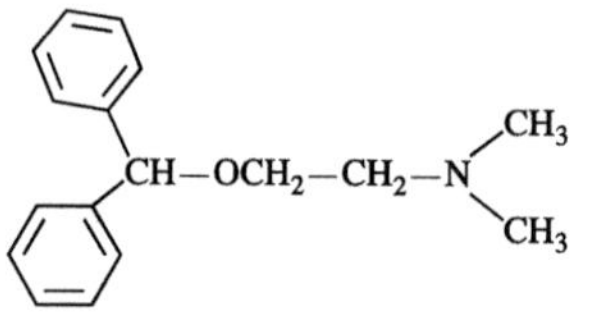

Diphenhydramine
2.110

A more rapid approach than the standard shake flask method[341] for determination of $\log P$ values that was described above has been reported[342] for neutral compounds. The reverse-phase HPLC method takes about 20 min per compound with a wide range of lipophilicities (6 $\log P$ units) with good accuracy and excellent reproducibility. The value obtained by this method is referred to as the $\text{ELog}\,P_{oct}$. Hydrophilic interaction chromatography,[343] and ultraperformance liquid chromatography[344] can be used to obtain $\log P$ data for the neutral form of basic compounds. A reverse-phase HPLC method[345] and a microfluidic liquid–liquid extraction method[346] for determination of $\log D$ values ($\text{ELog}\,D_{oct}$) have also been devised.

2.2.5.2.3. Computer Automation of log *P* Determination

The determination of $\log P$ values has become less of a chore as a result of computer automation of the method.[347] A nonlinear regression model for the estimation of partition coefficients was developed by Bodor et al.[348] using the following molecular descriptors: molecular surface, volume, weight, and charge densities. It was shown to have excellent predictive power for the estimation of the log *P* for complex molecules. A semiquantitative method for calculating log *P* values was developed by Moriguchi et al. (MLog *P*)[349] using a multiple regression analysis of 1230 organic molecules having a wide variety of structures; excellent correlation was observed between the observed log *P* and the calculated log *P*.

Probably the simplest way to get log *P* values for unknown compounds is with the use of one of the numerous software packages that are now commercially available, such as Biobyte (CLog *P*, developed by the medicinal chemistry group at Pomona College; also available within ChemDraw (Perkin Elmer Informatics)), Advanced Chemistry Development (ACD/log *P* DB), CTIS (AUTO-LOG™), Scivision (SciLogP), and Bio-Rad (PredictIt™ log *P* and log *D*). The problem with the software packages, however, is that the results can differ widely (2 or more log *P* units) and differ from the experimental value.[350] The reason is related to the fact that, as mentioned above, π values can be constitutive; depending on the structure of the compound, the π value can differ. Also, ionization of groups varies with concentration and counter ions. So, no software package can account for π values for substituents on *every* scaffold. Most software packages will always give an answer, but CLog *P* will not calculate a value when it does not have sufficient data (e.g., if an atom is poorly parameterized). Of course, it is frustrating when a computer tells you it cannot compute (which may be detrimental to the software company sales), but it may be the most honest approach. One way to obtain the most accurate predictive results for your particular family of compounds is to determine experimentally the actual log *P* value for one of the members of the family, then ask a variety of software packages to predict the log *P* value, and use the program that comes closest to the experimental value for that family of compounds.

2.2.5.2.4. Membrane Lipophilicity

Although the log *P* values determined from 1-octanol/water partitioning are excellent models for in vivo lipophilicity, it has been found that for a variety of aromatic compounds whose log *P* values are greater than 5.5 (very lipophilic) or whose molar volumes are greater than 230 cm^3/mol, there is a breakdown in the correlation of these values with those determined from partitioning between L-α-phosphatidylcholine dimyristoyl membrane vesicles and water.[351] Above a log *P* of 5.5 the solvent solubility for these molecules is greater than their membrane solubility. As the compound increases in size, more energy per unit volume is required to form a cavity in the structured membrane phase. This is consistent with observations that branched molecules have lower log *P* values than their straight-chain counterparts, and that this effect is even greater in membranes than in organic solvents.

It should be noted that although log *P* values are most commonly determined with 1-octanol/water mixtures, this is not universal because hydrophilicity resulting from acceptance of a hydrogen bond is not reflected well by partitioning in 1-octanol, which can accept hydrogen bonds almost as well as does water.[352] Consequently, this gives an apparently higher lipophilicity value than is reflected in membrane partitioning. Other nonhydroxylic solvents, such as cyclohexane, can provide insights into these processes. Because of this, Seiler[353] introduced a new additive constitutive substituent constant for solvents other than 1-octanol. Therefore, when using log *P* values, it is important to be aware of the solvent used to obtain the log *P* data.

2.2.5.3. Balancing Potency of Ionizable Compounds with Lipophilicity and Oral Bioavailability

Receptors are typically proteins, composed of amino acid residues having side chains with varying ionization states depending on the pH of the environment. For example, anionic groups in proteins include carboxylic acids (aspartic and glutamic acids, pK_a 4–4.5), phenols (tyrosine, pK_a 9.5–10), sulfhydryls (cysteine, pK_a 8.5–9) and hydroxyls (serine and threonine, pK_a 13.5–14). Cationic groups in proteins include imidazole (histidine, pK_a 6–6.5), amino (lysine, pK_a 10–10.5), and guanidino (arginine, pK_a 12–13) groups. At physiological pH (pH 7.4), even the mildly acidic groups, such as carboxylic acid groups, will be essentially completely in the carboxylate anionic form; phenolic hydroxyl groups may be partially ionized. Likewise, basic groups, such as amines, will be partially or completely protonated to give the cationic form. As discussed previously, the same is true for a drug; the ionization state of a drug will depend on the pK_a values of the ionizable groups and the pH of the medium with which it has to interact (see also discussion of effect of protein microenvironment later in this section). Thus, ionization can have a profound effect not only on a drug's interaction with a receptor but also on its lipophilicity.

What if the drug you are attempting to discover binds at an ionized site in the receptor, so ionization of your drug favors binding to the receptor (see Chapter 3, Section 3.2.2.2), but ionization of the drug also blocks its ability to cross various membranes prior to reaching the receptor? How is it possible to design a compound that is neutral when it needs to cross membranes, but ionized when it finally reaches the target receptor? This is possible because there is an equilibrium established between the neutral and ionized form of a molecule or group that depends on the pH of the medium and the pK_a of the ionizable group (Scheme 2.9). *It is important to note that when referring to the pK_a of a* ***basic*** *substance, for example, RNH_2 in the first equation of* Scheme 2.9, *in this text we are really referring to the pK_a of its conjugate acid (RNH_3^+) unless otherwise indicated.*

$$RNH_2 + H^+ \rightleftharpoons RNH_3^+$$

$$RCOOH \rightleftharpoons RCOO^- + H^+$$

SCHEME 2.9 Ionization equilibrium for an amine base and a carboxylic acid

For an ionizable substance with a given pK_a, the proportion of ionized to nonionized form is given by the Henderson–Hasselbalch equation (Eqn (2.14)). This equation is directly derivable from the well-known acid

$$pH = pK_a - \log \frac{[HA]}{[A^-]} \quad (2.14)$$

dissociation constant equation. According to Eqn (2.14), when the pH of the medium equals the pK of the molecule, then $\log[HA]/[A^-]=0$, which holds when $[HA]=[A^-]$. Therefore, when the pH of the medium equals the pK of the molecule, half the molecules are in the neutral form and half are in the ionized form. As another example, if the pH is 2 units lower than the pK_a, then $[HA]/[A^-]=100$, that is, the protonated form predominates over the unprotonated form by a factor of 100. The neutral form of a molecule is presumed to be best able to pass through a membrane, but once on the other side, the equilibrium with the ionized form is reestablished (the equilibrium mixture again will depend on the pH on the other side of the membrane), so there are now ionized molecules on the other side of the membrane to interact with the target receptor. The ionized molecules that did not cross the membrane also reestablish equilibrium and become a mixture of ionized and neutral molecules, so more neutral molecules can get across the membrane. If the equilibria could be reestablished indefinitely, eventually all the molecules would cross the membranes and bind to the target receptor. However, drugs get metabolized and excreted (see Chapter 8), so they may never get across the membrane before they are excreted. To adjust the ionization equilibrium of the lead compound, one can add electron-withdrawing or electron-donating groups to vary the pK_a of the molecule (electron-withdrawing groups will lower the pK_a, making acids more ionizable and bases less ionizable; the opposite holds for electron-donating groups).

The importance of ionization was recognized in 1924 when Stearn and Stearn[354] suggested that the antibacterial activity of stabilized triphenylmethane cationic dyes was related to an interaction of the cation with some anionic group in the bacterium. Increasing the pH of the medium also increased the antibacterial effect, presumably by increasing the ionization of the receptors in the bacterium or by decreasing the ionization of the compound outside the cell, thereby promoting higher permeability. Albert and coworkers[355] made the first rigorous proof that a correlation between ionization and biological activity existed. A series of 101 aminoacridines having a variety of pK_a values, including the antibacterial drug, 9-aminoacridine or aminacrine (**2.111**), were tested against 22 species of bacteria. A direct correlation was observed between ionization (formation of the cation) of the aminoacridines and antibacterial activity. However, at lower pH values, protons can compete with these cations for the receptor, and antibacterial activity is diminished. When this was realized, Albert[356] noted that the Australian Army during World War II was advised to pretreat wounds with sodium bicarbonate to neutralize any acidity prior to treatment with aminacrine. This, apparently, was quite effective in increasing the potency of the drug. The mechanism of action of aminoacridines is discussed in Chapter 6 (Section 6.3.1.3.1).

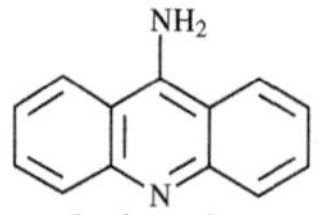

NH_2 N

Aminacrine
2.111

The great majority of alkaloids that act as neuroleptics, local anesthetics, and barbiturates have pK_a values between 6 and 8; consequently, both neutral and cationic forms are present at physiological pH. This may allow them to penetrate membranes in the neutral form and exert their biological action in the ionic form.

Antihistamines and antidepressants tend to have pK_a values of about 9. The uricosuric (increases urinary excretion of uric acid) drug phenylbutazone (**2.112,** Scheme 2.10; Butazolidine), $R=(CH_2)_3CH_3$) has a pK_a of 4.5 and is active as the anion (**2.113**). However, since the pH of urine is 4.6 or higher, suboptimal concentrations of the anion were found in the urinary system. Sulfinpyrazone (**2.112,** $R=CH_2CH_2SOPh$; Anturane) has a lower pK_a (2.8) and is about 20 times more potent than phenylbutazone; the anionic form blocks reabsorption of uric acid by renal tubule cells.[357]

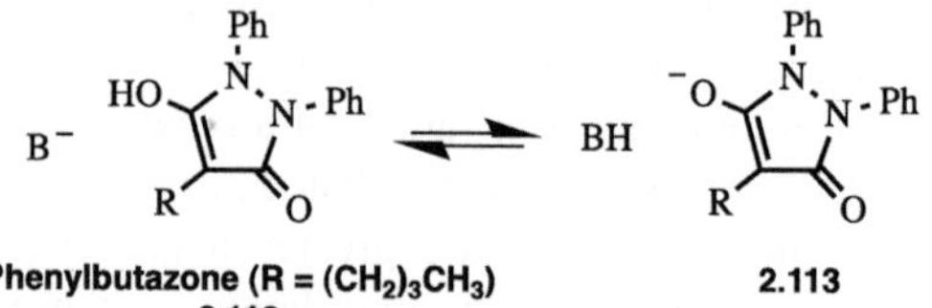

Ph HO N N-Ph B$^-$ R O ⇌ BH Ph $^-$O N N-Ph R O

Phenylbutazone (R = $(CH_2)_3CH_3$)
2.112

2.113

SCHEME 2.10 Ionization equilibrium for phenylbutazone

The antimalarial drug pyrimethamine (**2.114**; Daraprim) has a pK_a of 7.2 and is best absorbed from solutions of sufficient alkalinity such that it has a high proportion of molecules in the neutral form (to cross membranes). Its mode of action, the inhibition of the parasitic enzyme dihydrofolate reductase, however, requires that it be in the protonated cationic form.

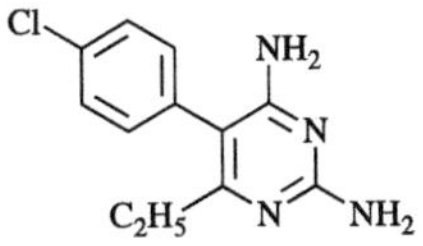

Pyrimethamine
2.114

The effect of ionization can be rationalized either from a pharmacokinetic or pharmacodynamic perspective. For example, if changing the pK_a increases its potency, it could be because the neutral form becomes more prevalent and, therefore, crossing membranes becomes favored (pharmacokinetic argument), or it could be because there is a hydrophobic pocket in the receptor that the neutral form prefers to bind into (pharmacodynamic argument). How can the relative importance of these two properties be determined? If the drugs act on microbial systems, one way is to compare results of assaying the test compounds in a cell-free system (in which there are no membranes to cross) and in an intact cell system (in which it is necessary to cross a membrane to get to the receptor). For example, the pharmacokinetics of the antibacterial agent sulfamethoxazole (**2.115**, Scheme 2.11; Bactrim) depends on their nonionized form (**2.115**), but the pharmacodynamics depends on the anionic form (**2.116**). In a cell-free system, the antibacterial activity of **2.115** and other sulfonamides was directly proportional to the degree of ionization, supporting the importance of ionization on pharmacodynamics, but in intact cells, where the drug must cross a membrane to get to the site of action, the antibacterial activity also depended on the neutral form[358] (supporting the notion that the neutral form is not important to pharmacodynamics, only to pharmacokinetics).

The structure and function of a receptor and of a drug can strongly depend on the pH of the medium, especially if an in vitro assay is being used. However, you must be careful when trying to assess pK_a values of groups within a binding site of a receptor, because these values can be quite variable and will depend on the microenvironment. On the basis of molecular dynamics simulations of several proteins in water, the interiors of these proteins were calculated to have dielectric constants of about 2–3,[359] which is comparable to the dielectric constant of nonpolar solvents such as benzene ($\varepsilon = 2.28$) or *p*-dioxane ($\varepsilon = 2.21$). This is quite different from the dielectric constant of water ($\varepsilon = 78.5$), which is a result of the strong dipole moment of the O–H bonds. If a carboxyl group is in a nonpolar region, its pK_a will rise because the anionic form will be destabilized (no polar groups to stabilize the charge). Glutamate-35 in the lysozyme–glycolchitin complex has a pK_a of 8.2[360]; the pK_a of glutamate in water is 4–4.5. The pK_a of Asp-99 in a nonpolar region of 3-oxo-Δ^5-steroid isomerase is a remarkable 9.5![361] That is, a change in equilibrium by a factor of 10^5 (remember, pK_a is a logarithm) in favor of the neutral form! If the carboxylate forms a salt bridge, it will be stabilized, and its pK_a will be lowered. If several carboxylic acid groups are near an essential active site carboxylic acid, the anionic form will be stabilized, and its pK_a also will be lowered. Likewise, an amino group buried in a nonpolar microenvironment will have a lower pK_a because protonation will be disfavored (to avoid the polar cationic character); the ε-amino group of the active site lysine residue in the enzyme acetoacetate decarboxylase has a pK_a of 5.9,[362] whereas in water, it is about 10–10.5. If the ammonium group of lysine forms a salt bridge, it will be stabilized, deprotonation will be inhibited, and the pK_a will rise. Given this large change in pK_a values in different microenvironments, it is worthwhile to make large changes in pK_a values of compounds in a lead modification library to see how the potency changes in both in vitro and in vivo assays. Once it is established whether the potency of a molecule is favored in the neutral or ionized form, then pK_a considerations can be employed in further lead modification approaches.

2.2.5.4. Properties that Influence Ability to Cross the Blood–Brain Barrier

Computational studies to determine what drug properties are important for crossing the blood–brain barrier and for CNS activity have been carried out by several groups.[363] CNS-active and CNS-inactive compounds were selected from the CMC and the MDDR databases (MDDR stands for MDL Drug Data Report; a database from MDL Inc., part of Accelrys, Inc. since 2010) by Ajay et al. Each molecule was described by seven one-dimensional descriptors (e.g., MW, number of hydrogen bond donors, and number of hydrogen bond acceptors) and 166 2D descriptors (e.g., log$P$, TPSA, and molar refractivity (MR) (see Section 2.2.6.2.1). Using all of these descriptors, 83% of the CNS-active compounds and 79% of the CNS-inactive compounds in these databases were correctly predicted. In general, they concluded that if the MW, the degree of branching, the number of

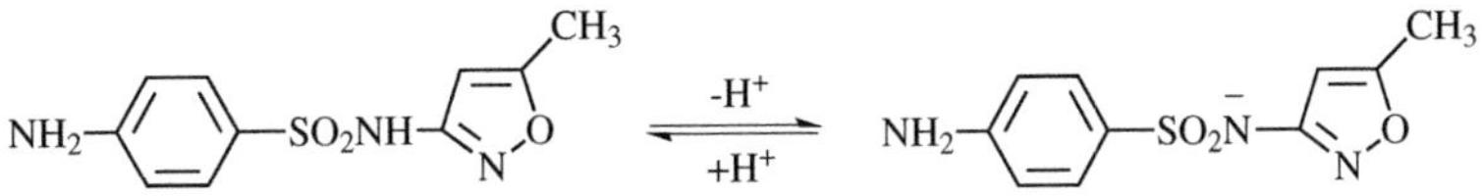

2.115 ⇌ (−H⁺ / +H⁺) **2.116**

SCHEME 2.11 Ionization equilibrium for sulfamethoxazole

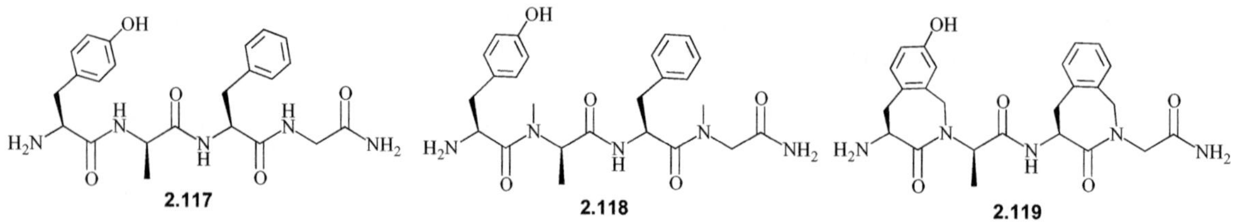

2.117 **2.118** **2.119**

rotatable bonds, or the number of hydrogen bond acceptors is increased, the compound will be *less* likely to be CNS active. If the aromatic density or log *P* is increased, the compound is *more* likely to be CNS active. A more recent study that analyzed the properties of 119 marketed CNS drugs and 108 CNS candidates from Pfizer considered not only calculable parameters, such as those above, but also other parameters that can be readily determined experimentally, such as LE and related values (see Section 2.1.2.3.5), stability to in vitro liver microsome preparations (liver microsomes are a major site of metabolism for many drugs), permeability through artificial membrane preparations, and susceptibility to the action of P-glycoprotein, a protein that promotes the transport of some drugs out of cells.[364] From this analysis, median values for physicochemical properties of favorable CNS drugs across multiple parameters were deduced (Table 2.16).

Another study was carried out to determine the effect of lipophilicity vs structural flexibility on blood–brain barrier penetration using **2.117–2.119**.[365] Peptide amide **2.117** is much more polar than the other two, both of which have two fewer NH bonds; **2.117** is not active in vivo by peripheral administration because of its inability to cross the blood–brain barrier. Compounds **2.118** and **2.119**, which are almost identical in CLog *P* values, PSA values, and number of H-bond donors/acceptors, were active by intravenous administration. However, there was little difference in antinociceptive properties between **2.118** and **2.119**, indicating that lipophilicity is more important than structural flexibility for blood–brain barrier penetration.

TABLE 2.16 Median Values of Calculated Experimental Properties for a Set of Marketed CNS Drugs

Parameter	How Determined	Median Value of Parameter
CLog *P*	Calculated	2.8
CLog *D* (pH 7.4)	Calculated	1.7
MW	Calculated	305.3
TPSA	Calculated	44.8
Hydrogen bond donors (HBD)	Calculated	1
pK_a	Calculated or by titration experiment	8.4
Passive permeability (Papp)	MDCK (canine kidney) cells	>10E-5 cm/s
Efflux to Influx Ratio	MDCK cells expressing MDR1	≤2.5
Liver microsome stability (CLint,u)	In vitro microsome preparations	≤100 mL/min/kg
Ligand efficiency (LE)	Potency assay, MW	0.46
LLE	Potency assay, CLog *P*	6.4
LELP	Calculated from LE and CLog *P*	5.9

2.2.5.5. Correlation of Lipophilicity with Promiscuity and Toxicity

Recent studies at Pfizer investigated the association of different physicochemical properties of compounds with toxic effects.[366] The studies drew upon data from in vivo toleration (IVT) studies in rats and dogs, a form of preliminary toxicology studies that involve administering the compound to animals repeatedly (typically once or twice a day) over several days (typically 4–7 days) and studying the animals for toxic effects. Because IVT studies are rather expensive (compared to in vitro assays) and may require multigram quantities of compound (depending on the potency of the compound), they are normally not carried out until later stages of lead optimization, before making the decision to advance a compound to more detailed toxicity studies requiring even larger quantities of compound and more animals. The study concluded that compounds with CLog *P*>3 and PSA<75 Å^2 were much more likely to show toxic effects. PSA, which was mentioned earlier in the context of properties associated with good oral bioavailability (see Section 2.1.2.3.2), represents the area on the surface of a molecule that is contributed by polar atoms.[367] High CLog *P* and low PSA also correlated with the propensity for a compound to be more promiscuous, that is, interact with several different types of biological target. It is likely that this greater promiscuity is at least in part responsible for the increased toxicity observed in the IVT studies. Moreover, higher log *P* is generally correlated with increased propensity for a molecule to undergo metabolic transformation to one or more new compounds (*metabolites*, see Chapter 8), some of which could be more toxic than the parent molecule.

2.2.6. Computational Methods in Lead Modification

2.2.6.1. Overview

Computational methods applied to lead discovery were discussed in Section 2.1.2.3.4. That discussion focused on various types of hypotheses (models) that could be proposed to predict the structural requirements for a desired activity, against which one or more databases of compounds could be compared (virtually screened) to identify new or improved lead compounds. In most cases, the compounds being considered in the virtual screen could be regarded as readily available for actual screening; for example, they may already exist in an in-house compound collection, they may be listed as commercially available, or, at the very least, they are believed to be rapidly synthesizable, perhaps using parallel methods to prepare many compounds simultaneously. The purpose of the virtual screen was to prioritize members of a large collection of compounds for inclusion in an actual biological screen.

Many of these models can also be useful during lead modification efforts. In many cases, the difference between the use of a given model to support lead modification rather than lead discovery lies not so much in the type of model than in how it is developed and used. One such difference is that during lead modification, the purpose of the computational support is much more often to help with prioritization of compounds for synthesis. In this context, there is likely to be a much smaller set of compounds being considered, for example, a set of 10 to 50 analogs of a lead series that have been identified as possible analogs for synthesis. There may also be differences in the proposed analogs from the viewpoint of ease or expense of synthesis or confidence in the likelihood that a given analog will show the desired activity. The more believable and predictive the model is, the more weight is likely to be placed on it as a prioritization tool, as opposed to reliance on other factors, such as ease of synthesis or obvious similarity to the lead molecule.

When applied to lead modification, the information used to develop a model is typically much more refined than the information used to develop a model for support of lead identification. For example, during lead discovery, there may be only one or a small number of active compounds around which to derive a model, and if there are multiple active molecules, proposed structural correspondence among them may be speculative. On the other hand, when used to support lead modification, frequently there are a number of closely related molecules possessing a range of activities against the target that can be used to develop and test a hypothesis.

The following sections will discuss computational methods and tools that have particular applicability to lead modification. Quantitative structure-activity relationships (QSARs) are especially useful for making predictions about proposed analogs when data already exist for a significant number of structurally related molecules. *Scaffold hopping* is really a special case of lead discovery, but is discussed under lead modification since it is an approach that is frequently carried out during the lead modification stage of a drug discovery project. *Molecular graphics* methods allow the researcher to view and manipulate molecular structures and shapes on a computer screen. Because of the interactive nature of such exercises, for example, to view the results of calculated predictions, molecular graphics tools find their broadest applicability in the more detailed assessments typically encountered during the lead modification phase of a project.

2.2.6.2. Quantitative Structure–Activity Relationships (QSARs)

2.2.6.2.1. Historical Overview. Steric Effects: The Taft Equation and Other Equations

In 1868, Crum-Brown and Fraser[368] predicted that some day a mathematical relationship between structure and activity would be expressed. It was not for almost 100 years that this prediction began to be realized, and a new era in drug design was born. In 1962, Corwin Hansch attempted to quantify the effects of particular substituent modifications, and from this, QSARs developed.[369] The concept of quantitative drug design is based on the fact that the biological properties of a compound are influenced by physicochemical parameters, that is, physical properties, such as solubility, lipophilicity (Section 2.2.5.2), electronic effects, ionization, stereochemistry, and so forth, that have a profound influence on the chemistry of the compounds. The first attempt to relate a physicochemical parameter to a pharmacological effect was reported in 1893 by Richet.[370] He observed that the narcotic action of a group of organic compounds was inversely related to their water solubility (*Richet's rule*). Overton[371] and Meyer[372] related tadpole narcosis, induced by a series of nonionized compounds added to the water in which the tadpoles were swimming, to the ability of the compounds to partition between oil and water. These early observations regarding the depressant action of structurally nonspecific drugs were rationalized by Ferguson.[373] He reasoned that when in a state of equilibrium, simple thermodynamic principles could be applied to drug activities, and that the important parameter for correlation of narcotic activities was the relative saturation (termed thermodynamic activity by Ferguson) of the drug in the external phase or extracellular fluids. This is known as *Ferguson's principle*, which is useful for the classification of the general mode of action of a drug and for predicting the degree of its biological effect. The numerical range of the thermodynamic activity for structurally nonspecific drugs is 0.01–1.0, indicating that they are active only at relatively high concentrations. Structurally specific drugs have thermodynamic activities considerably less than 0.01 and normally below 0.001.

In 1951, Hansch et al.[374] noted a correlation between the plant growth activity of phenoxyacetic acid derivatives and the electron density at the ortho position (lower electron density gave increased activity). They made an attempt to quantify this relationship by the application of the Hammett σ functions (see Section 2.2.5.1), but this was unsuccessful.

The crucial breakthrough in QSAR came when Hansch and coworkers[375] conceptualized the action of a drug as depending upon two processes. As already discussed (Section 2.2.5.2.1), the first process is the journey of the drug from its point of entry into the body to the site of action (pharmacokinetics), and the second process is the interaction of the drug with the specific site (pharmacodynamics). Because of the importance of pharmacokinetics to the success of a drug, he developed the octanol–water scale (log P) for lipophilicity (see Section 2.2.5.2.2) as a measurable physicochemical parameter to consider in addition to electronic effects that were developed by Hammett (see Section 2.2.5.1), which was the basis for the development of Hansch's lipophilicity equations.

Because interaction of a drug with a receptor involves the mutual approach of two molecules, another important parameter for QSAR is the steric effect. In much the same way that Hammett derived quantitative electronic effects (see Section 2.2.5.1), Taft[376] defined the steric parameter E_s (Eqn (2.15)). Taft used for the reference reaction the relative rates of the acid-catalyzed hydrolysis of α-substituted acetates (XCH_2CO_2Me). This parameter is

$$E_s = \log k_{XCO_2Me} - \log k_{CH_3CO_2Me} = \log k_X/k_o \quad (2.15)$$

normally standardized to the methyl group ($XCH_2=CH_3$) so that $E_s(CH_3)=0.0$; it is possible to standardize it to hydrogen by adding 1.24 to every methyl-based E_s value.[377] Hancock et al.[378] claimed that this model reaction was under the influence of hyperconjugative effects and, therefore, developed corrected E_s values for the hyperconjugation of α-hydrogen atoms (Eqn (2.16)), where E_s^c is the corrected E_s value and n is the number of α-hydrogen atoms.

$$E_s^c = E_s + 0.306\,(n-3) \quad (2.16)$$

Two other steric parameters worth mentioning are *molar refractivity* (*MR*) and the Verloop parameter. MR,[379] the molar volume corrected by the refractive index, which represents the size and polarizability of a fragment or molecule, is defined by the Lorentz–Lorenz equation (Eqn (2.17)), where n is the index of refraction at the sodium D line, MW is the

$$\mathrm{MR} = \frac{n^2-1}{n^2+2}\,\frac{\mathrm{MW}}{d} \quad (2.17)$$

molecular weight, and d is the density of the compound. The greater the positive MR value of a substituent, the larger is its steric or bulk effect. This parameter also measures the electronic effect and, therefore, may reflect dipole–dipole interactions at the receptor site.

The *Verloop steric parameters*[380] are used in a program called STERIMOL to calculate the steric substituent values from standard bond angles, van der Waals radii, bond lengths, and user-determined reasonable conformations. Five parameters are involved. One (L) is the length of the substituent along the axis of the bond between the substituent and the parent molecule. Four width parameters (B_1–B_4) are measured perpendicular to the bond axis. These five parameters describe the positions, relative to the point of attachment and the bond axis, of five planes that closely surround the group. In contrast to E_s values that, because of the reaction on which they are based, cannot be determined for many substituents, the Verloop parameters are available for any substituent.

2.2.6.2.2. Methods Used to Correlate Physicochemical Parameters with Biological Activity

Now that we can obtain numerous *physicochemical parameters* (*descriptors*) for any substituent, how do we use these parameters to gain information regarding what compound to synthesize next in an attempt to optimize the lead compound? First, several (usually, many) compounds related to the lead are synthesized, and the biological activities are determined in the assay of interest. These data, then, can be manipulated by a number of QSAR methods. The most common is Hansch analysis.

2.2.6.2.2.1. Hansch Analysis: A Linear Multiple Regression Analysis With the starting assumption of at least two considerations for biological activity, namely, lipophilicity (required for the journey of the drug to the site of action) and electronic factors (required for drug interaction with the site of action), and that lipophilicity is a parabolic function, Hansch and Fujita[381] expanded Eqn (2.7) to that shown in either Eqn (2.18a or b), known as the *Hansch equation*, where C is the molar concentration (or dose) that elicits a

$$\log 1/C = -k\pi^2 + k'\pi + \rho\sigma + k'' \quad (2.18a)$$

$$\log 1/C = -k(\log P)^2 + k'(\log P + \rho\sigma + k'' \quad (2.18b)$$

standard biological response (e.g., ED_{50}, the dose required for 50% of the maximal effect; IC_{50}, the concentration that gives 50% inhibition of an enzyme or antagonism of a receptor; LD_{50}, the lethal dose for 50% of the animal population); k, k', ρ, and k'' are the regression coefficients derived from statistical curve fitting, and π and σ are the lipophilicity and electronic substituent constants, respectively. The reciprocal of the concentration ($1/C$) reflects greater potency with a lower dose, and the negative sign for the π^2 (or $(\log P)^2$) term reflects the expectation of an optimum lipophilicity, i.e., the π_o or $\log P_o$.

Because of the importance of steric effects and other shape factors of molecules for receptor interactions, an E_s term and a variety of other shape, size, or topography terms (S) have been added to the Hansch equation (see Eqn (2.19)). The way these parameters are used is by the

$$\log 1/C = -a\pi^2 + b\pi + \rho\sigma + cE_s + dS + e \qquad (2.19)$$

application of the method of linear multiple regression analysis.[382] The best least squares fit of the dependent variable (the biological activity) to a linear combination of the independent variables (the descriptors) is determined. Hansch analysis, also called the *extrathermodynamic method*, then, is a linear free energy approach to drug design in *congeneric series* (a group of compounds in the same structural class), in which equations are set up involving different combinations of the physicochemical parameters; the statistical methodology allows the best equation to be selected and the statistical significance of the correlation to be assessed. Once this equation has been established, it can be used to predict the activities of untested compounds. Problems associated with the use of multiple regression analysis in QSAR studies have been discussed by Deardon.[383] Despite many years of modification of this approach, it is infrequently used today, because large numbers of compounds must be included in the data set, and it is restricted to simple substituent changes, generally at just a few positions.

2.2.6.2.2.2. Manual Stepwise Methods: Topliss Operational Schemes and Others Because of the lack of easy access to computers by chemists in the early 1970s, Topliss[384] developed a nonmathematical, nonstatistical, and noncomputerized (hence, manual) guide to the use of the Hansch principles. This method is most useful when the synthesis of large numbers of compounds is difficult and when biological testing of compounds is readily available. It is an approach for the efficient optimization of the potency of a lead compound with the minimization of the number of compounds that need to be synthesized. The only prerequisite for the technique is that the lead compound must contain an unfused benzene ring. However, according to literature surveys at the time that this method was published, 40% of all reported compounds[385] contained an unfused benzene ring and 50% of drug-oriented patents[386] were concerned with substituted benzenes. This approach relies heavily on π and σ values and, to a much lesser degree, on E_s values. The methodology will be outlined here; this approach has historical significance because it was the initial description of rational design for chemists using the Hansch principles. It is now ingrained in the minds of medicinal chemists to think routinely in terms of physicochemical parameters.

Consider that your lead compound is benzenesulfonamide (**2.120**, R=H) and its potency has been measured in whatever screen is being used. Because many systems are $+\pi$ dependent, that is, the potency increases with increasing π values, then a good choice for your first analog would be one with a substituent having a $+\pi$ value. Because $\pi_{4\text{-Cl}}=0.71$ and $\sigma_{4\text{-Cl}}=0.23$ (remember, $\pi_H=\sigma_H=0$), the 4-chloro analog (**2.120,** R=Cl) should be synthesized and tested. There are three possible outcomes of this effort, namely, the 4-chloro analog is more potent (M), equipotent (E), or less potent (L) than the parent compound. If it is more potent, then it can be attributed to a $+\pi$ effect, a $+\sigma$ effect, or to both. To determine which is important, one term could be held more or less constant and the other varied. For example, the 4-phenylthio analog ($\pi_{4\text{-PhS}}=2.32$, $\sigma_{4\text{-PhS}}=0.18$) would be a good test of the importance of lipophilicity, and the 4-trifluoromethyl analog ($\pi_{4\text{-}CF_3} = 0.88$, $\sigma_{4\text{-}CF_3} = 0.88$) would test the importance of electron withdrawal. If the 4-phenylthio analog is more potent than the 4-chloro analog, further increases in lipophilicity would be desirable. At this point, a potency tree, termed a *Topliss decision tree*, could be constructed (Figure 2.28), and additional analogs could be made.

R–C_6H_4–SO_2NH_2

2.120

What if the 4-chloro analog were equipotent with the parent compound? This could result from a favorable $+\pi$ effect counterbalanced by an unfavorable $+\sigma$ effect, or vice versa. If this is the case, then the 4-methyl analog ($\pi_{4\text{-Me}}=0.56$, $\sigma_{4\text{-Me}}=-0.17$) might show enhanced potency. Enhancement of potency by the 4-methyl analog would suggest that the synthesis of analogs with increasing π values and decreasing σ values would be propitious. If the 4-methyl analog is worse than the 4-chloro analog, perhaps the equipotency of the 4-chloro compound was the result of

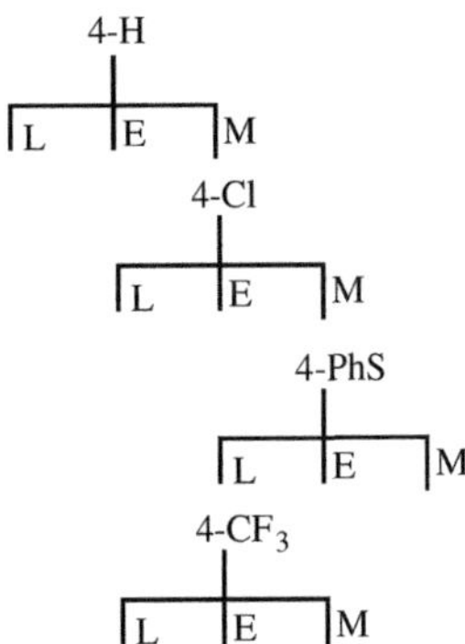

FIGURE 2.28 Topliss decision tree (M, more potent; E, equipotent; L, less potent).

a favorable σ effect and an unfavorable π effect. The 4-nitro analog ($\pi_{4\text{-}NO_2} = -0.28$, $\sigma_{4\text{-}NO_2} = 0.78$) would, then, be a wise next choice.

If the 4-chloro analog were less potent than the lead, then there may be a steric problem at the 4-position or increased potency depends on $-\pi$ and $-\sigma$ values. The 3-chloro analog ($\pi_{3\text{-}Cl}=0.71$, $\sigma_{3\text{-}Cl}=0.37$) could be synthesized to determine if a steric effect is the problem. Note that the σ constant for the 3-Cl substituent is different from that for the 4-Cl one because these descriptors are constitutive. If there is no steric effect, then the 4-methoxy compound ($\pi_{4\text{-}OMe}=-0.04$, $\sigma_{4\text{-}OMe}=-0.27$) could be prepared to investigate the effect of adding a $-\pi$ and $-\sigma$ substituent. Increased potency of the 4-OMe substituent would suggest that other substituents with more negative π and/or σ constants be tried.

Another way to increase both π and σ values would be by synthesizing the 3,4-dichloro analog ($\pi_{3,4\text{-}Cl_2} = -1.25$, $\sigma_{3,4\text{-}Cl_2} = 0.52$). Again, the 3,4-dichloro analog could be more potent, equipotent, or less potent than the 4-chloro compound. If it is more potent, then determination of whether $+\pi$ or $+\sigma$ is more important could be made by selection of appropriate substituents with higher π and/or σ values. If the 3,4-dichloro compound were less potent than the 4-chloro analog, it could be that the optimum values of π and σ were exceeded or that the 3-chloro group has an unfavorable steric effect. The latter hypothesis could be tested by the synthesis of the 4-trifluoromethyl analog ($\pi_{4\text{-}Cl_2} = 0.88$, $\sigma_{4\text{-}Cl_2} = 0.54$), which has no 3-substituent, but has a high σ and intermediate π value. It must be stressed that this analysis was based almost exclusively on π and σ values, and that many other descriptors have been neglected.

Topliss extended the operational scheme for side-chain problems when the group is adjacent to a carbonyl, amino, or amide functionality, i.e., –COR, –NHR, –CONHR, and –NHCOR, where R is the variable substituent. This approach is applicable to a variety of situations other than direct substitution on the aromatic nucleus. In this case, the parent molecule is the one where R=CH_3, and π, σ, and E_s parameters are used. Note that in the Topliss operational scheme, as in the other methods in this section, the procedure is stepwise, that is, the next compound is determined on the basis of the results obtained with the previous one.

Three other manual, stepwise methods will be mentioned only briefly: Craig plots,[387] Fibonacci search method,[388] and sequential simplex strategy.[389] The Topliss decision tree approach evolved from the work of Craig, who pointed out the utility of a simple graphical plot of π versus σ (or any two parameters) to guide the choice of a substituent (Figure 2.29). Once the Hansch equation has been expressed for an initial set of compounds, the sign and magnitude of the π and σ regression coefficients determine the particular quadrant of the Craig plot that is to be used to direct further synthesis. Therefore, if both the π and

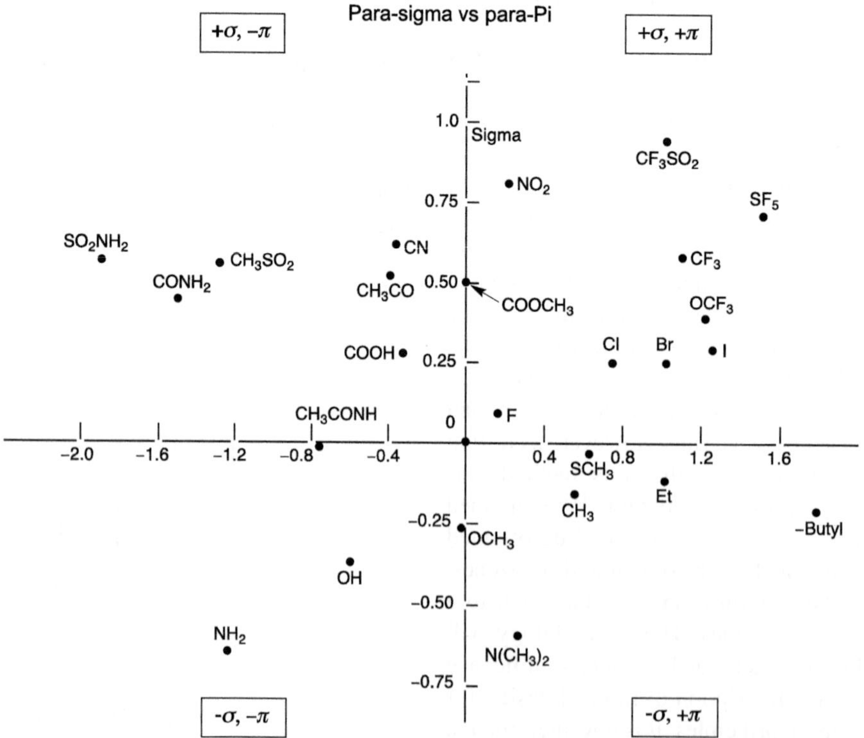

FIGURE 2.29 Craig plot of σ constants versus π values for aromatic substituents. *This material is reproduced with the permission of John Wiley & Sons, Inc. From Craig, P. N. (1980)* In *"Burger's Medicinal Chemistry," (M. E. Wolff, ed.), 4th ed., Part I, p. 343. Wiley, New York.*

σ terms have positive coefficients, then substituents in the upper right-hand quadrant of the plot (Figure 2.29) should be selected for future analogs.

The Fibonacci search technique is a manual method for the determination of the optimal properties of a parabolic function, such as potency versus $\log P$, using a minimum number of steps. Sequential simplex strategy is another stepwise technique suggested when potency depends on two physicochemical parameters such as π and σ.

2.2.6.2.2.3. Batch Selection Methods: Batchwise Topliss Operational Scheme, Cluster Analysis, and Others The inherent problem with the Topliss operational scheme described above is its stepwise nature. Provided that pharmacological results can be obtained quickly, this is probably not much of a problem; however, sometimes biological evaluation is slow. Topliss[390] proposed an alternative scheme that uses batchwise analysis of small groups of compounds. Substituents were grouped by Topliss according to π, σ, π^2, and a variety of $x\pi$– and $y\sigma$–weighted combinations. The approach starts with the synthesis of five derivatives, the unsubstituted (4-H), 4-chloro, 3,4-dichloro, 4-methyl, and 4-methoxyl compounds. After these five analogs have been screened, they are ranked in order of decreasing potency. The potency order determined for these analogs is then compared with the rankings in Table 2.17 to determine which parameter or combination of parameters is most dominant. If, for example, the potency order is 4-OCH_3>4-CH_3>H>4-Cl>3,4-Cl_2, then $-\sigma$ is the dominant parameter. Once the parameter dependency is determined, Table 2.18 is consulted to discover what substituents should be investigated next. In the above example, 4-$N(C_2H_5)_2$, 4-$N(CH_3)_2$, 4-NH_2, 4-NHC_4H_9, 4-OH, 4-$OCH(CH_3)_2$, 3-CH_3, and 4-OCH_3 would be suitable choices. The major weakness of this approach is that it is difficult to extend the method to additional parameters unless computers are used.

A computational batch selection method, known as cluster analysis, was introduced by Hansch et al.[391] Substituents were grouped into clusters with similar properties according to their σ, π, π^2, E_s, F (field constant), R (resonance constant), MR, and MW values. Some of the clusters are shown in Table 2.19.[392] One member of each cluster would be selected for substitution into the lead compound, and the compounds would be synthesized and tested. If a substituent showed dominant potency, then other substituents from that cluster would be selected for further investigation. The important advantage of the batch selection methods is that the initial batch of analogs prepared is derived from the widest range of parameters possible so that the dominant physicochemical property can be revealed early in the lead modification process. A variety of QSAR methods were applied to understand the activity of the

TABLE 2.18 New Substituent Selections

Probable Operative Parameters	New Substituent Selection
π, $\pi+\sigma$, σ	3-CF_3, 4-Cl; 3-CF_3, 4-NO_2; 4-CF_3, 2,4-Cl_2; 4-c-C_5H_9; 4-c-C_6H_{11}
π, $2\pi-\sigma$, $\pi-\sigma$	4-$CH(CH_3)_2$; 4-$C(CH_3)_3$; 3,4-$(CH_3)_2$; 4-$O(CH_2)_3CH_3$; 4-OCH_2Ph; 4-$N(C_2H_5)_2$
$\pi-2\sigma$, $\pi-3\sigma$, $-\sigma$	4-$N(C_2H_5)_2$; 4-$N(CH_3)_2$; 4-NH_2; 4-NHC_4H_9; 4-OH; 4-$OCH(CH_3)_2$; 3-CH_3, 4-OCH_3
$2\pi-\pi^2$	4-Br; 3-CF_3; 3,4-$(CH_3)_2$; 4-C_2H_5; 4-$O(CH_2)_2CH_3$; 3-CH_3, 4-cl; 3-cl; 3-CH_3; 3-OCH_3; 3-$N(CH_3)_2$; 3-CF_3; 3,5-Cl_2
Ortho effect	2-Cl; 2-CH_3; 2-OCH_3; 2-F
Other	4-F; 4-$NHCOCH_3$; 4-$NHSO_2CH_3$; 4-NO_2; 4-$COCH_3$; 4-SO_2CH_3; 4-$CONH_2$; 4-SO_2NH_2

TABLE 2.17 Potency Order for Various Parameter Dependencies

	Parameters									
Substituent	π	$2\pi-\pi^2$	σ	$-\sigma$	$\pi+\sigma$	$2\pi-\sigma$	$\pi-\sigma$	$\pi-2\sigma$	$\pi-3\sigma$	E_4[1]
3,4-Cl_2	1	1–2	1	5	1	1	1–2	3–4	5	2–5
4-Cl	2	1–2	2	4	2	2–3	3	3–4	3–4	2–5
4-CH_3	3	3	4	2	3	2–3	1–2	1	1	2–5
4-OCH_3	4–5	4–5	5	1	5	4	4	2	2	2–5
H	4–5	4–5	3	3	4	5	5	5	3–4	1

[1] *Unfavorable steric effect from 4-substitution.*

TABLE 2.19 Typical Members of Clusters Based on *σ*, *π*, *F*, *R*, *MR*, and *MW*

Cluster number[1]	Typical Members
1	Me, H, 3,4-(OCH_2O), CH_2CH_2COOH, $CH{=}CH_2$, Et, CH_2OH
2	CH=CHCOOH
3a	CN, NO_2, CHO, COOH, COMe
3b	C≡CH, CH_2Cl, Cl, NNN, SH, SMe, CH=NOH, CH_2CN, OCOMe, SCOMe, COOMe, SCN
4a	$CONH_2$, CONHMe, SO_2NH_2, SO_2Me, SOMe
4b	NHCHO, NHCOMe, $NHCONH_2$, $NHCSNH_2$, $NHSO_2Me$
5	F, OMe, NH_2, $NHNH_2$, OH, NHMe, NHEt, NMe_2
6	Br, OCF_3, CF_3, NCS, I, SF_5, SO_2F
7	CH_2Br, SeMe, $NHCO_2Et$, SO_2Ph, OSO_2Me
8	NHCOPh, $NHSO_2Ph$, OSO_2Ph, COPh, N=NPh, OCOPh, PO_2Ph
9	3,4-$(CH_2)_3$, 3,4-$(CH_2)_4$, Pr, *i*-Pr, 3,4-$(CH)_4$, NHBu, Ph, CH_2Ph, *t*-Bu, OPh
10	Ferrocenyl, adamantyl

[1]*Clusters three and four contain many of the common substituents used in medicinal chemistry; hence, these clusters are further subdivided according to their cluster membership when 20 clusters have been made.*
Reprinted with permission from Martin, Y. C. (1979). *In* "Drug Design" (E. J. Ariens, ed.), Vol. 8, p. 5. Academic Press, New York. Copyright© 1979 Academic Press, Inc.

anti-Alzheimer drug donepezil hydrochloride (**2.121**; Aricept) and its close analogs.[393]

Donepezil hydrochloride
2.121

A serious problem with the Hansch method is that most QSAR equations are overfit chance correlations as a result of testing many descriptors against small data sets and using few that actually correlate with activity.[394] Regression methods such as *PLS with cross-validation*,[395] useful when the number of descriptors is comparable to or greater than the number of compounds, allow many more variables to be tested without overfitting; perfect correlations involving descriptor subsets are not detected by PLS if the number of irrelevant descriptors is excessive.[396] The problem of overfitting, however, may return with programs like *Dragon* (http://www.talete.mi.it), which can compute nearly 5000 molecular descriptors.

2.2.6.2.2.4. Free and Wilson or de Novo Method Not long after Hansch proposed the extrathermodynamic approach, Free and Wilson[397] reported a general mathematical method for assessing the occurrence of additive substituent effects and for quantitatively estimating their magnitude. It is a method for the optimization of substituents within a given molecular framework that is based on the (tenuous) assumption that the introduction of a particular substituent at any one position in a molecule always changes the relative potency by the same amount, regardless of what other substituents are present in the molecule (additivity and independence of positions). A series of linear equations, constructed of the form shown in Eqn (2.20), where BA is the magnitude of the biological activity, X_i is the *i*th

$$BA = \sum a_i X_i + \mu \tag{2.20}$$

substituent with a value of 1 if present and 0 if not, a_i is the contribution of the *i*th substituent to the BA, and μ is the overall average activity of the parent skeleton, are solved by the method of least squares for the a_i and μ. All activity contributions at each position of substitution must sum to zero. The pros and cons of the Free–Wilson method have been discussed.[398] Fujita and Ban[399] suggested modifications of the Free–Wilson approach on the assumption that the effect on the activity of a certain substituent at a certain position in a compound is constant and additive.

As an example of the Free–Wilson approach, consider the hypothetical compound **2.122**.[400] If in one pair of analogs for which R^1, R^2, R^3, and R^4 are constant and R^5 is Cl or CH_3, the methyl compound is one-tenth as potent as the chloro analog, then the Free–Wilson method assumes that every R^5 methyl analog (where R^1–R^4 are varied) will be one-tenth as potent as the corresponding R^5 chloro analog. A requirement for this approach, then, is a series of compounds that have changes at more than one position. In addition, each type of substituent must occur more than once at each position in which it is found. The outcome is a table of the contribution to potency of each substituent at each position. If the free energy relationships of the extrathermodynamic method are linear or position specific, then Free–Wilson calculations will be successful. Chance correlations, as noted in the section above, do not occur with the Free and Wilson method because each substituent is treated as a significant variable; therefore, multiple regression analysis is not used to test the variables.

2.122

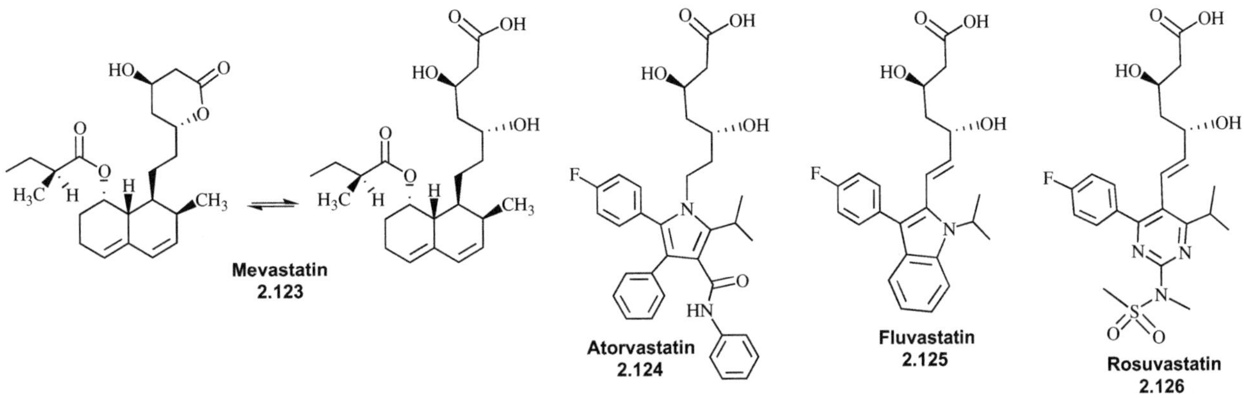

Mevastatin 2.123 Atorvastatin 2.124 Fluvastatin 2.125 Rosuvastatin 2.126

FIGURE 2.30 Example of scaffold hopping to identify new cholesterol-lowering statins from mevastatin

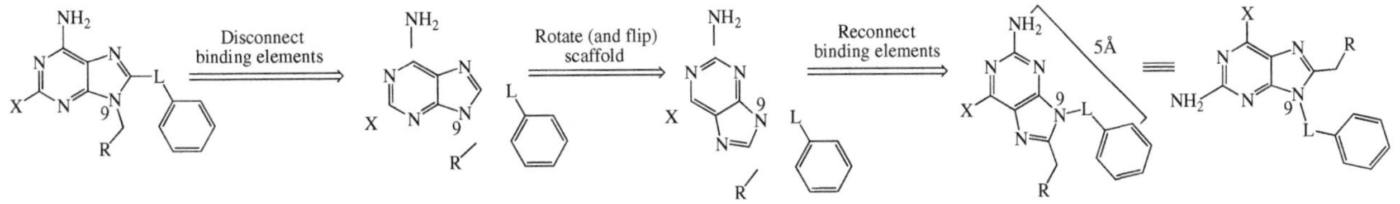

Disconnect binding elements Rotate (and flip) scaffold Reconnect binding elements 5Å

FIGURE 2.31 A variation of scaffold hopping that involves disconnection of bonds, rotation or flipping of the core, and reconnection of the pharmacophoric groups

2.2.6.2.2.5. Computational Methods for ADME Descriptors Because of the importance of physicochemical properties of molecules, such as log *P*, log *D*, and TPSA, to pharmacokinetics, a variety of computational methods to predict these parameters have become available, for example, ChemBioDraw (Cambridgesoft, Cambridge, MA, USA), Sybyl-X (Tripos, Inc., St Louis, MO, USA), Physicochemical & ADMET Prediction and ACD/ADME Suite (Advanced Chemistry Development, Inc., Toronto, Canada), and Discovery Studio ADME Descriptors (Accelyrs, San Diego, CA, USA).

In addition to simple physicochemical parameters, pharmacokinetic descriptors, such as cytochrome P450 inhibition and substrate activity (important to drug–drug interactions), blood–brain barrier penetration, P-glycoprotein activity (a measure of efflux from the brain), and oral bioavailability, can be estimated with commercial software. A few examples include *GastroPlus* (http://www.simulations-plus.com/), which simulates drug–drug interactions, preferred drug administration routes, effects of transporters, dissolution dependence of particle size, and metabolism. *ADMET Predictor* (http://www.simulations-plus.com/) predicts pK_a, membrane permeability, solubility, log *P*, log *D*, blood–brain barrier penetration, human plasma protein binding, volume of distribution, cytochrome P450 metabolism and inhibition, and intrinsic clearance. *Cloe PK* (https://www.cloegateway.com/) predicts human, rat, and mouse pharmacokinetics from early in vitro ADME and physicochemical data. A list of pharmacokinetics prediction software can be found at http://www.boomer.org/pkin/soft.html.

2.2.6.3. Scaffold Hopping

The concept of developing and utilizing models derived from 2D descriptors, 3D descriptors, or 3D-QSAR in the context of lead discovery was discussed earlier in this chapter (Section 2.1.2.3.4). The approach may also be applied at various points during the lead modification process. For example, the concept of *scaffold hopping*,[401] which could be considered a subset of the concept of bioisosterism (Section 2.2.4.3), might be applied during lead optimization to identify one or more new series of compounds having a fundamentally different core structure to improve, for example, its potency or selectivity profile, synthetic accessibility, intellectual property position, or pharmacokinetic properties. The goal of scaffold hopping is to attain novel structures by changing the central core framework (the scaffold) of active compounds without affecting its binding to the target receptor. In a scaffold hopping exercise, the information already gathered during the lead optimization work is used to develop a more advanced model, employing the methods described in the Lead Discovery section, against which a compound database can be virtually screened to identify new core variants of lead compounds. Examples of scaffold hopping can be found in the discovery of a family of 3-hydroxy-3-methylglutaryl-coenzyme A reductase inhibitors, the cholesterol-lowering statins (**2.124–2.126**), developed from the natural product mevastatin (**2.123,** Figure 2.30).[402]

A variant of scaffold hopping uses the same scaffold, but reorients it (Figure 2.31). In this example, the key pharmacophoric groups are disconnected, the core is rotated and/or flipped, and then the pharmacophoric groups are reconnected, trying to preserve the distance between the amino and phenyl groups.[403] By doing so, a novel purine series resulted having a different substitution pattern.

2.2.6.4. Molecular Graphics-Based Lead Modification

QSAR studies have relied heavily on the use of computers from the beginning for statistical calculations involving multiparameter equations. It was soon realized that drug design could be aided significantly if structures of receptors and drugs could be displayed on a computer terminal and molecular processes could be visualized. *Molecular graphics* is the visualization and manipulation of 3D representations of molecules on a graphics display device. The origins of molecular graphics have been traced by Hassall[404] to the project MAC (Multiple Access Computer),[405] which produced molecular graphics models of macromolecules for the first time. The potential to apply this technology to protein crystallography was quickly realized, and by the early 1970s, electron density data from X-ray diffraction studies could be presented and manipulated in stick or space-filling multicolored representations on a computer terminal.[406] Because the number of X-ray crystal structures available in the Protein Data Bank (PDB)[407] went from seven when it started in 1971 to about 200 in 1990 and to more than 100,000 in 2014, this lead modification tool has become even more powerful.

Medicinal chemists saw the potential of this approach in drug design as well. Stick (Dreiding) and space-filling (CPK) molecular models have been used extensively by organic chemists for years for small molecules, but these handheld models have major disadvantages.[408] Space-filling models often obscure the structure of the molecule, and wire or plastic models can give false impressions of molecular flexibility and often tend to change into unfavorable conformations at inopportune moments. Plastic models of proteins are much too cumbersome to work with. A 3D computer graphics representation of a protein that can be manipulated in three dimensions allows the operator to visualize the interactions of docked small molecules with biologically important macromolecules (see Section 2.1.2.3.4). The superimposition of structures, which is cumbersome at best with manual models, can be performed easily by molecular graphics. Also, some systems have the capability to graphically synthesize new structures by the assemblage of appropriate molecular fragments from a fragment file.

An early successful example of the application of molecular graphics to structure-based lead optimization is the discovery of zanamivir (**2.127**; Relenza), an antiviral agent used against influenza A and B infections.[409] The hemagglutinin at the surface of the virus binds to sialic acid (**2.128**) residues that are attached through glycoside linkages to receptors at the host cell surface. The virus enters the cell and replicates in the nucleus. The progeny virus particles escape the cell and stick to the sialic acid residues on the cell surface as well as to each other. *Neuraminidase* (also known as *sialidase*) is a key viral surface enzyme that catalyzes the cleavage of terminal sialic acid residues from the cell surface, which releases the virus particles to spread into the respiratory tract and infect new cells. The important feature of this enzyme that made it an attractive target for drug design is that its active site is lined with amino acids that are invariant in neuraminidases of all known strains of influenza A and B. Therefore, inhibition of this enzyme should be effective against all strains of influenza A and B. Random screening did not produce any potent inhibitors of the enzyme, although a nonselective neuraminidase inhibitor (**2.129**, R=OH) was identified. The breakthrough came when the crystal structure of the influenza A neuraminidase[410] with inhibitors bound[411] was obtained. The active site of the enzyme with **2.129** (R=OH) bound was probed computationally using Goodford's GRID program (discussed previously under Lead Discovery, Section 2.1.2.3.4).[412] Predictions by GRID of energetically favorable substitutions suggested replacement of the 4-hydroxyl group of **2.129** (R=OH) by an amino group (**2.129**, R=NH_2), which when protonated would form a favorable electrostatic interaction with Glu-119 (Figure 2.32(A)). It was predicted from the crystal structure that extension of the 4-ammonium group with a 4-guanidinium group, giving zanamivir (**2.127**), would produce an even tighter affinity because of the increase in basicity of the guanidinium group and also because it could interact with both Glu-119 and Glu-227 (Figure 2.32(B)).

From 1918 to 1920, a worldwide pandemic of influenza virus (a strain of the H_1N_1 swine flu that caused an epidemic in 2009) occurred that infected 500 million people, which was one-third of the world population at that time! Estimates were made that 50–100 million people died during

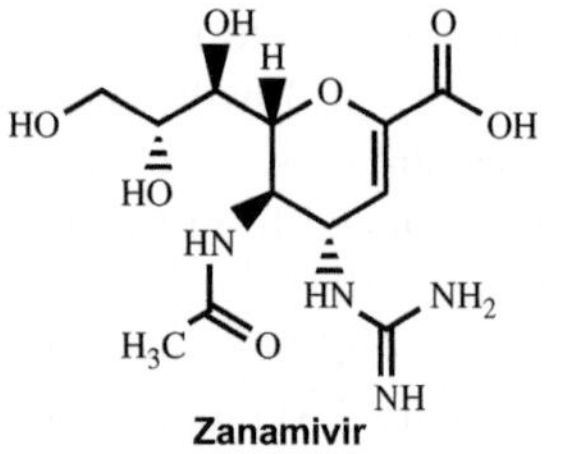

Zanamivir 2.127 **2.128** 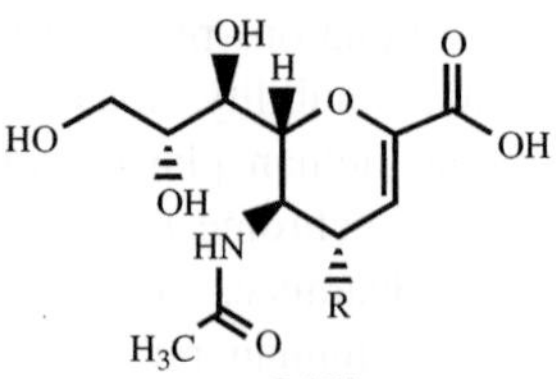 **2.129**

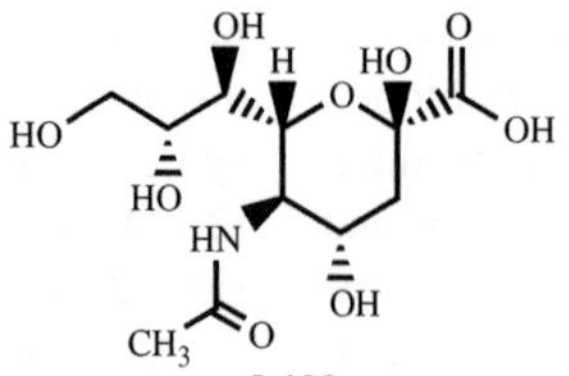

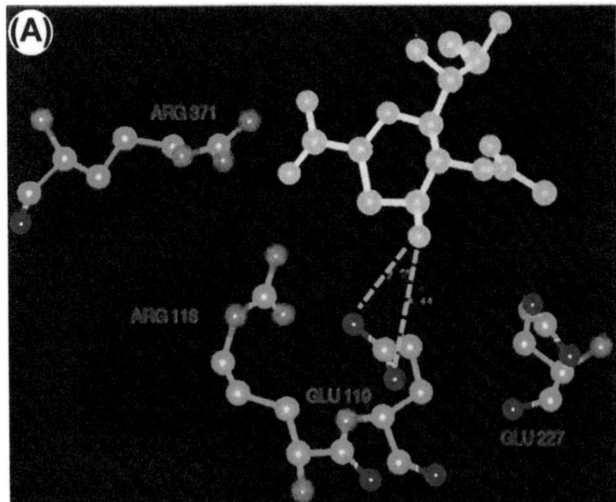

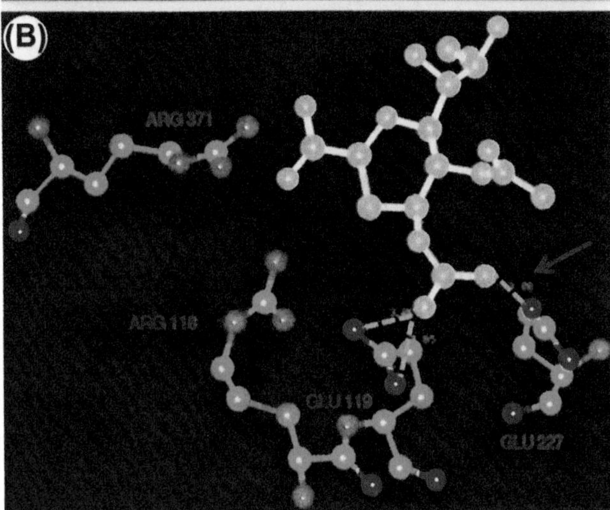

FIGURE 2.32 Crystal structure of neuraminidase active site with inhibitors bound. (A) Interaction of the protonated amino group of **2.129** ($R=NH_3^+$) with Glu-119. (B) Interaction of the protonated guanidinium group of **2.127** with Glu-119 and (red arrow) Glu-227. *Adapted with permission from Macmillan Publishers Ltd: Nature (von Itzstein, M., et al. Rational design of potent sialidase-based inhibitors of influence virus replication.* Nature ***1993***, *363, 418–423) Copyright 1993.*

that pandemic. Samples of tissue from those who died were frozen and were later used to determine that the neuraminidase of that flu was similar to that of current flu viruses; therefore, had zanamavir been available then, millions of people's lives could have been saved.[413]

Typically, the ideal compound is not realized so quickly. Rather, an idea for lead modification can be stimulated by viewing the crystal structure of the lead compound bound to the receptor. The new analog is synthesized and tested; maybe only a minor improvement in potency is produced (or maybe lower potency). If there is some improvement, a new crystal structure may be obtained for further refinement of ideas. In some lead optimization campaigns, this process is reiterated with further rounds of design, synthesis, testing, and crystal structure determination until higher potency analogs are obtained. Sometimes even a crystal structure with the ligand bound is not sufficient. A high-resolution crystal structure of thymidylate synthase with a ligand bound did not properly account for a ligand-induced enzyme conformational change during structure-based drug design.[414] As a result, the structure imparted an improper bias into the design of novel ligands.

Earlier in the chapter, the SAR of paclitaxel was described (**2.72**, Section 2.2.3). By overlaying the molecular graphics depiction of the crystal structure of paclitaxel with those of four other natural products also found to promote stabilization of microtubules in competition with paclitaxel (Figure 2.33), a common pharmacophore was proposed (Figure 2.34).[415] This gives a new perspective to lead modification, and permits the construction of new synthetic analogs having hybrid structures of each of the four unrelated scaffolds. On the basis of this pharmacophore model, **2.130** was synthesized and was shown to stabilize microtubules as well. Other 3D computer models of paclitaxel binding to microtubules have been promoted as well.[416] However, this pharmacophore model is particularly interesting because 5 years later the crystal structure of epothilone A (one of the four other natural products overlaid with paclitaxel) bound to α,β-tubulin was compared to that of paclitaxel, and it was found that there was *no common pharmacophore*; each compound binds in the pocket in a unique and independent manner.[417] This finding reminds us that caution must be taken with computer models, and the results should not be accepted as the gospel truth!

AcO O OH HO O O OH O OAc O Boc N H O

2.130

An exciting story of the design of analogs of the antitumor agent bryostatin 1 (**2.131**) having greater potency and exhibiting additional activities, such as in Alzheimer disease and HIV, evolved from an overlay of crystal structures of compounds that bind to protein kinase C.[418] Tumor promoters that activate protein kinase C are 1,2-diacyl-*sn*-glycerols (**2.132**), phorbol esters (**2.133**), and related diterpenes.[419] Other potent tumor promoters that activate protein kinase C include indole alkaloids, polyacetates, and a family of macrocyclic lactone marine natural products called bryostatins, having a highly complex architecture (as is apparent from **2.131**). Compounds with a synthetically more easily accessible structure than the bryostatins were sought; consequently, an overlay of the crystal structure of bryostatin 1 with crystal and calculated

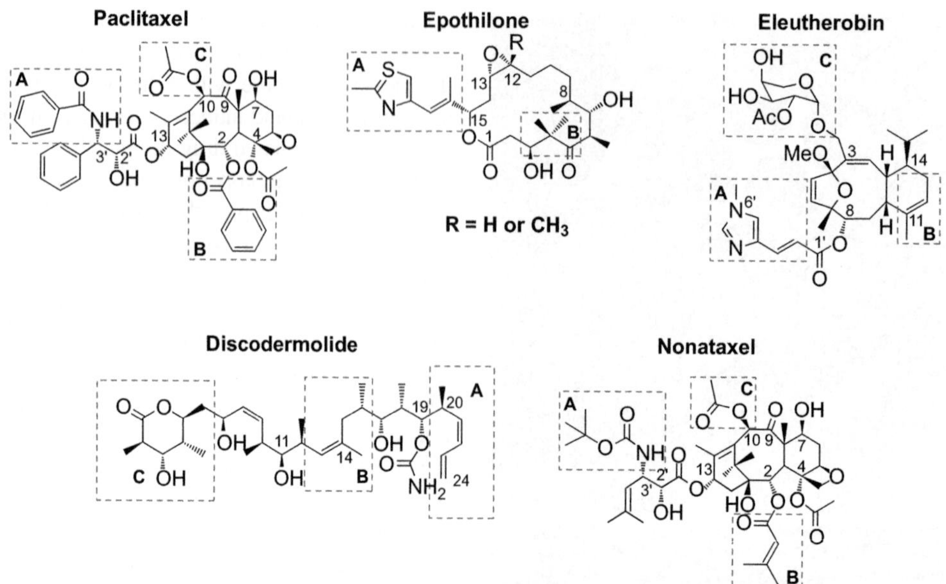

FIGURE 2.33 Five natural products found to promote stabilization of microtubules. The boxed sections were used to identify a common pharmacophore. *With permission for I. Ojima (1999). Reprinted with permission from Proc. Natl. Acad. Sci. USA* ***1999****, 96, 4256–4261. Copyright 1999 National Academy of Sciences, U.S.A.*

structures of the diterpene, indole alkaloid, and 1,2-diacyl-*sn*-glycerol classes of protein kinase C activators was performed to identify the relevant pharmacophore of these diverse compounds.[420] A marked similarity among all the classes of compounds was observed in the relative positions of various oxygen atoms and hydrophobic groups. Further molecular modeling indicated that the top part of bryostatin could be replaced by a simpler segment to give **2.134** ($R = CH_3$), which is much easier to synthesize, binds to protein kinase C with a K_i of 3.4 nM (K_i of bryostatin 1 is 1.4 nM), and inhibits the growth of several human tumor cell lines.[421] Later, it was found that **2.134** (R = H) could be synthesized more straightforwardly, had a K_i for protein kinase C of 250 pM, and is considerably more potent than bryostatin 1 in 24 of 35 human cancer cell lines.[422] Although the synthesis required 29 steps, because of its potency, an adequate amount of product could be obtained for clinical studies.

It is also known that activation of protein kinase C causes activation of α-secretases, which reduce the formation of amyloid beta, the main component of amyloid plaques found in the brains of patients with Alzheimer disease; therefore, activation of protein kinase C could be a new mechanism for the treatment of Alzheimer disease.[423] Both bryostatin 1 and **2.134** (R = H) were found to activate α-secretase activity at nanomolar to subnanomolar concentrations.[424] A peripheral Alzheimer disease biomarker, previously autopsy validated, was used to determine the therapeutic efficacy of these compounds. Both compounds converted the Alzheimer disease phenotype skin fibroblasts into those of the normal skin fibroblast control, demonstrating their potential in the treatment of Alzheimer disease.

Another important indication for these bryostatin analogs is in the treatment of AIDS. Although the current highly active antiretroviral therapeutics (HAART) fight the active virus by attacking numerous pathways simultaneously, these

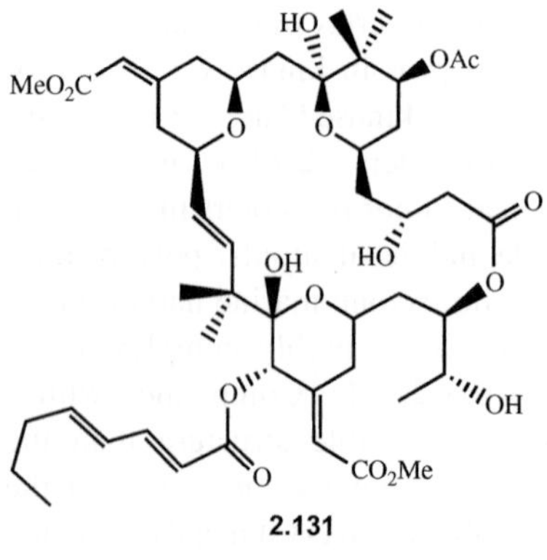

2.131 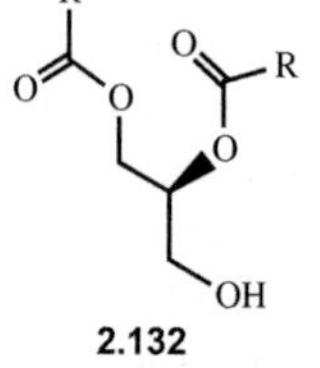 **2.132** 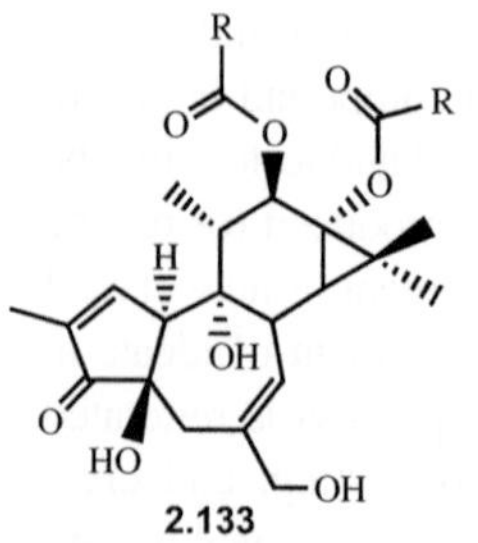 **2.133** 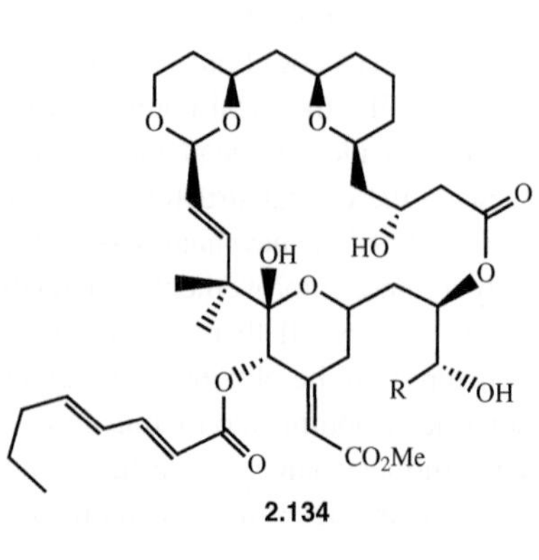 **2.134**

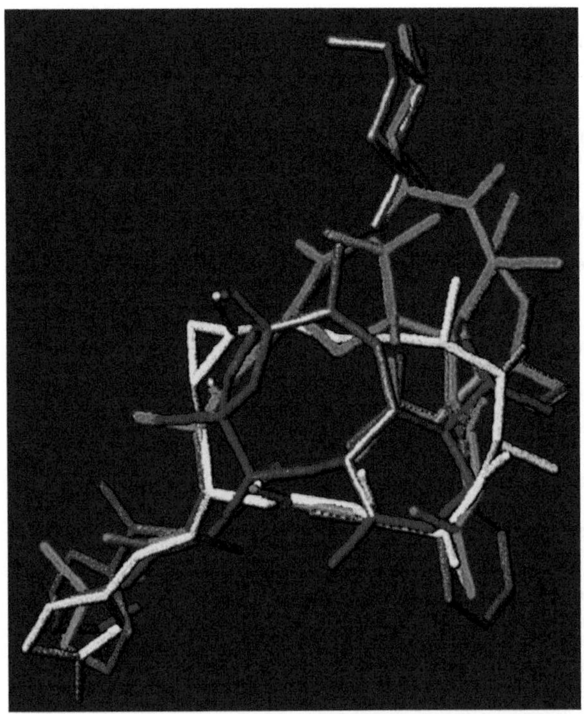

FIGURE 2.34 Common pharmacophore based on the composite of boxed sections in Figure 2.33. *With permission for I. Ojima (1999). Reprinted with permission from* Proc. Natl. Acad. Sci. USA ***1999***, 96, *4256–4261. Copyright 1999 National Academy of Sciences, U.S.A.*

drugs need to be taken chronically because of the latency of the virus; it can adopt a dormant provirus form in immune cells and then emerge at a later time in its active form. If the latency were eliminated, AIDS could be cured by these HIV therapeutics. Bryostatin 1 and its analogs were shown to bring the provirus out of dormancy; combination therapy with HAART would combine the purging of the latent virus and the destruction of the active virus, thereby eradicating the disease.[425]

The initial expectation for structure-based drug design, that potent receptor binders would be designed rapidly leading to the discovery of many new drugs, has not yet become a reality. Several problems with this approach may contribute to its less-than-optimal effectiveness. Table 2.20 lists various advantages to the use of molecular modeling approaches and its many limitations. Although the ease of visualization is appealing, the main problems are (1) that the structure of the molecular model may be completely different from the actual structure in the living organism; (2) even if the structure is correct, the resolution of the structure is insufficient to make an accurate assessment of ligand binding; and (3) until a crystal structure of a complex with bound ligand can be obtained, the bioactive conformation of a given ligand is not known, so the appropriate conformations may not be used in modeling experiments. Another important reason why there has not been a large increase in the number of new drugs being developed by molecular modeling techniques derives from the fact that pharmacokinetics is ignored by this method. Prior to the drug candidate interacting with a receptor, it must be properly absorbed, it must reach the receptor without metabolic or chemical degradation (unless it is a prodrug; see Chapter 9), excretion must be at an appropriate rate, and the drug candidate and metabolites must not be toxic or lead to undesirable side effects.

Because of all the uncertainty involved in this method, the process of molecular modeling, synthesis, testing, and molecular modeling again needs to undergo many iterations. Structure-based drug design has to be taken as yet another tool available to the medicinal chemist; it is not yet the answer to drug discovery, but it can be a very important part of the process.

2.2.7. Epilogue

On the basis of what was discussed in this chapter, it appears that even if you uncover a lead, it may be a fairly slow and random process to optimize its potency. Since 1950, in inflation-adjusted terms, the number of FDA-approved drugs per billion dollars spent in R&D has halved every 9 years.[426] However, the cost to get a drug in the market has increased from $4 million in 1962 to $350 million in 1996, $500 million in 2000,[427] $600–$800 million in 2002, and $1.8 billion in 2010.[428] Between 1960 and 1980, the time for development of a compound from synthesis to the market almost quadrupled, but the time has remained fairly constant since 1980 at about 12–15 years of research. An important cause for the increase in the length of time to bring a drug into the market that occurred in 1962 was the devastating effect of the drug thalidomide, a hypnotic drug shown to cause severe fetal limb abnormalities (phocomelia) when taken in the first trimester of pregnancy (see Chapter 3, Section 3.2.5.2). This tragedy led to the passage of the Harris-Kefauver Amendments to the Food, Drug, and Cosmetic Act in 1962, which required sufficient pharmacological and toxicological research in animals before a drug could be tested in humans; the data of the animal studies had to be submitted to the FDA in an application for approval of an investigational new drug (IND) before human testing could begin. After 1–5 years (average 2.6 years) of animal testing, three phases of clinical (human) trials were adopted (lasting from 4 to 10 years; see Chapter 1, Section 1.3.6.2) before a new drug application (NDA) could be submitted for commercial approval of a new drug.[429] Because of the declining approval rate of new drugs and increased costs, 84% of all prescriptions written in the US in 2012 were for generic drugs.[430]

It has been estimated that in 1950, 7000 compounds had to be isolated and tested for each one that made it to the market; by 1979, this number rose to 10,000 compounds, and now it is greater than 20,000 compounds. There are

TABLE 2.20 Advantages and Limitations to the Use of Molecular Modeling in Lead Modification

Advantages

- Proteins can be visualized in 3D, and every amino acid can be located.
- The structure can be manipulated so that it can be observed from any direction in 3D.
- Particular regions, e.g., the binding site, can be enlarged for better viewing.
- The physicochemical properties, e.g., hydrophobic, polar, positive or negative charge, etc., of each part of the receptor can be viewed.
- Distances between groups can be determined.
- Small molecules can be docked into various regions to determine their fit and interactions.
- Residues that are most suitable to mutate for mechanism studies can be determined.

Limitations

- The coordinates from an X-ray crystal structure or NMR solution structure are required.
- Crystals are obtained by crystallization of proteins under nonphysiological conditions, such as at low or high pH, well below 37 °C and in the presence of additives, such as buffers or detergents. Are the proteins really in the same conformation as in the living cell?
- Crystal structures represent the thermodynamically most stable conformation of the protein under these nonphysiological conditions. Therefore, the crystal structure may depict the protein in a conformation very different from that in a living cell.
- Often crystal structures with ligands bound are obtained by soaking the ligand into the preformed crystal. If binding of the ligand in solution results in a conformational change, it is highly unlikely that it will occur in the crystalline state because the crystal packing forces will favor the preexisting conformation.
- The protein structure is considered to be rigid, but small conformational changes of side chains can induce large changes in the size, shape, and interaction pattern of binding pockets. Typically, when a small molecule binds to a protein, there is some movement of side chains.
- Resolutions of crystal structures are generally in the range of 2–2.5 Å; some at 1.5–2.0 Å; rarely more resolved (although <1.0 Å are known).[1] Therefore, there is *much* uncertainty as to the exact position of each atom. A rule of thumb is that the positional error of atoms is about one-sixth of the resolution[2]; so a structure at 2.4 Å resolution has an uncertainty of every atom of 0.4 Å.
- Small molecules in the ground state are generally energy minimized to give the lowest energy conformers prior to docking them into the structure, but a ligand does not have to bind in the lowest energy conformation, and it can be quite different from the ground-state conformation. Also, solvent effects generally are not taken into account.
- For highly flexible molecules with several torsional angles, there may be many different geometries having the same conformational energy, but significantly different shapes.
- You tend to believe what you see in your molecular model and think it is accurate! This leads to many wrong assumptions.

[1]*For example, Betzel, C.; Gourinath, S.; Kumar, P.; Kaur, P.; Perbrandt, M.; Eschenburg, S.; Singh, T.P. Biochemistry* **2001**, *40, 3080.*
[2]*Böhm, H.-J.; Klebe, G. Angew. Chem. Int. Ed. Engl.* **1996**, *35, 2588.*

only about 6000 known drugs in the CMC database of the estimated 10^{60} possible compounds that could be drug-like, and these 6000 drugs interact with only about 400–500 different human targets or <1% of the human *proteome* (all proteins expressed by humans). It has been estimated that the number of potential drug targets may be between 5000 and 10,000.[431] Therefore, *genomics* (identifying and analyzing new gene targets from a genome) and *proteomics* (identifying and analyzing proteins expressed by the genes in the genome) have become very important aspects to drug discovery.[432] Once a new target from the proteome is identified, *bioinformatics*, in which databases of known proteins are scanned to find known proteins with structures similar to that of the new target, may be employed. When the similarities are known, inhibitors of the known protein can be tested with the new target protein. In addition to these biological methodologies, which appear to be increasing the rate of lead discovery, other rational approaches to lead discovery and lead optimization, based on chemical and biochemical principles, must be used. However, despite the significant advances in medicinal chemistry over the past decade, the failure rate in clinical trials has increased.[433] Between 2007 and 2011, the success rate in Phase II clinical trials dropped to just 18%; over half of the failures were the result of insufficient efficacy compared to placebo, 29% due to strategic miscalculations (drug not that different from existing drugs or going after existing targets), and 19% due to safety concerns.[434] In

Phase III, the failure rate was 50%, two-thirds resulting from lack of sufficient efficacy and 21% because of safety concerns.[435] Strategic issues in Phase III could involve advancing to Phase III despite marginal results in Phase II or attempting to use an existing drug for an indication that is different from what it was intended. For example, sunitinib (**2.135**, Sutent) is marketed for GI stromal tumors; it failed Phase III clinical trials for hepatic cancer. Maybe Thomas Edison said it best: "I have not failed. I've just found 10,000 ways that won't work."

Sunitinib
2.135

2.3. GENERAL REFERENCES

Drug Discovery

Drug Discovery and Development. Volume 2. Drug Development. Chorghade, M. S. (Ed.); John Wiley & Sons: Hoboken, NJ, 2007.

Drug Discovery Research. New Frontiers in the Post-Genomic Era. Huang, Z.; Wiley & Sons: Hoboken, NJ 2007.

Real World Drug Discovery: A Chemist's Guide to Biotech and Pharmaceutical Research; Rydzewski, R. M.; Elsevier: Amsterdam, 2008.

Lead Generation Approaches in Drug Discovery, Rankovic, Z.; Morphy, R., Eds.; Wiley & Sons: Hoboken, NJ, 2010.

Combinatorial Chemistry

Kshirsagar, T. (Ed.) *High-throughput Lead Optimization in Drug Discovery*; CRC Press: Boca Raton, FL, 2008.

Journals

Combinatorial Chemistry & High Throughput Screening

Journal of Combinatorial Chemistry

Molecular Diversity

webpages

http://www.5z.com

http://www.warr.com/ombichem.html (yes, "ombichem")

http://www.combi-web.com

Fragment-Based Drug Discovery

Jahnke, W.; Erlanson, D. A. *Fragment-based approaches in drug discovery*; Wiley: Weinheim, Germany, 2007.

Jhoti, H.; Leach, A. R. *Structure-based drug discovery*; Springer: Dordrecht, The Netherlands, 2007.

Ji, H.; Silverman, R. B. Case study 3: Fragment hopping to design highly potent and selective neuronal nitric oxide synthase inhibitors. *In* Scaffold Hopping in Medicinal Chemistry; Brown, N., ed.; Wiley-VCH Verlag: Weinheim, Germany; 2013, Chapter 18.

Zartler, E.; Shapiro, M., Eds. *Fragment-based drug discovery: a practical approach;* Wiley: Chichester, U.K., 2008.

Peptides and Peptidomimetics

Peptide Drug Discovery and Development: Translational Research in Academia and Industry; Castanho, M.; Santos, N., Eds.; Wiley-VCH: Weinheim, 2011.

Bursavich, M. G.; Rich, D. H. Designing non-peptide peptidomimetics in the twenty-first century: inhibitors targeting conformational ensembles. *J. Med. Chem.* **2002**, *45*, 541–558.

Ripka, A. S.; Rich, D. H. Peptidomimetic design. *Current Opin. Chem. Biol.* **1998**, *2*, 441–452.

Giannis, A.; Kolter, T. Peptidomimetics for receptor ligands – discovery, development and medical perspectives. *Angew. Chem. Int. Ed. Engl.* **1993**, *32*, 1244–1267.

Pharmacokinetics

Curry, S. H.; Whelpton, R. *Drug Disposition and Pharmacokinetics. From Principles to Applications*; Wiley-Blackwell: Chichester, U. K., 2011.

Hardman, J. G.; Limbird, L. E.; Gilman, A. G. *Goodman and Gilman's The Pharmacological Basis of Therapeutics*; 10th Edition; McGraw-Hill: New York, 2001.

Kerns, E. H.; Di, L. Drug-like properties: concepts, structure and design and methods: from ADME to toxicity optimization; Academic Press: Amsterdam, 2008.

Meanwell, N. A. Improving drug candidates by design: a focus on physicochemical properties as a means of improving compound disposition and safety. *Chem. Res. Toxicol.* **2011**, *24*, 1420–1456.

Smith, D. A.; van de Waterbeemd, H.; Walker, D. K.; Mannhold, R.; Kubinyi, H.; Timmerman, H. *Pharmacokinetics and Metabolism in Drug Design*; Wiley: New York, 2000.

QSAR.

Books

Reviews in Computational Chemistry; Lipkowitz, K. B.; Boyd, D. B., eds.; Wiley-VCH: New York. (Entire series of volumes)

Computational Drug Design: A Guide for Computational and Medicinal Chemists, D. C. Young, John Wiley & Sons: Hoboken, NJ, 2009.

Comparative QSAR, Edited by J. Devillers, Taylor and Francis: Washington, DC, 1998.

QSAR: Hansch Analysis and Related Approaches, H. Kubinyi, VCH: Weinheim, 1993.

C. Hansch; A. Leo, *Exploring QSAR*, Vol. 1 Fundamentals and Applications in Chemistry and Biology, ACS Publ.: Washington, DC, 1995.

A. Leo; C. Hansch; D. Hoekman, *Exploring QSAR*, Vol. 2 Hydrophobic, Electronic, and Steric Constants, ACS Publ.: Washington, DC, 1995.
3D-QSAR in Drug Design. Theory, Methods and Applications, Kubinyi, H., Ed., ESCOM, Leiden (NL), 1993.

Journals

Journal of Chemical Information and Computer Science

Computer Modeling

Höltje, H.-D.; Sippl, W.; Rognan, D.; Folkers, G. *Molecular Modeling: Basic Principles and Applications*; Wiley-VCH: Weinheim, Germany, 2008.
Young, D. C. Computational Drug Design: A Guide for Computational and Medicinal Chemists; John Wiley & Sons, 2009.
Bohacek, R. S.; McMartin, C.; Guida, W. C. The art and practice of structure-based drug design: a molecular modeling perspective. *Med. Res. Rev.* **1996**, *16*, 3.
Gubernator, K.; Böhm, H.-J. (Eds.) *Structure-Based Ligand Design*; Wiley-VCH: Weinheim, 1998.
Kirkpatrick, D. L.; Watson, S.; Ulhaq, S. Structure-based drug design: combination chemistry and molecular modeling. *Combinatorial Chem. High Throughput Screening* **1999**, *2*, 211–221.
Antel, J. Integration of combinational chemistry and structure-based drug design. *Curr. Opin. Drug Discovery Dev.* **1999**, *2*, 224–233.
Ooms, F. Molecular modeling and computer aided drug design. Examples of their applications in medicinal chemistry. *Curr. Med. Chem.* **2000**, *7*, 141–158.
Böhm, H.-J.; Stahl, M. Structure-based library design: molecular modeling merges with combinational chemistry. *Curr. Opin. Chem. Biol.* **2000**, *4*, 283–286.
Böhm, H.-J. Computational tools for structure-based ligand design. *Prog. Biophys. Mol. Biol.* **1996**, *66*, 197–210.
Kuntz, I. D., E. C. Meng, and Shoichet B. K. Structure-based molecular design. *Acc. Chem. Res.* **1994**, *27*, 117–123.
Lauri, G.; Bartlett, P. A. CAVEAT: A program to facilitate the design of organic molecules. *J. Comput. Aided Mol. Design* **1994**, *8*, 51–66.

Journals

Annual Reports in Combinatorial Chemistry and Molecular Design
Journal of Computer-Aided Molecular Design
Journal of Molecular Graphics

Software

Sybyl™ (Tripos, Inc.)
Insight II™ (MDL, Inc.)
Molecular Conceptor™ Courseware (Synergix, Ltd.)
CaChe™ Software (Fujitsu, Inc.)
http://www.netsci.org/Resources/Software/Modeling/CADD

References for Computer-Based Drug Design Methodologies.

Active Site Analysis.

MCSS

Caflish, A.; Miranker, A.; Karplus, M. Multiple copy simultaneous search and construction of ligands in binding sites: application to inhibitors of HIV-1 aspartic proteinase. *J. Med. Chem.* **1993**, *36*, 2142–2167.
Stultz, C. M.; Karplus, M. MCSS functionality maps for a flexible protein. *Proteins* **1999**, *37*, 512–529.

GRID

Goodford, P. J. A computational procedure for determining energetically favorable binding sites on biologically important macromolecules. *J. Med. Chem.* **1985**, *28*, 849–857.
Boobbyer, D. N. A.; Goodford, P. J.; McWhinnie, P. M.; Wade, R. C. New Hydrogen-bond potentials for use in determining energetically favorable binding sites on molecules of known structure. *J. Med. Chem.* **1989**, *32*, 1083–1094.
Wade, R. C.; Goodford, P. J. Further development of hydrogen bond functions for use in determining energetically favorable binding sites on molecules of known structure. 1. Ligand probe groups with the ability to form two hydrogen bonds. *J. Med. Chem.* **1993**, *36*, 140–147.
Wade, R. C.; Goodford, P. J. Further development of hydrogen bond functions for use in determining energetically favorable binding sites on molecules of known structure. 2. Ligand probe groups with the ability to form more than two hydrogen bonds. *J. Med. Chem.* **1993**, *36*, 148–156.

Applications of MCSS and GRID

Bitetti-Putzer, R.; Joseph-McCarthy, D.; Hogle, J. M.; Karplus, M. Functional group placement in protein binding sites: a comparison of GRID and MCSS. *J. Comput.-aided Mol. Des.* **2001**, *15*, 935–960.
Powers, R. A.; Shoichet, B. K. Structure-based approach for binding site identification on AmpC β-Lactamase. *J. Med. Chem.* **2002**, *45*, 3222–3234.

MCSS Applied to Structure-Based Ligand Design

Stultz, C. M.; Karplus, M. Dynamic ligand design and combinatorial optimization: designing inhibitors to endothiapepsin. *Proteins* **2000**, *40*, 258–289.
Joseph-McCarthy, D.; Tsang, S. K.; Filman, D. J.; Hogle, J. M.; Karplus, M. Use of MCSS to design small targeted libraries: application to picornavirus ligands. *J. Am. Chem. Soc.* **2001**, *123*, 12,758–12,769.

GRID Applied to 3D-QSAR

Davis, A. M.; Gensmantel, N. P.; Johansson, E.; Marriott, D. P. The use of the GRID program in the 3D-QSAR analysis of a series of calcium-channel agonists. *J. Med. Chem.* **1994**, *37*, 963–972.
Cruciani, G.; Watson, K. A. Comparative molecular field analysis using GRID force-field and GOLPE variable selection methods in a study of inhibitors of glycogen phosphorylase b. *J. Med. Chem.* **1994**, *37*, 2589–2601.

Bohm, M.; Klebe, G. Development of new hydrogen-bond descriptors and their application to comparative molecular field analysis. *J. Med. Chem.* **2002**, *45*, 1585–1597.

Pastor, M.; Cruciani, G.; Mclay, I.; Pickett, S.; Clemente, S. Grid-INdependent Descriptors (GRIND): a novel class of alignment-independent 3-D molecular descriptors. *J. Med. Chem.* **2000**, *43*, 3233–3243.

GRID Applied to Selectivity Analysis

Kastenholz, M. A.; Pastor, M.; Gruciani, G.; Haaksma, E. E.; Fox, T. GRID/CPCA: A new computational tool to design selective ligand. *J. Med. Chem.* **2000**, *43*, 3033–3044.

X-SITE

Laskowski, R. A.; Thornton, J.M.; Humblet, C.; Singh, J. X-SITE: use of empirically derived atomic packing preferences to identify favourable interaction regions in the binding sites of proteins. *J. Mol. Biol.* **1996**, *259*, 175–201.

Molecular Docking

DOCK, GOLD, Glide and FlexX are used as virtual screening tools when the 3-D structure of the binding site of the receptor is known.

Fast Shape Matching (DOCK)

Kuntz, I. D.; Blaney, J. M.; Oatley, S. J.; Langridge, R.; Ferrin, T. E. A geometric approach to macromolecule–ligand interactions. *J. Mol. Biol.* **1982**, *161*, 269–288.

Incremental Sonstruction (FlexX, Hammerhead, Surflex)

Rarey, M.; Kramer, B.; Lengauer, T.; Klebe, G. A fast flexible docking method using as incremental construction algorithm. *J. Mol. Biol.* **1996**, *261*, 470–489.

Welch, W.; Ruppert, J.; Jain, A. N. Hammerhead: fast fully automated docking of flexible ligands to protein binding sites. *Chem. Biol.* **1996**, *3*, 449–462.

Jain, A. N. Surflex: fully automatic flexible molecular docking using a molecular similarity-based search engine. *J. Med. Chem.* **2003**, 46, 499–511.

Tabu Search (Pro_Leads)

Baxter, C. A.; Murray, C. W.; Clark, D. E.; Westhead, D. R.; Eldridge, M. D. Flexible docking using Tabu search and an empirical estimate of binding affinity. *Proteins* **1998**, *33*, 367–382.

Genetic algorithm (GOLD, AutoDock 3.0, AutoDock 4.0)

Jones, G.; Wilett, P.; Glen, R. C.; Leach, A. R.; Taylor, R. Development and validation of a genetic algorithm for flexible docking. *J. Mol. Biol.* **1997**, *267*, 727–748.

Morris, G. M.; Goodsell, D. S.; Halliday, R.; Huey, R.; Hart, W. E.; Belew, R. K.; Olson, A. J. Automated docking using a Lamarckian genetic algorithm and an empirical binding free energy function. *J. Comput. Chem.* **1998**, *19*, 1639–1662.

Morris, G. M.; Huey, R.; Lindstrom, W.; Sanner, M. F.; Belew, R. K.; Goodsell, D. S.; Olson, A. J. AutoDock4 and AutoDockTools4: automated docking with selective receptor flexibility. *J. Comput. Chem.* **2009**, *30*, 2785–2791.

Monte Carlo Simulations (Glide, ICM, MCDOCK, QXP)

Abagyan, R. A.; Totrov, M. M. Biased probability Monte Carlo conformational searches and electrostatic calculations for peptides and proteins. *J. Mol. Biol.* **1994**, *235*, 983–1002.

Liu, M.; Wang, S. MCDOCK: a Monte Carlo simulation approach to the molecular docking problem. *J. Comput-Aided Mol. Des.* **1999**, *13*, 435–451.

McMartin, C.; Bohacek, R. S. QXP: powerful, rapid computer algorithms for structure-based drug design. *J. Comput-Aided Mol. Des.* **1997**, *11*, 333–344.

Friesner, R. A.; Banks, J. L.; Murphy, R. B.; Halgren, T. A.; Klicic, J. J.; Mainz, D. T.; Repasky, M. P.; Knoll, E. H.; Shelley, M.; Perry, J. K.; Shaw, D. E.; Francis, P.; Shenkin, P. S. Glide: a new approach for rapid, accurate docking and scoring. 1. Method and assessment of docking accuracy. *J. Med. Chem.* **2004**, *47(7)*, 1739–1749.

Simulated Annealing (AutoDock 2.4)

Goodsell, D. S.; Olson, A. J. Automated docking of substrates to proteins by simulated annealing. *Proteins* **1990**, *8*, 195–202.

Post-Docking Treatments

Free Energy Perturbation (FEP)

FEP is an accurate method, but it is very time consuming.

Kollman, P. A. Advances and continuing challenges in achieving realistic and predictive simulations of the properties of organic and biological molecules. *Acc. Chem. Res.* **1996**, *29*, 461–469.

Kollman, P. A. Free-energy calculations – applications to chemical and biochemical phenomena. *Chem. Rev.* **1993**, *93*, 2395–2417.

Toba, S.; Damodaran, K. V.; Merz, Jr., K. M. Binding preferences of hydroxamate inhibitors of the matrix metalloproteinase human fibroblast collagenase. *J. Med. Chem.* **1999**, *42*, 1225–1234.

OWFEG

OWFEG can be used to simplify the FEP method.

Pearlman, D. A. Free energy grids: a practical qualitative application of free energy perturbation to ligand design using the OWFEG method. *J. Med. Chem.* **1999**, *42*, 4313–4324.

Thermodynamic Integration (TI)

Reddy, M. R.; Viswanadhan, V. N.; Weinstein, J. N. Relative differences in the binding free energies of human immunodeficiency virus 1 protease inhibitors: a thermodynamic cycle-perturbation approach, *Proc. Natl. Acad. Sci. USA.* **1991**, *88*, 10,287–10,291.

McDonald, J. J.; Brooks, C. L. Theoretical approach to drug design. 2. Relative thermodynamics of inhibitor binding by chicken dihydrofolate reductase to ethyl derivatives of trimethoprim substituted at 3′-, 4′-, and 5′-positions. *J. Am. Chem. Soc.* **1991**, *113*, 2295–2301.

Guimaraes, C. R. W.; Bicca de Alencastro, R. Thermodynamic analysis of thrombin inhibition by benzamidine and p-methylbenzamidine via free-energy perturbations: inspection of intraperturbed-group contributions using the finite difference thermodynamic integration (FDTI) algorithm *J. Phys. Chem. B.* **2002**, *106*, 466–476.

Scoring Methods for Virtual Screening

Force Field Scoring Functions:
DOCK

Kuntz, I. D.; Blaney, J. M.; Oatley, S. J.; Langridge, R.; Ferrin, T. E. A geometric approach to macromolecule–ligand interactions. *J. Mol. Biol.* **1982**, *161*, 269–288.

GOLD

Jones, G.; Wilett, P.; Glen, R. C.; Leach, A. R.; Taylor, R. Development and validation of a genetic algorithm for flexible docking. *J. Mol. Biol.* **1997**, *267*, 727–748.

Validate

Head, R. D.; Smythe, M. L.; Opera, T. I.; Waller, C. L.; Green, S. M.; Marshall, G. R. VALIDATE: A new method for the receptor-based prediction of binding affinities of novel ligands. *J. Am. Chem. Soc.* **1996**, *118*, 3959–3969.

Empirical Free Energy Scoring Functions:
LUDI

Bohm, H. J. The development of a simple empirical scoring function to estimate the binding constant for a protein-ligand complex of known three-dimensional structure. *J. Comput-Aided Mol. Des.* **1994**, *8*, 243–256.

Chemscore

Eldrige, M.; Murray, C. W.; Auton, T. A.; Paolini, G. V.; Lee, R. P. Empirical scoring functions: I. the development of a fast empirical scoring function to estimate the binding affinity of ligands in receptor complexes. *J. Comput.-Aided Mol. Des.* **1997**, *11*, 425–445.

FlexX

Rarey, M.,; Kramer, B.; Lengauer, T.; Klebe, G. A fast flexible docking method using as incremental construction algorithm. *J. Mol. Biol.* **1996**, *261*, 470–489.

Score

Wang, R.; Liu, L.; Lai, L.; Tang, Y. Score: a new empirical method for estimating the binding affinity of ligands in receptor complexes. *J. Comput.-Aided Mol. Des.* **1997**, *11*, 425–445.

Fresno

Rognan, D.; Lauemoller, S. L.; Holm, A.; Buus, S.; Tschinke, V. Predicting binding affinities of protein ligands from three-dimensional models: application to peptide binding to class I major histocompatibility proteins. *J. Med. Chem.* **1999**, *42*, 4650–4658.

X-Score

Wang, R.; Lai, L.; Wang, S. Further development and validation of empirical scoring functions for structure-based binding affinity prediction. *J. Comput.-Aided Mol. Des.* **2002**, *16*, 11–26.

Knowledge-Based Scoring Functions.
PMF

Muegge, I.; Martin, Y. C. A general and fast scoring function for protein-ligand interactions: a simplified potential approach. *J. Med. Chem.* **1999**, *42*, 791–804.

DrugScore

Gohlke, H.; Hendlich, M.; Klebe, G. Knowledge-based scoring function to predict protein-ligand interaction. *J. Mol. Biol.* **2000**, *295*, 337–356.

SmoG2001

Ishchenko, A. V.; Shakhnovich, E. I.; Small Molecule Growth 2001 (SmoG2001): An improved knowledge-based scoring function for protein-ligand interactions. *J. Med. Chem.* **2002**, *45*, 2770–2780.

Gohlke, H.; Klebe, G. DrugScore meets CoMFA: Adaptation of fields for molecular comparison (AFMoC) or How to tailor knowledge-based pair-potentials to a particular protein. *J. Med. Chem.* **2002**, *45*, 4153–4170.

Consensus Scoring

Charifson, P. S.; Corkey, J. J. Murcko, M. A.; Walters, W. P. Consensus scoring: a method for obtaining improved hit rates from docking databases of three-dimensional structure into proteins. *J. Med. Chem.* **1999**, *42*, 5100–5109.

Terp, G. E.; Johansen, B. N.; Christensen, I. T.; Jorgensen, F. S. A new concept for multidimensional selection of ligand conformations (MultiSelect) and multidimensional scoring (MultiScore) of protein-ligand binding affinities. *J. Med. Chem.* **2001**, *44*, 2333–2343.

De novo Lead Design
LUDI

Bohm, H. J. LUDI: rule-based automatic design of new substituents for enzyme inhibitor leads. *J. Comput.-Aided Mol. Des.* **1992**, *6*, 593–606.

BOMB

Jorgensen, W. L. The many roles of computation in drug discovery. *Science* **2004**, *303*, 1813–1818.

LEGEND

Nishibata, Y.; Itai, A. Automatic creation of drug candidate structures based on receptor structure. Starting point for artificial lead generation. *Tetrahedron* **1991**, *47*, 8985–8990.

Nishibata, Y.; Itai, A. Confirmation of usefulness of a structure construction program based on three-dimensional receptor structure for rational lead generation. *J. Med. Chem.* **1993**, *36*, 2921–2928.

Honma, T.; Hayashi, K.; Aoyama, T.; Hashimoto, N.; Machida, T.; Fukasawa, K.; Iwama, T.; Ikeura, C.; Ikuta, M.; Suzuki-Takahashi, I.; Iwasawa, Y.; Hayama, T.; Nishimura, S.; Morishima, H. Structure-based generation of a new class of potent Cdk4 inhibitors: new de novo design strategy and library design. *J. Med. Chem.* **2001**, *44*, 4615–4627.

LeapFrog: From Tripos/SYBYL. www. Tripos.com.

GROW

Moon, J. B.; Howe, W. J. Computer design of bioactive molecules: a method for receptor-based de novo ligand design. *Proteins* **1991**, *11*, 314–328.

GROWMOL (more recent version is called AlleGrow; www.bostondenovo.com)

Bohacek, R. S.; McMartin, C. Multiple Highly diverse structures complementary to enzyme binding sites: results of extensive application of a de novo design method incorporating combinatorial growth. *J. Am. Chem. Soc.* **1994**, *116*, 5560–5571.

PRO_LIGAND

Clark, D. E.; Frenkel, D.; Levy, S. A.; Li, J.; Murray, C. W.; Robson, B.; Waszkowycz, B.; Westhead, D. R. PRO-LIGAND: an approach to de novo molecular design. 1. Application to the design of organic molecules. *J. Comput.-Aided Mol. Des.* **1995**, *9*, 13–32.

Waszkowycz, B.; Clark, D. E.; Frenkel, D.; Li, J.; Murray, C.W.; Robson, B.; Westhead, D. R. PRO_LIGAND: an approach to de novo molecular design. 2. design of novel molecules from molecular field analysis (MFA) models and pharmacophores. *J. Med. Chem.* **1994**, *37*, 3994–4002.

SPROUT

Gillet, V.; Johnson, A. P.; Mata, P.; Sike, S.; Williams, P. SPROUT: a program for structure generation. *J. Comput.-Aided Mol. Des.* **1993**, *7*, 127–153.

Gillet, V. J.; Newell, W.; Mata, P.; Myatt, G.; Sike, S.; Zsoldos, Z.; Johnson, A. P. SPROUT: recent developments in the de novo design of molecules. *J. Chem. Inf. Comput. Sci.* **1994**, *34*, 207–217.

MCSS/HOOK

Eisen, M. B.; Wiley, D. C.; Karplus, M.; Hubbard, R. E. HOOK: a program for finding novel molecular architectures that satisfy the chemical and steric requirements of a macromolecule binding site. *Proteins* **1994**, *19*, 199–221.

CAVEAT

Lauri, G.; Bartlett, P. A. CAVEAT: a program to facilitate the design of organic molecules. *J. Comput-Aided Mol. Des.* **1994**, *8*, 51–66.

PRO_LIGAND

NEWLEAD

Tschinke, V.; Cohen, N. C. The NEWLEAD program: a new method for the design of candidate structures from pharmacophoric hypotheses. *J. Med. Chem.* **1993**, *36*, 3863–3870.

3. Lattice-Based Sampling

CLIX

Lawrence, M. C.; Davis, P. C. CLIX: a search algorithm for finding novel ligands capable of binding proteins of known three-dimensional structure. *Proteins* **1992**, *12*, 31–41.

MCSS/DLD

Stultz, C. M.; Karplus, M. Dynamic ligand design and combinatorial optimization: designing inhibitors to endothiapepsin. *Proteins* **2000**, *40*, 258–289.

4. Molecular Dynamics Simulation

Concerts

Pearlman, D. A.; Murcko, M. A. CONCERTS: dynamic connection of fragments as an approach to de novo ligand design. *J. Med. Chem.* **1996**, *39*, 1651–1663.

Virtual Combinatorial Screening

Methods for virtual combinatorial screening

Legion: from SYBYL

CombiGilde: from Schroedinger

PRO_SELECT

Murray, C. W.; Clark, D. E.; Auton, T. R.; Firth, M. A.; Li, J.; Sykes, R. S.; Waszkowycz, B.; Westhead, D. R.; Young. S. C. PRO_SELECT: Combining structure-based drug design and combinatorial chemistry for rapid lead discovery. 1. Technology. *J. Comput.-Aided Mol. Des.* **1997**, *11*, 193–207.

Liebeschuetz, J. W.; Jones, S. D.; Morgan, P. J.; Murray, C. W.; Rimmer, A. D.; Roscoe, J. M.; Waszkowycz, B.; Welsh, P. M.; Wylie, W. A.; Young, S. C.; Martin, H.; Mahler, J.; Brady, L.; Wilkinson, K. PRO_SELECT: combining structure-based drug design and array-based chemistry for rapid lead discovery. 2. The development of a series of highly potent and selective factor Xa inhibitors. *J. Med. Chem.* **2002**, *45*, 1221–1232.

Profile-QSAR

Martin, E.; Mukherjee, P.; Sullivan, D.; Jansen, J. Profile-QSAR: a novel *meta*-QSAR method that combines activities across the kinase family to accurately predict affinity, selectivity, and cellular activity. *J. Chem. Inf. Model.* **2011**, *51*, 1942–1956.

Combinatorial library design

Drug Likeness

MoSELECT

Gillet, V. J.; Khatib, W.; Willett, P.; Fleming, P. J.; Green, D. V. Combinatorial library design using a multiobjective genetic algorithm. *J. Chem. Inf. Comput. Sci.* **2002**, *42*, 375–385.

Gillet, V. J.; Willett, P.; Fleming, P. J.; Green, D. V. Designing focused libraries using MoSELECT. *J. Mol. Graph Model* **2002**, *20*, 491–498.

REOS

Walters, W. P.; Murcko, M. A. Prediction of 'drug-likeness'. *Adv. Drug Deliv. Rev.* **2002**, *54*, 255–271.

Molecular Diversity

Martin, Y. C. Challenges and prospects for computational aids to molecular diversity. *Perspectives in Drug Discovery and Design.* **1997**, *7/8*, 159–172.

DiverseSolution: from SYBYL

Cramer, R. D.; Clark, R. D.; Petterson, D. E.; Ferguson, A. M. Bioisosterism as a molecular diversity descriptor: steric fields of single "Topomeric" conformers. *J. Med. Chem.* **1996**, *39*, 3060.

Patterson, D. E.; Cramer R. D.; Ferguson A. M.; Clark R. D.; Weinberger, L. E. Neighborhood behavior: a useful concept for validation of molecular diversity descriptors. *J. Med. Chem.* **1996**, *39*, 3049.

FlexSim-S

Briem, H.; Lessel, U. F. In vitro and in silico affinity fingerprints: finding similarities beyond structural classes. *Perspective in Drug Discovery and Design.* **2000**, *20*, 231–244.

LASSOO

Koehler, R. T.; Dixon, S. L.; Villar, H. O. LASSOO: A generalized directed diversity approach to the design and enrichment of chemical libraries. *J. Med. Chem.* **1999**, *42*, 4695–4704.

Others

Mount, J.; Ruppert, J.; Welch, W.; Jian, A. N. Ice Pick: A flexible surface-based system for molecular diversity. *J. Med. Chem.* **1999**, *42*, 60–66.

Makara, G. M. Measuring molecular similarity and diversity: total pharmacophore diversity. *J. Med. Chem.* **2001**, *44*, 3563–3571.

Andrews, K. M.; Cramer, R. D. Toward general methods of targeted library design: topomer shape similarity searching with diverse structures as queries. *J. Med. Chem.* **2000**, *43*, 1723–1740.

Mason, J. S.; Morize, I.; Menard, P. R.; Cheney, D. L.; Hulme, C.; Labaudiniere, R. F. New 4-point pharmacophore method for molecular similarity and diversity applications: overview of the method and applications, including a novel approach to the design of combinatorial libraries containing privileged substructures. *J. Med. Chem.* **1999**, *42*, 3251–3264.

Srinivasan, J.; Castellino, A.; Bradley E. K.; Eksterowicz, J. E.; Grootenhuis P. D. J.; Putta, S.; Stanton R. Evaluation of a novel shape-based computational filter for lead evolution: application to thrombin inhibitors. *J. Med. Chem.* **2002**, *45*, 2494–2500.

Synthetic accessibility

Lewell, X. Q.; Judd, D. B.; Watson, S. P.; Hann, M. M. RECAP–retrosynthetic combinatorial analysis procedure: a powerful new technique for identifying privileged molecular fragments with useful applications in combinatorial chemistry. *J. Chem. Inf. Comput. Sci.* **1998**, *38*, 511–522.

Gasteiger, J.; Pfortner, M.; Sitzmann, M.; Hollering, R.; Sacher, O.; Kostka, T.; Karg, N. Computer-assisted synthesis and reaction planning in combinatorial chemistry. *Perspective in Drug Discovery and Design.* **2000**, *20*, 245–264.

Gillet, V. J.; Nicolotti, O. Evaluation of reactant-based and product-based approaches to the design of combinatorial libraries. *Perspective in Drug Discovery and Design.* **2000**, *20*, 265–287.

Ligand optimization

ADME (Pharmacokinetics)

Property-based ligand design

van De Waterbeemd, H; Smith, D. A.; Beaumont, K.; Walker, D. K. Property-based design: optimization of drug absorption and pharmacokinetics. *J. Med. Chem.* **2001**, *44*, 1313–1333.

VolSurf

Cruciani, G.; Pastor, M.; Guba, W. VolSurf: a new tool for the pharmacokinetic optimization of lead compounds. *Eur. J. Pharm. Sci.* **2000**, *Suppl 2*, S29–39.

ChemGPS

Oprea, T. I.; Zamora, I.; Ungell, A. L. Pharmacokinetically based mapping device for chemical space navigation. *J. Comb. Chem.* **2002**, *4*, 258–266.

Ooms, F.; Weber, P.; Carrupt, P. A.; Testa, B. A simple model to predict blood–brain barrier permeation from 3D molecular fields. *Biochim. Biophys. Acta* **2002**, *1587*, 118–125.

Crivori, P.; Cruciani, G.; Carrupt, P. A.; Testa, B. Predicting blood–brain barrier permeation from three-dimensional molecular structure. *J. Med. Chem.* **2000**, *43*, 2204–2216.

Alifrangis, L. H.; Christensen, I. T.; Berglund, A.; Sandberg, M.; Hovgaard, L.; Frokjaer, S. Structure-property model for membrane partitioning of oligopeptides. *J. Med. Chem.* **2000**, *43*, 103–113.

Cruciani, G.; Pastor, M.; Mannhold, R. Suitability of molecular descriptors for database mining. A comparative analysis. *J. Med. Chem.* **2002**, *45*, 2685–2694.

Egan, W. J.; Merz Jr., K. M.; Baldwin, J. J. Prediction of drug absorption using multivariate statistics. *J. Med. Chem.* **2000**, *43*, 3867–3877.

QSAR

2D-QSAR

Hansch method is still used.

Manallack, D. T.; Ellis, D. D.; Livingstone D. J. Analysis of Linear and Nonlinear QSAR Data Using Neural Networks. *J. Med. Chem.* **1994**, *37*, 3639–3654.

Domine, D.; Devillers J.; Chastrette, M. A nonlinear map of substituent constants for selecting test series and deriving structure-activity relationships. 1. aromatic series. *J. Med. Chem.* **1994**, *37*, 973–980.

Domine, D.; Devillers, J.; Chastrette, M. A nonlinear map of substituent constants for selecting test series and deriving structure-activity relationships. 2. aliphatic series. *J. Med. Chem.* **1994**, *37*, 981–987.

3D-QSAR

CoMFA (Comparative Molecular Field Analysis)

Cramer III, R. D.; Paterson, D. E.; Bunce, J. D. Comparative molecular field analysis (CoMFA). I. Effect of shape on binding of steroids to carried proteins. *J. Am. Chem. Soc.* **1988**, *110*, 5959–5967.

Cramer, R. D. Topomer CoMFA: a design methodology for rapid lead optimization. *J. Med. Chem.* **2003**, *46*, 374–388.

Robinson, D. D.; Winn P. J.; Lyne P. D.; Richards, W. G. Self-organizing molecular field analysis: a tool for structure-acitivity studies. *J. Med. Chem.* **1999**, *42*, 573–583.

So, S.-S.; Karplus, M. Three-dimensional QSAR from molecular similarity matrices and genetic neural networks. 1. method and validations. *J. Med. Chem.* **1997**, *40*, 4347–4359.

So, S.-S.; Karplus, M. Three-dimensional QSAR from molecular similarity matrices and genetic neural networks. 2. applications. *J. Med. Chem.* **1997**, *40*, 4360–4371.

CoMSIA (Comparative Molecular Similarity Analysis)

Klebe, G.; Abraham, U.; Mietzner, T. Molecular similarity indices in a comparative analysis (CoMSIA) of drug molecules to correlate and predict their biological activity. *J. Med. Chem.* **1994**, *37*, 4130–4146.

CoMMA (Comparative Molecular Moment Analysis)

Silverman, B. D.; Platt, D. E. Comparative molecular moment analysis (CoMMA): 3D-QSAR without molecular superposition. *J. Med. Chem.* **1996**, *39*, 2129–2140.

COMBINE (Comparative Binding Energy)

Ortiz, A. R.; Pisabarro, M. T.; Gago, F.; Wade R. C. Prediction of drug binding affinities by comparative binding energy analysis. *J. Med. Chem.* **1995**, *38*, 2681–2691.

Perez, C.; Pastor, M.; Ortiz, A. R. Gago F. Comparative binding energy analysis of HIV-1 protease inhibitors: incorporation of solvent effects and validation as a powerful tool in receptor-based drug design. *J. Med. Chem.* **1998**, *41*, 836–852.

GRID/GOLPE

Cruciani, G.; Watson, K. A. Comparative molecular field analysis using GRID force-field and GOLPE variable selection methods in a study of inhibitors of glycogen phosphorylase b. *J. Med. Chem.* **1994**, *37*, 2589–2601.

Cho, S. J.; Tropsha A. Cross-validated R2-guided region selection for comparative molecular field analysis: a simple method to achieve consistent results. *J. Med. Chem.* **1995**, *38*, 1060–1066.

4D-QSAR

Klein, C. D.; Hopfinger, A. J. Pharmacological activity and membrane interactions of antiarrhythmics: 4D-QSAR/QSPR analysis. *Pharm. Res.* **1998**, *15*, 303–311.

5D-QSAR

Vedani, A.; Dobler, M. 5D-QSAR: the key for simulating induced fit? *J. Med. Chem.* **2002**, *45*, 2139–49.

6D-QSAR

Vedani, A.; Dobler, M.; Lill, M. A. Combining protein modeling and 6D-QSAR. Simulating the binding of structurally diverse ligands to the estrogen receptor. *J. Med. Chem.* **2005**, *48*, 3700–3703.

Molecular Superposition

FlexS

Lemmen, C.; Lengauer, T.; Klebe, G. FLEXS: a method for fast flexible ligand superposition. *J. Med. Chem.* **1998**, *41*, 4502–4520.

SQ

Miller, M. D.; Sheridan, R. P.; Kearsely, S. K. SQ: A program for rapidly producing pharmacophorically relevant molecular superpositions. *J. Med. Chem.* **1999**, *42*, 1505–1514.

QXP

McMartin, C.; Bohacek, R. S. QXP: powerful, rapid computer algorithms for structure-based drug design. *J. Comput.-Aided Mol. Des.* **1997**, *11*, 333–344.

Pharmacophore Elucidation and Pharmacophore-based database screening

Martin, Y. C. 3D database searching in drug design. *J. Med. Chem.* **1992**, *35*, 2145–2154.

DISCO from SYBYL

Martin, Y. C. DISCO: what we did right and what we missed. *IUL Biotechnology Series* **2000**, *2*, 49–68.

Martin, Y. C.; Bures, M. G.; Danaher, E. A.; DeLazzer, J.; Lico, I.; Pavlik, P. A. A fast new approach to pharmacophore mapping and its application to dopaminergic and benzodiazepine agonists. *J. Comput.-Aided Mol. Des.* **1993**, *7*, 83–102.

Marriott, D. P.; Dougall, I. G.; Meghani, P. Liu Y.-J.; Flower, D. R. Lead generation using pharmacophore mapping and three-dimensional database searching: application to muscarinic M3 receptor antagonists. *J. Med. Chem.* **1999**, *42*, 3210–3216.

GASP from SYBYL

Jones, G.; Willett, P.; Glen, R. C. GASP: genetic algorithm superimposition program. *IUL Biotechnology Series* **2000**, *2*, 85–106.

Holliday, J.D.; Willett, P. Using a genetic algorithm to identify common structural features in sets of ligands. *J. Mol. Graph. Model.* **1997**, *15*, 221–232.

RECEPTOR from SYBYL

Josien, H.; Convert, O.; Berlose, J.-P.; Sagan, S.; Brunissen, A.; Lavielle, S.; Chassaing, G. Topographic analysis of the S7 binding subsite of the tachykinin neurokinin-1 receptor. *Biopolymers* **1996**, *39*, 133–147.

UNITY from SYBYL

Catalyst

Kurogi, Y.; Guner, O. F. Pharmacophore modeling and three-dimensional database searching for drug design using catalyst. *Curr. Med. Chem.* **2001**, *8*, 1035–1055.

2.4. PROBLEMS (ANSWERS CAN BE FOUND IN THE APPENDIX AT THE END OF THE BOOK)

1. What are some of the advantages and disadvantages of random screening?
2. Name some situations in which the endogenous ligand of a new biological target may not serve as a good lead compound.
3. Cholecystokinin C-terminal tetrapeptide (CCK-4) is the smallest (molecular weight 596.7, CLog $P = -2.1$) active form of CCK found in the brain. Release of CCK in the brain is believed to promote anxiety.
   (a) What would be possible concerns with using CCK-4 as a lead compound for an orally active antianxiety drug?
   (b) Name some alternative approaches to identifying a lead compound.
4. Consider the earlier days, when drug screening was often done in whole animals, vs. today's more modern methods of drug discovery that often start with a high-throughput screening assay. Which of the following factors do you think are most influential in driving the change toward modern methods?
   a. Amount of compound needed for the first test.
   b. Expense of animals needed for the first test.
   c. Early information on in vivo pharmacokinetic properties.
   d. Ability to have a direct indication of pharmacodynamic properties.
5. Consider the data in Table 2.2 and then consider the assumption discussed in Section 2.1.2.3.1 that "similar compounds will have similar biological activities". What conclusions can you draw about this assumption?
6. "Semisynthesis" is a term used to describe preparation of analogs of a natural product using the natural product isolated from natural sources as the starting material. What are some likely pros and cons of the semisynthesis approach?
7. Nitric oxide synthase catalyzes the conversion of L-arginine to L-citrulline and nitric oxide. If you wanted to interfere with the production of NO, design some potential leads.

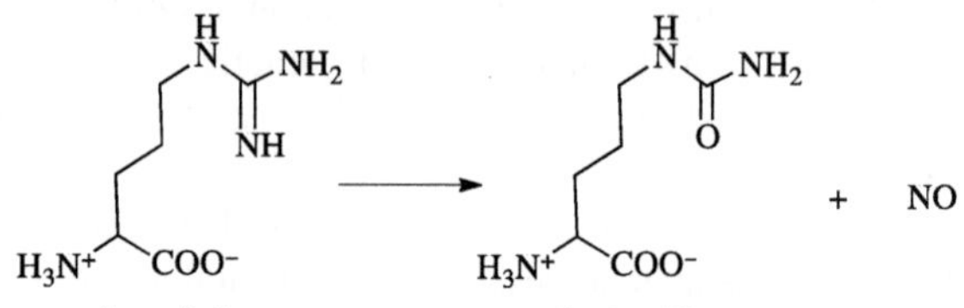

L-arginine → L-citrulline + NO

8. **a.** Compound **1** was found to be a lead compound at Pfizer for enhancement of cytotoxic effects of cancer drugs. (see Canan Koch, S. S. *et al.*, *J. Med. Chem.* **2002**, *45*, 4961–74). Suggest an approach you would take if you wanted to determine the pharmacophoric groups, the groups interfering with receptor (in this case an enzyme) binding, and those not involved at all.

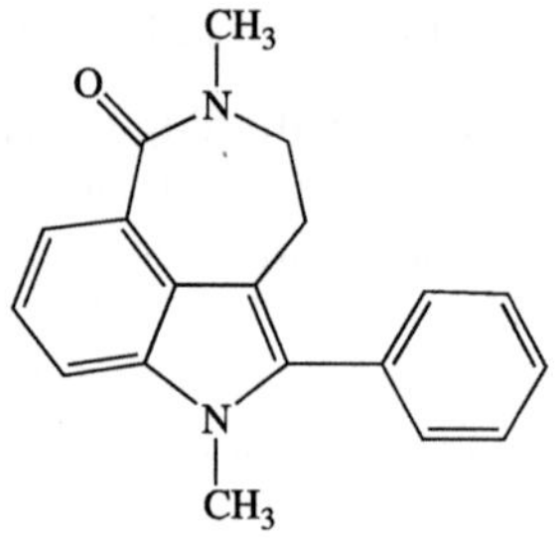

1

b. How would you interpret the following results (compounds 2–7)

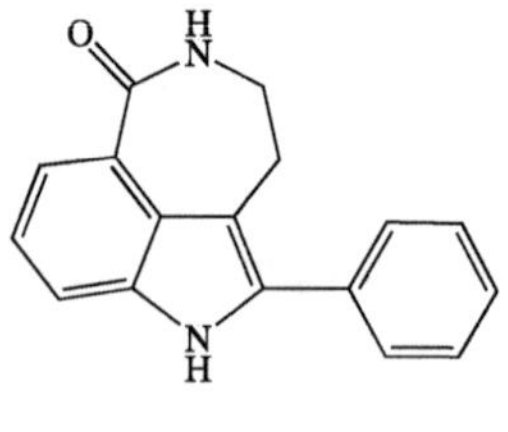

2
$> 10^3$ increase in potency over **1**

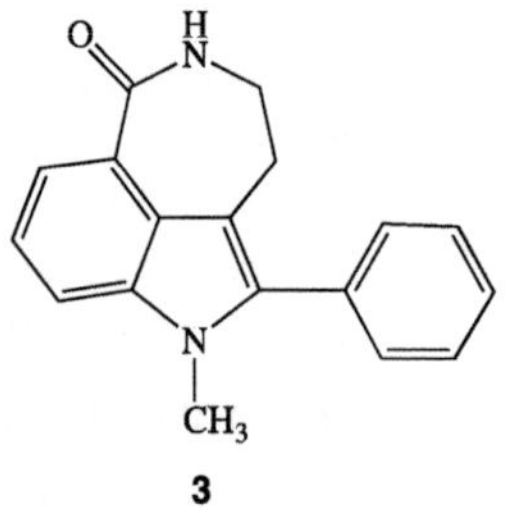

3
About same as **2**

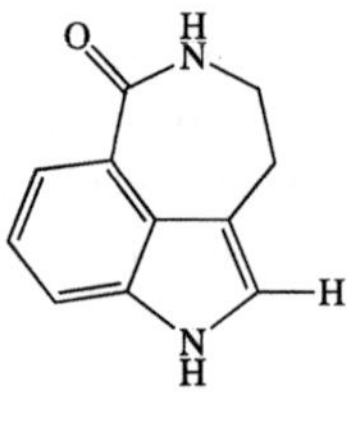

4
1/6 potency of **2**

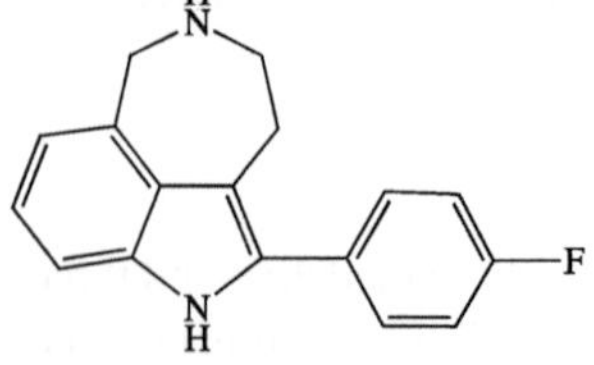

5
$> 10^3$ decrease in potency over **2**

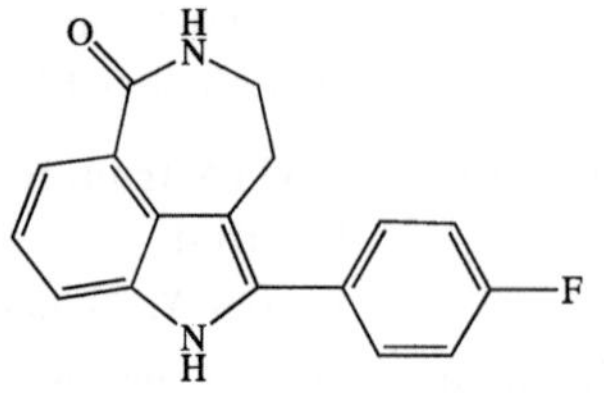

6
A little more potent than **2**

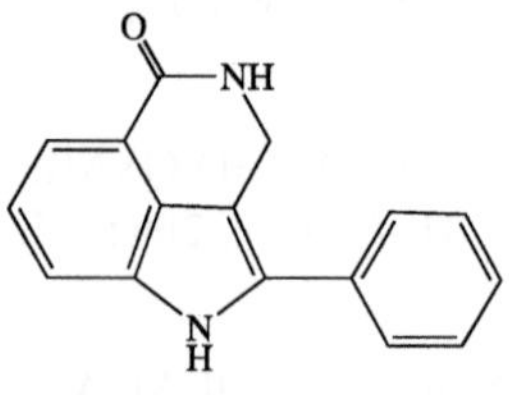

7
A little more potent than **2**

9. A new protein was identified that appears to be abundant in individuals prone to a certain type of cancer. You are trying to identify a molecule that will bind to this protein as a lead compound, but you do not know the structure or function of the protein. How would you proceed?
10. The LD_{50} for a potential antiobesity compound was found to be 10 mg/kg and the ED_{50} was 2 mg/kg. Is this an important drug candidate? Why or why not?
11. Candesartan cilexetil is an antihypertensive prodrug that antagonizes the AT_1 angiotensin receptor. Within its structure are four lead modification approaches with which you should be familiar. Working backward, draw a lead molecule from which this drug may have been derived and point out where the lead modifications occurred.

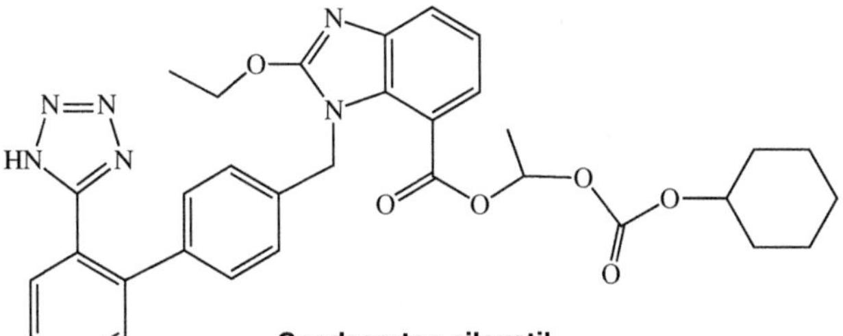

Candesartan cilexetil

12. What are some potential problems in using bioisosteric replacements for lead modification?
13. a. A claimed advantage of solid-phase chemistry was that because excess reagents could be used and then readily removed (by filtration), purer products should result. What is the fallacy in this argument?
    b. Running many reactions in parallel generally implies that they are all done in the same solvent, at the same temperature, for the same time period, and with the same method of agitation, and that they are worked up in the same way. The efficiency advantages in this approach are evident. Name at least one potential disadvantage.
    c. Combinatorial power is increased substantially when more points of diversity are used in a library, i.e., many more compounds can be made using a comparatively small number of monomers. Give two reasons why more compounds under this circumstance may not be an advantage.
14. In solid-phase chemistry, the starting material and product of a reaction are bound to a solid support and the reagents are in solution. In solution-phase chemistry, the starting material and product are in solution and either reagents or reactant scavengers are frequently bound to solid support. What are some advantages of the latter mode of operation?
15. (a) What is the rationale for considering the following when designing a screening collection: (1) drug- or lead-like properties, (2) privileged structures, and (3) toxicophores (b) What might be a disadvantage?
16. Briefly describe the distinction between the terms "ligand-based" and "structure-based".
17. Compounds **8** and **9** below were leads determined from an SAR by NMR study of a new receptor. Based on this analysis, **10** was synthesized, and *n* was varied, but all the compounds made had much lower potency than either **8** or **9**.

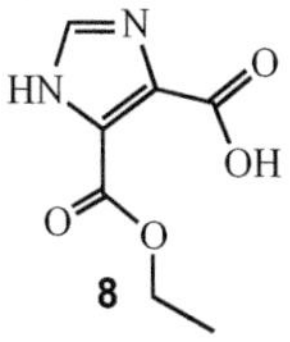

8 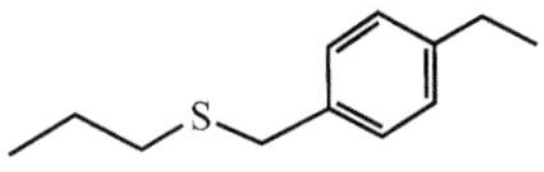 **9**

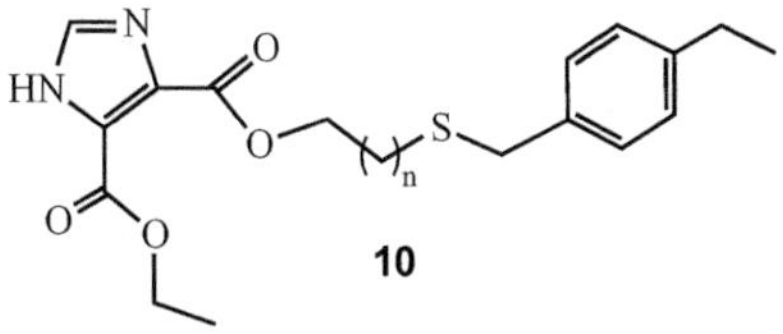

10

    a. What conclusions can you draw from this result (no, the experiment *was* done correctly)?
    b. What structure would you try next? Why?
18. A convenient variant of LE is the binding efficiency index (BEI; Abad-Zapatero, and Metz, *Drug Discovery Today* **2005**, *10*, 464), which can be defined as:

$$BEI = pIC_{50}/MW$$

where the IC_{50} is expressed in *M* and MW is expressed in kD.
    a. Calculate the BEI for the two compounds below:

Cpd	Structure	MW (g/mol)	IC_{50} (μM)	CLog*P*
1	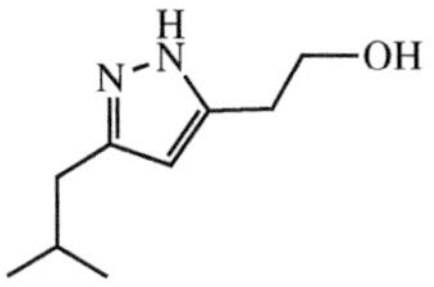	168.2	8	0.93
2	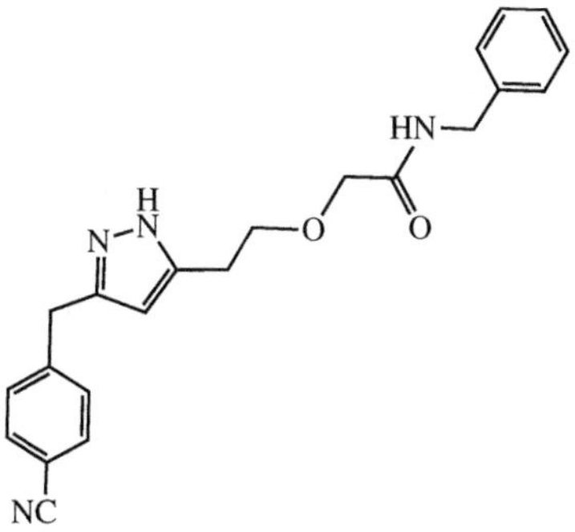	374.4	0.3	2.08

b. Calculate and new index, the BEI-LP, which is analogous to LELP but using BEI in place of LE
c. If only potency is taken into account, which is identified as the better starting point?
d. If BEI is used to prioritize the starting point, which is considered better?
e. If BEI-LP is used to prioritize the starting point, which is considered better?

19. Design six new analogs of the lead molecule compound **1** in problem 18 according to the different design principles specified:
a. Two analogs designed to help identify the pharmacophore.
b. Two analogs to test homologation.
c. Two analogs to test functional group modification.

20.a. Design three peptidomimetics for Glu-Tyr-Val, one using a ring–chain transformation, one a scaffold peptidomimetic, and one having at least one bioisosteric replacement.
b. Would you normally expect a bioisosteric replacement to improve at least one parameter (e.g., activity, safety, or pharmacokinetics) of your lead compound? Why?

21. Based on your knowledge of how the Hammett equation was developed (and basic organic mechanisms), show a mechanism and explain how a change in X will affect the rate of the following reaction.

$X\text{-}C_6H_4\text{-}COOH \xrightarrow[EtOH]{H_2SO_4} X\text{-}C_6H_4\text{-}CO_2Et$

22. Use the partial list of log *P* values at the end of the problem set to calculate the log *P* for the given molecules.
a.

b.

c. Why can there be many correct, but different answers to these calculations?

23.a. In general, why does the log *D* change with pH for some compounds, but not others?
b. Rationalize the change in log *D* vs pH for omeprazole shown below. Refer to the structure when discussing the log *D* changes with pH.

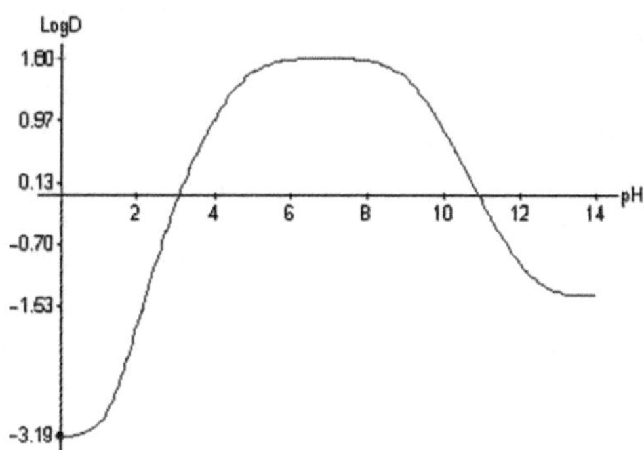

Omeprazole

24. Do you predict that compound **11** will have good oral bioavailability? Why or why not?

11

25. The following compound has potent antifungal activity in a cell-free system, but has poor activity in mice.

a. Why is it not effective in mice?
b. Suggest a structural modification that might increase antifungal activity in mice.

26. You just discovered a lead compound for the treatment of chemistry phobia with structure **12**.

Br

CN

CO_2Me CO_2Me CO_2Me

12 **13** **14**

a. Compound **13** was synthesized and found to be less potent than **12.** Offer explanations and suggest other compounds to synthesize to improve the potency.

b. Compound **14** was prepared and also was less potent than **12**, but more potent than **13.** Explain and suggest other compounds (with rationalizations) to synthesize.

27. *Briefly* describe the basis for Kuntz's DOCK program and Cramer's CoMFA.

28. For computer modeling approaches in drug design, what could be the problems associated with using a crystal structure of the target receptor without a small molecule bound to it?

29. Steric, electronic, lipophilic, and H-bonding effects are important parameters of molecules employed in computer-aided drug design. Why are each of these effects important in drug design?

Compound	log P_{oct}	Compound	log P_{oct}	Compound	log P_{oct}
CH_3OH	−0.66	CH_2=CHCOOH	0.43	▷-O-CH_3	1.20
CH_3NH_2	−0.57	CH_3CH_2CN	0.16	CH_2=CH–OCH_2CH_3	1.04
CCl_3COOH	1.49	H_3C–C(=O)–CH_3	−0.24	$CH_3CH_2CH_2COOH$	0.79
$BrCH_2COOH$	0.41	CH_2=$CHCH_2OH$	0.17	$CH_3CH_2CH_2CH_2OH$	0.83
$ClCH_2COOH$	0.47	CH_3CH_2CHO	0.38	$CH_3CH_2OCH_2CH_3$	0.77
FCH_2COOH	−0.12	CH_3CO_2Me	0.18	$CH_3CH_2OCH_2CH_2OH$	−0.54
ICH_2COOH	0.87	CH_3CH_2COOH	0.33	$CH_3CH_2NHCH_2CH_3$	0.57
CH_3CN	−0.34	CH_3OCH_2COOH	−0.55	piperidine (N–H)	0.85
CH_3CHO	0.43	$CH_3CH_2CH_2Br$	2.10	$CH_3CH_2CH_2CH_2CH_2F$	2.33
CH_3COOH	−0.17	$CH_3CH_2CH_2NO_2$	0.65	$PhCH_2OH$	1.10
$HOCH_2COOH$	−1.11	$CH_3OCH_2OCH_3$	0.00	$PhCH_2NH_2$	1.09
CH_3CH_2Br	1.74	$CH_3OCH_2CH_2OH$	−0.60	phthalimide (O, NH, O)	1.15
CH_3CH_2Cl	1.54	Me_3N	0.27	$PhCH_2COOH$	1.41
CH_3CH_2I	2.00	CH_3I	1.69	$PhOCH_2COOH$	1.26
CH_3CONH_2	−1.46	CH_3NO_2	−0.33	benzene	2.13
$CH_3CH_2NO_2$	0.18	uracil (O, HN, O, N–H)	−1.07	cyclohexane-OH, C≡CH	1.73

Continued

Compound	$\log P_{oct}$	Compound	$\log P_{oct}$	Compound	$\log P_{oct}$
CH_3CH_2OH	−0.32	HOOCCH=CHCOOH	0.28	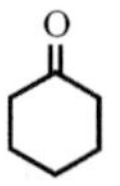	0.81
Me_2NH	−0.23	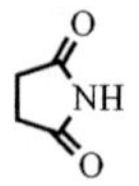	−1.21	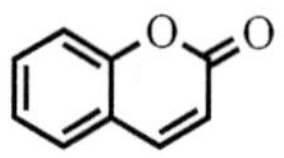	1.39
$CH_3CH_2NH_2$	−0.19	CH_2=CH–O–Ch=CH=CH_2	1.81	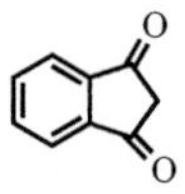	0.61
$HOCH_2CH_2NH_2$	−1.31	CH_3CH=CHCOOH	0.72	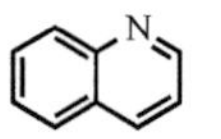	2.03
HC≡CCO_2H	0.46	$HOOCCH_2CH_2COOH$	−0.59	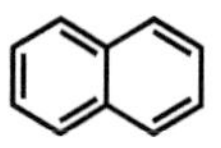	3.37
CH_2=CHCN	−0.92	CH_2=$CHCH_2OCH_3$	0.94	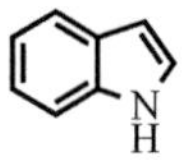	2.00

From Leo, A.; Hansch, C.; Elkins, D. *Chem Rev.* **1971**, *71*, 525.

REFERENCES

1. (a) Cheng, X.-M.; J. A. Bristol Chapter 34. To Market, To Market - 1995. *Annu. Rep. Med. Chem.*, **1996**, *31*, 337–355. (b) Moreau, S. C.; Murphy, W. A. et al. Comparison of somatuline (BIM-23014) and somatostatin on endocrine and exocrine activities in the rat. *Drug Dev. Res.* **1991**, *22*, 79–93.
2. Daws, L. C. Unfaithful neurotransmitter transporters: focus on serotonin uptake and implications for antidepressant efficacy. *Pharmacol. Ther.* **2009**, *121*, 89–99.
3. White, Jr., J. R. Apple trees to sodium glucose co-transporter inhibitors: a review of SGLT2 *inhibition. Clin. Diabetes* **2010**, *28*, 5–10.
4. (a) Proudfoot, J. R. Drugs, leads, and drug-likeness: an analysis of some recently launched drugs. *Bioorg. Med. Chem. Lett.* **2002**, *12*, 1647–50. (b) Oprea, T. I.; Davis, A. M.; Teague, S. J.; Leeson, P. D. Is there a difference between leads and drugs? A historical perspective. *J. Chem. Inf. Comput. Sci.* **2001**, *41*, 1308–1315.
5. (a) Boguski, M. S.; Mandl, K. D.; Sukhatme, V. P. Repurposing with a difference. *Science* (Washington, DC) **2009**, *324*, 1394–1395. (b) Doan, T. L.; Pollastri, M.; Walters, M. A.; Georg, G. I. The future of drug repositioning: old drugs, new opportunities. *Annu. Rep. Med. Chem.* **2011**, *46*, 385–401. (c) Aubé, J. Drug repurposing and the medicinal chemist. *ACS Med. Chem. Lett.* **2012**, *3*, 442–444.
6. Keiser, M. J.; Setola, V.; Irwin, J. J. et al. Predicting new molecular targets for known drugs. *Nature* **2009**, *462*, 175–182.
7. (a) Ritz, B.; Rhodes, S. L.; Qian, L.; Schernhammer, E.; Olsen, J. H.; Friis, S. L-type calcium channel blockers and Parkinson's disease in Denmark. *Ann. Neurol.* **2010**, *67*(5), 600–606. (b) Becker, C.; Jick, S. S.; Meier, C. R. Use of antihypertensives and the risk of Parkinson's disease. *Neurology* **2008** *70*(16, Pt. 2), 1438–1444. (c) Chan C. S.; Gertler T. S; Surmeier D. J. A molecular basis for the increased vulnerability of substantia nigra dopamine neurons in aging and Parkinson's disease. *Movement Disorders: official journal of the Movement Disorder Society* **2010**, *25* Suppl 1, S63–70.
8. (a) Skljarevski, V.; Desaiah, D.; Liu-Seifert, H.; Zhang, Q.; Chappell, A. S.; Detke, J. J.; Iyengar, S.; Atkinson, J. H.; Backonja, M. Efficacy and safety of duloxetine in patients with chronic back pain. *Spine* **2010**, *35*, E578–85. (b) Skljarevski, V.; Zhang, S.; Chappell, A. S.; Walker, D. J.; Murray, I.; Backonja, M. Maintenance of effect of duloxetine in patients with chronic low back pain: a 41-week uncontrolled, dose-blinded study, *Pain Med.* **2010**, *11*, 648–657.
9. Wong, K. K.; Kuo, D. W.; Chabin, R. M.; Founier, C.; Gegnas, L. D.; Waddell, S. T.; Marsilio, F.; Leiting, B.; Pompliano, D. L. Engineering a cell-free murein synthetic pathway: combinatorial enzymology in drug discovery. *J. Am. Chem. Soc.* **1998**, *120*, 13527–13528.
10. (a) Gao, J.; Cheng, X.; Chen, R.; Sigal, G. B.; Bruce, J. E.; Schwartz, B. L.; Hofstadler, S. A.; Anderson, G. A.; Smith, R. D.; Whitesides, G. M. Screening derivatized peptide libraries for tight binding inhibitors to carbonic anhydrase II by electrospray ionization-mass spectrometry. *J. Med. Chem.* **1996**, *39*, 1949–1955. (b) Rossi, D. T.; Sinz, M. W., Eds. *Mass Spectometry in Drug Discovery*, Marcel Dekker: New York, **2002**.
11. (a) Hajduk, P. J.; Olejniczak, E. T.; Fesik, S. W. One dimensional relaxation- and diffusion-edited NMR methods for screening compounds that bind to macromolecules. *J. Am. Chem. Soc.* **1997**, *119*, 12257–12261. (b) Hajduk, P. J.; Gerfin, T.; Boehlen, J.-M.; Häberli, M.; Marek, D.; Fesik, S. W. High-throughput nuclear magnetic resonance-based screening. *J. Med. Chem.* **1999**, *42*, 2315–2317.
12. (a) Mayr, L. M.; Bojanic, D. Novel trends in high-throughput screening. *Curr. Opin. Pharmacol.* **2009**, *9*(5), 580–588. (b) Dunn, D.; Orlowski, M.; McCoy, P.; Gastgeb, F.; Appell, K.; Ozgur, L.; Webb, M.; Burbaum, J. Ultra-high throughput screen of two-million-member combinatorial compound collection in miniaturized, 1536-well

assay format. *J. Biomol. Screen.* **2000**, *5*, 177–187. (c) Kenny, B. A.; Bushfield, M.; Parry-Smith, D. J.; Fogarty, S.; Treherne, J. M. The application of high-throughput screening to novel lead discovery. *Prog. Drug Res.* **1998**, *51*, 245–269.

13. Davis, A. M.; Keeling, D. J. Steel, J.; Tomkinson, N. P.; Tinker, A. C. Components of successful lead generation. *Curr. Top. Med. Chem.* **2005**, *5*, 421–439.
14. Drews, J. Drug discovery: a historical perspective. *Science* **2000**, *287*, 1960–1963.
15. Hann, M.M.; Oprea, T.I. Pursuing the leadlikeness concept in pharmaceutical research. *Curr. Opin. Chem. Biol.* **2004**, *8*(3), 255–263.
16. Agrestia, J. J.; Antipovc, E.; Abatea, A. R.; Ahna, K.; Rowata, A. C.; Barete, J. C.; Marquezf, M.; Klibanovc, A. M.; Griffiths, A. D.; Weitz, D. A. Ultrahigh-throughput screening in drop-based microfluidics for directed evolution. *Proc. Natl. Acad. Sci. U.S.A.* **2010**, *107*(9), 4004–4009.
17. Schonbrun, E.; Abate, A.R.; Steinvurzel, P. E.; Weitz, D. A.; Crozier, K. B. High-throughput fluorescence detection using an integrated zone-plate array. *Lab on a Chip* **2010**, *10*(7), 852–856.
18. Davis, A. M.; Keeling, D. J.; Steele, J.; Tomkinson, N. P.; Tinker, A. C. Components of successful lead generation. *Curr. Top. Med. Chem.* **2005**, *5*, 421–439.
19. (a) Harris, P. A.; Boloor, A.; Cheung, M.; Kumar, R.; Crosby, R. M.; Davis-Ward, R. G.; Epperly, A. H.; Hinkle, K. W.; Hunter III, R. N.; Johnson, J. H.; Knick, V. G.; Laudeman, C. P.; Luttrell, D. K.; Mook, R. A.; Nolte, R. T.; Rudolph, S. K.; Szewczyk, J. R.; Truesdale, A. T.; Veal, J. M.; Wang, L.; Stafford, J. A. Discovery of 5-[[4-[(2,3-Dimethyl-2H-indazol-6-yl)methylamino]-2-pyrimidinyl] amino]-2-methyl-benzenesulfonamide (pazopanib), a novel and potent vascular endothelial growth factor receptor inhibitor. *J. Med. Chem.* **2008**, *51*, 4632–4640. (b) Epple, R., C.; Cow, C. et al. Novel bisaryl substituted thiazoles and oxazoles as highly potent and selective peroxisome proliferator-activated receptor (agonists. *J. Med. Chem.* **2010**, *53*, 77–105. (c) Fritch, P. C.; McNaughton-Smith, G., et al. Novel KCNQ2/3 agonists as potential therapeutics for epilepsy and neuropathic pain. *J. Med. Chem.* **2010**, *53*, 887–896. (d) Morwick, T. F.; Büttner, H., et al. Hit to lead account of the discovery of bisbenzamide and related ureidobenzamide inhibitors of rho kinase. *J. Med. Chem.* **2010**, *53*, 759–777. (e) Yeh, J.-Y.; Coumar, M. S.; Horng, J.-T.; Shiao, H.-Y.; Kuo, F.-M.; Lee, H.-L.; Chen, I.-C.; Chang, C.-W.; Tang, W.-F.; Tseng, S.-N.; Chen, C.-J.; Shih, S.-R.; Hsu, J. T.-A.; Liao, C.-C.; Chao, Y.-S.; Hsieh, H.-P. Anti-influenza drug discovery: structure-activity relationship and mechanistic insight into novel angelicin derivatives. *J. Med. Chem.* **2010**, *53*, 1519–1533.
20. Dixon, St. L.; Villar, H. O. Bioactive diversity and screening library selection via affinity fingerprinting. *J. Chem. Inf. Comput. Sci.* **1998**, *38*, 1192–1203.
21. Harvey, A. L. Natural products in drug discovery. *Drug Discov. Today* **2008**, *13*(19/20), 894–901.
22. (a) Cragg, G. M.; Newman, D. J.; Snader, K. M. Natural products in drug discovery and development. *J. Nat. Prod.* **1997**, *60*, 52–60. (b) Newman, D. J.; Cragg, G. M. Natural products as sources of new drugs over the last 25 years. *J. Nat. Prod.* **2007**, *70*, 461–477.
23. Hung, D. T.; Jamison, T. F.; Schrieber, S. L. Understanding and controlling the cell cycle with natural products. *Chem. Biol.* **1996**, *3*, 623–639.
24. (a) Butler, M. S.. Natural products to drugs: Natural product-derived compounds in clinical trials. *Nat. Prod. Rep.* **2008**, *25*, 475–516. (b) Borzilleri, R. M.; Vite, G. D. Case history: Discovery of ixabepilone (IXEMPRATM), a first-in-class epothilone analog for treatment of metastatic breast cancer. *Annu. Rep. Med. Chem.* **2009**, *44*, 301–322.
25. Dancik, V.; Seiler, K. P.; Yound, D. W.; Schreiber, S. L.; Clemons, P. A. Distinct biological network properties between the targets of natural products and disease genes. *J. Am. Chem. Soc.* **2010**, *132*, 9259–9261.
26. Rosen, J.; Gottfires, J.; Muresan, S.; Backlund, A.; Oprea, T.I. Novel chemical space exploration via natural products. *J. Med. Chem.* **2009**, *52*, 1953–1962.
27. (a) Quinn, R. J.; Carroll, A. R. Pham, N. B.; Baron, P.; Palframan, M. E.; Suraweera, L.; Pierens, G. K.; Muresan, S. Developing a drug-like natural product library. *J. Nat. Prod.* **2008**, *71*, 464–468. (b) Ganesan, A. The impact of natural products upon modern drug discovery. *Curr. Opin. Chem. Biol.* **2008**, *12*, 306–317.
28. Dolle, R. E. Comprehensive survey of combinatorial library synthesis. *J. Comb. Chem.* **2004**, *6*(5), 623–679.
29. (a) Spandl, R. J.; Thomas, G. L.; Diaz-Gavilan, M.; O'Connell, K. M. G.; Spring, D. R. An introduction to diversity - oriented synthesis, In *Linker strategies in solid-phase organic synthesis*, P. Scott, Ed.: Wiley, 2009, pp. 241–262. (b) Peuchmaur, M.; Wong, Y.-S. Expanding the chemical space in practice: diversity - oriented synthesis. *Combinatorial Chemistry & High Throughput Screening* **2008**, *11*(8), 587–601.
30. Wermuth, C. G.; Ganellin, C. R.; Lindberg, P.; Mitscher, L. Glossary of terms used in medicinal chemistry. *Pure Appl. Chem.* **1998**, *70*(5), 1129–1143.
31. Furka, A. *Notariell Beglaubigtes Dokument Nr 36237/1982*, Budapest, Hungary, 1982.
32. Geysen, H. M.; Meloen, R. H.; Barteling, S. J. Use of peptide synthesis to probe viral antigens for epitopes to a resolution of a single amino acid. *Proc. Natl. Acad. Sci. U.S.A.* **1984**, *81*, 3998–4002.
33. Houghten, R. A. General method for the rapid solid-phase synthesis of large numbers of peptide: specificity of antigen-antibody reaction at the level of individual amino acids. *Proc. Natl. Acad. Sci. U.S.A.* **1985**, *82*, 5131–5135.
34. Zuckermann, R. N.; Martin, E. J.; Spellmeyer, D. C.; Stauber, G. B.; Shoemaker, K. R.; Karr, J. M.; Figliozzi, G. M.; Goff, D. A.; Siani, M. A.; Simon, R. J.; Banville, S. C.; Brown, E. G.; Wang, L.; Richter, L. S.; Moos, W. H. Discovery of nanomolar ligands for 7-transmembrane G-protein-coupled receptors from a diverse N-(substituted) glycine peptoid library. *J. Med. Chem.* **1994**, *37*, 2678–2685.
35. (a) Thompson, L. A.; Ellman, J. A. Synthesis and applications of small molecule libraries. *Chem. Rev.* **1996**, *96*, 555–600. (b) Ellman, J. A. Design, synthesis, and evaluation of small-molecule libraries. *Acc. Chem. Res.* **1996**, *29*, 132–143.
36. Terrett, N. K.; Gardner, M.; Gordon, D. W.; Kobylecki, R. J.; Steele, J. Combinational synthesis – the design of compound libraries and their application to drug discovery. *Tetrahedron* **1995**, *51*, 8135–8173.
37. Seneci, P., Taylor, J.B; Triggle, J.B. Combinatorial chemistry, In *Comprehensive Medicinal Chemistry II*. Elsevier: Oxford, 2007, pp. 697–760.
38. R. B. Merrifield, R. B. Solid phase peptide synthesis. I. The synthesis of a tetrapeptide. *J. Am. Chem. Soc.* **1963**, *85*(14), 2149–2154.
39. Dolle, R. E.; Le Bourdonnec, B.; Goodman, A. J.; Morales, G. A.; Thomas, C. J.; Zhang, W. Comprehensive survey of chemical libraries for drug discovery and chemical biology: 2007 *J. Comb. Chem.* **2008**, *10*(6), 753–802.

40. Plunkett, M. J.; Ellman, J. A. Solid-phase synthesis of structurally diverse 1,4-benzodiazepine derivatives using the Stille coupling reaction. *J. Am. Chem. Soc.* **1995**, *117*, 3306–3307.
41. Maclean, D.; Baldwin, J. J.; Ivanov, V. T.; Kato, Y.; Shaw, A.; Schneider, P.; Gordon, E. M. Glossary of terms used in combinatorial chemistry. *J. Comb. Chem.* **1999**, *2*, 562–578.
42. Boldi, A. M.; Dener, J. M.; Hopkins, T. P. Solid-phase library synthesis of alkoxyprolines. *J. Comb. Chem.* **2001**, *3*, 367–373.
43. (a) Furka, A.; Sebestyen, F.; Asgedom, M.; Dibo, G. General method for rapid synthesis of multicomponent peptide mixtures. *Int. J. Pept. Protein Res.* **1991**, *37*, 487–493. (b) Lam, K. S.; Salmon, S. E.; Hersh, E. M.; Hruby, V. J.; Kazmierski, W. M.; Knapp, R. J. A new type of synthetic peptide library for identifying ligand-binding activity. *Nature* **1991**, *354*, 82–84. (c) Zuckermann, R. N.; Kerr, J. M.; Siani, M. A.; Banville, S. C. Design, construction and application of a fully automated equimolar peptide mixture synthesizer. *Int. J. Pept. Protein Res.* **1992**, *40*, 497–506.
44. Houghten, R. A.; Pinilla, C.; Appel, J. R.; Blondelle, S. E.; Dooley, C. T.; Eichler, J.; Nefzi, A.; Ostresh, J. M. Mixture-based synthetic combinatorial libraries. Mixture-based synthetic combinatorial libraries. *J. Med. Chem.* **1999**, *42*, 3743–3778.
45. (a) Brummel, C L.; Lee, I. N. W.; Zhou, Y.; Benkovic, S. J.; Winograd, N. A mass spectrometric solution to the address problem of combinatorial libraries. *Science* **1994**, *264*, 399–402. (b) Zambias, R. A.; Boulton, D. A.; Griffin, P. R. Microchemical structure determination of a peptoid covalently bound to a polymeric bead by matrix-assisted laser desorption ionization time-of-flight mass spectometry. *Tetrahedron Lett.* **1994**, *35*, 4283. (c) Youngquist, R. S.; Fuentes, G. R.; Lacey, M. P.; Keough, T. Generational and screening of combinatorial peptide libraries designed for rapid sequencing by mass spectrometry. *J. Am. Chem. Soc.* **1995**, *117*, 3900–3906.
46. Brenner, S.; Lerner, R. A. Encoded combinatorial chemistry. *Proc. Natl. Acad. Sci. U.S.A.* **1992**, *89*, 5381–5383.
47. (a) Ohlmeyer, M. H.; Swanson, R. N.; Dillard, L. W.; Reader, J. C.; Asouline, G.; Kobayashi, R.; Wigler, M.; Still, W. C. Complex synthetic chemical libraries indexed with molecular tags. *Proc. Natl. Acad. Sci. U.S.A.* **1993**, *90*, 10922–10926. (b) Nestler, H. P.; Bartlett, P. A.; Still, W. C. A general method for molecular tagging of encoded combinatorial chemistry libraries. *J. Org. Chem.* **1994**, *59*, 4723–4724.
48. (a) Newman, D. J.; Cragg, G. M. Natural products as sources of new drugs over the last 25 years. *J. Nat. Prod.* **2007**, *70*, 461–477. (b) Newman, D. J. Natural products as leads to potential drugs: an old process of the new hope for drug discovery? *J. Med. Chem.* **2008**, *51*, 2589–2599.
49. Ashton, M.; Moloney, B. Solution phase parallel chemistry. In *Comprehensive Medicinal Chemistry II*; Elsevier: Oxford, 2007, p 761–790.
50. Krueger, E. B.; Hopkins, T. P.; Keaney, M. T.; Walters, M. A.; Boldi, A. M. Solution-phase library synthesis of furanoses. *J. Comb. Chem.* **2002**, *4*, 229–238.
51. Baek, K.-W.; Song, S.-H.; Kand, S.-H.; Rhee, Y.-W.; Lee, C.-S.; Lee, B.-J.; Hudson, S. Adsorption kinetics of boron by anion exchange resin in packed column bed. *J. Indust. Eng. Chem.* **2007**, *13*, 452–456.
52. Koppitz, M. Maximizing efficiency in the production of compound libraries. *J. Comb. Chem.* **2008**, *10*, 573–579.
53. Lipinski, C. A.; Lombardo, F.; Dominy, B.-W.; Feeney, P. J. Experimentational and computational approaches to estimate solubility and permeability in drug discovery and development settings. *Adv. Drug Deliv. Rev.* **1997**, *23*, 3–25.
54. Overington, J. P.; Al-Lazikani, B.; Hopkins, A. L. How many drug targets are there? *Nat. Rev. Drug Disc.* **2006**, *5*, 993–996.
55. Gleeson, M. P. Generation of a set of simple, interpretable ADMET rules of thumb. *J. Med. Chem.* **2008**, *51*, 817–834.
56. (a) Lipinski, C. A., personal communication. (b) Van de Waterbeemd, H.; Camenisch, G.; Folkers, G.; Chretien, J. R.; Raevsky, O. A. Estimation of blood-brain barrier crossing of drugs using molecular size and shape, and H-bonding descriptors. *J. Drug Target.* **1998**, *6*, 151–165.
57. (a) Wender, P. A.; Galliher, W. C.; Goun, E. A.; Jones, L. R.; Pillow, T. H. The design of guanidinium-rich transporters and their internationalization mechanisms. *Adv. Drug Deliv. Rev.* **2008**, *60*, 452–472. (b) Dubikovskaya, E. A. Thorne, S. H.; Pillow, T. H.; Contag, C. H.; Wender, P. A. Overcoming multidrug resistance of small-molecule therapeutics through conjugation with releasable octaarginine transporters. *Proc. Natl. Acad. Sci. U.S.A.* **2008**, *105*, 12128–12133. (c) Cooley, C. B.; Trantow, B. M.; Nederberg, F.; Kieseweter, M. K.; Hedrick J. L.; Waymouth, R. M.; Wender, P.A. Oligocarbonate molecular transporters: oligomerization-based syntheses and cell-penetrating studies. *J. Am. Chem. Soc.* **2009**, *131*, 16401–16403. (d) Wender, P. A.; Mitchell, D. J.; Pattabiraman, K.; Pelkey, E. T.; Steinman, L.; Rothbard, J. B. The design, synthesis, and evaluation of molecules that enable to enhance cellular uptake: peptoid molecular transporters. *Proc. Natl. Acad. Sci. U.S.A.* **2000**, *97*, 13003–13008. (e) Rothbard, J. B.; Garlington, S.; Lin, Q.; Kirschberg, T.; Kreider, E.; McGrane, L. P.; Wender, P. A.; Khavari, P. A. Conjugation or agrinine oligomers to cyclosporin A facilitates topical delivery and inhibition of inflammation. *Nat. Med.* **2000**, *6*, 1253–1257.
58. Veber, D. F.; Johnson, S. R.; Cheng, H.-Y.; Smith, B. R.; Ward, K. W.; Kopple, K. D. Molecular properties that influence the oral bioavailability of drug candidates. *J. Med. Chem.* **2002**, *45*, 2615–2623.
59. Ertl, P.; Rohde, B.; Selzer, P. Fast calculation of molecular polar surface area as a sum of fragment-based contributions and its application to the prediction of drug transport properties. *J. Med. Chem.* **2000**, *43*, 3714–3717.
60. Ertl, P. Polar Surface Area, In *Molecular Drug Properties*, R. Mannhold Ed.; Wiley-VCHL007, pp. 111–126.
61. Martin, Y. C. A bioavailability score. *J. Med. Chem.* **2005**, *48*, 3164–3170.
62. Yang, Y.; Engkvist, O.; Llinàs, A.; Chen, H. Beyond size, ionization state, and lipophilicity: influence of molecular topology on absorption, distribution, metabolism, excretion, and toxicity for druglike compounds. *J. Med. Chem.* **2012**, *55*, 3667–3677.
63. Yang, Y.; Chen, H.; Nilsson, I.; Muresan, S.; Engkvist, O. Investigation of the relationship between topology and selectivity for druglike molecules. *J. Med. Chem.* **2010**, *53*, 7709–7714.
64. Lovering, F.; Bikker, J.; Humblet, C. Escape from flatland: increasing saturation as an approach to improving clinical success. *J. Med. Chem.* **2009**, *52*, 6752–6756.
65. Thomas, V. H.; Bhattachar, S.; Hitchingham, L.; Zocharski, P.; Naath, M.; Surendran, N.; Stoner C. L.; El-Kattan, A. The road map to oral bioavailability: an industrial perspective. *Expert Opin. Drug Metab. Toxicol.* **2006**, *2*, 591–608.
66. Wager, T. T.; Hou, X.; Verhoest, P. R.; Villalobos, A. Moving beyond rules: the development of a central nervous system multiparameter optimization (CNS MPO) approach to enable alignment of druglike properties. *ACS Chem. Neurosci.* **2010**, *1*, 435–449.
67. (a) Vieth, M. Siegel, M. G.; Higgs, R. E.; Watson, I. A.; Robertson, D. H.; Savin, K. A.; Durst G. L.; Hipskind, P. A. Characteristic physical

properties and structural fragments of marketed oral drugs. *J. Med. Chem.* **2004**, *47*, 224–232. (b) Proudfoot, J. R. The evolution of synthetic oral drug properties. *Bioorg. Med. Chem. Lett.* **2005**, *15*, 1087–1090.

68. Leeson, P. S.; Davis, A. M. Time-related differences in the physical property profiles of oral drugs. *J. Med. Chem.* **2004**, *47*, 6338–6348.
69. Ajay; Walters, W. P.; Murcko, M. A. Can we learn to distinguish between "drug-like" and "nondrug-like" molecules? *J. Med. Chem.* **1998**, *41*, 3314–3324.
70. This is an electronic database of volume 6 of *Comprehensive Medicinal Chemistry* (Pergammon Press) available from Accelrys, Inc. San Diego, CA.
71. Sadowski, J.; Kubinyi, H. A scoring scheme for discriminating between drugs and nondrugs. *J. Med. Chem.* **1998**, *41*, 3325–3329.
72. The ACD is available from Accelrys, Inc. San Diego, CA and contains specialty and bulk commercially-available chemicals.
73. The WDI is from Derwent Information.
74. (a) Walters, W. P.; Stahl, M. T.; Murcko, M. A. Virtual screening – an overview. *Drug Discov. Today* **1998**, *3*, 160–178. (b) Walters, W. P.; Ajay; Murcko, M. A. Recognizing molecules with drug-like properties. *Curr. Opin. Chem. Biol.* **1999**, *3*, 384–387. (c) Teague, S. J.; Davis, A. M.; Leeson, P. D.; Oprea, T. The design of leadlike combinatorial libraries. *Angew. Chem. Int. Ed. Engl.* **1999**, *38*, 3743–3747. (d) Oprea, T. I. Property distribution of drug-related chemical databases. *J. Comput. Aided Mol. Des.* **2000**, *14*, 251–264. (e) Gillet, V. J.; Willett, P. L.; Bradshaw, J. Identification of biological activity profiles using substructural analysis and genetic algorithms. *J. Chem. Inf. Comput. Sci.* **1998**, *38*, 165–179. (f) Wagener, M.; vanGeerestein, V. J. Potential drugs and nondrugs: prediction and identification of important structural features. *J. Chem. Inf. Comput. Sci.* **2000**, *40*, 280–292. (g) Ghose, A. K.; Viswanadhan, V.N.; Wendoloski, J. J. A knowledge-based approach in designing combinatorial or medicinal chemistry libraries for drug discovery. 1. A qualitative and quantitative characterization of known drug databases. *J. Comb. Chem.* **1999**, *1*, 55–68. (h) Xu, J.; Stevenson, J. Drug-like index: a new approach to measure drug-like compounds and their diversity. *J. Chem. Inf. Comput. Sci.* **2000**, *40*, 1177–1187. (i) Muegge, I.; Heald, S. L.; Brittelli, D. Simple selection criteria for drug-like chemical matter. *J. Med. Chem.* **2001**, *44*, 1841–1846. (j) Anzali, S.; Barnickel, G.; Cezanne, B.; Krug, M.; Filimonov, D.; Poroikiv, V. Discriminating between drugs and nondrugs by prediction of activity spectra for substances (PASS). *J. Med. Chem.* **2001**, *44*, 2432–2437. (k) Brüstle, M.; Beck, B.; Schindler, T.; King, W.; Mitchell, T.; Clark, T. Descriptors, physical properties, and drug-likeness. *J. Med. Chem.* **2002**, *45*, 3345–3355.
75. Ohno, K.; Nagahara, Y.; Tsunoyama, K.; Orita, M. Are there differences between launched drugs, clinical candidates, and commercially available compounds? *J. Chem. Inf. Comput. Sci.* **2010**, *50*, 815–821.
76. Teague, S. J.; Davis, A. M.; Leeson, P. D.; Oprea, T. The design of leadlike combinatorial libraries. *Angew. Chem. Int. Ed. Engl.* **1999**, *38*, 3743–3747.
77. (a) Welsch, M. E.; Snyder, S. A.; Stockwell, B. R. Privileged scaffolds for library design and drug discovery. *Curr. Opin. Chem. Biol.* **2010**, *14*, 347–361. (b) Duarte, C. D.; Barreiro, E. J.; Fraga, C. A. M. Privileged structures: a useful concept for the rational design of new lead drug candidates. *Mini-Rev. Med. Chem.* **2007**, *7*, 1108–1119. (c) Horton, D. A.; Bourane, G. T.; Smythe, M. L. The combinatorial synthesis of bicyclic privileged structures or privileged structures. *Chem. Rev.* **2003**, *103*, 893–930.
78. Evans, B. E.; Rittle, K. E.; Bock, M. G.; DiPardo, R. M.; Freidinger, R. M.; Whitter, W. L.; Lundell, G. F.; Veber, D. F.; Anderson, P. S.; Chang, R. S. L.; Lotti, V. J.; Cerino, D. J.; Chen, T. B.; Kling, P. J.; Kunkel, K. A.; Springer, J. P.; Hirshfield, J. Methods for drug discovery: development of potent, selective, orally effective cholecystokinin antagonists. *J. Med. Chem.* **1988**, *31*, 2235–2246.
79. (a) Ariëns, E. J.; Beld, A. J.; Rodrigues de Miranda, J. F.; Simonis, A. M. in The Receptors: A Comprehensive Treatise; O'Brien, R. D., Ed.; Plenum Press: New York, 1979, p. 33. (b) Ariëns, E. J. Stereochemistry in the analysis of drug action. Part II. *Med. Res. Rev.* **1987**, *7*, 367–387.
80. Bemis, G. W.; Murcko, M. A. The properties of known drugs. 1. Molecular frameworks. *J. Med. Chem.* **1996**, *39*, 2887–2893.
81. D DeSimone, R. W.; Currie, K. S.; Mitchell, S. A.; Darrow, J. W.; Pippin, D. A. Privileged structures: applications in drug discovery. *Comb. Chem. High Throughput Screen.* **2004**, *7*, 473–493.
82. Bemis, G. W.; Murcko, M. A. Properties of known drugs. 2. Side chains. *J. Med. Chem.* **1999**, *42*, 5095–5099.
83. (a) Williams, D. P.; Naisbitt, D. J. Toxicophores: groups and metabolic routes associated with increased safety risk. *Curr. Opin. Drug Disc.* **2002**, *5*, 104–115. (b) Blagg, J. Structure-activity relationships for in vitro and in vivo toxicity. *Annu. Rep. Med. Chem.* **2006**, 41, 353–368. (c) Williams, D. P. Toxicophores: investigations in drug safety. *Toxicology* **2006**, *226*, 1–11. (d) Villar, H. O.; Hansen, M. R. Computational techniques in fragment based drug discovery. *Curr. Top. Med. Chem.* **2007**, *7*, 1509–1513.
84. Kazius, J.; McGuire, R.; Bursi, R. Derivation and validation of toxicophores for mutagenicity prediction. *J. Med. Chem.* **2005**, *48*, 312–320.
85. (a) Coan, K. E. D.; Maltby, D. A.; Burlingame, A. L.; Shoichet, B. K. Promiscuous aggregate-based inhibitors promote enzyme unfolding. *J. Med. Chem.* **2009**, *52*(7), 2067–2075. (b) Feng, B. Y.; Simeonov, A.; Jadhav, A.; Babaoglu, K.; Inglese, J.; Shoichet, B. K.; Austin, C. P. A high-throughput screen for aggregation-based inhibition in a large compound library. *J. Med. Chem.* **2007**, *50*, 2385–2390. (c) Doak, A. K.; Wille, H.; Prusiner, S. B.; Shoichet, B. K. Colloid formation by drugs in simulated intestinal fluid. *J. Med. Chem.* **2010**, *53*, 4259–4265.
86. (a) Ryan, A. J.; Gray, N. M.; Lowe, P. N.; Chung, C.-W. Effect of detergent on "promiscuous" inhibitors. *J. Med. Chem.* **2003**, *46*, 3448–3451. (b) Thorne, N.; Auld, D. S.; Inglese, J. Apparent activity in high-throughput screening: origins of compound-dependent assay interference. *Curr. Opin. Chem. Biol.* **2010**, *14*, 315–324.
87. LaPlante, S. R.; Carson, R.; Gillard, J.; Aubry, N.; Coulombe, R.; Bordeleau, S.; Bonneau, P.; Little, M.; O'Meara, J.; Beaulieu, P. L. Compound aggregation in drug discovery: implementing a practical NMR assay for medicinal chemists. *J. Med. Chem.* **2013**, *56*, 5142–5150.
88. (a) Hert, J.; Willett, P.; Wilton, D. J.; Acklin, P.; Azzaoui, K.; Jacoby, E.; Schuffenhauer, A. Enhancing the effectiveness of similarity-based virtual screening using nearest-neighbor information. *J. Med. Chem.* **2005**, *48*, 7049–7054. (b) Shanmugasundaram, V.; Maggiora, G. M.; Lajiness, M. S. Hit-directed nearest-neighbor searching. *J. Med. Chem.* **2005**, *48*, 240–248.
89. Houghten, R. A.; Pinilla, C.; Appel, J. R.; Blondelle, S. E.; Dooley, C. T.; Eichler, J.; Nefzi, A.; Ostresh, J. M. Mixture-based synthetic combinatorial libraries. *J. Med. Chem.* **1999**, *42*, 3743–3778.
90. Feng, B. Y.; Soichet, B. K. Synergy and antagonism of promiscuous inhibition in multiple-compound mixtures. *J. Med. Chem.* **2006**, *49*, 2151–2154.

91. (a) Rester, U. From virtuality to reality - Virtual screening in lead discovery and lead optimization: A medicinal chemistry perspective. *Curr. Opin. Drug Discov. Devel.* **2008**, *11*(4), 559–68. (b) Rollinger J. M.; Stuppner H.; Langer T. Virtual screening for the discovery of bioactive natural products. *Prog. Drug Res.* **2008**, *65*(211), 213–49. (c) Schneider, G. Virtual screening: an endless staircase? *Nat. Rev. Drug Discov.* **2010**, *9*, 273–276.
92. Saario, S. M.; Poso, A.; Juvonen, R. O.; Järvinen, T.; Salo-Ahen, O. M. H. Fatty acid amide hyrolase inhibitors from virtual screening of the endocannabinoid system. *J. Med. Chem.* **2006**, *49*, 4650–4656.
93. Irwin, J. J.; Shoichet, B. K. Zinc - a free database of commercially available compounds for virtual screening. *J. Chem. Inf. Model.* **2005**, *45*, 177–182. http://zinc.docking.org/(see subset1 for the compounds with different properties).
94. Boehm, M.; Wu, T.-Y.; Claussen, H. Similarity searching and scaffold hopping in synthetically accessible combinatorial library space. *J. Med. Chem.* **2008**, *51*, 2468–2480.
95. Dolle, R. E.; Le Bourdonnec, B.; Goodman, A. J.; Morales, G. A.; Thomas, C. J.; Zhang, W. Comprehensive survey of chemical libraries for drug discovery and chemical biology: 2008. *J. Comb. Chem.* **2009**, *11*, 739–790.
96. (a) Ajay; Walters, W. P.; Murcko, M. A. Can we learn to distinguish between "drug-like" and "nondrug-like" molecules? *J. Med. Chem.* **1998**, *41*, 3314–3324. (b) Martin, Y. C.; Kofron, J. L.; Traphagen, L. M. Do structurally similar molecules have similar biological activity? *J. Med. Chem.* **2002**, *45*, 4350–4358. (c) Sheridan, R. P.; Kearsley, S. K. Why do we need so many chemical similarity search methods? *Drug Discov. Today* **2002**, *7*, 903–911. (d) Willett, P. Similarity-based virtual screening using 20 fingerprints. *Drug Discov. Today* **2006**, *11*, 1046–1053.
97. Fligner, M. A.; Verducci, J. S.; Blower, P. E. The modification of the Jaccard-Tanimoto Similarity Index for diverse selection of chemical compounds using binary strings. *Technometrics* **2002**, *44*, 110–119.
98. Kubinyi, H. Success stories of computer-aided design. In *Computer Applications in Pharmaceutical Research and Development*; Ekins, S., Ed., Wiley-Interscience: New York, NY, 2006, pp. 377–424.
99. Rogers, D.; Hahn, M. Extended-connectivity fingerprints. *J. Chem. Inf. Model.* **2010**, *50*, 742–754.
100. Martin, Y. C.; Kofron, J. L.; Traphagen, L. M. Do structurally similar molecules have similar biological activity? *J. Med. Chem.* **2002**, *45*, 4350–4358.
101. Ng, H. P.; May, K.; Bauman, J. G.; Ghannam, A.; Islam, I.; Liang, M.; Horuk, R.; Hesselgesser, J.; Snider, R. M.; Perez, H. D.; Morrissey, M. M. Discovery of novel non-peptide CCR1 receptor antagonists. *J. Med. Chem.* **1999**, *42*, 4680–4694.
102. Srivastava, S.; Richardson, W. W.; Bradley, M. P.; Crippen, G. M. Three-dimensional receptor modeling using distance geometry and voronoi polyhedra. In *3D-QSAR in Drug Design: Theory, Methods, and Applications*; Kubinyi, H., Ed.; ESCOM, 1993, pp. 409–430.
103. (a) Crippen, G. M. Distance geometry approach to rationalizing binda data. *J. Med. Chem.* **1979**, *22*, 988–997. (b) Ghose, A. K.; Crippen, G. M. Quantitative structure-activity relationship by distance geometry *J. Med. Chem.* **1982**, *25*, 892. (c) Crippen, G. M. *Distance Geometry and Conformational Calculations*, Research Studies Press: New York, 1981.
104. (a) Sheridan, R. P.; Nilakantan, R.; Dixon, J. S.; Venkataraghavan, R. The ensemble approach to distance geometry: application to the nicotinic pharmacore. *J. Med. Chem.* **1986**, *29*, 899–206. (b) Sheridan, R. P.; Venkataraghaven, R. New methods in computer-aided drug design. *Acc. Chem. Res.* **1987**, *20*, 322–329.
105. (a) Nicklaus, M. C.; Neamati, N.; Hong, H.; Mazumder, A.; Sunder, S.; Chen, J.; Milne, G. W.; Pommier, Y. HIV-1 integrase pharmacore: discovery of inhibitors through three-dimensional database searching. *J. Med. Chem.* **1997**, *40*, 920–929. (b) Hong, H.; Neamati, N.; Wang, S.; Nicklaus, M. C.; Mazumder, A. Zhao, H.; Burke, T. R. J.; Pommier, Y.; Milne, G. W. Discovery of HIV-1 integrase inhibitors by pharmacore searching. *J. Med. Chem.* **1997**, *40*, 930–936.
106. (a) Hopfinger, A. J. A QSAR investigation of dihydrofolate reductase inhibition by Baker triazines based upon molecular shape analysis. *J. Am. Chem. Soc.* **1980**, *102*, 7196–7206. (b) Hopfinger, A. J. Inhibition of dihydrofolate reductase: structure-activity correlations of 2,4-diamino-5-benzylpyrimidines based upon molecular shape analysis. *J. Med. Chem.* **1981**, *24*, 818–822. (c) Hopfinger, A. J.Theory and application of molecular potential energy fields in molecular shape analysis: a quantitative structure-activity relationship study of 2,4-diamino-5-benzylpyrimidines as dihydrofolate reductase inhibitors. *J. Med. Chem.* **1983**, *26*, 990–996.
107. Carhart, R. E.; Smith, D. H.; Venkataraghavan, R. Atom pairs as molecular features in structure-activity studies: definition and applications. *J. Chem. Inf. Comput. Sci.* **1985**, *25*, 64–73.
108. Nilakantan, R.; Bauman, N.; Dixon, J. S.; Venkataraghavan, R. Topical torsion: a new molecular descriptor for SAR applications. Comparison with other descriptors. *J. Chem. Inf. Comput. Sci.* **1987**, *27*, 82–85.
109. Cramer, R. D. III; Patterson, D. E.; Bunce, J. D. Comparative molecular field analysis (CoMFA).1. Effect of shape on binding of steroids to carrier proteins. *J. Am. Chem. Soc.* **1988**, *110*, 5959–5967.
110. Kim, K.H.; Greco, G.; Novellino, E. In *3D QSAR in Drug Design*; Kybinyi, H.; Folkers, G.; Martin, Y. C., Eds.; Kluwer Academic Publ.: Dordrecht; 1998, Vol. 3, p. 257.
111. Green, S. M.; Marshall, G. R. 3D-QSAR: a current perspective. *TIPS* **1995**, *16*, 285–291.
112. Klebe, G.; Abraham, U.; Mietzner, T. Molecular similarity indices in a comparative analysis (CoMSIA) of drug molecules to correlate and predict their biological activity. *J. Med. Chem.* **1994**, *37*, 4130–4146.
113. Ortiz, A. R.; Pisabarro, M. T.; Gago, F.; Wade, R. C. Prediction of drug binding affinities by comparative binding energy analysis. *J. Med. Chem.* **1995**, *38*, 2681–2691.
114. Silverman, B. D.; Platt, D. E.; Pitman, M.; Rigoutsos, I. In *3D QSAR in Drug Design*; Kybinyi, H.; Folkers, G.; Martin, Y. C., Eds.; Kluwer Academic Publ.: Dordrecht; 1998, Vol. 3, p. 183.
115. Heritage, T. W.; Ferguson, A. M.; Turner, D. B.; Willett, P. In *3D QSAR in Drug Design*; Kybinyi, H.; Folkers, G.; Martin, Y. C., Eds.; Kluwer Academic Publ.: Dordrecht; 1998, Vol. 2, p. 381.
116. Todeschini, R.; Gramatica, P. In *3D QSAR in Drug Design*; Kybinyi, H.; Folkers, G.; Martin, Y. C., Eds.; Kluwer Academic Publ.: Dordrecht; 1998, Vol. 2, p. 355.
117. Pastor, M.; Cruciani, G.; McLay, I.; Pickett, S.; Clementi, S. GRid-INdependent descriptors (GRIND): a novel class of alignment-independent three-dimensional molecular descriptors. *J. Med. Chem.* **2000**, *43*(17), 3233–3243.
118. (a) Cramer, R. D. III; Patterson, D. E.; Bunce, J. D. Comparative molecular field analysis (CoMF). 1. Effect of shape on binding of steroids to carrier proteins. *J. Am. Chem. Soc.* **1988**, *110*, 5959–5967; (b) Cramer, R. D.; Jilek, R. J.; Guessregen, S.; Clark, S. J.; Wendt, B.; Clark, R. D. Lead hopping. Validation of topomer similarity as a superior predictor of similar biological activities. *J. Med. Chem.* **2004**, 47, 6777–6791.
119. Cramer, R. D.; Jilek, R. J.; Andrews, K. M. dbtop: Topomer similarity searching of conventional structure databases. *J. Mol. Graph. Model.* **2002**, *20*, 447–462.

120. Cramer, R. D. Topomer CoMFA: a design methodology for rapid lead optimization. *J. Med. Chem.* **2003**, *46*, 374–388.
121. (a) Cleves, A. E.; Jain, A. J. Robust ligand-based modeling of biological targets of known drugs. *J. Med. Chem.* **2006**, *49*, 2921–2938. (b) Giganti, D.; Guillemain, H.; Spadoni, J.-L.; Nilges, M.; Zagury, J.-F.; Montes, M. Comparative evaluation of 3D virtual ligand screening methods. *J. Chem. Inf. Model.* **2010**, *50*, 992–1004.
122. Heritage, T. W.; Lowis, D. R. Molecular hologram QSAR. In *Rational Drug Design, Novel Methodology and Practical Applications*, Parrill, A. L.; Reddy, M. R., Eds.; American Chemical Society: Washington, D. C.; Symposium Series, 1999, Vol. 719, Chapter 14, pp. 212–225.
123. Laurini, E.; Zampieri, D.; Mamolo, M. G.; Vio, L.; Zanette, C.; Florio, C.; Posocco, P.; Fermeglia, M.; Pricl, S. A. 3D-Pharmacophore model for sigma2 receptors based on a series of substituted benzo[d]oxazol-2(3h)-one derivatives. *Bioorg. Med. Chem. Lett.* **2010**, *20*, 2954–2957.
124. Guner, O. F. (Ed.) *Pharmacophore, Perception, Development, and Use in Drug Design*; IUL Biotechnology Series; International University Line: La Jolla, CA, 2000.
125. (a) Sufrin, J. R.; Dunn, D. A.; Marshall, G. R. Steric mapping of the L-methionine binding site of ATP: L-methionine S-adenosyltransferase. *Mol. Pharmacol.* **1981**, *19*, 307–313. (b) Humblet, C.; Marshall, G. R. Pharmacore identification and receptor mapping. *Annu. Rep. Med. Chem.* **1980**, *15*, 267–276. (c) Marshall, G. R. Structure-activity studies: a three-dimensional probe of receptor specificity. *Ann. N. Y. Acad. Sci.* **1985**, *439*, 162–169.
126. Marshall, G. R.; Barry, C. D.; Bosshard, H. E.; Dammkoehler, R. A.; Dunn, D. A. The conformational parameter in drug design: the active analog approach. *ACS Symp. Ser.*, Vol. 112 (1979), Computer-Assisted Drug Design, Chapter 9 pp 205–226. Chapter DOI: 10.1021/bk-1979-0112.ch009
127. (a) Brinkworth, R. I.; Fairlie, D. P.; Leung, D.; Young, P. R. Homology model of the dengue 2 virus NS3 protease: putative interactions with both substrate and NS2B cofactor. *J. Gen. Virol.* **1999**, *80*, 1167–1177. (b) Naus, J. L.; Reid, R. H.; Sadegh-Nasseri, S. Accuracy of a structural homology model for a class II histocompatibility protein, HLA-DR1: comparison to the crystal structure. *J. Biomol. Struct. Dyn.* **1995**, *12*, 1213–1233. (c) Carlson, G. M.; MacDonald, R. J.; Meyer, E. F., Jr. Computer aided prediction and evaluation of the tertiary structure for rat elastase II. *J. Theor. Biol.* **1986**, *119*, 107–124.
128. Kaczanowski S.; Zielenkiewicz P. Why similar protein sequences encode similar three-dimensional structures? *Theoret. Chem. Acc.* **2010**, *125*, 543–50.
129. (a) Chen, L. In pursuit of the high-resolution structure of nicotinic acetylcholine receptors. *J. Physiol.* **2010**, *588*, 557–564. (b) Congreve, M.; Marshall, F. The impact of GPCR structures on pharmacology and structure-based drug design. *Br. J. Pharmacol.* **2010**, *159*, 986–996. (c) Drahl, C. From picture to pill. *Chem. Engg. News* **2011**, *89*, 15–21. (d) Drahl, C. Dopamine show-and-tell: structural biology: First close-up of a dopamine receptor. *Chem. Engg. News* **2010**, *88*, 8. (e) Tse, M. T. Crystallizing how agonists bind. *Nat. Rev. Drug Discov.* **2011**, *10*, 97.
130. Goodford, P. J. A computational procedure for determining energetically favorable binding sites on biologically important macromolecules. *J. Med. Chem.* **1985**, *28*, 849–857.
131. Miranker, A.; Karplus, M. Functionality maps of binding sites: a multiple copy simultaneous search method. *Proteins* **1991**, *11*, 29–34.
132. Wade, R. C.; Goodford, P. J. Further development of hydrogen bond functions for use in determining energetically favorable binding sites on molecules of known structure. 2. Ligand probe groups with the ability to form more than two hydrogen bonds. *J. Med. Chem.* **1993**, *36*, 148–156.
133. (a) Kuntz, I. D.; Blaney, J. M.; Oatley, S. J.; Langridge, R.; Ferrin, T. E. A geometric approach to macromolecule-ligand interactions. *J. Mol. Biol.* **1982**, *161*, 269–288. (b) Ewing, T. J.; Makino, S.; Skillman, A. G.; Kuntz, I. D. Dock 4.0: search strategies for automated molecular docking of flexible molecule databases. *J. Comput. Aided Mol. Des.* **2001**, *15*, 411–428. (c) Lorber, D. A.; Shoichet, B. K. Flexible ligand docking using conformational ensembles. *Protein Sci.* **1998**, *7*, 938–950.
134. Morris, G. M.; Huey, R.; Lindstrom, W.; Sanner, M. F.; Belew, R. K.; Goodsell, D. S.; Olson, A. J. AutoDock4 and AutoDockTools4: automated docking with selective receptor flexibility. *J. Comput. Chem.* **2009**, *30*, 2785–2791.
135. Rarey, M.; Kramer, B.; Lengauer, T.; Klebe, G. A fast flexible docking method using an incremental construction algorithm. *J. Mol. Biol.* **1996**, *261*(3), 470–489.
136. Friesner, R. A.; Banks, J. L.; Murphy, R. B.; Halgren, T. A.; Klicic, J. J.; Mainz, D. T.; Repasky, M. P.; Knoll, E. H.; Shelley, M.; Perry, J. K.; Shaw, D. E.; Francis, P.; Shenkin, P. S. Glide: A new approach for rapid, accurate docking and scoring. 1. Method and assessment of docking accuracy. *J. Med. Chem.* **2004**, *47*, 1739–1749.
137. (a) Jones, G.; Willett, P.; Glen, R. C.; Leach, A. R.; Taylor, R. Development and validation of a genetic algorithm for flexible docking. *J. Mol. Biol.* **1997**, *267*, 727–48. (b) Fogel, G. B.; Cheung, M.; Pittman, E.; Hecht, D. Modeling the inhibition of quadruple mutant Plasmodium falciparum dihydrofo- late reductase by pyrimethamine derivatives. *J. Comput. Aided Mol. Des.* **2008**, *22*, 29–38.
138. (a) Jain, A.N. Surflex: fully automatic flexible molecular docking using a molecular similarity-based search engine. *J. Med. Chem.* **2003**, *46*(4), 499–511. (b) Jain, A. N. Effects of protein conformation in docking: improved pose prediction through protein pocket adaptation. *J. Comput. Aided Mol. Des.* **2009**, *23*(6), 355–374.
139. Thomsen, R.; Christensen, M. H. MolDock: A new technique for high-accuracy molecular docking. *J. Med. Chem.* **2006**, *49*, 3315–3321.
140. Cornell, W. D. Recent evaluations of high throughput docking methods for pharmaceutical lead finding - consensus and caveats. *Ann. Rep. Comput. Chem.* **2006**, *2*, 297–323.
141. (a) Klebe, G. Toward a more efficient handling of conformational flexibility in computer-assisted modeling of drug molecules. *Perspect. Drug Discov.* **1995**, *3*, 85–105. (b) Beusen, D. D.; Shands, E. F. B. Systematic search strategies in conformational analysis. *Drug Discov. Today* **1996**, *1*, 429–437.
142. (a) DesJarlais, R. L.; Sheridan, R. P.; Dixon, J. S.; Kuntz, I. D. Venkatarghavan, R. Docking flexible ligands to macromolecular receptors by molecular shape. *J. Med. Chem.* **1986**, *29*, 2149–2153. (b) DesJarlais, R. L.; Sheridan, R. P.; Seibel, G. L.; Dixon, J. S.; Kuntz, I. D.; Venkataraghavan, R. Using shape complementarity as an initial screen in designing ligands for a receptor binding site of known three-dimensional structure. *J. Med. Chem.* **1988**, *31*, 722–729. (c) Moustakas, D. T.; Lang, P. T.; Pegg, S.; Pettersen, E.; Kuntz, I. D.; Brooijmans, N.; Rizzo, R. C. Development and validation of a modular, extensible docking program: DOCK 5. *J. Comput. Aided Mol. Des.* **2006**, *20*(10–11), 601–619. (d) Lang, P. T.; Brozell, S. R.; Mukherjee, S.; Pettersen, E. F.; Meng, E. C.; Thomas, V.; Rizzo, R. C.; Case, D. A.; James, T. L.; Kuntz, I.D. DOCK 6: combining techniques to model RNA-small molecule complexes. *RNA* **2009**, *15*(6), 1219–30.

143. Kuntz, I. D. Structure-based strategies for drug design and discovery. *Science* **1992**, 257, 1078–1082.
144. DesJarlais, R. L.; Seibel, G. L.; Kuntz, I. D.; Furth, P. S.; Alvarez, J. C.; Ortiz de Montellano, P. R.; Decamp, D. L.;. Bab™, L. M.; Craik, C. S. Structure-based design of nonpeptide inhibitors specific for the human immunodeficiency virus 1 protease. *Proc. Natl. Acad. Sci. U.S.A.* **1990**, *87*, 6644–6648.
145. (a) Friedman, S. H.; DeCamp, D. L.; Sijbesma, R. P.; Srdanov, G.; Wudl, F.; Kenyon, G. L. Inhibition of the HIV-1 protease by fullerene derivatives: model building studies and experimental verification. *J. Am. Chem. Soc.* **1993**, *115*, 6506–6509. (b) Friedman, S. H.; Ganapathi, P. S.; Rubin, Y.; Kenyon, G. L. Optimizing the binding of fullerene inhibitors of the HIV-1 protease through predicted increases in hydrophobic desolvation. *J. Med. Chem.* **1998**, *41*, 2424–2429.
146. Kolb P., Rosenbaum D. M., Irwin J. J., Fung J. J., Kobilka B. K., Shoichet B. K Structure-based discovery of beta2-adrenergic receptor ligands. *Proc Natl. Acad. Sci. U.S.A.* **2009**, *106*(16), 6843–8.
147. (a) Cherezov, V; Rosenbaum, D. M.; Hanson, M. A.; Rasmussen, S. G.; Thian, F. S.; Kobilka, T. S.; Choi, H. J.; Kuhn, P.; Weis, W. I.; Kobilka, B. K.; Stevens, R. C. High-resolution crystal structure of an engineered human β2-adrenergic G protein-coupled receptor. *Science* **2007**, *318*, 1258–1265. (b) Rosenbaum, D. M.; Cherezov, V.; Hanson, M. A.; Rasmussen, S. G.; Thian, F. S.; Kobilka, T. S.; Choi, H. J.; Yao, X. J.; Weis, W. I.; Stevens, R. C.; Kobilka, B. K. GPCR engineering yields high-resolution structural insights into β2-adrenergic receptor function. *Science* **2007**, *318*, 1266–1273.
148. Wacker D.; Fenalti G.; Brown M. A.; Katritch V.; Abagyan R.; Cherezov V.; Stevens R. C. Conserved binding mode of human β2 adrenergic receptor inverse agonists and antagonist revealed by x-ray crystallography. *J. Am. Chem. Soc.* **2010**, *132*(33), 11443–11445.
149. (a) Zhang, Q.; Muegge, I. Scaffold hopping through virtual screening using 2D and 3D similarity descriptors: ranking, voting, and consensus scoring. *J. Med. Chem.* **2006**, *49*, 1536–1548. (b) Sheridan, R. P.; Kearsley, S. K. Why Do We Need So Many Chemical Similarity Search Methods? *Drug Discov. Today* **2002**, *7*, 903–911.
150. Ferreira, R. S.; Simeonov, A.; Jadhav, A.; Eidam, O.; Mott, B. T.; Keiser, M. J.; McKerrow, J. H.; Maloney, D. J.; Irwin, J. J.; Shoichet, B. K. Complementarity between a docking and a high-throughput screen in discovering new cruzain inhibitors. *J. Med. Chem.* **2010**, *53*, 4891–4905.
151. Schneider, G.; Fechner, U. Computer-based de novo design of drug-like molecules. *Nat. Rev. Drug Discov.* **2005**, *4*(8), 649–663
152. Böhm, H.-J. The computer program LUDI. A new method for the de novo design of enzyme inhibitors. *J. Comput. Aided Mol. Des.* **1992**, *6*, 61–78.
153. Nishibata, Y.; Itai, A. Confirmation of usefulness of a structure construction program based on three-dimensional receptor structure for rational lead generation. *J. Med. Chem.* **1993**, *36*, 2921–2928.
154. Barreiro, G.; Kim, J. T.; Guimara˜ es, C. R. W.; Bailey, C. M.; Domaoal, R. A.; Wang, L.; Anderson, K. S.; Jorgensen, W. L. From docking false-positive to active anti-HIV agent. *J. Med. Chem.* **2007**, *50*, 5324–5329.
155. Mauser, H.; Guba, W. Recent developments in de novo design and scaffold hopping. *Curr. Opin. Drug Discov. Devel.* **2008**, *11*(3), 365–374.
156. (a) Ji, H.; Stanton, B. Z.; Igarashi, J.; Li, H.; Martásek, P.; Roman, L. J.; Poulos, T. L.; Silverman, R. B. Minimal pharmacophoric elements and fragment hopping, an approach directed at molecular diversity and isozyme selectivity. Design of selective neuronal nitric oxide synthase inhibitors. *J. Am. Chem. Soc.* **2008**, *130*(12), 3900–3914. (b) Ji, H.; Silverman, R. B. Case study 3: Fragment hopping to design highly potent and selective neuronal nitric oxide synthase inhibitors. *In* Scaffold Hopping in Medicinal Chemistry; Brown, N. ed.; Wiley-VCH Verlag: Weinheim, Germany; 2013, Chapter 18.
157. Besnard, J.; Ruda, G. F.; Setola, V. et al. Automated design of ligands to polypharmacological profiles. *Nature* **2012**, *492*, 215–222.
158. (a) Gillespie, P.; Goodnow, R. A. The hit-to-lead process in drug discovery. *Annu. Rep. Med. Chem.* **2004**, *39*, 293–304. (b) Keserü, G. M.; Makara, G. M. Hit discovery and hit-to-lead approaches. *Drug Discov. Today* **2006**, *11*, 741–748.
159. Schnecke, V.; Boström, J. Computational chemistry-driven decision making in lead generation. *Drug Discov. Today* **2006**, *11*, 43–50.
160. Cheng, A.; Merz, Jr., K. M. Prediction of aqueous solubility of a diverse set of compounds using quantitative structure-property relationships. *J. Med. Chem.* **2003**, *46*, 3572–3580.
161. Yan, A.; Gasteiger, J. Prediction of aqueous solubility of organic compounds based on a 3D structure representation. *J. Chem. Inf. Comput. Sci.* **2003**, *43*, 429–434.
162. Lamanna, C.; Bellini, M.; Padova, A.; Westerberg, G.; Maccari, L. Straightforward recursive partitioning model for discarding insoluble compounds in the drug discovery process. *J. Med. Chem.* **2008**, *51*, 2891–2897.
163. Sun, H.; Chow, E. C. Y.; Liu, S.; Du, Y.; Pang, K. S.. The Caco - 2 cell monolayer: Usefulness and limitations. *Expert Opin. Drug Metab. Toxicol.* **2008**, *4*(4), 395–411.
164. (a) Lu, C.; Liao, M.; Cohen, L.; Xia, C. Q. Emerging in vitro tools to evaluate cytochrome P450 and transporter-mediated drug-drug interactions. *Curr. Drug Disc. Tech.* **2010**, *7*(3), 199–222. (b) Bambal, R. B.; Clarke, S. E. Cytochrome P450: Structure, function and application in drug discovery and development. In *Evaluation of Drug Candidates for Preclinical Development*, Chao, H.; Davis, C. G.; Wang, B., Eds. 2010, pp 55–107.
165. Galetin, A.; Gertz, M.; Houston, J. B. Contribution of intestinal cytochrome P450 –mediated metabolism to drug - drug inhibition and induction interactions. *Drug Metab. Pharmacol.* **2010**, *25*(1), 28–47.
166. (a) Riley, R. J.; Parker, A. J.; Trigg, S.; Manners, C. N. Development of a generalized quantitative physicochemical model for CYP3A4 inhibition for use in early drug discovery. *Pharm. Res.* **2001**, *18*, 652–655. (b) Zlokarnik, G.; Grootenhuis, P. D. J.; Watson, J. B. High throughput P450 inhibition screens in early drug discovery. *Drug Discov. Today* **2005**, *10*, 1443–1450.
167. (a) Vaz, R. J.; Li, Y.; Rampe, D. Human ether - a - go - go related gene (hERG): A chemist's perspective. *Prog. Med. Chem.* **2005**, *43*, 1–18. (b) Staudacher, I.; Schweizer, P. A.; Katus, H. A.; Thomas, D. hERG: protein trafficking and potential for therapy and drug side effects. *Curr. Opin. Drug Discov. Devel.* **2010**, *13*(1), 23–30. (c) Jamieson, C.; Moir, E. M.; Rankovic, Z.; Wishart, G. Medicinal chemistry of hERG optimizations: highlights and hang-ups. *J. Med. Chem.* **2006**, *49*, 5029–5046. (d) Aronov, A. M. Common pharmacophores for uncharged human ether-a-go-go-related gene (hERG) blockers. *J. Med. Chem.* **2006**, *49*, 6917–6921.
168. Fermini, B.; Fossa, A. A. The impact of drug-induced QT interval prolongation on drug discovery and development. *Nat. Rev. Drug Discov.* **2003**, *2*, 439–447.
169. Hopkins, A. L.; Groom, C, R.; Alex, A. Ligand efficiency: A useful metric for lead selection. *Drug Discov. Today* **2004**, *9*, 430–431.
170. Abad-Zapatero, C. and J. T. Metz. Ligand efficiency indices as guideposts for drug discovery. *Drug Discov. Today* **2005**, *10*, 464–469.
171. Perola, E. An analysis of the binding efficiencies of drugs and their leads in successful drug discovery programs. *J. Med. Chem.* **2010**, *53*, 2986–2997.

172. (a) Leeson, P. D.; Springthorpe, B. The influence of drug-like concepts on decision making in medicinal chemistry. *Nat. Rev. Drug Discov.* **2007**, *6*, 881–890. (b) Edwards, M. P.; Price, D. A. Role of physicochemical properties and ligand lipophilicity efficiency in addressing drug safety risks. *Annu. Rep. Med. Chem.* **2010**, *45*, 381–391.
173. Keserű, G. M.; Makara, G. M. The influence of lead discovery strategies on the properties of drug candidates. *Nat. Rev. Drug Discov.* **2009**, *8*, 203–212.
174. Tarcsay, Á.; Nyíri, K.; Keserű, G. M. Impact of lipophilic efficiency on compound quality. *J. Med. Chem.* **2012**, *55*, 1252–1260.
175. (a) Snowden, M.; Green, D. V. The impact of diversity-based, high-throughput screening on drug discovery: "Chance favours the prepared mind". *Curr. Opin. Drug Disc.* **2008**, *11*(4), 553–558; (b) Golebiowski, A.; Klopfenstein, S. R.; Portlock, D. E. Lead compounds discovered from libraries. *Curr. Opin. Chem. Biol.* **2001**, *5*, 273–284; (c) Golebiowski, A.; Klopfenstein, S. R.; Portlock, D. E. Lead compounds discovered from libraries: Part 2. *Curr. Opin. Chem. Biol.* **2003**, *7*, 308–325; (d) Fox, S.; Farr-Jones, S.; Sopchak, L.; Boggs, A.; Nicely, H. W.; Khoury, R.; Biros, M. High-throughput screening: update on practices and success. *J. Biomol. Screen.* **2006**, *11*(7), 864–869; (e) Posner, B. A. High-throughput screening-driven lead discovery: meeting the challenges of finding new therapeutics. *Curr. Opin. Drug Disc.* **2005**, *8*(4), 487–494.
176. (a) Bender, A.; Bojanic, D.; Davies, J. W.; Crisman, T. J.; Mikhailov, D.; Scheiber, J.; Jenkins, J. L.; Deng, Z.; Hill, W. A.; Popov, M.; Jacoby, E.; Glick, M. Which aspects of HTS are empirically correlated with downstream success? *Curr. Opin. Drug Disc.* **2008**, *11*(3), 327–337; (b) Lahana, R. How many leads from HTS? *Drug Discov. Today* **1999**, *4*(10), 447–448; (c) Shelat, A. A.; Guy, R. K. The interdependence between screening methods and screening libraries. *Curr. Opin. Chem. Biol.* **2007**, *11*(3), 244–251; (d) Bender, A.; Bojanic, D.; Davies, J. W.; Crisman, T. J.; Mikhailov, D.; Scheiber, J.; Jenkins, J. L.; Deng, Z.; Hill, W. A.; Popov, M.; Jacoby, E.; Glick, M. Which aspects of HTS are empirically correlated with downstream success? *Curr. Opin. Drug Discov. Devel.* **2008**, *11*(3), 327–337.
177. (a) Bohacek, R. S.; McMartin, C.; Guida, W. C. The art and practice of structure-based drug design: a molecular modeling perspective. *Med. Res. Rev.* **1996**, *16*(1), 3–50. (b) Martin, Y. C. Challenges and prospects for computational aids to molecular diversity. *Perspect. Drug Discov. Des.* **1997**, 7–8, 159–172.
178. Teague, S. J.; Davis, A. M.; Leeson, P. D. ; Oprea, T. The design of leadlike combinatorial libraries. *Angew. Chem. Int. Ed. Engl.* **1999**, *38*(24), 3743–3748.
179. (a) Hann, M. M.; Oprea, T. I. Pursuing the leadlikeness concept in pharmaceutical research. *Curr. Opin. Chem. Biol.* **2004**, *8*(3), 255–263. (b) Rishton, G. M. Molecular diversity in the context of leadlikeness: compound properties that enable effective biochemical screening. *Curr. Opin. Chem. Biol.* **2008**, *12*(3), 340–351.
180. (a) Wenlock, M. C.; Austin, R. P.; Barton, P.; Davis, A. M.; Leeson, P. D. A comparison of physiochemical property profiles of development and marketed oral drugs. *J. Med. Chem.* **2003**, *46*(7), 1250–1256; (b) Vieth, M.; Sutherland, J. J. Dependence of molecular properties on proteomic family for marketed oral drugs. *J. Med. Chem.* **2006**, *49*(12), 3451–3453.
181. Leeson, P. D.; Springthorpe, B. The influence of drug-like concepts on decision-making in medicinal chemistry. *Nat. Rev. Drug Discov.* **2007**, *6*(11), 881–890.
182. Hann, M. M.; Leach, A. R.; Harper, G. Molecular complexity and its impact on the probability of finding leads for drug discovery. *J. Chem. Inf. Comput. Sci.* **2001**, *41*(3), 856–864.
183. Hopkins, A. L.; Groom, C. R.; Alex, A. Ligand efficiency: a useful metric for lead selection. *Drug Discov. Today* **2004**, *9*, 430–431.
184. (a) Erlanson, D. A. Introduction to fragment-based drug discovery. *Top. Curr. Chem.* **2012**, *317*, 1–32. (b) Konteatis, Z. D. In silico fragment - based drug design. *Expert Opin. Drug Discov.* **2010**, *5*(11), 1047–1065. (c) Loving, K.; Alberts, I.; Sherman, W. Computational approaches for fragment - based and de novo design. *Curr. Topics in Med. Chem.* **2010**, *10*(1), 14–32. (d) Murray, C. W.; Blundell, T. L. *Curr. Opin. Struct. Biol.* **2010**, *20*, 497–507. (e) Feyfant, E.; Cross, J. B.; Paris, K.; Tsao, D. H. Fragment-based drug design. *Meth. Mol. Biol.* **2011**, *685*, 241–252. (f) Congreve, M.; Chessari, G.; Tisi, D.; Woodhead, A. J. Recent developments in fragment-based drug discovery. *J. Med. Chem.* **2008**, *51*, 3661–3680. (g) Schulz, M. N.; Hubbard, R. E. Recent progress in fragment-based lead discovery. *Curr. Opin. Pharmacol.* **2009**, *9*, 615–621. (h) Erlanson, D. A. Fragment-based lead discovery: a chemical update. *Curr. Opin. Biotechnol.* **2006**, *17*(6), 643–652. (i) Hajduk, P. J.; Greer, J. A decade of fragment-based drug design: strategic advances and lessons learned. *Nat. Rev. Drug Discov.* **2007**, *6*(3), 211–219. (j) Hubbard, R. E.; Davis, B.; Chen, I.; Drysdale, M. J. The SeeDs Approach: Integrating fragments into drug discovery. *Curr. Top. Med. Chem.* **2007**, *7*, 1568–1581.
185. (a) Pellecchia, M.; Bertini, I.; Cowburn, D.; Dalvit, C.; Giralt, E.; Jahnke, W.; James, T. L.; Homans, S. W.; Kessler, H.; Luchinat, C.; Meyer, B.; Oschkinat, H.; Peng, J.; Schwalbe, H.; Siegal, G. Perspectives on NMR in drug discovery: a technique comes of age. *Nat. Rev. Drug Discov.* **2008**, *7*(9), 738–45. (b) Lepre, C. A.; Moore, J. M.; Peng, J. W. Theory and application of NMR-based screening in pharmaceutical research. *Chem. Rev.* **2004**, *104*(8), 3641–3675. (c) Meyer, B.; Peters, T. NMR spectroscopy techniques for screening and identifying ligand binding to protein receptors. *Angew. Chem. Int. Ed. Engl.* **2003**, *42*(8), 864–890. (d) Pellecchia, M.; Sem, D. S.; Wüthrich, K. NMR in drug discovery. *Nat. Rev. Drug Discov.* **2002**, *1*(3), 211–219.
186. (a) Jhoti, H.; Cleasby, A.; Verdonk, M.; Williams, G. Fragment-based screening using X-ray crystallography and NMR spectroscopy. *Curr. Opin. Chem. Biol.* **2007**, *11*(5), 485–493; (b) Hartshorn, M. J.; Murray, C. W.; Cleasby, A.; Frederickson, M.; Tickle, I. J.; Jhoti, H. Fragment-based lead discovery using X-ray crystallography. *J. Med. Chem.* **2005**, *48*, 403–413. (c) Blundell, T. L.; Patel, S. High-throughput X-ray crystallography for drug discovery. *Curr. Opin. Pharmacol.* **2004**, *4*(5), 490–496; (d) Blundell, T. L.; Jhoti, H.; Abell, C. High-throughput crystallography for lead discovery in drug design. *Nat. Rev. Drug Discov.* **2002**, *1*(1), 45–54; (e) Kuhn, P.; Wilson, K.; Patch, M. G.; Stevens, R. C. The genesis of high-throughput structure-based drug discovery using protein crystallography. *Curr. Opin. Chem. Biol.* **2002**, *6*(5), 704–710.
187. (a) Erlanson, D. A.; Hansen, S. K. Making drugs on proteins: site-directed ligand discovery for fragment-based lead assembly. *Curr. Opin. Chem. Biol.* **2004**, *8*(4), 399–406; (b) Hofstadler, S. A.; Sannes-Lowery, K. A. Applications of ESI-MS in drug discovery: interrogation of noncovalent complexes. *Nat. Rev. Drug Discov.* **2006**, *5*(7), 585–595.
188. (a) Neumann, T.; Junker, H.-D.; Schmidt, K.; Sekul, R. SPR-based fragment screening: advantages and applications. *Curr. Top. Med. Chem.* **2007**, *7*(16), 1630–1642. (b) Navratilova, I.; Hopkins, A. L. Fragment screening by surface plasmon resonance. *ACS Med. Chem.*

Lett. **2010**, *1*, 44–48. (c) Kreatsoulas, C. and K. Narayan. Algorithms for the automated selection of fragment-like molecules using single-point surface plasmon resonance measurements. *Anal. Biochem.* **2010**, *402*, 179–184.
189. Hesterkamp, T.; Whittaker, M. Fragment-based activity space: smaller is better. *Curr. Opin. Chem. Biol.* **2008**, *12*(3), 260–268.
190. Fink, T.; Bruggesser, H.; Reymond, J. L. Virtual exploration of the small-molecule chemical universe below 160 D. *Angew. Chem. Int. Ed.* **2005**, *44*, 1504–1508.
191. Schuffenhauer, A.; Ruedisser, S.; Marzinzik, A. L.; Jahnke, W.; Blommers, M.; Selzer, P.; Jacoby, E. Library design for fragment based screening. *Curr. Top. Med. Chem.* **2005**, *5*(8), 751–762.
192. Hann, M. M; Leach, A. R.; Harper, G. Molecular complexity and its impact on the probability of finding leads for drug discovery. *J. Chem. Inf. Comput. Sci.* **2001**, *41*, 856–864.
193. Siegel, M. G.; Vieth, M. Drugs in other drugs: a new look at drugs as fragments. *Drug Discov. Today* **2007**, *12*(1–2), 71–79.
194. Congreve, M.; Carr, R.; Murray, C.; Jhoti, H. A "Rule of Three" for fragment-based lead discovery? *Drug Discov. Today* **2003**, *8*, 876–877.
195. Hajduk, P. J. Fragment-based drug design: how big is too big? *J. Med. Chem.* **2006**, *49*(24), 6972–6976.
196. Babaoglu, K.; Shoichet, B. K. Deconstructing fragment-based inhibitor discovery. *Nat. Chem. Biol.* **2006**, *2*(12), 720–723.
197. Siegal, G.; AB, E.; Schultz, J. Integration of fragment screening and library design. *Drug Discov. Today* **2007**, *12*(23–24), 1032–1039.
198. Bostrom, J.; Hogner, A.; Schmitt, S. Do structurally similar ligands bind in a similar fashion? *J. Med. Chem.* **2006**, *49*, 6716–6725.
199. Hajduk, P. J.; Huth, J. R.; Fesik, S. W. Druggability indices for protein targets derived from NMR-based screening data. *J. Med. Chem.* **2005**, *48*(7), 2518–2525.
200. Schuffenhauer, A.; Ruedisser, S.; Marzinzik, A. L.; Jahnke, W.; Blommers, M.; Selzer, P.; Jacoby, E. Library design for fragment based screening. *Curr. Top. Med. Chem.* **2005**, *5*(8), 751–762
201. Rees, D. C.; Congreve, M.; Murray, C. W.; Carr, R. Fragment-based lead discovery. *Nat. Rev. Drug Disc.* **2004**, *3*, 660–672.
202. Jencks, W. P. On the attribution and additivity of binding energies. *Proc. Natl. Acad. Sci. U.S.A.* **1981**, *78*, 4046–4050.
203. Nakamura, C. E.; Abeles, R. H. Mode of interaction of ß-hydroxy-ß-methylglutaryl coenzyme A reductase with strong binding inhibitors: compactin and related compounds. *Biochemistry* **1985**, *24*, 1364–1376.
204. Shuker, S. B.; Hajduk, P. J.; Meadows, R. P.; Fesik, S. W. Discovering high-affinity ligands for proteins: SAR by NMR. *Science* **1996**, *274*, 1531–1534.
205. (a) Whittaker, M.; Floyd, C. D.; Brown, P.; Gearing, J. H. Design and therapeutic application of matrix metalloproteinase inhibitors. *Chem. Rev.* **1999**, *99*, 2735–2776. (b) Woessner, J. F., Jr. Matrix metalloproteinases and their inhibitors in connective tissue remodeling. *FASEB J.* **1991**, *5*, 2145–2154.
206. (a) Hajduk, P. J.; Sheppard, G.; Nettesheim, D. G.; Olejniczak, E. T.; Shuker, S. B.; Meadows, R. P.; Steinman, D. H.; Carrera, G. M.; Marcotte, P. A.; Severin, J.; Walter, K.; Smith, H.; Gubbins, E.; Simmer, R.; Holtzman, T. F.; Morgan, D. W.; Davidsen, S. K.; Summers, J. B.; Fesik, S. W. Discovery of potent nonpeptide inhibitors of stromelysin using SAR by NMR *J. Am. Chem. Soc.* **1997**, *119*, 5818–5827. (b) Olejniczak, E. T.; Hajduk, P. J.; Marcotte, P. A.; Nettesheim, D. G.; Meadows, R. P.; Edalji, R.; Holtzman, T. F.; Fesik, S. W. Stromelysin inhibitors designed from weakly bound fragments: effects of linking and cooperativity. *J. Am. Chem. Soc.* **1997**, *119*, 5828–5832.
207. (a) Christin T., Shoemaker, A. R.; Adickes, J. et al. ABT-263: A potent and orally bioavailable Bcl-2 family inhibitor. *Cancer Res.* **2008**, *68*, 3421–3428. (b) Oltersdorf, T.; Elmore, S. W.; Shoemaker, A. R.; Armstrong, R. C.; et al. An inhibitor of Bcl-2 family proteins induces regression of solid tumors. *Nature* **2005**, *435*, 677–681.
208. Adams, J. M.; Cory, S. The Bcl-2 protein family: arbiters of cell survival. *Science* **1998**, *281*, 1322–1325.
209. Cory, S.; Huang, D. C. S.; Adams, J. M. The Bcl-2 family: roles in cell survival and oncogenesis. *Oncogene* **2003**, *22*, 8590–8607.
210. Sattler, M.; Liang, H.; Nettesheim, D.; et al. Structure of Bcl-xL-Bak peptide complex: recognition between regulators of apoptosis. *Science* **1997**, *275*, 983–986.
211. Petros, A. M.; Dinges, J.; Augeri, D. J.; Baumeister, S.A.; et al. Discovery of a potent inhibitor of the antiapoptotic protein Bcl-xL from NMR and parallel synthesis. *J. Med. Chem.* **2006**, *49*, 656–663.
212. Souers, A. J.; Leverson, J. D.; Boghaert, E. R. et al. ATT-199, a potent and selective BCL-2 inhibitor, achieves antitumor activity while sparing platelets. *Nat. Med.* **2013**, *19*, 202–208.
213. Baker, M. Fragment-based lead discovery grows up. *Nat. Rev. Drug Disc.* **2013**, *12*, 5–7.
214. Sprangers R; Kay L. E. Quantitative dynamics and binding studies of the 20S proteasome by NMR. *Nature* **2007**, *445*(7128), 618–22.
215. (a) Smales, C. M.; James, D. C. (Eds.) Therapeutic Proteins. Methods and Protocols; Humana Press: Totowa, NJ, 2005. (b) Petsko, G. A. For medicinal purposes. *Nature* **1996**, *384* (Supp. 7), 7–9. (c) Blundell, T. L. Structure-based drug design. *Nature* **1996**, *384*(Supp. 7), 23–26. (d) Martin, J. L. Protein crystallography and examples of its applications in medicinal chemistry. *Curr. Med. Chem.* **1996**, *3*, 419–436.
216. Hajduk, P. J.; Gerfin, T.; Böhlen, J.-M.; Häberli, M; Marek, D.; Fesik, S. W. High throughput nuclear magnetic resonance-based screening. *J. Med. Chem.* **1999**, *42*, 2315–2317.
217. Talamas, F. X.; Ao-leong, G.; Brameld, K. A.; et al. De novo fragment design: a medicinal chemistry approach to fragment-based lead generation. *J. Med. Chem.* **2013**, *56*, 3115–3119.
218. Maly, D. J.; Choong, I. C.; Ellman, J. A. Combinatorial target-guided ligand assembly: identification of a potent subtype-selective c-SRC inhibitors. *Proc. Natl. Acad. Sci. U.S.A.* **2000**, *97*, 2419–2424.
219. (a) Hofstadler, S. A.; Sannes-Lowery, K. A. Interrogation of noncovalent complexes by ESI- MS: a powerful platform for high throughput drug discovery. Methods and Principles in Medicinal Chemistry 2007, *36*(Mass Spectrometry in Medicinal Chemistry), 321–338. (b) Griffey, R. H.; Hofstadler, S. A.; Sannes-Lowery, K. A.; Ecker, D. J.; Crooke, S. T. Determinants of aminoglycoside-binding specificity for rRNA by mass spectrometry. *Proc. Natl. Acad. Sci. U.S.A.* **1999**, *96*, 10129–10133. (c) Hofstadler, S. A.; Sannes-Lowery, K. A.; Crooke, S. T.; Ecker, D. J.; Sasmor, H.; Manalili, S.; Griffey, R. H. Multiplexed screening of neutral mass-tagged RNA targets against ligand libraries with electrospray ionization FTICR MS: a paradigm for high-throughput affinity screening. *Anal. Chem.* **1999**, *71*, 3436–3440.
220. Seth, P. P.; Miyaji, A.; Jefferson, E. A.; Sannes-Lowery, K. A.; Osgood, S. A.; Propp, S. S.; Ranken, R.; Massire, C.; Sampath, R.; Ecker, D. J.; Swayze, E. E.; Griffey, R. H. SAR by MS: Discovery of a new class of RNA-binding small molecules for the hepatitis c virus: internal ribosome entry site iia subdomain. *J. Med. Chem.* **2005**, *48*(23), 7099–7102.
221. (a) Wood, W. J. L.; Patterson, A. W.; Tsuruoka, H.; Jain, R. K.; Ellman, J. A. Substrate activity screening: a fragment-based method for the rapid identification of nonpeptidic protease inhibitors. *J. Am.*

Chem. Soc. **2005**, *127*(44), 15521–15527; (b) Brak, K.; Doyle, P. S.; McKerrow, J. H.; Ellman, J. A. Identification of a new class of nonpeptidic inhibitors of cruzain. *J. Am. Chem. Soc.* **2008**, *130*(20), 6404–6410; (c) Soellner, M. B.; Rawls, K. A.; Grundner, C.; Alber, T.; Ellman, J. A. Fragment-based substrate activity screening method for the identification of potent inhibitors of the Mycobacterium tuberculosis phosphatase PtpB. *J. Am. Chem. Soc.* **2007**, *129*(31), 9613–9635.

222. (a) Frizler, M.; Stirnberg, M.; Sisay, M. T.; Guetschow, M. Development of nitrile-based peptidic inhibitors of cysteine cathepsins. *Curr. Top. Med. Chem.* **2010**, *10*(3), 294–322. (b) Gupta, S.; Singh, R. K.; Dastidar, S.; Ray, A. Cysteine cathepsin S as an immunomodulatory target: present and future trends. *Exp. Opin. Ther. Targets* **2008**, *12*(3), 291–299.
223. (a) Lewis W. G; Green L. G; Grynszpan F.; Radic Z.; Carlier P. R; Taylor P.; Finn M G; Sharpless K. B. Click chemistry in situ: acetylcholinesterase as a reaction vessel for the selective assembly of a femtomolar inhibitor from an array of building blocks. *Angew. Chem. Int. Ed.* **2002**, *41*(6), 1053–7. (b) Manetsch, R.; Krasinski, A.; Radic, Z.; Raushel, J.; Taylor, P.; Sharpless, K. B.; Kolb, H. C. In situ click chemistry: enzyme inhibitors made to their own specifications. *J. Am. Chem. Soc.* **2004**, *126*(40), 12809–18. (c) Krasinski, A.; Radic, Z.; Manetsch, R.; Raushel, J.; Taylor, P.; Sharpless, K. B.; Kolb, H. C. In situ selection of lead compounds by click chemistry: target-guided optimization of acetylcholinesterase inhibitors. *J. Am. Chem. Soc.* **2005**, *127*(18), 6686–92.
224. Ruda, G. F.; Campbell, G.; Aibu, V. P.; Barrett, M. P.; Brenk, R.; Gilbert, I. H. Virtual fragment screening for novel inhibitors of 6-phosphogluconate dehydrogenase. *Bioorg. Med. Chem.* **2010**, *18*, 5056–5062.
225. Zych, A. J.; Herr, R. J. Tetrazoles as carboxylic acid bioisosteres in drug discovery. *PharmaChem* **2007**, *6*, 21–24.
226. (a) Wermuth, C. G. Selective optimization of side activities: another way for drug discovery. *J. Med. Chem.* **2004**, *47*, 1303–1314. (b) Wermuth, C. G. The SOSA approach: an alternative to high-throughput screening. *Med. Chem. Res.* **2001**, *10*, 431–439.
227. Prestwick Chemical Library®; www.prestwickchemical.com.
228. Frederickson, M.; Callaghan, O.; Chessari, G.; Congreve, M.; Cowan, S. R.; Matthews, J. E.; McMenamin, R.; Smith, D-M.; Vinkovic, M.; Wallis N. G.Fragment-based discovery of mexiletine derivatives of bioavailable inhibitors of urokinase-type plasminogen activator. *J. Med. Chem.* **2008**, *51*, 183–186.
229. Wong, A. P.; Cortez S. L.; Baricos, W. H. Role of plasmin and gelatinase in extracellular matrix degradation by cultured rat mesangial cells. *Am. J. Physiol. Renal Physiol.* **1992**, *263*, 1112–1118.
230. Gveric, D.; Hanemaaijer, R.; Newcombe J.; van Lent, N. A.; Sier C. F.; Cuzner, M. L. Plasminogen activators in multiple sclerosis legions. *Brain* **2001**, *124*, 1978–1988.
231. Schweinitz, A. et al. Design of novel and selective inhibitors of urokinase-type of plasminogen activator with improved pharmacokinetic properties for use as antimetastatic agents. *J. Biol. Chem.* **2004**, *279*, 33613–33622.
232. Grewe, R. The problem of morphine synthesis. *Naturwissenschaften* **1946**, *33*, 333–336.
233. Schnider, O.; Grussner, A. Synthesis of hydroxymorphinans. *Helv. Chem. Acta*, **1949**, *32*, 821–828.
234. May, E. L.; Murphy, J. G. Structures related to morphine. III. Synthesis of an analog of N-methylmorphinan. *J. Org. Chem.* **1955**, *20*, 257–258.
235. Schaumann, O. Über eine neue Klasse von Verbindungen mit spasmolytischer und zentral analgetischer Wirksamkeit unter besonderer Berücksichtigung des 1-Methyl-4-phenyl-piperidin-4-äure-äthylesters (Dolantin). *Naunyn-Schmiedebergs Arch. Pharmacol. Exp. Pathol.* **1940**, *196*, 109.
236. (a) Bentley, K. W.; Hardy, D. G. Novel analgesics and molecular rearrangements in the morphine thebaine group. III. Alcohols of the 6,14-endo-ethenotetrahydrooripavine series and derived analogs of N-allylnormorphine and –norcodeine. *J. Am. Chem. Soc.* **1967**, *89*, 3281–3292. (b) Bentley, K. W.; Hardy, D. G. Novel analgesics and molecular rearrangements in the morphine thebaine group. II. Alcohols derived form 6,14-endo-etheno- and 6,14-endo-ethanotetrahydrothebaine. *J. Am. Chem. Soc.* **1967**, *89*, 3273–3280. (c) Bentley, K. W.; Hardy, D. G. Novel analgesics and molecular rearrangements in the morphine thebaine group. I. Ketones derived from 6,14-endo-ethenotetrahydrothebaine. *J. Am. Chem. Soc.* **1967**, *89*, 3267–73.
237. Andrews, P. R.; Craik, D. J.; Martin, J. L. Functional group contributions to drug-receptors interactions. *J. Med. Chem.* **1984**, *27*, 1648–1657.
238. Plobeck, N.; Delorme, D.; Wei, Z.-Y.; Yang, H.; Zhou, F.; Schwarz, P.; Gawell, L.; Gagnon, H.; Pelcman, B.; Schmidt, R.; Yue, S. Y.; Walpole, C.; Brown, W.; Zhou, E.; Labarre, M.; Payza, K.; St-Onge, S.; Kamassah, A.; Morin, P.-E.; Projean, D.; Ducharme, J.; Roberts, E. New diarylmethylpiperazines as potent and selective nonpeptidic δ opiod receptor agonists with increased in vitro metabolic stability. *J. Med. Chem.* **2000**, *43*, 3878–3894.
239. (a) Crum-Brown, A.; Fraser, T. R. On the connection between chemical constitution and physiological action. Part I. On the physiological action of the salts of the ammonium bases, derived from strychnine, brucia, thebaia, codeia, morphia, and nicotia. Part II. On the physiological action of the ammonium bases derived from atropia and conia. *Trans. Roy. Soc. Edinburgh* **1868**–1869, *25*, 151, 693. (b) Crum-Brown, A.; Fraser, T. R. On the connection between chemical constitution and physiological action-continued. On the physiological action of the salts of trimethylsulphin. *Proc. Roy. Soc. Edinburgh* **1872**, *7*, 663–665.
240. Richardson, B. W. Lectures on experimental and practical medicine. Physiological research on alcohols. *Med. Times Gaz.* **1869**, *2*, 703–706.
241. Wassermann, A. M.; Wawer, M.; Bajorath, J. Activity Landscape representations for structure–activity relationship analysis. *J. Med. Chem.* **2010**, *53*, 8209–8223.
242. (a) Stumpfe, D.; Bajorath, J. Exploring activity cliffs in medicinal chemistry. *J. Med. Chem.* **2012**, 55, 2932–2942. (b) Dimova, D.; Heikamp, K.; Stumpfe, D.; Bajorath, J. Do medicinal chemists learn from activity cliffs? A systematic evaluation of cliff progression in evolving compound data sets. *J. Med. Chem.* **2013**, *56*, 3339–3345.
243. Xia, G.; Benmohamed, R.; Kim, J.; Arvanites, A. C.; Morimoto, R. I.; Ferrante, R. J.; Kirsch, D. R.; Silverman, R. B. Pyrimidine-2,4,6-trione Derivatives and their inhibition of mutant SOD1-dependent protein aggregation. Toward a treatment for amyotrophic lateral sclerosis (ALS). *J. Med. Chem.* **2011**, *54*, 2409–2421.
244. Northey, E. H. *The Sulfonamides and Allied Compounds*, American Chemical Society Monograph Series, Reinhold: New York, 1948.
245. Loubatieres, A. In *Oral Hypoglycemic Agents*; Campbell, G. D. Ed.; Academic Press: New York, 1969.
246. High-ceiling diuretics are highly efficacious inhibitors of Na+K+2Cl-symport. Jackson, E. D. In *Goodman & Gilman's The Pharmacological Basis of Therapeutics*, 9th edition; Hardman, J. G.; Limbird, L. E.; Molinoff, P. B.; Ruddon, R. W.; Gilman, A. G., Eds.; McGraw-Hill: New York, 1996, p. 685.

247. Sprague, J. M. In *Topics in Medicinal Chemistry*; Robinowitz, J. L.; Myerson, R. M. Eds.; Wiley: New York, 1968; Vol. 2.
248. (a) He, L.; Jagtap, P. G.; Kingston, D. G. I.; Shen, H.-J.; Orr, G. A.; Horwitz, S. A common pharmacophore for taxol and the epothilones based on the biological activity of a taxane molecule lacking a C-13 side chain. *Biochemistry* **2000**, *39*, 3972–3978. (b) Díaz, J. F.; Strobe, R.; Engelborghs, Y.; Souto, A. A.; Andreu, J. M. Molecular recognition of taxol by microtubules. Kinetics and thermodynamics of binding of fluorescent taxol derivatives to an exposed site. *J. Biol. Chem.* **2000**, *275*, 26265–26276. (c) Yvon, A. C.; Wadsworth, P.; Jordan, M. Taxol suppresses dynamics of individual microtubules in living human tumor cells. *Mol. Biol. Cell* **2000**, *10*, 947–959.
249. (a) Kingston, D. Recent advances in the chemistry of taxol. *J. Nat. Prod.* **2000**, *63*, 726–734. (b) Dubois, J.; Thoret, S.; Guéritte, F.; Guénard, D. Synthesis of 5(20)deoxydocetaxel, a new active docetaxel analogue. *Tetrahedron Lett.* **2000**, *41*, 3331–3334. (c) Yuan, H.; Kingston, D. Synthesis and biological evaluation of C-1 and ring modified A-norpaclitaxels. *Tetrahedron* **1999**, *55*, 9089–9100.
250. Agrafiotis, D. K.; Shemanarev, M.; Connolly, P. J.; Farnum, M.; Lobanov, V. S. SAR maps: a new SAR visualization technique for medicinal chemists. *J. Med. Chem.* **2007**, *50*, 5926–5937.
251. Peltason, L.; Bajorath, J. SAR index: quantifying the nature of structure-activity relationships. *J. Med. Chem.* **2007**, *50*, 5571–5578.
252. Huang, Z.; Yang, G.; Lin, Z.; Huang, J. 2-[N1-2-pyrimidylamino-benzenesulfonamido]ethyl 4-bis(2-chloroethyl)aminophenylbutyrate A potent antitumor agent. *Bioorg. Med. Chem. Lett.* **2001**, *11*, 1099–1103.
253. Hodges, B.; Mazur, J. E. Intravenous ethanol for the treatment of alcohol withdrawal syndrome in critically ill patients. *Pharmacother.* **2004**, *24*, 1578–1585.
254. (a) Williams, D. P.; Naisbitt, D. J. Toxicophores: groups and metabolic routes associated with increased safety risk. *Curr. Opin. Drug Discov. Devel.* **2002**, *5*, 104–115. (b) Williams, D. P. Toxicophores: investigations in drug safety. *Toxicol.* **2006**, *226*, 1–11.
255. (a) Kazius, J.; McGuire, R.; Bursi, R. Derivation and validation of toxicophores for mutagenicity prediction. *J. Med. Chem.* **2005**, *48*, 312–320. (b) Hakimelahi, G. H.; Khodarahmi, G. A. The identification of toxicophores for the prediction of mutagenicity, hepatotoxicity and cardiotoxicity. *J. Iran. Chem. Soc.* **2005**, *2*, 244–267.
256. Stepan, A. F.; Walker, D. P.; Bauman, J.; Price, D. A.; Baillie, T. A.; Kalgutkar, A. S.; Aleo, M. D. Structural alert/reactive metabolite concept as applied in medicinal chemistry to mitigate the risk of idiosyncratic drug toxicity: a perspective based on the critical examination of trends in the top 200 drugs marketed in the United States. *Chem. Res. Toxicol.* **2011**, *24*, 1345–1410.
257. Richardson, B. W. Lectures on experimental and practical medicine. Physiological research on alcohols. *Med. Times Gaz.* **1869**, *2*, 703–706.
258. Dohme, A. R. L.; Cox, E. H.; Miller, E. Preparation of the acyl and alkyl derivatives of resorcinol. *J. Am. Chem. Soc.* **1926**, *48*, 1688–1693.
259. Funcke, A. B. H.; Ernsting, M. J. E.; Rekker, R. F.; Nauta, W. T. Spasmolytics I. Esters of mandelic acid. *Arzneimittel. Forsch.* **1953**, *3*, 503–506.
260. (a) Thornber, C. W. Isosterism and molecular modification in drug design. *Chem. Soc. Rev.* **1979**, *8*, 563–580. (b) Lima, L. M.; Barreiro, E. J. Bioisosterism: a useful strategy for molecular modification and drug design. *Curr. Med. Chem.* **2005**, *12*, 23–49. (c) Langdon, S. R.; Ertl, P.; Brown, N. Bioisosteric replacement and scaffold hopping in lead generation and optimization. *Mol. Inf.* **2010**, *29*, 366–385. (d) Meanwell, N. A. Synopsis of some recent tactical application of bioisosteres in drug design. *J. Med. Chem.* **2011**, *54*(8), 2529–2591.
261. Friedman, H. L., *National Academy of Sciences, National Research Council Publication No. 206*. Washington, DC, 1951, p. 295.
262. Burger, A. In *Medicinal Chemistry*, 3rd ed.; Burger, A. Ed.; Wiley: New York, 1970.
263. Korolkovas, A *Essentials of Molecular Pharmacology*: Background for Drug Design; Wiley: New York, 1970, pp 54–57.
264. Lipinski, C. A. Biososterism in drug design. *Annu. Rep. Med. Chem.* **1986**, *21*, 283–291.
265. Langmuir, I. Isomorphism, isosterism and covalence. *J. Am. Chem. Soc.* **1919**, *41*, 1543–1559.
266. (a) Grimm, H. G. Structure and size of the non-metallic hydrides. *Z. Elektrochem.* **1925**, *31*, 474–480. (b) Grimm, H. G. The different types of chemical union. *Z. Elektrochem.* **1928**, *34*, 430–437.
267. Erlenmeyer, H. Isosteric compounds and their chemical resemblance. *Bull. Soc. Chim. Biol.* **1948**, *30*, 792–805.
268. (a) Patani, G. A.; LaVoie, E. J. Biososterism: a rational approach in drug design. *Chem. Rev.* **1996**, *96*, 3147–3176. (b) Olsen, P. H. The use of bioisosteric groups in lead optimization. *Curr. Opin. Drug Disc.* **2001**, *4*(4), 471–478. (c) Chen, X.; Wang, W. The use of bioisosteric groups in lead optimization. *Annu. Rep. Med. Chem.* **2003**, *38*, 333–346. (d) Lima, L. M.; Barreiro, E.J. Bioisosterism: A useful strategy for molecular modification and drug design. *Curr. Med. Chem.* **2005**, *12*(1), 23–49. (e) Ertl, P. In silico identification of bioisosteric functional groups. *Curr. Opin. Drug Disc.* **2007** *10*(3), 281–288. (f) Wermuth, C. G.; Ciapetti, P.; Giethlen, B.; Bazzini, P. Bioisosterism. In *Comprehensive Medicinal Chemistry II* Taylor, J. B., Triggle, D. J., Eds.; Elsevier: 2006; pp 649–711. (g) Devereux, Mike; Popelier, Paul L. A. In silico techniques for the identification of bioisosteric replacements for drug design. *Curr. Top. Med. Chem.* (Sharjah, United Arab Emirates) **2010**, *10*(6), 657–668. (h) Langdon, S. R.; Ertl, P.; Brown, N. Bioisosteric replacement and scaffold hopping in lead generation and optimization. *Mol. Inf.* **2010**, *29*(5), 366–385. (i) Kalgutkar, A. S.; Daniels, J. S. Carboxylic acids and their bioisosteres. *RSC Drug Discovery Series* **2010**, 1(Metabolism, Pharmacokinetics and Toxicity of Functional Groups), 99–167. (j) Wuitschik, G.; Carreira, E. J.; Wagner, B.; Fischer, H.; Parrilla, I.; Schuler, F.; Roger-Evans, M.; Muller, K. Oxetanes in drug discovery: structural and synthetic insights. *J. Med. Chem.* **2010**, *53*, 3227–3246. (k) Bhatia, R.; Sharma, V.; Shrivastava, B.; Singla, R. K. A review on bioisosterism: A rational approach for drug design and molecular modification. *Pharmacologyonline* **2011**, *1*, 272–299.
269. Song, Y.; Connor, D. T.; Doubleday, R.; Sorenson, R. J.; Sercel, A. D.; Unangst, P. C.; Roth, B. D.; Gilbertsen, R. B.; Chan, K.; Schrier, D. J.; Guglietta, A.; Bornemeier, D. A.; Dyer, R. D. Synthesis, structure-activity relationships, and in vivo evaluations of substituted di-tert-butylphenols as a novel class of potent, selective, and orally activecyclooxygenase-2 inhibitors. 1. Thiazolone and oxazolone series. *J. Med. Chem.* **1999**, *42*, 1151–1160.
270. Böhm, H.-J.; Banner, D.; Bendels, S.; Kansy, M.; Kuhn, B.; Müller, K.; Obst-Sander, U.; Stahl, M. Fluorine in medicinal chemistry, *ChemBioChem.* **2004**, *5*, 637–643.
271. (a) Clift, M. D.; Ji, H.; Deniau, G. P.; O'Hagan, D.; Silverman, R. B. The enantiomers of 4-amino-3-fluorobutanoic acid as substrates for γ-aminobutyric acid aminotransferase. Conformational probes for GABA binding. *Biochemistry* **2007**, *46*(48), 13819–13828.

(b) Winkler, M.; Moraux, T.; Khairy, H. A.; Scott, R. H.; Slawin, A. M. Z.; O'Hagan, D. Synthesis and vanilloid receptor (TRPV1) activity of the enantiomers of a-fluorinated capsaicin. *ChemBioChem.* **2009**, *10*, 823–828.

272. Dalvit, C.; Vulpetti, A. Intermolecular and intramolecular hydrogen bonds involving fluorine atoms: implications for recognition, selectivity, and chemical properties. *ChemMedChem.* **2012**, *7*, 262–272.

273. Xue, F.; Li, H.; Delker, S.; Fang, J.; Martásek, P.; Roman, L. J.; Poulos, T. L.; Silverman, R. B. Potent, highly selective, and orally bioavailable gem-difluorinated monocationic inhibitors of neuronal nitric oxide synthase, *J. Am. Chem. Soc.* **2010**, *132*(40), 14229–14238.

274. Pinto, D. J. P.; Orwat, M. J.; Wang, S.; Fevig, J. M.; Quan, M. L. Amparo, E. et al. Discovery of 1-[3-(aminomethyl)phenyl]-N-[3-fluoro-2'-(methylsulfonyl)-[1,1'-biphenyl]-4-yl]-3-(trifluoromethyl)-1H-pyrazole-5-carboxamide (DPC423), a highly potent, selective, and orally bioavailable inhibitor of blood coagulation factor Xa. *J.Med.Chem.* **2001**, *44*, 566–578.

275. Stepan, A. F.; Subramanyam, C.; Efremov, I. V. et al. Application of the bicyclo[1.1.1]pentane motif as a nonclassical phenyl ring bioisostere in the design of a potent and orally active γ-secretase inhibitor. *J. Med. Chem.* **2012**, *55*, 3414–3424.

276. Burkhard, J. A.; Wuitschik, G.; Rogers-Evans, M.; Müller, K.; Carreira, E. M. Oxetanes as versatile elements in drug discovery and synthesis. *Angew. Chem. Int. Ed.* **2010**, *49*, 9052–9067.

277. Wuitschik, G.; Rogers-Evans, M.; Müller, K.; Fischer, H.; Wagner, B.; Schuler, F.; Polonchuk, L.; Carreira, E. M. Oxetanes as promising modules in drug discovery. *Angew. Chem. Int. Ed.* **2006**, *45*, 7736–7739.

278. Wuitschik, G.; Rogers-Evans, M.; Buckl, A.; Bernasconi, M.; Märki, M.; Godel, T.; Fischer, H.; Wagner, B.; Parrilla, I.; Schuler, F.; Schneider, J.; Alker, A.; Schweizer, W. B.; Müller, K.; Carreira, E. M.; Spirocyclic oxetanes: synthesis and properties. *Angew. Chem. Int. Ed.* **2008**, *47*, 4512–4515.

279. Stump. C. A.; Bell, I. A.; Bednar, R. A.; Fay, J. F.; Gallicchio, S. N.; Hershey, J. C.; Jelley, R.; Kreatsoulas, C.; Moore, E. L.; Mosser, S. D.; Quigley, A. G.; Roller, A. S.; Salvatore, C. A.; Sharik, S. A.; Theberge, A. R.; Zartman, C. G.; Kane, S. A.; Graham, S. L.; Selnick, H. G.; Vacca, J. P.; Williams, T. M. Identification of potent, highly constrained CGRP receptor antagonists. *Bioorg. Med. Chem. Lett.* **2010**, *20*, 2572–2576.

280. Brickner, S. J.; Barbachyn, M. R.; Hutchinson, D. K.; Manninen, P. R. Linezolid (ZYVOX), the first member of a completely new class of antibacterial agents for treatment of serious Gram-positive infections. *J. Med. Chem.* **2008**, *51*, 1981–1990.

281. (a) Tinoco, A. D.; Saghatellan, A. Investigating endogenous peptides and peptidases using peptidomics. *Biochemistry* **2011**, *50*, 7447–7461. (b) Gelman, J. S.; Fricker, L. D. Hemopressin and other bioactive peptides from cytosolic proteins: are these non-classical neuropeptides? *AAPS J.* **2010**, *12*(3), 279–289. (c) Swanson, H. H. Peptides. In *Brain Mechanism Psychotropic Drugs*; Baskys, A.; Remington, G., Eds.; CRC Press: Boca Raton, FL; 1996, pp 131–150.

282. Boman, H. G. Peptide antibiotics and their role in innate immunity. *Annu. Rev. Immunol.* **1995**, *13*, 61–92.

283. Harder, J.; Bartels, J.; Christophers, E.; Schröder, J.-M. A peptide antibiotic from human skin. *Nature* **1997**, *387*, 861.

284. (a) Schiller, P. W. Bi- or multifunctional opioid peptide drugs. Life Sci. **2010**, 86, 598–603. (b) Horvat, S. Opiod peptides and their glycoconjugates: structure activity relationships. *Curr. Med. Chem.: Cent. Nerv. Syst. Agents* **2001**, *1*, 133–154. (c) Vaccarino, A. L.; Kastin, A. J. Endogenous opiates: 1999. *Peptides* **2000**, *21*, 1975–2034. (d) Roques, B. P.; Noble, F.; Fournie-Zaluski, M.-C. Endogenous opioid peptides and analgesia. In *Opioids Pain Control*: Stein, C., Ed.; Cambridge University Press: Cambridge, UK, 1999; pp 21–45. (e) Stein, C.; Cabot, P. J.; Schafer, M. Peripheral opioid analgesia: mechanisms and clinical implications. In *Opioids Pain Contro*; Stein, C., Ed.; Cambridge University Press: Cambridge, UK, 1999; pp 96–108.

285. (a) Kohan, D. E.; Rossi, N. F.; Inscho, E. W.; Pollock, D. M. Regulation of blood pressure and salt homeostasis by endothelin. *Physiol. Rev.* **2011**, *91*(1), 1–77. (b) Sagnella, G. A. Atrial natriuretic peptide mimetics and vasopeptide inhibitors. *Cardivasc. Res.* **2001**, *51*, 416–428.

286. Boccardo, F.; Amoroso, D. Management of breast cancer: is there a role for somatostatin and its analogs? *Chemotherapy* **2001**, *47*(Supp. 2), 62–67.

287. (a) Cushman, D. W.; Ondetti, M. A. Inhibitors of angiotensin-converting enzyme *Prog. Med. Chem.* **1980**, *17*, 41–104. (b) Cushman, D. W.; Cheung, H. S.; Sabo, E. F.; Ondetti, M. A. Development and design of specific inhibitors of angiotensin-converting enzyme. *Am. J. Cardiol.* **1982**, *49*, 1390–1394.

288. Hughes, J.; Smith, T. W.; Kosterlitz, H. W.; Fothergill, L. A.; Morgan, B. A.; Morris, H. R. Identification of two related pentapeptides from the brain with potent opiate agonist activity. *Nature* **1975**, *258*, 577–579.

289. (a) Smith, G. S.; Griffin, J. F. Conformation of [Leu5]-enkephalin from x-ray diffraction: features important for recognition at opiate receptor. *Science* **1978**, *199*, 1214–1216. (b) Bradbury, A. F.; Smyth, D. G.; Snell, C. R. Biosynthetic origin and receptor conformation of methionine enkephalin. *Nature* **1976**, *260*, 165–166.

290. Farmer, P. S. In *Drug Design*; Ariens, E. J., Ed.; Academic Press: New York, 1980.

291. Giannis, A.; Kolter, T. Peptide mimetics for receptor ligands: discovery, development, and medicinal perspectives. *Angew. Chem. Int. Ed. Engl.* **1993**, *32*, 1244–1267.

292. (a) Hsieh, K.-H.; LaHann, T. R.; Speth, R. C. Topographic probes of angiotensin and receptor: potent angiotensin II agonist containing diphenylalanine and long-acting antagonists containing biphenylalanine and 2-indan amino acid in position 8. *J. Med. Chem.* **1989**, *32*, 898–903. (b) Corey, E. J.; Link, J. O. A general, catalytic, and enantioselective synthesis of a-amino acids. *J. Am. Chem. Soc.* **1992**, *114*, 1906–1908. (c) Schiller, P. W.; Weltrowka, G.; Nguyen, T. M. D.; Lemieux. C.; Chung, N. N.; Marken, B. J.; Wilke, B. C. Conformational restriction of the phenylalanine residue in a cyclic opioid peptide analog: effects on receptor selectivity or stereospecificity. *J. Med. Chem.* **1991**, *34*, 3125–3132. (d) Holladay, M. W.; Lin, C. W.; May, C.; Garvey, D.; Witte, D. G.; Miller, T. R.; Wolfram, C. A. W.; Nadzan, A. M. Trans-3-n-propyl-L-proline is a highly favorable, conformationally restricted replacement for methionine in the C-terminal tetrapeptide of cholecystokinin. Stereoselective synthesis of 3-allyl- and 3-n-propyl-L-proline derivatives from 4-hydroxy-L-proline. *J. Med. Chem.* **1991**, *34*, 455–457.

293. (a) Yanagisawa, H.; Ishihara, S.; Ando, A.; Kanazaki, T.; Miyamoto, S.; Koike, H.; Iijima, Y.; Oizumi, K.; Matsushita, Y.; Hata, T. Angiotensin-converting enzyme inhibitors. Perhydro-1,4-thiazepin-5-one derivatives. *J. Med. Chem.* **1987**, *30*, 1984–1991. (b) Giannis, A.; Kolter, T. Peptide mimetics for receptor ligands: discovery, development, and medicinal perspectives. *Angew. Chem. Int. Ed. Engl.* **1993**, *32*, 1244–1267.

294. Olson, G. L.; Bolin, D. R.; Bonner, M. P.; Bös, M.; Cook, C. M.; Fry, D. C.; Graves, B. J.; Hatada, M.; Hill, D. E.; Kahn, M.; Madison, V. S.; Rusiecki, V. K.; Sarabu, R.; Sepinwall, J.; Vincent, G. P.; Voss, M. E. Concept and progress in the development of peptide mimetics. *J. Med. Chem.* **1993**, *36*, 3039–3049.
295. (a) Nagai, U.; Sato, K.; Nakamura, R.; Kato, R. Bicyclic turned dipeptide (BTD) as a ß-turn mimetic: its design, synthesis and incorporation into bioactive peptides. *Tetrahedron* **1993**, 49, 3577–3592. (b) Sato, K.; Nagai, U. Synthesis and antibiotic activity of a gramicidin S analogue containing bicyclic ß-turn dipeptides. *J. Chem. Soc. Perkin Trans 1*, **1986**, 1231–1234.
296. (a) Brandmeier, V.; Sauer, W. H. B.; Feigel, M. Antiparallel ß-sheet conformation in cyclopeptides containing a pseudo-amino acid with a biphenyl moiety. *Helv. Chim. Acta* **1994**, *77*, 70–85. (b) Wagner, G.; Feigel, M. Parallel ß-sheet conformation in macrocycles. *Tetrahedron* **1993**, 49, 10831–10842.
297. (a) Kemp, D. S.; Curran, T. P.; Davis, W. M; Boyd, J. G.; Muendel, C. Studies of N-terminal templates for α-helix formation. Synthesis and conformational analysis of (2S, 5S, 8S,11S)-1-acetyl-1,4-diaza-3-keto-5-carboxy-10-thiatricyclo[2.8.1.04,8]tridecane (Ac-Hel1-OH). *J. Org. Chem.* **1991**, *56*, 6672–6682. (b) Kemp, D. S.; Curran, T. P.; Boyd, J. G.; Allen, T. J. Studies of N-terminal templates for α-helix formation. Synthesis and conformational analysis of peptide conjugates of (2S,5S,8S,11S)-1-acetyl-1,4-diaza-3-keto-5-carboxy-10-thiatricyclo[2.8.1.04,8]tridecane (Ac-Hel1-OH). *J. Org. Chem.* **1991**, *56*, 6683–6697.
298. Sarabu, R.; Lovey, K.; Madison, V. S.; Fry, D. C.; Greeley, D. W.; Cook, C. M.; Olson, G. L. Design, synthesis, and three-dimensional structural characterization of a constrained Ω-loop excised from interleukin-1a. *Tetrahedron* **1993**, *49*, 3629–3640.
299. Smith, A. B. III; Keenan, T. P.; Holcomb, R. C.; Sprengeler, P. A.; Guzman, M. C.; Wood, J. L.; Carroll, P. J.; Hirschmann, R. Design, synthesis, and crystal structure of a pyrrolinone-based peptidomimetic possessing the conformation of a ß-strand: potential application to the design of novel inhibitors of proteolytic enzymes. *J. Am. Chem. Soc.* **1992**, *114*, 10672–10674.
300. Lincoff, A. M.; Califf, R. M.; Topol, E. J. Platelet glycoprotein IIb/IIIa receptor blockade in coronary artery disease. *J. Am. Coll. Cardiol.* **2000**, *35*, 1103–1115.
301. Hirschmann, R.; Sprengeler, P. A.; Kawasaki, T.; Leahy, J. W.; Shakespeare, W. C.; Smith, A. B. III The first design and synthesis of a steroidal peptidomimetic. The potential value of peptidomimetics in elucidating the bioactive conformation of peptide ligands. *J. Am. Chem. Soc.* **1992**, *114*, 9699–9701.
302. Fisher, M. J.; Gunn, B.; Harms, C. S.; Kline, A. D.; Mullaney, J. T.; Nunes, A.; Scarborough, R. M.; Arfsten, A. E.; Skelton, M. A.; Um, S. L.; Utterback, B. G.; Jakubowski, J. A. Non-peptide RGD surrogates which mimic a gly-asp ß -turn: potent antagonists of platelet glycoprotein IIb-IIIa. *J. Med. Chem.* **1997**, *40*, 2085–2101.
303. Blackburn, B. K.; Lee, A.; Baier, M.; Kohl, B.; Olivero, A. G.; Matamoros, R.; Robarge, K. D.; McDowell, R. S. From peptide to non-peptide. E. Atropisomeric GPIIbIIIa antagonists containing the 3,4-dihydro-1H-1,4-benzodiazepine-2,5-dione nucleus. *J. Med. Chem.* **1997**, *40*, 717–729.
304. Hirschmann, R.; Nicolaou, K. C.; Pietranico, S.; Salvino, J.; Leahy, E. M.; Sprengeler, P. A.; Furst, G.; Smith, A. B. III.; Strader, C. D. Cascieri, M. A.; Candelore, M. R.; Donaldson, C.; Vale, W.; Maechler, L. Nonpeptidal peptidomimetics with β-D-glucose scaffolding. A partial somatostatin agonist bearing a close structural relationship to a potent, selective substance P antagonist. *J. Am. Chem. Soc.* **1992**, *114*, 9217–9218.
305. Miyamoto, M.; Yamazaki, N.; Nagaoka, A.; Nigawa, Y. Effects of TRH and its analogue, DN1417, on memory impairment in animal models. *Ann. N. Y. Acad. Sci.* **1989**, *553*, 508–510.
306. Olson, G. L.; Cheung, H.-C.; Chiang, E.; Madison, V. S.; Sepinwall, J.; Vincent, G. P.; Winokur A.; Gary, K. A. Peptide mimetics of thyrotropin-releasing hormone based on a cyclohexane framework: design, synthesis, and cognition-enhancing properties. *J. Med. Chem.* **1995**, *38*, 2866–2879.
307. Spatola, A. F. Peptide backbone modifications: a structure-activity analysis of peptides containing amid bond surrogates, conformational constraints, and related backbone replacements. *Chem. Biochem. Amino Acids Pept. Prot.* **1983**, *7*, 267.
308. Gante, J. Azapeptides. *Synthesis* **1989**, *6*, 405–413.
309. Han, H.; Janda, K. D. Azatides: solution and liquid phase syntheses of a new peptidomimetic. *J. Am. Chem. Soc.* **1996**, *118*, 2539–2544.
310. (a) Simon, R. J.; Kania, R. S.; Zuckermann, R. N.; Huebner, V. D.; Jewell, D. A.; Banville, S.; Ng, S.; Wang, L.; Rosenberg, S.; Marlowe, C. K.; Spellmeyer, D. C.; Tan, R. Y.; Frankel, A. D.; Santi, D. V.; Cohen, F. E.; Bartlett, P. Peptoids: a modular approach to drug discovery. A. *Proc. Natl. Acad. Sci. U.S.A.* **1992**, *89*, 9367–9371. (b) Zuckermann, R. N.; Kerr, J. M.; Kent, S. B. H.; Moos, W. H. Efficient method for the preparation of peptoids [oligo(N-substituted glycines)] by submonomer solid-phase synthesis. *J. Am. Chem. Soc.* **1992**, *114*, 10646–10647.
311. (a) Ripka, A. S.; Rich, D. H. Peptidomimetic design. *Curr. Opin. Chem. Biol.* **1998**, *2*, 441–452. (b) Estiarte, M. A.; Rich, D. H. In *Burger's Medicinal Chemistry and Drug Discovery*, 6th edition; Abraham, D., Ed.; John Wiley: New York, 2002, Volume I.
312. (a) Ren, S.; Lien, E. J. Development of HIV protease inhibitors. A survey. *Prog. Drug Res.* **1998**, *51*, 1–31. (b) Tomasselli, A. G.; Heinrikson, R. L Targeting the HIB-protease in AIDS therapy: a current clinical perspective. *Biochim. Biophys. Acta* **2000**, *1477*, 189–214. (c) Wensing, A. M. J.; van Maarseveen, N. M.; Nijhuis, M. Fifteen years of HIV protease inhibitors: raising the barrier to resistance. *Antiviral Res.* **2010**, *85*, 59–74.
313. Sakurai, M.; Higashida, S.; Sugano, M.; Handa, H.; Komai, T.; Yagi, R.; Nishigaki, T.; Yabe, Y. Studies of human immunodeficiency virus type 1 (HIV-1) protease inhibitors. iii. structure-activity relationship of HIV-1 protease inhibitors containing cyclohexylalanylalanine hydroxyethylene depeptide isostere. *Chem. Pharm. Bull.* **1994**, *42*, 534.
314. Vlieghe, P.; Lisowski, V.; Martinez, J.; Khrestchatisky. M. Synthetic therapeutic peptides: science and market. *Drug Discov. Today* **2010**, *15*, 40–56.
315. Wiley, R. A.; Rich, D. H. Peptidomimetics derived from natural products. *Med. Res. Rev.* **1993**, *13*, 327–384.
316. Meanwell, N. A. Improving drug candidates by design: a focus on physicochemical properties as a means of improving compound disposition and safety. *Chem. Res. Toxicol.* **2011**, *24*, 1420–1456.
317. Cole, M. J.; Janiszewski, J. S.; Fouda, H. G. In *Practical Spectroscopy*; Pramanik, B. N.; Ganguly, A. K.; Gross, M. L., Eds.; Marcel Dekker: New York; 2002; vol. 32, pp 211–249.
318. (a) Wessel, M. D.; Mente, S. ADME by computer. *Annu. Rep. Med. Chem.* **2001**, 36, 257–266. (b) Blake, J. F. Chemoinformatics – predicting the physicochemical properties of "drug-like" molecules. *Curr. Opin. Biotechnol.* **2000**, *11*, 104–107. (c) Clark, D. E.; Pickett, S. D. Computational methods for the prediction of "drug-likeness".

Drug Discov. Today **2000**, *5*, 49–58. (d) Kramer, S. D. Absorption prediction from physicochemical parameters. *Pharm. Sci. Technol. Today* **1999**, *2*, 373–380. (e) Norinder, U. Calculated molecular properties and multivariate statistical analysis. In *Methods and Principles in Medicinal Chemistry*; Wiley-VCH: Weinheim, 2009, Vol. 40, pp 375–408.

319. Mandagere, A.; Thompson, T. N.; Hwang, K.-K. Graphical model for estimating oral bioavailability of drugs in humans and other species from their caco-2 permeability and in vitro liver enzyme metabolic stability rates. *J. Med. Chem.* **2002**, *45*, 304–311.
320. Gorswant, C. V.; Thoren, P.; Engstrom, S. Triglyceride-based microemulsion for intravenous administration of sparingly soluble substances. *J. Pharm. Sci.* **1998**, *87*, 200–208.
321. (a) Yee, S. Experimental challenge of beef calves vaccinated intranasally with bovine myxovirus parainfluenza-3 vaccine. *Pharm. Res.* **1997**, *6*, 763–766. (b) Sugano, K.; Cucurull-Sanchez, L.; Bennett, J. Membrane permeability—measurement and prediction in drug discovery. In *Methods and Principles in Medicinal Chemistry*; Wiley-VCH: Weinheim, 2009, Vol. *43*, 117–143.
322. Smith, D. A.; Jones, B. C.; Walker, D. K. Design of drugs involving the concepts and theories of drug metabolism and pharmacokinetics. *Med. Res. Rev.* **1996**, *16*, 243–266.
323. (a) Hansch, C.; Maloney, P. P.; Fujita, T.; Muir, R. M. Correlation of biological activity of phenoxyacetic acids with Hammett substituent constants and partition coefficients. *Nature* **1962**, *194*, 178–180. (b) Fujita, T.; Iwasa, J.; Hansch, C. A new substituent constant, π, derived from partition coefficients. *J. Am. Chem. Soc.* **1964**, *86*, 5175–5180.
324. Hitchcock, S. A.; Pennington, L. D. Structure-brain exposure relationships. *J. Med. Chem.* **2006**, *49*, 7559–7583.
325. Singer, S. J.; Nicolson, G. L. Fluid mosaic model for the structure of cell membranes. *Science* **1972**, *175*, 720–731.
326. Richet, M. C. Sur le rapport entre la toxicité et les propriétés physiques des corps. *Compt. Rend. Soc. Biol.* **1893**, *45*, 775–776.
327. Overton, E. Z. About the osmotic characteristics of the cell in their meaning for the toxicology and the pharmacology (with special consideration of the ammonia and alkaloid). [machine translation]. *Phys. Chem.* **1897**, *22*, 189–209.
328. Meyer, H. Theory of narcosis. I. *Arch. Exp. Pathol. Pharmacol.* **1899**, *42*, 109–118.
329. Hansch, C.; Maloney, P. P.; Fujita, T.; Muir, R. M. Correlation of biological activity of the phenoxyacetic acids with Hammett substituent constants and partition coefficients. *Nature* **1962**, *194*, 178–180.
330. Fujita, T.; Iwasa, J.; Hansch, C. A new substituent constant, n, derive from partition coefficients. *J. Am. Chem. Soc.* **1964**, *86*, 5175–5180.
331. Sangster, J. *Octanol-Water Partition Coefficients*: Fundamentals and Physical Chemistry; Wiley: New York, 1997; pp 79–112.
332. (a) Collander, R. The permeability of Nitella cells to nonelectrolytes. *Physiol. Plant* **1954**, *7*, 420–445. (b) Collander, R. Partition of organic compounds between higher alcohols and water. *Acta Chem. Scand.* **1951**, *5*, 774–780.
333. Hansch, C.; Steward, A. R.; Anderson, S. M.; Bentley, D. Parabolic dependence of drug action upon lipophilic character as revealed by a study of hypnotics. *J. Med. Chem.* **1968**, *11*, 1–11.
334. Hansch, C.; Steward, A. R.; Anderson, S. M.; Bentley, D. Parabolic dependence of drug action upon lipophilic character as revealed by a study of hypnotics. *J. Med. Chem.* **1968**, *11*, 1–11.
335. (a) Scherrer, R. A.; Howard, S. M. Use of distribution coefficients in quantitative structure-activity relations. *J. Med. Chem.* **1977**, *20*, 53–58. (b) Stopher D.; McClean, S. An improved method for the determination of distribution coefficients. *J. Pharm. Pharmacol.* **1990**, *42*, 144–146. (c) Scherrer, R. A.; Leo, A. J. Multi-pH QSAR: a method to differentiate the activity of neutral and ionized species and obtain true correlations when both species are involved. *Mol. Inf.* **2010**, *29*, 687–693.
336. (a) Hansch, C.; Maloney, P. P.; Fujita, T.; Muir, R. M. Correlation of biological activity of phenoxyacetic acids with Hammett substituent constants and partition coefficients. *Nature* **1962**, *194*, 178–180. (b) Hansch, C.; Fujita, T. ρ- σ-π Analysis; a method for the correlation of biological activity and chemical structure. *J. Am. Chem. Soc.* **1964**, *86*, 1616–1626. (c) Fujita, T.; Iwasa, J.; Hansch, C. A new substituent constant, n, derived from partition coefficients. *J. Am. Chem. Soc.* **1964**, *86*, 5175–5180.
337. Hansch, C.; Leo, A. *Substituent Constants for Correlation Analysis in Chemistry and Biology*; Wiley: New York, 1979.
338. Leo, A.; Hansch, C.; Elkins, D. Partition coefficients and their uses. *Chem. Rev.* **1971**, *71*, 525–616.
339. Iwasa, J.; Fujita, T.; Hansch, C. Substituent constants for aliphatic functions obtained from partition coefficients. *J. Med. Chem.* **1965**, *8*, 150–153.
340. Leo, A.; Hansch, C.; Elkins, D. Partition coefficients and their uses. *Chem. Rev.* **1971**, *71*, 525–616.
341. Sangster, J. *Octanol-Water Partition Coefficients*: Fundamentals and Physical Chemistry; Wiley: New York, 1997; pp. 79–112.
342. Lombardo, F.; Shalaeva, M. Y.; Tupper, K. A.; Gao, F.; Abraham, M. H. ELogD_{oct}: a tool for lipophilicity determination in drug discovery. *J. Med. Chem.* **2000**, *43*, 2922–2928.
343. Bard, B.; Carrupt, P.-A.; Martel, S. Lipophilicity of basic drugs measured by hydrophilic interaction chromatography. *J. Med. Chem.* **2009**, *52*, 3416–3419.
344. Henchoz, Y.; Guillarme, D.; Rudaz, S.; Veuthey, J.-L.; Carrupt, P.-A. High-throughput log P determination by ultraperformance liquid chromatography: a convenient tool for medicinal chemists. *J. Med. Chem.* **2008**, *51*, 396–399.
345. Lombardo, F.; Shalaeva, M. Y.; Tupper, K. A.; Gao, F.; Abraham, M. H. ELogD_{oct}: a tool for lipophilicity determination in drug discovery. 2. Basic and neutral compounds. *J. Med. Chem.* **2001**, *44*, 2490–2497.
346. Alimuddin, M.; Grant, D.; Bulloch, D. ; Lee, N. ; Peacock, M. ; Dahl, R. Determination of log D via automated microfluidic liquid-liquid extraction. *J. Med. Chem.* **2008**, *51*, 5140–5142.
347. (a) Chou, J. T.; Jurs, P. C. Computer-assisted computation of partition coefficients from molecular structures using fragment constants. *J. Chem. Inf. Comput. Sci.* **1979**, *19*, 172–178. (b) Pomona College Medicinal Chemistry Project; see Hansch, C.; Björkroth, J. P.; Leo, A. Hydrophobicity and central nervous system agents: on the principle of minimal hydrophobicity in drug design. *J. Pharm. Sci.* **1987**, *76*, 663–687.
348. (a) Bodor, N.; Gabanyi, Z.; Wong, C.-K. A new method for the estimation of partition coefficient. *J. Am. Chem. Soc.* **1989**, *111*, 3783–3786. (b) Bodor, N.; Gabanyi, Z.; Wong, C.-K. A new method for the estimation of partition coefficient [erratum to document cited in CA110(25) :231045Z]. *J. Am. Chem. Soc.* **1989**, *111*, 8062.
349. Moriguchi, I.; Hirono, S.; Liu, Q.; Nakagome, I.; Matsushita, Y. Simple method of calculating octanol/water partition coefficient. *Chem. Pharm. Bull.* **1992**, *40*, 127–130.
350. (a) Moriguchi, I.; Hirono, S.; Liu, Q.; Nakagome, Y.; Matsushita, Y. Comparison of reliability of log P values for drugs calculated by

several methods. *Chem. Pharm. Bull.* **1994**, *42*, 976–978. (b) Leo, A. J. Critique of recent comparison of log P calculation methods. *Chem. Pharm. Bull.* **1995**, *43*, 512–513.

351. Gobas, F. A. P. C.; Lahittete, J. M.; Garofalo, G.; Shiu, W. Y.; Mackay, D. A novel method for measuring membrane-water partition coefficients of hydrophobic organic chemicals: comparison with 1-octanol-water partitioning. *J. Pharm. Sci.* **1988**, *77*, 265–272.
352. Abraham, M. H.; Lieb, W. R.; Franks, N. P. Role of hydrogen bonding in general anesthesia. *J. Pharm. Sci.* **1991**, *80*, 719–724.
353. Seiler, P. Interconversion of lipophilicites from hydrocarbon/water systems into the octanol/water system. Eur. *J. Med. Chem.* **1974**, *9*, 473–479.
354. Stearn, A.; Stearn, E. The chemical mechanism of bacterial behavior. III. The problem of bacteriostasis. *J. Bacteriol.* **1924**, *9*, 491–509.
355. (a) Albert, A.; Rubbo, S.; Goldacre, R. Correlation of basicity and antiseptic action in an acridine series. *Nature* (London) **1941**, *147*, 332–333. (b) Albert, A.; Rubbo, S.; Goldacre, R.; Davey, M.; Stone, J. The influence of chemical constitution on antibacterial activity. II. A general survey of the acridine series. *Br. J. Exp. Pathol.* **1945**, *26*, 160–192. (c) Albert, A.; Goldacre, R. Action of acridine antibacterials. *Nature* (London) **1948**, *161*, 95. (d) Albert, A. *The Acridines, Their Preparation, Properties, and Uses*, 2nd ed.; Edward Arnold: London, 1966.
356. Albert, A. *Selective Toxicity*, 7th ed.; Chapman and Hall: London, 1985; p 398.
357. Burns, J.; Yuü, T.; Dayton, P.; Gutman, A.; Brodie, B. Biochemical and pharmacological considerations of the phenylbutazone and its analogs. *Ann. N. Y. Acad. Sci.* **1960**, *86*, 253–262.
358. Miller, G.; Doukos, P.; Seydel, J. Sulfonamide structure-activity relation in a cell-free system. Correlation of inhibition of folate synthesis with antibacterial activity and physiochemical parameters. *J. Med. Chem.* **1972**, *15*, 700–706.
359. Simonson, T.; Brooks, C. L. III Charge screening and the dielectric constant of proteins: insights from molecular dynamics. *J. Am. Chem. Soc.* **1996**, *118*, 8452–8458.
360. (a) Parsons, S. M.; Raftery, M. A. Ionization behavior of the catalytic carboxyls of lysozyme. Effects of ionic strength. *Biochemistry* **1972**, *11*, 1623–1629. (b) Parsons, S. M.; Raftery, M. A. Ionization behavior of the catalytic carboxyls of lysozyme. Effects of temperature. *Biochemistry* **1972**, *11*, 1630–1633. (c) Parsons, S. M.; Raftery, M. A. Ionization behavior of the cleft carboxyls in lysozyme-substrate complexes. *Biochemistry* **1972**, *11*, 1633–1638.
361. (a) Wu, Z. R.; Ebrahimian, S.; Zawrotny, M. E.; Thornburg, L. D.; Perez-Alvarado, G. C.; Brothers, P.; Pollack, R. M.; Summers, M. F. Solution structure of 3-oxo-Δ5-steroid isomerase. *Science* **1997**, *276*, 415–418. (b) Cho, H.-S.; Choi, G.; Choi, K. Y.; Oh, B.-H. Crystal structure and enzyme mechanism of Δ5-3-ketosteroid isomerase from Pseudomonas testosteroni. *Biochemistry* **1998**, *37*, 8325–8330.
362. Schmidt, D. E.; Westheimer, F. H. pK of the lysine amino group at the active site of acetoacetate decarboxylase. *Biochemistry* **1971**, *10*, 1249–1253.
363. (a) Ajay; Bemis, G. W.; Murcko, M. A. Designing libraries with CNS activity. *J. Med. Chem.* **1999**, *42*, 4942–4951. (b) Mahar-Doan, K. M. Humphreys, J. E.; Webster, L. O. et al. Passive permeability and P-glycoprotein-mediated efflux differentiate central nervous system (CNS) and non-CNS marketed drugs. *J. Pharmacol. Exp. Ther.* **2002**, *303*, 1029–1037. (c) Hitchcock, S. A.; Pennington, L. D. Structure-brain exposure relationships. *J. Med. Chem.* **2006**, *49*, 7559–7583. (d) Ghose, A. K. ; Herbertz, T. ; Hudkins, R. L. ; Dorsey, B. D. ; Mallamo, J. P. Knowledge-based, central nervous system (CNS) lead selection and lead optimization for CNS drug discovery. *ACS Chem. Neurosci.* **2012**, *3*, 50–68. (e) Di, L. ; Rong, H. ; Feng, B. Demystifying brain penetration in central nervous system drug discovery. *J. Med. Chem.* **2012**, *56*, 2–12.
364. Wager, T. T; Chandrasekaran, R. Y.; Hou, X.; Troutman, M. D.; Verhoest, P. R.; Villalobos, A.; Will, Y. Defining desirable central nervous system drug space through the alignment of molecular properties, in vitro ADME, and safety attributes. *ACS Chem. Neurosci.* **2010**, *1*, 420–434.
365. Ballet, S.; Misicka, A.; Kosson, P.; Lemieux, C.; Chung, N.N.; Schiller, P. W.; Lipkowski, A. W.; Tourwe, D. Blood-brain barrier penetration by two dermorphin tetrapeptide analogues: role of lipophilicity vs structural flexibility. *J. Med. Chem.* **2008**, *51*, 2571–2574.
366. (a) Hughes, J. D.; Blagg, J.; Price, D. A.; Bailey, S.; DeCrescenzo, G. A.; Devraj, R. V.; Ellsworth, E.; Fobian, Y. M.; Gibbs, M. E.; Gilles, R.W.; Greene, N.; Huang, E.; Krieger-Burke, R.; Loesel, J.; Wager, T.; Whiteley, L.; Zhang, Y. Physicochemical drug properties associated with in vivo toxicological outcomes. *Bioorg. Med. Chem. Lett.* **2008**, *18*, 4872–4875. (b) Price, D. A. ; Blagg, J. ; Jones, L.; Greene, N.; Wager, T. Physicochemical drug properties associated with in vivo toxicological outcomes: a review. *Exp. Opin. Drug Metab. Toxicol.* **2009**, *5*, 921–931.
367. Ertl, P.; Rohde, B.; Selzer, P. Fast calculation of molecular polar surface area as a sum of fragment-based contributions and its application to the prediction of drug transport properties. *J. Med. Chem.* **2000**, *43*, 3714–3717.
368. (a) Crum-Brown, A.; Fraser, T. R. On the connection between chemical constitution and physiological action. Part I. On the physiological action of the salts of the ammonium bases, derived from strychnine, brucia, thebaia, codeia, morphia, and nicotia. Part II. On the physiological action of the ammonium bases derived from atropia and conia. *Trans. Roy. Soc. Edinburgh* **1868–1869**, *25*, 151 and 693. (b) Crum-Brown, A.; Fraser, T. R. On the connection between chemical constitution and physiological action: on the physiological action of the salts of ammonia, of tri-methylamine, and of tetra-methyl-ammonium ; of the salts of tropia, and of the ammonium bases derived from it; and of tropic atropic, and isotropic acids and their salts. With further details of the physiological action of the salts of methyl-strychnium and of ethyl-strychnium. *Proc. Roy. Soc. Edinburgh* **1869**, *6*, 556–561. (c) Crum-Brown, A.; Fraser, T. R. On the connection between chemical constitution and physiological action-continued. On the physiological action of the salts of trimethylsulphin. *Proc. Roy. Soc. Edinburgh* **1872**, *7*, 663–665.
369. Hansch, C.; Maloney, P. P.; Fujita, T.; Muir, R. M. Correlation of biological activity of phenoxyacetic acids with Hammett substituent constants and partition coefficients. *Nature* **1962**, *194*, 178–180.
370. Richet, M. C. Sur le rapport entre la toxicité et les propriétés physiques des corps. *Compt. Rend. Soc. Biol.* **1893**, *45*, 775–776.
371. Overton, E. About the osmotic characteristics of the cell in their meaning for the toxicology and the pharmacology (with special consideration of the ammonia and alkaloid). [machine translation] Z. *Phys. Chem.* **1897**, *22*, 189–209.
372. Meyer, H. Theory of narcosis. I. *Arch. Exp. Pathol. Pharmacol.* **1899**, *42*, 109–118.
373. Ferguson, J. Use of chemical potentials as indexes of toxicity. *Proc. Roy. Soc. London, Ser. B* **1939**, *127*, 387–404.
374. Hansch, C.; Muir, R. M.; Metzenberg, R. L., Jr. Further evidence for a chemical reaction between plant growth-regulators and a plant substrate. *Plant Physiol.* **1951**, *26*, 812–821.

375. (a) Hansch, C.; Maloney, P. P.; Fujita, T.; Muir, R. M. Correlation of biological activity of phenoxyacetic acids with Hammett substituent constants and partition coefficients. *Nature* **1962**, *194*, 178–180. (b) Hansch, C.; Fujita, T. ϱ-σ-nAnalysis; method for the correlation of biological activity and chemical structure. *J. Am. Chem. Soc.* **1964**, *86*, 1616–1626.
376. Taft, R. W. In *Steric Effects in Organic Chemistry*; Neuman, M. S. Ed.; Wiley: New York, 1956; pp 556–675.
377. Unger, S. H.; Hansch, C. Quantitative models of steric effects. *Prog. Phys. Org. Chem.* **1976**, *12*, 91–118.
378. Hancock, C. K.; Meyers, E. A.; Yager, B. J. Quantitative separation of hyperconjugation effects from stearic substituent constants. *J. Am. Chem. Soc.* **1961**, *83*, 4211–4213.
379. Hansch, C.; Leo, A.; Unger, S. H.; Kim, K. H.; Nikaitani, D.; Lien, E. J. Aromatic substituent constants for structure-activity correlations. *J. Med. Chem.* **1973**, *16*, 1207–1216.
380. (a) Verloop, A.; Hoogenstraaten, W.; Tipker, J. Development and application of new steric substituent parameters in drug design. In *Drug Design*; Ariens, E. J., Ed.; Academic Press: New York, 1976; Vol. VII, pp 165–207. (b) Draber, W. Sterimol and its role in drug research. *Z. Naturforsch., C: Biosci.* **1996**, *51*, 1–7.
381. Hansch, C.; Fujita, T. ϱ-σ-π Analysis; method for the correlation of biological activity and chemical structure. *J. Am. Chem. Soc.* **1964**, *86*, 1616–1626.
382. (a) Daniel, C.; Wood, F. S. *Fitting Equations to Data*; Wiley: New York, 1971. (b) Draper, N. R.; Smith, H. *Applied Regression Analysis*; Wiley: New York, 1966. (c) Snedecor, G. W.; Cochran, W. G. *Statistical Methods*; Iowa State University Press: Ames, 1967.
383. Deardon, J. C. In *Trends in Medicinal Chemistry*; Mutschler, E.; Winterfeldt, E., Eds.; VCH: Weinheim, 1987; pp 109–123.
384. Topliss, J. G. Utilization of operational schemes for analog synthesis in drug design. *J. Med. Chem.* **1972**, *15*, 1006–1011.
385. Granito, C. E.; Becker, G. T.; Roberts, S.; Wiswesser, W. J.; Windlinz, K. J. Computer-generated substructure codes (bit screens) *J. Chem. Doc.* **1971**, *11*, 106–110.
386. Goodford, P. J. Prediction of pharmacological activity b the method of physicochemical-activity relations. *Adv. Pharmacol. Chemother.* **1973**, *11*, 51–97.
387. (a) Craig, P. N. Interdependence between physical parameters and selection of substituent groups for correlation studies. *J. Med. Chem.* **1971**, *14*, 680–684. (b) Craig, P. N. Guidelines for drug and analog design. In *Burger's Medicinal Chemistry*, 4th ed.; Wolff, M. E., Ed.; Wiley: New York, 1980; Part I, Chapter 8.
388. (a) Bustard, T. M. Optimization of alkyl modifications by Fibonacci search. *J. Med. Chem.* **1974**, *17*, 777–778. (b) Santora, N. J.; Auyang, K. Noncomputer approach to structure-activity study. Expanded Fibonacci search applied to structurally diverse types of compounds. *J. Med. Chem.* **1975**, *18*, 959–963. (c) Deming, S. N. On the use of Fibonacci searches in structure-activity studies. *J. Med. Chem.* **1976**, *19*, 977–978.
389. Darvas, F. Application of the sequential simplex method in designing drug analogs. *J. Med. Chem.* **1974**, *17*, 799–804.
390. Topliss, J. G. A manual method for applying the Hansch approach to drug design. *J. Med. Chem.* **1977**, *20*, 463–469.
391. Hansch, C.; Unger, S. H.; Forsythe, A. B. Strategy in drug design. Cluster analysis as an aid in the selection of substituents. *J. Med. Chem.* **1973**, *16*, 1217–1222.
392. Martin, Y. C. In *Drug Design; Ariens*, E. J., Ed.; Academic Press: New York, 1979; Vol. VIII, p. 5.
393. Kawakami, Y.; Inoue, A.; Kawai, T.; Wakita, M.; Sugimoto, H.; Hopfinger, A. J. The rationale for E2020 as a potent acetylcholinesterase inhibitor. *Bioorg. Med. Chem.* **1996**, *4*, 1429–1446.
394. Topliss, J. G.; Costello, R. J. Chance correlations in structure-activity studies using multiple regression analysis. *J. Med. Chem.* **1972**, *15*, 1066–1068.
395. (a) Lindberg, W.; Persson, J-Å.; Wold, S. Partial least-squares method for spectrofluorimetric analysis of mixtures of humic acid and ligninsulfonate. *Anal. Chem.* **1983**, *55*, 643–648. (b) Cramer, R. D. III; Bunce, J. D. ; Patterson, D. E.; Frank, I. E. Cross-validation, bootstrapping, and partial least squares compared with multiple regression in conventional QSAR studies. *Mol. Informatics* **2006**, *7*, 18–25.
396. Clark, M.; Cramer, R. D. III. The probability of chance correlation using partial least square (PLS). *Quant. Struct. Act. Rel.* **1993**, *12*, 137–145.
397. (a) Free, S. M., Jr.; Wilson, J. W. A mathematical contribution to structure-activity studies. *J. Med. Chem.* **1964**, *7*, 395–399. (b) Tomic, S.; Nilsson, L.; Wade, R. C. Nuclear receptor-DNA binding specificity: a COMBINE and Free-Wilson QSAR analysis. *J. Med. Chem.* **2000**, *43*, 1780–1792.
398. (a) Blankley, C. J. In *Quantitative Structure-Activity Relationships of Drugs*; Topliss, J. G., Ed.; Academic Press: New York, 1983; Chapter 1. (b) Schaad, L. J.; Hess, B. A., Jr.; Purcell, W. P.; Cammarata, A.; Franke, R.; Kubinyi, H. Compatibility of the Free-Wilson and Hansch quantitative structure-activity relations. *J. Med. Chem.* **1981**, *24*, 900–901.
399. Fujita, T.; Ban, T. Structure-activity relation. 3. Structure-activity study of phenethylamines as substrates of biosynthetic enzymes of sympathetic transmitters. *J. Med. Chem.* **1971**, *14*, 148–152.
400. (a) Martin, Y. C. *Quantitative Drug Design: A Critical Introduction; Marcel Dekker*: New York, 1978; Chapter 2. (b) Tute, M. S. In *Physical Chemical Properties of Drugs; Yalkowsky*, S. H.; Sinkula, A. A.; Valvani, S. C., Eds.; Marcel Dekker: New York, 1980; p 141.
401. (a) Boehm, H.-J.; Flohr, A.; Stahl, M. Scaffold hopping. *Drug Discov. Today Technol.* **2004**, *1*, 217–224. (b) Zhao, H. Scaffold selection and scaffold hopping in lead generation: a medicinal chemistry perspective. *Drug Discov. Today* **2007**, *12*, 149–155. (c) Langdon, S. R.; Ertl, P.; Brown, N. Bioisosteric replacement and scaffold hopping in lead generation and optimization. *Mol. Inf.* **2010**, *29*, 366–385. (d) Jenkins, J. L.; Glick, M.; Davies, J. W. A 3D similarity method for scaffold hopping from known drugs or natural ligands to new chemotypes. *J. Med. Chem.* **2004**, *47*, 6144–6159.
402. Istvan, E. S.; Deisenhofer, J. Structural mechanism for statin inhibition of HMG-CoA reductase. *Science* **2001**, *292*, 1160–1164.
403. Kasinbhatla, S. R.; Hong, K. et al, Rationally designed high-affinity 2-amino-6-halopurine heat shock protein 90 inhibitors that exhibit potent antitumor activity *J. Med. Chem.* **2007**, *50*, 2767–2778.
404. Hassall, C. H. Computer graphics as an aid to drug design. *Chem. Brit.* **1985**, *21*(1), 39 41–43, 45–46.
405. Levinthal, C. Molecular model-building by computer. *Sci. Am.* **1966**, *214*, 42–52.
406. (a) Barry, C. D.; North, A. C. T. Use of a computer-controlled display system in the study of molecular conformations. *Cold Spring Harbor Quant. Biol.* **1971**, *36*, 577–584. (b) Barry, C. D. Quantitative determination of mononucleotide conformations in solution using lanthanide ion shift and broadening NMR probes. *Nature* (London) **1971**, *232*, 236–245.

407. (a) Kirchmair, J.; Markt, P.; Distinto, S.; Schuster, D.; Spitzer, G. M.; Liedl, K. R.; Langer, T.; Wolber, G. The Protein Data Bank (PDB), its related services and software tools as key components for in silico guided drug discovery. *J. Med. Chem.* **2008**, *51*, 7021–7040. (b) http://www.rcsb.org/pdb.
408. Tollenaere, J. P.; Janssen, P. A. J. Conformational analysis and computer graphics in drug research. *Med. Res. Rev.* **1988**, *8*, 1–25.
409. von Itzstein, M.; Wu, W.-Y.; Kok, G. B.; Pegg, M. S.; Dyason, J. C.; Jin, B.; van Phan, T.; Smythe, M. L.; White, H. F.; Oliver, S. W.; Colman, P. M.; Varghese, J. N.; Ryan, D. M.; Woods, J. M.; Bethell, R. C.; Hotham, V. J.; Cameron, J. M.; Penn, C. R. Rational design of potent sialidase-based inhibitors of influenza virus replication. *Nature* **1993**, *363*, 418–423.
410. (a) Varghese, J. N.; Laver, W. G.; Colman, P. M. Structure of the influenza virus glycoprotein antigen neuraminidase at 2.9 Å resolution. *Nature* **1983**, *303*, 35–40. (b) Colman, P. M.; Varghese, J. N.; Laver, W. G. Structure of the catalytic and antigenic sites in influenza virus neuraminidase. *Nature* **1983**, *303*, 41–44.
411. (a) Varghese, J. N.; Colman, P. M. Three-dimensional structure of the neuraminidase of influenza A/Tokyo/3/67 at 2.2 Å resolution. *J. Mol. Biol.* **1991**, *221*, 473–486. (b) Varghese, J. N.; McKimm-Breschkin, J. Caldwell, J. B.; Kortt, A. A.; Colman, P. M. The structure of the complex between influenza virus neuraminidase and sialic acid, the viral receptor. *Proteins* **1992**, *14*, 327–332.
412. von Itzstein, M.; Wu, W.-Y.; Kok, G. B.; Pegg, M. S.; Dyason, J. C.; Jin, B. et al. Rational design of potent sialidase-based inhibitors of influenza virus replication. *Nature* **1993**, *363*, 418–423.
413. (a) Taubenberger, J. K.; Palese, P. The origin and virulence of the 1918 "Spanish" influenza virus. In *Influenza Virology*, Kawaoka, Y., Ed.; 2006; pp 299–321. (b) Gibbs, M. J.; Armstrong, J. S.; Gibbs, A. J. The haemagglutinin gene, but not the neuraminidase gene, of "Spanish flu" was a recombinant. *Phil. Trans. R. Soc. Lond. B* **2001**, *356*, 1845–1855.
414. Anderson, A.; O'Neill, R.; Surti, T.; Stroud, R. Approaches to solving the rigid receptor problem by identifying a minimal set of flexible residues during ligand docking. *Chem. Biol.* **2001**, *8*, 445–457.
415. Ojima, I.; Chakravarty, S.; Inoue, T.; Lin, S.; He, L.; Horwitz, S. B.; Kuduk, S. D.; Danishefsky, S. J. A common pharmacophore for cytotoxic natural products that stabilize microtubules. *Proc. Natl. Acad. Sci. U.S.A.* **1999**, *96*, 4256–4261.
416. (a) Wang, M.; Xia, X.; Kim, Y.; Hwang, D.; Jansen, J. M.; Botta, M.; Liotta, D. C.; Snyder, J. P. A unified and quantitative receptor model for the microtubule binding of paclitaxel and epothilone. *Org. Lett.* **1999**; *1*; 43–46. (b) Metaferia, B. B.; Hoch, J.; Glass, T. E.; Bane, S. L.; Chatterjee, S. K.; Snyder, J. P.; Lakdawala, A.; Cornett, B.; Kingston, D. G. I. Synthesis and biological evaluation of novel macrocyclic paclitaxel analogues. *Org. Lett.* **2001**, *3*, 2461–2464.
417. Nettles, J. H.; Li, H.; Cornett, B.; Krahn, J. M.; Snyder, J. P.; Downing, K. H. The binding mode of epothilone A on a, β-tubulin by electron crystallography. *Science* **2004**, *305*, 866–869.
418. Wender, P. A.; Koehler, K. F.; Konrad, F.; Sharkey, N.A.; Dell'Aqila, M. L.; Blumberg, P. M. Analysis of the phorbol ester pharmacophore on protein kinase C as a guide to the rational design of new classes of analogs. *Proc. Natl. Acad. Sci. U.S.A.* **1986**, *83*, 4214–4218.
419. Blumberg, P. M. Protein kinase C as the receptor for the phorbol ester tumor promoters: Sixth Rhoads Memorial Award Lecture. *Cancer Res.* **1988**, *48*, 1–8.
420. Wender, P. A.;Cribbs, C. M.; Koehler, K. F.; Sharkey, N. A.; Herald, C. L.; Kamano, Y.; Pettit, G. ; Blumberg, P. M. Modeling of the bryostatins to the phorbol ester pharmacophore on protein kinase C. *Proc. Natl. Acad. Sci. U.S.A.* **1988**, *85*, 7197–7201.
421. Wender, P. A. De Brabander, J.; Harran P. G.; Jimenez, J-M.; Koehler, M. F. T.; Lippa, B.; Park, C-M.; Shiozaki, M. Synthesis of the first members of a new class of biologically active bryostatin analogs. *J. Am. Chem. Soc.* **1998**, *120*, 4534–4535.
422. Wender, P. A.; Baryza, J. L.; Bennett, C. E. Bi, F. C.; Brenner, S. E.; Clarke, M. O.; Horan, J. C.; Kan, C.; Lacote, E.; Lippa B.; et al. The practical synthesis of a novel and highly potent analogue of bryostatin. *J. Am. Chem. Soc.* **2002**, *124*, 13648–13649.
423. (a) Alkon, D. L.; Sun, M. K.; Nelson, T. J. PKC signaling deficits: a mechanistic hypothesis for the origins of Alzheimer's disease. *Trends Pharmacol. Sci.* **2007**, *28*, 51–60. (b) Choi, D. S. ; Wang, D. Yu, GQ.; Zhu, G.; Kharazia, V. N.; Paredes, J. P.; Chang, W. S.; Deitchman, J. K.; Mucke, L.; Messing, R. O. PKCε increases endothelin converting enzyme activity and reduces amyloid plaque pathology in transgenic mice. *Proc. Natl. Acad. Sci. U.S.A.* **2006**, *103*, 8215–8220.
424. Khan, T. K.; Nelson, T. J.; Verma, V. A.; Wender, P. A.; Alkon, D. L. A cellular model of Alzheimer's disease therapeutic efficacy: PKC activation reverses Aβ-induced biomarker abnormality on cultured fibroblasts. *Neurobiol. Dis.* **2009**, *34*, 332–339.
425. (a) DeChristopher, B. A.; Loy, B. A.; Marsden, M. D.; Schrier, A.; Zack, J. A.; Wender, P. A. Designed, synthetically accessible bryostatin analogues potently induce activation of latent HIV reservoirs in vitro. *Nat. Chem.* **2012**, *4*, 705–710. (b) Beans, E. J.; Fournogerakis, D.; Gauntlett, C.; Heumann, L. V.; Kramer, R.; Marsden, M. D.; Murray, D.; Chun, T.-W.; Zack, J. A.; Wender, P. A. Highly potent, synthetically accessible prostatin analogs induce latent HIV expression in vitro and ex vivo. *Proc. Natl. Acad. Sci. U.S.A.* **2013**, *110*, 11698–11703.
426. Scannell, J. W.; Blanckley, A.; Boldon, H.; Warrington, B. Diagnosing the decline in pharmaceutical R&D efficiency. *Nat. Rev. Drug Discov.* **2012**, *11*, 191–200.
427. Ooms, F. Molecular modeling and computer aided drug design. Examples of their applications in medicinal chemistry. *Curr. Med. Chem.* **2000**, *7*, 141–158.
428. Paul, S. M.; Mytelka, D. S.; Dunwiddie, C. T.; Persinger, C. C.; Munos, B. H.; Lindborg, S. R.; Schacht, A. L. How to improve R&D productivity: the pharmaceutical industry's grand challenge. *Nat. Rev. Drug Disc.* **2010**, *9*, 203–214.
429. Nies, A. S.; Spielberg, S. P. In *Goodman & Gilman's The Pharmacological Basis of Therapeutics*; 9th edition; Hardman, J. G.; Limbird, L. E.; Molinoff, P. B.; Ruddon, R. W.; Gilman, A. G., Eds.; McGraw Hill: New York, 1996; p 43.
430. IMS Institute for Healthcare Informatics. Declining medicine use and costs: For better or worse? A review of the use of medicines in the United States in 2012. http://www.imshealth.com/cds/ims/Global/Content/Insights/IMS%20Institute%20for%20Healthcare%20Informatics/2012%20U.S.%20Medicines%20Report/2012_U.S.Medicines_Report.pdf
431. Drews, J. In *Human Disease-from Genetic Causes to Biochemical Effects*, Drews, J.; Ryser, St., Eds.; Blackwell: Berlin, 1997; pp 5–9.
432. (a) Wang, J. H.; Hewick, R. M. Proteomics in drug discovery. *Drug Discov. Today* **1999**, *4*, 129–133. (b) Borman, S. Proteomics: taking over where the genomics leaves off. *Chem. Engg. News* **2000**, (July 31), *78*, 31–37.
433. Silberman, S. Placebos are getting more effective. Drugmakers are desperate to know why. *Wired Magazine* **2009**, Aug. 24.
434. Arrowsmith, J. Trial watch Phase II failures: 2008–2010. *Nat. Rev. Drug Disc.* **2011**, *10*, 328–329.
435. Arrowsmith, J. Trial watch Phase III and submission failures: 2007–2010. *Nat. Rev. Drug Disc.* **2011**, *10*, 87.

Chapter 3

Receptors

Chapter Outline

3.1. INTRODUCTION

Up to this point in our discussion it appears that a drug is taken, it travels through the body to a target site, and it elicits a pharmacological effect. The site of drug action, which is ultimately responsible for the pharmaceutical effect, is a *receptor*, any biological molecule with which the drug interacts. Allusions were made in Chapter 2 to the binding of a drug to a receptor, which constitutes *pharmacodynamics*. In this chapter, the emphasis will be on pharmacodynamics of general receptors; in Chapter 4, a special class of proteins that have catalytic properties, called enzymes, will be discussed; and in Chapter 6, a nonprotein receptor, deoxyribonucleic acid (*DNA*), will be the topic of discussion. The drug–receptor properties discussed in this chapter will also apply to drug–enzyme and drug–DNA complexes. The receptors discussed in this chapter include some of the major drug targets, such as *guanine nucleotide-binding regulatory protein* (*G protein*)*-coupled receptors* (*GPCRs*), ion channels, nuclear receptors, and receptor tyrosine kinases (RTKs). GPCRs are the largest class of receptors known; about 800 different human genes (~4% of the human genome) are predicted to be members of the GPCR superfamily.[1] Over 80% of hormones use GPCRs for signaling. These seven-transmembrane proteins are activated by a variety of ligands such as peptides, hormones, neurotransmitters, chemokines, lipids, glycoproteins, divalent cations, and light.[2] Binding of these ligands causes a conformational change in the structure of these cell-surface receptors to facilitate interaction of the receptor with a member of the G protein family. G protein activation by the receptor results in the activation of intracellular signal transduction cascades, which leads to a change in the activity of ion channels and enzymes, thereby causing an alteration in the rate of production of intracellular second messengers.[3] Therefore, the GPCRs are involved in the control of every aspect of our behavior and physiology and are linked to numerous diseases, including cardiovascular problems, mental disorders, retinal degeneration, cancer, and acquired immunodeficiency syndrome. Almost half of all drugs target GPCRs by either activating or inactivating them. An *ion channel* is a transmembrane pore that is composed of

The Organic Chemistry of Drug Design and Drug Action. http://dx.doi.org/10.1016/B978-0-12-382030-3.00003-9

the following elements: a *pore*, which is responsible for the transit of the ion, and one or more *gates* that open and close in response to specific stimuli that are received by the *sensors*. Conformational mobility is an integral component of the function of ion channels; the three states of a channel, closed, open, and inactivated, are all believed to be regulated by conformational changes. Ligands can gain access to the channel either by membrane permeation or through an open channel state. *Nuclear receptors* are ligand-dependent transcription factors responsible for sensing steroid and thyroid hormones, bile acids, fatty acids, and certain vitamins and prostaglandins.[4] In response to ligand binding, these protein receptors work with other proteins to regulate gene expression, thereby controlling the development, differentiation, metabolism, and reproduction of the organism. Ligand binding to a nuclear receptor results in a conformational change in the receptor, which activates the receptor, resulting in up- or downregulation of gene expression. RTKs are a subclass of cell-surface growth factor receptors having a ligand-dependent enzymatic activity (kinase activity) of catalyzing the transfer of the γ-phosphate group from a nucleoside triphosphate donor, such as adenosine triphosphate, to hydroxyl groups of tyrosine residues of target proteins.[5] The ligand-binding domain, which is usually glycosylated, is connected to the cytoplasmic domain by a single transmembrane helix. RTKs play an important role in the control of most fundamental cellular processes, including the cell cycle, cell migration, cell metabolism and survival, as well as cell proliferation and differentiation, and, therefore, also play a crucial role in carcinogenesis. RTKs function in many signal transduction cascades by which extracellular signals are transmitted through the cell membrane (transmembrane) to the cytoplasm and often to the nucleus, where gene expression may be modified.

In 1878 John N. Langley, a physiology graduate student at Cambridge University, while studying the mutually antagonistic action of the alkaloids atropine (**3.1**; now used as a smooth muscle relaxant in a variety of drugs, such as Prosed) and pilocarpine (**3.2**; Salagen; causes sweating and salivation) on cat salivary flow, suggested that both of these chemicals interacted with some yet unknown substance (no mention of "receptors" was made) in the nerve endings of the gland cells.[6] Langley, however, did not follow-up this notion for over 25 years.

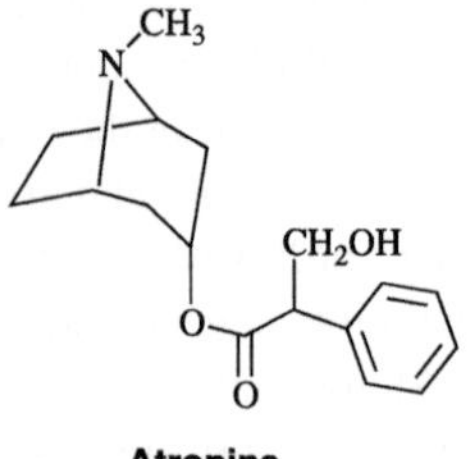

Atropine
3.1

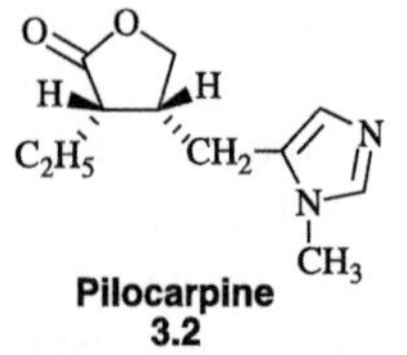

Pilocarpine
3.2

Paul Ehrlich worked for a dye manufacturing company and was fascinated by the observation that dyes could attach so tightly to fabrics that they could not be removed by washing. He also was intrigued by why different bacteria caused different diseases and thought that the toxins generated by bacteria might produce their effects by attaching tightly to specific sites in the cells of the body, just as dyes attach to fabrics. In 1897, Ehrlich suggested his *side chain theory*.[7] According to this hypothesis, cells have side chains attached to them that contain specific groups capable of combining with a particular group of a toxin. Ehrlich termed these side chains *receptors*. Another ground-breaking facet of this hypothesis was that when toxins combined with the side chains, excess side chains are produced and released into the bloodstream. In today's biochemical vernacular, these excess side chains would be called *antibodies*, and they can combine with macromolecular toxins stoichiometrically.

In 1905, Langley[8] (at this time Chair of the Department of Physiology at Cambridge, where he did his graduate studies, and editor of the *Journal of Physiology*, where he published his work on cat salivary flow as a graduate student) studied the antagonistic effects of curare (see Figure 2.16) on nicotine stimulation of skeletal muscle. He concluded that there was a *receptive substance* that received the stimulus and, by transmitting it, caused muscle contraction. This was the first time that attention was drawn to the two fundamental characteristics of a receptor, namely, a *recognition capacity* for specific molecules and an *amplification component*, the ability of the complex between the molecule and the receptor to initiate a biological response.

Receptors are mostly membrane-bound proteins that selectively bind small molecules (*ligands*) and elicit a physiological response. Many receptors are integral proteins that are embedded in the phospholipid bilayer of cell membranes (see Figure 2.26). Since such receptors typically function in the membrane environment, their properties and mechanisms of action depend on the phospholipid milieu. Vigorous treatment of cells with detergents is required to dissociate these proteins from the membrane. Once they become dissociated, however, they generally lose their integrity. Because they usually exist in minute quantities and can be unstable, few membrane-bound receptors have been purified and little structural information is known about most of them. Advances in molecular biology have permitted the isolation, cloning, and sequencing of receptors,[9] and this is leading to further approaches to the molecular characterization of these proteins. However, these receptors, unlike many enzymes, are still typically characterized in terms of their function rather than by their structural properties. The two functional components of receptors, the recognition component and the amplification component, may represent the same or different sites on the same protein. Various hypotheses regarding the mechanisms by which drugs may initiate a biological response are discussed in Section 3.2.4.

3.2. DRUG–RECEPTOR INTERACTIONS

3.2.1. General Considerations

To appreciate the mechanisms of drug action, it is important to understand the forces of interaction that bind drugs to their receptors. Because of the low concentration of drugs and receptors in the bloodstream and other biological fluids, the law of mass action alone cannot account for the ability of small doses of structurally specific drugs to elicit a total response by combination with all, or practically all, of the appropriate receptors. The enlightening calculation shown below supports the notion that something more than mass action is required to get the desired drug–receptor interaction.[10] One mole of a drug contains 6.02×10^{23} molecules (Avogadro's number). If the molecular weight of an average drug is 300 g/mol, then 15 mg (an effective dose for many drugs) will contain $6.02 \times 10^{23}(15 \times 10^{-3})/300 = 3 \times 10^{19}$ molecules of drug. The human organism is composed of about 3×10^{13} cells. Therefore, each cell will be acted upon by $3 \times 10^{19}/3 \times 10^{13} = 10^{6}$ drug molecules. One erythrocyte cell contains about 10^{10} molecules. On the assumption that the same number of molecules is found uniformly in all cells, then for each drug molecule, there are $10^{10}/10^{6} = 10^{4}$ molecules of the human body! With this ratio of human molecules to drug molecules, Le Chatelier would have a difficult time explaining how the drug could interact and form a stable complex with the desired receptor.

The driving force for the drug–receptor interaction can be considered as a low energy state of the drug–receptor complex (Scheme 3.1), where k_{on} is the rate constant for formation of the drug–receptor complex, which depends on the concentrations of the drug and the receptor, and k_{off} is the rate constant for breakdown of the complex, which depends on the concentration of the drug–receptor complex as well as other forces. The biological activity of a drug is related to its affinity for the receptor, i.e., the stability of the drug–receptor complex. This stability is commonly measured by how difficult it is for the complex to dissociate, which is represented by its K_d, the dissociation constant for the drug–receptor complex at equilibrium (Eqn (3.1)). Note that

$$K_d = \frac{[\text{drug}]\,[\text{receptor}]}{[\text{drug - receptor complex}]} \tag{3.1}$$

because K_d is a *dissociation* constant, the smaller the K_d, the larger the concentration of the drug–receptor complex, the more stable is that complex, and the greater is the affinity of the drug for the receptor. K_d roughly represents the concentration of the drug required to reach an equilibrium of 50% in the drug–receptor complex. To give you an idea of the affinity of a typical drug for its target, it has been estimated that the median K_d for enzyme inhibitor drugs on the market is about 20 nM[11] (at 20 nM concentration of drug, the enzyme is 50% in the drug–enzyme complex). Formation of the drug–receptor complex involves an elaborate equilibrium. Solvated ligands (such as drugs) and solvated proteins (such as receptors) generally exist as an equilibrium mixture of several conformers each. To form a complex, solvent molecules that occupy the binding site of the receptor must be displaced by the drug to produce a solvated complex; interactions between the drug and the receptor are stronger than the interactions between the drug and receptor with the solvent molecules.[12] Drug–receptor complex formation is also entropically unfavorable; it causes a loss in conformational degrees of freedom for both the protein and the ligand, as well as the loss of three rotational and three translational degrees of freedom.[13] Therefore, highly favorable enthalpic contacts (interactions) between the receptor and the drug must compensate for the entropic loss.

3.2.2. Important Interactions (Forces) Involved in the Drug–Receptor Complex

Interactions involved in the drug–receptor complex[14] are the same forces experienced by all interacting organic molecules and include covalent bonding, ionic (electrostatic) interactions, ion–dipole and dipole–dipole interactions, hydrogen bonding, charge-transfer interactions, hydrophobic interactions, cation–π interactions, halogen bonding, and van der Waals interactions. Weak interactions usually are possible only when molecular surfaces are close and complementary, that is, bond strength is distance dependent. The spontaneous formation of a bond between atoms occurs with a decrease in free energy, that is, a noncovalent bond will occur only when there is a negative ΔG, which is the sum of an enthalpic term (ΔH) and an entropic term ($-T\Delta S$). The change in free energy (*binding energy*) is related to the binding equilibrium constant (K_{eq}) according to Eqn (3.2). Therefore, at physiological temperature (37 °C), changes in free energy of a

$$\Delta G^0 = -\text{RT} \ln K_{eq} \tag{3.2}$$

few kilocalories per mole can have a major effect on the establishment of good secondary interactions. In fact, if the K_{eq} were only 0.01 (i.e., 1% of the equilibrium mixture in the form of the drug–receptor complex), then a ΔG^0 of interaction of −5.45 kcal/mol would shift the binding equilibrium

$$\text{drug} + \text{receptor} \underset{k_{off}}{\overset{k_{on}}{\rightleftharpoons}} \text{drug-receptor complex}$$

SCHEME 3.1 Equilibrium between a drug, a receptor, and a drug–receptor complex constant to 100 (i.e., 99% in the form of the drug–receptor complex). It would be desirable for observed interactions to be additive; however, molecular interactions tend to behave in a highly nonadditive fashion.[15] A particular interaction may be worth different amounts of free energy depending on the specific molecular structure involved. The multiplicity of interactions in one protein–ligand complex is a compromise of attractive and repulsive forces; solvation processes, long-range interactions, and conformational changes are often neglected. Also, it is very easy to be misled by drug–receptor interactions in crystal structures, which present static views of interactions and do not take into account the energy cost of displacement of water molecules from the binding site.[16]

Generally, the bonds formed between a drug and a receptor are weak noncovalent interactions; consequently, the effects produced are reversible. Because of this, a drug becomes inactive as soon as its concentration in the extracellular fluids decreases, generally by metabolism (see Chapter 8). Often it is desirable for the drug effect to last only for a limited time so that the pharmacological action can be terminated. In the case of central nervous system (CNS) stimulants and depressants, for example, a prolonged action could be harmful. Sometimes, however, the effect produced by a drug should persist, and even be irreversible. For example, it is most desirable for a *chemotherapeutic agent*, a drug that acts selectively on a foreign organism or tumor cell, to form an irreversible complex with its receptor so that the drug can exert its toxic action for a prolonged period.[17] In this case, a covalent bond would be desirable.

In the following subsections, the various types of drug–receptor interactions are discussed briefly. These interactions are applicable to all types of receptors, including enzymes and DNA, that are described in this book.

3.2.2.1. Covalent Bonds

The *covalent bond* is the strongest bond, generally worth anywhere from −40 to −110 kcal/mol in stability. It is seldom formed by a drug–receptor interaction, except with enzymes and DNA. These bonds will be discussed further in Chapters 5 and 6.

3.2.2.2. Ionic (or Electrostatic) Interactions

For protein receptors at *physiological pH* (generally taken to mean pH 7.4, the pH of blood), basic groups such as the amino side chains of arginine, lysine, and, to a much lesser extent, histidine, are protonated and, therefore, provide a cationic environment. Acidic groups, such as the carboxylic acid side chains of aspartic acid and glutamic acid, are deprotonated to give anionic groups.

Drug and receptor groups will be mutually attracted provided they have opposite charges. This *ionic interaction* can be effective at distances farther than those required for other types of interactions, and they can persist longer. A simple ionic interaction can provide a $\Delta G^0 = -5$ kcal/mol, which declines by the square of the distance between the charges. If this interaction is reinforced by other simultaneous interactions, the ionic interaction becomes stronger ($\Delta G^0 = -10$ kcal/mol) and persists longer. The antidepressant drug pivagabine (Tonerg) is used as an example of a molecule that can hypothetically participate in an ionic interaction with an arginine residue (Figure 3.1).

3.2.2.3. Ion–Dipole and Dipole–Dipole Interactions

As a result of the greater electronegativity of atoms such as oxygen, nitrogen, sulfur, and halogens relative to that of carbon, C–X bonds in drugs and receptors, where X is an electronegative atom, will have an asymmetric distribution of electrons; this produces electronic dipoles. These dipoles in a drug molecule can be attracted by ions (*ion–dipole interaction*) or by other dipoles (*dipole–dipole interaction*) in the receptor, provided charges of opposite sign are properly aligned. Because the charge of a dipole is less than that of an ion, a dipole–dipole interaction is weaker than an ion–dipole interaction. In Figure 3.2, the insomnia drug zaleplon (Sonata) is used to demonstrate these interactions, which can provide a $\Delta G^0 = -1$ to -7 kcal/mol.

3.2.2.4. Hydrogen Bonds

Hydrogen bonds are a type of dipole–dipole interaction formed between the proton of a group X–H, where X is an electronegative atom, and one or more other electronegative atoms (Y) containing a pair of nonbonded electrons. The most significant hydrogen bonds occur in molecules where

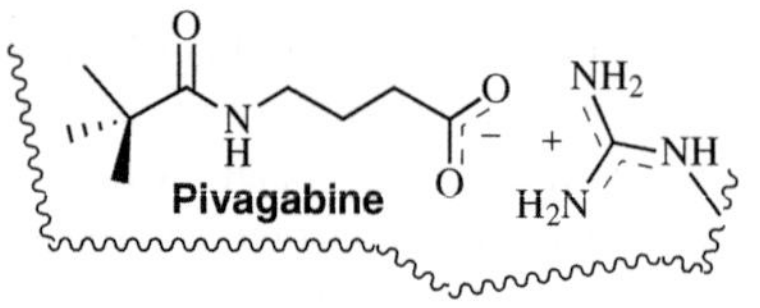

FIGURE 3.1 Example of an electrostatic (ionic) interaction. Wavy line represents the receptor cavity.

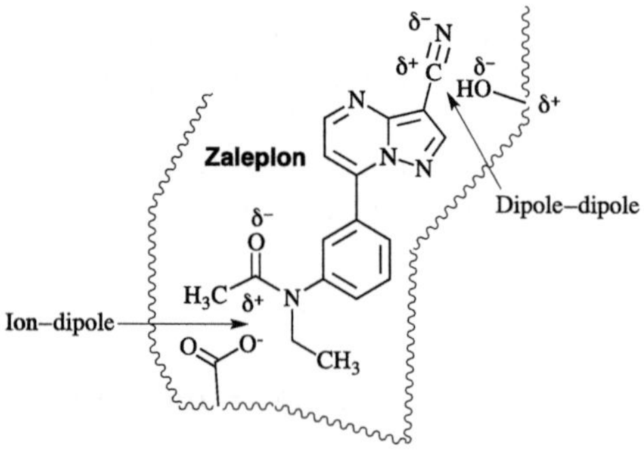

FIGURE 3.2 Examples of ion–dipole and dipole–dipole interactions. Wavy line represents the receptor cavity.

X and Y are N and O and, to a lesser extent, F[18]; interesting special cases of weak hydrogen bonding for X=C also have been described.[19] X removes electron density from the hydrogen so it has a partial positive charge, which is strongly attracted to the nonbonded electrons of Y. The interaction is denoted as a dotted line, –X–H···Y–, to indicate that a covalent bond between X and H still exists, but that an interaction between H and Y also occurs. In this depiction, X is referred to as the *hydrogen bond donor* and Y is the *hydrogen bond acceptor*. When X and Y are equivalent in electronegativity and degree of ionization, the proton can be shared equally between the two groups, i.e. –X···H···Y–, referred to as a *low-barrier hydrogen bond*.[20] On average, the hydrogen bond between a carbonyl oxygen and an alcohol proton is 2.75 Å long and that between a carbonyl and an NH proton is 2.90 Å.[21]

The hydrogen bond is unique to hydrogen because it is the only atom that can carry a positive charge at physiological pH while remaining covalently bonded in a molecule, and which also is small enough to allow close approach of a second electronegative atom. The strength of the hydrogen bond is related to the Hammett σ constants.[22]

There are *intramolecular* and *intermolecular* hydrogen bonds; the former are stronger (see salicylic acid used in wart removal remedies, in Figure 3.3). Intramolecular hydrogen bonding is an important property of molecules that may have a significant effect on lead modification approaches.[23] As discussed in Chapter 2 (Sections 2.1.2.3.4 and 2.2.4.4), the bioactive conformation of a molecule is the optimal conformation of the molecule when bound to its receptor. When there are hydrogen bond donor and acceptor groups in a compound that have the possibility of interacting to form a five- to seven-membered intramolecular ring, those interactions will produce a stable conformation that may or may not approximate the bioactive conformation. The order of stability for intramolecular hydrogen bond rings is six-membered ring>>five-membered ring>seven-membered ring with acceptor strength (carbonyl>heterocyclic N acceptor>sulfoxide>alkoxyl[24]) enhancing the probability of the intramolecular hydrogen bond. Intramolecular hydrogen bonding becomes increasingly important if a bioisosteric replacement of an oxygen atom in an ether (capable of forming strong hydrogen bonds) is replaced by a sulfur atom in a thioether (which forms very weak or no hydrogen bonds); this could have a major impact on the potency and even activity of the compound if an intramolecular hydrogen bond changes the conformation of the molecule. This same difference between oxygen or nitrogen and sulfur also becomes important in intermolecular bonding between the drug molecule and the receptor. Intramolecular hydrogen bonding also may mask the binding of a pharmacophoric group. For example, methyl salicylate (**3.3**, wintergreen oil), an active ingredient in many muscle pain remedies and antiseptics, is a weak antibacterial agent. The corresponding para-isomer, methyl *p*-hydroxybenzoate (**3.4**), however, is considerably more potent as an antibacterial agent and is used as a food preservative. It is believed that the antibacterial activity of **3.4** is derived from the phenolic hydroxyl group. In **3.3**, this group is masked by intramolecular hydrogen bonding.[25]

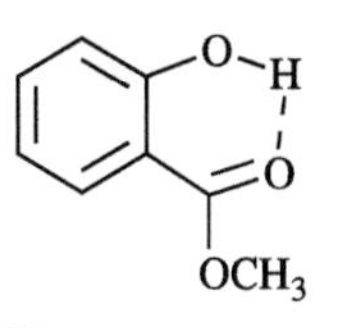

Methyl salicylate
3.3

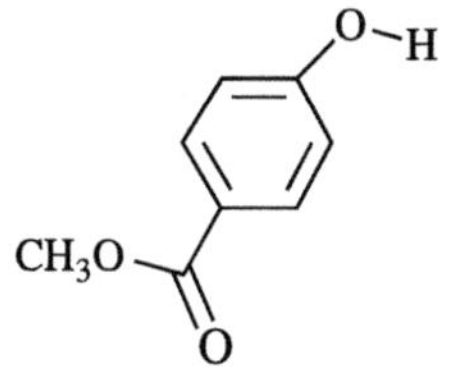

3.4

Intramolecular hydrogen bonding produces structures that can be thought of as bioisosteres of bicyclic compounds, a type of scaffold hopping (see Chapter 2, Section 2.2.6.3) (Figure 3.4).[26] Some intramolecular hydrogen bonds are strong enough to persist in water.[27]

Because intramolecular hydrogen bonding removes one donor and one acceptor moiety from the molecule, it increases its lipophilicity and membrane permeability and decreases its aqueous solubility. This can have a significant impact on pharmacokinetics. For example, the increased brain penetration and pharmacological activity of neurokinin 1 receptor antagonists were attributed to increased

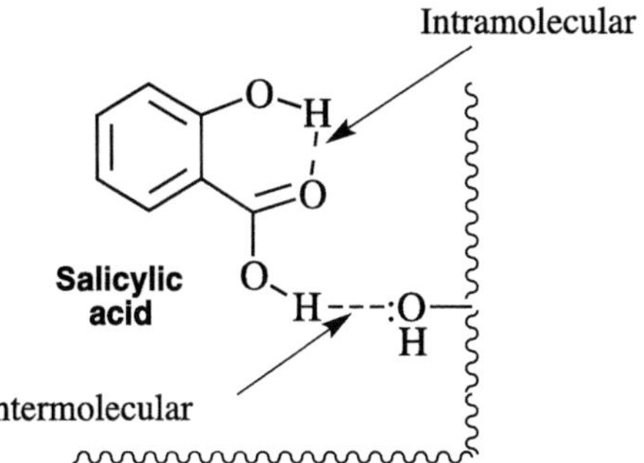

FIGURE 3.3 Examples of hydrogen bonds. Wavy line represents the receptor cavity.

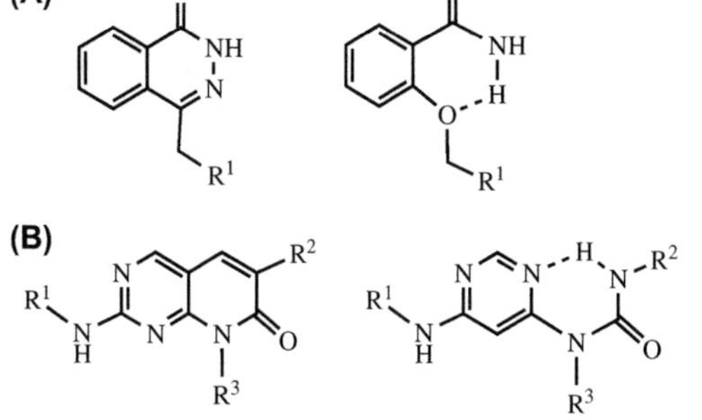

FIGURE 3.4 Two examples (A and B) of how intramolecular hydrogen bonding can mimic a bioisosteric heterocycle.

lipophilicity resulting from intramolecular hydrogen bonding.[28] Also, CLog *P* calculations (see Chapter 2, Sections 2.2.5.2.2 and 2.2.5.2.3) typically underestimate the lipophilicity of molecules that can undergo intramolecular hydrogen bonding; on average, the CLog *P* values should be increased by 0.4 for each intramolecular hydrogen bond in the molecule.[29]

Hydrogen bonds are essential to maintain the structural integrity of α-helix (**3.5**) and β-sheet (**3.6**) conformations of peptides and proteins and the double helix of DNA (**3.7**) (Figure 3.5). As is discussed in Chapter 6, many antitumor agents act by alkylation of the DNA bases, thereby preventing hydrogen bonding. This disrupts the double helix and destroys the DNA.

The ΔG^0 for hydrogen bonding can be between −1 and −7 kcal/mol, but usually is in the range −3 to −5 kcal/mol. Binding affinities increase by about one order of magnitude per hydrogen bond.

3.2.2.5. Charge–Transfer Complexes

When a molecule (or group) that is a good electron donor comes into contact with a molecule (or group) that is a good electron acceptor, the donor may transfer some of its charge

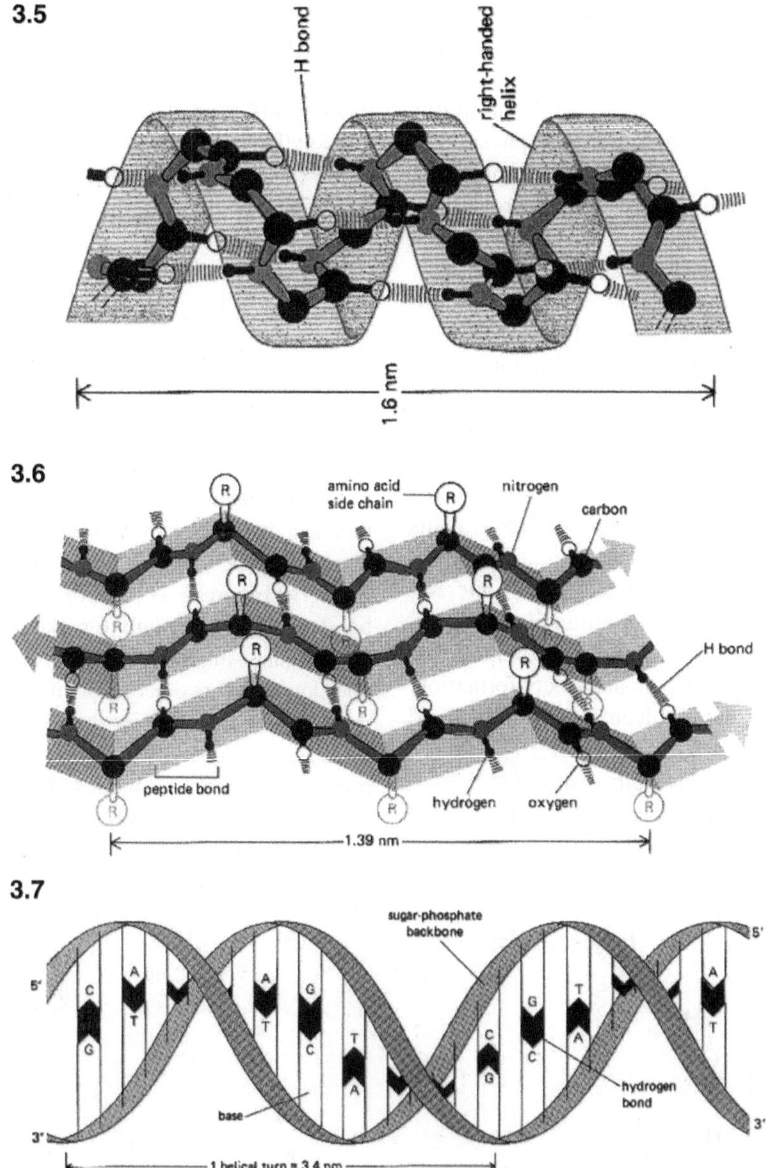

FIGURE 3.5 **3.5** is an example of an α-helix in a protein—Copyright 2007 from Molecular Biology of the Cell, Fifth Edition by Alberts, et al. Reproduced by permission of Garland Science/Taylor & Francis LLC. **3.6** is an example of a β-sheet in a protein—Copyright 2007 from Molecular Biology of the Cell, Fifth Edition by Alberts, et al. Reproduced by permission of Garland Science/Taylor & Francis LLC. **3.7** is an example of a double helix in DNA—Copyright 2007 from Molecular Biology of the Cell, Fifth Edition by Alberts, et al. *Reproduced by permission of Garland Science/Taylor & Francis LLC.*

to the acceptor. This forms a *charge-transfer complex*, which, in effect, is a molecular dipole–dipole interaction. The potential energy of this interaction is proportional to the difference between the ionization potential of the donor and the electron affinity of the acceptor.

Donor groups contain π-electrons, such as alkenes, alkynes, and aromatic moieties with electron-donating substituents, or groups that contain a pair of nonbonded electrons, such as oxygen, nitrogen, and sulfur moieties. Acceptor groups contain electron-deficient π-orbitals, such as alkenes, alkynes, and aromatic moieties having electron-withdrawing substituents, and weakly acidic protons. There are groups on receptors that can act as electron donors, such as the aromatic ring of tyrosine or the carboxylate group of aspartate.

Charge-transfer interactions are believed to provide the energy for intercalation of certain planar aromatic antimalarial drugs, such as chloroquine (**3.8,** Aralen), into parasitic DNA (see Chapter 6). The fungicide, chlorothalonil (Bravo), is used in Figure 3.6 as a hypothetical example for a charge-transfer interaction with a tyrosine. The ΔG^o for charge-transfer interactions also can range from −1 to −7 kcal/mol.

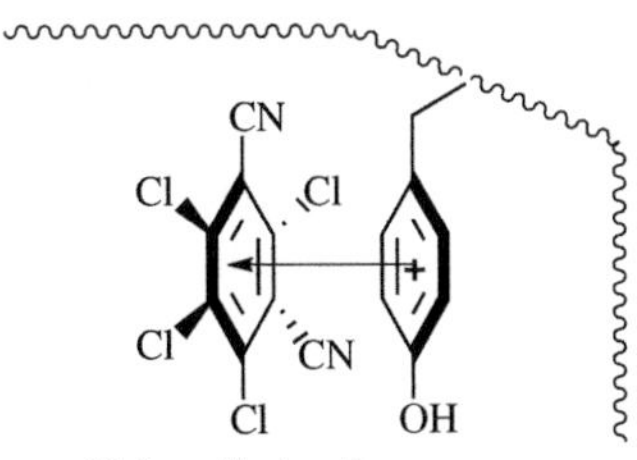

Chloroquine
3.8

Chlorothalonil

FIGURE 3.6 Example of a charge-transfer interaction. Wavy line represents the receptor cavity.

3.2.2.6. Hydrophobic Interactions

In the presence of a nonpolar molecule or region of a molecule, the surrounding water molecules orient themselves and, therefore, are in a higher energy state than when only other water molecules are around. When two nonpolar groups, such as a lipophilic group on a drug and a nonpolar receptor group, each surrounded by ordered water molecules, approach each other, these water molecules become disordered in an attempt to associate with each other. This increase in entropy, therefore, results in a decrease in the free energy ($\Delta G = \Delta H - T\Delta S$), which stabilizes the drug–receptor complex. This stabilization is known as a *hydrophobic interaction* (see Figure 3.7). Consequently, this is not an attractive force of two nonpolar groups "dissolving" in one another, but, rather, is the decreased free energy of the nonpolar group because of the increased entropy of the surrounding water molecules. Jencks[30] has suggested that hydrophobic forces may be the most important single factor responsible for noncovalent intermolecular interactions in aqueous solution. Hildebrand,[31] on the other hand, is convinced that there is no hydrophobia between water and alkanes; instead, he believes that there just is not enough hydrophilicity to break the hydrogen bonds of water and allow alkanes to go into solution without the assistance of other polar groups. Addition of a single methyl group that can occupy a receptor-binding pocket improves binding by −1.5 kcal/mol or by about a factor of 12.[32] This hydrophobic interaction has been referred to as a "magic methyl" interaction.[33] In Figure 3.8 the topical anesthetic butamben is depicted in a hypothetical hydrophobic interaction with an isoleucine group.

Another type of hydrophobic interaction, called a π–π interaction, involves two aryl groups.[34] The more common π–π interactions involve a parallel arrangement of aromatic rings,[35] in which the π-electrons interact in a face-to-face arrangement, known as π-stacking.[36] In Figure 3.9, the phenyl ring of the anticonvulsant drug lacosamide (Vimpat) is shown in a hypothetical π-stacking interaction with a receptor phenylalanine. Alternatively, a T-shaped arrangement (edge-to-face interaction) is possible, in which the edge of one aromatic ring forms a T-shape with the face

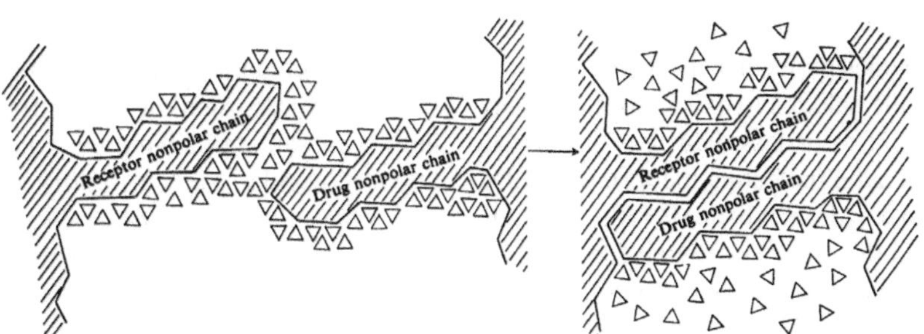

FIGURE 3.7 Formation of hydrophobic interactions. *From Korolkovas, A. (1970).* Essentials of Molecular Pharmacology, *p. 172. Wiley, New York. This material is reproduced with permission of John Wiley & Sons, Inc. and by permission of Kopple, K. D. 1966.* Peptides and Amino Acids. *Addison-Wesley, Reading, MA.*

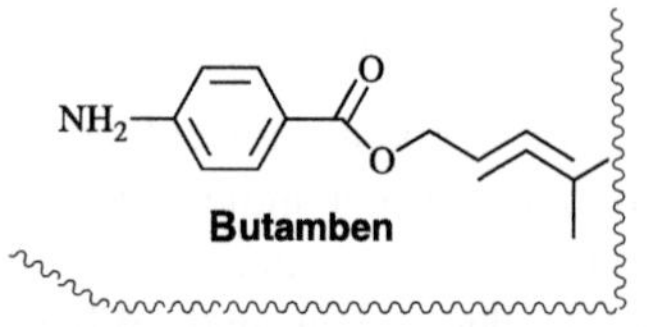

FIGURE 3.8 **Example of hydrophobic interactions.** The wavy line represents the receptor cavity.

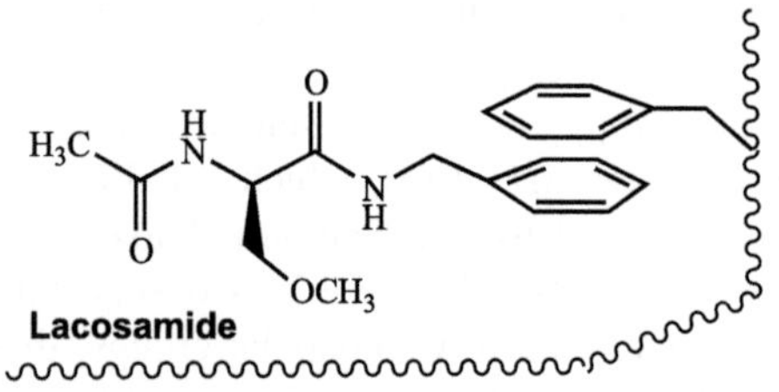

FIGURE 3.9 **Example of π–π stacking.** The wavy line represents the receptor cavity.

of the other aromatic ring. When one ring is electron deficient and the other is electron rich, then charge-transfer processes become important as well.

3.2.2.7. Cation–π Interaction

These interactions are very common in protein structure and also can be exploited for drug–receptor interactions.[37] In proteins, the most common aromatic group involved in a cation–π interaction is tryptophan (although phenylalanine, tyrosine, and histidine also participate), and the most common cation is arginine (although lysine is also important). A cationic group on a drug can undergo a cation–π interaction with an aromatic group on the receptor, or vice versa. Figure 3.10 is an example of a hypothetical cation–π interaction between the ammonium ion of lisdexamfetamine (Vyvanse), a drug for attention-deficit hyperactivity disorder, and a tryptophan residue. The ΔG^0 for cation–π interactions can be between −0.5 and −7 kcal/mol, but usually is in the range −1 to −5 kcal/mol.

3.2.2.8. Halogen Bonding

It has now been well established that a covalently bonded halogen atom can act as an electron acceptor (Lewis acid) to undergo *halogen bonding* with an electron-rich donor atom, such as O, N, or S.[38] On the basis of crystal structure and quantum mechanics/molecular mechanics data, it was found that many of the halogen to oxygen (or nitrogen) bond distances were equal to or less than the sum of the respective van der Waals radii, indicating the formation of a halogen bond (similar to a hydrogen bond).[39] The strength of these interactions is in the order H≈I>Br>Cl>>F. The interaction is caused by anisotropy of electron density on the halogen, resulting from the σ-hole, a positively charged region on the back side of the halogen atom along the R–X bond axis. These interactions can govern the conformation

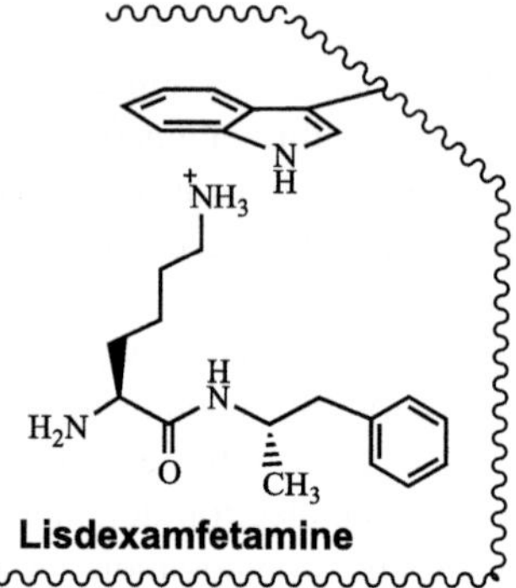

FIGURE 3.10 **Example of a cation–π interaction.** The wavy line represents the receptor cavity.

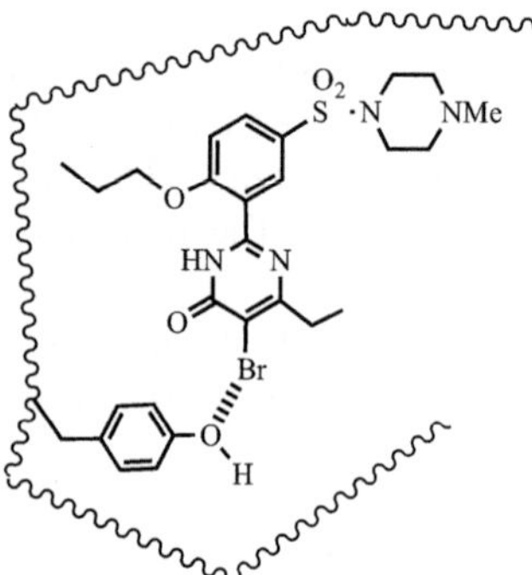

FIGURE 3.11 **Example of halogen bonding.** A compound bound into phosphodiesterase 5. The wavy line represents the enzyme cavity.

of molecules in the binding site of proteins. A series of phosphodiesterase type 5 inhibitors was designed with F, Cl, Br, and I atoms incorporated.[40] The potential halogen bond strengths were calculated, the molecules were synthesized, and then they were assayed; a good correlation was observed between the calculated binding energies and the activity of the molecules. The predicted interactions between the halogen atom and the phenolic oxygen atom of Tyr-612 were validated by X-ray crystallography, as shown in Figure 3.11. The ΔG^0 for halogen bonding can be between −1 and −15 kcal/mol,[41] but usually is in the range −1 to −5 kcal/mol.

3.2.2.9. van der Waals or London Dispersion Forces

Atoms in nonpolar molecules may have a temporary nonsymmetrical distribution of electron density, which results in the generation of a temporary dipole. As atoms from different molecules (such as a drug and a receptor) approach each other, the temporary dipoles of one molecule induce opposite dipoles in the approaching molecule. Consequently, intermolecular attractions, known as *van der Waals forces*, result. These weak universal forces only become significant when there is a close surface contact of the atoms; however, when there is molecular complementarity, numerous atomic interactions result (each interaction contributing about −0.5 kcal/mol to the ΔG^0), which can add up to a significant overall drug–receptor binding component. Other weak interactions may contribute to receptor–ligand binding as well.[42]

3.2.2.10. Conclusion

Because noncovalent interactions are generally weak, cooperativity by several types of interactions is critical. Once the first interaction has taken place, translational entropy is lost. This results in a much lower entropy loss in the formation of the second interaction. The effect of this cooperativity is that several rather weak interactions may combine to produce a strong interaction. This phenomenon is the basis for why the SAR by NMR approach to lead modification (see Chapter 2, Section 2.1.2.3.6) can produce such high-affinity ligands from two moderate- or poor-affinity ligands. Because several different types of interactions are involved, selectivity in drug–receptor interactions can result. In Figure 3.12, the local anesthetic dibucaine is used as an example to show the variety of interactions that are possible.

The binding constants for 200 drugs and potent enzyme inhibitors were used to calculate the average strength of noncovalent bonds (i.e., the binding energy) associated with 10 common functional groups in an average drug–receptor environment.[43] As suggested above, charged groups bind more strongly than polar groups, which bind more tightly than nonpolar groups; ammonium ions form the best electrostatic interactions (11.5 kcal/mol), then phosphate (10.0 kcal/mol), and then carboxylate (8.2 kcal/mol). For loss of rotational and translational entropy, 14 kcal/mol of binding energy has to be subtracted and 0.7 kcal/mol of energy is subtracted for each degree of conformational freedom restricted.[44] Compounds that bind to a receptor exceptionally well have measured binding energies that exceed the calculated average binding energy, and those whose binding energy is less than the average calculated value fit poorly into the receptor.

3.2.3. Determination of Drug–Receptor Interactions

Hormones and neurotransmitters are important endogenous molecules that are responsible for the regulation of a myriad of physiological functions. These molecules interact with a specific receptor in a tissue and elicit a specific characteristic response. For example, the activation of a muscle by the CNS is mediated by release of the excitatory neurotransmitter acetylcholine (ACh; **3.9**). If the logarithm of the concentration of ACh added to a muscle tissue preparation is plotted against the percentage of total muscle contraction, the graph shown in Figure 3.13 may result. This is known as a *dose–response* or *concentration–response curve*. The low concentration part of the curve results from too few neurotransmitter molecules available for collision with the receptor. As the concentration increases, it reaches a point where a linear relationship is observed between the logarithm of the neurotransmitter concentration and the biological response. As most of the receptors become occupied, the probability of additional drug-receptor interactions diminishes, and the curve deviates from linearity (the high concentration end). Concentration–response curves are a means of measuring drug–receptor interactions by showing the relationship between drug concentration, usually plotted on the *X*-axis, and a biological response, usually plotted on the *Y*-axis. As shown in Figure 3.13, key parameters, such as K_d (the concentration of test compound that gives half-maximal binding) or EC_{50}, (the [effective] concentration of drug that elicits 50% of the total biological response) can be determined from the concentration–response data. EC_{50} is a common standard measure for comparing potencies of compounds that interact with a receptor and elicit a particular biological response. When the experiment is conducted in a whole animal, then *dose* (rather than concentration) is the variable normally plotted on the *X*-axis; in analogy to the terminology discussed above, the plot is called a *dose–*response curve, and ED_{50} (the [effective] *dose* that elicits 50% of the total response) is the parameter that is typically used to compare potencies across compounds.

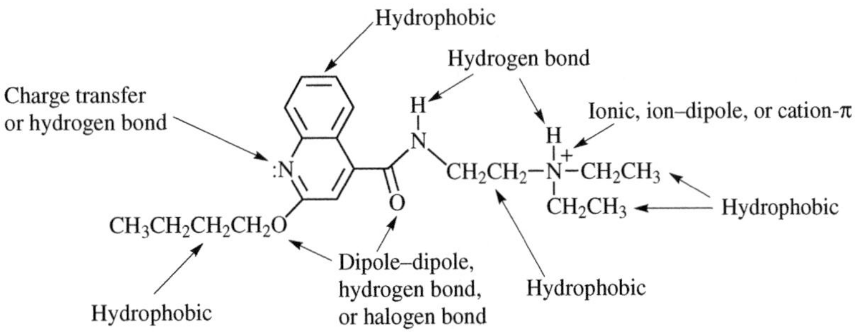

Hydrophobic
Hydrogen bond
Charge transfer or hydrogen bond
Ionic, ion–dipole, or cation-π
$CH_2CH_2-\overset{+}{N}H-CH_2CH_3$
CH_2CH_3
Hydrophobic
$CH_3CH_2CH_2CH_2O$
Dipole–dipole, hydrogen bond, or halogen bond
Hydrophobic
Hydrophobic
Dibucaine

FIGURE 3.12 **Example of potential multiple drug–receptor interactions. The van der Waals interactions are excluded.**

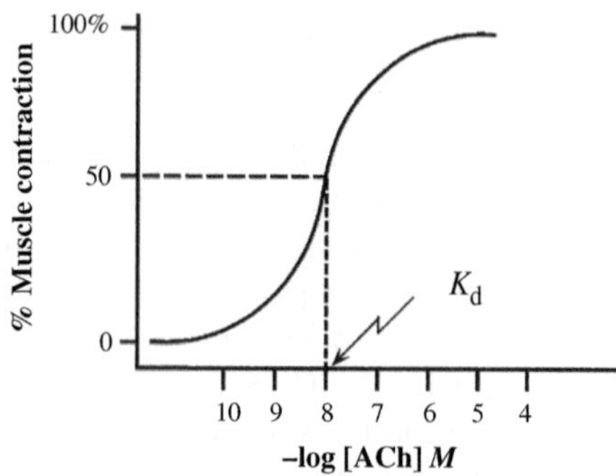

FIGURE 3.13 Effect of increasing the concentration of a neurotransmitter (ACh) on muscle contraction. The K_d is measured as the concentration of neurotransmitter that gives 50% of the maximal activity.

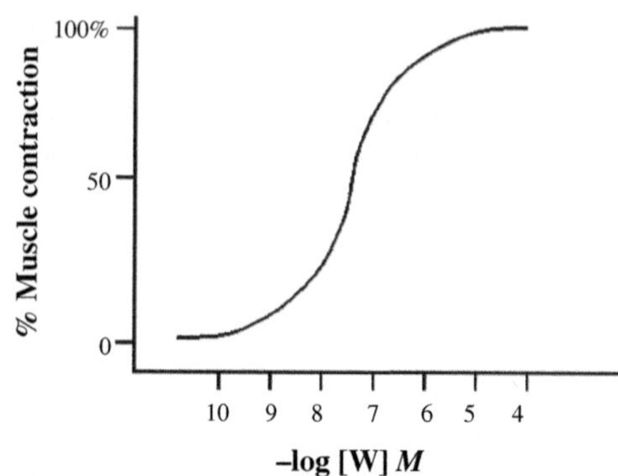

FIGURE 3.14 Dose–response curve for a full agonist (W).

3.9

If another compound (W) is added in increasing amounts to the same tissue preparation, and the curve shown in Figure 3.14 results, the compound, which produces the same maximal response as the neurotransmitter, is called a *full agonist.*

A second compound (X) added to the tissue preparation shows no response at all (Figure 3.15A); however, if it is added to the neurotransmitter, and the effect of the neurotransmitter is blocked until a higher concentration of the neurotransmitter is added (Figure 3.15B), compound X is called a *competitive antagonist.* There are two general types of antagonists, competitive antagonists and noncompetitive antagonists. The former, which is the larger category, is one in which the degree of antagonism depends on the relative concentrations of the agonist and the antagonist; both bind to the same site on the receptor, or, at least, the antagonist directly interferes with the binding of the agonist. The most common assessment of the potency of competitive antagonists to establish a SAR is by determination of their IC_{50} values (the concentration of compound that inhibits the response of a given agonist by 50%) (Figure 3.15C). This allows for a direct comparison of different antagonists. The degree of blocking of a *noncompetitive antagonist* (X′) is independent of the amount of agonist present, so the EC_{50} does not change with increasing neurotransmitter (Figure 3.15D). Two different binding sites may be involved; when the noncompetitive antagonist binds to its *allosteric* binding site, a site to which the endogenous ligand normally does not bind, it may cause a conformational change in the protein, which affects binding of the endogenous molecule. Only competitive antagonists will be discussed further in this text.

If a compound Y is added to the tissue preparation and some response is elicited, but not a full response, regardless of how high a concentration of Y is used, then Y is called a *partial agonist* (see Figure 3.16A). A partial agonist has properties of both an agonist and an antagonist.

When Y is added to low concentrations of a neurotransmitter sufficient to give a response less than the maximal response of the partial agonist (for example, 15%, as shown in Figure 3.16B), additive effects are observed as Y is increased, but the maximum response does not exceed that produced by Y alone. Under these conditions, the partial agonist has an agonistic effect. However, if Y is added to high concentrations of a neurotransmitter sufficient to give full response of the neurotransmitter, then antagonistic effects are observed; as Y increases, the response decreases to the point of maximum response of the partial agonist (Figure 3.16C). If this same experiment is done starting with higher concentrations of the neurotransmitter, the same results are obtained except that the dose–response curves shift to the right resembling the situation of adding an antagonist to the neurotransmitter (Figure 3.16C).

In a hypothetical situation, compound Z is added to the tissue preparation and muscle relaxation occurs (the opposite effect of the agonist). This would be a *full inverse agonist*, a compound that binds to the receptor, but displays an effect opposite to that of the natural ligand (Figure 3.17A). Valium (see **1.17**), for example, binds to a γ-aminobutyric acid (GABA) receptor and has an anticonvulsant effect, similar to that of the natural ligand GABA, and is thus an agonist; β-carbolines (**3.10**) bind to the same receptor, but act as convulsants, and are inverse agonists.[45] Just as an antagonist can displace an agonist or natural ligand (Figure 3.15B), it also can displace an inverse agonist (Figure 3.17B). A *partial inverse agonist* (Z′) is one that, at any concentration, does not give 100% of the effect of a full inverse agonist (Figure 3.17C).

3.10

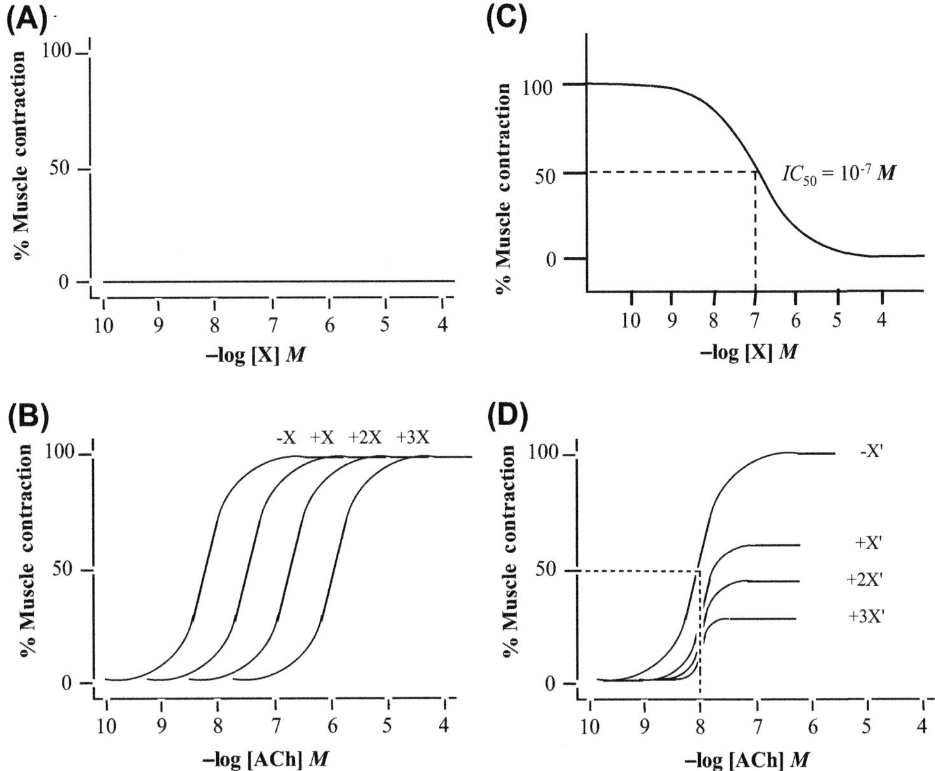

FIGURE 3.15 (A) Dose-response curve for an antagonist (X); (B) effect of a competitive antagonist (X) on the response of a neurotransmitter (acetylcholine; ACh); (C) effect of varying concentration of a competitive antagonist X in the presence of a fixed, maximally effective concentration of agonist (ACh); and (D) effect of various concentrations of a noncompetitive antagonist (X') on the response of the neurotransmitter (ACh).

On the basis of the above discussion, if you need a drug to effect a certain response of a receptor, an agonist would be desired; if you need a drug to prevent a particular response of a natural ligand, an antagonist would be required; if you need a drug that causes the opposite effect of the natural ligand, then an inverse agonist is what you want.

Sometimes, there are great structural similarities among a series of agonists, but little structural similarity in a series of competitive antagonists. For example, Table 3.1 shows some agonists and antagonists for histamine and epinephrine; a more detailed list of agonists and antagonists for specific receptors has been reported.[46] The differences in the structures of the antagonists is not surprising because a receptor can be blocked by an antagonist simply by its binding to a site near enough to the binding site for the agonist that it physically blocks the agonist from reaching its binding site. This may explain why antagonists are frequently much more bulky than the corresponding agonists. It is easier to design a molecule that blocks a receptor site than one that interacts with it in the specific way required to elicit a response. An agonist can be transformed into an antagonist by appropriate structural modifications, sometimes by relatively minor modifications. For example, both **3.11a** and **3.11b** (Table 3.1) bind to the progesterone receptor; however, **3.11a** is an antagonist (IC_{50} 5.0 nM) and **3.11b** is an agonist (EC_{50} 1.3 nM). Compound **3.11a** exhibited contraceptive activity in rats and monkeys.[47]

How is it possible for an antagonist to bind to the same site as an agonist and not elicit a biological response? There are several ways that this may occur. In Figure 3.18, panel A shows an agonist with appropriate groups interacting with three receptor-binding sites and eliciting a response. In panel B of Figure 3.18, the compound has two groups that can interact with the receptor, but one essential group is missing. In the case of enantiomers (panel C shows the enantiomer of the compound in panel A), only two groups are able to interact with the proper receptor sites. If appropriate groups must interact with all three binding sites in order for a response to be elicited, then the compounds depicted by panels B and C would be antagonists.

There are two general categories of compounds that interact with receptors: (1) compounds that occur naturally within the body, such as hormones, neurotransmitters, and other agents that modify cellular activity (*autocoids*) and (2) *xenobiotics*, compounds that are foreign to the body. Receptor selectivity is very important, but often difficult to attain because receptor structures are often unknown. Many current drugs are pharmacologically active at multiple receptors, some of which are not associated with the illness that is being treated. This can lead to side effects. For example, the clinical effect of neuroleptics (an early class of antipsychotic drugs with tranquilizing properties) is believed to result from their antagonism of dopamine receptors.[48] In general, this class of drugs also blocks cholinergic and

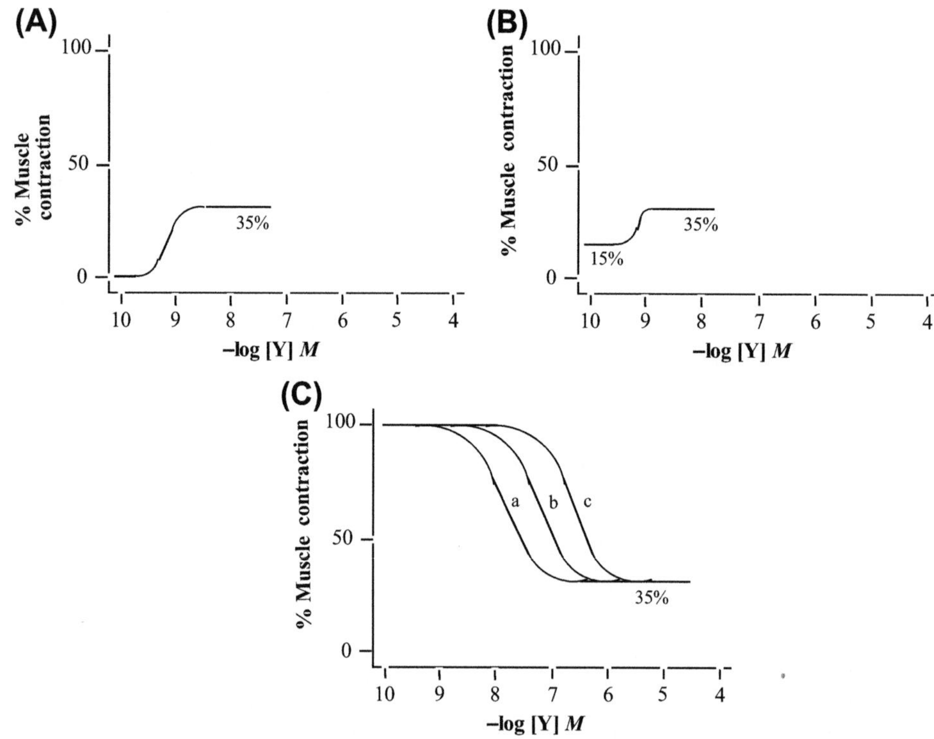

FIGURE 3.16 (A) Dose–response curve for a partial agonist (Y); (B) effect of a low concentration of neurotransmitter on the response of a partial agonist (Y); and (C) effect of a high concentration of neurotransmitter on the response of a partial agonist (Y). In (C), the concentration of the neurotransmitter (*a,b,c*) is $c>b>a$.

α-adrenergic receptors, and this results in side effects such as sedation and hypotension.

3.2.4. Theories for Drug–Receptor Interactions

Over the years a number of hypotheses have been proposed to account for the ability of a drug to interact with a receptor and elicit a biological response. Several of the more important proposals are discussed here, starting from the earliest hypothesis (the occupancy theory) to the current one (the multistate model).

3.2.4.1. Occupancy Theory

The *occupancy theory* of Gaddum[49] and Clark[50] states that the intensity of the pharmacological effect is directly proportional to the number of receptors occupied by the drug. The response ceases when the drug–receptor complex dissociates. However, as discussed in Section 3.2.3, not all agonists produce a maximal response. Therefore, this theory does not rationalize partial agonists, and it does not explain inverse agonists.

Ariëns[51] and Stephenson[52] modified the occupancy theory to account for partial agonists, a term coined by Stephenson. These authors utilized the original Langley concept of a receptor, namely, that drug–receptor interactions involve two stages: first, there is a complexation of the drug with the receptor, which they both termed the *affinity*; second, there is an initiation of a biological effect, which Ariëns termed the *intrinsic activity* and Stephenson called the *efficacy*. Affinity, then, is a measure of the capacity of a drug to bind to the receptor, and depends on the molecular complementarity of the drug and the receptor. Intrinsic activity (α) now refers to the maximum response induced by a compound relative to that of a given reference compound, and efficacy is the property of a compound that produces the maximum response or the ability of the drug–receptor complex to initiate a response.[53] Because of the slight change in definitions, we will use the term efficacy to refer to the ability of a compound to initiate a biological response. In the original theory, this latter property was considered to be constant. Examples of affinity and efficacy are given in Figure 3.19. Figure 3.19A shows the theoretical dose–response curves for five drugs with the same affinity for the receptor ($pK_d = 8$), but having efficacies varying from 100% of the maximum to 20% of the maximum. The drug with 100% efficacy is a full agonist; the others are partial agonists. Figure 3.19B shows dose–response curves for four drugs with the same efficacy (all full agonists), but having different affinities varying from a pK_d of 9 to 6.

Antagonists can bind tightly to a receptor (great affinity), but be devoid of activity (no efficacy). Potent agonists may have less affinity for their receptors than partial

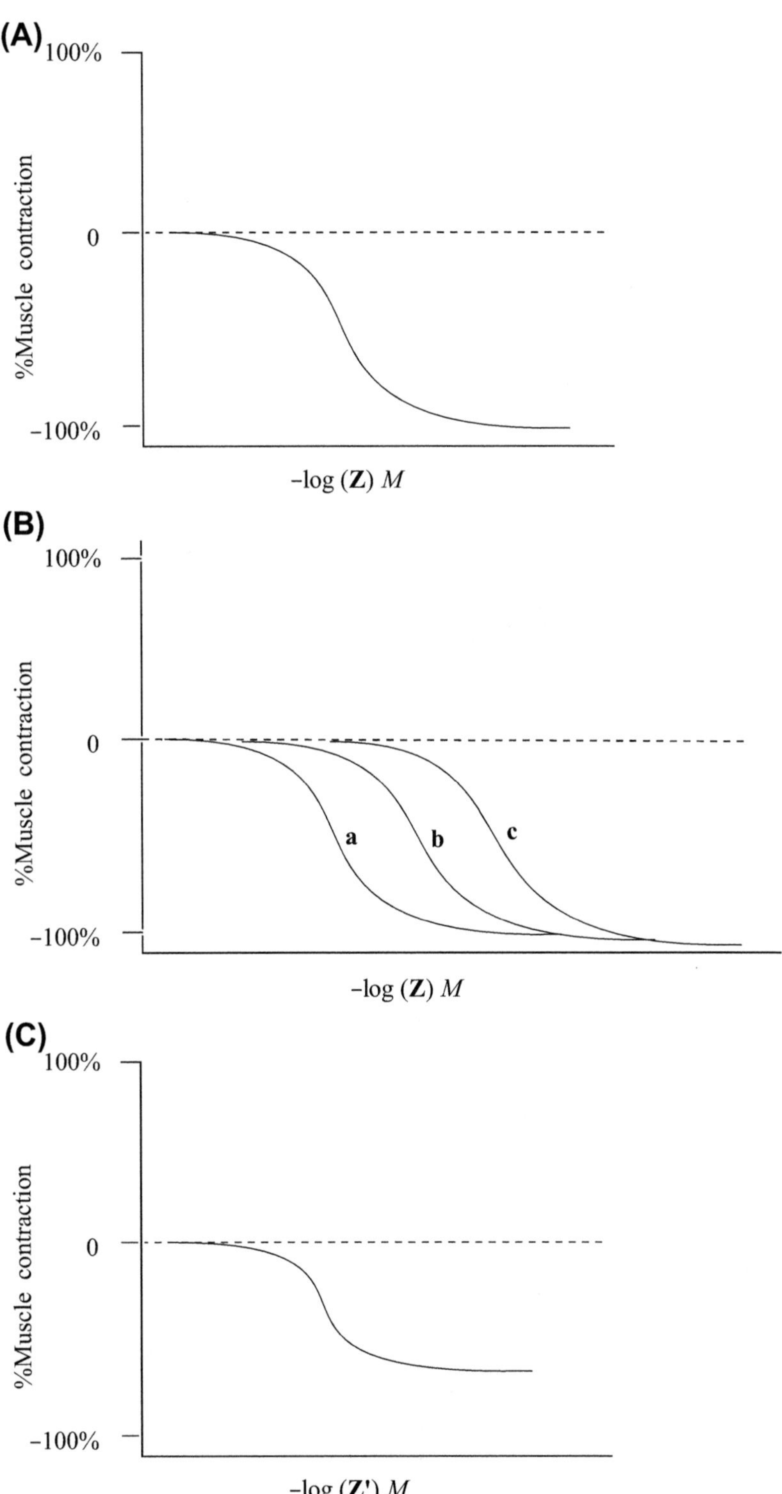

FIGURE 3.17 (A) Dose–response curve for a full inverse agonist (Z); (B) effect of a competitive antagonist on the response of a full inverse agonist (a, b, and c represent increasing concentrations of the added antagonist or natural ligand to Z); and (C) dose–response curve for a partial inverse agonist (Z′).

TABLE 3.1 Agonists and Antagonists

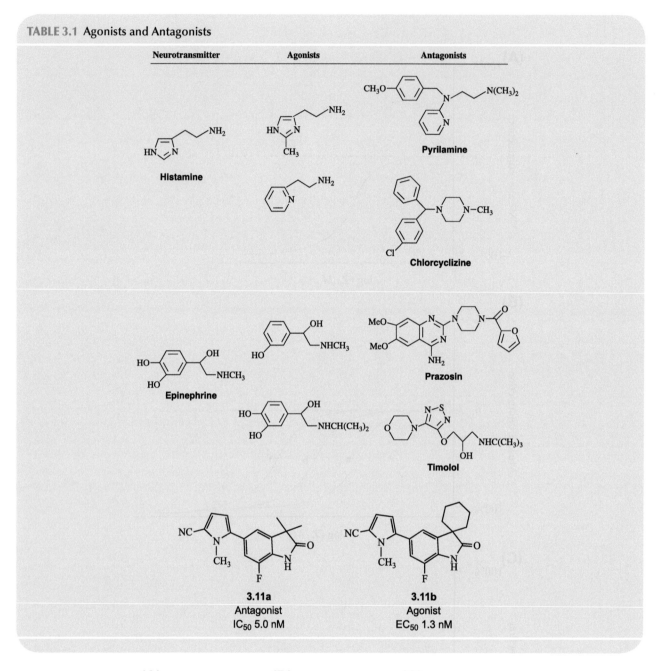

Neurotransmitter	Agonists	Antagonists
Histamine		Pyrilamine
		Chlorcyclizine
Epinephrine		Prazosin
		Timolol
	3.11b Agonist EC_{50} 1.3 nM	3.11a Antagonist IC_{50} 5.0 nM

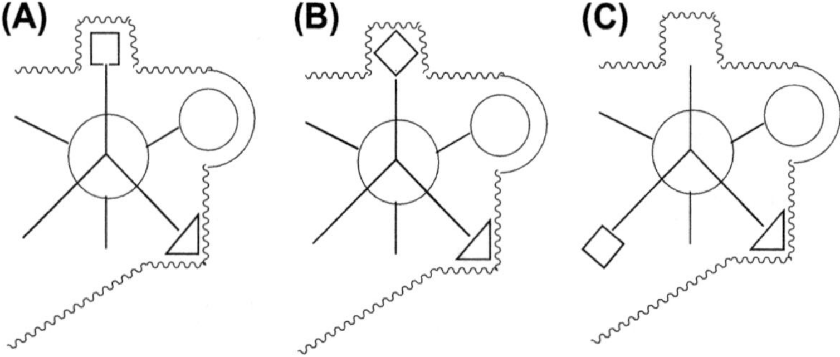

FIGURE 3.18 Inability of an antagonist to elicit a biological response. The wavy line is the receptor cavity. *Adapted with permission from W. O. Foye (Ed.), 1989 "Principles of Medicinal Chemistry," 3rd ed., p. 63. Copyright © 1989 Lea & Febiger, Philadelphia, Pennsylvania (Lippincott Williams & Wilkins/Wolters Kluwer).*

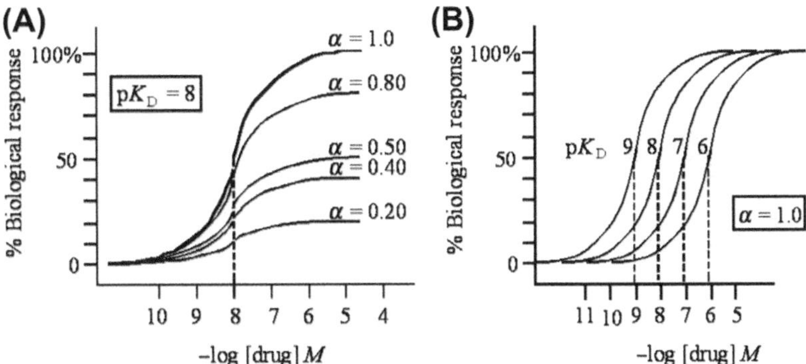

FIGURE 3.19 Theoretical dose–response curves illustrate (A) drugs with equal affinities and different efficacies (the top compound is a full agonist, and the others are partial agonists) and (B) drugs with equal efficacies (all full agonists) but different affinities.

agonists or antagonists. Therefore, these two properties, affinity and efficacy, are uncoupled. Also, the terms agonist, partial agonist, antagonist, and inverse agonist are biological system dependent and not necessarily properties of drugs. A compound that is an agonist for one receptor may be an antagonist or inverse agonist for another receptor. A particular receptor is considered to have an intrinsic *maximum response*; this is the largest magnitude of response that the receptor is capable of producing by any ligand. A compound that elicits the maximum response is a full agonist; a particular compound may be capable of exceeding the maximum response of a tissue, but the observed response can only be the maximum response of that particular tissue. A drug that is not capable of eliciting the maximum response of the tissue, which depends on the structure of the drug, is a partial agonist. A full agonist or partial agonist is said to display *positive efficacy*, an antagonist displays zero efficacy, and a full or partial inverse agonist displays *negative efficacy* (depresses basal tissue response).

The modified occupancy theory accounts for the existence of partial agonists and antagonists, but it does not account for why two drugs that can occupy the same receptor can act differently, i.e., one as an agonist, the other as an antagonist.

3.2.4.2. Rate Theory

As an alternative to the occupancy theory, Paton[54] proposed that the activation of receptors is proportional to the total number of encounters of the drug with its receptor per unit time. Therefore, the *rate theory* suggests that the pharmacological activity is a function of the rate of association and dissociation of the drug with the receptor and not the number of occupied receptors. Each association would produce a quantum of stimulus. In the case of agonists, the rates of both association and dissociation would be fast (the latter faster than the former). The rate of association of an antagonist with a receptor would be fast, but the dissociation would be slow. Partial agonists would have intermediate drug–receptor complex dissociation rates. At equilibrium, the occupancy and rate theories are mathematically equivalent. As in the case of the occupancy theory, the rate theory does not rationalize why the different types of compounds exhibit the characteristics that they do.

3.2.4.3. Induced-Fit Theory

The *induced-fit theory* of Koshland[55] was originally proposed for the action of substrates with enzymes, but it could apply to drug–receptor interactions as well. According to this theory, the receptor need not necessarily exist in the appropriate conformation required to bind the drug. As the drug approaches the receptor, a *conformational change* is induced, which orients the essential binding sites (Figure 3.20). The conformational change in the receptor could be responsible for the initiation of the biological response (movement of residues to interact with the substrate). The receptor (enzyme) was suggested to be elastic, and could return to its original conformation after the drug (product) was released. The conformational change need not occur only in the receptor (enzyme); the drug (substrate) also could undergo deformation, even if this resulted in strain in the drug (substrate). According to this theory, an agonist would induce a conformational change and elicit a response, an antagonist would bind without a conformational change, and a partial agonist would cause a partial conformational change. The induced-fit theory can be adapted to the rate theory. An agonist would induce a conformational change in the receptor, resulting in a conformation to which the agonist binds less tightly and from which it can dissociate more easily. If drug–receptor complexation does not cause a conformational change in the receptor, then the drug–receptor complex will be stable, and an antagonist will result.

Other theories evolved from the induced-fit theory, such as the macromolecular perturbation theory, the activation–aggregation theory, and multistate models.

3.2.4.4. Macromolecular Perturbation Theory

Having considered the conformational flexibility of receptors, Belleau[56] suggested that in the interaction of a drug with a receptor two general types of *macromolecular*

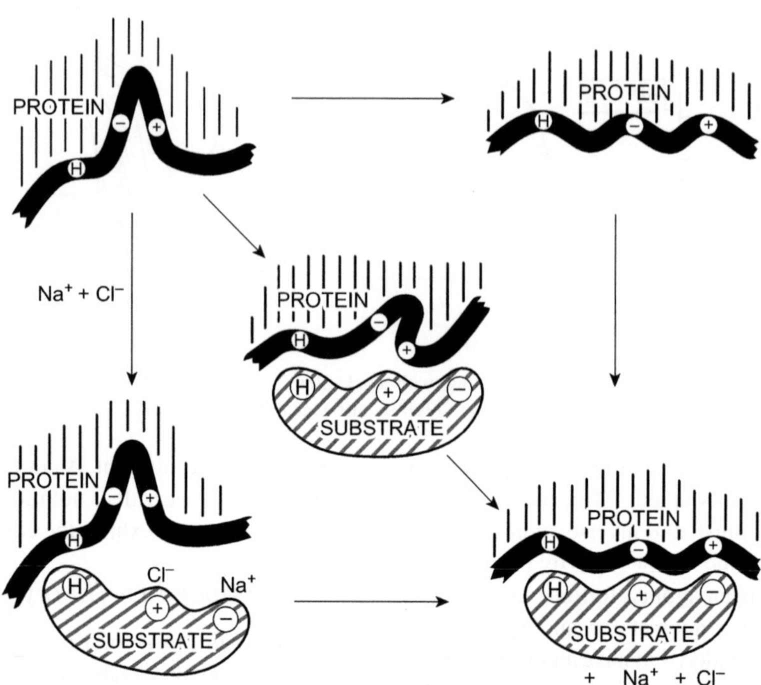

FIGURE 3.20 Schematic of the induced-fit theory. *Koshland, Jr., D. E., and Neet, K. E., Annu. Rev. Biochem., Vol. 37, 1968. Annual Review of Biochemistry by Annual Reviews. Reproduced with permission of Annual Reviews via Copyright Clearance Center, 2013.*

perturbations could result: *a specific conformational perturbation* makes possible the binding of certain molecules that produce a biological response (an agonist) and *a nonspecific conformational perturbation* accommodates other types of molecules that do not elicit a response (e.g., an antagonist). If the drug contributes to both macromolecular perturbations, a mixture of two complexes will result (a partial agonist). This theory offers a physicochemical basis for the rationalization of molecular phenomena that involve receptors, but does not address the concept of inverse agonism.

3.2.4.5. Activation–Aggregation Theory

An extension of the macromolecular perturbation theory (which also is based on the induced-fit theory) is the *activation–aggregation theory* of Monad, Wyman, and Changeux[57] and Karlin.[58] According to this theory, even in the absence of drugs, a receptor is in a state of dynamic equilibrium between an activated form (R_0), which is responsible for the biological response, and an inactive form (T_0). Using this theory, agonists bind to the R_0 form and shift the equilibrium to the activated form, antagonists bind to the inactive form (T_0), and partial agonists bind to both conformations. In this model, the agonist binding site in the R_0 conformation can be different from the antagonist binding site in the T_0 conformation. If there are two different binding sites and conformations, then this could account for the structural differences in these classes of compounds and could rationalize why an agonist elicits a biological response but an antagonist does not. This theory can explain the ability of partial agonists to possess both the agonistic and antagonistic properties as depicted in Figure 3.16. In Figure 3.16B, as the partial agonist interacts with the remaining unoccupied receptors, there is an increase in the response up to the maximal response for the partial agonist interaction. In Figure 3.16C, the partial agonist competes with the neurotransmitter for the receptor sites. As the partial agonist displaces the neurotransmitter, it changes the amount of R_0 and T_0 receptor forms (T_0 increases and, therefore, the response decreases) until all the receptors have the partial agonist bound. This theory, however, also does not address inverse agonists.[59]

3.2.4.6. The Two-State (Multistate) Model of Receptor Activation

The concept of a conformational change in a receptor inducing a change in its activity has been viable for many years.[60] The Monod–Wyman–Changeux idea described above involves a two-state model of receptor activation, but it does not go far enough. This model was revised based mostly on observations with GPCRs (see Section 3.1).[61]

The revised *two-state model of receptor activation* proposes that, in the absence of the natural ligand or agonist, receptors exist in equilibrium (defined by equilibrium constant *L*; Figure 3.21) between an active state (R^*), which is able to initiate a biological response, and a resting state (R), which cannot. In the absence of a natural ligand or agonist,

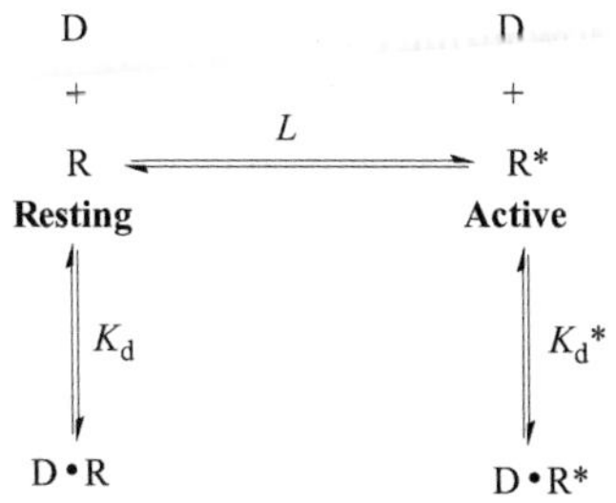

$$\begin{array}{ccc} D & & D \\ + & & + \\ R & \overset{L}{\rightleftharpoons} & R^* \\ \textbf{Resting} & & \textbf{Active} \\ \Big\updownarrow K_d & & \Big\updownarrow K_d^* \\ D\bullet R & & D\bullet R^* \end{array}$$

FIGURE 3.21 **Two-state model of receptor activation.** D is the drug, R is the receptor, and L is the equilibrium between the resting (R) and the active (R^*) state of the receptor.

the equilibrium between R^* and R defines the basal activity of the receptor. A drug can bind to one or both of these conformational states, according to equilibrium constants K_d and K_d^* for formation of the drug–receptor complex with the resting (D·R) and active (D·R^*) states, respectively. Full agonists alter the equilibrium fully to the active state by binding to the active state and causing maximum response; partial agonists preferentially bind to the active state, but not to the extent that a full agonist does, so maximum response is not attained; full inverse agonists alter the equilibrium fully to the resting state by binding to the resting state, causing a negative efficacy (a decrease in the basal activity); partial inverse agonists preferentially bind to the resting state, but not to the extent that a full inverse agonist does; and antagonists have equal affinities for both states (i.e., have no effect on the equilibrium or basal activity, and, therefore, exhibit neither positive nor negative efficacy).[62] A competitive antagonist is able to displace either an agonist or an inverse agonist from the receptor.

Leff and coworkers further extended the two-state receptor model to a *three-state receptor model*.[63] In this model, there are two active conformations (this becomes a multistate model by extension to more than two active conformations) and an inactive conformation. This accommodates experimental findings regarding variable agonist and inverse agonist behavior (both affinities and efficacies) in different systems containing the same receptor type (called *receptor promiscuity*). According to this hypothesis, the basis for differential agonist efficacies among different agonists is their different affinities for the different active states.

3.2.5. Topographical and Stereochemical Considerations

Up to this point in our discussion of drug–receptor interactions, we have been concerned with what stabilizes a drug–receptor complex, how drug–receptor interactions are measured, and possible ways that the drug–receptor complex may form. In this section, we turn our attention to molecular aspects and examine the topography and stereochemistry of drug–receptor complexes.

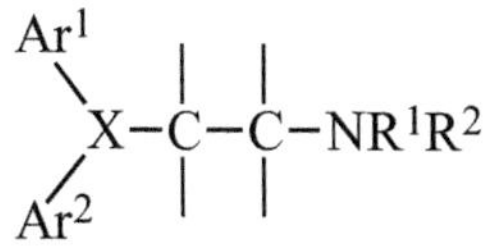

$$(Ar^1)(Ar^2)X{-}\overset{|}{\underset{|}{C}}{-}\overset{|}{\underset{|}{C}}{-}NR^1R^2$$

FIGURE 3.22 General structure of antihistamines

3.2.5.1. Spatial Arrangement of Atoms

It was indicated in the discussion of bioisosterism (Chapter 2, Section 2.2.4.3) and from SAR studies that many antihistamines have a common pharmacophore (Figure 3.22).[64] In Figure 3.22, Ar^1 is aryl, such as phenyl, substituted phenyl, or heteroaryl (2-pyridyl or thienyl) and Ar^2 is aryl or arylmethyl. The two aryl groups can also be connected through a bridge (as in phenothiazines, **2.75**), and the C-C-N moiety can be part of another ring (as in chlorcyclizine (Di-Paralene), Table 3.1). X is CH–O–, N–, or CH–; C–C is a short carbon chain (two or three atoms), which may be saturated, branched, contain a double bond, or be part of a ring system. These compounds are called *antihistamines* because they are antagonists of a histamine receptor known as the *H_1 histamine receptor*. When a sensitized person is exposed to an allergen, an antibody is produced, an antigen–antibody reaction occurs, and histamine is released. Histamine binding to the H_1 receptor can cause stimulation of smooth muscle and produce allergic and hypersensitivity reactions such as hay fever, pruritus (itching), contact and atopic dermatitis, drug rashes, urticaria (edematous patches of skin), and anaphylactic shock. Antihistamines are used widely to treat these symptoms. Unlike histamine (see Table 3.1 for structure), most H_1 antagonists contain tertiary amino groups, usually a dimethylamino or pyrrolidino group. At physiological pH, then, this group will be protonated, and it is believed that an ionic interaction with the receptor is a key binding contributor. The commonality of structures of antihistamines suggests that there are specific binding sites on the H_1 histamine receptor that have an appropriate topography for interaction with certain groups on the antihistamine, which are arranged in a similar configuration (see Section 3.2.2). It must be reiterated, however, that although the antihistamines are competitive antagonists of histamine for the H_1 receptor, the same set of atoms on the receptor need not interact with both histamine and the antagonists.[65] Consequently, it is difficult to make conclusions regarding the receptor structure on the basis of antihistamine structure–activity relationships. Because of the essentiality of various parts of antihistamine molecules, it is likely that the minimum binding requirements include a negative charge or π system on the receptor to interact with the ammonium cation (electrostatic or cation–π, respectively) and hydrophobic (van der Waals) interactions with the aryl groups. Obviously, many other interactions are involved.

From the very simplistic view of drug–receptor interactions discussed above, it is not possible to rationalize the fact that enantiomers, i.e., compounds that are nonsuperimposable mirror images of each other, can have quite different binding properties to receptors. This phenomenon is discussed in more detail in the next section.

3.2.5.2. Drug and Receptor Chirality

Histamine is an achiral molecule, but many of the H_1 receptor antagonists are chiral molecules. Proteins are polyamino acid macromolecules, and amino acids are chiral molecules (in the case of mammalian proteins, they are almost all L isomers); consequently, proteins (receptors) also are chiral substances. Complexes formed between a receptor and two enantiomers are diastereomers, not enantiomers, and, as a result, they have different energies and chemical properties. This suggests that dissociation constants for drug–receptor complexes of enantiomeric drugs may differ and may even involve different binding sites. The chiral antihistamine dexchlorpheniramine (**3.12**, Polaramine) is highly stereoselective (one stereoisomer is more potent than the other); the *S*-(+)-isomer is about 200 times more potent than the *R*-(−)-isomer.[66] According to the nomenclature of Ariëns,[67] when there is isomeric stereoselectivity, the more potent isomer is termed the *eutomer* and the less potent isomer is the *distomer*. The ratio of the potency of the more potent (higher affinity) enantiomer to the potency of the less potent enantiomer is termed the *eudismic ratio*. The in vivo eudismic ratio (−/+) for etorphine (Immobilon), a highly potent analgesic agent used to immobilize large nondomestic animals (see **2.63**, R=CH_3, R′=C_3H_7), is greater than 6666.[68]

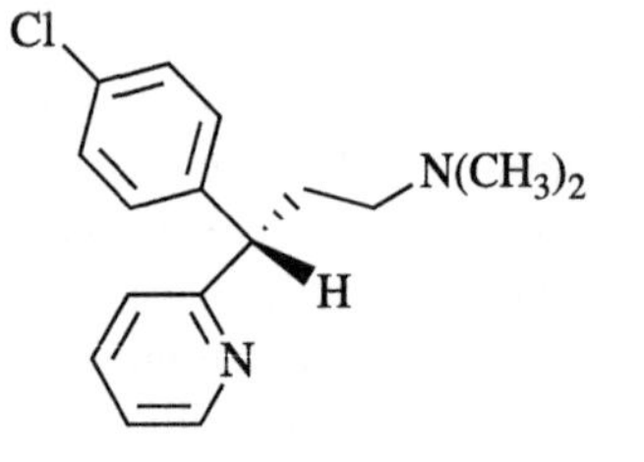

Cl N(CH$_3$)$_2$ H N

S-(+)-Dexchlorpheniramine
3.12

High-potency antagonists are those having a high degree of complementarity with the receptor. When the antagonist contains a stereogenic center in the pharmacophore (see Chapter 2, Section 2.2.1), a high eudismic ratio is generally observed for the stereoisomers because the receptor complementarity would not be retained for the distomer. This increase in eudismic ratio with an increase in potency of the eutomer is *Pfeiffer's rule*.[69] Small eudismic ratios are typically observed when the stereogenic center lies outside of the region critically involved in receptor binding, i.e., is not part of the pharmacophore, or when both the eutomer and distomer have low affinity for the receptor (poor molecular complementarity).

The distomer actually should be considered as an impurity in the mixture, or, in the terminology of Ariëns, the *isomeric ballast*. It, however, may contribute to undesirable side effects and toxicity; in that case, the distomer for the biological activity may be the eutomer for the side effects. For example, *d*-ketamine (**3.13**; the asterisk marks the chiral carbon) is a hypnotic and analgesic agent; the *l*-isomer is responsible for the undesired side effects[70] (note that *d* is synonymous with (+) and *l* is synonymous with (−)). Probably the most horrendous example of toxicity by a distomer is that of thalidomide (**3.14**, Contergan), a drug used in the late 1950s and the early 1960s as a sedative and to prevent morning sickness during pregnancy, which was shown to cause severe fetal limb abnormalities (phocomelia, shortening of limbs, and amelia, absence of limbs) when taken in the first trimester of pregnancy. This tragedy led to the development of three phases of clinical trials and the requirement for Food and Drug Administration (FDA) approval of drugs (see Chapter 1, Section 1.3.6.2 and 1.3.6.3). Later, it was thought that the teratogenicity (birth defect) of thalidomide was caused by the (*S*)-isomer only[71]; however, then it was found that the (*R*)-isomer was converted into the (*S*)-isomer in vivo.[72] Despite the potential danger of this drug, it is back on the market (as the racemate, Thalomid) for the treatment of moderate or severe erythema nodosum leprosum in leprosy patients and for the treatment of multiple myeloma, but it is not administered to pregnant women and preferably only to those women beyond child-bearing age. The target for thalidomide seems to be the protein cereblon (CRBN), which is involved in limb outgrowth.[73]

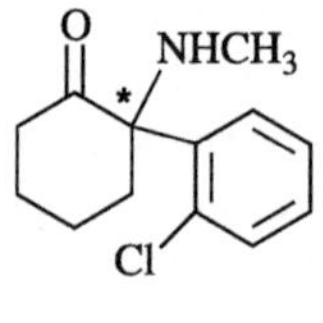

O NHCH$_3$ * Cl

3.13

O O NH N O O

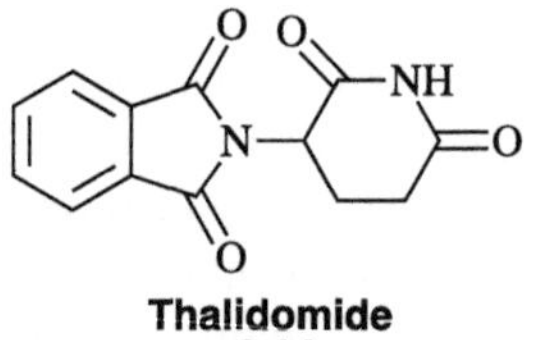

Thalidomide
3.14

It also is possible that both isomers are biologically active, but only one contributes to the toxicity, such as the local anesthetic prilocaine (**3.15**, Citanest).[74]

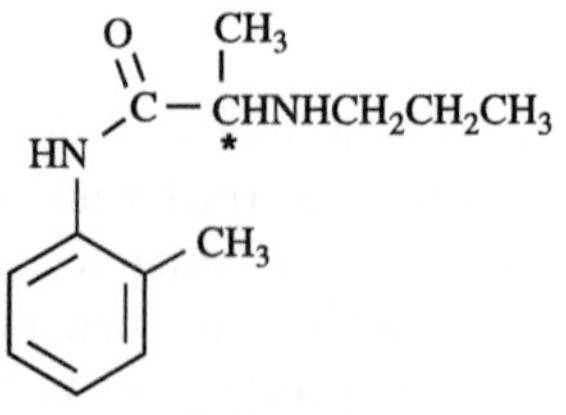

O CH$_3$ C − CHNHCH$_2$CH$_2$CH$_3$ * HN CH$_3$

Prilocaine
3.15

In some cases it is desirable to have both isomers present. Both isomers of bupivacaine (**3.16**, Sensorcaine) are local anesthetics, but only the *l*-isomer shows vasoconstrictive activity.[75] The experimental diuretic (increases water excretion) drug indacrinone (**3.17**) has a uric acid retention side effect. The *d*-isomer of **3.17** is responsible (i.e., is the eutomer) for both the diuretic activity and the uric acid retention side effect. Interestingly, however, the *l*-isomer acts as a uricosuric agent (reduces uric acid levels). Unfortunately, the ratio that gives the optimal therapeutic index (see Chapter 2; Section 2.2.4) is 1:8 (*d*:*l*), not 1:1 as is present in the racemic mixture.[76]

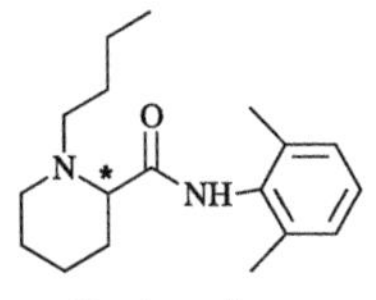

Bupivacaine
3.16

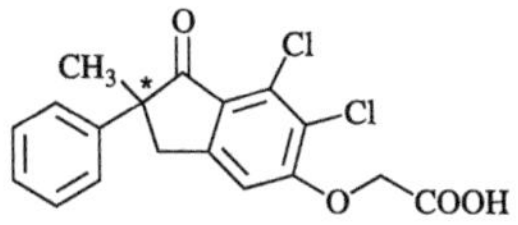

Indacrinone
3.17

Enantiomers may have different therapeutic activities as well.[77] Darvon (**3.18**), 2*R*,3*S*-(+)-dextropropoxyphene, is an analgesic drug and its enantiomer, Novrad (**3.19**), 2*S*,3*R*-(−)-levopropoxyphene, is an antitussive (anticough) agent, an activity that is not compatible with analgesic action. Consequently, these enantiomers are marketed separately. You may have noticed that the trade names are enantiomeric as well! The (*S*)-(+)-enantiomer of the antiinflammatory/analgesic drug ketoprofen (**3.20**, Orudis) is the eutomer; the (*R*)-(−)-isomer shows activity against bone loss in periodontal disease.

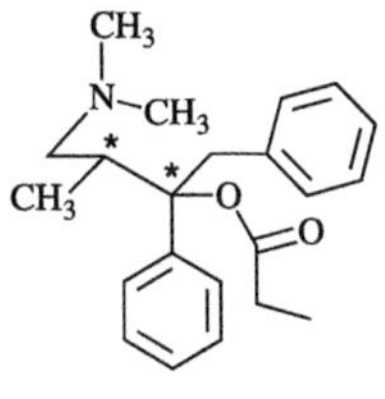

Dextropropoxyphene
3.18

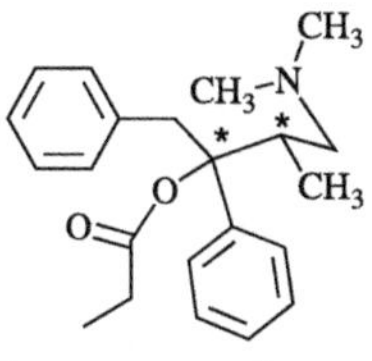

Levopropoxyphene
3.19

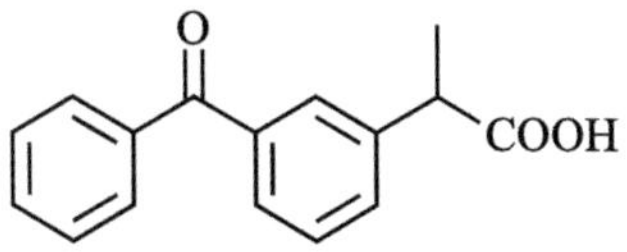

Ketoprofen
3.20

It, also, is possible for enantiomers to have opposite effects.[78] The (*R*)-(−)-enantiomer of 1-methyl-5-phenyl-5-propylbarbituric acid (**3.21**) is a narcotic, and the (*S*)-(+)-enantiomer is a convulsant![79] The (+)-isomer of the experimental narcotic analgesic picenadol (**3.22**) is an opiate agonist, the (−)-isomer is a narcotic antagonist, and the racemate is a partial agonist.[80] This suggests a potential danger in studying racemic mixtures; one enantiomer may antagonize the other, and no effect will be observed. For example, the racemate of UH-301 (**3.23**) exhibits no serotonergic activity; (*R*)-UH-301 is an agonist of the 5-hydroxytryptamine 1A (5-HT_{1A}) receptor, but (*S*)-UH-301 is an antagonist of the same receptor.[81] Consequently, no activity is observed with the racemate.

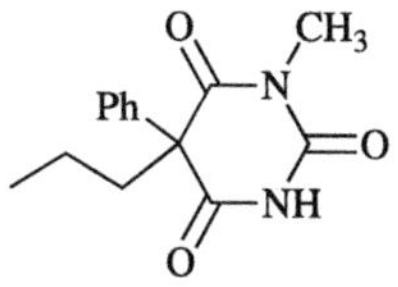

3.21

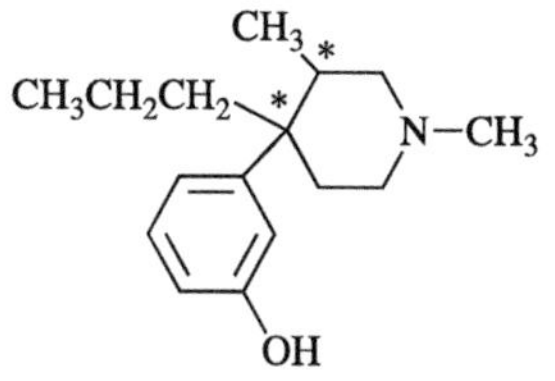

3.22

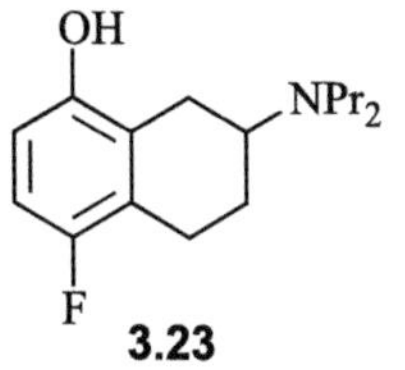

3.23

It is quite common for chiral compounds to show stereoselectivity with receptor action, and the stereoselectivity of one compound can vary for different receptors. For example, (+)-butaclamol (**3.24**) is a potent antipsychotic, but the (−)-isomer is essentially inactive; the eudismic ratio (+/−) is 1250 for the D_2-dopaminergic, 160 for the D_1-dopaminergic, and 73 for the α-adrenergic receptors. (−)-Baclofen (**3.25**, Lioresal) is a muscle relaxant that binds to the $GABA_B$ receptor; the eudismic ratio (−/+) is 800.[82]

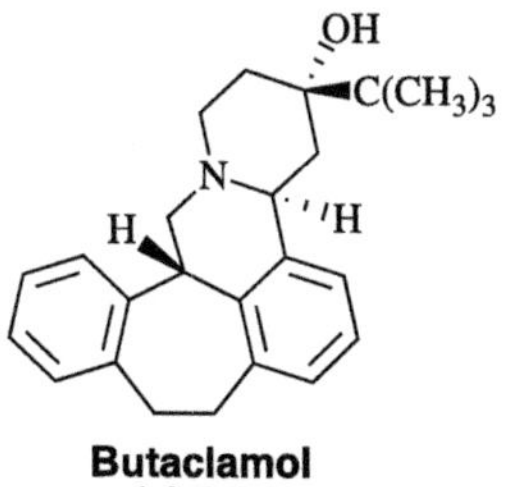

Butaclamol
(+)-3.24

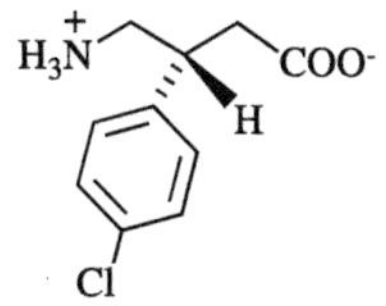

Baclofen
(-)-3.25

Remember that the (+)- and (−)-nomenclature refers to the effect of the compound on the direction of rotation of plane polarized light and has nothing to do with the stereochemical configuration of the molecule. The stereochemistry about a stereogenic carbon atom is noted by the *R*,*S* convention of Cahn et al.[83] Because the *R*,*S* convention is determined by the atomic numbers of the substituents about the stereogenic center, two compounds having the same stereochemistry, but a different substituent can have opposite chiral nomenclatures. For example, the eutomer of the antihypertensive agent, propranolol (Inderal) is the *S*-(−)-isomer (**3.26**, X = $NHCH(CH_3)_2$).[84] If X is varied so that the attached atom has an atomic number greater than that of oxygen, such as F, Cl, Br, or S, then the nomenclature rules dictate that the molecule is designated as an *R* isomer, even though there is no change in the stereochemistry. Note, however, that even though the absolute configuration about the stereogenic carbon remains unchanged after variation of the X group in **3.26**, the effect on plane polarized light cannot necessarily be predicted; the compound with a different substituent X can be either + or −. The most common examples of this phenomenon in nature are some of the amino acids. (*S*)-Alanine, for example, is the (+)-isomer and (*S*)-serine (same absolute stereochemistry) is the (−)-isomer; the only difference is a CH_3 group for alanine and a CH_2OH group for serine.

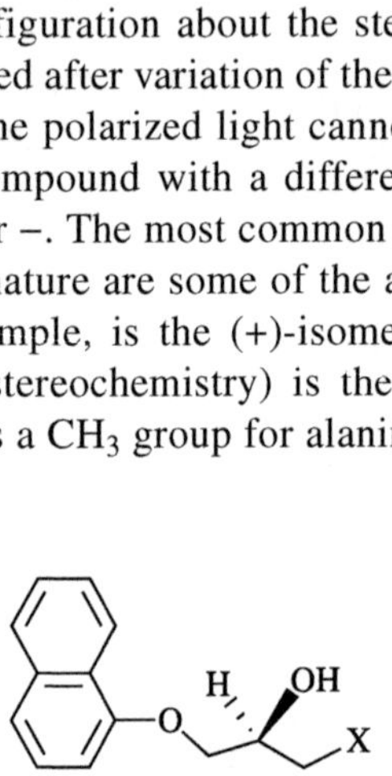

Propranolol
(-)-3.26

Propranolol (**3.26**, Inderal, X = $NHCH(CH_3)_2$), the first member of a family of drugs known as *β-blockers* (Sir James W. Black shared a Nobel Prize in Medicine in 1988 for this discovery), is a competitive antagonist (blocker) of the β-adrenergic receptor, which triggers a decrease in blood pressure and regulates cardiac rhythm and oxygen consumption for those with cardiovascular disease. The β_1- and β_2-adrenergic receptors are important to cardiac and bronchial vasodilation, respectively; propranolol is nonselective in its antagonism for these two receptors. The eudismic ratio (−/+) for propranolol is about 100; however, propranolol also exhibits local anesthetic activity for which the eudismic ratio is 1. The latter activity apparently is derived from some mechanism other than β-adrenergic receptor blockade. A compound of this type that has two separate mechanisms of action and, therefore, different therapeutic activities has been called a *hybrid drug* by Ariëns.[85] (+)-Butaclamol (**3.24**), which interacts with a variety of receptors, is another hybrid drug. However, butaclamol has three chiral centers and, therefore, has eight possible isomeric forms. When multiple isomeric forms are involved in the biological activity, the drug is called a *pseudohybrid drug*. Another important example of this type of drug is the antihypertensive agent, labetalol (Figure 3.23, Normodyne), which has two stereogenic centers and therefore exists in four stereoisomeric forms (two diastereomeric pairs of enantiomers), having the stereochemistries (*RR*), (*SS*), (*RS*), and (*SR*). In the drug form, labetalol contains a mixture of all four stereoisomers and has α- and β-adrenergic blocking properties (note that although labetalol has two stereogenic centers, all four isomers do not have to be included in the formulation, but they are in this case). The (*RR*)-isomer is predominantly the β-blocker (the eutomer for β-adrenergic blocking action), and the (*SR*)-isomer is mostly the α-blocker (the eutomer for α-adrenergic blocking); the other 50% of the isomers, the (*SS*)- and (*RS*)-isomers, are almost inactive (the isomeric ballast). Labetalol, then, is a pseudohybrid drug, a mixture of isomers having different receptor-binding properties.

Labetalol also is an example of how relatively minor structural modifications of an agonist can lead to transformation into an antagonist. *l*-Epinephrine (**3.27**) is a natural hybrid molecule that induces both α- and β-adrenergic effects. Introduction of the phenylalkyl substituent on the nitrogen transforms the α-adrenergic activity of the agonist *l*-epinephrine into the α-adrenergic antagonist labetalol. The modification of one of the catechol hydroxyl groups of *l*-epinephrine to a carbamyl group of labetalol changes the β-adrenergic action (agonist) to a β-adrenergic blocking action (antagonist).

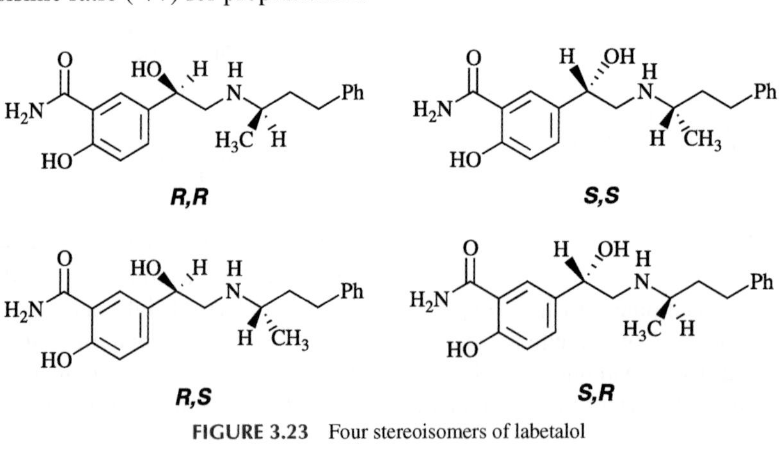

R,R **S,S** **R,S** **S,R**

FIGURE 3.23 Four stereoisomers of labetalol

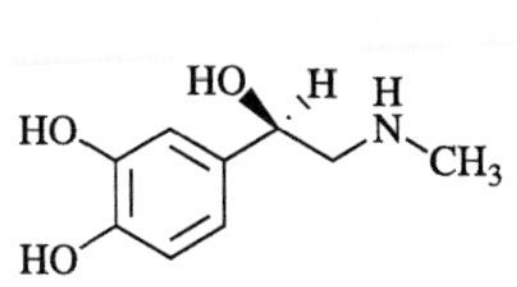

l-epinephrine
3.27

As pointed out by Ariëns[86] and by Simonyi,[87] it is quite common for mixtures of isomers, particularly racemates, to be marketed as a single drug, even though at least half of the mixture not only may be inactive for the desired biological activity but also may, in fact, be responsible for various side effects. In the early 1980s, only 58 of the 1200 drugs available were single enantiomers;[88] however, this has changed dramatically. In 2004, for the first time, all new chiral drugs introduced in the market (13 of them) were single enantiomers. This was motivated by guidelines issued by various regulatory agencies in the late 1980s and early 1990s (the FDA in 1992[89]), which allowed drug companies to choose whether to develop chiral drugs as racemates or single enantiomers, but required applicants with racemic drugs to submit rigorous scientific evidence why the racemate was developed rather than a single enantiomer.[90] Racemates were developed if it was discovered that the single enantiomer racemized easily in vitro and/or in vivo, if the enantiomers had similar pharmacological and toxicological profiles, or when the use of racemates resulted in synergistic effects (see Chapter 7), leading to better pharmacological or toxicological properties. The challenges, and sometimes prohibitive expense, in the synthesis of single enantiomer drugs have been assuaged by advancements in asymmetric synthesis and chiral separation technologies.

To further encourage companies to prepare and market single-entity drugs, the concept of a *racemic switch* (also called *chiral switch*) was introduced. This is the redevelopment in single enantiomer form of a drug that is being marketed as a racemate (the racemate is switched for the eutomer). Even if the racemate is currently covered by an active patent, the patent office would allow a new patent to a second company for the eutomer of the racemate. Of course, the same company can be awarded a patent for a racemic switch as well, which is an interesting strategy to extend the life of exclusivity for a drug. For example, AstraZeneca markets the antiulcer drug omeprazole (**3.28**, Prilosec) as a racemate, but shortly before the patent expired, a new patent was issued to the same company for the active (*S*)-isomer, which was approved for marketing as esomeprazole (Nexium). Because the racemate had already been approved by the FDA, less testing was needed for the active enantiomer. Interestingly, the (*R*)-isomer is more potent than either the (*S*)-isomer or racemate in rats; the two enantiomers are equipotent in dogs, and the (*S*)-isomer is most potent in humans (apparently because of the higher bioavailability and consistent pharmacokinetics compared with the other enantiomer).[91]

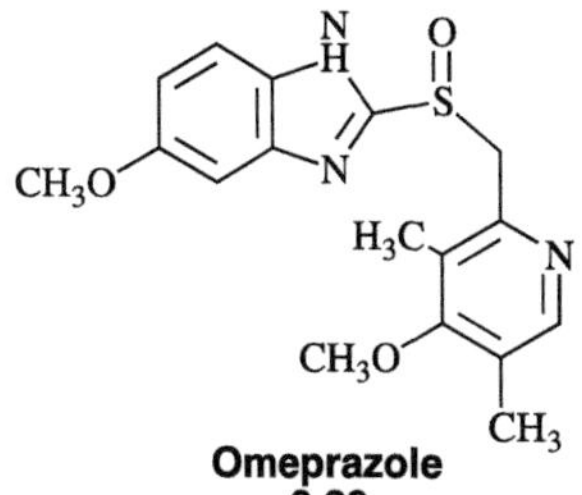

Omeprazole
3.28

The use of a single enantiomer is generally expected to lower side effects and toxicity. For example, the antiasthma drug albuterol (**3.29**, Ventolin/Proventil) is an agonist for β_2-adrenergic receptors on airway smooth muscle, leading to bronchodilation. The racemic switch, levalbuterol (the *R*-(−)-isomer, Xopenex), appears to be solely responsible for the therapeutic effect. The (*S*)-isomer seems to produce side effects such as pulse rate increases, tremors, and decreases in blood glucose and potassium levels. Because of this advantage, single isomer drug sales have been steadily increasing worldwide; in 1996, they accounted for 27% of the market, and in 2002, 39% of drug sales were of single enantiomers.[92]

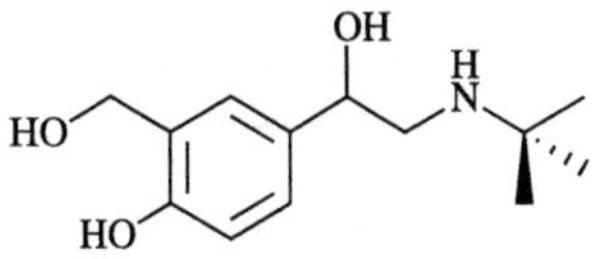

Albuterol
3.29

However, it is not always best to use the single enantiomer of the drug. The antidepressant drug fluoxetine (**3.30**, Prozac) is marketed as the racemate (in this case both isomers are active as selective serotonin reuptake inhibitors). Clinical trials with just the (*R*)-isomer at a higher dosage, however, produced a cardiac side effect. Another unusual problem associated with the use of a single enantiomer may occur if the two enantiomers have synergistic pharmacological activities. For example, the (+)-isomer of the antihypertensive drug nebivolol (**3.31**, Nebilet) is a β-blocker (see above); the (−)-isomer is not a β-blocker, but it is still a vasodilating agent (via the nitric oxide pathway), so the drug is sold as a racemate to take advantage of two different antihypertensive mechanisms. Sometimes an unexpected side benefit is associated with the use of a racemic mixture. The racemic calcium ion channel blocker (see Section 3.2.6) verapamil (**3.32**, Calan) has long been used as an antihypertensive drug. The (*S*)-isomer is the eutomer, but the (*R*)-isomer has been found to inhibit the resistance of cancer cells to anticancer drugs.[93]

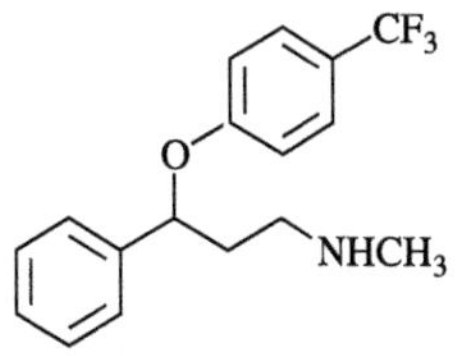

Fluoxetine
3.30

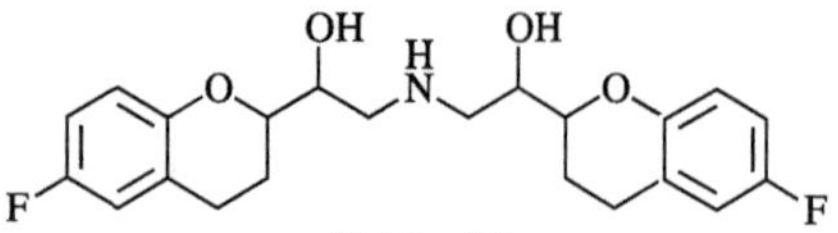

Nebivolol
3.31

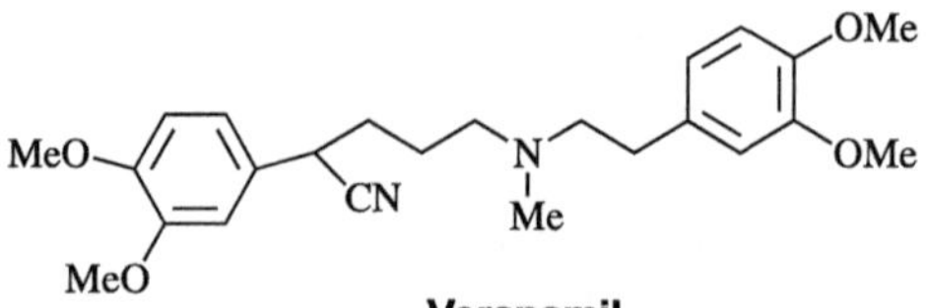

Verapamil
3.32

For cases in which the enantiomers are readily interconvertible in vivo, there is no reason to go to the expense of marketing a single enantiomer. Enantiomers of the antidiabetes drug rosiglitazone (**3.33**, Avandia) spontaneously racemize in solution, so it is sold as a racemate. Because of the reasons noted above for continuing the use of racemates, about 10% of annual drug approvals (13% for FDA and 9% worldwide) are still racemates.[94]

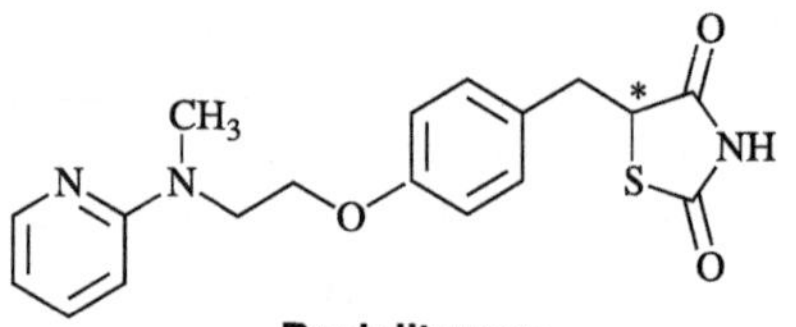

Rosiglitazone
3.33

Because of the potential vast differences in activities of two enantiomers, caution should be used when applying quantitative structure-activity relationship (QSAR) methods such as Hansch analyses (see Chapter 2; Section 2.2.6.2.2.1) to racemic mixtures. These methods really should be applied to the separate isomers.[95]

It is quite apparent from the above discussion that receptors are capable of recognizing and selectively binding optical isomers. Cushny[96] was the first to suggest that enantiomers could have different biological activities because one isomer could fit into a receptor much better than the other. How are they able to accomplish this? If you consider two enantiomers, such as epinephrine, interacting with a receptor that has only two binding sites (Figure 3.24), it becomes apparent that the receptor cannot distinguish between them. However, if there are at least three binding sites (Figure 3.25), the receptor easily can differentiate them. The *R*-(−)-isomer has three points of interaction and is held in the conformation shown to maximize molecular complementarity. The *S*-(+)-isomer can have only two sites of interaction (the hydroxyl group cannot interact with the hydroxyl binding site and may even have an adverse steric interaction); consequently it has a lower binding energy. Easson and Stedman[97] were the first to recognize this *three-point attachment* concept: a receptor can differentiate enantiomers if there are as few as three binding sites. As in the case of the β-adrenergic receptors discussed above, the structure of α-adrenergic receptors to which epinephrine

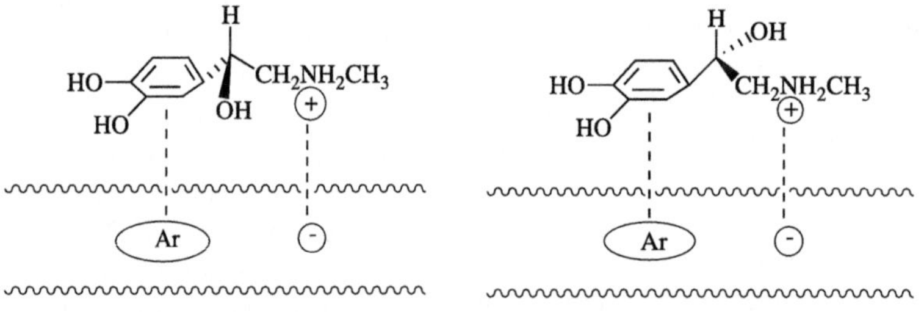

R-(-)-Epinephrine **S-(+)-Epinephrine**

FIGURE 3.24 Binding of epinephrine enantiomers to a two-site receptor. The wavy lines are the receptor surfaces.

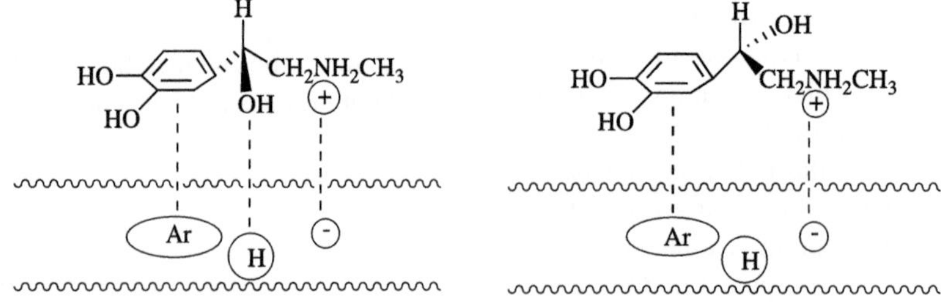

FIGURE 3.25 Binding of epinephrine enantiomers to a three-site receptor. The wavy lines are the receptor surfaces.

binds is unknown. α-Adrenergic receptors appear to mediate vasoconstrictive effects of catecholamines in bronchial, intestinal, and uterine smooth muscle. The eudismic ratio (*R/S*) for vasoconstrictor activity of epinephrine is only 12–20,[98] indicating that there is relatively little difference in binding energy for the two isomers to the α-adrenergic receptor. Although the above discussion was directed at the enantioselectivity of receptor interactions, it should be noted that there is also enantioselectivity with respect to pharmacokinetics, i.e., absorption, distribution, metabolism, and excretion, which will be discussed in Chapter 8.[99]

As noted in Chapter 2 (Section 2.1.2.3.1), a relatively large percentage of antiinfectives and antitumor compounds are natural products or are analogs of natural products. The above discussion would suggest that the chirality of natural products should be very important to their biological activities. This is true; however, the nonnatural enantiomer of the natural product could be even more potent than the natural product. For example, *ent*-(−)-roseophilin (**3.34**), the unnatural enantiomer of the natural antitumor antibiotic, is 2–10 times more potent than the natural (+)-isomer in cytotoxicity assays,[100] and *ent*-fredericamycin A (**3.35**) is as cytotoxic as its natural enantiomer.[101] Why should that be so? The organisms that produce these natural products may not be producing them for the purpose of protecting themselves from the disease state we have in mind for these compounds. After all, are these organisms really concerned with developing cancer? There are many possible mechanisms of action for antitumor agents, and the *ent*-natural product may bind to the relevant receptor better than the natural product does.

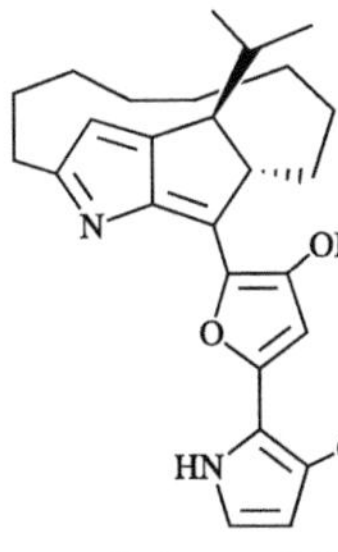

***Ent*-roseophilin**
3.34

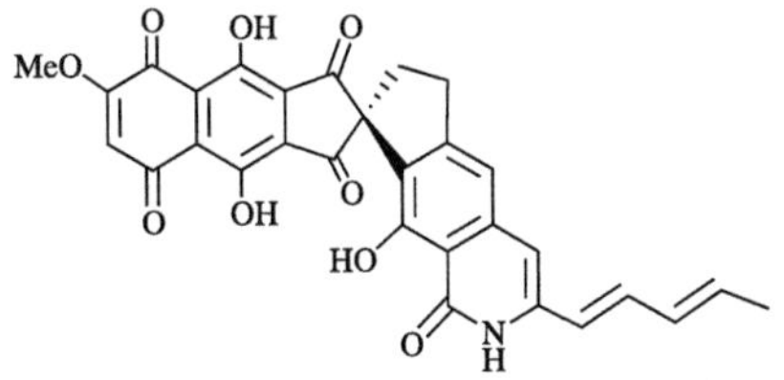

***Ent*-fredericamycin A**
3.35

3.2.5.3. Diastereomers

Two (or more) compounds having different spatial arrangements (i.e., are stereoisomers) that are not mirror images of each other (i.e., are not enantiomers of each other) are diastereomers. Geometric isomers (*E*- and *Z*-isomers[102]) are a special case of diastereomers. Epimers (a pair of compounds with multiple stereogenic centers that have opposite configuration at only one stereocenter) are another special case. Diastereomers are different compounds, having different energies and stabilities. As a result of their different configurations, receptor interactions will be different. Unlike enantiomers, which are relatively difficult to separate, diastereomers often can be easily separated by chromatography or recrystallization, so they should be tested separately. The antihistamine activity of *E*-triprolidine (**3.36a**, found in cold remedies, such as Actifed) was found to be 1000-fold greater than the corresponding *Z*-isomer (**3.35b**).[103] Likewise, the neuroleptic potency of the *Z*-isomer of the antipsychotic drug chlorprothixene (**3.37a**, Taractan) is more than 12 times greater than that of the corresponding *E*-isomer (**3.37b**).[104] On the other hand, the *E*-isomer of the anticancer drug diethylstilbestrol (**3.38a**) has 14 times greater estrogenic activity than the *Z*-isomer (**3.38b**), possibly because its overall structure and the interatomic distance between the two hydroxyls in the *E*-isomer are similar to that of estradiol (**3.39**).

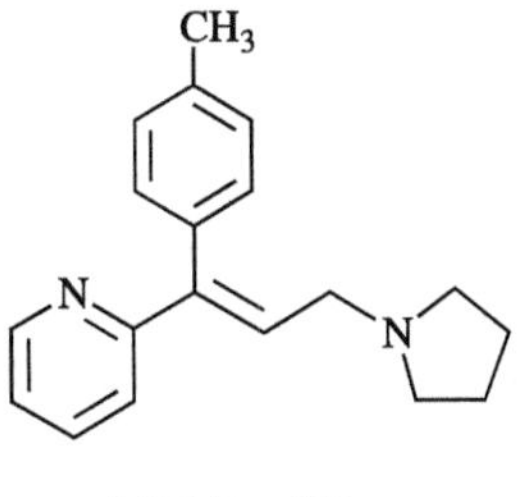

(*E*)-Triprolidine
3.36a

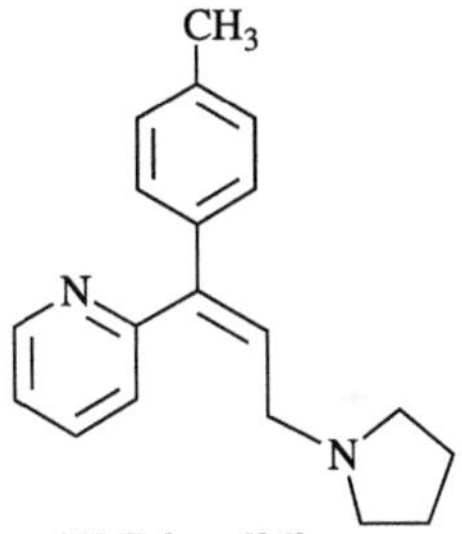

(*Z*)-Triprolidine
3.36b

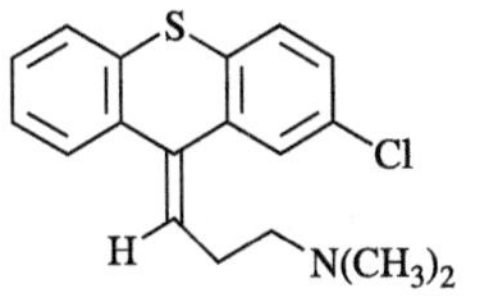

(*Z*)-Chlorprothixene
3.37a

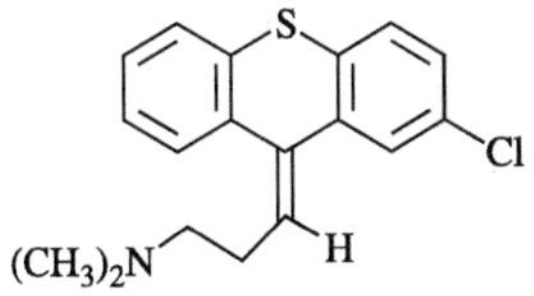

(*E*)-Chlorprothixene
3.37b

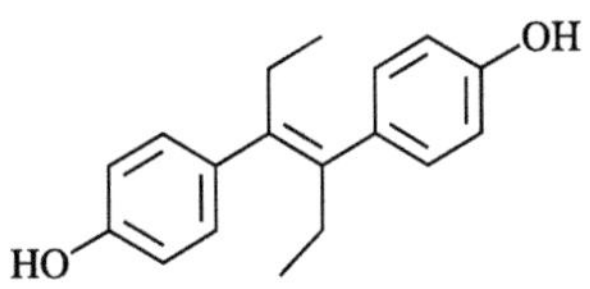

Diethylstilbesterol
3.38a

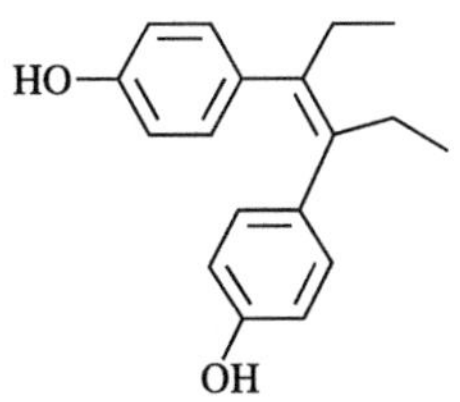

3.38b

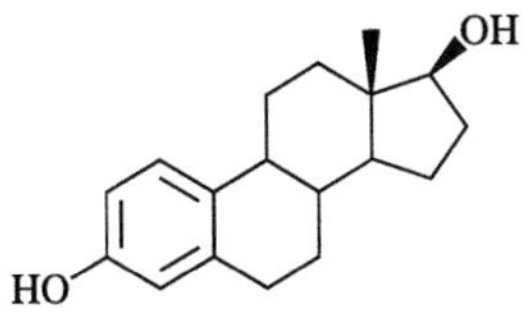

3.39

Although, in some cases, the *cis*- and *trans*-nomenclature does correspond with *Z*- and *E*-, respectively, it should be kept in mind that these terminologies are based on different conventions, so there may be confusion. The *Z,E* nomenclature is unambiguous, and should be used.

3.2.5.4. Conformational Isomers

Diastereomers and enantiomers can be separated, isolated, and screened individually. There are isomers, however, that typically cannot be separated, namely *conformational isomers* or *conformers* (isomers generated by a change in conformation). As a result of free rotation about single bonds in acyclic molecules and conformational flexibility in many cyclic compounds, a drug molecule can assume a variety of *conformations*, i.e., the location of the atoms in space without breakage of bonds. The pharmacophore of a molecule is defined not only by the configuration of a set of atoms but also by the bioactive conformation in relation to the receptor-binding site. A receptor may bind only one of the conformers. As was pointed out in Chapter 2 (Section 2.2.4.4), the conformer that binds to a receptor need *not* be the lowest energy conformer observed in the crystalline state, as determined by X-ray crystallography, or found in solution, as determined by NMR spectrometry, or determined theoretically by molecular mechanics calculations. The binding energy to the receptor may overcome the barrier to the formation of a higher energy conformation. In order for drug design to be efficient, it is extremely helpful to know the *bioactive conformation* (the active conformation when bound to the receptor) in the drug–receptor complex. Figure 3.26 is a crystal structure of the antidiabetic drug rosiglitazone (see **3.33**, Avandia) bound to the peroxisome proliferator-activated receptor gamma (PPARγ), a transcription factor.[105] Note that the bioactive conformation of the drug bound to the receptor is an inverted U shape, rather than an extended conformation. Compounds that cannot attain this inverted U structure will not be able to bind to that site.

If a lead compound has low potency, it may only be because the population of the active conformer in solution is low (higher in energy); for example, with PPARγ shown in Figure 3.26, an inverted U conformation of the compound is essential for high potency. The energy of a conformer will determine the relative population of that conformer in the equilibrium mixture of conformers. A higher energy conformer will be in lower concentration in the equilibrium mixture of conformers. Therefore, if the bioactive conformation is a high-energy conformer, the K_d for the molecule will appear high (poor affinity), not because the structure of the compound is incorrect, but because the population of the ideal conformation is so low. If the conformation of the ideal conformer were in higher concentration, the K_d would be much lower. To give you a simple example of conformational populations from organic chemistry, consider 1-*tert*-butylcyclohexane (Scheme 3.2). Cyclohexanes can exist in numerous conformations, including a chair form with the substituent in the equatorial position (**a**), a half-chair (**b**), a boat (**c**) (including twist-boat), a different half-chair (**d**), and a chair conformer with the substituent axial (**e**). The difference in free energy for the two chair conformers with the *tert*-butyl group either equatorial (**a**) or axial (**e**), which in a receptor-binding site would make an enormous difference on binding effectiveness, is −5.4 kcal/mol; this translates into an equilibrium mixture ([equatorial]/[axial]) at 37 °C of 6619 ($\Delta G^0 = -RT \ln K$). If the axial conformer were the bioactive conformation, and the mixture were 1 μM in 1-*tert*-butylcyclohexane, it would only be 0.00015 μM (i.e., 150 pM) in the axial conformer of 1-*tert*-butylcyclohexane. This would lead to the conclusion that 1-*tert*-butylcyclohexane was inactive, whereas, if only the axial conformer existed in solution, it could be the most potent binder ever observed for that receptor.

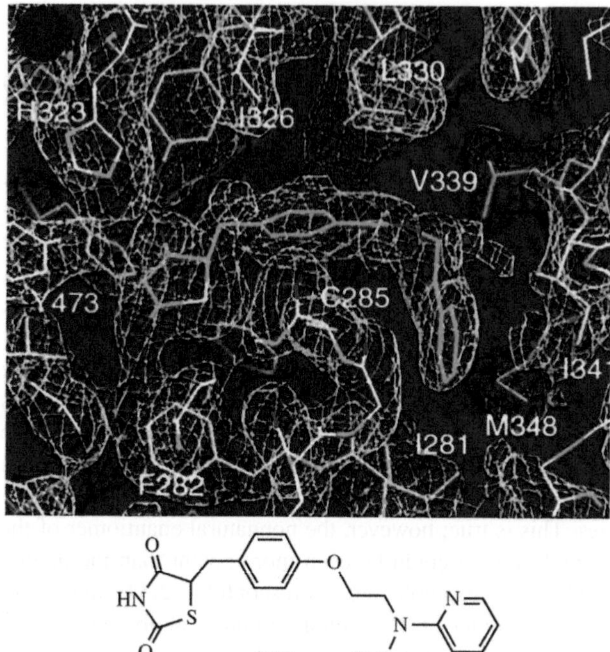

FIGURE 3.26 An example of a bioactive conformation. Rosiglitazone (**3.33**), an antidiabetic drug (green structure in the middle), bound to PPARγ. Note the sickle-shaped conformation in the binding site to accommodate the shape formed by the active site residues. *Reprinted from* Mol. Cell, *Vol. 5, "Asymmetry in the PPARγ/RXRα Crystal Structure Reveals the Molecular Basis of Heterodimerization among Nuclear Receptors", pp. 545–555. Reprinted with permission of Elsevier.*

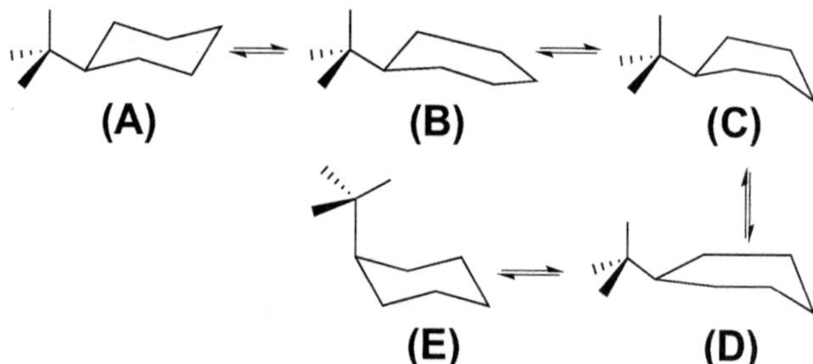

SCHEME 3.2 Cyclohexane conformations. **a**, chair (substituent equatorial); **b**, half-chair; **c**, boat; **d**, half-chair; **e**, chair (substituent axial).

A unique approach has been taken to determine the bioactive conformation of a drug molecule in the drug–receptor complex. This approach involves the synthesis of *conformationally rigid analogs* of flexible drug molecules. The potential pharmacophore becomes locked in various configurations by judicious incorporation of cyclic or unsaturated moieties into the drug molecule. These conformationally rigid analogs are, then, tested, and the analog with the optimal activity (or potency) can be used as the prototype for further structural modification. Conformationally rigid analogs are propitious because key functional groups, presumably part of the pharmacophore, are constrained in one position, thereby permitting the determination of the *pharmacophoric conformation*. The major drawback to this approach is that in order to construct a rigid analog of a flexible molecule, usually additional atoms and/or bonds must be attached to the original compound, and these can affect the chemical and physical properties. Consequently, it is imperative that the conformationally rigid analog and the drug molecule be as similar as possible in size, shape, and mass.

First we will look at the conformationally rigid analog approach to determine the bioactive conformation of a natural ligand (a neurotransmitter); then we will apply this methodology to lead modification. An example of the use of conformationally rigid analogs for the elucidation of receptor-binding site topography is the study of the interaction of the neurotransmitter, acetylcholine (ACh), with its receptors. There are at least two important receptors for ACh, one activated by the alkaloid muscarine (**3.40**) and the other by the alkaloid nicotine (**3.41**; presumably in the protonated pyrrolidine form); binding of nicotine to the ACh receptor is stabilized by a strong cation–π interaction between the ammonium ion of nicotine and a tryptophan residue in the receptor.[106] ACh has a myriad of conformations; four of the more stable possible conformers (groups staggered) are **3.42a–3.42d**. There are also conformers with groups eclipsed that are higher in energy. Four different *trans*-decalin stereoisomers were synthesized[107] (**3.43a–d**) corresponding to the four ACh conformers shown as **3.42**. All four isomers exhibited low muscarinic receptor activity; however, **3.43a** (which corresponds to the most stable conformer, the *anti*-conformer) was the most potent (0.06 times the potency of ACh). The low potency of **3.43a** is believed to be the result of the additional atoms present in the *trans*-decalin moiety. A comparison of *erythro*-(**3.44**) and *threo*-2,3-dimethylacetylcholine (**3.45**) gave the startling result that **3.44** was 14 times *more* potent than ACh and **3.45** was only 0.036 times as potent as ACh; in one case, the additional methyl groups enhanced the potency, and in the other case, they decreased the potency. Compound **3.43a** corresponds to *threo*-isomer **3.45**, and, therefore, is expected to have low potency. The corresponding *erythro* analog (**3.44**) does not have a *trans*-decalin analogy, so it could not be tested. To minimize the number of extra atoms added to ACh, *trans*-(**3.46**) and *cis*-1-acetoxy-2-trimethylammoniocyclopropanes (**3.47**) were synthesized and tested[108] for *cholinomimetic properties*, i.e., production of a response resembling that of ACh. The (+)-*trans*-isomer (shown in **3.46**)[109] has about the same muscarinic activity as does ACh, thus indicating the importance of minimizing additional atoms; the (−)-*trans*-isomer is about 1/500th the potency of ACh. This strongly supports the *anti*-conformer (**3.42a**) as the bioactive conformer. Unfortunately, the other conformers cannot be modeled by substituted cyclopropane analogs; the *cis*-isomer (**3.47**) models an eclipsed conformer of ACh. Nonetheless, the racemic *cis*-isomer has negligible activity. The (+)-*trans*-isomer has the same absolute configuration as the active enantiomers of the two muscarinic receptor agonists muscarine and acetyl β-methylcholine. These results suggest that ACh binds in an extended form (**3.42a**). However, both the *cis*- and the *trans*-cyclopropyl isomers, as well as all of the *trans*-decalin stereoisomers (**3.43a–d**), were only weakly active with the nicotinic cholinergic receptor. Because **3.46**, the lowest energy conformer, corresponding to ACh, is not active with the nicotinic ACh receptor, it can be supposed that a higher energy conformer is. This supports the concept that the lowest energy conformer does not have to be the one that binds to a receptor. In general, for this type of analysis to be convincing, it is best to identify a conformationally rigid analog that exhibits comparable or improved potency to that of the flexible compound being mimicked; otherwise, there could be many alternative interpretations for the results.

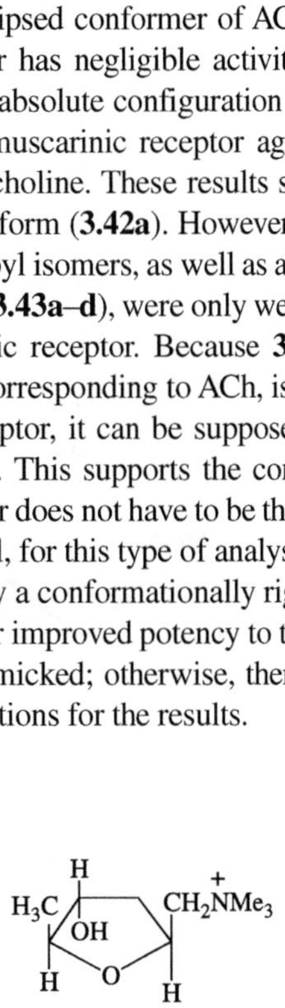

Muscarine
3.40

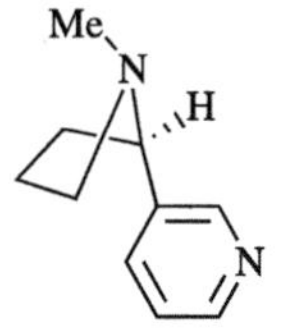

Nicotine
3.41

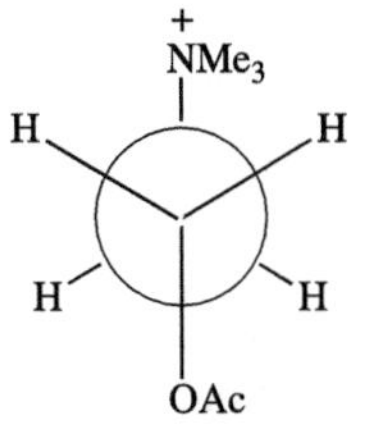

3.42a

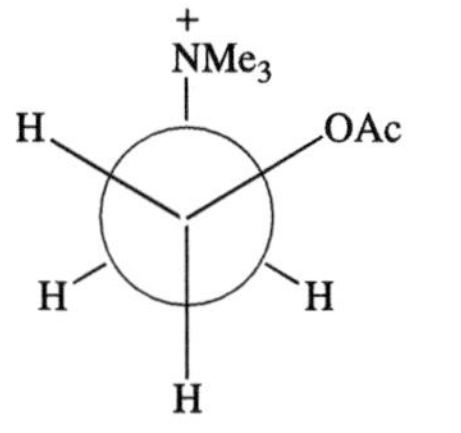

3.42b

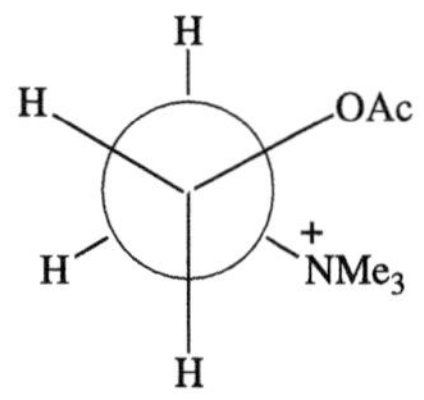

3.42c

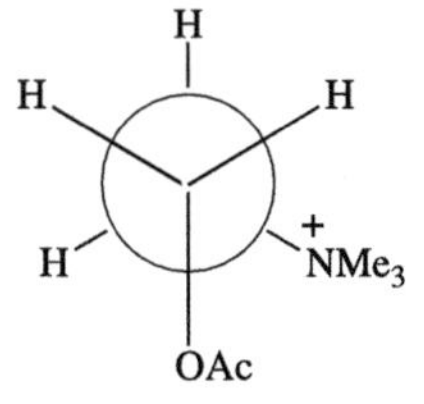

3.42d

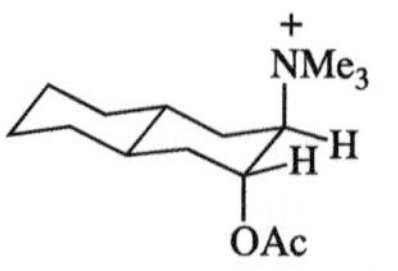

3.43a

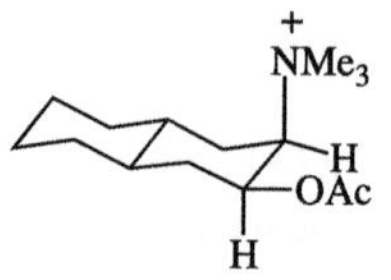

3.43b

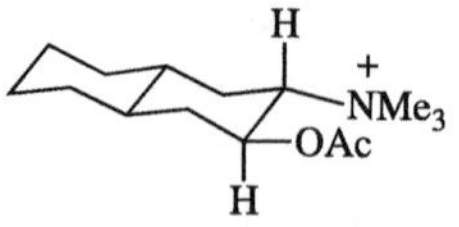

3.43c

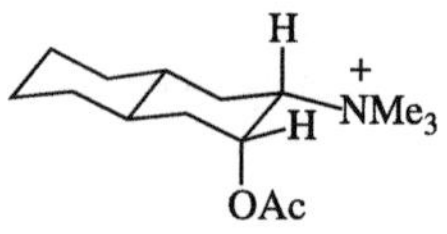

3.43d

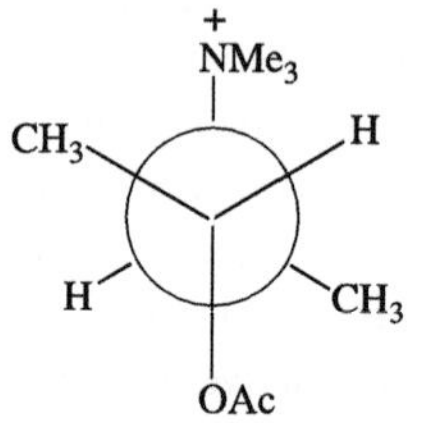

3.44

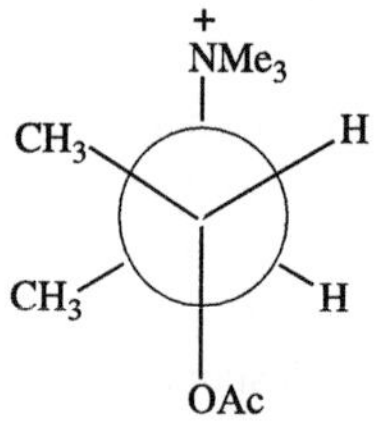

3.45

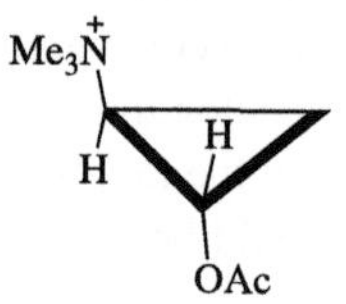

3.46

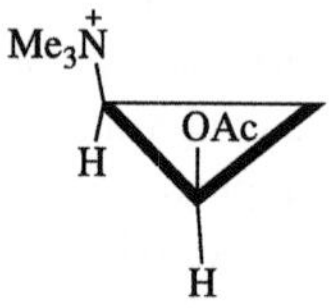

3.47

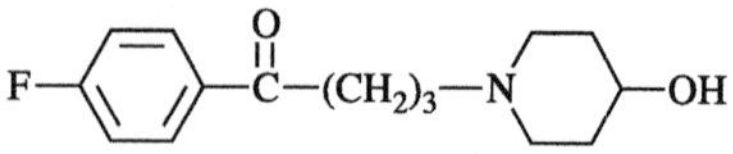

3.48

An example of the use of conformationally rigid analogs in drug design was reported by Li and Biel.[110] 4-(4-Hydroxypiperidino)-4′-fluorobutyrophenone (**3.48**) was found to have moderate tranquilizing activity in lower animals and man; however, unlike the majority of antipsychotic butyrophenone-type compounds, it only had minimal antiemetic (prevents vomiting) activity. The piperidino ring can exist in various conformations (**3.49a–d**, R=F-$C_6H_4CO(CH_2)_3$-), including two chair forms (**3.49a** and **3.49d**) and two twist-boat forms (**3.49b** and **3.49c**). The difference in free energy between the axial and equatorial hydroxyl conformers of the related compound, *N*-methyl-4-piperidinol (**3.49**, R=Me) is 0.94±0.05 kcal/mol at 40 °C (the equatorial conformer is favored by a factor of 4.56 over the axial conformer).[111] Energies for the twist-boat conformers are about 6 kcal/mol higher, but because of hydrogen bonding, **3.49b** should be more stable than **3.49c**. On the assumption that the chair conformers are more likely, three conformationally rigid chair analogs, **3.50–3.52**, were synthesized to determine the effect on receptor binding of the hydroxyl group in the equatorial (**3.50**), axial (**3.51**), and both (**3.52**) positions. Of course, with **3.52**, it must be assumed that, if the hydroxyl group is involved in hydrogen bonding, it is as an acceptor, not as a donor. Also, for synthetic reasons, the conformationally rigid analog of **3.49d** could not be made; instead, the diastereomer (with the R group still equatorial, but the hydroxyl group axial) was synthesized. This study, then, provides data for the preference of the position of the hydroxyl group, not strictly for the conformer preference. When subjected to muscle relaxation tests, the order of potency was **3.51**>**3.52**>**3.50**, indicating again that the conformationally less stable compound with the axial hydroxyl group has better molecular complementarity with the receptor than does the more stable compound with the equatorial hydroxyl group. This suggests that further analogs should be prepared where the axial hydroxyl is the more stable conformer or where it can be held in that configuration.

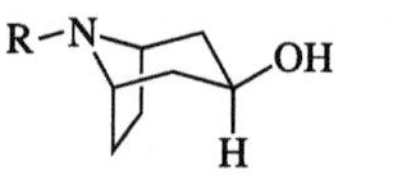

3.50

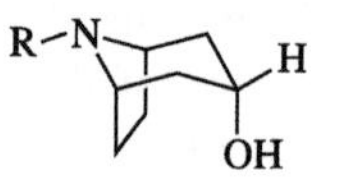

3.51

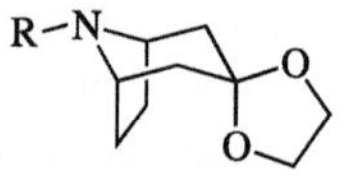

3.52

Another use of conformationally rigid analogs is to determine the appropriate orientation of pharmacophoric groups for binding to related receptors of unknown structure. The *N*-methyl-D-aspartate (NMDA) subclass of glutamate receptors is composed of an ion channel with multiple binding sites, including one for phencyclidine (PCP, **3.53**, Semylan, Figure 3.27). PCP analogs can bind to the PCP site of the NMDA receptor, the σ receptor, and the dopamine-D_2 receptor.[112] Neither the physical nature nor endogenous ligands for the σ receptor has been identified, but several structurally unrelated ligands are known.[113] PCP is a flexible molecule that can undergo conformational ring inversion of both the cyclohexyl and piperidinyl rings as well as rotation of the phenyl group. The various conformations place the ammonium ion and the phenyl ring in different spatial orientations, which may be responsible for binding to the various receptor sites. Conformationally rigid analogs of PCP were synthesized that fixed the orientation of the ammonium center of the PCP with

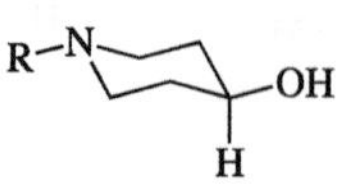

3.49a

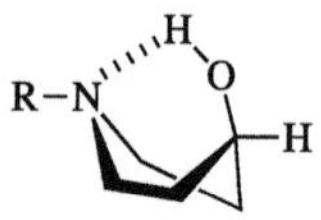

3.49b

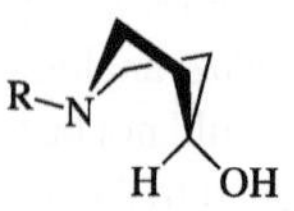

3.49c

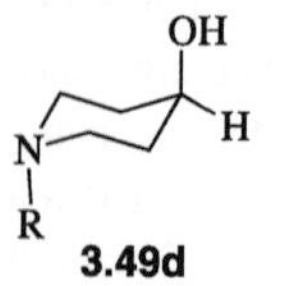

3.49d

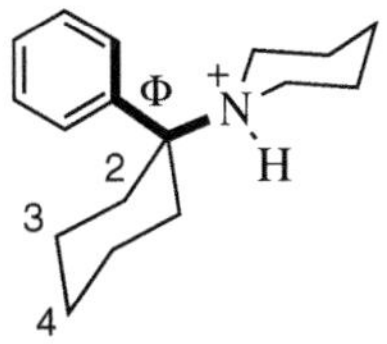

Phencyclidine (PCP)
3.53

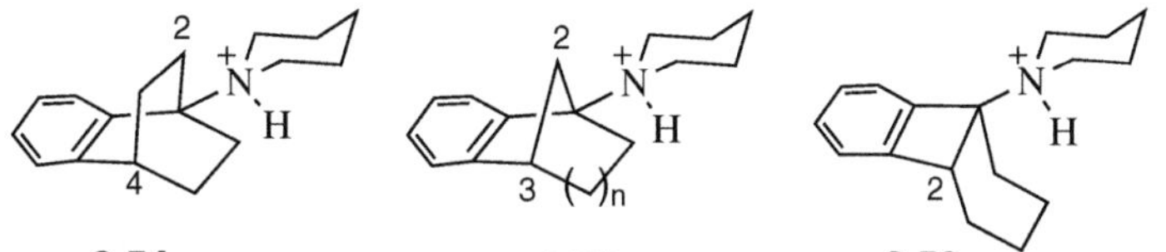

3.54 3.55 3.56

FIGURE 3.27 PCP, 3.53 and three conformationally rigid analogs of PCP

respect to the centrum of the phenyl ring to determine the importance of conformation on selectivity between the PCP and σ sites (Figure 3.27; ϕ is the angle defined by the darkened bonds in **3.53**).[114] The designed analogs incorporated a new bond connecting the *ortho* position of the phenyl ring to the 4-, 3-, or 2-position of the cyclohexane ring to give, respectively, **3.54**, **3.55** (n=2), and **3.56**; for synthetic reasons, **3.55** (n=1) was actually prepared and tested in place of **3.55** (n=2). In **3.54**, ϕ is 0°; in **3.55,** ϕ is 30°; and in **3.56,** ϕ is 60°. As the rigidity increases (**3.56**→**3.55**→**3.54**), the affinity for the PCP site of the NMDA receptor is diminished, and none binds well to this site (the best, **3.56**, only has 2% of the affinity of PCP). However, all three bind well to the σ site, almost twice as well as does PCP itself and fit a pharmacophore model for the σ receptor.[115]

As described in Chapter 2 (Section 2.2.4.5) peptides are unfavorable structures for drug discovery because they are too polar and flexible. Polarity, and some flexibility, can be handled with the use of peptidomimetics and bioisosteres (Chapter 2, Section 2.2.4.3). A conformationally rigid analog approach was taken to lock in the amide bond conformation, which generally favors the trans-conformation, and give either the cis- or trans-conformation using a 1,4- or 1,5-disubstituted 1,2,3-triazole, respectively, as an amide bond bioisostere (Figure 3.28).[116]

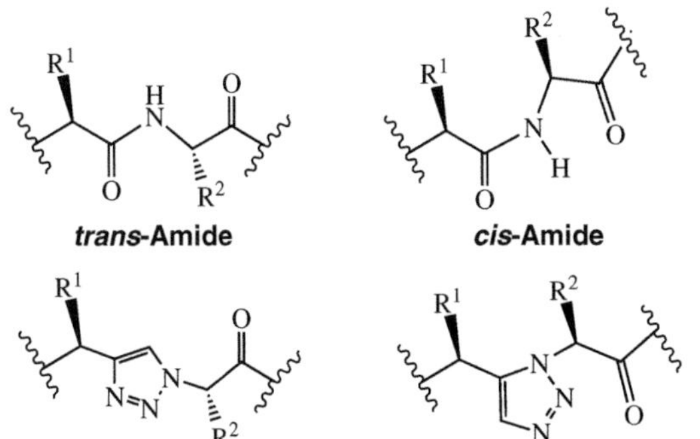

FIGURE 3.28 Use of a triazole as a conformationally rigid bioisostere to lock in an amide bond conformation

3.2.5.5. Atropisomers

Another type of conformational chirality, called *atropisomerism*, occurs when there is hindered rotation about a single bond as a result of steric or electronic constraints, causing slow interconversion of two conformers (a rule of thumb is having a half-life >1000 s) (Figure 3.29).[117] Slow interconversion can lead to two conformers, thereby giving a single chiral compound, a mixture of two chiral compounds (diastereomers), or a racemate, depending on whether there are stereogenic centers in the molecule and the rate of interconversion. This can be problematic in drug design, if you are assuming that the molecule exists as a single structure.[118] Energy barriers to rotation were calculated using quantum mechanics, and it was estimated that 20 kcal/mol was the minimum bond rotation energy to distinguish between atropisomers and nonatropisomers with a prediction accuracy of 86%.[119] Figure 3.30 shows an example of three analogs having different bond rotation energies that are (**a**) nonatropisomeric, (**b**) unstably atropisomeric, and (**c**) stably atropisomeric. The stable atropisomers can be separated and treated as two individual compounds, as in the case of enantiomers. The utility of this calculation is found in the prediction and validation of atropisomerism during the hit-to-lead (Chapter 2, Section 2.1.2.3.5) and lead modification (Chapter 2, Section 2.2) stages of drug discovery. Telenzepine (**3.57**, Figure 3.31), an anticholinergic compound, is atropisomeric, and the enantiomers have been resolved; the (+)-isomer is

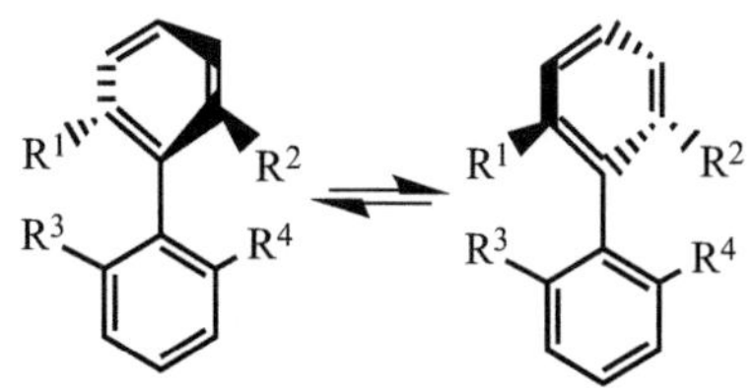

FIGURE 3.29 General example of atropisomerization

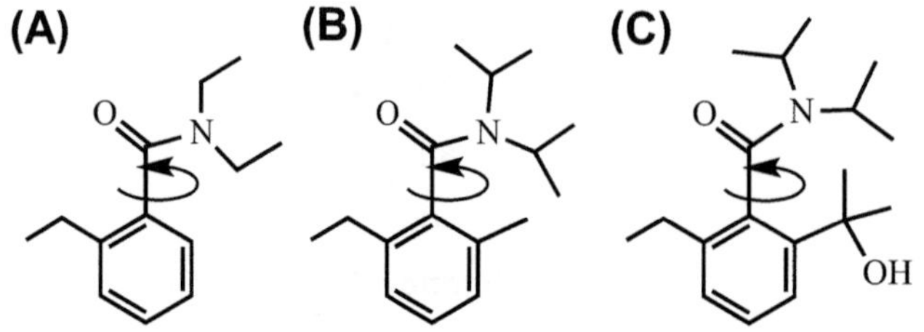

	(A)	(B)	(C)
$E_{rotation}$ (kcal/mol^{-1})	14.2	25.7	30.0
$t_{1/2}$ (rotation)	0.002 second	10 days	>10 years

FIGURE 3.30 Example of a nonatropisomer, an unstable atropisomer, and a stable atropisomer

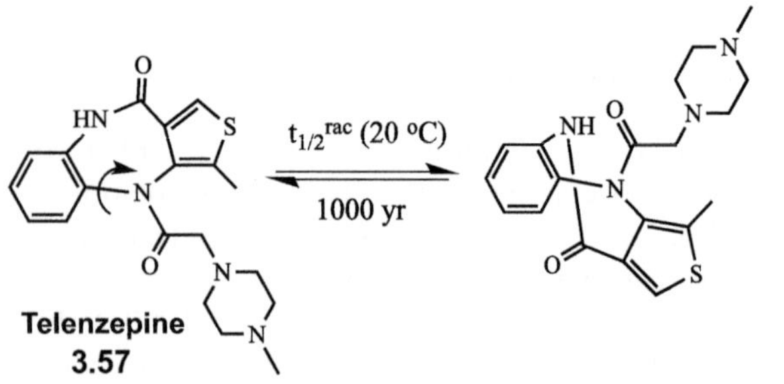

Telenzepine 3.57

$t_{1/2}^{rac}$ (20 °C) 1000 yr

FIGURE 3.31 Exceedingly slow isomerization of atropisomers of telenzepine (**3.57**)

atropisomeric or nonatropisomeric. For example, a neurokinin 1 (NK_1) antagonist (**3.58**) was identified at AstraZeneca for the treatment of depression, but it was shown to be a composite of four atropisomers because of restriction about two amide single bonds.[120] The active atropisomer had conformation **3.59**. A conformationally rigid analog of **3.59** was designed (**3.60**), which, unfortunately, was found to exist as two diastereomeric atropisomers, but **3.61** existed as a single conformer (the additional methyl group hinders rotation), which had excellent potency in vitro and in vivo.

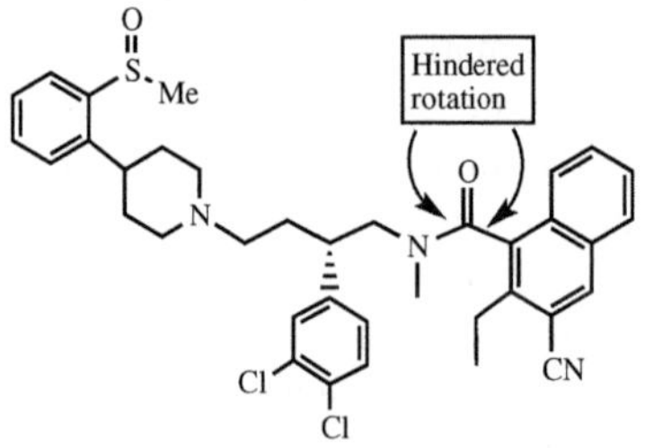

Hindered rotation

3.58 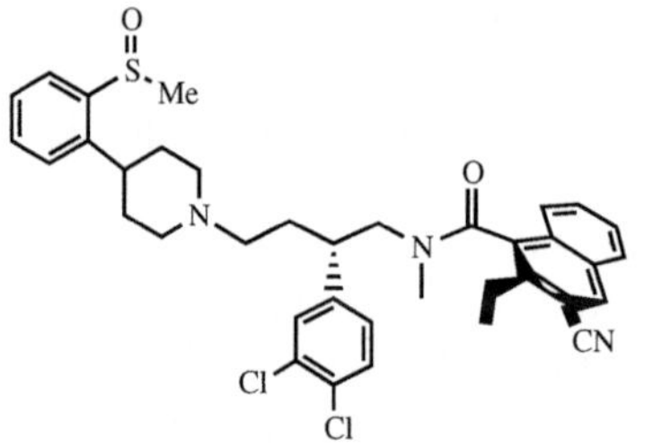 **3.59** 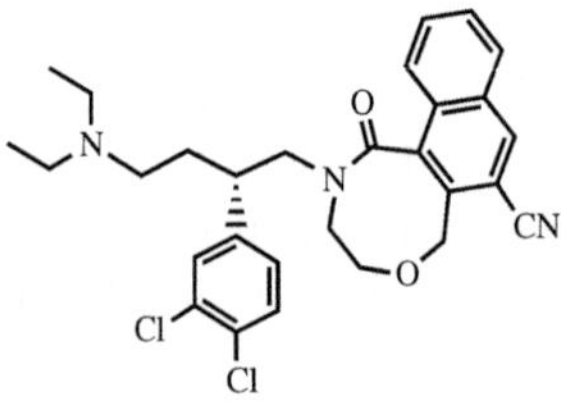 **3.60** 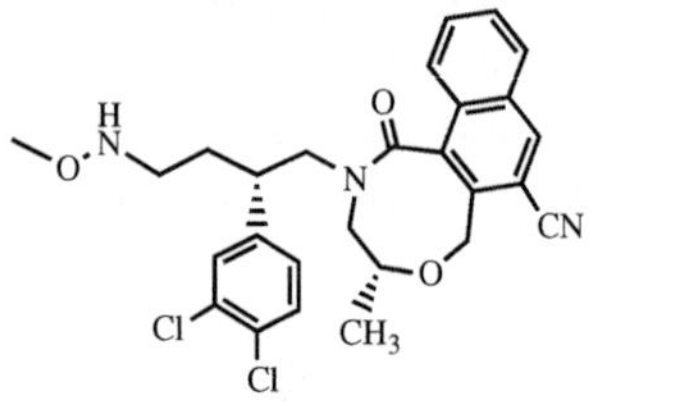 **3.61**

500 times more potent than the (−)-isomer at muscarinic receptors.

It is important to recognize when your compounds are potentially atropisomeric so you can either be sure that they are chiral or not and can ascertain how stable they are. If they are in the pseudo-atropisomer regime, you may need modification to make them either stably

Another way to deal with atropisomerism is to engineer the molecule so that it has faster bond rotation, making the conformers interconvertible. Compound **3.62** was an effective monocarboxylate transporter 1 antagonist having immunomodulatory activity but existed in four atropisomeric forms.[121] Modification to **3.63** allowed all conformers to readily interconvert.

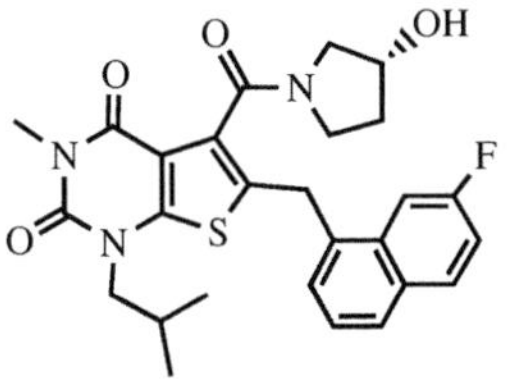

3.62

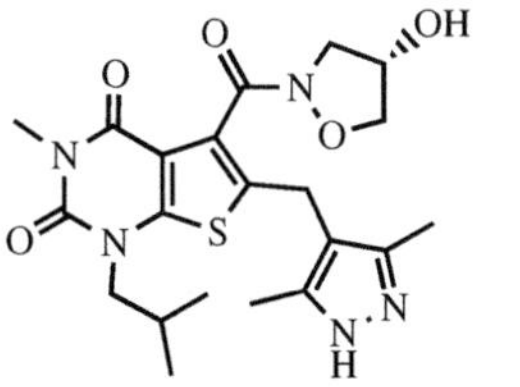

3.63

A third way to avoid atropisomerism is by symmetrization. A group at Schering-Plough (now Merck) was interested in developing a C–C chemokine receptor type 5 (CCR5) antagonist, which inhibits human immunodeficiency virus (HIV) entry into host cells.[122] A clinical candidate (**3.64**, SCH 351125) reduced levels of HIV-1 RNA in infected patients, but it existed in four atropisomeric forms (a pair of diastereomeric enantiomers). Symmetric isomer **3.65** eliminated the two diastereomeric conformers, and the two remaining enantiomeric conformers were rapidly interconverted.[123]

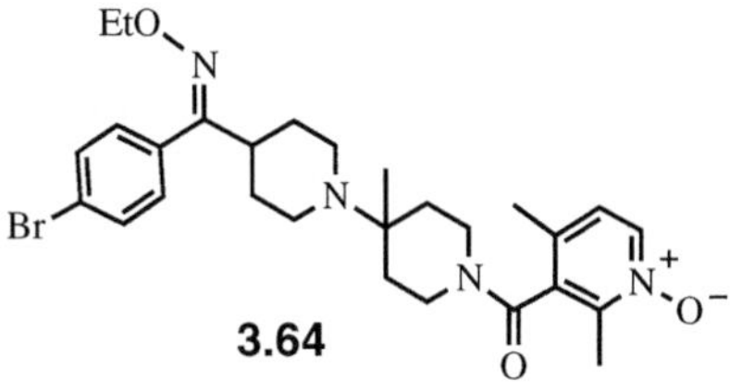

3.64

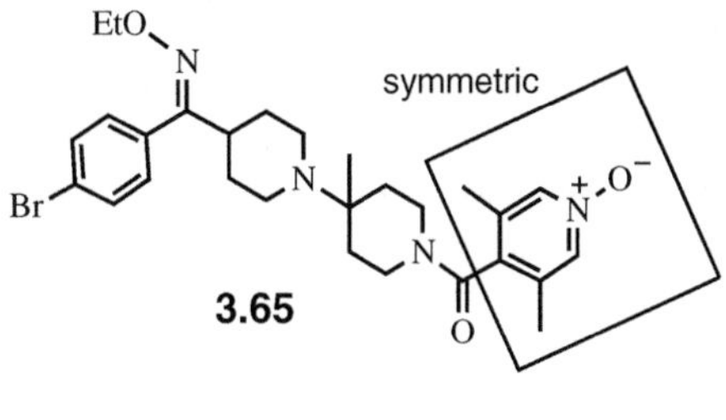

symmetric

3.65

3.2.5.6. *Ring Topology*

Tricyclic psychomimetic drugs show an almost continuous transition of activity in going from structures such as the tranquilizer chlorpromazine (**3.66**, Thorazine) through the antidepressant amitriptyline (**3.67**, Elavil), which has a tranquilizing side effect, to the pure antidepressant agent imipramine (**3.68**, Tofranil).[124] Stereoelectronic effects seem to be the key factor, even though tranquilizers and antidepressants have different molecular mechanisms. Three angles can be drawn to define the positions of the two aromatic rings in these compounds (Figure 3.32). The angle α (**3.69**) describes the bending of the ring planes; β (**3.70**) is the annellation angle of the ring axes that passes through carbon 1 and 4 of each aromatic ring; γ (**3.71**) is the torsional angle of the aromatic rings as viewed from the side of the molecule. In general, the tranquilizers have essentially only a bending angle α and little or no β and γ angles. The mixed tranquilizer–antidepressants have both bending (α) and annellation angles (β), but no γ angle. The pure antidepressants exhibit all three angles. The activities arise from the binding of the compounds to different receptors; these angles determine the overall three-dimensional structure of the pharmacophoric groups of the compound, which dictate the binding affinities for various receptors.

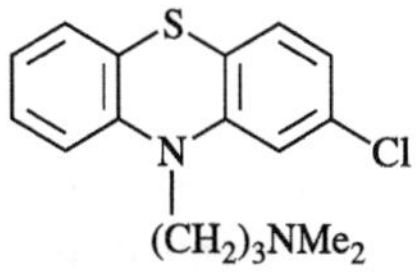

$(CH_2)_3NMe_2$
Chlorpromazine
3.66

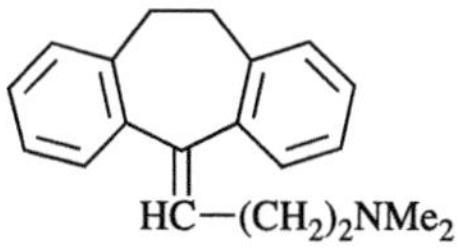

$HC{-}(CH_2)_2NMe_2$
Amitriptyline
3.67

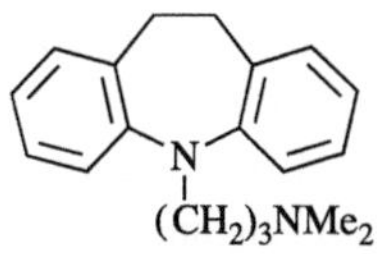

$(CH_2)_3NMe_2$
Imipramine
3.68

3.2.6. Case History of the Pharmacodynamically Driven Design of a Receptor Antagonist: Cimetidine

There are many drugs on the market that were discovered by rational design using the application of physical organic chemical principles. The antiulcer drug cimetidine (**3.72**, Tagamet) is a truly elegant early example of a pharmacodynamically driven approach in drug discovery, utilizing various lead modification methods discussed in Chapter 2, to uncover the first histamine H_2 receptor antagonist and an entirely new class of drugs. Cimetidine is one of the first drugs discovered by a rational approach, thanks to the valiant efforts of medicinal chemists C. Robin Ganellin and Graham Durant and pharmacologist James Black at Smith, Kline, & French Laboratories (now GlaxoSmithKline; Sir James W. Black shared the 1988 Nobel Prize in Physiology or Medicine for the discovery of propranolol and is also credited for the discovery of this drug; actually, the medicinal chemists would have made the discovery). This is a case, however, where neither QSAR nor molecular graphics approaches were utilized. As described in Section 3.2.5.1, histamine binds to the H_1 receptor and causes allergic and hypersensitivity reactions, which antihistamines antagonize. Black found that another action of histamine is the stimulation of gastric acid secretion.[125] However, antihistamines have no effect on this activity; consequently, it was suggested that there was a

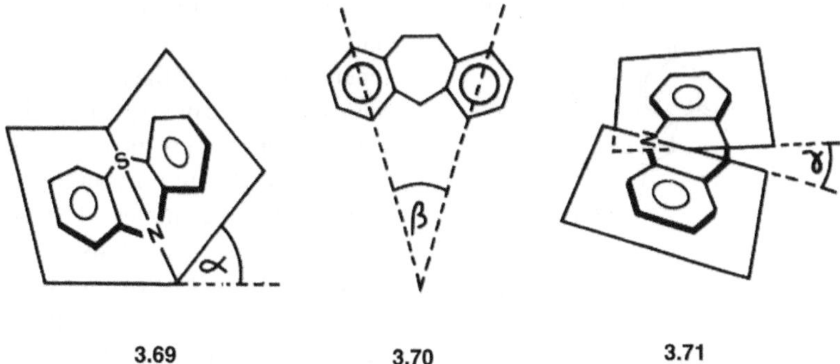

3.69 3.70 3.71

FIGURE 3.32 Ring topology of tricyclic psychomimetic drugs. *Reproduced with permission from Nogrady, T. (1985). In* "Medicinal Chemistry: A Biochemical Approach," *p. 29. Oxford University Press, New York. By permission of Oxford University Press, USA.*

second histamine receptor, which was termed the H_2 *receptor*. The H_1 and H_2 receptors can be differentiated by agonists and antagonists. 2-Methylhistamine (**3.73**) preferentially elicits H_1 receptor responses, and 4-methylhistamine (**3.74**) has the corresponding preferential effect on H_2 receptors. An antagonist of the histamine H_2 receptor would be beneficial for the treatment of hypersecretory conditions such as duodenal and gastric ulcers (peptic ulcers). Consequently, in 1964, Smith, Kline & French Laboratories in England initiated a search for a lead compound that would antagonize the H_2 receptor.[126] Actually, now there are four different histamine receptors known, each one responsible for a different physiological function.[127] The critically important challenge in drug design is to get selectivity of action of molecules.

Cimetidine
3.72

3.73 **3.74**

The first requirement for initiation of a lead discovery program for the H_2 receptor is an efficient bioassay (screen). Unfortunately, there were no high-throughput screens at that time. In fact, no in vitro screen was possible, so a tedious in vivo screen was developed: histamine was infused into anesthetized rats to stimulate gastric acid secretion, the stomach was perfused, and then the pH of the perfusate from the lumen of the stomach was measured before and after administration of the compound. Needless to say, this is a highly time-consuming and variable assay.

The lead discovery approach that was taken involved the use of the endogenous ligand of the receptor as the lead, as described in Chapter 2 (Section 2.1.2.1). Histamine analogs were synthesized on the assumption that the receptor would recognize that general backbone structure. However, the structure had to be sufficiently different so as not to stimulate a response (i.e., act as an agonist) and defeat the purpose. After 4 years, none of the 200 or so compounds made showed any H_2 receptor antagonistic activity. Then a new, more sensitive assay was developed, and some of the same compounds were retested, which identified the first lead compound, N^{α}-guanylhistamine (**3.75**). This compound was only very weakly active as an inhibitor of histamine stimulation; later it was determined to be a partial agonist, not an antagonist. An isostere, isothiourea **3.76**, was made, which was found to be more potent. The corresponding conformationally rigid analog **3.77** (a ring–chain transformation), however, was less potent than **3.76**; consequently, it was thought that flexibility in the side chain was important. Many additional compounds were synthesized, but they acted as partial agonists. They could block histamine binding, but they could not fully prevent acid secretion.

3.75

3.76 **3.77**

It, therefore, became necessary to separate the agonist and antagonist activities. The reason for their agonistic activity, apparently, was their structural similarity to histamine. Not only were these compounds imidazoles, but at physiological pH, the side chains were protonated and positively charged, just like histamine. Consequently, it was reasoned that the imidazole ring should be retained for receptor recognition, but the side chain could be modified to eliminate the positive charge. After numerous substitutions, the neutral thiourea analog (**3.78**) was prepared, having weak antagonistic activity without stimulatory activity. Homologation of the side chain gave a purely competitive antagonist (**3.79**, R = H); no agonist effects were observed. Methylation

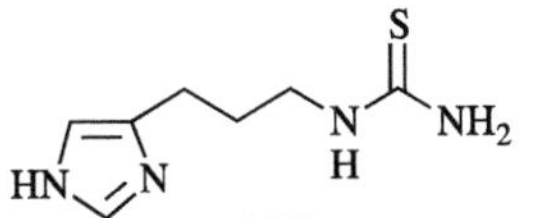

3.78 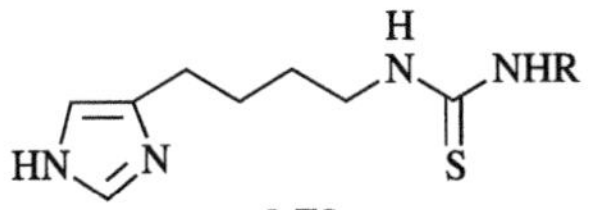 **3.79**

and further homologation on the thiourea nitrogen were carried out; the *N*-methyl analog (**3.79**, R = CH_3) called burimamide, was found to be highly specific as a competitive antagonist of histamine at the H_2 receptor. It was shown to be moderately effective in the inhibition of histamine-stimulated gastric acid secretion in rat, cat, dog, and man. Burimamide was the first H_2 receptor antagonist tested in humans,[128] but it lacked adequate oral activity, so the search for analogs with improved activity continued.

The poor oral potency of burimamide could be a pharmacokinetic problem (poor ability for the drug to reach its target) or a pharmacodynamic problem (suboptimal interaction of the drug with the target). The Smith, Kline, and French group decided to consider the latter. In aqueous solution at physiological pH, the imidazole ring can exist in three main forms (**3.80a–3.80c**, Figure 3.33; R is the rest of burimamide). The thioureido group can exist as four conformers (**3.81a–3.81d**, Figure 3.34; R is the remainder of burimamide). The side chain can exist in a myriad of conformers. Therefore, it is possible that only a very small fraction of the molecules in equilibrium would

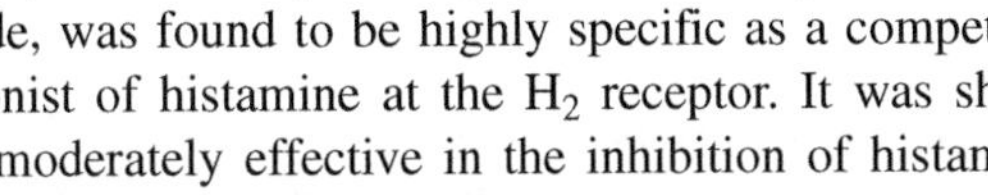

3.80a **3.80b** **3.80c**

FIGURE 3.33 Three principal forms of 5-substituted imidazoles at physiological pH

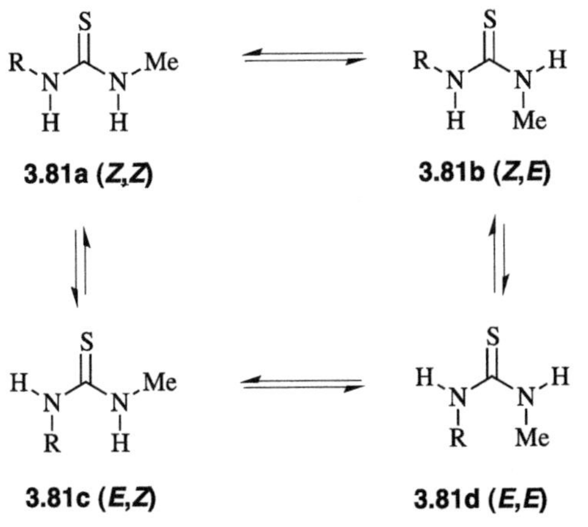

3.81a (*Z,Z*) **3.81b (*Z,E*)**

3.81c (*E,Z*) **3.81d (*E,E*)**

FIGURE 3.34 Four conformers of the thioureido group

have the bioactive conformation, and this could account for the low potency.

One approach taken to increase the potency of burimamide was to compare the population of the imidazole form in burimamide at physiological pH to that in histamine.[129] The population can be estimated from the electronic influence of the side chain, which alters the electron densities at the ring nitrogen atoms, and, therefore, affects the proton acidity. This effect is more important at the nearer nitrogen atom, so if R is electron donating, it would make the adjacent nitrogen more basic, and **3.80c** (Figure 3.33) should predominate; if R is electron withdrawing, it would make the adjacent nitrogen less basic, and **3.80a** should be favored. The fraction present as **3.80b** can be determined from the ring pK_a and the pH of the solution. The electronic effect of R can be calculated from the measured ring pK_a with the use of the Hammett equation (Eqn (3.3)), where pK_a^R is the pK_a of the substituted imidazole, pK_a^H is that of

$$pK_a^R = pK_a^H + \rho\sigma_m \quad (3.3)$$

imidazole (R = H), σ_m is the meta electronic substituent constant, and ρ is the reaction constant (see Chapter 2, Section 2.2.5.1). Imidazole has a pK_a of 6.80, and at physiological temperature and pH, 20% of the molecules are in the protonated form. The imidazole in histamine, under these conditions, has a pK_a of 5.90. This indicates that the side chain in histamine is electron withdrawing, thus favoring tautomer **3.80a** (to the extent of 80%), and only 3% of the molecules are in the cationic form (**3.80b**). The pK_a of the imidazole in burimamide, however, is 7.25, indicating an electron-donating side chain, which favors tautomer **3.80c**. The cation is one of the principal species, about 40% of the molecules. Therefore, even though the side chains in histamine and burimamide appear to be similar, they have opposite electronic effects on the imidazole ring. On the assumption that the desired form of the imidazole should resemble that in histamine, the Smith, Kline & French group decided to increase the electron-withdrawing effect of the side chain of burimamide; however, they did not want to make a major structural modification. Incorporation of an electron-withdrawing atom into the side chain near the imidazole ring was contemplated, and the isosteric replacement of a methylene by a sulfur atom to give thiaburimamide (**3.82**, R = H) was carried out. A comparison of the physical properties of the two compounds (**3.79**, R = CH_3 and **3.82**, R = H) shows that they have similar van der Waals radii and bond angles, although the C–S bond is slightly longer than the C–C bond and is more flexible. A sulfur atom is also slightly more hydrophilic than a methylene group; the log *P* for thiaburimamide is 0.16 and for burimamide is 0.39. The pK_a of the imidazole in thiaburimamide was determined to be 6.25, indicating that the electron-withdrawing effect of the side chain increased, and more of the favored tautomeric form was the same as that in histamine (**3.80a**).

Thiaburimamide is about three times more potent as a histamine H_2 receptor antagonist in vitro than burimamide.

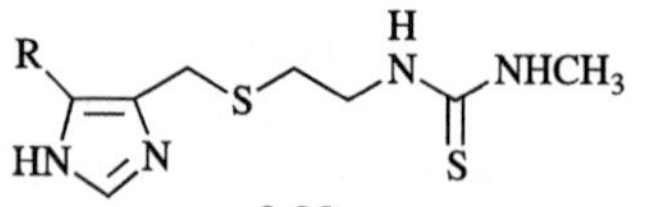

3.82

A second way to increase the population of tautomer **3.80a** would be to introduce an electron-donating substituent at the 4-position of the ring, because electron-donating groups increase the basicity of the adjacent nitrogen, which is chemically equivalent to putting an electron-withdrawing group at the side chain position in thiaburimamide. Because 4-methylhistamine (**3.74**) is a known H_2 receptor agonist, there should be no steric problem with a 4-methyl group. However, the addition of an electron-donating group should increase the pK_a of the ring, thereby increasing the population of the cation (**3.80b**). Although the increase in tautomer **3.80a** is somewhat offset by the decrease in the total uncharged population, the overall effect was favorable. Metiamide (**3.82**, R=CH_3) has a pK_a identical to that of imidazole, indicating that the effect of the electron-withdrawing side chain exactly balanced the effect of the electron-donating 4-methyl group; the percentage of molecules in the charged form was 20%. The important result, however, is that metiamide is eight to nine times more potent than burimamide.

You would think that the tautomeric form would be shifted even more favorably toward **3.80a** by substitution of the side chain with a more electronegative oxygen atom instead of a sulfur atom. Theoretically, that should be the case. This compound, oxaburimamide, was synthesized, but it was *less* potent than burimamide! The explanation for this unexpected result is that intramolecular hydrogen bonding between the oxygen atom and the thiourea NH produces an unfavorable *conformationally restricted analog* (**3.83**). This is one of the problems associated with isosteric replacements; although CH_2, NH, O, and S can have similar biological activity, NH and O can participate in intramolecular (and intermolecular) hydrogen bonding, which changes the shape of the compound and may disfavor (although in other cases, it may favor) the bioactive conformation.

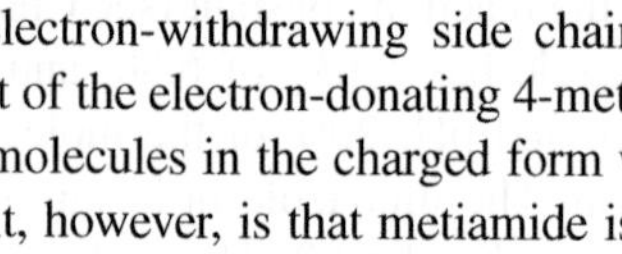

3.83

Metiamide was tested on 700 patients with duodenal ulcers and was found to produce a significant increase in the healing rate with marked symptomatic relief. However, a few cases of granulocytopenia (reduction of the number of white blood cells in the blood) developed. Even though this was a reversible side effect, it was undesirable (compromises the immune system), and it halted further clinical work with this compound.

The Smith, Kline & French group conjectured that the granulocytopenia associated with metiamide was caused by the thiourea group; consequently, alternative substituents were sought. An isosteric replacement approach was taken. The corresponding urea (**3.84**, X=O) and guanidino (**3.84**, X=NH) analogs were synthesized and found to be 20 times less potent than metiamide. Of course, the guanidino analog would be positively charged at physiological pH, and that could be the cause of the lower potency. Charton[130] found a Hammett relationship between the σ and pK_a values for *N*-substituted guanidines; consequently, if guanidino basicity were the problem, then substitution of the guanidino nitrogen with electron-withdrawing groups could lower the pK_a. In fact, cyanoguanidine and nitroguanidine have pK_a values of −0.4 and −0.9, respectively (compared with −1.2 for thiourea), a drop of about 14 pK_a units from that of guanidine. The corresponding cyanoguanidine (**3.84**, X=N–CN; cimetidine, Tagamet) and nitroguanidine (**3.84**, X=N–NO_2) were synthesized in 1972, and both were potent H_2 antagonists, comparable in potency to metiamide, but without the granulocytopenia (cimetidine was slightly more potent than **3.84**, X=NNO_2).

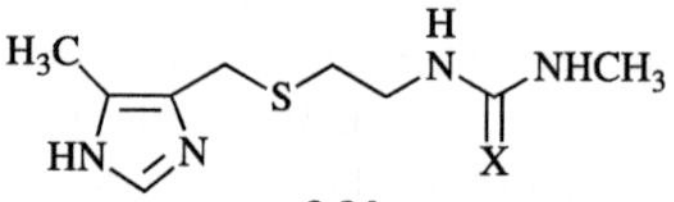

3.84

Because strong electron-withdrawing substituents on the guanidino group favor the imino tautomer, the cyanoguanidino and nitroguanidino groups correspond to the thiourea structure (**3.84**, X=NCN, NNO_2, and S, respectively). These three groups are actually bioisosteres; they are all planar structures of similar geometries, are weakly amphoteric (weakly basic and acidic); being unionized in the pH range 4–11, are very polar; and are hydrophilic. The crystal structures of metiamide (**3.82**, R=CH_3) and cimetidine (**3.84**, X=NCN) are almost identical. The major difference in the two groups is that whereas *N,N'*-disubstituted thioureas assume three stable conformers (see Figure 3.34; *Z,Z*, *Z,E*, and *E,Z*), *N,N'*-disubstituted cyanoguanidines appear to assume only two stable conformers (*Z,E* and *E,Z*). This suggests that the most stable conformer of metiamide, the *Z,Z* conformer, is not the bioactive conformation. An isocytosine analog (**3.85**) also was prepared (pK_a 4.0), which can exist only in the *Z,Z* and *E,Z* conformations. It was only about one-sixth as potent as cimetidine. However, the isocytosino group has a lower log *P* (more hydrophilic) than that of the *N*-methylcyanoguanidino group, and it was thought that lipophilicity may be an important physicochemical parameter. There was, indeed, a correlation found between the H_2 receptor antagonist activity in vitro and the octanol–water partition coefficient of the corresponding

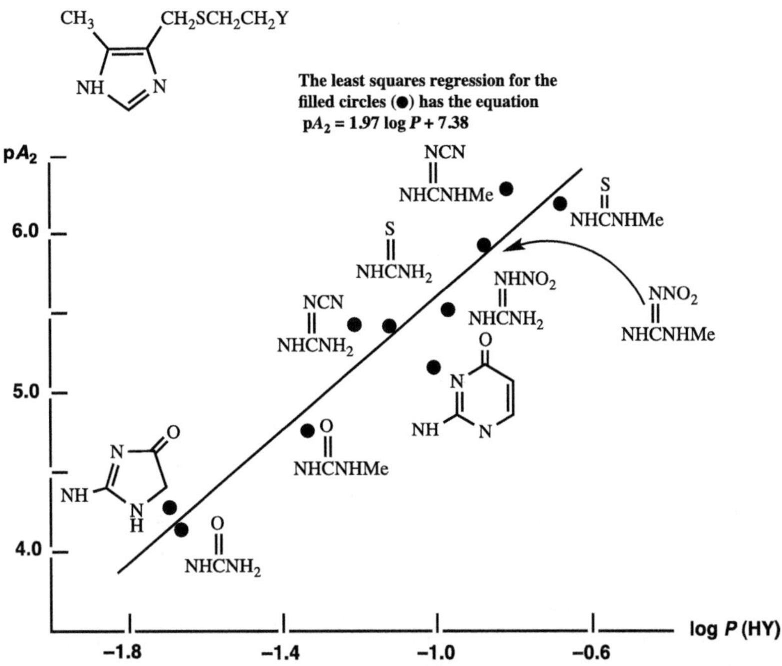

FIGURE 3.35 Linear free energy relationship between H_2 receptor antagonist activity (pA_2) and the partition coefficient. *Reprinted with Permission of Elsevier. This article was published in Pharmacology of Histamine Receptors, Ganellin, C. R., and Parsons, M. E. (1982), p. 83, Wright-PSG, Bristol.*

acid of the substituent Y (Figure 3.35). Although increased potency correlates with increased lipophilicity, all these compounds are fairly hydrophilic. Because the correlation was determined in an in vitro assay, membrane transport is not a concern; consequently, these results probably reflect a property involved with receptor interaction, not with transport. Therefore, it is not clear if the lower potency of the isocytosine analog is structure or hydrophilicity dependent.

3.85

Cimetidine was first marketed in the United Kingdom in 1976; therefore, it took only 12 years from initiation of the H_2 receptor antagonist program to commercialization. Subsequent to the introduction of cimetidine onto the US drug market, three other H_2 receptor antagonists were approved, ranitidine (**3.86**, Zantac, Glaxo Laboratories), which rapidly became the largest selling drug worldwide, famotidine (**3.87**, Pepcid, Yamanouchi/Merck), and nizatidine (**3.88**, Axid, Eli Lilly); the only difference in structure between ranitidine and nizatidine is the heterocyclic ring incorporated. It is apparent that an imidazole ring is not essential for H_2 receptor recognition and that a positive charge near the heterocyclic ring (the Me_2N– of **3.86** and **3.88** and the guanidino group of **3.87** will be protonated at physiological pH) is not unfavorable.

Ranitidine 3.86

Famotidine 3.87

Nizatidine 3.88

Cimetidine became the first drug ever to achieve more than $1 billion a year in sales, thereby having the distinction of being the first *blockbuster drug*. The discovery of cimetidine is one of the many examples now of how the judicious use of physical organic chemistry can result in lead discovery, if not in drug discovery.

One approach for combination therapy with H_2 receptor antagonists stems from the discovery that a bacterium, *Helicobacter pylori*, is found in the stomach and is associated with peptic ulcers.[131] The organism protects itself from the acid in the stomach partly because it lives within the layer of mucus that the stomach secretes to protect itself against the acid and partly because the bacterium produces the enzyme urease, which converts urea in the blood into ammonia to neutralize the acid.[132] This bacterium was discovered in 1983 by Drs Barry J. Marshall and J. Robin Warren at the Royal Perth Hospital in Australia, who were trying to grow mysterious cells taken from the stomach. The culture was left much longer to grow than normal because of the four-day Easter weekend that year, and upon their return, they noticed the growth of a bacterium with spiral, helix-shaped cells, which they called *Helicobacter*. Although it took more than a decade to convince others that this bacterium was really living in the stomach and that could cause ulcers, it is now widely accepted. In fact, Marshall and Warren jointly received the 2005 Nobel Prize in Physiology or Medicine for this discovery. Because there are many people who have this bacterium, but do not have ulcers, there must be additional factors, such as stress, that are needed for ulcer formation.[133] Treatment with antibacterial agents can kill these bacteria, but generally other drugs that can lower stomach acid are needed in combination.[134]

3.2.7. Case History of the Pharmacokinetically Driven Design of Suvorexant

Over 50% of American adults polled reported at least one symptom of insomnia in the previous year, and one-third reported symptoms almost every night;[135] about one-fourth of American adults take sleep medication.[136] Orexins A and B are neuropeptides that have been shown to affect the sleep/wake cycle by binding to the orphan G protein-coupled receptors orexin receptor 1 (OX_1R) and orexin receptor 2 (OX_2R).[137] Mice deficient in these neuropeptides exhibit excessive sleepiness;[138] intracerebroventricular infusion of these neuropeptides into rat cerebrospinal fluid leads to an increase in wakefulness.[139] Therefore, an antagonist of these receptors could be an important new mechanism for the treatment of insomnia.

A high-throughput screen of the Merck sample collection revealed four active scaffolds, including *N,N*-disubstituted 1,4-diazepanes, such as **3.89**, having IC_{50} values for OX_1R and OX_2R of 630 and 98 nM, respectively.[140] The western heterocycle was optimized to **3.90**, and the eastern heterocycle to **3.91**; the composite structure (**3.92**) had IC_{50} values for OX_1R and OX_2R of 29 and 27 nM, respectively. Compound **3.92** promoted sleep in rats when orally dosed at 100 mg/kg. The problem with **3.92**, however, was its low oral bioavailability (16% in dogs; 2% in rats) and rapid oxidative metabolism (see Chapter 8) of all three rings. The most detrimental oxidation was that of the 1,4-diazepane ring adjacent to the eastern nitrogen (**3.93**, Scheme 3.3).

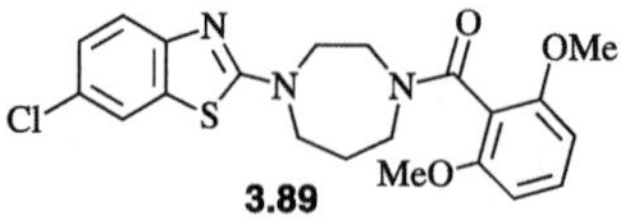

3.89 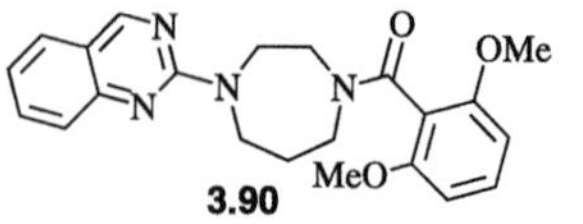 3.90 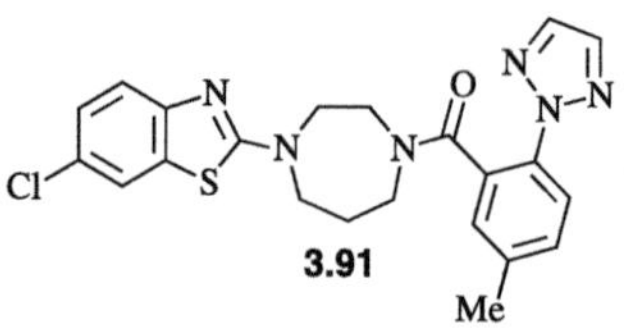 3.91 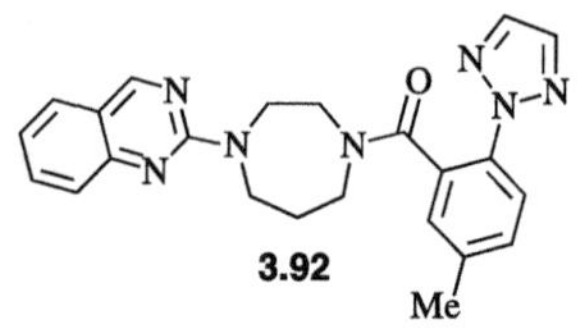 3.92

This is in equilibrium with the corresponding aldehyde, which was trapped by added semicarbazide to detect its presence (**3.94**), indicating its potential as an undesirable reactive metabolite. It was found that an effective way to block that metabolism was by methylation, and the (*R*)-antipode had superior potency (**3.95**). Although the clearance rate decreased, the oral bioavailability decreased until the western heterocycle was metabolically protected with a fluorine and the methyl group on the eastern heterocycle was removed (**3.96**), leading to comparable potency as **3.92** and an oral bioavailability of 37% in dogs, which is still only moderate. To further decrease the clearance rate and increase the potency and oral bioavailability, several additional modifications were made to the western heterocycle on the basis of earlier studies in dog, which identified benzoxazole as having reduced clearance properties. Installation of the unsubstituted benzoxazole gave **3.97**, which had the lowest clearance rate to that date, but the potency dropped 20-fold. From their SAR, they knew that increased lipophilicity on that heterocycle was highly beneficial for potency, and addition of a chlorine atom (**3.98**, suvorexant) increased the potency 10-fold with excellent brain penetration and an oral bioavailability of 56% in dogs. Suvorexant has completed Phase III clinical trials.

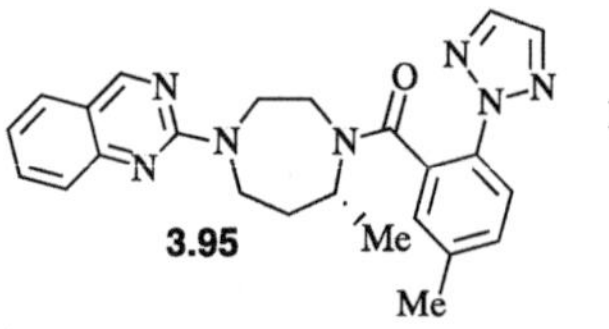

3.95 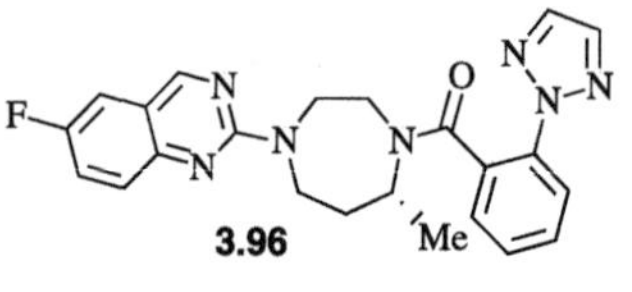 3.96 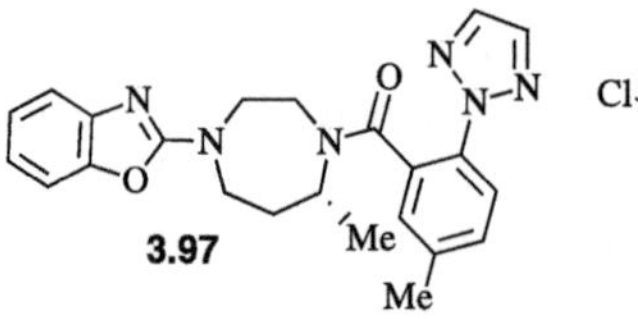 3.97 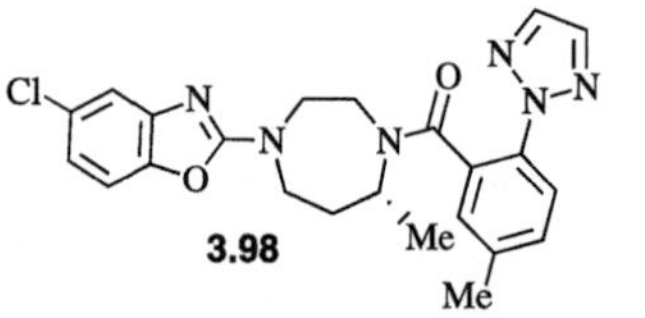 3.98

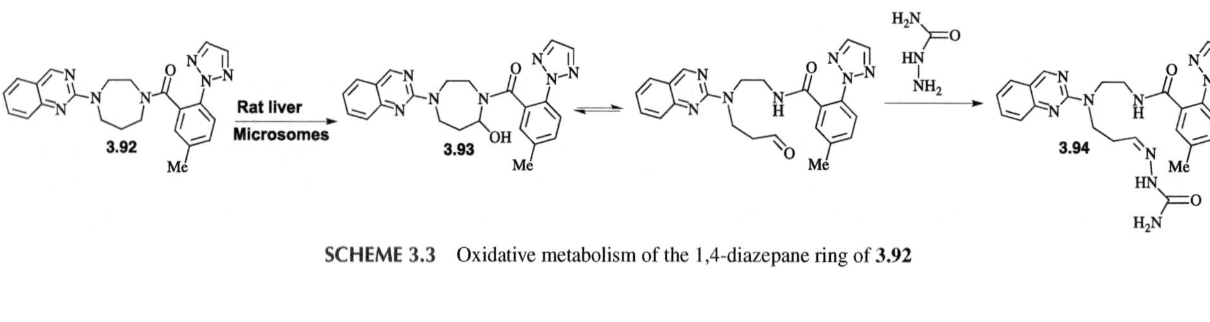

3.92 Rat liver Microsomes 3.93 3.94

SCHEME 3.3 Oxidative metabolism of the 1,4-diazepane ring of **3.92**

A backup compound for suvorexant was designed from an understanding of the conformational properties believed to favor high orexin receptor binding. Conformational studies with 1,4-diazepane carboxamides were the rationale for the design of molecules with a piperidine core that could permit intramolecular aryl–aryl interactions. Methylation of the piperidine core adjacent to the amide nitrogen atom gave analogs with a piperidine in the chair conformation having a 2,5-*trans*-diaxial conformation (**3.99**), which promotes the desired aryl–aryl interaction. This compound was found to be more potent than suvorexant in vivo and is in Phase II clinical trials.

3.99

Next, we turn our attention to a special class of receptors called enzymes, which also are very important targets for drug design.

3.3. GENERAL REFERENCES

Membranes and Receptors; G Protein-Coupled Receptors

Foreman, J. C.; Johansen, T.; Gibb, A. J. (Eds.) *Textbook of Receptor Pharmacology*, 3rd ed., CRC Press, Boca Raton, FL, 2011.

Luckey, M. *Membrane Structural Biology*, Cambridge University Press, Cambridge, U.K., 2008.

Poyner, D.; Wheatley, M. *G Protein-Coupled Receptors: Essential Methods*, Wiley-Blackwell, Chichester, U.K., 2010.

Neve, K. A. (Ed.) *Functional Selectivity of G Protein-Coupled Receptor Ligands. New Opportunities for Drug Discovery*, Humana Press: New York, NY, 2009.

Siehler, S.; Milligan, G. (Eds.) *G Protein-Coupled Receptors. Structure, Signaling, and Physiology*, Cambridge University Press, Cambridge, U.K., 2011.

Drug–Receptor Interactions

Cannon, J. G., *Pharmacology for Chemists*; Oxford University Press, 1999.

Kenakin, T. *A Pharmacology Primer*, 3rd ed., Elsevier, Amsterdam, 2009.

Francotte, E.; Lindner, W. (Eds.) *Chirality in Drug Research*, Wiley-VCH, Weinheim, 2007.

Huang, Z. (Ed.) *Drug Discovery Research*, Wiley, Hoboken, NJ, 2007.

Albert, A. *Selective Toxicity*, 7th ed., Chapman and Hall, London, 1985.

Drug–Receptor Theories

Kenakin, T. *A Pharmacology Primer*, 3rd ed., Elsevier, Amsterdam, 2009.

O'Brien, R. D. (Ed.) *The Receptors*, Plenum, New York, 1979.

Prull, C.-R.; Maehle, A.-H.; Halliwell, R. F. *A Short History of the Drug Receptor Concept*, Palgrave Macmillan, Hampshire, U.K., 2009.

Stereochemical Considerations

Smith, D. F., (Ed.) *CRC Handbook of Stereoisomers: Therapeutic Drugs*, CRC Press, Boca Raton, 1989.

Francotte, E.; Lindner, W. (Eds.) *Chirality in Drug Research*, Wiley-VCH, Weinheim, 2007.

Histamine Receptors and Antagonists

de Esch, I.; Timmerman, H.; Leurs, R. Histamine receptors. In *Textbook of Drug Design and Discovery*, 4th ed., CRC Press, Boca Raton, FL, 2010, pp. 283–297.

Hill, S. J.; Ganellin, C. R.; Timmerman, H.; Schwartz, J. C.; Shankley, N. P.; Young, J. M.; Schunack, W.; Levi, R.; Haas, H. L. International Union of Pharmacology. XIII. Classification of histamine receptors. *Pharmacol. Rev.* **1997**, *49*, 253–278.

Blaya, B.; Nicolau-Galmes, F.; Jangi, S. M.; Ortega-Martinez, I.; Alonso-Tejerina, E.; Burgos-Bretones, J.; Perez-Yarza, G.; Asumendi, A.; Boyano, M. D. Histamine and histamine receptor antagonists in cancer biology. Histamine and histamine receptor antagonists in cancer biology. *Inflammation & Allergy: Drug Targets* **2010**, *9*(3), 146–157.

Jadidi-Niaragh, F.; Mirshafiey, A. Histamine and histamine receptors in pathogenesis and treatment of multiple sclerosis. *Neuropharmacology* **2010**, *59*(3), 180–189.

3.4. PROBLEMS (ANSWERS CAN BE FOUND IN THE APPENDIX AT THE END OF THE BOOK)

1. Indicate what drug–receptor interactions are involved at every arrow shown. More than one kind of interaction is possible for each letter.

a

b

c

d

e

2. A receptor has lysine and histidine residues important to binding, which do not interact with each other.

The pK_a of the lysine residue is 6.4 (pK_a in solution is 10.5), and the pK_a of the histidine residue is 9.4 (pK_a in solution is 6.5). On the basis of the discussion in Chapter 2 about pK_a variabilities as a result of the environment, what can you say about possible other residues in the binding site to rationalize these observations.

3. Draw a dose–response curve for:
a. a full agonist
b. a mixture of a full agonist and a competitive antagonist

4. Draw dose–response curves (place on same plot) for a series of three compounds with the following properties:

	K_d (*M*)	α
1	10^{-6}	1.0
2	10^{-9}	0.8
3	10^{-9}	0.4

5. A series of dopamine analogs was synthesized and assayed for their effect on the D_2 dopamine receptor. The results are shown in Table 3.2.

TABLE 3.2 Effect of dopamine analogs on D_2 dopamine receptor

Compound	K_d (nM)	% Change in Basal Activity	% Change When 100 μM Dopamine Added
1	180	−12.1	-6.2
2	37	0.1	0.8
3	0.46	19	19
4	14.9	12.2	18.4
Dopamine	1.9	19	

a. Compare the affinities of **1–4** to that of dopamine.
b. Compare the efficacies of **1–4** to dopamine.
c. What type of effect is produced by **1–4**?

6. a. What problems are associated with administration of racemates?
b. How can you increase the eudismic ratio?

7. Design conformationally-rigid analogs for:
a. 4-aminobutyric acid (GABA) 
b. Epinephrine 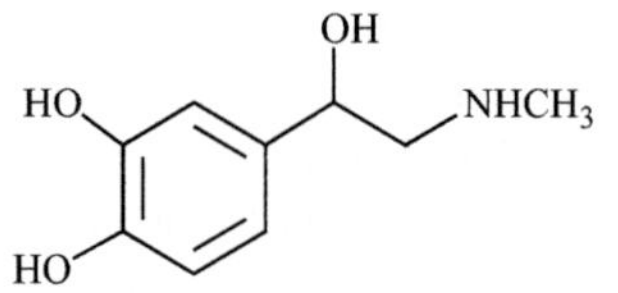
c. Nicotine

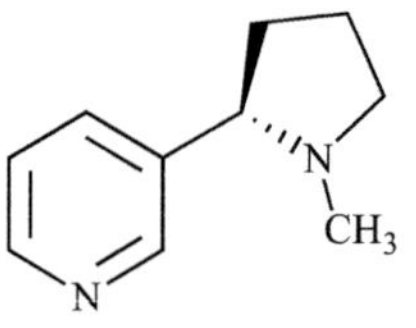

8. On the basis of generalizations about ring topology discussed in the chapter, would you expect the compound below to be a tranquilizer, have both antidepressant and tranquilizing properties, or be an antidepressant agent? Why?

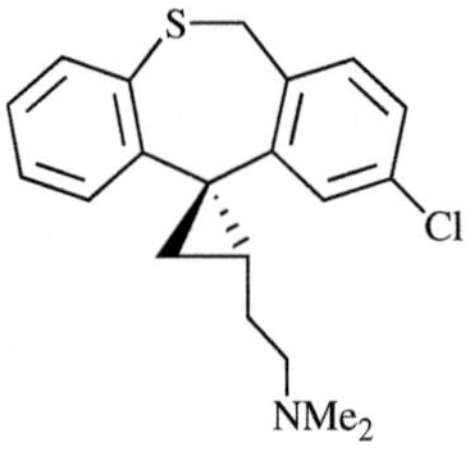

9. An isosteric series of compounds shown below, where X=CH_2, NH, O, S, was synthesized. The order of potency was X=NH>O>S>CH_2. How can you rationalize these results (you need to consider the three-dimensional structure)?

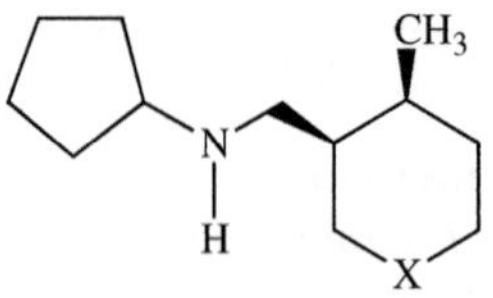

10. Tyramine binds to a receptor that triggers the release of norepinephrine, which can raise the blood pressure. If the tyramine receptor was isolated, and you wanted to design a new antihypertensive agent, discuss what you would do in terms of lead discovery and modification.

HO–(benzene ring)–$CH_2CH_2NH_2$

Tyramine

11. A receptor was isolated, a crystal structure was obtained with the natural ligand bound, and it was found that the binding site displayed C_2 symmetry. Computer modeling was done, and a C_2 symmetric antagonist (**5**) was designed. However, it exhibited very low potency. What could be the problem? Show it.

H_2N OH NH_2

5

REFERENCES

1. (a) Ren, H.; Yu, D.; Ge, B.; Cook, B.; Xu, Z.; Zhang, S. High-level production, solubilization and purification of synthetic human GPCR chemokine receptors CCR5, CCR3, CXCR4 and CX3CR1. *PLoS ONE* **2009**, *4*, e4509. (b) Costanzi, S.; Siegel, J.; Tikhonova, I. G.; Jacobson, K. A. Rhodopsin and the others: a historical perspective on structural studies of G protein-coupled receptors. *Curr. Pharm. Des.* **2009**, *15*(35), 3994–4002.
2. (a) Milligan, G.; Rees, S. Chapter 24. Oligomerisation of G protein-coupled receptors. *Annu. Rep. Med. Chem.* **2000**, *35*, 271–279. (b) Hamm, H. E. How activated receptors couple to G proteins. *Proc. Natl. Acad. Sci. U.S.A.* **2001**, *98*, 4819–4821.
3. (a) Schoneberg, T.; Schulz, A.; Gudermann, T. The structural basis of G-protein-coupled receptor function and dysfunction in human diseases. *Rev. Physiol. Biochem. Pharmacol.* **2002**, *144*, 143–227. (b) Lombardi, M. S.; Kavelaars, A.; Heijnen, C. J. Role and modulation of G protein-coupled receptor signaling in inflammatory processes. *Crit. Rev. Immunol.* **2002**, *22*, 141–163.
4. (a) Elfaki, D. A. H.; Bjornsson, E.; Lindor, K. D. Review article: nuclear receptors and liver disease-current understanding and new therapeutic implications. *Aliment. Pharmacol. Therap.* **2009**, *30*, 816–825. (b) Evans, R. M. The steroid and thyroid hormone receptor superfamily. *Science* **1988**, *240*(4854), 889–895. (c) McEwan, I. J. Nuclear receptors: one big family. *Meth. Mol. Biol.* **2009**, *505*, 3–18. (d) Tobin, J. F.; Freedman, L. P. Nuclear receptors as drug targets in metabolic diseases: new approaches to therapy. *Trends Endocrinol. Metab.* **2006**, *17*, 284–290.
5. (a) Schlessinger, J. Cell signaling by receptor tyrosine kinases. *Cell* **2000**, *103*, 211–225. (b) Gschwind, A.; Fischer, O. M.; Ullrich, A. The discovery of receptor tyrosine kinases: targets for cancer therapy. *Nat. Rev. Cancer* **2004** *4*, 361–370. (c) Pyne, N. J.; Pyne, S. Receptor tyrosine kinase-G-protein-coupled receptor signaling platforms: out of the shadow? *Trends Pharmacol. Sci.* **2011**, *32*(8), 443–450. (d) Preis, M.; Korc, M. Signaling pathways in pancreatic cancer. *Crit. Rev. Eukaryot. Gene Express.* **2011**, *21*(2), 115–129.
6. Langley, J. N. On the physiology of the salivary secretion: Part II. On the mutual antagonism of atropine and pilocarpin, having especial reference to their relations in the sub-maxillary gland of the cat. *J. Physiol. (Lond.)* **1878**, *1*, 339–369.
7. Ehrlich, P. Die Wertbemessung des Diphterieheilserums und deren theoretische Grundlagen. *Klin. Jahr.* **1897**, *6*, 299–326.
8. Langley, J. N. On the reaction of cells and of nerve-endings to certain poisons, chiefly as regards the reaction of striated muscle to nicotine and to curare. *J. Physiol. (Lond.)* **1905**, *33*, 374–413.
9. (a) Kenakin, T. *Molecular Pharmacology: A Short Course*, Blackwell Science, Cambridge, MA, 1997. (b) Lindstrom, J. In *Neurotransmitter Receptor Binding*, Yamamura, H. I.; Enna, S. J.; Kuhar, M. J. (Eds.), Raven Press, New York, 1985, p. 123. (c) Douglass, J.; Civelli, O.; Herbert, E. Polyprotein gene expression: generation of diversity of neuroendrocrine peptides. *Annu. Rev. Biochem.* **1984**, *53*, 665–715.
10. Litter, M. *Farmacologia*, 2nd ed., El Ateneo, Buenos Aires, 1961.
11. Overington, J. P.; Al-Lazikani, B.; Hopkins, A. L. How many drug targets are there? *Nat. Rev. Drug Discov.* **2006**, *5*, 993–996.
12. (a) Ringe, D. What makes a binding site a binding site? *Curr. Opin. Struct. Biol.* **1995**, *5*, 825–829. (b) Karplus, P. A.; Faerman, C. Ordered water in macromolecular structure. *Curr. Biol.* **1994**, *4*, 770–776.
13. (a) Searle, M. S.; Williams, D. H. The cost of conformational order: entropy changes in molecular associations. *J. Am. Chem. Soc.* **1992**, *114*, 10690–10697. (b) Babine, R. E.; Bender, S. L. Molecular recognition of protein-ligand complexes: applications to drug design. *Chem. Rev.* **1997**, *97*, 1359–1472.
14. Bissantz, C.; Kuhn, B.; Stahl, M. A medicinal chemist's guide to molecular interactions. *J. Med. Chem.* **2010**, *53*, 5061–5084.
15. (a) Dill, K. A. Additive principles in biochemistry. *J. Biol. Chem.* **1997**, *272*, 701–704. (b) Mark, A. E.; van Gunsteren, W. F. Decomposition of the free energy of a system in terms of specific interactions. Implications for theoretical and experimental studies. *J. Mol. Biol.* **1994**, *240*, 167–176.
16. Levitt, M.; Park, B. H. Water: now you see it, now you don't. *Structure* **1993**, *1*, 223–226.
17. Albert, A. *Selective Toxicity*, 7th ed., Chapman and Hall, London, 1985, p. 206.
18. (a) Carosati, E.; Sciabola, S.; Cruciani, G. Hydrogen bonding interactions of covalently bonded fluorine atoms: from crystallographic data to a new angular function in the GRID force field. *J. Med. Chem.* **2004**, *47*, 5114–5125.
19. Pierce, A.C.; ter Haar, E.; Binch, H. M.; Kay, D. P.; Patel, S. R.; Li, P. CH···O and CH···N hydrogen bonds in ligand design: a novel quinazolin-4-ylthiazol-2-ylamine protein kinase inhibitor. *J. Med. Chem.* **2005**, *48*, 1278–1281.
20. (a) Cleland, W. W.; Kreevoy, M. M. Low-barrier hydrogen bonds and enzymic catalysis. *Science* **1994**, *264*, 1887–1890. (b) Cleland, W. W.; Frey, P. A.; Gerlt, J. A. The low barrier hydrogen bond in enzymic catalysis. *J. Biol. Chem.* **1998**, *273*, 25529–25532.
21. Bissantz, C.; Kuhn, B.; Stahl, M. A medicinal chemist's guide to molecular interactions. *J. Med. Chem.* **2010**, *53*, 5061–5084.
22. Jencks, W. P. *Catalysis in Chemistry and Enzymology*, McGraw-Hill, New York, 1969, p. 340.
23. Kuhn, B.; Mohr, P.; Stahl, M. Intramolecular hydrogen bonding in medicinal chemistry. *J. Med. Chem.* **2010**, *53*, 2601–2611.
24. Laurence, C.; Brameld, K. A.; Graton, J.; Le Questel, J.-Y.; Renault, E. The pKBHX database: toward a better understanding of hydrogen-bond basicity for medicinal chemists. *J. Med. Chem.* **2009**, *52*, 4073–4086.

25. Korolkovas, A. *Essentials of Molecular Pharmacology*, Wiley, New York, 1970, p. 159.
26. (a) Menear, K. A.; Adcock, C.; Alonso, F. C.; Blackburn, K.; Copsey, L.; Drzewiecki, J.; Fundo, A.; Le Gall, A.; Gomez, S.; Javaid, H.; Lence C. F.; Martin, N. M. B.; Mydlowski, C.; Smith, G. C. M. Novel alkoxybenzamide inhibitors of poly (ADP-ribose) polymerase. *Bioorg. Med. Chem. Lett.* **2008**, *18*, 3942–3945. (b) Furet, P.; Caravatti, G.; Guagnano, V.; Lang, M.; Meyer, T.; Schoepfer, J. Entry into a new class of protein kinase inhibitors by pseudo ring design. *Bioorg. Med. Chem. Lett.* **2008**, *18*, 897–900.
27. Jansma, A.; Zhang, Q.; Li, B.; Ding, Q.; Uno, T.; Bursulaya, B; Liu, Y.; Furet, P.; Gray, N. S.; Geierstanger, B. H. Verification of a designed intramolecular hydrogen bond in a drug scaffold by nuclear magnetic resonance spectroscopy. *J. Med. Chem.* **2007**, *50*, 5875–5877.
28. Ashwood, V.A.; Field, M. J. et al. Utilization of an intramolecular hydrogen bond to increase the CNS penetration of an NK1 receptor antagonist. *J. Med. Chem.* **2001**, *44*, 2276–2285.
29. Kuhn, B.; Mohr, P.; Stahl, M. Intramolecular hydrogen bonding in medicinal chemistry. *J. Med. Chem.* **2010**, *53*, 2601–2611.
30. Jencks, W. P. *Catalysis in Chemistry and Enzymology*, McGraw-Hill, New York, 1969, p. 393.
31. Hildebrand, J. H. Is there a "hydrophobic effect"? *Proc. Natl. Acad. Sci. U.S.A.* **1979**, *76*, 194.
32. Kuntz, I. D.; Chen, K.; Sharp, K. A.; Kollman, P. A. The maximal affinity of ligands. *Proc. Natl. Acad. Sci. U.S.A.* **1999**, *96*, 9997–10002.
33. Smith, D. This is your captain speaking. *Drug Disc. Today* **2002**, *7*, 705–706.
34. (a) Burley, S. K.; Petsko, G. A. Aromatic-aromatic interaction: a mechanism of protein structure stabilization. *Science*. **1985**, *229*, 23–28. (b) Salonen, L. M.; Ellermann, M.; Diederich, F. Aromatic rings in chemical and biological recognition: energetics and structures. *Angew. Chem. Int. Ed.* **2011**, 50, 4808–4842. (c) Raju, R. K.; Bloom, J. W. G.; Wheeler, S. E. Substituent effects on non-covalent interactions with aromatic rings: insights from computational chemistry. *ChemPhysChem* **2011**, *12*, 3116–3130.
35. McGaughey, G. B.; Gagne, M.; Rappe, A. K. π-stacking interactions. Alive and well in proteins. *J. Biol. Chem.* **1998**, *273*, 15458–15463.
36. Tsuzuki, S.; Honda, K.; Uchimaru, T.; Mikami, M.; Tanabe, K. Origin of attraction and directionality of the π/π interaction: model chemistry calculations of benzene dimer interaction. *J. Am. Chem. Soc.* **2002**, *124*, 104–112.
37. (a) Mecozzi, S.; West, A. P., Jr.; Dougherty, D. A. Cation–π interactions in aromatics of biological and medicinal interest: electrostatic potential surfaces as a useful qualitative guide. *Proc. Natl. Acad. Sci. U.S.A.* **1996**, *93*, 10566–10571. (b) Gallivan, J. P.; Dougherty, D. A. Cation–π interactions in structural biology. *Proc. Natl. Acad. Sci. U.S.A.* **1999**, *96*, 9459–9464. (c) Dougherty, D. A. The Cation–π interaction. *Chem. Rev.* **1997**, *97*, 1303–1324. (d) Frontera, A.; Quinonero, D.; Deya, P. M. Cation–π and anion–π interactions. *Comput. Mol. Sci.* **2011**, *1*, 440–459.
38. (a) Politzer, P.; Lane, P.; Concha, M. C.; Ma, Y.; Murray, J. S. An overview of halogen bonding. *J. Mol Model.* **2007**, *13*, 305–311. (b) Auffinger, P.; Hays, F. A.; Westhof, E.; Ho, P. S. Halogen bonds in biological molecules. *Proc. Natl. Acad. Sci. U.S.A.* **2004**, *101*, 16789–16794. (c) Ibrahim, M. A. A. Molecular mechanical study of halogen bonding in drug discovery. *J. Comput. Chem.* **2011**, *32*, 2564–2574. (d) Hardegger, L. A.; Kuhn, B.; Spinnler, B.; Anselm, L.; Ecabert, R.; Stihle, M.; Gsell, B.; Thoma, R.; Diez, J.; Benz, J.; Plancher, J.-M.; Hartmann, G.; Banner, D. W.; Haap, W.; Diederich, G. Systematic investigation of halogen bonding in protein-ligand interactions. *Angew. Chem. Int. Ed.* **2011**, *50*, 314–318. (e) Wilcken, R.; Zimmermann, M. O.; Lange, A.; Joerger, A. C.; Boeckler, F. M. Principles and applications of halogen bonding in medicinal chemistry and chemical biology. *J. Med. Chem.* **2013**, *56*, 1363–1388.
39. Lu, Y.; Shi, T.; Wang, Y.; Yang H.; Yan, X.; Luo, X.; Jiang, H.; Zhu, W. Halogen bonding—a novel interaction of rational drug design? *J. Med. Chem.* **2009**, *52*, 2854–2862.
40. Xu, Z.; Liu, Z. et al. Utilization of halogen bond in lead optimization: A case study of rational design of potent phosphodiesterase type 5 (PDE5) inhibitors. *J. Med. Chem.* **2011**, *54*, 5607–5611.
41. Lu, Y.-X.; Zou, J.-W.; Wang, Y.-H.; Jiang, Y.-J.; Yu, Q.-S. Ab initio investigation of the complexes between bromobenzene and several electron donors: some insights into the magnitude and nature of halogen bonding interactions. *J. Phys. Chem.* **2007**, *111*, 10781–10788.
42. (a) Paulini, R.; Muller, K.; Diederich, F. Orthogonal multipolar interactions in structural chemistry and biology. *Angew. Chem. Int. Ed.* **2005**, *44*, 1788–1805. (b) Bartlett, G. J.; Choudhary, A.; Raines, R. T.; Woolfson, D. N. $n \rightarrow \pi^*$ interactions in proteins. *Nat. Chem. Biol.* **2010**, *6*, 615–620.
43. Andrews, P. R.; Craik, D. J.; Martin, J. L. Functional group contributions to drug-receptor interactions. *J. Med. Chem.* **1984**, *27*, 1648–1657.
44. (a) Page, M. I. Entropy, binding energy, and enzymic catalysis. *Angew. Chem. Int. Ed.* **1977**, *16*, 449–459. (b) Page, M. I. In *Quantitative Approaches to Drug Design*, Dearden, J. C. (Ed.), Elsevier, Amsterdam, 1983, p. 109.
45. Allen, M. S.; Tan, Y.-C.; Trudell, M. L.; Narayanan, K.; Schindler, L. R.; Martin, M. J.; Schultz, C.; Hagen, T. J.; Koehler, K. F.; Codding, P. W.; Skolnick, P.; Cook, J. M. Synthetic and computer-assisted analyses of the pharmacophore for the benzodiazepine receptor inverse agonist site. *J. Med. Chem.* **1990**, *33*, 2343–2357.
46. (a) Williams, M.; Enna, S. J. The receptor: from concept to function. *Annu. Rep. Med. Chem.* **1986**, *21*, 211–235. (b) Kuo, C. L.; Wang, R. B.; Shen, L. J.; Lien, L. L.; Lien, E. J. G-protein coupled receptors: SAR analyses of neurotransmitters and antagonists. *J. Clin. Pharm. Ther.* **2004**, *29*(3), 279–298.
47. Fensome, A.; Adams, W. R.; Adams, A. L.; Berrodin, T. J.; Cohen, J.; Huselton, C.; Illenberger, A.; Kern, J. C.; Hudak, V. A.; Marella, M. A.; Melenski, E. G.; McComas, C. C.; Mugford, C. A.; Slayden, O. D.; Yudt, M.; Zhang, Z.; Zhang, P.; Zhu, Y.; Winneker, R. C.; Wrobel, J. E. Design, synthesis, and SAR of new pyrrole-oxindole progesterone receptor modulators leading to 5-(7-fluoro-3,3-dimethyl-2-oxo-2,3-dihydro-1H-indol-5-yl)-1-methyl-1H-pyrrole-2-carbonitrile (WAY-255348). *J. Med. Chem.* **2008**, *51*, 1861–1873.
48. Costall, B.; Naylor, R. J. The hypotheses of different dopamine receptor mechanisms. *Life Sci.* **1981**, *28*, 215–259.
49. Gaddum, J. H. The action of adrenaline and ergotamine on the uterus of the rabbit. *J. Physiol. (Lond.)* **1926**, *61*, 141–150.
50. Clark, A. J. The reaction between acetylcholine and muscle cells. *J. Physiol. (Lond.)* **1926**, *61*, 530–546.
51. (a) Ariëns, E. J. Affinity and intrinsic activity in the theory of competitive inhibition. I. Problems and theory. *Arch. Intern. Pharmacodyn. Thérap.* **1954**, *99*, 32–49. (b) van Rossum, J. M.; Ariëns, E. J. Receptor-reserve and threshold phenomena. II. Theories of drug action and a quantitative approach to spare receptors and threshold values. *Arch. Intern. Pharmacodyn. Thérap.* **1962**, *136*, 385–413. (c) van Rossum, J. M. The relation between chemical structure and biological activity. Discussion of possibilities, pitfalls and limitations. *J. Pharm. Pharmacol.* **1963**, *15*, 285–316.

52. Stephenson, R. P. A modification of receptor theory. *Brit. J. Pharmacol. Chemother.* **1956**, *11*, 379–393.
53. Wermuth, C.-G.; Ganellin, C. R.; Lindberg, P.; Mitscher, L. A. Glossary of terms used in medicinal chemistry (IUPAC recommendations 1997). *Annu. Rep. Med. Chem.* **1998**, *33*, 385–395.
54. Paton, W. D. M. A theory of drug action based on the rate of drug-receptor combination. *Proc. Roy. Soc. London, Ser. B* **1961**, *154*, 21–69.
55. (a) Koshland, D. E., Jr. An application of a theory of enzyme specificity to protein synthesis. *Proc. Natl. Acad. Sci. U.S.A.* **1958**, *44*, 98–105. (b) Koshland, D. E., Jr. Lipid biosynthesis. *Biochem. Pharmacol.* **1961**, *8*, 57–124. (c) Koshland, D. E., Jr.; Neet, K. E. The catalytic and regulatory properties of enzymes. *Annu. Rev. Biochem.* **1968**, *37*, 359–410.
56. (a) Belleau, B. The chemical basis for cholinomimetic and cholinolytic activity. IV. A molecular theory of drug action based on induced conformational perturbations of receptors. *J. Med. Chem.* **1964**, *7*, 776–784. (b) Belleau, B. Conformational perturbation in relation to the regulation of enzyme and receptor behavior. *Adv. Drug Res.* **1965**, *2*, 89–126.
57. Monad, J.; Wyman, J.; Changeux, J.-P. On the nature of allosteric transitions: a plausible model. *J. Mol. Biol.* **1965**, *12*, 88–118.
58. Karlin, A. Application of "a plausible model" of allosteric proteins to the receptor for acetylcholine. *J. Theor. Biol.* **1967**, *16*, 306–320.
59. Milligan, G.; Bond, R. A.; Lee, M. Inverse agonism: pharmacological curiosity or potential therapeutic strategy? *Trends Pharmacol. Sci.* **1995**, *16*, 10–13.
60. del Castillo, J.; Katz, B. Interaction at end-plate receptors between different choline derivatives. *Proc. Roy. Soc. London Ser. B* **1957**, *146*, 369–381.
61. (a) Leff, P. The two-state model of receptor activation. *Trends Pharmacol. Sci.* **1995**, *16*, 89–97. (b) Kenakin, T. P. Agonist-receptor efficacy II: agonist trafficking of receptor signals. *Trends Pharmacol. Sci.* **1995**, *16*, 232–238. (c) Bond, R.; Milligan, G.; Bouvier, M. Inverse agonism. *Handb. Exp. Pharmacol.* **2000**, *148*, 167–182. (d) De Ligt, R. A. F.; Kourounakis, A. P.; Ijzerman, A. P. Inverse agonism at G protein-coupled receptors: (patho)physiological relevance and implications for drug discovery. *Br. J. Pharmacol.* **2000**, *130*, 1–12. (e) Perez, D. M.; Karnik, S. S. Multiple signaling states of G-protein-coupled receptors. *Pharmacol. Rev.* **2005**, *57*, 147–161.
62. Samama, P.; Cotecchia, S.; Costa, T.; Lefkowitz, R. J. A mutation-induced activated state of the ß2-adrenergic receptor. Extending the ternary complex model. *J. Biol. Chem.* **1993**, *268*, 4625–4636.
63. Leff, P.; Scaramellini, C.; Law. C.; McKechnie, K. A three-state receptor model of agonist action. *Trends Pharmacol. Sci.* **1997**, *18*, 355–362.
64. Ganellin, C. R. *Pharmacology of Histamine Receptors*; Ganellin, C. R.; Parsons, M. E., (Eds.), Wright-PSG, Britol, 1982, Chapter 2.
65. Ariëns, E. J.; Simonis, A. M.; van Rossum, J. M. In *Molecular Pharmacology*, Ariëns, E. J. (Ed.), Academic, New York, 1964; Vol. I, pp. 212 and 225.
66. Roth, F. E.; Govier, W. M. Comparative pharmacology of chlorpheniramine (Chlor-Trimeton) and its optical isomers. *J. Pharmacol. Exp. Ther.* **1958**, *124*, 347–349.
67. (a) Ariëns, E. J. Stereochemistry: a source of problems in medicinal chemistry. *Med. Res. Rev.* **1986**, *6*, 451–466. (b) Ariëns, E. J. Stereochemistry in the analysis of drug action. Part II. *Med. Res. Rev.* **1987**, *7*, 367.
68. Jacobson, A. E. In *Problems of Drug Dependence 1989*, Harris, L. S. (Ed.), U. S. Printing Office, Washington, DC, p. 556.
69. Pfeiffer, C. Optical isomerism and pharmacological action, a generalization. *Science* **1956**, *124*, 29–41.
70. White, P.; Ham, J.; Way, W.; Trevor, A. Pharmacology of ketamine isomers in surgical patients. *Anesthesiology* **1980**, *52*, 231–239.
71. Mason, S. The left hand of nature. *New Scientist* **1984**, *101*, 10–14.
72. Winter, W.; Frankus, E. Thalidomide enantiomers. *The Lancet* **1992**, *339*, 365.
73. Ito, T.; Ando, H.; Suzuki, T.; Ogura, T.; Hotta, K.; Imamura, Y.; Yamaguchi, Y.; Handa, H. Identification of a primary target of thalidomide teratogenicity. *Science* **2010**, *327*, 1345–1350.
74. Takada, T.; Tada, M.; Kiyomoto, A. Pharmacological studies of a new local anesthetic, 2′-methyl-2-methyl-2-n-propylaminopropionanilide hydrochloride (LA-012). *Nippon Yakurigaku Zasshi* **1966**, *62*, 64–74; Takada, T.; Tada, M.; Kiyomoto, A. Pharmacological studies of a new local anesthetic, 2′-methyl-2-methyl-2-n-propylaminopropionanilide hydrochloride (LA-012). *Chem. Abstr.* **1967**, *67*, 72326s.
75. Aps, C.; Reynolds, F. An intradermal study of the local anesthetic and vascular effects of the isomers of bupivacaine. *Br. J. Clin. Pharmacol.* **1978**, *6*, 63–68.
76. Tobert, J.; Cirillo, V.; Hitzenberger, G.; James, I.; Pryor, J.; Cook, T.; Buntinx, A.; Holmes, I.; Lutterbeck, P. Enhancement of uricosuric properties of inKdacrinone by manipulation of the enantiomer ratio. *Clin. Pharmacol. Ther.* **1981**, *29*, 344–350.
77. Drayer, D. E. Pharmacodynamic and pharmacokinetic differences between drug enantiomers in humans: an overview. *Clin. Pharmacol. Ther.* **1986**, *40*, 125–133.
78. Knabe, J. In *Chirality and Biological Activity*, Alan R. Liss, New York, 1990, pp. 237–246.
79. Knabe, J.; Rummel, W.; Büch, H. P.; Franz, N. Optically active barbiturates. Synthesis, configuration and pharmacological effects. *Arzneim. Forsch.* **1978**, *28*, 1048–1056.
80. Zimmerman, D.; Gesellchen, P. Analgesics (peripheral and central), endogenous opioids and their receptors. *Annu. Rep. Med. Chem.* **1982**, *17*, 21–30.
81. Hillver, S. E.; Björk, I.; Li, Y.-L.; Svensson, B.; Ross, S.; Andén, N.-E.; Hacksell, U. (S)-5-fluoro-8-hydroxy-2-(dipropylamino)tetralin: a putative 5-HT1A-receptor antagonist. *J. Med. Chem.* **1990**, *33*, 1541–1544.
82. Hill, D. R.; Bowery, N. G. 3H-baclofen and 3H-GABA bind to bicuculline-insensitive GABAB sites in rat brain. *Nature* (*Lond.*) **1981**, *290*, 149–152.
83. Cahn, R. S.; Ingold, C. K.; Prelog, V. Specification of molecular chirality. *Angew. Chem. Int. Ed. Engl.* **1966**, *5*, 385–415.
84. Dukes, M.; Smith, L. H. ß-Adrenergic blocking agents. 9. Absolute configuration of propranolol and of a number of related aryloxpropanolamines and arylethanolamines. *J. Med. Chem.* **1971**, *14*, 326–328.
85. Ariëns, E. J. Stereochemical implication of hybrid and pseudo-hybrid drugs. Part III. *Med. Res. Rev.* **1988**, *8*, 309–320.
86. Ariëns, E. J. Stereochemistry in the analysis of drug-action. Part II. *Med. Res. Rev.* **1987**, *7*, 367–387.
87. Simonyi, M. On chiral drug action. *Med. Res. Rev.* **1984**, *4*, 359–413.
88. Ariëns, E. J.; Wuis, E. W.; Veringa, E. J. Stereoselectivity of bioactive xenobiotics. A pre-pasteur attitude in medicinal chemistry, pharmacokinetics and clinical pharmacology. *Biochem. Pharmacol.* **1988**, *37*, 9–18.
89. Food and Drug Administration policy statement for the development of new stereoisomeric drugs. Available at: www.fda.gov/cder/guidance/stereo.htm. *Fed. Reg.* **1992**, *22*, 249.

90. Murakami, H. From racemates to single enantiomers-chiral synthetic drugs over the last 20 years. *Top. Curr. Chem.* **2007**, *269*, 273–299.
91. Lindberg, P.; Keeling, D.; Fryklund, J.; Andersson, T.; Lundborg, P.; Carlsson, E. *Aliment. Pharmacol. Ther.* **2003**, *17*, 481–488.
92. (a) Stinson S. C. Counting on chirality. *Chem. Eng. News* **2000**, *76*(38), 83–104. (b) Rouhi, A. M. Chiral business. *Chem. Eng. News* **2003**, *81*(18), 45–55.
93. Stinson, S. C. Chiral drugs: new single-isomer products on the chiral drug market create demand for enantiomeric intermediates and enantioselective technologies. *Chem. Eng. News* **Sept. 1994**, *19*, pp. 38, 40.
94. Agranat, I.; Wainschtein, S. R.; Zusman, E. Z. The predicted demise of racemic new molecular entities is an exaggeration. *Nat. Rev. Drug Disc.* **2012**, *11*, 972–973.
95. Lien, E. J.; Rodrigues de Miranda, J. F.; Ariëns, E. J. Quantitative structure-activity correlation of optical isomers: a molecular basis for Pfeiffer's rule. *Mol. Pharmacol.* **1976**, *12*, 598–604.
96. Cushny, A. *Biological Relations of Optically Isomeric Substances*, Williams and Wilkins, Baltimore, 1926.
97. Easson, L. H.; Stedman, E. Studies on the relationship between chemical constitution and physiological action. V. Molecular dissymmetry and physiological activity. *Biochem. J.* **1933**, *27*, 1257–1266.
98. Blaschko, H. Action of local hormones: remarks on chemical specificity. *Proc. Roy. Soc. B* **1950**, *137*, 307–311.
99. Jamali, F.; Mehvar, R.; Pasutto, F. M. Enantioselective aspects of drug action and disposition: therapeutic pitfalls. *J. Pharm. Sci.* **1989**, *78*, 695–715.
100. Boger, D. L.; Hong, J. Asymmetric total synthesis of ent-(−)-roseophilin: assignment of absolute configuration. *J. Am. Chem. Soc.* **2001**, *123*, 8515–8519.
101. Boger, D. L.; Hüter, O.; Mbiya, K.; Zhang, M. Total synthesis of natural and ent-Fredericamycin A. *J. Am. Chem. Soc.* **1995**, *117*, 11839–11849.
102. Cross, L. C.; Klyne, W. Rules for the nomenclature of organic chemistry. Section E: stereochemistry (recommendations 1974). *Pure Appl. Chem.* **1976**, *45*, 11–30.
103. Casy, A. F.; Ganellin, C. R.; Mercer, A. D.; Upton, C. Analogs of triprolidine: structural influences upon antihistamine activity. *J. Pharm. Pharmacol.* **1992**, *44*, 791–795.
104. Kaiser, C.; Setler, P. E. In *Burger's Medicinal Chemistry*, 4th ed., Wolff, M. E. (Ed.), Wiley, New York, 1981, Part III, Chapter 56.
105. Gampe, R. T.; Montana, V. G.; Lambert, M. H.; Miller, A. B.; Bledsoe, R. K.; Milburn, M. V.; Kliewer, S. A.; Willson, T. M.; Xu, H. E. Asymmetry in the PPARγ/RXRa crustal structure reveals the molecular basis of heterodimerization among nuclear receptors. *Mol. Cell* **2000**, *5*, 545–555.
106. Xiu, X.; Puskar, N. L.; Shanata, J. A. P.; Lester, H. A.; Dougherty, D. A. Nicotine binding to brain receptors requires a strong cation–π interaction. *Nature* **2009**, *458*, 534–538.
107. Smissman, E. E.; Nelson, W. L.; LaPidus, J. B.; Day, J. L. Conformational aspects of acetylcholine receptor sites. The isomeric 3-trimethylammonium-2-acetoxy-transdecalin halides and the isomeric a,ß-dimethylacetylcholine halides. *J. Med. Chem.* **1966**, *9*, 458–465.
108. (a) Armstrong, P. D.; Cannon, J. G.; Long, J. P. Conformationally rigid analogs of acetylcholine. *Nature* (*Lond.*) 1968, 220, 65–66. (b) Chiou, C. Y.; Long, J. P.; Cannon, J. G.; Armstrong, P. D. Cholinergic effects and rates of hydrolysis of conformationally rigid analogs of acetylcholine. *J. Pharmacol. Exp. Ther.* **1969**, *166*, 243–248.
109. Armstrong, P. D.; Cannon, J. G. Small ring analogs of acetylcholine. Synthesis and absolute configurations of cyclopropane derivatives. *J. Med. Chem.* **1970**, *13*, 1037–1039.
110. Li, J. P.; Biel, J. H. Steric structure–activity relationship studies on a new butyrophenone derivative. *J. Med. Chem.* **1969**, *12*, 917–919.
111. Chen, C.-Y.; LeFèvre, R. J. W. Free energy difference between the axial and equatorial dispositions of the hydroxyl group in N-methyl-4-piperidinol by proton magnetic resonance spectroscopy. *Tetrahedron Lett.* **1965**, 4057–4063.
112. (a) Garey, R. E.; Heath, R. G. The effects of phencyclidine on the uptake of 3H-catecholamines by rat striatal and hypothalamic synaptosomes. *Life Sci.* **1976**, *18*, 1105–1110. (b) Smith, R. C.; Meltzer, H. Y.; Arora, R. C.; Davis, J. M. Effects of phencyclidine on [3H]catecholamine and [3H]serotonin uptake in synaptosomal preparations from rat brain. *Biochem. Pharmacol.* **1977**, *26*, 1435–1439.
113. (a) Maurice, T.; Lockhart, B. P. Neuroprotective and anti-amnesic potentials of sigma (σ) receptor ligands. *Prog. Neuro-Psychopharmacol. Biol. Psychiat.* **1997**, *21*, 69–102. (b) Su, T. P.; London, E. D.; Jaffe, J. H. Steroid binding at (σ) receptors suggests a link between endocrine, nervous and immune systems. *Science* **1988**, *240*, 219–221. (c) Ross, S. B.; Gawaell, L.; Hall, H. Stereoselective high-affinity binding of 3H-alaproclate to membranes from rat cerebral cortex. *Pharmacol. Toxicol.* **1987**, *61*, 288–292.
114. Moriarity, R. M.; Enache, L. A.; Zhao, L.; Gilardi, R.; Mattson, M. V.; Prakash, O. Rigid phencyclidine analogs. Binding to the phencyclidine and σ1 receptors. *J. Med. Chem.* **1998**, *41*, 468–477.
115. Hudkins, R. L.; Mailman, R. B.; DeHaven-Hudkins, D. L. Novel (4-phenylpiperidinyl)- and (4-phenylpiperazinyl)alkyl-spaced esters of 1-phenylcyclopentanecarboxylic acids as potent σ-selective compounds. *J. Med. Chem.* **1994**, *37*, 1964–1970.
116. Tischler, M.; Nasu, D.; Empting, M.; Schmelz, S.; Heinz, D. W.; Rottmann, P.; Kolmar, H.; Buntkowsky, G.; Tietze, D.; Avrutina, O. Braces for the peptide backbone: Insights into structure-activity relationships of protease inhibitor mimics with locked amide conformations. *Angew. Chem. Int. Ed.* **2012**, *51*, 3708–3712.
117. Oki, M. Recent advances in atropisomerism. *Top. Stereochem.* **1983**, *14*, 1–81.
118. (a) Clayden, J.; Moran, W. J.; Edwards, P. J.; LaPlante, S. R. The challenge of atropisomerism in drug discovery. *Angew. Chem. Int. Ed.* **2009**, *48*, 6398–6401. (b) LaPlante, S. R.; Fader, L. D.; Fandrick, K. R.; Fandrick, D. R.; Hucke, O.; Kemper, R.; Miller, S. P. F.; Edwards, P. J. Assessing atropisomer axial chirality in drug discovery and development. *J. Med. Chem.* **2011**, *54*, 7005–7022.
119. LaPlante, S. R.; Edwards, P. J.; Fader, L. D.; Jakalian, A.; Hucke, O. Revealing atropisomer axial chirality in drug discovery. *ChemMedChem* **2011**, *6*, 505–513.
120. Albert, J. S.; Aharony, D.; Andisik, H.; Barthlow, P. R. et al. Design, synthesis, and SAR of tachykinin antagonists: modulation of balance in NK_1/NK_2 receptor antagonist activity. *J. Med. Chem.* **2002**, *45*, 3972–3983.
121. Guile, S. D.; Bantick, J. R.; Cooper, M. E.; Donald, D. K. et al. Optimization of monocarboxylate transporter 1 blockers through analysis and modulation of atropisomer interconversion properties. *J. Med. Chem.* **2007**, *50*, 254–263.
122. Palani, A.; Shapiro, S.; Clader, J. W.; Greenlee, W. J. et al. Biological evaluation and interconversion of studies of rotamers of SCH 351125, an orally bioavailable CCR5 antagonist. *Bioorg. Med. Chem. Lett.* **2003**, *13*, 705–708.
123. Palani, A.; Shapiro, S.; Clader, J. W.; Greenlee, W. J. et al. Oximino-piperidino-piperidine-based CCR5 antagonists. Part 2: synthesis, SAR and biological evaluation of symmetrical heteroaryl carboxamides. *Bioorg. Med. Chem. Lett.* **2003**, *13*, 709–712.

124. Nogrady, T. *Medicinal Chemistry*, Oxford, New York, 1985, p. 28.
125. Black, J. W.; Duncan, W. A. M.; Durant, C. J.; Ganellin, C. R.; Parsons, E. M. Definition and antagonism of histamine H2-receptors. *Nature* (*Lond.*) **1972**, *236*, 385–390.
126. (a) Ganellin, C. R. The 1980 award in medicinal chemistry. Medicinal chemistry and dynamic structure-activity analysis in the delivery of drugs acting at histamine H2 receptors. *J. Med. Chem.* **1981**, *24*, 913–920. (b) Ganellin, C. R.; Durant, G. J. In *Burger's Medicinal Chemistry*, 4th ed., Wolff, M. E., Ed.; Wiley, New York, 1981, Part III, Chapter 48.
127. (a) Hill, S. J.; Ganellin, C. R.; Timmerman, H.; Schwartz, J. C.; Shankley, N.P.; Young, J. M.; Schunack, W.; Levi, R.; Haas, H. L. International union of pharmacology. XIII. Classification of histamine receptors. *Pharmacol. Rev.* **1997**, *49*, 253–278. (b) O'Mahony, L.; Akdis, M.; Akdis, C. A. Regulation of the immune response and inflammation by histamine and histamine receptors. *J. Allergy Clin. Immunol.* **2011**, 128, 1153–1162.
128. Wyllie, J. H.; Hesselbo, T.; Black, J. W. Effects in man of histamine H2-receptro blockade by burmamide. *Lancet* **1972**, 2(7787), 1117–1120.
129. Black, J. W.; Durant, G. J.; Emmett, J. C.; Ganellin, C. R. Sulfurmethylene isosterism in the development of metiamide, a new histamine H2-receptor antagonist. *Nature* (*Lond.*) **1974**, *248*, 65–67.
130. Charton, M. The application of the Hammett equation to amidines. *J. Org. Chem.* **1965**, *30*, 969–973.
131. Marshall, B. J. In *Helicobacter pylori*, Mobley, H. L. T.; Mendz, G. L.; Hazell, S. L. (Eds.), ASM Press, Herndon, VA, 2001; pp. 19–24.
132. Montecucco, C.; Rappuoli, R. Living dangerously: how *Helicobacter pylori* survives in the human stomach. *Nat. Rev. Mol. Cell Biol.* **2001**, *2*, 457–466.
133. Israel, D. A.; Peek, R. M. Review article: Pathogenesis of *Helicobacter pylori*-induced gastric inflammation. *Aliment. Pharmacol. Ther.* **2001**, *15*, 1271–1290.
134. (a) Xia, H. H.-X.; Wong, B. C. Y.; Talley, N. J.; Lam, S. K. *Helicobacter pylori* infection - current treatment practice. *Exp. Opin. Pharmacother.* **2001**, *2*, 253–266. (b)Williamson, J. S. *Helicobacter pylori*: current chemotherapy and new targets for drug design. *Curr. Pharm. Des.* **2001**, *7*, 355–392.
135. Passarella, S.; Duong, M.-T. Diagnosis and treatment of insomnia. *Am. J. Health-Syst. Pharm.* **2008**, *65*, 927–934.
136. Sullivan, S. S.; Guilleminault, C. Emerging drugs for insomnia: new frontiers for old and new targets. *Exp. Opin. Emerging Drugs* **2009**, *14*, 411–422.
137. (a) De Lecea, L.; Kilduff, T. S.; Peyron, C.; Gao, X.-B.; Foye, P. E.; Danielson, P. E.; et al. Th hypocretins: hypothalamus-specific peptides with neuroexcitatory activity. *Proc. Natl. Acad. Sci. U.S.A.* **1998**, *95*, 322–327. (b) Sakurai, T.; Amemiya, A.; Ishii, M.; Matsuzaki, I. et al. Orexins and orexin receptors: a family of hypothalamic neuropeptides and G-protein receptors that regulate feeding behavior. *Cell* **1998**, *92*, 573–585.
138. Chemelli, R. M.; Willie, J. T.; Sinton, C. M.; Elmquist, J. K. et al. Narcolepsy in orexin knockout mice: molecular genetics of sleep regulation. *Cell* **1999**, *98*, 437–451.
139. Akanmu, M. A.; Honda, K. Selective stimulation of orexin receptor type 2 promotes wakefulness in freely behaving rats. *Brain Res.* **2005**, *1048*, 138–145.
140. Whitman, D. B.; Cox, C. D.; Breslin, M. J.; Brashear, K. M.; Schreier, J. D. et al. Discovery of a potent, CNS-penetrant orexin receptor antagonist based on an N, N-disubstituted-1,4-diazepane scaffold that promotes sleep in rats. *ChemMedChem* **2009**, *4*, 1069–1074.

Enzymes

Chapter Outline

4.1. ENZYMES AS CATALYSTS

Enzymes are special types of receptors. The receptors discussed in Chapter 3 are mostly membrane-bound proteins that interact with natural ligands and agonists to form complexes that then elicit a biological response. Subsequent to the response, the ligand is released intact. Enzymes, most of which are soluble and found in the cytosol of cells, interact with substrates to form complexes, but, unlike receptors, it is from these *enzyme–substrate complexes* that enzymes catalyze reactions, thereby transforming the substrates into products that are released. Therefore, the two characteristics of enzymes are their ability to recognize a substrate and to catalyze a reaction with it.

4.1.1. What are Enzymes?

Enzymes are natural proteins that catalyze chemical reactions; ribonucleic acids (RNA) also can catalyze chemical reactions.[1] The first enzyme to be recognized as a protein was jack bean urease, which was crystallized in 1926 by Sumner[2] and was shown to catalyze the hydrolysis of urea to CO_2 and NH_3. It took almost 70 years more, however, before its crystal structure would be obtained by Andrew Karplus (for the enzyme from *Klebsiella aerogenes*).[3] By the 1950s, hundreds of enzymes had been discovered, and many were purified to homogeneity and crystallized. In 1960, Hirs, Moore, and Stein[4] were the first to sequence an enzyme, namely, ribonuclease A, having only 124 amino acids (molecular mass: 13,680 Da). This was an elegant piece of work, and William H. Stein and Stanford Moore shared the Nobel Prize in chemistry in 1972 for the methodology of protein sequencing, which was developed to determine the ribonuclease A sequence.

Enzymes can have molecular masses of several thousand to several million daltons, yet catalyze transformations on molecules as small as carbon dioxide or nitrogen. Carbonic anhydrase from human erythrocytes, for example, has a molecular mass of about 31,000 Da and each enzyme molecule can catalyze the hydration of 1,400,000 molecules of CO_2 to H_2CO_3 per second! This is almost 10^8 times faster than the uncatalyzed reaction, which is actually on the low side of rate enhancements for an enzyme. Orotidine 5′-phosphate decarboxylase, for example, catalyzes the decarboxylation of orotidine 5′-phosphate to uridine 5′-phosphate 10^{17} times faster than the nonenzymatic rate.[5]

The Organic Chemistry of Drug Design and Drug Action. http://dx.doi.org/10.1016/B978-0-12-382030-3.00004-0

4.1.2. How do Enzymes Work?

Scheme 4.1 shows a generalized scheme for an enzyme-catalyzed reaction, where S is the starting reactant (substrate), E is the enzyme, TS is the transition state, and P is the product. In general, enzymes function by lowering transition state energies and energetic intermediates and by raising the ground state energy (*ground state destabilization*). The transition state for an enzyme-catalyzed reaction, just as in the case of a chemical reaction, is a high energy state having a lifetime of about 10^{-14} s, the time for one bond vibration.[6] There is no spectroscopic method that can detect the transition state structure in an enzyme.

At least 21 different hypotheses for how enzymes catalyze reactions have been proposed.[7] The one common link between all these proposals, however, is that an enzyme-catalyzed reaction always is initiated by the formation of an *enzyme–substrate (or E·S) complex*, from which the catalysis takes place. The concept of an E·S complex was originally proposed independently in 1902 by Brown[8] and Henri[9]; this idea is an extension of the 1894 *lock and key hypothesis* of Fischer[10] in which it was proposed that an enzyme is the lock into which the substrate (the key) fits. This interaction of the enzyme and substrate would account for the high degree of specificity of enzymes, but the lock and key hypothesis does not rationalize certain observed phenomena. For example, compounds whose structures are related to that of the substrate, but have *less* bulky substituents, often fail to be substrates even though they should have fit into the enzyme. Some compounds with *more* bulky substituents are observed to bind *more* tightly to the enzyme than does the substrate. If the lock and key hypothesis were correct, it would be thought that a more bulky compound would not fit into the lock. Some enzymes that catalyze reactions between two substrates do not bind one substrate until the other one is already bound to the enzyme. These curiosities led Koshland[11] in 1958 to propose the *induced-fit hypothesis* (discussed in relationship to receptors in Chapter 3, Section 3.2.4.3), namely, that when a substrate begins to bind to an enzyme, interactions of various groups on the substrate with particular enzyme functional groups are initiated, and these mutual interactions induce a *conformational change* in the enzyme. This results in a change of the enzyme from a low catalytic form to a high catalytic form by destabilization of the enzyme and/or by inducing proper alignment of the groups involved in catalysis. The conformational change could serve as a basis for substrate specificity. Compounds resembling the substrate except with smaller or larger substituents may bind to the enzyme but may not induce the conformational change necessary for catalysis. Unlike the lock and key hypothesis, which implies a rigid active site, the induced-fit hypothesis requires a flexible active site to accommodate different binding modes and conformational changes in the enzyme. Actually, the concept of a flexible active site was stated earlier by Pauling[12] who hypothesized that an enzyme is a flexible template that is most complementary to substrates at the transition state rather than at the ground state. This flexible model is consistent with many observations regarding enzyme action.

In 1930, Haldane[13] suggested that an E·S complex requires additional activation energy prior to enzyme catalysis, and this energy may be derived from substrate strain energy on the enzyme. As early as 1921, Polanyi proposed that transition state binding is essential for catalysis,[14] and he and his student Eyring developed transition state theory,[15] which is the basis for the above-mentioned hypothesis of Pauling. According to this hypothesis, the substrate does not bind most effectively in the E·S complex; as the reaction proceeds, the enzyme conforms to the transition state structure, leading to the tightest interactions (increased binding energy) with the transition state structure.[16] This increased binding energy, known as *transition state stabilization*, results in rate enhancement. Schowen has suggested[17] that all of the above-mentioned 21 hypotheses of enzyme catalysis (as well as other correct hypotheses) are just alternative expressions of transition state stabilization.

Similar to the case of noncatalytic receptors, in which the pharmacophore of the drug interacts with a relatively small part of the total receptor, the substrate likewise binds to only a small part of the enzyme known as the *active site* of the enzyme. There may be only a dozen or so amino acid residues that comprise the active site, and of these, only a few may be involved directly in substrate binding and/or catalysis. Because all of the catalysis takes place in the active site of the enzyme, you may wonder why it is necessary for enzymes to be so large. There are several hypotheses regarding the function of the remainder of the enzyme. One suggestion[18] is that the most effective binding of the substrate to the enzyme (the largest binding energy) results from close packing of the atoms within the protein; possibly, the remainder of the enzyme outside of the active site is required to maintain the integrity of the active site for effective catalysis. The protein may also serve the function of channeling the substrate into the active site.

Enzyme catalysis is characterized by two features: *specificity* and *rate acceleration*. The active site contains moieties that are responsible for both of these properties of an enzyme, namely, amino acid residues and, in the case of some enzymes, cofactors. A *cofactor*, also called a *coenzyme*, is an organic molecule or a metal ion that binds to the active site, in some cases covalently and in others noncovalently, and is essential for the catalytic action of those enzymes that require cofactors.

$$\mathrm{E+S} \underset{}{\overset{K_s}{\rightleftharpoons}} \mathrm{E{\bullet}S} \overset{k_{cat}}{\rightleftharpoons} [\mathrm{E{\bullet}TS}] \rightleftharpoons \mathrm{E{\bullet}P} \rightleftharpoons \mathrm{E+P}$$

SCHEME 4.1 Generalized enzyme-catalyzed reaction. The overall rate constant for conversion for E·S to E·P is called k_{cat}.

4.1.2.1. Specificity of Enzyme-Catalyzed Reactions

Specificity refers to both specificity of binding and specificity of reaction. Certain active-site constituents are involved in these binding interactions, which are responsible for the binding specificity. These interactions are the same as those discussed in Chapter 3 (Section 3.2.2) for the interaction of an agonist with a noncatalytic receptor and include covalent, electrostatic, ion–dipole, dipole–dipole, hydrogen bonding, charge–transfer, hydrophobic, cation–π, halogen bonding, and van der Waals interactions.

4.1.2.1.1. Binding Specificity

As indicated above, maximum binding interactions at the active site occur at the transition state of the reaction. An enzyme binds the transition state structure on average about 10^{12} times more tightly than it binds the substrate or products (the enzyme orotidine 5′-monophosphate decarboxylase binds the transition state 10^{17} times greater than substrate![19]). Therefore, it is important that an enzyme does not bind to intermediate states excessively or this will increase the free energy difference between the intermediate and transition states. The binding interactions set up the substrate for the reaction that the enzyme catalyzes (Scheme 4.1).

At one end of the spectrum, binding specificity can be absolute, that is, essentially only one substrate forms an E·S complex with a particular enzyme, which then leads to product formation. At the other end of the spectrum, binding specificity can be very broad, in which case many molecules of related structure can bind and be converted to product, such as the family of enzymes known as *cytochrome P450*, which protects us from the toxins we eat and breathe.[20] Specificity may involve E·S complex formation with only one enantiomer of a racemate or E·S complex formation with both enantiomers, but only one is converted to product. The reason enzymes can accomplish this enantiomeric specificity, just as receptors do with enantiomers (Section 3.2.5.2), is because they are chiral molecules (mammalian enzymes are comprised of only L-amino acids); therefore, interactions of an enzyme with a racemic mixture results in the formation of two diastereomeric complexes. This is analogous to the principle behind the resolution of racemic mixtures with chiral reagents. If, for example (Scheme 4.2), a pure *R*-isomer of a chiral amine such as (*R*)-2-methylbenzylamine (**4.1**) is mixed with a racemic mixture of a carboxylic acid such as the nonsteroidal antiinflammatory drug ibuprofen (**4.2**; Advil), then two diastereomeric salts, (*R·R*) and (*R·S*), will be formed. Because these salts are no longer enantiomers, they will have different properties and can be separated by physical means. When an enzyme is exposed to a racemic mixture of a substrate, the binding energy for E·S complex formation with one enantiomer may be much higher than that with the other enantiomer either because of differential binding interactions as noted above or for steric reasons. For example (Figure 4.1), after the ammonium and carboxylate substituents of phenylalanine have interacted with active-site groups, a third substituent at the stereogenic center (the benzyl group) has two possible orientations; in the case of the *S*-isomer, there is a binding pocket (**A**), but the benzyl group of the *R*-isomer (**B**) points in the other direction toward the leucine side chain and causes steric hindrance (i.e., the *R*-isomer does not bind to the active site). If the binding energies for the two complexes are significantly different, then only one E·S complex may form (as would be the case in Figure 4.1). Alternatively, both E·S complexes may form, but for steric or electronic reasons, only one E·S complex may lead to product formation. The enantiomer that forms the E·S complex that is not *turned over* (i.e., converted by the enzyme to product) is said to undergo *nonproductive binding* to the enzyme. Enzymes also can demonstrate complete stereospecificity with geometric isomers since these are diastereomers already.

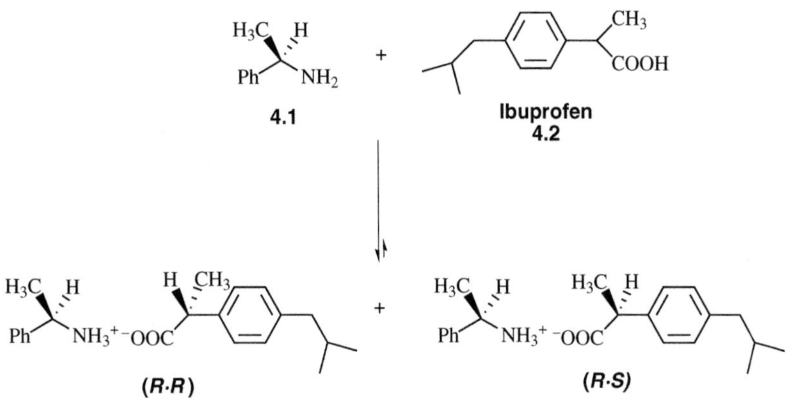

H_3C H Ph NH_2 **4.1** + CH_3 COOH **Ibuprofen 4.2**

H_3C H Ph NH_3^+ ^-OOC H CH_3 **(*R·R*)** + H_3C H Ph NH_3^+ ^-OOC H_3C H **(*R·S*)**

SCHEME 4.2 Resolution of a racemic mixture

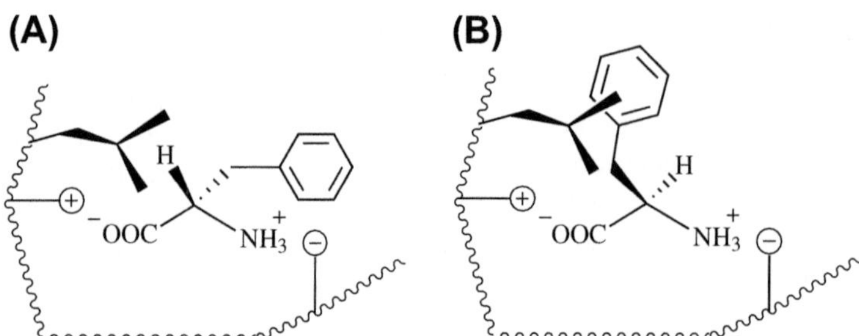

FIGURE 4.1 Differential binding interactions by enantiomers. (A) binding pocket for the (*S*)-isomer; (B) stertic hindrance with the (*R*)-isomer.

4.1.2.1.2. Reaction Specificity

Reaction specificity also arises from constituents of the active site, namely, specific acid, base, and nucleophilic functional groups of amino acids (Section 4.2) and from cofactors (Section 4.3). Unlike most reactions in solution, enzymes can show specificity for chemically identical protons (Figure 4.2). If there are specific binding sites for R and R′ at the active site of the enzyme, and a base (B^-) of an amino acid side chain is juxtaposed so that it can only reach proton H_a, then abstraction of H_a will occur stereospecifically, even though in most nonenzymatic reactions H_a and H_b would be chemically equivalent and, therefore, would have equal probability to be abstracted. The approach taken by synthetic chemists in designing chiral reagents for stereospecific reactions is modeled after this.

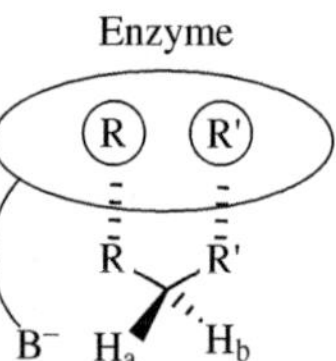

FIGURE 4.2 Enzyme specificity for chemically identical protons. R and R′ on the enzyme are groups that interact specifically with R and R′, respectively, on the substrate.

4.1.2.2. *Rate Acceleration*

In general, catalysts stabilize the transition state relative to the ground state, and this decrease in activation energy is responsible for the rate acceleration that results (Figure 4.3A).

Jencks proposed that the fundamental feature that distinguishes enzymes from simple chemical catalysts is the ability of enzymes to utilize binding interactions away from the site of catalysis.[21] These binding interactions facilitate reactions by positioning substrates with respect to one another and with respect to the catalytic groups at the active site. Because an enzyme has numerous opportunities to invoke catalysis, for example, by stabilization of the transition states (thereby lowering the transition state energy), by destabilization of the E·S complex (thereby raising the ground state energy), by destabilization of intermediates, and during product release, multiple steps, each having small activation energies, may be involved (Figure 4.3B). As a result of these multiple catalytic steps, rate accelerations of 10^{10}–10^{14} over the corresponding nonenzymatic reactions are common. Wolfenden hypothesized that the rate acceleration produced by an enzyme is proportional to the affinity of the enzyme for the transition state structure of the bound substrate[22]; the reaction rate is proportional to the amount of substrate that is in the transition state complex. The enzyme has to be able to bind tightly only to the unstable transition state structure (with a lifetime of one bond vibration) and not to either the substrate or the products. A conformational change in the protein structure plays an important role in this operation.

Enzyme catalysis does not alter the equilibrium of a reversible reaction. If an enzyme accelerates the rate of the forward reaction, it must accelerate the rate of the corresponding back reaction by the same amount; its effect is to accelerate the attainment of the equilibrium, but not the relative concentrations of substrates and products at equilibrium.

Typically, enzymes have *turnover numbers* (also termed k_{cat}), that is, the number of molecules of substrate converted to product (i.e., turned over) per unit of time per molecule of enzyme active site, on the order of $10^3\ s^{-1}$ (about 1000 molecules of substrate are converted by a single enzyme molecule to product every second!). The enzyme catalase is one of the most efficient enzymes,[23] having a turnover number of $10^7\ s^{-1}$. Because there are two other important steps to enzyme catalysis, namely, substrate binding and product release, high turnover numbers are only useful if these two physical steps occur at faster rates. This is not always the case.

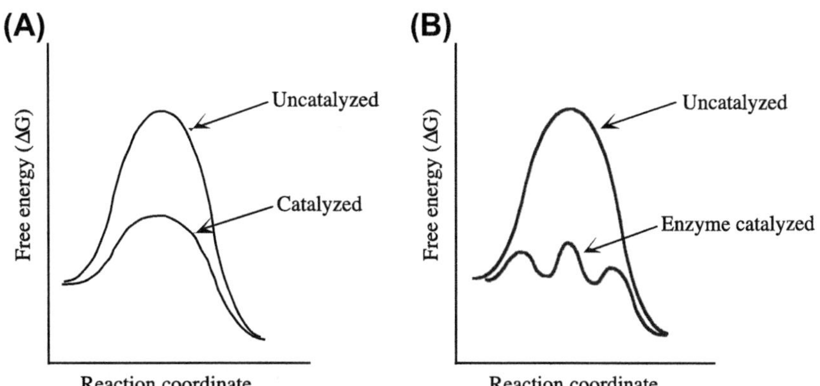

FIGURE 4.3 Examples of the effect of (A) a chemical catalyst and (B) an enzyme on the activation energy.

4.2. MECHANISMS OF ENZYME CATALYSIS

Once the substrate binds to the active site of the enzyme via the interactions noted in Chapter 3 (Section 3.2.2), there are a variety of mechanisms that the enzyme can utilize to catalyze the conversion of the substrate to product. The most common mechanisms[24–26] are approximation, covalent catalysis, general acid–base catalysis, electrostatic catalysis, desolvation, and strain or distortion. All of these act by stabilizing the transition state energy or destabilizing the ground state (which is generally not as important as transition state stabilization).

4.2.1. Approximation

Approximation is the rate enhancement by proximity, that is, the enzyme serves as a template to bind a substrate so that it is close to reactive groups of the enzyme (or to bind multiple substrates so that they are close to each other) in the reaction center. This results in a loss of rotational and translational entropies of the substrate upon binding to the enzyme; however, this entropic loss is offset by a favorable binding energy of the substrate, which provides the driving force for catalysis. Furthermore, since the catalytic groups are now an integral part of the same molecule, the reaction of an enzyme-bound substrate becomes first order, whereas the reaction would be second order if the substrate were free in solution. Holding the reaction centers in close proximity and in the correct geometry for reaction is equivalent to increasing the concentration of the reacting groups. This phenomenon can be exemplified with nonenzymatic model studies. For example, consider the second-order reaction of acetate with an aryl acetate (Scheme 4.3). If the rate constant k for this reaction is set equal to $1.0\,M^{-1}\,s^{-1}$, and then the effect of decreasing rotational and translational entropy is determined by measuring the corresponding first-order rate constants

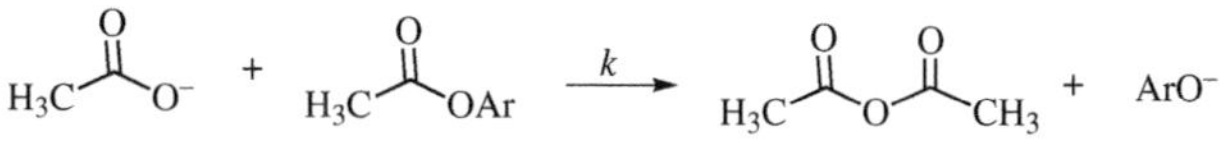

$$H_3C{-}C(=O){-}O^- + H_3C{-}C(=O){-}OAr \xrightarrow{k} H_3C{-}C(=O){-}O{-}C(=O){-}CH_3 + ArO^-$$

SCHEME 4.3 Second-order reaction of acetate with aryl acetate

for related molecules that can undergo the corresponding intramolecular reactions, it is apparent from Table 4.1 that forcing the reacting groups to be closer to each other increases the reaction rate.[27,28] Thirty-six years after the original experimental study of the effect of restricted rotation on rate acceleration, a theoretical investigation using MM3 calculations showed that when the nucleophile and electrophile are closely arranged, and the van der Waals surfaces are properly juxtaposed, the activation energy is lowered as a result of a decrease in the enthalpy of the reaction (ΔH^0), and the rate of the reaction really should increase, thereby supporting the earlier experimental observations.[29]

Although first- and second-order rate constants cannot be compared directly, the efficiency of an intramolecular reaction can be defined in terms of its *effective molarity* (EM),[30] the concentration of the reactant (or catalytic group) required to cause the intermolecular reaction to proceed at the observed rate of the intramolecular reaction. The EM is calculated by dividing the first-order rate constant for the intramolecular reaction by the second-order rate constant for the corresponding intermolecular reaction (Table 4.1). What this indicates is that the acetate ion would have to be at a concentration of, for example, 220 M ($220\,s^{-1}/1\,M^{-1}\,s^{-1}$) for the intermolecular reaction of acetate and aryl acetate to proceed at a rate comparable to that of the glutarate monoester reaction. Of course, 220 M acetate ion is an imaginary number (pure water is only 55 M), so the effect of decreasing the entropy is quite significant. EMs for a wide range of intramolecular reactions have been measured, and the conclusion is that the efficiency of intramolecular catalysis varies with structure and can be as high as 10^{16} M for

TABLE 4.1 Effect of Approximation on Reaction Rate

	k_{obs} (30 °C)	Relative Rate, k	EM
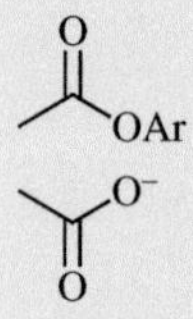 Ethyl ester (OAr) + acetate (O^-)	$3.36\times10^{-7}M^{-1}s^{-1}$	$1.0M^{-1}s^{-1}$	
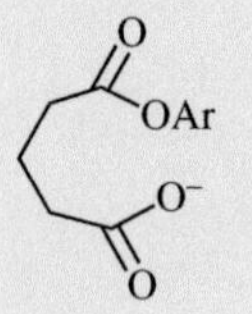 Glutarate monoester (OAr, O^-)	$7.39\times10^{-5}s^{-1}$	$220s^{-1}$	220M
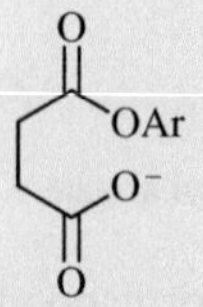 Succinate monoester (OAr, O^-)	$1.71\times10^{-2}s^{-1}$	$5.1\times10^{4}s^{-1}$	$5.1\times10^{4}M$
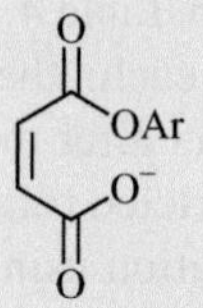 Maleate monoester (OAr, O^-)	$7.61\times10^{-1}s^{-1}$	$2.3\times10^{6}s^{-1}$	$2.3\times10^{6}M$
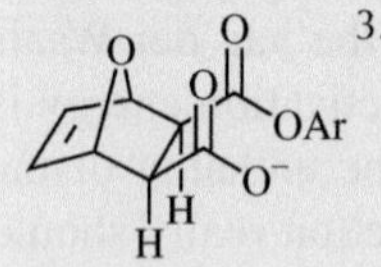 Bicyclic (bridged) monoester (OAr, O^-, H, H)	$3.93s^{-1}$	$1.2\times10^{7}s^{-1}$	$1.2\times10^{7}M$

reactive systems. Therefore, holding groups proximal to each other, particularly when the orbitals of the reacting moieties in an enzyme–substrate complex are aligned correctly for reaction, can be an important contributor to catalysis.

4.2.2. Covalent Catalysis

Some enzymes use nucleophilic amino acid side chains or cofactors in the active site to form covalent bonds to the substrate; in some cases, a second substrate then can react with this enzyme–substrate intermediate to generate the product. This is known as *nucleophilic catalysis* (Scheme 4.4), a subclass of *covalent catalysis* that involves covalent bond formation as a result of attack by an enzyme nucleophile at an electrophilic site on the substrate.[31] For example, if Y in Scheme 4.4 is an amino acid or peptide and Z^- is a hydroxide ion, then the enzyme would be a peptidase (or protease). For nucleophilic catalysis to be most effective, Y should be converted into a better leaving group than X, and the covalent intermediate (**4.3**, Scheme 4.4) should be more reactive than the substrate. The most common active-site nucleophiles are the thiol group of cysteine, the hydroxyl group of serine or threonine, the imidazole of histidine, the amino group of lysine, and the carboxylate group of aspartate or glutamate. These active-site nucleophiles are generally activated by deprotonation, often by a neighboring histidine imidazole or by a water molecule, which is deprotonated in a general base reaction (Section 4.2.3). Amide bonds of peptides have low reactivity, but the nucleophile in the active site is made more nucleophilic in a nonpolar environment; also, alkoxides (e.g., ionized serine) and thiolates (ionized cysteine) are better nucleophiles than hydroxide ion. Once **4.3** is generated, the carbonyl becomes much more reactive (if X is the hydroxyl of serine or threonine, then **4.3** is an ester; if X is the thiol of cysteine, then it is a thioester).

Nucleophilic catalysis is the enzymatic analogy to anchimeric assistance by neighboring groups in organic reaction mechanisms. *Anchimeric assistance* is the process by which a neighboring functional group assists in the expulsion of a leaving group by intermediate covalent bond formation.[32] This results in accelerated reaction rates. Scheme 4.5 shows how a neighboring sulfur atom makes the displacement of a β-chlorine a much more facile reaction than it would be without the sulfur atom. If the sulfur atom were part of an active-site nucleophile, such as a methionine, and the C–Cl bond were part of a substrate, and RO^- were HO^- generated by enzyme-catalyzed deprotonation of water, this would represent covalent catalysis in an enzyme-catalyzed reaction, where the covalent adduct is represented by the episulfonium intermediate.

Typical enzymatic reactions where nucleophilic catalysis is important include many of the proteolytic enzymes,[33] including serine, threonine, and cysteine proteases. Serine proteases[34] utilize a serine residue at the active site as the nucleophile and include enzymes such as elastase (degrades elastin, a connective tissue prevalent in the lung) and plasmin (lyses blood clots). Threonine proteases[35] utilize an active-site threonine residue as the nucleophile and include the protease component of the proteasome (multiprotein complex that degrades cellular proteins through specific regulatory mechanisms).[36] Cysteine proteases[37] utilize an active-site cysteine residue as the nucleophile and include enzymes such as papain (found in papaya fruit and used in digestion) and caspases[38] (involved in programmed cell death).

4.2.3. General Acid–Base Catalysis

In any reaction where proton transfer occurs, *general acid catalysis* and/or *general base catalysis* can be an important mechanism for specificity and rate enhancement.

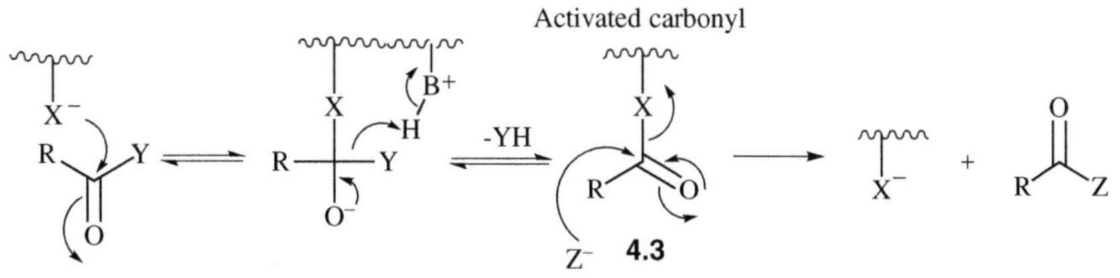

Activated carbonyl

4.3

SCHEME 4.4 Nucleophilic catalysis

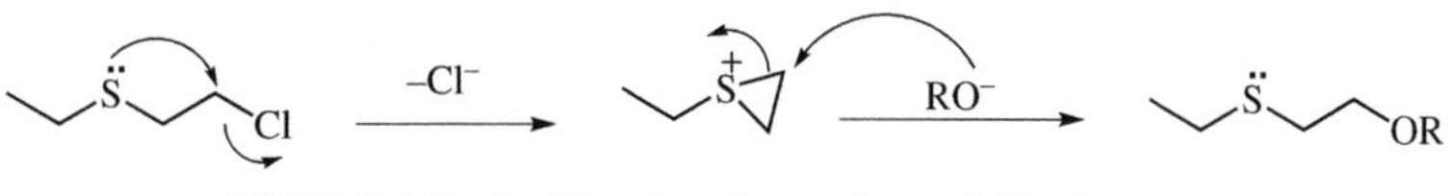

SCHEME 4.5 Anchimeric assistance by a neighboring group

$$H_3C-\overset{O}{\overset{\|}{C}}-OC_2H_5 + H_2O \rightleftharpoons H_3C-\overset{O}{\overset{\|}{C}}-OH + C_2H_5OH$$

SCHEME 4.6 Hydrolysis of ethyl acetate

$$H_3C-\overset{O}{\overset{\|}{C}}-OC_2H_5 \xrightarrow{HO^-} H_3C-\overset{O}{\overset{\|}{C}}-OH + C_2H_5O^- \rightleftharpoons H_3C-\overset{O}{\overset{\|}{C}}-O^- + C_2H_5OH$$

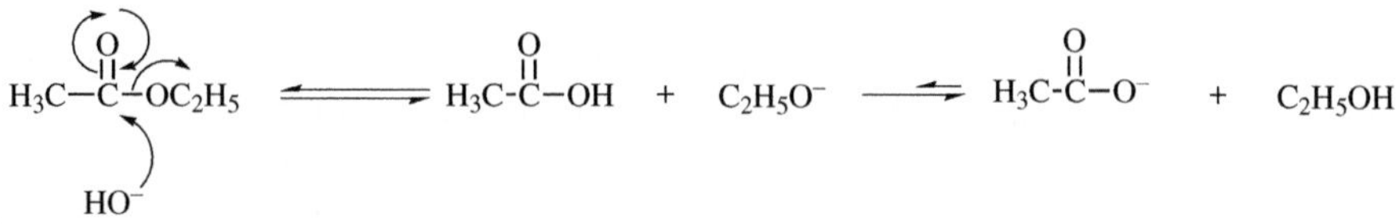

SCHEME 4.7 Alkaline hydrolysis of ethyl acetate

$$H_3C-\overset{O}{\overset{\|}{C}}-OC_2H_5 + H_3O^+ \rightleftharpoons \left[H_3C-\overset{\overset{+}{O}H}{\overset{\|}{C}}-OC_2H_5 \longleftrightarrow H_3C-\overset{OH}{\overset{|}{\underset{+}{C}}}-OC_2H_5 \right] \xrightarrow{H_2O} H_3C-\overset{O}{\overset{\|}{C}}-OH + C_2H_5OH$$

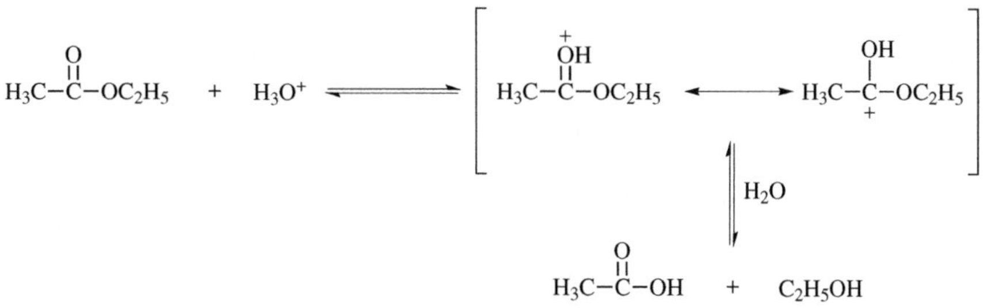

SCHEME 4.8 Acid hydrolysis of ethyl acetate

There are two kinds of acid–base catalysis: general catalysis and specific catalysis. If catalysis occurs by a hydronium (H_3O^+) or hydroxide (HO^-) ion and is determined only by the pH, not the buffer concentration, it is referred to as *specific acid or specific base catalysis*, respectively. As an example of how specific acid–base catalysis works, consider the hydrolysis of ethyl acetate (Scheme 4.6). This is an exceedingly slow reaction at neutral pH because both the nucleophile (H_2O) and the electrophile (the carbonyl of ethyl acetate) are unreactive. The reaction rate could be accelerated, however, if the reactivity of either the nucleophile or the electrophile could be enhanced. An increase in the pH increases the concentration of hydroxide ion, which is a much better nucleophile than is water, and, in fact, the rate of hydrolysis at higher pH increases (Scheme 4.7). Likewise, a decrease in the pH increases the concentration of the hydronium ion, which can protonate the ester carbonyl, thereby increasing its electrophilicity, and this also increases the hydrolysis rate (Scheme 4.8). That being the case, then the hydrolysis rate should be doubly increased if base *and* acid are added together, right? Of course not. Addition of an acid to a base in solution would only result in neutralization and loss of any catalytic effect.

Unlike reactions in solution, however, an enzyme *can* utilize acid and base catalysis simultaneously (Scheme 4.9) for even greater catalysis. The protonated base in Scheme 4.9 is either an acidic amino acid side chain or a basic side chain in the protonated form, and the free base is either a basic residue or an acidic residue in the deprotonated form.

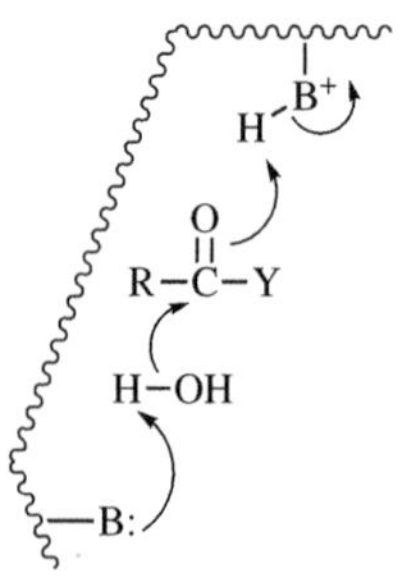

SCHEME 4.9 Simultaneous acid and base enzyme catalysis

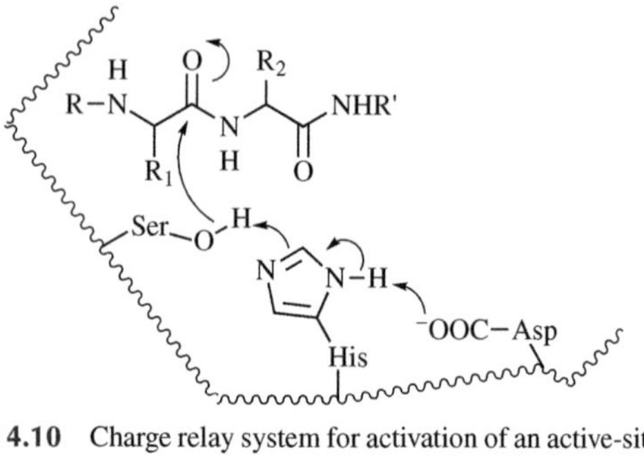

SCHEME 4.10 Charge relay system for activation of an active-site serine residue

As was discussed in Chapter 2 (Section 2.2.5.3), it is important to appreciate the fact that the pK_a values of amino acid side chain groups within the active site of enzymes are not necessarily the same as those measured in solution. Also, pK_a values can change drastically in hydrophobic environments. Therefore, removal of seemingly higher pK_a protons from substrates by active-site bases may not be as unreasonable as would appear if only solution chemistry were taken into consideration.

General acid catalysis occurs when acids other than hydronium ion accelerate the reaction rate. Similarly, *general base catalysis* occurs when bases other than hydroxide ion accelerate the rate. In solution, general acid–base catalysis can be demonstrated when the reaction rate increases with increasing buffer concentration at a constant pH and ionic strength, and shows a larger increase with a buffer that contains a more concentrated acid or base component. Because the hydronium or hydroxide ion concentration is not increasing (the pH is constant), it must be the buffer that is catalyzing the reaction. In enzyme-mediated catalysis, general acid–base catalysis occurs when an acidic or basic residue at the active site is used to facilitate proton transfers in the reaction.

As an example of general acid–base (and covalent) catalysis, consider the enzyme *α-chymotrypsin*, another member of the serine protease family discussed in the previous section. The generally accepted mechanism of action of this enzyme, which features nucleophilic attack by a serine hydroxyl group onto the carbonyl group of a peptide bond, is depicted in Scheme 4.10. Although the serine hydroxyl group is not normally regarded as a potent nucleophile, aspartic acid and histidine residues nearby have been implicated in the conversion of the serine to an alkoxide by a mechanism called the *charge relay system* by Blow and coworkers,[39] the discoverers of the existence of the hydrogen bonding network involving Asp-102, His-57, and Ser-195. This *catalytic triad* involves the aspartate carboxylate (pK_a of the acid is 3.9 in solution) removing a proton from the histidine imidazole (pK_a 6.1 in solution) which, in turn, removes a proton from the serine hydroxyl group (pK_a 14 in solution). Any respectable organic chemist would find that suggestion absurd; how can a base such as aspartate, whose conjugate acid is two pK_a units lower than that of the histidine imidazole, remove the imidazole proton efficiently, and, then, how can this imidazole remove the proton from the hydroxyl group of serine, which is eight pK_a units higher than the (protonated) imidazole of histidine? The equilibrium is only 1% in favor of the first proton transfer, and the equilibrium for the second proton transfer favors the back direction by a factor of 10^8! One explanation (see above) could be that the pK_a values of some of these acids and bases at the active site are different from those in solution. Furthermore, a full deprotonation is likely not occurring; rather, because these groups are held close together at the active site, as the proton is beginning to be removed from the serine hydroxyl group, the charge density proceeds to the next step (attack of the oxygen electron density at the peptide carbonyl), thereby driving the equilibrium in the forward direction. This is the beauty of enzyme-catalyzed reactions; the approximation of the groups and the fluidity of the active-site residues working in concert permit reactions to occur that would be nearly impossible in solution.

4.2.4. Electrostatic Catalysis

An enzyme catalyzes a reaction by stabilization of the transition state and by destabilization of the ground state. Stabilization of the transition state may involve the presence of an ionic charge or partial ionic charge at the active site to interact with an opposite charge developing on the substrate at the transition state of the reaction (Scheme 4.11). In the case of the tetrahedral intermediate shown in Scheme 4.11, the site in the enzyme that leads to this stabilization is referred to as the *oxyanion hole*.[40] Electrostatic interactions may not be as pronounced as is shown in Scheme 4.11; instead of a full positive charge on the enzyme, there may be one or more local dipoles having partial positive charges directed at the incipient transition state anion or a protonated group available for hydrogen bonding. In the case of the serine protease subtilisin, it has been suggested that the lowering of the free energy of the activated complex is the result of hydrogen bonding of the developing oxyanion with protein residues.[41] When the suspected active-site proton donor was replaced by a leucine residue

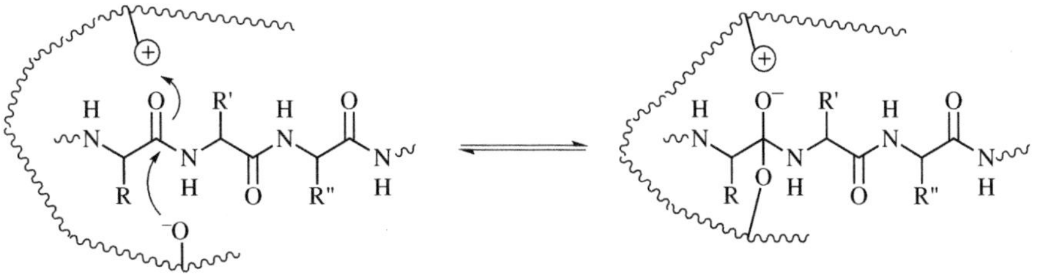

SCHEME 4.11 Electrostatic stabilization of the transition state

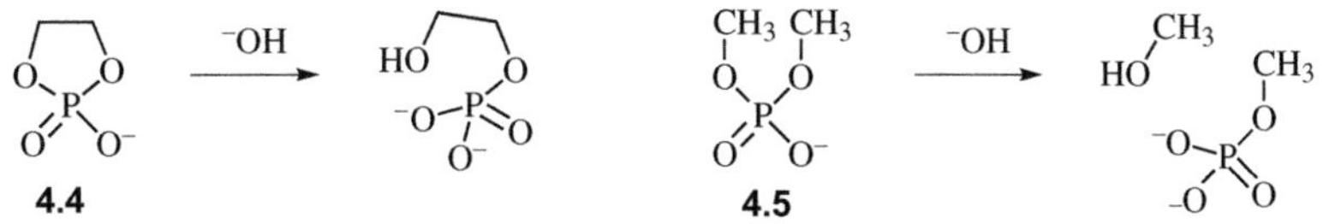

SCHEME 4.12 Alkaline hydrolysis of phosphodiesters as an example of strain energy

using site-directed mutagenesis, the k_{cat} greatly diminished, but the K_m remained the same, indicating the importance of hydrogen bonding to catalysis. Furthermore, a *mutant* (i.e., a form of the enzyme that has one or more of its amino acid residues changed, generally by site-directed mutagenesis) of the protease subtilisin, in which all three of the catalytic triad residues (serine-221, histidine-64, and aspartate-32) were replaced by alanine residues, still was able to hydrolyze amides 10^3 times faster than the uncatalyzed hydrolysis rate, albeit 2×10^6 times slower than the *wild-type enzyme* (i.e., the nonmutated form).[42] This suggests that factors other than nucleophilic and general base catalysis must also be important.

4.2.5. Desolvation

Many reactions are much faster in the gas phase than in water. The desolvation hypothesis posits that an enzyme active site, which is largely or completely devoid of water, can mimic the reaction environment found in the gas phase. Thus, when the substrate enters the active site, water molecules are removed from polar or charged groups on the reactants (i.e., substrate(s), enzyme, or both), which can result in ground state destabilization, since the charged or polar groups on the reactants are no longer stabilized by the water, but instead are exposed to a lower dielectric constant environment.[43] It has been argued, however, that rather than viewing the rate acceleration as being analogous to that in the gas phase, it is more appropriate to view the mechanism of the enzyme as involving "solvent substitution" rather than desolvation.[44] In this view, enzyme active sites provide specific polar environments (substituting the enzyme polar groups for solvent water) that are designed for electrostatic stabilization of ionic transition states, such as seen in Scheme 4.11; these polar groups stabilize ("solvate") these transition states more than water does.

4.2.6. Strain or Distortion

In organic chemistry, *strain* and *distortion* play an important role in the reactivity of molecules. The much higher reactivity of epoxides relative to other ethers demonstrates this phenomenon and is another type of ground state destabilization. Cyclic phosphate ester hydrolysis is another example. Considerable ring strain in **4.4** (Scheme 4.12) is released upon alkaline hydrolysis; the rate of hydrolysis of **4.4** is 10^8 times greater than that for the corresponding acyclic phosphodiester **4.5**.[45] Therefore, if strain or distortion could be induced during enzyme catalysis, then the enzymatic reaction rate would be enhanced. This effect could be induced either in the enzyme, thereby converting it to a high-activity state, or in the substrate, thereby raising the ground state energy (destabilization) of the substrate and making it more reactive. In Section 3.2.4.3 of Chapter 3, the induced-fit hypothesis of Koshland[46] was mentioned. This hypothesis suggests that the enzyme need not necessarily exist in the appropriate conformation required to bind the substrate. As the substrate approaches the enzyme, various groups on the substrate interact with particular active-site functional groups, and this mutual interaction induces a conformational change in the active site of the enzyme. This can result in a change of the enzyme from a low catalytic form to a high catalytic form by destabilization of the enzyme (strain or distortion) or by inducing proper alignment of active-site groups involved in catalysis, which could be responsible for the initiation of catalysis. Inspection of substrate binding sites in protein crystallographic databases indicates that most enzymes have at least a portion of the active site in a structure that is complementary to the substrate to permit binding on the first collisional event. The conformational change need not occur only in the enzyme; the substrate also could undergo deformation, which would lead to strain (destabilization; higher ground state energy) in the substrate. The enzyme was suggested to be elastic and could return to its original conformation after the product was released. This rationalizes how high-energy conformations of substrates are able to bind to enzymes.

According to Jencks,[47] strain or distortion of the bound substrate is essential for catalysis. Since ground state stabilization of the substrate occurs concomitant with transition state stabilization, the $\Delta G^{\ddagger}$ is no different from that of the uncatalyzed reaction, only displaced downward (Figure 4.4B). To lower the $\Delta G^{\ddagger}$ for the catalytic reaction, the E·S complex must be destabilized by strain, desolvation, or loss of entropy upon binding, thereby raising the $\Delta G^{\ddagger}$ of the E·S and E·P complexes (Figure 4.4C). As the reaction proceeds, the $\Delta G^{\ddagger}$ can be lowered by release of strain energy or by other mechanisms described above.

4.2.7. Example of the Mechanisms of Enzyme Catalysis

A very important bacterial enzyme in medicinal chemistry is the peptidoglycan transpeptidase, the enzyme that catalyzes the cross-linking of peptidoglycan strands to make the bacterial cell wall.[48] Most bacterial peptidoglycan transpeptidases are embedded in the cell membrane, making them difficult to isolate in their active form, but a strain of *Streptomyces* was found to secrete a soluble enzyme, D-alanine–D-alanine carboxypeptidase, which catalyzes reactions analogous to those catalyzed by peptidoglycan transpeptidase.[49] This soluble D-alanine–D-alanine carboxypeptidase has been isolated, and the three-dimensional structure has been solved by X-ray crystallography. On the basis of detailed studies of this carboxypeptidase, the hypothetical mechanism shown in Scheme 4.13 can be proposed for the peptidoglycan transpeptidases. Consistent with the principles discussed above, the E–S complex (**A**) formed from the two strands of peptidoglycan with transpeptidase would be stabilized by the appropriate noncovalent binding interactions (Chapter 3, Section 3.2.2). These interactions could place the peptide carbonyl, which is ultimately the site of transpeptidation, very close (approximation) to the serine residue that is involved in covalent catalysis. Base catalysis (via a catalytic triad) could be utilized to activate the active-site serine and electrostatic catalysis (for example, by a positively charged arginine residue) could stabilize the oxyanion intermediate (**B**). Alternatively, an active-site acidic group could donate a proton (acid catalysis) to the incipient oxyanion to lower the transition state energy for the formation of the tetrahedral intermediate. Breakdown of this intermediate could be facilitated by proton donation (acid catalysis) to the leaving D-alanine residue in addition to appropriate strain energy in the sp^3 tetrahedral carbon produced by a conformational change that favors sp^2 hybridization of the product (the covalent intermediate). If a proton-donating mechanism instead of electrostatic mechanism were used to activate the initial covalent reaction, then the active-site conjugate base could remove the proton to facilitate tetrahedral intermediate breakdown. The third step, the cross-linking of the second peptidoglycan strand with the newly formed ester linkage of the activated initial peptidoglycan strand, could be catalyzed by approximation

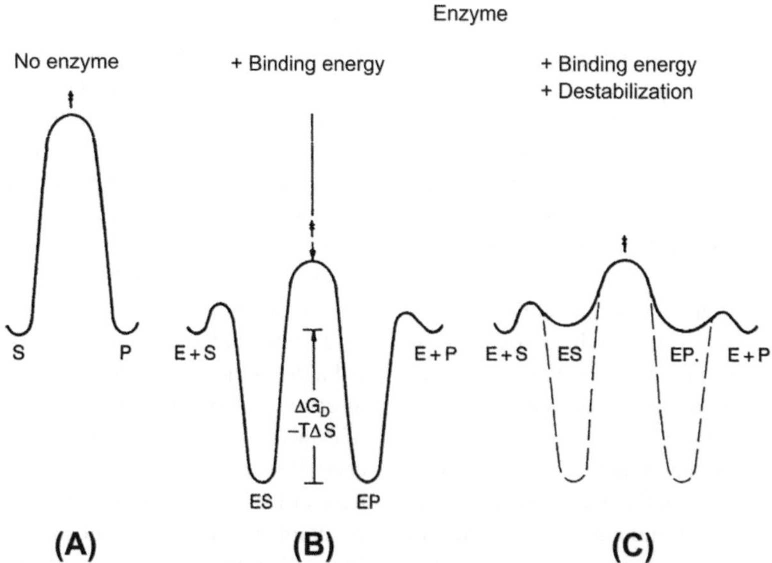

FIGURE 4.4 Energetic effect of enzyme catalysis. (A) no enzyme catalysis; (B) concomitant transition state and ground state stabilization; (C) ground state destabilization. *Reproduced with permission from Jencks, W. P. (1987) Cold Spring Harbor* Symp. Quant. Biol. *52, 65. Copyright © 1987 Cold Spring Harbor Laboratory Press.*

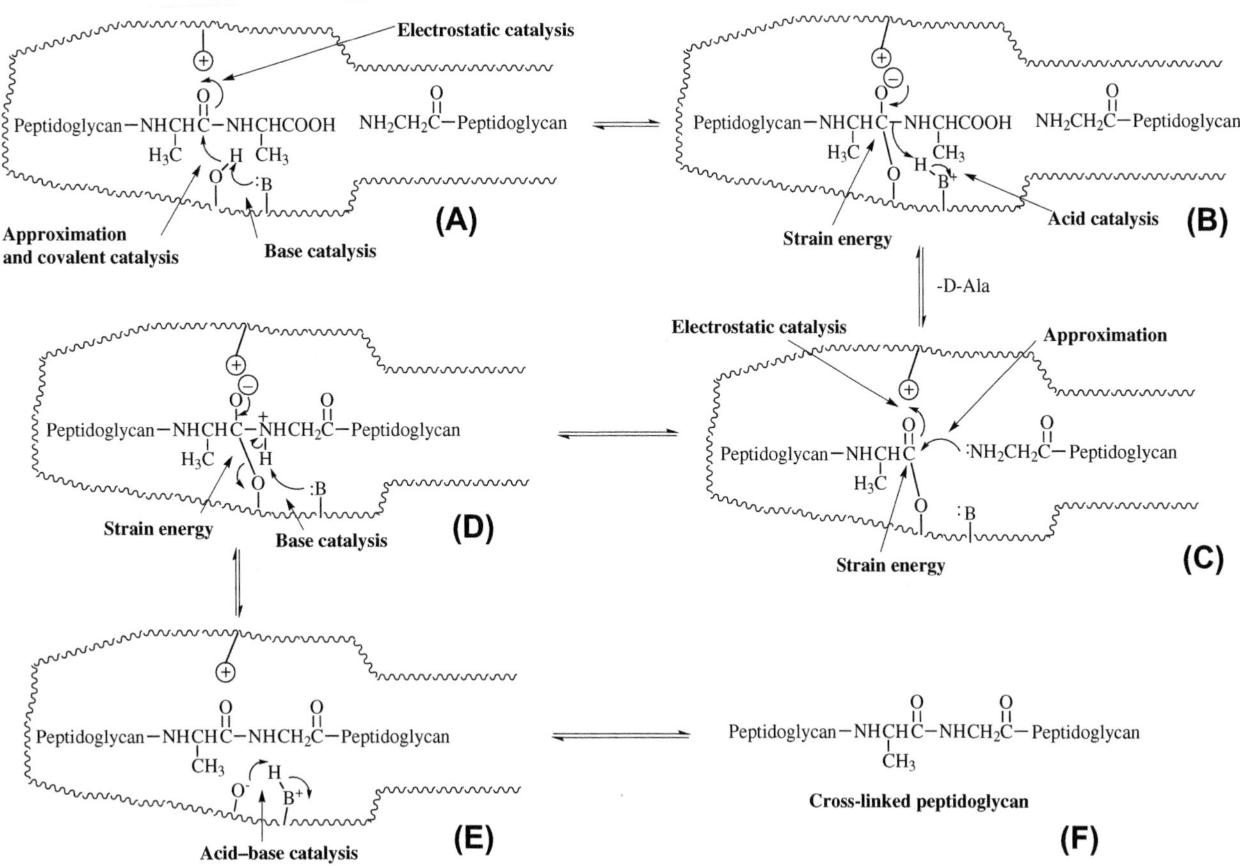

SCHEME 4.13 Hypothetical mechanism for peptidoglycan transpeptidase.

of the two reacting centers, by electrostatic catalysis as in the first step, and by another conformational change in the enzyme to distort the sp^2 ester carbonyl toward the second sp^3 tetrahedral intermediate (**C**). Breakdown of that tetrahedral intermediate (**D**) could be catalyzed again by strain energy (sp^3 back to sp^2) as well as by base catalysis. Release of product (**F**) may occur more readily with the enzyme in the charged form shown in (**E**). Proton transfer after product release would return the enzyme to its normal energy state.

It must be emphasized that the mechanism shown in Scheme 4.13 is hypothetical. However, on the basis of the mechanism of enzyme catalysis described above, it is not an unreasonable hypothesis.

4.3. COENZYME CATALYSIS

A *coenzyme*, or *cofactor*, is any organic molecule or metal ion that is essential for the catalytic action of the enzyme. The usual organic coenzymes (**4.6–4.13b**) are generally derived as products of the metabolism of vitamins that we consume (Table 4.2). Other organic molecules that are involved in essential enzyme functioning, but which are not derived from vitamins, include coenzyme A (**4.13b**), which is used to make the CoA thioesters of carboxylic acid substrates (and is not related to vitamin A); heme (**4.14**; protoporphyrin IX) and the tripeptide glutathione (**4.15**; GSH), which are very important to enzymes involved in drug metabolism (Chapter 8); adenosine triphosphate (ATP) (**4.16**), which supplies the energy required to activate certain substrates during enzyme-catalyzed reactions; and lipoic acid (**4.17**) and ascorbic acid (**4.18**, vitamin C), involved in oxidation and reduction reactions, respectively.

4.6

4.7

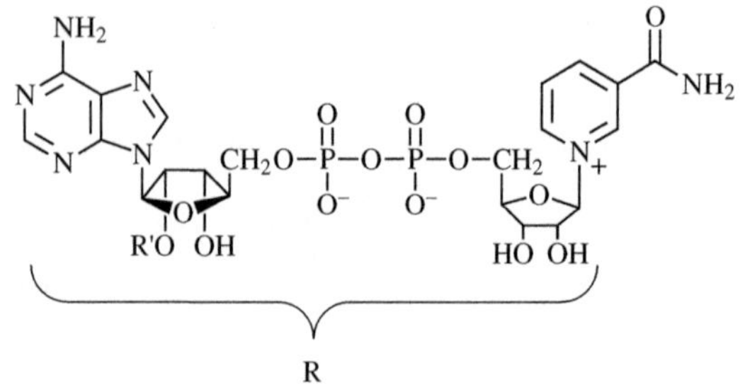

R

4.8a

R

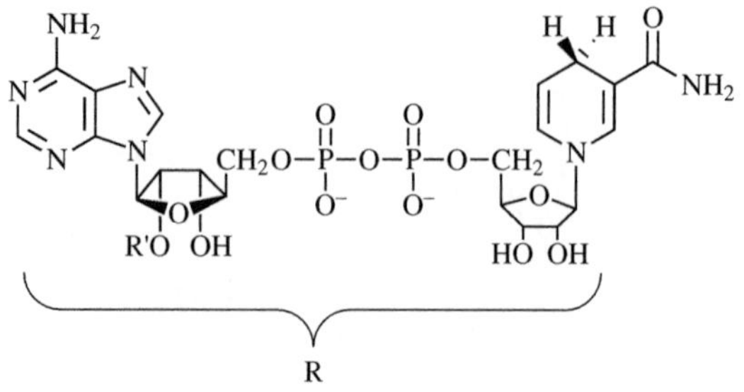

4.8b

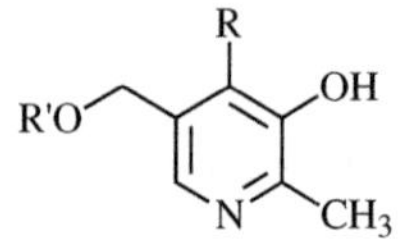

4.9

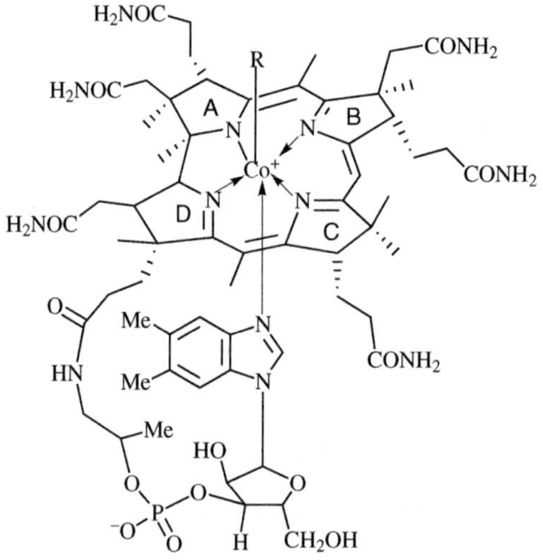

4.10

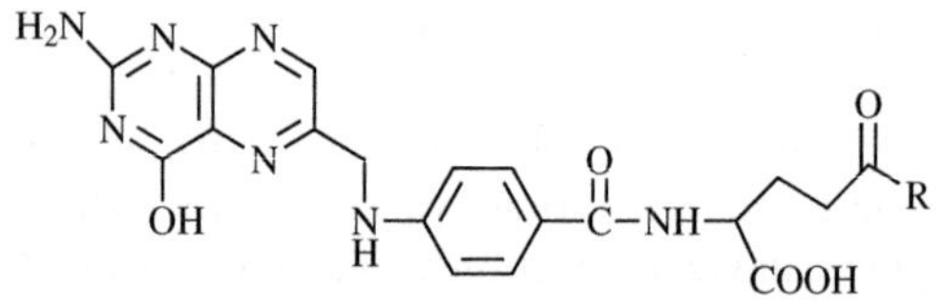

4.11a

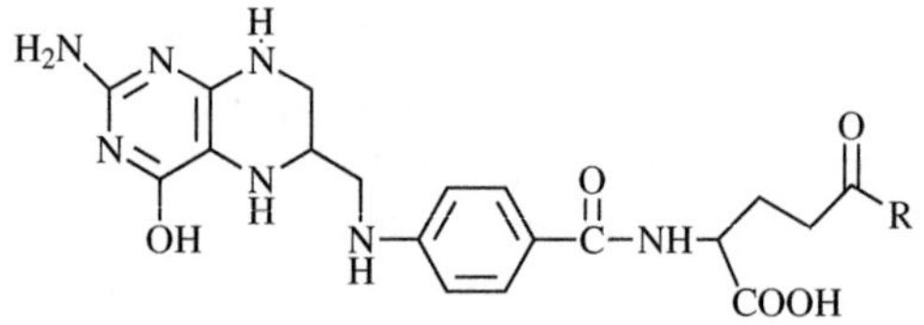

4.11b

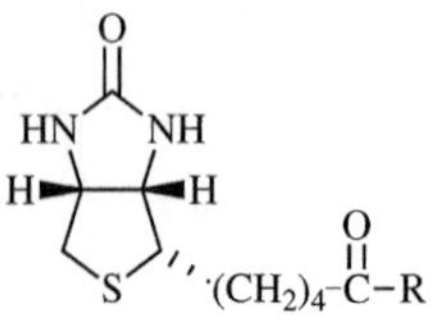

4.12

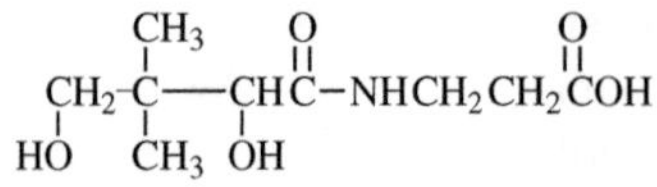

4.13a

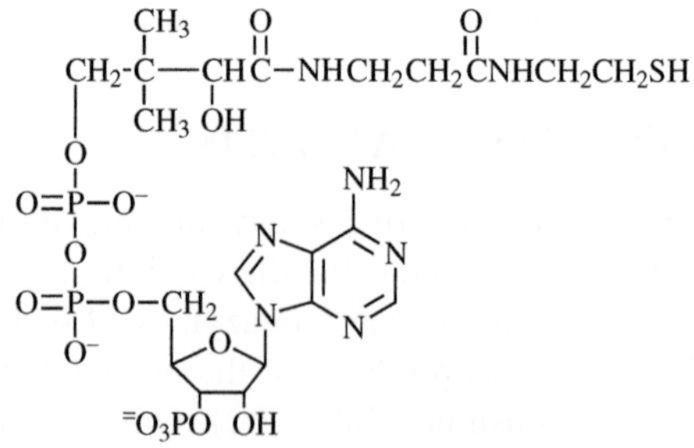

4.13b

TABLE 4.2 Coenzymes Derived from Vitamins

Vitamin	Structure	Coenzyme Form	Structure	Coenzyme Acronym
Vitamin B_1 (thiamin)	**4.6** (R=H)	Thiamin diphosphate	**4.6** $\left(R = {}^-O_2POPO_3^{2-}\right)$	TPP (TDP)
Vitamin B_2 (riboflavin)	**4.7** (R=H)	Flavin mononucleotide	**4.7** $\left(R = PO_3^{2-}\right)$	FMN
		Flavin adenine dinucleotide	**4.7** $\left(R = {}^-O_2POPO_3^- - 5' - Ado\right)$	FAD
Vitamin B_3 (niacinamide)	**4.8a** (no R)	Nicotinamide adenine dinucleotide	**4.8a** (R′=H)	NAD^+
(Niacin)	Corresponding carboxylic acid	Nicotinamide dinucleotide phosphate	**4.8a** $\left(R' = PO_3^{2-}\right)$	$NADP^+$
		Reduced nicotinamide adenine dinucleotide	**4.8b** (R′=H)	NADH
		Reduced nicotinamide adenine dinucleotide phosphate	**4.8b** $\left(R' = PO_3^{2-}\right)$	NADPH
Vitamin B_6 (pyridoxine)	**4.9** (R=CH_2OH, R′=H)	Pyridoxal 5′-phosphate	**4.9** $\left(R = CHO,\ R' = PO_3^{2-}\right)$	PLP
		Pyridoxamine 5′-phosphate	**4.9** $\left(R = CH_2NH_2,\ R' = PO_3^{2-}\right)$	PMP
Vitamin B_9 (folic acid)	**4.11a** (R=OH or poly-γ-glutamyl)	Tetrahydrofolate	**4.11b** (R=OH or poly-γ-glutamyl)	THF
Vitamin B_{12} (cyanocobalamin)	**4.10** (R=CN)	Adenosylcobalamin	**4.10** (R=5′-Ado)	CoB_{12}
Biotin	**4.12** (R=OH)	Covalently bound to enzyme as amide	**4.12** (R=lysine residue of enzyme)	–
Pantothenic acid	**4.13a**	Coenzyme A (no relation to vitamin A)	**4.13b**	CoASH

Fe^{3+} COOH COOH

4.14

Cysteine $\overset{+}{N}H_3$ ^-OOC NH O NH COO^- O SH γ-Glutamyl Glycine

4.15

NH_2 O O O $^-O-P-O-P-O-P-O-CH_2$ O^- O^- O^- N N N N O HO OH

4.16

S–S H OH O

4.17

OH H HO O O HO OH

4.18

TABLE 4.3 Symptoms of Vitamin Deficiency

Deficient Vitamin	Disease State
Thiamin	Beriberi
Riboflavin	Dermatitis and anemia
Niacin	Pellagra
Pyridoxine	Dermatitis and convulsions
Vitamin B12	Pernicious anemia
Folic acid	Pernicious anemia
Biotin	Dermatitis
Pantothenic acid	Neuromuscular effects
Ascorbic acid	Scurvy
Vitamin D	Rickets
Vitamin K	Hemorrhage
Vitamin A	Night blindness

Vitamins are, by definition, essential nutrients; human metabolism is incapable of producing them. Deficiency in a vitamin results in the shutting down of the catalytic activity of various enzymes that require the coenzyme made from the vitamin or results from the lack of other activities of the vitamins. This leads to certain disease states (Table 4.3).[50]

The remainder of this chapter is devoted to the chemistry of coenzyme catalysis. The discussion is limited to only those coenzymes whose chemistry of action will be important to the mechanisms of drug action that are described in later chapters and is not an exhaustive treatment.

4.3.1. Pyridoxal 5′-Phosphate

Enzymes dependent on pyridoxal 5′-phosphate (PLP) catalyze several different reactions of amino acids which, at first glance, appear to be unrelated; however, when the mechanisms are discussed, the relationships will become apparent.[51] The overall reactions are summarized in Table 4.4. PLP is the most versatile coenzyme but all of these reactions are very specific depending on the enzyme to which the PLP is bound. Although these reactions can be catalyzed by PLP nonenzymatically, in this case, typically several of the possible types of reactions take place simultaneously.[52,53] For a given enzyme, only one type of reaction occurs almost exclusively.

Although there are several noncovalent binding interactions responsible for holding the PLP in the active site, the major interaction is a covalent one (**4.19a**, Figure 4.5). To make the chemistry easier to visualize, this coenzyme will be abbreviated as shown in **4.19b** when the chemistry is discussed. The aldehyde group of the PLP is held tightly at the active site by a Schiff base (iminium) linkage to a lysine residue. In addition to securing the PLP in the optimal position at the active site, Schiff base formation activates the carbonyl for nucleophilic attack. This is very important to the catalysis because the first step in *all* PLP-dependent enzyme transformations of amino acids (Table 4.4) is a transimination reaction, i.e., the conversion of the lysine–PLP imine (**4.20**, Scheme 4.14) to the substrate–PLP imine (**4.21**). In Scheme 4.14, two different bases are shown to be involved in acid–base catalysis; a similar mechanism could be drawn with a single base. It is from **4.21** that all of the reactions shown in Table 4.4 occur. The property of **4.21** that links all of these reactions is that the pyridinium group can act as an electron sink to stabilize electrons by resonance from the C–H, C–COO^-, or C–R bonds. This could account for why all three of these bonds can be broken nonenzymatically. An important question is how can an enzyme catalyze the *regiospecific* cleavage of only one of these three bonds?

The bond that breaks must lie in a plane perpendicular to the plane of the PLP–imine π-electron system. In Figure 4.6, the C–H bond is the one perpendicular to the plane of the π-system, i.e., parallel with the p-orbitals. This configuration results in maximum π-electron overlap (the sp^3 σ-orbital of the C–H bond and the interacting p-orbitals of the conjugated system) and, therefore, minimizes the transition state energy for bond breakage of the C–H bond. The problem for the enzyme to solve, then, is how to control the conformation about the C_α–N bond so that only the bond that is to be cleaved is perpendicular to the plane of the π-system at the active site of the enzyme. The *Dunathan hypothesis*[54] gives a rational explanation for how an enzyme could control the C_α–N bond rotation (Figure 4.7). A positively charged residue at the active site could form a salt bridge with the carboxylate group of the amino acid bound to the PLP. This would make it possible for an enzyme to restrict rotation about the C_α–N bond and hold the H (**A**), the COO^- (**B**), or the R group (**C**) perpendicular to the plane of the conjugated system. If the Dunathan hypothesis is accepted, then all of the PLP-dependent enzyme reactions can be readily understood. In fact, crystal structures of PLP-dependent enzymes generally show a group at the active site, often an arginine residue, positioned to bind to the substrate-bound carboxylate group.[55]

In the next four subsections, mechanisms for only the classes of PLP-dependent enzymes that will be relevant to later chapters are described. The mechanism for α-cleavage will be discussed in Section 4.3.2. PLP-dependent enzymes are important targets for therapeutic agents.[56]

4.3.1.1. Racemases

D-Amino acids are commonly found in bacteria for use in the assembly of their cell wall.[57] D-Serine, D-aspartate,

TABLE 4.4 Reactions Catalyzed by Pyridoxal 5′-Phosphate-Dependent Enzymes

Substrate	Product	Reaction
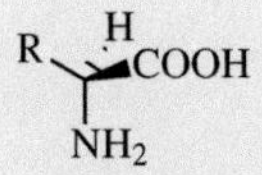 R, H, COOH, NH_2	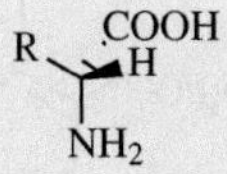 R, COOH, H, NH_2	Racemization
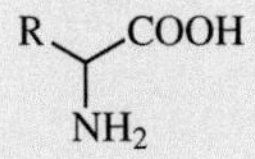 R, COOH, NH_2	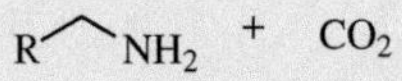 R, NH_2 + CO_2	Decarboxylation
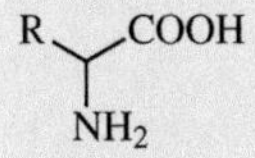 R, COOH, NH_2	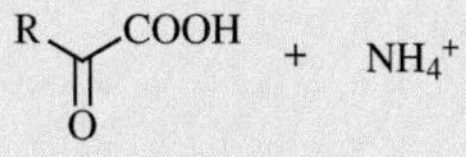 R, COOH, O + NH_4^+	Transamination
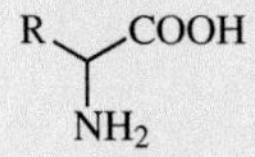 R, COOH, NH_2	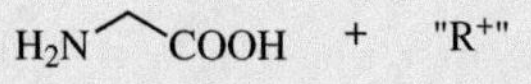 H_2N, COOH + "R^+"	α-cleavage
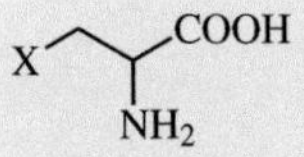 X, COOH, NH_2	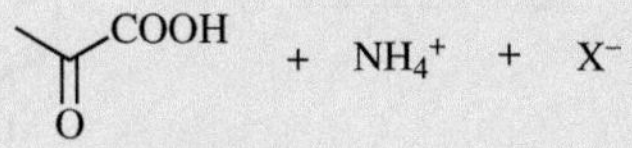 COOH, O + NH_4^+ + X^-	β-elimination
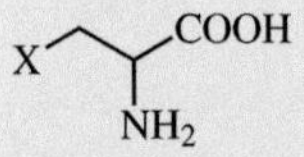 X, COOH, NH_2	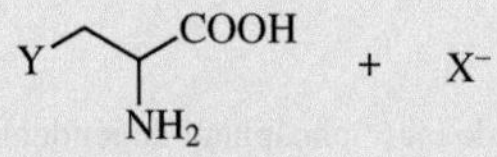 Y, COOH, NH_2 + X^-	β-replacement
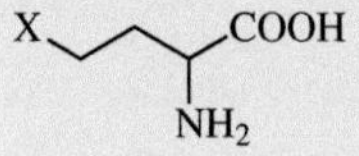 X, COOH, NH_2	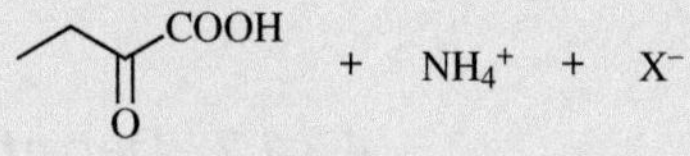 COOH, O + NH_4^+ + X^-	γ-elimination
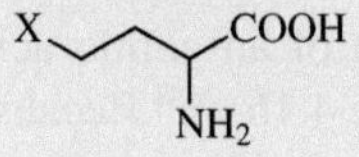 X, COOH, NH_2	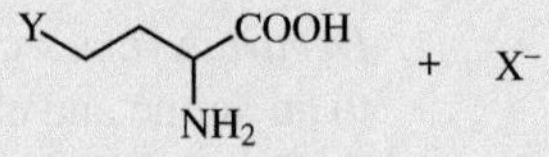 Y, COOH, NH_2 + X^-	γ-replacement

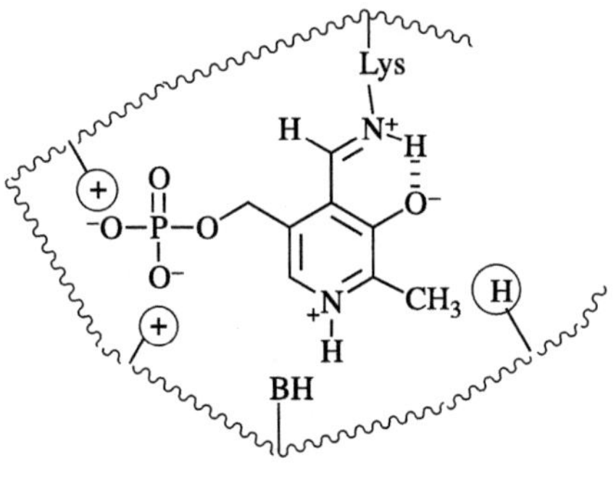

4.19a

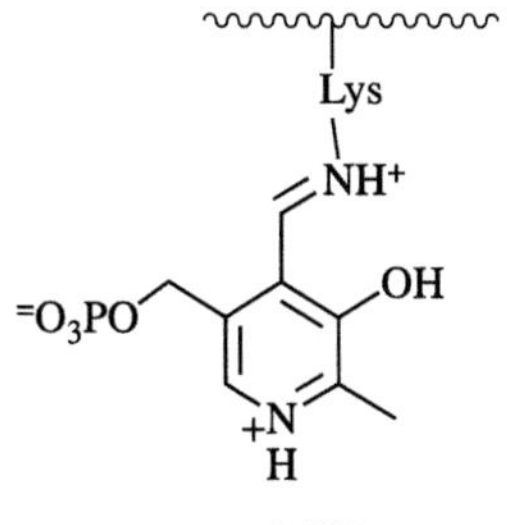

4.19b

FIGURE 4.5 Pyridoxal 5′-phosphate covalently bound to the active site of an enzyme.

and other D-amino acids, however, have also been detected in mammalian brain[58]; D-serine activates glutamate/NMDA receptors in neurotransmission.[59] Because the amino acids in plant and animal proteins have the L-configuration, enzymes that convert the L-amino acids into their enantiomers must be available for the organisms to get the D-isomers. These enzymes are part of the family of enzymes known as *racemases*,[60] which are highly specific for the amino acid used as its substrate; alanine racemase, for example, does not catalyze the racemization of glutamate. Also (and at least as fascinating as the substrate specificity), a decarboxylase does not also catalyze racemization of its substrate, so histidine decarboxylase converts L-histidine to histamine without formation of D-histidine, although there are some PLP-dependent enzymes that do catalyze a slow side reaction different

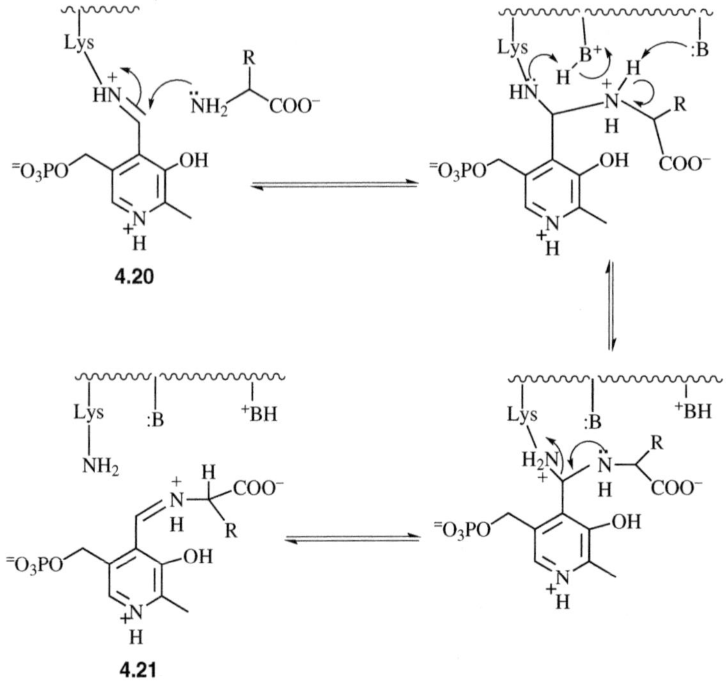

SCHEME 4.14 First step in all pyridoxal 5′-phosphate-dependent enzyme reactions

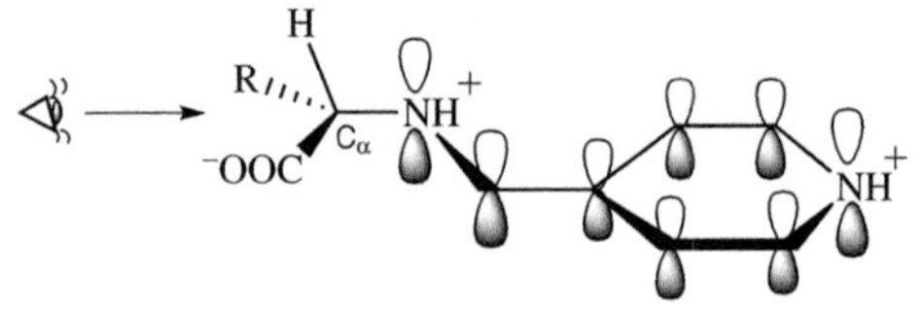

FIGURE 4.6 π-Electron system of the PLP-imine

from its normal function. Consider alanine racemase as an example of this type of reaction. The reaction catalyzed by this enzyme is the reversible interconversion of L- and D-alanine with an equilibrium constant of 1, typical for racemases, so it does not matter which enantiomer is used as the substrate, the racemate is produced. The mechanism shown in Scheme 4.15 is typical for PLP-dependent racemases (for alanine racemase, $R = CH_3$). Because the carbanion produced by α-proton removal from the PLP-imine is so highly delocalized, the pK_a of the α-proton is much lower than that in the parent amino acid. Deprotonation of the bound amino acid could occur with either one base for both enantiomers or two bases.[61] Alanine racemase uses two different active-site bases to deprotonate bound D-alanine (Lys-39) or L-alanine (Tyr-265) because the protons being removed are facing in different directions in the active site.[62] Because of the importance of D-alanine to the construction of the bacterial cell wall, inhibition of this enzyme (and prevention of the formation of D-alanine) has been an approach for potential antibacterial drug design.

4.3.1.2. Decarboxylases

Decarboxylases catalyze the conversion of an amino acid to an amine and carbon dioxide (Scheme 4.16).[63] Because the loss of CO_2 is irreversible, the amine cannot be converted back to the amino acid. However, the last step in Scheme 4.16 is reversible, so it is possible to catalyze the exchange of the α-proton of the amine with solvent protons, for example, when the isolated enzyme reaction is carried out in ^2H_2O.

Specific decarboxylases are known for more than 10 of the common amino acids. L-Aromatic amino acid decarboxylase (sometimes referred to as dopa decarboxylase) plays a vital role in the conversion of the antiparkinsonian drug L-dopa to dopamine in the brain; low dopamine levels is a characteristic of parkinsonism.[64] Many of the bacterial amino acid decarboxylases have low pH optima. It is possible that at least one of the functions of these decarboxylases, which generate amine bases from neutral amino acids, is to neutralize acidic conditions in the cell. Another function may be to control the bacterial intracellular CO_2 pressure.

4.3.1.3. Aminotransferases (Formerly Transaminases)

Aminotransferases are the most complicated of the PLP-dependent reactions; they involve two substrates going to two products in two half reactions. Various PLP-dependent aminotransferases are known for α-, β-, and γ-amino acids. As the name implies, the amino group of the substrate

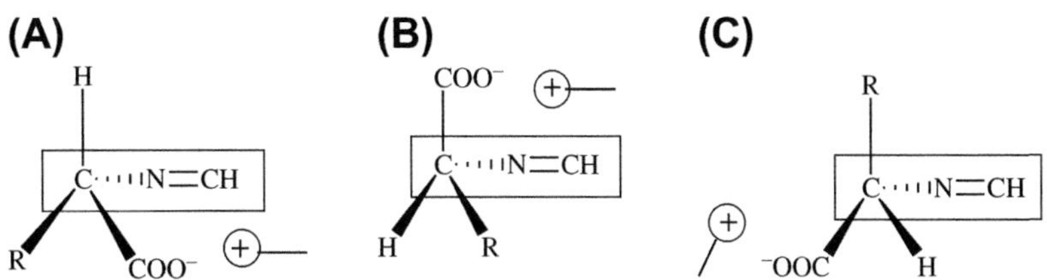

FIGURE 4.7 Dunathan hypothesis for PLP activation of the C_α–N bond. The rectangle represents the plane of the pyridine ring of the PLP. The angle of sight is that shown by the eye in Figure 4.6. *From Dunathan, H. C. (1971) In "Advances in Enzymology," Vol. 35, p. 79, Meister, A., Ed. Copyright © 1971. This material is reproduced with permission of John Wiley & Sons, Inc.*

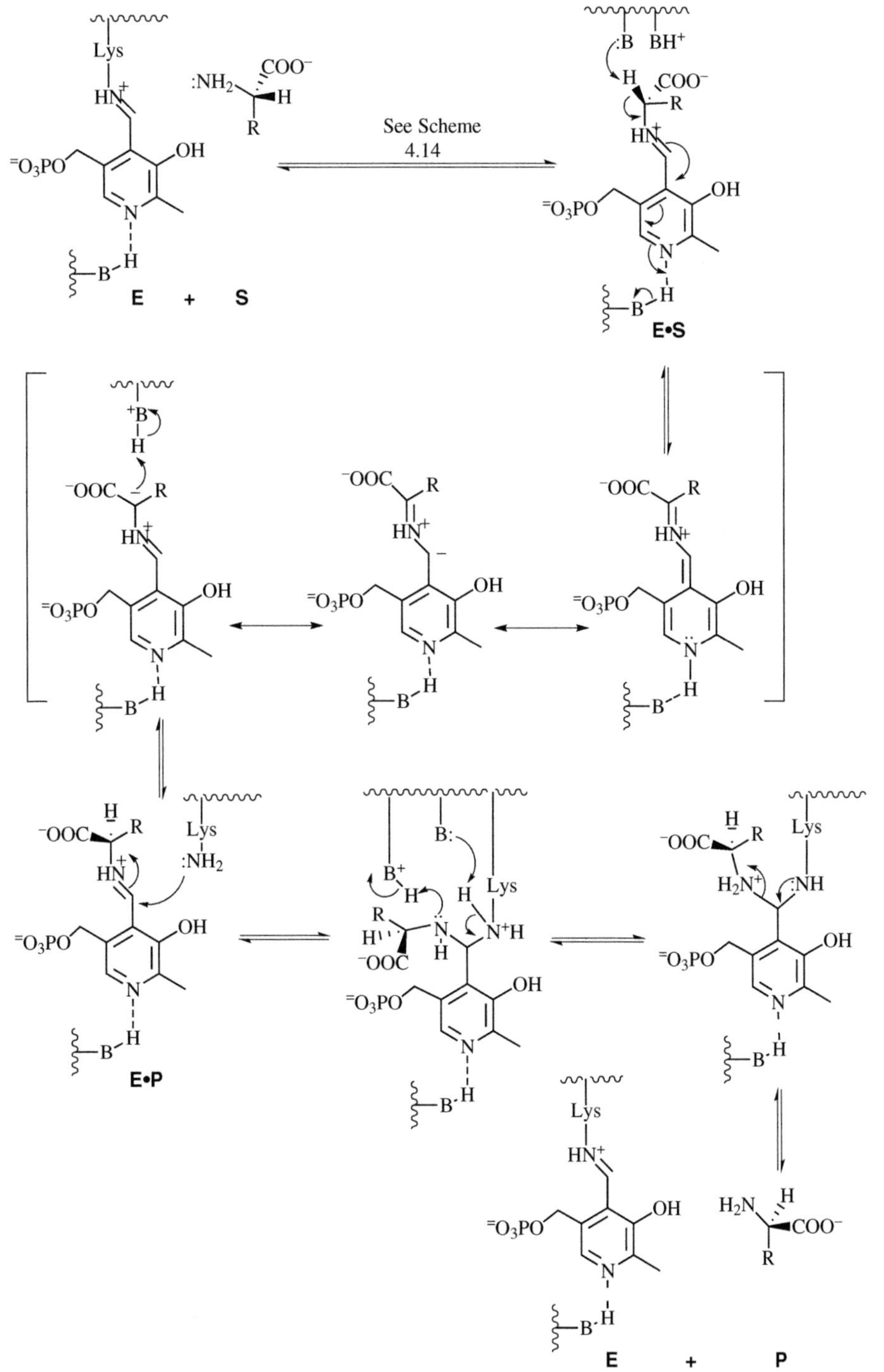

SCHEME 4.15 Mechanism for PLP-dependent racemases

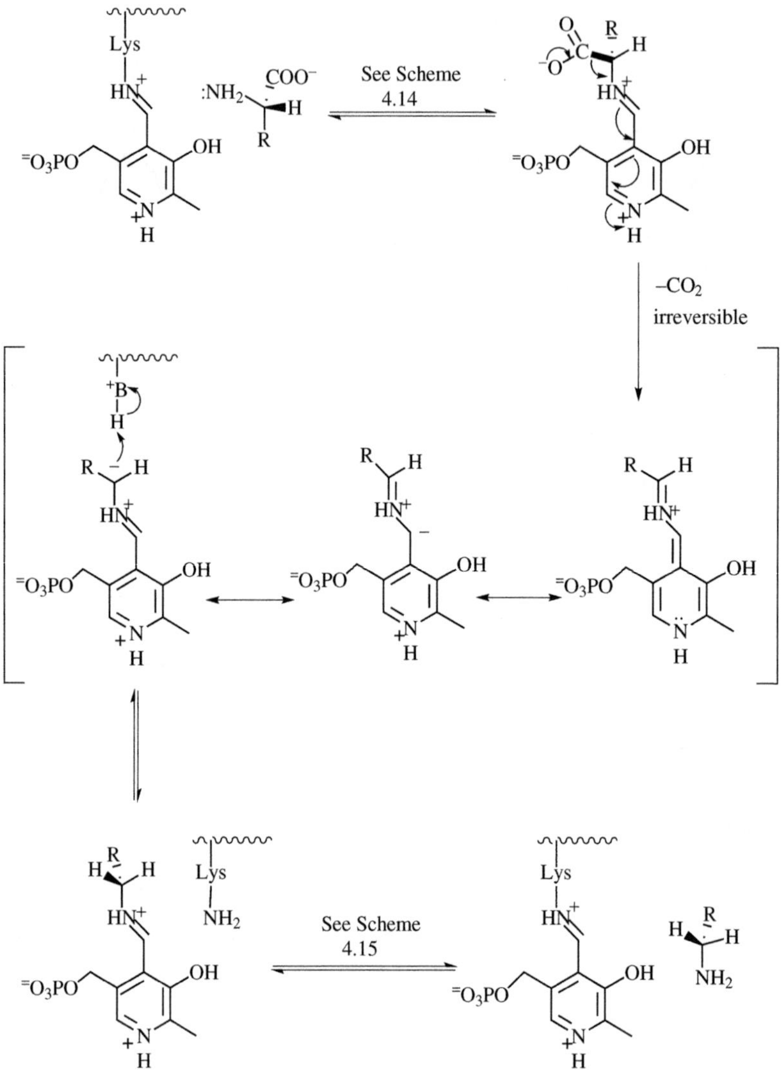

SCHEME 4.16 Mechanism for PLP-dependent decarboxylases

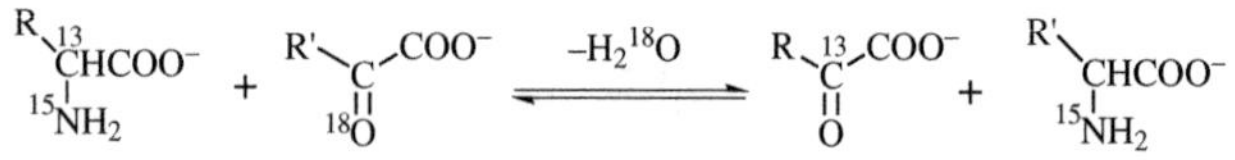

SCHEME 4.17 Overall reaction catalyzed by PLP-dependent aminotransferases

amino acid is transferred to another molecule, an α-keto acid (Scheme 4.17). The labeling pattern given in Scheme 4.17 indicates that the amino group of **4.22** is transferred to the α-keto acid (**4.23**), which loses its oxygen as a molecule of water. The initial amino acid (**4.22**), therefore, is converted into an α-keto acid (**4.24**), and the starting α-keto acid (**4.23**) is converted into an amino acid (**4.25**). A mechanism consistent with these observations is given in Schemes 4.18 and 4.19. In Scheme 4.18, the ^{15}N substrate is transferred to the coenzyme, which is converted into ^{15}N-containing pyridoxamine 5′-phosphate (PMP) (**4.28**). This is an internal redox reaction (a tautomerization): the amino acid substrate is oxidized to an α-keto acid at the expense of the reduction of the PLP to PMP. Note that the first two steps (up to the three resonance structures) are identical to the mechanism for PLP-dependent racemization (Scheme 4.15), yet racemization does not occur. So how does the enzyme alter this racemization pathway in favor of transamination? It does so simply by the placement of an appropriate acidic residue closer to the PMP carbon (**4.26b**, Scheme 4.18) than the substrate α-carbon (**4.26c**). This leads to the PMP-imine Schiff base (**4.27**), which can be hydrolyzed to PMP and the α-keto acid. At this point, the coenzyme is no longer in the proper oxidation state (i.e., an imine) for reaction with another amino acid substrate molecule (remember, the first step is Schiff base formation with an amino acid).

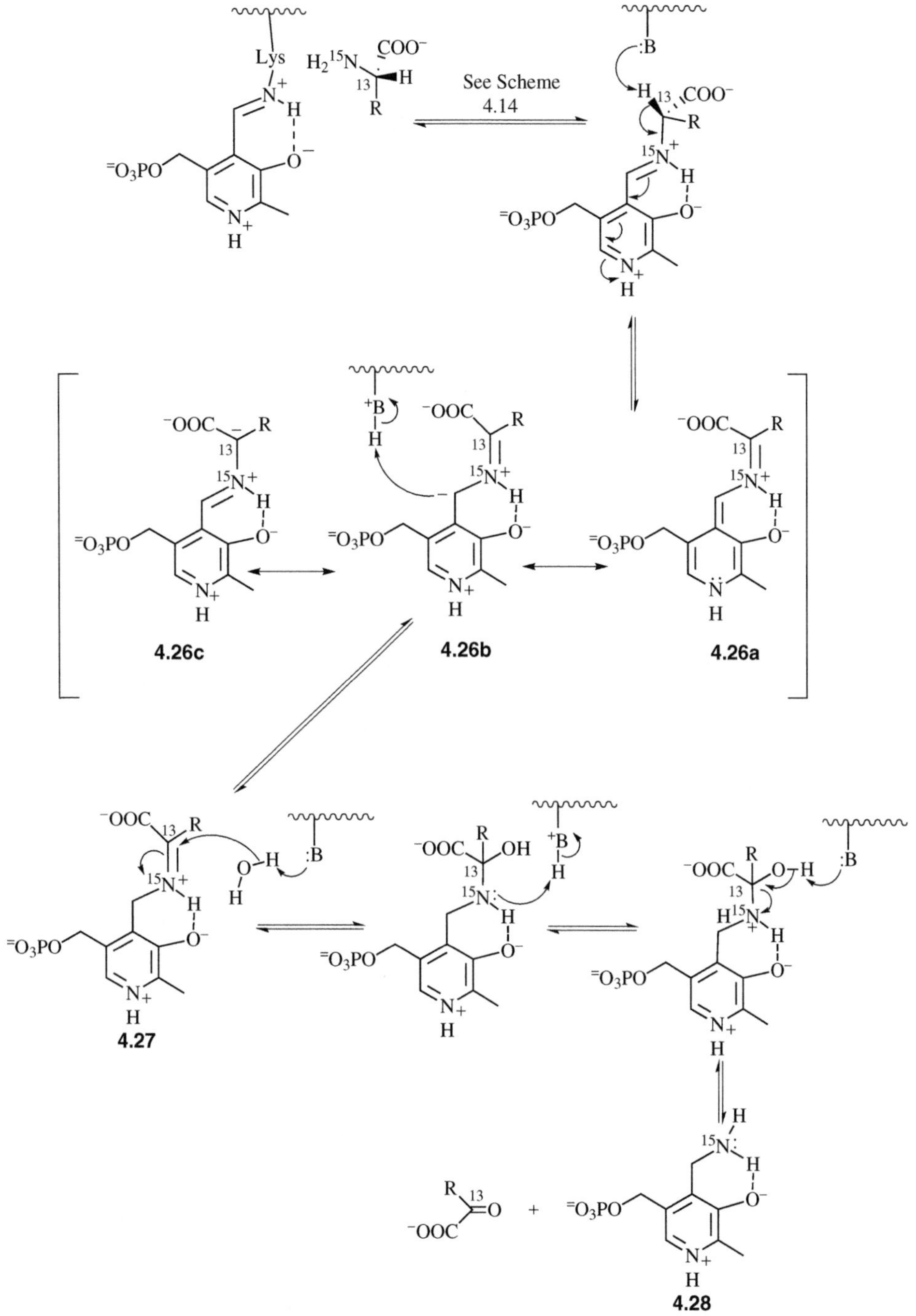

SCHEME 4.18 First half-reaction for the mechanism of PLP-dependent aminotransferases

If this were the end of the enzyme reaction, then the enzyme would be a reagent, not a catalyst, because only one turnover has taken place, and the enzyme is inactive. To return the coenzyme in its PMP form to the requisite oxidation state of the PLP form for continued substrate turnover, a second substrate, another ketone, generally an α-keto acid (not the product α-keto acid), binds to the active site of the enzyme, which may have undergone a conformational change, and the exact reverse of the reaction shown in Scheme 4.18 occurs (Scheme 4.19). In this process, the amino group on the PMP, which originally came from the substrate amino acid, is then transferred to the second substrate (the α-keto acid), thereby producing a new amino acid and PLP. In the overall reaction (Schemes 4.18 and 4.19),

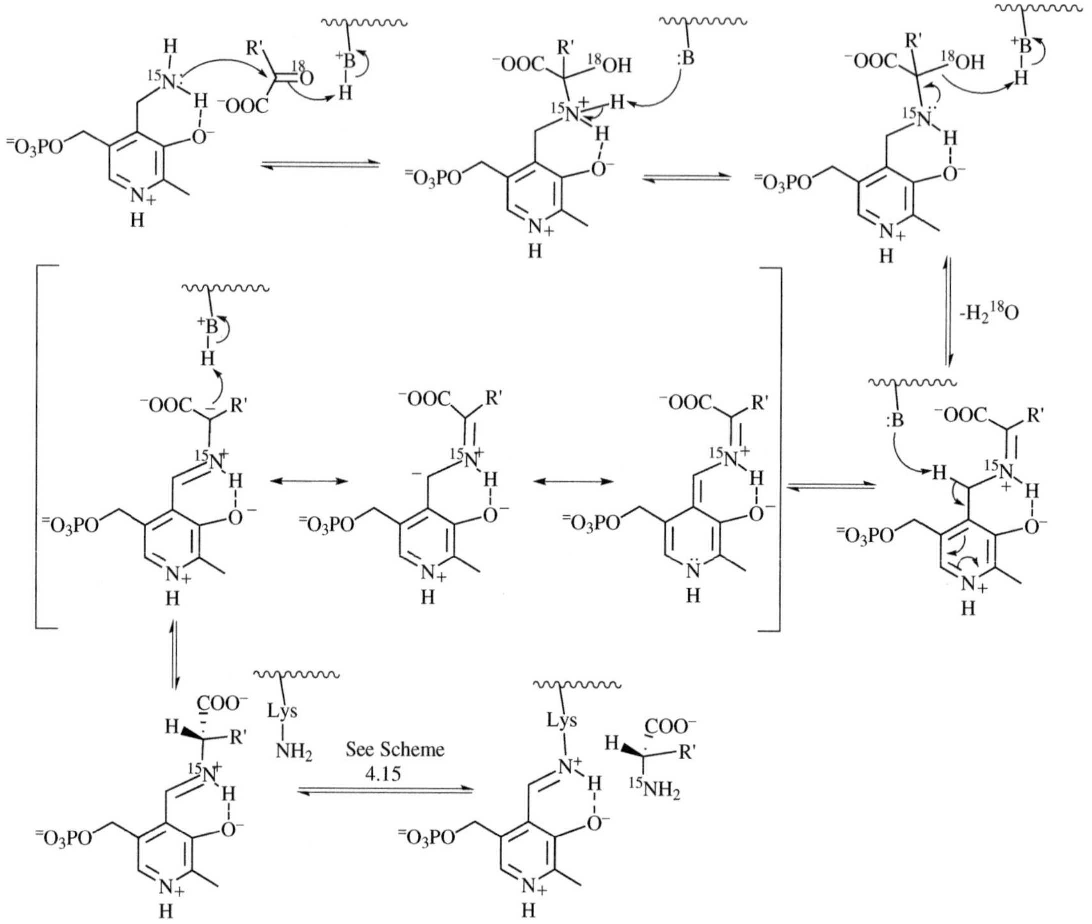

SCHEME 4.19 Second half-reaction for the mechanism of PLP-dependent aminotransferases

the second half reaction (Scheme 4.19) is only utilized to convert the enzyme back to the active form. There are some aminotransferases, however, in which the converse is true, namely, that the first half reaction is to convert the enzyme to the PMP form so that the intended substrate (an α-keto acid) can be converted to the product amino acid.

4.3.1.4. PLP-Dependent β-Elimination

Some PLP-dependent enzymes catalyze the elimination of H–X from substrate molecules that contain a β-leaving group. Although these enzymes are not directly relevant to the later discussions, this sort of mechanism will be an important alternative pathway for certain inactivators of other PLP-dependent enzymes (Chapter 5, Section 5.3.3.3). The mechanism is shown in Scheme 4.20.

4.3.2. Tetrahydrofolate and Pyridine Nucleotides

The most important way for groups containing one carbon to be added to molecules involves enzymes that utilize coenzymes derived from human vitamin B_9 or folic acid (**4.29**, Scheme 4.21). Although folic acid is a monoglutamate, the coenzyme forms actually can contain an oligomer of as many as twelve glutamate residues, depending on the enzyme. The C–N double bonds in folic acid are reduced to tetrahydrofolate (**4.33**) by another coenzyme, reduced nicotinamide adenine dinucleotide (NADH)[65] (**4.30**; see **4.8b** for the entire structure), in the enzyme dihydrofolate reductase.[66] Because the sugar phosphate part of the pyridine nucleotide coenzymes is not involved in the chemistry, it is abbreviated as R; likewise, the part of folic acid not involved in the chemistry is abbreviated by R′. NADH and reduced nicotinamide adenine dinucleotide phosphate (NADPH, **4.8b**, $R' = PO_3^{2-}$) can be thought of as Mother Nature's sodium borohydride, a reagent used in organic chemistry to reduce active carbonyl and imine functional groups. As in the case of sodium borohydride, the reduced forms of the pyridine nucleotide coenzymes are believed to transfer their reducing equivalents as hydride ions[67] (Scheme 4.21).

The carbon atom at the C-4 position of NADH and NADPH is prochiral. An atom is *prochiral* if, by changing one of its substituents, it is converted from achiral to chiral.

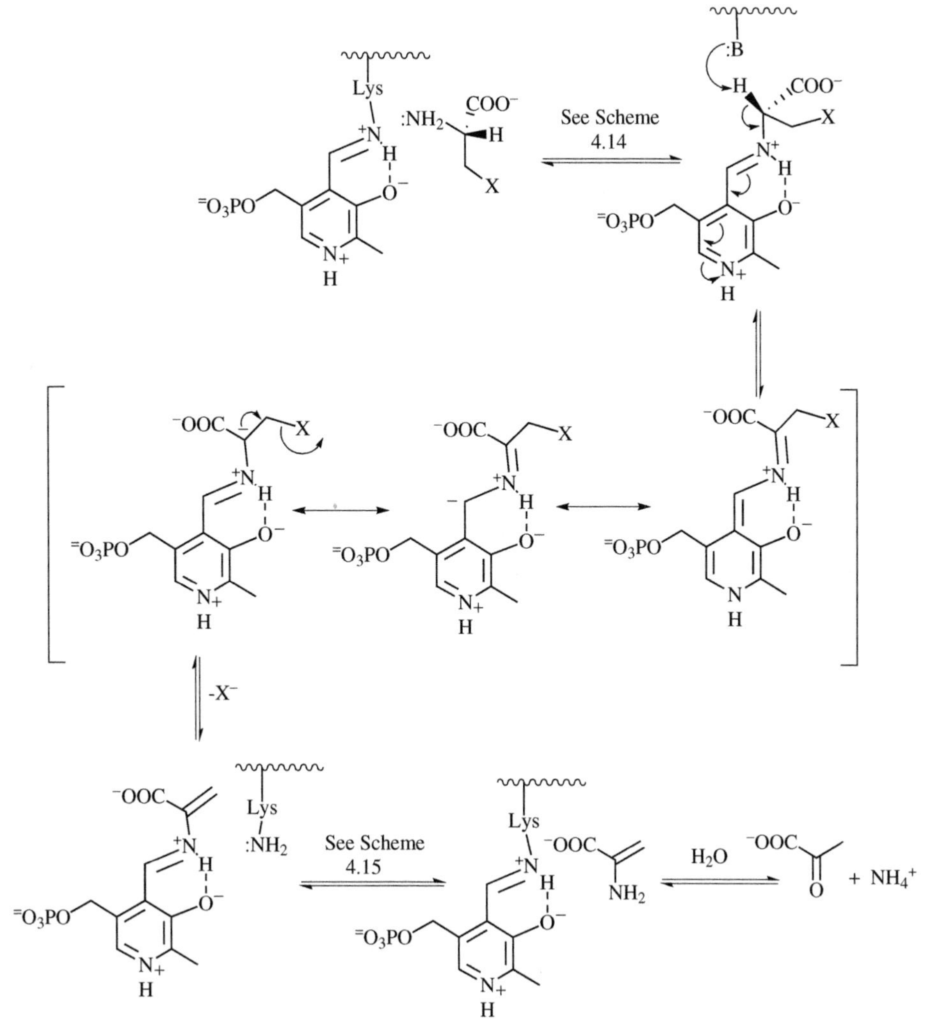

SCHEME 4.20 Mechanism for PLP-dependent β-elimination reactions. X is a leaving group.

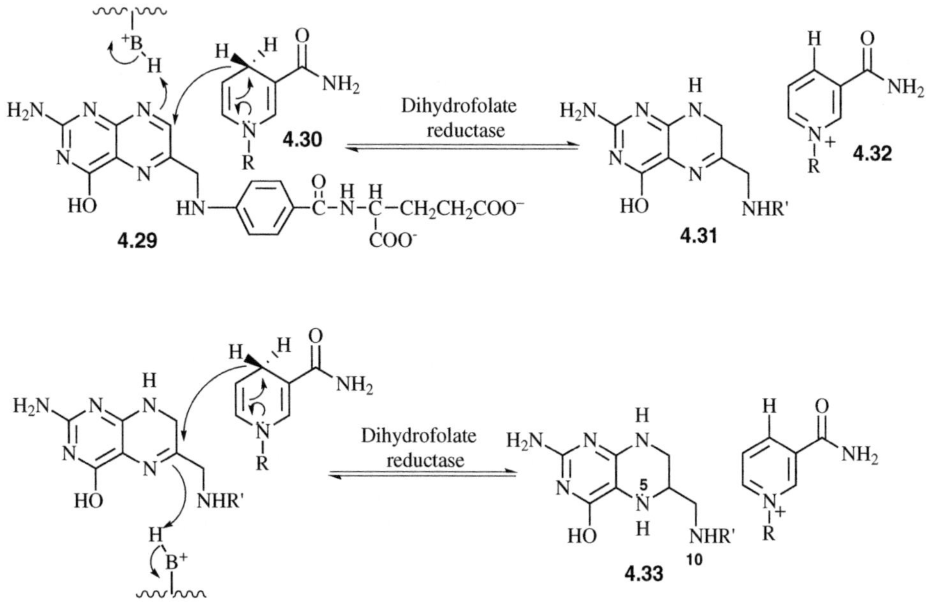

SCHEME 4.21 Pyridine nucleotide-dependent reduction of folic acid to tetrahydrofolate

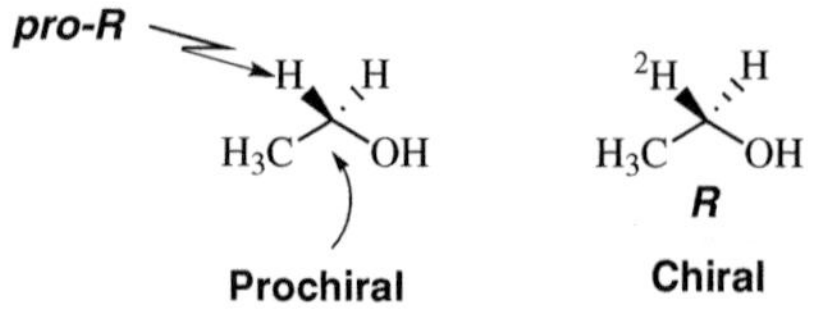

pro-R

Prochiral Chiral

R

FIGURE 4.8 Determination of prochirality

The C-4 carbon of NADH has two hydrogen atoms attached to it; consequently, it is achiral. If one of the hydrogens is replaced by a deuterium, then the carbon becomes chiral. If the chiral center that is generated by replacement of hydrogen by deuterium has the *S*-configuration, then the hydrogen that was replaced is called the *pro-S* hydrogen; if the *R*-configuration is produced, then the hydrogen replaced is the *pro-R* hydrogen. This is demonstrated for ethanol and deuterated ethanol in Figure 4.8. The *pro-R* and *pro-S* hydrogen of the reduced pyridine nucleotide coenzymes are noted in **4.34**. As indicated in Section 4.1.2.1.1, because enzymes are chiral and they bind molecules in specific orientations, any achiral compound bound to an enzyme can become prochiral, that is, its hydrogens can be differentiated by the enzyme. Some enzymes, such as dihydrofolate reductase, utilize the *pro-R* hydrogen (also called the A-side hydrogen) of the reduced pyridine nucleotides, others use the *pro-S*, or B-side, hydrogen. However, the reaction always is stereospecific; if an enzyme uses the *pro-R* hydrogen of NADH, then this is the only hydrogen that it transfers.

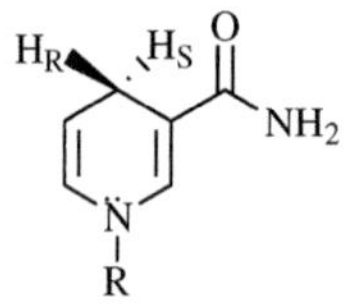

4.34

Although folate is converted into a reduced form (tetrahydrofolate), this coenzyme is generally not involved in redox reactions (one exception is thymidylate synthetase, which will be discussed in Section 5.3.3.3.5). Actually, tetrahydrofolate is not the full coenzyme; the complete coenzyme form contains an additional carbon atom between the N^5 and N^{10} positions (see **4.33** in Scheme 4.21 for nomenclature), which is transferred to other molecules. That additional carbon atom is derived from the methylene group of L-serine in a reaction catalyzed by the PLP-dependent enzyme serine hydroxymethyltransferase.[68] The reaction catalyzed by this PLP-dependent enzyme is one that we have not yet discussed, namely, α-cleavage (Scheme 4.22; the carbon atom that will be transferred to the tetrahydrofolate is marked with an asterisk). The hydroxymethyl group is held perpendicular to the plane of the PLP aromatic system (**4.35**), as in the case of the group cleaved during all of the other PLP-dependent reactions, so that deprotonation and, in this case, a carbon–carbon bond cleavage (a retroaldol reaction) can occur readily. This α-cleavage is an uncommon PLP-dependent reaction, but the cation formed from C–C bond cleavage is stabilized by the adjacent oxygen atom. The serine methylene is converted to formaldehyde as a result of this reaction.

Because of the potential toxicity of released formaldehyde (it is highly electrophilic and reacts readily with amino groups), serine hydroxymethyltransferase does not catalyze the degradation of serine (to an appreciable extent) until *after* the acceptor for formaldehyde, namely, tetrahydrofolate, is already bound at an adjacent site.[69] Once the tetrahydrofolate binds, the degradation of serine is triggered, and the formaldehyde generated reacts directly with the tetrahydrofolate to give, initially, the carbinolamine **4.36**, as shown in Scheme 4.23. Kallen and Jencks showed that the more basic nitrogen of tetrahydrofolate is the N^5 nitrogen,[70] which attacks the formaldehyde first;[71] the N^{10} nitrogen is attached to the aromatic ring of R′ (see structures **4.11**) *para* to a carbonyl group, and, therefore, this nitrogen has amide-like character, and is not very basic. Carbinolamine **4.36** is dehydrated to N^5-methylenetetrahydrofolate (**4.37**), which is in equilibrium with N^5,N^{10}-methylenetetrahydrofolate (**4.38**) and N^{10}-methylenetetrahydrofolate (**4.39**). The equilibrium in solution strongly favors the cyclic form (**4.38**) ($K_{eq} = 3.2 \times 10^4$ at pH 7.2).

The coenzyme forms of tetrahydrofolate are responsible for the transfer of one-carbon units in three different oxidation states: formate (transferred to a nucleophile as a formyl group), formaldehyde (transferred to a nucleophile as a hydroxymethyl group), and methanol (transferred to a nucleophile as a methyl group). Methylenetetrahydrofolate (**4.37–4.39**) contains the transferrable carbon atom in the formaldehyde oxidation state, which is the correct oxidation state to transfer a hydroxymethyl group. What if the cell needs a one-carbon unit at the formate oxidation state? This is one oxidation state higher than the formaldehyde oxidation state. An $NADP^+$-dependent dehydrogenase (N^5,N^{10}-methylenetetrahydrofolate dehydrogenase) is involved in the oxidation of methylenetetrahydrofolate to give N^5,N^{10}-methenyltetrahydrofolate (**4.40**, Scheme 4.24). Hydrolysis of **4.40**, catalyzed by the enzyme N^5,N^{10}-methenyltetrahydrofolate cyclohydrolase, gives the carbinolamine **4.41**, which can break down to give either N^5-formyltetrahydrofolate (**4.42**, pathway a) or N^{10}-formyltetrahydrofolate (**4.43**, pathway b). Some enzymes utilize **4.42**, and others use **4.43** as their coenzyme.

If the cell needs a one-carbon unit at the methanol oxidation state, then N^5,N^{10}-methylenetetrahydrofolate (**4.37–4.39**) is reduced by N^5,N^{10}-methylenetetrahydrofolate reductase[72] to N^5-methyltetrahydrofolate (**4.44**, Scheme 4.25). This enzyme also requires another coenzyme, flavin adenine dinucleotide (FAD),[73] discussed in the next section.

All of these forms of the coenzyme are involved in enzyme-catalyzed transfers of one-carbon units at the

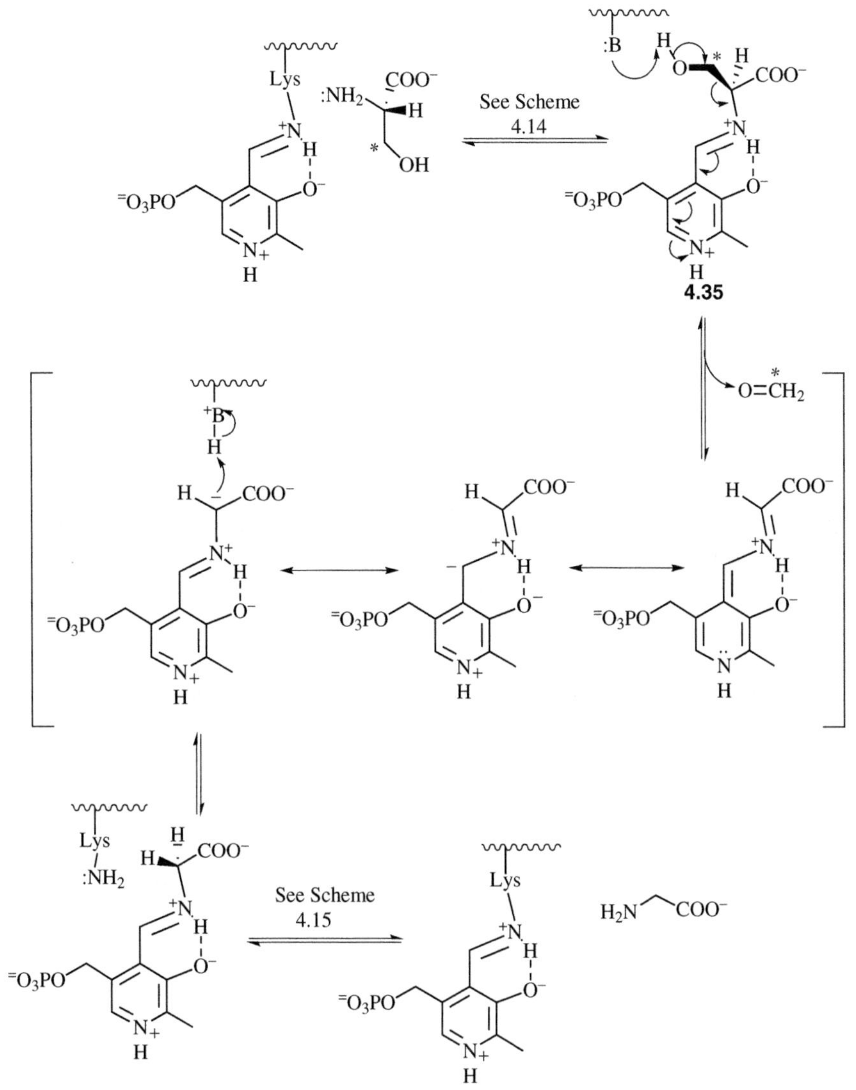

See Scheme 4.14

4.35

See Scheme 4.15

SCHEME 4.22 First half-reaction for serine hydroxymethylase-catalyzed transfer of formaldehyde from serine to tetrahydrofolate

4.36

4.39 4.38 4.37

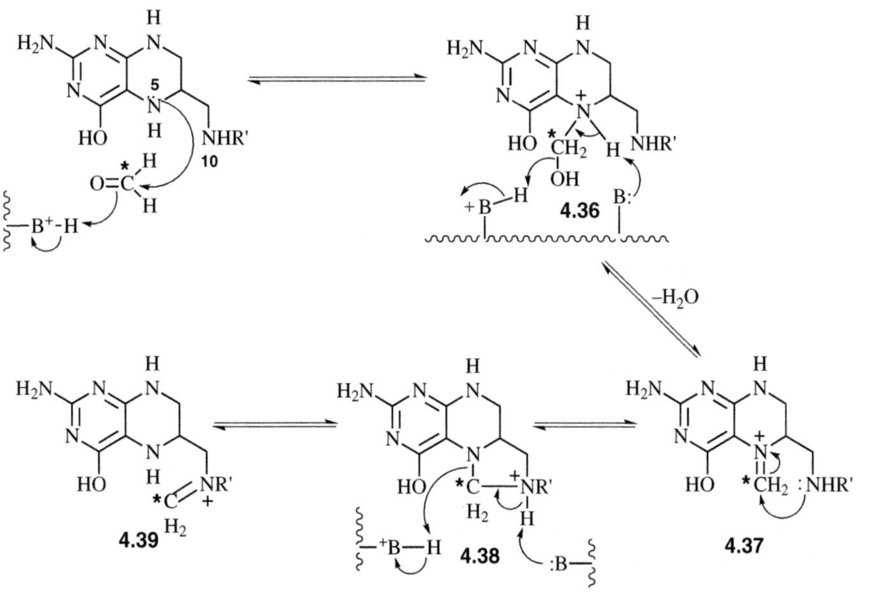

SCHEME 4.23 Second half-reaction for serine hydroxymethylase-catalyzed transfer of formaldehyde from serine to tetrahydrofolate

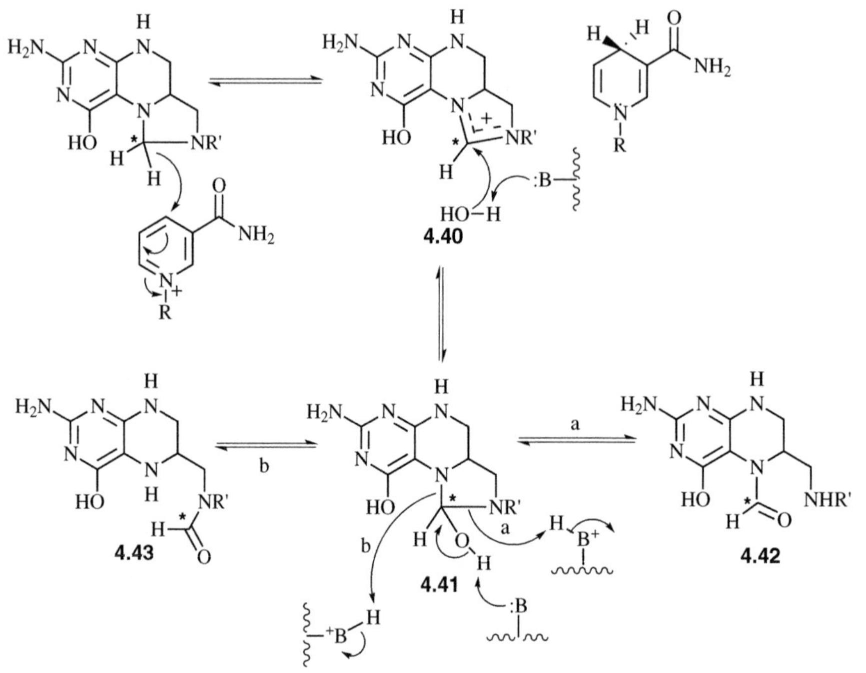

SCHEME 4.24 Oxidation of N^5,N^{10}-methylenetetrahydrofolate to N^5,N^{10}-methenyltetrahydrofolate and to N^5- and N^{10}-formyltetrahydrofolate

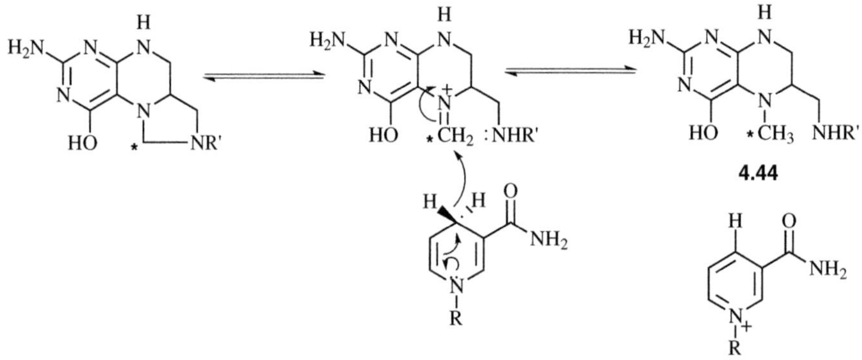

SCHEME 4.25 Reduction of N^5,N^{10}-methylenetetrahydrofolate to N^5-methyltetrahydrofolate

formate (**4.40**, **4.42**, or **4.43**), formaldehyde (**4.37-4.39**), or methanol (**4.44**) oxidation states. An example of an enzyme that catalyzes the transfer of a one-carbon unit at the formate oxidation state (i.e., as a formyl group) is 5-aminoimidazole-4-carboxamide-5′-ribonucleotide transformylase (AICAR, Scheme 4.26).[74] This is the penultimate enzyme in the de novo biosynthetic pathway to purines, and it catalyzes the conversion of 5′-phosphoribosyl-4-carboxamide-5-aminoimidazole (**4.45**) to 5′-phosphoribosyl-4-carboxamide-5-formamidoimidazole (**4.47**), presumably via the tetrahedral intermediate **4.46**. The formyl group is transferred from 10-formyltetrahydrofolate, which is converted to tetrahydrofolate. The product is enzymatically dehydrated by inosine monophosphate cyclohydrolase (IMPCH) to inosine monophosphate (**4.48**), which is further converted to adenosine monophosphate (AMP) and guanosine monophosphate by other enzymes. The carbon in between the two nitrogens in the five-membered ring of purines also was derived from a methenyltetrahydrofolate-dependent reaction earlier in the biosynthetic sequence.

A one-carbon transfer at the formaldehyde oxidation state occurs in the biosynthesis of the pyrimidine DNA precursor, thymidylate, by the enzyme thymidylate synthase,[75] which is discussed in more detail in Chapter 5 (Section 5.3.3.3.5).

N^5-Methyltetrahydrofolate is involved in the enzyme-catalyzed transfer of a one-carbon unit at the methanol oxidation state of homocysteine, to give methionine.[76] Methyl transfer in general, however, is not carried out by methyltetrahydrofolate; another methyl transfer agent, *S*-adenosylmethionine,[77] usually is implicated.

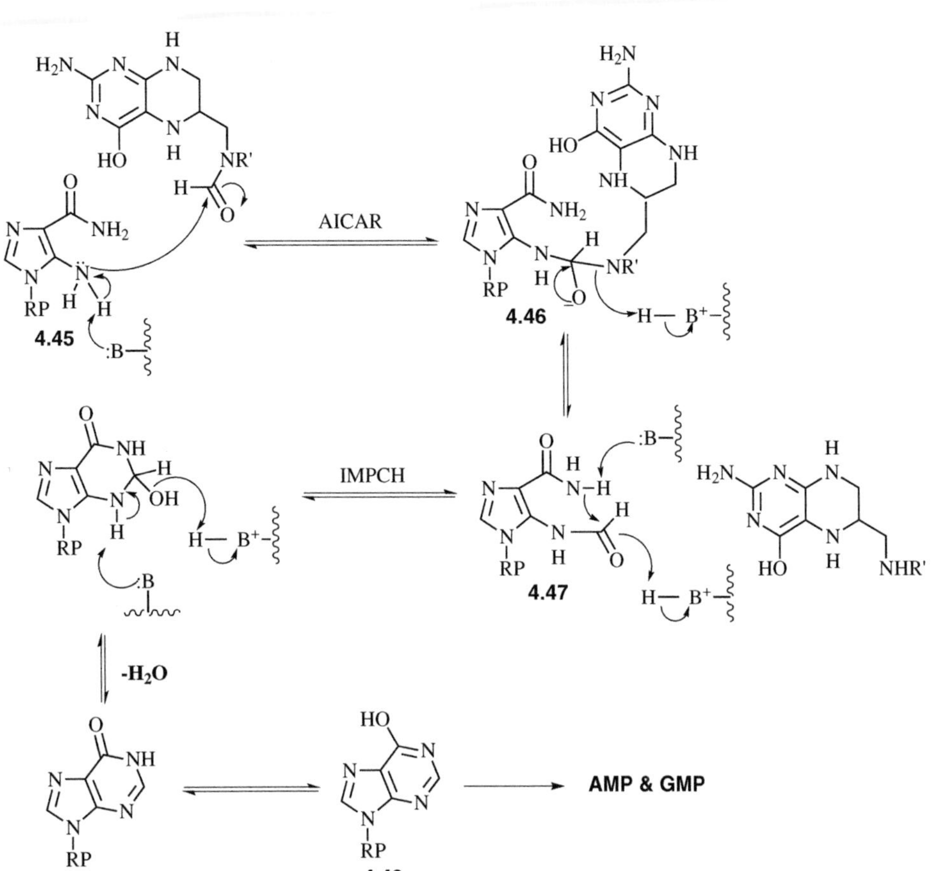

SCHEME 4.26 N^{10}-Formyltetrahydrofolate in the biosynthesis of purines. RP, ribose phosphate

4.3.3. Flavin

As in the case of the pyridine nucleotide coenzymes (see the section above), flavin coenzymes exist in several different forms.[78] All of the forms are derived from riboflavin (vitamin B_2, **4.49**), which is enzymatically converted to two other forms, flavin mononucleotide (FMN, **4.50**) and flavin adenine dinucleotide (FAD, **4.51**). Both **4.50** and **4.51** appear to be functionally equivalent, but some enzymes use one form, some the other, and some use one of each.

Unlike PLP, which catalyzes reactions of amino acids, the flavin coenzymes catalyze a wide variety of *redox* and *monooxygenation reactions* on diverse classes of compounds (Table 4.5).

The highly conjugated isoalloxazine tricyclic ring system of the flavins is an excellent electron acceptor, and this is responsible for its strong redox properties. Flavins can accept either one electron at a time or two electrons

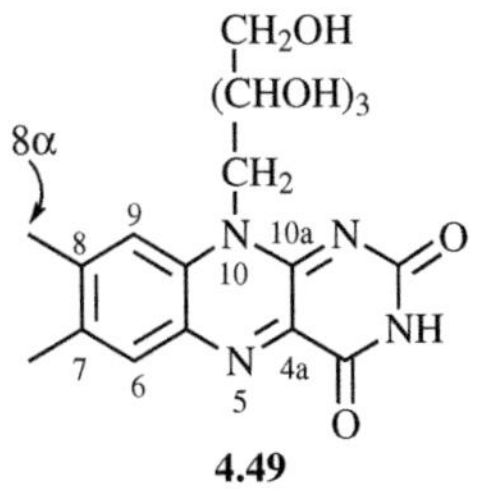

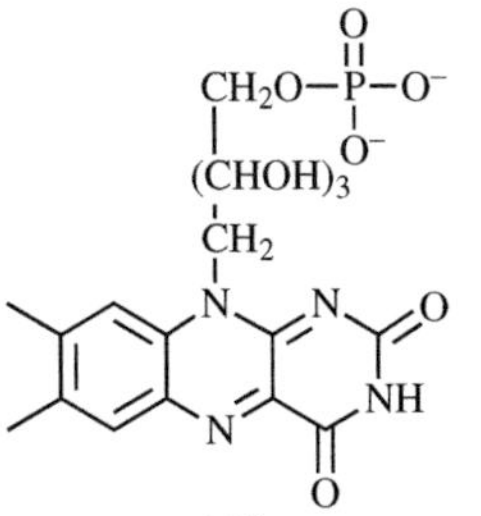

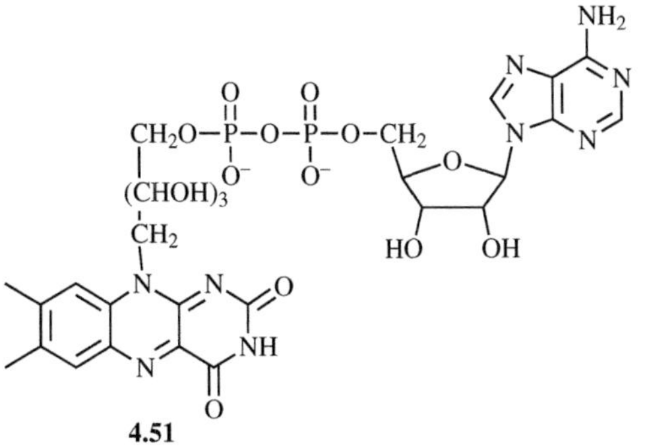

TABLE 4.5 Reactions Catalyzed by Flavin-Dependent Enzymes

Substrate	Product
R–CH(OH)–R'	R–C(=O)–R'
R–CH(NH_2)–R' (amines and amino acids)	[R–C(=NH^+)–R'] $\xrightarrow{H_2O}$ R–C(=O)–R' + NH_4^+
RR'CH–CH(R'')–C(=O)–R'''	RR'C=C(R'')–C(=O)R'''
RCH(SH)–$(CH_2)_n$–CH(SH)R'	RCH(S–S)$(CH_2)_n$CHR' (disulfide)
R–C_6H_5	R–C_6H_4–OH
R–C_6H_4–OH	R–C_6H_3(OH)–OH
R–C(=O)–R'	R–C(=O)–OR'
R–C(=O)H	R–C(=O)–OH
R–CH(OH)COOH	RCOOH + CO_2

simultaneously (Scheme 4.27); in many overall two electron oxidations, it is not clear if the reaction proceeds by a single two-electron reaction or by two one-electron transfer steps. The three forms of the coenzyme are the oxidized form (Fl_{ox}), the semiquinone form ($Fl^{-}\cdot$), and the reduced form (FlH^{-}). Although most flavin-dependent enzymes (also called *flavoenzymes*) bind the flavin with noncovalent interactions, some enzymes have covalently bound flavins, in which the flavin is attached at its 8α-position or 6-position (see **4.49** for numbering) to either an active-site histidine[79] or cysteine[80] residue. Once the flavin has been reduced, the enzyme requires a second substrate to return the flavin to the oxidized form so that it can accept electrons from another substrate molecule. This is reminiscent of the requirement of the second substrate (an α-keto acid) that is required to return pyridoxamine phosphate to PLP after PLP-dependent enzyme transamination. In the case of flavoenzymes, however, there are two mechanisms for conversion of reduced flavin back to oxidized flavin.

Some flavoenzymes are called *oxidases* and others *dehydrogenases*. The distinction between these names refers to the way in which the reduced form of the coenzyme is reoxidized, so the catalytic cycle can continue. Those enzymes that utilize electron transfer proteins, such as ubiquinone or cytochrome b_5, to accept electrons from the reduced flavin and proceed by two one-electron transfers (Scheme 4.28) are called *dehydrogenases*.[81] Oxidases use molecular oxygen to oxidize the coenzyme with concomitant formation of hydrogen peroxide; Scheme 4.29 shows possible mechanisms for this oxidation. Pathway a depicts the reaction with triplet oxygen, leading first to the caged radical pair of **4.52** and superoxide by electron transfer, which, after spin inversion, can undergo either radical combination via pathway c to give **4.53**[82] or second electron transfer from the flavin semiquinone (pathway d) to go directly to the oxidized flavin. Pathway b shows the analogous reaction with singlet oxygen to give the flavin hydroperoxide (**4.53**) directly, which will only occur if there is a mechanism for spin inversion from the normal triplet oxygen, such as with a metal ion. Loss of hydrogen peroxide gives oxidized flavin.

Each of the reactions shown in Table 4.5 is specific for a particular enzyme. Some of the reactions are difficult to carry out nonenzymatically; therefore, although the flavin is essential for the redox reaction, the enzyme is responsible for catalyzing these reactions, and for both substrate and reaction specificity. Four types of mechanisms can be considered—one involving a carbanion intermediate, one with a carbanion and a radical intermediate, one with radical intermediates, and one with a hydride intermediate. As we will see, there is no definitive mechanism for flavin-dependent enzymes; each of these mechanisms may be applicable to different flavoenzymes and/or different substrates.

4.3.3.1. Two-Electron (Carbanion) Mechanism

Evidence for a carbanion mechanism (Scheme 4.30) has been provided for the flavoenzyme D-amino acid oxidase,[83] which catalyzes the oxidation of D-amino acids to α-keto acids and ammonia (the second reaction in Table 4.5).

4.3.3.2. Carbanion Followed by Two One-Electron Transfers

Some enzymes may initiate a carbanion mechanism but then proceed by an electron transfer (radical) pathway. Enzymatic evidence for this kind of mechanism is available from studies with general acyl-CoA dehydrogenase,[84] a family of enzymes that catalyzes the oxidation of fatty acid acyl-CoA derivatives (**4.54**, Scheme 4.31) to the corresponding α,β-unsaturated acyl-CoA compound (**4.55**).

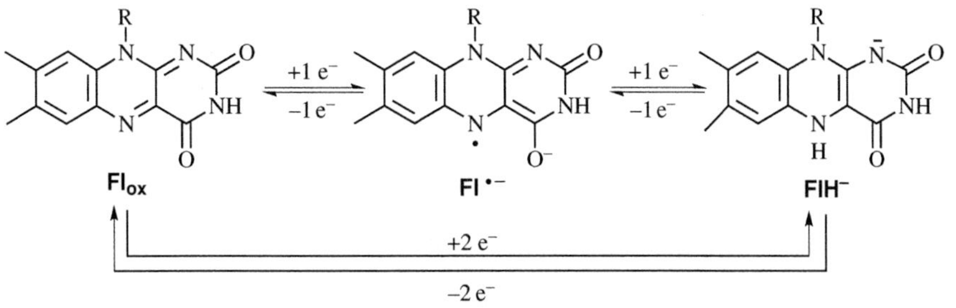

Fl_{ox} ⇌ (+1 e⁻ / −1 e⁻) $Fl^{\bullet -}$ ⇌ (+1 e⁻ / −1 e⁻) FlH^-

+2 e⁻ / −2 e⁻

SCHEME 4.27 One- and two-electron reductions of flavins

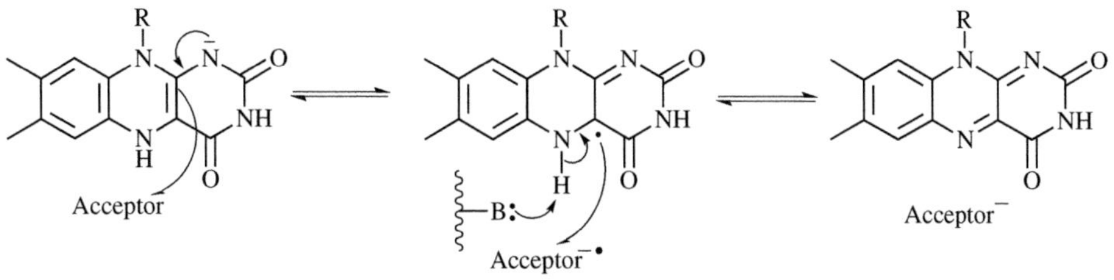

Acceptor, Acceptor$^{-\bullet}$, Acceptor$^-$

SCHEME 4.28 Mechanism for dehydrogenase-catalyzed flavin oxidation

a, b: O=O, $+H^+$; 4.53; $-H_2O_2$; Fl_{ox}; Radical combination c; d: 2nd e⁻ transfer $+H^+$; 4.52

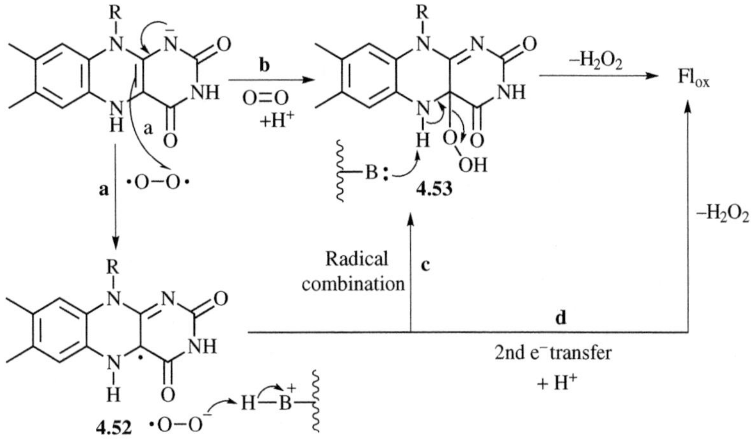

SCHEME 4.29 Mechanism for oxidase-catalyzed flavin oxidation

$-H^+$; $-FlH^-$; H_2O, $-NH_4^+$

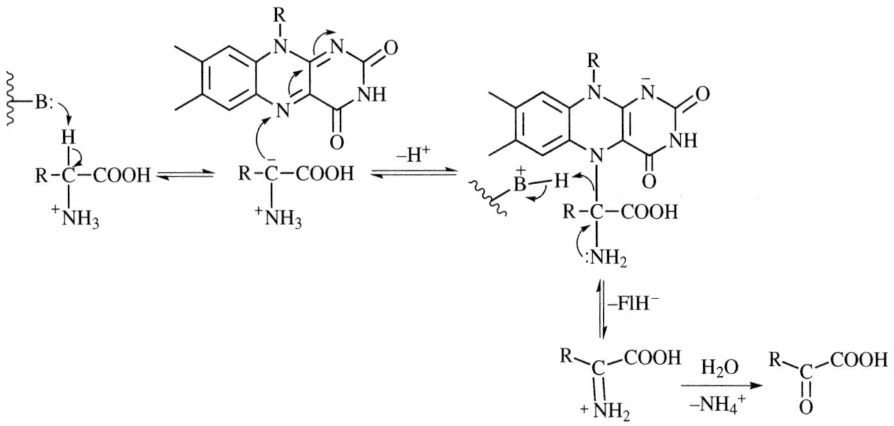

SCHEME 4.30 Two-electron mechanism for flavin-dependent D-amino acid oxidase-catalyzed oxidation of D-amino acids

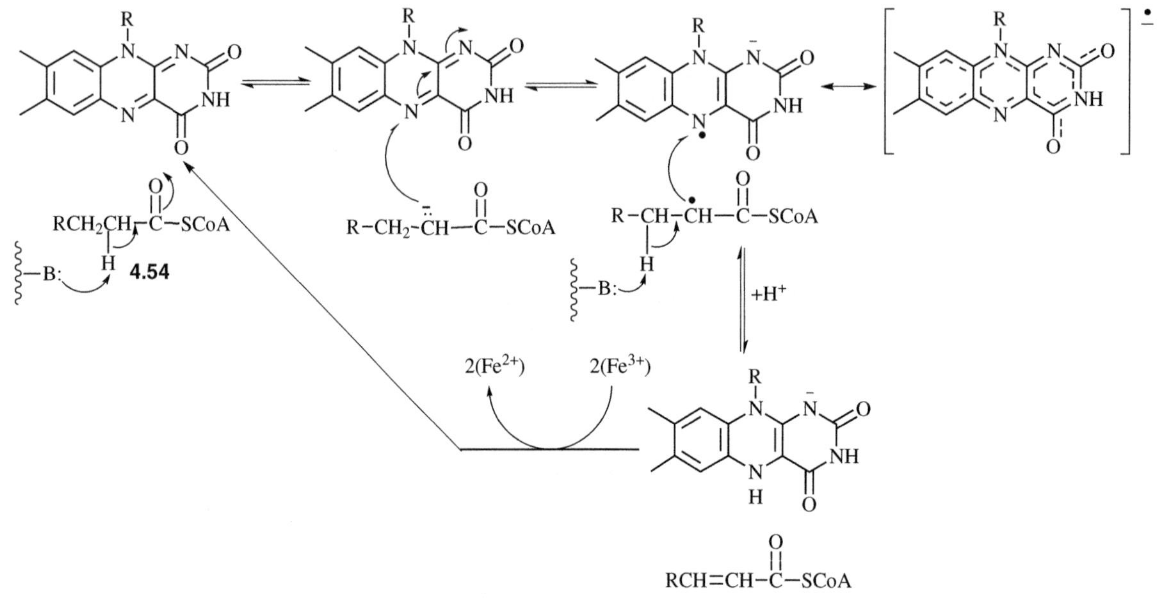

SCHEME 4.31 Two-electron followed by one-electron mechanism for general acyl-CoA dehydrogenase

4.3.3.3. One-Electron Mechanism

Evidence for a one-electron (radical) flavin mechanism comes from a variety of experiments with monoamine oxidase (MAO),[85] a flavoenzyme important in medicinal chemistry (see Chapter 5, Sections 5.3.3.3.3 and 5.3.3.3.4). This is one of the enzymes responsible for the catabolism of various biogenic amine neurotransmitters, such as norepinephrine and dopamine. Possible mechanisms for MAO, which exist in two isozymic forms called MAO A and MAO B, and which catalyze the degradation of biogenic amine neurotransmitters, such as norepinephrine and dopamine (**4.56**) to their corresponding aldehydes (**4.57**), are summarized in Scheme 4.32.

4.3.3.4. Hydride Mechanism

The second step in bacterial cell wall peptidoglycan biosynthesis, catalyzed by uridine diphosphate-*N*-acetylenolpyruvylglucosamine reductase (or MurB), is the reduction of enolpyruvyl uridine diphosphate-*N*-acetylglucosamine (**4.58**) to give uridine diphosphate-*N*-acetylmuramic acid (**4.59**) (Scheme 4.33). The proposed reduction mechanism[86] involves initial reduction of the FAD cofactor by NADPH (because a reduced flavin is highly prone to oxidation, it is typically formed at the active site of the enzyme by in situ reduction with a pyridine nucleotide cofactor, at the time it is needed), followed by hydride transfer from reduced FAD via a Michael addition to the double bond of the α,β-unsaturated carboxylate (**4.58**).

Flavin monooxygenases are members of the class of liver microsomal mixed function oxygenases that are important in the oxygenation of *xenobiotics* (foreign substances) that enter the body, including drugs (Chapter 8).[87] Unlike flavin oxidases and dehydrogenases, flavin monooxygenases incorporate an oxygen atom from molecular oxygen into the substrate. It is believed that a flavin C^{4a}-hydroperoxide is an important intermediate in this process,[88] which can be formed by the same mechanisms suggested for the oxidation of reduced flavin (Scheme 4.29, pathways a and b/c). This then requires the flavin monooxygenase to have the flavin in its reduced form. As noted in the preceding paragraph, the way this is accomplished is with a stable reducing agent, NADH or NADPH, which converts the oxidized flavin to its reduced form and initiates the oxygenation reaction (Scheme 4.34). On the basis of stopped-flow spectroscopic evidence for the formation and decay of C^{4a}-flavin hydroperoxide anion, C^{4a}-flavin hydroperoxide, and C^{4a}-flavin hydroxide intermediates in a flavoenzyme that converts a phenol into a catechol (reaction 6 in Table 4.5), the simplest mechanism that can be drawn involves nucleophilic attack by the substrate at the distal oxygen of the flavin hydroperoxide.[89] In this case, the electron pair on X in Scheme 4.34 is a pair of π-electrons from the phenyl ring of the phenol.

4.3.4. Heme

Heme, or protoporphyrin IX, is an iron(III)-containing porphyrin cofactor (**4.14**) for a large number of liver microsomal mixed function oxygenases principally in the cytochrome P450 family of enzymes. These enzymes, like flavin monooxygenases, are important in the metabolism of xenobiotics, including drugs (Chapter 8).[90] As in

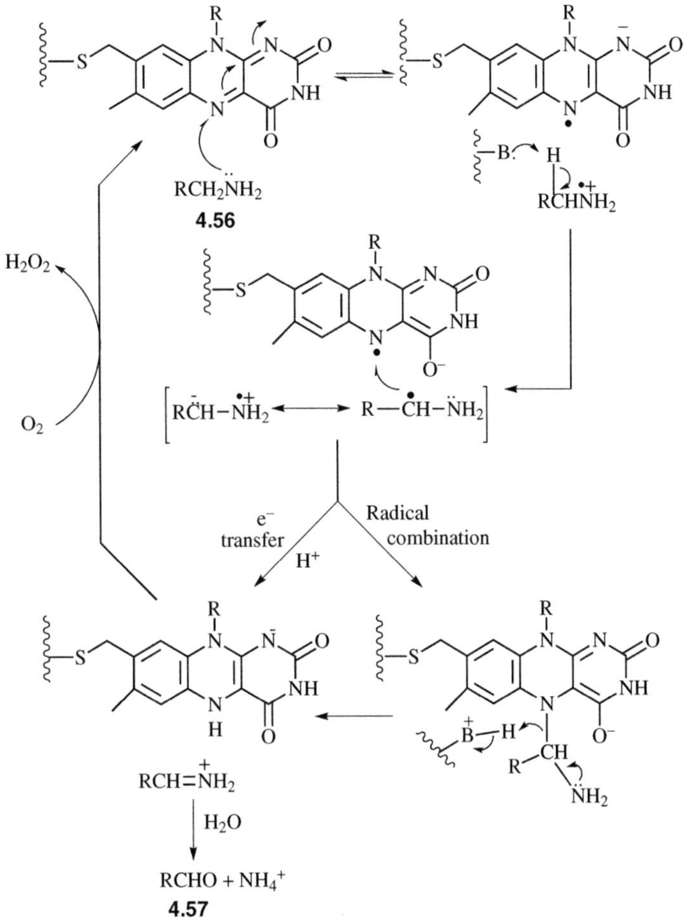

SCHEME 4.32 One-electron mechanism for monoamine oxidase

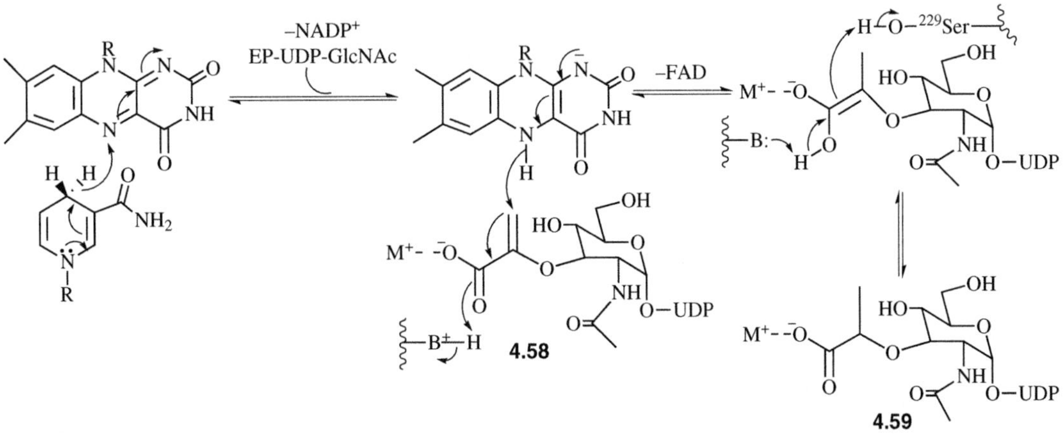

SCHEME 4.33 Hydride reduction mechanism for uridine diphosphate-*N*-acetylenolpyruvylglucosamine reductase (MurB)

the case of the flavin monooxygenases, molecular oxygen binds to the heme cofactor (after Fe^{3+} has been reduced to Fe^{2+}) and is converted into a reactive form, which is used in a variety of oxygenation reactions, especially hydroxylation and epoxidation reactions. The hydroxylation reactions often occur at seemingly unactivated carbon atoms. The mechanism for this class of enzymes is still in debate, but a high-energy iron-oxo species has always been suspected.[91] Formally, this species can be written as any one of the resonance structures **4.61a–4.61d** (Scheme 4.35). Direct rapid freeze-quench techniques with spectroscopic (EPR, UV–vis, and Mössbauer) and kinetic characterization

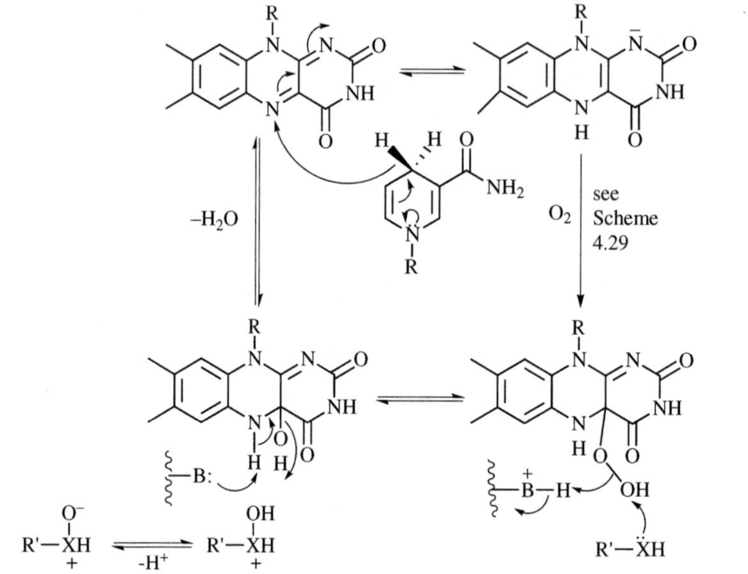

SCHEME 4.34 Mechanism for oxygenation of substrate R′XH by flavin monooxygenase

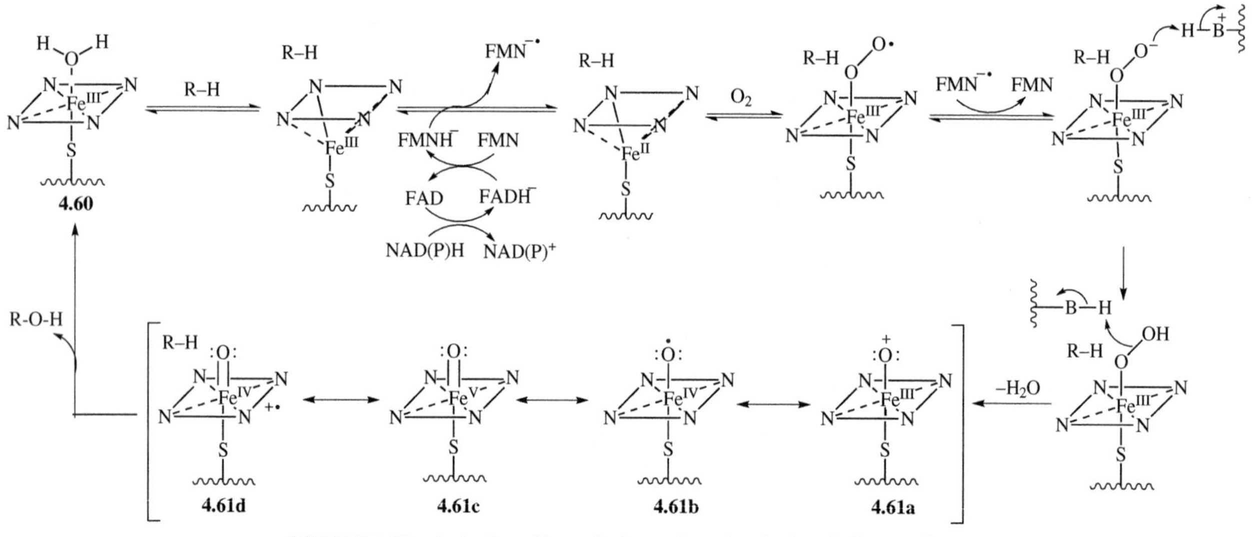

SCHEME 4.35 Activation of heme for heme-dependent hydroxylation reactions

of this intermediate was obtained, and the favored depiction appears to be **4.61d**.[92] The heme is abbreviated as **4.60**, where the peripheral nitrogens represent the four pyrrole nitrogens. The axial ligands in the case of cytochrome P450 are water and a cysteine thiolate from the protein. The electrons for reduction of the heme of cytochrome P450 (the second and fourth steps of Scheme 4.35) come from an enzyme complexed with cytochrome P450 called NADPH-cytochrome P450 reductase,[93] which contains NADPH and two different flavin coenzymes (FAD and FMN).

Heme-dependent enzymes catalyze reactions on a wide variety of substrates. Our discussion will be limited to reactions of cytochrome P450 because of its importance to drug metabolism (Chapter 8). Despite the substrate variability, there appears to be a common mechanism (with variations) for each involving initial transfer of one electron from the substrate to the high-energy iron-oxo species **4.61**. Below are specific mechanisms for hydroxylation of alkanes (Scheme 4.36), epoxidation of alkenes (Scheme 4.37), and hydroxylation of molecules containing heteroatoms, such as sulfur (Scheme 4.38). In Scheme 4.36, the high-energy iron-oxo species (**4.61d**) abstracts a hydrogen atom from the alkane to give a very short-lived carbon radical (**4.62**),[94] which accepts a hydroxyl radical from the heme species (known as *oxygen rebound*) to give the alcohol. Alkene epoxidation (Scheme 4.37) is similar

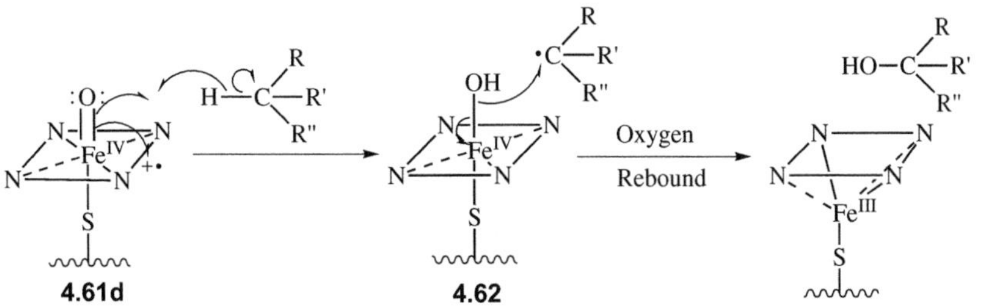

SCHEME 4.36 Mechanism for heme-dependent hydroxylation reactions

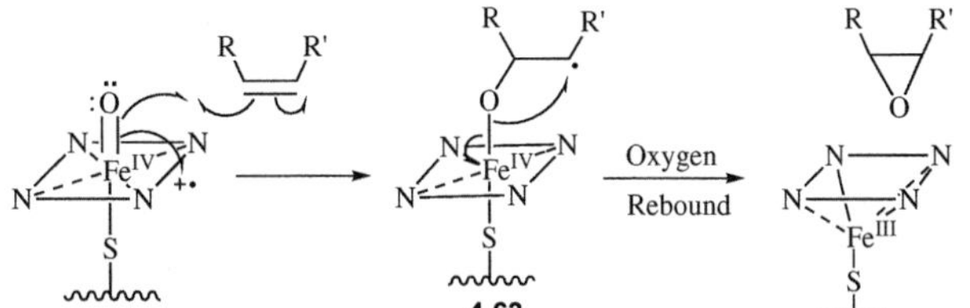

SCHEME 4.37 Mechanism for heme-dependent epoxidation reactions

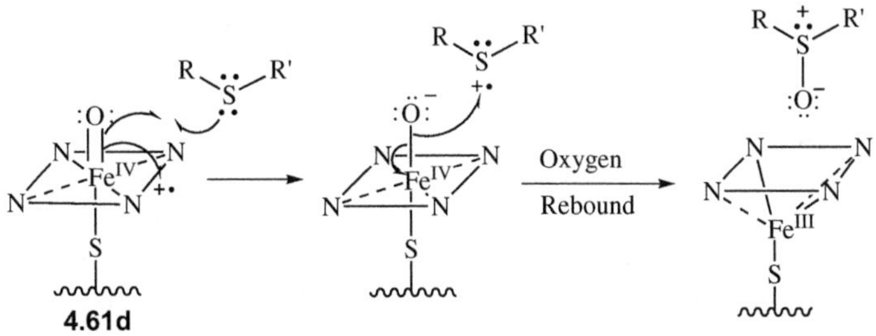

SCHEME 4.38 Mechanism for heme-dependent heteroatom oxygenation reactions

$$PhCO_2H + NH_3 \longrightarrow PhCO_2^- NH_4^+$$

SCHEME 4.39 Reaction of an amine with a carboxylic acid

to alkane hydroxylation, but instead of a hydrogen atom abstracted from the substrate, an electron is "abstracted" from the double bond to give the alkylated iron-oxo species (**4.63**), followed by Fe–O bond homolysis. Chemical model studies,[95] however, suggest that alkene epoxidation may involve an initial charge–transfer complex between the iron-oxo species and the alkene followed by a concerted process; if **4.63** is formed, it is very short-lived (as in the case of alkane hydroxylation). Oxygenation of sulfides (Scheme 4.38) also is similar to hydroxylation of alkanes, except instead of initial hydrogen atom abstraction, the iron-oxo species "abstracts" an electron from the readily oxidizable nonbonded electron pair of the sulfur atom, followed by oxygen rebound. This electron-transfer mechanism occurs because of the low oxidation potential of sulfur.

4.3.5. Adenosine Triphosphate and Coenzyme A

Adenosine triphosphate (ATP) provides energy to those molecules that are not sufficiently reactive to allow them to undergo chemical reactions. Consider the nonenzymatic reaction of a carboxylic acid with ammonia. If you mix benzoic acid with a base such as ammonia, what are you going to get? You will not get an amide (unless you heat it excessively); you get the salt, ammonium benzoate (Scheme 4.39). What if you wanted to get the amide; how would you do it? There are numerous ways to make amides, but in general, the carboxylic acid must first be activated, because once the proton is removed, the carboxylate usually does not react with nucleophiles. Typically, dehydrating agents such as thionyl chloride or ethyl chloroformate are used first to activate the carboxylic acid to an acid chloride (**4.64**, Scheme 4.40) or anhydride (**4.65**), respectively. Acid chlorides and anhydrides are very reactive toward nucleophiles, so ammonia can displace the good leaving groups chloride ion and monoethyl carbonate (which decomposes to ethanol and carbon dioxide), respectively,

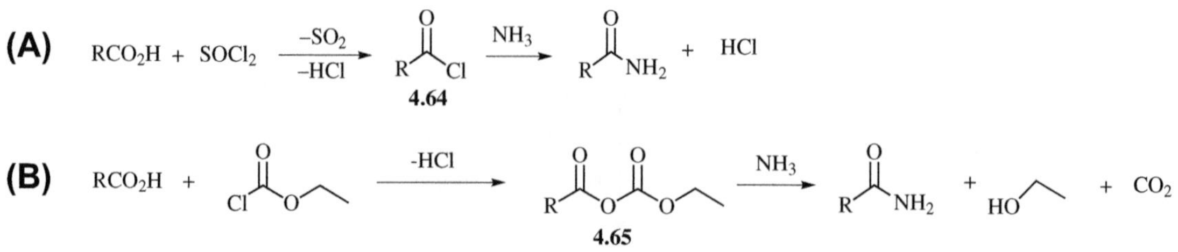

(A) $RCO_2H + SOCl_2 \xrightarrow[-HCl]{-SO_2}$ RCOCl (4.64) $\xrightarrow{NH_3}$ $RCONH_2$ + HCl

(B) RCO_2H + ClCO$_2$Et $\xrightarrow{-HCl}$ RCOOCO$_2$Et (4.65) $\xrightarrow{NH_3}$ $RCONH_2$ + EtOH + CO_2

SCHEME 4.40 Activation of carboxylic acids via acid chloride (A) and anhydride (B) intermediates

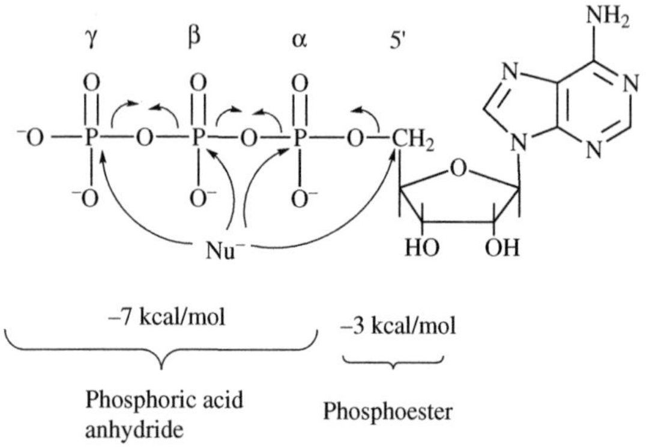

−7 kcal/mol: Phosphoric acid anhydride; −3 kcal/mol: Phosphoester

ATP

FIGURE 4.9 Electrophilic sites on ATP

and produce amides. That, in fact, is what enzymes do also, except thionyl chloride would be a bit harsh on our tissues, so instead, ATP is the substitute. Think of ATP, then, as the endogenous form of thionyl chloride or ethyl chloroformate.

The structure of ATP is shown in Figure 4.9. There are four electrophilic sites where nucleophiles can attack: at the γ-phosphate, the β-phosphate, the α-phosphate, and the 5′-methylene group. Attacks at the γ- and α-positions are the most common because those reactions are thermodynamically favored. Attack at the 5′-methylene is rare. For the most part, after the complete enzyme-catalyzed reaction sequence, ATP is converted to adenosine diphosphate + P_i (inorganic phosphate) or to AMP + PP_i (inorganic diphosphate), depending upon what site is attacked.

In the case of the conversion of fatty acids to the corresponding fatty acyl-coenzyme A derivatives (**4.67**), which are the substrates for general acyl-CoA dehydrogenase (see Section 4.3.3.2; Scheme 4.31), ATP is used to activate the carboxylic acid as an AMP anhydride (**4.66**) so that it is reactive enough for conversion to the CoA thioester (Scheme 4.41). PPi is generated as a by-product in this reaction. Then the AMP anhydride intermediate undergoes rapid reaction with the highly nucleophilic thiol group of coenzyme A (**4.13b**) to form the desired fatty acid CoA thioester; AMP is the by-product of the second step.

Coenzyme A thioesters serve three important functions for different enzymes. With general acyl-CoA dehydrogenase, the thioester group makes the α-proton much more acidic than that of a carboxylate, and therefore easier to remove. The pK_a of the α-proton of a thioester (about 21) is lower than that of an ester (about 25–26) and considerably lower than the α-proton of a carboxylate anion (about 33–34).[96] Second, thioesters are much more reactive toward acylation of nucleophilic substrates than are carboxylic acids and oxygen esters, but not so reactive, as in the case of the phosphoryl anhydrides made from carboxylic acids with ATP, that they cannot be functional in aqueous media. Third, coenzyme A esters are important in the transport of molecules and in their binding to enzymes.

4.4. ENZYME CATALYSIS IN DRUG DISCOVERY

4.4.1. Enzymatic Synthesis of Chiral Drug Intermediates

As we have seen, many drugs are chiral, which makes their syntheses much more complicated. However, enzymes are highly stereoselective, if not stereospecific, so process chemistry groups are using enzymes, either natural or engineered, to carry out critical stereogenic synthetic steps.

The first chiral synthesis of pregabalin (**4.68**, Lyrica), a drug for the treatment of neuropathic pain, fibromyalgia, and epilepsy, used Evans' chiral oxazolidinone chemistry,[97] but there were difficulties in scale up with this method. Another synthesis utilized a chiral resolution of racemic 4-amino-3-isobutylbutanoic acid using (*S*)-(+)-mandelic acid to give pregabalin in 95% enantiomeric excess (ee) after one recrystallization; after three crystallizations, 100% ee.[98] Another chiral route (Scheme 4.42) involved the asymmetric hydrogenation of 3-cyano-5-methylhex-3-enoic acid (**4.69**, Scheme 4.42) using a rhodium catalyst.[99] More recently, however, the process group at Pfizer developed a chemoenzymatic method to obtain a chiral intermediate, which was converted to pregabalin.[100] The enzymatic step uses *Thermomyces lanuginosus* lipase (Lipolase), a commercially available enzyme that catalyzes the hydrolysis of

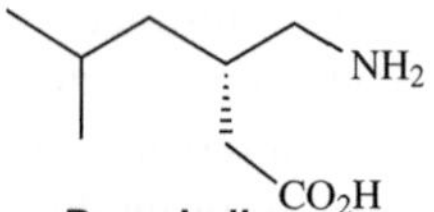

Pregabalin
4.68

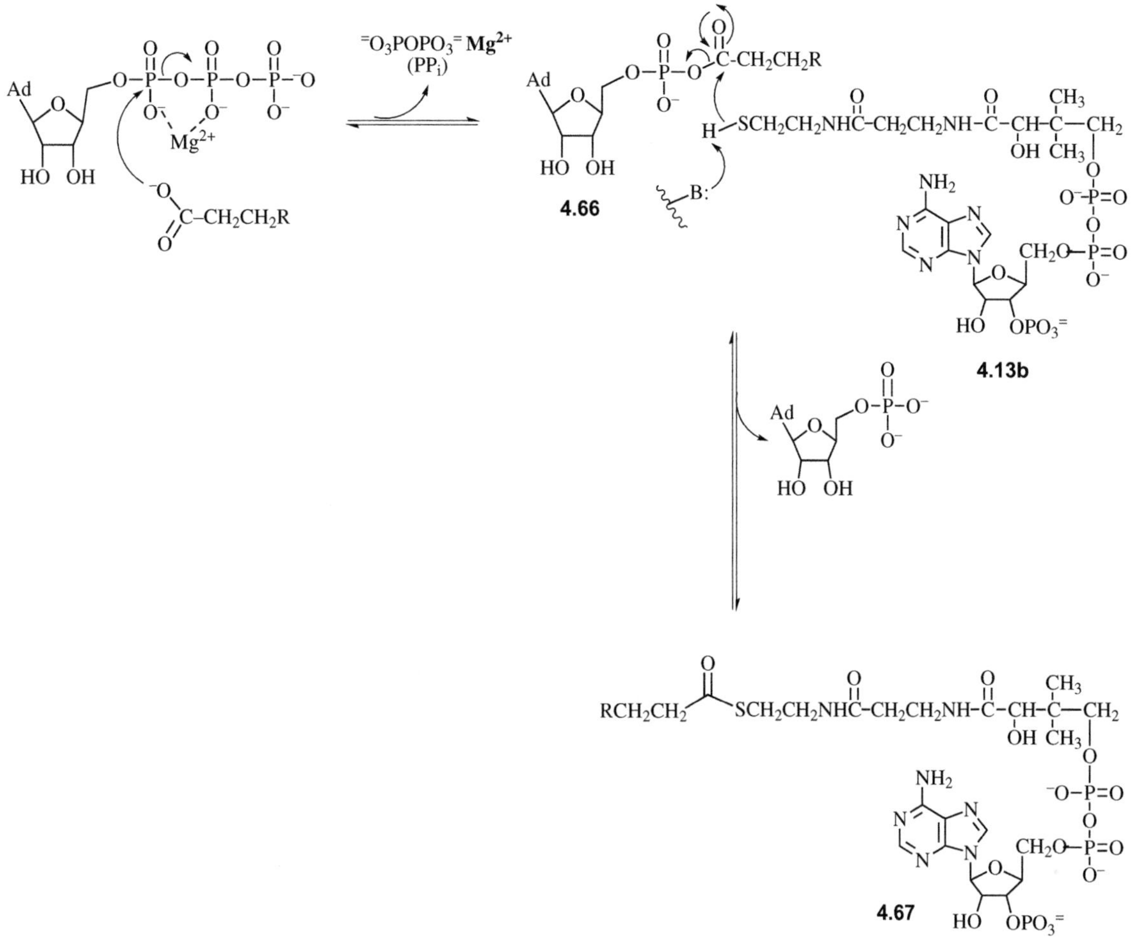

SCHEME 4.41 ATP-activation of fatty acids in the biosynthesis of fatty acid CoA thioesters

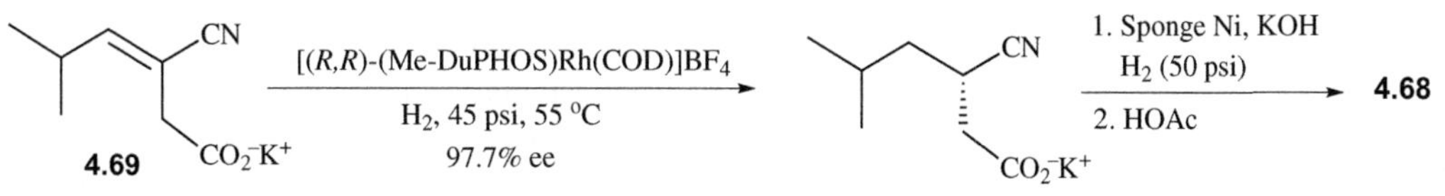

SCHEME 4.42 Asymmetric synthesis of pregabalin

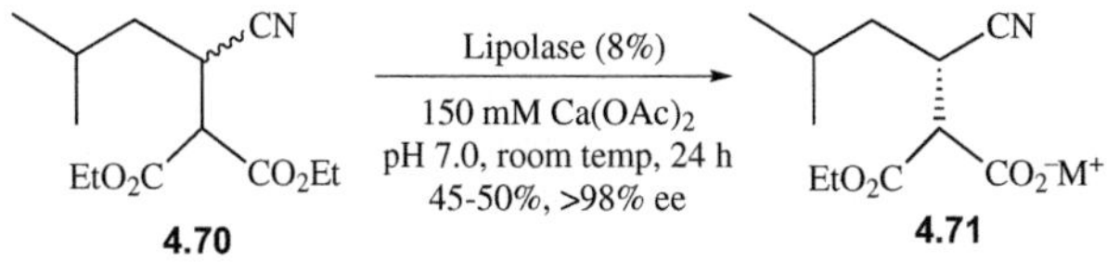

SCHEME 4.43 Enzymatic step in the chemoenzymatic synthesis of pregabalin

one of the two ethyl esters of **4.70** to give the chiral intermediate (**4.71**, Scheme 4.43) in >98% ee at 45–50% conversion (remember, the best you can do with conversion of a racemate to a single enantiomer is 50%). And this is done in water at pH 7.0 and room temperature. Decarboxylation of **4.71** with heat, and saponification of the ethyl ester, gives the same intermediate shown in Scheme 4.42 that is hydrogenated in the presence of Sponge Ni to **4.68**.

Another type of biocatalysis[101] used in drug synthesis involves protein engineering, the modification of specific amino acids in an enzyme to accommodate a particular substrate structure. As we have found, enzymes typically bind specific molecules and catalyze reactions on specific molecules. In a drug synthesis, the intermediates generally will not have the same structure as natural substrates for an enzyme. However, by site-directed mutagenesis, any amino acid at the active site can be modified to allow for binding of nonnatural molecules. This is the approach that was taken at Merck in their biocatalytic synthesis of the antidiabetes drug sitagliptin phosphate (**4.72**, Januvia).[102] The chemical synthesis involved a high-pressure rhodium-based chiral catalytic hydrogenation of an enamine as the last step, giving sitagliptin with a 97% ee. A group at Codexis, Inc. and the process group at Merck teamed up to find a biocatalytic

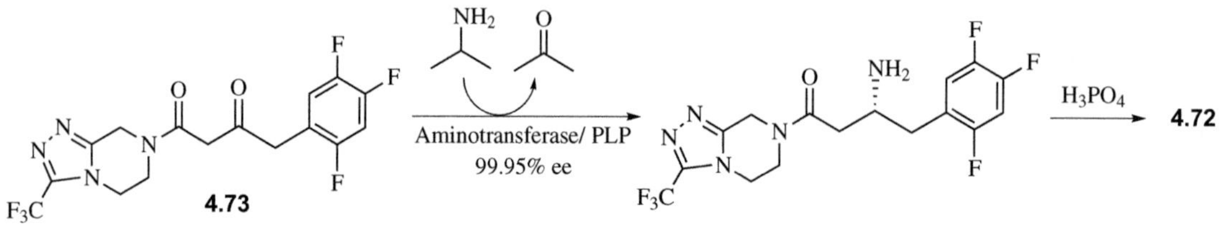

SCHEME 4.44 Directed evolution of an aminotransferase for the synthesis of sitagliptin

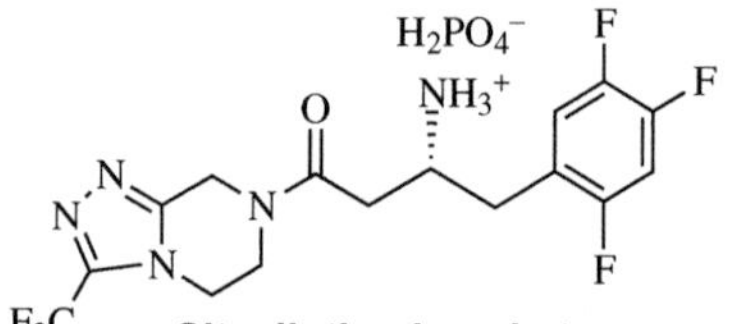

Sitagliptin phosphate
4.72

chiral synthesis for **4.72** and decided to screen aminotransferases to convert a ketone to the chiral amino group of sitagliptin. Aminotransferases typically have a limited substrate range; usually a methyl ketone is the largest group that can be accommodated in the active site. They started with an (*R*)-selective PLP-dependent aminotransferase (ATA-117), built a structural homology model (a model of a protein derived by inserting the amino acid sequence of the desired protein into an experimental three-dimensional structure of a related homologous protein),[103] then did docking studies to see what amino acid(s) needed to be mutated to get the desired ketone substrate (**4.73**, Scheme 4.44) to fit. The initial analysis suggested that four residues interacting with the trifluorophenyl group should be mutated; several mutations at each position were performed, and the engineered enzyme was screened for activity with **4.73**. This gave the first enzyme with activity. A second round of directed evolution[104] was carried out with 12 mutations; after 11 rounds of evolution, they obtained an aminotransferase with 27 mutations of ATA-117 that gave sitagliptin with 99.95% ee (Scheme 4.44).

4.4.2. Enzyme Therapy

Throughout this chapter, small organic and some inorganic molecules are discussed in terms of their design and mechanism of action. In the next chapter, drugs that inhibit the catalytic action of enzymes are described. Some enzymes, however, are themselves useful as drugs. For the most part, the enzymes that have therapeutic utility catalyze hydrolytic reactions. For example, amylase, ligase, cellulase, trypsin, papain, and pepsin are proteolytic or lipolytic digestive enzymes used for gastrointestinal disorders resulting from poor digestion. Trypsin also is used for degrading necrotic tissue from wounds. Collagenase hydrolyzes collagen in necrotic tissue, but does not attack collagen in healthy tissue. Lactase (β-D-galactosidase) is taken by those having low lactase activity for the hydrolysis of lactose into glucose and galactose. Deoxyribonuclease and fibrinolysin are used to dissolve the DNA and fibrinous material, respectively, in purulent exudates and in blood clots. Because malignant leukemia cells are dependent on an exogenous source of asparagine for survival, whereas normal cells are able to synthesize asparagine, the enzyme asparaginase (L-asparagine amidohydrolase), which hydrolyzes the exogenous asparagine to aspartate, has been successful in the treatment of acute lymphocytic leukemia. Three different enzymes, urokinase, streptokinase, and tissue plasminogen activator, convert the inactive protein plasminogen into plasmin, a proteolytic enzyme that digests fibrin clots; therefore, these enzymes are effective in the treatment of myocardial infarction, venous and arterial thrombosis, and pulmonary embolism.[105] Excessive bilirubin in the blood is the cause for neonatal jaundice. A blood filter containing immobilized bilirubin oxidase can degrade more than 90% of the bilirubin in the blood in a single pass through the filter.[106] Recombinant human *N*-acetylgalactosamine-6-sulfatase completed Phase III clinical trials for the treatment of mucopolysaccharidosis type IVA (Morquio A syndrome), a rare inherited lysosomal storage disorder resulting in the accumulation of complex carbohydrates in the body.[107] The use of genetic engineering techniques to produce altered active enzymes that have enhanced stability should lead to increased use of enzymes as drugs. The major drawbacks to the use of enzymes in therapy are enzyme instability (other proteases degrade them), poor bioavailability by most routes of administration, and allergic responses.

In the next chapter, we examine the design and mechanism of action of enzyme inhibitors as drugs.

4.5. GENERAL REFERENCES

Enzyme Catalysis

Boyer, P. D. (Ed.) *The Enzymes*, 3rd ed., Academic, New York, 1970–1987.

Fersht, A. *Enzyme Structure and Mechanism*, 2nd ed., W. H. Freeman, New York, 1985.

Jencks, W. P. *Catalysis in Chemistry and Enzymology*, McGraw-Hill, New York, 1969.
Jencks, W. P. Binding energy, specificity, and enzymic catalysis: the circe effect. *Adv. Enzymol.* **1975**, *43*, 219–410.
McGrath, B. M.; Walsh, G. (Eds.) *Directory of Therapeutic Enzymes*, CRC Taylor & Francis, Boca Raton, FL, 2006.
Schomburg, D.; Schomburg, I. (Eds.) *Springer Handbook of Enzymes*, 2nd ed., (39 volumes covering all classes of enzymes), Springer, Berlin, 2001–2009, after 2009, supplement volumes.
Segel, I. H. *Enzyme Kinetics*, Wiley, New York, 1975.
Silverman, R. B. *The Organic Chemistry of Enzyme-Catalyzed Reactions*, Academic Press, San Diego, 2002.
Stein, R. L. *Kinetics of Enzyme Action. Essential Principles for Drug Hunters*; Wiley: Hoboken, NJ, 2011.

Pyridoxal 5′-Phosphate (PLP)

Dolphin, D.; Poulson, R.; Avramovic, O. (Eds.) *Vitamin B_6 Pyridoxal Phosphate*, Wiley, New York, 1986, Parts A and B.
Eliot, A. C.; Kirsch, J. F. Pyridoxal phosphate enzymes: mechanistic, structural, and evolutionary considerations. *Ann. Rev. Biochem.* **2004**, *73*, 383–415.

Pyridine Nucleotides (NADH and NADPH)

Dolphin, D.; Poulson, R.; Avramovic O. (Eds.) *Pyridine Nucleotide Coenzymes*, Wiley, New York, 1987; Parts A and B.
Chen, L.; Petrelli, R.; Felczak, K.; Gao, G.; Bonnac, L.; Yu, J. S.; Bennett, E. M.; Pankiewicz, K. W. Nicotinamide adenine dinucleotide based therapeutics. *Curr. Med. Chem.* **2008**, *15*(7), 650–670.

Heme

Ortiz de Montellano, P. R. (Ed.) *Cytochrome P450: Structure, Mechanism, and Biochemistry*, 3rd ed., Kluwer Academic/ Plenum, New York, 2005.

Enzyme Therapy

Blohm, D.; Bollschweiler, C.; Hillen, H. Pharmaceutical proteins. *Angew. Chem. Int. Ed. Engl.* **1988**, *27*, 207–308.
Circhoke, A. J. *Enzymes and Enzyme Therapy*, 2nd ed., Keats Publishing, Glenview, IL, 2000.
Bohager, T. *Everything You Need to Know About Enzymes: a Simple Guide to Using Enzyme to Treat Everything from Digestive Problems and Allergies to Migraines and Arthritis*, Greenleaf Book Group Press, Austin, TX, 2008.

4.6. PROBLEMS (ANSWERS CAN BE FOUND IN THE APPENDIX AT THE END OF THE BOOK)

1. What are the two most important characteristics of an enzyme?

2. The (3*S*,4*S*)-isomer of deuterated compound **1** is a substrate for a dephosphorylase that gives an anti-elimination to the *Z*-isomer (**2**). The corresponding (3*R*,4*S*)-isomer is not a substrate. How do you explain that?

$^{-}$OOC … $OPO_3^{=}$ → $^{-}$OOC … H, D

1 **2**

3. Compound **3** is a substrate for an enzyme that catalyzes the removal of the H_S proton exclusively. If compound **4** is used as the substrate, the deprotonation occurs mostly at H_S, but a small amount of products are observed from removal of H_R. This second reaction was shown to be catalyzed by the same enzyme as well. How could this be rationalized?

H_S H_R **3** O_2N H_S H_R **4**

4. The following reaction is catalyzed by an enzyme at pH 7 and 25 °C, whereas nonenzymatically, this reaction does not occur under these conditions. Explain how the enzyme can easily catalyze this reaction.

OH $\underset{+H_2O}{\overset{-H_2O}{\rightleftharpoons}}$

5. Porphobilinogen synthase is a zinc-dependent enzyme (contains two zinc ions) that catalyzes the condensation of two molecules of 5-aminolevulinic acid (**5**) to give porphobilinogen (**6**). A condensed mechanism is shown below. Indicate with an arrow every place possible that the enzyme catalyzes this reaction, and name the catalytic mechanisms.

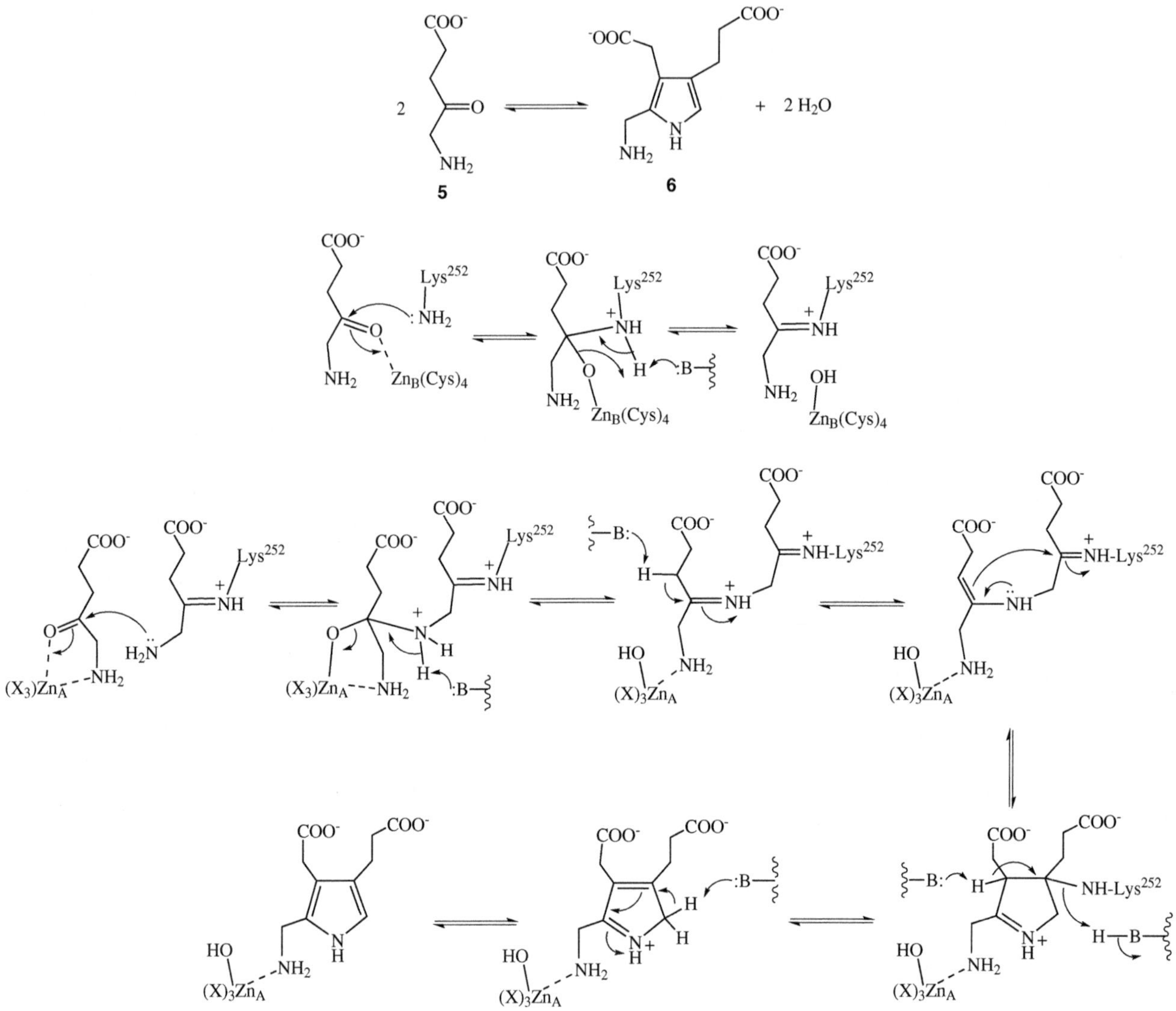

6. Draw a mechanism for the enzyme-catalyzed enolization of a ketone utilizing an electrostatic interaction.
7. Transition state stabilization accounts for a large portion of the stabilization energy of an enzyme-catalyzed reaction. Haloalkane dehalogenase catalyzes an S_N2 reaction of an active site carboxylate with an alkyl halide followed by hydrolysis to the alcohol. What transition state stabilization processes could the enzyme utilize (draw the transition state for guidance and note the processes)? 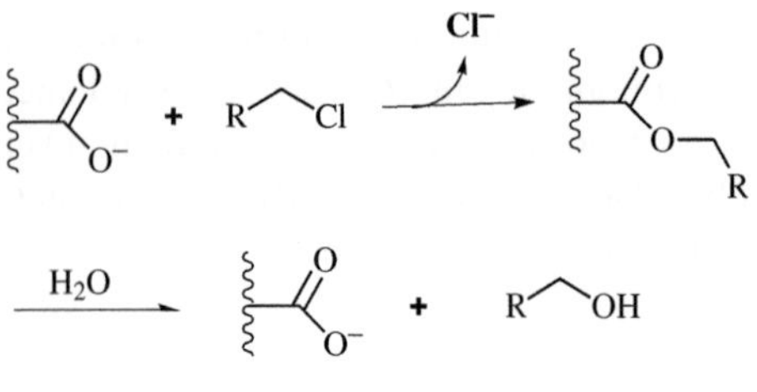
8. If you were using aspartate aminotransferase, and the enzyme stopped catalyzing the reaction, what *two* components of the enzyme reaction would you check?
9. Dopa (**7**, Ar = 3,4-dihydroxyphenyl) is converted into norepinephrine (**8**, Ar = 3,4-dihydroxyphenyl; R = H) in two enzyme-catalyzed reactions, both of which require coenzymes. If the first enzyme-catalyzed reaction is run in ^2H_2O, the norepinephrine is stereospecifically labeled with deuterium (R = ^{2}H). Draw mechanisms for the two enzyme-catalyzed reactions.

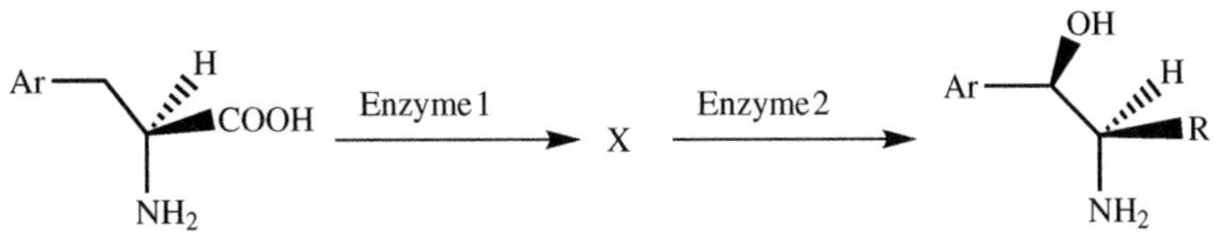

Ar, H, COOH, NH_2 **7** —Enzyme 1→ X —Enzyme 2→ OH, Ar, H, R, NH_2 **8**

10. γ-Aminobutyric acid (GABA) aminotransferase catalyzes a PLP-dependent conversion of GABA to succinic semialdehyde (SSA).

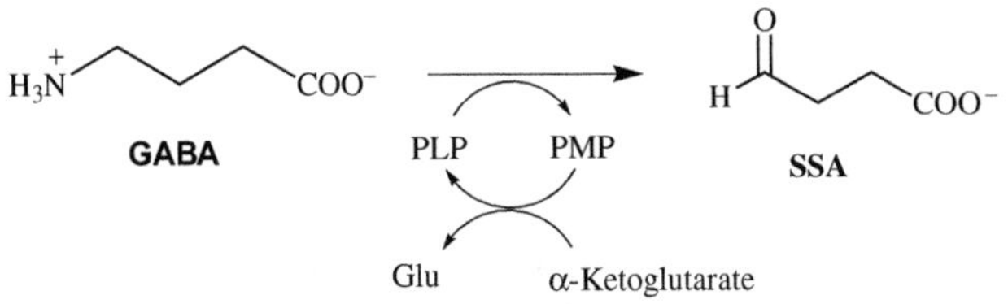

$H_3\overset{+}{N}$–COO^- (**GABA**) → (PLP, PMP; Glu, α-Ketoglutarate) H(C=O)–COO^- (**SSA**)

a. Draw a mechanism for this reaction.

b. Why is only one molecule of succinic semialdehyde formed in the absence of α-ketoglutarate

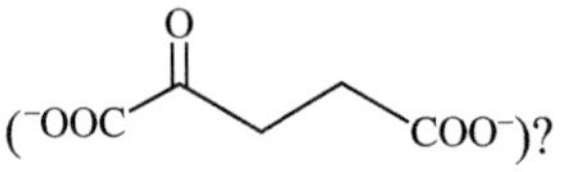

$(^-OOC$–C(=O)–$COO^-)$?

c. If the reaction were carried out in $^2H_2{}^{18}O$, what would the products be?

11. *p*-Hydroxyphenylacetate 3-hydroxylase is a flavin adenine dinucleotide enzyme that catalyzes the reaction below. Draw a mechanism.

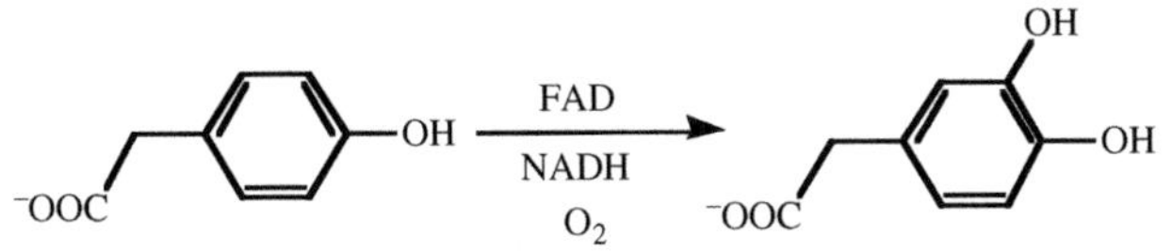

^-OOC–CH_2–C_6H_4–OH —(FAD, NADH, O_2)→ ^-OOC–CH_2–C_6H_3(OH)–OH

Hint: The mechanism is related to that in Scheme 4.34.

12. Draw a mechanism for the formation of N^5,N^{10}-methylenetetrahydrofolate and the transfer of a hydroxymethyl group to uridylate.

Hint: The first step in the transfer involves a Michael addition of an enzyme cysteine residue to uridylate.

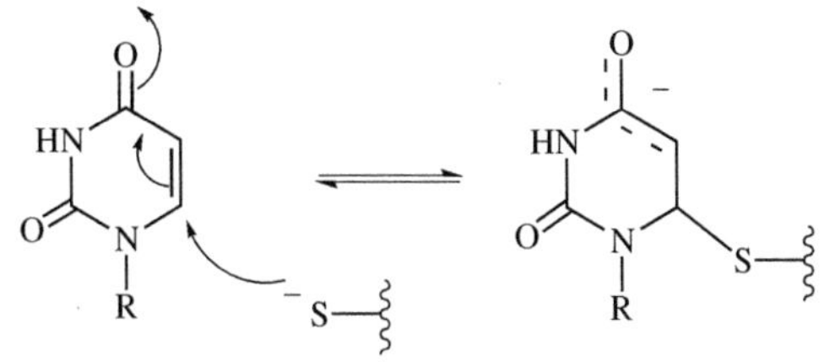

13. Cytochrome P450, a heme-dependent enzyme, catalyzes the oxidation of a wide variety of xenobiotics. Draw a mechanism for the conversion of propranolol (**9**) to **10** and acetone.

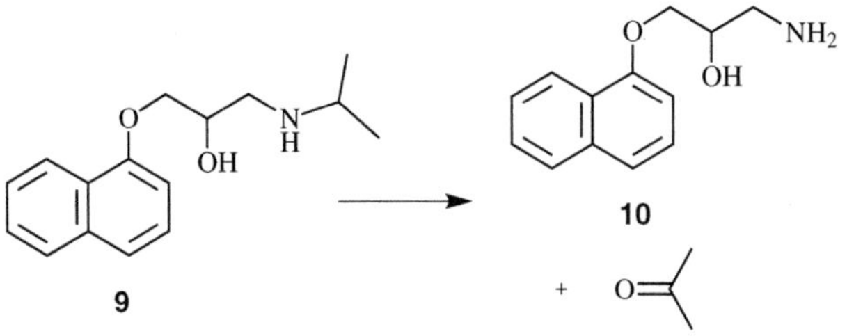

9 → **10** + acetone

14. Vitamin B_6 (pyridoxine; **11**) is converted to pyridoxal 5′-phosphate by a two-enzyme sequence, one that phosphorylates the 5′-hydroxyl group with ATP and one that oxidizes the 4′-hydroxyl group with NAD^+. Draw mechanisms for the biosynthesis of PLP.

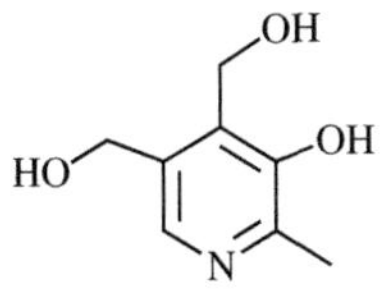

11

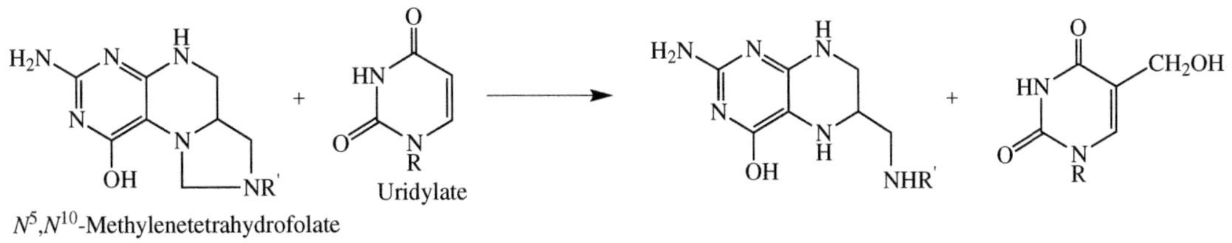

N^5,N^{10}-Methylenetetrahydrofolate + Uridylate →

15. The formyl group (CHO) of *N*-formylmethionine (**13**) can be added to methionine (**12**) by incubation of **12** with three enzymes plus another amino acid and coenzymes. You need **13** with a ^{14}C label at the formyl carbon atom, which can be done biosynthetically starting from a ^{14}C-labeled amino acid.

SCH_3 H_2N COOH **12** —3, Enzymes→ SCH_3 O H 14 N H COOH **13**

Show the starting amino acid with the ^{14}C label, the radioactive products of the first two enzymatic reactions (**13** is the product of the third enzyme), and any coenzymes involved in each enzyme reaction (indicate which reaction the coenzyme belongs to).

16. Draw a mechanism for the aminotransferase-catalyzed reaction described in Scheme 4.44.

REFERENCES

1. (a) Joyce, G. F. Nucleic acid enzymes: playing with a fuller deck. *Proc. Natl. Acad. Sci. U.S.A.* **1998**, *95*, 5845–5847. (b) Wilcox, J. L.; Ahluwalia, A. K.; Bevilacqua, P. C. Charged nucleobases and their potential for RNA catalysis. *Acc. Chem. Res.* **2011**, *44*(12), 1270–1279. (c) Mulhbacher, J.; St-Pierre, P.; Lafontaine, D. A. Therapeutic applications of ribozymes and riboswitches. *Curr. Opin. Pharmacol.* **2010**, *10*(5), 551–556. (d) Mastroyiannopoulos, N. P.; Uney, J. B.; Phylactou, L. A. The application of ribozymes and DNAzymes in muscle and brain. *Molecules.* **2010**, *15*, 5460–5472.
2. Sumner, J. B. The isolation and crystallization of the enzyme ureas. Preliminary paper. *J. Biol. Chem.* **1926**, *69*, 435–441.
3. (a) Jabri, E.; Carr, M. B.; Hausinger, R. P.; Karplus, P. A. The crystal structure of urease from Klebsiella aerogenes. *Science.* **1995**, *268*, 998–1004. (b) Jabri, E.; Karplus, P. A. Structures of the Klebsiella aerogenes urease apoenzyme and two active-site mutants. *Biochemistry* **1996**, *35*, 10616–10626.
4. Hirs, C. H. W.; Moore, S.; Stein, W. H. Sequence of the amino acid residues in performic acid-oxidized ribonuclease. *J. Biol. Chem.* **1960**, *235*, 633–647.
5. Radzicka, A.; Wolfenden, R. A proficient enzyme. *Science.* **1995**, *267*, 90–93.
6. Schramm, V. L. Enzymatic transition states, transition-state analogs, dynamics, thermodynamics, and lifetimes. *Annu. Rev. Biochem.* **2011**, *80*, 703–732.
7. Page, M. I. In *Enzyme Mechanisms*, Page, M. I.; Williams, A. (Eds.), Royal Society of Chemistry, London, 1987, p. 1.
8. Brown, A. J. Enzyme action. *Trans. Chem. Soc. (Lond.)* **1902**, *81*, 373–388.
9. Henri, V. General theory of the effect of some enzymes. *Acad. Sci. Paris.* **1902**, *135*, 916–919.
10. Fischer, E. Influence of configuration on the action of enzymes. *Berichte* **1894**, *27*, 2985–2993.
11. (a) Koshland, D. E., Jr. Application of a theory of enzyme specificity to protein synthesis. *Proc. Natl. Acad. Sci. U.S.A.* **1958**, *44*, 98–104. (b) Koshland, D. E.; Neet, K. E. The catalytic and regulatory properties of enzymes. *Annu. Rev. Biochem.* **1968**, *37*, 359–410.
12. (a) Pauling, L. Molecular architecture and biological reactions. *Chem. Eng. News.* **1946**, *24*, 1375–1377. (b) Pauling, L. Chemical achievement and hope for the future. *Am Sci.* **1948**, *36*, 51–58.
13. Haldane, J. B. S. *Enzymes*, Longmans and Green, London, 1930 (reprinted by MIT Press: Cambridge, 1965).
14. Polanyi, M. Adsorption catalysis. *Z. Elektrochem.* **1921**, *27*, 142–150.
15. Eyring, H. Activated complex in chemical reactions. *J. Chem. Phys.* **1935**, *3*, 107–115.
16. Hackney, D. D. *The Enzymes*, 3rd ed., Sigman, D. S.; Boyer, P.D. (Eds.); Academic Press, San Diego, 1990; vol. 19, pp. 1–37.
17. Schowen, R. L. In *Transition States of Biochemical Processes*, Gandour, R. D.; Schowen, R. L. (Eds.), Plenum, New York, 1978, p. 77.
18. Richards, F. M. Areas, volumes, packing and protein structure. *Annu. Rev. Biophys. Bioeng.* **1977**, *6*, 151–176.
19. Miller, B. G.; Wolfenden, R. Catalytic proficiency: the unusual case of OMP decarboxylase. *Annu. Rev. Biochem.* **2002**, *71*, 847–885.
20. (a) Guengerich, F. P. Cytochrome P450 and chemical toxicology. *Chem. Res. Toxicol.* **2008**, *21*(1), 70–83. (b) Neve, E. P. A.; Ingelman-Sundberg, M. Cytochrome P450 proteins: retention and distribution from the endoplasmic reticulum. *Curr. Opin. Drug Disc. Develop.* **2010**, *13*(1), 78–85.
21. Jencks, W. P. Binding energy, specificity, and enzymic catalysis: the circe effect. *Adv. Enzymol.* **1975**, *43*, 219–410.
22. Wolfenden, R. Analog approaches to the structure of the transition state in enzyme reactions. *Acc. Chem. Res.* **1972**, *5*, 10–18.
23. Eigen, M.; Hammes, G. G. Elementary steps in enzyme reactions (as studied by relaxation spectrometry). *Adv. Enzymol.* **1963**, *25*, 1–38.
24. Jencks, W. P. *Catalysis in Chemistry and Enzymology*. McGraw-Hill, New York, 1969, Chaps. 1, 2, 3, and 5.
25. Jencks, W. P. Binding energy, specificity, and enzymic catalysis: the circe effect. *Adv. Enzymol.* **1975**, *43*, 219–410.
26. Wolfenden, R.; Frick, L. In *Enzyme Mechanisms* (Page, M. I.; Williams, A., (Ed.)) Roy. Soc. Chem., London, 1987, p. 97.
27. Bruice, T. C.; Pandit, U. K. The effect of geminal substitution, ring size, and rotamer distribution on the intramolecular nucleophilic catalysis of the hydrolysis of monophenyl esters of dibasic acids and the solvolysis of the intermediate anhydrides. *J. Am. Chem. Soc.* **1960**, *82*, 5858–5865.
28. Bruice, T. C.; Pandit, U. K. Intramolecular models depicting the kinetic importance of fit in enzymic catalysis. *Proc. Natl. Acad. Sci. U.S.A.* **1960**, *46*, 402–404.
29. (a) Lightstone, F. C.; Bruice, T. C. Ground state conformations and entropic and enthalpic factors in the efficiency of intramolecular and enzymic reactions. 1. Cyclic anhydride formation by substituted glutarates, succinate, and 3,6-Endoxo-Δ 4-tetrahydrophthalate monophenyl esters. *J. Am. Chem. Soc.* **1996**, *118*, 2595–2605. (b) Bruice, T. C.; Lightstone, F. C. Ground state and transition state contributions to the rate of intramolecular and enzymic reactions. *Acc. Chem. Res.* **1999**, *32*, 127–136.
30. Kirby, A. J. Effective molarities for intramolecular reactions. *Adv. Phys. Org. Chem.* **1980**, *17*, 183–278.
31. Kenyon, G. L. Covalent nucleophilic catalysis. *Encyclopedia of Life Sciences* [Online]. http://dx.doi.org/10.1038/npg.els.0000603. Published Online, Apr. 19, **2001**.

32. March, J. *Advanced Organic Chemistry*, 3rd ed., Wiley, New York, 1985, p. 268.
33. (a) Fersht, A. *Enzyme Structure and Mechanism*, 2nd ed., W. H. Freeman, New York, 1985, pp. 405–426. (b) Polgár, L. *Mechanisms of Protease Action*; CRC Press, Boca Raton, 1989.
34. Maryanoff, B. E. Inhibitors of serine proteases as potential therapeutic agents. *J. Med. Chem.* **2004**, *47*, 769–787.
35. Tschan, S.; Mordmueller, B.; Kun, J. F. J. Threonine peptidases as drug targets against malaria. *Exp. Opin. Ther. Targ.* **2011**, *15*, 365–378.
36. Coux, O.; Tanaka, K.; Goldberg, A. L. Structure and functions of the 20S and 26S proteasomes. *Annu. Rev. Biochem.* **1996**, *65*, 801–847.
37. (a) Chorna, V. I.; Lyanna, O. L. Inhibitors of lysosomal cysteine proteases. *Biopolymers Cell.* **2011**, *27*(3), 181–192. (b) Toh, E. C. Y.; Huq, N. L.; Dashper, S. G.; Reynolds, E. C. Cysteine protease inhibitors: from evolutionary relationships to modern chemotherapeutic design for the treatment of infectious diseases. *Curr. Prot. Pept. Sci.* **2010**, *11*(8), 725–743.
38. Zhenodarova, S. M. Small molecule caspase inhibitors. *Russ. Chem. Rev.* **2010**, *79*(2), 119–143.
39. Blow, D. M.; Birktoft, J.; Hartley, B. S. Role of a buried acid group in the mechanism of action of chymotrypsin. *Nature.* **1969**, *221*, 337–340.
40. Kraut, J. Serine proteases: structure and mechanism of catalysis. *Annu. Rev. Biochem.***1977**, *46*, 331–358.
41. Bryan, P.; Pantoliano, M. W.; Quill, S. G.; Hsaio, H.-Y.; Poulos, T. L. Site-directed mutagenesis and the role of the oxyanion hole in subtilisin. *Proc. Natl. Acad. Sci. U.S.A.* **1986**, *83*, 3743–3745.
42. Carter, P.; Wells, J. A. Dissecting the catalytic triad of a serine protease. *Nature.* **1988**, *332*, 564–568.
43. Dewar, M. J. S.; Storch, D. M. Alternative view of enzyme reactions. *Proc. Natl. Acad. Sci. U.S.A.* **1985**, *82*, 2225–2229.
44. Warshel, A.; Aqvist, J.; Creighton, S. Enzymes work by solvation substitution rather than by desolvation. *Proc. Natl. Acad. Sci. U.S.A.* **1989**, *86*, 5820–5824.
45. Covitz, T.; Westheimer, F. H. The hydrolysis of methyl ethylene phosphate: steric hindrance in general base catalysis. *J. Am. Chem. Soc.* **1963**, *85*, 1773–1777.
46. (a) Koshland, D. E., Jr. Application of a theory of enzyme specificity to protein synthesis. *Proc. Natl. Acad. Sci. U.S.A.* **1958**, *44*, 98–105. (b) Koshland, D. E., Jr. Biological specificity in protein small molecule interactions. *Biochem. Pharmacol.* **1961**, *8*, 57. (c) Koshland, D. E., Jr.; Neet, K. E. The catalytic and regulatory properties of enzymes. *Annu. Rev. Biochem.* **1968**, *37*, 359–410.
47. Jencks, W. P. Economics of enzyme catalysis. *Cold Spring Harbor Symp. Quant. Biol.* **1987**, *52*, 65–73.
48. Waxman, D. J.; Strominger, J. L. Penicillin-binding proteins and the mechanism of action of β-lactam antibiotics. *Annu. Rev. Biochem.* **1983**, *52*, 825–869.
49. (a) Frère, J.-M. *Streptomyces* R61 D-Ala-D-Ala carboxypeptidase. In *Handbook of Proteolytic Enzymes*, 2nd ed., Barrett A. J.; Rawlings N. D.; Woessner J. F. (Eds.), Elsevier, 2004, p. 1959. (b) Rhazi, N.; Delmarcelle, M.; Sauvage, E; Jacquemotte, F.; Devriendt, K; Tallon, V.; Ghosez, L.; Frère, J.-M. Specificity and reversibility of the transpeptidation reaction catalyzed by the *Streptomyces* R61 D-Ala-D-Ala peptidase. *Protein Sci.* **2005**, *14*, 2922–2928.
50. Marcus, R.; Coulston, A. M. In *Goodman and Gilman's The Pharmacological Basis of Therapeutics*, 7th ed., Gilman, A. G.; Goodman, L. S.; Rall, T. W.; Murad, F. (Eds.), Macmillan, New York, 1985; p. 1551.
51. (a) Toney, M. D. Controlling reaction specificity in pyridoxal phosphate enzymes. *Biochim. Biophys. Acta.* **2011**, *1814*(11), 1407–1418. (b) Toney, M. D. Pyridoxal phosphate enzymology. *Biochim. Biophys. Acta.* **2011**, *1814*(11), 1405–1406. (c) Toney, M. D. PLP-Dependent Enzymes, Chemistry of. *Wiley Encyclopedia of Chemical Biology.* 2009, *3*, 731–735. (d) Mozzarelli, A.; Bettati, S. Exploring the pyridoxal 5′-phosphate-dependent enzymes. *Chem. Rec.* **2006**, *6*(5), 275–287.
52. Martell, A. E. Vitamin B_6-catalyzed reactions of α-amino and α-keto acids: model systems. *Acc. Chem. Res.* **1989**, *22*, 115–124.
53. Leussing, D. L. In *Vitamin B_6 Pyridoxal Phosphate*, Dolphin, D.; Poulson, R.; Avramovic, O. (Eds.), Wiley, New York, 1986; Part A, p. 69.
54. Dunathan, H. C. Stereochemical aspects of pyridoxal phosphate catalysis. *Adv. Enzymol.* **1971**, *35*, 79–134.
55. John, R. A. Pyridoxal phosphate-dependent enzymes. *Biochim. Biophys. Acta.* **1995**, *1248*, 81–96.
56. Amadasi, A.; Bertoldi, M.; Contestabile, R.; Bettati, S.; Cellini, B.; di Salvo, M. L.; Borri-Voltattorni, C.; Bossa, F.; Mozzarelli, A. Pyridoxal 5′-phosphate enzymes as targets for therapeutic agents. *Curr. Med. Chem.* **2007**, *14*(12), 1291–1324.
57. Walsh, C. T. Enzymes in the D-alanine branch of bacterial cell wall peptidoglycan assembly. *J. Biol. Chem.* **1989**, *264*, 2393–2396.
58. (a) Hashimoto, A.; Nishikawa, T.; Oka, T.; Takahashi, K. Endogenous D-serine in rat brain: N-methyl-D-aspartate receptor-related distribution and aging. *J. Neurochem.* **1993**, *60*, 783–786. (b) Hashimoto, A.; Oka, T. Free D-aspartate and D-serine in the mammalian brain and periphery. *Prog. Neurobiol.* **1997**, *52*, 325–353. (c) Ohide, H.; Miyoshi, Y.; Maruyama, R.; Hamase, K.; Konno, R. D-Amino acid metabolism in mammals: Biosynthesis, degradation and analytical aspects of the metabolic study *J. Chromatog. B* **2011**, *879*(29), 3162–3168. (d) Jiraskova-Vanickova, J.; Ettrich, R.; Vorlova, B.; Hoffman, H. E.; Lepsik, M.; Jansa, P.; Konvalinka, J. Inhibition of human serine racemase, an emerging target for medicinal chemistry. *Curr. Drug Targets.* **2011**, *12*(7), 1037–1055.
59. Wolosker, H.; Blackshaw, S.; Snyder, S. H. Serine racemase: a glial enzyme synthesizing D-serine to regulate glutamate-N-methyl-D-aspartate neurotransmission. *Proc. Natl. Acad. Sci. U.S.A.* **1999**, *96*, 13409–13414.
60. Yoshimura, T.; Esaki, N. Amino acid racemases: functions and mechanisms. *J. Biosci. Bioeng.* **2003**, *96*(2), 103–109.
61. Soda, K.; Tanaka, H.; Tanizawa, K. In *Vitamin B_6 Pyridoxal Phosphate*, Dolphin, D.; Poulson, R.; Avramovic, O. (Eds.), Wiley, New York, 1986, part B, p. 223.
62. Watanabe, A.; Yoshimura, T.; Mikami, B.; Hayashi, H.; Kagamiyama, H.; Esaki, N. Reaction mechanism of alanine racemase from Bacillus stearothermophilus: X-ray crystallographic studies of the enzyme bound with N-(5′-phosphopyridoxyl)alanine. *J. Biol. Chem.* **2002**, *277*, 19166–19172.
63. Sukhareva, B. S. In *Vitamin B_6 Pyridoxal Phosphate*, Dolphin, D.; Poulson, R.; Avramovic, O. (Eds.), Wiley, New York, 1986; Part B, p. 325.
64. Allen, G. F. G.; Land, J. M.; Heales, S. J. R. A new perspective on the treatment of aromatic L-amino acid decarboxylase deficiency. *Mol. Genet. Metab.* **2009**, *97*(1), 6–14.
65. Houtkooper, R. H.; Canto, C.; Wanders, R. J.; Auwerx, J. The Secret Life of NAD+: An old metabolite controlling new metabolic signaling pathways. *Endocrine Rev.* **2010**, *31*(2), 194–223.

66. Bertino, J. R.; Booth, B. A.; Bieber, A. L.; Cashmore, A.; Sartorelli, A. C. The inhibition of dihydrofolate reductase by the folate antagonists. *J. Biol. Chem.* **1964**, *239*, 479–485.
67. Westheimer, F. H. In *Pyridine Nucleotide Coenzymes*, Dolphin, D.; Poulson, R.; Avramovic, O. (Eds.), Wiley: New York, 1987; Part A, p. 253.
68. (a) Schirch, L. G. In *Folates and Pterins*; Blakely, R. L.; Benkovic, S. J. (Eds.), Wiley: New York, 1984, Vol. 1, p. 399. (b) Matthews, R. G.; Drummond, J. T. Providing one-carbon units for biological methylations: mechanistic studies on serine hydroxymethyltransferase, methylenetetrahydrofolate reductase, and methyltetrahydrofolate-homocysteine methyltransferase. *Chem. Rev.* **1990**, *90*, 1275–1290. (c) Florio, R.; di Salvo, M. L.; Vivoli, M.; Contestabile, R. Serine hydroxymethyltransferase: A model enzyme for mechanistic, structural, and evolutionary studies. *Biochim. Biophys. Acta.* **2011**, *1814*(11), 1489–1496.
69. (a) Benkovic, S. J.; Bullard, W. P. In *Progress in Bioorganic Chemistry*, Kaiser, E. T.; Kedzy, F. (Eds.), Wiley, New York; 1973, vol. 2, p. 133. (b) Jordan, P. M.; Akhtar, M. Mechanism of action of serine transhydroxymethylase. *Biochem. J.* **1970**, *116*, 277–286.
70. Kallen, R. G.; Jencks, W. P. Mechanism of the condensation of formaldehyde with tetrahydrofolic acid. *J. Biol. Chem.* **1966**, *241*, 5845–5863.
71. Kallen, R. G.; Jencks, W. P. Mechanism of the condensation of formaldehyde with tetrahydrofolic acid. *J. Biol. Chem.* **1966**, *241*, 5851–5863.
72. Ballou, D. P.; Yamada, K.; Pejchal, R.; Ludwig, M. L.; Matthews, R. G.; Trimmer, E. E. In *Flavins and Flavoproteins 2002, Proceedings of the International Symposium, 14th, Cambridge, United Kingdom, July 14–18, 2002* (2002), pp. 13–22, Chapman, S. K.; Perham, R. N.; Scrutton, N. S. (Eds.), *Studies of the mechanisms of bacterial and mammalian methylenetetrahydrofolate reductases*.
73. (a) Daubner, S. C.; Matthews, R. G. Purification and properties of methylenetetrahydrofolate reductase from pig liver. *J. Biol. Chem.* **1982**, *257*, 140–145. (b) Matthews, R. G.; Drummond, J. T. Providing one-carbon units for biological methylations: mechanistic studies on serine hydroxymethyltransferase, methylenetetrahydrofolate reductase, and methyltetrahydrofolate-homocysteine methyltransferase. *Chem. Rev.* **1990**, *90*, 1275–1290.
74. (a) Wolan, D. W.; Greasley, S. E.; Beardsley, G. P.; Wilson, I. A. Structural insights into the avian AICAR transformylase mechanism. *Biochemistry.* **2002**, *41*, 15505–15513. (b)Baggott, J. E.; Tamura, T. Evidence for the hypothesis that 10-formyldihydrofolate is the in vivo substrate for aminoimidazolecarboxamide ribotide transformylase *Exp. Biol. Med. (Lond., United Kingdom)*. **2010**. *235*(3), 271–277.
75. Carreras, C. W.; Santi, D. V. The catalytic mechanism and structure of thymidylate synthase. *Annu. Rev. Biochem.* **1995**, *64*, 721–762.
76. (a) González, J. C.; Peariso, K.; Penner-Hahn, J. E.; Matthews, R. G. Cobalamin-Independent Methionine Synthase from *Escherichia coli*: A Zinc Metalloenzyme. *Biochemistry* **1996**, *35*, 12228–12234. (b) Matthews, R. G.; Goulding, C. W. Enzyme-catalyzed methyl transfers to thiols: the role of zinc. *Curr. Opin. Chem. Biol.* **1997**, *1*, 332–339.
77. Martinez-Lopez, N.; Varela-Rey, M.; Ariz, U.; Embade, N.; Vazquez-Chantada, M.; Fernandez-Ramos, D.; Gomez-Santos, L.; Lu, S. C.; Mato, J. M.; Martinez-Chantar, M. L. S-adenosylmethionine and proliferation: new pathways, new targets. *Biochem. Soc. Trans.* **2008**, *36*(5), 848–852.
78. (a) Walsh, C. Flavin coenzymes: at the crossroads of biological redox chemistry. *Acc. Chem. Res.* **1980**, *13*, 148–155. (b) Bruice, T. C. Mechanisms of flavin catalysis. *Acc. Chem. Res.* **1980**, *13*, 256–262. (c) Ghisla, S.; Massey, V. Mechanisms of flavoprotein-catalyzed reactions. *Eur. J. Biochem.* **1989**, *181*, 1–17. (c) Macheroux, P.; Kappes, B.; Ealick, S. E. Flavogenomics—a genomic and structural view of flavin-dependent proteins. *FEBS J.* **2011**, *278*(15), 2625–2634. (d) Joosten, V.; Van Berkel, W. jh. Flavoenzymes. *Curr. Opin. Chem. Biol.* **2007**, *11*(2), 195–202.
79. (a) Chlumsky, L. J.; Sturgess, A. W.; Nieves, E.; Jorns, M. S. Identification of the covalent flavin attachment site in sarcosine oxidase. *Biochemistry* **1998**, *37*, 2089–2095. (b)Singer, T. P.; Edmondson, D. E. Structure, properties, and determination of covalently bound flavins. *Methods Enzymol.* **1980**, *66*, 253–264. (c) Edmondson, D. E.; Kenney, W. C.; Singer, T. P. Synthesis and isolation of 8α-substituted flavins and flavin peptides. *Methods Enzymol.* **1978**, *53*, 449–465.
80. (a) Kearney, E. B.; Salach, J. I.; Walker, W. H.; Seng, R. L.; Kenney, W.; Zeszotek, E.; Singer, T. P. Covalently-bound flavin of hepatic monoamine oxidase. I. Isolation and sequence of a flavin peptide and evidence for binding at the 8α position. *Eur. J. Biochem.* **1971**, *24*, 321. (b) Steenkamp, D. J.; Denney, W. C.; Singer, T. P. A novel type of covalently bound coenzyme in trimethylamine dehydrogenase. *J. Biol. Chem.* **1978**, *253*, 2812–2817.
81. Massey, V.; Müller, F.; Feldberg, R.; Schuman, M.; Sullivan, P. A.; Howell, L. G.; Mayhew, S. G.; Matthews, R. G.; Foust, G. P. Reactivity of flavoproteins with sulfite. Possible relevance to the problem of oxygen reactivity. *J. Biol. Chem.* **1969**, *244*, 3999–4006.
82. Bruice, T. C. Oxygen-flavin chemistry. *Israel J. Chem.* **1984**, *24*, 54–61.
83. (a) Neims, A. H.; DeLuca, D. C.; Hellerman, L. Crystalline D-amino acid oxidase. III. Substrate specificity and σ–ρ relations. *Biochemistry* **1966**, *5*, 203–213. (b) Walsh, C. T.; Schonbrunn, A.; Abeles, R. H. Mechanism of action of D-amino acid oxidase. Evidence for removal of substrate α-hydrogen as a proton. *J. Biol. Chem.* **1971**, *246*, 6855–6866. (c) Todone, F.; Vanoni, M. A.; Mozzarelli, A.; Bolognesi, M.; Coda, A.; Curti, B.; Mattevi, A. Active site plasticity in D-amino acid oxidase: A crystallographic analysis. *Biochemistry* **1997**, *36*, 5853–5860. (d) Fitzpatrick, P. F. Oxidation of amines by flavoproteins. *Arch. Biochem. Biophys.* **2010**, *493*(1), 13–25.
84. (a) Lenn, N. D.; Shih, Y.; Stankovich, M. T.; Liu, H.-W. Studies of the inactivation of general acyl-CoA dehydrogenase by racemic (methylenecyclopropane)acetyl-CoA-new evidence suggesting a radical mechanism of this enzyme catalyzed reaction. *J. Am. Chem. Soc.* **1989**, *111*, 3065–3067. (b) Lai, M.-T.; Liu, L.-D.; Liu, H.-W. Mechanistic study on the inactivation of general acyl-CoA dehydrogenase by a metabolite of hypoglycin A. *J. Am. Chem. Soc.* **1991**, *113*, 7388–7397. (c) Lai, M.-T.; Oh, E.; Liu, H.-W. Inactivation of medium-chain acyl-CoA dehydrogenase by a metabolite of hypoglycin: characterization of the major turnover product and evidence suggesting an alternative flavin modification pathway. *J. Am. Chem. Soc.* **1993**, *115*, 1619–1628.
85. (a) Silverman, R. B. In *Advances in Electron Transfer Chemistry*, vol. 2, Mariano, P. S. (Ed.), JAI Press, Greenwich, CT, 1992, pp. 177–213. (b) Silverman, R. B. Radical ideas about monoamine oxidase. *Acc. Chem. Res.* **1995**, *28*, 335–342. (c) Lu, X.; Rodriguez, M.; Ji, H.; Silverman, R. B.; Vintém, A. P. B.; Ramsay, R. R. In *Flavins and Flavoproteins 2002: Proceedings of the 14th International Symposium*, Chapman, S. K.; Perham, R. N.; Scrutton, N. S. (Eds.), Rudolf Weber, Berlin, Germany, 2002; pp. 817–830.

86. Benson, T. E.; Marquardt, J. L.; Marquardt, A. C.; Etzkorn, F. A.; Walsh, C. T. Overexpression, purification, and mechanistic study of UDP-N-acetylenolpyruvylglucosamine reductase. *Biochemistry* **1993**, *32*, 2024–2030.
87. (a) Ellis, H. R. The FMN-dependent two-component monooxygenase systems. *Arch. Biochem. Biophys.* **2010**, *497*(1–2), 1–12. (b) Palfey, B. A.; McDonald, C. A. Control of catalysis in flavin-dependent monooxygenases. *Arch. Biochem. Biophys.* **2010**, *493*(1), 26–36. (c) Cashman, J. R. Role of flavin-containing monooxygenase in drug development. *Exp. Opin. Drug Metab. Toxicol.* **2008**, *4*(12), 1507–1521.
88. Ghisla, S.; Massey, V. Mechanisms of flavoprotein-catalyzed reactions. *Eur. J. Biochem.* **1989**, *181*, 1–17.
89. Massey, V. Activation of molecular oxygen by flavins and flavoproteins. *J. Biol. Chem.* **1994**, *269*, 22459–22462.
90. (a) Guengerich, F. P. (Ed.) *Mammalian Cytochromes P-450*, CRC Press, Boca Raton, FL, 1987; vol. I and II. (b) Ortiz de Montellano, P. R. (Ed.), *Cytochrome P450: Structure, Mechanism, and Biochemistry*, 3rd ed., Kluwer Academic/Plenum, New York, 2005.
91. Akhtar, M.; Wright, J. N. A unified mechanistic view of oxidative reactions catalyzed by P-450 and related iron-containing enzymes. *Nat. Prod. Rep.* **1991**, *8*, 527–551.
92. Rittle, J.; Geen, M. T. Cytochrome P450 compound I: capture, characterization, and C–H bond activation kinetics. *Science* **2010**, *330*, 933–937.
93. (a) Kim, J.-J. P.; Roberts, D. L.; Djordjevic, S.; Wang, M.; Shea, T. M.; Masters, B. S. S. Crystallization studies of NADPH-cytochrome P450 reductase. *Methods Enzymol.* **1996**, *272*, 368–377. (b) Strobel, H. W.; Dignam, J. D.; Gum, J. R. NADPH cytochrome P-450 reductase and its role in the mixed function oxidase reaction. *Pharmacol. Ther.* **1980**, *8*, 525–537. (c) Laursen, T.; Jensen, K.; Moller, B. L. Conformational changes of the NADPH-dependent cytochrome P 450 reductase in the course of electron transfer to cytochromes P 450. *Biochim. Biophys. Acta.* **2011**, *1814*(1), 132–138. (d) Jensen, K.; Moller, B. L. Plant NADPH-cytochrome P450 oxidoreductases. *Phytochemistry* **2010**, *71*(2–3), 132–141.
94. Newcomb, M.; Le Tadic-Biadatti, M.-H.; Chestney, D. L.; Roberts, E. S.; Hollenberg, P. F. A nonsynchronous concerted mechanism for cytochrome P-450 catalyzed hydroxylation. *J. Am. Chem. Soc.* **1995**, *117*, 12085–12091.
95. (a) Ostovic, D.; Bruice, T. C. Intermediates in the epoxidation of alkenes by cytochrome P-450 models. 5. Epoxidation of alkenes catalyzed by a sterically hindered (meso-tetrakis(2,6-dibromophenyl) porphinato)iron(III) chloride. *J. Am. Chem. Soc.* **1989**, *111*, 6511–6517. (b) Ostovic, D.; Bruice, T. C. Mechanism of alkene epoxidation by iron, chromium, and manganese higher valent oxo-metalloporphyrins. *Acc. Chem. Res.* **1992**, *25*, 314–320.
96. Richard, J. P.; Williams, G.; O'Donoghue, A. C.; Amyes, T. L. Formation and stability of enolates of acetamide and acetate anion: an Eigen plot for proton transfer at α-carbonyl carbon. *J. Am. Chem. Soc.* **2002**, *124*, 2957–2968.
97. Yuen, P. W.; Kanter, G. D.; Taylor, C. P.; Vartanian, M. G. Enantioselective synthesis of PD144723: a potent stereospecific anticonvulsant. *Bioorg. Med. Chem. Lett.* **1994**, *4*(6), 823–826.
98. Hoekstra, M. S.; Sobieray, D. M.; Schwindt, M. A.; Mulhern, T. A.; Grote, T. M.; Huckabee, B. K.; Hendrickson, V. S.; Franklin, L. C.; Granger, E. J.; Karrick, G. L. Chemical development of C;-1008, an enantiomerically pure anticonvulsant. *Org. Process Res. Devel.* **1997**, *1*, 26–38.
99. Burk, M. J.; de Koning, P. D.; Grote, T. M.; Hoekstra, M. S.; Hoge, G.; Jennings, R. A.; Kissel, W. S.; Le, T. V.; Lennon, I. C.; Mulhern, T. A.; Ramsden, J. A.; Wade, R. A. An enantioselective synthesis of (S)-(+)-3-aminomethyl-5-methylhexanoic acid via asymmetric hydrogenation. *J. Org. Chem.* **2003**, *68*, 5731–5734.
100. Martinez, C. A.; Hu, S.; Dumond, Y.; Tao, J.; Kelleher, P.; Tully, L. Development of a chemoenzymatic manufacturing process for pregabalin. *Org. Process Res. Devel.* **2008**, *12*, 392–398.
101. (a) Patel, R. N. Synthesis of chiral pharmaceutical intermediates by biocatalysis. *Coord. Chem. Rev.* **2008**, *252*, 659–701. (b) Patel, R. N. Enzymatic synthesis of chiral drug intermediates. *Encyclopedia of Industrial Biotechnology: Bioprocess, Bioseparation, and Cell Technology*, Flickinger, M. C. (Ed.), John Wiley & Sons, Hoboken, NJ, 2010, pp. 1–18.
102. Savile, C. K.; Janey, J. M.; Mundorff, E. C.; Moore, J. C.; Tan, S.; Jarvis, W. R.; Colbeck, J. C.; Krebber, A.; Fleitz, F. J.; Brands, J.; Devine, P. N.; Huisman, G. W.; Hughes, G. J. Biocatalytic asymmetric synthesis of chiral amines from ketones applied to sitagliptin manufacture. *Science (Washington, DC)* **2010**, *329*, 305–309.
103. Cavasotto, C. N.; Phatak, S. S. Homology modeling in drug discovery: current trends and applications. *Drug Disc. Today.* **2009**, *14*, 676–683.
104. (a) Dalby, P. A. Strategy and success for the directed evolution of enzymes. *Curr. Opin. Struct. Biol.* **2011**, *21*, 473–480. (b) Turner, N. J. Directed evolution drives the next generation of biocatalysts. *Nature Chem. Biol.* **2009**, *5*, 567–573.
105. Haber, E.; Quertermous, T.; Matsueda, G. R.; Runge, M. S. Innovative approaches to plasminogen activator therapy. *Science* **1989**, *243*, 51–56.
106. Lavin, A.; Sung, C.; Klibanov, A. M.; Langer, R. Enzymic removal of bilirubin from blood: a potential treatment for neonatal jaundice. *Science* **1985**, *230*, 543–545.
107. Crunkhorn, S. Enzyme replacement success in Phase III trial for rare metabolic disorder. *Nat. Rev. Drug Disc.* **2013**, *12*, 12.

Chapter 5

Enzyme Inhibition and Inactivation

Chapter Outline

The Organic Chemistry of Drug Design and Drug Action. http://dx.doi.org/10.1016/B978-0-12-382030-3.00005-2

5.1. WHY INHIBIT AN ENZYME?

Many diseases, or at least the symptoms of diseases, arise from a deficiency or excess of a specific metabolite in the body, from an infestation of a foreign organism, or from aberrant cell growth. If the metabolite deficiency or excess can be normalized, and if the foreign organisms and aberrant cells can be destroyed, then these disease states will be remedied. Many of these problems can be addressed by specific enzyme inhibition.

Any compound that slows down or blocks enzyme catalysis is an *enzyme inhibitor*. If the interaction with the *target enzyme* (specific enzyme for inhibition) is irreversible (usually covalent), then the compound is a special type of enzyme inhibitor referred to as an *enzyme inactivator*. Of the 149 new small molecule and biological drugs that went on the world drug market in the 6 years 2006–2011, one-third (51/149) were enzyme inhibitors; if only small molecules (including peptides) are considered, then 41% (51/123) were enzyme inhibitors.[1] Of the 51 small molecule new drugs that are enzyme inhibitors, 12 inhibit kinase enzymes (10 of which are cancer treatments), 14 inhibit protease enzymes (six for cardiovascular disease, five for diabetes, and three for viral infection), and 11 inhibit enzymes involved with deoxyribonucleic acid (DNA) or ribonucleic acid (RNA) synthesis or function (eight for viral or bacterial infection and three for cancer or proliferative syndrome treatments). Representative inhibitors of these enzyme classes are shown in Table 5.1.

Consider what happens when an enzyme activity is blocked. The substrates for that enzyme cannot be metabolized, and the metabolic products are not generated (that is, unless there is another enzyme that can metabolize the substrate, and unless there is another metabolic pathway that generates the same product). Why should these two outcomes be important to drug design? If a cell has a deficiency of the substrate for the target enzyme and as a result of that deficiency, a disease state results, then inhibition of that enzyme would prevent the degradation of the substrate, thereby increasing its concentration. An example of this is the onset of seizures that arises from diminished γ-aminobutyric acid (GABA) levels in the brain. Inhibition of the enzyme that degrades GABA, namely, GABA aminotransferase (GABA-AT), leads to an anticonvulsant effect. If there is an excess of a particular metabolite that produces a disease state, then inhibition of an enzyme on the pathway for biosynthesis of that metabolite should diminish its concentration. Excess uric acid can lead to gout. Inhibition of xanthine oxidase, the enzyme that catalyzes the conversion of xanthine to uric acid, decreases the uric acid levels, and results in an antihyperuricemic effect. If the product of an enzyme reaction is required to carry out an important physiological function that the drug is supposed to block, then inhibition of that enzyme decreases the concentration of that product and can interfere with the physiological effect. Prostaglandins (PGs) are important hormones that are involved in the pathogenesis of inflammation and fever. Inhibition of PG synthase results in antiinflammatory, antipyretic, and analgesic effects. In the case of foreign organisms such as bacteria and parasites, or in the case of tumor cells, inhibition of one of their essential enzymes can prevent important metabolic processes from taking place, resulting in inhibition of growth or replication of the organism or aberrant cell. Inhibition of the bacterial alanine racemase, the enzyme that makes D-alanine for incorporation into peptidoglycan strands, for example, would prevent the biosynthesis of the peptidoglycan and, therefore, the biosynthesis of the bacterial cell wall. These compounds possess antibacterial activity. The use of drugs to combat foreign organisms and aberrant cells is called *chemotherapy*.

Enzyme inhibition is a promising approach for the rational discovery of new leads or drugs. Although there are numerous drugs that exert their therapeutic action by inhibiting specific enzymes, the mechanisms of action of some of these drugs were determined subsequent to the discovery of the therapeutic properties of the drugs. Target enzymes selected for rational drug design are those whose inhibition in vivo would lead to the desired therapeutic effect.

There are two general categories of target enzymes. In most cases, a potential drug is designed for an enzyme whose inhibition is known to produce a specific pharmacological effect, but existing inhibitors have certain undesirable properties such as lack of potency or specificity or exhibit side effects. A more daring approach is to design inhibitors of enzymes whose inhibition has not yet been established to lead to a desired therapeutic effect. This category of enzyme targets requires knowledge of the pathophysiology of disease processes and the ability to identify important metabolites whose function or dysfunction results in a disease state. While preliminary proof of concept may be obtained through such methods as decreasing enzyme production through inhibitory RNA or inhibition of the enzyme with an antibody, not until a small molecule inhibitor is obtained will it be possible to determine the real effect of inhibition of that enzyme on the metabolism of the organism under the conditions anticipated for the actual therapy. Once an enzyme target is identified, lead compounds must be prepared that can inhibit it completely and specifically.

Of all the protein targets for potential therapeutic use, including hormone and neurotransmitter receptors and transporter proteins, enzymes are the most promising for rational inhibitor design. Enzyme purification is generally a much simpler task than receptor purification; a homogeneous enzyme preparation can be obtained for preliminary screening purposes, and in some cases, may be used to elucidate the active site structure, which is useful for computer-based drug design approaches (see Chapter 2, Section 2.2.6). Furthermore, whereas effective receptor antagonists often bear little structural similarity to agonists, enzyme inhibitors are

TABLE 5.1 Representative Commercially Marketed Enzyme Inhibitors Launched between 2006 and 2010

Drug	Structure	Indication	Enzyme Inhibited	Enzyme Class
Dasatinib	5.1	Anticancer	bcr-abl (including mutants)	Kinase
Pazopanib	5.2	Anticancer	Vascular endothelial growth factor receptor kinase	Kinase
Aliskiren	5.3	Antihypertensive	Renin	Aspartyl protease
Saxagliptin	5.4	Antidiabetic	DPP-4	Serine protease
Rivaroxaban	5.5	Anticoagulant	Factor Xa	Serine protease
Besifloxacin	5.6	Antibacterial	DNA gyrase/ topoisomerase	Enzymes modulating DNA/ RNA synthesis/ function
Clevudine	5.7	Antiviral	DNA polymerase	Enzymes modulating DNA/ RNA synthesis/ function
Vorinostat	5.8	Anticancer	Histone deacetylase	Enzymes modulating DNA/ RNA synthesis/ function

frequently able to be designed starting from substrates or products of the target enzyme. In addition, knowledge of enzyme mechanisms can be used in the design of transition state analogs and multisubstrate inhibitors (Section 5.2.3), slow tight-binding inhibitors (Section 5.2.4), and mechanism-based enzyme inactivators (Section 5.3.3).

To minimize side effects, there are certain properties that ideal enzyme inhibitors and/or enzyme targets should possess. An ideal enzyme inhibitor should be totally specific for the one target enzyme. Because this is rare, if attained at all, highly selective inhibition is a more realistic objective. By adjustment of the dose administered, essentially specific inhibition may be possible. In some cases, such as infectious diseases, enzyme targets can be identified because of biochemical differences in essential metabolic pathways between foreign organisms and their hosts.[2] In other instances, there are substrate specificity differences between enzymes from the two sources that can be utilized in the design of selective enzyme inhibitor drugs. Unfortunately, when dealing with some organisms, and especially with tumor cells, the enzymes that are essential for their growth also are vital to human health. Inhibition of these enzymes can destroy human cells as well. Nonetheless, this approach is taken in various types of chemotherapy. The reason this approach can be effective is that foreign organisms and tumor cells replicate at a much faster rate than do most normal human cells (those in the gut, the bone marrow, and the mucosa are exceptions). Consequently, rapidly proliferating cells have an elevated requirement for essential metabolites. *Antimetabolites*, compounds whose structures are similar to those of essential metabolites and which disrupt metabolic processes (typically by blocking or acting as an alternative substrate for an enzyme in a metabolic pathway), are taken up by the rapidly replicating cells and, therefore, these cells are selectively inhibited. The *selective toxicity* in this case derives from a kinetic difference rather than a qualitative difference in the metabolism.

An ideal enzyme target in a foreign organism or aberrant cell would be one that is essential for its growth, but which is either nonessential for human health, or, even better (in the case of foreign organisms), not even present in humans. This type of selective toxicity would destroy only the foreign organism or aberrant cell, and would not require the careful administration of a drug that is necessary when the inhibited enzyme is important to human metabolism as well. The penicillins, for example, which inhibit the bacterial peptidoglycan transpeptidase essential for the biosynthesis of the bacterial cell wall, are nontoxic to humans, since human cells do not have cell walls (nor do viruses, so penicillins are nontoxic to them as well).

Enzyme inhibitors can be grouped into two general categories: reversible and irreversible inhibitors. As the name implies, inhibition of enzyme activity by a *reversible inhibitor* is reversible, suggesting that noncovalent interactions are involved. This is not strictly the case; there also can be reversible covalent interactions. An *irreversible enzyme inhibitor*, also called an *enzyme inactivator*, is one that prevents the return of enzyme activity for an extended period, suggesting the involvement of a covalent bond. This also is not strictly the case; it is possible for noncovalent interactions to be so effective that the enzyme–inhibitor complex is, for all intents and purposes, irreversibly formed.

5.2. REVERSIBLE ENZYME INHIBITORS

5.2.1. Mechanism of Reversible Inhibition

The most common enzyme inhibitor drugs are the reversible type, particularly ones that compete with the substrate for active site binding. These are known as *competitive reversible inhibitors*, typically compounds that have structures similar to those of the substrates or products of the target enzymes and which bind at the substrate binding sites, thereby blocking substrate binding. Typically, these inhibitors establish their binding equilibria with the enzyme rapidly, so that inhibition is observed as soon as the enzyme is assayed for activity, although there are some cases (see Sections 5.2.4, 5.3.2, and 5.3.3) in which inhibition can be relatively slow.

As in the case of the interaction of a substrate with an enzyme, an inhibitor (I) also can form a complex with an enzyme (E) (Scheme 5.1). The equilibrium constant K_i (k_{off}/k_{on}) is a *dissociation* constant for breakdown of the E·I complex; therefore, as discussed for the K_d of drug–receptor complexes (see Chapter 3, Section 3.2.1), the *smaller* the K_i value for an inhibitor I, the more potent the inhibitor. Another common measurement for inhibition, which is quicker to determine, but is more of an estimate, is the IC_{50}, the inhibitor concentration that produces 50% enzyme inhibition. An IC_{50} value is not a constant, but rather depends on the substrate concentration used in the assay. Therefore, the IC_{50} is useful for comparing the potency of inhibitors so long as all the inhibitors were tested in the presence of the same concentration of substrate. The IC_{50} value can be converted into an approximate K_i value by Eqn (5.1).[3]

$$IC_{50} = \left(1 + \frac{[S]}{K_m}\right) K_i \tag{5.1}$$

When the inhibitor binds at the active site (the substrate binding site), then it is a competitive inhibitor. Formation of the E·I complex prevents the binding of substrate and,

$$\begin{array}{ccccc} E + I & \underset{k_{off}}{\overset{k_{on}}{\rightleftharpoons}} & E{\bullet}I & & \\ {\scriptstyle -S}\big\uparrow\big\downarrow{\scriptstyle +S} & & & & \\ E{\bullet}S & \rightleftharpoons & E{\bullet}P & \rightleftharpoons & E + P \end{array}$$

SCHEME 5.1 Kinetic scheme for competitive enzyme inhibition therefore, blocks the catalytic conversion of the substrate to product. In some cases, the inhibitor may also act as a substrate, whereby it may be converted to a metabolically useless product. In the context of drug design, this is generally not a favorable process because the product formed may be toxic or may lead to other toxic metabolites. Nonetheless, there are drugs that function by this mechanism (see Section 5.2.2.3).

Interaction of the inhibitor with the enzyme can occur at a site other than the substrate-binding site (i.e., at an *allosteric binding site*) and still result in inhibition of substrate turnover. When this occurs, often as a result of an inhibitor-induced conformational change in the enzyme to give a form of the enzyme that does not bind the substrate properly, then the inhibitor is a *noncompetitive reversible inhibitor*. The discussion in this chapter will focus on the design and mechanism of action of competitive enzyme inhibitors.

The equilibrium shown in Scheme 5.1, and therefore the EI concentration, will depend on the concentrations of the inhibitor and the substrate, as well as the K_i for the inhibitor and the K_m for the substrate. As the concentration of the inhibitor is increased, it drives the equilibrium toward the E·I complex. Because the substrate and the competitive inhibitor bind to the enzyme at the same site, they both cannot interact with the enzyme simultaneously. When the inhibitor concentration diminishes, the E·I complex concentration diminishes, and the effect of the inhibitor can be overcome by the substrate.

If the enzyme inhibitor is a drug, the maximal pharmacological effect will occur when the drug concentration is maintained at a saturating level at the target enzyme active site. As the drug is metabolized (see Chapter 8), and the concentration of I diminishes, additional administration of the drug is required to maintain the prevalence of the E·I complex. Drugs that are rapidly cleared from the circulation by metabolism or other means may need to be taken several times a day. To increase the potency of reversible inhibitors, and thereby reduce the dosage of the drug, the binding interactions with the target enzyme must be optimized (that is, the inhibitor should have a low K_i value).

When a drug is designed to be an enzyme inhibitor, it frequently will be a competitive inhibitor because the lead compound often will be the substrate for the target enzyme. Because an enzyme is just a specific type of receptor, an analogy can be made between agonists, partial agonists, and antagonists with good substrates, poor substrates, and competitive inhibitors, respectively.

5.2.2. Selected Examples of Competitive Reversible Inhibitor Drugs

In this section, we will take a look at five different approaches to the design of competitive reversible inhibitors: simple competitive inhibition, stabilization of an inactive enzyme conformation, alternative substrate inhibition, transition state analog inhibition, and slow, tight binding inhibition.

5.2.2.1. Simple Competitive Inhibition

The most common type of inhibition is simple competitive inhibition, in which the inhibitor rapidly and reversibly binds to the active site of the enzyme. Frequently the structure of the inhibitor resembles that of the substrate or product of the target enzyme.

5.2.2.1.1. Epidermal Growth Factor Receptor Tyrosine Kinase as a Target for Cancer

For many years, various combinations of surgery, radiation, and chemotherapy have constituted mainstream treatments for cancer. Chemotherapy is generally designed to inhibit processes related to cell division, based on the premise that rapidly dividing cells, i.e., those in tumors, will be selectively affected by inhibitors of cell division (see discussions in Section 5.3.3.3.5 and in Chapter 6, Section 6.1.1). While this strategy can be effective, normal cells, and particular normal cells that divide relatively rapidly, such as hair cells and cells in the gastrointestinal tract, rarely escape the effects of the treatment, and side effects such as hair loss and gastrointestinal distress (for example, nausea) frequently occur.

Since the early 2000s, much emphasis has been placed on *targeted* therapy. In the context of cancer, this term implies the ability to identify and inhibit a biological target that exhibits aberrant behavior that specifically promotes the disease. Several prominent examples of targeted therapies have emerged in the past decade, resulting in lifesaving therapy for patients with certain types of cancers. A number of these examples are protein kinase inhibitors, and in particular, inhibitors of protein kinases that are abnormal, and as a result of this abnormality cause uncontrolled proliferation of certain types of cells. Recall from Chapter 2, Scheme 2.1 and associated discussion, that protein kinases catalyze the transfer of the terminal phosphate group of adenosine triphosphate (ATP) to the hydroxyl group on the side chain of serine, threonine, or tyrosine residues of proteins. The reason that protein kinases often appear in the context of uncontrolled cell proliferation is that in normal cells, these kinases are usually present in the biological pathways that communicate to cells when to grow and divide. In normal cells, these pathways are tightly controlled so that signals to proliferate are only turned on at specified times. However, it has been found that aberrations in specific kinases in these pathways can circumvent the regulatory control and the signal to proliferate is therefore enhanced, leading to uncontrolled cell proliferation.

Epidermal growth factor (EGF) is a polypeptide that binds to receptors on the surface of targeted cells to stimulate the growth, proliferation, and survival of these cells. This mechanism is used, for example, to regulate the growth of new cells in the lining of the intestine. The epidermal growth factor receptor (EGFR) is a transmembrane receptor wherein the intracellular domain functions as a kinase enzyme (Figure 5.1).

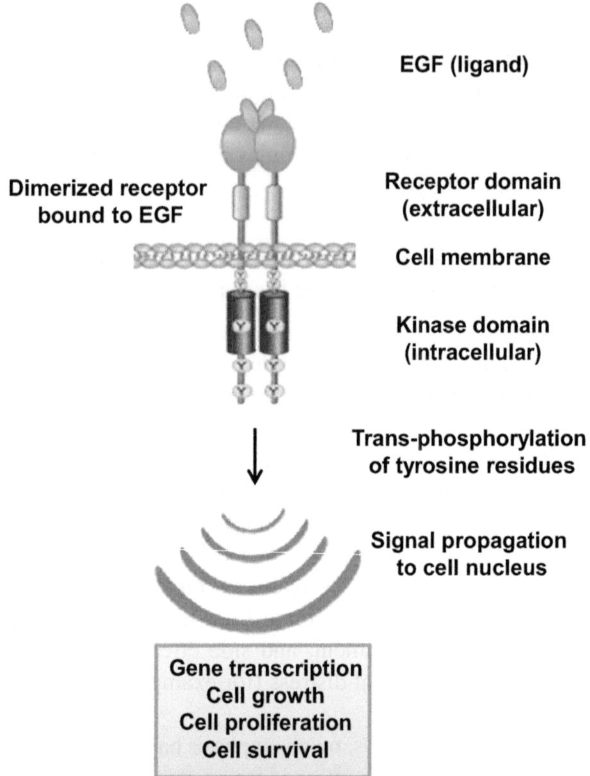

FIGURE 5.1 Schematic depiction of epidermal growth factor (EGF) and its receptor, activation of which leads to cell growth and proliferation. In its activated state, the EGFR normally functions in a dimeric form, as shown. *Adapted from Pawson, T.; Jorgensen, C. Signal Transduction by Growth Factor Receptors (Chapter 11) In The Molecular Basis of Cancer, 3rd ed. Mendelsohn, J.; Howley, P.M.; Israel, M. A.; Gray, J. W.; Thompson, C. B. (Eds.), Elsevier, **2008**, pp. 155–168.*

Thus, interaction of EGF with the extracellular domain of the receptor results in activation of the kinase domain inside the cell and represents one mechanism by which signals outside a cell can lead to specific responses inside the cell (this is one example of how a receptor, in this case EGFR, propagates the response to an agonist, in this case EGF). Note that EGFR is a member of the family of *receptor tyrosine kinases*, which function as receptors to mediate the transmission, via the kinase domain, of signals of certain growth factors. Here we will be primarily concerned with the kinase activity of EGFR, that is, the enzymatic activity of the intracellular kinase domain of the receptor (see Figure 5.1).

Analysis of cells from a number of different tumor types, for example, from lung or colon cancer, revealed abnormalities in EGF or EGFR.[4] In some cases, signals for growth and proliferation were being transmitted into the nucleus of these cells even in the absence of EGF, whereas in normal cells EGF is required for initiation of these signals. Thus, inhibition of EGFR signaling was recognized as a possible approach to slow the rate of growth of tumors harboring aberrant signaling via EGFR. Possible approaches include antagonists at the receptor-binding site or inhibition of the intracellular kinase activity. The natural ligand at the receptor and the natural substrate(s) of the kinase are polypeptides and, therefore, difficult to use as leads, particularly toward orally administered compounds (antibodies that block the extracellular receptor domain have found clinical use but are administered by injection). However, kinases have a second substrate, namely, ATP, which is the source of the phosphoryl group that is transferred to the protein substrate. Since there are many human kinases (and at least 500 human protein kinases) that all use ATP as a substrate, the design of a kinase inhibitor that is specific for a single kinase based on the structure of ATP as a lead has been a significant challenge. However, there are sufficient differences in binding regions in and around the ATP-binding site of different kinases that specificity for individual kinases, or at least for a comparatively small number of kinases, has indeed proven possible.

5.2.2.1.2. Discovery and Optimization of EGFR Inhibitors

Scientists at Zeneca (now AstraZeneca) and other pharmaceutical companies set out to discover inhibitors of EGFR kinase.[5] Mechanistic studies indicated that catalysis was effected through a ternary complex of enzyme, ATP, and the peptide substrate, and that phosphate transfer occurred through direct nucleophilic attack of the tyrosine hydroxyl group on the phosphorus atom of the γ-phosphate group of ATP (see Chapter 4, Section 4.3.5). The lead discovery approach at Zeneca for this target was through virtual screening (see Chapter 2, Section 2.1.2.3.4), which searched for compounds with the following characteristics (recall that one of the main requirements for virtual screening was a hypothesis of features required for interaction with the target):

1. A six-membered aromatic ring to mimic the phenolic ring of tyrosine in the substrate.
2. An oxygen or nitrogen atom at a suitable distance from the above aromatic ring to mimic the phenolic hydroxyl group of tyrosine in the substrate.
3. Any oxygen or nitrogen atom at a suitable distance from the tyrosine hydroxyl group to mimic at least two of the nonbridging oxygens of the triphosphate group of ATP.

The screening compounds were a 1500-member subset of the Zeneca corporate collection, which totaled about 250,000 compounds at that time. The screening subset was designed to be representative of the range of chemotypes in the broader collection. A single low-energy conformation was predicted for each compound in the screening set using Concord software, and Aladdin software was used to computationally compare the test compounds with the requirements of the hypothesis (the search query). The search query matched 152 compounds, three of which in fact inhibited EGFR kinase with reasonable potency. Additional compounds from the corporate collection were then tested on the basis of their similarity to these inhibitors. In this way, compound **5.9** was discovered, which exhibited an IC_{50} of 40 nM. Kinetic studies showed that this inhibitor is not competitive with peptide substrate (in contrast to the hypothesis used in the virtual screen) but is competitive with ATP. At about the same time, Parke-Davis Pharmaceuticals (subsequently acquired by Pfizer, then closed) reported that research starting with a mass screening campaign had yielded compounds **5.10** and **5.11** as EGFR kinase inhibitors, which exhibited IC_{50} values of 0.31 and 0.025 nM, respectively.[6] While differences in assay conditions used by the two groups do not allow a direct comparison of the potencies of **5.9** and **5.10**, it is apparent that the 6,7-dimethoxyl functionality of **5.10** contributes substantially to increased potency. This effect may, at least in part, be the result of the electron-donating properties of the methoxyl groups leading to increased electron density on the quinazoline nitrogen atoms. Subsequent lead optimization studies focused on varying the substituents on the benzo portion of the quinazoline ring as well as on the phenyl ring of the anilino appendage. These efforts led to two marketed drugs to treat lung and other cancers: gefitinib (**5.12**, Iressa) from the Zeneca efforts and erlotinib (**5.13**, Tarceva) from the Parke-Davis lead compound. Note that in both cases, lead optimization led to incorporation of polar functionality in the substituents on the benzo portion of the quinazoline ring, no doubt intended to increase solubility of the compounds for improved pharmacokinetic and pharmaceutical properties. The fact that inhibition of EGFR kinase by **5.13** is competitive with ATP (IC_{50} = 2 nM), in analogy to its earlier analogs, was confirmed by kinetic studies.[7]

An X-ray crystal structure of erlotinib in complex with EGFR kinase shows a likely hydrogen bonding interaction between the quinazoline N^1 and the backbone NH of Met-793 of the enzyme (Figure 5.2).[8] An X-ray crystal structure has also been solved for the complex of EGFR with ATP analog adenosine-5′[β,γ-imido]triphosphate (AMP-PNP, **5.14**, Figure 5.3). AMP-PNP is identical to ATP, except that the distal phosphoric anhydride oxygen has been replaced by NH (ionized to N^- at physiological pH) to stabilize the terminal phosphate linkage. As shown in Figure 5.3, AMP-PNP also forms a hydrogen bond with Met-793 of EGFR; a comparison of Figures 5.2 and 5.3 suggests that the erlotinib-binding region, at least partially, overlaps with the ATP-binding site, which strongly supports the conclusion deduced from enzyme kinetic studies.

5.2.2.2. Stabilization of an Inactive Conformation: Imatinib, an Antileukemia Drug

5.2.2.2.1. The Target: Bcr-Abl, a Constitutively Active Kinase

Chronic myelogenous leukemia (CML) is a disease characterized by excessive proliferation in the bone marrow of specific types of white blood cells and their precursors. It was found that most cases of CML were associated with the Philadelphia chromosome, a chromosomal abnormality that led to the expression of an abnormal protein called Bcr-Abl. Abl, short for Abelson (Abl) kinase, is a tyrosine kinase normally expressed in many cells, and which like other tyrosine kinases catalyzes the phosphorylation of phenolic hydroxyl groups on the side chain of tyrosine residues of other proteins. Bcr (short for breakpoint cluster region) is a polypeptide fragment that is fused to Abl as coded by the

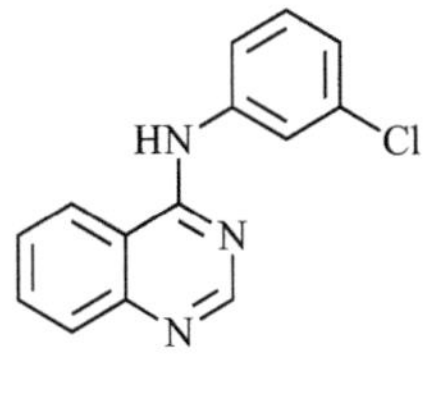

5.9

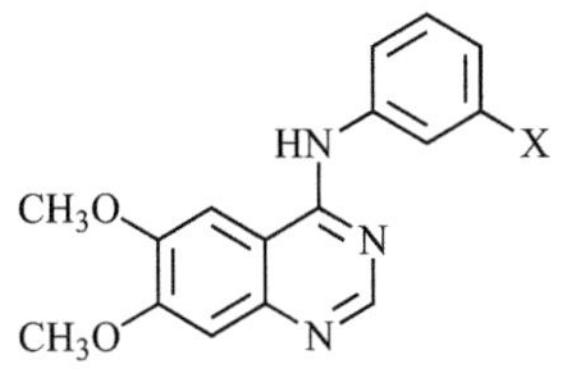

5.10, X = Cl
5.11, X = Br

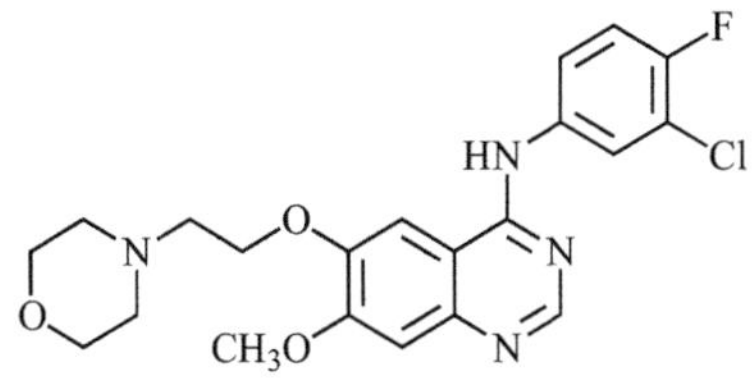

Gefitinib
5.12

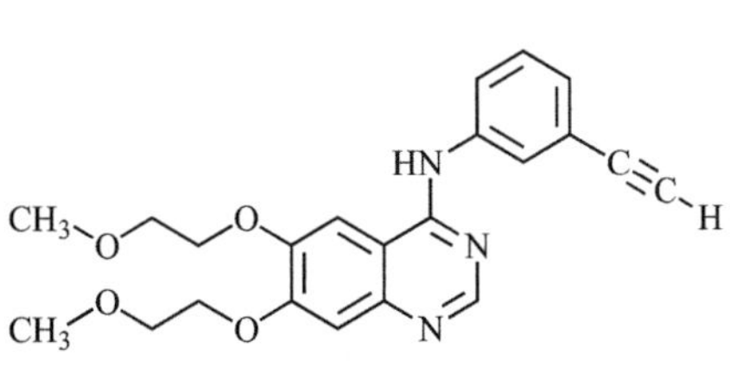

Erlotinib
5.13

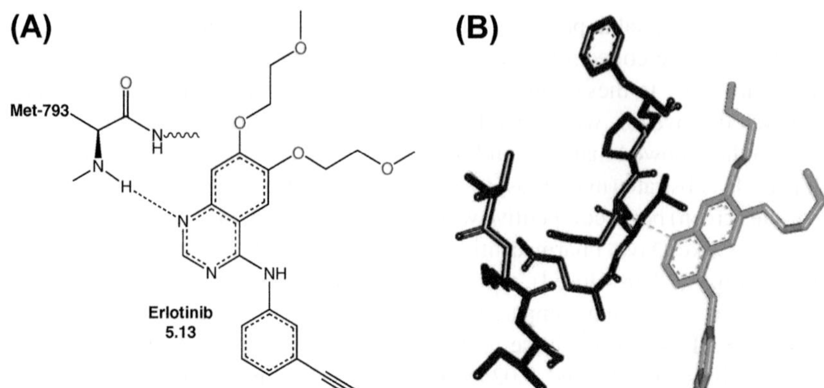

FIGURE 5.2 (A) Schematic drawing of the interactions of erlotinib (**5.13**) with the active site of EGFR kinase. Modified image from PoseView (University of Hamburg, Germany) created from Protein Data Bank ID 1M17. EGFR residue numbering corresponds to that in PDB 2ITX. (B) Three-dimensional structure of erlotinib (green) bound to EGFR kinase (selected residues shown in blue). Data from Protein Data Bank ID 1M17 modified using Accelrys DS Viewer software. (PDB ID:1M17). *Stamos, J., Sliwkowski, M.X., Eigenbrot, C. Structure of the epidermal growth factor receptor kinase domain alone and in complex with a 4-anilinoquinazoline inhibitor. J. Biol. Chem.* ***2002****, 277, 46265–46272.*

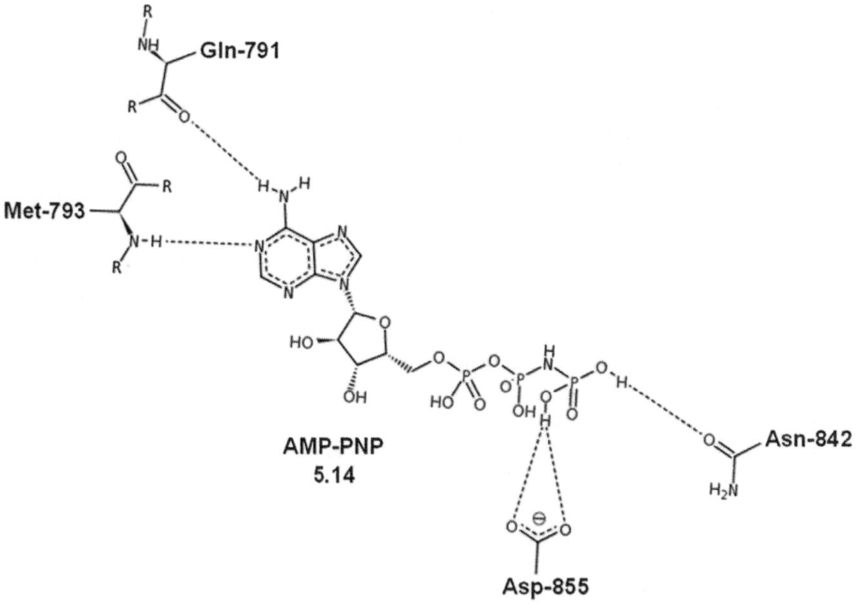

FIGURE 5.3 Schematic drawing of the interactions of stable ATP analog AMP-PNP (**5.14**) with the active site of EGFR kinase. *Modified image from PoseView (University of Hamburg, Germany) created from Protein Data Bank ID 2ITX. PDB ID: 2ITX (Yun, C.-H., Boggon, T.J., Li, Y., Woo, S., Greulich, H., Meyerson, M., Eck, M.J. Structures of lung cancer-derived EGFR mutants and inhibitor complexes: mechanism of activation and insights into differential inhibitor sensitivity. Cancer Cell* ***2007****, 11, 217–227.)*

Philadelphia chromosome. The effect of the fused Bcr segment is to significantly enhance the kinase activity of Abl. In 1990, three different groups reported that mice transfected with the *bcr-abl* gene developed symptoms closely resembling CML in humans.[9]

5.2.2.2.2. Lead Discovery and Modification

With validation of Bcr-Abl as a target for CML at hand, scientists at Ciba–Geigy (now Novartis since the Ciba-Geigy merger with Sandoz in 1996) set out to discover an inhibitor.[10] Like EGFR kinase, Abl is a tyrosine kinase. However, in contrast to EGFR, Abl is a soluble kinase found in the cytosol and, therefore, does not have a transmembrane receptor domain. On the basis of the interactions of compounds like AMP-PNP to kinases (cf. Figure 5.3), many aminopyrimidine-like molecules designed to mimic one or more of these key interactions were synthesized and tested against various kinases as potential inhibitor leads.

The Novartis team had originally identified anilinopyrimidine **5.15** as a lead structure for inhibition of a different kinase, platelet-derived growth factor receptor (PDGFR) kinase.[11] Key objectives of that effort were to enhance potency for

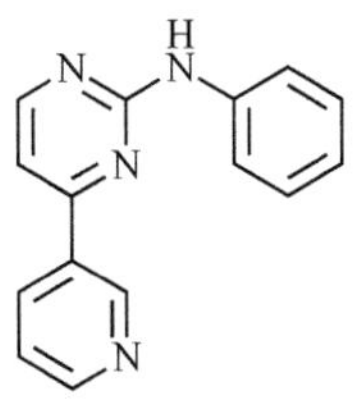

5.15 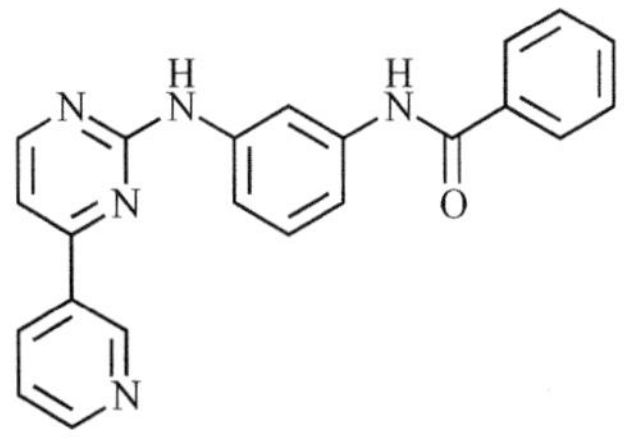 **5.16** **5.17**

5.16	IC50 (μM)
PDGFR	5
PKC	1.2

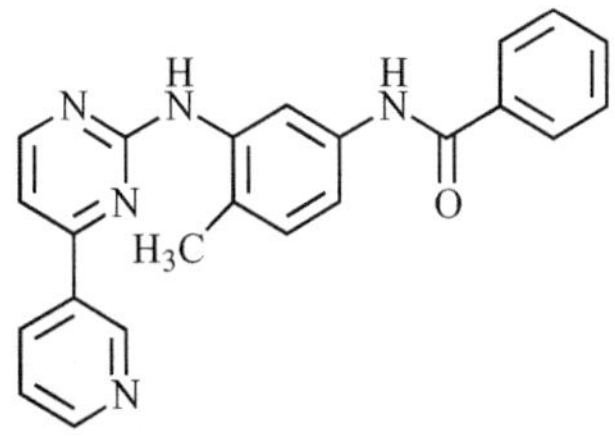

5.17	IC50 (μM)
PDGFR	0.1
PKC	72

PDGFR kinase as well as increase selectivity of PDGFR kinase over protein kinase C (PKC). Addition of a benzamido moiety gave **5.16** having low micromolar potency against PDGFR kinase, but poor selectivity against PKC. Remarkably, among a number of analogs synthesized, compound **5.17**, bearing one additional methyl group on the central phenyl ring, resulted in a 50-fold increase in potency against PDGFR kinase and a 60-fold decrease in potency against PKC! A possible role for this methyl group (called the "flag methyl" by the Ciba-Geigy team) is discussed in Section 5.2.2.2.3.

After identification of Abl kinase as an important therapeutic target, **5.17** was found to be a potent inhibitor of the enzyme, but it was not yet an ideal drug since it exhibited poor bioavailability. Since the aqueous solubility of **5.17** was poor, it was hypothesized that increasing the solubility would also lead to improved bioavailability, on the basis of the following rationale. When a compound is poorly soluble, poor absorption from the gastrointestinal tract may result, since the compound must be in solution prior to passing through gut membranes. In addition, once a compound reaches the bloodstream from the gut, it must next pass through the liver, where metabolic enzymes target hydrophobic compounds for metabolic conversion to more soluble derivatives to promote elimination in the urine. Therefore, the next objective was to increase the solubility of **5.17**. Appending hydrophilic groups such as *N*-methylpiperazine, in this case through a methylene linker, proved successful in improving bioavailability leading to imatinib (**5.18**, Gleevec).

5.2.2.2.3. Binding Mode of Imatinib to Abl Kinase

Shortly after imatinib was approved for treatment of CML in 2001, a crystal structure of the drug bound to Abl kinase was published.[12] In the bioactive conformation of imatinib, the central phenyl ring is positioned approximately 90° out of the plane of the neighboring pyrimidine ring; the flag methyl group promotes this conformation by destabilizing the planar conformation shown in **5.18** because of steric hindrance.

In the same paper, the complex of Abl kinase with another inhibitor, PD173955 (**5.19**), was also reported. Interestingly, the conformation of the enzyme was shown to be markedly different in the two structures (Figure 5.4). In Figure 5.4, the position of the activation loop (red in (A); blue in (B)) prominently defines the two different conformations. Only the conformation in (A) is able to bind ATP (when PD173955 is absent), and is therefore called the "active conformation". On the other hand, imatinib cannot bind to the active conformation, since the extended portion of the molecule (see arrow in Figure 5.4) would clash with a portion of the activation loop. In contrast, the inactive conformation in (B) contains an extended region that not only accommodates but also forms specific interactions with the extended portion of imatinib. Therefore, imatinib binds to the inactive conformation and stabilizes it, thereby preventing ATP from binding and causing inhibition of the enzyme.

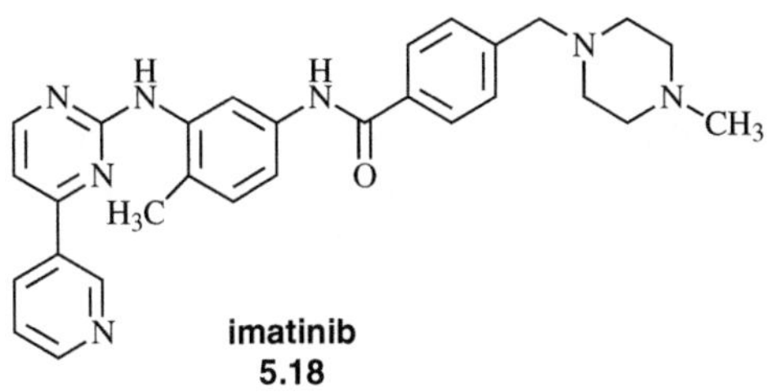

imatinib
5.18

	IC50 (μM)
Abl	0.1 - 0.3
PDGFR	0.1
c-KIT	0.1

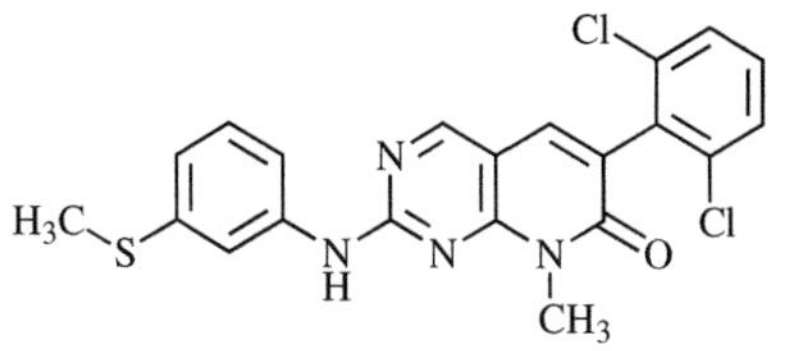

PD173955
5.19

It is assumed that the different conformational states of the apoenzyme (e.g., Abl_{active} and $Abl_{inactive}$ in Scheme 5.2) are present and in equilibrium in a cell at rest (see

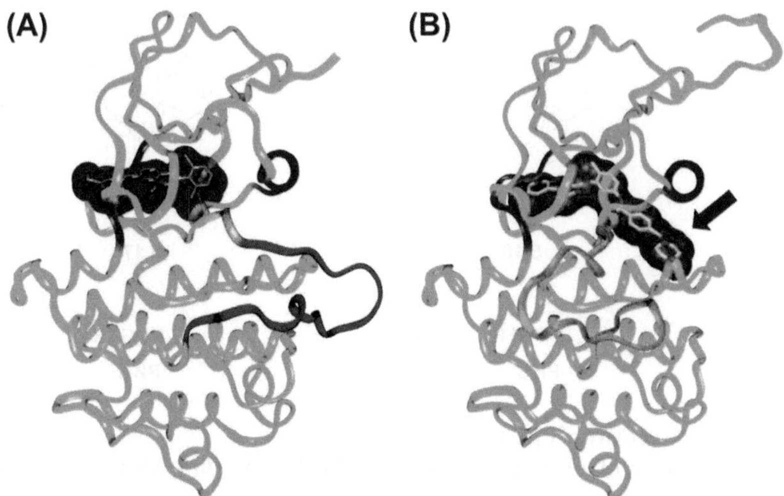

FIGURE 5.4 Ribbon representations of the structure of the Abl kinase domain (green) in complex with (A) PD173955 and (B) imatinib (originally known as STI-571). The protein strand colored red in (A) and blue in (B) is the "activation loop" of the kinase; the arrow points to the *N*-methylpiperazine portion of imatinib. *Adapted by permission from the American Association for Cancer Research: Nagar, B. et al. Crystal structures of the kinase domain of c-Abl in complex with the small molecule inhibitors PD173955 and imatinib (STI-571). Cancer Res.* ***2002****, 62, 4236–4243.*

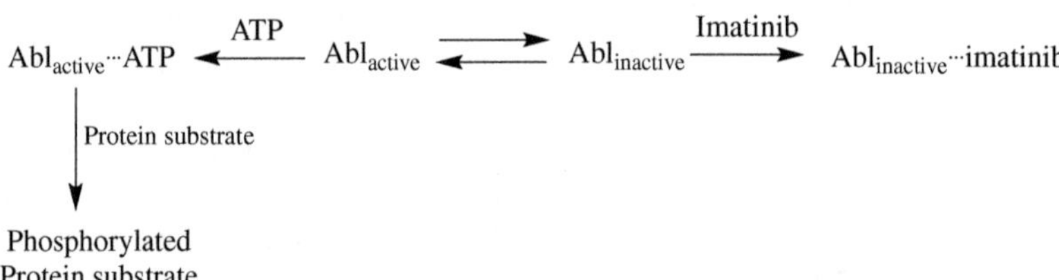

SCHEME 5.2 Equilibrium between active and inactive conformations of Abl kinase, where the active conformation binds to ATP, leading to catalysis of the normal enzymatic reaction; the inhibitor imatinib binds to and stabilizes the inactive form, pulling the equilibrium to the right.

Chapter 3, Section 3.1). Normal cells have regulatory mechanisms that influence the position of this equilibrium. For example, phosphorylation of one or more amino acid hydroxyl groups of Abl$_{inactive}$ by another kinase serves to shift the equilibrium in Scheme 5.2 away from Abl$_{inactive}$ and toward Abl$_{active}$, resulting in increased enzymatic activity. Phosphorylation by kinases (counterbalanced by dephosphorylation by phosphatases) has been established as a key mechanism for cellular regulation (in simple terms, serving as on and off switches for many cellular processes).[13] The Bcr fusion in Bcr-Abl is an abnormal mechanism for shifting the equilibrium in Scheme 5.2 toward Abl$_{active}$.

The presence of multiple conformations, as characterized by the position of the activation loop, is a common theme among protein kinases. It has been hypothesized that the design of inhibitors that preferentially interact with an inactive conformation, for example, in a manner similar to imatinib, offers an approach to achieve selectivity of the inhibitor among different kinases.[14] Support for the hypothesis comes from the observation that the extended region present in the inactive conformation is more dissimilar among kinases than is the site in the active conformation, which was designed by nature to interact with ATP. A tangible example in support of this hypothesis is that PD173955, which binds to the active conformation of Abl, is also a potent inhibitor of the related kinase Src, whereas imatinib is not, despite the high sequence and structural similarity of the two enzymes.

5.2.2.2.4. Inhibition of Other Kinases by Imatinib

Despite the foregoing discussion, imatinib does inhibit other enzymes in addition to Abl, including platelet-derived growth factor receptor (PDGFR) and KIT. Both of these enzymes have been found to be overactive in certain types of tumors. Overactivated KIT is a primary driver of gastrointestinal stromal tumors (GIST), and imatinib has also been approved as a new treatment for this disease. Since the approval of imatinib for chronic myelogenous leukemia (CML) and GIST, targeting kinases for cancer has been a significant focus in the pharmaceutical industry. More recently, research on kinase inhibitors to treat inflammation and autoimmune diseases,[15] which involve overproliferation of certain cells of the immune system, also has been initiated. The challenge of this area of research is that a higher margin of safety is generally required for such diseases, which may be debilitating, but usually are not life threatening.

5.2.2.3. Alternative Substrate Inhibition: Sulfonamide Antibacterial Agents (Sulfa Drugs)

Competitive inhibition of an enzyme also can be attained with a molecule that not only binds to the enzyme but also acts as a substrate. In this case, however, the product produced is not a compound that is useful to the organism. While the alternative substrate is being turned over, it is preventing the actual substrate from being converted to the product that the organism needs. The principal disadvantage to this approach is that the product generated could be toxic or cause an unwanted side effect. In the example below, this is not the case.

5.2.2.3.1. Lead Discovery

At the beginning of the twentieth century, Paul Ehrlich showed that various azo dyes were effective agents against trypanosomiasis in mice; however, none was effective in man. In the early 1930s, Gerhard Domagk, head of bacteriological and pathological research at the Bayer Company in Germany, who was trying to find agents against streptococcal infections, tested a variety of azo dyes. One of the dyes, Prontosil (**5.20**), showed dramatically positive results, and successfully protected mice against streptococcal infections.[16] However, Bayer was unwilling to move rapidly on getting Prontosil onto the drug market. As Albert[17] tells it, when, in late 1935, Domagk's daughter cut her hand and was about to die of a streptococcal infection, her father gave her Prontosil! Although she turned bright red from the dye, her recovery was rapid,[18] and the effectiveness of the drug became quite credible. In 1939, Domagk was awarded the Nobel Prize in Medicine for this achievement.

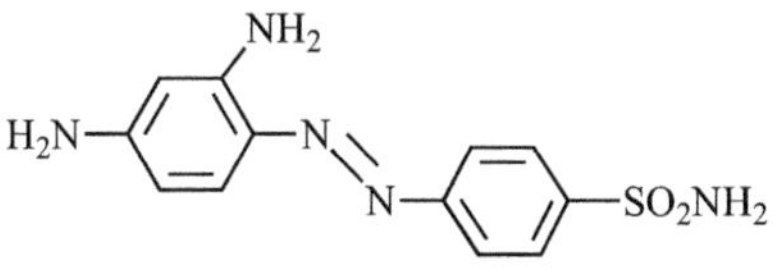

NH_2 H_2N N N SO_2NH_2

Prontosil
5.20

An unexpected property of Prontosil, however, was that it had no activity against bacteria in vitro. Tréfouël and coworkers[19] found that if a reducing agent was added to Prontosil, then it *was* effective in vitro. They suggested that the reason for the lack of in vitro activity, but high in vivo activity, was that Prontosil was metabolized by reduction to the active antibacterial agent, namely, *p*-aminobenzenesulfonamide (also called sulfanilamide) (**5.21**; AVC); therefore, Prontosil is a prodrug (see Chapter 9), a compound that requires metabolic activation to be effective. Furthermore, they demonstrated that sulfanilamide was as effective as Prontosil in protecting mice against streptococcal infections, and that it exerted a *bacteriostatic effect* in vitro. Unlike a *bacteriocidal* agent, which kills bacteria, a bacteriostatic drug inhibits further growth of the bacteria, thus allowing the host defenses to catch up in their fight against the bacteria. Because microorganisms replicate rapidly (with *Escherichia coli*, for example, the number of cells can double every 20–30 min), a bacteriostatic agent will interrupt this rapid growth and allow the immune system to destroy the organism. Of course, with immunocompromised individuals, who are unable to contribute natural body defenses to fight their disease, a bacteriocidal agent is necessary to prevent continuation of growth of the organism when the drug is withdrawn.

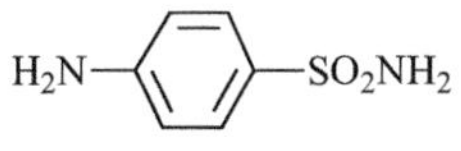

H_2N SO_2NH_2

Sulfanilamide
5.21

5.2.2.3.2. Lead Modification

The discovery of Prontosil marks the beginning of modern chemotherapy. During the next decade thousands of sulfonamides were synthesized and tested as antibacterial agents. These were the first structure–activity relationship studies (see Chapter 2, Section 2.2.3), which demonstrated the importance of molecular modification in drug design. Also, this was one of the first examples where new lead compounds for other diseases were revealed from side effects observed during pharmacological and clinical studies (see Chapter 1, Sections 1.2.4 and 1.2.5; Chapter 2, Section 2.1.2.2). These early studies led to the development of new antidiabetic and diuretic agents. Another important scientific advance that was derived from work with sulfonamides was a simple method for the assay of these compounds in body fluids and tissues.[20] This method involved diazotization of the aromatic amino group followed by diazo coupling with dimethyl-α-naphthylamine to produce a dye that could be measured colorimetrically. Furthermore, it was shown that the antibacterial effect of sulfanilamide was proportional to its concentration in the blood, and that at a given dose this varied from patient to patient. This was the beginning of the monitoring of blood drug levels during chemotherapy treatment, which led to the initiation of the routine use of *pharmacokinetics*, the study of the absorption, distribution, and excretion of drugs, in drug development programs. Proper drug dosage requirements could now be calculated.

5.2.2.3.3. Mechanism of Action

On the basis of the work by Stamp,[21] who showed that bacteria and other organisms contained a heat-stable substance that counteracted the antibacterial action of sulfonamides, Woods[22] in 1940 reported a breakthrough in the determination of the mechanism of action of this class of drugs. He hypothesized that because enzymes are inhibited by compounds whose structures resemble those of their substrates, the counteractive substance should be a substrate for an essential enzyme, and the substrate should have a structure similar to that of sulfanilamide. After various chemical tests, and a vague notion of the possible structure of this counteractive substance, he deduced that it must be *p*-aminobenzoic acid (**5.22**) and proceeded to show that **5.22** potently counteracted sulfanilamide-induced bacteriostasis. The results of his experiments showed that sulfanilamide was competitive with *p*-aminobenzoic acid for microbial growth. To maintain growth with increasing concentrations of sulfanilamide, it is also necessary to increase the concentration of *p*-aminobenzoic acid. Selbie[23] found that coadministration of *p*-aminobenzoic acid and sulfanilamide into streptococcal-infected mice prevented the antibacterial action of the drug.

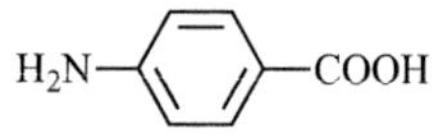

H_2N—C_6H_4—COOH

5.22

The observation of competitive inhibition by sulfanilamide was the basis for Fildes[24] to propose his theory of *antimetabolites*. An antimetabolite is a compound that disrupts metabolic processes, typically by blocking or acting as an alternative substrate for an enzyme in a metabolic pathway. He proposed a rational approach to chemotherapy, namely, enzyme inhibitor design, and suggested that the molecular basis for enzyme inhibition was that either the inhibitor combines with the enzyme and displaces its substrate or coenzyme or it combines directly with the substrate or coenzyme.

In the mid-1940s, Miller and coworkers[25] demonstrated that sulfanilamide inhibited folic acid biosynthesis, and in 1948, Nimmo-Smith et al.[26] showed that the inhibition of folic acid biosynthesis by sulfonamides was competitively reversed by *p*-aminobenzoic acid. Two enzymes from *E. coli* were purified by Richey and Brown,[27] one that catalyzed the diphosphorylation of 2-amino-4-hydroxy-6-hydroxymethyl-7,8-dihydropteridine (**5.23**) and the other (dihydropteroate synthase), which catalyzed the synthesis of dihydrofolate (**5.25**) from diphosphate **5.24** and *p*-aminobenzoic acid (Scheme 5.3). The name of the enzyme stems from the fact that folic acid is a derivative of the pterin ring system (**5.26**, R = H). Because of the structural similarity of sulfanilamide to *p*-aminobenzoic acid, it is a potent competitive inhibitor of the second enzyme. The reversibility of the inhibition

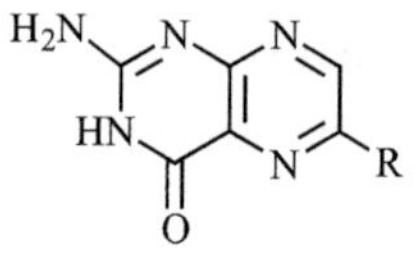

5.26

was demonstrated by Weisman and Brown,[28] who suggested that sulfonamides were incorporated into the dihydrofolate. This was verified by Bock et al.,[29] who incubated dihydropteroate synthase with diphosphate **5.24** and [^{35}S]-sulfamethoxazole (**5.27**) and identified the product as **5.28** (Scheme 5.4). Therefore, this is an example of competitive reversible inhibition in which the inhibitor also is a substrate. However, the product (**5.28**) cannot produce dihydrofolate, and, therefore, the organism cannot get the tetrahydrofolate needed as a coenzyme to make purines (see Chapter 4, Section 4.3.2), which are needed for DNA biosynthesis. This is why the sulfonamides are bacteriostatic, not bacteriocidal. Inhibition of tetrahydrofolate biosynthesis only inhibits replication; it does not kill the existing bacteria.

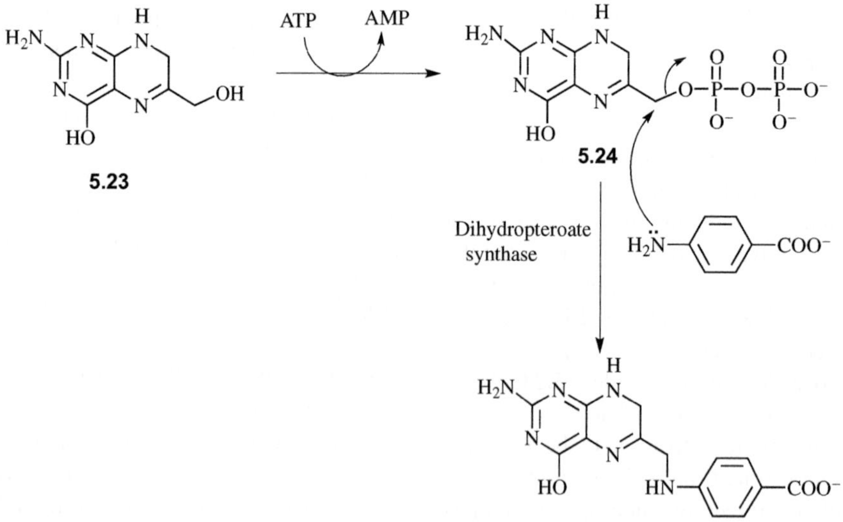

5.23 —(ATP → AMP)→ **5.24**

Dihydropteroate synthase, H_2N—C_6H_4—COO^-

5.25

SCHEME 5.3 Biosynthesis of bacterial dihydrofolic acid

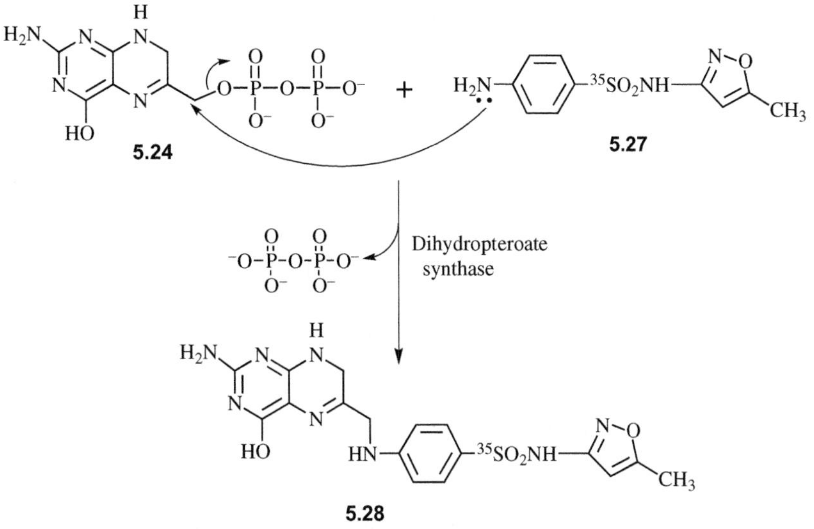

SCHEME 5.4 Dihydropteroate synthase use of sulfamethoxazole in place of *para*-aminobenzoic acid

Inhibitors of dihydropteroate synthase, however, have no effect on humans, because we do not biosynthesize folic acid and, therefore, do not have that enzyme. Folic acid is a vitamin and must be eaten by humans. Furthermore, because bacteria biosynthesize their folic acid, they do not have a transport system for it.[30] Consequently, we can eat all the folic acid we want, and the bacteria cannot utilize it. This is another example of *selective toxicity*, inhibition of the growth of a foreign organism without affecting the host, and falls into the category of an ideal enzyme inhibitor (see Section 5.1). It is interesting to note that sulfonamides are not effective with pus-forming infections because pus contains many compounds that are the end products of tetrahydrofolate-dependent reactions, such as purines, methionine, and thymidine. Therefore, inhibition of folate biosynthesis is unimportant, and pus can contribute to bacterial sustenance.

A major deterrent to the use of sulfa drugs is their central nervous system (CNS) side effect profile, including headaches, tremors, nausea, vomiting, and insomnia. The side effects are believed to derive from the off-target inhibition of the enzyme sepiapterin reductase, an enzyme involved in the biosynthesis of the coenzyme tetrahydrobiopterin (**5.26** doubly reduced; R=(1*R*,2*S*)-dihydroxypropyl), which is essential for producing several neurotransmitters, such as serotonin, dopamine, norepinephrine, and epinephrine; depletion of these neurotransmitters could cause the neurological side effects.[31]

Dihydropteroate synthase satisfies at least three of four important criteria for a good antimicrobial drug target: (1) the target is essential to the survival of the microorganism; (2) the target is unique to the microbe, so its inhibition does not harm humans; and (3) the structure and function of the target is highly conserved across a variety of species of that microbe so that inhibitors are broad-spectrum agents.[32] The fourth criterion may be the most difficult to attain, namely, that resistance to inhibitors of the target not be easily acquired. Typically, it takes between 1 and 4 years for resistance to an antibacterial drug to emerge; in the case of the sulfonamides, it was almost 7 years.[33] There are several mechanisms of resistance to sulfonamide antibacterial drugs; these are discussed in Chapter 7.

5.2.3. Transition State Analogs and Multisubstrate Analogs

5.2.3.1. Theoretical Basis

As discussed in Chapter 4 (Section 4.1.2.2), an enzyme accelerates the rate of a reaction by stabilizing the transition state, which lowers the free energy of activation. The enzyme achieves this rate enhancement by changing its conformation so that the strongest interactions occur between the substrate and enzyme active site *at the transition state* of the reaction. Some enzymes act by straining or distorting the substrate toward the transition state. The catalysis-by-strain hypothesis led to early observations that some enzyme inhibitors owe their effectiveness to a resemblance to the strained species. Bernhard and Orgel[34] theorized that inhibitor molecules resembling the transition state species would be much more tightly bound to the enzyme than would be the substrate; 11 years prior to that Pauling had mentioned that the best inhibitor of an enzyme would be one that resembled the "activated complex".[35] Therefore, why should we design inhibitors on the basis of the ground state substrate structure? A potent enzyme inhibitor should be a stable compound whose structure resembles that of the substrate at a *postulated transition state* (or transient intermediate) of the reaction rather than that at the ground state. A compound of this type would bind much more tightly to the enzyme, and is called a *transition state analog inhibitor*. Jencks[36] was the first to suggest the existence of transition state analog inhibitors, and cited several possible literature examples; Wolfenden[37] and Lienhard[38] developed the concept further. Values for dissociation constants (K_i) of 10^{-15} M for enzyme–transition state complexes may not be unreasonable given the normal range of 10^{-3}–10^{-5} M for dissociation constants of enzyme–substrate complexes (K_m).

For the design of a transition state inhibitor to be effective, the mechanism of the enzyme reaction must be understood, so that a theoretical structure for the substrate at the transition state can be hypothesized. Because many enzyme-catalyzed reactions have similar transition states (for example, the different serine proteases), the basic structure of a transition state analog for one enzyme can be modified to meet the specificity requirements of another enzyme in the same mechanistic class and, thereby, generate a transition state analog for the other enzyme. This modification may be as simple as changing an amino acid in a peptidyl transition state analog inhibitor for one protease so that it conforms to the peptide specificity requirement of another protease. Christianson and Lipscomb[39] have renamed reversible inhibitors that undergo a bond-forming reaction with the enzyme prior to the observation of the enzyme–inhibitor complex as *reaction coordinate analogs*.

Schramm[40] argues that transition states only have a lifetime of a few femtoseconds,[41] whereas enzymes have turnover numbers (k_{cat}) on the millisecond timescale; therefore, the transition state occupies only about 10^{-12} of the enzymatic reaction cycle, and the transition state complex will occupy the enzyme about one-trillionth of the time and the remainder will be in the ground state. Therefore, it is not reasonable to expect a thermodynamic equilibrium explanation, such as binding, to account for this short-lived chemical event. Rather, to determine transition states, kinetic isotope effect (KIE) measurements of the chemical steps,[42] which perturb bond vibrational states and influence the transition state, should be used[43]; binding isotope effects would only give insight into ground state interactions.[44]

For those of you who have not studied *KIEs*, a brief description follows. For any bond that is broken *in the rate-determining step*, the rate of the reaction will be affected by incorporation of an isotope for one of the atoms connected to the bond that breaks (called a *primary KIE*). The degree of this effect will be determined by the mass of the atom that is substituted because that is related to the energy of the bond that breaks: the greater the mass, the lower the vibrational frequency, which means the lower the zero-point energy and, therefore, the more energy needed to break the bond. A C-^{2}H bond, then, is a little stronger than a C-^{1}H bond and requires more energy to break. The substitution of a ^{2}H for a ^{1}H (mass difference of 100%) will have a much greater effect than the substitution of a ^{13}C atom for ^{12}C (mass difference of only 1/12 or about 8%). The rate of a reaction that breaks a bond to a ^{1}H can be seven times greater than that of the same reaction with a ^{2}H, whereas the corresponding reaction with a ^{12}C will only be 1.04 times that with ^{13}C.

KIEs provide information regarding bonding and geometric differences between reactants and transition state structures. If KIEs are combined with computational methods that correlate experimental values of van der Waals geometry and electrostatic potential surfaces with predicted chemical models, a three-dimensional depiction of the transition state of an enzyme-catalyzed reaction can be obtained, which permits the design of transition state analog inhibitors.[45] Because geometry and electrostatics are the dominant properties for ligand binding, the more similar these properties are to those of the transition state,[46] the more tightly the molecule will bind. The ideal enzyme target for transition state analog inhibitor design, therefore, would catalyze reactions that have altered geometry and charge on the conversion of reactants to the transition state. An example of drug design that uses this approach is given in Section 5.2.3.2.3.

When more than one substrate is involved in the enzyme reaction, a single stable compound can be designed that has a structure similar to that of the two or more substrates at the transition state of the reaction. This special case of a transition state analog is termed a *multisubstrate analog inhibitor*. Because these compounds are a combination of two or more substrates, their structures are unique, and they are often highly specific. Their great binding affinity for the target enzyme arises because the free energy of binding is roughly the product of the free energies of each independent substrate that it mimics.

5.2.3.2. Transition State Analogs

5.2.3.2.1. Enalaprilat

Enalaprilat (**5.29**, Figure 5.5) is a very potent slow, tight-binding inhibitor of angiotensin-converting enzyme (ACE) (see Section 5.2.4.2). The rationalization for its binding effectiveness was that it had multiple binding interactions with the substrate- and product-binding sites. The resemblance of enalaprilat to the substrate and products of the enzyme reaction (Figure 5.5) supports this notion.

Both of these rationalizations are ground state arguments, but transition state theory suggests that the most effective interactions occur at the transition state of the reaction. With that in mind, let's consider a potential mechanism for ACE-catalyzed substrate hydrolysis (Scheme 5.5) and see if the transition state structure is relevant. It is not known if a general base mechanism, as shown in Scheme 5.5, or a covalent catalytic mechanism is involved. Enalaprilat has been drawn beneath transition states 1 and 2 ($\ddagger_1$ and $\ddagger_2$) to show how the structures are related. An enzyme conformational change at the transition state (shown as a blocked enzyme instead of a rounded enzyme in Scheme 5.5) could increase the binding interactions.

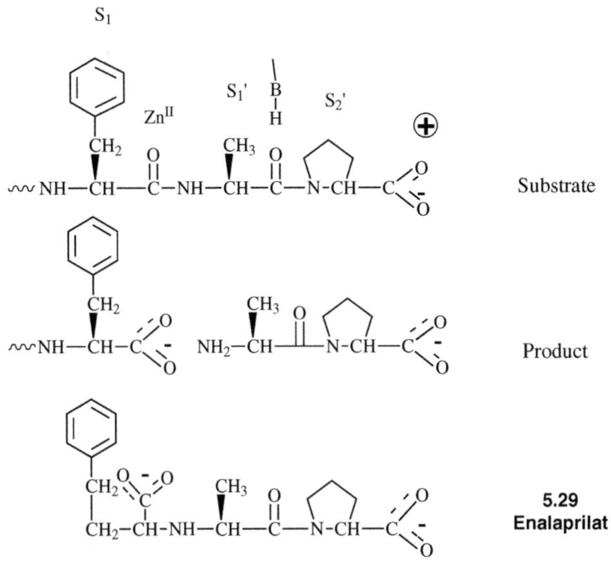

Substrate

Product

5.29
Enalaprilat

FIGURE 5.5 Hypothetical interactions of enalaprilat with ACE

5.2.3.2.2. Pentostatin

2′-Deoxycoformycin (pentostatin, **5.30**; Nipent), an antineoplastic agent isolated from fermentation broths of the bacterium *Streptomyces antibioticus*, is a potent inhibitor of the enzyme adenosine deaminase (adenosine aminohydrolase).[47] The K_i is 2.5×10^{-12} M which is 10^7 times lower than the K_m for adenosine! As in the case of enalaprilat (Sections 5.2.3.2.1 and 5.2.4.2), pentostatin is a slow, tight-binding inhibitor (see Section 5.2.4).[48] The k_{on} with human erythrocyte adenosine deaminase is 2.6×10^6 $m^{-1}s^{-1}$, and the k_{off} is 6.6×10^{-6} s^{-1}. The very small k_{off} value is reflected in the very low K_i value ($K_i = k_{off}/k_{on}$).

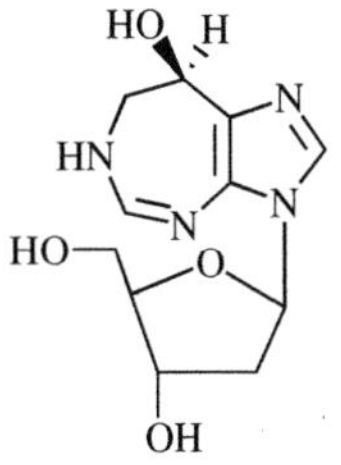

Pentostatin
5.30

Pentostatin is an analog of the natural nucleoside 2′-deoxyinosine (**5.31**, Scheme 5.6), in which the purine is modified to contain a seven-membered ring with two sp^3 carbon atoms. It is believed that this compound mimics the transition state structure of the substrates adenosine and 2′-deoxyadenosine (**5.31**, Scheme 5.6) during their hydrolysis to inosine and 2′-deoxyinosine (**5.34**), respectively, by adenosine deaminase.

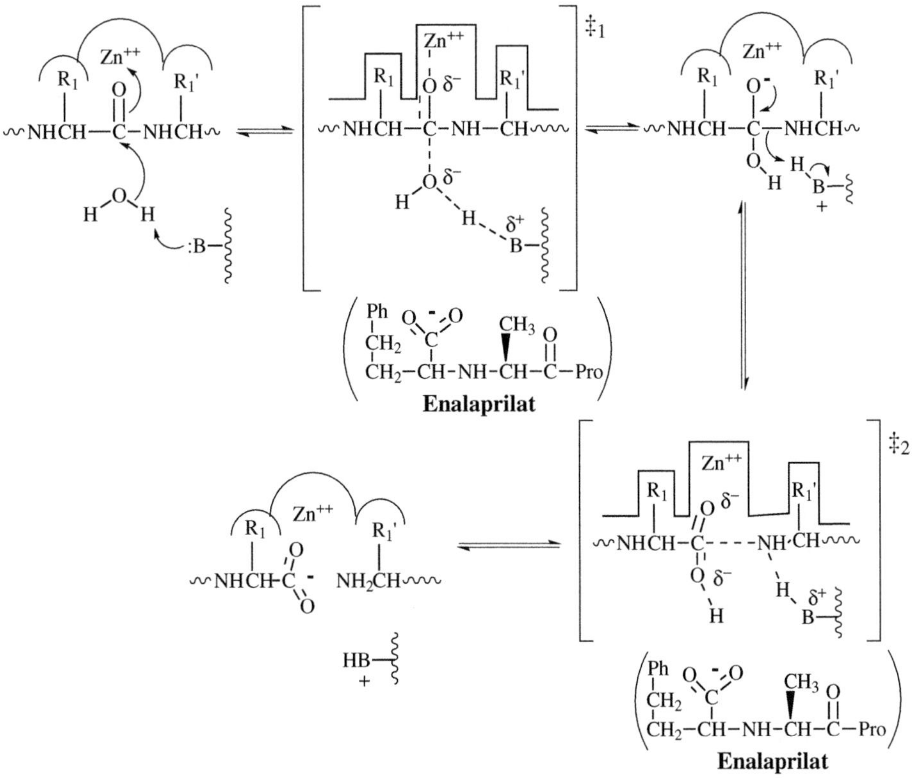

Enalaprilat

Enalaprilat

SCHEME 5.5 Hypothetical mechanism for ACE-catalyzed peptide hydrolysis

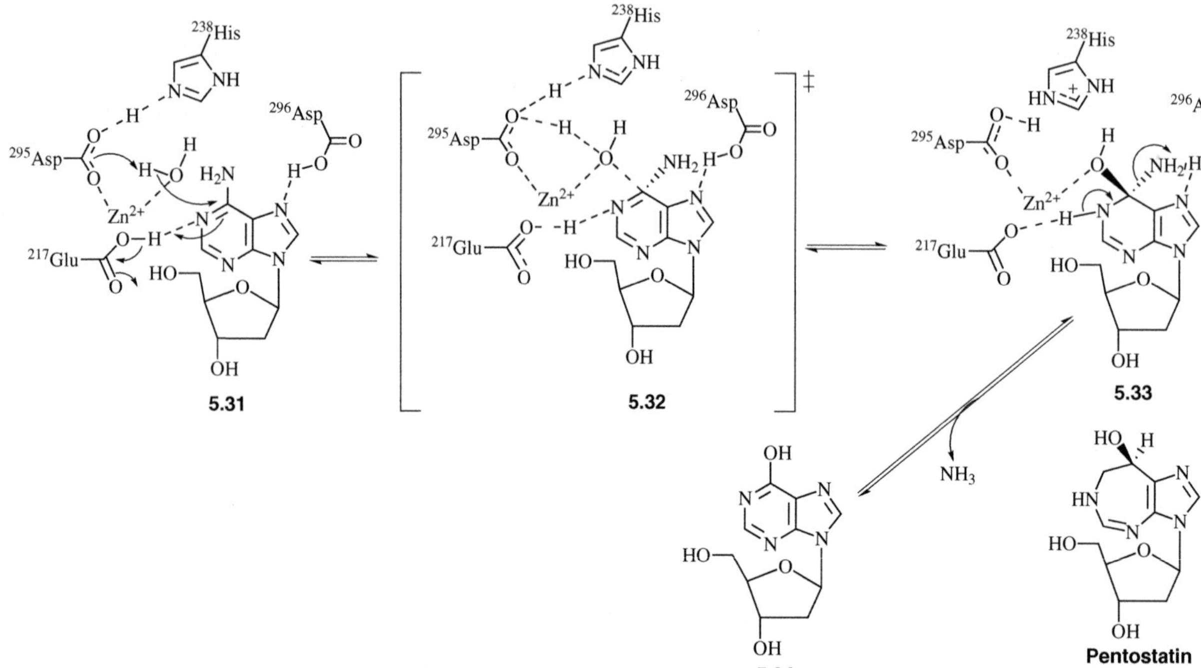

SCHEME 5.6 Hypothetical mechanism for adenosine deaminase-catalyzed hydrolysis of 2′-deoxyadenosine

A crystal structure to 2.4 Å resolution of the transition state analog 6*R*-hydroxy-1,6-dihydropurine ribonucleoside (**5.35**), complexed to adenosine deaminase, revealed a zinc ion cofactor and suggested a possible mechanism for the enzyme.[49] In this case, the resemblance of pentostatin (**5.30**) to the intermediate **5.33** (Scheme 5.6) is clearer than its similarity to the transition state shown (**5.32**); however, a late transition state would look more like **5.33**.

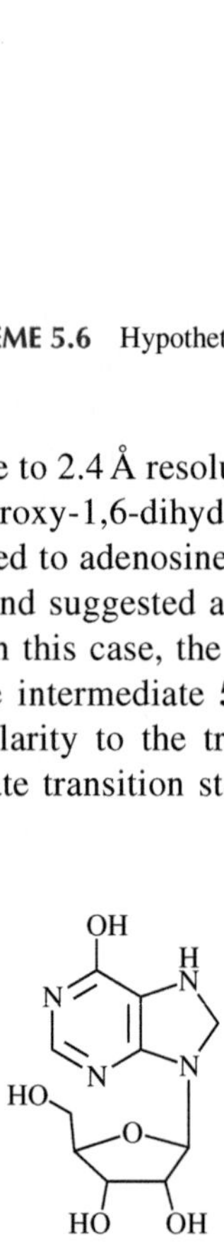

5.35

Pentostatin is most effective in lymphocytic disease, for example, hairy cell leukemia and chronic lymphocytic leukemia.[50] It is known that adenosine deaminase levels are elevated in T-lymphocytes (white blood cells that mature in the thymus).[51] Since inhibition of an enzyme can lead to accumulation of its substrate(s), one hypothesis for the action of pentostatin is that it induces lymphotoxicity via accumulation of 2′-deoxyadenosine, which inhibits ribonucleotide reductase and *S*-Adenosylhomocysteine hydrolase.[52] Ribonucleotide reductase catalyzes the conversion of ribonucleotides to the corresponding 2′-deoxyribonucleotides and is essential for DNA biosynthesis. Thus, inhibition of this enzyme leads to inhibition of DNA biosynthesis. *S*-Adenosylhomocysteine competitively inhibits most of the methyltransferases that utilize *S*-adenosylmethionine (SAM) as the methyl-donating agent. This, apparently, is a mechanism for the regulation of these methyltransferases. Inhibition of *S*-adenosylhomocysteine hydrolase, the enzyme that degrades *S*-adenosylhomocysteine, results in accumulation of *S*-adenosylhomocysteine, which inhibits the growth and replication of various tumors (and viruses), particularly those requiring a methylated 5′-cap structure on their messenger ribonucleic acids. Furthermore, various lymphocytic functions are suppressed by the accumulation of extracellular adenosine. Resistance to antipurines is discussed in Chapter 7.

5.2.3.2.3. Forodesine and DADMe-ImmH

Examples of transition state analog inhibitor design combining KIEs and computational methods are immucillin-H (ImmH; forodesine; **5.36**), in clinical trials for T-cell malignancies,[53] and 4′-deaza-1′-aza-2′-deoxy-1′-(9-methylene)immucillin-H (DADMe-ImmH; BCX4208; **5.37**), in clinical trials for gout,[54] both potent inhibitors of purine nucleoside phosphorylase (PNP), which catalyzes the phosphate-dependent conversion of purine nucleosides or deoxynucleosides to α-D-(deoxy)ribose-1-phosphate and the purine base (Scheme 5.7).[55]

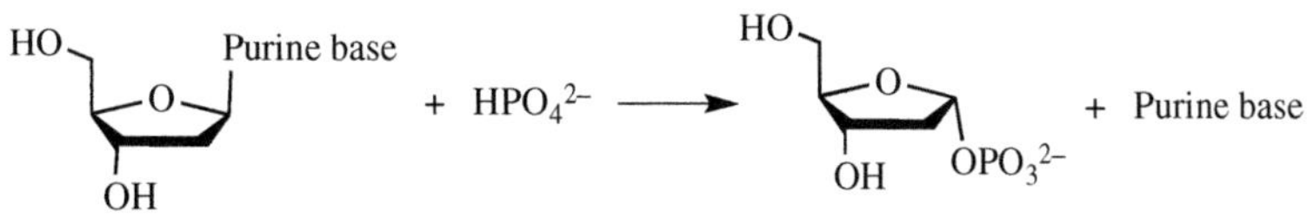

SCHEME 5.7 Reaction catalyzed by PNP

Inhibition of PNP causes elevated deoxyguanosine,[56] which is triphosphorylated by lymphocytes,[57] leading to the accumulation of deoxyguanosine triphosphate, known to be toxic to T cells[58]; T-cell deficiency induces apoptosis (cell death).[59] Normally, there is no measurable deoxyguanosine in human plasma or urine because of highly active PNP, so deoxyguanosine triphosphate does not accumulate. Diseases that could be impacted by controlling T-cell function by inhibiting PNP include T-cell leukemia and lymphoma, tissue transplant rejection, and T-cell autoimmune diseases, such as psoriasis, inflammatory bowel disease, rheumatoid arthritis, and insulin-dependent juvenile diabetes.[60]

Using KIE studies with bovine PNP (assumed to be similar to human PNP because of 87% amino acid sequence identity) and inosine as the substrate, a transition state structure shown in Figure 5.6 was deduced.[61] The computer-aided comparison of the electrostatic image of the transition state and a series of analogs led to the design of forodesine (**5.36**, Figure 5.6) as a transition state inhibitor; the K_d was found to be 23 pM![62] A crystal structure of PNP with **5.36** and phosphate bound showed the presence of at least six new hydrogen bonds at the active site relative to the substrate, which could account for its 10^6 times enhancement in binding.[63] However, **5.36** has a K_d for human PNP of only 56 pM (which, of course, is still phenomenal), indicating that the transition state for human PNP is a little different from that for bovine PNP,[64] and the aim was to treat humans, not cows. KIE studies with human PNP indicated a more dissociated transition state, with the positive charge on C1′ rather than at the sugar oxygen and the purine leaving group about 3 Å away (Figure 5.7). The inhibitor that was designed for this transition state had the positive charge on the sugar ring moved over to the C1′ position with the purine leaving group moved farther from the sugar ring by the addition of a methylene group; in addition, as the target substrate is 2′-deoxyguanosine, a 2′-deoxysugar ring mimic was incorporated. This led to DADMe-ImmunH (BCX4208; **5.37**); the K_d for **5.37** with human PNP was 9 pM![65] This insightful use of KIEs and computer modeling is a general approach that should be applicable to the design of any transition state inhibitor.

Transition state structure for bovine PNP

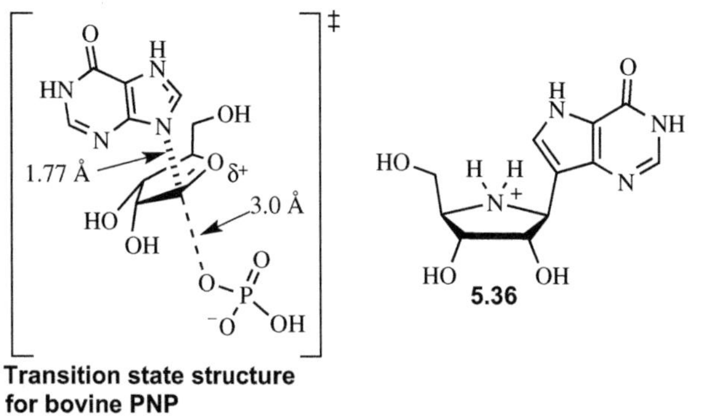

FIGURE 5.6 Hypothesized transition state structure for the reaction catalyzed by bovine PNP; structure of forodesine (**5.36**).

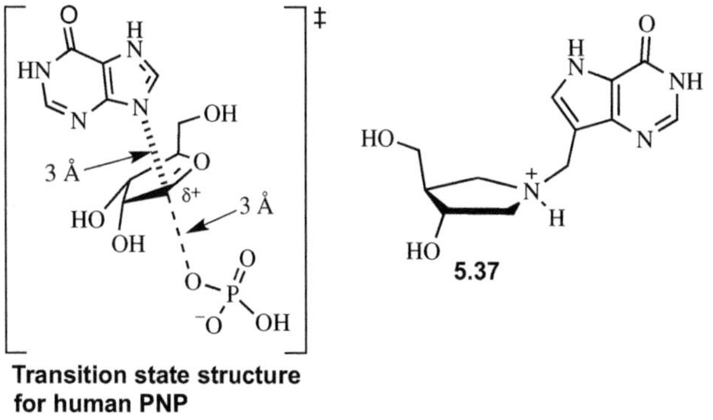

Transition state structure for human PNP

FIGURE 5.7 Hypothesized transition state structure for the reaction catalyzed by human PNP; structure of DADMe-ImmunH; BCX4208; **5.37**.

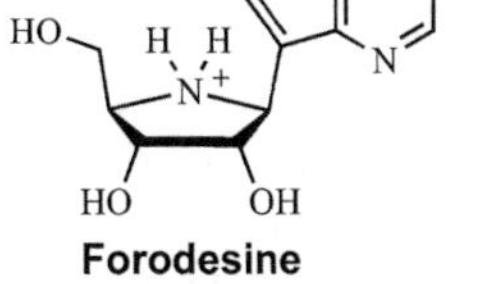

Forodesine **5.36**

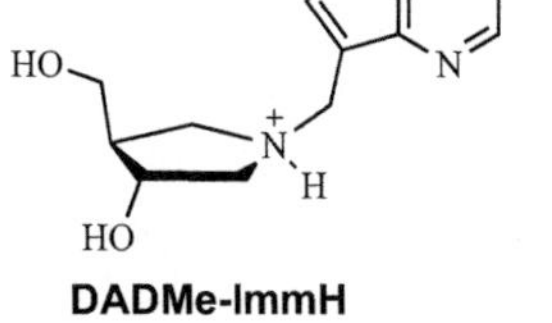

DADMe-ImmH **5.37**

5.2.3.2.4. Multisubstrate Analogs

One of the first steps in the de novo biosynthesis of pyrimidines is the condensation of carbamoyl phosphate (**5.38**) and L-aspartic acid, catalyzed by aspartate transcarbamylase, which produces *N*-carbamoyl-L-aspartate (**5.39**, Scheme 5.8). Below the transition state structure is drawn *N*-phosphonoacetyl-L-aspartate (**5.40**, PALA), which is a stable compound (the isosteric exchange of a CH_2 for the O prohibits loss of the PO_4^{3-} moiety) that resembles the transition state for condensation of the two substrates.[66] Although PALA is an effective inhibitor of aspartate transcarbamylase, it was ineffective in clinical trials as a result of tumor resistance. The proposed mechanism of resistance is discussed in Chapter 7.

Design of multisubstrate analogs has not yet led to many clinically successful drugs.[67] This is possibly because many

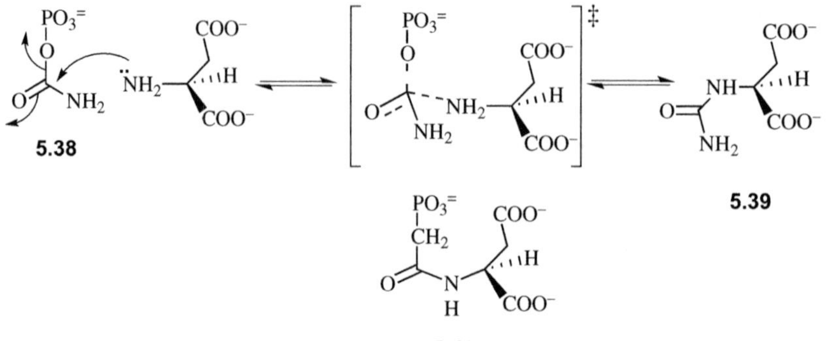

SCHEME 5.8 Hypothetical mechanism for the reaction catalyzed by aspartate transcarbamylase

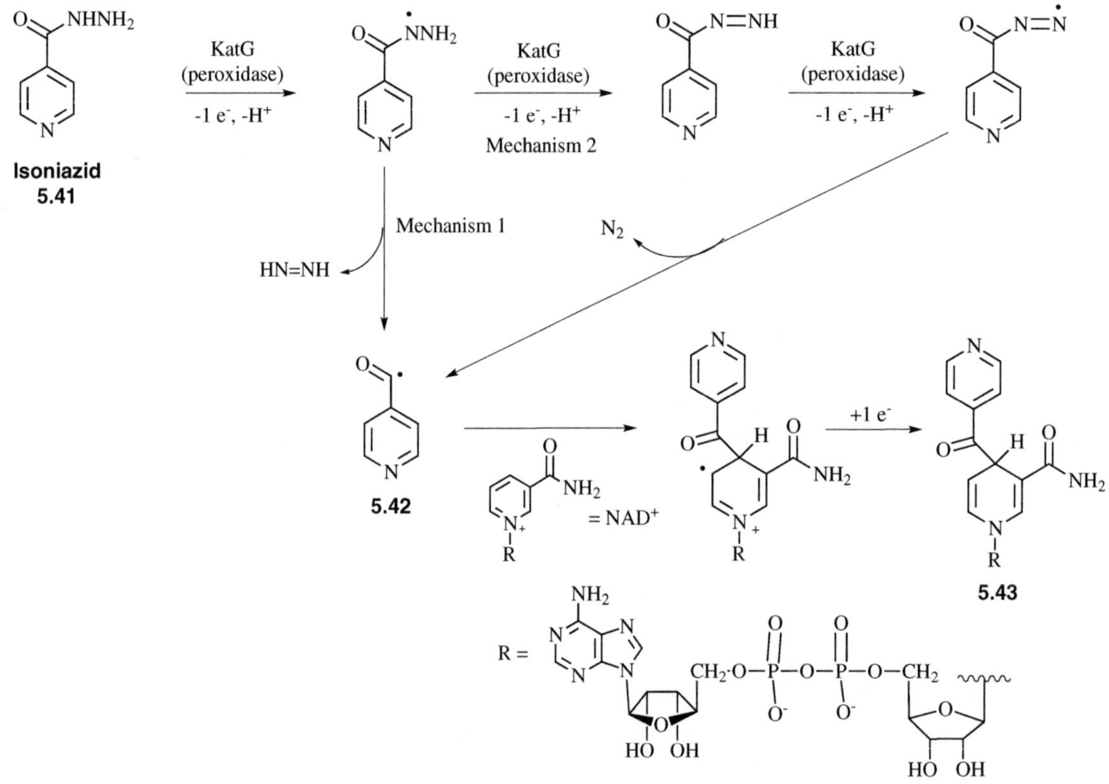

SCHEME 5.9 Proposed mechanism of action of isoniazid

of the multisubstrate inhibitors designed to date have had high molecular weights or are charged molecules and therefore do not penetrate cell membranes to reach their intended targets. However, there are some intriguing examples of bisubstrate analogs that are formed in vivo from smaller precursors that do have suitable druglike properties. For example, isoniazid (**5.41**, Scheme 5.9) is an antituberculosis drug that was used clinically for many years before its target was elucidated. Current evidence supports the hypothesis that the target is a reduced nicotinamide adenine dinucleotide (NADH)-dependent enoyl reductase, called *InhA*, involved in fatty acid metabolism, which is essential for cell wall synthesis in *Mycobacterium tuberculosis*. However, isoniazid itself is not active against this enzyme. The presently accepted mechanism of action of isoniazid (Scheme 5.9) proposes the oxidation of **5.41** by the heme-containing peroxidase enzyme *KatG* to form acyl radical **5.42**.[68] As shown in Scheme 5.9, two possible pathways have been proposed for peroxidase-catalyzed conversion of aryl hydrazides to acyl radicals, one involving a one-electron oxidation followed by loss of diimide (mechanism 1), and the other involving three sequential one-electron oxidations followed by loss of molecular nitrogen (mechanism 2).[69] Subsequent reaction of **5.42** with NAD^+ followed by a one-electron reduction by superoxide (possibly generated by KatG-catalyzed reduction of O_2)[70] produces the bisubstrate analog **5.43**, which is a potent inhibitor of *InhA*. A subsequently reported X-ray crystal structure of **5.43** bound to a close relative of *InhA* showed that **5.43** occupied the binding site for NADH and the binding region for the fatty acid substrates of the enzyme.

5.2.4. Slow, Tight-Binding Inhibitors

5.2.4.1. Theoretical Basis

As indicated above, the equilibrium between an enzyme and a reversible inhibitor is typically established rapidly. With *slow-binding inhibitors*, however, the equilibrium between enzyme and inhibitor is reached slowly, and inhibition is time-dependent, reminiscent of the kinetics for irreversible inhibition (see Section 5.3.2.1). *Tight-binding inhibitors* are those inhibitors for which substantial inhibition occurs when the concentrations of inhibitor and enzyme are comparable.[71] *Slow, tight-binding inhibitors* have both properties. These inhibitors can bind noncovalently[72] or covalently[73]; when a covalent bond is formed, a slowly reversible adduct may be involved. Noncovalent slow, tight-binding inhibitors are the bridge between rapidly reversible and covalent irreversible inhibitors. Depending on the tightness of binding, these inhibitors can become functionally equivalent to covalent, irreversible inhibitors with half-lives (time for half of the E·I complex to break down) of hours, days, or even months!

The reason for slow-binding inhibition is not definitively known. One possibility is that these inhibitors are such good analogs of the substrate that they induce a conformational change in the enzyme that resembles the conformation associated with the transition state[74]; typically, these compounds are transition state analogs (see Section 5.2.3.1). If this is the case, then inhibitor binding would be slow because it does not have all the essential structural features of the substrate transition state geometry. The dissociation would be even slower because the dissociation rate is not enhanced by product formation. The conformational change may result from a change in the protonation state of the enzyme[75] or from the displacement of an essential water molecule by the inhibitor.[76] The tightness of binding is often measured by the *residence time*, the duration of the inhibition, or the dissociative half-life of the drug–protein complex.[77] The longer the residence time, the longer is the drug's effect. Assuming adequate pharmacokinetic properties, this should translate into greater in vivo effects.[78] To differentiate simple competitive inhibition from slow-binding inhibition, it is necessary to carry out kinetic studies. Many simple competitive inhibitors may turn out to be slow- or slow, tight-binding inhibitors after the kinetic analysis.

5.2.4.2. Captopril, Enalapril, Lisinopril, and Other Antihypertensive Drugs

5.2.4.2.1. Humoral Mechanism for Hypertension

The elucidation of the molecular details of the *renin-angiotensin system*, one of the humoral mechanisms for blood pressure control, began over 60 years ago.[79] Angiotensinogen, an α-globulin produced by the liver,[80] is hydrolyzed by the proteolytic enzyme renin to a decapeptide, angiotensin I (Scheme 5.10), which has little, if any, biological activity. The C-terminal histidylleucine dipeptide is cleaved from angiotensin I by angiotensin-converting enzyme (ACE or dipeptidyl carboxypeptidase I) mainly in the lungs and blood vessels to give the octapeptide angiotensin II. This peptide is responsible for the increase in blood pressure by acting as a very potent vasoconstrictor[81] and by triggering release of a steroid hormone, aldosterone (**5.44**), which regulates the electrolyte balance of body fluids by promoting excretion of potassium ions and retention of sodium ions and water. Both vasoconstriction and sodium ion/water retention lead to an increase in blood pressure. Angiotensin II is converted to another peptide hormone, angiotensin III, by aminopeptidase A[82]; angiotensin III is also made by aminopeptidase A hydrolysis of angiotensin I to **5.45** (Scheme 5.10), then by ACE-catalyzed hydrolysis of **5.45** to angiotensin III.[83] This hormone also stimulates the secretion of aldosterone and causes vasoconstriction.[84] Angiotensin II also is hydrolyzed by aminopeptidase N to a hexapeptide, angiotensin IV, which may be involved in memory retention and neuronal development, but it is unclear if it is involved in vasopressin release; angiotensin III is also converted to angiotensin IV by aminopeptidase N.[85] To make matters even worse, in addition to cleaving angiotensin I to angiotensin II, and **5.45** to angiotensin III, ACE catalyzes the hydrolysis of the two C-terminal amino acids from the potent vasodilator nonapeptide bradykinin (Arg-Pro-Pro-Gly-Phe-Ser-Pro-Phe-Arg), thereby destroying its vasodilation activity. Consequently, the action of ACE results in the generation of potent hypertensive agents (angiotensin II and angiotensin III), which also stimulate the release of another hypertensive agent (aldosterone), and destroys a potent antihypertensive agent (bradykinin). All these outcomes of ACE action result in hypertension, an increase in blood pressure. There are compensatory pathways that decrease the blood pressure as well (e.g., see Section 5.2.4.2.4), but when there is an imbalance in these systems where the net result is high blood pressure, an antihypertensive drug is needed. ACE, therefore, is an important target for the design of antihypertensive agents; inhibition of ACE would shut down its three hypertensive mechanisms.[86]

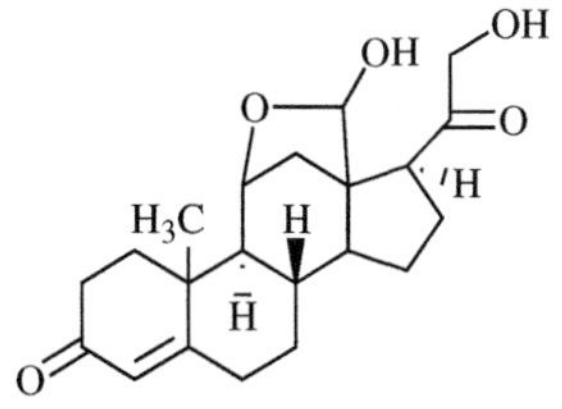

Aldosterone
5.44

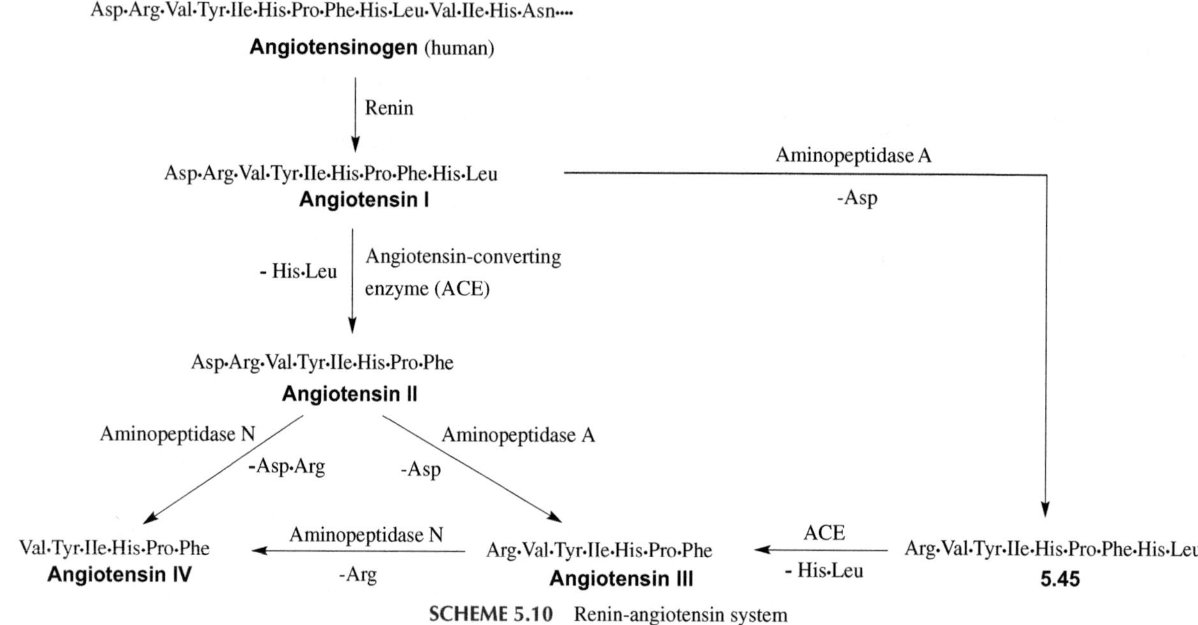

SCHEME 5.10 Renin-angiotensin system

5.2.4.2.2. Lead Discovery

In 1965, Ferreira[87] reported that a mixture of peptides in the venom of the South American pit viper *Bothrops jararaca* potentiated the action of bradykinin by inhibition of some bradykininase activity. Bakhle and coworkers[88] subsequently showed that these peptides also inhibited the conversion of angiotensin I to angiotensin II. Nine active peptides were isolated from this venom; the structure of a pentapeptide (Pyr-Lys-Trp-Ala-Pro, where Pyr is L-pyroglutamate) was identified.[89] This peptide was shown to inhibit the conversion of angiotensin I to II and bradykinin degradation in vitro[90] and in vivo.[91] The structures of six more of the peptides were determined by Ondetti and coworkers.[92] The peptide with the greatest in vitro activity was the pentapeptide,[93] but a nonapeptide (Pyr-Trp-Pro-Arg-Pro-Gln-Ile-Pro-Pro) called teprotide had the greatest in vivo potency[94] and was effective in lowering blood pressure.[95] Five other active peptides were isolated from the venom of the Japanese pit viper *Agkistrodon halys blomhoffii*.[96] Because these compounds were peptides, they were not effective when administered orally, but they laid the foundation for the design of orally active ACE inhibitors.

5.2.4.2.3. Lead Modification and Mechanism of Action

The fact that *N*-acylated tripeptides are substrates of ACE indicated that it may be possible to prepare a small orally active ACE inhibitor. After testing numerous peptides as competitive inhibitors of ACE, it was concluded that proline was best in the C-terminal position and alanine was best in the penultimate position. An aromatic amino acid is preferred in the antepenultimate position.

When the search for a potent inhibitor of ACE was initiated at Squibb (now Bristol-Myers Squibb (BMS)) and Merck pharmaceutical companies, the enzyme had not yet been purified. Because the enzyme was inhibited by ethylenediaminetetraacetic acid and other chelating agents, particularly bidentate ligands, it was believed to be a metalloenzyme. In fact, ACE purified to homogeneity from rabbit lung[97] was shown to contain 1 gram-atom of zinc ion per mole of protein. The zinc ion is believed to be a cofactor that assists in the catalytic hydrolysis of the peptide bond by both coordination to the carbonyl oxygen, making the carbonyl more electrophilic, and by coordination to a water molecule, making the water more nucleophilic. Coordination of both molecules to the zinc ion lowers the activation energy for attack of the water on the scissile peptide bond (Figure 5.8). Because the structure of the enzyme was not known, it was not obvious what peptidelike structures would be the best inhibitors. It was hypothesized that the mechanism and active site of ACE may resemble those of carboxypeptidase A, another zinc-containing peptidase whose X-ray structure was known.[98] Three important binding interactions between carboxypeptidase A and peptides are a carboxylate-binding group, a group that binds the C-terminal amino acid side chain, and the zinc ion that

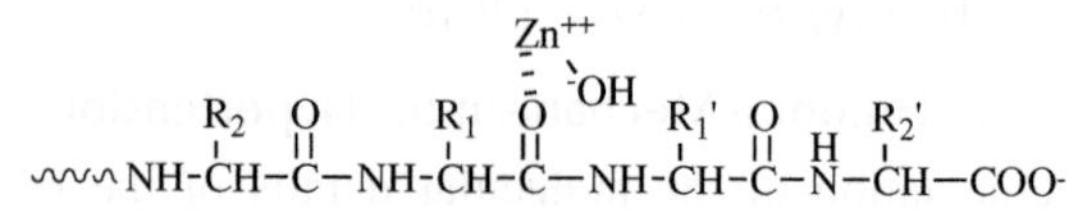

Zn^{++} ... OH

~~~NH-CH(R_2)-C(=O)-NH-CH(R_1)-C(=O)-NH-CH(R_1')-C(=O)-N(H)-CH(R_2')-COO⁻

FIGURE 5.8 Function of the Zn(II) cofactor in ACE catalysis
~~~

coordinates to the carbonyl of the penultimate (the scissile) peptide bond (Figure 5.9).[99] (*R*)-2-Benzylsuccinic acid, which can bind at all three of these sites, is a potent inhibitor of carboxypeptidase A.[100] The extreme potency of inhibition of carboxypeptidase A by (*R*)-2-benzylsuccinic acid was suggested to be derived from the resemblance of this inhibitor to the *collected products* (Figure 5.10) of hydrolysis of the substrate, and, therefore, it combines all of their individual binding characteristics into a single molecule. With this as a model, and the known effectiveness of a C-terminal proline for ACE inhibition, a series of peptidomimetic carboxyalkanoylproline derivatives (**5.46**) were tested as inhibitors of ACE. Note that to increase stability and decrease peptide-like character, the N-terminal amino group was substituted by an isosteric CH_2 group to which the Zn(II)-coordinating carboxylate was attached. Although the results were encouraging, all of these compounds were only weak inhibitors of ACE. To increase the potency of the compounds, a better Zn(II)-coordinating ligand, a thiol group, was substituted for the carboxylate (**5.47**).[66] These compounds were very potent inhibitors of ACE. Figure 5.11 shows a hypothesized depiction of the interaction of **5.46** and **5.47** with ACE. Note that carboxypeptidase A is a C-terminal *exopeptidase* (it cleaves

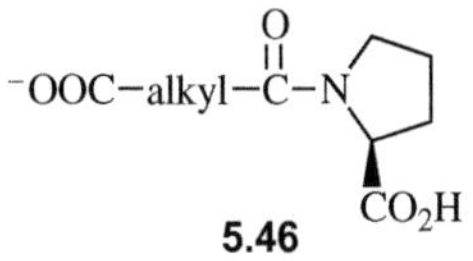

5.46 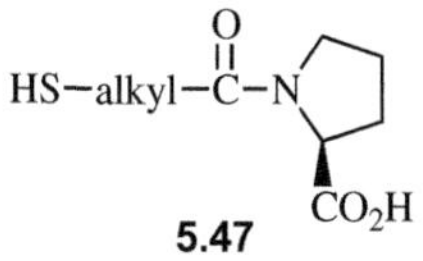 **5.47**

the C-terminal amino acid), whereas ACE is a C-terminal *endopeptidase* or, more precisely, a *dipeptidyl*

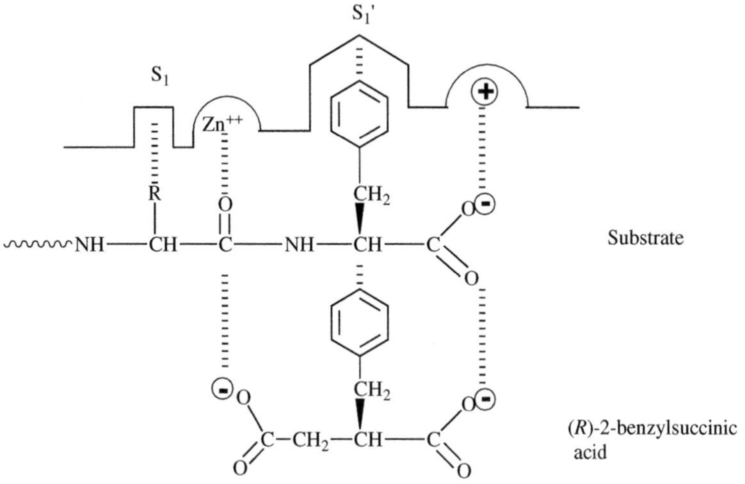

FIGURE 5.9 Hypothetical active site of carboxypeptidase A. *Adapted with permission from Cushman, D. W., Cheung, H. S., Sabo, E. F., Ondetti, M. A.* Biochemistry *1977* 16, *5484. Copyright© 1977 American Chemical Society*

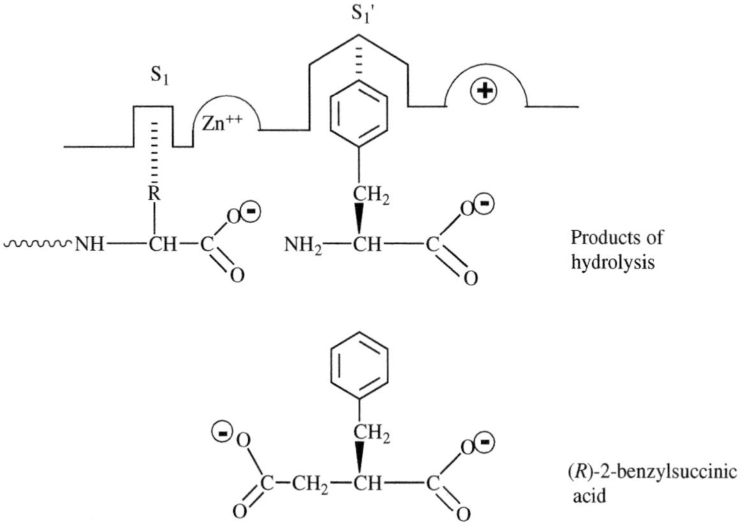

FIGURE 5.10 The collected products hypothesis of enzyme inhibition using inhibition of carboxypeptidase A as an example

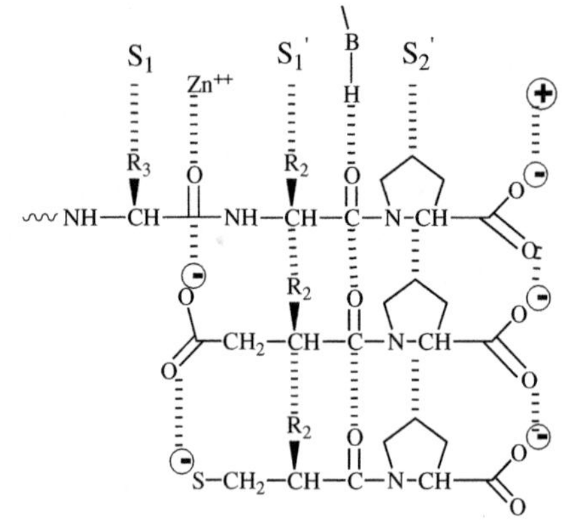

Substrate

5.46

5.47

FIGURE 5.11 Hypothetical binding of carboxyalkanoylproline and mercaptoalkanoylproline derivatives to ACE. *Adapted with permission from Cushman, D. W., Cheung, H. S., Sabo, E. F., Ondetti, M. A.* Biochemistry *1977* 16, *5484. Copyright© 1977 American Chemical Society*

carboxypeptidase (it cleaves a C-terminal dipeptide). Therefore, the active site of ACE (Figure 5.11) has two additional binding sites than carboxypeptidase A (Figure 5.9) has between the Zn(II) and the group that interacts with the C-terminal carboxylate group. The compound that had the best binding properties was **5.48** (captopril), a competitive inhibitor of ACE with a K_i of 1.7×10^{-9}M under standard assay conditions. Furthermore, captopril is highly specific for ACE; the K_i values for captopril with carboxypeptidase A and carboxypeptidase B, two other Zn(II)-containing peptidases, are 6.2×10^{-4}M and 2.5×10^{-4}M, respectively.[101] Presumably, the reason for the specificity is that there are many functional groups in **5.48** that can regio- and stereospecifically interact with groups at the active site of ACE, but they cannot interact to the same degree or perhaps at all with groups in other peptidases (compare Figures 5.9 and 5.11). The carboxylate group of the inhibitor can be stabilized by an electrostatic interaction with a cationic group on the enzyme, the amide carbonyl can be hydrogen bonded to a hydrogen donor group, the sulfhydryl can be ligated to the zinc ion, and the proline and (*S*)-methyl group can be involved in stereospecific hydrophobic and van der Waals interactions.

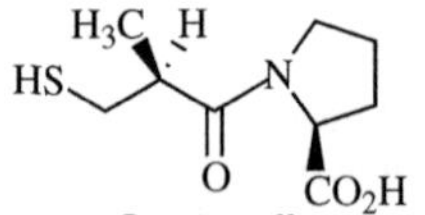

Captopril
5.48

All of these interactions must be important because deletion or alteration of any of these groups raises the K_i considerably (Table 5.2). A myriad of analogs of this basic structure, including compounds with Zn(II)-coordinating ligands other than carboxylate and thiol groups, have been synthesized and tested as ACE inhibitors (see the general references).

TABLE 5.2 Effect on K_i of Structural Modification of Captopril

Analog	Relative K_i
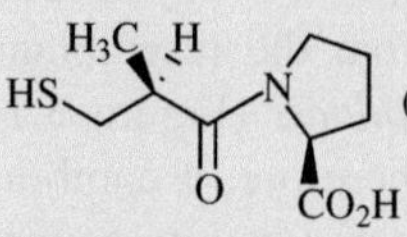 (Captopril)	1.0
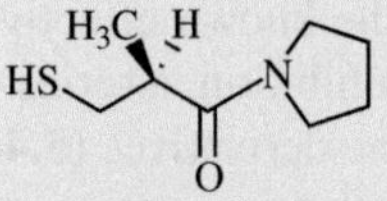	12,500
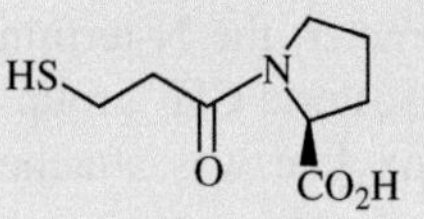	10
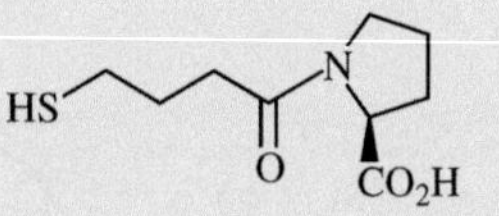	12,000
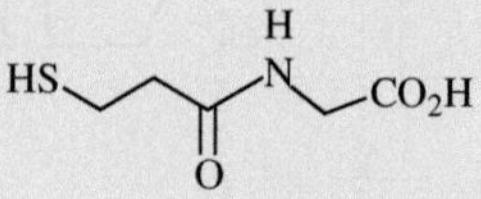	120
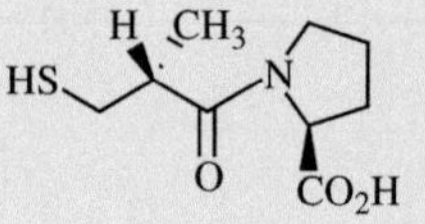	120
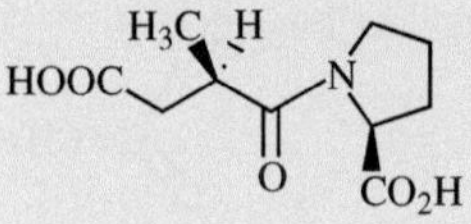	1100

Captopril was the first ACE inhibitor in the drug market, and it was shown to be effective for the treatment of both hypertension and congestive heart failure. Given alone, captopril can normalize the blood pressure of about 50% of the hypertensive population. When given in combination with a diuretic, such as hydrochlorothiazide (**5.49**; Aldoril) (remember, angiotensin II releases aldosterone, which causes water retention), this can be extended to 90% of the hypertensive population. In more severe cases, an antagonist for the β-adrenergic receptor (a *β-blocker*), which triggers vasodilation, may be used in a triple therapy with captopril and a diuretic.

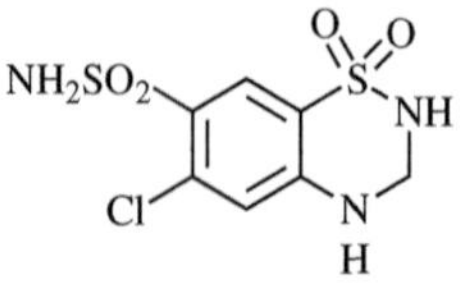

Hydrochlorothiazide
5.49

Two side effects were observed in some patients during the early usage of captopril, namely, rashes and loss of taste. Both of these side effects were reversible on drug withdrawal or reduction of the dose.[102] Considering the potential lethality of hypertension, a minor rash or loss of taste would seem insignificant to the benefits of the therapy. However, hypertension is a disease without a symptom (that is, until it is too late); generally, a patient discovers he has this disease when his physician takes his blood pressure. Because of this lack of immediate discomfort, there may be difficulties getting the patient to comply with the therapy, especially if unpleasant side effects arise when the drug is taken.

Consequently, a Merck group investigated the cause for the side effects. Because similar side effects arise when penicillamine is administered, it was hypothesized that the thiol group may be responsible.[103] Furthermore, deletion of this functional group should give inhibitors greater metabolic stability because thiols undergo facile in vivo oxidation to disulfides. The approach taken was to attempt to increase the previously found weak potency of the carboxyalkanoylproline analogs by adding groups that can interact with additional sites on the enzyme, that is, to increase the pharmacophore.

If the carboxyalkanoylproline derivatives are collected product inhibitors (see Figure 5.10), then there are two features that can be built into these analogs to make them look more productlike. One is to make them structurally more similar to product dipeptides by substituting an NH for a CH_2 such as **5.50** (R=R′=H). Disappointingly, however, this compound had less than twice the potency of the isostere with a CH_2 in place of NH. The reason for this could be compensatory factors. An NH (or its protonated form) is much more hydrophilic than a CH_2 group and might make the molecule too hydrophilic. If that occurs, then an additional hydrophobic group could be added to counterbalance this hydrophilic effect. When a methyl group was appended (**5.50**, R=CH_3, R′=H), the potency increased about 55-fold. Because the other feature that could make these compounds structurally more similar to the collected products would be to append a group that might interact with the substrate S_1 subsite, the R group of **5.50** was modified further, and **5.50** (R=(*S*)-$PhCH_2CH_2$, R′=H), called enalaprilat (see Section 5.2.3.2.1), emerged as the viable drug candidate (note, in the early studies with peptides, the antepenultimate site favored an aromatic group). The IC_{50} for enalaprilat under standard assay conditions is 19 times lower than that for captopril, suggesting that there are increased interactions of enalaprilat with ACE. These may be hydrophobic interactions of the phenylethyl group with the S_1 subsite (see Figure 5.5).

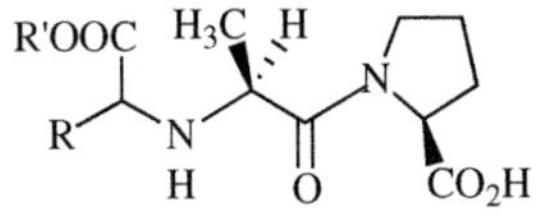

5.50

The interactions of captopril and enalaprilat with purified rabbit lung ACE was studied in detail, and the kinetics indicated that these compound are slow, tight-binding inhibitors.[104] For example, the rate constant for the formation of the E·I complex (k_{on}) with enalaprilat at pH 7.5 was determined to be 2×10^6 $m^{-1}s^{-1}$, which is at least two orders of magnitude smaller than expected for a diffusion-controlled reaction. Steady-state kinetics gave the value of the K_i as 1.8×10^{-10}M. Since $K_i=k_{off}/k_{on}$, the k_{off} should be 3.6×10^{-4} s^{-1}, which is in satisfactory agreement with the measured k_{off} value of 1.6×10^{-4} s^{-1}. The small k_{off} value for this noncovalent E·I complex emphasizes the strong affinity of enalaprilat for ACE.

Enalaprilat, however, is poorly absorbed orally and, therefore, must be given by intravenous (IV) injection. This problem was remedied simply by conversion of the carboxyl group to an ethyl ester (**5.50**, R=(*S*)-$PhCH_2CH_2$, R′=CH_3CH_2), giving enalapril, which has excellent oral activity. Because the in vitro IC_{50} for enalapril is 10^3 times higher than that for enalaprilat, the ethyl ester group must be hydrolyzed by esterases in the body to liberate the active form of the drug, namely, enalaprilat. Enalapril, then, is an example of a *prodrug*, a compound that requires metabolic activation for activity (see Chapter 9). The effect of esterification is to lower the pK_a of the NH group (pK_a 5.5 in enalapril, but 7.6 in enalaprilat)[105] and to remove the charge of the carboxylate, both of which would increase membrane permeability. A prodrug related to enalapril is trandolapril (**5.51**, Mavik), which is highly lipophilic with prolonged ACE inhibitory activity, even at a dose of 2 mg once a day.

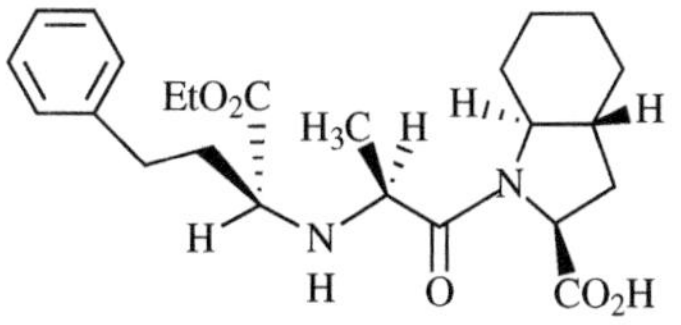

Trandolapril
5.51

Another compound, prepared by the Merck chemists as an alternative to enalaprilat, was the lysylproline analog called lisinopril (**5.52**; Prinivil). Note the stereochemistry shown (*S,S,S*) is that found in the most potent isomer of lisinopril, and is also the stereochemistry of enalaprilat. Lisinopril is more slowly and less completely absorbed than enalapril, but its longer oral duration of action, and the fact that it does not require metabolic activation, makes it an attractive alternative. The crystal structure of lisinopril bound to ACE showed that it bears little similarity to the structure of carboxypeptidase A, which was the protein used as the basis for the design of captopril.[106]

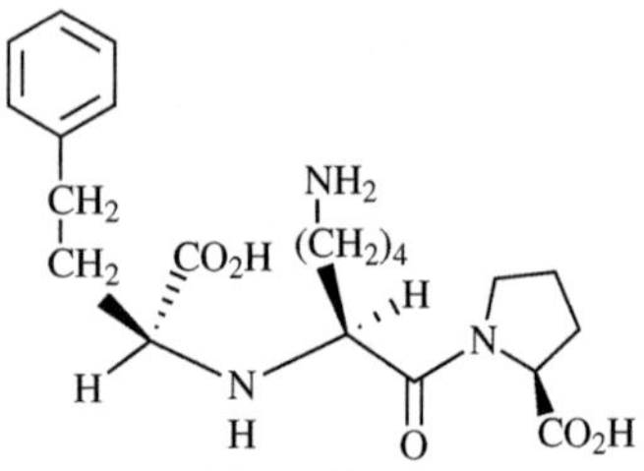

Lisinopril
5.52

5.2.4.2.4. Dual-Acting Drugs: Dual-Acting Enzyme Inhibitors

When there are two related enzymes whose inhibition would give an enhanced effect compared to the sum of the effects of inhibiting either enzyme alone (see Drug Synergism, Chapter 7), it may be beneficial to design a single inhibitor of both enzymes, a compound known as a *dual-acting enzyme inhibitor*. There are several advantages to the design of one compound that inhibits two different enzymes rather than two compounds, one for each enzyme: (1) with two drugs, two separate syntheses, two formulations, and two different metabolism studies (see Chapter 8) have to be developed; (2) two drugs will have different pharmacokinetic rates and metabolic profiles, making it difficult for both to be optimal in the same time frame; (3) the likelihood that both drugs would progress to the clinic at the same rate is small; (4) the cost for three sets of safety studies and three separate clinical trials (one for each drug plus one with the combination) would be enormous; and (5) the odds for a single drug just starting clinical trials to be approved for the drug market is 1 in 10 and that for two drugs entering the market would be 1 in 100! With a single, dual-acting drug, none of these problems exists.

An example of dual-acting enzyme inhibition is related to our discussion above of the "pril" antihypertensive agents. Neutral endopeptidase (NEP) is another Zn^{++}-containing endopeptidase. It degrades and deactivates atrial natriuretic peptide (ANP), a 28-amino acid vasoactive peptide hormone produced by the heart that causes vasodilation and inhibition of the formation of aldosterone, the steroid hormone that regulates the electrolyte balance and leads to retention of sodium ions and water. Therefore, ANP acts to *lower* the blood pressure; the hormone actions of ANP and angiotensin II, therefore, are functionally opposite. If the formation of angiotensin II is *blocked* by the inhibition of ACE, and the concentration of ANP is *increased* by the inhibition of NEP, there should be a synergistic (or at least additive) antihypertensive effect.[107]

The lead compound devised at BMS (**5.53**), which had a structure composed of part of captopril and enalapril, was found to have IC_{50} values for ACE and NEP of 30 and 400 nM, respectively, which is excellent for a lead compound.[108] The thiol group was retained because subsequent to the launch of captopril it was found that the loss of taste and rash side effects initially observed were related to high dosing, not to the sulfur atom hypothesized at Merck; at a lower dose, captopril was just as effective, but without those side effects. Homologation gave a compound that was potent in vitro for both enzymes, but not very potent in vivo (**5.54**). The research group turned to conformationally restricted analogs to increase potency. Compound **5.55** was potent both in vitro and in vivo, and **5.56** had IC_{50} values of 5 and 17 nM for ACE and NEP, respectively, with oral activity greater than that of captopril in rats. A 7,6-fused bicyclic thiazepinone analog (**5.57**, omapatrilat, Vanlev) with IC_{50} values of 5 and 8 nM for ACE and NEP, respectively, was advanced to clinical trials.[109] The key advantages of the dual-acting enzyme inhibitor omapatrilat over the ACE inhibitors is its ability to lower both diastolic and systolic blood pressure better and its effectiveness in controlling subpopulations of patients, such as African-Americans and diabetics, where traditional drugs have been less effective. Unfortunately, omapatrilat was not approved by the Food and Drug Administration (FDA) because of a larger incidence of angioedema (swelling of the eyelids, lips, face, tongue, and throat) than ACE inhibitors, which is caused by the increased bradykinin levels that occurs more with these inhibitors than ACE inhibitors.[110]

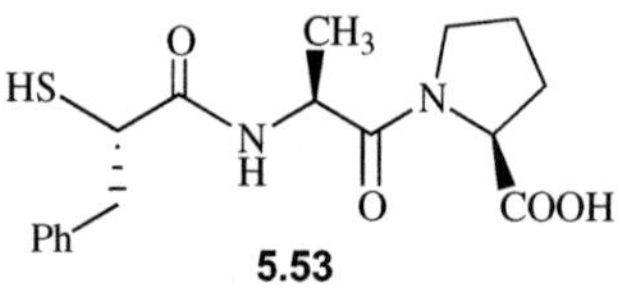

5.53

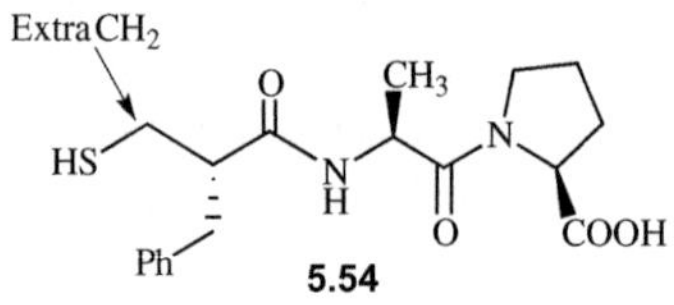

Extra CH_2

5.54

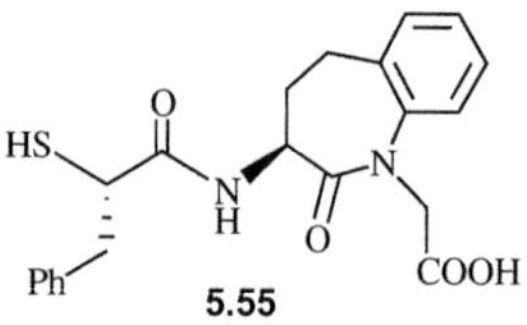

5.55

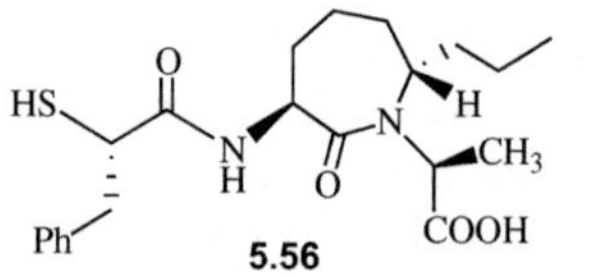

5.56

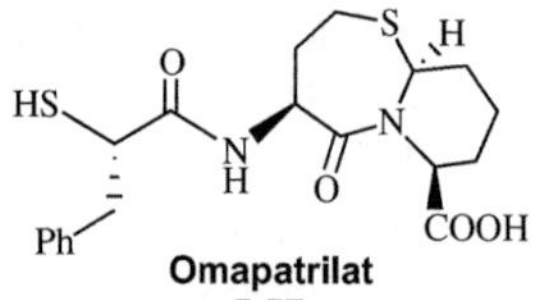

Omapatrilat
5.57

This dual-acting inhibitor approach for the design of new antihypertensive agents has been extended further; there is a third zinc metalloprotease, endothelin-converting enzyme (ECE), which displays high amino acid sequence identity with NEP, especially in the active site,[111] and hydrolyzes a 38-amino acid inactive peptide into endothelin-1, a 21-amino acid peptide that is the most potent vasoconstrictor (constricts blood vessels, thereby raising blood pressure) known.[112] Several classes of compounds that act as triple-acting enzyme inhibitors for ACE, NEP, and ECE have been identified.[113]

A *dual-acting drug* does not have to be limited to compounds that inhibit two different enzymes. It could also be a compound that inhibits one enzyme and acts as an antagonist for a receptor, a compound that is an antagonist for two different receptors, an agonist for two different receptors, or any combination thereof.

An example of a dual-acting drug that acts as an inhibitor of an enzyme and as an antagonist for a receptor is Z-350 (**5.58**),[114] an inhibitor of steroid 5α-reductase and an antagonist for the α_1-adrenoceptor.[115] Benign prostatic hyperplasia (BPH) is a progressive enlargement of the prostate gland, leading to bladder outlet obstruction. This obstruction consists of a static component related to prostatic tissue mass and a dynamic component related to excessive contraction of the prostate and urethra.[116] Antagonists of the α_1-adrenoceptor, such as terazosin HCl (**5.59**, Hytrin), are used to relax the smooth muscle of the prostate and urethra.[117] Because dihydrotestosterone is known to be a dominant factor in prostatic growth, inhibitors of steroid 5α-reductase, the enzyme that converts testosterone into dihydrotestosterone, such as finasteride (**5.60**, Proscar), are also used to treat BPH. A dual-acting agent was designed[118] on the basis of the α_1-adrenoceptor antagonist **5.61** (the structure is drawn in a conformation to resemble **5.58**) and the steroid 5α-reductase inhibitor **5.62** by combining features of both molecules into one structure (**5.63**), a common approach for the design of dual-acting agents. This compound was a potent antagonist for the α_1-adrenoceptor but needed increased potency against steroid 5α-reductase. It was well established that the lipophilic part and the butanoic acid moiety are essential for steroid 5α-reductase activity, but the benzanilide moiety could be replaced by an acyl indole, so the next structures included **5.64**, which is as potent as **5.63** as an antagonist for α_1-adrenoceptor and more potent in steroid 5α-reductase inhibition.[119] Merging the structures of **5.63** and **5.64** led to the design of **5.58**, which significantly reduces prostatic growth in rabbits and rats.

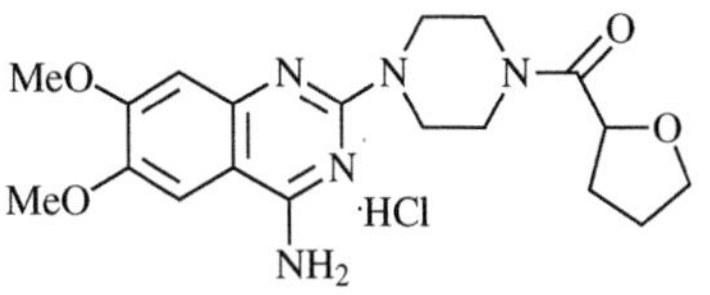

Terazosin HCl
5.59

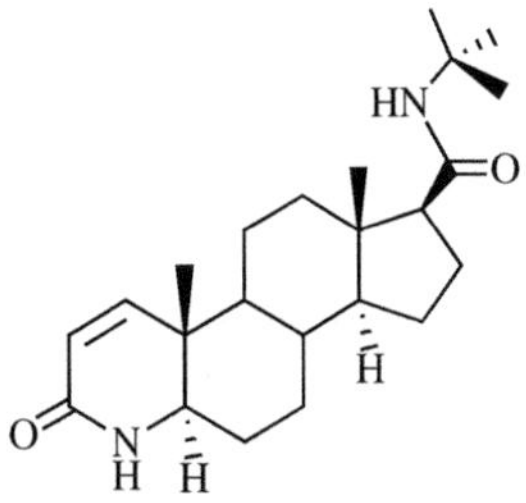

Finasteride
5.60

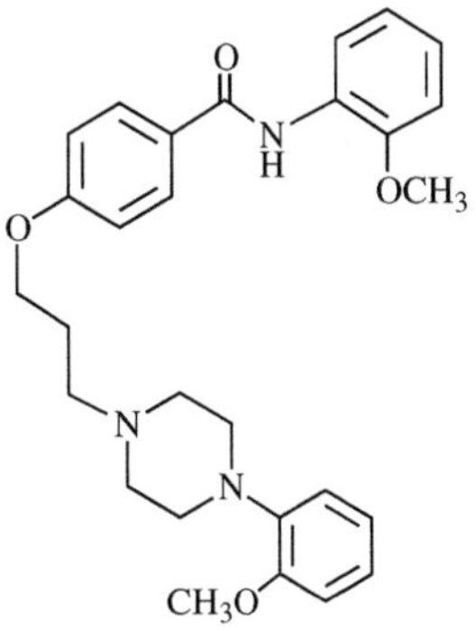

5.61

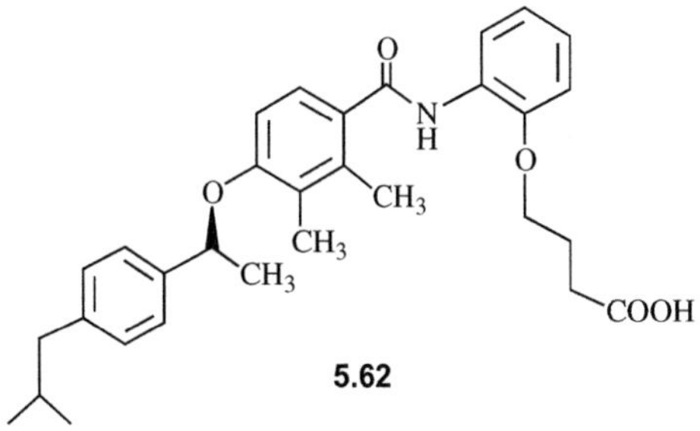

5.62

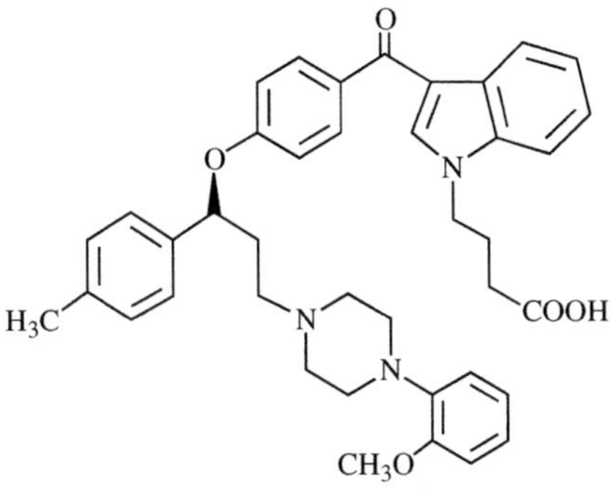

Z-350
5.58

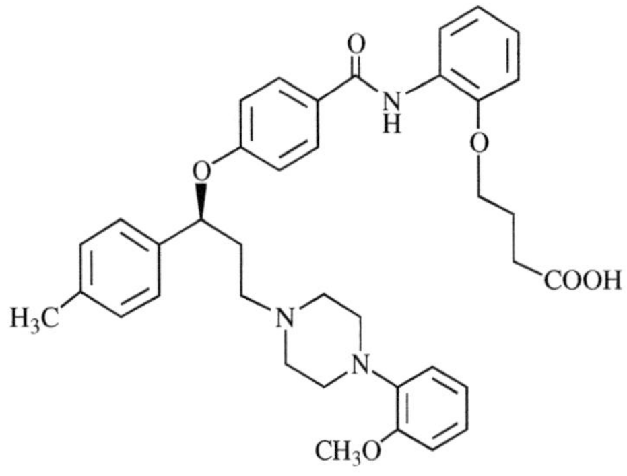

5.63

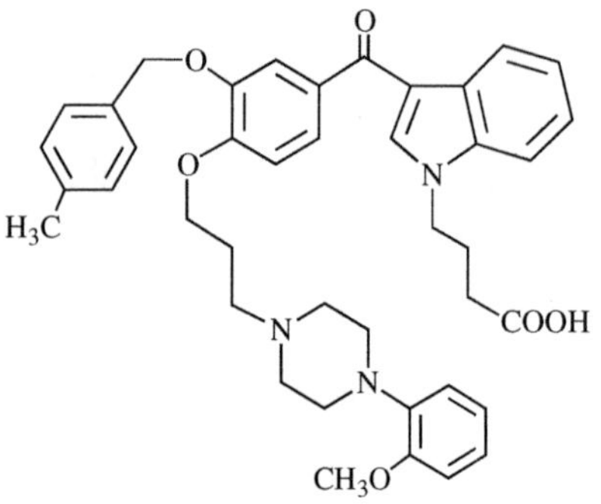

5.64

An example of a dual-acting receptor agonist is a compound that acts as an agonist for both the D_2-receptor and the β_2-adrenoceptor for the treatment of airway diseases, such as chronic obstructive pulmonary disease (COPD) and asthma. A D_2-receptor agonist reduces reflex bronchoconstriction, dyspnea, cough, and mucus production, but less likely diminishes bronchoconstrictor activity of locally released mediators of bronchoconstriction.[120] β_2-Adrenoceptor agonists are the most commonly used antibronchoconstrictor agents,[121] but they have little effect on dyspnea, cough, and mucus production. A dual-acting agent for these receptors should combine all the desired features of this class of drugs. The structures of a weak β_2-adrenoceptor agonist (**5.65**) and a potent D_2-receptor agonist (**5.66**) were hybridized to give **5.67**. Modification of the side chain gave **5.68**,[122] which went into clinical trials for the treatment of the symptoms of COPD.

5.2.4.3. *Lovastatin (Mevinolin) and Simvastatin, Antihypercholesterolemic Drugs*

An example of noncovalent slow, tight-binding inhibitors is a family of drugs used to treat high cholesterol levels.

5.2.4.3.1. Cholesterol and Its Effects

Coronary heart disease is the leading cause of death in the United States and other Western countries; about one-half of all deaths in the United States can be attributed to atherosclerosis,[123] which results from the buildup of fatty deposits called plaque on the inner walls of arteries. The major component of atherosclerotic plaque is cholesterol. In humans, more than one-half of the total body cholesterol is derived from its de novo biosynthesis in the liver.[124] Cholesterol biosynthesis requires more than 20 enzymatic steps starting from acetyl CoA. The rate-determining step is the conversion of 3-hydroxy-3-methylglutaryl coenzyme A (HMG-CoA; **5.69**) to mevalonic acid (**5.70**), catalyzed by HMG-CoA reductase (Scheme 5.11). For the structure of coenzyme A, see Chapter 4, **4.13b**. Because hypercholesterolemia is a primary risk factor for coronary heart disease,[125] and the overall rate of cholesterol biosynthesis is a function of this enzyme, efforts were initiated to inhibit HMG-CoA reductase as a means of lowering plasma cholesterol levels.[126]

5.2.4.3.2. Lead Discovery

Endo and coworkers[127] at the Sankyo Company in Tokyo tested 8000 strains of microorganisms for metabolites that inhibited sterol biosynthesis in vitro and discovered three active compounds in the culture broths of the fungus *Penicillium citrinum*. The most potent compound, called mevastatin, was also isolated from broths of *Penicillium brevicompactum* by Brown and coworkers[128] at Beecham Pharmaceuticals in England, who named it compactin (**5.72**, R=H). A second, more potent, compound was isolated by Endo[129] from the fungus *Monascus ruber*, which he named monacolin K; the same compound was isolated by a group at Merck from *Aspergillus terreus*,[130] which they named mevinolin (**5.72**, R=CH_3). Mevinolin is now known as lovastatin (Mevacor). Several related metabolites were also isolated from cultures of these fungi,[131] including dihydrocompactin from *P. citrinum* (**5.73**, R=H, R′=(*S*)-$CH_3CH_2CH(CH_3)CO_2^-$), dihydromevinolin from *A. terreus* (**5.73**, R=CH_3, R′=(*S*)-$CH_3CH_2CH(CH_3)CO_2^-$), and dihydromonacolin L from a mutant strain of *M. ruber* (**5.73**, R=CH_3, R′=H).

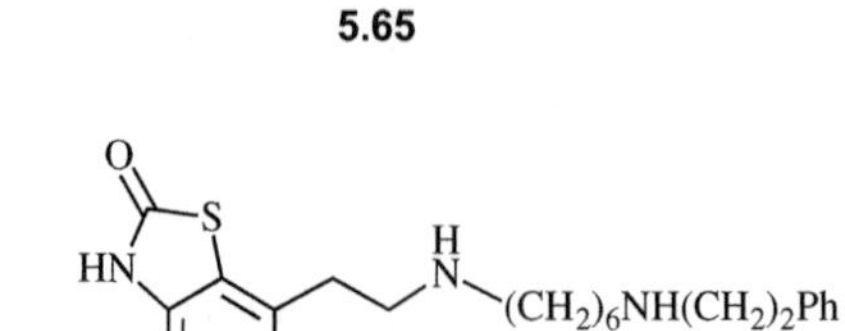

5.65

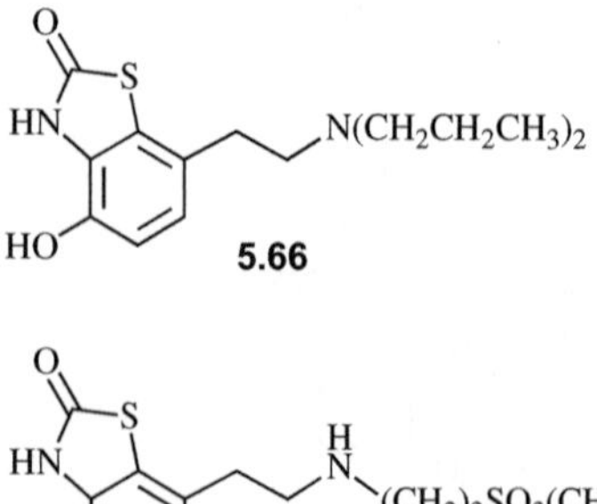

5.66

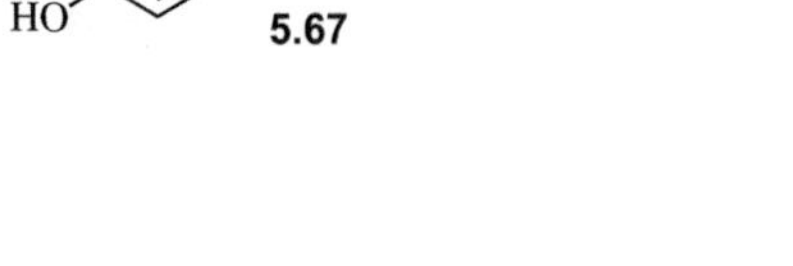

5.67

5.68

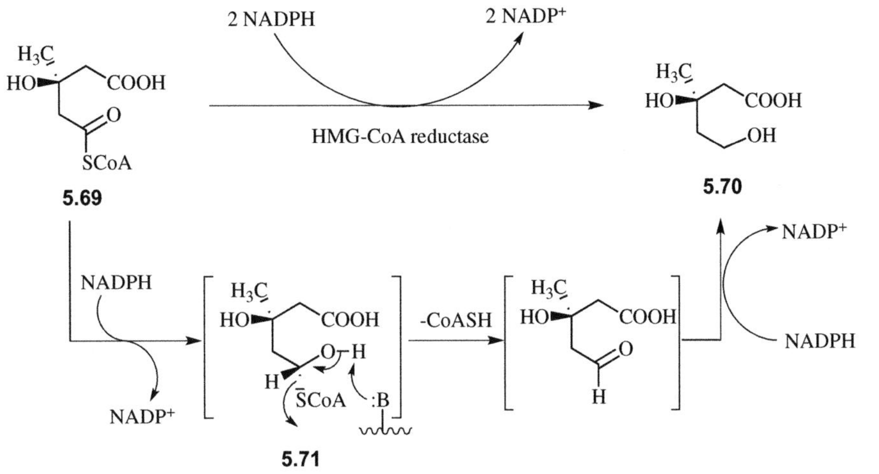

SCHEME 5.11 HMG-CoA reductase, the rate-determining enzyme in de novo cholesterol biosynthesis

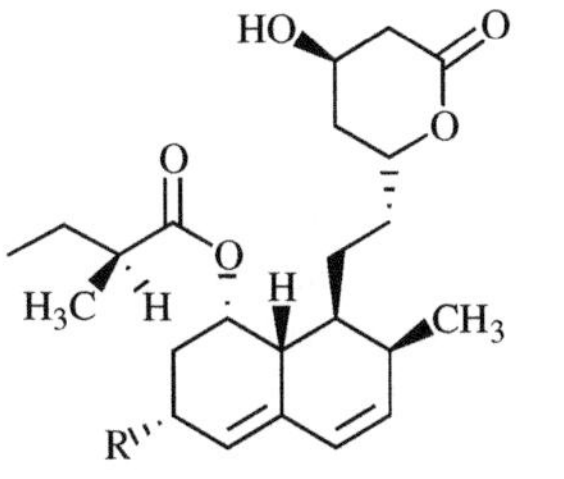

Mevastatin or compactin (R = H)
Monacolin K, mevinolin, or lovastatin (R = CH_3)
5.72

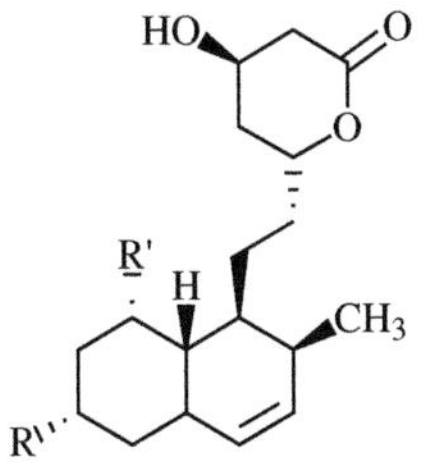

5.73

5.2.4.3.3. Mechanism of Action

Compactin[132] and lovastatin are potent competitive reversible inhibitors of HMG-CoA reductase. The K_i for compactin is 1.4×10^{-9} M and for lovastatin is 6.4×10^{-10} M (rat liver enzyme); for comparison, the K_m for HMG-CoA is about 10^{-5} M. Therefore, the affinity of HMG-CoA reductase for compactin and lovastatin is 7140 and 16,700 times, respectively, *greater* than that for its substrate. Compactin and lovastatin do not affect any other enzyme in cholesterol synthesis except HMG-CoA reductase.

It may not be immediately obvious why lovastatin is a competitive reversible inhibitor of HMG-CoA reductase, given that it does not closely resemble the structure of the substrate or product, which might be expected of a competitive inhibitor. The reason is that the active form is not that shown in **5.72**, but rather, the hydrolysis product, that is, the open chain 3,5-dihydroxyvaleric acid form (**5.74**). This form mimics the structure of proposed intermediate **5.71** (Scheme 5.11) in the reduction of HMG-CoA by HMG-CoA reductase, that is, if the structure of CoA (see Chapter 4, **4.13b**) is taken into account. Enzyme studies with compactin and analogs indicate that there are two important binding domains at the active site, the HMG binding domain, to which the upper part of **5.74** binds and a hydrophobic pocket located adjacent to the active site, to which CoA and the decalin (lower) part of **5.74** bind.[133] The high affinity of compactin and its analogs to HMG-CoA reductase derives from the simultaneous interactions of the two parts of these inhibitors, having an ethylene linker, with the two binding domains on the enzyme. This is the same phenomenon as was responsible for the increased binding of two inhibitors attached by a linker in SAR by NMR (see Chapter 2, Section 2.1.2.3.6). As a result of the interactions in two adjacent binding pockets, dissociation of the E·I complex is very slow. A kinetic analysis of the on and off rate constants (k_{on} and k_{off}, respectively; see Scheme 5.1) with yeast HMG-CoA reductase were 1.9×10^5 $m^{-1}s^{-1}$ and 0.11 s^{-1} for HMG-CoA and 2.7×10^7 $m^{-1}s^{-1}$ and 6.5×10^{-3} s^{-1} for compactin, respectively. Therefore, compactin binds faster and dissociates more slowly than does HMG-CoA, which accounts for the difference in their K_m and K_i (k_{off}/k_{on}) values. It is also possible to classify these inhibitors as transition state analogs (see Section 5.2.3).

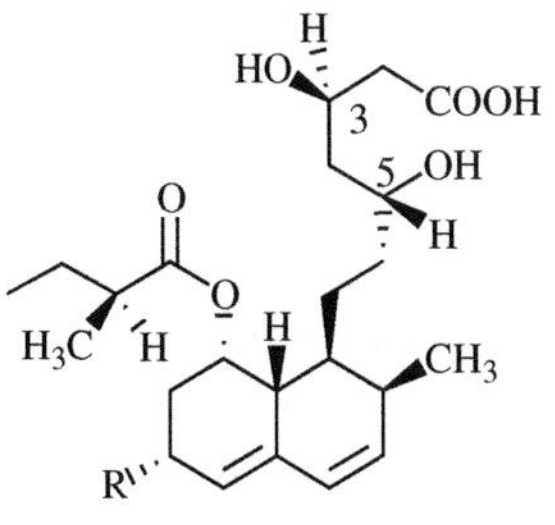

5.74

5.2.4.3.4. Lead Modification

Numerous structural modifications were made to compactin and lovastatin to determine the importance of the lactone moiety and its stereochemistry, the ability of the lactone moiety to be opened to the dihydroxy acid, the optimal length and structure of the moiety bridging the lactone and the lipophilic groups, and the size and shape of the lipophilic group. It was found that potency was greatly reduced[134] unless a carboxylate anion could be formed and the hydroxyl groups were left unsubstituted in an erythro relationship. Insertion of a bridging unit other than ethylene or (*E*)-ethenyl between the 5-carbinol moiety and the lipophilic moiety also diminishes the potency. Modifications of the lower (lipophilic) part of compactin, in most cases, led to compounds with considerably lower potencies, except for certain substituted biphenyl analogs.[135] If the substituted biphenyl rings were constrained as fluorenylidene moieties, the corresponding potencies decreased.[136] When the hydroxyl group in the lactone ring was replaced by an amino or thiol group, diminished potencies were observed.[137]

Modification of the 2(*S*)-methylbutyryl ester side chain ($CH_3CH_2CH\ (CH_3)\ CO_2^-$) of lovastatin indicated that introduction of an additional aliphatic group on the carbon α to the carbonyl group increased the potency of lovastatin.[138] To block ester hydrolysis and produce a compound with a much longer plasma half-life, a second methyl group was added, by synthesis,[139] to the side chain of the lower half of lovastatin (**5.72** ($R = CH_3$), but with ester side chain $CH_3CH_2C(CH_3)_2\ CO_2^-$) called simvastatin (Zocor), which has a potency about 2.5 times greater than that of lovastatin. The lactone epimer of lovastatin (the epimer of the carbon adjacent to the lactone oxygen in **5.72**) has less than 10^{-4} times the potency of lovastatin.[140] Modifications in the 3,5-dihydroxyvaleric acid moiety of analogs of **5.74** resulted in lower potencies, except when the 5-hydroxyl group was replaced by a 5-keto group, in which case potencies comparable to the parent compounds were observed. Presumably, the 5-keto group becomes reduced, but it is not known if this occurs by HMG-CoA reductase or by cytochrome (see Chapter 8; Section 8.4.2.2.1) and, if it does, whether it occurs prior to inhibition or whether it is the cause for inhibition.

The generic names of this family of drugs end in the suffix "-statin", so they are known collectively as the *statins*. The early analogs, such as compactin, lovastatin, and simvastatin, are classified as *type 1 statins*; the newer analogs, known as *type 2 statins*, retain the HMG mimic of the molecule, but the CoA mimic has been modified. Cerivastatin (**5.75**, Baycol), a very potent anticholesterol drug (0.1–0.15 mg taken once a day!), was launched by Bayer in 1999 but had to be recalled from the drug market in 2001 because of its association with deaths of some patients taking it who developed rhabdomyolysis, a side effect involving muscular weakness.

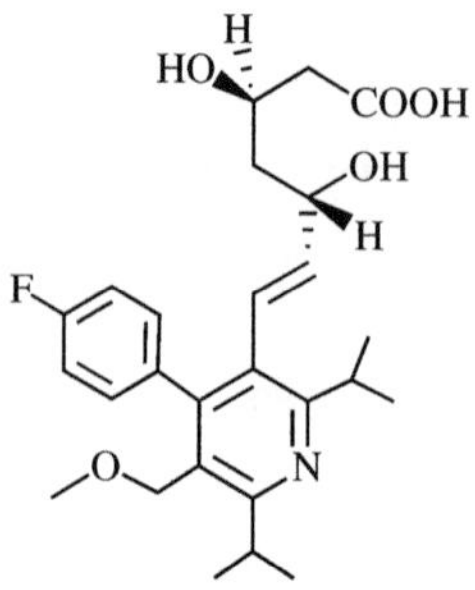

Cerivastatin
5.75

Although side effects generally have a negative connotation, another "side effect" of both the type 1 and type 2 statins is an enhancement in new bone formation in rodents.[141] This new bone growth is associated with an increased expression of the bone morphogenetic protein-2 gene in bone cells, which may have implications in the treatment of osteoporosis.

The X-ray crystal structures of several of the statins bound to HMG-CoA reductase confirm the biochemical hypothesis and show that they occupy a portion of the HMG-CoA binding site, thereby blocking access of the substrate to the active site.[142]

5.2.4.4. Saxagliptin, a Dipeptidyl Peptidase-4 Inhibitor and Antidiabetes Drug

Diabetes is a family of diseases characterized by high blood glucose levels resulting from defects in the body's ability to produce and/or use insulin, a peptide hormone produced in the pancreas that reduces blood sugar. The most common form, type 2 diabetes, characterized by elevated plasma glucose and glycosylated hemoglobin, is the leading cause of blindness, kidney failure, and limb amputation worldwide.[143] As of 2011, over 8% of the US population had diabetes (with an increasing number of children having the disease). This is a growing chronic problem, which has been a major focus of drug discovery research since the mid-1990s.

Glucagon-like peptide-1 (GLP-1) and *glucose-dependent insulinotropic polypeptide* are two *incretins*, gastrointestinal hormones that increase insulin by increasing insulin secretion from the pancreas (an *insulin secretagogue*), decreasing glucagon secretion from the liver (a *glucagon suppressor*), increasing insulin gene expression, decreasing food intake, and promoting insulin sensitivity. These polypeptides are the natural safeguards against diabetes; however, the half-life of GLP-1 is only about 2 min because of *dipeptidyl peptidase-4* (DPP-4), the serine protease that degrades it. Therefore, considerable effort by numerous laboratories to identify inhibitors of DPP-4 was initiated.[144]

BMS took as their starting point cyanopyrrolidine amide **5.76**, a compound reported by Ferring Research as a potent inhibitor of DPP-4,[145] which had moderate pharmacokinetic duration and chemical instability as a result of an intramolecular cyclization that occurred (Scheme 5.12). To retard

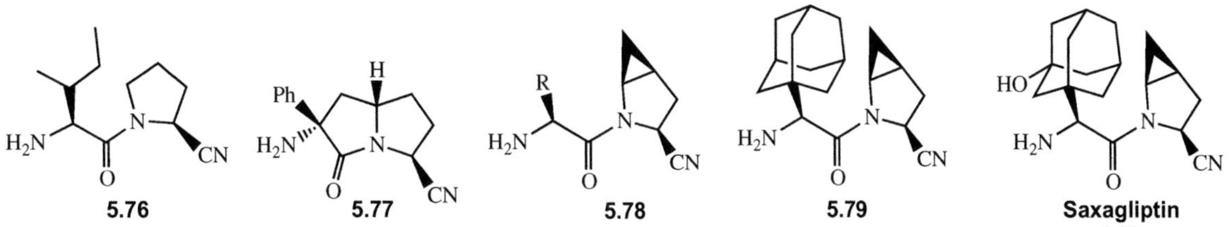

5.76 5.77 5.78 5.79 Saxagliptin 5.80

this cyclization, conformationally restricted analogs were required. A potent DPP-4 inhibitor from another company, Phenomix, had just such an active scaffold (**5.77**), which indicated that the cyclization could be prevented by conformational effects. However, to skirt the intellectual property issue as well as to combat the cyclization problem, BMS incorporated a fused cyclopropane ring onto each position of the pyrrolidine ring in both stereochemistries to impact the conformation of the pyrrolidine ring; the one that had the best pharmacodynamic and pharmacokinetic properties and solution stability was **5.78** (R=1-methylpropyl). It was then found that increasing the steric bulk (**5.78**, R=*tert*-butyl) further enhanced solution stability and in vitro potency (K_i=7 nM) with excellent pharmacokinetic properties (including >50% oral bioavailability in rats). Given the enhanced potency with increasing bulkiness of the R group, larger groups were added, and adamantyl was found to be most potent (**5.79**, K_i=0.9 nM). However, its in vivo stability was low with rapid metabolism and low bioavailability. The problem was found to be microsomal oxidation (see Chapter 8, Section 8.4.2.1.4) of the adamantyl ring to give **5.80** (saxagliptin, Onglyza), which was found to be almost as potent as **5.79** but with superior duration in vivo, slow metabolism with microsomes, and with >60% bioavailability in rats. Kinetic analysis of **5.80** showed that it was a slow, tight-binding inhibitor of DPP-4, which was dependent on two structural moieties: (1) the bulky adamantyl substituent to displace a water molecule at the active site, driving the slow entropic on rate and preventing enzymatic hydrolysis and (2) the nitrile, which formed a reversible, covalent bond with the active site serine, driving the slow enthalpic off rate; both of these properties resulted in the low K_i value. An X-ray crystal structure of **5.80** bound to DPP-4 revealed the covalent bond between Ser-630 and the nitrile carbon atom; however, upon dialysis, complete enzyme activity was recovered, indicating the reversibility of the bond.[146]

Before we discuss covalent irreversible inhibitors, let's look in detail at another case history for the design of a

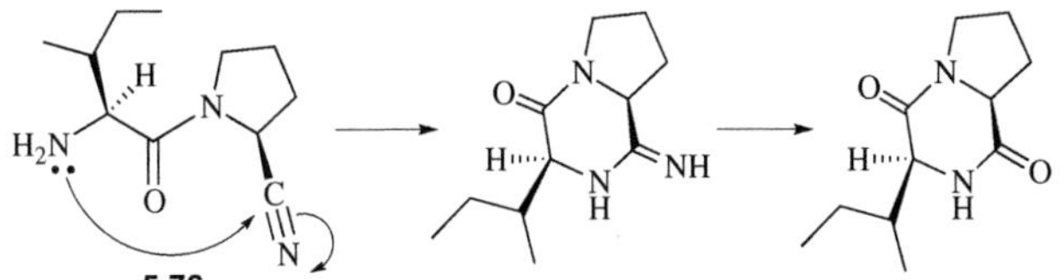

5.76

SCHEME 5.12 Intramolecular cyclization of **5.76**, which causes its short pharmacokinetic duration

competitive reversible inhibitor, one in which pharmacokinetics was of primary concern.

5.2.5. Case History of Rational Drug Design of an Enzyme Inhibitor: Ritonavir

The genome of the human immunodeficiency virus-1 (HIV-1) encodes an aspartate protease (HIV-1 protease), which proteolytically processes the *gag* and *gag-pol* gene products into mature, functional proteins.[147] If these processing steps are blocked, the progeny virions are immature and noninfectious. Consequently, HIV-1 protease should be an important target for design of inhibitors to act as potential anti-acquired immune deficiency syndrome (AIDS) drugs. Several companies discovered very potent and effective inhibitors by various lead discovery/modification approaches. What follows is the approach taken at Abbott Laboratories leading to ritonavir, a potent HIV-1 protease inhibitor with high oral bioavailability that was shown to be an effective drug for the treatment of AIDS. Similar approaches were taken by other pharmaceutical companies as well.

The general approach is first to find molecules that have good potency (preferably in the nanomolar range), and then use those molecules as starting points for solving pharmacokinetic problems. There has to be constant monitoring of potency while the pharmacokinetic issues are being addressed.

5.2.5.1. Lead Discovery

HIV protease is an unusual enzyme because the *homodimer* (two identical polypeptides come together to form the active enzyme) has C_2 *symmetry* (a 180° rotation about an axis through the center gives the same structure). This symmetry element was used as a key starting point in the design of novel inhibitor structures.[148] The initial plan was to design C_2 symmetric compounds that should show selectivity for HIV protease over other mammalian aspartate proteases because of the lack of symmetry with other aspartate proteases. The other design element was to make a transition state analog (or, more accurately, an intermediate analog). The proposed tetrahedral intermediate structure for the hydrolysis of a good asymmetric substrate, such as -Phe-Pro-, was bisected either through the scissile carbon (**5.81**, Figure 5.12) or adjacent to it (**5.82**). However, to make a C_2 symmetric compound, the two amino acid residues must be the same. Because the *P region* (the *N*-terminal side of the scissile bond) had been shown to be more important than the *P' region* (the *C*-terminal side of the scissile bond), the P′ region was deleted and the proline was substituted by phenylalanine. A C_2 symmetry operation was performed on the remainder of the substrate, generating two possible lead compounds, **5.83a** and **5.84a** (Figure 5.12), respectively. The amino groups were acylated with a variety of *N*-protecting groups to increase lipophilicity, including Ac (**b**), Boc (**c**), and Cbz (**d**). Compound **5.83b** was very weakly inhibitory; the stereoisomers of **5.84c**, however, were good inhibitors.

5.2.5.2. Lead Modification

To increase the pharmacophore of these inhibitors, P_2/P_2' residues were added. The Cbz-Val analogs, **5.83e** and **5.84e** (Figure 5.12), were low nanomolar and subnanomolar inhibitors, respectively; those analogs with the **5.84** structure were generally 10 times more potent than **5.83** analogs. Varying the stereochemistry of the hydroxyl groups in **5.84e** had little or no effect on inhibition. Compounds **5.84e** (different stereochemistries) were potent in vitro inhibitors of HIV-1 protease in H9 cells (IC_{50} values 20–150 nM). The therapeutic indices were in the range 500–5000, which is much better than is generally required for agents designed to treat life-threatening diseases.

At this point in the lead modification process, relatively potent inhibitors have been identified, so efforts can concentrate on pharmacokinetic problems. Because of generally poor pharmacokinetic behavior of peptides, peptidomimetics are generally sought (see Chapter 2, Section 2.2.4.5). However, peptidomimetics also can suffer from low oral/intestinal absorption and rapid hepatic elimination.[149] The major causes for poor peptidomimetic pharmacokinetic profiles are high molecular weight, low aqueous solubility, susceptibility to proteolytic degradation, hepatic metabolism, and biliary extraction.

The aqueous solubilities of **5.83e** and **5.84e** were very poor. The crystal structure of HIV-1 protease with **5.83e** bound[150] indicated that the Cbz groups were auxophoric, that is, they were not interacting with the protein as part of the pharmacophore and were also not interfering with protein binding. As discussed in Chapter 2 (Section 2.2.1), these are ideal groups to modify, because it is likely that modification will not affect potency. The crystal structure showed that there was room for structural modification, so the terminal Cbz phenyl groups were modified with polar, heterocyclic bioisosteres, such as pyridinyl (at the 2-, 3-, and 4-positions of the pyridinyl ring) and thiazolyl (at the 2- and 4-positions) groups.[151] The P_1 phenyl (from Phe) and P_2 isopropyl (from Val) groups were embedded in lipophilic pockets, so modification of these groups would not be fruitful. Conversion of one of the Cbz phenyls to a pyridinyl group increased the aqueous solubility by 20-fold without a change in the inhibitory potency, but the water solubility was still less than 1 μg/mL. More basic nonaromatic heterocycles had much greater water solubilities, but the potencies dropped. Substitution of both Cbz phenyl groups by heteroaromatic groups had little effect on the inhibitory potencies, but gave dramatic increases in water solubilities. Some of the compounds showed oral bioavailabilities in the range of 10–20%, but these bioavailabilities did not exceed concentrations required for effective anti-HIV activity in vitro. Compound **5.85** had

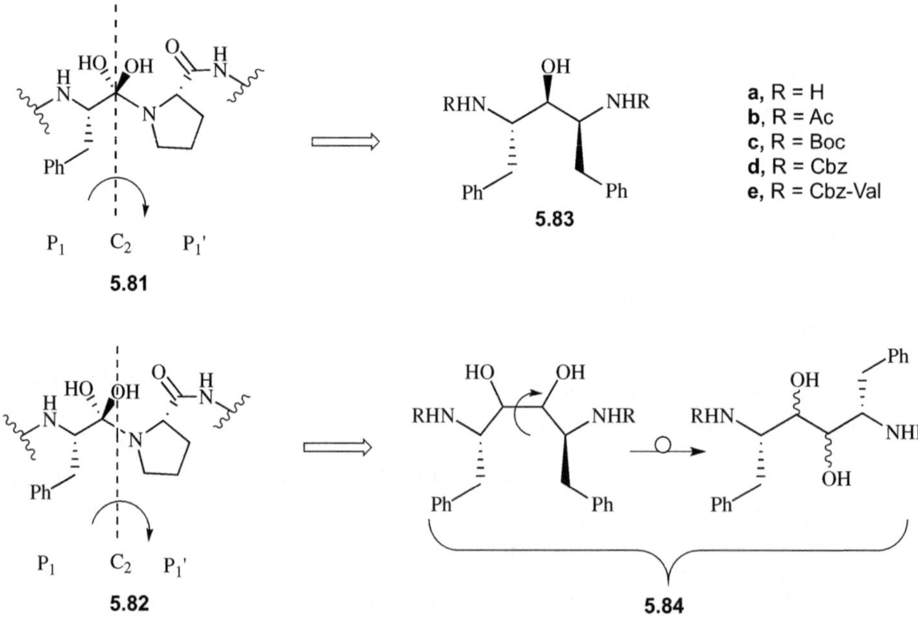

a, R = H
b, R = Ac
c, R = Boc
d, R = Cbz
e, R = Cbz-Val

FIGURE 5.12 Lead design for C_2 symmetric inhibitors of HIV protease based on the structure of the tetrahedral intermediate during hydrolysis of HIV protease

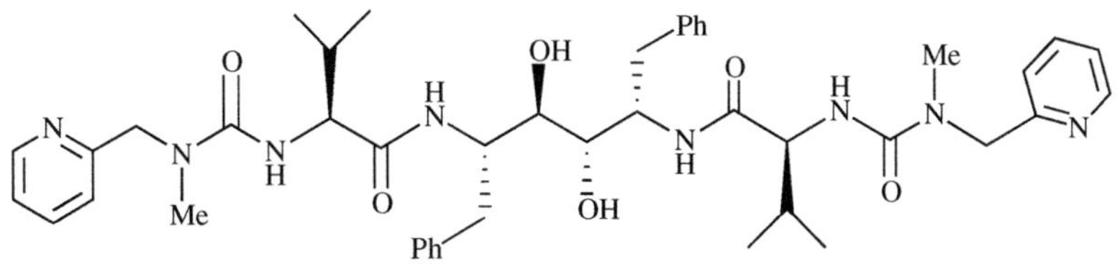

5.85

the best combination of inhibitory potency and aqueous solubility; unfortunately, it showed no oral bioavailability in rats. In general, the compounds with the **5.83** framework, although less potent than the compounds with the **5.84** framework, had consistently superior oral bioavailabilities in animals. To take advantage of both of these properties, the next series of analogs investigated had general structure **5.86**, which combined the extended framework of **5.84** with the mono alcohol structure of **5.83**.[152]

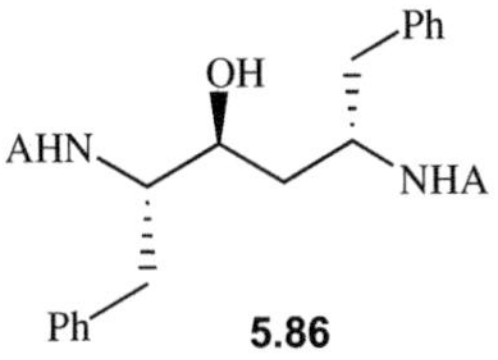

5.86

Although these were still poorly bioavailable in rats, the **5.86** series was more potent than the **5.84** series (IC_{50} values were generally subnanomolar). With all of these modifications and still no oral bioavailability, it is time to consider Lipinski's Rule of 5 (see Chapter 2, Section 2.1.2.3.2). One of his rules for oral bioavailability is that the molecular weight should not exceed about 500; the molecular weight of **5.85** is 794. To increase oral bioavailability, then, it was reasonable to try to decrease the size of the molecules. But to do that, and retain the C_2 symmetry, both ends of **5.85** would have to be deleted, but that brings the molecule back to the **5.83** and **5.84** series. That was when the C_2 symmetry design had to be put aside, and unsymmetrically substituted derivatives (different A groups at the *N*- and *C*-termini) of **5.84** and **5.86** were investigated.[153] Several SAR observations were made that were incorporated into the later designs: (1) incorporation of a carbamate linkage (ROCONHR′) resulted in improved potency over the use of an *N*-alkylurea linkage (RNR′CONHR″); (2) an *N*-ethylurea linkage was less tolerated than *N*-methylurea; (3) there was no difference between 2-pyridinyl- and 3-pyridinyl groups at P_3; and (4) methyl substitution on the P_3 pyridinyl group did not diminish potency and often enhanced it. With regard to the pharmacokinetics, there was little correlation between aqueous solubility and oral bioavailability! Therefore, although aqueous solubility is a necessary (or at least helpful) parameter for oral bioavailability, it certainly is not always a sufficient one. This makes sense. Aqueous solubility serves an important role in helping to increase the exposure of the drug to the gut membranes through which it must pass; however, solubility can have less of an influence than other factors on how well the drug will pass through those membranes or how well the drug will survive metabolism by the liver, which is one of the first hurdles that a drug faces on entering the bloodstream by the oral route (see Chapter 8, Section 8.1). Compounds with an *N*-methylurea linkage between the pyridinyl group and the P_2 aminoacyl residue generally showed greater oral bioavailability and solubility than the ones with a carbamate linkage. Also, compounds in the **5.86** series exhibited greater oral bioavailability than those in the **5.84** series. This is presumed to be because of improved absorption in the gut, but a decreased liability to liver metabolism or biliary excretion are other possible contributors.

A good measure of the overall potential of HIV-1 protease inhibitors is the ratio of the maximum plasma levels achieved in vivo (C_{max}) to the concentration required for anti-HIV activity in vitro (ED_{50}). The compound that emerged from this study, **5.87**, has a C_{max}/ED_{50} of 21.6 (4.11 μM/0.19 μM). The aqueous solubility is 3.2 μg/mL, the oral bioavailability is 32%, and the plasma half-life after a 5 mg/kg IV dose was 2.3 h. This molecule is appreciably smaller than those in the **5.83** and **5.84** series (but the molecular weight is 653, still quite high), yet it maintained the submicromolar in vivo antiviral activity. However, the relatively short plasma half-life prohibits the maintenance of the plasma levels sufficiently in excess of the ED_{95} for viral replication observed in vitro. A pharmacokinetic study of the metabolism of **5.87** indicated that the probable cause for the short plasma half-life was the production of three metabolites: the *N*-oxide

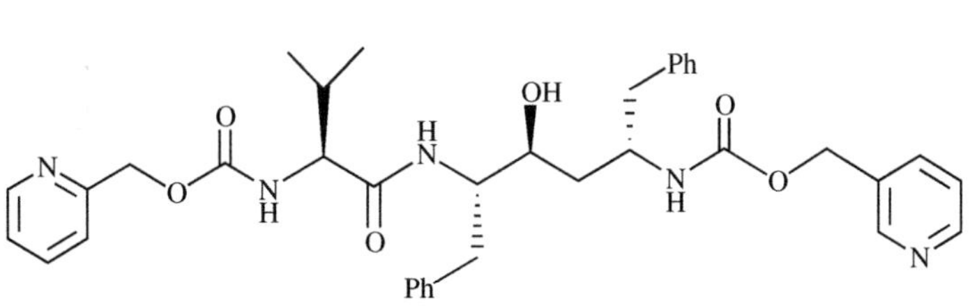

5.87

of the 2-pyridinyl group, the *N*-oxide of the 3-pyridinyl group, and the bis(pyridine *N*-oxide). Using ^{14}C-labeled **5.87**, these metabolites corresponded to 92–95% of the total bile radioactivity. Obviously, the next modification had to be to minimize this metabolism. Attempts to hinder oxidative metabolism sterically by the addition of substituents at the 6-position of the pyridinyl group generally yielded compounds that showed lower C_{max} values and oral bioavailability, although they did have greater potency than **5.87**. Attempts to modify the electronic nature of the P_3 pyridinyl group by the addition of electron-donating groups, such as methoxyl or amino, led to more potent and more soluble analogs, but they had poorer pharmacokinetic profiles.

By diminishing the oxidation potential of the electron-rich pyridinyl groups, drug metabolism should be diminished; consequently, the pyridinyl groups were replaced with other heteroaromatic bioisosteres.[154] The more electron-deficient (less basic) heterocycles, however, also had lower aqueous solubility and oral absorption. 5-Pyrimidinyl substitution gave inhibitors of equal potencies to the pyridinyl compounds, but with lower bioavailability; the furanyl analogs were more potent, but showed even lower bioavailability. However, 5-thiazolyl analogs, which generally increase aqueous solubility, showed excellent pharmacokinetic properties, although their solubilities were still low.

Alkyl substitution on the P_3 heterocycle led to increased antiviral potency, but substitution on the P_2' heterocycle gave decreased potency. Later, the crystal structure of ritonavir bound to the enzyme showed that the alkyl group participated in a hydrophobic interaction with Val-82; this interaction was optimized with an isopropyl group. As noted above, though, an *N*-methylurea linkage between the P_3 and P_2' groups provided much higher solubility than the corresponding carbamate linkages. Consequently, P_3 *N*-methylurea analogs were made. Furthermore, it was found that in the *N*-methylurea series, but not the carbamate series, the regioisomeric position of the hydroxyl group became significant; compounds in which the hydroxyl group was distal to the P_2 valine residue (e.g., **5.88**) were 10-fold more potent than the corresponding analogs with the hydroxyl

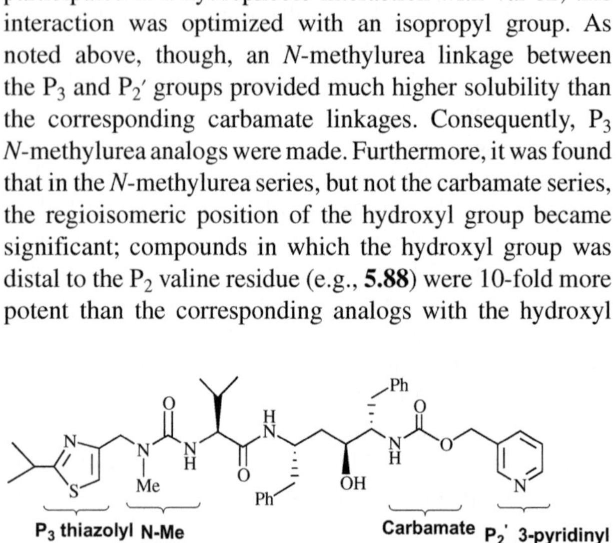

5.88

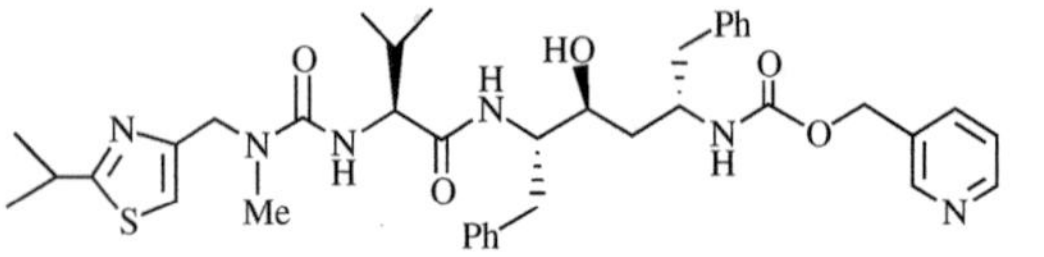

5.89

group proximal to the P_2 valine (**5.89**). By combining all of these characteristics, the best analog was found to be **5.90** (ritonavir, Norvir), which has a solubility of 6.9 μg/mL at pH 4, a C_{max}/ED_{50} of 105 (2.62 μM/0.025 μM), and an oral

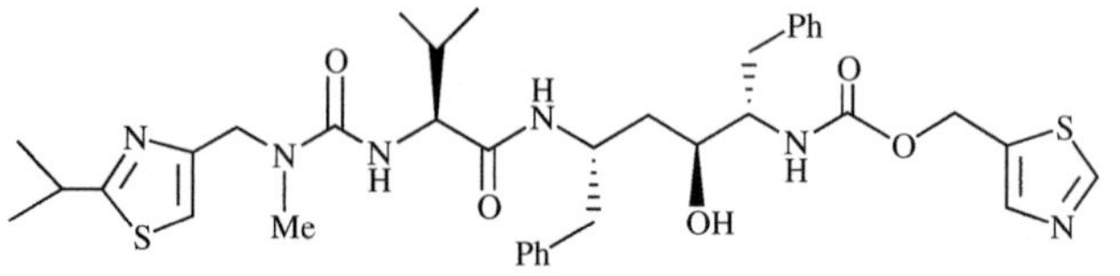

Ritonavir
5.90

bioavailability of 78%. The in vitro K_i for HIV-1 protease is 15 pM! The metabolic reactivity of pyridinyl groups versus thiazolyl groups was elucidated by determining the metabolism of **5.87** vs **5.88** vs **5.90**. The relative rates of metabolism are **5.87** (1.0)>**5.88** (0.2)>**5.90** (0.05), as predicted. Oxazolyl and isoxazolyl analogs had profiles similar to that of ritonavir, both in potency and pharmacokinetic behavior, but plasma concentrations were maintained for shorter periods than ritonavir. It was later found that one additional reason for the improved pharmacokinetic properties of ritonavir is that it is a potent inhibitor of the 3A4 isozyme of cytochrome P450, the enzyme responsible for the oxidative metabolism of ritonavir.[155] Inhibition results from the binding of the unhindered nitrogen atom on the P_2' 5-thiazolyl group to the heme cofactor in the active site of CYP3A4.[156]

The general medicinal chemistry approach taken is to start by modifying the lead to increase potency for binding to the appropriate receptor (pharmacodynamics). Once this is accomplished, modifications to improve the pharmacokinetic behavior need to be carried out in conjunction with studies to determine if these modifications affect the pharmacodynamics. The discovery of ritonavir demonstrates that these approaches can work very well[157] and also shows that peptidomimetic analogs can be capable of high absorption, oral bioavailability, and slow hepatic clearance.

5.3. IRREVERSIBLE ENZYME INHIBITORS

5.3.1. Potential of Irreversible Inhibition

A reversible enzyme inhibitor is effective as long as a suitable concentration of the inhibitor is present to drive the equilibrium $E+I \rightleftharpoons E \cdot I$ to the right (see Section 5.2.1). Therefore, a reversible inhibitor drug is effective only while the drug concentration is maintained at a high enough level to sustain the enzyme–drug complex. Because of drug metabolism and excretion (see Chapter 8), repetitive administration of the drug is required.

A competitive *irreversible enzyme inhibitor*, also known as an *active-site directed irreversible inhibitor* or an *enzyme inactivator*, is a compound whose structure is similar to that of the substrate or product of the target enzyme and which generally forms a covalent bond to an active site residue (a slow, tight-binding inhibitor, however, often is a noncovalent inhibitor that can be *functionally* irreversibly bound). In the case of irreversible inhibition, it is not necessary to sustain the inhibitor concentration to retain the enzyme–inhibitor interaction because this is an irreversible reaction (usually covalent), and once the target enzyme has reacted with the irreversible inhibitor, the complex cannot dissociate (again, there are exceptions). The enzyme, therefore, remains inactive, even in the absence of additional inhibitor. This effect could translate into a requirement for smaller and fewer doses of the drug. It could also lead to lower toxicity because once the enzyme is inactivated, drug metabolism can rapidly clear the body of the unbound drug, lowering potential toxicity from the drug, but retaining the pharmacological activity of the drug over an extended period.[158] In fact, metabolic stability becomes less of an issue with an irreversible inhibitor because rapid clearance may be advantageous, especially if covalent bond formation occurs rapidly.[159] Vigabatrin (Section 5.3.3.3.1) is a good example of this phenomenon.[160] In rats[161] and humans,[162] vigabatrin has a metabolic half-life of 1–3 and 5 h, respectively, but a binding half-life to GABA-AT, the target of this drug, of several days. The return of GABA-AT activity after vigabatrin is administered corresponds to the rate of genetic synthesis of new GABA-AT.[163] In some cases, however, particularly where genetic translation of the target enzyme is extremely slow, it may be safer to design reversible inhibitors whose effects can be controlled more effectively by termination of their administration. For diseases that require extended, high target occupancy, such as cancer and microbial infections, irreversible inhibition might be the most effective therapy.[164]

Even though the target enzyme is destroyed by an irreversible inhibitor, it does not mean that only one dose of the drug would be sufficient to destroy the enzyme permanently. Yes, it destroys that copy of the enzyme permanently, but our genes are constantly encoding more copies of proteins that diminish. As the enzyme loses activity, additional copies of the enzyme are synthesized, but this process can take hours or even days (as is the case above with GABA-AT).[165] One may wonder what the effect on metabolism would be if a particular enzyme activity were completely inhibited for an extended period. Consider the case of aspirin, an irreversible inhibitor of PG synthase (See Section 5.3.2.2.2). If the quantity of aspirin consumed in the United States (about 80 billion tablets a year in 1994)[166] were averaged over the entire population, then every man, woman, and child would be taking about 240 mg of aspirin every day, enough to shut down human PG biosynthesis for the entire country permanently! Suffice it to say that there are many irreversible enzyme inhibitor drugs in medical use. In fact, there is at least one example of an approved irreversible inhibitor drug for about one-third of all enzyme drug targets.[167] Three of the 10 top-selling drugs in 2009 (clopidogrel, **5.91** [Plavix]; esomeprazole, **5.92** [Nexium]; and lansoprazole, **5.93** [Prevacid]) are covalent inhibitors of their biological targets. However, most of the covalent drugs were not designed as such but their covalent mechanisms were determined after their efficacy was discovered, generally through a screening campaign.

The anxiety about the use of covalent drugs stems from the fear of potential off-target reactions, resulting from reactive functional groups, and immunogenic effects, which may have derived from studies in the 1970s that showed hepatotoxicity from compounds, such as acetaminophen, that underwent metabolism to highly reactive intermediates, which covalently bonded to liver proteins.[168] Modification of proteins, either by direct attack by the drug or by a reactive metabolite, produces a hapten–protein complex, which can elicit an immune response.[169] These early studies may not be relevant to many covalent drugs.

The term irreversible is a loose one; either a very stable covalent bond or a labile covalent bond may be formed between the drug and the enzyme active site. A reversible covalent inhibitor, which dissociates at a rate that is faster than the physiological degradation of the protein, has the advantage of longer lived inhibition than noncovalent inhibitors and without the potential immunological effects of irreversible inhibitors. As pointed out earlier, some tight-binding reversible inhibitors are also functionally irreversible. As long as return of enzyme activity is slower than degradation of the enzyme, the enzyme is considered irreversibly inhibited. The two principal types of enzyme inactivators are reactive compounds called affinity labeling agents and unreactive compounds that are activated by the target enzyme, known as mechanism-based enzyme inactivators.

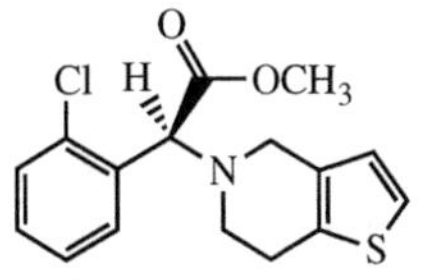

Cl H O OCH_3 N S

Clopidogrel
5.91

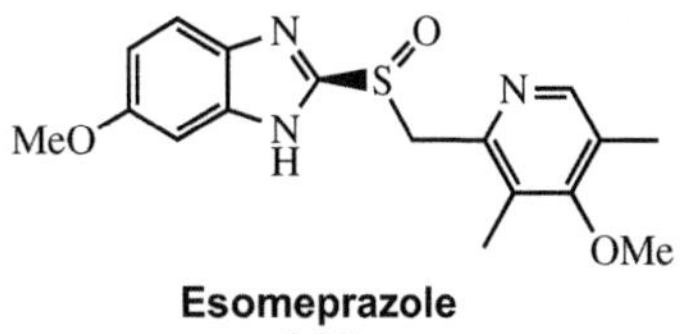

MeO N N H S O N OMe

Esomeprazole
5.92

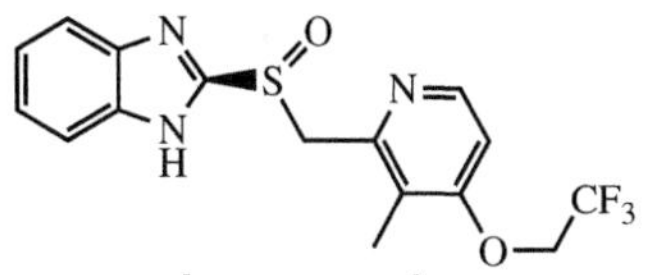

N N H S O N O CF_3

Lansoprazole
5.93

5.3.2. Affinity Labeling Agents

5.3.2.1. Mechanism of Action

An *affinity labeling agent* is a reactive compound that has a structure similar to that of the substrate for a target enzyme. Subsequent to reversible E·I complex formation, it reacts with active site nucleophiles (usually amino acid side chains), generally by acylation or alkylation (S_N2) mechanisms, thereby forming a stable covalent bond to the enzyme (Scheme 5.13). Note that this reaction scheme is similar to that for the conversion of a substrate to a product (see Chapter 4, Section 4.1.2); instead of a k_{cat}, the catalytic rate constant for product formation, there is a k_{inact}, the inactivation rate constant for enzyme inactivation. On the assumption that the equilibrium for reversible E·I complex formation (K_I) is rapid, and the rate of dissociation of the E·I complex (k_{off}) is fast relative to that of the covalent bond forming reaction, then k_{inact} will be the rate-determining step. In this case, unlike simple reversible inhibition, there will be a time-dependent loss of enzyme activity (as is the case with slow, tight-binding reversible inhibitors because of the relatively small k_{off}).

The rate of inactivation is proportional to low concentrations of inhibitor, but becomes independent at high concentrations. As is the case with substrates, the inhibitor can also reach *enzyme saturation* when the k_{inact} is slow relative to the k_{off}. Once all the enzyme molecules are tied up in an E·I complex, the addition of more inhibitor will have no effect on the rate of inactivation.

Because an affinity labeling agent contains a reactive functional group, it can react not only with the active site of the target enzyme but also can react with thousands of nucleophiles associated with many other enzymes and biomolecules in the body. Consequently, these inactivators are potentially quite toxic. In fact, many cancer chemotherapy drugs (see Chapter 6) are affinity labeling agents, and they are quite toxic.

There are several principal reasons why these reactive molecules, nonetheless, can be effective drugs. First, once the inactivator forms an E·I complex, a unimolecular reaction ensues (the E·I complex is now a single molecule), which can be many orders of magnitude (10^8 times; see Chapter 4, Section 4.2.1) more rapid than nonspecific bimolecular reactions with nucleophiles on other proteins.

$$E + I \underset{}{\overset{K_I}{\rightleftharpoons}} E \cdot I \xrightarrow{k_{inact}} E - I$$

SCHEME 5.13 Basic kinetic scheme for an affinity labeling agent

Furthermore, the inactivator may form an E·I complex with other enzymes, but if there is no nucleophile near the reactive functional group, no reaction will take place. Third, in the case of antitumor agents, mimics of DNA precursors are rapidly transported to the appropriate site and, therefore, they are preferentially concentrated at the desired target.

The keys to the effective design of affinity labeling agents as drugs are specificity of binding and moderate reactivity of the reactive moiety. If the molecule has a very low K_i for the target enzyme, then E·I complex formation with the target enzyme will be favored, and the selective reactivity will be enhanced. The effectiveness of modulating the activity of the reactive functional group can be seen by comparing the relatively moderate reactivity of the functional groups in the nontoxic affinity labeling agents described in Section 5.3.2.2 with the highly reactive functional groups in some of the cancer chemotherapeutic drugs in Chapter 6 (Section 6.3.2).

One approach that takes advantage of modulated reactivity of an affinity labeling agent has been termed by Krantz[170] *quiescent affinity labeling*. In this case, the inactivator reactivity is so low that reactions with nucleophiles in solution at physiological pH and temperature are exceedingly slow or nonexistent. However, because of the exceptional nucleophilicity of groups in some enzymes that use covalent catalysis as a catalytic mechanism (see Chapter 4, Section 4.2.2), the poorly electrophilic sites in the quiescent affinity labeling agents are reactive enough for nucleophilic reaction, but only at the active site of the enzyme. High selectivity for inactivation of a particular enzyme can be built into a molecule if the structure of the inactivator is designed so that it binds selectively to the target enzyme. For example, peptidyl acyloxymethyl ketones (**5.94**, Scheme 5.14) have low chemical reactivity (the acyloxyl group is a weak leaving group), but are potent and highly selective inactivators of cathepsin B,[171] a cysteine protease that has been implicated in osteoclastic bone resorption,[172] tumor metastasis,[173] and muscle wasting in Duchenne muscular dystrophy.[174]

A third feature that would increase the potential effectiveness of an affinity labeling agent can be built into the molecule, if something is known about the location of the active site nucleophiles. In cases where a nucleophile is known to be at a particular position relative to the bound substrate, the reactive functional group can be incorporated into the affinity labeling agent so that it is near that site when the inactivator is bound to the target enzyme. This increases the probability for reaction with the target enzyme by approximation (see Chapter 4, Section 4.2.1).

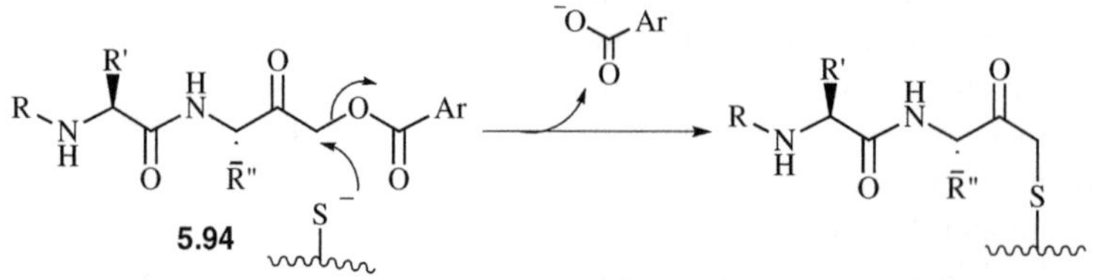

5.94

SCHEME 5.14 Peptidyl acyloxymethyl ketones as an example of a quiescent affinity labeling agent

Because many enzymes involved in DNA biosynthesis utilize substrates with similar structures, high concentrations of reversible inhibitors can block multiple enzymes. A lower concentration of an irreversible inhibitor may be more selective, if it only reacts with those enzymes having appropriately juxtaposed active site nucleophiles.

Even when all these design factors are taken into consideration, nonspecific reactions still can take place that may result in side effects. Judicious use of low electrophilic moieties can result in highly effective potential drug candidates. Another approach, known as targeted covalent inhibition,[175] targets a noncatalytic nucleophilic residue that is poorly conserved across the target protein family, thereby giving selectivity for the desired target protein. Of course, in all cases, enhanced potency for the target protein will lead to favorable selectivity and safety; the majority of drugs withdrawn from the market because of safety issues, known as *idiosyncratic drug toxicities*,[176] which only occur in humans and are of unknown causes (although reactive metabolites are suspected), were administered at doses greater than 100 mg/day, whereas drugs taken at <10 mg/day seem to have generally good safety profiles.[177]

5.3.2.2. Selected Affinity Labeling Agents

5.3.2.2.1. Penicillins and Cephalosporins/Cephamycins

Penicillins have the general structure **5.95**; for example, **5.95a** is penicillin G, **5.95b** is penicillin V, **5.95c** (R′=H) is oxacillin (Baetocill), **5.95c** (R′=Cl) is cloxacillin (Cloxapen), **5.95d** (R′=H) is ampicillin (Principen), and **5.95d** (R′=OH) is amoxicillin (Amoxil). The differences in these derivatives (other than structure) are related to absorption properties, susceptibility to deactivation by penicillinases, and specificity for organisms for which they are most effective.[178]

The structures of some cephalosporins and cephamycins are shown in **5.96**; for example, **5.96a** is cefazolin (injectable; Ancef), **5.96b** is cefoxitin (injectable; Mefoxin), **5.96c** is cefaclor (oral; Ceclor), and **5.96d** is ceftizoxime (injectable; Cefizox). The analogs where X=H are cephalosporins and those where X=OCH_3, such as cefoxitin, **5.96b**, are called cephamycins. The penicillins, cephalosporins, and cephamycins all have in common the β-lactam ring and are known collectively as *β-lactam antibiotics*. Cephalosporins and cephamycins are classified by generations, which are based on general features of antimicrobial activity.[179] Cefazolin is a first-generation cephalosporin, cefoxitin and cefaclor are second-generation cephalosporins, and ceftizoxime is a third-generation cephalosporin. Modifications in the structure of these antibiotics have been extensive,[180] so much so that essentially every atom excluding the lactam nitrogen has been replaced or modified in the search for improved antibiotics.

The discovery of penicillin was described in Chapter 1, Section 1.2.2.[181] Penicillins and cephalosporins/cephamycins are *bacteriocidal* (they kill existing bacteria), unlike the sulfonamides (see Section 5.2.2.3), which are bacteriostatic. They are ideal drugs in that they inactivate an enzyme that is essential for bacterial growth, but which does not exist in animals, namely, the peptidoglycan transpeptidase. As discussed in Chapter 4 (Section 4.2.7.), this enzyme catalyzes the cross-linking of the peptidoglycan to form the bacterial cell wall. Scheme 4.13 shows that one (the *acyl donor*) of the

5.95

a, R = $PhCH_2$–

b, R = $PhOCH_2$–

c, R = (R′-substituted phenyl)-methylisoxazolyl

d, R = (R′-substituted phenyl)

5.95d

5.96

a, R = tetrazolyl-CH_2– X = H Y = $-CH_2S$-(methylthiadiazolyl)

b, R = thienyl-CH_2– X = OCH_3 Y = $-CH_2OC(=O)NH_2$

c, R = phenyl-CH(NH_2)– X = H Y = Cl

d, R = (iminothiazolinyl)-C=N-OCH_3 X = H Y = H

two substrates of the peptidoglycan transpeptidase contains a peptide segment having D-alanyl-D-alanine at the C-terminus (Scheme 4.13, graphic **A**). According to Scheme 4.13, the enzyme initially acts analogously to a serine protease to carry out the first half of the transpeptidase reaction. That is, nucleophilic attack of an enzyme serine residue on the carbonyl group of the first D-Ala residue of the C-terminal D-Ala-D-Ala-OH segment results in formation of a serine ester linkage with the release of the C-terminal D-Ala-OH (Scheme 4.13, graphics **A-C**). The active site serine involved in catalysis has been identified.[182] Scheme 4.13 (graphics **C-F**) further shows the reaction of the amino terminus of a second peptidoglycan unit (the *acyl acceptor*) with the serine ester to form the cross-linked peptidoglycan. In the case of a serine protease, a water molecule would take the place of the second peptidoglycan segment to result in simple hydrolysis of the serine ester. As discussed below, penicillins interfere with these peptidoglycan cross-linking processes.

By comparison of a molecular model of penicillin with that of D-alanyl-D-alanine, Tipper and Strominger[183] suggested that penicillin could mimic the structure of the terminus of the peptidoglycan, and bind at the active site of the transpeptidase (Figure 5.13). The N^a to N^b distances (3.3 Å) and the N^b to carboxylate carbon distances (2.5 Å) in both molecules are identical. The N^a to carboxylate carbon distance is 5.4 Å in the penicillins and 5.7 Å in D-alanyl-D-alanine. The β-lactam carbonyl may be further activated by torsional effects of the thiazolidine ring in the penicillins. This carbonyl corresponds to the carbonyl in the acyl D-alanyl-D-alanine that acylates the active site serine and, therefore, penicillins also could acylate the transpeptidase serine residue[184] (Scheme 5.15). This hypothesis was supported many years later in a crystal structure obtained to 1.2 Å resolution of a cephalosporin bound to a bifunctional serine type D-alanyl-D-alanine carboxypeptidase/transpeptidase.[185] The bulk of the penicillin molecule, when it is attached to the active site, precludes hydrolysis or transamidation either for steric reasons or because it induces a conformational change in the enzyme that prevents these processes from occurring.[186] Covalent binding at the active site prevents the peptidoglycan acyl donor substrate from binding. Similar arguments could be made for cephalosporins. The double bond in the dihydrothiazine ring also may activate the β-lactam carbonyl (Scheme 5.16).[187] It should be noted that β-lactam antibiotics, as well as all bactericidal antibiotics, have been shown to produce highly toxic hydroxyl radicals, which may be the ultimate mechanism of bacterial cellular death.[188]

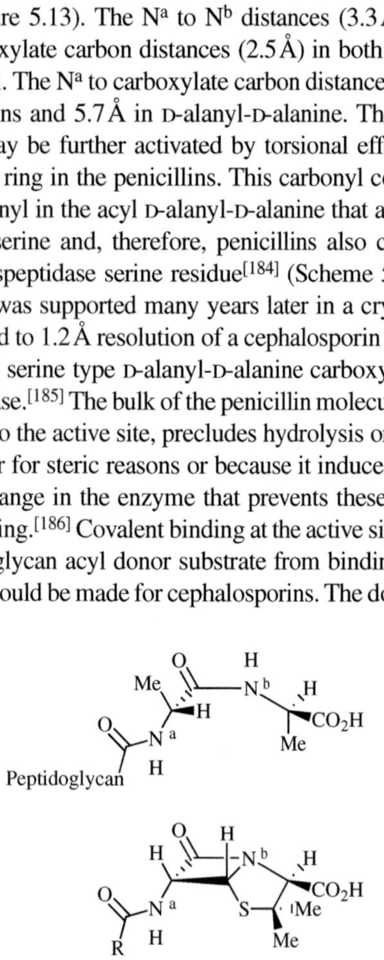

On the basis of the structural similarity of penicillins to acyl D-alanyl-D-alanine (see Figure 5.13), Tipper and Strominger predicted that 6-methylpenicillin (a methyl on the sp^3 carbon adjacent to N^a) would be a more potent inhibitor than the parent molecule. However, both 6-methylpenicillin and 7-methylcephalosporin (the numbering is different for cephalosporins because it has one more carbon in the ring, but the methyl in 7-methylcephalosporin corresponds to that in 6-methylpenicillin) were synthesized and were shown to be inactive.[189] Because the corresponding 7-methoxycephalosporins (that is, cephamycins) are better inhibitors than the parent cephalosporins, it is not clear why the methyl analogs are poor inhibitors.

The beauty of the penicillins (and cephalosporins) is that they are not exceedingly reactive; consequently, few nonspecific acylation reactions occur. Their modulated reactivity and nontoxicity make them ideal drugs. If it were not for allergic responses and problems associated with digestion and drug resistance (see Chapter 7), penicillins might be considered nutritious foods, composed of various carboxylic acid derivatives (the RCO side chains), cysteine, and the essential amino acid valine!

Because of the discovery of highly effective antibiotics, such as the penicillins and the sulfa drugs, and the realization of even more effective synthetic analogs of these natural products, Nobel laureate immunologist F. Macfarlane Burnet noted in 1962 that by the late twentieth century "the virtual elimination of infectious disease as a significant factor in social life" should be anticipated.[190] William Stewart, the

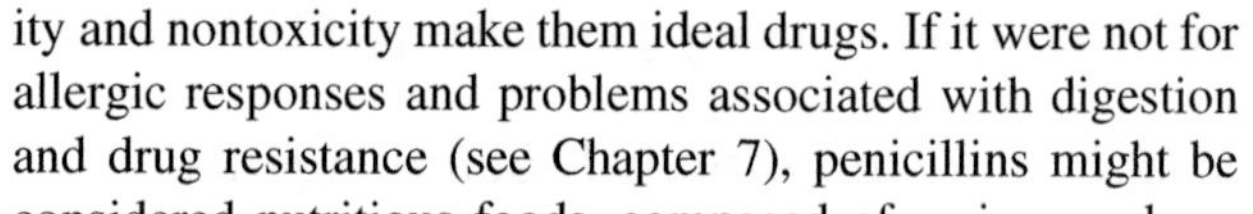

FIGURE 5.13 Comparison of the structure of penicillins with acyl D-alanyl-D-alanine

SCHEME 5.15 Acylation of peptidoglycan transpeptidase by penicillins

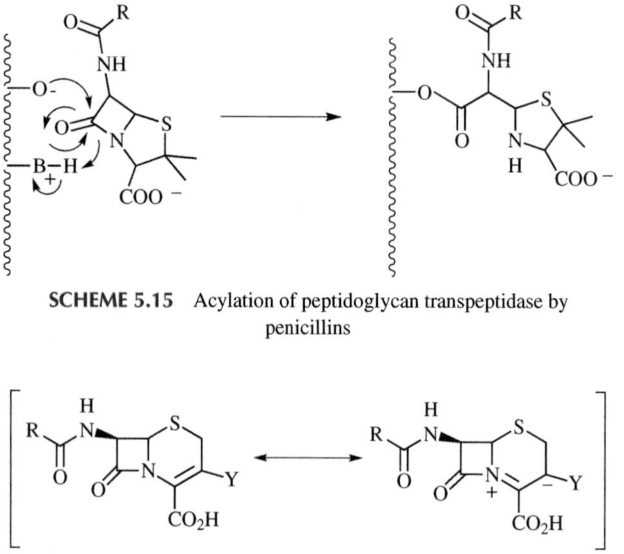

SCHEME 5.16 Activation of the β-lactam carbonyl of cephalosporins

former Surgeon General of the United States, testified before Congress in 1967 that it was time to "close the book on infectious diseases". Apparently, rampant resistance (Chapter 7) to these and all antibacterial drugs was not foreseen.

5.3.2.2.2. Aspirin

It is stated in the *Papyrus Ebens* (c. 1550 B.C.) that dried leaves of myrtle provide a remedy for rheumatic pain. Hippocrates (460–377 B.C.) recommended the use of willow bark for pain during childbirth.[191] A boiled vinegar extract of willow leaves was suggested by Aulus Cornelius Celsus (30 A.D.) for relief of pain.[192] In 1763, Reverend Edmund Stone of England announced his findings that the bark of the willow (*Salix alba vulgaris*) provided an excellent substitute for Peruvian bark (*Cinchona* bark, a source of quinine) in the treatment of fevers.[193] The connection between the two barks was discovered by his tasting them and making the observation that they both had a similar bitter taste. The bitter active ingredient with antipyretic activity in the willow bark was called salicin, which was first isolated in 1829 by Leroux. Upon hydrolysis, salicin produced glucose and salicylic alcohol, which was metabolized to salicylic acid. Sodium salicylate was first used for the treatment of rheumatic fever and as an antipyretic agent in 1875. Toward the end of the nineteenth century, the father of Felix Hoffmann, a chemist employed by Bayer Company, suffered from severe rheumatoid arthritis and pleaded with his son to search for a less irritating drug than the sodium salicylate he was using (salicylic acid not only tastes awful but also causes ulcerations of the mouth and stomach linings with prolonged use). Hoffmann synthesized many salicylate derivatives, and acetylsalicylic acid (aspirin, **5.97**) was the best. He gave it to his father, who responded well, and in 1899 Bayer introduced aspirin as an antipyretic (fevers), antiinflammatory (arthritis), and analgesic (pain) agent.[194] Aspirin became the first drug ever to be tested in clinical trials before registration. Also, because of its insolubility in water, it had to be sold in solid form, and, therefore, became the first major medicine to be sold in the form of tablets. The trade name aspirin was coined by adding an "a" for acetyl to spirin for *Spiraea*, the plant species from which salicylic acid was once prepared. In the first 100 years since its introduction on the market, it was estimated that one trillion (10^{12}) aspirin tablets had been consumed.[195]

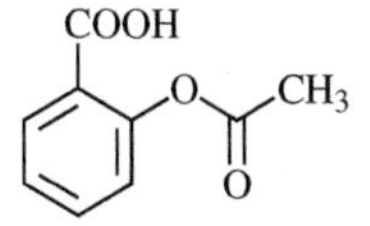

COOH O CH_3 O

Aspirin
5.97

The mechanism of action was initially reported by Vane,[196] who shared the 1982 Nobel Prize in Medicine (with Sune Bergström and Bengt Samuelsson) for this discovery, and by Smith and Willis,[197] to be the result of inhibition of prostaglandin (PG) biosynthesis. PGs are derived from arachidonic acid (**5.98**) by the action of PG synthase (also known as cyclooxygenase (COX)) (Scheme 5.17). At the time of the initial report there was evidence that various PGs were involved in the pathogenesis of inflammation

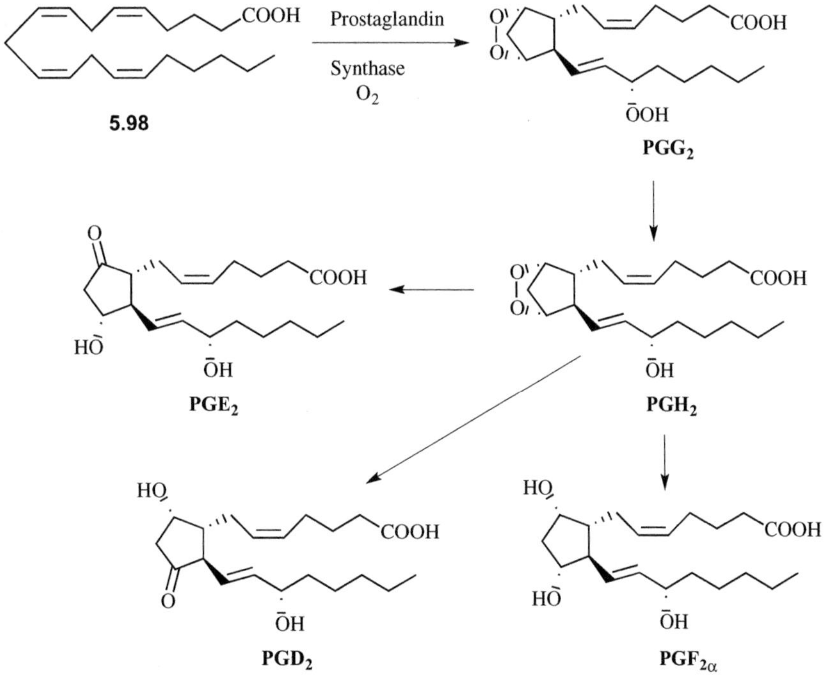

COOH **5.98** — Prostaglandin Synthase O_2 → COOH OOH **PGG$_2$** → COOH OH **PGH$_2$** → COOH HO OH **PGE$_2$**; **PGH$_2$** → HO COOH O OH **PGD$_2$**; **PGH$_2$** → HO COOH HO OH **PGF$_{2\alpha}$**

SCHEME 5.17 Biosynthesis of prostaglandins (PGs) from arachidonic acid and fever. It is now known that all mammalian cells (except erythrocytes) have microsomal enzymes that catalyze the biosynthesis of PGs, and that PGs are always released when cells are damaged, and are in increased concentrations in inflammatory exudates. When PGs are injected into animals, the effects are reminiscent of those observed during inflammatory responses, namely, redness of the skin (erythema) and increased local blood flow. A long-lasting vasodilatory action is also prevalent. PGs can cause headache and vascular pain when infused in man. Elevation of body temperature during infection is also mediated by the release of PGs.

From the above discussion, it is apparent that inhibition of PG synthase, the enzyme responsible for the biosynthesis of all the PGs and related compounds (PGH_2 also is converted by prostacyclin synthase to prostacyclin (**5.99**) and by thromboxane synthase to thromboxane A_2 (**5.100**)), would be a desirable approach for the design of antiinflammatory, analgesic, and antipyretic drugs. Inhibition of platelet COX is particularly effective in blocking PG biosynthesis because, unlike most other cells, platelets cannot regenerate the enzyme, because they have little or no capacity for protein biosynthesis. Therefore, a single dose of 40 mg/day of aspirin is sufficient to destroy the COX for the life of the platelet! When PG synthesis is blocked, there is less stimulation of the pain-sensitive nerve endings, resulting in less aching of muscles and joints, as well as less relaxation of blood vessels in the head, so fewer headaches; fewer PGs in the hypothalamus means that the body temperature set point is not changed, and no fever is induced.

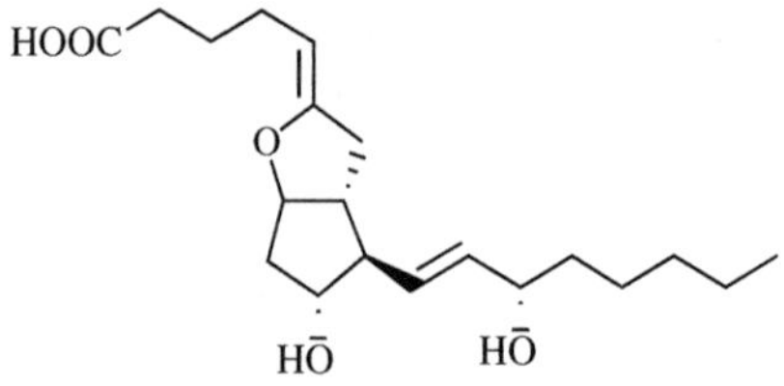

5.99

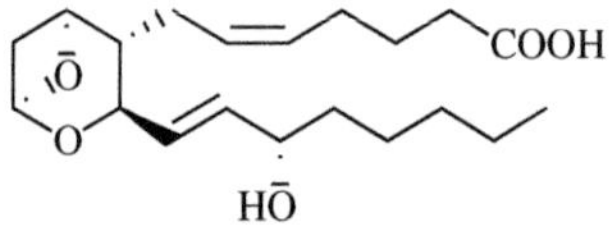

5.100

Sheep vesicular gland PG synthase was shown to be irreversibly inactivated by aspirin.[198] When microsomes of sheep seminal vesicles were treated with aspirin tritiated in the methyl of the acetyl group, acetylation of a single protein was observed. The same experiment carried out with aspirin tritiated in the benzene ring resulted in no tritium incorporation, suggesting that acetylation was occurring. Incubation of purified PG synthase with [^{3}H-acetyl]aspirin led to irreversible inactivation with incorporation of one acetyl group per enzyme molecule.[199] Pepsin digestion of the tritiated enzyme gave a 22-amino-acid-labeled peptide; tryptic digestion of this peptide gave a tritiated decapeptide in which the serine residue was acetylated.[200] Thermolysin digestion of the tritiated enzyme gave the labeled dipeptide Phe–Ser, where the hydroxyl group of Ser-530 had become acetylated.[201] The most straightforward mechanism for acetylation would be a transesterification mechanism by aspirin acting as an affinity labeling agent; on the basis of site-directed mutagenesis studies, it was proposed by Marnett and coworkers[202] that Tyr-385 and Tyr-348 in the active site hydrogen bond to the acetyl carbonyl of aspirin, which directs it specifically to Ser-530 for acetylation (Scheme 5.18).

The principal side effect of aspirin and all nonsteroidal antiinflammatory drugs (NSAIDs) is that their chronic use leads to ulceration of the stomach lining. In the late 1980s, it was found that COX activity was increased in certain inflammatory states and could be induced by inflammatory cytokines.[203] This suggested the existence of two different forms (isozymes) of COX, now known as COX-1 and COX-2. COX-1, the *constitutive form* of the enzyme (that is, present in the cells all the time), is responsible for the physiological production of PGs that are important in maintaining tissues in the stomach lining and kidneys. COX-2,[204] induced by cytokines in inflammatory cells, is responsible for the elevated production of PGs during inflammation and is associated with inflammation, pain, and fevers. The discovery of a second COX enzyme

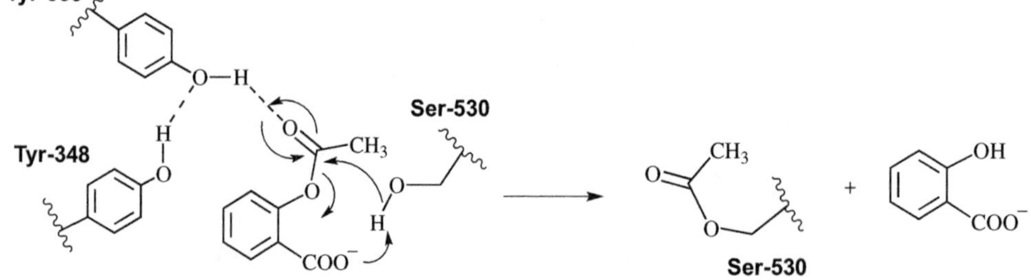

SCHEME 5.18 Hypothetical mechanism for acetylation of prostaglandin synthase by aspirin

suggested that aspirin and other NSAIDs may inhibit both COX-1 and COX-2, leading to stomach irritation (and, in some cases, ulcers); ideally, inhibition of only COX-2 would produce the desired antiinflammatory effect without stomach lining irritation.[205]

Both G. D. Searle (now defunct) and Merck initiated programs to discover a COX-2-selective (reversible) inhibitor. At Searle, more than 2500 compounds were synthesized and almost 2000 screened of which 280 were potent. Almost 350 compounds were tested for oral activity, of which seven candidates were selected. Compound **5.101** has an IC_{50} of 60 nM and a selectivity of >1700 against COX-2 vs COX-1.[206] Series **5.102** has a range of IC_{50} values of 10–100 nM and in vitro selectivities of 10^3–10^4 for COX-2 vs COX-1.[207] However, celecoxib (**5.103**, Celebrex), with an IC_{50} of 40 nM and in vitro selectivity of only 375 in favor of COX-2 inhibition, emerged as the commercial compound because of its overall favorable properties.[208] Its in vivo selectivity (inflammatory effect/stomach irritation; a therapeutic index) was >1000. Merck also discovered a COX-2-selective inhibitor, rofecoxib (**5.104**, Vioxx),[209] which is about 5.5 times more selective than celecoxib and was a once-a-day treatment for osteoarthritis and rheumatoid arthritis. Searle then discovered valdecoxib (**5.105**; Bextra), having an IC_{50} for COX-1 of 140 μM and for COX-2 of 5 nM or 28,000-fold selectivity for COX-2 over COX-1.[210] In 2003, the global sales of Vioxx reached $2.5 billion, but in 2004, Merck voluntarily took the drug off the market after a study found an increased incidence of heart attacks and strokes among patients taking the drug compared to those on placebo.[211] This event engendered congressional hearings and much finger-pointing as questions were asked about how much and when the FDA and Merck knew about the potentially fatal side effects of the drug. Whether celecoxib and valdecoxib might share this property was also questioned, and sales of valdecoxib were discontinued in 2005, while as of 2014, celecoxib was still on the market. It appears that the lower selectivity of celecoxib relative to rofecoxib and valdecoxib may be key to its safer properties.[212] Inhibition of COX-2 reduces production of prostacyclin (**5.99**), and thereby attenuates prostacyclin's athero-protective properties. Inhibition of COX-1 decreases production of thromboxane A2 and thereby attenuates thromboxane A2's prothrombotic properties. A less selective inhibition may lead to a more favorable balance between these opposing effects.[213]

5.101

5.102

Celecoxib
5.103

Rofecoxib
5.104

Valdecoxib
5.105

In addition to its oxygenation of arachidonic acid (**5.98**, Scheme 5.17), COX-2 is also responsible for oxygenation of the endocannabinoids, which exert analgesic and anti-inflammatory effects at cannabinoid receptors CB1 and CB2; oxygenation of endocannabinoids turns off their pain-relieving effects. Therefore, inhibition of the oxygenation of endocannabinoids should be a treatment for neuropathic pain.[214] The (*S*)-enantiomers of chiral NSAIDs ibuprofen (**5.106**, Advil) and naproxen (**5.107**, Aleve) are the eutomers for inhibition of the oxygenation of arachidonic acid by COX-2; the (*R*)-enantiomers are inactive. However, it was found that the (*R*)-enantiomers are potent substrate-selective inhibitors of the endocannabinoid oxygenation activity of COX-2,[215] suggesting their potential role in pain management.

A third isozyme of COX called COX-3 (actually a variant of COX-1), found principally in the cerebral cortex, was shown to be selectively inhibited by NSAIDs that have good analgesic and antipyretic activity, but low antiinflammatory activity, such as acetaminophen (**5.108**, Tylenol).[216] Analgesic/antipyretic drugs penetrate the blood–brain barrier well and may accumulate in the brain where they inhibit COX-3. NSAIDs that contain a carboxylate group, such as aspirin (**5.97**) and ibuprofen (**5.106**), cross the blood–brain barrier poorly, but are more potent inhibitors of COX-3, so they also exhibit analgesic and antipyretic activities. COX-2 selective inhibitor drugs inhibit inflammatory pain,[217] whereas COX-1-selective inhibitor drugs, such as aspirin, ibuprofen, and naproxen,[218] are superior to COX-2 inhibitors against chemical pain stimulators.[219] The analgesic effect of COX-1-selective inhibitors occurs at lower doses than those needed to inhibit inflammation.[220]

X-ray crystal structures of COX-1 and COX-2 are known, and there is very little difference in the active sites of these two isozymes. COX-1[221] has an active site isoleucine at position 523 (Ile-523) and COX-2[222] has a valine (Val-523) at that position. Because isoleucine is larger than valine (by one methylene group), when COX-2 selective inhibitors bind to COX-1, a substituent on the inhibitor has a repulsive interaction with the Ile, but not with the smaller Val in COX-2, resulting in preferential binding to COX-2. Site-directed mutagenesis of these amino acids demonstrated that they were important to the selective inhibitor binding.[223] This small difference in size between Ile and Val is not sufficient for structure-based drug design approaches (see Chapter 2, Section 2.2.6), and in fact the COX-2-selective inhibitors were discovered without the aid of crystal structures. Note that these COX-2-selective inhibitors are reversible inhibitors. They are being discussed in this section of irreversible inhibitors only because they follow from our discussion of aspirin.

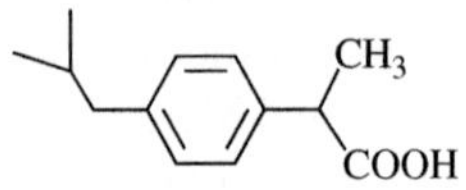

Ibuprofen
5.106

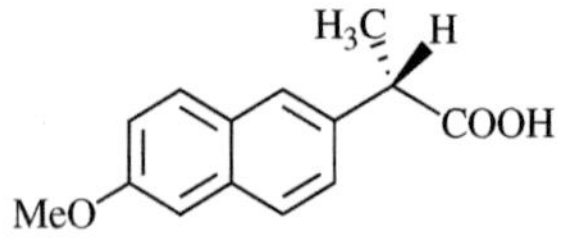

Naproxen
5.107

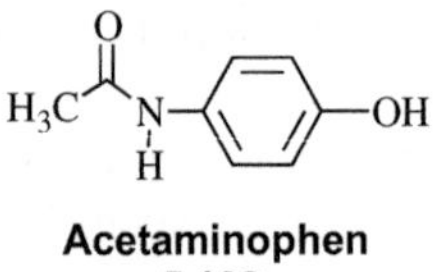

Acetaminophen
5.108

Both penicillin and aspirin are examples of affinity labeling agents that involve acylation mechanisms. Other affinity labeling agents function by alkylation or arylation mechanisms. An interesting example involves selective, covalent modification of β-tubulin, leading to disruption of microtubule polymerization and cytotoxicity against multidrug-resistant tumors (multidrug resistance is discussed in Chapter 7).[224] The pentafluorobenzene analog, **5.109**, produces time-dependent binding to β-tubulin; tritium-labeled **5.109** incorporates tritium into a protein that comigrates on sodium dodecyl sulfate/polyacrylamide gel electrophoresis with β-tubulin. Covalent modification occurs at a conserved cysteine residue (Cys-239) shared by β_1, β_2, and β_4 tubulin isoforms. Replacement of the *para*-fluoro group with other halogens causes a substantial decrease in potency; replacement with a hydrogen abolishes the cytotoxicity of the compound.[225] These results are consistent with a nucleophilic aromatic substitution mechanism for covalent inactivation (Scheme 5.19).

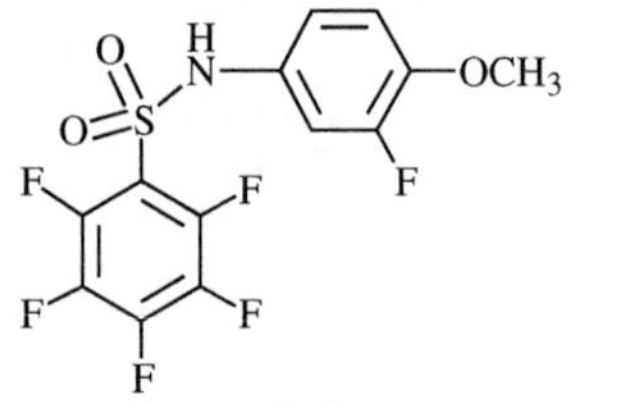

5.109

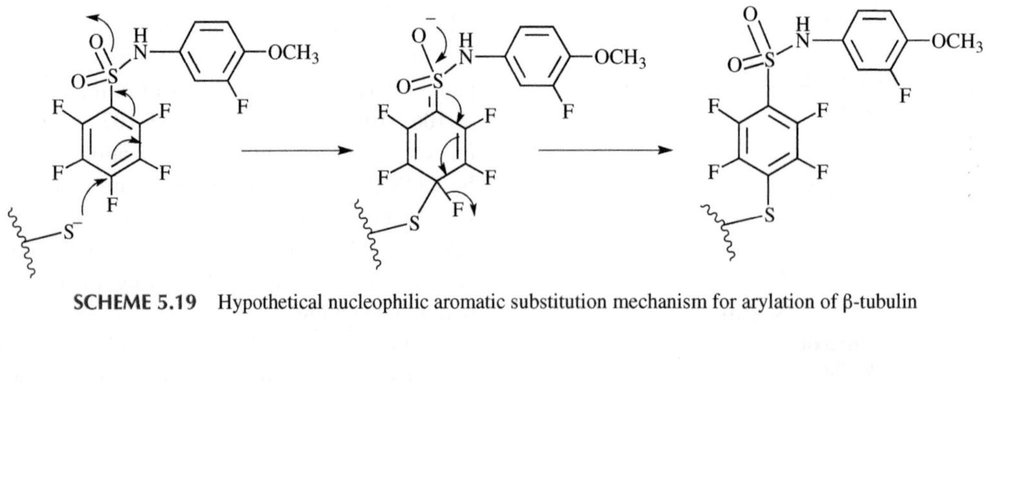

SCHEME 5.19 Hypothetical nucleophilic aromatic substitution mechanism for arylation of β-tubulin

5.3.3. Mechanism-Based Enzyme Inactivators

5.3.3.1. Theoretical Aspects

A *mechanism-based enzyme inactivator* is an unreactive compound that bears a structural similarity to the substrate for a specific enzyme. Once the inactivator binds to the active site, the target enzyme, via its normal catalytic mechanism, converts the compound into a product that inactivates the enzyme prior to escape from the active site. Therefore, these inactivators are acting initially as substrates for the target enzyme. Although the product generally forms a covalent bond to the target enzyme, it is not essential (a tight-binding inhibitor may form). However, the target enzyme must transform the inactivator into the actual inactivating species, and inactivation must occur prior to the release of this species from the active site. Because there is an additional step in the inactivation process relative to that for an affinity labeling agent (see Section 5.3.2), the kinetic scheme for mechanism-based inactivation differs from that for affinity labeling (Scheme 5.20). Provided k_4 is a fast step and the equilibrium k_1/k_{-1} is set up rapidly, k_2 is the inactivation rate constant (k_{inact}) that determines the rate of the inactivation process. The two key features of this type of inactivator that differentiate it from an affinity labeling agent are its initial unreactivity and the requirement for the enzyme to catalyze a reaction on it, thereby converting it into a product, that is, the actual inactivator species. Often this converted inactivator species is quite reactive, and therefore acts as an affinity labeling agent that already is at the active site of the target enzyme. Inactivation (k_4 in Scheme 5.20) does not necessarily occur every time the inactivator is transformed into the inactivating species; sometimes this species escapes from the active site (k_3 in Scheme 5.20). The ratio of the number of turnovers that gives a released product per inactivation event, k_3/k_4, is called the *partition ratio*. Ideally, the partition ratio should be zero to avoid release of potentially reactive species, but this is rare.[226]

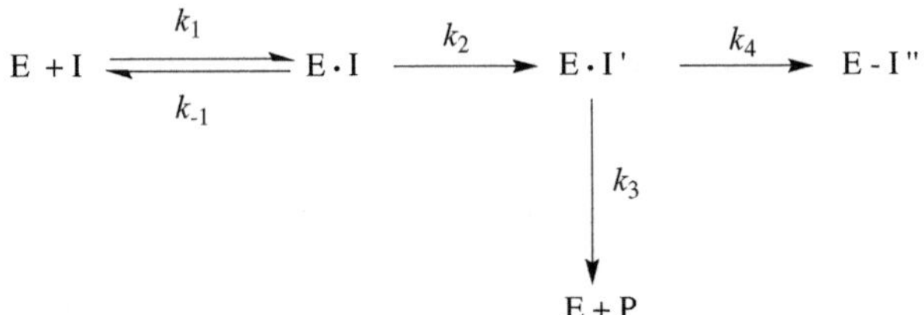

SCHEME 5.20 Kinetic scheme for simple mechanism-based enzyme inactivation

5.3.3.2. Potential Advantages in Drug Design Relative to Affinity Labeling Agents

Because of the generally high reactivity of affinity labeling agents, they can react with enzymes and biomolecules other than the target enzyme. When this occurs, toxicity and side effects can arise. However, mechanism-based enzyme inactivators are unreactive compounds, and this is the key feature that makes them so amenable to drug design. Consequently, nonspecific alkylations and acylations of other proteins should not be a problem. In the ideal case, only the target enzyme will be capable of catalyzing the appropriate chemistry for conversion of the inactivator to the activated species, and inactivation will result with every turnover, that is, the partition ratio will be zero. This may be quite important for potential drug use. If the partition ratio is greater than zero, the released activated species may react with other proteins, possibly resulting in a toxic effect. In this case, the inactivator is called a *metabolically activated inactivator*.[227] Alternatively, the released species may be hydrolyzed by the aqueous medium or scavenged by a scavenger molecule such as glutathione prior to reaction with other biomolecules, but the product formed may be toxic or may be metabolized further to other toxic substances. Under the ideal conditions mentioned above, the inactivator would be a strong drug candidate because it would be highly enzyme specific and low in toxicity. In fact, α-difluoromethylornithine (eflornithine), a specific mechanism-based inactivator of ornithine decarboxylase (see Section 5.3.3.3.2), has been administered to patients in amounts of 30 g/day for several weeks with only minor side effects![228]

There are quite a few drugs in current medical use that are mechanism-based inactivators, but most were determined ex post facto, rather than being designed, to be mechanism-based inactivators. A few known mechanism-based inactivators not discussed in the section below include the antidepressant drug phenelzine sulfate (**5.110**, Nardil) and the antihypertensive drug, hydralazine hydrochloride (**5.111**; Apresoline), both of which inactivate monoamine oxidase (MAO); clavulanic acid (**5.112**; see Chapter 7, Section 7.2.2.1; in combination with amoxicillin called Augmentin), a compound used to protect penicillins and cephalosporins against bacterial degradation (inactivates β-lactamases); the antiviral agent, trifluridine (**5.113**, Viroptic), which inactivates thymidylate synthase; gemcitabine-HCl (**5.114**, Gemzar), an antitumor drug that inactivates ribonucleotide reductase; the antihyperuricemic agent, allopurinol (**5.115**, inactivates xanthine oxidase; Aloprim); and the antithyroid drug, methimazole (**5.116**, Tapazole), which inactivates thyroid peroxidase. There are quite a few mechanism-based inactivators of cytochrome P450; however, those inactivators do not derive their pharmacological effects as a result of that inactivation.

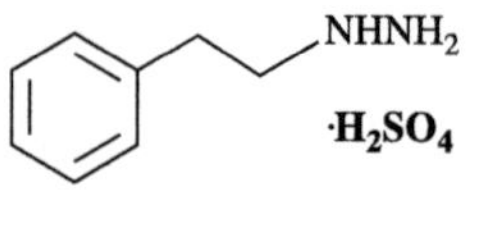

Phenelzine sulfate 5.110 | **Hydralazine hydrochloride 5.111** | **Clavulanate 5.112** | **Trifluridine 5.113**

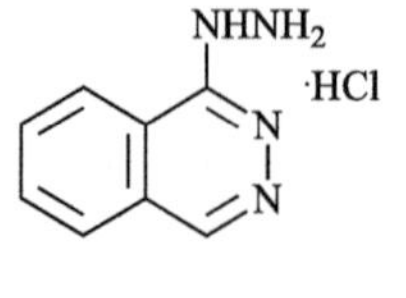

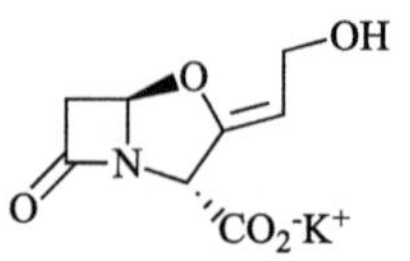

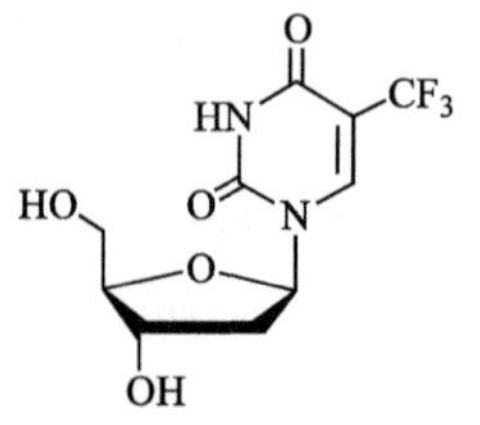

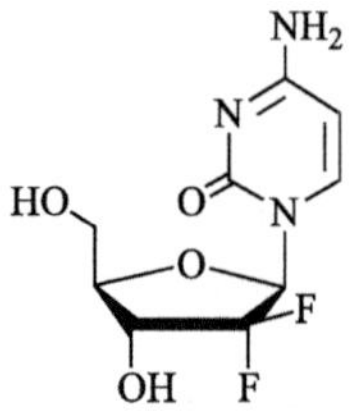

Gemcitabine HCl 5.114 | **Allopurinol 5.115** | **Methimazole 5.116**

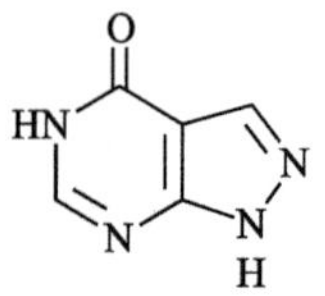

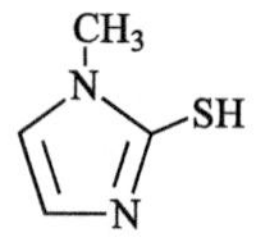

Because the activation of mechanism-based inactivators depends on the catalytic mechanism of their target enzyme, these inactivators can be designed by a rational organic mechanistic approach. However, this approach to drug design was infrequently used for several reasons. First, knowledge of the catalytic mechanism of the target enzyme is required to use the mechanism in the design of the inactivator. Also, this approach requires that a specific molecule be synthesized, which may be time consuming to do, just to test the mechanistic hypothesis. Even if the compound inactivates the enzyme, issues of potency, specificity, and pharmacokinetics have to be managed while maintaining the core structure for catalytic turnover. Because of all of these complications, relatively few mechanism-based inactivator drugs are commercial. Furthermore, by the last decade of the twentieth century, high-throughput screening (see Chapter 2, Section 2.1.2.3) and combinatorial chemistry (see Chapter 2, Section 2.1.2.3.1.3.1) made random and focused screens popular again, leaving the more rational approaches behind. However, with the low success of the combinatorial chemistry approaches, the more rational approaches are finding renewed life. In the next section, several mechanism-based inactivators are discussed, first in terms of the medicinal relevance of their target enzyme, and then from a mechanistic point of view. The reason for the relatively large number of examples of this type of enzyme inhibitor discussed below, despite the lesser prominence of this approach in drug design, derives from our interest in organic mechanisms and the history of these drugs.

5.3.3.3. Selected Examples of Mechanism-Based Enzyme Inactivators

5.3.3.3.1. Vigabatrin, an Anticonvulsant Drug

Epilepsy is a disease that was described over 4000 years ago in early Babylonian and Hebrew writings. Full clinical descriptions were written in Hippocrates' monograph *On the Sacred Disease* in about 400 BC.[229] Epilepsy is not a single disease, but a family of CNS diseases characterized by recurring convulsive seizures. By that definition then 1–2% of the world population has some form of epilepsy.[230] The disease is categorized as *primary* or *idiopathic* when no cause for the seizure is known, and *secondary* or *symptomatic* when the etiology has been identified. Symptomatic epilepsy can result from specific physiological phenomena such as brain tumors, syphilis, cerebral arteriosclerosis, multiple sclerosis, Buerger disease, Pick disease, Alzheimer disease, sunstroke or heat stroke, acute intoxication, lead poisoning, head trauma, vitamin B_6 deficiency, hypoglycemia, and labor, inter alia. The biochemical mechanism leading to CNS electrical discharges and epilepsy is unknown, but there may be multiple mechanisms involved. However, it has been shown that convulsions arise when there is an imbalance in two principal neurotransmitters in the brain, L-glutamic acid, an excitatory neurotransmitter, and GABA, an inhibitory neurotransmitter. The concentrations of L-glutamic acid and GABA are regulated by two PLP-dependent enzymes, L-glutamic acid decarboxylase (GAD), which converts glutamate to GABA, and GABA-AT, which

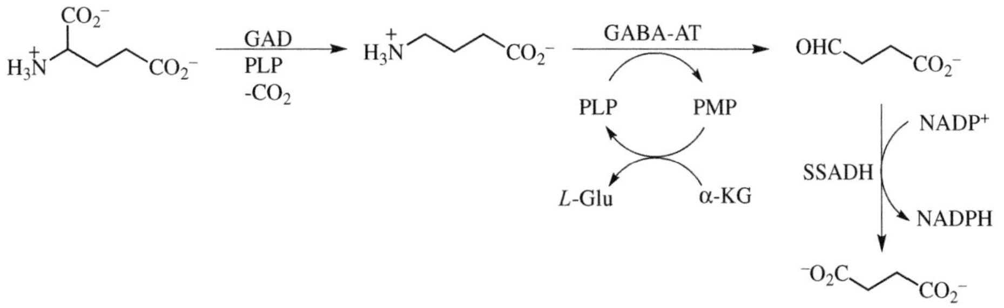

SCHEME 5.21 Metabolism of L-glutamic acid. SSADH, is semialdehyde dehydrogenase

degrades GABA to succinic semialdehyde with the regeneration of glutamate (Scheme 5.21). Although succinic semialdehyde is toxic to cells, there is no buildup of this metabolite, because it is efficiently oxidized to succinic acid by the enzyme succinic semialdehyde dehydrogenase. GABA system dysfunction has been implicated in the symptoms associated with epilepsy, Huntington's disease, Parkinson's disease, and tardive dyskinesia. When the concentration of GABA diminishes below a threshold level in the brain, convulsions begin. If a convulsion is induced in an animal, and GABA is injected directly into the brain, the convulsions cease. It would seem, then, that an ideal anticonvulsant agent would be GABA; however, peripheral administration of GABA produces no anticonvulsant effect. This was shown to be the result of the failure of GABA, under normal circumstances, to cross the *blood–brain barrier*, a membrane that surrounds the capillaries of the circulatory system in the brain and protects it from passive diffusion of undesirable (generally hydrophilic) chemicals from the bloodstream. Another approach for increasing the brain GABA concentration, however, would be to design a compound capable of permeating the blood–brain barrier that subsequently inactivates GABA-AT, the enzyme that catalyzes the degradation of GABA. Provided that GAD is also not inhibited, GABA concentrations should rise. This, in fact, has been shown to be an effective approach to the design of anticonvulsant agents.[231] Compounds that both cross the blood–brain barrier and inhibit GABA-AT in vitro have been reported to increase whole brain GABA levels in vivo and possess anticonvulsant activity. The anticonvulsant effect does not correlate with whole brain GABA levels, but it does correlate with an increase in the GABA concentration at the nerve terminals of the *substantia nigra*.[232]

The mechanism for the PLP-dependent aminotransferase reactions was discussed earlier (see Chapter 4, Section 4.3.1.3). On the basis of this mechanism, researchers at the former Merrell Dow Pharmaceuticals (now Sanofi)[233] designed 4-amino-5-hexenoic acid (vigabatrin; **5.117**, Scheme 5.22; Sabril). This is the first rationally designed mechanism-based inactivator drug. The proposed inactivation mechanism is shown in Scheme 5.22. By comparison of Scheme 5.22 with Scheme 4.18 in Chapter 4 (in Scheme 4.18 an α-amino acid is used, but here a γ-amino acid is the example), it is apparent that identical mechanisms are proposed up to compound **5.119** (Scheme 5.22) and compound **4.27** (Scheme 4.18). In the case of normal substrate turnover, hydrolysis of **4.27** gives pyridoxamine 5′-phosphate (PMP) and the α-keto acid (Scheme 4.18). The same hydrolysis could occur with **5.119** to give the corresponding products, PMP and **5.121**. However, **5.119** is a potent electrophile, a Michael acceptor, which can undergo conjugate addition by an active site nucleophile (X^-) and produce inactivated enzyme (**5.120a** or **5.120b**). The active site nucleophile was identified as Lys-329, the lysine residue that holds the PLP at the active site.[234] The mechanism in Scheme 5.22 (pathway a) appears to be relevant for about 70% of the inactivation.[235] The other 30% of the inactivation is accounted for by an allylic isomerization and enamine rearrangement leading to **5.122** (pathway b). Note that vigabatrin is an unreactive compound that is converted by the normal catalytic mechanism of the target enzyme into a reactive compound (**5.119**), which attaches to the enzyme. This is the typical course of events for a mechanism-based inactivator.

It may seem strange that GABA does not cross the blood–brain barrier, but vigabatrin, which is also a small charged molecule, *can* diffuse through that lipophilic membrane (albeit poorly). The attachment of a vinyl substituent to GABA, apparently, has two effects that permit this compound to cross the blood–brain barrier. First, the vinyl substituent increases the lipophilicity of the molecule. Second, it is an electron-withdrawing substituent that would have the effect of lowering the pK_a of the amino group. This would increase the concentration of the nonzwitterionic form (**5.123b**, Scheme 5.23), which is more lipophilic than the zwitterionic form (**5.123a**) because it is uncharged.

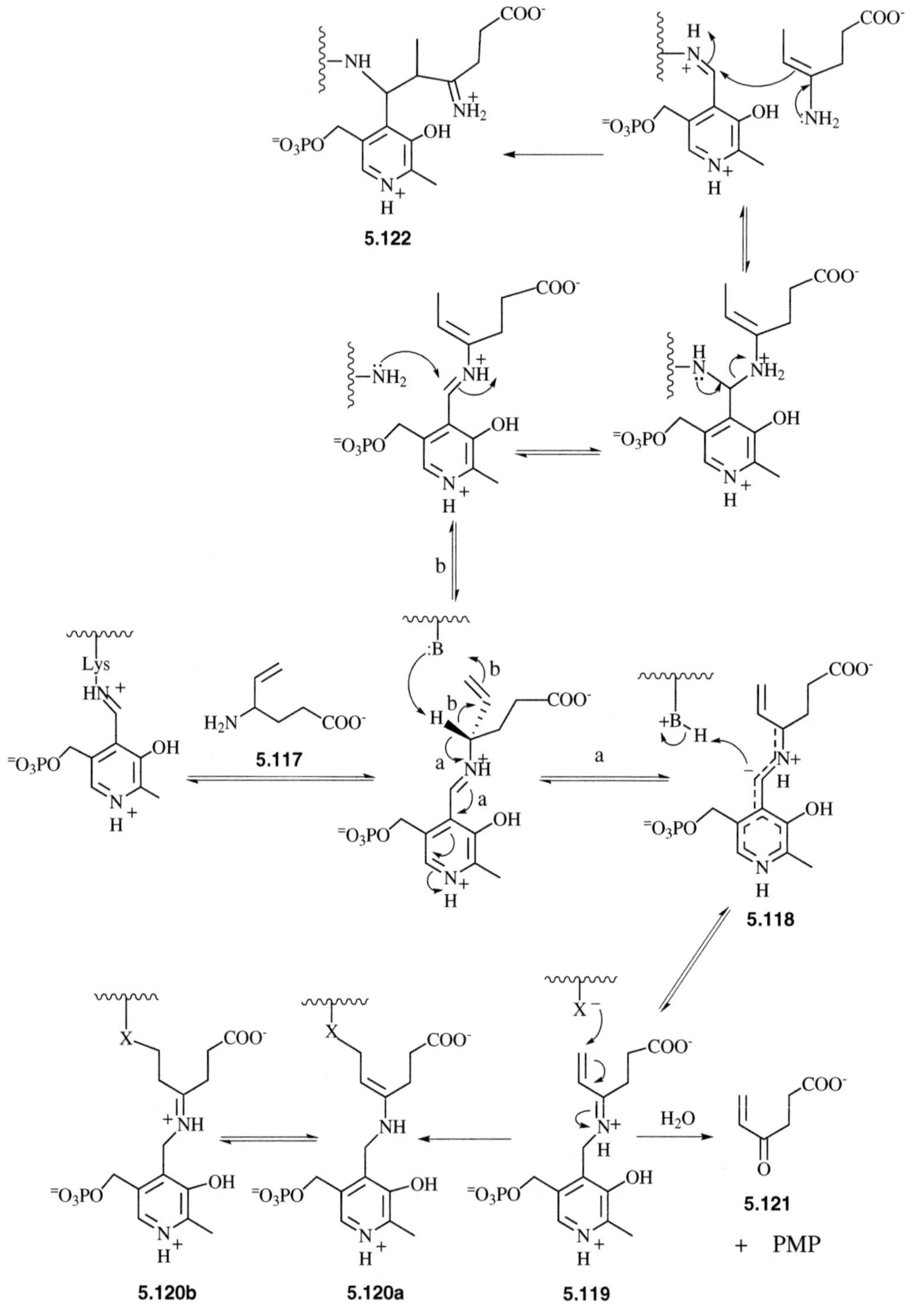

SCHEME 5.22 Hypothetical mechanism for the inactivation of GABA aminotransferase by vigabatrin

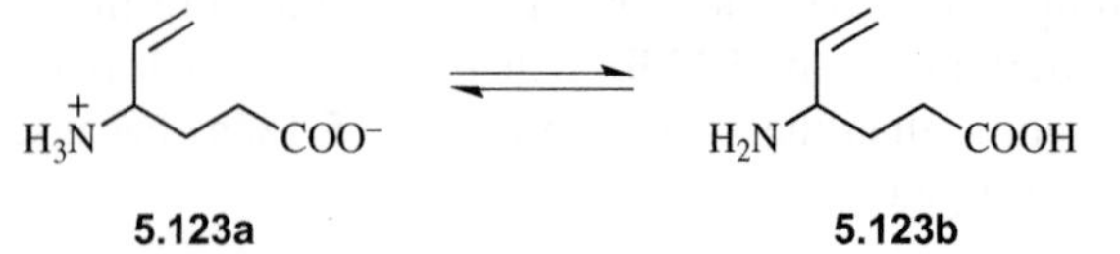

SCHEME 5.23 Zwitterionic and nonzwitterionic forms of vigabatrin

5.3.3.3.2. Eflornithine, an Antiprotozoal Drug and Beyond

The *polyamines*, spermidine (**5.124**) and spermine (**5.125**), and their precursor, putrescine (**5.126**) are important regulators of cell growth, division, and differentiation. The mechanisms by which they do this are unclear, but they appear to be

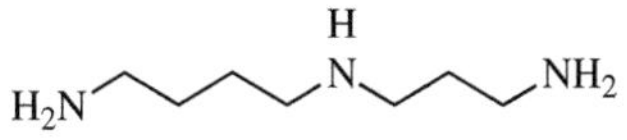

5.124

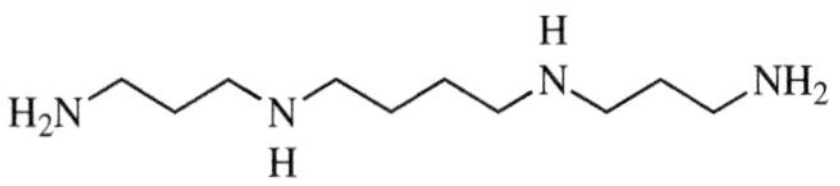

5.125

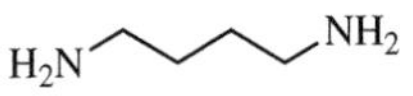

5.126

required for DNA synthesis. Rapidly growing cells have much higher levels of polyamines (and ornithine decarboxylase; see below) than do slowly growing or quiescent cells. When quiescent cells are stimulated, the polyamine and ornithine decarboxylase levels increase prior to an increase in the levels of DNA, RNA, and protein. The polycationic nature of the polyamines may be responsible for their interaction with cellular structures that have negatively charged groups such as DNA.

Ornithine decarboxylase, the enzyme that catalyzes the conversion of ornithine to putrescine (**5.126**), is the rate-limiting step in polyamine biosynthesis (Scheme 5.24). Spermidine (**5.124**) is produced by the spermidine synthase-catalyzed reaction of putrescine with *S*-adenosylhomocysteamine (**5.128**), which is derived from the decarboxylation of SAM (**5.127**) in a reaction catalyzed by SAM decarboxylase. Another aminopropyltransferase, namely, spermine synthase, catalyzes the reaction of spermidine with **5.128** to produce spermine (**5.125**).

Because polyamines are important for rapid cell growth, inhibition of polyamine biosynthesis should be an effective approach for the design of antitumor and antimicrobial agents. In fact, inactivation of ornithine decarboxylase by the potent mechanism-based inactivator α-difluoromethylornithine eflornithine, **5.129**, Ornidyl), results in virtually complete reduction in the putrescine and spermidine content; however, the spermine concentration is only slightly affected. There are at least two reasons for this latter observation. Inactivation of ornithine decarboxylase activity by eflornithine in vivo may not be complete because gene synthesis of ornithine decarboxylase has a half-life of only 30 min, and, also, the endogenous ornithine that is present competitively protects the enzyme from the inactivator. Furthermore, inactivation of ornithine decarboxylase induces an increase in the levels of *S*-adenosylmethionine decarboxylase, which leads to higher levels of **5.128**. This can be efficiently used to drive the existing putrescine and spermidine to spermine.[236] The

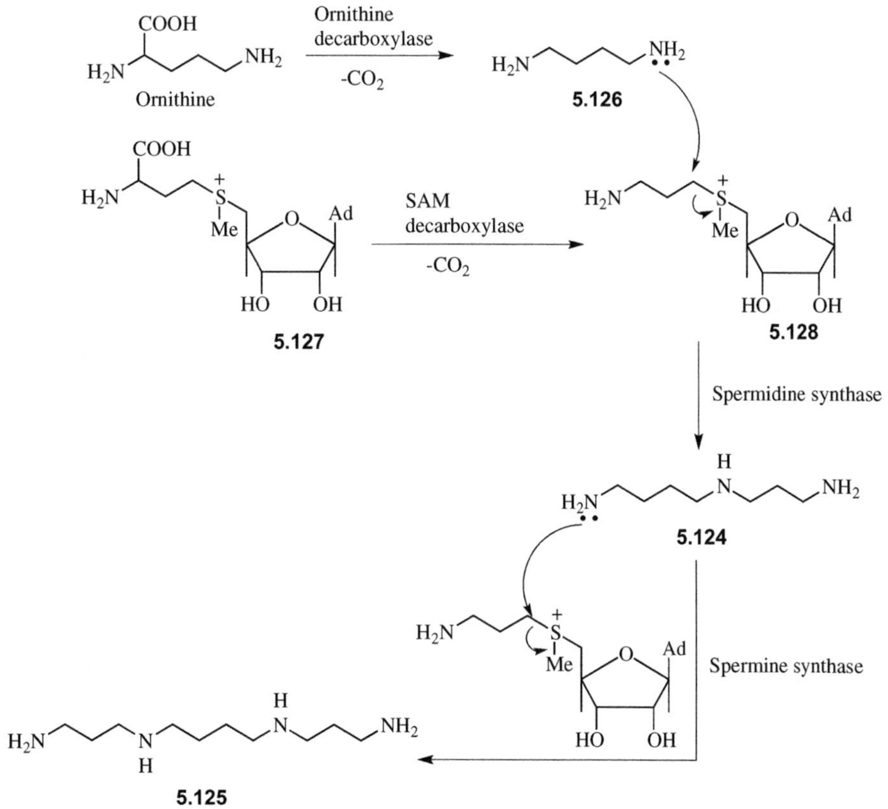

Ornithine

Ornithine decarboxylase

$-CO_2$

5.126

SAM decarboxylase

$-CO_2$

5.127

5.128

Spermidine synthase

5.124

Spermine synthase

5.125

SCHEME 5.24 Polyamine biosynthesis

inability of eflornithine to shut down all polyamine biosynthesis may be responsible for its discouragingly poor antitumor effects observed in clinical trials.[237] However, eflornithine has been found to have great value in the treatment of certain protozoal infestations such as *Trypanosoma brucei gambiense* (the more virulent strain, *Trypanosoma brucei rhodesiense*, is resistant), which causes African sleeping sickness (300,000 deaths per year)[238] and *Pneumocystis carinii*, the microorganism that produces pneumonia in AIDS patients.[239] Unfortunately, large amounts of the compound (150 mg/kg) are needed every 4–6 h, which makes it very expensive, especially in Third World countries, where the disease is prevalent.

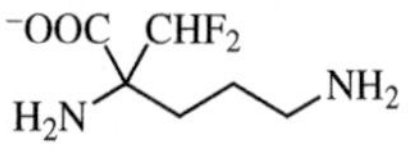

Eflornithine
5.129

Another indication has been discovered for eflornithine hydrochloride, namely, as a topical cream for the reduction of unwanted facial hair in women (SkinMedica in the United States and Vaniqa in Europe). The mechanism of action is not known, but it could involve inhibition of polyamine biosynthesis important to hair growth.

The mechanism for the PLP-dependent decarboxylases was discussed in Chapter 4 (see Section 4.3.1.2). Ornithine decarboxylase catalyzes the decarboxylation of eflornithine (**5.129**), producing a reactive product that inactivates the enzyme; a possible inactivation mechanism is shown in Scheme 5.25.

According to the mechanism for PLP-dependent decarboxylases (see Chapter 4, Scheme 4.16), the only irreversible step is the one where CO_2 is released. Therefore, if you incubate the decarboxylase with the product amine, it catalyzes the reverse reaction up to the loss of CO_2, that is, the removal of the α-proton. This is the microscopic reverse of the reaction that occurs once the CO_2 is released. The *principle of microscopic reversibility* states that for a reversible reaction, the

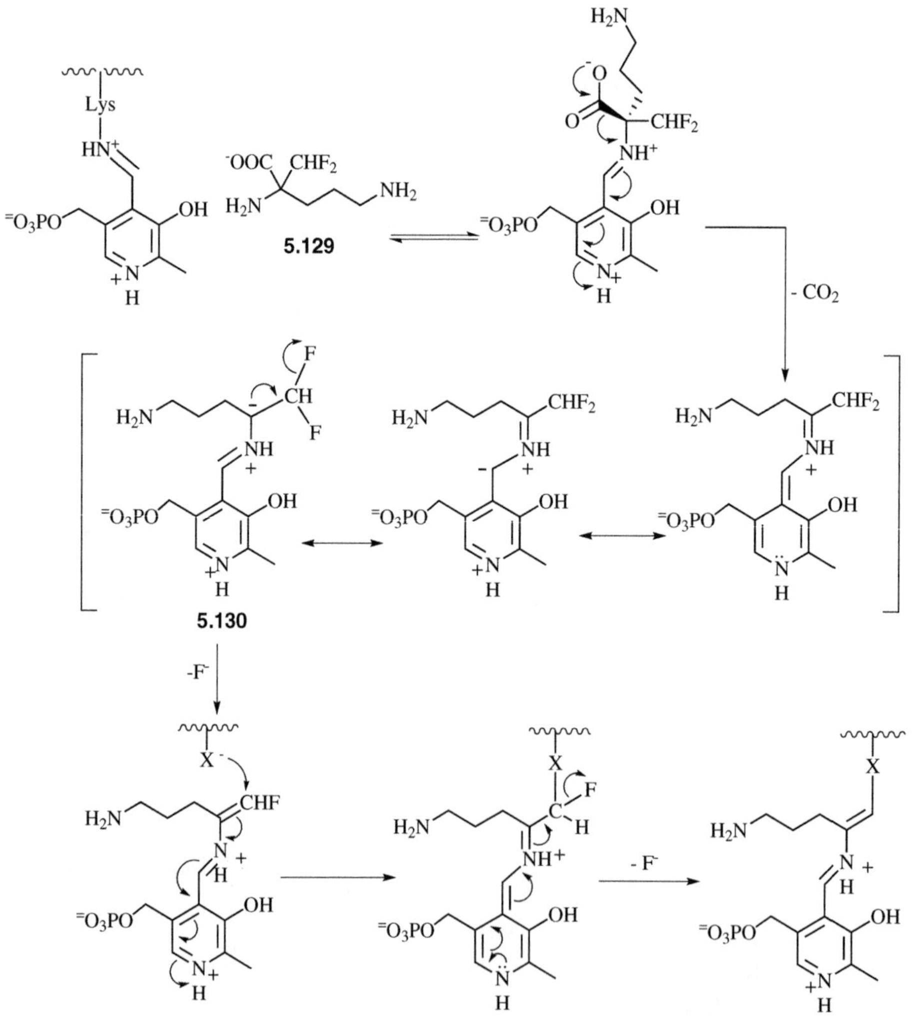

SCHEME 5.25 Hypothetical mechanism for inactivation of ornithine decarboxylase by eflornithine

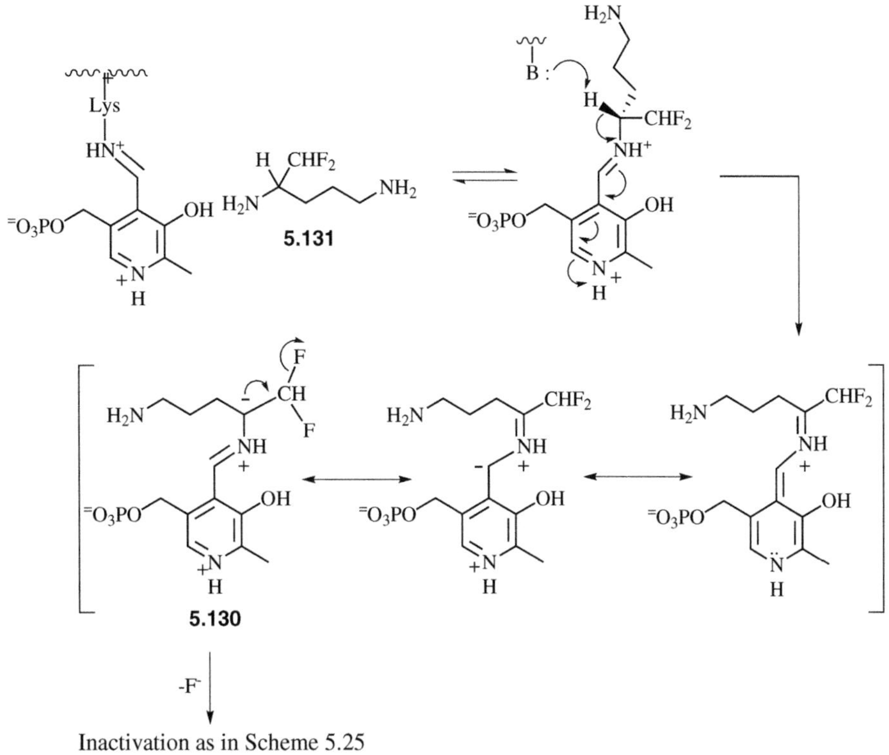

Inactivation as in Scheme 5.25

SCHEME 5.26 Hypothetical mechanism for inactivation of ornithine decarboxylase by α-difluoromethylputrescine

same mechanistic pathway will be followed in the forward and reverse reactions. Therefore, a mechanism-based inactivator that has a structure similar to that of the product of the ornithine decarboxylase reaction (the amine) should undergo catalytic α-deprotonation, which then would produce the same set of resonance structures produced by decarboxylation of the ornithine analog. α-Difluoromethylputrescine (**5.131**, Scheme 5.26) was shown to inactivate ornithine decarboxylase,[240] presumably because deprotonation gives the same intermediate that is obtained by decarboxylation of eflornithine (compare **5.130** in Scheme 5.25 with **5.130** in Scheme 5.26).

For any target enzyme that is reversible, mechanism-based inactivators could be designed for both the forward (substratelike) and back (productlike) reactions. Depending on the metabolic pathway involved, enzyme selectivity may be more favorable for one over the other.

5.3.3.3.3. Tranylcypromine, an Antidepressant Drug

The modern era of therapeutics for the treatment of depression began in the late 1950s with the introduction of both the monoamine oxidase (MAO) inhibitors and the tricyclic antidepressants. The first MAO inhibitor was iproniazid (**5.132**), which initially was used as an antituberculosis drug until it was observed that patients taking it exhibited excitement and euphoria.[241] In 1952, Zeller et al.[242] showed that iproniazid was a potent inhibitor of MAO, and clinical studies were underway in the late 1950s.[243]

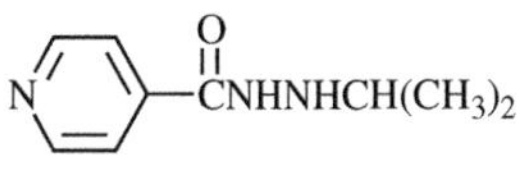

Iproniazid
5.132

The brain concentrations of various biogenic (pressor) amines such as norepinephrine, serotonin, and dopamine were found to be depleted in chronically depressed individuals. A correlation was observed between an increase in the concentrations of these brain biogenic amines and the onset of an antidepressant effect.[244] This was believed to be the result of MAO inhibition, because MAO is one of the enzymes responsible for the catabolism of these biogenic amines. By the early 1960s several MAO inactivators were being used clinically for the treatment of depression. Unfortunately, it was found that in some cases there was a cardiovascular side effect that led to the deaths of several patients. Consequently, these drugs were withdrawn from the drug market until the cause of death could be ascertained. Within a few months the problem was understood. It was determined that all of those who died while taking an MAO inhibitor had two things in common: they had all died from a hypertensive crisis and, prior to their deaths, they had eaten foods containing high tyramine content (e.g.,

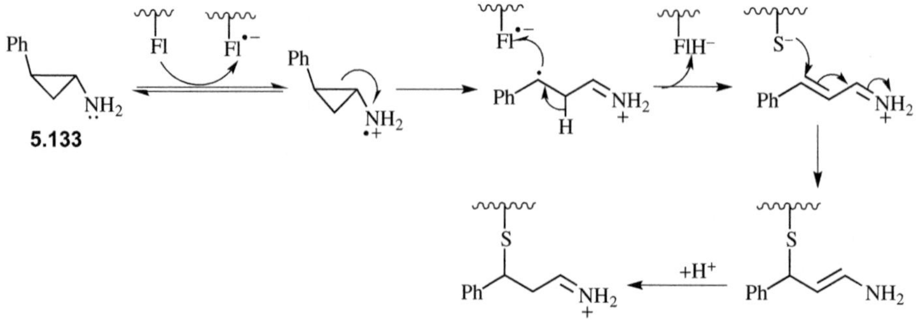

SCHEME 5.27 Hypothetical mechanism for the inactivation of monoamine oxidase by tranylcypromine

cheese, wine, beer, and chocolate). The connection between these observations is that the ingested tyramine triggers the release of norepinephrine, a potent vasoconstrictor, which raises the blood pressure. Under normal conditions, the excess norepinephrine is degraded by MAO and catecholamine *O*-methyltransferase. If the MAO is inactivated, then the norepinephrine is not degraded fast enough, the blood pressure keeps rising, and this can lead to a hypertensive crisis. This series of events has been termed the *cheese effect* because of the high tyramine content found in many cheeses. Because the MAO inhibitors were not toxic, except when taken with certain foods, these drugs were allowed to return to the drug market, but they were prescribed with strict dietary regulations. Because of this inconvenience, and the discovery of the tricyclic antidepressants (which block the reuptake of biogenic amines at the nerve terminals), MAO inhibitors are not the drugs of choice, except in those types of depression that do not respond to tricyclic antidepressants or when treating phobic anxiety disorders, which respond well to MAO inhibitors.

The resurgence of interest of the pharmaceutical industry in MAO inhibitors in the late 1980 and 1990s[245] came because MAO exists in at least two isozymic forms,[246] termed MAO A and MAO B. The main difference in these two isozymes is their selectivity for the oxidation of the various biogenic amines. Because the antidepressant effect is related to increased concentrations of brain serotonin and norepinephrine, both of which are MAO A substrates, compounds that selectively inhibit MAO A possess antidepressant activity; selective inhibitors of MAO B show potent antiparkinsonian properties.[247] To have an antidepressant drug without the cheese effect, however, it is necessary to inhibit brain MAO A selectively without inhibition of peripheral MAO A, particularly MAO A in the gastrointestinal tract and sympathetic nerve terminals. Inhibition of brain MAO A increases the brain serotonin and norepinephrine concentrations, which leads to the antidepressant effect; the peripheral MAO A must remain active to degrade the peripheral tyramine and norepinephrine that cause the undesirable cardiovascular effects.

Tranylcypromine (**5.133**; Parnate), a nonselective MAO A/B inactivator that exhibits a cheese effect, was one of the first MAO inactivators approved for clinical use; many other cyclopropylamine analogs show antidepressant activity.[248] The one-electron mechanism for MAO, a flavoenzyme, was discussed in Chapter 4 (Section 4.3.3.3; Scheme 4.32). If tranylcypromine acts as a substrate, and one-electron transfer from the amino group to the flavin occurs, the resulting cyclopropylaminyl radical will undergo rapid cleavage (Scheme 5.27),[249] and the benzylic radical produced could combine with an active site radical[250]. However, it was suggested[251] that an adduct with an active site cysteine may have formed. In Scheme 5.27, a mechanism for formation of a cysteine adduct is shown.

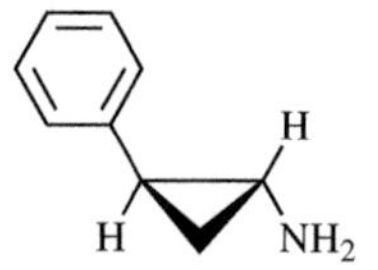

Tranylcypromine
5.133

5.3.3.3.4. Selegiline (L-Deprenyl) and Rasagiline: Antiparkinsonian Drugs

Parkinson's disease, the second most common neurodegenerative disease, afflicting more than one-half million people in the United States, is characterized by chronic, progressive motor dysfunction resulting in severe tremors, rigidity, and akinesia. The symptoms of Parkinson's disease arise from the degeneration of dopaminergic neurons in the *substantia nigra* and a marked reduction in the concentration of the pyridoxal 5′-phosphate-dependent aromatic L-amino acid decarboxylase, the enzyme that catalyzes the conversion of L-dopa to the inhibitory neurotransmitter dopamine. Because dopamine is metabolized primarily by MAO B in man (see Section 5.3.3.3.3), and Parkinson's disease is characterized by a reduction in the brain dopamine concentration (see Chapter 9, Section 9.2.2.10), selective inactivation of MAO B has been shown to be an effective approach to increase the dopamine concentration and, thereby, treat this disease. Actually, an MAO B-selective inactivator is used in combination with the antiparkinsonian drug L-dopa (see Chapter 9, Section 9.2.2.10). Selective inhibition of MAO B does not interfere with the MAO A-catalyzed degradation of tyramine and norepinephrine; therefore, no cardiovascular side effects (the cheese effect; see previous section) are observed with the use of an MAO B-selective inactivator. The earliest MAO B-selective inactivator, selegiline (**5.134**; Eldepryl),[252] was approved in 1989 by the FDA for the treatment of Parkinson's disease in the United States; in 2006, another MAO B-selective inactivator, rasagiline (**5.135**; Azilect),[253] was approved.

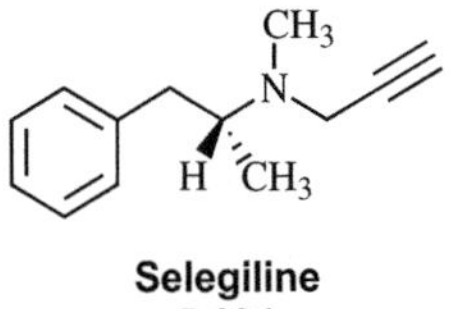

Selegiline
5.134

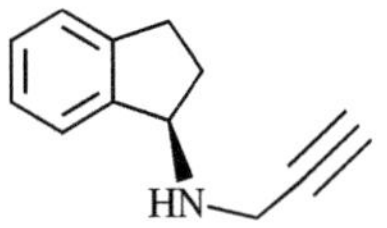

Rasagiline
5.135

Little was known about the cause for this disease until 1976, when a previously healthy 23-year-old man, who was a chronic street drug user, was referred to the National Institute of Mental Health for investigation of symptoms of what appeared to be Parkinson's disease.[254] Although the patient responded favorably to the usual treatment for Parkinson's disease (L-dopa/carbidopa; see Chapter 9), the speed with which, and the age at which, the symptoms developed were inconsistent with the paradigm of Parkinson's disease as a regressive geriatric disease. When this man later died of a drug overdose, an autopsy showed the same extensive destruction of the *substantia nigra* that is found in idiopathic parkinsonism. In 1982, seven young Californians who had tried some "synthetic heroin" also developed symptoms of an advanced case of Parkinson's disease, including near total immobility.[255] It was found that the drug they had been taking was a *designer drug*, a synthetic narcotic that has a structure designed to be a variation of an existing controlled drug; at that time, because the new structure was not listed as a controlled substance, it was not illegal to possess it (this law has since been changed to include variants of controlled substances).

In this case, the "designers" of the street drug were using the controlled analgesic meperidine (**5.136**; Demerol) as the basis for their structure modification. The compound synthesized, 1-methyl-4-phenyl-4-propionoxypiperidine (**5.137**), was referred to as a "reverse ester" of meperidine (note the ester oxygen in **5.137** is on the opposite side of the carbonyl from that in **5.136**). However, by "reversing" the ester, the drug designers converted a stable ethoxycarbonyl group of meperidine to a propionoxy group, a good leaving group, in **5.137**. In fact, **5.137** decomposes on heating or in the presence of acids with elimination of propionate to give 1-methyl-4-phenyl-1,2,5,6-tetrahydropyridine (MPTP; **5.138**). Analysis of several samples of the designer drug (called "new heroin") revealed the MPTP contamination. This same contamination was identified in the samples of drugs taken by the 23-year-old in 1976, but the drug samples at that time were found to exhibit no neurotoxicity in rats, so it was thought that MPTP was not responsible for the neurological effects. However, when the young California drug addicts were observed to have similar symptoms, tests in primates[256] and mice[257] showed that MPTP produced the same neurological symptoms and histological changes in the *substantia nigra* as those observed with idiopathic parkinsonism. Rats, however, it is now known, are remarkably resistant to the neurotoxic effects of MPTP, which explains why the earlier tests in rats were negative. The observation that an industrial chemist developed Parkinson's disease after synthesizing large amounts of MPTP as a starting material led to the suggestion that cutaneous absorption or vapor inhalation may be significant pathways for introduction of the neurotoxin. Because of this, it can be hypothesized that Parkinson's disease is, at least partially, an environmental disease arising from long-term slow degeneration of dopaminergic neurons by ingested or inhaled neurotoxins similar to MPTP. Because the symptoms of Parkinson's disease do not appear until 60–80% of the dopaminergic neurons are destroyed, it is reasonable that this disease is associated with the elderly. Opponents of this hypothesis note that the interregional and subregional patterns of striatal dopamine loss by MPTP differ from those of idiopathic Parkinson's disease.[258] However, results of surveys taken throughout the world suggest that environmental toxins may have an important role in the etiology of Parkinson's disease.[259] Furthermore, it is interesting that many frequently used medicines have structures related to MPTP; some, particularly neuroleptics, produce parkinsonian side effects.[260] Also, genetic factors do not appear to be important in most cases of Parkinson's disease.

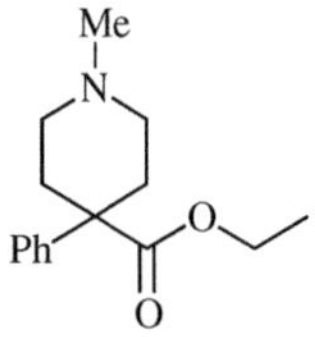

Meperidine
5.136

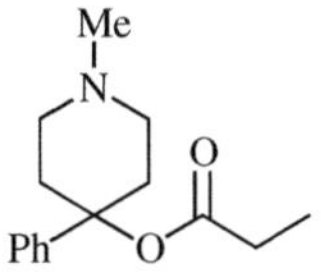

5.137

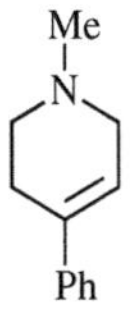

5.138

Once a connection was made between MPTP and Parkinson's disease, a vast amount of research with MPTP was initiated, and it soon was realized that the neurotoxic agent was actually a metabolite of MPTP. Pretreatment of animals with the MAO B-selective inactivator selegiline was shown to protect the animals from the neurotoxic effects (both the disease symptoms and the damage to the *substantia nigra*) of MPTP, whereas the MAO-A selective agent clorgyline (**5.139**) did not.[261] This indicated that the neurotoxic metabolite was generated by MAO B oxidation of MPTP; the lower levels of MAO B in rat brain could explain the negative effect on rats when the street drug was initially tested.[262] The two metabolites produced by MAO B are 1-methyl-4-phenyl-2,3-dihydropyridinium ion (**5.140**) and 1-methyl-4-phenylpyridinium ion (MPP^+; **5.141**).[263] The latter compound (**5.141**) accumulates in selected areas in the brain, and therefore, is believed to be the actual neurotoxic agent.[264] Because MPTP is a neutral molecule, it can cross the blood–brain barrier and enter the brain; once it is oxidized by MAO B, the pyridinium ion (**5.141**) cannot diffuse out of the brain. Selective toxicity of MPTP appears to be the result of the transport of MPP^+, but not MPTP, into dopamine neurons via an amine uptake system.[265] These studies suggest that Parkinson's disease may be more than just a geriatric disease; it may be caused by molecules in the environment, possibly pesticides, that are consumed and metabolized to molecules that are neurotoxic.[266]

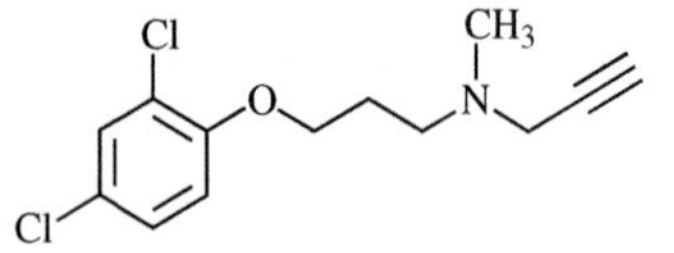

5.139

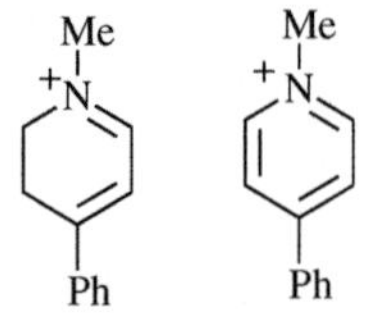

5.140 **5.141**

It is apparent, then, that, by inactivation of MAO B, selegiline is important both for the prevention of the oxidation of neurotoxin precursors such as MPTP and for preventing the degradation of dopamine. The mechanism of inactivation of MAO by a simpler analog of selegiline and rasagiline, 3-dimethylamino-1-propyne (**5.142**), has been studied; this compound becomes attached to the N^5 position of the flavin.[267] Because of the evidence for a one-electron mechanism for MAO-catalyzed oxidations (see Chapter 4, Section 4.3.3.3), the inactivation mechanisms for selegiline and rasagiline,[268] shown in Scheme 5.28 for rasagiline, seem most reasonable. The covalent N^5-adduct depicted in Scheme 5.28 was confirmed by X-ray crystallography.[269] Rasagaline, which has a very similar pharmacological action to that of selegiline,[270] is 10 times more potent than selegiline and is not metabolized to L-amphetamine or L-methampethamine as is selegiline, which could account for selegiline's sympathomimetic activity.[271]

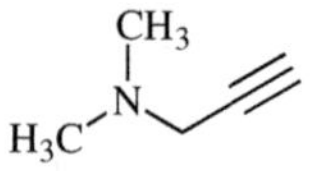

5.142

5.3.3.3.5. 5-Fluoro-2′-deoxyuridylate, Floxuridine, and 5-Fluorouracil: Antitumor Drugs

Cancer is a family of diseases characterized by abnormal and uncontrolled cell division. Neither the etiology nor the way in which it causes death is fully understood in most cases. One important approach to *antineoplastic* (antitumor) agents is the design of an antimetabolite (see Section 5.2.2.3.3), whose structure is related to those of pyrimidines and purines

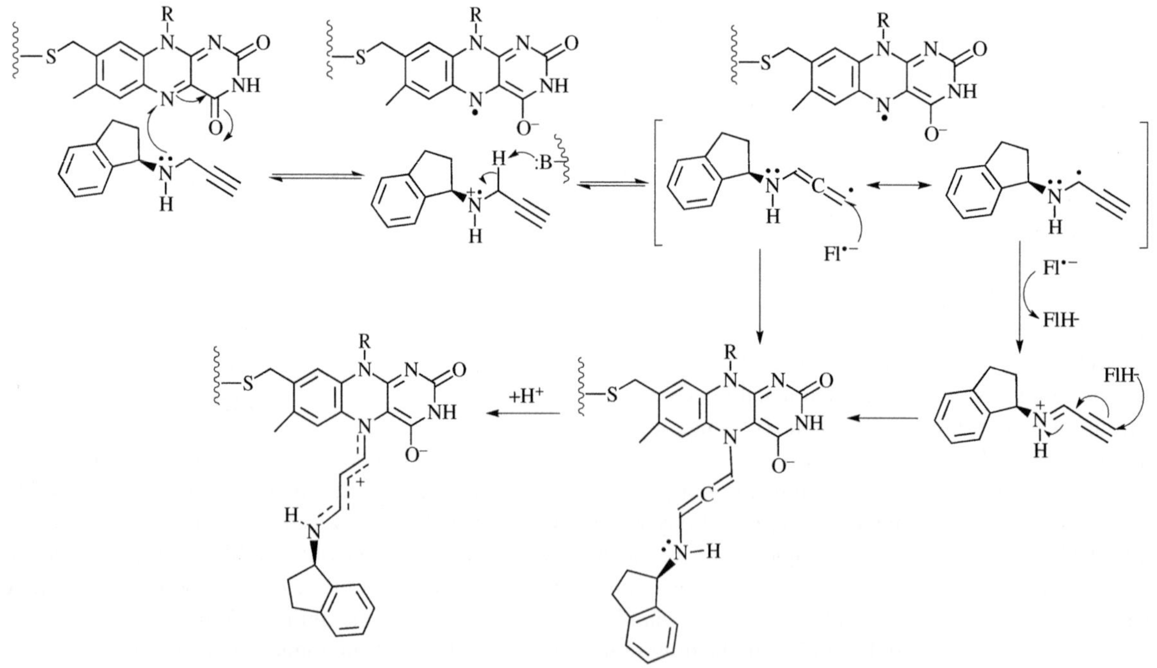

SCHEME 5.28 Hypothetical mechanism for the inactivation of monoamine oxidase by selegiline

involved in the biosynthesis of DNA. These compounds interfere with the formation or utilization of one of these essential normal cellular metabolites. This interference generally results from the inhibition of an enzyme in the biosynthetic pathway of the metabolite or from incorporation, as a false building block, into vital macromolecules such as proteins and polynucleotides. Antimetabolites usually are obtained by making a small structural change in the metabolite, such as a bioisosteric interchange (see Chapter 2, Section 2.2.4.3).

5-Fluorouracil (**5.143**, one name for systemic use is Adrucil; for topical use is Efudex), its 2′-deoxyribonucleoside, floxuridine (**5.144**, FUDR), and its 2′-deoxyribonucleotide, 5-fluoro-2′-deoxyuridylate (**5.145**), are potent antimetabolites of uracil and its congeners, and are also potent antineoplastic agents. 5-Fluorouracil itself is not active, but it is converted in vivo to the 2′-deoxynucleotide (**5.145**), which is the active form. Fluorine is often used as a replacement for hydrogen in medicinal chemistry[272] (the van der Waals radius of fluorine is 1.35 Å vs 1.20 Å for hydrogen), and 5-fluorouracil and its metabolites are recognized by enzymes that act on uracil and its metabolites. There are several pathways for this in vivo activation (Scheme 5.29). A minor pathway to intermediate 5-fluoruouridylate (**5.147**) begins with the conversion of 5-fluorouracil (**5.143**) to 5-fluorouridine (**5.146**), catalyzed by ribose-1-phosphate uridine phosphorylase, followed by further conversion of **5.146** to **5.147**, catalyzed by uridine kinase. The major pathway to **5.147** is the direct conversion of 5-fluorouracil, which is catalyzed by orotate phosphoribosyltransferase. 5-Fluoro-2′-deoxyuridylate (**5.145**) is produced from **5.147** by the circuitous route shown in Scheme 5.29 or by direct conversion of **5.143** to its 2′-deoxyribonucleoside, floxuridine (**5.144**), catalyzed by uridine phosphorylase, followed by 5′-phosphorylation, which is catalyzed by thymidine kinase (Scheme 5.29). However, when **5.144** is administered rapidly, it is converted back to **5.143** faster than it is phosphorylated to **5.145**. Under these circumstances, attempts to use floxuridine to bypass the long metabolic route for conversion of **5.143** to **5.145** are unsuccessful. Continuous intra-arterial infusion of floxuridine, however, enhances the direct conversion of **5.144** to **5.145**.

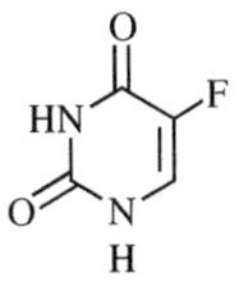

5.143 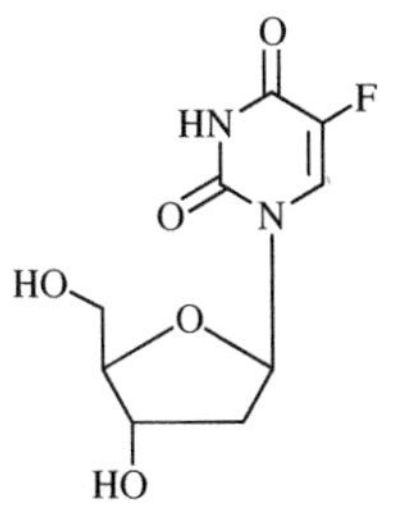 **5.144** 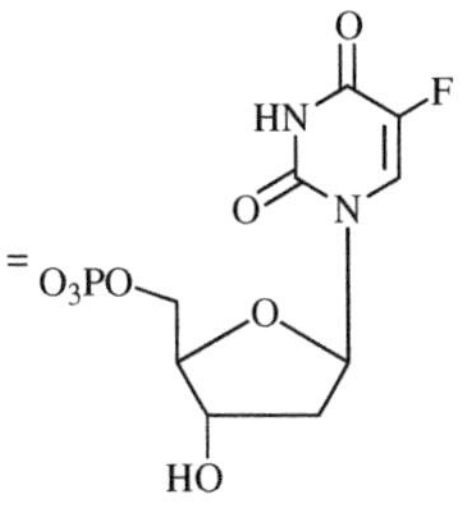 **5.145**

The principal site of action of **5.145** is thymidylate synthase, the enzyme that catalyzes the last step in de novo biosynthesis of thymidylate, namely, the conversion of 2′-deoxyuridylate to 2′-deoxythymidylate (referred to as just thymidylate). The reaction catalyzed by thymidylate synthase is the only de novo source of thymidylate, which is an essential constituent of the DNA. Therefore, inhibition of thymidylate synthase in tumor cells inhibits DNA biosynthesis and produces what is known as *thymineless death* of the cell.[273] Unfortunately, normal cells also require thymidylate synthase for de novo synthesis of their thymidylate. Nonetheless, inhibitors of thymidylate synthase are effective antineoplastic agents. There are several reasons for this selective toxicity against tumor cells; all are related to the difference in the rates of cell division for normal and abnormal cells. Because aberrant cells replicate much more rapidly than do most normal cells, the rapidly proliferating tumor cells have a higher requirement for their DNA (and DNA precursors) than do the slower proliferating normal cells. This means that the activity of thymidylate synthase is elevated in tumor cells relative to normal cells. Because uracil is one of the precursors of thymidylate, it *and* 5-fluorouracil are taken up into tumor cells much more efficiently than into normal cells. Finally, and possibly most importantly, enzymes that degrade uracil in normal cells also degrade 5-fluorouracil, and these degradation processes do not take place in cancer cells.[274] The adverse side effects accompanying the use of 5-fluorouracil in humans generally arise from the inhibition of thymidylate synthase and destruction of the rapidly proliferating normal cells of the intestines, the bone marrow, and the mucosa. The effects of anticancer drugs are discussed in more detail in Chapter 6.

Unlike other tetrahydrofolate-dependent enzymes (see Chapter 4, Section 4.3.2), thymidylate synthase utilizes methylenetetrahydrofolate as both a one-carbon donor *and* as a reducing agent (Scheme 5.30).[275] An active site cysteine residue undergoes Michael addition to the 6-position of 2′-deoxyuridylate (**5.148**, dRP is deoxyribose phosphate) to give an enolate (**5.149**), which attacks 5,10-methylenetetrahydrofolate (more likely, in one of the more reactive open, iminium forms; see Chapter 4, Scheme 4.23), and forms a ternary complex (**5.150**) of the enzyme, the substrate, and the coenzyme. Enzyme-catalyzed removal of the C-5 proton leads to β-elimination of tetrahydrofolate (**5.151**). Oxidation of the tetrahydrofolate (a hydride mechanism is shown, but a one-electron mechanism, that is, first transfer of an N-5 nitrogen nonbonded electron to the alkene followed by hydrogen atom transfer, is possible) gives dihydrofolate (**5.152**) and the enzyme-bound thymidylate enolate (**5.153**). Reversal of the first step releases the active site cysteine residue and produces thymidylate (**5.154**). This reaction changes the oxidation state of the coenzyme. Because of this, another enzyme, dihydrofolate reductase, is required to reduce the dihydrofolate back to tetrahydrofolate (see Chapter 4, Scheme 4.21).

5-Fluoro-2′-deoxyuridylate (**5.145**) inactivates thymidylate synthase because once the ternary complex (**5.155**)

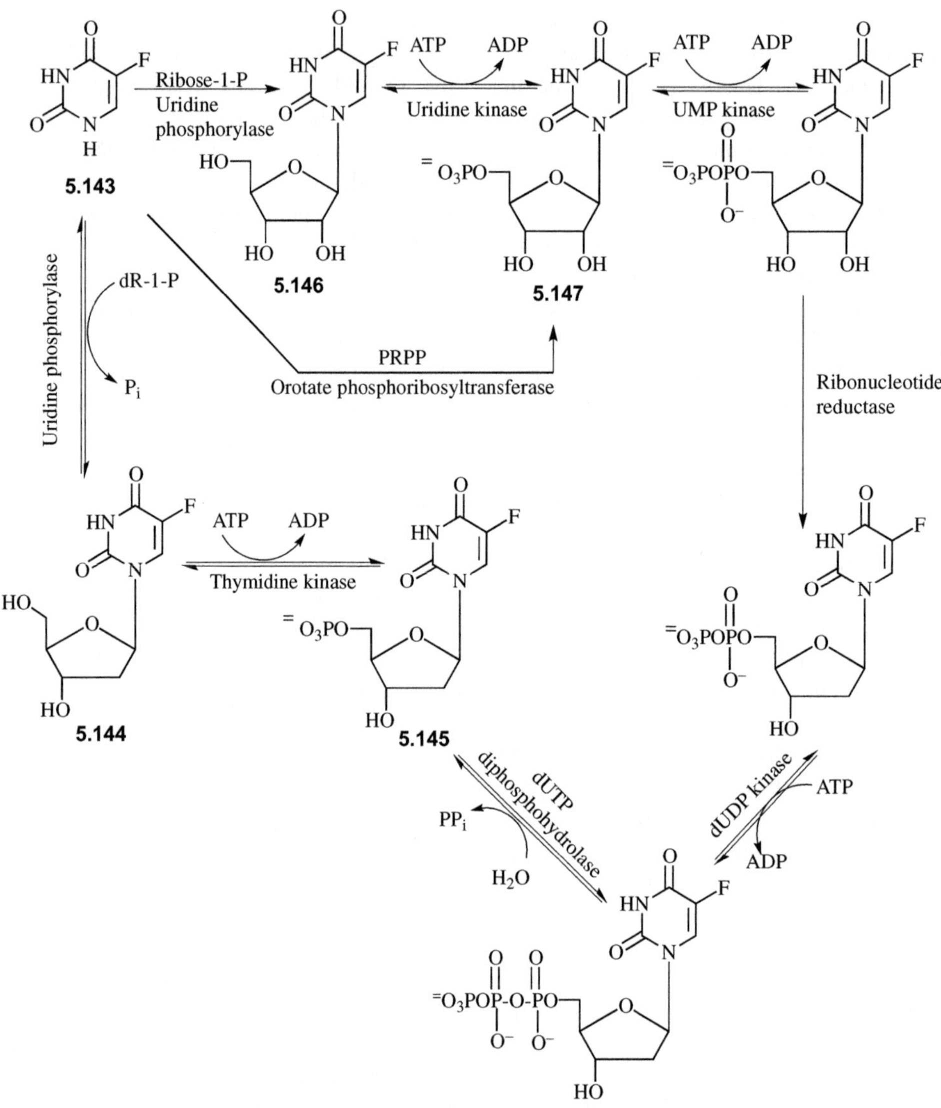

SCHEME 5.29 Metabolism of 5-fluorouracil

forms, there is no C-5 proton that the enzyme can remove to eliminate the tetrahydrofolate (Scheme 5.31).[276] Consequently, the enzyme remains as the ternary complex.

Note that in this mechanism the inactivator is not converted into a reactive compound that attaches to the enzyme. Instead, it first attaches to the enzyme, then requires condensation with 5,10-methylenetetrahydrofolate to generate a stable complex. A mechanism-based inactivator, therefore, does not necessarily require formation of a reactive species. It only requires the enzyme to catalyze a reaction on it that leads to inactivation prior to release of the product. If the enzyme were inactivated without the requirement of 5,10-methylenetetrahydrofolate, that is, by simple Michael addition of the active site cysteine to the C-6 position of the inactivator, then **5.145** would be an affinity labeling agent.

Next, in Chapter 6, we will consider drug interactions with another type of receptor, DNA.

5.4. GENERAL REFERENCES

Copeland, R. A. *Evaluation of Enzyme Inhibitors in Drug Discovery: A Guide for Medicinal Chemists and Pharmacologists*, 2nd Edition, John Wiley & Sons: Hoboken, NJ, 2013.

Lu, C.; Li, A. P. *Enzyme Inhibition in Drug Discovery and Development. The Good and the Bad*, John Wiley & Sons: Hoboken, NJ, 2010.

Sulfonamides

Weidner-Wells, M. A.; Macielag, M. J. Sulfonamides, *Kirk-Othmer Encyclopedia of Chemical Technology*, 5th ed., 2007, *23*, 493–513.

Dibbern, D. A., Jr.; Montanaro, A. Allergies to sulfonamide antibiotics and sulfur-containing drugs. *Ann. Allergy, Asthma, Immunol.* **2008**, *100*, 91–100.

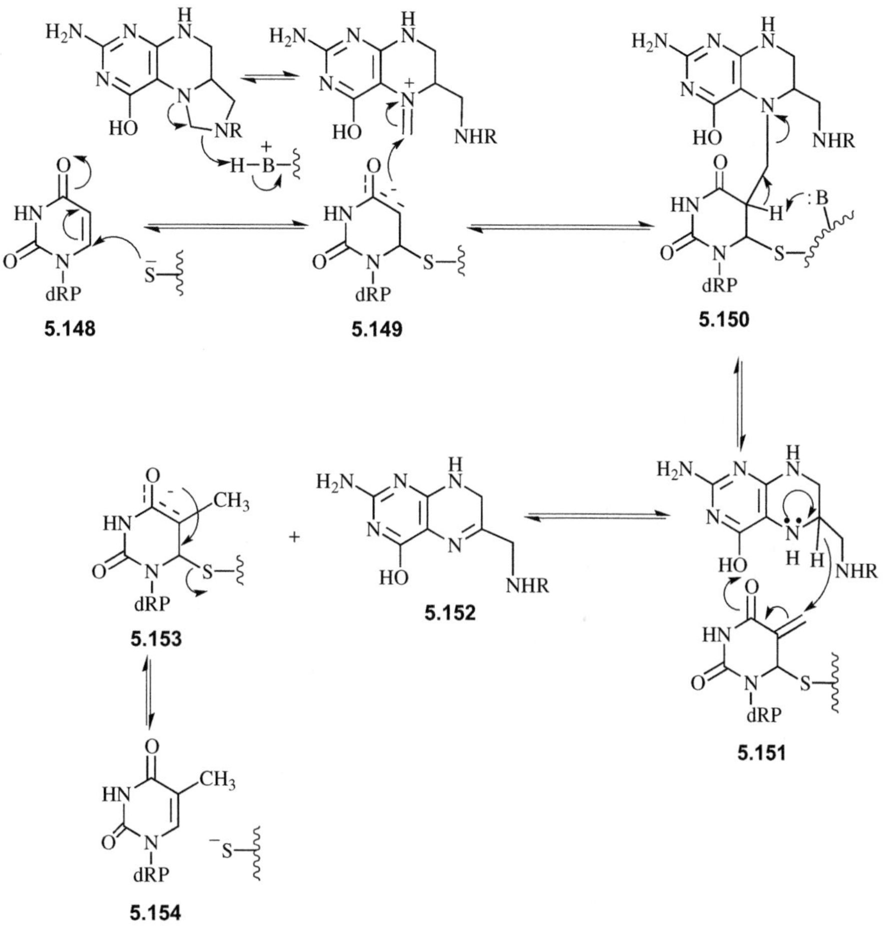

SCHEME 5.30 Hypothetical mechanism for thymidylate synthase (dRP is deoxyribose phosphate)

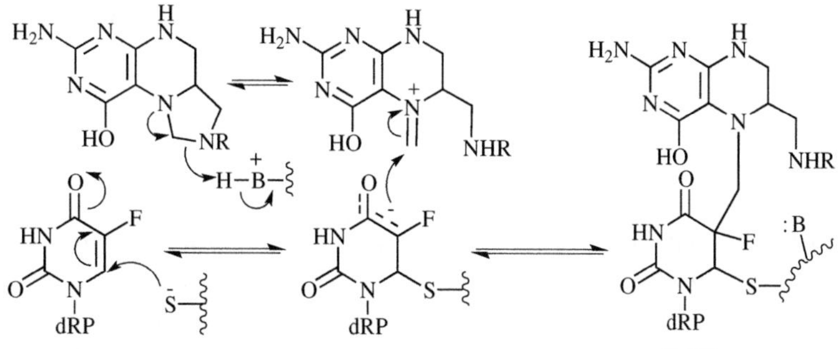

SCHEME 5.31 Hypothetical mechanism for the inactivation of thymidylate synthase by 5-fluoro-2′-deoxyuridylate

Supuran, C. T. Diuretics: from classical carbonic anhydrase inhibitors to novel applications of the sulfonamides. *Curr. Pharmaceut. Des.* **2008**, *14*, 641–648.

Kalgutkar, A. S.; Jones, R.; Sawant, A. Sulfonamide as an essential functional group in drug design. *RSC Drug Discovery Series* **2010**, *1*, 210–274.

Chen, X.; Hussain, S.; Parveen, S.; Zhang, S.; Yang, Y.; Zhu, C. Sulfonyl group-containing compounds in the design of potential drugs for the treatment of diabetes and its complications. *Curr. Med. Chem.* **2012**, *19*, 3578–3604.

Slow, Tight-Binding Inhibitors

Silverman, R. B. Enzyme inhibition. *Wiley Encyclopedia Chem. Biol.* **2009**, *1*, 663–681.

Szedlacsek, S. E.; Duggleby, R. G. Kinetics of slow and tight-binding inhibitors. *Meth. Enzymol.* **1995**, *249*, (Enzyme Kinetics and Mechanism, Part D), 144–180.

Morrison, J. F.; Walsh, C. T. The behavior and significance of slow-binding enzyme inhibitors. *Adv. Enzymol.* **1988**, *61*, 201–301.

Schloss, J. V. Significance of slow-binding enzyme inhibition and its relationship to reaction-intermediate analogs. *Acc. Chem. Res.* **1988**, *21*, 348–353.

Sculley, M. J.; Morrison, J. F. The determination of kinetic constants governing the slow, tight-binding inhibition of enzyme-catalysed reactions. *Biochim. Biophys. Acta* **1986**, *874*, 44–53.

Transition State Analogues

Schramm, V. L.; Tyler, P. C. Transition state analogue inhibitors of N-ribosyltransferases. In *Iminosugars*, Compain, P.; Martin, O. R. (Eds.), Wiley, Southern Gate, Chichester, 2007, pp. 177–208.

Smyth, T. P. Substrate variants versus transition state analogues as noncovalent reversible enzyme inhibitors. *Bioorg. Med. Chem.* **2004**, *12*(15), 4081–4088.

Schramm, V. L. Enzymic transition states and transition state analog design. *Annu. Rev. Biochem.* **1998**, *67*, 693–720.

Andrews, P. R.; Winkler, D. A. In *Drug Design: Fact or Fantasy?* Jolles, G.; Wooldridge, K. R. H. (Eds.), Academic: London, 1984, p. 145.

Wolfenden, R. Transition state analog inhibitors and enzyme catalysis. *Annu. Rev. Biophys. Bioeng.* **1976**, *5*, 271–306.

Wolfenden, R. Transition state analogs as potential affinity labeling reagents. *Meth. Enzymol.* **1977**, *46*, 15–28.

Multisubstrate Analogues

Le Calvez, P. B.; Scott, C. J.; Migaud, M. E. Multisubstrate adduct inhibitors: drug design and biological tools. *J. Enz. Inhib. Med. Chem.* **2009**, *24*(6), 1291–1318.

Radzicka, A.; Wolfenden, R. Transition state and multisubstrate analog inhibitors. *Meth. Enzymol.* **1995**, *249* (Enzyme Kinetics and Mechanism, Part D), 284–312.

Page, M. I. Enzyme inhibition In *Comprehensive Medicinal Chemistry*, Hansch, C.; Sammes, P. G.; Taylor, J. B. (Eds.), Pergamon Press, Oxford, 1990, Vol. 2, pp. 61–87.

Broom, A. D. Rational design of enzyme inhibitors: multisubstrate analogue inhibitors. *J. Med. Chem.* **1989**, *32*, 2–7.

Covalent Inhibitors

Mehdi, S. Covalent enzyme inhibition in drug discovery and development, In *Enzyme Technologies: Pluripotent Players in Discovering Therapeutic Agents*. Yang, H.-C.; Yeh, W.-K.; McCarthy, J. R. (Eds.), John Wiley & Sons, Hoboken, NJ, 2014.

Penicillins and Cephalosporins

Bush, K.; Mobashery, S. How beta-lactamases have driven pharmaceutical drug discovery. From mechanistic knowledge to clinical circumvention. *Adv. Exp. Med. Biol.* **1998**, *456*, 71–98.

Mandell, G. L.; Sande, M. A. In *Goodman and Gilman's The Pharmacological Basis of Therapeutics*, 7th ed., Gilman, A. G.; Goodman, L. S.; Rall, T. W.; Murad, F., Eds.; Macmillan, New York, 1985; p. 1115.

Morin, R. B.; Gorman, M. (Eds.), *Chemistry and Biology of β-Lactam Antibiotics*; Academic: New York, 1982.

Aspirin

De Caterina, R.; Renda, G. Clinical use of aspirin in ischemic heart disease: past, present, and future. *Curr. Pharmaceut. Des.* **2012**, *18*, 5215–5223.

Ugurlucan, M.; Caglar, I. M.; Caglar, F. N. T.; Ziyade, S.; Karatepe, O.; Yildiz, Y.; Zencirci, E.; Ugurlucan, F. G.; Arslan, A. H.; Korkmaz, S.; et al. Aspirin: from a historical perspective. *Recent Patents Cardiovasc. Drug Disc.* **2012**, *7*, 71–76.

Thun, M. J.; Jacobs, E. J.; Patrono, C. The role of aspirin in cancer prevention. *Nature Rev. Clin. Oncol.* **2012**, *9*, 259–267.

Lordkipanidze, M. Advances in monitoring aspirin therapy. *Platelets* **2012**, *23*, 526–536.

Choubey, A. K. Aspirin: a wonder drug. *J. Pharm. Res.* **2011**, *4*, 3803–3805.

Flower, R. J.; Moncada, S.; Vane, J. R. In *Goodman and Gilman's The Pharmacological Basis of Therapeutics*, 7th ed., Gilman, A. G.; Goodman, L. S.; Rall, T. W.; Murad, F. (Eds.), Macmillan, New York, 1985; p. 674.

Mechanism-Based Enzyme Inactivators

Silverman, R. B. *Mechanism-Based Enzyme Inactivation: Chemistry and Enzymology*; CRC: Boca Raton, FL, 1988, Vol. 1–2.

Silverman, R.B. The Potential Use of Mechanism-Based Enzyme Inactivators in Medicine. *J. Enz. Inhib.* **1988**, *2*, 73–90.

Silverman, R. B. Mechanism-Based Enzyme Inactivators. *Methods Enzymol.* **1995**, *249*, 240–283.

Polyamines

Bachmann, A. S.; Levin, V. A. Clinical applications of polyamine-based therapeutics. *RSC Drug Disc. Series* **2012**, *17* (Polyamine Drug Discovery), 257–276.

Huang, Y.; Marton, L. J.; Woster, P. M. The design and development of polyamine-based analogs with epigenetic targets. *RSC Drug Disc. Series* **2012**, *17* (Polyamine Drug Discovery), 238–256.

Goodwin, A. C.; Murray-Stewart, T. R.; Casero, R. A., Jr. Targeting the polyamine catabolic enzymes spermine oxidase, N1-acetylpolyamine oxidase and spermidine/spermine N1-acetyltransferase. *RSC Drug Disc. Series* **2012**, *17* (Polyamine Drug Discovery), 135–161.

Pegg, A. E. Inhibitors of polyamine biosynthetic enzymes. *RSC Drug Disc. Series* **2012**, *17* (Polyamine Drug Discovery), 78–103.

Clark, K.; Niemand, J.; Reeksting, S.; Smit, S.; Brummelen, A. C.; Williams, M.; Louw, A. I.; Birkholtz, L.

Functional consequences of perturbing polyamine metabolism in the malaria parasite, *Plasmodium falciparum*. *Amino Acids* **2010**, *38*, 633–644.

Wallace, H. M. Targeting polyamine metabolism: a viable therapeutic/preventative solution for cancer? *Exp. Opin. Pharmacother.* **2007**, *8*, 2109–2116.

Heby, O.; Persson, L.; Rentala, M. Targeting the polyamine biosynthetic enzymes: a promising approach to therapy of African sleeping sickness, Chagas' disease, and leishmaniasis. *Amino Acids* **2007**, *33*, 359–366.

Tabor, C. W.; Tabor, H. Polyamines. *Annu. Rev. Biochem.* **1984**, *53*, 749–790.

Monoamine Oxidase Inhibitors

Poewe, W.; Mahlknecht, P.; Jankovic, J. Emerging therapies for Parkinson's disease. *Curr. Opin. Neurol.* **2012**, *25*, 448–459.

Jankovic, J.; Poewe, W. Therapies in Parkinson's disease. *Curr. Opin. Neurol.* **2012**, *25*, 433–447.

Carradori, S.; Secci, D.; Bolasco, A.; Chimenti, P.; D'Ascenzio, M. Patent-related survey on new monoamine oxidase inhibitors and therapeutic potential. *Exp. Opin. Therapeut. Patents* **2012**, *22*(7), 759–801.

Helguera, A. M.; Perez-Machado, G.; Cordeiro, M. N. D. S.; Borges, F. Discovery of MAO-B inhibitors-present status and future directions part I: oxygen heterocycles and analogs. *Mini-Rev. Med. Chem.* **2012**, *12*(10), 907–919.

Flockhart, D. A. Dietary restrictions and drug interactions with monoamine oxidase inhibitors: an update. *J. Clin. Psych.* **2012**, *73* (Suppl. 1), 17–24.

Hoy, S. M.; Keating, G. M. Rasagiline: a review of its use in the treatment of idiopathic Parkinson's disease. *Drugs* **2012**, *72*, 643–669.

Schapira, A. H. V. Monoamine oxidase B inhibitors for the treatment of Parkinson's disease: a review of symptomatic and potential disease-modifying effects. *CNS Drugs* **2011**, *25*, 1061–1071.

***β*-Lactamase Inhibitors**

Biondi, S.; Long, S.; Panunzio, M.; Qin, W. L. Current trends in β-lactam based β-lactamase inhibitors. *Curr. Med. Chem.* **2011**, *18*, 4223–4236.

Bush, K.; Macielag, M. J. New β-lactam antibiotics and β-lactamase inhibitors. *Exp. Opin. Therapeut. Patents* **2010**, *20*, 1277–1293.

Cartwright, S. J.; Waley, S. G. Beta-lactamase inhibitors. *Med. Res. Rev.* **1983**, *3*, 341–382.

Knowles, J. R. Penicillin resistance: the chemistry of β-lactamase inhibition. *Acc. Chem. Res.* **1985**, *18*, 97–104.

5.5. PROBLEMS (ANSWERS CAN BE FOUND IN THE APPENDIX AT THE END OF THE BOOK)

1. Why would you want to design a drug that is an enzyme inhibitor?
2. If you wanted to inhibit an enzyme in a microorganism that is also in humans, what approaches would you take in your research?
3. *S*-Adenosylmethionine (SAM) is biosynthesized from methionine and ATP, catalyzed by methionine adenosyl transferase. The mechanism is shown below.

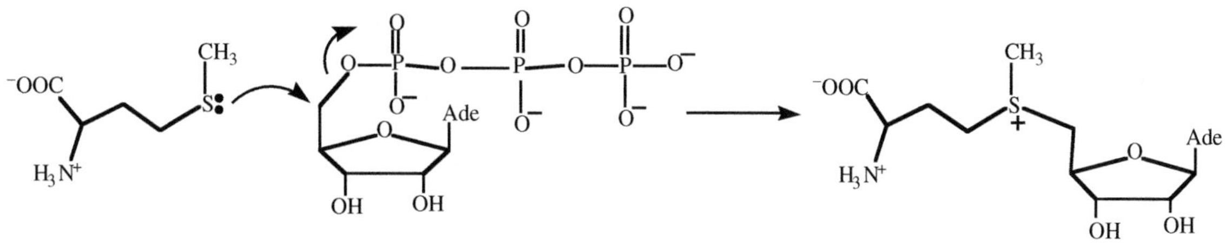

   a. Design a competitive reversible inhibitor for this enzyme.
   b. Design a multi-substrate analogue inhibitor, and show the basis for your design.
4. What advantage does a slow, tight-binding inhibitor have over a simple reversible inhibitor?
5. Thromboxane A_2 (TXA_2) is biosynthesized in humans from prostaglandin H_2 (PGH_2) by the enzyme thromboxane A_2 synthase. Binding of TXA_2 to a receptor causes vasoconstriction (raises the blood pressure) and platelet aggregation; therefore, TXA_2 has been implicated as a causative factor in a number of cardiovascular and renal diseases.
   a. Without looking at part b of this question, briefly describe three approaches you could take regarding TXA_2 to design a new antihypertensive agent.
   b. Thromboxane A_2 synthase inhibitors, such as **2**, have not been successful, possibly because of the buildup of PGH_2, the substrate for thromboxane

A_2 synthase, which is also a potent agonist at the TXA_2 receptor. PGH_2 is converted to PGI_2, a potent vasodilator and platelet inhibitory agent, by PGI_2 synthase. TXA_2 receptor antagonists, such as **3**, block the binding of both PGH_2 and TXA_2 to the TXA_2 receptor, but because TXA_2 synthase is not inhibited, there is no buildup of PGH_2 and therefore no conversion to PGI_2. What approach would you take to get around these problems?

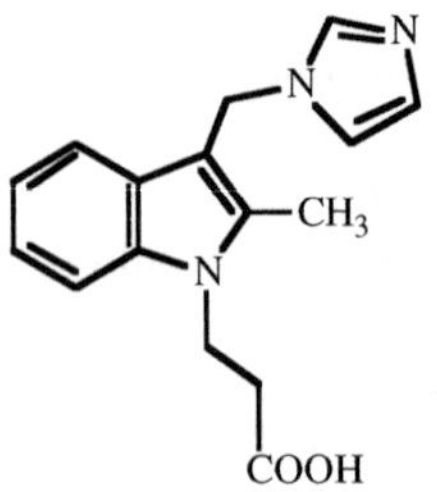

2 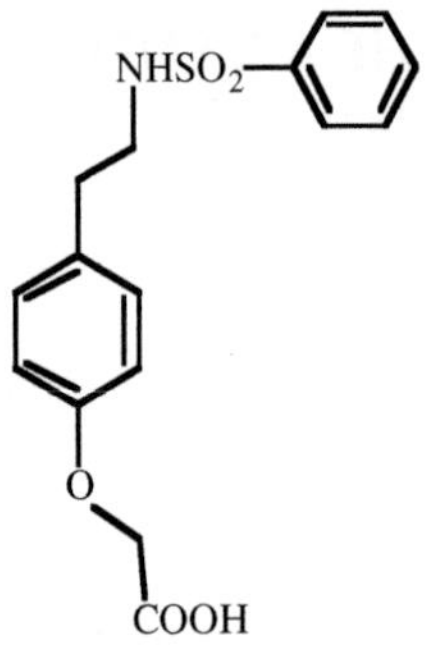 **3**

c. Design a molecule that uses the principle you propose in part b.

6. Show the transition state for the reaction below, and draw a reasonable transition state analog inhibitor.

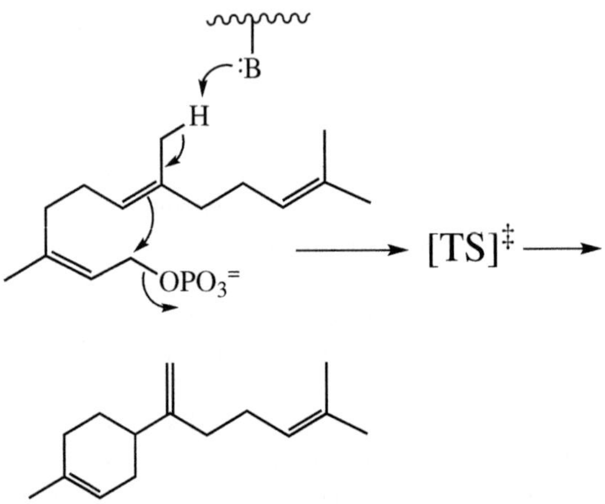

$\longrightarrow [TS]^{\ddagger} \longrightarrow$

7. 5-Fluoro-2′-deoxyuridylate (5-FdUMP) inactivates thymidylate synthase and is an antitumor agent. What would be a good choice for drug combination with 5-FdUMP? Why?

8. 5-Lipoxygenase requires a ferric ion as a cofactor. Show how **4** acts as an inhibitor of 5-lipoxygenase.

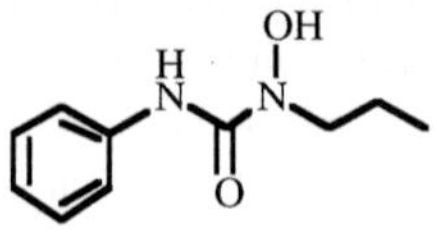

4

9. AG 7088 was designed to inhibit a picornavirus protease and is a potent antiviral agent.

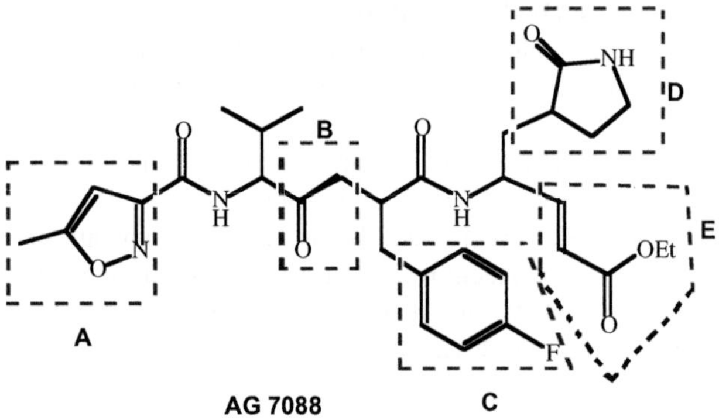

AG 7088

Why do you think moieties A-E were incorporated into the structure?

10. **a.** Two isoforms of an enzyme were discovered; isoform-1 produces a hormone that causes muscle spasms and isoform-2 makes another hormone from the same substrate that lowers cholesterol levels. What would you do to prevent muscle spasms without raising cholesterol levels?

b. If the active sites of isoform-1 and -2 are the same except isoform-1 has a cysteine residue and isoform-2 has a phenylalanine residue at that same position, what two approaches would you take for a muscle spasm drug without a cholesterol level increase side effect?

11. Fosfomycin (**5**) is a potent antibacterial agent (see Chapter 7, Figure 7.3) that interferes with cell wall biosynthesis at the enzyme called MurA.

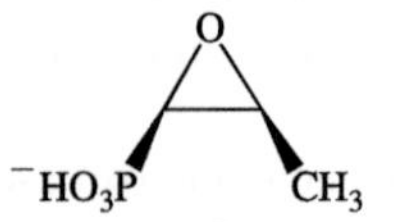

5

Draw a potential chemical mechanism for how fosfomycin acts.

12. Carfilzomib (**5a**, Kyprolis) is a proteasome inhibitor for the treatment of multiple myeloma and solid tumors. It acts by irreversible inhibition of the 20S proteasome, an enzyme that degrades unwanted proteins. Inhibition of this enzyme in the tumor cell leads to a build-up of ubiquitinated proteins, which causes apoptosis and inhibition of tumor cell growth. The N-terminal threonine of the 20S proteasome reacts to form a morpholine adduct (**5b**). Draw a reasonable inactivation mechanism that rationalizes why acid- or base-catalysis would be favored in each step.

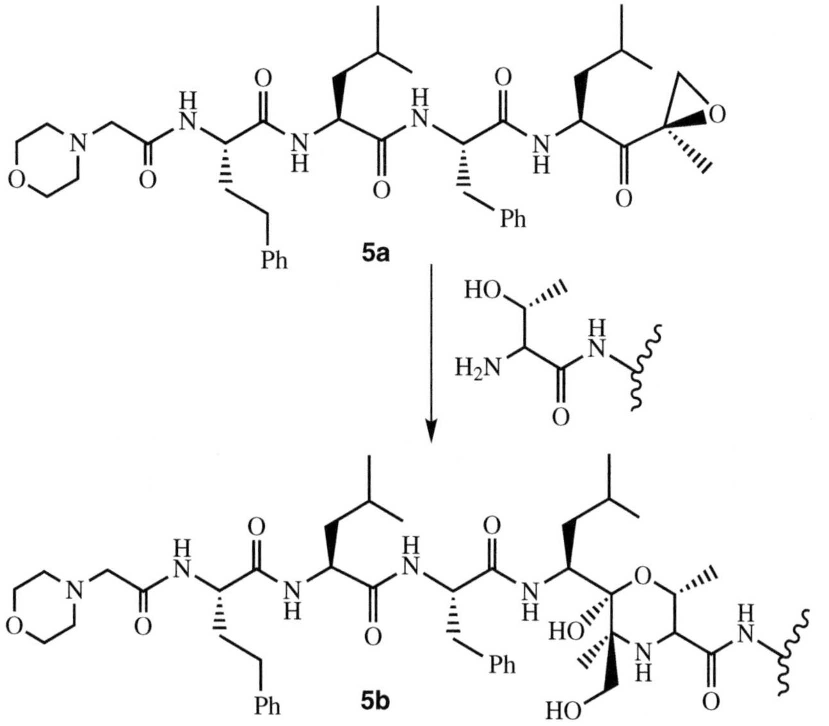

5a

5b

13. Note if any of the following drugs have the potential to be irreversible inhibitors. If so, draw the most likely reaction that they undergo.

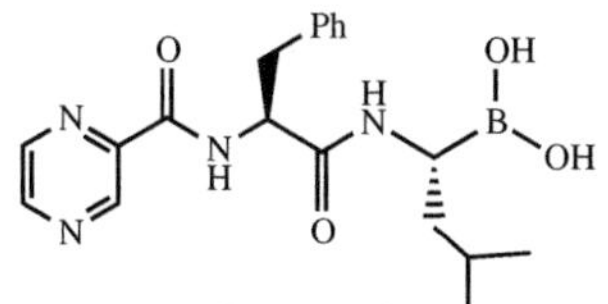

Bortezomib 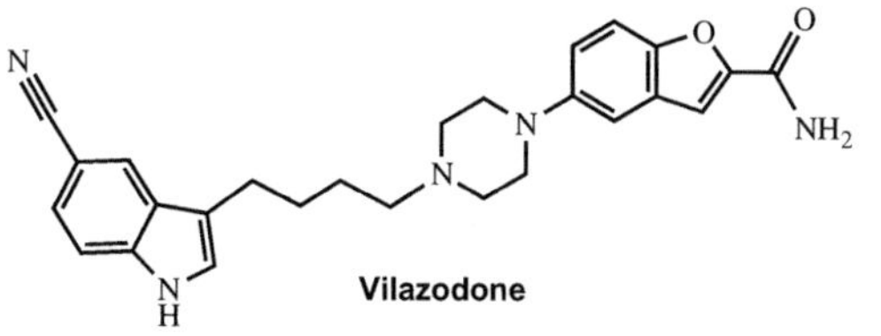 **Vilazodone** 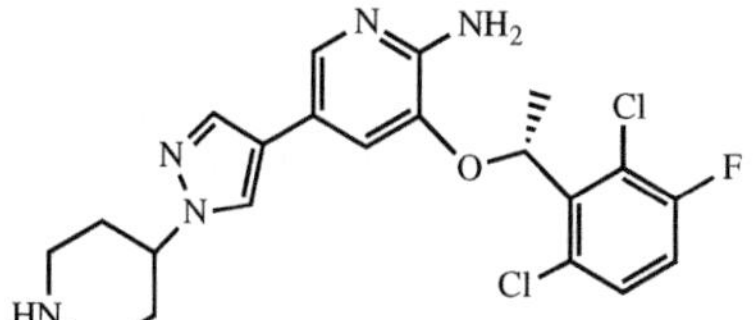 **Crizotinib**

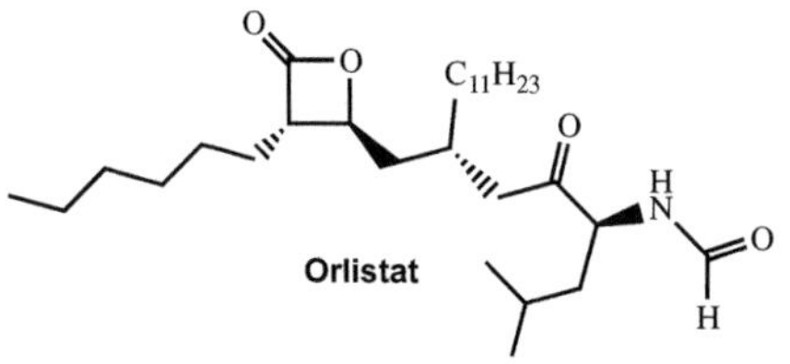

Orlistat 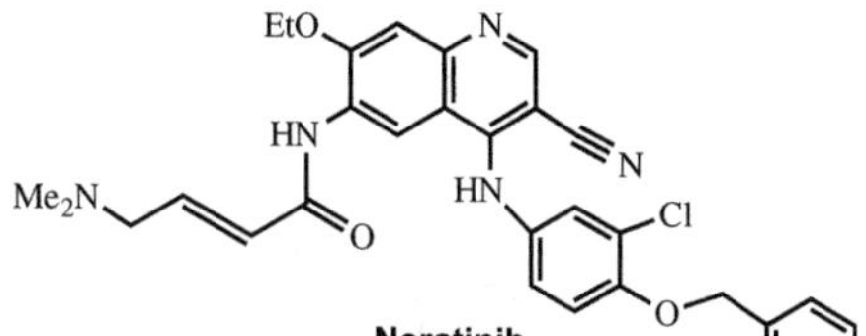 **Neratinib**

14. GABA aminotransferase catalyzes a PLP-dependent conversion of GABA to succinic semialdehyde (see Section 5.3.3.3.1).

a. Draw a mechanism for how **6** inactivates this enzyme.

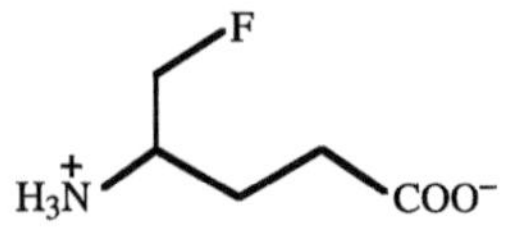

6

b. A hypothetical anticonvulsant drug that inhibits GABA aminotransferase was given to a patient in overdose quantities. Not only did the patient stop convulsing, but he went into a coma. If the problem was that the GABA concentration became too high, mention two possible solutions to the problem.

15. An excess of androgenic hormones such as testosterone can cause benign prostatic hypertrophy (enlarged prostate). Most of the androgenic activity appears to be caused by a metabolite of testosterone (**7**), namely, 5α-dihydrotestosterone (**8**), produced from testosterone in a NADPH-dependent reaction catalyzed by steroid 5α-reductase. The mechanism for this reductase is shown below.

Finasteride (**9**, Proscar) is a potent inhibitor of steroid 5α-reductase. It has been proposed to be a mechanism-based inhibitor. Draw a mechanism consistent with this proposal.

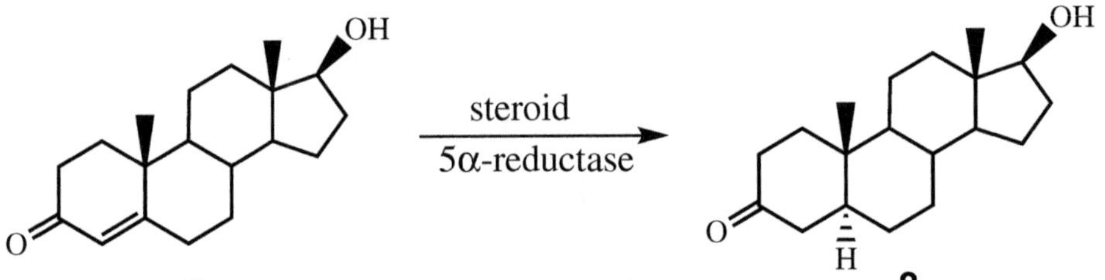

steroid 5α-reductase

7 **8**

8

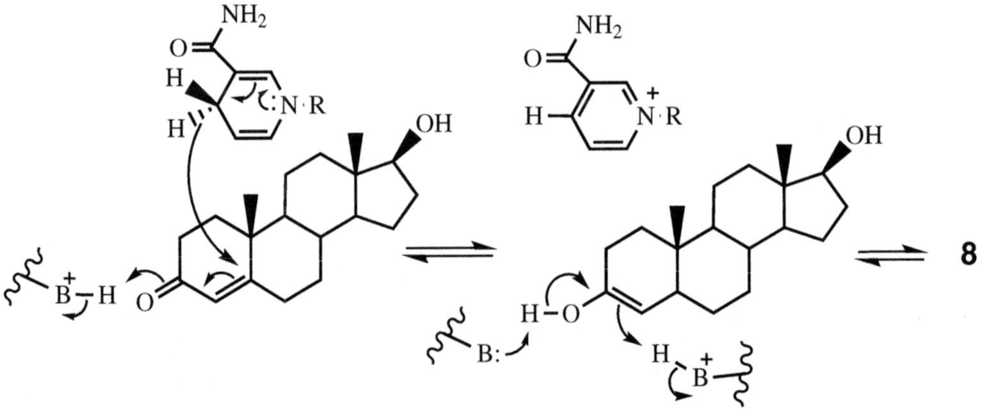

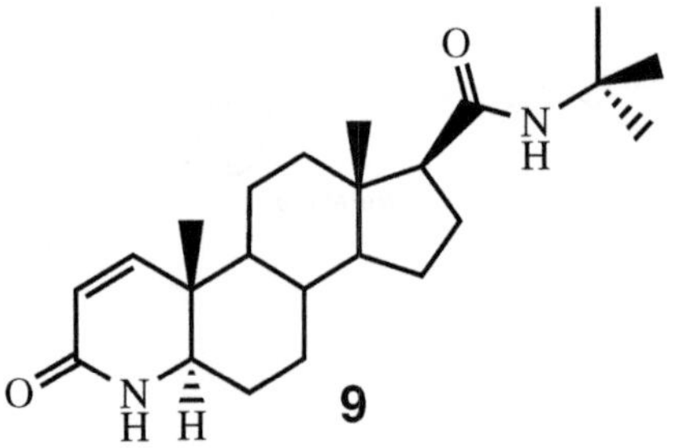

9

REFERENCES

1. *Statistics taken from Annual Reports in Medicinal Chemistry*, Academic Press, San Diego, CA, Vol. 42–47, 2007–2012.
2. Cohen, S. S. Comparative biochemistry and drug design for infectious disease. *Science*, **1979**, *205*, 964–971.
3. (a) Segal, I. H. *Enzyme Kinetics*, John Wiley & Sons, New York, 1975, p. 106. (b) Burlingham, B. T.; Widlanski, T. S. An intuitive look at the relationship of Ki and IC_{50}: a more general use of the Dixon plot. *J. Chem. Educ.* **2003**, *80*, 214–218.
4. Normano, N.; De Luca, A.; Bianco, C.; Strozzo. L.; Mancino, M.; Maiello, M. R.; Carotenuto, A.; De Feo, G.; Caponigro, F.; Salomon, D. S. Epidermal growth factor receptor (EGFR) signaling in cancer. *Gene* **2006**, *366*, 2–16.
5. Ward, W. H. J.; Cook, P. N.; Slate, A. M.; Davies, D. H.; Holdgate, G. A.; Green, L. R. Epidermal growth factor receptor tyrosine kinase: investigation of catalytic mechanism, structure-based searching and discovery of a potent inhibitor. *Biochem. Pharmacol.* **1994**, *48*, 659–666.
6. (a) Fry, D. W.; Kraker, A. J.; McMichael, A.; Ambroso, L. A.; Nelson, J. M.; Leopold, W. R.; Connors, R. W.; Bridges, A. J. A specific inhibitor of the epidermal growth factor receptor tyrosine kinase. *Science* **1994**, *265*, 1093–1095. (b) Rewcastle, G. W.; Denny, W. A.; Bridges, A. J.; Zhou, H.; Cody, D. R.; McMichael, A.; Fry, D. W. Tyrosine kinase inhibitors. 5. Synthesis and structure-activity relationships for 4-[(phenylmethyl)amino]- and 4-phenylamino)quinazolines as potent adenosine 5'-triphosphate binding site inhibitors of the tyrosine kinase domain of the epidermal growth factor receptor. *J. Med. Chem.* **1995**, *38*, 3482–3487.
7. Moyer, J. D.; Barbacci, E. G.; Iwata, K. K; Arnold, L.; Boman, B.; Cunningham, A.; DiOrio, C.; Doty, J.; Morin, M. J.; Moyer, M.; Neveu, M.; Pollack, V. A.; Pustilnik, L. R.; Reynolds, M. M.; Sloan, D.; Theleman, A.; Miller, P. Induction of apoptosis and cell cycle arrest by CP-358,774, an inhibitor of epidermal growth factor receptor tyrosine kinase. *Cancer Res.* **1997**, *57*, 4838–4848.
8. Stamos, J.; Sliwkowski, M. X.; Eigenbrot, C. Structure of epidermal growth factor receptor kinase domain alone and in complex with a 4-anilinoquinazoline inhibitor. *J. Biol. Chem.* **2002**, *277*, 46265–46272.
9. (a) Daley, G. O.; Van Etten, R. A.; Baltimore, D. Induction of chronic myelogenous leukemia in mice by the P210bcr/abl gene of the Philadelphia chromosome. *Science* **1990**, *247*, 824–830. (b) Kelliher, M. A.; McLaughlin, J.; Witte, O. N.; Rosenberg, N. Induction of a chronic myelogenous leukemia-like syndrome in mice with v-abl and BCR/ABL. *Proc. Natl. Acad. Sci. U.S.A.* **1990**, *87*, 6649–6653. (c) Heisterkamp, N.; Jenster, G.; ten Hoeve, J.; Zovich, D.; Pattengale, P. K.; Groffen, J. Acute leukaemia in bcr/abl transgenic mice. *Nature* **1990**, *344*, 251–253.
10. Capdeville, R.; Buchdunger, E.; Zimmerman, J.; Matter, A.C. Glivec (STI571, imatinib), a rationally developed, targeted anticancer drug. *Nat. Rev. Drug Disc.* **2002**, *1*, 493–502.
11. Zimmerman, J.; Buchdunger, E.; Mett, H.; Meyer, T.; Lydon, N.; Traxler, P. Phenylamino-pyrimidine (PAP) – derivatives: a new class of potent and highly selective PDGF-receptor autophosphorylation inhibitors. *Bioorg. Med. Chem. Lett.* **1996**, *6*, 1221–1226.
12. Nagar, B.; Bornmann, W. G.; Pellicena, P.; Schindler, T.; Veach, D. R.; Miller, W. T.; Clarkson, B.; Kuriyan, J. Crystal structure of the kinase domain of c-abl in complex with the small molecule inhibitors PD173955 and imatinib (STI-571). *Cancer Res.* **2002**, *62*, 4236–4243.
13. Krebs, E. G. Historical perspectives on protein phosphorylation and a classification system for protein kinases. *Philos. Trans. R. Soc. London B Biol. Sci.* **1983**, *302*, 3–11.
14. Liu, Y.; Gray, N. S. Rational design of inhibitors that bind to inactive kinase conformations. *Nature Chem. Biol.* **2006**, *2*, 358–364.
15. (a) Bhagwat, S. S. Kinase inhibitors for the treatment of inflammatory and autoimmune disorders. *Purinergic Signal.* **2009**, *5*, 107–115. (b) Rokosz, L. L.; Beasley, J. R.; Carroll, C. D.; Lin, T.; Zhao, J.; Appell, K. C.; Webb, M. L. Kinase inhibitors as drugs for chronic inflammatory and immunological diseases: progress and challenges. *Exp. Opin. Ther. Targets* **2008**, *12*, 883–903.
16. Domagk, G. Chemotherapy of bacterial infections. *Deut. Med. Wschr.* **1935**, *61*, 250–253.
17. Albert, A. *Selective Toxicity*, 7th ed., Chapman and Hall, London, 1985, p. 220.
18. Domagk, G. Chemotherapy for streptococcus infection. *Klin. Wochenschr.* **1936**, *15*, 1585–1590.
19. Tréfouël, J.; Tréfouël, Mme. J.; Nitti, F.; Bovet, D. Action of p-aminophenylsulfamide in experimental streptococcus infections of mice and rabbits. *Compt. Rend. Soc. Biol.* Paris, **1935**, *120*, 756–758.
20. (a) Marshall, E. K., Jr. Determination of sulfanilamide in blood and urine. *J. Biol. Chem.* **1937**, *122*, 263–273. (b) Bratton, A. C.; Marshall, E. K., Jr. A new coupling component for sulfanilamide determination. *J. Biol. Chem.* **1939**, *128*, 537–550.
21. Stamp, T. C. Bacteriostatic action of sulfanilamide in vitro. Influence of fractions isolated from hemolytic streptococci. *Lancet* **1939**, *ii*, 10–17.
22. Woods, D. D. The relation of p-aminobenzoic acid to the mechanism of the action of sulfanilamide. *Br. J. Exper. Pathol.* **1940**, *21*, 74–90.
23. Selbie, F. R. The inhibition of the action of sulfanilamide in mice by p-aminobenzoic acid. *Br. J. Exper. Pathol.* **1940**, *21*, 90–93.
24. Fildes, P. A rational approach to research in chemotherapy. *Lancet*, **1940**, *i*, 955–957.
25. Miller, A. K. Folic acid and biotin synthesis by sulfonamide-sensitive and sulfonamide-resistant strains of *E. coli*. *Proc. Soc. Exper. Pathol. Med.* **1944**, *57*, 151–153. Miller, A. K.; Bruno, P.; Berglund, R. M. The effect of sulfathiazole on the in vitro synthesis of certain vitamins by *Escherichia coli*. *J. Bacteriol.* **1947**, *54*, 9 (G20).
26. Nimmo-Smith, R. H.; Lascelles, J.; Woods, D. D. The synthesis of "folic acid" by *Streptobacterium plantarum* and its inhibition by sulfonamides. *Br. J. Exper. Pathol.* **1948**, *29*, 264–281.
27. Richey, D. P.; Brown, G. M. Biosynthesis of folic acid. IX. Purification and properties of the enzymes required for the formation of dihydropteroic acid. *J. Biol. Chem.* **1969**, *244*, 1582–1592.
28. Weisman, R. A.; Brown, G. M. The biosynthesis of folic acid. V. Characteristics of the enzyme system that catalyzes the synthesis of dihydropteroic acid. *J. Biol. Chem.* **1964**, *239*, 326–331.
29. Bock, L.; Miller, G. H.; Schaper, K.-J.; Seydel, J. K. Sulfonamide structure-activity relations in a cell-free system. 2. Proof for the formation of a sulfonamide-containing folate analog. *J. Med. Chem.* **1974**, *17*, 23–28.
30. Wood, R. C.; Ferone, R.; Hitchings, G. H. Relation of cellular permeability to the degree of inhibition by amethopterin and pyrimethamine in several species of bacteria. *Biochem. Pharmacol.* **1961**, *6*, 113–124.
31. Haruki, H.; Pedersen, M. G.; Gorska, K. E.; Pojer, F.; Johnsson, K. Tetrahydrobiopterin biosynthesis as an off-target of sulfa drugs. *Science* **2013**, *340*, 987–991.

32. Wong, K. K.; Pompliano, D. L. In *Resolving the Antibiotic Paradox*, Rosen, B. P.; Mobashery, S. (Eds.), Plenum Publ., New York, 1998, pp. 197–217.
33. Davies, J. E. In *Antibiotic Resistance: Origins, Evolution and Spread*, Chadwick, D. J.; Goode, J. (Eds.), Wiley, New York, 1997.
34. Bernhard, S. A.; Orgel, L. E. Mechanism of enzyme inhibition by phosphate esters. *Science* **1959**, *130*, 625–626.
35. Pauling, L. Chemical achievement and hope for the future. *Am. Sci.* **1948**, *36*, 51–58.
36. Jencks, W. P. In *Current Aspects of Biochemical Energetics*; Kennedy, E. P., Ed., Academic, New York, 1966, p. 273.
37. Wolfenden, R. Transition state analog inhibitors and enzyme catalysis. *Annu. Rev. Biophys. Bioeng.* **1976**, *5*, 271–306. Wolfenden, R. Transition state analogues for enzyme catalysis. *Nature* **1969**, *223*, 704–705. Wolfenden, R. Transition state analogs as potential affinity labeling regents. *Meth. Enzymol.* **1977**, *46*, 15–28.
38. Lienhard, G. E. Enzymic catalysis and transition-state theory. *Science* **1973**, *180*, 149–154. Lienhard, G. E. Transition state analogs as enzyme inhibitors. *Annu. Rep. Med. Chem.* **1972**, *7*, 249–258.
39. Christianson, D. W.; Lipscomb, W. N. Carboxypeptidase A. *Acc. Chem. Res.* **1989**, *22*, 62–69.
40. Schramm, V. L. Enzymatic transition states: thermodynamics, dynamics and analogue design. *Arch. Biochem. Biophys.* **2005**, *433*, 13–26.
41. Saen-Oon, S.; Quaytman-Machleder, S.; Schramm, V. L.; Schwartz, S. D. Atomic detail of chemical transformation at the transition state of an enzymatic reaction. *Proc. Natl. Acad. Sci.* **2008**, *105*, 16543–16548.
42. (a) Cleland, W. W. Isotope effects: Determination of enzyme transition state structure. *Meth. Enzymol.* **1995**, *249*, 341–373. (b) Schramm, V. L. Enzymatic transition state poise and transition state analogues. *Acc. Chem. Res.* **2003**, *36*, 588–596. (c) Northrup, D. B. The expression of isotope effects on enzyme-catalyzed reactions. *Annu. Rev. Biochem.* **1981**, *50*, 103–131.
43. Cleland, W. W. The use of isotope effects to determine the transition-state structure for enzymatic reactions. *Meth. Enzymol.* **1982**, *87*, 625–641.
44. Lewis, B. E.; Schramm, V. L. Enzymatic binding isotope effects and the interaction of glucose with hexokinase. *Isotope Effects in Chemistry and Biology*; Kohen, A.; Limbach, H.-H. (Eds.), CRC Press, Boca Raton, FL, 2006; pp. 1019–1053.
45. Schramm, V. L. Enzymatic transition states, transition-state analogs, dynamics, thermodynamics, and lifetimes. *Annu. Rev. Biochem.* **2011**, *80*, 703–732.
46. (a) Bagdassarian, C. K.; Schramm, V. L.: Schwartz, S. D. Molecular electrostatic potential analysis for enzymatic substrates, competitive inhibitors and transition state inhibitors. *J. Am. Chem. Soc.* **1996**, *118*, 8825–8836. (b) Bagdassarian, C. K.; Braunheim, B. B.; Schramm, V. L.; Schwartz, S. D. Quantitative measures of molecular similarity: measures to analyze transition-state analogs for enzymic reactions. *Int. J. Quantum. Chem.: Quantum. Biol. Symp.* **1996**, *60*, 73–80.
47. Loo, T. L.; Nelson, J. A. In *Cancer Medicine*, 2nd ed., Holland, J. F.; Frei, E. III. (Eds.), Lea & Febiger, Philadelphia, 1982, p. 790. McCormack, J. J.; Johns, D. G. In *Pharmacologic Principles of Cancer Treatment*, Chabner, B. A. (Ed.), W. B. Saunders, Philadelphia, **1982**, p. 213.
48. Agarwal, R. P.; Spector, T.; Parks, R. E., Jr. Tight-binding inhibitors. IV. Inhibition of adenosine deaminases by various inhibitors. *Biochem. Pharmacol.* **1977**, *26*, 359–367.
49. (a) Wilson, D. K.; Rudolph, F. B.; Quiocho, F. A. Atomic structure of adenosine deaminase complexed with a transition-state analog: understanding catalysis and immunodeficiency mutations. *Science* **1991**, *252*, 1278–1284. (b) See PDB ID 2ADA (pdb.org) for a correction to reference 49a.
50. Sauter, C.; Lamanna, N.; Weiss, M. A. Pentostatin in chronic lymphocytic leukemia. *Expert Opin. Drug Metab. Toxicol.* **2008**, *4*, 1217–1222.
51. Daddona, P. E.; Kelley, W. N. Control of adenosine deaminase levels in human lymphoblasts. *Adv. Enzyme Regul.* **1982**, *20*, 153–163.
52. Berne, R. M.; Rall, T. W.; Rubio, R. (Eds.), Regulatory Functions of Adenosine; Martinus Nijhoff, Boston, 1983.
53. Balakrishnan, K.; Verma, D.; O'Brien, S.; Kilpatrick, M.; Chen, Y.; Tyler, B.; Bickel, S.; Santia, S.; Keating, M. Jabtarhuabm H. Phase 2 and pharmacodynamic study of oral forodesine in patients with advanced, fludarabine-treated chronic lymphocytic leukemia. *Blood* **2010**, *116*, 886–892.
54. Bantia, S. Parker, C.; Upshaw, R.; Cunningham, A.; Kotian, P.; Kilpatrick, M.; Morris, P.; Chand, P.; Babu, Y. Potent orally bioavailable purine nucleoside phosphorylase inhibitor BCX4208 induces apoptosis in B- and T-lymphocytes–a novel approach for autoimmune diseases, organ transplantation and hematologic malignancies. *Int. Immunopharmacol.* **2010**, *10*, 784–790.
55. Kline, P. C.; Schramm, V. L. Pre-steady-state transition-state analysis of the hydrolytic reaction catalyzed by purine nucleoside phosphorylase. *Biochemistry* **1995**, *34*, 1153–1162.
56. Cohen, A.; Gudas, L. J.; Ammann, A. J.; Staal, G. E. J.; Martin, Jr., D. W. Deoxyguanosine triphosphate as a possible toxic metabolite in the immunodeficiency associated with purine nucleoside phosphorylase deficiency. *J. Clin. Invest.* **1978**, *61*, 1405–1409.
57. Tattersall, M. H.; Ganeshaguru, K.; Hoffbrand, A. V. The effect of external deoxyribonucleosides on deoxyribonucleoside triphosphate concentrations in human lymphocytes. *Biochem. Pharmacol.* **1975**, *24*, 1495–1498.
58. Mitchell, B. S.; Mejias, E.; Daddona, P. E.; Kelley, W. N. Purinogenic immunodeficiency diseases: selective toxicity of deoxyribonucleosides for T cells. *Proc. Natl. Acad. Sci. U.S.A.* **1978**, *75*, 5011–5014.
59. Bantia, S.; Ananth, S. L.; Parker, C. D.; Horn, L.L.; Upshaw, R. Mechanism of inhibition of T-acute lymphoblastic leukemia cells by PNP inhibitor-BCX-1777. *Int. Immunopharmcol.* **2003**, *3*, 879–887.
60. Morris, P. E., Jr.; Montgomery, J. A. Inhibitors of the enzyme purine nucleoside phosphorylase. *Expert Opin. Ther. Pat.* **1998**, *8*, 283–299.
61. Kline, P. C.; Schramm, V. L. Purine nucleoside phosphorylase. Catalytic mechanism and transition state analysis of the arsenolysis reaction. *Biochemistry* **1993**, *32*, 13212–13219.
62. Miles, R. W.; Tyler, P. C.; Furneaux, R. H.; Bagdassarian, C. K.; Schramm, V. L. One-third-the-sites transition state inhibitors for purine nucleoside phosphorylase. *Biochemistry* **1998**, *37*, 8615–8621.
63. Fedorov, A; Shi, W.; Kicska, G.; Federov, E.; Tyler, P. C.; Furneaux, R. H.; Hanson, J. C.; Gainsford, G. J.; Larese, J. Z.; Schramm, V. L. Transition state structure of purine nucleoside phosphorylase and principles of atomic motion in enzymatic catalysis. *Biochemistry* **2001**, *40*, 853–860.
64. Lewandowicz, A.; Schramm, V. L. Transition state analysis for human and Plasmodium falciparum purine nucleoside phosphorylases. *Biochemistry* **2004**, *43*, 1458–1468.
65. Lewandowicz, A.; Tyler, P. C. ; Tyler, P.; Evans, G., Furneaux, R.; Schramm, V. Achieving the ultimate physiological goal in transition state analogue inhibitors for purine nucleoside phosphorylase. *J. Biol. Chem.* **2003**, *278*, 31465–31468.

66. (a) Stark, G. R.; Bartlett, P. A. Design and use of potent, specific enzyme inhibitors. *Pharmacol. Ther.* **1983**, *23*, 45–78. (b) Collins, K. D.; Stark, G. R. Aspartate transcarbamylase. Interaction with the transition state analog N-(phosphonacetyl)-L-aspartate. *J. Biol. Chem.* **1971**, *246*, 6599–6605.
67. Frantom, P. A.; Blanchard, J. S. Bisubstrate analog inhibitors. In *Comprehensive Natural Products II Chemistry and Biology*, Mander, L.; Liu, H.-W. (Eds.), Elsevier: Amsterdam, 2010, Vol. 8; pp. 689–717.
68. (a) Johnsson, K.; Schultz, P. G. Mechanistic studies of the oxidation of isoniazid by the catalase peroxidase from Mycobacterium tuberculosis. *J. Am. Chem. Soc.* **1994**, *116*, 7425–7426. (b) Zhao, X.; Yu, H.; Yu, S.; Wang, F.; Sacchettini, J. C.; Magliozzo, R. S. Hydrogen peroxide-mediated isoniazid activation catalyzed by Mycobacterium tuberculosis catalase-peroxidase (KatG) and its S315T mutant. *Biochemistry* **2006**, *45*, 4131–4140.
69. Aitken, S. M.; Ouellet, M.; Percival, M. D.; English, A. M. Mechanism of horseradish peroxidase inactivation by benzhydrazide: a critical evaluation of arylhydrazides as peroxidase inhibitors. *Biochem. J.* **2003**, *375*, 613–621.
70. Wiseman, B.; Carpena, X.; Feliz, M.; Donald, L. J.; Pons, M.; Fita, I.; Loewen, P. C. Isonicotinic acid hydrazide conversion to isonicotinyl-NAD by catalase-peroxidases. *J. Biol. Chem.* **2010**, *285*, 26662–26673.
71. (a) Waley, S. G. The kinetics of slow-binding and slow, tight-binding inhibition: The effects of substrate depletion. *Biochem. J.* **1993**, *294*, 195–200. (b) Sculley, M. J.; Morrison, J. F. The determination of kinetic constants governing the slow, tight-binding inhibition of enzyme-catalysed reactions. *Biochim. Biophys. Acta* **1986**, *874*, 44–53. (c) Schloss, J. V. Significance of slow-binding enzyme inhibition and its relationship to reaction-intermediate analogs. *Acc. Chem. Res.* **1988**, *21*, 348–353. (d) Morrison, J. F.; Walsh, C. T. The behavior and significance of slow-binding enzyme inhibitors. *Adv. Enzymol.* **1988**, *61*, 201–301.
72. (a) Rich, D. H. Pepstatin-derived inhibitors of aspartic proteinases. A close look at an apparent transition-state analog inhibitor. *J. Med. Chem.* **1985**, *28*, 263–273. (b) Morrison, J. F.; Walsh, C. T. The behavior and significance of slow-binding enzyme inhibitors. *Adv. Enzymol.* **1988**, *61*, 201–301.
73. (a) Imperiali, B.; Abeles, R. H. Inhibition of serine proteases by peptidyl fluoromethyl ketones. *Biochemistry* **1986**, *25*, 3760. (b) Stein, R. L.; Strimpler, A. M.; Edwards, P. D.; Lewis, J. J.; Mauger, R. C.; Schwartz, J. A.; Stein, M. M.; Trainor, D. A.; Wildonger, R. A.; Zottola, M. A. Mechanism of slow-binding inhibition of human leukocyte elastase by trifluoromethyl ketones. *Biochemistry* **1987**, *26*, 2682–2689.
74. Morrison, J. F. The slow-binding and slow, tight-binding inhibition of enzyme-catalyzed reactions. *Trends Biochem. Sci.* **1982**, *7*, 102–105.
75. Bartlett, P. A.; Marlowe, C. K. Evaluation of intrinsic binding energy from a hydrogen bonding group in an enzyme inhibitor. *Science (Washington, D. C.)* **1987**, *235*, 569–571.
76. Rich, D. H.; Pepstatin-derived inhibitors of aspartic proteinases. A close look at an apparent transition-state analog inhibitor. *J. Med. Chem.* **1985**, *28*, 263–273.
77. Copeland, R. A.; Pompliano, D. L.; Meek, T. D. Drug-target residence time and its implications for lead optimization. *Nat. Rev. Drug Disc.* **2006**, *5*, 730–739.
78. Lu, H.; England, K.; am End, C.; Truglio, J.; Luckner, S.; Reddy B. G.; Marlenee N.; Knudson, S.; Knudson, D.; Bowen, R. Slow-onset inhibition of the FabI enoyl reductase from *Francisella tularensis*: Residence time and in vivo activity. *ACS Chem. Biol.* **2009**, *4*, 221–231.
79. Espiner, E. A.; Nicholls, M. G. In *The Renin-Angiotensin System*, Robertson, J. I. S.; Nicholls, M. G. (Eds.), Gower Medical Publishing, London, 1993; pp. 33.1–33.24.
80. Tewksbury, D. A.; Dart, R. A.; Travis, J. The amino terminal amino acid sequence of human angiotensinogen. *Biochem. Biophys. Res. Commun.* **1981**, *99*, 1311–1315; Tewksbury, D. In Biochemical Regulation of Blood Pressure; Soffer, R. L. (Ed.), Wiley, New York, 1981; p. 95.
81. Moeller, I.; Allen, A. M.; Chai, S.-Y.; Zhuo, J.; Mendelsohn, F. A. O. Bioactive angiotensin peptides. *J. Hum. Hypertens.* **1998**, *12*, 289–293.
82. Wilk, D.; Healy, D. P. Glutamyl aminopeptidase (aminopeptidase A), the BP-1/6C3 antigen. *Adv. Neuroimmunol.* **1993**, *3*, 195–207.
83. Larner, A.; Vaughan, E. D., Jr.; Tsai, B.-S.; Peach, M. J. Role of converting enzyme in the cardiovascular and adrenal cortical responses to (des-Asp1)-angiotensin I. *Proc. Soc. Exp. Biol. Med.* **1976**, *152*, 631–634.
84. (a) Zini, S.; Fournie-Zaluski, M.-C.; Chauvel, E.; Roques, B. P.; Corvol, P.; Llorens-Cortes, C. Identification of metabolic pathways of brain angiotensin II and III using specific aminopeptidase inhibitors: Predominant role of angiotensin III in the control of vasopressin release. *Proc. Natl. Acad. Sci. U.S.A.* **1996**, *93*, 11968–11973. (b) Blair-West, J. R.; Coghlan, J. P.; Denton, D. A.; Funder, J. W.; Scoggins, B. A.; Wright, R. D. Effect of the heptapeptide (2-8) and hexapeptide (3-8) fragments of angiotensin II on aldosterone secretion. *J. Clin. Endocrinol. Metab.* **1971**, *32*, 575–578. (c) Caldicott, W. J. H.; Taub, K. J.; Hollenberg, N. K. Identical mesenteric, femoral and renal vascular responses to angiotensins II and III in the dog. *Life Sci.* **1977**, *20*, 517.
85. Palmieri, F. E.; Bausback, H. H.; Ward, P. E. Metabolism of vasoactive peptides by vascular endothelium and smooth muscle aminopeptidase M. *Biochem. Pharmacol.* **1989**, *38*, 173–180.
86. (a) Powers B.; Greene L.; Balfe L. M. Updates on the treatment of essential hypertension: a summary of AHRQ's comparative effectiveness review of angiotensin-converting enzyme inhibitors, angiotensin II receptor blockers, and direct renin inhibitors. *J. Manag. Care Pharm.* **2011**, *17*(Suppl. 8), S1–S4. (b) White C. M.; Greene L. Summary of AHRQ's comparative effectiveness review of angiotensin-converting enzyme inhibitors, angiotensin II receptor blockers added to standard medical therapy for treating stable ischemic heart disease. *J. Manag. Care Pharm.* **2011**, *17*(Suppl. 5), S1–S15. (c) Karthikeyan V. J.; Lip G. Y. H. Review: Angiotensin converting enzyme inhibitors and angiotensin receptor blockers prevent atrial fibrillation. *Evidence-based Med.* **2006**, *11*(1), 15. (d) Sica, D. A. Pharmacotherapy review: angiotensin-converting enzyme inhibitors. *J. Clin. Hypertens.* (Greenwich, Conn.) **2005**, *7*(8), 485–488.
87. Ferreira, S. H. A bradykinin-potentiating factor (BPF) present in the venom of *Bothrops jararaca*. *Br. J. Pharmacol. Chemother.* **1965**, *24*, 163–169.
88. (a) Bakhle, Y. S. Conversion of angiotensin I to angiotensin II by cell-free extracts of dog lung. *Nature (London)* **1968**, *220*, 919–921. (b) Bakhle, Y. S.; Reynard, A. M.; Vane, J. R. Metabolism of the angiotensins in isolated perfused tissues. *Nature (London)* **1969**, *222*, 956–959.

89. Ferreira, S. H.; Bartelt, D. C.; Greene, L. J. Isolation of bradykinin-potentiating peptides from *Bothrops jararaca* venom. *Biochemistry* **1970**, *9*, 2583–2593.
90. Ferreira, S. H.; Greene, L. J.; Alabaster, V. A.; Bakhle, Y. S.; Vane, J. R. Activity of various fractions of bradykinin potentiating factor against angiotensin I converting enzyme. *Nature (London)* **1970**, *225*, 379–380.
91. Stewart, J. M.; Ferreira, S. H.; Greene, L. J. Bradykinin potentiating peptide pyrrolidonecarbonyl-Lys-Trp-Ala-Pro. Inhibitor of the pulmonary inactivation of bradykinin and conversion of angiotensin I to II. *Biochem. Pharmacol.* **1971**, *20*, 1557–1567.
92. Ondetti, M. A.; Williams, N. J.; Sabo, E. F.; Pluscec, J.; Weaver, E. R.; Kocy, O. Angiotensin-converting enzyme inhibitors from the venom of *Bothrops jararaca*. Isolation, elucidation of structure, and synthesis. *Biochemistry* **1971**, *10*, 4033–4039.
93. Cheung, H. S.; Cushman, D. W. Inhibition of homogeneous angiotensin-converting enzyme of rabbit lung by synthetic venom peptides of *Bothrops jararaca*. *Biochim. Biophys. Acta* **1973**, *293*, 451–463.
94. Cushman, D. W.; Cheung, H. S. In *Hypertension*, Genest, J.; Koiw, E. (Eds.), Springer: Berlin, 1972, p. 532.
95. Ondetti, M. A.; Cushman, D. W. In *Biochemical Regulation of Blood Pressure*, Soffer, R. L. (Ed.), Wiley, New York, 1981, p. 165.
96. Kato, H.; Suzuki, T. Bradykinin-potentiating peptides from the venom of *Agkistrodon halys*. Isolation of five bradykinin potentiators and the amino acid sequences of two of them, potentiators B and C. *Biochemistry* **1971**, *10*, 972–980.
97. Das, M.; Soffer, R. L. Pulmonary angiotensin-converting enzyme. II. Structural and catalytic properties. *J. Biol. Chem.* **1975**, *250*, 6762–6768.
98. Quiocho, F. A.; Lipscomb, W. N. Carboxypeptidase A: a protein and an enzyme. *Adv. Protein Chem.* **1971**, *25*, 1–78.
99. Cushman, D. W.; Cheung, H. S.; Sabo, E. F.; Ondetti, M. A. Design of potent competitive inhibitors of angiotensin-converting enzyme. Carboxyalkanoyl and mercaptoalkanoyl amino acids. *Biochemistry*, **1977**, *16*, 5484–5491.
100. Byers, L. D.; Wolfenden, R. Potent reversible inhibitor of carboxypeptidase A. *J. Biol. Chem.* **1972**, *247*, 606–608. Byers, L. D.; Wolfenden, R. Binding of the by-product analog benzylsuccinic acid by carboxypeptidase A. *Biochemistry* **1973**, *12*, 2070–2078.
101. Ondetti, M. A.; Cushman, D. W.; Sabo, E. F.; Cheung, H. S. In *Drug Action and Design: Mechanism-Based Enzyme Inhibitors*. Kalman, T. I. (Ed.), Elsevier/North Holland, New York, 1979, p. 271.
102. Atkinson, A. B.; Robertson, J. I. S. Captopril in the treatment of clinical hypertension and cardiac failure. *Lancet* **1979**, *ii*, 836–839.
103. Patchett, A. A.; Harris, E.; Tristram, E. W.; Wyvratt, M. J.; Wu, M. T.; Taub, D.; Peterson, E. R.; Ikeler, T. J.; ten Broeke, J.; Payne, L. G.; Ondeyka, D. L.; Thorsett, E. D.; Greenlee, W. J.; Lohr, N. S.; Hoffsommer, R. D.; Joshua, H.; Ruyle, W. V.; Rothrock, J. W.; Aster, S. D.; Maycock, A. L.; Robinson, F. M.; Hirschmann, R.; Sweet, C. S.; Ulm, E. H.; Gross, D. M.; Vassil, T. C.; Stone, C. A. A new class of angiotensin-converting enzyme inhibitors. *Nature (London)***1980**, *288*, 280–283.
104. (a) Shapiro, R.; Riordan, J. F. Inhibition of angiotensin converting enzyme: mechanism and substrate dependence. *Biochemistry* **1984**, *23*, 5225–5233. (b) Bull, H. G.; Thornberry, N. A.; Cordes, M. H. J.; Patchett, A. A.; Cordes, E. H. Inhibition of rabbit lung angiotensin-converting enzyme by Nα[(S)-1-carboxy-3-phenylpropyl]L-alanyl-L-proline and Nα -[(S)-1-carboxy-3-phenylpropyl]L-lysyl-L-proline. *J. Biol. Chem.* **1985**, *260*, 2952–2962.
105. Wyvratt, M. J.; Patchett, A. A. Recent developments in the design of angiotensin-converting enzyme inhibitors. *Med. Res. Rev.* **1985**, *4*, 483–531.
106. Natesh, R.; Schwager, S. L. U.; Sturrock, E. D.; Acharya, K. R. Crystal structure of the human angiotensin-converting enzyme-lisinopril complex. *Nature* **2003**, 421, 551–554.
107. (a) De Lombaert, S.; Chatelain, R. E.; Fink, C. A.; Trapani, A. J. Design and pharmacology of dual angiotensin-converting enzyme and neutral endopeptidase inhibitors. *Curr. Pharm. Des.* **1996**, *2*, 443–462. (b) Fink, C. A. Recent advances in the development of dual angiotensin-converting enzyme and neutral endopeptidase inhibitors. *Exp. Opin. Ther. Pat.* 1996, 6, 1147–1164. (c) Seymour, A. A.; Asaad, M. M.; Lanoce, V. M.; Langenbacher, K. M.; Fennell, S. A.; Rogers, W. L. Systemic hemodynamics, renal function and hormonal levels during inhibition of neutral endopeptidase 3.4.24.11 and angiotensin-converting enzyme in conscious dogs with pacing-induced heart failure. *J. Pharmacol. Exp. Ther.* **1993**, *266*, 872–883. (d) Pham, I.; Gonzalez, W.; El Amraani, A. I.; Fournie-Zaluski, M. C.; Philippe, M.; Laboulandine, I.; Roques, B. P.; Michel, J. B. Effects of converting enzyme inhibitor and neutral endopeptidase inhibitor on blood pressure and renal function in experimental hypertension. *J. Pharmacol. Exp. Ther.* **1993**, *265*, 1339–1347.
108. Robl, J. A.; Cimarusti, M. P.; Simpkins, L. M.; Brown, B.; Ryono, D. E.; Bird, J. E.; Asaad, M. M.; Schaeffer, T. R.; Trippodo, N. C. Dual metalloprotease inhibitors. 6. incorporation of bicyclic and substituted monocyclic azepinones as dipeptide surrogates in angiotensin-converting enzyme (ACE)/neutral endopeptidase (NEP) inhibitors. *J. Med. Chem.* **1996**, *39*, 494–502.
109. Robl, J. A.; Sun, C. Q.; Stevenson, J.; Ryono, D. E.; Simpkins, L. M.; Cimarusti, M. P.; Dejneka, T.; Slusarchyk, W. A.; Chao, S.; Stratton, L.; Misra, R. N.; Bednarz, M. S.; Asaad, M. M.; Cheung, H. S.; AbboaOffei, B. E.; Smith, P. L.; Mathers, P. D.; Fox, M.; Schaeffer, T. R.; Seymour, A. A.; Trippodo, N. C. Dual metalloprotease inhibitors: Mercaptoacetyl-based fused heterocyclic dipeptide mimetics as inhibitors of angiotensin-converting enzyme and neutral endopeptidase. *J. Med. Chem.* **1997**, *40*, 1570–1577.
110. (a) Pickering, T. G. The rise and fall of omapatrilat. *Medscape News Today* **2002**; http://www.medscape.com/viewarticle/443224. (b) Venugopal, J. Pharmacological modulation of the natriuretic peptide system. *Exp. Opin. Ther. Pat.* **2003**, *13*, 1389–1409.
111. Turner, A. J.; Tanzawa, K. Mammalian membrane metallopeptidases: NEP, ECE, KELL, and PEX. *FASEB J.* **1997**, *11*, 355–364.
112. (a) Rubanyi, G. M.; Polokoff, M. A. Endothelins: molecular biology, biochemistry, pharmacology, physiology, and pathophysiology. *Pharmacol. Rev.* **1994**, *46*, 325–415. (b)Patel, T. R. Therapeutic potential of endothelin receptor antagonists in cerebrovascular disease. *CNS Drugs* **1996**, *5*, 293–310. (c) Benigni, A.; Remuzzi, G. The renoprotective potential of endothelin receptor antagonists. *Exp. Opin. Ther. Pat.* **1997**, *7*, 139–149.
113. (a) Loffler, B.-M. The renoprotective potential of endothelin receptor antagonists. *J. Cardiovasc. Pharmacol.* **2000**, *35*(Suppl. 2), S79–S82. (b) Vemulapalli, S.; Chintala, M.; Stamford, A.; Watkins, R.; Chiu, P.; Sybertz, E.; Fawzi, A. B. Renal effects of SCH 54470: a triple inhibitor of ECE, ACE, and NEP. *Cardiovasc. Drug Rev.* **1997**, *15*, 260–272.
114. Fukuda, Y.; Fukuta, Y.; Higashino, R.; Ogishima, M.; Yoshida, K.; Tamaki, H.; Takei, M. Z-350, a new chimera compound possessing α 1-adrenoceptor antagonistic and steroid 5α -reductase inhibitory actions. *Naunyn Schmiedebergs Arch. Pharmacol.* **1999**, *359*, 433–438.

115. (a) Furuta, S.; Fukuda, Y.; Sugimoto, T.; Miyahara, H.; Kamada, E.; Sano, H.; Fukuta, Y.; Takei, M.; Kurimoto, T. Pharmacodynamic analysis of steroid 5α -reductase inhibitory actions of Z-350 in rat prostate. *Eur. J. Pharmacol.* **2001**, *426*, 105–111. (b) Fukuta, Y.; Fukuda, Y.; Higashino, R.; Yoshida, K.; Ogishima, M.; Tamaki, H.; Takei, M. Z-350, a novel compound with α 1-adrenoceptor antagonistic and steroid 5α -reductase inhibitory actions: pharmacological properties in vivo. *J. Pharmacol. Exp. Ther.* **1999**, *290*, 1013–1018.
116. Kenny, B.; Ballard, S.; Blagg, J.; Fox, D. Pharmacological options in the treatment of benign prostatic hyperplasia. *J. Med. Chem.* **1997**, *40*, 1293–1315.
117. Lepor, H.; Knap-Maloney, G.; Sunshine, H. A dose titration study evaluating terazosin, a selective, once-a-day alpha 1-blocker for the treatment of symptomatic benign prostatic hyperplasia. *J. Urol.* **1990**, *144*, 1393–1397.
118. Yoshida, K.; Horikoshi, Y.; Eta, M.; Chikazawa, J.; Ogishima, M.; Fukuda, Y.; Sato, H. Synthesis of benzanilide derivatives as dual acting agents with α 1-adrenoceptor antagonistic action and steroid 5-α reductase inhibitory activity. *Bioorg. Med. Chem. Lett.* **1998**, *8*, 2967–2972.
119. Sato, H.; Kitagawa, O.; Aida, Y.; Chikazawa, J.; Kurimoto, T.; Takei, M.; Fukuta, Y.; Yoshida, K. Dual-acting agents with α 1-adrenoceptor antagonistic and steroid 5α -reductase inhibitory activities. synthesis and evaluation of arylpiperazine derivatives. *Bioorg. Med. Chem. Lett.* **1999**, *9*, 1553–1558.
120. (a) Missale, C.; Nash, S. R.; Robinson, S. W.; Jaber, M.; Caron, M. G. Dopamine receptors: from structure to function. *Physiol. Rev.* **1998**, *78*, 189–225. (b) Strange, P. G. In *Advances in Drug Research*, Testa, B., Meyer, U. A. (Eds.),Academic Press, London, 1996, Vol. 28, pp. 313–352.
121. Andersen, G. P. In *New Drugs for Asthma Therapy*. Agents and Actions (Suppl. 34) Anderson, G. P., Chapman, I. D., Morley, J. (Eds.), Birkhauser Verlag, Basel, 1991, pp. 97–115.
122. Bonnert, R. V.; Brown, R. C.; Chapman, D.; Cheshire, D. R.; Dixon, J.; Ince, F.; Kinchin, E. C.; Lyons, A. J.; Davis, A. M.; Hallam, C.; Harper, S. T.; Unitt, J. F.; Dougall, I. G.; Jackson, D. M.; McKechnie, K.; Young, A.; Simpson, W. T. Dual D2-receptor and β 2-adrenoceptor agonists for the treatment of airway diseases. 1. Discovery and biological evaluation of some 7-(2-aminoethyl)-4-hydroxybenzothiazol-2(3H)-one analogs. *J. Med. Chem.* **1998**, *41*, 4915–4917.
123. Witztum, J. L. In Goodman and Gilman's The Pharmacological Basis of Therapeutics, 9th ed., Hardman, J. G.; Limbird, L. E.; Molinoff, P. B.; Ruddon, R. W.; Gilman, A. G. (Eds.), McGraw-Hill, New York, 1996, p. 875.
124. Grundy, S. M. Cholesterol metabolism in man. *West. J. Med.* **1978**, *128*, 13–25.
125. (a) Stamler, J., Dietary and serum lipids in the multifactorial etiology of atherosclerosis. *Arch. Surg.* **1978**, *113*, 21. (b) Havel, R. J.; Goldstein, J. L.; Brown, M. S. In *Metabolic Control and Disease*, Bundy, P. K.; Rosenberg, L. E. (Eds.), W. B. Saunders, Philadelphia, 1980, p. 393.
126. Vigna, G. B.; Fellin, R. Pharmacotherapy of dyslipidemias in the adult population. *Exp. Opin. Pharmacother.* **2010**, *11(18)*, 3041–3052.
127. (a) Endo, A.; Kuroda, M.; Tsujita, Y. ML-236A, ML-236B, and ML-236C, new inhibitors of cholesterogenesis produced by *Penicillium citrinum*. *J. Antibiot.* **1976**, *29*, 1346–1348. (b) Endo, A.; Tsujita, Y.; Kuroda, M.; Tanzawa, K. Inhibition of cholesterol synthesis in vitro and in vivo by ML-236A and ML-236B, competitive inhibitors of 3-hydroxy-3-methylglutaryl-coenzyme A reductase. *Eur. J. Biochem.* **1977**, *77*, 31.
128. Brown, A. G.; Smale, T. C.; King, T. J.; Hasenkamp, R.; Thompson, R. H. Crystal and molecular structure of compactin, a new antifungal metabolite from *Penicillium brevicompactum*. *J. Chem. Soc. Perkin Trans. I* **1976**, 1165–1170.
129. (a) Endo, A. Monacolin K, a new hypocholesterolemic agent produced by a *Monascus* species. *J. Antibiot.* **1979**, *32*, 852–854. (b) Endo, A. Monacolin K, a new hypocholesterolemic agent that specifically inhibits 3-hydroxy-3-methylglutaryl coenzyme A reductase. *J. Antibiot.* **1980**, *33*, 334–336.
130. Alberts, A. W.; Chen, J.; Kuron, G.; Hunt, V.; Huff, J.; Hoffman, C.; Rothrock, J.; Lopez, M.; Joshua, H.; Harris, E.; Patchett, A.; Monaghan, R.; Currie, S.; Stapley, E.; Albers-Schonberg, G.; Hensens, O.; Hirschfield, J.; Hoogsteen, K.; Liesch, J.; Springer, J. Mevinolin: A highly potent competitive inhibitor of hydroxymethylglutaryl-coenzyme A reductase and a cholesterol-lowering agent. *Proc. Natl. Acad. Sci. U.S.A.* **1980**, *77*, 3957–3961.
131. Endo, A. Compactin (ML-236B) and related compounds as potential cholesterol-lowering agents that inhibit HMG-CoA reductase. *J. Med. Chem.* **1985**, *28*, 401–405.
132. Tanzawa, K.; Endo, A. Kinetic analysis of the reaction catalyzed by rat liver 3-hydroxy-3-methylglutaryl-coenzyme-A reductase using two specific inhibitors. *Eur. J. Biochem.* **1979**, *98*, 195–201.
133. Nakamura, C. E.; Abeles, R. H. Mode of interaction of β -hydroxy-β -methylglutaryl coenzyme A reductase with strong binding inhibitors: compactin and related compounds. *Biochemistry* **1985**, *24*, 1364–1376.
134. Stokker, G. E.; Hoffman, W. F.; Alberts, A. W.; Cragoe, E. J., Jr.; Deana, A. A.; Gilfillan, J. L.; Huff, J. W.; Novello, F. C.; Prugh, J. D.; Smith, R. L.; Willard, A. K. 3-Hydroxy-3-methylglutaryl-coenzyme A reductase inhibitors. I. Structural modification of 5-substituted 3,5-dihydroxypentanoic acids and their lactone derivatives. *J. Med. Chem.* **1985**, *28*, 347–358.
135. (a) Hoffman, W. F.; Alberts, A. W.; Cragoe, E. J., Jr.; Deana, A. A.; Evans, B. E.; Gilfillan, J. L.; Gould, N. P.; Huff, J. W.; Novello, F. C.; Prugh, J. D.; Rittle, K. E.; Smith, R. L.; Stokker, G. E.; Willard, A. K. 3-Hydroxy-3-methylglutaryl-coenzyme A reductase inhibitors. 2. Structural modification of 7-(substituted aryl)-3,5-dihydroxy-6-heptenoic acids and their lactone derivatives. *J. Med. Chem.* **1986**, *29*, 159–169. (b)Stokker, G. E.; Alberts, A. W.; Anderson, P. S.; Cragoe, E. J., Jr.; Deana, A. A.; Gilfillan, J. L.; Hirschfield, J.; Holtz, W. J.; Hoffman, W. F.; Huff, J. W.; Lee, T. J.; Novello, F. C.; Prugh, J. D.; Rooney, C. S.; Smith, R. L.; Willard, A. K. *J. Med. Chem.* **1986**, *29*, 170.
136. Stokker, G. E.; Alberts, A. W.; Gilfillan, J. L.; Huff, J. W.; Smith, R. L. 3-Hydroxy-3-methylglutaryl-coenzyme A reductase inhibitors. 5. 6-(Fluoren-9-yl)- and 6-(fluoren-9-ylidene)-3,5-dihydroxyhexanoic acids and their lactone derivatives. *J. Med. Chem.* **1986**, *29*, 852–855.
137. Bartmann, W.; Beck, G.; Granzer, E.; Jendralla, H.; Kerekjarto, B. V.; Wess, G. Convenient two-step stereospecific hydroxy-substitution with retention in β -hydroxy-δ-lactones. 4(R)-Heterosubstituted mevinolin and analogs. *Tetrahedron Lett.* **1986**, *27*, 4709–4712.
138. Hoffman, W. F.; Alberts, A. W.; Anderson, P. S.; Chen, J. S.; Smith, R. L.; Willard, A. K. 3-Hydroxy-3-methylglutaryl-coenzyme A reductase inhibitors. 4. Side-chain ester derivatives of mevinolin. *J. Med. Chem.* **1986**, *29*, 849–852.

139. Thaper, R. K.; Kumar, Y.; Kumar, S. M. D.; Misra, S.; Khanna, J. M. A cost-efficient synthesis of simvastatin via high-conversion methylation of an alkoxide ester enolate. *Org. Process Res. Dev.* **1999**, *3*, 476–479.
140. (a) Stokker, G. E.; Rooney, C. S.; Wiggins, J. M.; Hirschfield, J. Synthesis and x-ray characterization of 6(S)-epimevinolin, a lactone epimer. *J. Org. Chem.* **1986**, *51*, 4931–4934. (b) Heathcock, C. H.; Hadley, C. R.; Rosen, T.; Theisen, P. D.; Hecker, S. J. *J. Med. Chem.* **1987**, *30*, 1858.
141. Mundy, G.; Garrett, R.; Harris, S.; Chan, J.; Chen, D.; Rossini, G.; Boyce, B.; Zhao, M.; Gutierrez, G. Stimulation of bone formation in vitro and in rodents by statins. *Science* **1999**, *286*, 1946–1949.
142. Istvan, E. S.; Deisenhofer, J. Structural mechanism for statin inhibition of HMG-CoA reductase. *Science* **2001**, *292*, 1160–1164.
143. American Diabetes Association. http://www.diabetes. Org/diabetes-basics/; World Health Organization. http://www.who.int/diabetes/facts/en/, National Diabetes Education Program. http://ndep.nih.gov/diabetes-facts/index.aspx.
144. (a) van Genugten, R. E.; Raalte, D. H.; Diamant, M. Dipeptidyl peptidase-4 inhibitors and preservation of pancreatic islet-cell function: a critical appraisal of the evidence and Rasagiline. *Diabetes Obes. Metab.* **2012**, *14*, 101–111. (b) Matteucci, E.; Giampietro, O. Dipetidyl peptidase-4 inhibition: linking chemical properties to clinical safety. *Curr. Med. Chem.* **2011**, *18*, 4753–4760. (c) Ahren, B. Inhibition of dipetidyl peptidase-4 (DPP-4): a target to treat type 2 diabetes. *Curr. Enzym. Inhib.* **2011**, *7*, 205–217.
145. Ashworth, D. M.; Atrash, B.; Baker, G. R.; Baxter, A. J.; Jenkins, P, D.; Jones, D. M.; Szelke, M. 2-Cyanopyrrolidides as potent, stable inhibitors of dipeptidyl peptidase IV. *Bioorg. Med. Chem. Lett.* **1996**, *6*, 1163–1166.
146. Robl, J. A.; Hamman, L. H. The discovery of the dipeptidyl dipeptidase (DPP4) inhibitor Onglyza: from concept to market. In *Accounts in Drug Discovery: Case Studies in Medicinal Chemistry*, Barrish, J. C.; Carter, P. C.; Cheng, P. T. W.; Zahler, R. (Ed.), RSC Publishing, Cambridge, 2011, Chapter 1, pp. 1–24, http://dx.doi.org/10.1039/9781849731980-00001.
147. (a) Kramer, R. A.; Schaber, M. S.; Skalka, A. M.; Ganguly, K.; Wong-Staal, F.; Reedy, E. P. HTLV-III gag protein is processed in yeast cells by the virus pol-protease. *Science* **1986**, *231*, 1580–1584. (b) Debouck, C.; Gorniak, J. G.; Strickler, J. E.; Meek, T. D.; Metcalf, B.W.; Rosenberg, M. Human immunodeficiency virus protease expressed in *Escherichia coli* exhibits autoprocessing and specific maturation of the gag precursor. *Proc. Natl. Acad. Sci.* **1987**, *84*, 8903–8906. (c) Kohl, N. E.; Emini, E. A.; Schleif, W. A.; Davis, L. J.; Heimbach, J. C.; Dixon, R. A. F.; Scolnick, E. M.; Sigal, I. S. Active human immunodeficiency virus protease is required for viral infectivity. *Proc. Natl. Acad. Sci.* **1988**, *85*, 4686–4690.
148. Kempf, D. J.; Norbeck, D. W.; Codacovi, L. M.; Wang, X. C.; Kohlbrenner, W. E.; Wideburg, N. E.; Paul, D. A.; Knigge, M. F.; Vasavanonda, S.; Craigkennard, A.; Saldivar, A.; Rosenbrook, W.; Clement, J. J.; Plattner, J. J.; Erickson, J. Structure-based, C2 symmetric inhibitors of HIV protease. *J. Med. Chem.* **1990**, *33*, 2687–2689.
149. Plattner, J. J.; Norbeck, D. W. In Drug Discovery Technologies; Clark, R.; Moos, W. H., Eds.; Ellis Horwood Ltd: Chichester, 1990; pp. 92–126.
150. Erickson, J.; Neidhart, D. J.; Vandrie, J.; Kempf, D. J.; Wang, X. C.; Norbeck, D. W.; Plattner, J. J.; Rittenhouse, J. W.; Turon, M.; Wideburg, N.; Kohlbrenner, W. E.; Simmer, R.; Helfrich, R.; Paul, D. A.; Knigge, M. Design, activity, and 2.8. ANG. crystal structure of a C2 symmetric inhibitor complexed to HIV-1 protease. *Science* **1990**, *249*, 527–533.
151. Kempf, D. J.; Codacovi, L.; Wang, X. C.; Kohlbrenner, W. E.; Wideburg, N. E.; Saldivar, A.; Vasavanonda, S.; Marsh, K. C.; Bryant, P.; Sham, H. L.; Green, B. E.; Betebenner, D. A.; Erickson, J.; Norbeck, D. W. Symmetry-based inhibitors of HIV protease. Structure-activity studies of acylated 2,4-diamino-1,5-diphenyl-3-hydroxypentane and 2,5-diamino-1,6-diphenylhexane-3,4-diol. *J. Med. Chem.* **1993**, *36*, 320–330.
152. Kempf, D. J.; Norbeck, D. W.; Codacovi, L.; Wang, X. C.; Kohlbrenner, W. F.; Wideburg, N. E.; Saldivar, A.; Craig-Kennard, A.; Vasavanonda, S.; Clement, J. J.; Erickson, J. Recent Advances in the Chemistry of Anti-Infective Agents.Bentley, P. H.; Ponsford, R. (Eds.), Royal Society of Chemistry, Cambridge, 1993; pp. 297–313.
153. Kempf, D. J.; Marsh, K. C.; Fino, L. C.; Bryant, P.; Craig-Kennard, A.; Sham, H. L.; Zhao, C.; Vasavanonda, S.; Kohlbrenner, W. E. Design of orally bioavailable, symmetry-based inhibitors of HIV protease. *Bioorg. Med. Chem.* **1994**, *2*, 847–858.
154. Kempf, D. J.; Marsh, K. C.; Denissen, J. F.; McDonald, E.; Vasavanonda, S.; Flentge, C. A.; Green, B. E.; Fino, L.; Park, C. H. ABT-538 is a potent inhibitor of human immunodeficiency virus protease and has high oral bioavailability in humans. *Proc. Natl. Acad. Sci. U.S.A.* **1995**, *92*, 2484–2488.
155. Kumar, G. N.; Gravowski, B.; Lee, R.; Denissen, J. F. Hepatic drug-metabolizing activities in rats after 14 days of oral administration of the human immunodeficiency virus-type 1 protease inhibitor ritonavir (ABT-538). *Drug Metab. Dispos.* **1996**, *24*, 615–617.
156. Kempf, D. J.; Marsh, K. C.; Kumar, G.; Rodrigues, A. D.; Denissen, J. F.; McDonald, E.; Kukulka, M. J.; Hsu, A.; Granneman, G. R.; Baroldi, P. A.; Sun, E.; Pizzuti, D.; Plattner, J. J.; Norbeck, D. W.; Leonard, J. M. Pharmacokinetic enhancement of inhibitors of the human immunodeficiency virus protease by coadministration with ritonavir. *Antimicrob. Agents Chemother.* **1997**, *41*, 654–660.
157. Kempf, D. J.; Sham, H. L.; Marsch, K. C.; Flentge, C. A.; Betebenner, D.; Green, B. E.; McDonald, E.; Vasavanonda, S.; Saldivar, A.; Wideburg, N. E.; Kati, W. M.; Ruiz, L.; Zhao, C.; Fino, L.; Patterson, J.; Molla, A.; Plattner, J. J.; Norbeck, D. W. Discovery of ritonavir, a potent inhibitor of HIV protease with high oral bioavailability and clinical efficacy. *J. Med. Chem.* **1998**, *41*, 602–617.
158. Smith, A. J. T.; Zhang, X.; Leach, A. G.; Houk, K. N. Beyond picomolar affinities: quantitative aspects of noncovalent and covalent binding of drugs to proteins. *J. Med. Chem.* **2009**, *52*, 225.
159. Copeland, R.A. Irreversible enzyme inactivators In *Evaluation of Enzyme Inhibitors in Drug Discovery: A Guide for Medicinal Chemists and Pharmacologists*, 2nd Edition, John Wiley & Sons: Hoboken, N.J., 2013, Chapter 9, pp. 345–382.
160. Lewis, P. J.; Richen, E. Vigabatrin: a new antiepileptic drug. *Br. J. Clin. Pharmacol.* **1989**, *27*(Suppl 1), 1S–12S.
161. Gram, L.; Larsson, O. M.; Johnsen, A.; Schoesboe, A. Experimental studies of the influence of vigabatrin on the GABA system. *Br. J. Clin. Pharmacol.* **1989**, *27*(Suppl 1), 13S–17S.
162. Messenheimer, J. A. Lamotrigine. *Clin. Neuropharmacol.* **1994**, *17*, 548–559.
163. Browne, T. R; Mattson, R. H.; Penry, J. K.; Smith, D. B.; Wilder, B.J.; Treiman, D. M.; Ben-Menachem, E.; Miketta, R. M.; Sherry, K. M.; Szabo, G. K. A multicenter study of vigabatrin for drug resistant epilepsy. *Br. J. Clin. Pharmacol.* **1989**, *27*(Suppl. 1), 95S–100S.
164. Barf, T.; Kaptein, A. Irreversible protein kinase inhibitors: balancing the benefits and risks. *J. Med. Chem.* **2012**, *55*, 6243–6262.

165. Lippert, B.; Jung, M. J.; Metcalf, B. W. Biochemical consequences of reactions catalyzed by GAD and GABA-T. *Brain Res. Bull.* **1980**, *5*(Supp. 2), 375–379.
166. Jeffreys, D. *Aspirin: The remarkable story of a wonder drug*, Bloomsbury, New York, 2004.
167. Robertson, J. G. Mechanistic basis of enzyme-targeted drugs. *Biochemistry* **2005**, *44*, 5561–5571.
168. (a) Erve, J. C. Chemical toxicology: reactive intermediates and their role in pharmacology and toxicology. *Exp. Opin. Drug Metab.* **2006**, *2*, 923–946. (b) Baillie, T. C. Future of toxicology-metabolic activation and drug design: challenges and opportunities in chemical toxicology. *Chem. Res. Toxicol.* **2006**, *19*, 889–893.
169. (a) Uetrecht, J. Immune-mediated adverse drug reactions. *Chem. Res. Toxicol.* **2009**, *22*, 24–34. (b) Naisbitt, D. J.; Gordon, S. F.; Pirmohamed, M.; Park, B. K. Immunological principles of adverse drug reactions. *Drug Saf.* **2000**, *23*, 483–507.
170. Krantz, A. In *Advances in Medicinal Chemistry*, JAI Press, London, 1992, Vol. 1, pp. 235–261.
171. Smith, R. A.; Copp, L. J.; Coles, P. J.; Pauls, H. W.; Robinson, V. J.; Spencer, R. W.; Heard, S. B.; Krantz, A. New inhibitors of cysteine proteinases. Peptidyl acyloxymethyl ketones and the quiescent nucleofuge strategy. *J. Am. Chem. Soc.* **1988**, *110*, 4429–4431.
172. Kominami, E.; Tsukahara, T.; Bando, Y.; Katunuma, N. Distribution of cathepsins B and H in rat tissues and peripheral blood cells. *J. Biochem.* **1985**, *98*, 87–93.
173. Sloane, B. F.; Lah, T. T.; Day, N. A.; Rozhin, J.; Bando, Y.; Honn, K. V. In *Cysteine Proteinases and Their Inhibitors*, Turk, V. (Ed.), Walter de Gruyter: New York, 1986, pp. 729–749.
174. Prous, J. R. (Ed.), EST. *Drugs Future* 1986, *11*, 927–930.
175. Singh, J.; Petter, R. C.; Baillie, T. A.; Whitty, A. The resurgence of covalent drugs. *Nature Rev. Drug Discov.* **2011**, *10*, 307–317.
176. (a) Ulrich, R. Idiosyncratic toxicity: a convergence of risk factors. *Annu. Rev. Med.* **2007**, *58*, 17–34. (b) Uetrecht, J. Idiosyncratic drug reactions: past, present, and future. *Chem. Res. Toxicol.* **2008**, *21*, 84–92.
177. (a) Lammert, C.; Einarsson, S.; Saha, C.; Niklasson, A.; Bjornsson, E.; Chalasani, N. Relationship between daily dose of oral medications and idiosyncratic drug-induced liver injury: search for signals. *Hepatology* **2008**, *47*, 2003–2009. (b) Kalgutkar, A. S.; Gardner, I.; Obach, R. S.; Shaffer, C. L.; Callegari, E.; Henne, K. R.; Mutlib, A. E.; Dalvie, D. K.; Lee, J. S.; Nakai, Y. A comprehensive listing of bioactivation pathways of organic functional groups. *Curr. Drug Metab.* **2005**, *6*, 161–225. (c) Nakayama, S. A zone classification system for risk assessment of idiosyncratic drug toxicity using daily dose and covalent binding. *Drug Metab. Dispos.* **2009**, *37*, 1970–1977.
178. Neuhaus, F. C.; Georgopapadakou, N. H. In *Emerging Targets for Antibacterial and Antifungal Chemotherapy*, Sutcliffe, J.; Georgopapadakou, N. H. (Eds.), Chapman and Hall, New York, 1992, pp. 206–273.
179. Mandell, G. L. In *Principles and Practice of Infectious Diseases*, 2nd ed., Mandell, G. L.; Douglas, R. G., Jr.; Bennett, J. E. (Eds.), Wiley: New York, 1985; p. 180.
180. (a) Sammes, P. G. (Ed.), *Topics in Antibiotic Chemistry*, Ellis Horwood, Chichester, 1980; Vol. 4. (b) Brown, A. G.; Roberts, S. M. (Eds.), *Recent Advances in the Chemistry of β-Lactam Antibiotics*, Royal Society of Chemistry, London, 1985.
181. Bush, K.; Mobashery, S. How β -lactamases have driven pharmaceutical drug discovery: From mechanistic knowledge to clinical circumvention. *Adv. Exp. Med. Biol.* **1998**, *456*, 71–98.
182. Yocum, R. R.; Rasmussen, J. R.; Strominger, J. L. The mechanism of action of penicillin. Penicillin acylates the active site of *Bacillus stearothermophilus* D-alanine carboxypeptidase. *J. Biol. Chem.* **1980**, *255*, 3977–3986.
183. Tipper, D. J.; Strominger, J. L. Mechanism of action of penicillins; a proposal based on their structural similarity to acyl-D-alanyl-D-alanine. *Proc. Natl. Acad. Sci. U.S.A.* **1965**, *54*, 1133–1141.
184. Izaki, K.; Matsuhashi, M.; Strominger, J. L. Biosynthesis of the peptidoglycan of bacterial cell walls. xiii. peptidoglycan transpeptidase and d-alanine carboxypeptidase: penicil. *J. Biol. Chem.* **1968**, *243*, 3180–3192.
185. Lee, W.; McDonough, M. A.; Kotra, L. P.; Li, Z.-H.; Silvaggi, N. R.; Takeda, Y.; Kelly, J. A.; Mobashery, S. A 1.2-.ANG. snapshot of the final step of bacterial cell wall biosynthesis. *Proc. Natl. Acad. Sci. U.S.A.* **2001**, *98*, 1427–1431.
186. Kuzin, A.; Liu, H., Kelly, J. A.; Knox, J. R. Binding of cephalothin and cefotaxime to D-Ala-D-Ala-peptidase reveals a functional basis of a natural mutation in a low-affinity penicillin-binding protein and in extended-spectrum β -lactamases. *Biochemistry* **1995**, *34*, 9532–9540.
187. Sweet, R. M.; Dahl, K. F. Molecular architecture of the cephalosporins. Insights into biological activity based on structural investigations. *J. Am. Chem. Soc.* **1970**, *92*, 5489–5507.
188. Kohanski, M. A.; Dwyer, D. J.; Hayete, B.; Lawrence, C. A.; Collins, J. J. A common mechanism of cellular death induced by bactericidal antibiotics. *Cell* **2007**, *130*, 797–810.
189. Böhme, E. H. W.; Applegate, H. E.; Toeplitz, B.; Dolfini, J. E.; Gougoutas, J. Z. 6-methyl penicillins and 7-methyl cephalosporins. *J. Am. Chem. Soc.* **1971**, *93*, 4324–4326.
190. Kotra, L. P.; Golemi, D.; Vakulenko, S.; Mobashery, S. Bacteria fight back. *Chem. Ind.* 22 May **2000**, 341–344.
191. Gross, M.; Greenberg, L. A. The Salicylates. A Critical Bibliographic Review, Hillhouse, New Haven, 1948.
192. Margotta, R., In *An Illustrated History of Medicine*, Lewis, P. (Ed.), Paul Hamlyn, London, 1968.
193. Stone, E. An account of the success of the bark of the willow in the cure of ages. In a letter to the Right Honourable George Earl of Macclesfield, President of R. S. from the Rev. Mr. Edmund Stone, of Chipping-Norton in Oxfordshire. *Philos. Trans. R. Soc. London* **1963**, *53*, 195–200.
194. Martin, B. K., In *Salicylates, An International Symposium*, Dixon, A. St. J.; Martin, B. K.; Smith, M. V. H.; Wood, R. H. N. (Eds.), Little, Brown and Company, Boston, 1963, p. 6.
195. Jourdier, S. A miracle drug. *Chem. Br.* **1999**, *35*, 33–35.
196. Vane, J. R. Inhibition of prostaglandin synthesis as a mechanism of action for aspirin-like drugs. *Nature New Biol.* **1971**, *231*, 232–235.
197. Smith, J. B.; Willis, A. L. Aspirin selectively inhibits prostaglandin production in human platelets. *Nature New Biol.* **1971**, *231*, 235–237.
198. Roth, G. J.; Stanford, N.; Majerus, P. W. Acetylation of prostaglandin synthase by aspirin. *Proc. Natl. Acad. Sci. U.S.A.* **1975**, *72*, 3073–3076.
199. (a) Hemler, M.; Lands, W. E. M.; Smith, W. L. Purification of the cyclooxygenase that forms prostaglandins. Demonstration of two forms of iron in the holoenzyme. *J. Biol. Chem.* **1976**, *251*, 5575–5579. (b) Van der Ouderaa, F. J.; Buytenhek, M.; Nugteren, D. H.; Van Dorp, D. A. Acetylation of prostaglandin endoperoxide synthetase with acetylsalicylic acid. *Eur. J. Biochem.* **1980**, *109*, 1–8.

200. Roth, G. J.; Machuga, E. T.; Ozols, J. Isolation and covalent structure of the aspirin-modified, active-site region of prostaglandin synthetase. *Biochemistry* **1983**, *22*, 4672–4675.
201. (a) Van der Ouderaa, F. J.; Buytenhek, M.; Nugteren, D. H.; Van Dorp, D. A. Acetylation of prostaglandin endoperoxide synthetase with acetylsalicylic acid. *Eur. J. Biochem.* **1980**, *109*, 1–8. (b) DeWitt, D. L.; El-Harith, E. A.; Kraemer, S. A.; Andrews, M. J. Yao, E. F.; Armstrong, R. L.; Smith, W. L. The aspirin and heme-binding sites of ovine and murine prostaglandin endoperoxide synthases. *J. Biol. Chem.* **1990**, *265*, 5192–5198.
202. Hochgesang, G. P., Jr.; Rowlinson, S. W.; Marnett, L. J. Tyrosine-385 is critical for acetylation of cyclooxygenase-2 by aspirin. *J. Am. Chem. Soc.* **2000**, *122*, 6514–6515.
203. (a) Raz, A.; Wyche, A.; Siegel, N.; Needleman, P. Regulation of fibroblast cyclooxygenase synthesis by interleukin-1. *J. Biol. Chem.* **1988**, *263*, 3022. (b) Masferrer, J. L.; Zweifel, B. S.; Seibert, K.; Needleman, P. Selective regulation of cellular cyclooxygenase by dexamethasone and endotoxin in mice. *J. Clin. Invest.* **1990**, *86*, 1375–1379.
204. (a) Xie, W.; Chipman, J. G.; Robertson, D. L.; Erikson, R. L.; Simmons, D. L. Expression of a mitogen-responsive gene encoding prostaglandin synthase is regulated by mRNA splicing. *Proc. Natl. Acad. Sci. U.S.A.* **1991**, *88*, 2692–2696. (b) Kujubu, D. A.; Fletcher, B. S.; Varnum, C. R.; Lim, W.; Herschman, H. TIS10, a phorbol ester tumor promoter-inducible mRNA from Swiss 3T3 cells, encodes a novel prostaglandin synthase/cyclooxygenase homolog. *J. Biol. Chem.* **1991**, *266*, 12866–2872.
205. (a) Tally, J. J. Selective Inhibitors of cyclooxygenase-2. *Exp. Opin. Ther. Pat.* **1997**, *7*, 55–62. (b) Bjorkman, D. J. Nonsteroidal anti-inflammatory drug-induced gastrointestinal injury. *Am. J. Med.* **1996**, *101*(Suppl. A), 25S–32S. (c)Seibert, K.; Zhang, Y.; Leahy, K.; Hauser, S.; Masferrer, J.; Perkins, W.; Lee, L.; Isakson, P. Pharmacological and biochemical demonstration of the role of cyclooxygenase 2 in inflammation and pain. *Proc. Natl. Acad. Sci. U.S.A.* **1994**, *91*, 12013–12017.
206. Khanna, I. K.; Weier, R. M.; Yu, Y.; Collins, P. W.; Miyashiro, J. M.; Koboldt, C. M.; Veenhuizen, A. W.; Currie, J. L.; Seibert, K.; Isakson, P. C. 1,2-Diarylpyrroles as potent and selective inhibitors of cyclooxygenase-2. *J. Med. Chem.* **1997**, *40*, 1619–1633.
207. Khanna, I. K.; Weier, R. M.; Yu, Y.; Xu, X. D.; Koszyk, F. J.; Collins, P. W.; Koboldt, C. M.; Veenhuizen, A. W.; Perkins, W. E.; Casler, J. J.; Masferrer, J. L.; Zhang, Y. Y.; Gregory, S. A.; Seibert, K.; Isakson, P. C. 1,2-Diarylimidazoles as potent, cyclooxygenase-2 selective and orally active antiinflammatory agents. *J. Med. Chem.* **1997**, *40*, 1634–1647.
208. Penning, T. D.; Talley, J. J.; Bertenshaw, S. R.; Carter, J. S.; Collins, P. W.; Docter, S.; Graneto, M. J.; Lee, L. F.; Malecha, J. W.; Miyashiro, J. M.; Rogers, R. S.; Rogier, D. J.; Yu, S. S.; Anderson, G. D.; Burton, E. G.; Cogburn, J. N.; Gregory, S. A.; Koboldt, C. M.; Perkins, W. E.; Seibert, K.; Veenhuizen, A. W.; Zhang, Y. Y.; Isakson, P. C. Synthesis and biological evaluation of the 1,5-diarylpyrazole class of cyclooxygenase-2 inhibitors: Identification of 4-[5-(4-methylphenyl)-3-(trifluoromethyl)-1H-pyrazol-1-yl]benzenesulfonamide (SC-58635, celecoxib). *J. Med. Chem.* **1997**, *40*, 1347–1365.
209. (a) Prasit, P.; Wang, Z.; Brideau, C.; Chan, C. C.; Charleson, S.; Cromlish, W.; Ethier, D.; Evans, J. F.; Ford-Hutchinson, A. W.; Gauthier, J. Y.; Gordon, R.; Guay, J.; Gresser, M.; Kargman, S.; Kennedy, B.; Leblanc, Y.; Leger, S.; Mancini, J.; O'Neill, G. P.; Ouellet, M.; Percival, M. D.; Perrier, H.; Riendeau, D.; Rodger, I.; Tagari, P.; Therien, M.; Vickers, P.; Wong, E.; Xu, L. J.; Young, R. N.; Zamboni, R.; Boyce, S.; Rupniak, N.; Forrest, N.; Visco, D.; Patrick, D. The discovery of rofecoxib, (MK 966, Vioxx, 4-(4'-methylsulfonylphenyl)-3-phenyl-2(5H)-furanone), an orally active cyclooxygenase-2 inhibitor. *Bioorg. Med. Chem. Lett.* **1999**, *9*, 1773–1778. (b) Chan, C.-C.; Boyce, S.; Brideau, C.; Charleson, S.; Cromlish, W.; Ethier, D.; Evans, J.; Ford-Hutchinson, A. W.; Forrest, M. J.; Gauthier, J. Y.; Gordon, R.; Gresser, M.; Guay, J.; Kargman, S.; Kennedy, B.; Leblanc, Y.; Leger, S.; Mancini, J.; O'Neill, G. P.; Ouellet, M.; Patrick, D.; Percival, M. D.; Perrier, H.; Prasit, P.; Rodger, I.; Tagari, P.; Therien, M.; Vickers, P.; Visco, D.; Wang, Z.; Webb, J.; Wong, E.; Xu, L.-J.; Young, R. N.; Zamboni, R.; Riendeau, D. Rofecoxib (Vioxx, MK-0966; 4-(4'-methylsulfonylphenyl)-3-phenyl-2-(5H)-furanone): a potent and orally active cyclooxygenase-2 inhibitor. Pharmacological and biochemical profiles. *J. Pharmacol. Exp. Ther.* **1999**, *290*, 551–560.
210. Talley, J. J.; Brown, D. L.; Carter, J. S.; Graneto, M. J.; Koboldt, C. M.; Masferrer, J. L.; Perkins, W. E.; Rogers, R. S.; Shaffer, A. F.; Zhang, Y. Y.; Zweifel, B. S.; Seibert, K. 4-[5-methyl-3-phenylisoxazol-4-yl]- benzenesulfonamide, valdecoxib: A potent and selective inhibitor of COX-2. *J. Med. Chem.* **2000**, *43*, 775–777.
211. Rubin, R.; How the Vioxx debacle happen? USA Today, October 12, 2004, http://www.usatoday.com/news/health/2004-10-12-vioxx-cover_x.htm.
212. Egan, K. M.; Lawson, J. A.; Fries, S.; Koller, B.; Rader, D. J.; Smyth, E. M.; FitzGerald, G. A. COX-2-derived prostacyclin confers atheroprotection on female mice. *Science* **2004**, *306*, 1954–1957.
213. Dogné, J.-M.; Supuran, C. T.; Pratico, D. Adverse cardiovascular effects of the coxibs. *J. Med. Chem.* **2005**, *48*, 2251–2257.
214. Piomelli, D.; Giuffrida, A.; Calignano, A.; Rodriguez de Fonseca, F. The endocannabinoid system as a target for therapeutic drugs. *Trends Pharmacol. Sci.* **2000**, *21*, 218–224.
215. Duggan, K. C.; Hermanson, D. J.; Musse, J.; Prusakiewics, J. J.; Scheib, J. L.; Carter, B. Bannerjee, S.; Oates, J. A.; Marnett, L. H. (R)-Profens are substrate-selective inhibitors of endocannabinoid oxygenation by COX-2. *Nature Chem. Biol.* **2011**, *7*, 803–809.
216. (a) Chandrasekharan, N. V.; Dai, H.; Turepu Roos, K. L.; Evanson, N. K.; Tomsik, J.; Elton, T. S.; Simmons, D. L. COX-3, a cyclooxygenase-1 variant inhibited by acetaminophen and other analgesic/antipyretic drugs: cloning, structure, and expression. *Proc. Natl. Acad. Sci. U.S.A.* **2002**, *99*, 13926–13931. (b) Botting, R. M. Mechanism of action of acetaminophen: Is there a cyclooxygenase 3? *Clin. Infect. Dis.* **2000**, *31*(Suppl. 5), S202–S210. (c) Willoughby, D. A.; Moore, A. R.; Colville-Nash, P. R. COX-1, COX-2, and COX-3 and the future treatment of chronic inflammatory disease. *Lancet* **2000**, *355*, 646–648.
217. Matheson, A. J.; Figgitt, D. P. Rofecoxib: a review of its use in the management of osteoarthritis, acute pain and rheumatoid arthritis. *Drugs* **2001**, *61*, 833–865.
218. Warner, T. D.; Giuliano, F.; Vojnovic, I.; Bukasa, A.; Mitchell, J. A.; Vane, J. R. Nonsteroid drug selectivities for cyclo-oxygenase-1 rather than cyclo-oxygenase-2 are associated with human gastrointestinal toxicity: a full in vitro analysis. *Proc. Natl. Acad. Sci. U.S.A.* **1999**, *96*, 7563–7568.
219. Ochi, T.; Motoyama, Y.; Goto, T. The analgesic effect profile of FR122047, a selective cyclooxygenase-1 inhibitor, in chemical nociceptive models. *Eur. J. Pharmacol.* **2000**, *391*, 49–54.
220. Buckley, M. M.; Brogden, R. N. Ketorolac. A review of its pharmacodynamic and pharmacokinetic properties, and therapeutic potential. *Drugs* **1990**, *39*, 86–109.

221. Picot, D.; Loll, P. J.; Garavito, R. M. The X-ray crystal structure of the membrane protein prostaglandin H2 synthase-1. *Nature* **1994**, *367*, 243–249.
222. Kurumbail, R. G.; Stevens, A. M.; Gierse, J. K.; McDonald, J. J.; Stegeman, R. A.; Pak, J. Y.; Gildehaus, D.; Miyashiro, J. M.; Penning, T. D.; Seibert, K.; Isakson, P. C.; Stallings, W. C. Structural basis for selective inhibition of cyclooxygenase-2 by anti-inflammatory agents. *Nature* **1996**, *384*, 644–648.
223. Gierse, J. K.; McDonald, J. J.; Hauser, S. D.; Rangwala, S. H.; Koboldt, C. M.; Seibert, K. A single amino acid difference between cyclooxygenase-1 (COX-1) and -2 (COX-2) reverses the selectivity of COX-2 specific inhibitors. *J. Biol. Chem.* **1996**, *271*, 15810–15814.
224. Shan, B.; Medina, J. C.; Santha, E.; Frankmoelle, W. P.; Chou, T. C.; Learned, R. M.; Narbut, M. R.; Stott, D.; Wu, P. G.; Jaen, J. C.; Rosen, T.; Timmermans, P. B. M. W. M.; Beckman, H. Selective, covalent modification of β-tubulin residue Cys-239 by T138067, an antitumor agent with in vivo efficacy against multidrug-resistant tumors. *Proc. Natl. Acad. Sci. U.S.A.* **1999**, *96*, 5686–5691.
225. Medina, J. C.; Roche, D.; Shan, B.; Learned, R. M.; Frankmoelle, W. P.; Clark, D. L.; Rosen, T.; Jaen, J. C. Novel halogenated sulfonamides inhibit the growth of multidrug resistant MCF-7/ADR cancer cells. *Bioorg. Med. Chem. Lett.* **1999**, *9*, 1843–1846.
226. Silverman, R.B.; Invergo, B.J. Mechanism of inactivation of γ-aminobutyrate aminotransferase by 4-amino-5-fluoropentanoic acid. First example of an enamine mechanism for a γ-amino acid with a partition ratio of 0. *Biochemistry* **1986**, *25*, 6817–6820.
227. Nelson, S. D. Metabolic activation and drug toxicity. *J. Med. Chem.* **1982**, *25*, 753–765.
228. Schechter, P. J.; Barlow, J. L. R.; Sjoerdsma, A. In *Inhibition of Polyamine Metabolism. Biological Significance and Basis for New Therapies*. McCann, P. P.; Pegg, A. E.; Sjoerdsma, A. (Eds.), Academic, Orlando, FL, 1987, p. 345.
229. Isaacson, E. I.; Delgado, J. N. In Burger's Medicinal Chemistry. 4th ed.; Wolff, M. E. (Ed.), Wiley, New York, 1981, Part III; p. 829.
230. Houser, W. A. In *Epilepsy. A Comprehensive Textbook*, Engel, J.; Pedley, T. A. (Eds.), Lippincott-Raven, Philadelphia, 1997, Vol. 1, Section 1.
231. Nanavati, S. M.; Silverman, R. B. Design of potential anticonvulsant agents: mechanistic classification of GABA aminotransferase inactivators. *J. Med. Chem.* **1989**, *32*, 2413–2421.
232. Iadarola, M. J.; Gale, K. Substantia nigra: site of anticonvulsant activity mediated by γ-aminobutyric acid. *Science* **1982**, *218*, 1237–1240.
233. Lippert, B.; Metcalf, B. W.; Jung, M. J.; Casara, P. 4-Amino-hex-5-enoic acid, a selective catalytic inhibitor of 4-aminobutyric-acid aminotransferase in mammalian brain. *Eur. J. Biochem.* **1977**, *74*, 441–445.
234. De Biase, D.; Barra, D.; Bossa, F.; Pucci, P.; John, R. A. Chemistry of the inactivation of 4-aminobutyrate aminotransferase by the antiepileptic drug vigabatrin. *J. Biol. Chem.* **1991**, *266*, 20056–20061.
235. Nanavati, S. M.; Silverman, R. B. Mechanisms of inactivation of γ-aminobutyric acid aminotransferase by the antiepilepsy drug γ-vinyl GABA (vigabatrin). *J. Am. Chem. Soc.* **1991**, *113*, 9341–9349.
236. Pegg, A. E. Polyamine metabolism and its importance in neoplastic growth and as a target for chemotherapy. *Cancer Res.* **1988**, *48*, 759–754.
237. Schechter, P. J.; Barlow, J. L. R.; Sjoerdsma, A. In Inhibition of Polyamine Metabolism. Biological Significance and Basis for New Therapies; McCann, P. P.; Pegg, A. E.; Sjoerdsma, A. (Eds.), Academic, Orlando, FL, 1987, p. 345.
238. (a) Kuzoe, F. A. S. Current situation of African trypanosomiasis. *Acta Trop.* **1993**, *54*, 153–162. (b) Pegg, A. E.; Shantz, L. M.; Coleman, C. S. Ornithine decarboxylase as a target for chemoprevention. *J. Cell Biochem.* **1995**, *22*, 132–138. (c) Wang, C. C. In *Burger's Medicinal Chemistry and Drug Discovery*, 5th ed., Wolff, M. E. (Ed.), John Wiley & Sons, New York, 1997; Vol. 4, p. 459.
239. McCann, P. P.; Pegg, A. E. Ornithine decarboxylase as an enzyme target for therapy. *Pharmacol. Ther.* **1992**, *54*, 195–215.
240. Danzin, C.; Bey, P.; Schirlin, D.; Claverie, N. α -Monofluoromethyl and α -difluoromethyl putrescine as ornithine decarboxylase inhibitors: in vitro and in vivo biochemical properties. *Biochem. Pharmacol.* **1982**, *31*, 3871–3878.
241. Selikoff, I. J.; Robitzek, E. H.; Ornstein, G. G. Toxicity of hydrazine derivatives of isonicotinic acid in the chemotherapy of human tuberculosis. *Quart. Bull. Seaview Hosp.* **1952**, *13*, 17–27.
242. Zeller, E. A.; Barsky, J.; Fouts, J. P.; Kirchheimer, W. F.; Van Orden, L. S. Influence of isonicotinic acid hydrazide and 1-isonicotinoyl-2-isopropylhydrazine on bacterial and mammalian enzymes. *Experientia* **1952**, *8*, 349–350.
243. (a) Kline, N. S. Clinical experience with iproniazid (marsilid). *J. Clin. Exp. Psychopathol. Quart. Rev. Psychiat. Neurol.* **1958**, *19*(Suppl. 1), 72–78. (b)Zeller, E. A. (Ed.), *In vitro and in vivo inhibition of amine oxidases. Ann. N. Y. Acad. Sci.* **1959**, *80*, 583–589.
244. Ganrot, P. O.; Rosengren, E.; Gottfries, C. G. Effect of iproniazid on monoamines and monamine oxidase in human brain. *Experientia* **1962**, *18*, 260–261.
245. Dostert, P. L.; Strolin Benedetti, M.; Tipton, K. F. Interactions of monoamine oxidase with substrates and inhibitors. *Med. Res. Rev.* **1989**, *9*, 45–89.
246. (a) Squires, R. F.; Lassen, J. B. Pharmacological and biochemical properties of γ-morpholinobutyrophenone (NSD 2023), a new monoamine oxidase inhibitor. *Biochem. Pharmacol.* **1968**, *17*, 369–384. (b) Squires, R. F. Additional evidence for the existence of several forms of mitochondrial monoamine oxidase in the mouse. *Biochem. Pharmacol.* **1968**, *17*, 1401. (c) Johnston, J. P. Some observations on a new inhibitor of monoamine oxidase in brain tissue. *Biochem. Pharmacol.* **1968**, *17*, 1285–1297.
247. (a) Palfreyman, M. G.; McDonald, I. A.; Bey, P.; Schechter, P. J.; Sjoerdsma, A. Design and early clinical evaluation of selective inhibitors of monoamine oxidase. *Prog. Neuropsychopharmacol. Biol. Psychiat.* **1988**, *12*, 967–987. (b) McDonald, I. A.; Bey, P.; Palfreyman, M. G. In *Design of Enzyme Inhibitors as Drugs*, Sandler, M.; Smith, H. J. (Eds.), Oxford University Press: Oxford, 1989; p. 227.
248. Green, L. D.; Dawkins, K. In *Burger's Medicinal Chemistry and Drug Discovery*, 5th ed., Wolff, M. E., Ed.; Wiley: New York, 1997; Vol. 5, p. 121.
249. Maeda, Y.; Ingold, K. U. Kinetic applications of electron paramagnetic resonance spectroscopy. 35. The search for a dialkylaminyl rearrangement. Ring opening of N-cyclobutyl-N-n-propylaminyl. *J. Am. Chem. Soc.* **1980**, *102*, 328–331.
250. Silverman, R. B. Mechanism of inactivation of monoamine oxidase by trans-2-phenylcyclopropylamine and the structure of the enzyme-inactivator adduct. *J. Biol. Chem.* **1983**, *258*, 14766–14769.

251. Paech, C.; Salach, J. I.; Singer, T. P. Suicide inactivation of monoamine oxidase by trans-phenylcyclopropylamine. *J. Biol. Chem.* **1980**, *255*, 2700.
252. In *MAO-B Inhibitor Selegiline (R-(-)-Deprenyl)*. Riederer, P.; Przuntek, H. (Eds.), Springer-Verlag: Wein, 1987.
253. Weinreb, O.; Amit, T.; Riederer, P.; Youdim, M. B. H.; Mandel, S. A. Neuroprotective profile of the multitarget drug rasagiline in Parkinson's disease. *Int. Rev. Neurobiol.* **2011**, *100*, 127–149.
254. Davis, G. C.; Williams, A. C.; Markey, S. P.; Ebert, M. H.; Caine, E. D.; Reichert, C. M.; Kopin, I. J. Chronic parkinsonism secondary to intravenous injection of meperidine analogs. *Psychiat. Res.* **1979**, *1*, 249–254.
255. Langston, J. W.; Ballard, P.; Tetrud, J. W.; Irwin, I. Chronic Parkinsonism in humans due to a product of meperidine-analog synthesis. *Science (Washington, D.C.)* **1983**, *219*, 979–980.
256. (a) Burns, R. S.; Chiueh, C. C.; Markey, S. P.; Ebert, M. H.; Jacobowitz, D. M.; Kopin, I. J. A primate model of parkinsonism: Selective destruction of dopaminergic neurons in the pars compacta of the substantia nigra by N-methyl-4-phenyl-1,2,3,6-tetrahydropyridine. *Proc. Natl. Acad. Sci. U.S.A.* **1983**, *80*, 4546–4550. (b) Langston, J. W.; Forno, L. S.; Robert, C. J.; Irwin, I. Selective nigral toxicity after systemic administration of 1-methyl-4-phenyl-1,2,5,6-tetrahydropyridine (MPTP) in the squirrel monkey. *Brain Res.* **1984**, *292*, 390–394.
257. Heikkila, R. E.; Hess, A.; Duvoisin, R. C. Dopaminergic neurotoxicity of 1-methyl-4-phenyl-1,2,5,6-tetrahydropyridine in mice. *Science (Washington, D.C.)* **1984**, *224*, 1451–1453.
258. Hornykiewicz, O. Aging and neurotoxins as causative factors in idiopathic Parkinson's disease - a critical analysis of the neurochemical evidence. *Prog. Neuropsychopharmacol. Biol. Psychiat.* **1989**, *13*, 319–328.
259. Tanner, C. M. The role of environmental toxins in the etiology of Parkinson's disease. *Trends Neurosci.* **1989**, *12*, 49–54.
260. Markey, S. P.; Schmuff, N. R. The pharmacology of the Parkinsonian syndrome producing neurotoxin MPTP (1-methyl-4-phenyl-1,2,3,6-tetrahydropyridine) and structurally related compounds. *Med. Res. Rev.* **1986**, *6*, 389–429.
261. (a) Langston, W. B.; Irwin, I.; Langston, E. B.; Forno, L. S. Pargyline prevents MPTP-induced Parkinsonism in primates. *Science (Washington, D.C.)* **1984**, *225*, 1480–1482. (b) Heikkila, R. E.; Manzino, L.; Cabbat, F. S.; Duvoisin, R. C. Protection against the dopaminergic neurotoxicity of 1-methyl-4-phenyl-1,2,5,6-tetrahydropyridine by monoamine oxidase inhibitors. *Nature* **1984**, *311*, 467–469.
262. Langston, J. W. In Factor, S. A.; Weiner, W. J. *Parkinson's Disease. Diagnosis and Clinical Management*. Demos Medical Publishing, New York, 2002, Chap. 30.
263. Chiba, K.; Trevor, A.; Castagnoli, N. Jr. Metabolism of the neurotoxic tertiary amine, MPTP, by brain monoamine oxidase. *Biochem. Biophys. Res. Commun.* **1984**, *120*, 574–578.
264. Markey, S. P.; Johannessen, J. N.; Chiueh, C. C.; Burns, R. S.; Herkenham, M. A. Intraneuronal generation of a pyridinium metabolite may cause drug-induced parkinsonism. *Nature* **1984**, *311*, 464.
265. Javitch, J. A.; D'Amato, R. J.; Strittmater, S. M.; Snyder, S. H. Parkinsonism-inducing neurotoxin, N-methyl-4-phenyl-1,2,3,6-tetrahydropyridine: uptake of the metabolite N-methyl-4-phenylpyridine by dopamine neurons explains selective toxicity. *Proc. Natl. Acad. Sci. U.S.A.* **1985**, *82*, 2173–2177.
266. (a) Priyadarshi, A.; Khuder, S. A.; Schaub, E. A.; Priyadarshi, S. S. Environmental Risk Factors and Parkinson's Disease: A Metaanalysis. *Environ. Res.* **2001**, *86*, 122–127. (b) Priyadarshi, A.; Khuder, S. A.; Schaub, E. A.; Shrivastava, S. A meta-analysis of Parkinson's disease and exposure to pesticides. *Neurotoxicology* **2000**, *21*, 435–440. (c) Le Couteur, D. G.; McLean, A. J.; Taylor, M. C.; Woodham, B. L.; Board, P. G. A meta-analysis of Parkinson's disease and exposure to pesticides. *Biomed. Pharmacother.* **1999**, *53*, 122–130.
267. Maycock, A. L.; Abeles, R. H.; Salach, J. I.; Singer, T. P. The structure of the covalent adduct formed by the interaction of 3-dimethylamino-1-propyne and the flavine of mitochondrial amine oxidase. *Biochemistry* **1976**, *15*, 114–25.
268. Hubálek, F.; Binda, C.; Li, M.; Herzig, Y.; Sterling, J.; Youdim, M. B.H.; Mattevi, A.; Edmondson, D.E. Inactivation of purified human recombinant monoamine oxidase A and B by rasagiline and its analogues. *J. Med. Chem.* **2004**, *47*, 1760–1766.
269. Binda, C.; Hubálek, F.; Li, M.; Herzig, Y.; Sterling, J.; Edmondson, D. E.; Mattevi, A. Crystal structures of monoamine oxidase B in complex with four inhibitors of the *N*-propargylaminoindan class. *J. Med. Chem.* **2004**, *47*, 1767–1774.
270. Chen, J.J.; Swope, D.M.; Dashtipour, K. Comprehensive review of rasagiline, a second-generation monoamine oxidase inhibitor, for the treatment of Parkinson's disease. *Clin. Ther.* **2007**, *29*, 1825–1849.
271. Bar-Am, O.; Amit, T.; Sagi, Y.;Y, M.B.H. Contrasting neuroprotective and neurotoxic actions of respective metabolites of anti-Parkinson drugs rasagiline and selegiline. *Neurosci. Lett.* **2004**, *355*, 169–172.
272. Müller, K.; Faeh, C; Diederich, F. Fluorine in pharmaceuticals: looking beyond intuition. *Science* **2007**, *317*, 1881–1886 and supplementary material.
273. Cohen, S. S. Nature of thymineless death. *Ann. N. Y. Acad. Sci.* **1971**, *186*, 292–301.
274. Mukherjee, K. L.; Heidelberger, C. Fluorinated pyrimidines. IX. Degradation of 5-fluorouracil-6-C14. *J. Biol. Chem.* **1960**, *235*, 433–437.
275. (a) Douglas, K. T. The thymidylate synthesis cycle and anticancer drugs. *Med. Res. Rev.* **1987**, *4*, 441–475. (b) Benkovic, S. J. On the mechanism of action of folate- and biopterin-requiring enzymes. *Annu. Rev. Biochem.* **1980**, *49*, 227–251.
276. (a) Santi, D. V.; McHenry, C. S.; Raines, R. T.; Ivanetich, K. M. Kinetics and thermodynamics of the interaction of 5-fluoro-2'-deoxyuridylate with thymidylate synthase. *Biochemistry* **1987**, *26*, 8606–8613. (b) Silverman, R. B. Mechanism-Based Enzyme Inactivation: Chemistry and Enzymology, CRC, Boca Raton, FL, 1988; Vol. 1, p. 59.

Chapter 6

DNA-Interactive Agents

Chapter Outline

6.1. INTRODUCTION

6.1.1. Basis for DNA-Interactive Drugs

Another receptor (broadly defined) with which drugs can interact is deoxyribonucleic acid (DNA), the polynucleotide that carries the genetic information in cells. Because this receptor is so vital to human functioning, and from the perspective of a medicinal chemist the overall shape and chemical structure of DNA found in normal and abnormal cells is nearly indistinguishable, drugs that interact with this receptor (*DNA-interactive drugs*) are generally very toxic to normal cells. Therefore, these drugs are reserved only for life-threatening diseases such as cancers and microbial infections. Because the medical term for cancer is *neoplasm*, anticancer drugs may be referred to as *antineoplastic agents*. Unlike the design of drugs that act on enzymes in a foreign organism, there is little that is useful to direct the design of selective agents against abnormal DNA. One feature of cancer cells that differentiates them from that of most normal cells is that cancer cells undergo a rapid, abnormal, and uncontrolled cell division. Genes coding for differentiation in cancer cells appear to be shut off or inadequately expressed, while genes coding for cell proliferation are expressed when they should not be. Because these cells are continually undergoing mitosis, there is a constant need for rapid production of DNA (and its precursors). One difference, then, is quantitative rather than qualitative. Because of the correspondence of normal and abnormal DNA, a compound that reacts with a cancer cell will react with a normal cell as well. However, because of rapid cell division,

The Organic Chemistry of Drug Design and Drug Action. http://dx.doi.org/10.1016/B978-0-12-382030-3.00006-4

cancer cell mitosis can be halted preferentially to that found in normal cells where there is sufficient time for the triggering of repair mechanisms.[1] This quantitative difference is not the only difference. DNA damage in a cell is sensed by several as yet poorly defined mechanisms involving a number of proteins, especially p53.[2] Activation of p53 in response to DNA damage in normal cells can result in several other possible cellular responses, including upregulation of DNA repair systems, cell cycle arrest (to allow time for DNA repair to occur), or programmed cell death (*apoptosis*). Tumor cells, however, are defective in their ability to undergo cell cycle arrest or apoptosis in response to DNA damage. Cancer cells that cannot undergo cell cycle arrest are thus more sensitive to DNA-damaging agents.[3]

Because DNA is constantly being damaged, which leads to 80–90% of human cancers,[4] these DNA lesions must be excised, generally by DNA repair enzymes.[5] Repair systems in humans protect the genome by a variety of mechanisms that repair modified bases, DNA adducts, cross-links, and double-strand breaks, such as direct reversal, base excision, nucleotide excision, and recombination. Some nucleotides may be altered by UV light; enzymes known as *photolyases* can reverse the lesions in the presence of visible light.[6] Base excision repair, discovered in 1964,[7] eliminates single damaged base residues by the action of various *DNA glycosylases*, and then the abasic sugar is excised by *apurinic/apyrimidinic (AP) endonucleases*. Nucleotide excision repair enzymes hydrolyze two phosphodiester bonds, one on either side of the lesion. They can excise damage within oligomers that are 25–32 nucleotides long as well as other types of modified nucleotides. *DNA polymerases* and *ligases* are used to reinsert nucleotides and complete the repairs. Double-strand breaks in DNA are repaired by mechanisms that involve *DNA protein kinases* and recombination enzymes.

In general, anticancer drugs that target DNA are most effective against malignant tumors with a large proportion of rapidly dividing cells, such as leukemias and lymphomas. Unfortunately, the most common tumors are solid tumors, which have a small proportion of rapidly dividing cells.

This is not a chapter on antitumor agents, but rather on the organic chemistry of DNA-interactive drugs and the ways in which DNA damage relates to cancer chemotherapy. Therefore, only relatively few drugs have been selected as representative examples to demonstrate the organic chemistry involved. Some of the principles of antitumor drug design were discussed in Chapter 5 (Section 5.3.3.3.5) and will also be discussed in Chapter 9 (Sections 9.2.2.3, 9.2.2.5.2, 9.2.2.5.3, 9.2.2.6.2, 9.2.2.6.5, 9.2.2.7).

6.1.2. Toxicity of DNA-Interactive Drugs

The toxicity associated with cancer drugs usually is observed in those parts of the body where rapid cell division normally occurs, such as in the bone marrow, the gastrointestinal (GI) tract, the mucosa, and the hair. It is not known what causes chemotherapy-induced hair loss.[8] Some possibilities include ablation of proliferating epithelium, thereby blocking the normal maturation of precursor epithelial cells to the hair strand[9]; apoptosis of hair follicle dermal papilla cells by reactive oxygen species resulting from a downregulation of the antiapoptotic protein Bcl-2[10]; and upregulation of p53,[11] a transcription factor that regulates the cell cycle, and, therefore, acts as a tumor suppressor protein.

The clinical effectiveness of a cancer drug requires that it generally be administered at doses in the toxic range so that it kills tumor cells but allows enough normal cells in the critical tissues, such as the bone marrow and GI tract, to survive, thereby allowing recovery to be possible. There is some evidence that the nausea and vomiting that often occurs from these toxic agents are triggered by the central nervous system rather than as a result of destruction of cells in the GI tract.[12]

Even though cancer drugs are very cytotoxic, they must be administered repeatedly over a relatively long period of time to be assured that all of the malignant cells have been eradicated. According to the *fractional cell kill hypothesis*,[13] a given drug concentration that is applied for a defined time period will kill a constant fraction of the cell population, independent of the absolute number of cells. Therefore, each cycle of treatment will kill a specific fraction of the remaining cells, and the effectiveness of the treatment is a direct function of the dose of the drug administered and the frequency of repetition. Furthermore, it is now known that single-drug treatments are only partially effective and produce responses of short duration. When complete remission is obtained with these drugs, it is only short lived, and relapse is associated with resistance to the original drug (Chapter 7). Because of this phenomenon, combination chemotherapy was adopted.

6.1.3. Combination Chemotherapy

The introduction of cyclic *combination chemotherapy* for acute childhood lymphatic leukemia in the late 1950s marked a turning point in effective treatment of neoplastic disease. The improved effectiveness of combination chemotherapy compared to single-agent treatment is derived from a variety of reasons: initial resistance to any single agent is frequent; initially responsive tumors rapidly acquire resistance after drug administration, probably because of selection of the preexisting resistant tumor cells in the cell population; anticancer drugs themselves increase the rate of mutation of cells into resistant forms; multiple drugs having different mechanisms of action allow independent cell killing by each agent; cells resistant to one drug may be sensitive to another; if drugs have nonoverlapping toxicities, each can be used at full dosage and the effectiveness of each drug will be maintained in combination. Also, unlike enzymes, which require gene-encoded resynthesis to restore activity after inactivation, covalent modification of DNA can be reversed by repair enzymes. In repair-proficient tumor cells, it is possible to potentiate the cytotoxic effects of DNA-reactive drugs with a combination of alkylating drugs and inhibitors of DNA repair.[14]

6.1.4. Drug Interactions

The most significant problem associated with the use of combination chemotherapy is drug interactions; overlapping toxicities are of primary concern. For example, drugs that cause renal toxicity must be used cautiously or not at all with other drugs that depend on renal elimination as their primary mechanism of excretion. The order of administration also is important. An example (unrelated to DNA-interactive drugs) is the synergistic effects that are obtained when methotrexate (an inhibitor of dihydrofolate reductase) precedes 5-fluorouracil (an inhibitor of thymidylate synthase; see Chapter 5, Section 5.3.3.3.5).[15] Inhibition of dihydrofolate reductase leads to decreased production of 5,10-methylenetetrahydrofolate, which is used as a cofactor by thymidylate synthase (Scheme 5.30). Thus, the above order of administration first decreases the availability of this cofactor of thymidylate synthase and then directly inactivates the enzyme itself, leading to a synergistic effect. The opposite order of administration leads to initial inactivation of thymidylate synthase, after which decreasing the availability of one of its cofactors becomes inconsequential.

6.1.5. Drug Resistance

As indicated earlier, the prime reason for the utilization of combination chemotherapy is to avoid *drug resistance*. Mechanisms underlying drug resistance are discussed in more detail in Chapter 7. Suffice it to say here that tumor cells, like microbial cells, have multiple possible mechanisms for counteracting the effects of drugs, including mutating the drug target, excluding a drug from entering or staying in the cell, and mechanisms for destroying the drug once it enters the cell. After treatment with a drug, the cells may activate (upregulate) such mechanisms in an attempt to ensure their own survival. Mechanisms for activation or upregulation frequently involve gene amplification or increased expression of messenger RNA (mRNA) and proteins coded by the genes. Alternatively, cells may downregulate certain mechanisms, for example, key enzymes that are responsible for activating a prodrug to the active compound intended to kill the cell.

6.2. DNA STRUCTURE AND PROPERTIES

6.2.1. Basis for the Structure of DNA

The elucidation of the structure of DNA by Watson and Crick[16] was the culmination and synthesis of experimental results reported by a large number of scientists over several years.[17] Todd and co-workers[18] established that the four deoxyribonucleotides containing the two purine bases—adenine (A) and guanine (G)—and the two pyrimidine bases—cytosine (C) and thymine (T)—are linked by bonds joining the 5′-phosphate group of one nucleotide to a 3′-hydroxyl group on the sugar of the adjacent nucleotide to form 3′,5′-phosphodiester linkages (**6.1**). The phosphodiester bonds are stable because they are negatively charged, thereby repelling nucleophilic attack. Chargaff and co-workers[19] showed that for any duplex DNA molecule, the ratios of A/T and G/C are always equal to one regardless of the base composition of the DNA. They also noted that the number of adenines and thymines relative to the number of guanines and cytosines is characteristic of a given species but varies from species to species (in humans, for example, 60.4% of DNA is composed of adenine and thymine bases). Astbury reported the first X-ray photographs of fibrous DNA, which exhibited a very strong meridional reflection at 3.4 Å distance, suggesting that the bases are stacked upon each other.[20] On the basis of electrotitrimetric studies,

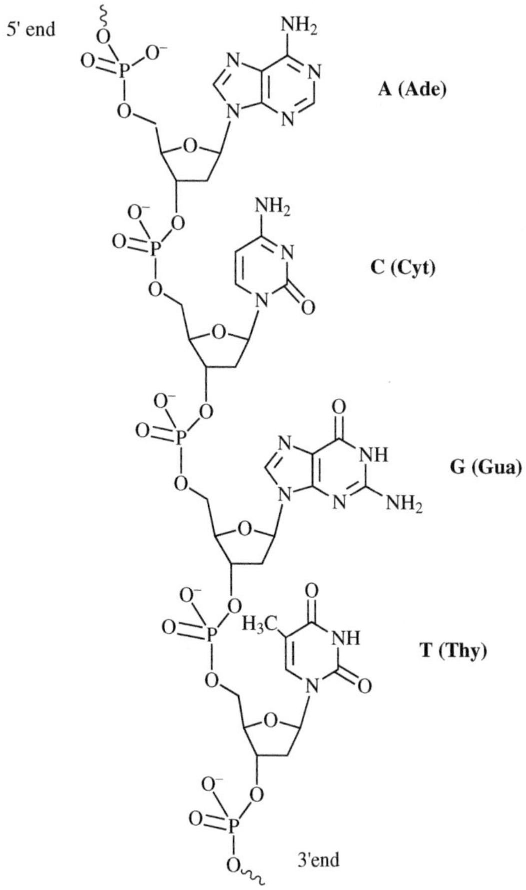

5' end
A (Ade)
C (Cyt)
G (Gua)
T (Thy)
3'end

6.1

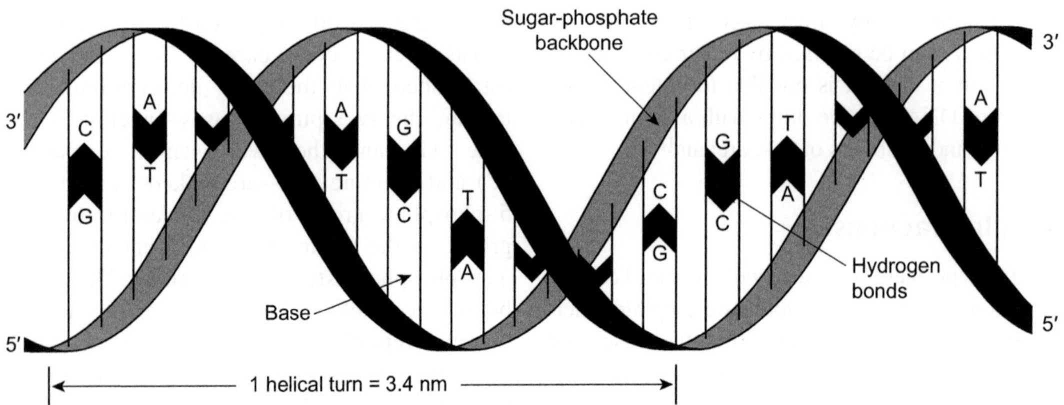

FIGURE 6.1 **DNA structure.** *Reproduced with permission from Alberts, B., Bray, D., Lewis, J., Raff, M., Roberts, K., and Watson, J.D. (1989). Molecular Biology of the Cell, 2nd ed., p. 99. Garland Publishing, New York. Copyright 1989 Garland Publishing.*

Gulland[21] concluded that the nucleotide bases were linked by hydrogen bonding. X-ray data by Wilkins[22] and Franklin[23] indicated that DNA was a helical molecule that was able to adopt a variety of conformations. All of these data were digested by Watson and Crick, who then proposed that two strands of DNA are intertwined into a helical duplex, which is held together by specific hydrogen bonding between base pairs of adenine with thymine (**6.2**) and guanine with cytosine (**6.3**) to explain the results of Gulland and Chargaff and that these base pairs were stacked at 3.4 Å distance, as observed in the X-ray photographs. Furthermore, right-handed rotation between adjacent base pairs by about 36° produces a double helix with 10 base pairs (bp) per turn. A model of the helix was constructed using dimensions and conformations of the individual nucleotides based on the structure of cytidine.[24] The bases are located along the axis of the helix with sugar phosphate backbones winding in an antiparallel orientation along the periphery (Figure 6.1). Because the sugar and phosphate groups are always linked together by 3′,5′-phosphodiester bonds, this part of DNA is very regular; however, the order (or sequence) of the nucleotides along the chain varies from one DNA molecule to another. The purine and pyrimidine bases are flat and tend to stack above each other approximately perpendicular to the helical axis; this base stacking is stabilized mainly by London dispersion forces[25] and by hydrophobic effects.[26] The two chains of the double helix are held together by hydrogen bonds between the bases.

6.2

6.3

All the bases of the DNA are on the inside of the double helix, and the sugar phosphates are on the outside; therefore, the bases on one strand are close to those on the other. Because of this fit, specific base pairings between a large purine base (either A or G) on one chain and a smaller pyrimidine base (T or C) on the other chain are essential. Base pairing between two purines would occupy too much space to allow a regular helix, and base pairing between two pyrimidines would occupy too little space. In fact, hydrogen bonds between guanine and cytosine or adenine and thymine are more effective than any other combination. Therefore, *complementary base pairs* (also called *Watson–Crick base pairs*) form between guanines and cytosines or adenines and thymines only, resulting in a complementary relation between sequences of bases on the two polynucleotide strands of the double helix. For example, if one strand has the sequence 5′-TGCATG-3′, then the complementary strand must have the sequence 3′-ACGTAC-5′ (note that the chains are antiparallel). As you might predict, because there are three hydrogen bonds between G and C base pairs and only two hydrogen bonds between A and T base pairs, the former are more stable.

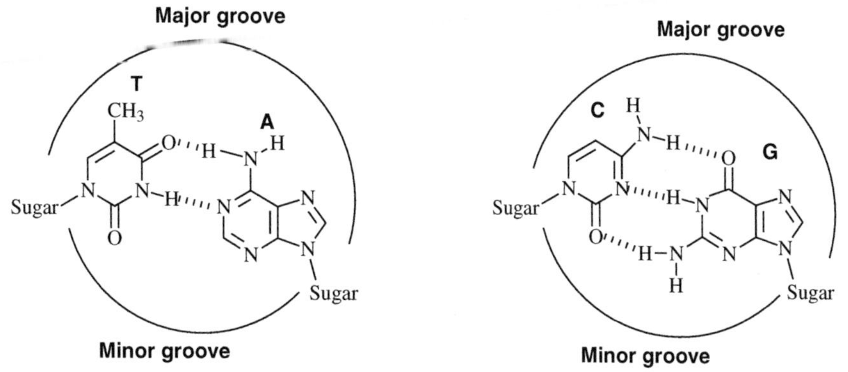

FIGURE 6.2 **Characteristic of DNA base pairs that causes formation of major and minor grooves**

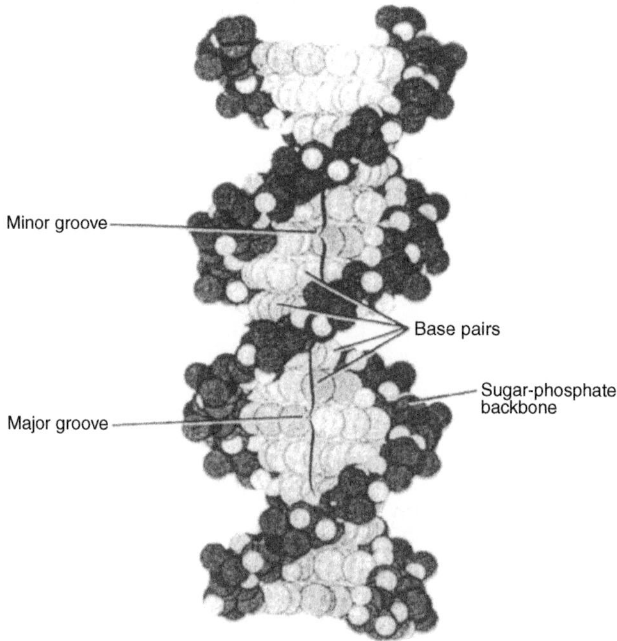

FIGURE 6.3 **Major and minor grooves of DNA.** *With permission from Kornberg, A. (1980); From DNA Replication by Arthur Kornberg. Copyright ©1980 by W. H. Freeman and Company. Used with permission.*

The two glycosidic bonds that connect the base pair to its sugar rings are not directly opposite to each other, and, therefore, the two sugar-phosphate backbones of the double helix are not equally spaced along the helical axis (Figure 6.2). As a result, the grooves that are formed between the backbones are not of equal size; the larger groove is called the *major groove* and the smaller one is called the *minor groove* (Figures 6.2 and 6.3). One side of every base pair faces into the major groove, and the other side faces into the minor groove. The floor of the major groove is filled with base pair nitrogen and oxygen atoms that project inward from their sugar-phosphate backbones toward the center of the DNA. The floor of the minor groove is filled with nitrogen and oxygen atoms of base pairs that project outward from their sugar-phosphate backbones toward the outer edge of the DNA.

6.2.2. Base Tautomerization

Because of the importance of hydrogen bonding to the structure of DNA, we need to consider the tautomerism of the different heterocyclic bases (Figure 6.4), which depends largely on the dielectric constant of the medium and on the pK of the respective heteroatoms.[27] As shown in Figure 6.4, a change in the tautomeric form would have disastrous consequences with regard to hydrogen bonding because groups that are hydrogen bond donors in one tautomeric form become hydrogen bond acceptors in another form, and protons are moved to different positions on the heterocyclic ring. At physiological pH, the more stable tautomeric form for the bases having an amino substituent (A, C, and G) is, by far (>99.99%), the amino form, not the imino form; the oxygen atoms of guanine and thymine also strongly prefer (>99.99%) to be in the keto form rather than the enol form.[28] Apparently, these four bases are ideal for maximizing the population of the appropriate tautomeric forms for complementary base recognition. Note that the donor and acceptor arrows for the tautomers in the box are complementary for C and G and for A and T, but not for C and A or G and T. Simple modifications of the bases, such as replacement of the carbonyl group in purines by a thiocarbonyl group, increase the enol population to about 7%[29]; this would have a significant effect on base pairing.

It has now become clear that hydrogen bonding is not the only factor that controls the specificity of base pairing; shape may also play a key role. Kool has synthesized several nonpolar nucleoside isosteres that lack hydrogen bonding functionality to determine the importance of hydrogen bonding for DNA (**6.4** and **6.5**, Figure 6.5).[30] The difluorotoluene isostere nucleoside (**6.4**, a thymidine isostere; log $P=+1.39$) makes a nearly perfect mimic of thymidine (log $P=-1.27$) in the crystalline state and in solution.[31] When substituted in a DNA in which it is paired opposite to adenine, it adopts a structure identical to that of a T–A base pair.[32] The benzimidazole isostere of deoxyadenosine (**6.5**) is less perfect in shape but is still a good adenosine mimic in DNAs.[33] When **6.4** was incorporated into a template strand of DNA, common polymerases could selectively

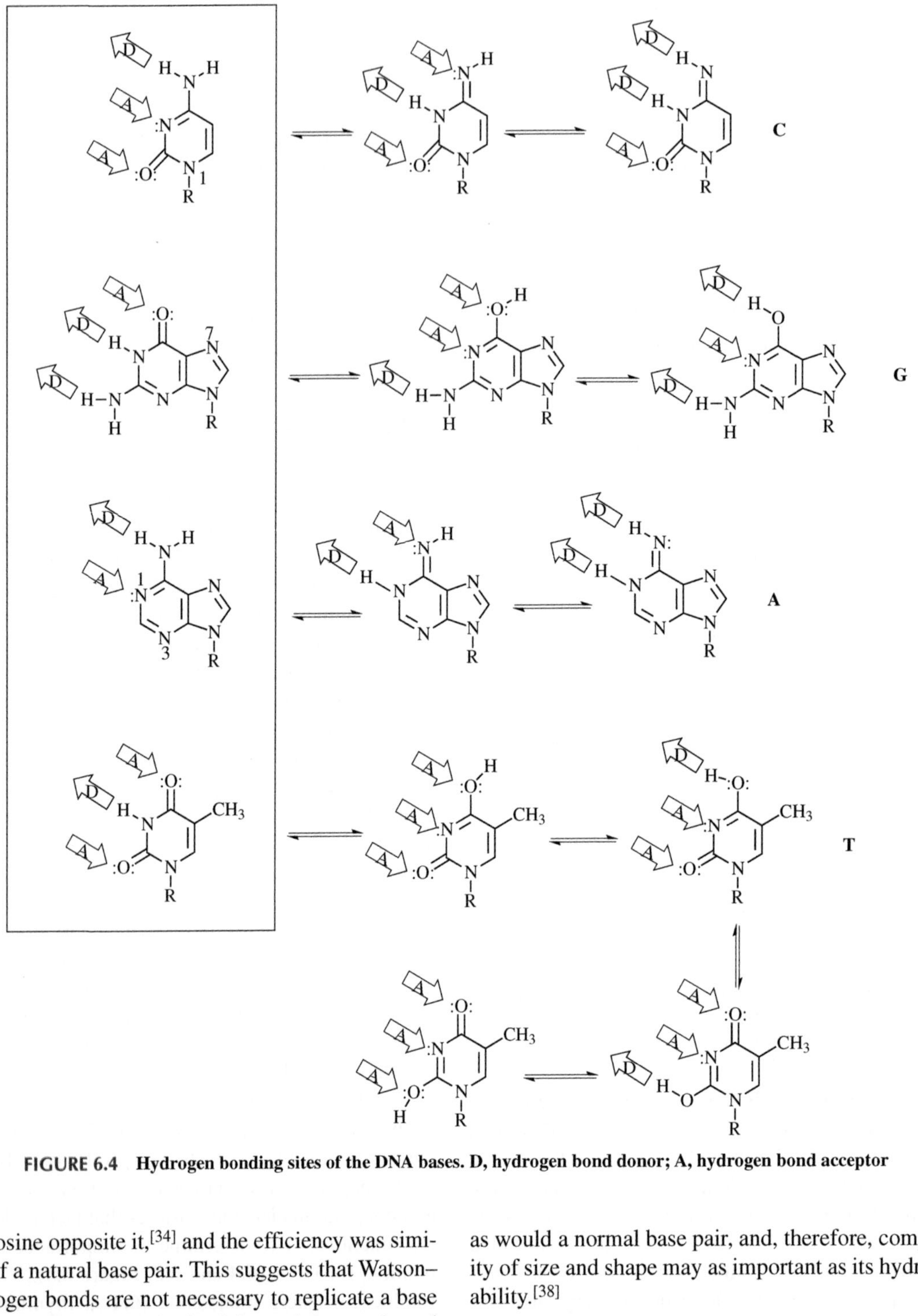

FIGURE 6.4 Hydrogen bonding sites of the DNA bases. D, hydrogen bond donor; A, hydrogen bond acceptor

insert adenosine opposite it,[34] and the efficiency was similar to that of a natural base pair. This suggests that Watson–Crick hydrogen bonds are not necessary to replicate a base pair with high efficiency and selectivity and that steric and geometric factors may be at least as important in the polymerase active site.[35] The nucleoside triphosphate derivative of **6.4** was made, and it was shown to insert selectively opposite to an A in the template strand.[36] Isostere **6.5** also was a substrate for polymerases, and a pair between **6.4** and **6.5** also was replicated well.[37] The most important property of bases for successful base pairing may be that they fit as snugly into the tight, rigid active site of DNA polymerase as would a normal base pair, and, therefore, complementarity of size and shape may as important as its hydrogen bond ability.[38]

6.2.3. DNA Shapes

DNA exists in a variety of sizes and shapes. The length of the DNA that an organism contains varies from micrometers to several centimeters in size. In the nucleus of human somatic cells, each of the 46 chromosomes consists of a single DNA duplex molecule about 4 cm long. If the chromosomes in each somatic cell were placed end to end, the DNA

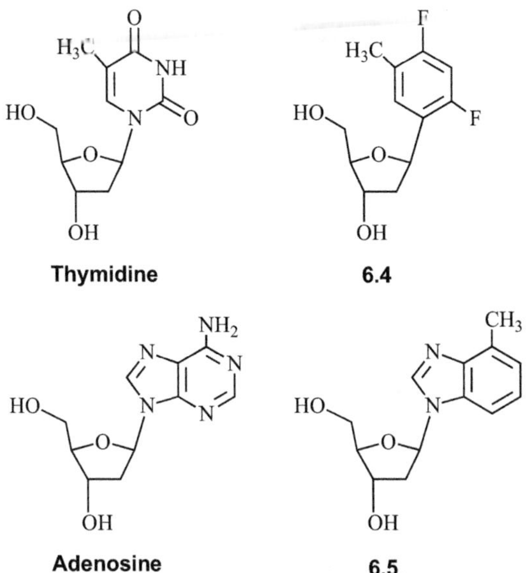

FIGURE 6.5 **Nonpolar nucleoside isosteres (6.4 and 6.5) of thymidine and adenosine, respectively, that base pair by non-hydrogen-bond interactions**

would stretch almost 2 m long! How is it possible for such a large quantity of DNA to be crammed into each nucleus of a cell, given that the nucleus is only 5 μm in diameter? It is accomplished by the packaging of DNA into *chromatin* (Figures 6.6 and 6.7). Formation of chromatin starts with *nucleosomes*, particles of DNA coiled around small, richly basic proteins called *histones* at regular intervals of about 200 bp.[39] The nucleosomes, held together by the electrostatic interactions between the positively charged lysine and arginine residues of the histone protein and the negatively charged phosphates of DNA, are then packed into chromatin fibers,[40] which then associate with the chromosome scaffold, and are packed further to form the metaphase chromosome. The latter steps in this process are not clear, but Figures 6.6 and 6.7 give an artist's rendition of what they could be.

Some DNA is single stranded or triple stranded (*triplex*), but mostly it is in the double-stranded (duplex) form. Some DNA molecules are linear and others (in bacteria) are circular (known as *plasmids*). Linear DNA can freely rotate until the ends become covalently linked to form circular DNA; then, the absolute number of times the DNA chains twist about each other (called the *linkage number*) cannot change. To accommodate further changes in the number of base pairs per turn of the duplex DNA, the circular DNA must twist, like when a rubber band is twisted, into *supercoiled DNA* (Figure 6.8).

Untwisting of the double helix prior to rejoining the ends in circular DNA usually leads to *negative supercoiling* (left-handed direction); overtwisting results in *positive supercoiling* (right-handed direction). Virtually all duplex DNA within cells exists as chromatin in the negative supercoiled state, which is the direction opposite to that of the

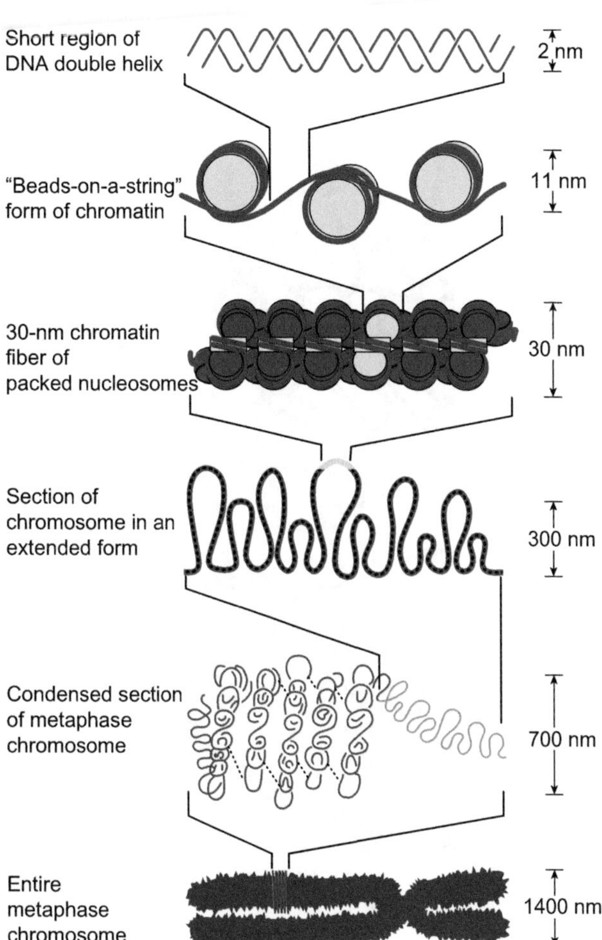

FIGURE 6.6 **Stages in the formation of the entire metaphase chromosome starting from duplex DNA.** *With permission from Alberts, B., (1994). Copyright ©1994 from* Molecular Biology of the Cell, *3rd ed. By Bruce Alberts, Dennis Bray, Julian Lewis, Martin Raff, Keith Roberts and James D. Watson. Reproduced by permission of Routledge, Inc., part of The Taylor & Francis Group.*

twist of the double helix (Section 6.2.4). Because supercoiled DNA is a higher energy state than uncoiled DNA, the cutting (called *nicking*) of one of the DNA strands of supercoiled DNA converts it into relaxed DNA.

Nicking of supercoiled DNA is catalyzed by a family of enzymes called *DNA topoisomerases*.[41] These ubiquitous nuclear enzymes catalyze the conversion of one topological isomer of DNA into another and also function to resolve topological problems in DNA, leading to overwinding and underwinding, knotting and unknotting and catenation and decatenation (Figures 6.9 and 6.10), which normally arise during replication, transcription, recombination, repair, and other DNA processes.[42] There are a number of ways for a long DNA molecule to lose or gain a few turns of twist; these excess or deficient turns of twist need to be corrected by topoisomerases.[43] DNA also has to untwist during several of its normal functions, for example, when it is copied into the mRNA (*transcription*), which is responsible for making proteins in the cell, near start sites of all genes

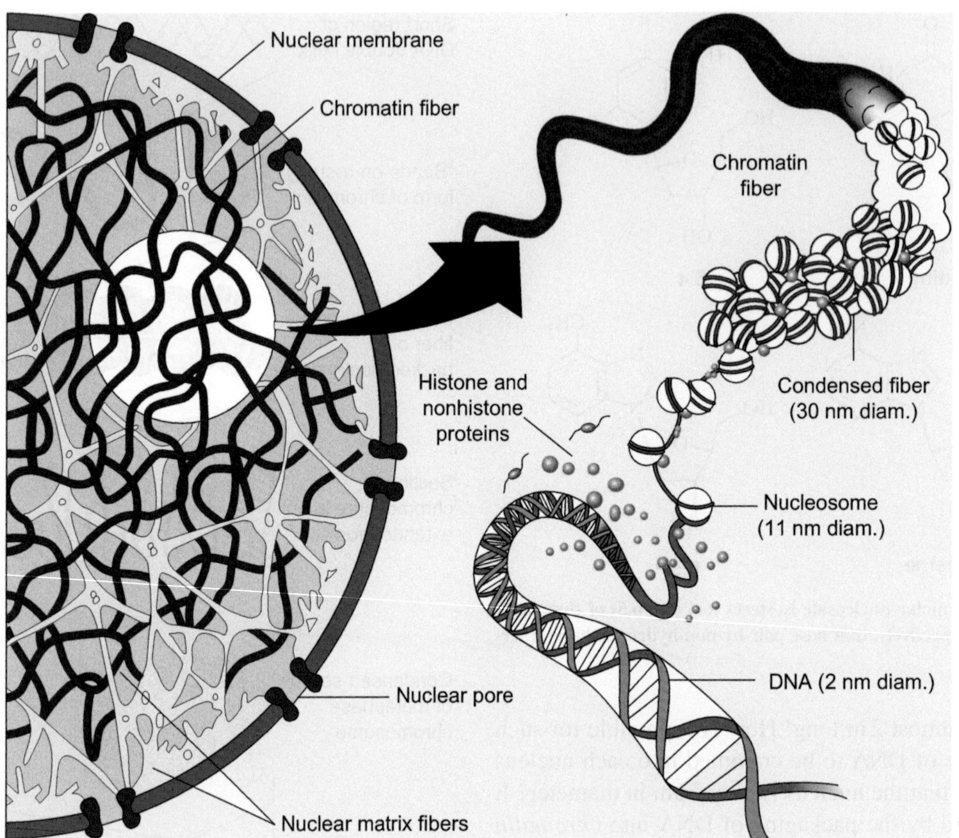

FIGURE 6.7 Artist rendition of the conversion of duplex DNA into chromatin fiber

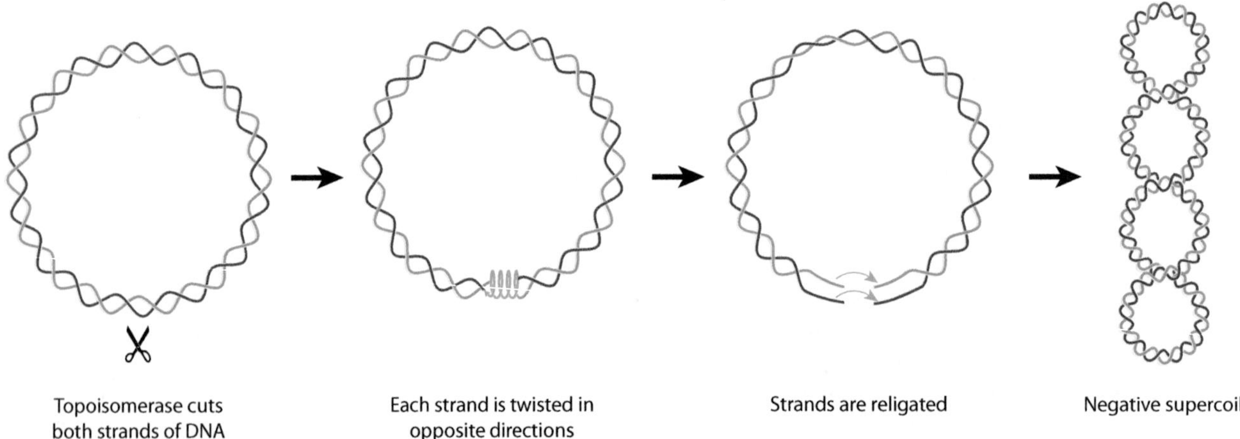

FIGURE 6.8 Conversion of duplex DNA into supercoiled DNA

so that RNA polymerase can construct new RNA strands, and when DNA is copied into another DNA strand by DNA polymerase just before a single cell divides into two cells (*replication*). During DNA replication, the two strands of the DNA must be unlinked by topoisomerases, and during transcription, the translocating RNA polymerase generates supercoiling tension in the DNA that must be relaxed. The association of DNA with histones and other proteins also introduces supercoiling that requires relaxation by topoisomerases. Transcription from some promoters in bacteria requires a minimal level of negative supercoiling, but too much supercoiling of either sign is disastrous. In all cells, completely replicated chromosomes must be untangled by DNA topoisomerases before partitioning and cell division can occur. Topoisomerases are known that relax only negative supercoils, that relax supercoils of both signs, or that introduce either negative (bacterial *DNA gyrase*) or positive supercoils into the DNA (*reverse gyrase*).

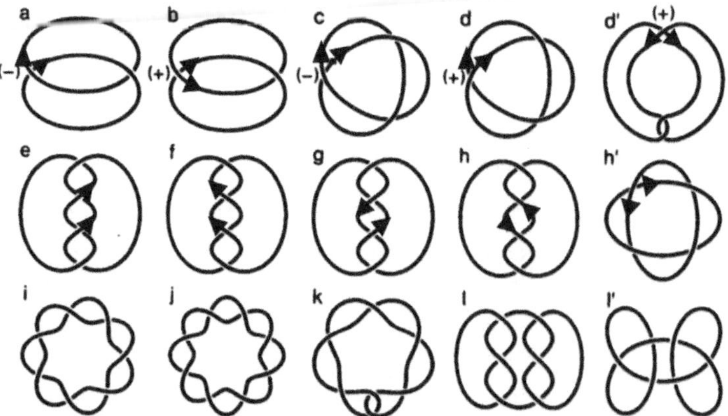

FIGURE 6.9 Catenane and knot catalog. Arrows indicate the orientation of the DNA primary sequence: a and b, singly linked catenanes; c and d, simplest knot, the trefoil; e–h, multiply interwound torus catenanes; i, right-handed torus knot with seven nodes; j, right-handed torus catenane with eight notes; k, right-handed twist knot with seven nodes; l, 6-noded knots composed of two trefoils. *Adapted with permission from Wasserman, S. A. and Cozzarelli, N. R. Biochemical topology: applications to DNA recombination and replication.* Science, ***1986***, 232, *952. Reprinted with permission from AAAS.*

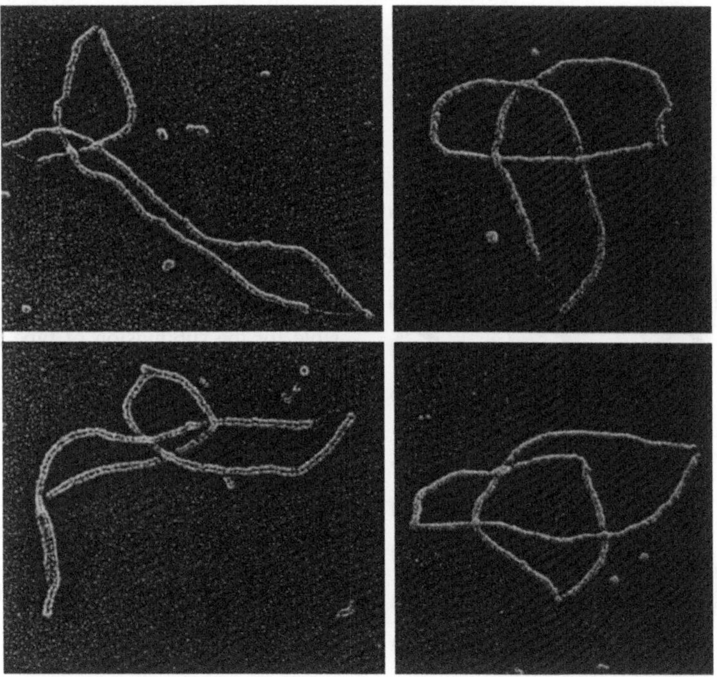

FIGURE 6.10 Visualization of trefoil DNA by electron microscopy. *Reproduced with permission from Griffith J.D., Nash, H.A.,* Proc. Natl. Acad. Sci. USA ***1985***, 82, *3124.*

There are two general types of topoisomerases into which at least six different topoisomerases have been classified[44]; these topoisomerases regulate the state of supercoiling of intracellular DNA. One type, known as *DNA topoisomerase I*,[45] removes positive and negative supercoils by catalyzing a transient break of one strand of duplex DNA and allowing the unbroken, complementary strand to pass through the enzyme-linked strand, thereby resulting in DNA relaxation by one positive turn. The other type is called *DNA topoisomerase II* (or, in the case of the bacterial enzyme, *DNA gyrase*),[46] which catalyzes the transient breakage of both strands of the duplex DNA, with a 4 bp stagger between the nicks. This generates a gate through which another region of DNA can be passed prior to *religating* (reattaching) the strands. The outcome of this process is the supercoiling of the DNA in the negative direction or relaxation of positively supercoiled DNA, which changes the linkage number by negative two. The type I enzymes are further classified into type IA,[47] type IB,[48] or type IC[49] subfamilies (Table 6.1; Figure 6.11).

Type II topoisomerases also are divided into type IIA and type IIB subfamilies by the same criteria. DNA topoisomerase

TABLE 6.1 Characteristics of Types IA, IB, and IC DNA Topoisomerases

Type IA	Type IB	Type IC
Relax negatively supercoiled DNA	Relax negatively and positively supercoiled DNA	Relax negatively and positively supercoiled DNA
Require divalent cations (Mg^{2+})	Do not require divalent cations (Mg^{2+})	Do not require divalent cations (Mg^{2+})
Require single-stranded DNA for activity	Require double-stranded DNA for activity	Require double-stranded DNA for activity
Covalent linkage to 5′-end of the broken strand	Covalent linkage to 3′-end of the broken strand	Covalent linkage to 3′-end of the broken strand
Enzyme-bridged strand passage mechanism[a]	Controlled (swiveling or strand) rotation mechanism[a]	Controlled (swiveling or strand) rotation mechanism[a]
Present in bacteria, eukarya, and archaea	Present in bacteria and eukarya	Present in archaea
Sequence similarities among all members	Sequence similarities among all members	At present only one member[b]
Structural similarities among all members	Structural similarities among all members	At present only one member[b]

[a]See Figure 6.11.
[b]9/13.

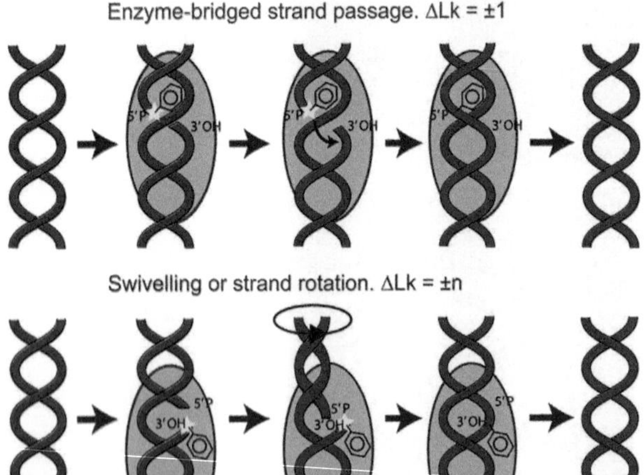

FIGURE 6.11 Mechanisms of DNA topoisomerase-catalyzed reactions. Drawings produced by Professor Alfonso Mondragón, Department of Molecular Biosciences, Northwestern University.

I has no requirement for an energy cofactor to complete the religation process, but DNA topoisomerase II requires ATP and Mg^{2+} for the "strand passing" activity.[50] *Topoisomerase III* (which actually is a member of the type IA subfamily of topoisomerases) catalyzes the removal of negative, but not positive, supercoils.[51] *Topoisomerase IV* (subfamily IIA), a target for quinolone antimicrobial agents, is required for the terminal stages of unlinking of DNA during replication; its inhibition causes accumulation of replication catenanes.[52] DNA *topoisomerase V* (subfamily type IC)[53] relaxes both negatively and positively supercoiled DNA in hyperthermophilic organisms in the temperature range from 60 to 122 °C and salt concentrations from 0 to 0.65 M KCl or NaCl or 0 to 3.1 M of potassium glutamate.[54] The discovery of DNA *topoisomerase VI* in hyperthermophilic archaea was responsible for the subdivision of type II topoisomerases into type IIA and IIB subfamilies.[55] *Topoisomerase VI*, a type IIB topoisomerase, is able to decatenate intertwined DNA and to relax either positively or negatively supercoiled DNA in the presence of ATP and divalent cations.

The mechanisms for DNA strand cleavage by DNA topoisomerase I and II are different, but a common feature is the involvement of an active-site tyrosine residue that catalyzes a covalent catalytic cleavage mechanism. Scheme 6.1 shows general mechanisms for topoisomerase IA and IB.[56] A tyrosyl group on the enzymes attacks the phosphodiester bond of DNA, giving one of the two possible covalent adducts, known as *cleavable complexes*, depending upon whether the enzyme is in subfamily type IA (Scheme 6.1, pathway a) or type IB (Scheme 6.1, pathway b). Figure 6.12 depicts a possible mechanism for a topoisomerase I reaction.

According to this mechanism, a single strand of DNA is cleaved by attack of an active-site tyrosine residue, which becomes covalently bound to the 5′-phosphate (this is a topoisomerase type IA) on one end of the cleaved strand (Figure 6.12, structure B). The enzyme holds onto both ends at the site of the break (the released 3′-end is held noncovalently), producing a bridge across the gap through which the intact strand passes (structure C). The enzyme religates the two ends (structure D), and releases the relaxed DNA

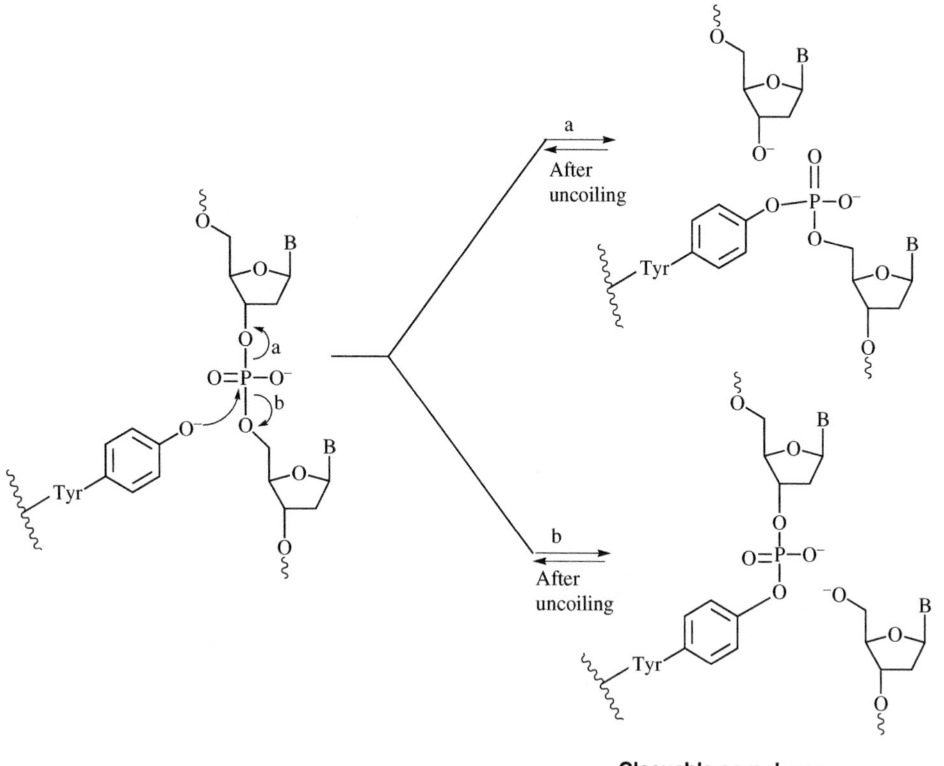

a
After uncoiling

b
After uncoiling

Cleavable complexes

SCHEME 6.1 **DNA topoisomerase-catalyzed strand cleavage to cleavable complexes**

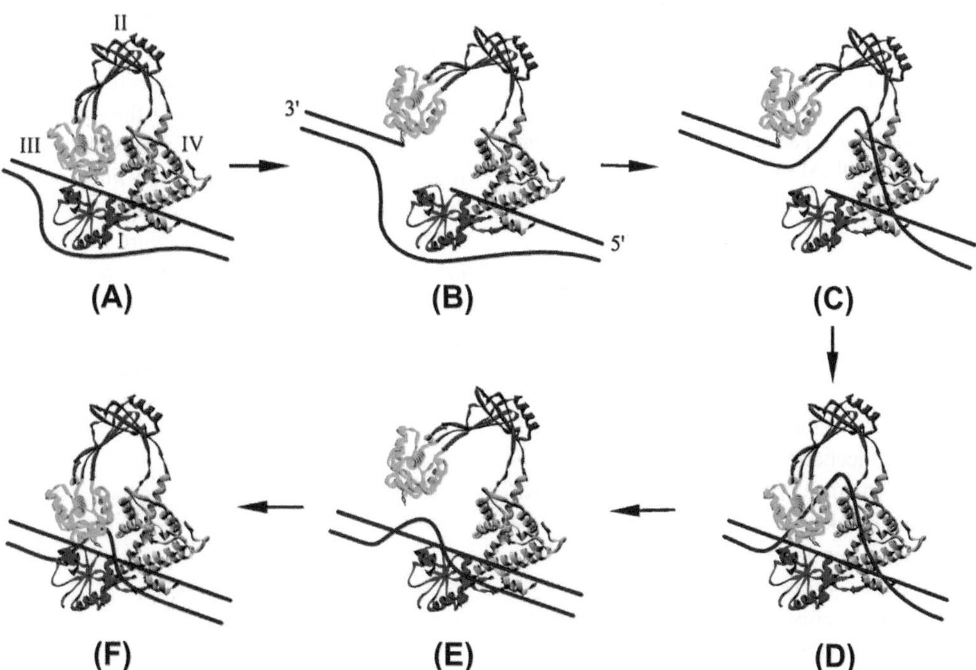

FIGURE 6.12 **Artist rendition of a possible mechanism for a topoisomerase I reaction. The colored sections are the topoisomerase, and the black lines are the double-stranded DNA.** *With permission from Champoux, J.J. (2010). With permission from the* Annual Review of Biochemistry. *Volume 70 ©2001 by* Annual Reviews. *www.annualreview.org.*

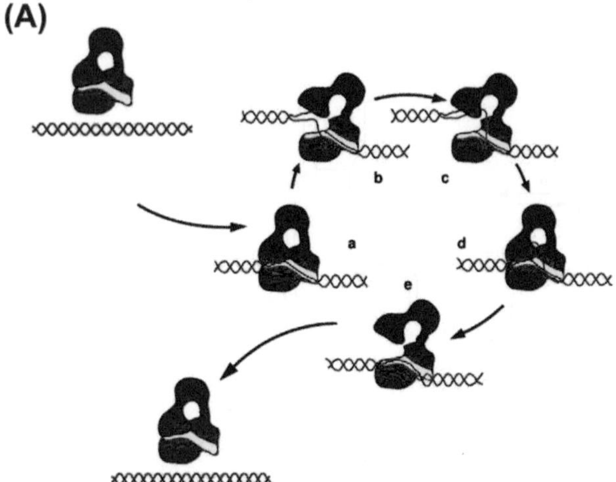

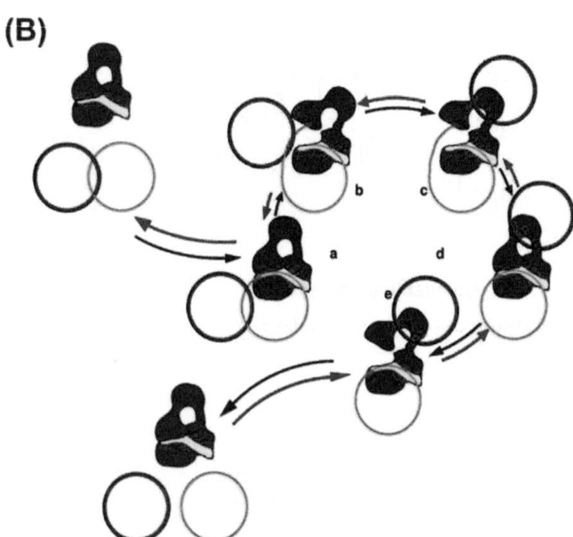

FIGURE 6.13 Artist rendition of possible mechanisms of topoisomerase IA-catalyzed relaxation of (A) supercoiled DNA and (B) decatenation of a DNA catenane. *From Li, Z.; Mondragon, A.; DiGate, R. J. The mechanism of IA topoisomerase-mediated DNA topological transformations.* Mol. Cell ***2001****, 7, 301.*

(structure E).[57] Figure 6.13 depicts a cartoon view of how topoisomerase IA may relax supercoiled DNA (A) or catalyze decatenation (B).[58]

A crystal structure of a fragment of the *Escherichia coli* enzyme[59] supports the mechanism in Figures 6.12 and 6.13. In the case of a topoisomerase type IB, nucleophilic attack by the tyrosine residue breaks the DNA strand to generate a phosphodiester link between the tyrosine and the 3′-phosphate, releasing a 5′-hydroxyl end (Scheme 6.1, pathway b).

In the case of topoisomerase II enzymes, two tyrosine residues cleave phosphodiester bonds on each of the DNA strands to form cleavable complexes. The mechanism for how these enzymes coil and uncoil DNA is even more in debate than the topoisomerase I enzymes and would be difficult to show in a pictorial form anyway, so no mechanism is given. As we will see (Section 6.3.1.3), one important mechanism of action of antitumor and antibacterial agents is the stabilization of the cleavable complex after it forms so that religation of the two DNA fragments does not occur, leaving the DNA cleaved.[60]

6.2.4. DNA Conformations

There are three general helical conformations of DNA; two are right-handed DNA (*A-DNA* and *B-DNA*) and one is the left-handed DNA (*Z-DNA*). Each conformation involves a helix made up of two antiparallel polynucleotide strands with the bases paired through Watson–Crick hydrogen bonding, but the overall shapes of the helices are quite different (Figure 6.14).

The right-handed forms differ in the distance required to make a complete helical turn (called the *pitch*), in the way their sugar groups are bent or puckered, in the angle of tilt that the base pairs make with the helical axis, and in the dimensions of the grooves. The predominant form by far is B-DNA, but in environments with low hydration, A-DNA occurs. In contrast to B-DNA, individual residues in A-DNA display uniform structural features; nucleotides in A-DNA have more narrowly confined conformations. Whereas there are 11 nucleotides in one helical turn in A-DNA, there are only 10 bp/pitch in B-DNA. Therefore, A-DNA is shorter and squatter than B-DNA.

Z-DNA, a minor component of the DNA of a cell, is a left-handed double helix having 12 bp per helical turn. It has only a minor groove because the major groove is filled with cytosine C-5 and guanine N-7 and C-8 atoms. In A- and B-DNA, the glycosyl bond is always oriented *anti* (**6.6**). In Z-DNA, the glycosyl bond connecting the base to the deoxyribose group is oriented *anti* at the pyrimidine residues but *syn* at the purine residues (**6.7**). This alternating *anti-syn* configuration gives the backbone (a line connecting the phosphorus atoms) an overall zigzag appearance, hence, the name Z-DNA. An X-ray crystal structure revealed that at the junction between B-DNA and Z-DNA, a single base pair is broken and is flipped out of the double helix.[61] This suggested that transient regions of Z-DNA, generated by transcription or unwrapping of DNA, form in our chromosomes and may play a role in virus pathogenesis and gene expression.

6.6

6.7

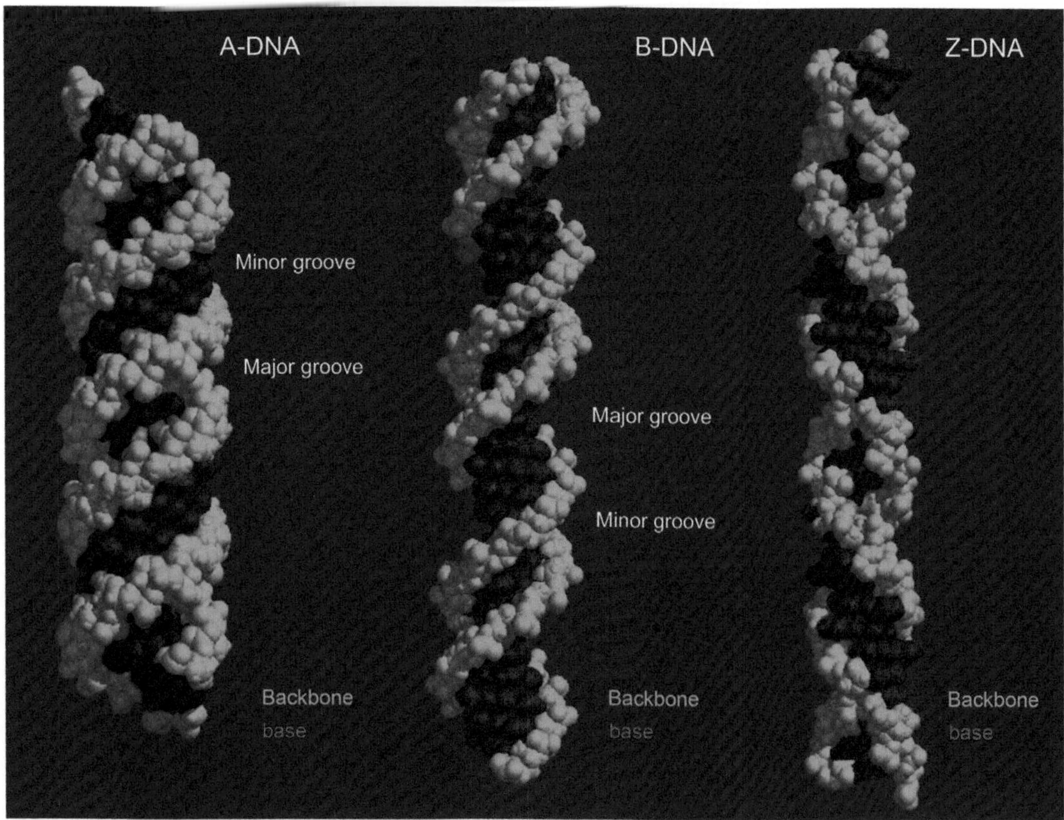

FIGURE 6.14 **Computer graphics depictions of A-DNA, B-DNA, and Z-DNA.** *Reproduced with permission from the Jena Library of Biological Macromolecules, Institute of Molecular Biotechnology (IMB), Jena, Germany; http://jenalib.fli-leibniz.de/ Hühne R., Koch F. T., Sühnel, J. A comparative view at comprehensive information resources on three-dimensional structures of biological macromolecules.* Brief Funct. Genomic Proteomic ***2007****, 6(3), 220–239.*

With this brief introduction to the structure of DNA, we can now explore the different mechanisms by which drugs interact with DNA.

6.3. CLASSES OF DRUGS THAT INTERACT WITH DNA

In general, there are three major classes of clinically important DNA-interactive drugs: *reversible binders*, which interact with DNA through the reversible formation of noncovalent interactions; *alkylators*, which react covalently with DNA bases; and *DNA strand breakers*, which generate reactive radicals that produce cleavage of the polynucleotide strands. The ideal DNA-interactive drug may turn out to be a nonpeptide molecule that is targeted for a specific sequence and site size.[62] However, it is not yet clear what DNA sequences (genes) should be targeted. Also, in traditional cancer chemotherapy, significant amounts of DNA damage (rather than small amounts of sequence-selective DNA damage) are required to elicit the cell killing that is necessary for effective anticancer drugs. Proteins are examples of molecules that exhibit unambiguous DNA sequence recognition. The primary DNA sequence recognition by proteins results from complementary hydrogen bonding between amino acid residues on the protein and nucleic acid bases in the major and minor grooves of DNA.[63] Proteins generally use major groove interactions with B-DNA because there are more donor and acceptor sites for hydrogen bonding than in the minor groove.[64]

You may be wondering how drugs can interact with DNA at all, given that essentially all the DNA in the cell is packed as chromatin (Section 6.2.3). A close-up view of a computer model of a nucleosome (Figure 6.15)[65] shows that the outer surface of the DNA is directly accessible to small molecules. Larger molecules, however, also can interact with DNA because in vitro studies have demonstrated that the nucleosomes are in dynamic equilibrium with uncoiled DNA (Figure 6.16),[66] so the drug can bind after uncoiling, which interferes with the binding of the DNA to the histone.

As this chapter is related to drugs that directly interact with DNA, we will not be discussing the many important antitumor, antibacterial, and antiviral drugs that inhibit the various DNA topoisomerases[67]; those would be applicable to the topics discussed in Chapter 5.

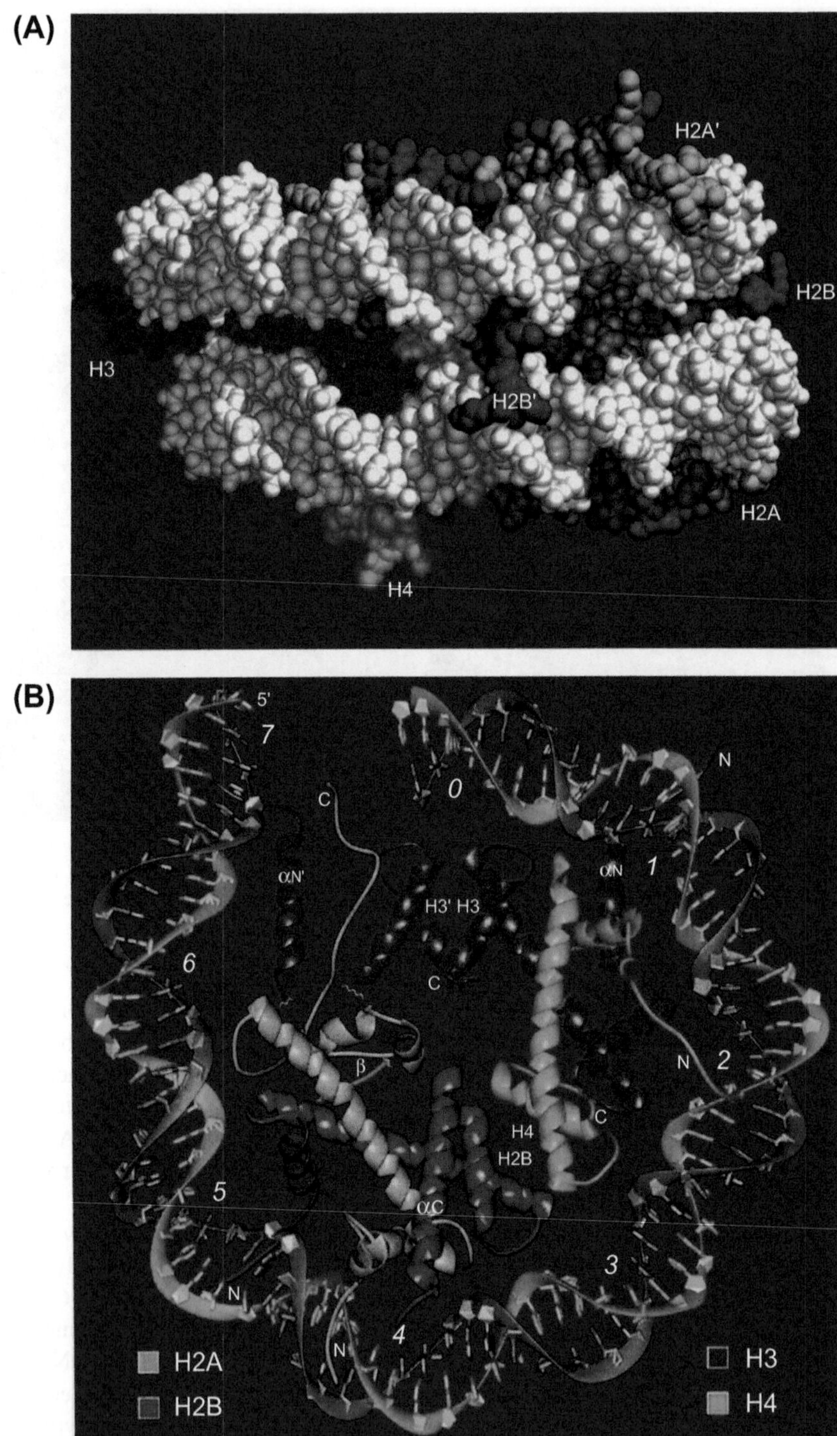

FIGURE 6.15 **(A) Molecular model of a nucleosome. (B) Cutaway view of the nucleosome with the histones in the center and duplex DNA wrapped around them.** *With permission from Luger, K. (1977). Reprinted with permission from Macmillan Publishers Ltd:* Nature ***1993***, 398, *251–260.*

6.3.1. Reversible DNA Binders

Nucleic acids inside the cell interact with a variety of small molecules, including water, metal cations, small organic molecules, and proteins, all of which are essential for stabilization of the nucleic acid structure.[68] Interference with these interactions can disrupt the DNA structure. There are three important ways that small molecules can reversibly bind to duplex DNA and lead to interference of DNA function: (1) by electrostatic binding along the exterior of the helix; (2) by interaction with the edges of the base pairs in either the major or minor

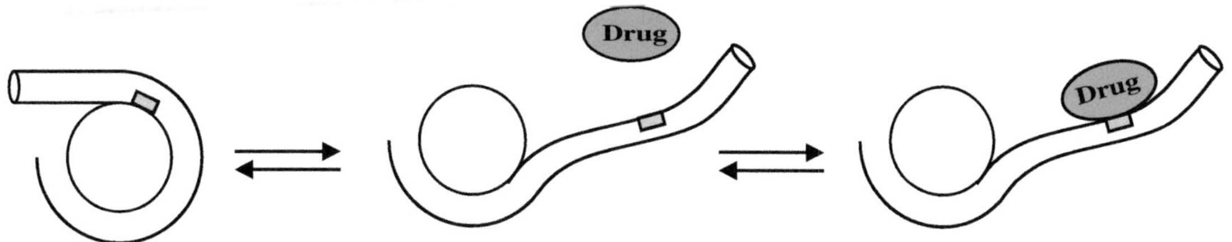

FIGURE 6.16 **Schematic of how a drug could bind to DNA wrapped around histones in the nucleosome.** *Polach K. J. Mechanism of protein access to specific DNA sequences in chromatin: A dynamic equilibrium model for gene regulation.* J. Mol Biol. ***1995***, 254, *130.*

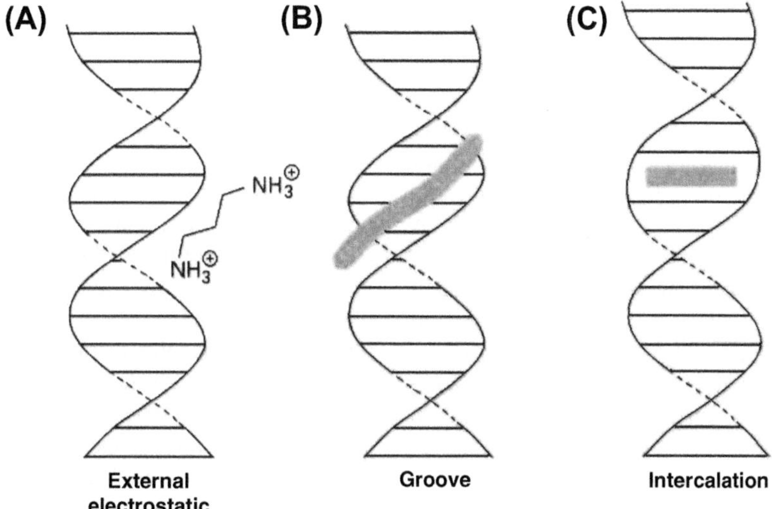

FIGURE 6.17 **Schematic of three types of reversible DNA binders. A**, external electrostatic binder; **B**, groove binder; **C**, intercalator. In **B** and **C**, the pink bar represents the drug. *Reproduced with permission from Blackburn G. M., Gait M. J., Eds.* Nucleic Acids in Chemistry and Biology, *2nd ed., 1996; p. 332. By permission of Oxford University Press.*

groove; and (3) by intercalation between the base pairs (Figure 6.17).

6.3.1.1. External Electrostatic Binding

Duplex DNA contains a negatively charged sugar-phosphate backbone; this polyanionic nature strongly affects both the structure and function of DNA. Cations and water molecules bind to DNA and allow it to exist in various secondary structures described in Section 6.2.4. Release of bound cationic counter ions of the negative phosphate groups upon binding of specific cation ligands can provide both favorable (increased entropy) and unfavorable (decreased enthalpy because of loss of specific ionic interactions) contributions to the overall free energy, leading to disruption of the DNA structure. These types of interactions are generally not dependent upon DNA sequence.

6.3.1.2. Groove Binding

The major and minor grooves have significant differences in their electrostatic potentials, hydrogen bonding characteristics, steric effects, and degree of hydration. Proteins exhibit binding specificity primarily through major groove interactions, but small molecules prefer minor groove binding. Minor groove binding molecules generally have aromatic rings connected by single bonds that allow for torsional rotation in order to fit into the helical curvature of the groove with displacement of water molecules.[69] The minor groove is generally not as wide in A–T-rich regions relative to G–C-rich regions; therefore, A–T regions might be more amenable to flat aromatic molecule binding than are G–C regions. The more narrow A–T regions produce a more snug fit of molecules into the minor groove and lead to van der Waals interactions with the DNA functional groups that define the groove. Binding also arises from interactions with the edges of the base pairs on the floor of the groove. Groove binders do not significantly unwind DNA base pairs. Hydrogen bonding from the C-2 carbonyl oxygen of T or the N-3 nitrogen of A to minor groove binders is very important. Similar groups also are present in the G–C base pairs, but the amino group of G sterically hinders hydrogen bond formation at N-3 of G and at the C-2 carbonyl oxygen of C. Also, the hydrogen bonds between the amino groups of G and carbonyl oxygen of C in G–C base pairs lie in the minor groove, and these sterically inhibit penetration of molecules into G–C-rich regions of this groove. Because of greater negative electrostatic potential in the A–T regions of the minor groove relative to the G–C regions,[70] there is a higher selectivity for cationic molecules in A–T regions; it is possible to enhance G–C region binding by the design of molecules that can accept hydrogen bonds from the amino group of G. Groove binders can be elongated to extend the interactions within the groove, which leads to high sequence-specific recognition by these molecules.

Molecules that bind in the A–T regions of the minor groove typically are crescent shaped with hydrogen bonding NH groups on the interior of the crescent. The NH groups hydrogen bond with the A–T base pairs in the minor groove, but are excluded from these interactions with G–C base pairs by the amino group of G. Cationic groups undergo electrostatic interactions with the negative electrostatic potential in the minor groove. A typical minor groove binder is the antitumor, antibacterial, and antiviral agent netropsin (**6.8**, Figure 6.18).[71] The refined 2.2 Å crystal structure of **6.8** bound to a B-DNA dodecamer shows that the drug displaces the water molecules so it is centered in the AATT region of the minor groove and forms three good bifurcated hydrogen bonds with N-3 of adenine and the C-2 carbonyl oxygen of thymine along the floor of the groove. The pyrrole rings of netropsin are packed against the C-2 positions of adenines, which leaves no room for the amino group of guanine, thereby providing a structural rationale for the A–T specificity of netropsin. Binding of netropsin neither unwinds nor elongates the double helix, but it causes a widening of the minor groove (0.5–2.0 Å) in the AATT region and a bending of the helix axis (8°) away from the site of binding.[72] The result of this widening is to interfere with the interaction of topoisomerase II (Section 6.3.1.3), leading to DNA damage.

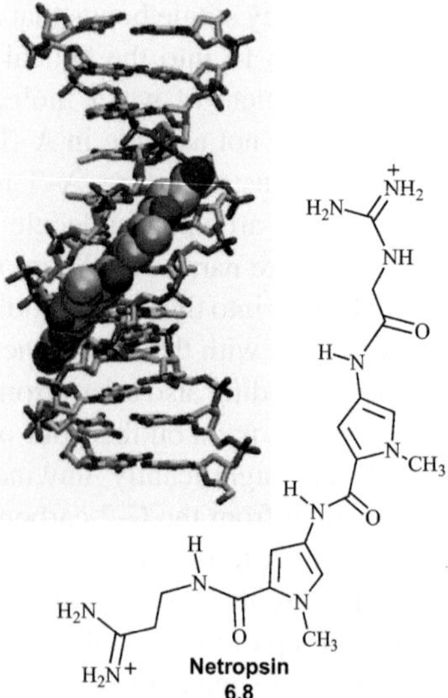

FIGURE 6.18 Model showing interaction of netropsin (colored ball model) with double helical DNA (colored stick model). The 2D structure of netropsin (**6.8**) is also shown. *Image created by JanLipfert from crystallographic coordinates deposited in the Protein Data Bank, accession code 101D.*

6.3.1.3. Intercalation and Topoisomerase-Induced DNA Damage

Flat, generally aromatic or heteroaromatic molecules bind to DNA by inserting (i.e., *intercalating*) and stacking between the base pairs of the double helix. The principal driving forces for intercalation are stacking and charge-transfer interactions, but hydrogen bonding and electrostatic forces also play a role in stabilization.[73] Intercalation, first described in 1961 by Lerman,[74] is a noncovalent interaction in which the drug is held rigidly perpendicular to the helix axis. This causes the base pairs to separate vertically, thereby distorting the sugar-phosphate backbone and decreasing the pitch of the helix (Figure 6.19 shows the intercalation of ethidium bromide into B-DNA). Intercalation, apparently, is an energetically favorable process, because it occurs so readily. Presumably, the van der Waals forces that hold the intercalated molecules to the base pairs are stronger than those found between the stacked base pairs. Much of the binding energy is the result of the removal of the drug molecule from the aqueous medium and a hydrophobic effect. Intercalation occurs preferentially (by 7–13 kcal/mol) into pyrimidine-3′,5′-purine sequences rather than into purine-3′,5′-pyrimidine sequences.[75] Intercalators do not bind between every base pair. The *neighbor exclusion principle* states that intercalators can, at most, bind at alternate possible base-pair sites on DNA because saturation is reached at a maximum of one intercalator between every second site.[76] One explanation for this principle is that binding in one site causes a conformational change in the adjacent site, which prevents binding of the intercalator in that adjacent site (known as *negative cooperativity*).

In general, intercalation does not disrupt the Watson–Crick hydrogen bonding, but it does destroy the regular helical structure, unwinds the DNA at the site of binding, and as a result of this, interferes with the action of DNA-binding enzymes such as DNA topoisomerases and DNA polymerases. Interference with topoisomerases alters the degree of supercoiling of DNA; interference with DNA polymerases inhibits the elongation of the DNA chain in the 5′ to 3′ direction and also prevents the correction of mistakes in the DNA by inhibiting the clipping out (via hydrolysis of the phosphodiester bond) of mismatched residues at the terminus. Most intercalators display either no sequence preferences in their binding or a slight G–C preference, which contrasts with the A–T binding preference of groove binders. Furthermore, groove binders, in general, exhibit significantly greater binding selectivity than intercalators. Groove binders interact with more base pairs than intercalators as they lie along the groove.

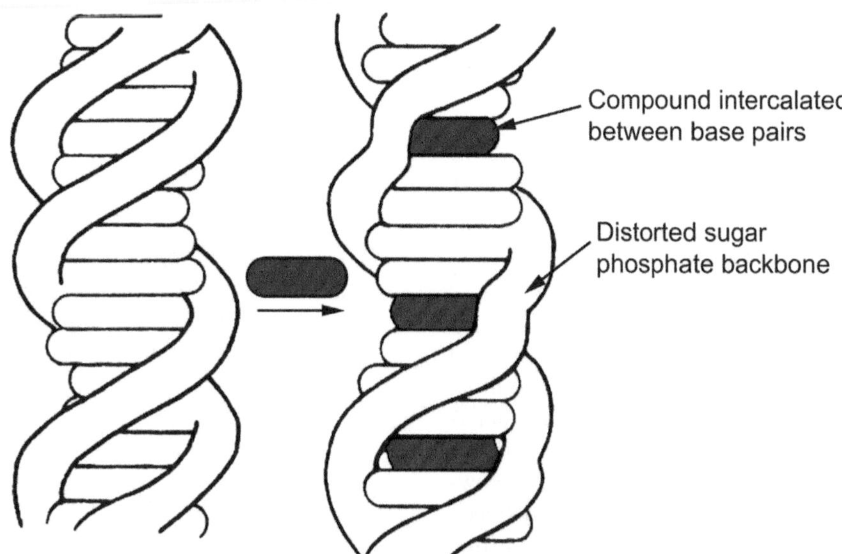

FIGURE 6.19 **Intercalation of ethidium bromide into B-DNA**

Also, grooves are distinct in A–T and G–C regions, which adds to the potential for specificity in groove binding (Section 6.3.1.2).

For simple cationic intercalators, intercalation involves two steps. First, the cation interacts with the negatively charged DNA sugar-phosphate backbone. Then the intercalator diffuses along the surface of the helix until it encounters gaps between base pairs that have separated because of thermal motion, thereby creating a cavity for intercalation.

Although the drugs in this section are categorized as being intercalators, it is now believed that intercalation of a drug into DNA is only the first step in the events that eventually lead to DNA damage by other mechanisms. For many classes of antitumor agents, there is an involvement of the DNA topoisomerases (Section 6.2.3) subsequent to intercalation.[77] Mammalian DNA topoisomerase I is involved in the action of the antitumor agent topotecan hydrochloride (**6.9**; Hycamtin),[78] whereas DNA topoisomerase II is involved in the action of a variety of classes of antitumor drugs, such as anthracyclines, anthracenediones, acridines, actinomycins, and ellipticines.[79] The action of quinolone antibacterial drugs, such as nalidixic acid (**6.10**; NegGram), also involves bacterial DNA topoisomerase II. The evidence to date suggests that intercalators that promote topoisomerase-induced DNA damage interfere with the breakage-rejoining reaction (Scheme 6.1) by trapping a key covalent reaction intermediate (presumably the tyrosine adducts shown in Scheme 6.1), termed the *cleavable complex*.[80] The cleavable complex may be stabilized by the formation of a reversible nonproductive (noncleavable) drug–DNA–topoisomerase ternary complex. It has been hypothesized that this ternary complex may collide with transcription and replication complexes; upon collision, the ternary complex may lose its reversibility and generate lethal double-strand DNA breaks.[81] It is not clear if the drug binds to DNA first, then topoisomerase II forms the ternary complex, or the drug binds to a topoisomerase II–DNA complex. An example of this phenomenon is TAS-103 (**6.11**), an antineoplastic agent that first intercalates into DNA, then kills cells by increasing the amount of DNA cleavage mediated by topoisomerase II as a result of its inhibition of the religation of the DNA from the cleavable complex.[82]

Topotecan hydrochloride
6.9

Nalidixic acid
6.10

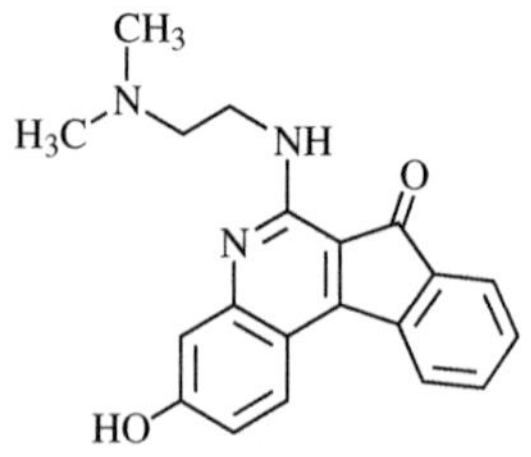

TAS-103
6.11

6.13

Actinomycin D
6.14

A study of a series of anthracycline analogs showed that DNA intercalation of these compounds is required, but not sufficient for topoisomerase II-targeted activity.[83] There was a strong correlation between the potency of intercalation and cleavable complex formation. However, there does not appear to be a correlation between DNA intercalation and antitumor activity. Some strong intercalators do not induce cleavable complexes, possibly because of certain structural requirements for the binding of the intercalated drug to the topoisomerase. Also, epipodophyllotoxins, such as the anticancer drug etoposide (**6.12**; Etopophos), are nonintercalating DNA topoisomerase II poisons.

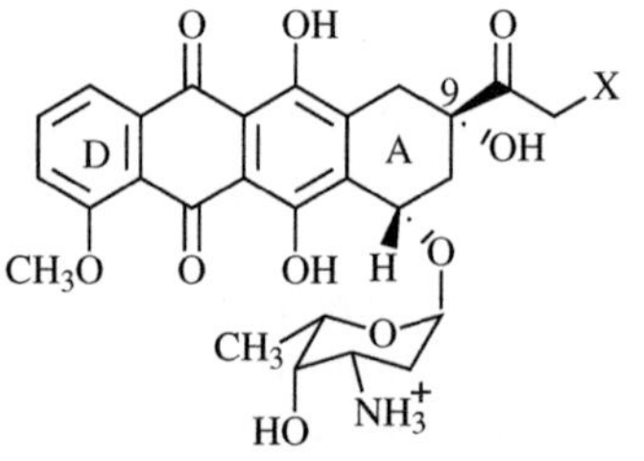

Doxorubicin (X = OH)
Daunorubicin (X = H)
6.15

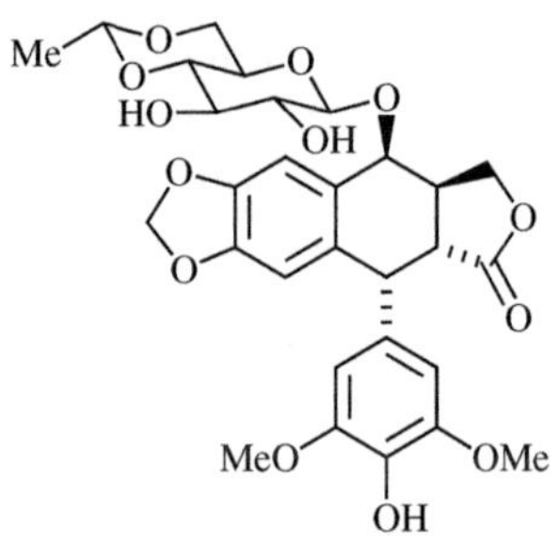

Etoposide
6.12

Although the mechanism of topoisomerase II-induced DNA damage is not clear, the ternary complex appears to be lethal to proliferating cells. Selective sensitivity of proliferating tumor cells to the cytotoxic effects of DNA topoisomerase II poisons may be the result of the high levels of DNA topoisomerase II found in proliferating cells and the very low levels found in quiescent cells.

Intercalation may not be the direct cause for DNA damage, but it does produce a conformational change (unwinding) in the double helix. This, then, can result in the positioning of the drug in the DNA appropriately for binding with the topoisomerase in the ternary complex, or it can position the drug for subsequent reactions, as will be discussed in Sections 6.3.2 and 6.3.3.

Three classes of drug molecules that have been well characterized as intercalators of DNA are the acridines (**6.13**), the actinomycins (**6.14**), and the anthracyclines (**6.15**).

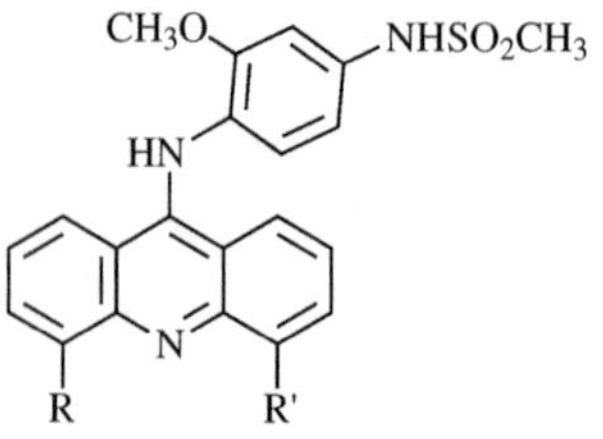

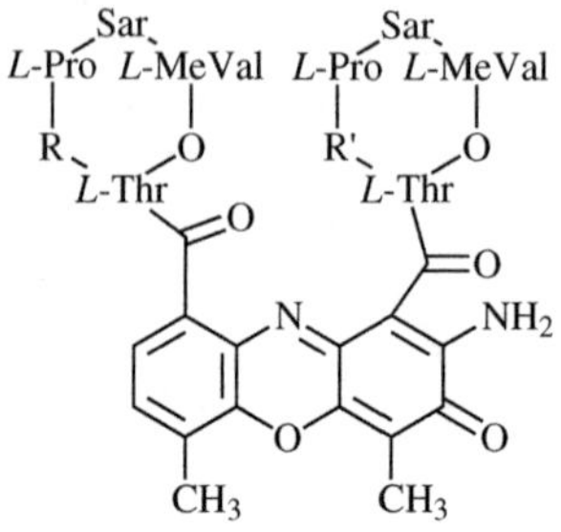

6.3.1.3.1. Amsacrine, an Acridine Analog

Acridine compounds, which were by-products of aniline dye manufacture, were first used in clinical medicine in the late nineteenth century against malaria.[84] By the First World War, acridine derivatives such as proflavine (**6.16**) were in widespread use as local antibacterial agents. After the Second World War, another acridine derivative, aminacrine (**6.17**; Monacrin), was the principal acridine antibacterial agent used.[85]

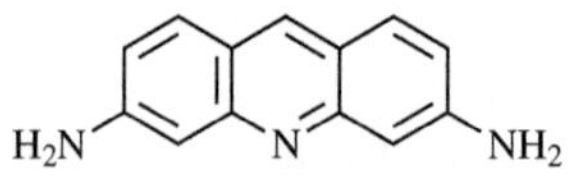

Proflavine
6.16

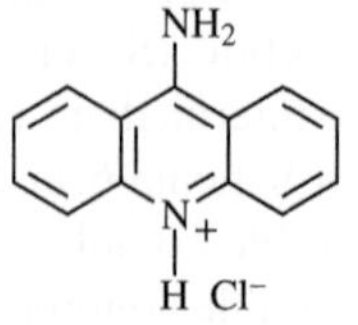

Aminacrine
6.17

In the 1960s and early 1970s, a variety of anilino-substituted analogs of 9-anilinoacridine (**6.18**) were prepared and tested for antitumor activity[86] on the basis of the reported antitumor activity of **6.18** (R = H, R′ = Me_2N).[87] Although the 3,4-diamino analog had good antitumor activity, it was unstable to air oxidation. On the basis of structure–activity relationships, it was reasoned that an electron donor group was needed on the anilino ring; consequently, a sulfonamide group, which would be partially anionic at physiological pH, was selected. In fact, **6.18** (R = H, R′ = $NHSO_2Me$) was as potent as the 3,4-diamino analog.[88] This was used as a lead compound, and it was found that the most potent analogs had other electron-donating substituents in addition to the sulfonamide group. Amsacrine (**6.18**; R = OMe, R′ = $NHSO_2Me$; Amsidyl) was the most potent of those tested;[89] its main use is now in the treatment of leukemia.[90]

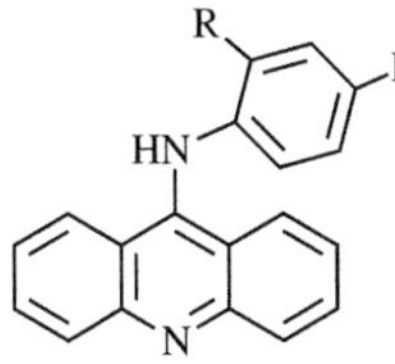

Amsacrine (R = OMe, R' = $NHSO_2Me$)
6.18

Paradoxically, although a very wide variation of structures of 9-anilinoacridines can be tolerated with retention of antitumor activity, among the active derivatives, large differences in potency are observed with small changes in structure.[91] The antitumor activity was parabolically related to drug lipophilicity as measured by log *P* values (Chapter 2, Section 2.2.5.2.2); compounds with log *P* values close to that of amsacrine were most potent. There also is a close correlation between the electronic properties (σ constant; see Chapter 2, Section 2.2.5.1) of groups at the *para*-position of the anilino ring and acridine pK_a values. Furthermore, when lipophilic and electronic effects of a series of bulky substituents at various positions on the 9-anilinoacridine framework are taken into account, the steric effects of the group play a dominant role.

These results are consistent with the mode of action of 9-anilinoacridines as intercalators of double-stranded (duplex) DNA.[92] Earlier studies showed that these compounds unwound closed circular duplex DNA.[93] By analogy with the crystal structure of 9-aminoacridine bound to a dinucleotide,[94] Denny et al.[95] hypothesized that the anilino ring, which lies almost at right angles to the plane of the acridine chromophore, is lodged in the minor groove with the 1′-substituent (the sulfonamide group) oriented at a 90° angle to the helical axis. The sulfonamide may interact with a second macromolecule, such as a regulatory protein, and this ternary complex could mediate the biological effects of the 9-anilinoacridines. More recent studies of the rates of dissociation of amsacrine from DNA suggest that the anilino group may bind in the major groove.[96]

Amsacrine lacks broad spectrum clinical activity and is difficult to formulate because of its low aqueous solubility. It was thought that the relatively high pK_a of the compound (8.02 for the acridine nitrogen) was important in limiting in vivo distribution; consequently, analogs with improved solubility and high DNA binding, but with lower pK_a, were sought. A compound was found (**6.13**, R = Me, R′ = CONHMe) that had all of the desirable physicochemical properties, showed superior antileukemic activity compared with amsacrine, and was broader in its spectrum of action.[97]

6.3.1.3.2. Dactinomycin, the Parent Actinomycin Analog

Actinomycin D (now called dactinomycin; **6.14**, R = R′ = D-Val; Cosmegen) was the first of a family of chromopeptide antibiotics isolated from a culture of a *Streptomyces* strain in 1940.[98] In 1952, these compounds were found to have antitumor activity and were used clinically.[99] Dactinomycin binds to double-stranded DNA and, depending upon its concentration, inhibits DNA-directed RNA synthesis or DNA synthesis. RNA chain initiation is not prevented, but chain elongation is blocked.[100] The phenoxazone chromophore intercalates between bases in the DNA.[101] Binding depends on the presence of guanine; the 2-amino group of guanine is important for the formation of a stable drug–DNA complex.[102] X-ray crystal structures of a 1:2 complex of dactinomycin with deoxyguanosine,[103] deoxyguanylyl-3′,5′-deoxycytidine (Figure 6.20),[104] and a complex of dactinomycin with d(ATGCAT)[105] are models for the intercalation of dactinomycin into DNA. These structures suggest that the phenoxazone ring can intercalate between deoxyguanosines and that the cyclic pentapeptide substituents can be involved in strong hydrogen bonding and hydrophobic interactions with DNA.[106] In particular, there are two crucial hydrogen bonds that stabilize the DNA binding complex. One strong hydrogen bond exists between neighboring cyclic pentapeptide chains connecting the N–H of one D-valine residue with the C=O of the other D-valine residue. Another strong hydrogen bond connects the guanine 2-amino group with the carbonyl oxygen of the L-threonine residue. A weaker hydrogen bond connects the guanine N-3 ring nitrogen with the NH group on this same L-threonine residue. Stacking forces are primarily responsible for the recognition and preferential binding of a guanine base to dactinomycin.[107] The biological activity appears to depend on the very slow rate of DNA–dactinomycin dissociation, which reflects the intermolecular hydrogen bonds, the planar interactions between

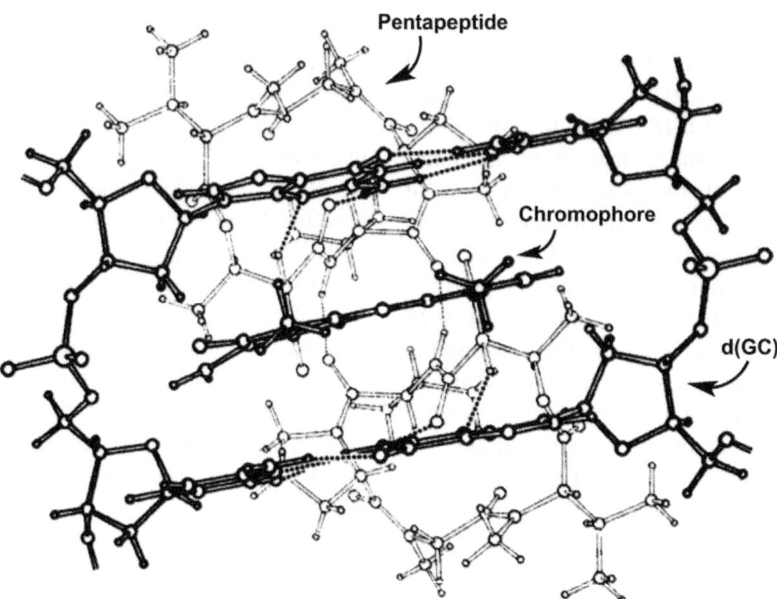

FIGURE 6.20 X-ray structure of a 1:2 complex of dactinomycin with d(GC). *Reprinted from* Journal of Molecular Biology, *Vol. 68, "Stereochemistry of actinomycin binding to DNA. II. Detailed molecular model of actinomycin DNA complex and its implications", pp. 26–34. Copyright. 1972 Academic Press, with permission from Elsevier.*

the purine rings and the chromophore, and the numerous van der Waals interactions between the polypeptide side chains and the DNA. The peptide substituents, which lie in the minor groove, may block the progression of the RNA polymerase along the DNA. It is clear, then, that intercalators can cause cytotoxicity by interfering with the normal cellular processing of DNA, such as replication and transcription.

6.3.1.3.3. Doxorubicin (Adriamycin) and Daunorubicin (Daunomycin), Anthracycline Antitumor Antibiotics

The anthracycline class of antitumor antibiotics exemplified by doxorubicin (**6.15**, X=OH; previously called adriamycin; Doxil) and daunorubicin (**6.15**, X=H; also called daunomycin; Cerubidine) are isolated from different species of *Streptomyces*. Although these two compounds differ by only one oxygen atom, there is a major difference in their antitumor activity. Whereas daunorubicin is active only against leukemia, doxorubicin is active against leukemia as well as a broad spectrum of solid tumors.

There is some controversy as to whether the mechanism of action of these compounds is related to their ability to intercalate into the DNA or to cause DNA strand breakage.[108] The vast majority of the intracellular drug is in the nucleus, where it intercalates into the DNA double helix (and forms the ternary complex with DNA topoisomerase II), with consequent inhibition of replication and transcription. X-ray[109] and nuclear magnetic resonance[110] studies of model daunorubicin–oligonucleotide complexes show that the oligonucleotides form a 6 bp right-handed double helix with two daunorubicin molecules intercalated in the d(CpG) sequences. The tetracyclic chromophore is oriented orthogonal to the long dimension of the DNA base pairs, and the ring that has the amino sugar substituent (A ring) protrudes into the minor groove (Figure 6.21). Substituents on the A ring hydrogen bond to base pairs above and below the intercalation site. The amino sugar nearly covers the minor groove, but without bonding to the DNA. Ring D protrudes into the major groove. The complex is stabilized by stacking energies and by hydrogen bonding of the hydroxyl and carbonyl groups at C-9 of the A ring. X-ray diffraction analysis of the DNA–daunorubicin complex indicates that there is no interaction between the ionized amino group of the drug and any part of the double helix; it sits in the center of

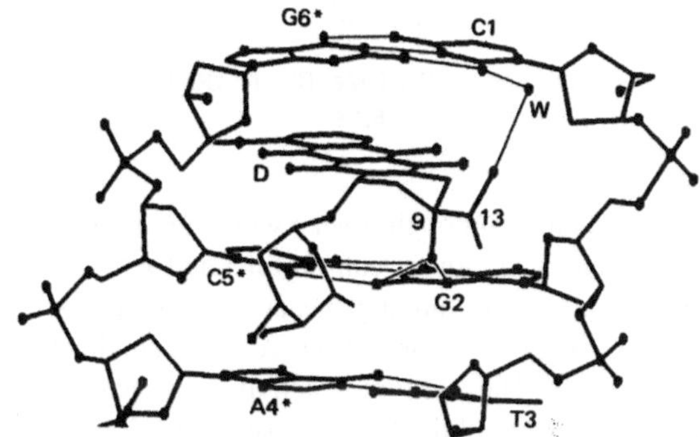

FIGURE 6.21 X-ray structure of daunorubicin intercalated into an oligonucleotide. Quigley, G. S.;Wang, A.; Ughetto, G.; Van der Marel, G.; Van Boom, J. H.; Rich, A. Molecular structure of an anticancer drug-DNA complex: Daunomycin plus d(CpGpTpApCpG). *Proc. Natl. Acad. Sci. USA* **1980**, *77*, p. 7206. *Reprinted with permission from Dr. C. J. Quigley.*

the minor groove. This is consistent with structure–activity studies,[111] which indicate that modification of the amino group does not necessarily affect biological activity.

In addition to intercalation and topoisomerase II-induced DNA damage, another mechanism of action of the anthracycline antitumor antibiotics involves radical-induced DNA strand breakage; this mechanism is discussed in Section 6.3.3.1.

6.3.1.3.4. *Bis*-intercalating Agents

Once success with intercalating agents was realized, it was thought that *bifunctional intercalating agents* (also called *bis-intercalating agents*), in which there are two potential intercalating molecules tethered together so that each could intercalate into different DNA strands, would have enhanced affinity for DNA and slower dissociation rates. In general, the DNA affinity of *bis*-intercalating agents is greater than that of their monointercalating counterparts, in some cases approaching values typical of those observed with natural repressor proteins. However, this affinity may result from the polycationic nature of the high-affinity *bis*-intercalators and not necessarily because of the intercalating group. The specific nature of the linker between the two quinoxaline groups of quinoxaline antibiotics such as triostin A (**6.19**, R=CH_2S–SCH_2) and echinomycin (**6.19**, R=$CH(SCH_3)$–SCH_2) appears to be important to their activity and there is no binding to DNA without the cyclic depsipeptide.[112] This may be the result of the lack of rigidity to maintain the two quinoxaline groups in the correct orientation for *bis*-intercalation with two base pairs between each intercalation site (consistent with the neighbor exclusion principle discussed in this section). Both the rigidity and length of the linker chain between the two intercalator molecules are important to constrain the *bis*-intercalator in the ideal configuration to form a sandwich with the two base pairs.[113] A general structure of *bis*-quinoxaline intercalators is shown in Figure 6.22. The synthetic diacridines with flexible linker chains are ineffective.

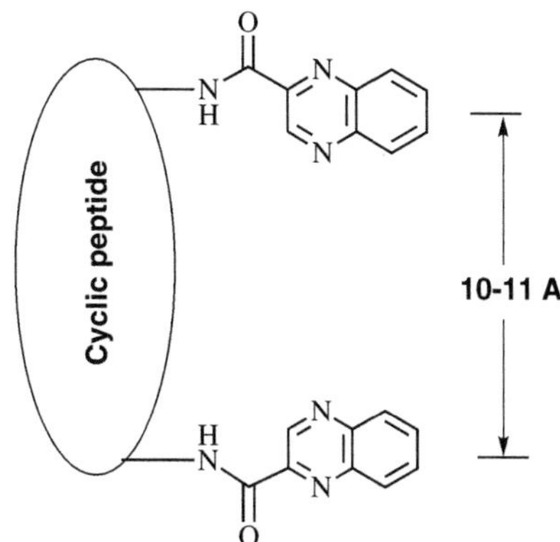

FIGURE 6.22 General structure of *bis*-quinoxaline intercalators

For the most part, much of the expected enhancement in free energy of *bis*-intercalation is not observed.[114] This is probably the result of the unfavorable entropy associated with loss of rotational freedom. In general, these agents do not show a remarkable improvement in specificity compared to that found for the monofunctional intercalators, although there are effective bis-intercalators that utilize flexible linkers.[115]

6.3.2. DNA Alkylators

The difference between the DNA alkylators and the DNA intercalators is akin to the difference between irreversible and reversible enzyme inhibitors (Chapter 5). The intercalators (reversible enzyme inhibitors) bind to the DNA (enzyme) with noncovalent interactions. The DNA alkylators (irreversible inhibitors) react with the DNA (enzyme) to form covalent bonds. Several of the important classes of alkylating drugs having different alkylation mechanisms are discussed below.

6.3.2.1. Nitrogen Mustards

6.3.2.1.1. Lead Discovery

6.20

Sulfur mustard (**6.20**) is a highly cytotoxic and vesicant chemical warfare agent that was used in World Wars I and II (as well as in later conflicts). Autopsies of soldiers killed

Triostin A (R = CH_2S-SCH_2)
Echinomycin (R = $CH(SCH_3)$-SCH_2)
6.19

in World War I by sulfur mustard revealed leukopenia (low white blood cell count), bone marrow aplasia (defective development), dissolution of lymphoid tissues, and ulceration of the GI tract.[116] All of these lesions indicated that sulfur mustard has a profound effect on rapidly dividing cells and suggested that related compounds may be effective as antitumor agents. In fact, in 1931, sulfur mustard was injected directly into tumors in humans,[117] but this procedure turned out to be too toxic for systemic use. Because of the potential antitumor effects of sulfur mustards, a less toxic form was sought. Gilman and others examined the antitumor effects of the isosteric nitrogen mustards (**6.21**), less reactive alkylating agents, and in 1942, the first clinical trials of a nitrogen mustard were initiated. However, this research was classified during World War II, so the usefulness of nitrogen mustard in the treatment of cancer was not known until 1946, when these studies became declassified and Gilman published a review of his findings.[118] Soon thereafter several other summaries of clinical research carried out during the war appeared.[119] This work marks the beginning of modern cancer chemotherapy.

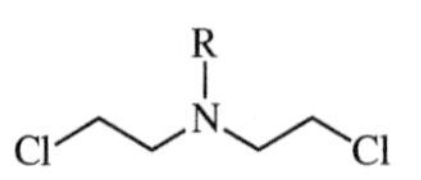

Mechlorethamine (R = CH_3)
6.21

Prior to a discussion of lead modification and other classes of alkylating agents, let us take a brief excursion into the chemistry of alkylating agents in general.

6.3.2.1.2. Chemistry of Alkylating Agents

According to Ross,[120] a biological alkylating agent is a compound that can replace a hydrogen atom with an alkyl group under physiological conditions (pH 7.4, 37 °C, aqueous solution). These alkylation reactions are generally described in terms of substitution reactions by N, O, and S heteroatomic nucleophiles with the electrophilic alkylating agent, although Michael addition reactions also are important. The two most common types of nucleophilic substitution reactions are S_N1 (Scheme 6.2A), a stepwise reaction via an intermediate carbenium ion, and S_N2 (Scheme 6.2B), a concerted reaction. In general, the relative rates of nucleophilic substitution at physiological pH are in the order thiolate > amino > phosphate > carboxylate.[121] For DNA, the most reactive nucleophilic sites are N-7 of guanine > N-3 of adenine > N-7 of adenine > N-3 of guanine > N-1 of adenine > N-1 of cytosine[122] (see **6.22** and **6.23** for the numbering system of purines (A and G) and pyrimidines (C and T), respectively). The N-3 of cytosine, the O-6 of guanine, and the phosphate groups also can be alkylated. Quantum mechanical calculations[123] confirm that the N-7 position of guanine is the most nucleophilic site. The reactivity of various nucleophilic sites on DNA is strongly controlled by steric, electronic, and hydrogen bonding effects. For example, some of the nucleophilic sites are in the interior of the DNA double helix and are sterically blocked. Also, only nucleophilic centers in the major and minor grooves or in the walls of the double helix are readily accessible to alkylating agents. In addition to these steric effects, the nucleophilicity of various sites on the purine and pyrimidine bases of the DNA is diminished because of their involvement in Watson–Crick hydrogen bonding.

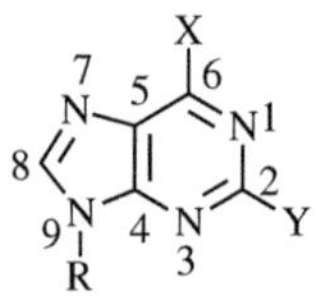

6.22 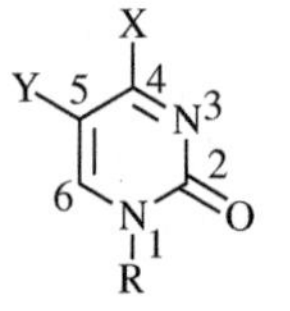 **6.23**

The reaction order for nucleophilic substitution depends on the chemical structure of the alkylating agent. Simple alkylating agents such as ethylenimines and methanesulfonates undergo S_N2 reactions. Alkylating agents such as nitrogen mustards, which have a nucleophile capable of *anchimeric assistance* (neighboring group participation), can undergo S_N1- or S_N2-type reactions, depending on the relative rates of the aziridinium ion formation and the nucleophilic attack on the aziridinium ion (Scheme 6.3).[124]

When aziridinium ion formation is fast, the overall reaction rate is second order (S_N2), but when aziridinium ion formation is slower than nucleophilic attack, the overall reaction is first order (S_N1).

$$\text{(A)}\quad \text{Alkyl—X} \underset{}{\overset{S_N1}{\rightleftharpoons}} \text{Alkyl}^+\ X^- \xrightarrow{Nu^-} \text{Alkyl—Nu}$$

$$\text{(B)}\quad \text{Alkyl—X} \underset{S_N2}{\overset{Nu^-}{\longrightarrow}} \text{Alkyl—Nu} + X^-$$

SCHEME 6.2 Nucleophilic substitution mechanisms

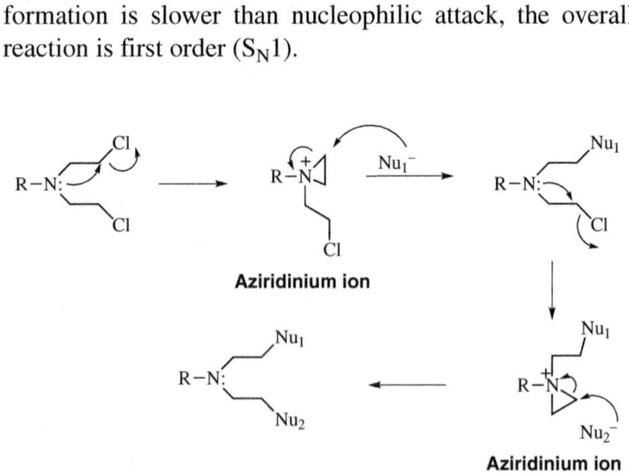

Aziridinium ion

Aziridinium ion

SCHEME 6.3 Alkylations by nitrogen mustards

In the case of the nitrogen mustards, which are *bifunctional alkylating agents* (i.e., they have two electrophilic sites), the DNA undergoes intrastrand and interstrand cross-linking.[125] Although there does not appear to be a direct correlation between the chemical reactivity of the alkylating agent and the therapeutic or toxic effects,[126] the compounds that are able to cross-link DNA are much more effective than monofunctional alkylating agents.[127] There is a rough correlation between antitumor efficacy and ability to induce mutations and inhibit DNA synthesis and a relationship between the rate of solvolysis of the alkylating agent and its cytotoxic effect on tumor cells in vitro. The differences in the effectiveness of the various alkylating agents probably result from differences in pharmacokinetic factors, lipid solubility, ability to penetrate the central nervous system, membrane transport properties, detoxification reactions (Chapter 8), and specific enzymatic reactions capable of repairing alkylated sites on the DNA.[128]

6.3.2.1.3. Lead Modification

The prototype of the nitrogen mustards is mechlorethamine (**6.21**, R=CH_3; Mustargen),[129] which is still used in the treatment of advanced Hodgkin's disease. Mechlorethamine is a bifunctional alkylating agent that reacts with the N-7 of two different guanines in DNA,[130] producing an interstrand cross-link (**6.24**) by the mechanism shown in Scheme 6.3. The formation of the N-7 ammonium ion makes the guanine more acidic and, therefore, appears to shift the equilibrium in favor of the enol tautomer.[131] Because guanine in this tautomeric form can make base pairs with thymine residues instead of cytosine (**6.25**), this leads to miscoding during replication. Furthermore, the N-7 alkylated guanosine is susceptible to hydrolysis to produce an abasic site (**6.26**, Scheme 6.4), a site on DNA in which the heterocyclic nucleobase has been displaced; this is the most common type of damage suffered by genomic DNA.[132] Anticancer drugs, particularly alkylating agents that modify guanine residues, can cause abasic sites.[133] This results in the destruction of the purine nucleus and in DNA strand scission (Scheme 6.4). Deglycosylation (loss of the sugar) is a very slow reaction.[134] In addition to interstrand cross-links, it also is possible for the second chloroethyl group of the nitrogen mustard to react with a thiol or amino group of a protein, resulting in a DNA–protein cross-link. Any of these reactions could explain both the mutagenic and cytotoxic effects of the nitrogen mustards. The generation of abasic sites also leads to interstrand cross-links in duplex DNA, especially if a guanine residue is nearby (Scheme 6.5).[135]

Mechlorethamine is quite unstable to hydrolysis. In fact, it is so reactive with water that it is marketed as a dry solid (HCl salt), and aqueous solutions are prepared immediately prior to injection; within minutes after administration, mechlorethamine reacts completely in the body. Because of this reactivity, a more stable analog was sought. Substitution of the methyl group of mechlorethamine with an electron withdrawing group, such as an aryl substituent (**6.27**), makes the nitrogen less nucleophilic, less able to participate in anchimeric assistance, and therefore slows down the rate of aziridinium ion formation (Scheme 6.3); this decreases the reactivity of the nitrogen mustard.[136] As a result of this stabilization, some of these compounds could be administered orally, and they would be able to undergo absorption and distribution before extensive alkylation occurred. Simple aryl substituted nitrogen mustards were not water soluble enough for intravenous administration, but the solubility problem was solved with the use of carboxylate-containing aryl substituents. Direct substitution on the phenyl with a carboxylate (**6.27**, R=CO_2H), however, gave a compound that was too stable because the nonbonded electrons of the nitrogen could be delocalized into the carboxylate carbonyl; this compound was not very active. To increase the reactivity, the electron-withdrawing effect of the carboxylate was attenuated by the insertion of methylenes

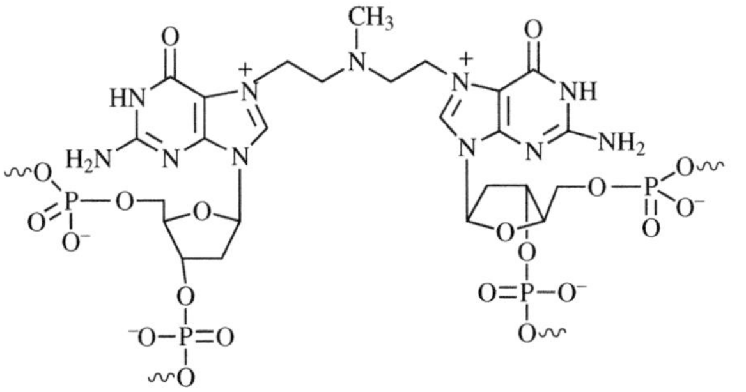

6.24

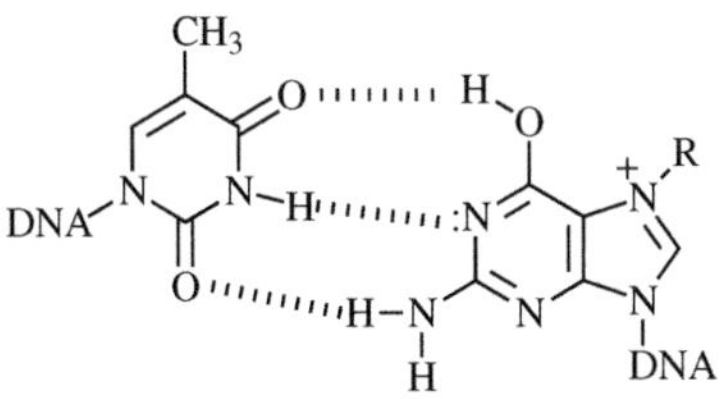

6.25

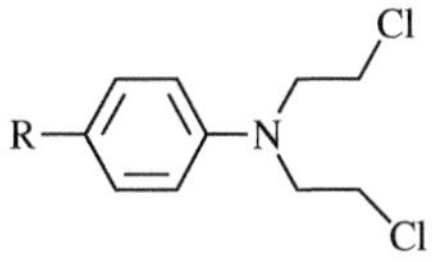

Chlorambucil (R = $(CH_2)_3CO_2H$)
Melphalan (R = $CH_2CH(NH_2)CO_2H$)
6.27

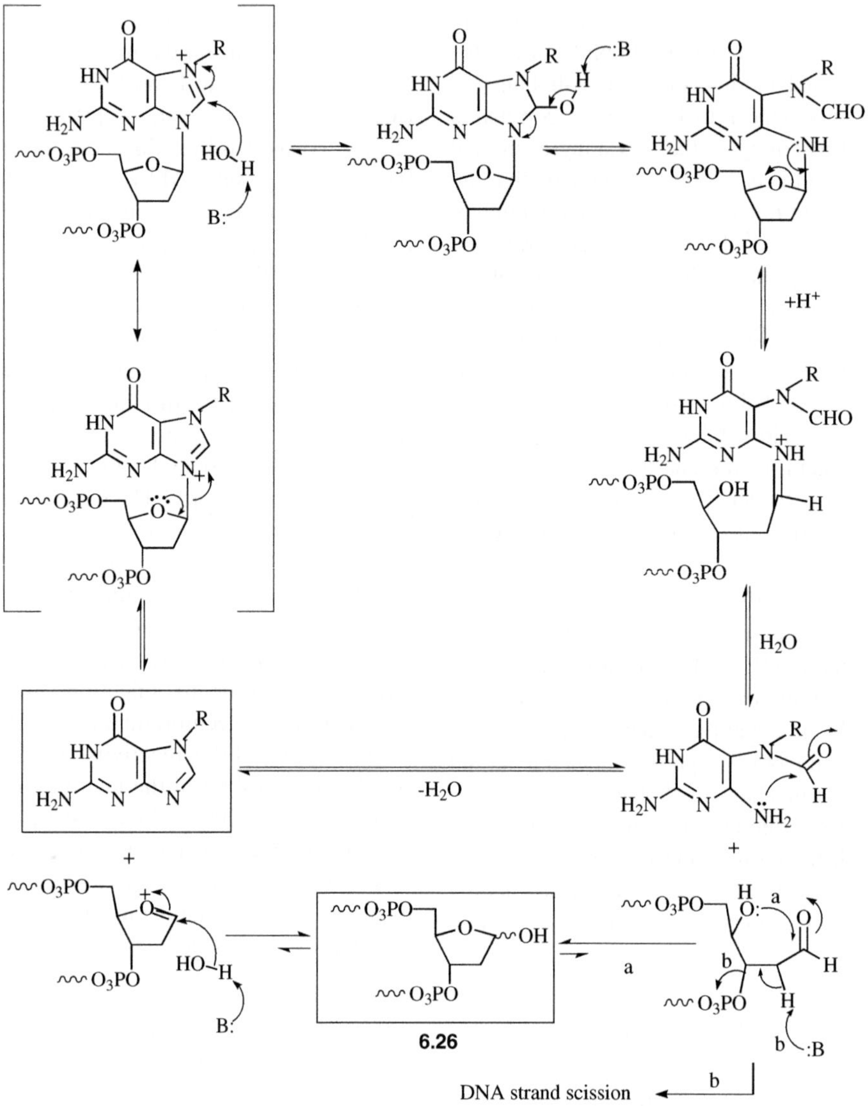

6.26

DNA strand scission

SCHEME 6.4 Depurination of N-7 alkylated guanines in DNA

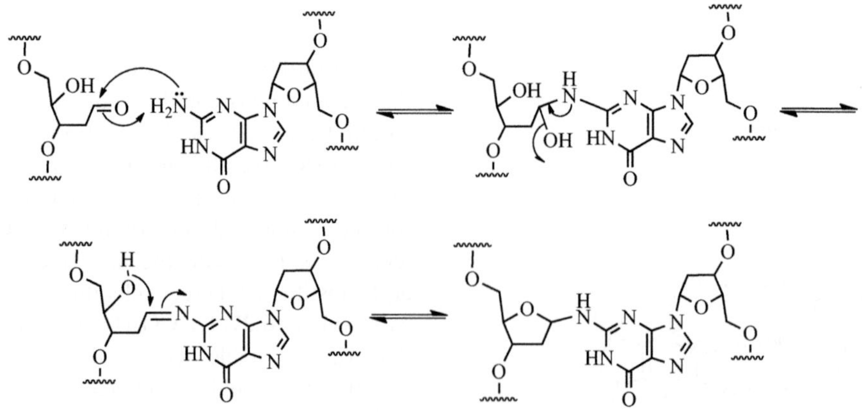

SCHEME 6.5 Interstrand cross-links of abasic sites in duplex DNA by reaction with guanine

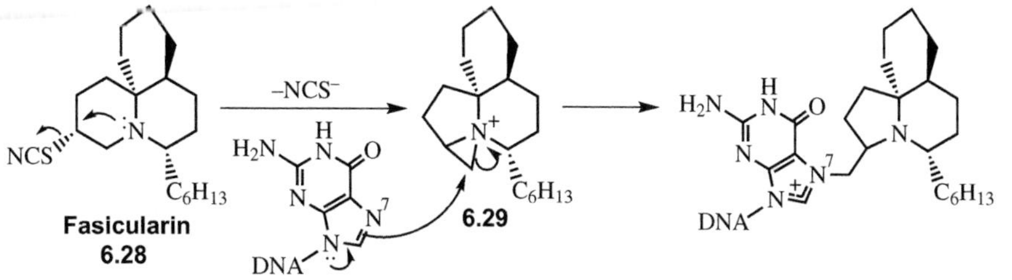

SCHEME 6.6 Proposed mechanism for DNA alkylation by fasicularin

between the phenyl and carboxylate groups. The optimal number of methylenes was found to be three, giving the antitumor drug chlorambucil (**6.27**, R = $(CH_2)_3CO_2H$; Leukeran).[137] This approach retained the water solubilizing effect of the carboxylate group, but the nitrogen lone pair could not be resonance stabilized, so its reactivity increased relative to **6.27**, R = CO_2H. Other nitrogen mustard analogs were prepared in an attempt to obtain an anticancer drug that would be targeted for a particular tissue. Because L-phenylalanine is a precursor to melanin, it was thought that L-phenylalanine nitrogen mustard (**6.27**, R = $CH_2CH(NH_2)CO_2H$; melphalan; Alkeran) might accumulate in melanomas. Although this analog is an effective, orally active anticancer drug (for multiple myelomas), it is not active against melanomas. The L-isomer (melphalan), the D-isomer (medphalan), and the racemic mixture (merphalan) have approximately equal potencies.[138] This lack of enantiospecificity suggests that there is no active transport of these compounds into cancer cells; however, a leucine carrier system appears to be involved in melphalan transport.[139]

An interesting variant of a nitrogen mustard is the natural product fasicularin (**6.28**, Scheme 6.6), an unusual thiocyanate containing alkaloid[140]; a mechanism was proposed that involved an intermediate aziridinium ion (**6.29**, Scheme 6.6).[141] Alternative oxidative mechanisms, e.g. by thiocyanate-derived radicals, were excluded, as was direct alkylation of DNA.

6.3.2.2. Ethylenimines

Because the reactive intermediate involved in DNA alkylation by nitrogen mustards (Scheme 6.3) is an aziridinium ion, an obvious extension of the nitrogen mustards is the use of aziridines (ethylenimines). Protonated ethylenimines are highly reactive (they are aziridinium ions), and would not be effective drugs. When electron-withdrawing groups are substituted on the aziridine nitrogen, however, the pK_a of the nitrogen is lowered to a point where the aziridine is not protonated at physiological pH. These aziridines are much less reactive. In general, two ethylenimine groups per molecule are required for antitumor activity to allow for cross-linking of DNA; compounds with three or four aziridines are not significantly more potent.[142] Examples of antitumor ethylenimines include triethylenemelamine (**6.30**), carboquone (**6.31**), and diaziquone (**6.32**). By appropriate addition of lipophilic substituents to the benzoquinone ethylenimines, antitumor activity in the central nervous system can be achieved.[143]

6.3.2.3. Methanesulfonates

Methanesulfonate is an excellent leaving group. The most prominent example of this class of alkylating agents is the bifunctional anticancer drug busulfan[144] (**6.33**, $n = 4$; Myleran).

Compounds with 1–8 methylene groups (**6.33**, $n = 1$–8) have antitumor activity, but maximum activity is obtained with four methylenes.[145] Alkylation of the N-7 position of guanine was demonstrated.[146] Unlike the nitrogen mustards (Section 6.3.2.1), however, intrastrand, not interstrand, cross-links form.[147]

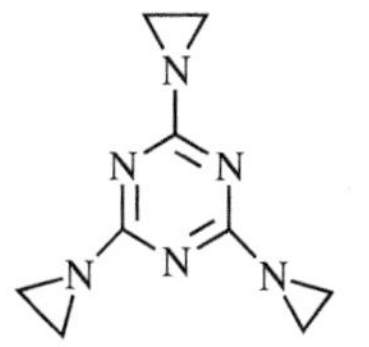

Triethylenemelamine
6.30

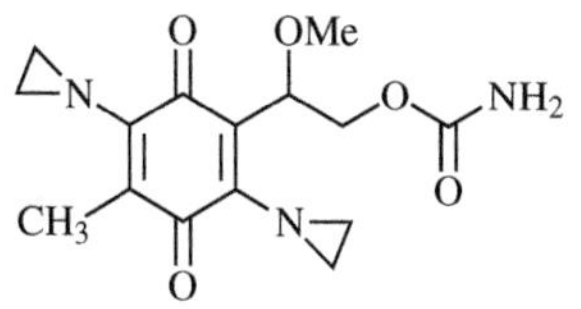

Carboquone
6.31

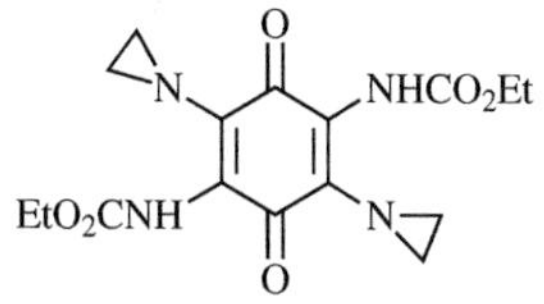

Diaziquone
6.32

$CH_3O_2SO—(CH_2)_n—OSO_2CH_3$

Busulfan (n = 4)
6.33

6.3.2.4. (+)-CC-1065 and Duocarmycins

(+)-CC-1065 (**6.34**),[148] (+)-duocarmycin A (**6.35**),[149] and (+)-duocarmycin SA (**6.36**),[150] natural products isolated from various strains of *Streptomyces*, are sequence-selective DNA alkylating agents.[151] Within each of these molecules is embedded a 4-spirocyclopropylcyclohexadienone (**6.37**), which is quite electrophilic because nucleophilic attack at the cyclopropane releases the electrons of the strained cyclopropane ring for delocalization into the enone system of the cyclohexadienone (Scheme 6.7).[152] However, all of these natural products are stable toward nucleophiles at neutral pH because they also contain a nitrogen atom that is conjugated with the enone system, thereby sharply decreasing its electrophilicity (Scheme 6.8). Nonetheless, these molecules alkylate the N-3 atom of adenine bases selectively within the AT-rich regions of the minor groove.[153] The reason for the selectivity of attachment appears to be because of a forced adoption of these molecules into helical conformations upon binding at the narrower, deeper AT-rich regions in the minor groove, which causes a greater degree of conformational change of the molecules. This conformational change twists the carbon–nitrogen bond of these molecules out of conjugation with the enone system, thereby decreasing the nitrogen stabilization of the enone; as a result, they become more like the spirocyclopropylcyclohexadienone system (**6.37**), which does not have a nitrogen in conjugation with the enone. Once the nitrogen is out of conjugation, nucleophilic attack at the cyclopropane is enhanced[154] (Scheme 6.9), and because this activation occurs most favorably in AT-rich regions of the minor groove, that is where alkylation selectively occurs. The presence of the nitrogen atom increases the stability of the spirocyclopropyl group 10^3–10^4 times at pH 7.[155] The rate

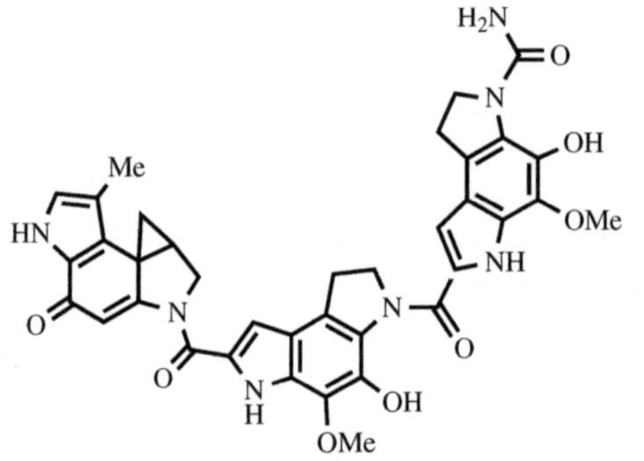

CC-1065
6.34

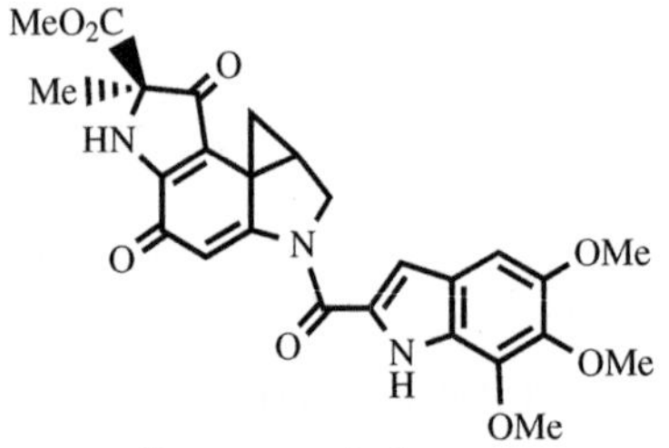

Duocarmycin A
6.35

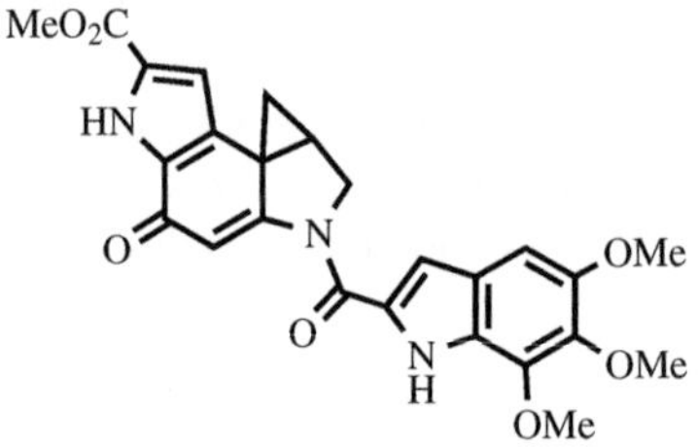

Duocarmycin SA
6.36

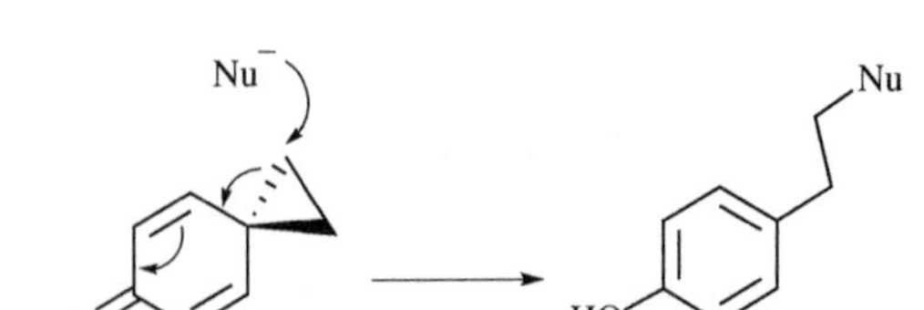

6.37

SCHEME 6.7 Reaction of nucleophiles with 4-spirocyclopropylcyclohexadienone

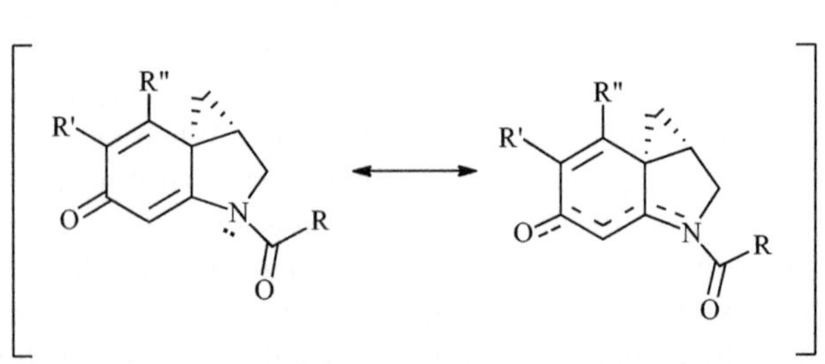

SCHEME 6.8 Stabilization of the spirocyclopropylcyclohexadienone by nitrogen conjugation

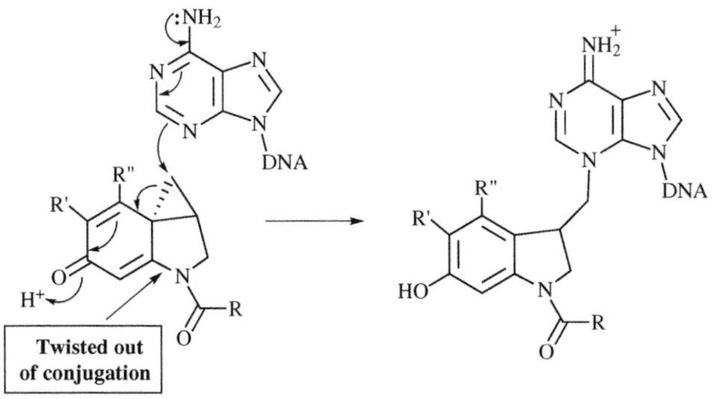

Twisted out of conjugation

SCHEME 6.9 N-3 adenine alkylation by CC-1065 and related compounds

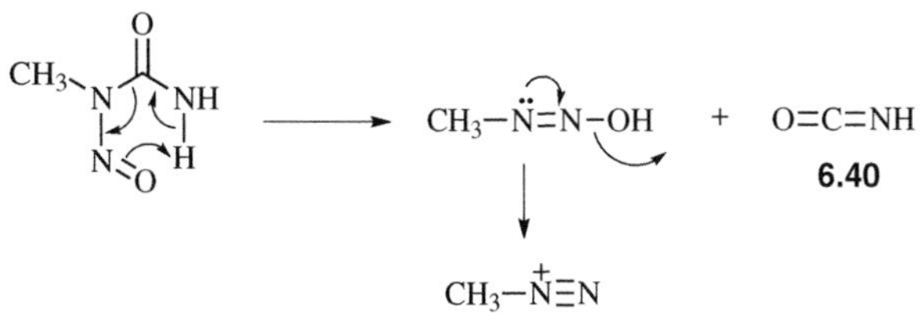

of DNA alkylation changes by less than a factor of two over a physiologically relevant range spanning two pH units.[156] This indicates that the alkylation reaction is not acid catalyzed, supporting an S_N2 type reaction.

Therefore, these compounds start out as relatively unreactive minor groove binders, but as a result of binding at a particular region (AT-rich regions), a conformational change activates these molecules for alkylation. This is the DNA equivalent of mechanism-based enzyme inactivation discussed in Chapter 5 (Section 5.3.3).[157] This class of alkylating agents is activated by binding to DNA, but there are many DNA alkylating agents that are stable until some biological species, such as one or more enzymes or a thiol, converts them into alkylating agents. Those compounds are called *metabolically activated alkylating agents.*

6.3.2.5. Metabolically Activated Alkylating Agents

6.3.2.5.1. Nitrosoureas

The nitrosoureas (**6.38**) were developed from the lead compound *N*-methylnitrosourea (**6.38**, R=CH_3, R′=H), which

$CH_3-\overset{+}{N}\equiv N$ **6.39**; O=C=NH **6.40**

SCHEME 6.10 Decomposition of *N*-methyl-*N*-nitrosourea

exhibited modest antitumor activity in animal tumor models.[158] Analogs with 2-chloroethyl substituents, such as carmustine (**6.38**, R=R′=CH_2CH_2Cl; BCNU; Gliadel) and lomustine (**6.38**, R=CH_2CH_2Cl, R′=cyclohexyl; CCNU; CeeNU), were found to possess much greater antitumor activity.[159] Because of their lipophilicity, the 2-chloroethyl analogs were able to cross the blood–brain barrier, and, consequently, have been used in the treatment of brain tumors. Despite the potency of these antitumor drugs, they are less desirable than others because of severe problems of delayed and cumulative bone marrow toxicity.

Extensive mechanistic studies have been carried out on nitrosoureas. Decomposition of the first active anticancer nitrosourea, **6.38** (R=CH_3, R′=H), produces methyl diazonium ion (**6.39**, Scheme 6.10), a potent methylating agent, and the ketenimine isocyanic acid (**6.40**), a carbamoylating agent.[160] Evidence that diazomethane is not the alkylating agent was provided by a model study[161] showing that under physiological conditions, 1-trideuteriomethyl-3-nitro-1-nitrosoguanidine (**6.41**, Scheme 6.11) alkylates nucleophiles with the trideuteriomethyl group intact; if diazomethane were the alkylating agent, dideuteriomethyl groups would have resulted. It is now known that *N*-nitrosamides (**6.42**) and *N*-nitrosourethanes (**6.43**), which produce alkylating, but not carbamoylating species, do have anticancer activity.[162] Furthermore, certain nitrosoureas that have little carbamoylating activity also are quite active, but nitrosoureas with no

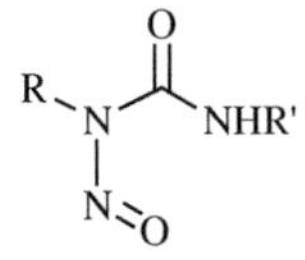

Carmustine (R = R′ = CH_2CH_2Cl)
Lomustine (R = CH_2CH_2Cl, R′ = cyclohexyl)
6.38

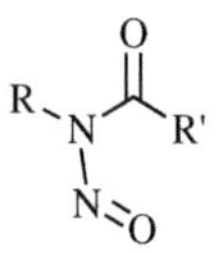

6.42 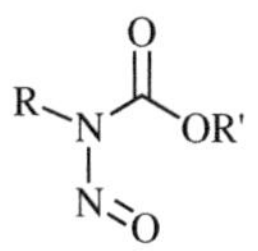 **6.43**

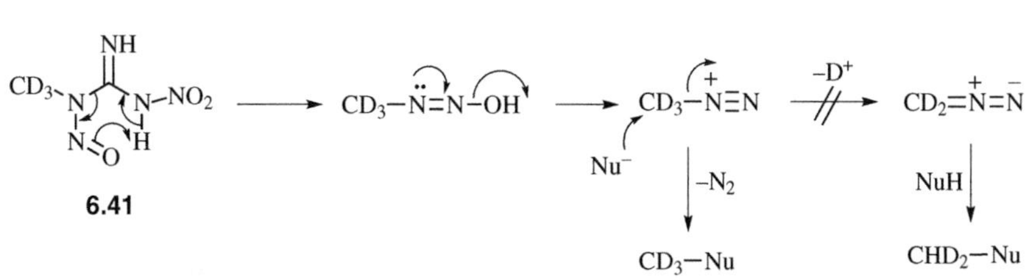

6.41

SCHEME 6.11 Deuterium labeling experiment to determine mechanism of activation of nitrosoureas detectable alkylating activity are either very weakly active or inactive. Therefore, the alkylating (diazonium), not carbamoylating (ketenimine), product appears to be the principal species responsible for the anticancer activity.

The carbamoylating isocyanate that is generated (e.g., **6.40** in Scheme 6.10) does not appear to be directly involved in the antitumor effects of nitrosoureas, but it does react with amines in proteins.[163] More importantly, it inhibits DNA polymerase[164] and other enzymes involved in the repair of DNA lesions,[165] such as O^6-alkylguanine-DNA alkyltransferase,[166] DNA nucleotidyl transferase, and DNA glycosylases.[167] It also inhibits RNA synthesis and processing[168] and plays a role in the toxicity of the nitrosoureas.[169]

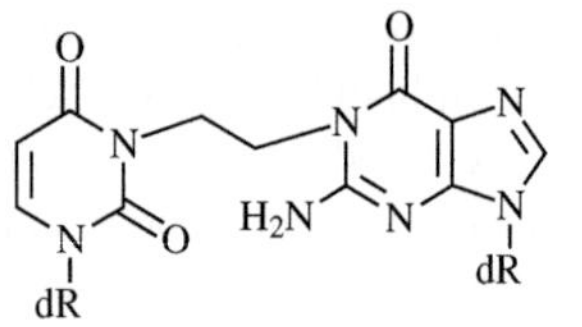

6.44

The 2-chloroethyl-substituted analogs (**6.38**, $R=CH_2CH_2Cl$) react with DNA and produce an interstrand cross-link between a guanine on one strand and a cytosine residue on another.[170] 1-[N^3-Deoxycytidyl]-2-[N^1-deoxyguanosinyl]ethane (**6.44**) was isolated from the reaction of *N,N'*-bis(2-chloroethyl)-N-nitrosourea (carmustine; **6.38**, $R=R'=CH_2CH_2Cl$). Because the same cross-link occurs with the mono-2-chloroethyl-substituted analog **6.38** ($R=CH_2CH_2Cl$, $R'=$cyclohexyl), the mechanism shown in Scheme 6.12 was proposed.

The same mechanism was proposed for fotemustine (**6.48**; Muphoran), and the reactive 1-(2-chloroethyl)-2-hydroxydiazene (**6.45**) was detected by electrospray ionization mass spectrometry.[171] To rationalize the regioselectivity of these alkylating agents, a kinetic analysis of the reaction was carried out, and an alternative reaction mechanism was suggested[172] (Scheme 6.13). The principal difference in these mechanisms is that in Scheme 6.13 a nucleoside on

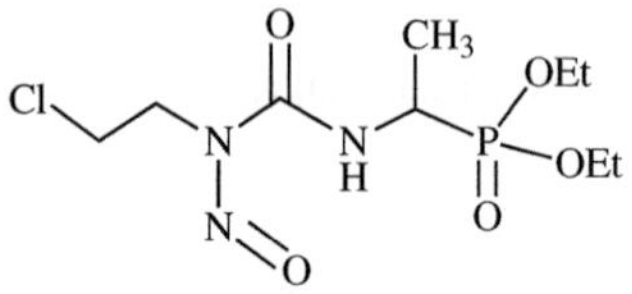

Fotemustine
6.48

the DNA reacts with the intact drug to form a tetrahedral intermediate (**6.49**), which after cyclization to **6.50**, undergoes reaction with the O-6 of a guanine to give **6.51**, the precursor to the diazonium ion that leads to interstrand cross-linking (Scheme 6.12). Evidence for the cyclization of **6.46** to **6.47** (Scheme 6.12) and nucleophilic attack to give interstrand cross-linking comes from model chemistry for this reaction.[173]

Evidence for the intermediacy of an O-6 guanine adduct such as **6.46** (Scheme 6.12) or **6.51** (Scheme 6.13) is based on the observation that cell lines capable of excising O-6 guanine adducts were resistant to cross-link formation[174] and that the addition of rat liver O^6-alkylguanine-DNA alkyltransferase,

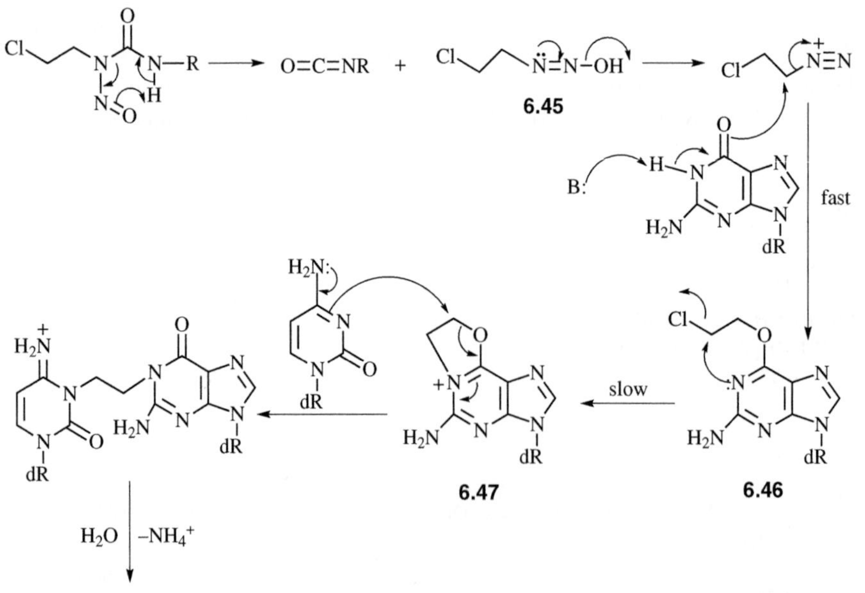

SCHEME 6.12 Mechanism proposed for cross-linking of DNA by (2-chloroethyl)nitrosoureas

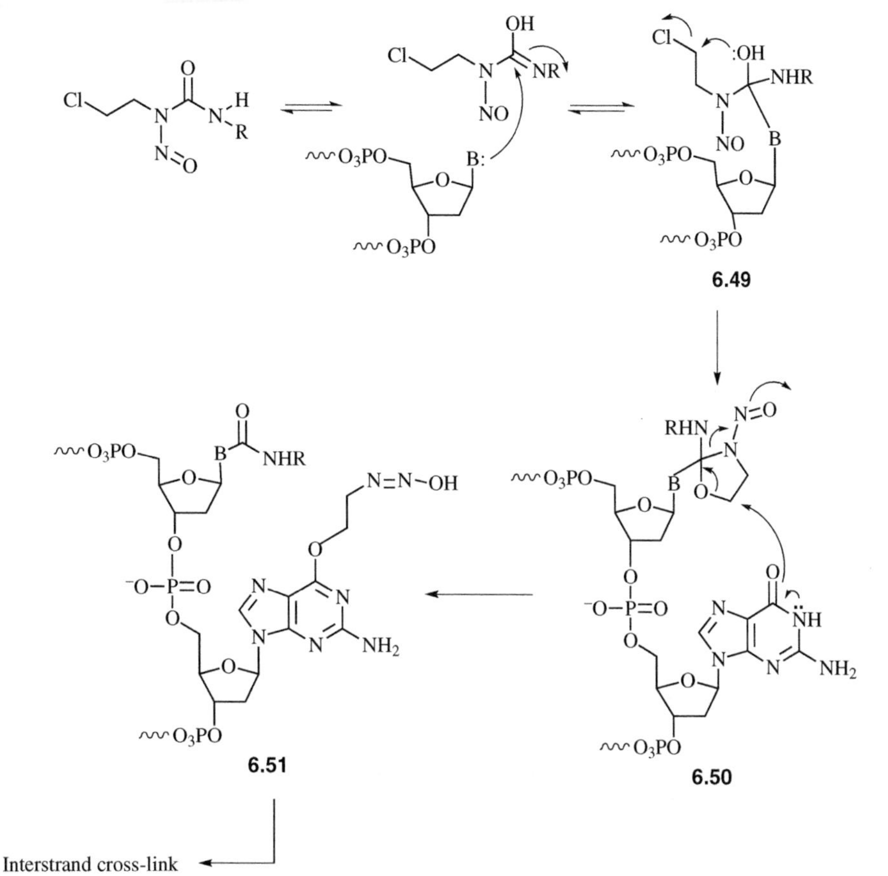

SCHEME 6.13 Alternative mechanism for the cross-linking of DNA by (2-chloroethyl)nitrosoureas

the enzyme that excises O-6 guanine adducts, prevents formation of the cross-links.[175] *O*^{6}-Alkylguanine adducts are among the most important DNA modifications that are responsible for the induction of cancer, mutation, and cell death.[176]

6.3.2.5.2. Triazene Antitumor Drugs

Other antitumor drugs are metabolically converted into the methyldiazonium ion, which methylates DNA. An example is the triazenoimidazoles, such as 5-(3,3-dimethyl-1-triazenyl)-1*H*-imidazole-4-carboxamide (**6.52**, Scheme 6.14 dacarbazine; DTIC-Dome), which is active against a broad range of cancers but is used preferably for the treatment of melanotic melanoma.[177] Although dacarbazine is a structural analog of 5-aminoimidazole-4-carboxamide, an intermediate in purine biosynthesis, the cytotoxicity of **6.52** is a result of its conversion into an alkylating agent, not its structural similarity to the metabolic intermediate. With the use of [^{14}C-*methyl*]dacarbazine, it was shown that formaldehyde is generated and that the DNA becomes methylated at the 7-position of guanine.[178] A mechanism that rationalizes these results is shown in Scheme 6.14.

6.3.2.5.3. Mitomycin C

Bioreductive alkylation is a process by which an inactive compound is metabolically reduced to an alkylating agent.[179] The prototype for antitumor antibiotics that act as bioreductive alkylating agents of DNA is mitomycin C (**6.53**, Scheme 6.15; Mutamycin) which contains three important carcinostatic functional groups—the quinone, the aziridine, and the carbamate group.[180] The mechanism proposed by Iyer and Szybalski[181] is shown in Scheme 6.15. Reduction of the quinone (a common initiation step for bioreductive alkylation) by one-electron reductants, such as cytochrome c reductase or cytochrome b_5 reductase, to the semiquinone (**6.54**, R = electron) or by the more common two-electron reductant, such as DT-diaphorase,[182] to the hydroquinone (**6.54**, R = H) converts the heterocyclic nitrogen from a vinylogous amide nitrogen (the nonbonded electrons of the nitrogen are in conjugation with the quinone carbonyl via one of the quinone double bonds), which is not nucleophilic, into an amine nitrogen, which can promote elimination of the β-methoxide ion (**6.54**). Tautomerization of the resultant immonium ion (**6.55**) gives **6.56**, which is set up for aziridine ring opening. This activates the drug by unmasking the electrophilic site at C-1, which

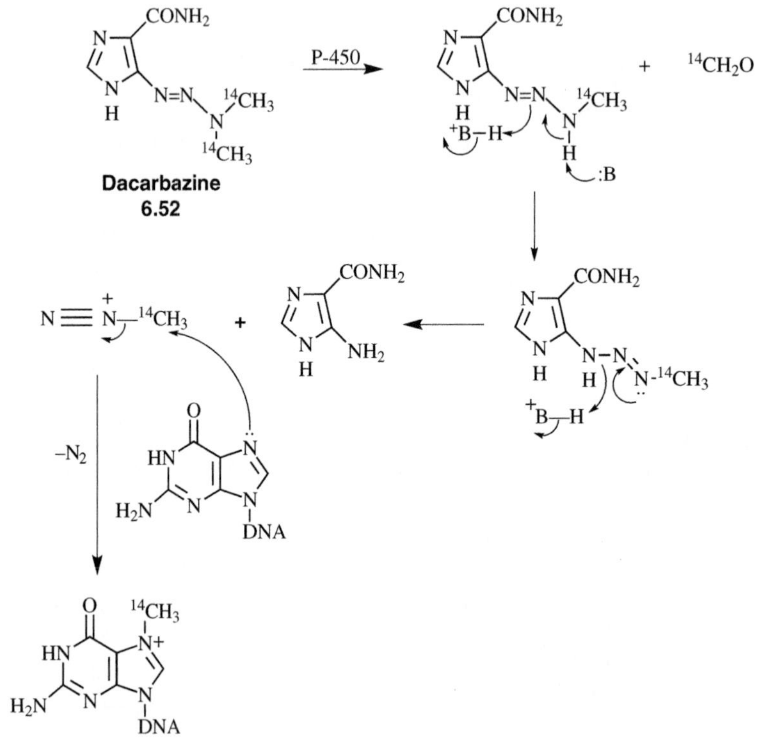

SCHEME 6.14 Mechanism for the methylation of DNA by dacarbazine

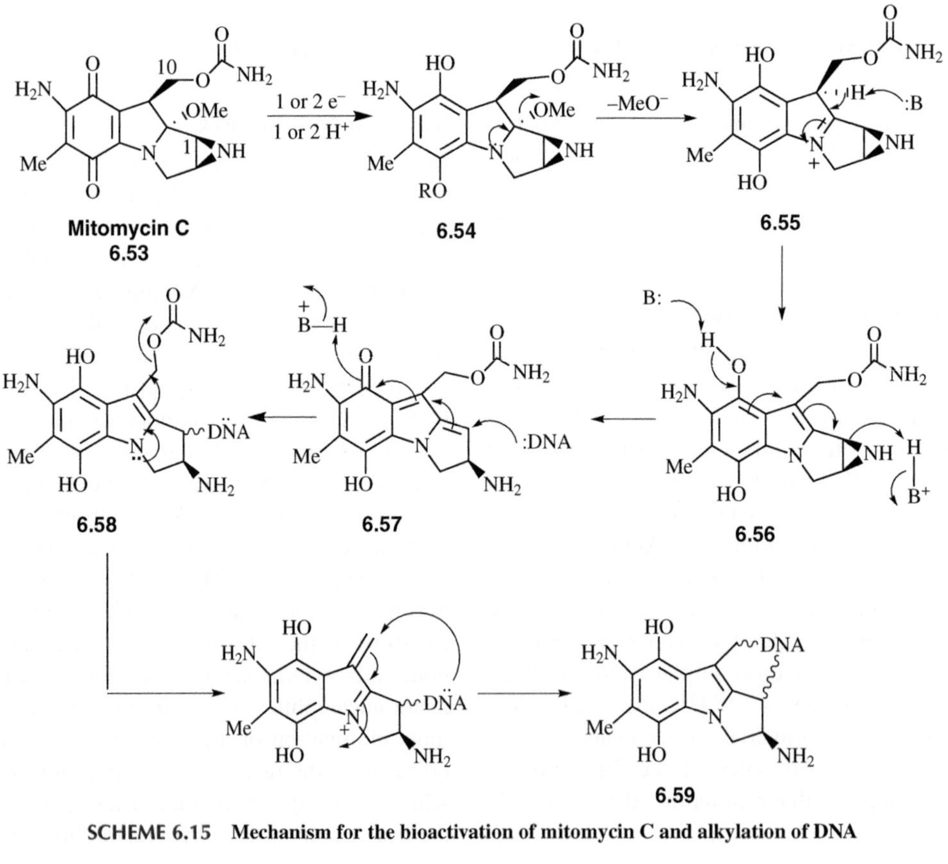

SCHEME 6.15 Mechanism for the bioactivation of mitomycin C and alkylation of DNA

alkylates the DNA (**6.57**). A subsequent reaction at C-10 (**6.58**) results in the cross-linking of the DNA (**6.59**). Bean and Kohn[183] showed in chemical models that nucleophiles react most rapidly at C-1; the reaction at C-10 to displace the carbamate also occurs, but at a slower rate. Reduction of the quinone is necessary for the covalent reaction of **6.53** to DNA, but controversy exists as to whether the semiquinone (**6.54**, R=electron) or hydroquinone (**6.54**, R=H) is the viable intermediate.[184] Chemical model studies on the mechanism of action of mitomycin C indicate that the conversion of **6.53** to **6.58** can occur at the semiquinone stage,[185] and the conversion of **6.58** to **6.59** occurs at the hydroquinone oxidation state.[186] Both monoalkylated[187] and *bis*-alkylated DNA adducts have been identified; the extent of mono- and *bis*-alkylation increases with increasing guanine base composition of the DNA.[188] Cross-links in the minor groove between two guanines at their C-2 amino groups form[189] with preferential interstrand cross-linking at 5′-CG rather than 5′-GC sequences.[190]

The bioreductive alkylation approach also was directed toward the design of new antineoplastic agents that may be selective for hypoxic (O_2-deficient) cells in solid tumors.[191] These cells are remote from blood vessels and are located at the center of the solid tumors. Hypoxia protects the tumor cells from radiation therapy, and because these cells are buried deep inside the tumor, appropriate concentrations of antitumor drugs may not reach them prior to drug metabolism. Because these cells might have a more efficient reducing environment, bioreductive alkylation seemed to be well suited. The bioreductive alkylation approach based on reduction of a quinone to the corresponding hydroquinone was utilized in the design of both mono- (Scheme 6.16)[192] and *bis*-alkylating agents (Scheme 6.17).[193] Electron-rich substituents lower the reduction potential of the quinones and make them more reactive.[194]

6.3.2.5.4. Leinamycin

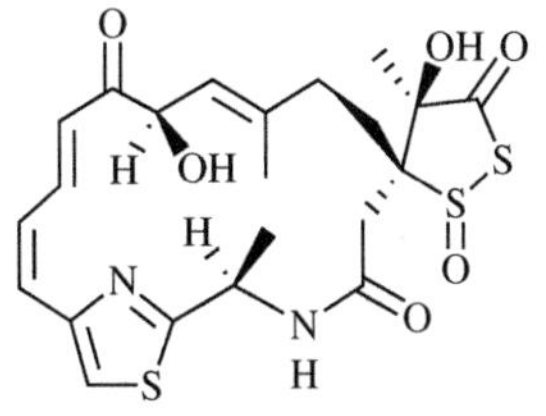

Leinamycin
6.60

Leinamycin (**6.60**) is a potent antitumor agent that was isolated from a strain of *Streptomyces*, and its structure was elucidated by spectroscopy and X-ray crystallography.[195] The cytotoxic activity of this compound is initiated by a reaction with thiols.[196] Gates and coworkers,[197] using simple chemical model compounds containing 1,2-dithiolan-3-one 1-oxide groups (the

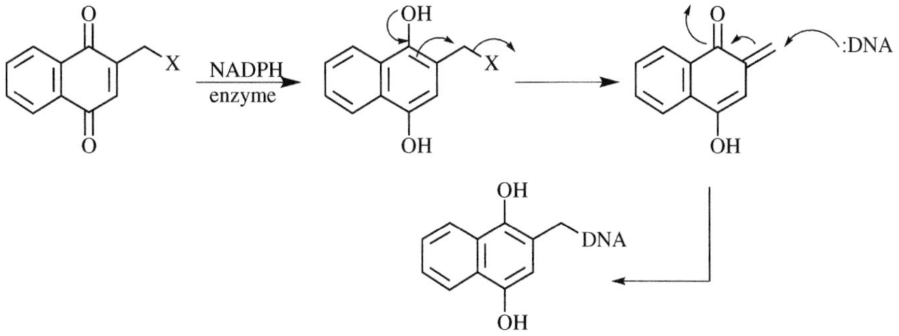

SCHEME 6.16 **Bioreductive monoalkylating agents**

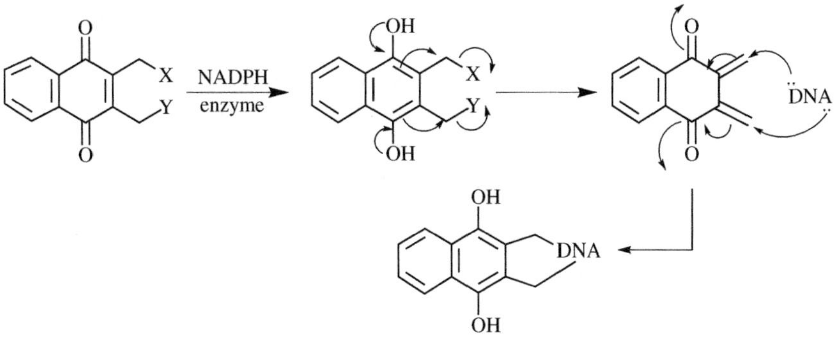

SCHEME 6.17 **Bioreductive *bis*-alkylating agents**

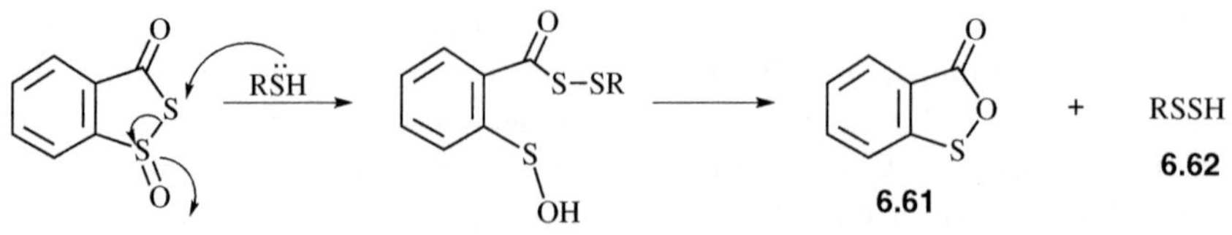

SCHEME 6.18 **Model reaction for the mechanism of activation of leinamycin**

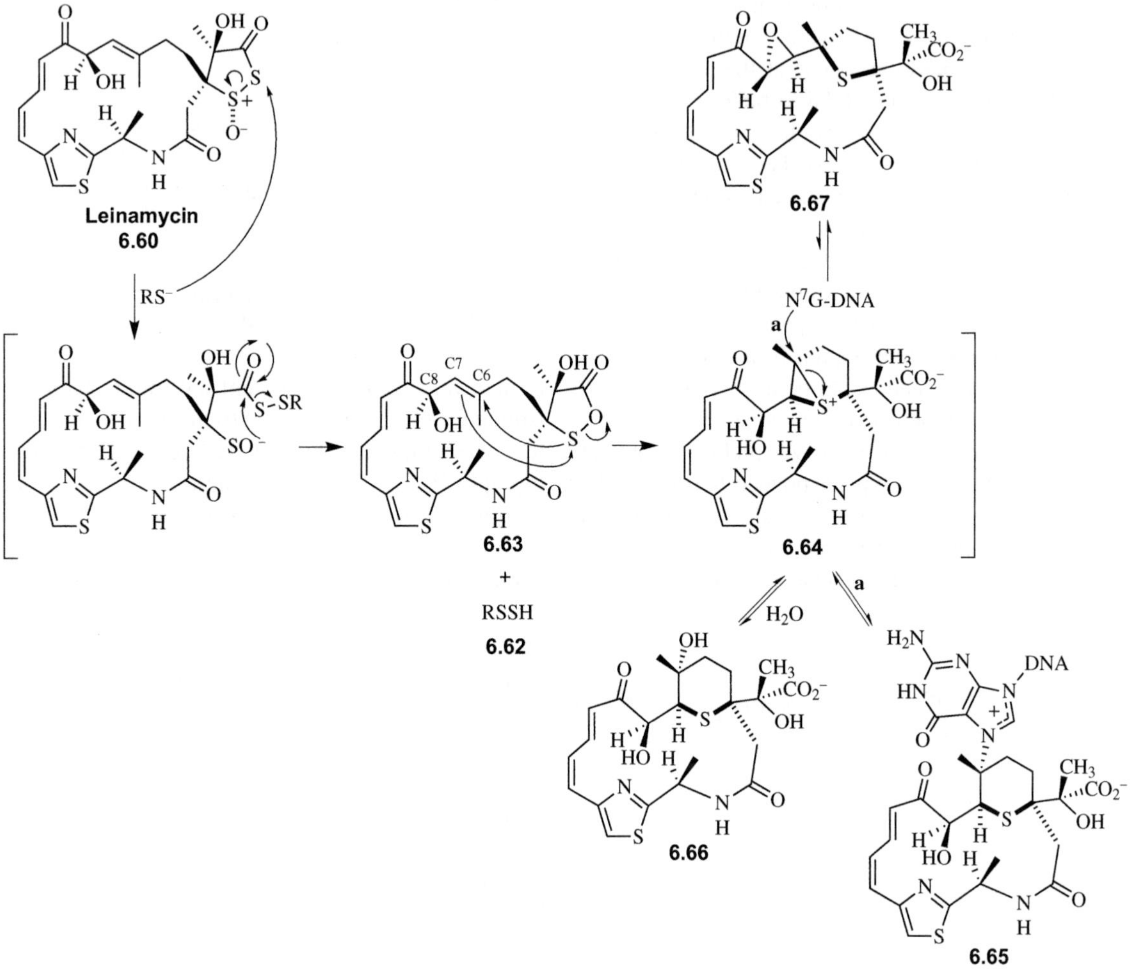

SCHEME 6.19 **Mechanism for DNA alkylation by leinamycin**

five-membered ring heterocycle found in leinamycin), proposed that two of the intermediates generated by thiol addition to these model compounds are an oxathiolanone (**6.61**) and a hydrodisulfide (**6.62**). These intermediates would arise from thiol attack at the 1,2-dithiolan-3-one 1-oxide functional group to give a sulfenic acid intermediate, which cyclizes to the oxathiolanone (Scheme 6.18).[198] Asai and coworkers[199] elegantly proved that the product of thiol addition to leinamycin in the presence of calf thymus DNA was an alkylated N-7 guanine adduct (**6.65**, Scheme 6.19), proposed to be derived from thiol attack on the 1,2-dithiolan-3-one 1-oxide to give oxathiolanone **6.63** (based on the above-cited model studies of Gates and coworkers), which rearranges to an episulfonium intermediate (**6.64**), a highly electrophilic species that alkylates the DNA. Although the episulfonium intermediate was too reactive to detect, epoxide **6.67** could be isolated and characterized and was shown to alkylate DNA; however, the product obtained is the one derived from attack on the episulfonium intermediate (**6.64**), indicating that the epoxide does not directly alkylate DNA, but is in equilibrium with **6.64**, which is the actual alkylating species.[200] In model studies, hydrolyzed leinamycin (**6.66**) from the leinamycin-modified DNA was observed, suggesting that leinamycin may be a reversible DNA alkylating agent.[201]

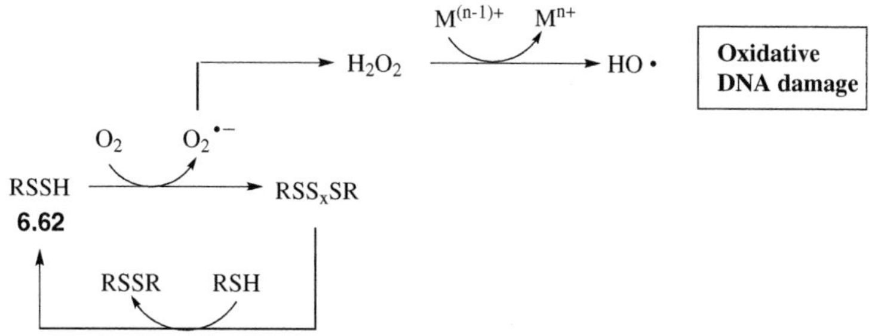

$M^{(n-1)+}$ M^{n+}

H_2O_2 → $HO\cdot$ → Oxidative DNA damage

O_2 $O_2^{\bullet-}$

RSSH → RSS_xSR

6.62

RSSR RSH

SCHEME 6.20 **Mechanism for hydrodisulfide activation of molecular oxygen to cause oxidative DNA damage**

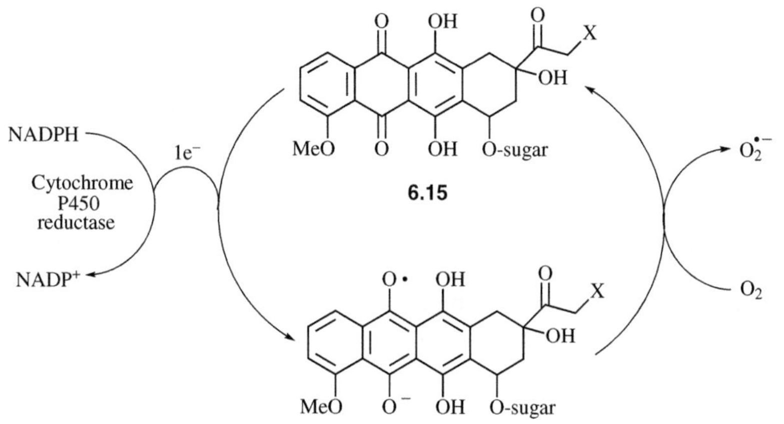

NADPH, Cytochrome P450 reductase, $NADP^+$, $1e^-$, 6.15, $O_2^{\bullet-}$, O_2, 6.68

SCHEME 6.21 **Electron transfer mechanism for DNA damage by anthracyclines**

This alkylation mechanism is not the only way that leinamycin modifies DNA. The other pathway involves further reactions of hydrodisulfide **6.62**, which can be oxidized by O_2 to give polysulfides, shown[202] to cause further thiol-dependent oxidative DNA damage, presumably via radical intermediates such as hydroxyl radicals (Scheme 6.20). This radical-induced DNA damage characterizes the third class of DNA-interactive agents, the DNA strand breakers (next section).

A major problem that limits the effectiveness of alkylating agents in general is resistance in tumor cells. Possible mechanisms based on in vitro or animal models are discussed in Chapter 7.

6.3.3. DNA Strand Breakers

Some DNA-interactive drugs initially intercalate into DNA, but then, under certain conditions, react in such a way as to generate radicals. These radicals typically abstract hydrogen atoms from the DNA sugar-phosphate backbone or from the DNA bases, leading to DNA strand scission. Therefore, these DNA-interactive compounds are metabolically activated radical generators. As examples of this mode of action of DNA-interactive drugs, we will consider the anthracycline antitumor antibiotics, bleomycin, tirapazamine (this one does not initially intercalate), and the enediyne antitumor antibiotics. It should be kept in mind that some of the compounds that lead to strand breakage act via the topoisomerase-induced mechanisms discussed in Section 6.3.1.3.

6.3.3.1. Anthracycline Antitumor Antibiotics

Doxorubicin (**6.15**, X=OH) and daunorubicin (**6.15**, X=H) are anthracyclines that were discussed in the section on DNA intercalators (Section 6.3.1.3); however, these drugs also cause an oxygen-dependent DNA damage.[203] Several mechanisms have been proposed to account for this destruction of DNA. Anthracyclines cause protein-associated breaks that correlate with their cytotoxicity,[204] and these breaks may be caused by the reaction of anthracyclines on topoisomerase II, an enzyme that promotes DNA strand cleavage and reannealing (Section 6.2.3).

Another mechanism for DNA damage involves electron transfer chemistry. A one-electron reduction of the anthracyclines, probably catalyzed by flavoenzymes such as NADPH cytochrome P-450 reductase,[205] produces the anthracycline semiquinone radical (**6.68**, Scheme 6.21), which can transfer an electron to molecular oxygen to regenerate the anthracycline and produce superoxide ($O_2\cdot^-$). Both the superoxide and anthracycline semiquinone radical anions can generate hydroxyl radical ($HO\cdot$) (Scheme 6.22), which is known to cause DNA strand breaks.[206]

$$2\,O_2^{\bullet -} + 2\,H^+ \longrightarrow H_2O_2 + O_2$$

$$O_2^{\bullet -} + H_2O_2 \xrightarrow{\text{slow}} HO\bullet + HO^- + O_2$$

or

$$H_2O_2 + \mathbf{6.68} \longrightarrow \mathbf{6.15} + HO\bullet + HO^-$$

or

$$O_2^{\bullet -} + Fe^{3+}\bullet\mathbf{6.15} \longrightarrow O_2 + Fe^{2+}\bullet\mathbf{6.15}$$

$$Fe^{2+}\bullet\mathbf{6.15} + H_2O_2 \longrightarrow Fe^{3+}\bullet\mathbf{6.15} + HO\bullet + HO^-$$

$$HO\bullet + \text{DNA} \longrightarrow \text{strand scission}$$

SCHEME 6.22 **Anthracycline semiquinone generation of hydroxyl radicals**

A third possibility for the mechanism of DNA damage by anthracyclines is the formation of a ferric complex, which binds to DNA by a mechanism different from intercalation and significantly tighter.[207] The binding constant for the doxorubicin–ferric complex (**6.69**) is 10^{33}.

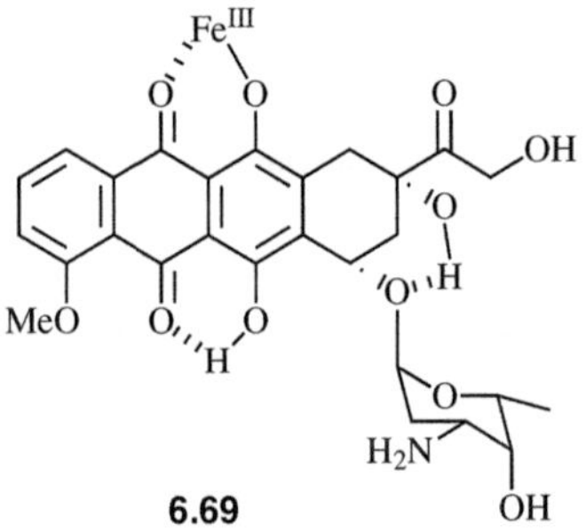

6.69

This ferric complex could react with superoxide to give oxygen and the corresponding ferrous complex (Scheme 6.22). The reaction of ferrous ions with hydrogen peroxide is known as the *Fenton reaction*,[208] which is used in the standard method for the generation of hydroxyl radicals (not Fe(IV)-oxo species, as some had suspected[209]) when doing *DNA footprinting*, a technique that indiscriminately cleaves the DNA to determine where protein–DNA interactions occur.[210]

Because the generation of the hydroxyl radicals occurs adjacent to DNA, it is unlikely that radical scavengers would be an effective method of cell protection, as was shown with anthracycline antibiotic-induced cardiac toxicity. However, an iron chelator (**6.70**) does prevent doxorubicin-induced cardiac toxicity in humans.[211] This iron chelator actually is a prodrug (Chapter 9) that, because of its nonpolar nature, enters the cell. Once inside the cell it is hydrolyzed to the active iron chelator **6.71** (Scheme 6.23), which is structurally related to the well-known iron chelator EDTA.

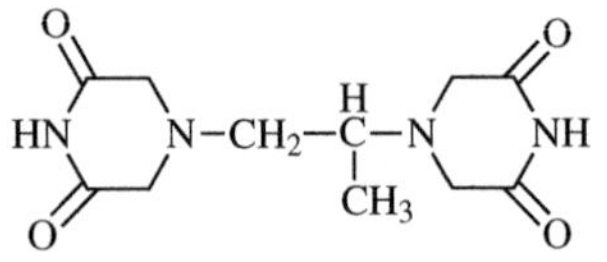

6.70

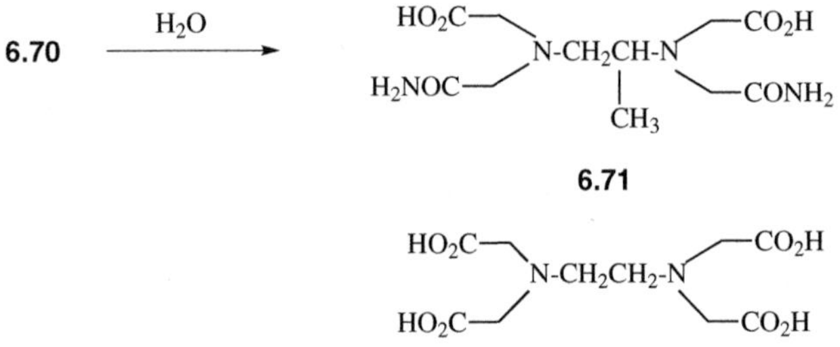

6.70 $\xrightarrow{H_2O}$ **6.71** + **EDTA**

SCHEME 6.23 **Conversion of iron chelator prodrug 6.70 into iron chelator 6.71**

Gianni et al.[212] showed that the iron–doxorubicin (**6.15**, X=OH) complex is more reactive than the iron–daunorubicin (**6.15**, X=H) complex because the hydroxymethyl ketone side chain of doxorubicin reacts spontaneously with iron to produce Fe^{2+}, $HO\cdot$, and H_2O_2.

6.3.3.2. Bleomycin

The anticancer drug bleomycin (**6.72**; BLM; Blenoxane) is actually a mixture of several glycopeptide antibiotics isolated from a strain of the fungus *Streptomyces verticellus*; the major component is bleomycin A_2 (**6.72**).[213] It cleaves double-stranded DNA selectively at 5′-GC and 5′-GT sites in the minor groove by a process that is both metal ion and oxygen dependent.[214] There are three principal domains in bleomycin: a domain that targets cancer cells, a domain that intercalates into the cancer cell DNA,

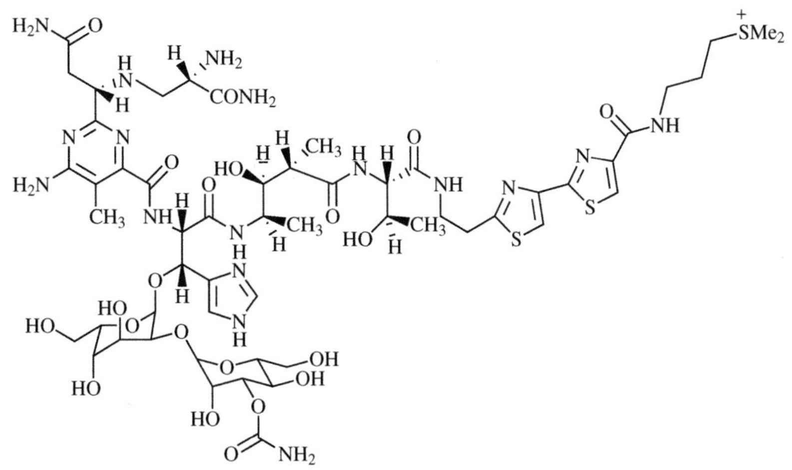

Bleomycin
6.72

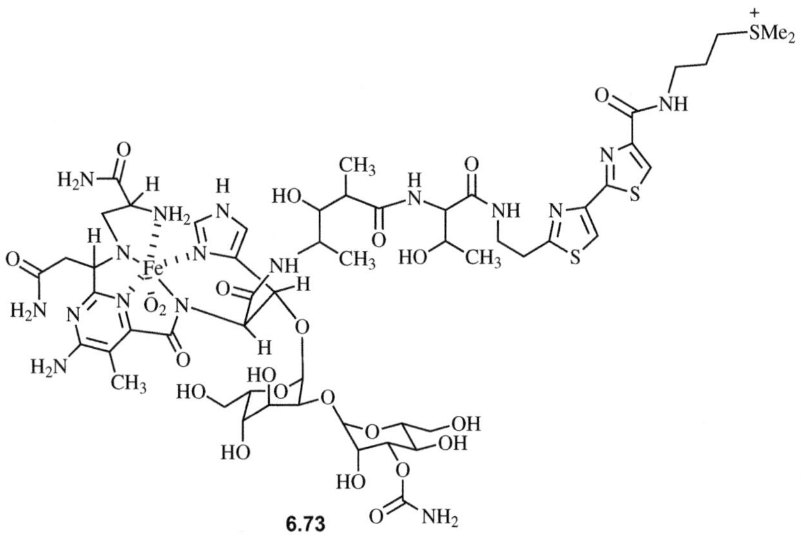

6.73

and a domain that damages the cancer cell DNA.[215] The targeting domain could be the gulose and carbamoylated mannose disaccharide moiety, as cancer cells have a strong requirement for carbohydrates (as well as proteins). The intercalating domain is composed of the bithiazole moiety (the five-membered N and S heterocycles) and the attached sulfonium ion-containing side chain. The bithiazole is important for sequence selectivity, presumably because of its intercalation properties with DNA;[216] possibly, the sulfonium ion is attracted electrostatically to a sugar-phosphate group.[217] The pyrimidine, the β-aminoalanine, and the α-alkoxyimidazole moieties make up the DNA-damaging domain. Bleomycin forms a stable complex with iron (II), which interacts with O_2 to give a ternary complex (**6.73**), believed to be responsible for the DNA cleaving activity.[218] The primary mechanism of action of bleomycin is the generation of single- and double-strand breaks in DNA. This results from the production of radicals by a 1:1:1 ternary complex of bleomycin, Fe(II), and O_2 (Scheme 6.24). Activation of this ternary complex may be self-initiated by the transfer of an electron from a second unit of the ternary complex or activation may be initiated by a microsomal NAD(P)H-cytochrome P450 reductase-catalyzed reduction.[219] The activated bleomycin binds tightly to guanine bases in DNA, principally via the tripeptide (called the *tripeptide S*) containing the bithiazole unit.[220] Binding by the bithiazole to G–T and G–C sequences is favored. Evidence for intercalation comes from the observation that when bleomycin, its tripeptide S moiety, or just bithiazole is mixed with DNA, a lengthening of the linear DNA and an unwinding of supercoiled circular DNA result.[221]

Activation of the bound ternary complex is believed to occur by a reaction related to that for heme-dependent enzyme activation (Chapter 4, Section 4.3.4) because it catalyzes the same reactions that are observed with these

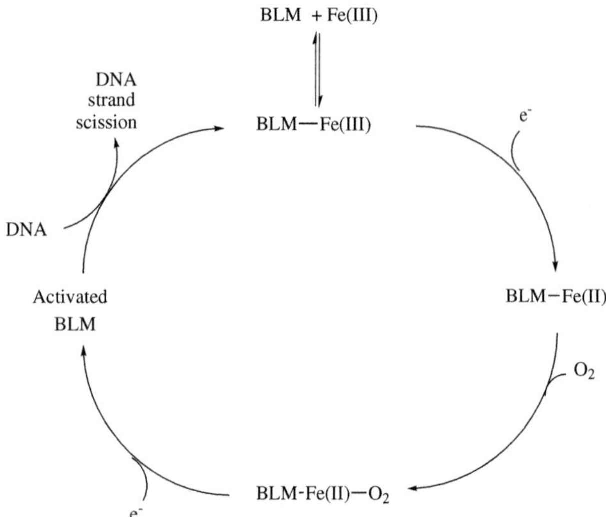

SCHEME 6.24 Cycle of events involved in DNA cleavage by bleomycin (BLM)

enzymes. As in the case of heme-dependent reactions, addition of an electron to the bleomycin–Fe(II)–O_2 complex would give a bleomycin–Fe(III)–OOH complex. This ferric hydroperoxo complex has been detected by UV–visible spectroscopy,[222] Electron paramagnetic resonance (EPR) spectrometry,[223] Mössbauer spectroscopy,[224] electrospray ionization mass spectrometry,[225] and X-ray absorption spectroscopy.[226] There are at least three possible mechanisms for the formation of an activated bleomycin from the bleomycin–Fe(III)–OOH complex that leads to DNA strand scission: (1) analogous to the case of the heme-dependent enzymes, addition of two protons would activate the O–O bond for heterolytic cleavage, giving a bleomycin–Fe(V)=O species, which could abstract a hydrogen atom from DNA; (2) the O–O bond could break homolytically to give a bleomycin–Fe(IV)=O species and hydroxyl radical, both of which could abstract a hydrogen atom from DNA; (3) Solomon and coworkers[227] favor a mechanism in which the bleomycin–Fe(III)–OOH complex undergoes a concerted reaction with DNA with concomitant O–O bond homolysis to give Fe(IV)=O, water, and a DNA radical. There is no experimental evidence that differentiates these, but density-functional theory calculations[228] show that heterolysis of the O–O bond of bleomycin–Fe(III)–OOH (mechanism 1) is predicted to be unfavorable by at least 40 kcal/mol, which is more than 150 kcal/mol less favorable than heterolysis of heme–OOH; direct reaction of bleomycin–Fe(III)–OOH with DNA (mechanism 3), however, is predicted to be close to thermoneutral. Mechanism 2 could be questioned because hydroxyl radicals are less selective than what is observed for bleomycin,[229] and homolytic cleavage of the O–O bond may have a large activation energy.[230]

The two major monomeric products formed when activated bleomycin reacts with DNA are nucleic base propenals (**6.77**, Scheme 6.25) and nucleic acid bases. Nucleic base propenal formation consumes an equivalent of O_2 in addition to that required for bleomycin activation and is accompanied by DNA-strand scission with the production of 3′-phosphoglycolate (**6.78**) and 5′-phosphate-modified DNA fragments (**6.76**).[231] DNA base formation does not require additional O_2 and results in destabilization of the DNA sugar-phosphate backbone. Evidence for the C-4′ radical (**6.74**) and the peroxy radical (**6.75**) comes from model studies of Giese and coworkers who used chemical methods to generate a C-4′ radical in a single-stranded oligonucleotide.[232] They detected the C-4′ radical and the peroxy radical and obtained similar (as well as additional) products as those observed from the reaction of bleomycin and DNA. On the basis of isotope studies with $^{18}O_2$ and $H_2{}^{18}O$ and other chemical precedence, two modified mechanisms are presented in Scheme 6.25. Mechanism A invokes a modified Criegee rearrangement to account for the ^{18}O labeling results, in which hydroxide is released rather than being added back to the adjacent carbon.[233] Mechanism B involves a cleavage of the sugar-phosphate bond early in the mechanism on the basis of other chemical precedence, and uses a Grob fragmentation step to break the hydroperoxide bond.[234]

DNA-strand scission by bleomycin is sequence selective, occurring most frequently at 5′-GC-3′ and 5′-GT-3′ sequences.[235] The specificity for cleavage of DNA at a residue located at the 3′ side of G appears to be absolute. Preference for cleavage at 5′-GC and 5′-GT instead of the corresponding 5′-AC or 5′-AT sites can be attributed to reduced binding affinity of bleomycin at adenine because of one less hydrogen bond relative to that to guanine.[236]

6.3.3.3. *Tirapazamine*

Tirapazamine (**6.79**) is a bioreductively activated antitumor agent that selectively kills oxygen-poor (hypoxic) cells in solid tumors.[237] One-electron reduction, possibly by enzymes such as NADPH-cytochrome P450 reductase or xanthine oxidase, produces the key radical intermediate (**6.80**), which can undergo homolytic cleavage to **6.81** and hydroxyl radicals, highly reactive radicals that readily degrade DNA (Scheme 6.26).[238] The DNA damage caused by reduction of tirapazamine is typical of hydroxyl radicals, namely, damage to both the DNA backbone and to the heterocyclic bases.[239] The activated form (**6.80**) is rapidly destroyed in normally oxygenated cells by reaction with O_2,[240] but under hypoxic conditions, this radical causes oxidative cleavage of the DNA backbone.[241] The tirapazamine

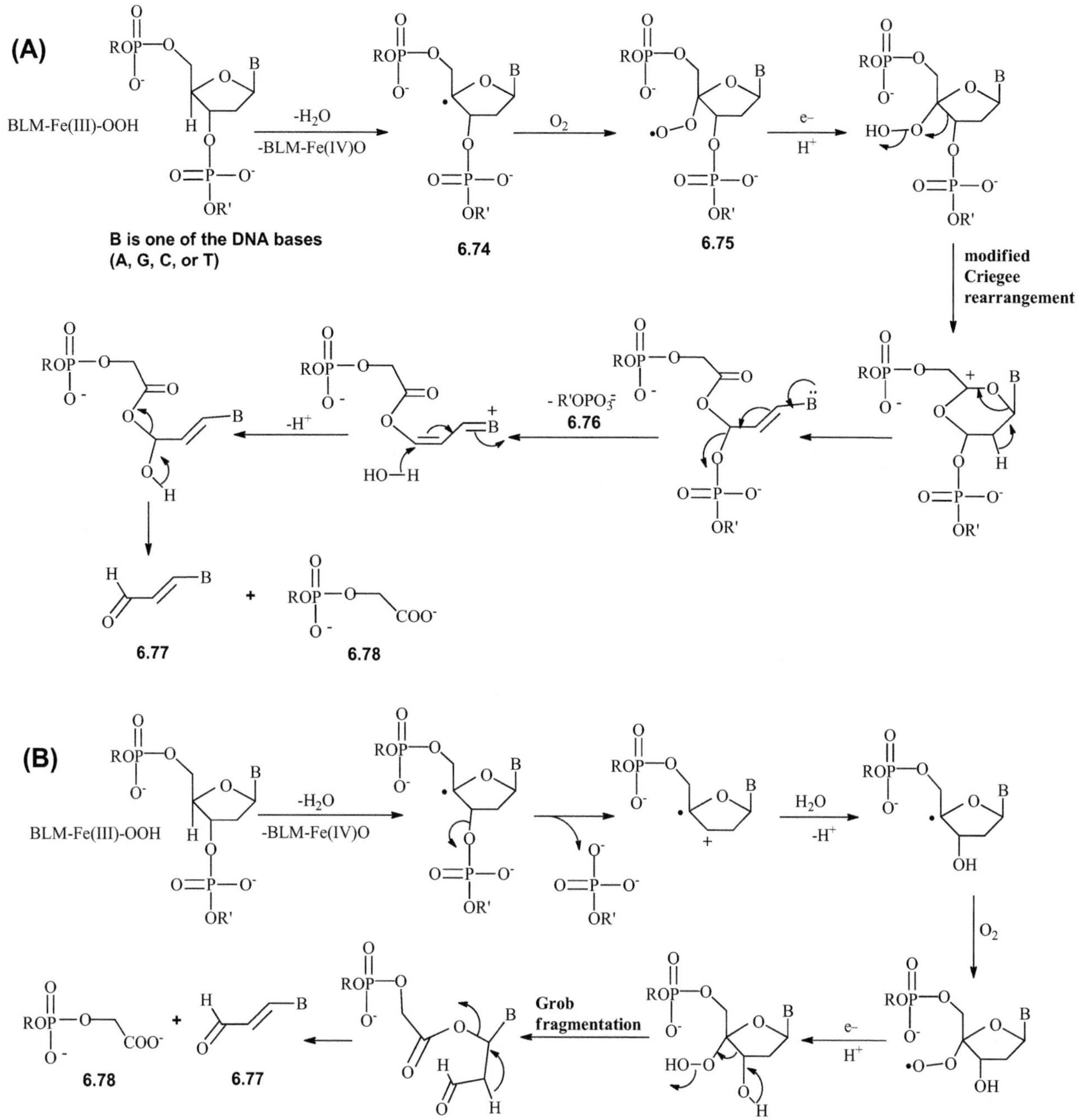

SCHEME 6.25 **Alternative mechanisms for base propenal formation and DNA strand scission by activated bleomycin: (A) Modified Criegee mechanism and (B) Grob fragmentation mechanism**

radical also causes extensive damage to the heterocyclic bases of DNA under hypoxic conditions.[242] The reason tirapazamine can efficiently cause DNA-strand cleavage under hypoxic conditions is that not only does it initiate the formation of deoxyribose radicals by hydroxyl radical abstraction of a hydrogen atom, but it also reacts with these DNA radicals and converts them into strand breaks, thereby serving as a surrogate for molecular oxygen (Scheme 6.27).[243]

6.3.3.4. Enediyne Antitumor Antibiotics

Except for zinostatin (**6.82**),[244] which was isolated in 1965, other members of the enediyne antibiotic class of antitumor agents, such as the esperamicins (**6.83**),[245] the calicheamicins (**6.84**),[246] dynemicin A (**6.85**),[247] kedarcidin (**6.86**),[248] C-1027 (**6.87**),[249] N1999A2 (**6.88**),[250] and maduropeptin chromophore[251] (**6.89**) were isolated from various microorganisms since the mid-1980s. Because

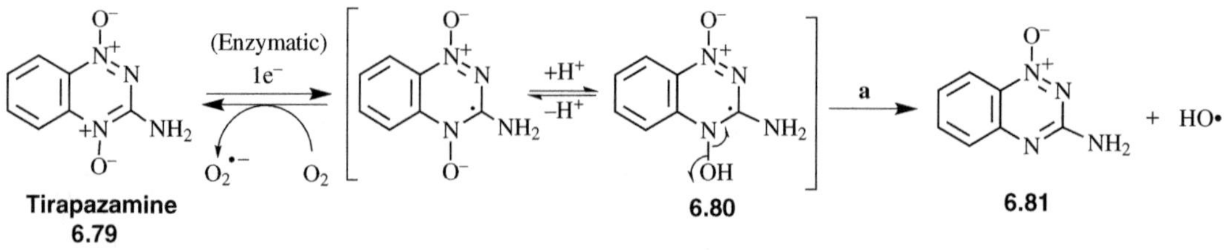

SCHEME 6.26 **Mechanism for formation of hydroxyl radicals by tirapazamine**

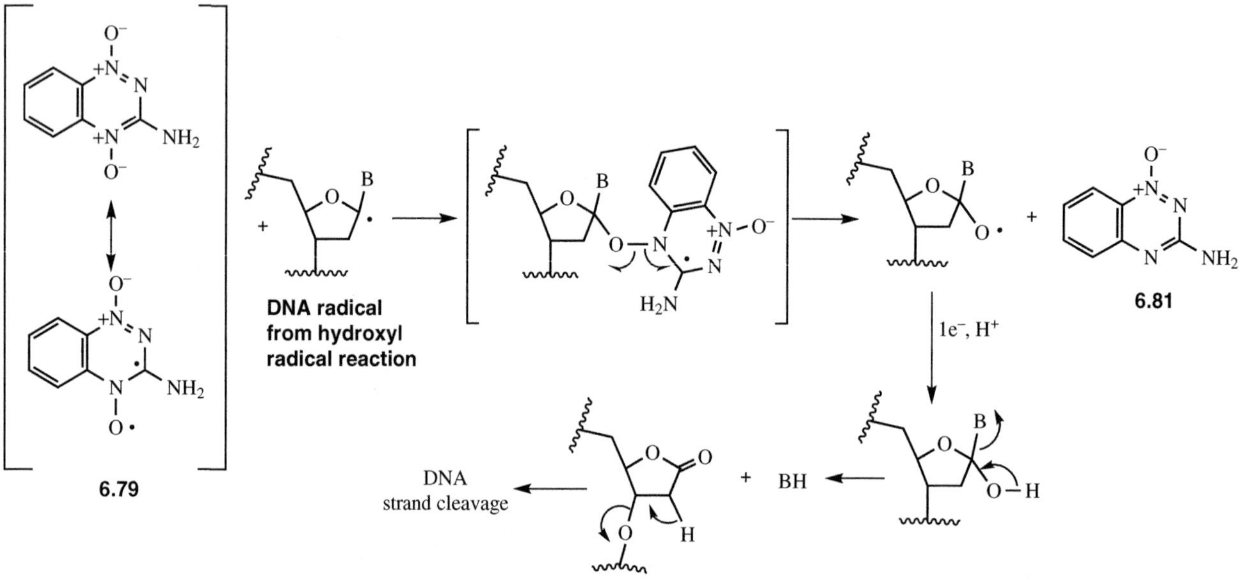

SCHEME 6.27 **Mechanism for DNA-strand cleavage by tirapazamine**

their common structural feature is a macrocyclic ring containing at least one double bond and two triple bonds, they are referred to as *enediyne antitumor antibiotics*.[252] Several of the enediyne antibiotics, such as zinostatin, kedarcidin, C-1027, and maduropeptin, are stabilized in nature as a noncovalent complex with a protein and are referred to as *chromoprotein enediyne antibiotics* because of the chromophoric properties of the compounds embedded in the protein. The protein associated with kedarcidin and zinostatin (and possibly all of the chromoprotein enediyne antibiotics) not only stabilizes the enediyne chromophore but also has been shown to selectively proteolyze basic peptides and proteins, such as histones, which, as we saw earlier (Section 6.2.3), are essential for packing of the DNA into chromatin.[253] Therefore, these compounds are dual DNA cleaving agents and histone proteolytic agents.

All of these compounds appear to share two modes of action, namely intercalation of part of the molecule into the minor groove of DNA[254] and reduction of the molecule by either thiol or NADPH, which triggers a reaction that leads to the generation of radicals, cleaving the DNA. It is not clear if DNA binding must precede the activation process.[255] Although knowledge of the chemistry involved in the latter process is only beginning to surface, we will be able to draw reasonable mechanisms for DNA damage by each of these classes of compounds.

Generally, there are two tests that are used to demonstrate minor groove binding to B-DNA. One indication is an asymmetric cleavage pattern on the 3′-side of the opposite strand of the DNA helix.[256] The other test is inhibition of DNA cleavage by the known minor groove binders distamycin A and netropsin (similar to substrate protection of enzymes from inhibition; see Chapter 5, Section 5.2.1).

There are two phases to the mechanism of DNA degradation by the enediyne antibiotics after binding. First, there is the activation of the antitumor agent (except for C-1027,[257] activation requires addition of a thiol or reducing agent), then there is the action of the activated antitumor agent on DNA. This is much akin to the process of DNA degradation that we discussed for bleomycin (Section 6.3.3.2) and tirapazamine (Section 6.3.3.3). Most of what is now known about the mechanism of the reaction of activated enediynes with DNA comes from studies with zinostatin, which will be discussed last. Because the chemistry of the activation of esperamicins and calicheamicins is virtually identical, we will first take a look at these compounds.

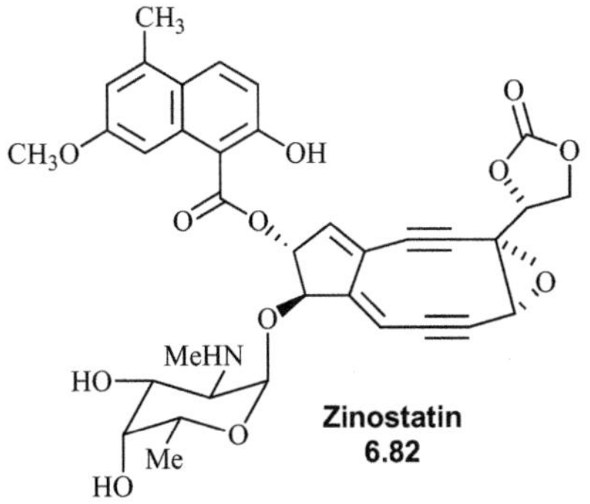

Zinostatin
6.82

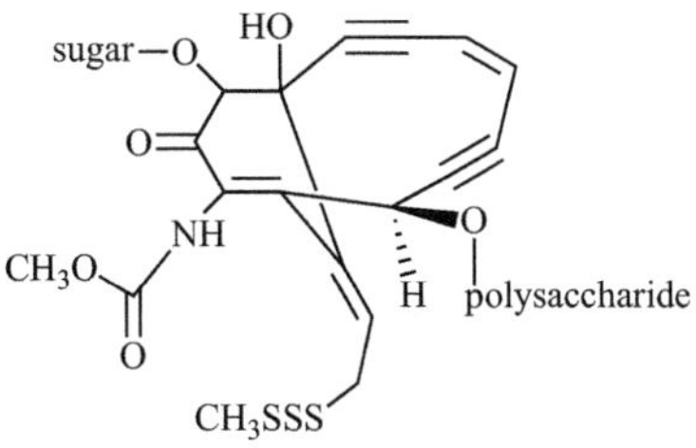

Esperamicins
6.83

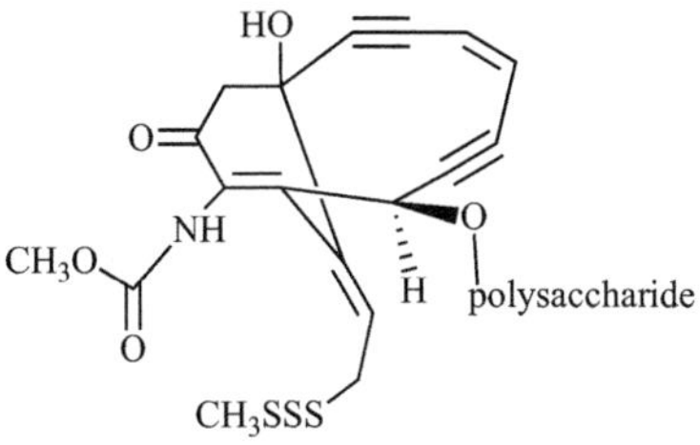

Calicheamicins
6.84

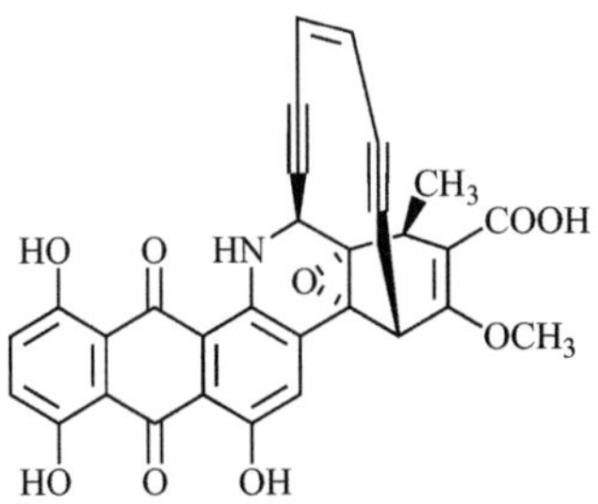

Dynemicin A
6.85

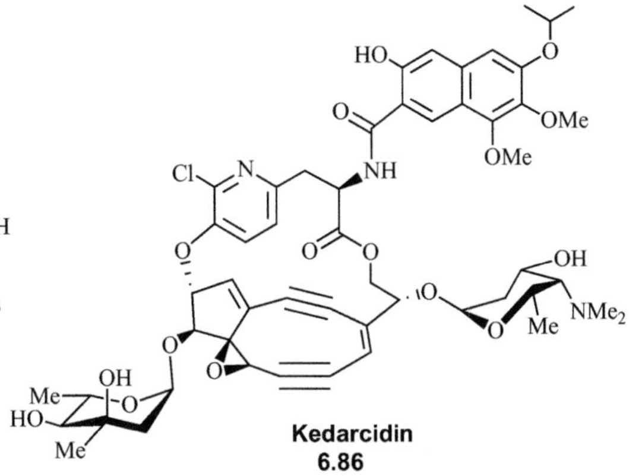

Kedarcidin
6.86

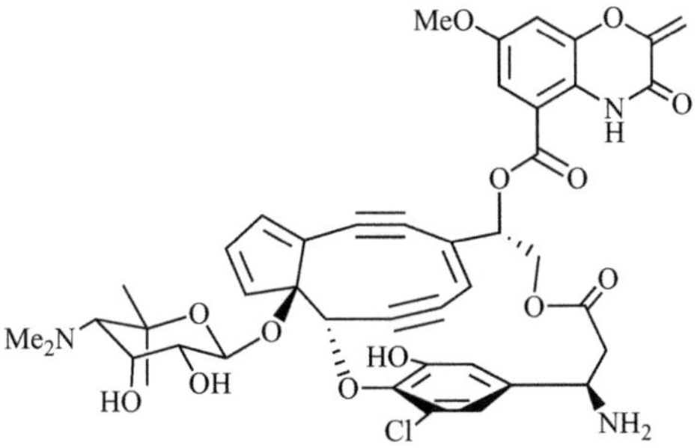

C-1027
6.87

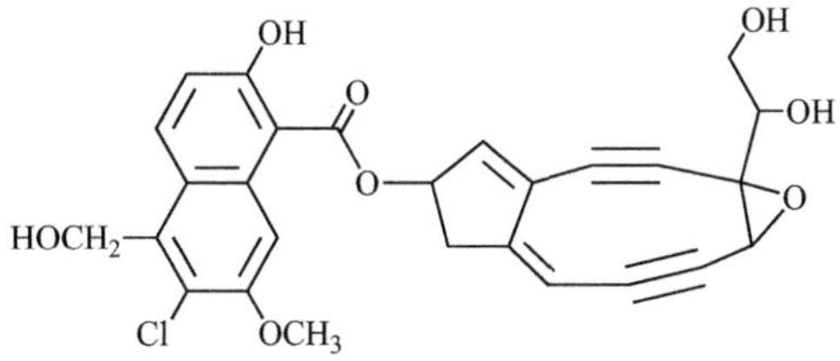

N1999A2
6.88

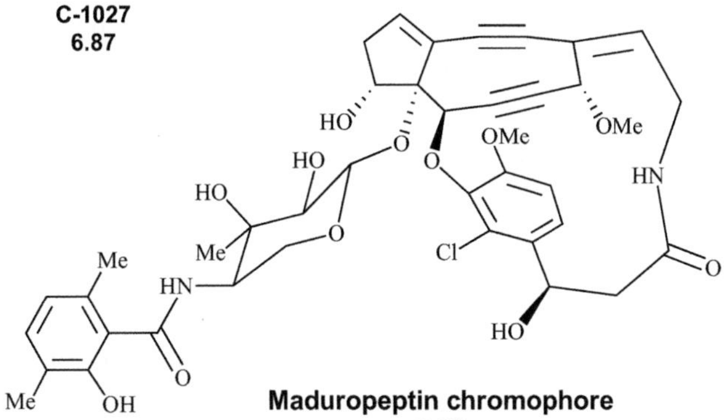

Maduropeptin chromophore
6.89

6.3.3.4.1. Esperamicins and Calicheamicins

The important structural features of these molecules (**6.83** and **6.84**) are a bicyclo[7.3.1] ring system, an allylic trisulfide attached to the bridgehead carbon, a 3-ene-1,5-diyne as part of the macrocycle, and an α,β-unsaturated ketone in which the double bond is at the bridgehead of the bicyclic system. It is believed that the enediyne moiety partially inserts into the minor groove, and then undergoes a reaction

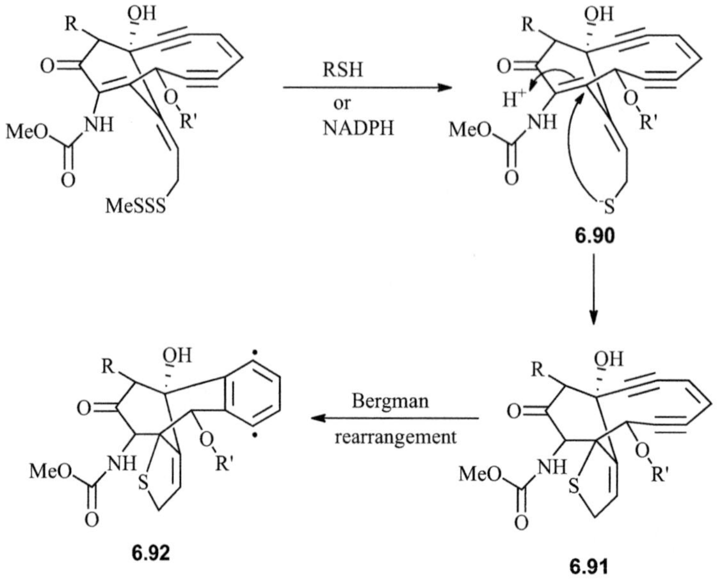

SCHEME 6.28 Activation of esperamicins and calicheamicins

with either a thiol or NADPH, which reduces the trisulfide to the corresponding thiolate (**6.90**, Scheme 6.28).[258]

Michael addition of this thiolate into the α,β-unsaturated ketone gives the dihydrothiophene (**6.91**), in which the bridgehead carbon hybridization has changed from sp^2 in **6.90** to sp^3 in **6.91**. This change in geometry at the bridgehead may be sufficient to allow the two triple bonds to interact with each other and to trigger a Bergman rearrangement,[259] giving the 1,4-dehydrobenzene biradical (**6.92**), which is the activated esperamicin or calicheamicin. The reaction of this biradical with DNA is, presumably, identical to that for the biradicals produced from all of the enediyne antibiotics.

6.3.3.4.2. Dynemicin A

Another member of the enediyne class of antitumor antibiotics is dynemicin A (**6.85**), which combines the structural features of both the anthracycline antitumor agents and the enediynes. The anthraquinone part of dynemicin A intercalates into the minor groove of DNA,[260] and then, depending upon whether activation is initiated by NADPH or a thiol, a reductive mechanism (Scheme 6.29) or nucleophilic mechanism (Scheme 6.30) of activation, respectively, is possible.[261] In the reductive activation mechanism (Scheme 6.29), as we have seen earlier (Section 6.3.2.5.3), either NADPH or a thiol reduces the quinone to a hydroquinone (**6.93**), which leads to opening of the epoxide. Following protonation to **6.94**, the geometry of the triple bonds becomes more favorable for the Bergman reaction to take place, leading to the generation of the 1,4-dehydrobenzene biradical (**6.95**). In the nucleophilic mechanism (Scheme 6.30), reaction of a thiol gives **6.96**, which is very similar in structure to that of the reduced intermediate **6.94** (Scheme 6.29). Bergman reaction of **6.96** gives the 1,4-dehydrobenzene biradical **6.97**. Either biradical (**6.95** or **6.97**) could be responsible for DNA degradation by the mechanism described in detail for zinostatin (Section 6.3.3.4.3).

6.3.3.4.3. Neocarzinostatin (Zinostatin)

The oldest known member of the enediyne family of antibiotics is zinostatin (**6.82**; drug component of Smancs), the only one of the enediynes (not in prodrug form) currently approved for clinical use, although not in the U.S. More mechanistic studies have been performed with this compound than any of the more recent additions to the enediyne family. The naphthoate ester moiety is believed to intercalate into DNA, thereby positioning the epoxybicyclo[7.3.0] dodecadiendiyne portion of the chromophore in the minor groove.[262] Activation by a thiol generates an intermediate (**6.98**, Scheme 6.31) capable of undergoing a Bergman rearrangement to biradical **6.99**.[263] This biradical differs from the biradicals generated by activation of esperamicin and calicheamicin (**6.92**, Scheme 6.28) and dynemicin A (**6.95**, Scheme 6.29 or **6.97**, Scheme 6.30) in that it is not a 1,4-dehydrobenzene biradical, but it is very similar.

Surprisingly, the reaction of small thiol reducing agents with zinostatin still complexed to its protein (see introduction to Section 6.3.3.4) leads to *deactivation* of the chromophore rather than activation.[264] Myers and coworkers[265] proposed that the mechanism for deactivation involves the same initial attack of the chromophore by the thiol, but they suggest that the epoxide is not disposed for nucleophilic opening because it is sequestered within a hydrophobic pocket of the protein. Therefore, the reaction proceeds with

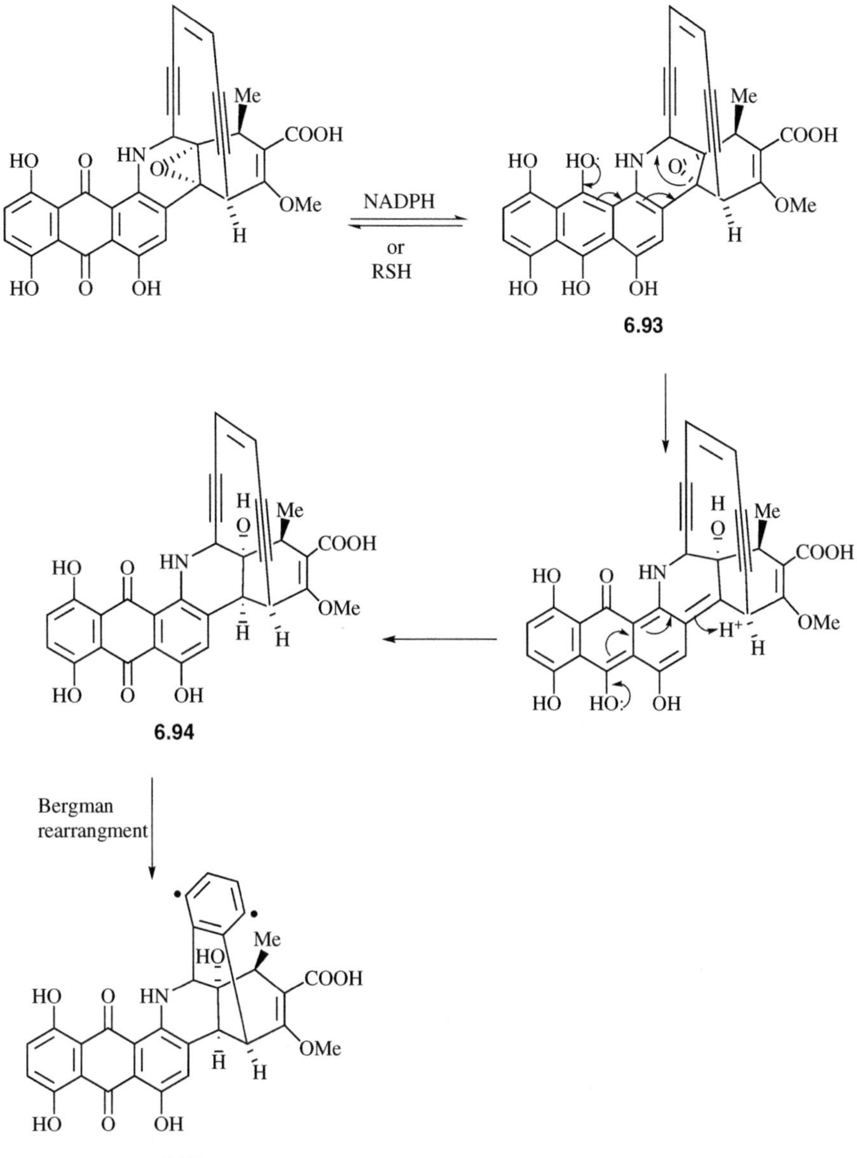

SCHEME 6.29 Reductive mechanism for activation of dynemicin A

homolytic cleavage of the epoxide ring, as shown in Scheme 6.32, followed by rapid reaction of the proposed polar resonance structure of biradical intermediate **6.100** with water to give **6.101**. σ,π-Biradicals, such as **6.100**, in contrast to σ,σ-biradicals, such as **6.99** (Scheme 6.31), are known to undergo polar addition reactions, so this may account for the deactivation process.[266]

Highly reactive biradical **6.99** (or **6.100**) is responsible for DNA-strand scission, which consumes one equivalent of O_2 per strand break. Both the C-4′ and the C-5′ hydrogens of DNA sugar phosphate residues are accessible to the biradical within the minor groove, and either can be abstracted. In the presence of O_2, two different mechanisms of DNA cleavage can result (Scheme 6.33).[267] About 80% of the cytotoxic lesions are the result of single-strand cleavages, mostly caused by C-5′ hydrogen atom abstraction from thymidine or deoxyadenosine residues (pathway a). This leads to the formation of the nucleoside 5′-aldehyde (**6.102**) and the 3′-phosphate (**6.103**). Abstraction of the C-4′ hydrogen atom (pathway b) gives a radical (**6.104**) that partitions between a modified basic carbohydrate terminus (**6.105**, pathway c) and a 3′-phosphoglycolate terminus (**6.106**, pathway d), as was observed with bleomycin (Section 6.3.3.2). It appears that single-strand breaks at AGT sequences occur with both C-4′ and C-5′ oxidation, but ACT breaks occur only with C-5′ oxidation.

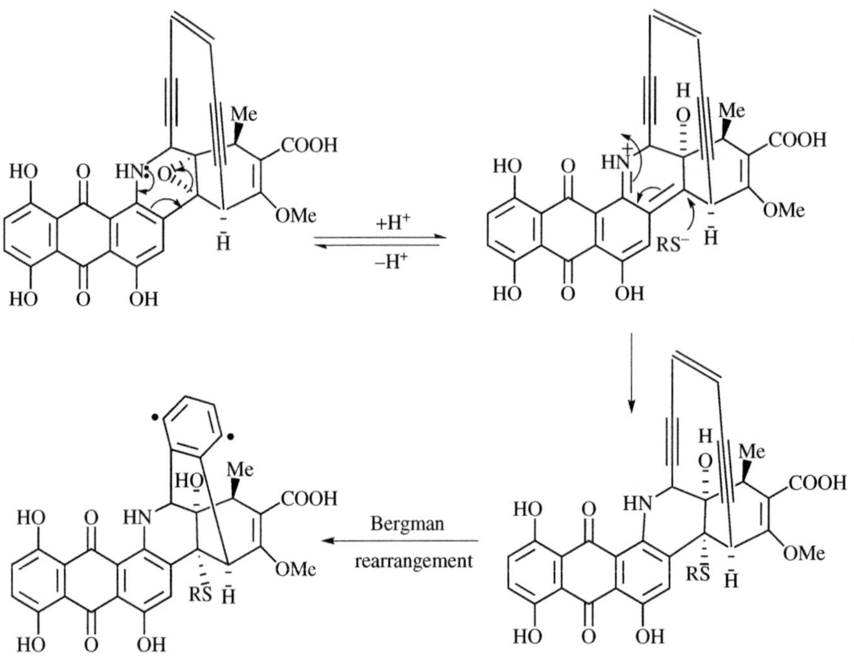

6.97 6.96

SCHEME 6.30 **Nucleophilic mechanism for activation of dynemicin A**

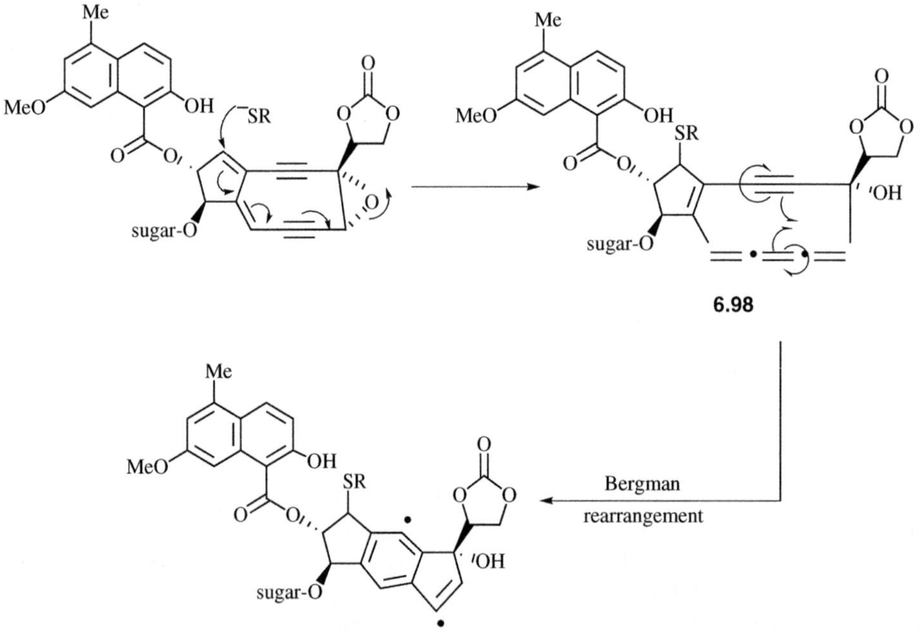

6.98

6.99

SCHEME 6.31 **Activation of zinostatin by thiols**

A catalytic antibody[268] was raised that catalyzes a Bergman cyclization of an enediyne (**6.107**), but instead of the expected tetralin (**6.108**), the biradical reacted with O_2, producing the corresponding quinone (**6.109**, Scheme 6.34).[269] Quinones also are known to be highly cytotoxic,[270] so this suggests that there may be more than one mechanism for cytotoxicity produced by enediyne antibiotics. The enediynes, in general, have been too toxic for clinical use;

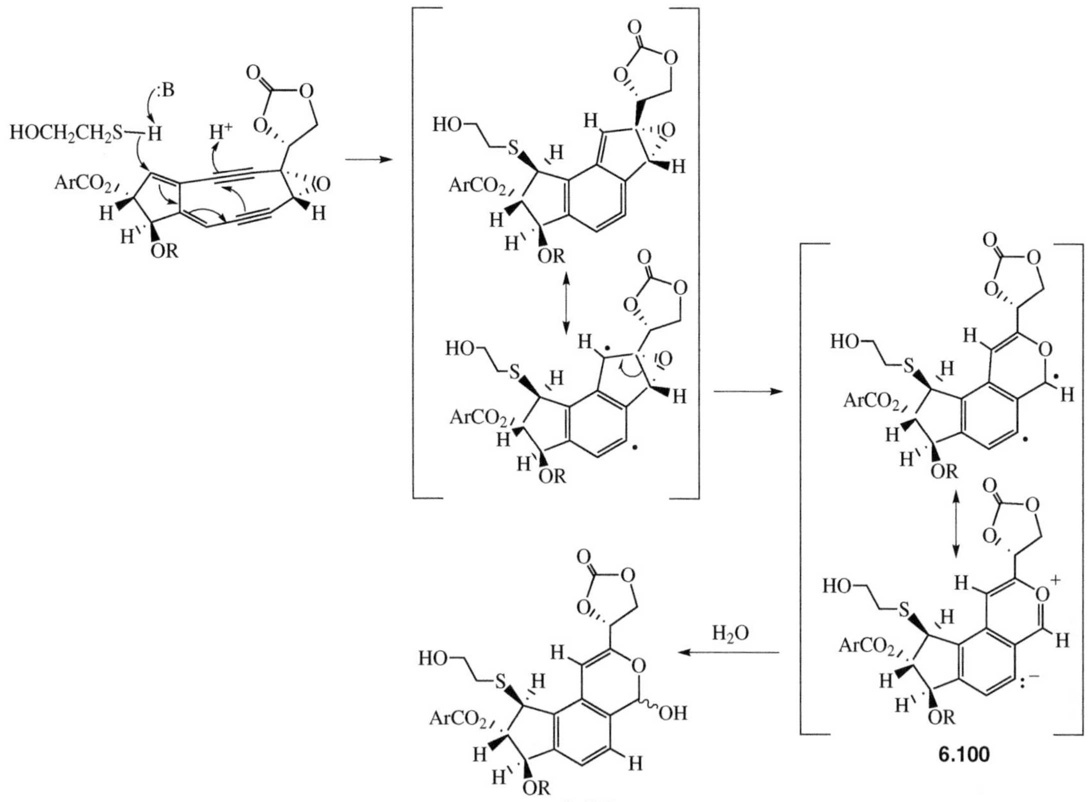

6.100

6.101

SCHEME 6.32 **Polar addition reaction to deactivate zinostatin**

a conjugate of an antibody-calicheamicin derivative (gemtuzumab ozogamicin, Mylotarg) was withdrawn from the market in 2010 for the treatment of acute myeloid leukemia (Chapter 9, Section 9.2.1.3.4).

All of these highly cytotoxic antibiotics were isolated from various microorganisms, all of which also contain reducing agents to activate the cytotoxins, and DNA that can be degraded. So why do these organisms produce these compounds, and why are the organisms not destroyed by them? Generally, these compounds are produced for protection against other microorganisms. The organism that produces the antibiotic protects itself in at least two ways. First, as was noted for the chromoprotein enediyne antibiotics, they produce proteins that sequester, stabilize, and transport the compounds out of the cell, thereby protecting the bacteria that produced them.[271] Another mechanism for protection is the production of enzymes that catalyze reactions to destroy the antibiotic in case it is accidentally released within the cell. A protein was identified that is responsible for self-resistance of the organism from its own production of calicheamicin.[272] As was noted above, the zinostatin–chromoprotein complex is deactivated by small thiol reducing agents, but not by larger thiols. Many prokaryotes, such as bacteria, use small thiols (cysteine or hydrogen sulfide) for metabolism in contrast to eukaryotes, which utilize the tripeptide thiol, glutathione.[273] The susceptibility of the zinostatin chromoprotein to small thiols may be a mechanism for protection by the organism if the zinostatin complex is not transported quickly enough out of the cell.

6.3.3.5. Sequence Specificity for DNA-Strand Scission

Many studies have been carried out to determine cleavage site preferences in the minor groove for each of the strand breakers.[274] The preferences are rationalized as occurring because of selective binding at particular sites on the basis of the structure of the compound. Evidence from in vitro and in vivo work suggest that adduct formation (alkylation)[275] and radical-mediated strand breakage[276] by small organic molecules is not markedly altered by chromatin structure, probably because of the accessibility of the outer surface of DNA in chromatin to small molecules. Therefore, in vitro studies of DNA do provide a reasonable model for the in vivo sequence preferences of DNA-damaging agents.

Although many of the DNA-interactive drugs react with the constituents of the minor groove, it is not yet clear if there is sufficient sequence specificity information in the minor groove of DNA to allow for the *design* of minor groove-selective agents. Nature has opted, for example, with protein–DNA interactions, to utilize major groove interactions.

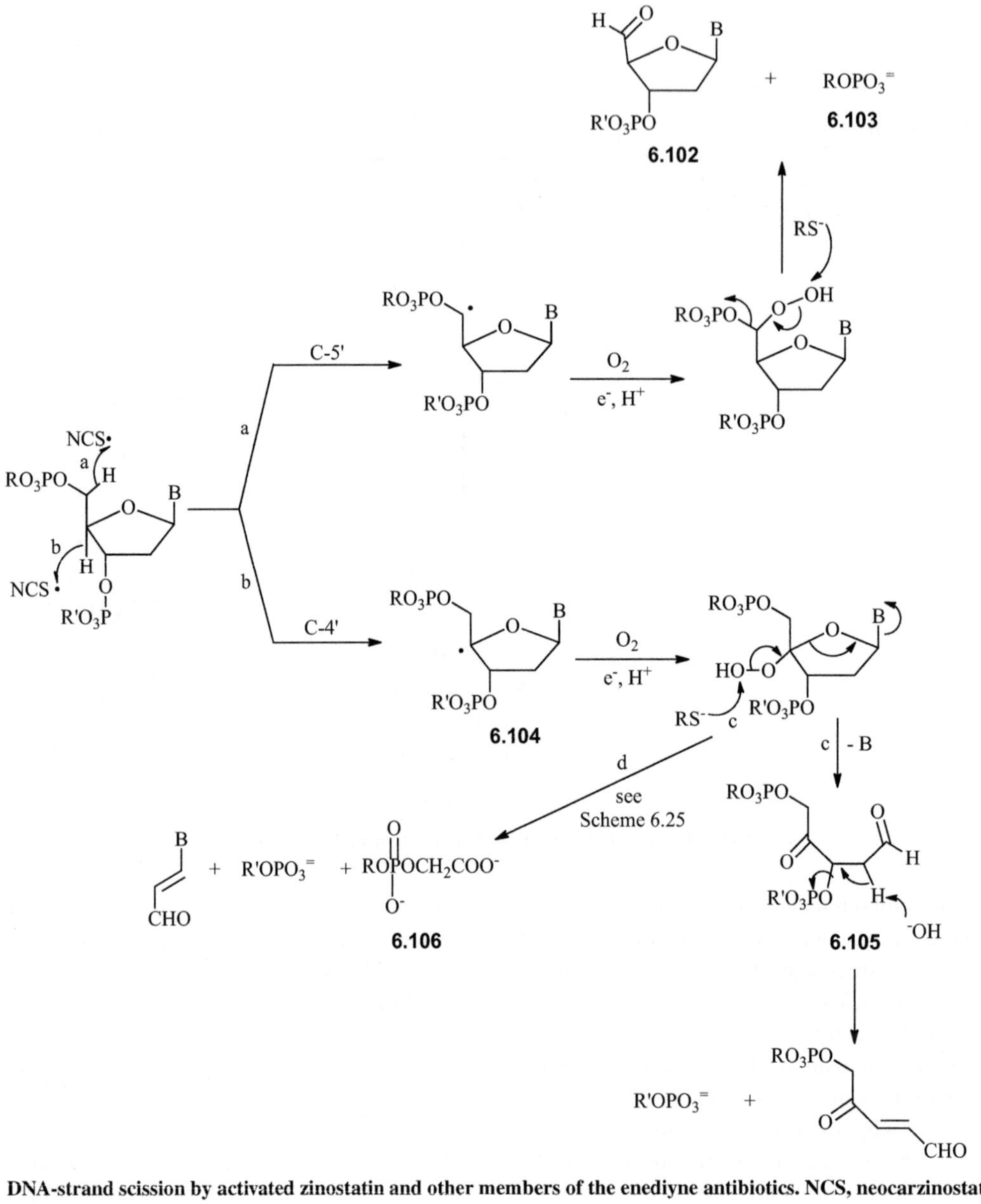

SCHEME 6.33 DNA-strand scission by activated zinostatin and other members of the enediyne antibiotics. NCS, neocarzinostatin (Zinostatin)

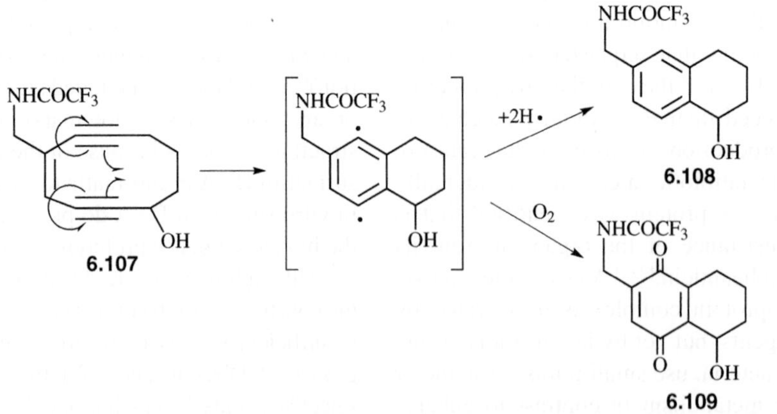

SCHEME 6.34 Catalytic antibody-catalyzed conversion of an enediyne into a quinone via oxygenation of the corresponding benzene biradical

6.4. GENERAL REFERENCES

DNA Structure and Function

Alberts, B.; Johnson, A.; Lewis, J.; Raff, M. *Molecular Biology of the Cell*, 5th ed., Garland Publishing, New York, 2008.

Blackburn, G. M.; Gait, M. J.; Loakes, D.; Williams, D. (Eds.) *Nucleic Acids in Chemistry and Biology*, 3rd ed., RSC Publishing, Cambridge, U.K., 2006.

Lodish, H.; Berk, A.; Kaiser, C. A.; Krieger, M.; Bretscher, A.; Ploegh, H.; Amon, A.; Scott, M. P. *Molecular Cell Biology*, 7th ed.; W. H. Freeman, New York, 2012.

Neidle, S. *Principles of Nucleic Acid Structure*, Academic Press, London, U.K., 2008.

Stormo, G. D. *Introduction to Protein-DNA Interactions: Structure, Thermodynamics, and Bioinformatics*, Cold Spring Harbor Laboratory Press, Cold Spring Harbor, NY, 2013.

Watson, J. D. *DNA the Secret of Life*, Alfred A. Knopf, New York, 2004.

Watson, J. D.; Baker, T. A.; Bell, S. P.; Gann, A.; Levine, M.; Losick, R. *Molecular Biology of the Gene*, 7th ed., Benjamin/Cummings Publishing, Menlo Park, CA, 2013.

DNA Repair

Doetsch, P. (Ed.) Mechanisms of DNA repair, Vol. 110. In *Progress in Molecular Biology and Translational Science*; Academic Press, San Diego, CA, 2012.

Friedberg, E. C.; Walker, G. C.; Siede, W.; Wood, R. D.; Schultz, R. A.; Ellenberger, T. *DNA Repair and Mutagenesis*, 2nd ed.; ASM Press, Washington, DC, 2006.

Kelley, M. R. (Ed.) *DNA Repair in Cancer Therapy: Molecular Targets and Clinical Applications*, Academic Press, San Diego, CA, 2012.

Madhusudan, S.; Wilson III, D. M. *DNA Repair and Cancer: From Bench to Clinic*, CRC Press, Boca Raton, FL, 2013.

Panasci, L.; Aloyz, R.; Alaoui-Jamali, M. (Eds.) *Advances in DNA Repair in Cancer Therapy*, Humana Press, New York, NY, 2013.

Von Sonntag, C. *Free-Radical-Induced DNA Damage and its Repair, A Chemical Perspective*, Springer, Berlin, 2006.

Topoisomerases

Andoh, T. *DNA Topoisomerases in Cancer Therapy: Present and Future*, Springer, New York, 2012.

Bates, A. D.; Maxwell, A. *DNA Topology*, Oxford University Press, Oxford, U.K., 2005.

Champoux, J. J. DNA topoisomerases: structure, function, and mechanism. *Annu. Rev. Biochem.* **2001**, *70*, 369–413.

Pommier, Y. (Ed.) *DNA Topoisomerases and Cancer*, Humana Press, New York, 2012.

Wang, J. C. *Untangling the Double Helix: DNA Entanglement and the Action of the DNA Topoisomerases*, Cold Spring Harbor Laboratory Press, Cold Spring Harbor, NY, 2009.

Antitumor Drugs

Chabner, B. A.; Longo, D. L. *Cancer Chemotherapy and Biotherapy: Principles and Practice*, 5th ed., Lippincott Williams & Wilkins, Philadelphia, 2011.

Neidle, S.; Waring, M. J. (Eds.) *Molecular Aspects of Anticancer Drug DNA Interaction*, Vol. 1, CRC Press, Boca Raton, FL, 1993.

Spencer, P.; Holt, W. (Eds.) *Anticancer Drugs: Design, Delivery and Pharmacology*, Nova Science Publishers, Hauppauge, NY, 2011.

6.5. PROBLEMS (ANSWERS CAN BE FOUND IN THE APPENDIX AT THE END OF THE BOOK)

1. Draw a mechanism for the topoisomerase-catalyzed cleavage of DNA that gives the free 5′-end.
2. Show how isoguanosine (**1**) could form a Watson-Crick base pair with isocytidine (**2**).

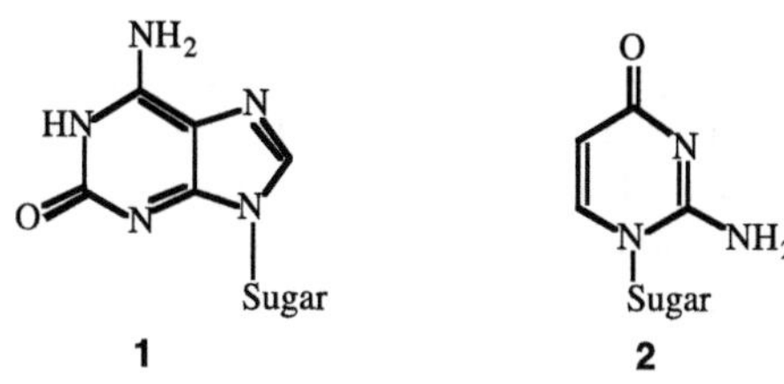

3. What properties would you incorporate into the design of DNA intercalating agents?
4. Alkylating agents are used in cancer chemotherapy to interfere with DNA biosynthesis or replication. Why can these be useful considering that normal cells also need DNA?
5. Draw a mechanism for the intrastrand cross linkage of two guanines by busulfan (**6.33**, n = 4).
6. Draw a mechanism from **6.51** to **6.44** (Scheme 6.13).
7. Amino-*seco*-CBI-TMI (**3**) is a minor groove alkylating agent. Show a mechanism for alkylation of DNA that does *not* involve direct S_N2 displacement of the chloride by DNA.

8. *N,N*-Diethylnitrosoamine (**4**) is carcinogenic, resulting in ethylation of DNA. Two other products formed are acetaldehyde and N_2. Under anaerobic conditions or in the absence of NADPH no carcinogenicity is observed. Draw a mechanism for DNA ethylation by this compound.

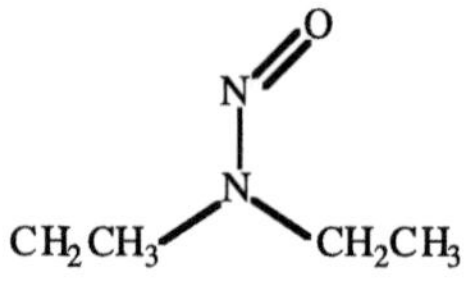

4

9. Compounds **6.31** and **6.32** are stable until some sort of activation occurs for antitumor activity. Show what is necessary for activation and explain why.
10. Why are enediyne antitumor agents stable until a thiol or NADPH reduction of some other part of the molecule occurs?
11. Draw a mechanism for the formation of a 1,4-benzene diradical from an activated enediyne (Bergman rearrangement).
12. Based on your knowledge of intercalating agents and strand breakers, design a new class of enediyne antitumor agents that utilizes both concepts.
13. A new antitumor agent was discovered from a random screen of soil microbes having structure **5**. The compound is not active until it undergoes one electron reduction, after which it was found to cause DNA strand breakage. Draw a possible mechanism for its activity.

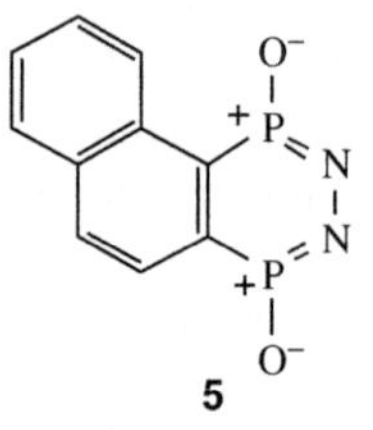

5

14. Draw a mechanism for how kedarcidin (**6**) might act as an antitumor agent (a reducing agent is required).

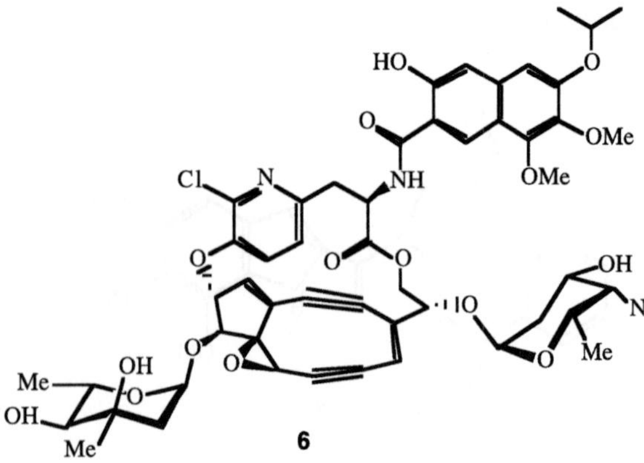

6

15. It was proposed that a common mechanism for antibiotics might be hydroxyl radical oxidation of guanine to 7,8-dihydro-8-oxoguanine (**7**), which is enzymatically excised from DNA and replaced by guanine; however, when too many guanines are oxidized, DNA strand breakage occurs, killing the cell. Draw a mechanism for how guanine could be oxidized by hydroxyl radicals to **7**.

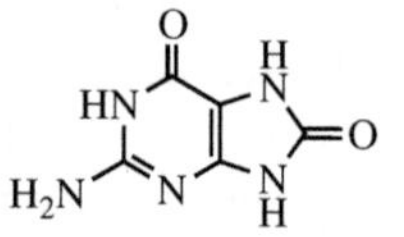

7

REFERENCES

1. (a) Gullotta, F.; De Marinis, E.; Ascenzi, P.; di Masi, A. Targeting the DNA double strand breaks repair for cancer therapy. *Curr. Med. Chem.* **2010**, *17*, 2017–2048. (b) Sancar, A. Structure and function of photolyase and in vivo enzymology: 50th anniversary. *J. Biol. Chem.* **2008**, *283*, 32153–32157. (c) Sedgwick, B.; Bates, P. A.; Paik, J.; Jacobs, S. C.; Lindahl, T. Repair of alkylated DNA: recent advances. *DNA Repair* **2007**, *6*, 429–442. (d) Anderson, J. P.; Loeb, L. A. DNA repair enzymes. In *Gene-Environment Interactions*; Costa, L. G.; Eaton, D. L. (Eds.), John Wiley & Sons, Hoboken, NJ, 2006.
2. (a) Sperka, T.; Wang, J.; Rudolph, K. L. DNA damage checkpoints in stem cells, ageing and cancer. *Nat. Rev. Mol. Cell Biol.* **2012**, *13*(9), 579–590. (b) Suvorova, I. I.; Katolikova, N. V.; Pospelov, V. A. New insights into cell cycle regulation and DNA damage response in embryonic stem cells. *Int. Rev. Cell Mol. Biol.* **2012**, *299*, 161–198. (c) Hoepker, K.; Hagmann, H.; Khurshid, S.; Chen, S.; Schermer, B.; Benzing, T.; Reinhardt, H. C. Putting the brakes on p53-driven apoptosis. *Cell Cycle.* **2012**, *11*(22), 4122–4128. (d) Stewart, Z. A.; Pietenpol, J. A. P53 signaling and cell cycle checkpoints. *Chem. Res. Toxicol.* **2001**, *14*, 243–263.
3. (a) Morandell, S.; Yaffe, M. B. Exploiting synthetic lethal interactions between DNA damage signaling, checkpoint control, and p53 for targeted cancer therapy. *Prog. Mol. Biol. Trans. Sci.* **2012**, *110* (Mechanisms of DNA Repair), 289–314. (b) Waldman, T.; Zhang, Y; Dillehay, L.; Yu, J.; Kinzler, K.; Vogelstein, B.; Williams, J. Cell-cycle arrest versus cell death in cancer therapy. *Nat. Med.* **1997**, *3*, 1034–1036.
4. Doll, R.; Peto, R. The causes of cancer: quantitative estimates of avoidable risks of cancer in the United States today. *J. Natl. Cancer Inst.* **1981**, *66*, 1191–1308.
5. (a) Kemp, M. G.; Sancar, A. DNA excision repair. Where do all the dimers go? *Cell Cycle* **2012**, *11*, 2997–3002. (b) Liu, Yuan; Wilson, Samuel H. DNA base excision repair: a mechanism of trinucleotide repeat expansion. *Trends Biochem. Sci.* **2012**, *37*, 162–172. (c) Rechkunova, N. I.; Krasikova, Yu. S.; Lavrik, O. I. Nucleotide excision repair: DNA damage recognition and preincision complex assembly. *Biochemistry (Moscow).* **2011**, *76*, 24–35. (d) Pascucci, B.; D'Errico, M.; Parlanti, E.; Giovannini, S.; Dogliotti, E. Role of nucleotide excision repair proteins in oxidative DNA damage repair: an updating. *Biochemistry (Moscow).* **2011**, *76*, 4–15. (e) Tuteja, N.; Tuteja, R. Unraveling DNA repair in human: molecular mechanisms and consequences of repair defect. *Crit. Rev. Biochem. Mol. Biol.* **2001**, *36*, 261–290.

6. (a) Yi, C.; He, C. DNA repair by reversal of DNA damage. *Cold Spring Harbor Perspect. Biol.* **2013**, *5*, a012575/1–a012575/18. (b) Benjdia, A. DNA photolyases and SP lyase: structure and mechanism of light-dependent and independent DNA lyases. *Curr. Opin. Struct. Biol.* **2012**, *22*, 711–720. (c) Brettel, K.; Byrdin, M. Reaction mechanisms of DNA photolyase. *Curr. Opin. Struct. Biol.* **2010**, *20*, 693–701. (d) Sancar, A. Structure and function of photolyase and in vivo enzymology: 50th anniversary. *J. Biol. Chem.* **2008**, *283*, 32153–32157. (e) Carell, T.; Burgdorf, L. T.; Kundu, L. M.; Cichon, M. The mechanism of action of DNA photolyases. *Curr. Opin. Chem. Biol.* **2001**, *5*, 491–498.
7. (a) Setlow, R. B.; Carrier, W. The disappearance of thymine dimers from DNA: an error-correcting mechanism. *Proc. Natl. Acad. Sci. U.S.A.* **1964**, *51*, 226–231. (b) Boyce, R.; Howard-Flanders, P. Release of ultraviolet light-induced thymine dimers from DNA in e. coli K-12. *Proc. Natl. Acad. Sci. U.S.A.* **1964**, *51*, 293–300. (c) Pettijohn, D.; Hanawalt, P. C. Evidence for repair-replication of ultraviolet damaged DNA in bacteria. *J. Mol. Biol.* **1964**, *9*, 395–410.
8. (a) Chon S. Y.; Champion R. W; Geddes E. R; Rashid R. M. Chemotherapy-induced alopecia. *J. Am. Acad. Dermatol.* **2012**, *67*, e37–47. (b) Paus, R.; Haslam, I. S.; Sharov, A. A.; Botchkarev, V. A. Pathobiology of chemotherapy-induced hair loss. *Lancet Oncol.* **2013**, *14*, e50–e59.
9. Paus, R.; Cotsarelis, G. The biology of hair follicles. *N. Engl. J. Med.* **1999**, *341*, 491–497.
10. Luanpitpong, S.; Nimmannit, U.; Chanvorachote, P.; Leonard, S. S.; Pongrakhananon, V.; Wang, L.; Rojanasakul, Y. Hydroxy radical mediates cisplatin-induced apoptosis in human hair follicle dermal papilla cells and keratinocytes through Bcl-2 dependent mechanism. *Apoptosis* **2011**, *16*, 769–782.
11. Botchkarev, V. A. Molecular mechanisms of chemotherapy-induced hair loss. *J. Invest. Dermatol. Symp. Proc.* **2003**, *8*, 72–75.
12. Borison, H. L.; Brand, E. D.; Orkand, R. K. Emetic action of nitrogen mustard (mechlorethamine hydrochloride) in dogs and cats. *Am. J. Physiol.* **1958**, *192*, 410–416.
13. Pittillo, R. F.; Schabel, F. M., Jr.; Wilcox, W. S.; Skipper, H. E. Experimental evaluation of potential anticancer agents. XVI. Basic study of effects of certain anticancer agents on kinetic behavior of model bacterial cell populations. *Cancer Chemother. Rep.* **1965**, *47*, 1–26.
14. (a) Illuzzi, J. L.; Wilson, D. M., III Base excision repair: contribution to tumorigenesis and target in anticancer treatment paradigms. *Curr. Med. Chem.* **2012**, *19*, 3922–3936. (b) Ralhan, R.; Kaur, J. Alkylating agents and cancer therapy. *Exp. Opin. Therap. Pat.* **2007**, *17*, 1061–1075. (c) Collins, A. R. S.; Squires, S.; Johnson, R. T. Inhibitors of repair DNA synthesis. *Nucleic Acids Res.* **1982**, *10*, 1203–1213.
15. Cadman, E.; Heimer, R.; Davis, L. Enhanced 5-fluorouracil nucleotide formation after methotrexate administration: explanation for drug synergism. *Science* **1979**, *205*, 1135–1137.
16. (a) Watson, J. D.; Crick, F. H. C. Molecular structure of nucleic acids. A structure for deoxyribose nucleic acid. *Nature* **1953**, *171*, 737–738. (b) Crick, F. H. C.; Watson, J. D. The complementary structure of deoxyribonucleic acid (DNA). *Proc. Roy. Soc. (Lond.) Ser. A* **1954**, *223*, 80–96.
17. Watson, J. D. *The Double Helix*, Weidenfeld and Nicholson, London, 1968.
18. Dekker, C. A.; Michelson, A. M.; Todd, A. R. Nucleotides. XIX. Pyrimidine deoxyribonucleoside diphosphates. *J. Chem. Soc.* **1953**, 947–951.
19. Zamenhof, S.; Braverman, G.; Chargaff, E. The deoxypentosenucleic acids from several micro-organisms. *Biochim. Biophys. Acta* **1952**, *9*, 402–405.
20. Astbury, W. T. X-ray studies of nucleic acids. *Symp. Soc. Exp. Biol. I. Nucleic Acid* **1947**, 66–76.
21. Gulland, J. M. The structures of nucleic acids. *Cold Spring Harbor Symp. Quant. Biol.* **1947**, *12*, 95–103.
22. Wilkins, M. H. F. Molecular configuration of nucleic acids. *Science* **1963**, *140*, 941–950.
23. Maddox, B. *Rosalind Franklin Dark Lady of DNA*, HarperCollins, London and New York, 2002.
24. Furberg, S. The crystal structure of cytidine. *Acta Crystallogr.* **1950**, *3*, 325–333.
25. Hanlon, S. The importance of London dispersion forces in the maintenance of the deoxyribonucleic acid helix. *Biochem. Biophys. Res. Commun.* **1966**, *23*, 861–867.
26. Herskovits, T. T. Nonaqueous solutions deoxyribonucleic acid (DNA). Factors determining the stability of the helical configuration in solution. *Arch. Biochem. Biophys.* **1962**, *97*, 474–484.
27. Beak. P. Energies and alkylations of tautomeric heterocyclic compounds: old problems—new answers. *Acc. Chem. Res.* **1977**, *10*, 186–192.
28. Wolfenden, R. V. Tautomeric equilibria in inosine and adenosine. *J. Mol. Biol.* **1969**, *40*, 307–310.
29. Chenon, M.-T.; Pugmire, R. J.; Grant, D. M.; Panzica, R. P.; Townsend, L. B. Carbon-13 magnetic resonance. XXVI. Quantitative determination of the tautomeric populations of certain purines. *J. Am. Chem. Soc.* **1975**, *97*, 4636–4642.
30. (a) Kool, E. T. Replacing the nucleobases in DNA with designer molecules. *Acc. Chem. Res.* **2002**, *35*, 936–943. (b) Schweitzer, B. A.; Kool, E. T. Aromatic nonpolar nucleosides as hydrophobic isosteres of pyrimidines and purine nucleosides. *J. Org. Chem.* **1994**, *59*, 7238–7242. (c) Khakshoor, O.; Wheeler, S. E.; Houk, K. N.; Kool, E. T. Measurement and theory of hydrogen bonding contribution to isosteric DNA base pairs. *J. Am. Chem. Soc.* **2012**, *134*, 3154–3163.
31. Guckian, K. M.; Kool, E. T. Highly precise shape mimicry by a difluoro-toluene deoxynucleoside, a replication-competent substitute for thymidine. *Angew. Chem. Int. Ed. Engl.* **1998**, *36*, 2825–2828.
32. Guckian, K. M.; Krugh, T. R.; Kool, E. T. Solution structure of a DNA duplex containing a replicable difluorotoluene-adenine pair. *Nat. Struct. Biol.* **1998**, *5*, 954–959.
33. Guckian, K. M.; Krugh, T. R.; Kool, E. T. Solution structure of a nonpolar, non-hydrogen-bonded base pair surrogate in DNA. *J. Am. Chem. Soc.* **2000**, *122*, 6841–6847.
34. Moran, S.; Ren, R. X.-F.; Rumney, S.; Kool, E. T. Difluorotoluene, a nonpolar isostere for thymine, codes specifically and efficiently for adenine in DNA replication. *J. Am. Chem. Soc.* **1997**, *119*, 2056–2057.
35. (a) Kunkel, T. A.; Bebenek, K. DNA replication fidelity. *Annu. Rev. Biochem.* **2000**, *69*, 497–529. (b) Goodman, M. F. Hydrogen bonding revisited: geometric selection as a principal determinant of DNA replication fidelity. *Proc. Natl. Acad. Sci. U.S.A.* **1997**, *94*, 10493–10495.
36. Moran, S.; Ren, R. X.-F.; Kool, E. T. A thymidine triphosphate shape analog lacking Watson-Crick pairing ability is replicated with high sequence selectivity. *Proc. Natl. Acad. Sci. U.S.A.* **1997**, *94*, 10506–10511.

37. Morales, J. C.; Kool, E. T. Efficient replication between non-hydrogen-bonded nucleoside shape analogs. *Nat. Struct. Biol.* **1998**, *5*, 950–954.
38. (a) Kool, E. T. Hydrogen bonding, base stacking, and steric effects in DNA replication. *Annu. Rev. Biophys. Biomol. Struct.* **2001**, *30*, 1–22. (b) Kool, E. T. Active site tightness and substrate fit in DNA replication. *Annu. Rev. Biochem.* **2002**, *71*, 191–219.
39. (a) Richmond, T. J. The structural basis of gene regulation for DNA organized as chromatin. *Chimia* **2001**, *55*, 487–491. (b) Widom, J. Chromatin: the nucleosome unwrapped. *Curr. Biol.* **1997**, *7*, R653–R655. (c) Ramakrishnan, V. Histone structure and the organization of the nucleosome. *Annu. Rev. Biophys. Biomol. Struct.* **1997**, *26*, 83–112.
40. Price, B. D.; D'Andrea, A. D. Chromatin remodeling at DNA double-strand breaks. *Cell* **2013**, *152*, 1344–1354.
41. (a) Wang, J. C. DNA topoisomerases. *Annu. Rev. Biochem.***1996**, *65*, 635–692. (b) Ghilarov, D. A.; Shkundina, I. S. DNA topoisomerases and their function in a cell. *Mol. Biol. (Moscow)* **2012**, *46*, 47–57. (c) Pommier, Y.; Leo, E.; Zhang, H.; Marchand, C. DNA topoisomerases and their poisoning by anticancer and antibacterial drugs. *Chem. Biol.* **2010**, *17*, 421–433. (d) Schoeffler, A. J.; Berger, J. M. DNA topoisomerases: harnessing and constraining energy to govern chromosome topology. *Quart. Rev. Biophys.* **2008**, *41*, 41–101. (e) Dong, K. C.; Berger, J. M. Structure and function of DNA topoisomerases. *Protein-Nucl. Acid Interact.* **2008**, 234–269.
42. (a) Pogorelcnik, B.; Perdih, A.; Solmajer, T. Recent advances in the development of catalytic inhibitors of human DNA topoisomerase IIa as novel anticancer agents. *Curr. Med. Chem.* **2013**, *20*, 694–709. (b) Lopez, V.; Martinez-Robles, M.-L.; Hernandez, P.; Krimer, D. B.; Schvartzman, J. B. Topo IV is the topoisomerase that knots and unknots sister duplexes during DNA replication. *Nucleic Acid Res.* **2012**, *40*, 3563–3573. (c) Witz, G.; Stasiak, A. DNA supercoiling and its role in DNA decatenation and unknotting. *Nucleic Acid Res.* **2010**, *38*, 2119–2133. (d) Schvartzman, J. B.; Stasiak, A. A topological view of the replicon. *EMBO Rep.* **2004**, *5*, 256–261.
43. Wang, J. C. Moving one DNA double helix through another by a type II DNA topoisomerase: the story of a simple molecular machine. *Q. Rev. Biophys.* **1998**, *31*, 107–144.
44. (a) Pommier, Y. Drugging topoisomerases: lessons and challenges. *ACS Chem. Biol.* **2013**, *8*, 82–95. (b) Champoux, J. J. DNA topoisomerases: structure, function, and mechanism. *Annu. Rev. Biochem.* **2001**, *70*, 369–413.
45. (a) Baker, N. M.; Rajan, R.; Mondragón, A. Structural studies of type I topoisomerases. *Nucleic Acids Res.* **2009**, *37*(3), 693–701. (b) Pommier, Y. DNA topoisomerase I inhibitors: chemistry, biology, and interfacial inhibition. *Chem. Rev.* **2009**, *109*, 2894–2902. (c) Leppard, J. B.; Champoux, J. J. Human DNA topoisomerase I: relaxation, roles, and damage control. *Chromosoma* **2005**, *114*, 75–85.
46. (a) Nitiss, J. L. DNA topoisomerase II and its growing repertoire of biological functions. *Nat. Rev. Cancer* **2009**, *9*, 327–337. (b) Roca, J. Topoisomerase II: a fitted mechanism for the chromatin landscape. *Nucleic Acid Res.* **2009**, *37*, 721–730. (c) Dal Ben, D.; Palumbo, M.; Zagotto, G.; Capranico, G.; Moro, S. DNA topoisomerase II structures and anthracycline activity: insights into ternary complex formation. *Curr. Pharmaceut. Des.* **2007**, *13*(27), 2766–2780. (d) McClendon, A. K.; Osheroff, N. DNA topoisomerase II, genotoxicity, and cancer. *Mutation Res.* **2007**, *623*, 83–97.
47. (a) Viard, T.; Bouthier de la Tour, C. Type IA topoisomerases: a simple puzzle? *Biochimie* **2007**, *89*, 456–467. (b) Bugreev, D. V.; Nevinsky, G. A. Structure and mechanism of action of type IA DNA topoisomerases. *Biochemistry (Moscow)* **2009**, *74*(13), 1467–1481. Champoux, J. J. Type IA DNA topoisomerases: strictly one step at a time. *Proc. Natl. Acad. Sci. U.S.A.* **2002**, *99*, 11998–12000.
48. (a) Pommier, Y.; Leo, E.; Zhang, H.; Marchand, C. DNA topoisomerases and their poisoning by anticancer and antibacterial drugs. *Chem. Biol.* **2010**, *17*, 421–433. (b) Reguera, R. M.; Diaz-Gonzalez, R.; Perez-Pertejo, Y.; Balana-Fouce, R. *Curr. Drug Targets.* **2008**, *9*(11), 966–978. (c) Redinbo, Matthew R.; Champoux, James J.; Hol, Wim G. J. Structural insights into the function of type IB topoisomerases. *Curr. Opin. Struct. Biol.* **1999**, *9*(1), 29–36.
49. Forterre, P.; Gribaldo, S.; Gadelle, D.; Serre, M.-C. Origin and evolution of DNA topoisomerases. *Biochimie* **2007**, *89*(4), 427–446.
50. Bates, A. D.; Maxwell, A. Energy coupling in type II topoisomerases: Why do they hydrolyze ATP? *Biochemistry* **2007**, *46*, 7929–7941.
51. (a) Rothstein, R.; Shor, E. DNA topoisomerases: type III-recQ helicase systems. *Encycl. Biol. Chem.* **2004**, *1*, 812–816. (b) Hanai, R.; Caron, P. R.; Wang, J. C. Human TOP3: a single-copy gene encoding DNA topoisomerase III. *Proc. Natl. Acad. Sci. U.S.A.* **1996**, *93*, 3653–3657.
52. (a) Drlica, K.; Zhao, X. DNA topoisomerase IV as a quinolone target. *Curr. Opin. Anti-Infect. Invest. Drugs* **1999**, *1*, 435–442. (b) Khodursky, A. B.; Zechiedrich, E L.; Cozzarelli, N. R. Topoisomerase IV is a target of quinolones in Escherichia coli. *Proc. Natl. Acad. Sci. U.S.A.* **1995**, *92*, 11801–11805.
53. Forterre, P. DNA topoisomerase V: a new fold of mysterious origin. *Trends Biotech.* **2006**, *24*, 245–247.
54. (a) Slesarev, A. I.; Belova. G. I.; Lake, J. A.; Kozyavkin, S. A. Topoisomerase V from methanopyrus kandleri. *Meth. Enzymol.* **2001**, *334*, 179–192. (b) Belova, G. I.; Prasad, R.; Nazimov, I. V.; Wilson, S. H.; Slesarev, A. I. The domain organization and properties of individual domains of DNA topoisomerase V, a type 1B topoisomerase with DNA repair activities. *J. Biol. Chem.* **2002**, *277*, 4959–4965.
55. (a) Buhler, C.; Lebbink, J. H. G.; Bocs, C.; Ladenstein, R.; Forterre, P. DNA topoisomerase VI generates ATP-dependent double-strand breaks with two-nucleotide overhangs. *J. Biol. Chem.* **2001**, *276*, 37215–37222. (b) Bocs, C.; Buhler, C.; Forterre, P.; Bergerat, A. DNA topoisomerases VI from hyperthermophilic archaea. *Meth. Enzymol.* **2001**, *334*, 172–179.
56. (a) Wang, J. C. Interaction between DNA and an Escherichia coli proteiñ *J. Mol. Biol.* **1971**, *55*, 523–533. (b) Tse, Y.-C.; Kirkegaard, K.; Wang, J. C. Covalent bonds between protein and DNA. Formation of phosphotyrosine linkage between certain DNA topoisomerases and DNA. *J. Biol. Chem.* **1980**, *255*, 5560–5565. (c) Champoux, J. J. DNA is linked to the rat liver DNA nicking-closing enzyme by a phosphodiester bond to tyrosine. *J. Biol. Chem.* **1981**, *256*, 4805–4809.
57. (a) Brown, P. O.; Cozzarelli, N. R. Catenation and knotting of duplex DNA by type 1 topoisomerases: A mechanistic parallel with type 2 topoisomerases. *Proc. Natl. Acad. Sci. U.S.A.* **1981**, *78*, 843–847. (b) Tse, Y.; Wang, J. C. Escherichia coli and Micrococcus luteus DNA topoisomerase I can catalyze catenation or decantenation of double-stranded DNA rings. *Cell* **1980**, *22*, 269–276.
58. Li, Z.; Mondragón, A.; DiGate, R. J. The mechanism of type IA topoisomerase-mediated DNA topological transformations. *Mol. Cell* **2001**, *7*, 301–307.
59. Lima, C. D.; Wang, J. C.; Mondragón, A. Three-dimensional structure of the 67K N-terminal fragment of E. coli DNA topoisomerase I. *Nature* **1994**, *367*, 138–146.

60. (a) Beretta, G. L.; Zuco, V.; De Cesare, M.; Perego, P.; Zaffaroni, N. Namitecan: a hydrophilic camptothecin with a promising preclinical profile. *Curr. Med. Chem.* **2012**, *19*(21), 3488–3501. (b) Beretta, G. L.; Petrangolini, G.; De Cesare, M.; Pratesi, G.; Perego, P.; Tinelli, S.; Tortoreto, M.; Zucchetti, M.; Frapolli, R.; Bello, E.; et al. Biological properties of IDN5174, a new synthetic camptothecin with the open lactone ring. *Cancer Res.* **2006**, *66*, 10975–10982. (c) Constantinou, A. I.; Husband, A. Phenoxodiol (2H-1-benzopyran-7-ol, 3-(4-hydroxyphenyl)), a novel isoflavone derivative, inhibits DNA topoisomerase II by stabilizing the cleavable complex. *Anticancer Res.* **2002**, *22*, 2581–2585.
61. Ha, S. C.; Lowenhaupt, K.; Rich, A.; Kim, Y.-G.; Kim, K. K. Crystal structure of a junction between B-DNA and Z-DNA reveals two extruded bases. *Nature* **2005**, *437*, 1183–1186.
62. (a) Dervan, P. B. Design of sequence-specific DNA-binding molecules. *Science* **1986**, *232*, 464–471. (b) Hurley, L. H.; Boyd, F. L. Approaches toward the design of sequence-specific drugs for DNA. *Annu. Rep. Med. Chem.* **1987**, *22*, 259–268. (c) Hurley, L. H. DNA and associated targets for drug design. *J. Med. Chem.* **1989**, *32*, 2027–2033.
63. (a) Meysman, P.; Marchal, K.; Engelen, K. DNA structural properties in the classification of genomic transcription regulation elements. *Bioinform. Biol. Insights* **2012**, *6*, 155–168. (b) Rohs, Remo; W., Sean M.; Sosinsky, A.; Liu, P.; Mann, R. S.; Honig, B. The role of DNA shape in protein-DNA recognition. *Nature* **2009**, *461*, 1248–1253. (c) Badis, G.; Chan, E. T.; van Bakel, H.; Pena-Castillo, L.; et al. *Mol. Cell* **2008**, *32*, 878–887. (d) Grigorescu, A. A.; Rosenberg, J. M. DNA sequence recognition by proteins. *Encycl. Biol. Chem.* **2004**, *1*, 788–793.
64. Branden, C; Tooze, J. *Introduction to Protein Structure*, Garland Press, New York, 1991, p. 83.
65. Luger, K.; Mäder, A. W.; Richmond, R. K.; Sargent, D. F.; Richmond, T. J. Crystal structure of the nucleosome core particle at 2.8. ANG. resolution. *Nature* **1997**, *389*, 251–260.
66. Polach, K.J.; Widom, J. Mechanism of protein access to specific DNA sequences in chromatin: A dynamic equilibrium model for gene regulation. *J. Mol. Biol.* **1995**, *254*, 130–149.
67. (a) Pommier, Y.; Leo, E.; Zhang, H.; Marchand, C. DNA topoisomerases and their poisoning by anticancer and antibacterial drugs. *Chem. Biol.* **2010**, *17*, 421–433. (b) Pommier, Y. DNA topoisomerase I inhibitors: chemistry, biology, and interfacial inhibition. *Chem. Rev.* **2009**, *109*, 2894–2902. (c) Denny, W. A. Deoxyribonucleic acid topoisomerase inhibitors. *Comprehensive Med. Chem. II* **2006**, *7*, 111–128. (d) Liu, W. M.; te Poele, R. H. Recent advances and developments in the inhibitors of DNA topoisomerases. *Curr. Enz. Inhib.* **2007**, *3*, 161–174.
68. (a) Ceron-Carrasco, J. P.; Jacquemin, D. Influence of Mg^{2+} on the guanine-cytosine tautomeric equilibrium: simulations of the induced intermolecular proton transfer. *ChemPhysChem* **2011**, *12*(14), 2615–2623. (b) Zarytova, V. F.; Levina, A. S. Polyamine-containing DNA fragments. *Adv. Chem. Res.* **2010**, *4*, 1–54. (c) Haq, I. Thermodynamics of drug-DNA interactions. *Arch. Biochem. Biophys.* **2002**, *403*, 1–15.
69. Bando, T.; Sugiyama, H. Synthesis and biological properties of sequence-specific DNA-alkylating pyrrole-imidazole polyamides. *Acc. Chem. Res.* **2006**, *39*, 935–944.
70. (a) Burridge, J. M.; Quarendon, P.; Reynolds, C. A.; Goodford, P. J. Electrostatic potential and binding of drugs to the minor groove of DNA. *J. Mol. Graphics* **1987**, *5*, 165–166. (b) Zakrzewska, K.; Lavery, R.; Pullman, B. Theoretical studies of the selective binding to DNA of two nonintercalating ligands: netropsin and SN 18071. *Nucleic Acids Res.* **1983**, *11*, 8825–8839.
71. (a) Goodsell, D. S.; Kopka, M. L.; Dickerson, R. E. Refinement of netropsin bound to DNA: bias and feedback in electron density map interpretation. *Biochemistry* **1995**, *34*, 4983–4993. (b) Nunn, C. M.; Garman, E.; Neidle, S. Crystal structure of the DNA decamer d(CGCAATTGCG) complexed with the minor groove binding drug netropsin. *Biochemistry* **1997**, *36*, 4792–4799.
72. Kopka, M. L.; Yoon, C.; Goodsell, D.; Pjura, P.; Dickerson, R. E. The molecular origin of DNA-drug specificity in netropsin and distamycin. *Proc. Natl. Acad. Sci. U.S.A.* **1985**, *82*, 1376–1380.
73. (a) Strekowski, L.; Wilson, B. Noncovalent interactions with DNA: an overview. *Mutat. Res., Fund. Mol. Mech. Mutagen.* **2007**, *623*(1–2), 3–13. (b) Eriksson, M.; Norden, B. Linear and circular dichroism of drug-nucleic acid complexes. *Meth. Enzymol.* **2001**, *340*, 68–98. (c) Chaires, J. B. Energetics of drug-DNA interactions. *Biopolymers* **1998**, *44*, 201–215. (d) Neidle, S.; Abraham, Z. Structural and sequence-dependent aspects of drug intercalation into nucleic acids. *CRC Crit. Rev. Biochem.* **1984**, *17*, 73–121.
74. Lerman, L. S. Structural considerations in the interaction of deoxyribonucleic acid and acridines. *J. Mol. Biol.* **1961**, *3*, 18–30.
75. (a) Krugh, T. R.; Reinhardt, C. G. Evidence for sequence preferences in the intercalative binding of ethidium bromide to dinucleoside monophosphates. *J. Mol. Biol.* **1975**, *97*, 133–162. (b) Nuss, M. E.; Marsh, F. J.; Kollman, P. A. Theoretical studies of drug-dinucleotide interactions. Empirical energy function calculations on the interaction of ethidium, 9-aminoacridine, and proflavin cations with the base-paired dinucleotides GpC and CpG. *J. Am. Chem. Soc.* **1979**, *101*, 825–833.
76. (a) Kapur, A.; Beck, J. L.; Sheil, M. M. Observation of daunomycin and nogalamycin complexes with duplex DNA using electrospray ionization mass spectrometry. *Rapid Commun. Mass Spectrom.* **1999**, *13*, 2489–2497. (b) Rao, S. N.; Kollman, P. A. Molecular mechanical simulations on double intercalation of 9-amino acridine into d(CGCGCGC)·d(GCGCGCG): analysis of the physical basis for the neighbor-exclusion principle. *Proc. Natl. Acad. Sci. U.S.A.* **1987**, *84*, 5735–5739.
77. (a) Topcu, Z. DNA topoisomerases as targets for anticancer drugs. *J. Clin. Pharm. Therap.* **2001**, *26*, 405–416. (b) Toonen, T. R.; Hande, K. R. Topoisomerase II inhibitors. *Cancer Chemother. Biol. Response Mod.* **2001**, *19*, 129–147. (c) Stewart, C. F. Topoisomerase I interactive agents. *Cancer Chemother. Biol. Response Mod.* **2001**, *19*, 85–128. (d) Holden, J. A. DNA topoisomerases as anticancer drug targets: from the laboratory to the clinic. *Curr. Med. Chem. Anti-Cancer Ag.* **2001**, *1*, 1–25. (e) Froelich-Ammon, S. J.; Osheroff, N. Topoisomerase poisons: harnessing the dark side of enzyme mechanism. *J. Biol. Chem.* **1995**, *270*, 21429–21432. (f) Entire issue of *Biochim. Biophys. Acta* **1998**, 1400.
78. Ulukan, H.; Swaan, P. W. Camptothecins: a review of their chemotherapeutic potential. *Drugs* **2002**, *62*, 2039–2057.
79. Liu, L. F. DNA topoisomerase poisons as antitumor drugs. *Annu. Rev. Biochem.* **1989**, *58*, 351–375.
80. Nelson, E. M.; Tewey, K. M.; Liu, L. F. Mechanism of antitumor drug action: poisoning of mammalian DNA topoisomerase II on DNA by 4′-(9-acridinylamino)methanesulfon-m-anisidide. *Proc. Natl. Acad. Sci. U.S.A.* **1984**, *81*, 1361–1365.
81. Zhang, H.; D'Arpa, P.; Liu, L. F. A model for tumor cell killing by topoisomerase poisons. *Cancer Cells* **1990**, *2*, 23–27.

82. (a) Byl, J. A. W.; Fortune, J. M.; Burden, D. A.; Nitiss, J. L.; Utsugi, T.; Yamada, Y.; Osheroff, N. DNA topoisomerases as targets for the anticancer drug TAS-103: primary cellular target and DNA cleavage enhancement. *Biochemistry* **1999**, *38*, 15573–15579. (b) Fortune, J. M.; Velea, L.; Graves, D. E.; Utsugi, T.; Yamada, Y.; Osheroff, N. DNA topoisomerases as targets for the anticancer drug TAS-103: DNA interactions and topoisomerase catalytic inhibition. *Biochemistry* **1999**, *38*, 15580–15586.
83. (a) Bodley, A.; Liu, L. F.; Israel, M.; Seshadri, R.; Koseki, Y.; Giuliani, F. C.; Kirschenbaum, S.; Silber, R.; Potmesil, M. DNA topoisomerase II-mediated interaction of doxorubicin and daunorubicin congeners with DNA. *Cancer Res.*, **1989**, *49*, 5969–5978. (b) D'Arpa, P.; Liu, L. F. Topoisomerase-targeting antitumor drugs. *Biochim. Biophys. Acta* **1989**, *989*, 163–167.
84. Mannaberg, J. Ueber die wirkung von chininderivaten und phosphinen bei malariafiebern. *Arch. Klin. Med.* **1897**, *59*, 185.
85. Albert, A. *The Acridines*, 2nd ed., Edward Arnold, London, 1966.
86. Cain, B. F.; Atwell, G. J.; Seelye, R. N. Potential antitumor agents. 11. 9-Anilinoacridines. *J. Med. Chem.* **1971**, *14*, 311–315.
87. Goldin, A.; Serpick, A. A.; Mantel, N. Experimental screening procedures and clinical predictability value. *Cancer Chemother. Rep.* **1966**, *50*, 173–218.
88. Atwell, G. J.; Cain, B. F.; Seelye, R. N. Potential antitumor agents. 12. 9-Anilinoacridines. *J. Med. Chem.* **1972**, *15*, 611–615.
89. Denny, W. A.; Cain, B. F.; Atwell, G. J.; Hansch, C.; Panthananickal, A.; Leo, A. Potential antitumor agents. 36. Quantitative relationships between experimental antitumor activity, toxicity, and structure for the general class of 9-anilinoacridine antitumor agents. *J. Med. Chem.* **1982**, *25*, 276–315.
90. Zittoun, R. Amsacrine with high-dose cytarabine in acute leukemia. *Cancer Treat. Rep.* **1985**, *69*, 1447–1448.
91. Cain, B. F.; Atwell, G. J.; Denny, W. A. Potential antitumor agents. 16. 4′-(Acridin-9-ylamino)methanesulfonanilides. *J. Med. Chem.* **1975**, *18*, 1110–1117.
92. Waring, M. J. DNA-binding characteristics of acridinylmethanesulfonanilide drugs: comparison with antitumor properties. *Eur. J. Cancer* **1976**, *12*, 995–1001.
93. Braithwaite, A. W.; Baguley, B. C. Existence of an extended series of antitumor compounds which bind to deoxyribonucleic acid by nonintercalative means. *Biochemistry* **1980**, *19*, 1101–1106.
94. Sakore, T. D.; Reddy, B. S.; Sobell, H. M. Visualization of drug-nucleic acid interactions at atomic resolution. IV. Structure of an aminoacridine-dinucleoside monophosphate crystalline complex, 9-aminoacridine-5-iodocytidylyl (3′–5′) guanosine. *J. Mol. Biol.* **1979**, *135*, 763–785.
95. Denny, W. A.; Baguley, B. C.; Cain, B. F.; Waring, M. J. In *Molecular Aspects of Anti-cancer Drug Action*; Neidle, S.; Waring, M. J. (Eds.), Verlag Chemie, Weinheim, 1983, pp. 1–34.
96. Denny, W. A.; Wakelin, L. P. G. Kinetic and equilibrium studies of the interaction of amsacrine and anilino ring-substituted analogs with DNA. *Cancer Res.* **1986**, *46*, 1717–1721.
97. Baguley, B. C.; Denny, W. A.; Atwell, G. J.; Finlay, G. J.; Rewcastle, G. W.; Twigden, S. J.; Wilson, W. R. Synthesis, antitumor activity, and DNA binding properties of a new derivative of amsacrine, N-5-dimethyl-9-[(2-methoxy-4-methylsulfonylamino)phenylamino]-4-acridinecarboxamide. *Cancer Res.* **1984**, *44*, 3245–3251.
98. Waksman, S. A.; Woodruff, H. B. Bacteriostatic and bactericidal substances produced by a soil Actinomyces. *Proc. Soc. Exp. Biol. Med.* **1940**, *45*, 609–614.
99. Schulte, G. Z. New cytostatic agents in [cases of] hemoblastosis and carcinoma and comparison of their effects to X-ray therapy. *Krebsforsch.* **1952**, *58*, 500–503.
100. Sobell, H. M. Actinomycin and DNA transcription. *Proc. Natl. Acad. Sci. U.S.A.* **1985**, *82*, 5328–5331.
101. Müller, W.; Crothers, D. M. Studies of the binding of actinomycin and related compounds to DNA. *J. Mol. Biol.* **1968**, *35*, 251–290.
102. Cerami, A.; Reich, E.; Ward, D. C.; Goldberg, I. H. The interaction of actinomycin with DNA: requirement for the 2-amino group of purines. *Proc. Natl. Acad. Sci. U.S.A.* **1967**, *57*, 1036–1042.
103. Sobell, H. M.; Jain, S. C. Stereochemistry of actinomycin binding to DNA. II. Detailed molecular model of actinomycin-DNA complex and its implications. *J. Mol. Biol.* **1972**, *68*, 21–34.
104. Takusagawa, F.; Dabrow, M.; Neidle, S.; Berman, H. M. The structure of a pseudointercalated complex between actinomycin and the DNA binding sequence d(GpC). *Nature* **1982**, *296*, 466–469.
105. Takusagawa, F.; Goldstein, B. M.; Youngster, S.; Jones, R. A.; Berman, H. M. Crystallization and preliminary X-ray study of a complex between d(ATGCAT) and actinomycin D. *J. Biol. Chem.* **1984**, *259*, 4714–4715.
106. Takusagawa, F. The role of the cyclic depsipeptide rings in antibiotics. *J. Antibiot.* **1985**, *38*, 1596–1604.
107. Chiao, Y.-C. C.; Krugh, T. R. Actinomycin D complexes with oligonucleotides as models for the binding of the drug to DNA. Paramagnetic induced relaxation experiments on drug-nucleic acid complexes. *Biochemistry* **1977**, *16*, 747–755.
108. Gewirtz, D. A. A critical evaluation of the mechanisms of action proposed for the antitumor effects of the anthracycline antibiotics Adriamycin and daunorubicin. *Biochem. Pharmacol.* **1999**, *57*, 727–741.
109. Quigley, G. J.; Wang, A. H.-J.; Ughetto, G.; van der Marel, G.; van Boom, J. H.; Rich, A. Molecular structure of an anticancer drug-DNA complex: Daunomycin plus d(CpGpTpApCpG). *Proc. Natl. Acad. Sci. U.S.A.* **1980**, *77*, 7204–7208.
110. Patel, D. J.; Kozlowski, S. A.; Rice, J. A. Hydrogen bonding, overlap geometry, and sequence specificity in anthracycline antitumor antibiotic. DNA complexes in solution. *Proc. Natl. Acad. Sci. U.S.A.* **1981**, *78*, 3333–3337.
111. Henry, D. W. Structure-activity relationships among daunorubicin and adriamycin analogs. *Cancer Treat. Rep.* **1979**, *63*, 845–854.
112. (a) Zolova, O. E.; Mady, A. S. A.; Garneau-Tsodikova, S. Recent developments in bisintercalator natural products. *Biopolymers* **2010**, *93*, 777–790. (b) Dawson, S.; Malkinson, J. P.; Paumier, D.; Searcey, M. Bisintercalator natural products with potential therapeutic applications: isolation, structure determination, synthetic and biological studies. *Nat. Prod. Rep.* **2007**, *24*, 109–126. (c) Ughetto, G.; Wang, A. H. J.; Quigley, G. J.; Van der Marel, G. A.; Van Boom, J. H.; Rich, A. A comparison of the structure of echinomycin and triostin A complexed to a DNA fragment. *Nucleic Acids Res.* **1985**, *13*, 2305–2323.
113. Wright, R. G.; Wakelin, L. P.; Fieldes, A.; Acheson, R. M.; Waring, M. J. Effects of ring substituents and linker chains on the bifunctional intercalation of diacridines into deoxyribonucleic acid. *Biochemistry* **1980**, *19*, 5825–5836.
114. Wakelin, L. P. G. Polyfunctional DNA intercalating agents. *Med. Res. Rev.* **1986**, *6*, 275–340.

115. (a) Guelev, V.; Lee, J.; Ward, J.; Sorey, S.; Hoffman, D. W.; Iverson, B. L. Peptide bis-intercalator binds DNA via threading mode with sequence specific contacts in the major groove. *Chem. Biol.* **2001**, *8*, 415–425. (b) Leng, F.; Priebe, W.; Chaires, J. B. Ultratight DNA binding of a new bisintercalating anthracycline antibiotic. *Biochemistry* **1998**, *37*, 1743–1753.
116. Krumbhaar, E. B.; Krumbhaar, H. D. The blood and bone marrow in yellow cross gas (mustard gas) poisoning. Changes produced in the bone marrow of fatal cases. *J. Med. Res.* **1919**, *40*, 497–507.
117. Adair, F. E.; Bagg, H. J. Experimental and clinical studies on the treatment of cancer by dichloroethyl sulfide (mustard gas). *Ann. Surgery* **1931**, *93*, 190–199.
118. Gilman, A.; Philips, F. S. The biological actions and therapeutic applications of β-chloroethylamines and sulfides. *Science* **1946**, *103*, 409–415.
119. (a) Rhoads, C. P. Nitrogen mustards in the treatment of neoplastic disease; official statement. *J. Am. Med. Assoc.* **1946**, *131*, 656–658. (b) Goodman, L. S.; Wintrobe, M. M.; Dameshek, W.; Goodman, M. J.; Gilman, A.; McLennan, M. Nitrogen mustard therapy. Use of methylbis(2-chloroethyl)amine hydrochloride and tris(2-chloroethyl)amine hydrochloride for Hodgkin's disease, lymphosarcoma, leukemia and certain allied and miscellaneous disorders. *J. Am. Med. Assoc.* **1946**, *132*, 126–132.
120. Ross, W. C. J. In *Biological Alkylating Agents*, Butterworths, London, 1962.
121. Montgomery, J. A.; Johnston, T. P.; Shealy, Y. F. In *Burger's Medicinal Chemistry*, 4th ed.; Wolff, M. E. (Ed.), Wiley: New York, 1979, Part II, p. 595.
122. (a) Beranek, D. T. Distribution of methyl and ethyl adducts following alkylation with monofunctional alkylating agents. *Mutation Res.* **1990**, *231*, 11–30. (b) Lawley, P. D.; Phillips, D. H. DNA adducts from chemotherapeutic agents. *Mutation Res.* **1996**, *355*, 13–40. (c) Lawley, P. D.; Brookes, P. Alkylation of nucleic acids and their constituent nucleotides. *Biochem. J.* **1963**, *89*, 127–138.
123. Pullman, A.; Pullman, B. Electrostatic effect of the macromolecular structure on the biochemical reactivity of the nucleic acids. Significance for chemical carcinogenesis. *Int. J. Quant. Chem., Quant. Biol. Symp.* **1980**, *7*, 245–259.
124. Price, C. C. In *Handbook of Experimental Pharmacology*, Sartorelli, A. C.; Johns, D. J. (Eds.), Springer-Verlag, Berlin, 1974; Vol. 38, Part 2, p. 4.
125. (a) Kohn, K. W.; Spears, C. L.; Doty, P. Interstrand crosslinking of DNA by nitrogen mustard. *J. Mol. Biol.* **1966**, *19*, 266–288. (b) Lawley, P. D.; Brookes, P. Interstrand cross-linking of DNA by difunctional alkylating agents. *J. Mol. Biol.* **1967**, *25*, 143–160.
126. (a) Colvin, M.; Chabner, B. A. In *Cancer Chemotherapy: Principles and Practice*, Chabner, B. A.; Collins, S. M. (Eds.), Lippincott, Philadelphia, 1990, p. 276. (b) Bardos, T. J.; Chmielewicz, Z. F.; Hebborn, P. Structure-activity relations of alkylating agents in cancer chemotherapy. *Ann. N. Y. Acad. Sci.* **1969**, *163*, 1006–1025.
127. (a) Niculescu-Duvaz, I.; Baracu, I.; Balaban, A. T. In *The Chemistry of Antitumour Agents*, Wilman, D. E. V. (Ed.), Blackie: Glasgow, 1990, p. 63. (b) Kohn, K. W.; Erickson, L. C.; Laurent, G.; Ducore, J. M.; Sharkey, N. A.; Ewig, R. A. G. In *Nitrosoureas, Current Status and New Developments*, Prestayo, W.; Crooke, S. T.; Karter, S. K.; Schein, P. S. (Eds.), Academic Press, New York, 1981, p. 69.
128. (a) Guza, R.; Pegg, A. E.; Tretyakova, N. Effects of sequence context on O^6-alkylguanine DNA alkyltransferase repair of O^6-alkyldeoxyguanosine adducts. *ACS Symp. Ser.* **2010**, *1041*, 73–101. (b) Dahlmann, H. A.; Vaidyanathan, V. G.; Sturla, S. J. Investigating the biochemical impact of DNA damage with structure-based probes: abasic sites, photodimers, alkylation adducts, and oxidative lesions. *Biochemistry* **2009**, *48*, 9347–9359. (c) Harris, A. L. DNA repair and resistance to chemotherapy. *Cancer Surv.* **1985**, *4*, 601–624.
129. Prelog, V.; Stepan, V. Bis(β-haloethyl)amines. VII. A new synthesis of N-monoalkylpiperazines. *Coll. Czech. Chem. Commun.* **1935**, *7*, 93–102.
130. Brookes, P.; Lawley, P. D. Reaction of mono- and difunctional alkylating agents with nucleic acids. *Biochem. J.* **1961**, *80*, 496–503.
131. (a) Oida, T.; Humphreys, W. G.; Guengerich, F. P. Preparation and characterization of oligonucleotides containing S-[2-(N7-guanyl)ethyl] glutathione. *Biochemistry* **1991**, *30*, 10513–10522. (b) Persmark, M.; Guengerich, F. P. Spectroscopic and thermodynamic characterization of the interaction of N7-guanyl thioether derivatives of d(TGCTG*CAAG) with potential complements. *Biochemistry* **1994**, *33*, 8662–8672.
132. (a) Gates, K. S. An overview of chemical processes that damage cellular DNA: Spontaneous hydrolysis, alkylation, and reactions with radicals. *Chem. Res. Toxicol.* **2009**, *22*, 1747–1760. (b) Kim, Y.-J.; Wilson, D. M. III. Overview of base excision repair biochemistry. *Curr. Mol. Pharmacol.* **2012**, *5*, 3–13.
133. Gates, K. S.; Nooner, T.; Dutta, S. Biologically relevant chemical reactions of N7-alkylguanine residues in DNA. *Chem. Res. Toxicol.* **2004**, *17*, 839–856.
134. Greenberg, M. M.; Hantosi, Z.; Wiederholt, C. J.; Rithner, C. D. Studies on N4-(2-Deoxy-D-pentofuranosyl)-4,6-diamino-5-formamidopyrimidine (Fapy.dA) and N6-(2-Deoxy-D-pentofuranosyl)-6-diamino-5-formamido-4-hydroxypyrimidine (Fapy.dG). *Biochemistry* **2001**, *40*, 15856–15861.
135. (a) Dutta, S.; Chowdhury, G.; Gates, K. S. Interstrand cross-links generated by abasic sites in duplex DNA. *J. Am. Chem. Soc.* **2007**, *129*, 1852–1853. (b) Gates, K. S. *J. Am. Chem. Soc.* **2013**, *135*, 1015–1025.
136. Haddow, A.; Kon, G. A. R.; Ross, W. C. J. Effects upon tumours of various haloalkylarylamines. *Nature* **1948**, *162*, 824.
137. Everett, J. L.; Roberts, J. J.; Ross, W. C. J. Aryl-2-haloalkylamines. XII. Some carboxylic derivatives of N, N-bis(2-chloroethyl)aniline. *J. Chem. Soc.* **1953**, 2386–2392.
138. Schmidt, L. H.; Fradkin, R.; Sullivan, R.; Flowers, A. Comparative pharmacology of alkylating agents. II. Therapeutic activities of alkylating agents and reference compounds against various tumor systems. *Cancer Chemother. Rep.* **1965**, (Suppl. 2), 1–1528.
139. Vistica, D. T. Cellular pharmacokinetics of the phenylalanine mustards. *Pharmacol. Ther.* **1983**, *22*, 379–406.
140. Patil, A. D.; et al. *Tetrahedron Lett.* **1997**, *38*, 363–364.
141. Dutta, S.; Abe, H.; Aoyagi, S.; Kibayashi, C.; Gates, K. S. DNA damage by fasicularin. *J. Am. Chem. Soc.* **2005**, *127*, 15004–15005.
142. Goldin, A.; Wood, H. B., Jr. Preclinical investigation of alkylating agents in cancer chemotherapy. *Ann. N. Y. Acad. Sci.* **1969**, *163*, 954–1005.
143. Khan, A. H.; Driscoll, J. S. Potential central nervous system antitumor agents. Aziridinylbenzoquinones. *J. Med. Chem.* **1976**, *19*, 313–317.
144. Haddow, A.; Timmis, G. M. Myleran in chronic myeloid leukemia. Chemical constitution and biological action. *Lancet* **1953**, *1*, 207–208.

145. Timmis, G. M.; Hudson, R. F. in Discussion: Part 1 in Comparative clinical and biological effects of alkylating agents. *Ann. N. Y. Acad. Sci.* **1958**, *68*, 727–730.
146. Brookes, P.; Lawley, P. D. The alkylation of guanosine and guanylic acid. *J. Chem. Soc.* **1961**, 3923–3928.
147. Tong, W. P.; Ludlam, D. B. Crosslinking of DNA by busulfan. Formation of diguanyl derivatives. *Biochim. Biophys. Acta* **1980**, *608*, 174–181.
148. Hanka, L. J.; Dietz, A.; Gerpheide, S. A.; Kuentzel, S. L.; Martin, D. G. CC-1065 (NSC-298223), a new antitumor antibiotic. Production, in vitro biological activity, microbiological assays and taxonomy of the producing microorganism. *J. Antibiot.* **1978**, *31*, 1211–1217.
149. Takahashi, I.; Takahashi, K.; Ichimura, M.; Morimoto, M.; Asano, K.; Kawamoto, I.; Tomita, F.; Nakano, H. Duocarmycin A, a new antitumor antibiotic from Streptomyces. *J. Antibiot.* **1988**, *41*, 1915–1917.
150. Ichimura, M.; Ogawa, T.; Takahashi, K.; Kobayashi, E.; Kawamoto, I.; Yasuzawa, T.; Takahashi, I.; Nakano, H. Duocarmycin SA, a new antitumor antibiotic from Streptomyces sp. *J. Antibiot.* **1990**, *43*, 1037–1038.
151. (a) Boger, D. L.; Garbaccio, R. M. Shape-dependent catalysis: insights into the source of catalysis for the CC-1065 and duocarmycin DNA alkylation reaction. *Acc. Chem. Res.* **1999**, *32*, 1043–1052. (b) Boger, D. L.; Johnson, D. S. CC-1065 and the duocarmycins: unraveling the keys to a new class of naturally derived DNA alkylating agents. *Proc. Natl. Acad. Sci. U.S.A.* **1995**, *92*, 3642–3649. (c) Searcey, M. Duocarmycins—nature's prodrugs? *Curr. Pharm. Design* **2002**, *8*, 1375–1389.
152. Baird, R.; Winstein, S. Neighboring carbon and hydrogen. LI. Dienones from Ar1θ-3 participation. Isolation and behavior of spiro[2,5] octa-1,4-dien-3-one. *J. Am. Chem. Soc.* **1963**, *85*, 567–578.
153. (a) Hurley, L. H.; Reynolds, V. L.; Swenson, D. H.; Petzold, G. L.; Scahill, T. A. Reaction of the antitumor antibiotic CC-1065 with DNA: structure of a DNA adduct with DNA sequence specificity. *Science* **1984**, *226*, 843–844. (b) Hurley, L. H.; Lee, C.-S.; McGovren, J. O.; Mitchell, M.; Warpehoski, M. A.; Kelly, R. C.; Aristoff, P. A. Molecular basis for sequence-specific DNA alkylation by CC-1065. *Biochemistry* **1988**, *27*, 3886–3892.
154. Lin, C. H.; Beale, J. M.; Hurley, L. H. Structure of the (+)-CC-1065-DNA adduct: critical role of ordered water molecules and implications for involvement of phosphate catalysis in the covalent reaction. *Biochemistry* **1991**, *30*, 3597–3602.
155. Boger, D. L.; Turnbull, P. Synthesis and evaluation of a carbocyclic analog of the CC-1065 and duocarmycin alkylation subunits: role of the vinylogous amide and implications on DNA alkylation catalysis. *J. Org. Chem.* **1998**, *63*, 8004–8011.
156. (a) Boger, D. L.; Boyce, C. W.; Johnson, D. S. pH dependence of the rate of DNA alkylation for (+)-duocarmycin SA and (+)-CCBI-TMI. *Bioorg. Med. Chem. Lett.* **1997**, *7*, 233–238. (b) Boger, D. L.; Garbaccio, R. M. Are the duocarmycin and CC-1065 DNA alkylation reactions acid-catalyzed? Solvolysis pH-rate profiles suggest they are not. *J. Org. Chem.* **1999**, *64*, 5666–5669.
157. Warpehoski, M. A.; Harper, D. E. Enzyme-like rate acceleration in the DNA minor groove. Cyclopropylpyrroloindoles as mechanism-based inactivators of DNA. *J. Am. Chem. Soc.* **1995**, *117*, 2951–2952.
158. Skinner, W. A.; Gram, H. F.; Greene, M. O. Potential anticancer agents. XXXI. Relationship of chemical structure to antileukemic activity with analogs of 1-methyl-3-nitro-1-nitrosoguanidine (NSC-9369). *J. Med. Pharm. Chem.* **1960**, *2*, 299–333.
159. Schabel, F. M., Jr.; Johnston, T. P.; McCaleb, G. S.; Montgomery, J. A.; Laster, W. R.; Skipper, H. E. Experimental evaluation of potential anticancer agents. VIII. Effects of certain nitrosoureas on intracerebral L1210 leukemia. *Cancer Res.* **1963**, *23*, 725–733.
160. Montgomery, J. A.; James, R.; McCaleb, G. S.; Johnston, T. P. Modes of decomposition of 1,3-bis(2-chloroethyl)-1-nitrosourea and related compounds. *J. Med. Chem.* **1967**, *10*, 668–674.
161. Wheeler, G. P. In *Handbook of Experimental Pharmacology*, Sartorelli, A. C.; Johns, D. G. (Eds.), Springer-Verlag, Berlin, 1974, Vol. 38, Part 2, p. 7.
162. Johnston, T. P.; Montgomery, J. A. Relationship of structure to anticancer activity and toxicity of the nitrosoureas in animal systems. *Cancer Treat. Rep.* **1986**, *70*, 13–30.
163. Schmall, B.; Cheng, C. J.; Fujimura, S.; Gersten, N.; Grunberger, D.; Weinstein, I. B. Modification of proteins by 1-(2-chloroethyl)3-cyclohexyl-1-nitrosourea (NSC 79037) in vitro. *Cancer Res.* **1973**, *33*, 1921–1924.
164. Baril, B. B.; Baril, E. F.; Laszlo, J.; Wheeler, G. P. Inhibition of rat liver DNA polymerase by nitrosoureas and isocyanates. *Cancer Res.* **1975**, *35*, 1–5.
165. (a) Kann, H. E., Jr.; Blumenstein, B. A.; Petkas, A.; Schott, M. A. Radiation synergism by repair-inhibiting nitrosoureas in L1210 cells. *Cancer Res.* **1980**, *40*, 771–775. (b) Robins, P.; Harris, A. L.; Goldsmith, I.; Lindahl, T. Crosslinking of DNA induced by chloroethylnitrosourea is prevented by O^6-methylguanine-DNA methyltransferase. *Nucleic Acids Res.* **1983**, *11*, 7743–7758.
166. (a) Tubbs, J. L.; Pegg, A. E.; Tainer, J. A. DNA binding, nucleotide flipping, and helix-turn-helix motif in base repair by O^6-alkylguanine-DNA alkyltransferase and its implications for cancer chemotherapy. *DNA Repair* **2007**, *6*, 1100–1115. (b) Hou, Q.; Du, L.; Gao, J.; Liu, Y.; Liu, C. QM/MM study on the reaction mechanism of O^6-alkylguanine-DNA alkyltransferase. *J. Phys. Chem. B* **2010**, *114*, 15296–15300.
167. Hang, B. Base excision repair. In *DNA Repair, Genetic Instability, and Cancer*, Wei, Q.; Li, L.; Chen, D. J. (Eds.), World Scientific Publishing, Hackensack, NJ, 2007.
168. Kann, H. E., Jr.; Kohn, K. W.; Widerlite, L.; Gullion, D. Effects of 1,3-bis(2-chloroethyl)-1-nitrosourea and related compounds on nuclear RNA metabolism. *Cancer Res.* **1974**, *34*, 1982–1988.
169. Panasci, L. C.; Green, D.; Nagourney, R.; Fox, P.; Schein, P. S. A structure-activity analysis of chemical and biological parameters of chloroethylnitrosoureas in mice. *Cancer Res.* **1977**, *37*, 2615–2618.
170. (a) Tong, W. P.; Kirk, M. C.; Ludlum, D. B. Formation of the crosslink 1-[N^3-deoxycytidyl]-2-[N^1-deoxyguanosinyl]ethane in DNA treated with N, N′-bis(2-chloroethyl)-N-nitrosourea. *Cancer Res.* **1982**, *42*, 3102–3105. (b) Tong, W. P.; Kirk, M. C.; Ludlum, D. B. Mechanism of action of the nitrosoureas. V. Formation of O^6-(2-fluoroethyl)guanine and its probable role in the crosslinking of deoxyribonucleic acid. *Biochem. Pharmacol.* **1983**, *32*, 2011–2015. (c) Lown, J. W.; McLaughlin, L. W.; Chang, Y.-M. Mechanism of action of 2-haloethylnitrosoureas on DNA and its relation to their antileukemic properties. *Bioorg. Chem.* **1978**, *7*, 97–110.
171. Hayes, M. T.; Bartley, J.; Parsons, P. G.; Eaglesham, G.K.; Prakash, A. S. Mechanism of action of fotemustine, a new chloroethylnitrosourea anticancer agent: evidence for the formation of two reactive intermediates contributing to cytotoxicity. *Biochemistry* **1997**, *36*, 10646–10654.

172. Buckley, N.; Brent, T. P. Structure-activity relations of (2-chloroethyl) nitrosoureas. 2. Kinetic evidence of a novel mechanism for the cytotoxically important DNA cross-linking reactions of (2-chloroethyl) nitrosoureas. *J. Am. Chem. Soc.* **1988**, *110*, 7520–7529.
173. Piper, J. R.; Laseter, A. G.; Johnston, T. P.; Montgomery, J. A. Synthesis of potential inhibitors of hypoxanthine-guanine phosphoribosyltransferase for testing as antiprotozoal agents. 2. 1-Substituted hypoxanthines. *J. Med. Chem.* **1980**, *23*, 1136–1139.
174. (a) Erickson, L. C.; Sharkey, N. A.; Kohn, K. W. DNA crosslinking and monoadduct repair in nitrosourea-treated human tumor cells. *Nature* **1980**, *288*, 727–729. (b) Brent, T. P.; Houghton, P. J.; Houghton, J. A. O^6-Alkylguanine-DNA alkyltransferase activity correlates with the therapeutic response of human rhabdomyosarcoma xenografts to 1-(2-chloroethyl)-3-(trans-4-methylcyclohexyl)-1-nitrosourea. *Proc. Natl. Acad. Sci. U.S.A.* **1985**, *82*, 2985–2989.
175. Ludlum, D. B.; Mehta, J. R.; Tong, W. P. Prevention of 1-(3-deoxycytidyl),2-(1-deoxyguanosinyl)ethane crosslink formation in DNA by rat liver O^6-alkylguanine-DNA alkyltransferase. *Cancer Res.* **1986**, *46*, 3353–3357.
176. (a) Dolan, M. E.; Pegg, A. E. O^6-Benzylguanine and its role in chemotherapy. *Clin. Cancer Res.* **1997**, *3*, 837–847. (b) Karran, P. Mechanisms of tolerance to DNA damaging therapeutic drugs. *Carcinogenesis* **2001**, *22*, 1931–1937.
177. Comis, R. L. DTIC (NSC-45388) in malignant melanoma: a perspective. *Cancer Treat. Rep.* **1976**, *60*, 165–176.
178. (a) Skibba, J. L.; Ramirez, G.; Beal, D. D.; Bryan, G. T. Metabolism of 4(5)-(3,3-dimethyl-1-triazeno)-imidazole-5(4)-carboxamide to 4(5)-aminoimidazole-5(4)-carboxamide in man. *Biochem. Pharmacol.* **1970**, *19*, 2043–2051. (b) Mizuno, N. S.; Humphrey, E. W. Metabolism of 5-(3,3-dimethyl-1-triazeno)imidazole-4-carboxamide (NSC-45388) in human and animal tumor tissue. *Cancer Chemother. Rep. (Part 1)* **1972**, *56*, 465–472.
179. (a) Vanelle, P.; Terme, T.; Giraud, L.; Crozet, M. P. Progress in electron transfer reactions of new quinone bioreductive alkylating agents. *Recent Res. Develop. Org. Chem.* **2000**, *4*(Part 1), 1–28. (b) Beall, H. D.; Winski, S. L. Mechanisms of action of quinone-containing alkylating agents. I. NQO1-directed drug development. *Front. Biosci. (Electronic)* **2000**, *5*, D639–D648. (c) Moore, H. W.; Czerniak, R. Naturally occurring quinones as potential bioreductive alkylating agents. *Med. Res. Rev.* **1981**, *1*, 249–280.
180. Tomasz, M.; Palom, Y. The mitomycin bioreductive antitumor agents: crosslinking and alkylation of DNA as the molecular basis of their activity. *Pharmacol. Therap.* **1997**, *76*, 73–87.
181. Iyer, V. N.; Szybalski, W. A. Mitomycins and porfiromycin: chemical mechanism of activation and cross-linking of DNA. *Science* **1964**, *145*, 55–58.
182. (a) Fitzsimmons, S. A.; Workman, P.; Grever, M.; Paull, K.; Camalier, R.; Lewis, A. D. Reductase enzyme expression across the National Cancer Institute tumor cell line panel: Correlation with sensitivity to mitomycin C and EO9. *J. Natl. Cancer Inst.* **1996**, *88*, 259–269. (b) Kumar, G. S.; Lipman, R.; Cummings, J.; Tomasz, M. Mitomycin C-DNA adducts generated by DT-diaphorase. Revised mechanism of the enzymic reductive activation of mitomycin C. *Biochemistry* **1997**, *36*, 14128–14136.
183. Bean, M.; Kohn, H. Studies on the reaction of mitomycin C with potassium thiobenzoate under reductive conditions. *J. Org. Chem.* **1985**, *50*, 293–298.
184. (a) Franck, R. W.; Tomasz, M. In *The Chemistry of Antitumor Agents*; Wilman, D. E. V. (Ed.), Blackie and Son, Glasgow, 1990, p. 379. (b) Remers, W. A. *The Chemistry of Antitumor Antibiotics*, Vol. 1, Wiley: New York, 1979, p. 271.
185. (a) Kohn, H.; Zein, N.; Lin, X. Q.; Ding, J.-Q.; Kadish, K. M. Mechanistic studies on the mode of reaction of mitomycin C under catalytic and electrochemical reductive conditions. *J. Am. Chem. Soc.* **1987**, *109*, 1833–1840. (b) Danishefsky, S. J.; Egbertson, M. The characterization of intermediates in the mitomycin activation cascade: a practical synthesis of an aziridinomitosene. *J. Am. Chem. Soc.* **1986**, *108*, 4648–4650. (c) Andrews, P. A.; Pan, S.-S.; Bachur, N. R. Electrochemical reductive activation of mitomycin C. *J. Am. Chem. Soc.* **1986**, *108*, 4158–4166.
186. Kohn, H.; Hong, Y. P. Observations on the activation of mitomycin C. Requirements for C-10 functionalization. *J. Am. Chem. Soc.* **1990**, *112*, 4596–4598.
187. Tomasz, M.; Chowdary, D.; Lipman, R.; Shimotakahara, S.; Veiro, D.; Walker, V.; Verdine, G. L. Reaction of DNA with chemically or enzymically activated mitomycin C: isolation and structure of the major covalent adduct. *Proc. Natl. Acad. Sci. U.S.A.* **1986**, *83*, 6702–6706.
188. (a) Borowy-Borowski, H.; Lipman, R.; Chowdary, D.; Tomasz, M. Duplex oligodeoxyribonucleotides crosslinked by mitomycin C at a single site: synthesis, properties, and crosslink. *Biochemistry* **1990**, *29*, 2992–2999. (b) Borowy-Borowski, H.; Lipman, R.; Tomasz, M. Recognition between mitomycin C and specific DNA sequences for cross-link formation. *Biochemistry* **1990**, *29*, 2999–3006.
189. (a) Tomasz, M.; Lipman, R.; Chowdary, D.; Pawlak, J.; Verdine, G. L.; Nakanishi, K. Isolation and structure of a covalent cross-link adduct between mitomycin C and DNA. *Science* **1987**, *235*, 1204–1208. (b) Tomasz, M.; Lipman, R.; McGuinness, B. F.; Nakanishi, K. Isolation and characterization of a major adduct between mitomycin C and DNA. *J. Am. Chem. Soc.* **1988**, *110*, 5892–5896.
190. Millard, J. T.; Weidner, M. F.; Raucher, S.; Hopkins, P. B. Determination of the DNA crosslinking sequence specificity of reductively activated mitomycin C at single-nucleotide resolution: deoxyguanosine residues at CpG are crosslinked preferentially. *J. Am. Chem. Soc.* **1990**, *112*, 3637–3641.
191. Kennedy, K. A.; Teicher, B. A.; Rockwell, S.; Sartorelli, A. C. The hypoxic tumor cell: a target for selective cancer chemotherapy. *Biochem. Pharmacol.* **1980**, *29*, 1–8.
192. Antonini, I.; Lin, T.-S.; Cosby, L. A.; Dai, Y.-R.; Sartorelli, A. C. 2- and 6-methyl-1,4-naphthoquinone derivatives and potential bioreductive alkylating agents. *J. Med. Chem.* **1982**, *25*, 730–735.
193. Lin, A. J.; Lillis, B. J.; Sartorelli, A. C. Potential bioreductive alkylating agents. 5. Antineoplastic activity of quinoline-5,8-diones, naphthazarins, and naphthoquinones. *J. Med. Chem.* **1975**, *18*, 917–921.
194. (a) Lin, A. J.; Sartorelli, A. C. Potential bioreductive alkylating agents. VI. Determination of the relationship between oxidation-reduction potential and antineoplastic activity. *Biochem. Pharmacol.* **1976**, *25*, 206–207. (b) Prakash, G.; Hodnett, E. M. Discriminant analysis and structure-activity relationships. 1. Naphthoquinones. *J. Med. Chem.* **1978**, *21*, 369–374.
195. Hara, M.; Takahashi, I.; Yoshida, M.; Kawamoto, I.; Morimoto, M.; Nakano, H. DC 107, a novel antitumor antibiotic produced by a Streptomyces sp. *J. Antibiot.* **1989**, *42*, 333–335.

196. Hara, M.; Saitoh, Y.; Nakano, H. DNA strand scission by the novel antitumor antibiotic leinamycin. *Biochemistry* **1990**, *29*, 5676–5681.
197. Behroozi, S. J.; Kim, W.; Gates, K. S. Reaction of n-propanethiol with 3H-1,2-benzodithiol-3-one 1-oxide and 5,5-dimethyl-1,2-dithiolan-3-one 1-oxide: studies related to the reaction of antitumor antibiotic leinamycin with DNA. *J. Org. Chem.* **1995**, *60*, 3964–3966.
198. (a) Gates, K. S. Mechanisms of DNA damage by leinamycin. *Chem. Res. Toxicol.* **2000**, *13*, 953–956. (b) Mitra, K.; Gates, K. S. Chemistry of thiol-dependent DNA damage by the antitumor antibiotic leinamycin. *Recent Res. Devel. Org. Chem.* **1999**, *3*, 311–317.
199. Asai, A.; Hara, M.; Kakita, S.; Kanda, Y.; Yoshida, M.; Saito, H.; Saitoh, Y. Thiol-mediated DNA alkylation by the novel antitumor antibiotic leinamycin. *J. Am. Chem. Soc.* **1996**, *118*, 6802–6803.
200. Asai, A.; Saito, H.; Saitoh, Y. Thiol-independent DNA cleavage by a leinamycin degradation product. *Bioorg. Med. Chem.* **1997**, *5*, 723–729.
201. Nooner, T.; Dutta, S.; Gates, K. S. Chemical properties of the leinamycin-guanine adduct in DNA. *Chem. Res. Toxicol.* **2004**, *17*, 942–949.
202. (a) Mitra, K.; Kim, W.; Daniels, J. S.; Gates, K. S. Oxidative DNA cleavage by the antitumor antibiotic leinamycin and simple 1,2-dithiolan-3-one 1-oxides: Evidence for thiol-dependent conversion of molecular oxygen to DNA-cleaving oxygen radicals mediated by polysulfides. *J. Am. Chem. Soc.* **1997**, *119*, 11691–11692. (b) Breydo, L.; Gates, K. S. Thiol-dependent DNA cleavage by 3H-1,2-benzodithiol-3-one 1,1-dioxide. *Bioorg. Med. Chem. Lett.* **2000**, *10*, 885–889. (c) Behroozi, S. B.; Kim, W.; Gates, K. S. 1,2-dithiolan-3-one 1-oxides: a class of thiol-activated DNA-cleaving agents that are structurally related to the natural product leinamycin. *Biochemistry* **1996**, *35*, 1768–1774.
203. (a) Lown, J. W.; Sim, S.-K.; Majumdar, K. C.; Chang, R.-Y. Studies related to antitumor antibiotics. Part XI. Strand scission of DNA by bound adriamycin and daunorubicin in the presence of reducing agents. *Biochem. Biophys. Res. Commun.* **1977**, *76*, 705–710. (b) Bachur, N. R.; Gordon, S. L.; Gee, M. V. A general mechanism for microsomal activation of quinone anticancer agents to free radicals. *Cancer Res.* **1978**, *38*, 1745–1750.
204. Ross, W. A.; Glaubiger, D. L.; Kohn, K. W. Protein-associated DNA breaks in cells treated with adriamycin or ellipticine. *Biochim. Biophys. Acta* **1978**, *519*, 23–30.
205. Pan, S.-S.; Pedersen, L.; Bachur, N. R. Comparative flavoprotein catalysis of anthracycline antibiotic. Reductive cleavage and oxygen consumption. *Mol. Pharmacol.* **1981**, *19*, 184–186.
206. Hertzberg, R. P.; Dervan, P. B. Cleavage of DNA with methidium-propyl-EDTA-iron(II): reaction conditions and product analyses. *Biochemistry* **1984**, *23*, 3934–3945.
207. Garnier-Suillerot, A. In Lown, J. W. (Ed.), *Anthracycline and Anthracenedione-based Anticancer Agents*, Elsevier, Amsterdam, 1988, pp. 129–157.
208. Walling, C. Fenton's reagent revisited. *Acc. Chem. Res.* **1975**, *8*, 125–131.
209. Pestovsky, O.; Stoian, S.; Bominaar, E.L.; Shan, X.; Münck, E.; Que, L., Jr.; Bakac, A. Aqueous Fe^{iv}=O: Spectroscopic identification and oxo-group exchange. *Angew. Chem. Int. Ed.* **2005**, *44*, 6871–6874.
210. Tullius, T. D.; Dombrowski, B. A.; Churchill, M. E. A.; Kam, L. Hydroxyl radical footprinting: a high-resolution method for mapping protein-DNA contacts. *Methods Enzymol.* **1987**, *155*, 537–538.
211. Speyer, J. L.; Green, M. D.; Kramer, E.; Rey, M.; Sanger, J.; Ward, C.; Dubin, N.; Ferran, V.; Stecy, P.; Zeleniuch-Jaquotte, A.; Wernz, J.; Feit, F.; Slater, W.; Blum, R.; Mugia, F. Protective effect of the bispiperazinedione ICRF-187 against doxorubicin-induced cardiac toxicity in women with advanced breast cancer. *N. Engl. J. Med.* **1988**, *319*, 745–752.
212. Gianni, L.; Vigano, L.; Lanzi, C.; Niggeler, M.; Malatesta, V. Role of daunosamine and hydroxyacetyl side chain in reaction with iron and lipid peroxidation by anthracyclines. *J. Natl. Cancer Inst.* **1988**, *80*, 1104–1111.
213. Umezawa, H.; Suhara, Y.; Takita, T.; Maeda, K. Purification of bleomycins. *J. Antibiot. Ser. A.* **1966**, *19*, 210–215.
214. (a) Stubbe, J.; Kozarich, J. W.; Wu, W.; Vanderwall, D. E. Bleomycins: a structural model for specificity, binding, and double strand cleavage. *Acc. Chem. Res.* **1996**, *29*, 322. (b) Povirk, L. F.; Han, Y.-H.; Steighner, R. J. Structure of bleomycin-induced DNA double-strand breaks: predominance of blunt ends and single-base 5′ extensions. *Biochemistry* **1989**, *28*, 5808–5814. (c) Steighner, R. J.; Povirk, L. F. Bleomycin-induced DNA lesions at mutational hot spots: implications for the mechanism of double-strand cleavage. *Proc. Natl. Acad. Sci. U.S.A.* **1990**, *87*, 8350–8354.
215. Stubbe, J.; Kozarich, J. W. Mechanisms of bleomycin-induced DNA degradation. *Chem. Rev.* **1987**, *87*, 1107–1136.
216. Hamamichi, N.; Hecht, S. M. On the role of individual bleomycin thiazoles in oxygen activation and DNA cleavage. *J. Am. Chem. Soc.* **1992**, *114*, 6278–6291.
217. Fisher, L. M.; Kuroda, R.; Sakai, T. T. Interaction of bleomycin A2 with deoxyribonucleic acid: DNA unwinding and inhibition of bleomycin-induced DNA breakage by cationic thiazole amides related to bleomycin A2. *Biochemistry* **1985**, *24*, 3199–3207.
218. (a) Sausville, E. A.; Stein, R. W.; Peisach, J.; Horwitz, S. B. Properties and products of the degradation of DNA by bleomycin and iron(II). *Biochemistry* **1978**, *17*, 2746–2754. (b) Hecht, S. M. The chemistry of activated bleomycin. *Acc. Chem. Res.* **1986**, *19*, 383–391.
219. (a) Ciriolo, M. R.; Magliozzo, R. S.; Peisach, J. Microsome-stimulated activation of ferrous bleomycin in the presence of DNA. *J. Biol. Chem.* **1987**, *262*, 6290–6295. (b) Mahmutoglu, I.; Kappus, H. Redox cycling of bleomycin-iron(III) by an NADH-dependent enzyme, and DNA damage in isolated rat liver nuclei. *Biochem. Pharmacol.* **1987**, *36*, 3677–3681.
220. Umezawa, H.; Takita, T.; Sugiura, Y.; Otsuka, M.; Kobayashi, S.;Ohno, M. DNA-bleomycin interaction. Nucleotide sequence-specific binding and cleavage of DNA by bleomycin. *Tetrahedron* **1984**, *40*, 501–509.
221. Povirk, L. F.; Hogan, M.; Dattagupta, N. Binding of bleomycin to DNA: intercalation of the bithiazole rings. *Biochemistry* **1979**, *18*, 96–101.
222. Burger, R. M.; Peisach, J.; Horwitz, S. B. Activated bleomycin. A transient complex of drug, iron, and oxygen that degrades DNA. *J. Biol. Chem.* **1981**, *256*, 11636–11644.
223. Heimbrook, D. C.; Mulholland, R. L., Jr.; Hecht, S. M. Multiple pathways in the oxidation of cis-stilbene by iron-bleomycin. *J. Am. Chem. Soc.* **1986**, *108*, 7839–7840.
224. Burger, R. M.; Kent, T. A.; Horwitz, S. B.; Münck, E.; Peisach, J. Moessbauer study of iron bleomycin and its activation intermediates. *J. Biol. Chem.* **1983**, *258*, 1559–1564.
225. Sam, J. W.; Tang, X.-J.; Peisach, J. Electrospray mass spectrometry of iron bleomycin: demonstration that activated bleomycin is a ferric peroxide complex. *J. Am. Chem. Soc.* **1994**, *116*, 5250–5256.
226. Westre, T. E.; Loeb, K. E.; Zaleski, J. M.; Hedman, B.; Hodgson, K. O.; Solomon, E. I. Determination of the geometric and electronic structure of activated bleomycin using X-ray absorption spectroscopy. *J. Am. Chem. Soc.* **1995**, *117*, 1309–1313.

227. Solomon, E. I.; Brunold, T. C.; Davis, M. I.; Kemsley, J. N.; Lee, S.-K.; Lehnert, N.; Neese, F.; Skulan, A. J.; Yang, Y.-S.; Zhou, J. Geometric and electronic structure/function correlations in non-heme iron enzymes. *Chem. Rev.* **2000**, *100*, 235–349.
228. Neese, F.; Zaleski, J. M.; Loeb, K. E.; Solomon, E. I. Electronic structure of activated bleomycin: oxygen intermediates in heme versus non-heme iron. *J. Am. Chem. Soc.* **2000**, *122*, 11703–11724.
229. Burger, R. M. Cleavage of nucleic acids by bleomycin. *Chem. Rev.* **1998**, *98*, 1153–1169.
230. Solomon, E. I.; Sundaram, U. M.; Machonkin, T. E. Multicopper oxidases and oxygenases. *Chem. Rev.* **1996**, *96*, 2563–2605.
231. (a) Giloni, L.; Takeshita, M.; Johnson, F.; Iden, C.; Grollman, A. P. Bleomycin-induced strand-scission of DNA. Mechanism of deoxyribose cleavage. *J. Biol. Chem.* **1981**, *256*, 8608–8615. (b) Murugesan, N.; Xu, C.; Ehrenfeld, G. M.; Sugiyama, H.; Kilkuskie, R. E.; Rodriguez, L. O.; Chang, L.-H.; Hecht, S. M. Analysis of products formed during bleomycin-mediated DNA degradation. *Biochemistry* **1985**, *24*, 5735–5744.
232. (a) Giese, B.; Beyrich-Graf, X.; Erdmann, P.; Petretta, M.; Schwitter, U. The chemistry of single-stranded 4'-DNA radicals: influence of the radical precursor on anaerobic and aerobic strand cleavage. *Chem. Biol.* **1995**, *2*, 367–375. (b) Giese, B.; Erdmann, P.; Giraud, L.; Göbel, T.; Petretta, M.; Schäfer, T. Heterolytic C, O-bond cleavage of 4′-nucleotide radicals. *Tetrahedron Lett.* **1994**, *35*, 2683–2686.
233. McGall, G. H.; Rabow, L. E.; Ashley, G. W.; Wu, S. H.; Kozarich, J. W.; Stubbe, J. New insight into the mechanism of base propenal formation during bleomycin-mediated DNA degradation. *J. Am. Chem. Soc.* **1992**, *114*, 4958–4967.
234. Giese, B.; Beyrich-Graf, X.; Erdmann, P.; Giraud, L.; Imwinkelried, P.; Muller, S. N.; Schwitter, U. Cleavage of single-stranded 4′-oligonucleotide radicals in the presence of O_2. *J. Am. Chem. Soc.* **1995**, *117*, 6146–6147.
235. (a) D'Andrea, A. D.; Haseltine, W. A. Sequence specific cleavage of DNA by the antitumor antibiotics neocarzinostatin and bleomycin. *Proc. Natl. Acad. Sci. U.S.A.* **1978**, *75*, 3608–3612. (b) Takeshita, M.; Grollman, A. P.; Ohtsubo, E.; Ohtsubo, H. Interaction of bleomycin with DNA. *Proc. Natl. Acad. Sci. U.S.A.* **1978**, *75*, 5983–5987.
236. (a) Boger, D. L.; Ramsey, T. M.; Cai, H.; Hoehn, S. T.; Kozarich, J. W.; Stubbe, J. Assessment of the role of the bleomycin A2 pyrimidoblamic acid C4 amino group. *J. Am. Chem. Soc.* **1998**, *120*, 53–65. (b) Bailly, C.; Waring, M. J. The purine 2-amino group as a critical recognition element for specific DNA cleavage by bleomycin and calicheamicin. *J. Am. Chem. Soc.* **1995**, *117*, 7311–7316.
237. Brown, J. M. The hypoxic cell: a target for selective cancer therapy-eighteenth Bruce F. Cain memorial award lecture. *Cancer Res.* **1999**, *59*, 5863–5870.
238. (a) Junnotula, V.; Sarkar, U.; Sinha, S.; Gates, K. S. Initiation of DNA strand cleavage by 1,2,4-benzotriazine 1,4-dioxide antitumor agents: mechanistic insight from studies of 3-methyl-1,2,4-benzotriazine 1,4-dioxide. *J. Am. Chem. Soc.* **2009**, *131*, 1015–1024. (b) Daniels, J. S.; Gates, K. S. DNA cleavage by the antitumor agent 3-amino-1,2,4-benzotriazine 1,4-dioxide (SR4233): Evidence for involvement of hydroxyl radical. *J. Am. Chem. Soc.* **1996**, *118*, 3380–3385. (c) Patterson, L. H.; Taiwo, F. A. Electron paramagnetic resonance spectrometry evidence for bioreduction of tirapazamine to oxidising free radicals under anaerobic conditions. *Biochem. Pharmacol.* **2000**, *60*, 1933–1935.
239. Kotandeniya, D.; Ganley, B.; Gates, K. S. Oxidative DNA base damage by the antitumor agent 3-amino-1,2,4-benzotriazine 1,4-dioxide (tirapazamine). *Bioorg. Med. Chem. Lett.* **2002**, *12*, 2325–2329.
240. (a) Wardman, P.; Priyadarsini, K. I.; Dennis, M. F.; Everett, S. A.; Naylor, M. A.; Patel, K. B.; Stratford, I. J.; Stratford, M. R. L.; Tracy, M. Chemical properties which control selectivity and efficacy of aromatic N-oxide bioreductive drugs. *Br. J. Cancer* **1996**, *74*, S70–S74. (b) Lloyd, R. V.; Duling, D. R.; Rumyantseva, G. V.; Mason, R. P.; Bridson, P. K. Microsomal reduction of 3-amino-1,2,4-benzotriazine 1,4-dioxide to a free radical. *Mol. Pharmacol.* **1991**, *40*, 440–445.
241. (a) Laderoute, K. L.; Wardman, P.; Rauth, M. Molecular mechanisms for the hypoxia-dependent activation of 3-amino-1,2,4-benzotriazine-1,4-dioxide (SR 4233). *Biochem. Pharmacol.* **1988**, *37*, 1487–1495. (b) Fitzsimmons, S. A.; Lewis, A. D.; Riley, R. J.; Workman, P. Reduction of 3-amino-1,2,4-benzotriazine-1,4-di-N-oxide (tirapazamine, WIN 59075, SR 4233) to a DNA-damaging species: a direct role for NADPH:cytochrome P450 oxidoreductase. *Carcinogenesis* **1994**, *15*, 1503–1510.
242. Birincioglu, M.; Jaruga, P.; Chowdhury, G.; Rodriguez, H.; Dizdaroglu, M.; Gates, K. S. DNA base damage by the antitumor agent 3-amino-1,2,4-benzotriazine 1,4-dioxide (tirapazamine). *J. Am. Chem. Soc.* **2003**, *125*, 11607–11615.
243. (a) Hwang, J.-T.; Greenberg, M. M.; Fuchs, T.; Gates, K. S. Reaction of the hypoxia-selective antitumor agent tirapazamine with a C1′-radical in single-stranded and double-stranded DNA: The drug and its metabolites can serve as surrogates for molecular oxygen in radical-mediated DNA damage reactions. *Biochemistry* **1999**, *38*, 14248–14255. (b) Daniels, J. S.; Gates, K. S.; Tronche, C.; Greenberg, M. M. Direct evidence for bimodal DNA damage induced by tirapazamine. *Chem. Res. Toxicol.* **1998**, *11*, 1254–1257. (c) Jones, G. D. D.; Weinfeld, M. Dual action of tirapazamine in the induction of DNA strand breaks. *Cancer Res.* **1996**, *56*, 1584–1590.
244. Ishida, N.; Miyazaki, K.; Kumagai, K.; Rikimaru, M. Neocarzinostatin, antitumor antibiotic of high molecular weight; isolation, physicochemical properties, and biological activities. *J. Antibiot.* **1965**, *18*, 68–76.
245. Konishi, M.; Ohkuma, H.; Saitoh, K.; Kawaguchi, H.; Golik, J.; Dubay, G.; Groenewold, G.; Krishnan, B.; Doyle, T. W. Esperamicins, a novel class of potent antitumor antibiotics. I. Physicochemical data and partial structure. *J. Antibiot.* **1985**, *38*, 1605–1609.
246. (a) Lee, M. D.; Ellestad, G. A.; Borders, D. B. Calicheamicins: discovery, structure, chemistry, and interaction with DNA. *Acc. Chem. Res.* **1991**, *24*, 235–243. (b) Lee, M. D.; Dunne, T. S.; Siegel, M. M.; Chang, C. C.; Morton, G. O.; Borders, D. B. Calichemicins, a novel family of antitumor antibiotics. 1. Chemistry and partial structure of calichemicin γ 1I. *J. Am. Chem. Soc.* **1987**, *109*, 3464–3466.
247. Konishi, M.; Ohkuma, H.; Matsumoto, K.; Tsuno, T.; Kamei, H.; Miyaki, T.; Oki, T.; Kawaguchi, H.; Van Duyne, G. D.; Clardy, J. Dynemicin A, a novel antibiotic with the anthraquinone and 1,5-diyn-3-ene subunit. *J. Antibiot.* **1989**, *42*, 1449–1452.
248. (a) Lam, K. S.; Hesler, G. A.; Gustavson, D. R.; Crosswell, A. R.; Veitch, J. M.; Forenza, S.; Tomita, K. Kedarcidin, a new chromoprotein antitumor antibiotic. I. Taxonomy of producing organism, fermentation and biological activity. *J. Antibiot.* **1991**, *44*, 472–478. (b) Leet, J. E.; Schroeder, D. R.; Langley, D. R.; Colson, K. L.; Huang, S.; Klohr, S. E.; Lee, M. S.; Golik, J.; Hofstead, S. J.; Doyle, T. W.; Matson, J. A. Chemistry and structure elucidation of the kedarcidin chromophore. *J. Am. Chem. Soc.* **1993**, *115*, 8432–8443.

249. (a) Hu, J.; Xue, Y.-C.; Xie, M.-Y.; Zhang, R.; Otani, T.; Minami, Y.; Yamada, Y.; Marunaka, T. A new macromolecular antitumor antibiotic, C-1027. I. Discovery, taxonomy of producing organism, fermentation and biological activity. *J. Antibiot.* **1988**, *41*, 1575–1579. (b) Minami, Y.; Yoshida, K.-i.; Azuma, R.; Saeki, M.; Otani, T. Structure of an aromatization product of C-1027 chromophore. *Tetrahedron Lett.* **1993**, *34*, 2633–2636. (c) Yoshida, K.-i.; Minami, Y.; Azuma, R.; Saeki, M.; Otani, T. Structure and cycloaromatization of a novel enediyne, C-1027 chromophore. *Tetrahedron Lett.* **1993**, *34*, 2637–2640.

250. Ando, T.; Ishii, M.; Kajiura, T.; Kameyama, T.; Miwa, K.; Sugiura, Y. A new non-protein enediyne antibiotic N1999A2: unique enediyne chromophore similar to neocarzinostatin and DNA cleavage feature. *Tetrahedron Lett.* **1998**, *39*, 6495–6498.

251. (a) Ling, J.; Horsman, G. P.; Huang, S.-X.; Luo, Y.; Lin, S.; Shen, B. Enediyne antitumor antibiotic maduropeptin biosynthesis featuring a C-methyltransferase that acts on a CoA-tethered aromatic substrate. *J. Am. Chem. Soc.* **2010**, *132*, 12534–12536. (b) Komano, K.; Shimamura, S.; Norizuki, Y.; Zhao, D.; Kabuto, C.; Sato, I.; Hirama, M. Total synthesis and structure revision of the (–)-maduropeptin chromophore. *J. Am. Chem. Soc.* **2009**, *131*, 12072–12073.

252. (a) Joshi, Mukesh C.; Rawat, Diwan S. Recent developments in enediyne chemistry. *Chem. Biodivers.* **2012**, *9*, 459–498. (b) Shao, R.-G. Pharmacology and therapeutic applications of enediyne antitumor antibiotics. *Curr. Mol. Pharmacol.* **2008**, *1*, 50–60. (c) Galm, U.; Hager, M. H.; Van Lanen, S. G.; Ju, J.; Thorson, J. S.; Shen, B. Antitumor antibiotics: bleomycin, enediynes, and mitomycin. *Chem. Rev.* **2005**, *105*, 739–758. (d) Smith, A. L.; Nicolaou, K. C. The enediyne antibiotics. *J. Med. Chem.* **1996**, *39*, 2103–2017.

253. Zein, N.; Casazza, A. M.; Doyle, T. W.; Leet, J. E.; Schroeder, D. R.; Solomon, W.; Nadler, S. G. Selective proteolytic activity of the antitumor agent kedarcidin. *Proc. Natl. Acad. Sci. U.S.A.* **1993**, *90*, 8009–8012.

254. (a) Kumar, R. A.; Ikemoto, N.; Patel, D. J. Solution structure of the esperamicin A1-DNA complex. *J. Mol. Biol.* **1997**, *265*, 173–186. (b) Kumar, R. A.; Ikemoto, N.; Patel, D. J. Solution structure of the calicheamicin γ 1I-DNA complex. *J. Mol. Biol.* **1997**, *265*, 187–201. (c) Gao, X.; Stassinopoulos, A.; Gu, J.; Goldberg, I. H. NMR studies of the post-activated neocarzinostatin chromophore-DNA complex. Conformational changes induced in drug and DNA. *Bioorg. Med. Chem.* **1995**, *3*, 795–809. (d) Gao, X.; Stassinopoulos, A.; Rice, J. S.; Goldberg, I. H. Basis for the sequence-specific DNA strand cleavage by the enediyne neocarzinostatin chromophore. Structure of the postactivated chromophore-DNA complex. *Biochemistry* **1995**, *34*, 40–49.

255. (a) Myers, A. G.; Cohen, S. B.; Tom, N. J.; Madar, D. J.; Fraley, M. E. Insights into the mechanism of DNA cleavage by dynemicin A as revealed by DNA-binding and -cleavage studies of synthetic analogs. *J. Am. Chem. Soc.* **1995**, *117*, 7574–7575. (b) Myers, A. G.; Cohen, S. B.; Kwon, B. M. A study of the reaction of calicheamicin γ 1 with glutathione in the presence of double-stranded DNA. *J. Am. Chem. Soc.* **1994**, *116*, 1255–1271.

256. Sluka, J. P.; Horvath, S. J.; Bruist, M. F.; Simon, M. I.; Dervan, P. B. Synthesis of a sequence-specific DNA-cleaving peptide. *Science* **1987**, *238*, 1129–1132.

257. Xu, Y.; Zhen, Y.; Goldberg, I. H. C1027 chromophore, a potent new enediyne antitumor antibiotic, induces sequence-specific double-strand DNA cleavage. *Biochemistry* **1994**, *33*, 5947–5954.

258. (a) Long, B. H.; Golik, J.; Forenza, S.; Ward, B.; Rehfuss, R.; Dabrowiak, J. C.; Catino, J. J.; Musial, S. T.; Brookshire, K. W.; Doyle, T. W. Esperamicins, a class of potent antitumor antibiotics: mechanism of action. *Proc. Natl. Acad. Sci. U.S.A.* **1989**, *86*, 2–6. (b) Zein, N.; Sinha, A. M.; McGahren, W. J.; Ellestad, G. A. Calicheamicin γ 1I: an antitumor antibiotic that cleaves double-stranded DNA site specifically. *Science* **1988**, *240*, 1198–1201.

259. (a) Kerrigan, N. J. Bergman cyclization. In *Name Reactions for Carbocyclic Ring Formations*, Li, J. J. (Ed.), Wiley & Sons, Hoboken, NJ, 2010. (b) Lockhart, T. P.; Comita, P. B.; Bergman, R. G. Kinetic evidence for the formation of discrete 1,4-dehydrobenzene intermediates. Trapping by inter- and intramolecular hydrogen atom transfer and observation of high-temperature CIDNP. *J. Am. Chem. Soc.* **1981**, *103*, 4082–4090.

260. Sugiura, Y.; Shiraki, T.; Konishi, M.; Oki, T. DNA intercalation and cleavage of an antitumor antibiotic dynemicin that contains anthracycline and enediyne cores. *Proc. Natl. Acad. Sci. U.S.A.* **1990**, *87*, 3831–3835.

261. Sugiura, Y.; Arawaka, T.; Uesugi, M.; Siraki, T.; Ohkuma, H.; Konishi, M. Reductive and nucleophilic activation products of dynemicin A with methyl thioglycolate. A rational mechanism for DNA cleavage of the thiol-activated dynemicin A. *Biochemistry* **1991**, *30*, 2989–2992.

262. (a) Povirk, L. F.; Dattagupta, N.; Warf, B. C.; Goldberg, I. H. Neocarzinostatin chromophore binds to deoxyribonucleic acid by intercalation. *Biochemistry* **1981**, *20*, 4007–4014. (b) Lee, S. H.; Goldberg, I. H. Sequence-specific, strand-selective, and directional binding of neocarzinostatin chromophore to oligodeoxyribonucleotides. *Biochemistry* **1989**, *28*, 1019–1026.

263. Myers, A. G.; Cohen, S. B.; Kwon, B.-M. DNA cleavage by neocarzinostatin chromophore. Establishing the intermediacy of chromophore-derived cumulene and biradical species and their role in sequence-specific cleavage. *J. Am. Chem. Soc.* **1994**, *116*, 1670–1682.

264. Sugiyama, H.; Yamashita, K.; Fujiwara, T.; Saito, I. Apoprotein-assisted unusual cyclization of neocarzinostatin chromophore. *Tetrahedron* **1994**, *50*, 1311–1325.

265. Myers, A. G.; Arvedson, S. P.; Lee, R. W. A new and unusual pathway for the reaction of neocarzinostatin chromophore with thiols. Revised structure of the protein-directed thiol adduct. *J. Am. Chem. Soc.* **1996**, *118*, 4725–4726.

266. Myers, A. G.; Dragovich, P. S.; Kuo, E. Y. Studies on the thermal generation and reactivity of a class of (σ,π)-1,4-biradicals. *J. Am. Chem. Soc.* **1992**, *114*, 9369–9386.

267. (a) Frank, B. L.; Worth, L., Jr.; Christner, D. F.; Kozarich, J. W.; Stubbe, J.; Kappen, L. S.; Goldberg, I. H. Isotope effects on the sequence-specific cleavage of DNA by neocarzinostatin: kinetic partitioning between 4′- and 5′-hydrogen abstraction at unique thymidine sites. *J. Am. Chem. Soc.* **1991**, *113*, 2271–2275. (b) Kappen, L. S.; Goldberg, I. H. Activation of neocarzinostatin chromophore and formation of nascent DNA damage do not require molecular oxygen. *Nucleic Acids Res.* **1985**, *13*, 1637–1648. (c) Kappen, L. S.; Goldberg, I. H.; Frank, B. L.; Worth, L., Jr.; Christner, D. F.; Kozarich, J. W.; Stubbe, J. Neocarzinostatin-induced hydrogen atom abstraction from C-4′ and C-5′ of the T residue at a d(GT) step in oligonucleotides: shuttling between deoxyribose attack sites based on isotope selection effects. *Biochemistry* **1991**, *30*, 2034–2042.

268. (a) Blackburn, G. M.; Datta, A.; Denham, H.; Wentworth, P., Jr. Catalytic antibodies. *Adv. Phys. Org. Chem.* **1998**, *31*, 249–392. (b) Wentworth, P., Jr.; Janda, K. D. Catalytic antibodies. *Curr. Opin. Chem. Biol.* **1998**, *2*, 138–144.
269. Jones, L. H.; Harwig, C. W.; Wentworth, P. Jr.; Simeonov, A.; Wentworth, A. D.; Py, S.; Ashley, J. A.; Lerner, R. A.; Janda, K. D. Conversion of enediynes into quinones by antibody catalysis and in aqueous buffers: implications for an alternative enediyne therapeutic mechanism. *J. Am. Chem. Soc.* **2001**, *123*, 3607–3608.
270. Lin, A. J.; Pardini, R. S.; Cosby, L. A.; Lillis, B. J.; Shansky, C. W.; Sartorelli, A. C. Potential bioreductive alkylating agents. 2. Antitumor effect and biochemical studies of naphthoquinone derivatives. *J. Med. Chem.* **1973**, *16*, 1268–1271.
271. Thorson, J. S.; Shen, B.; Whitwam, R. E.; Liu, W.; Li, Y.; Ahlert, J. Enediyne biosynthesis and self-resistance: a progress report. *Bioorg. Chem.* **1999**, *27*, 172–188.
272. Whitwam, R. E.; Ahlert, J.; Holman, T. R.; Ruppen, M.; Thorson, J. S. The gene calC encodes for a non-heme iron metalloprotein responsible for calicheamicin self-resistance in micromonospora. *J. Am. Chem. Soc.* **2000**, *122*, 1556–1557.
273. Newton, G. L.; Fahey, R. C. In *Glutathione: Metabolism and Physiological Functions*, Viña, J. (Ed.), CRC Press, Boca Raton, FL, 1990, pp 69–77.
274. (a) Reddy, B. S. P.; Sharma, S. K.; Lown, J. W. Recent developments in sequence selective minor groove DNA effectors. *Curr. Med. Chem.* **2001**, *8*, 475–508. (b) Goldberg, I. H.; Kappen, L. S.; Xu, Y. J.; Stassinopoulos, A.; Zeng, X.; Xi, Z.; Yang, C. F. Enediynes as probes of nucleic acid structure. *NATO ASI Series, Series C* **1996**, *479*, 1–21. (c) Stubbe, J.; Kozarich, J. W.; Wu, W.; Vanderwall, D. E. Bleomycins: a structural model for specificity, binding, and double strand cleavage. *Acc. Chem. Res.* **1996**, *29*, 322–330. (d) Fox, K.R.; Nightingale, K.P. Bleomycin-DNA interactions. *Nucleic Acids Mol. Biol.* **1994**, *8*, 167–183. (e) Smith, A. L.; Nicolaou, K. C. The enediyne antibiotics. *J. Med. Chem.* **1996**, *39*, 2103–2117. (f) Sugiura, Y.; Shiraki, T.; Konishi, M.; Oki, T. DNA intercalation and cleavage of an antitumor antibiotic dynemicin that contains anthracycline and enediyne cores. *Proc. Natl. Acad. Sci. U.S.A.* **1990**, *87*, 3831–3835.
275. Cloutier, J.-F.; Drouin, R.; Castonguay, A. Treatment of human cells with N-nitroso(acetoxymethyl)methylamine: distribution patterns of piperidine-sensitive DNA damage at the nucleotide level of resolution are related to the sequence context. *Chem. Res. Toxicol.* **1999**, *12*, 840–849.
276. Wu, J.; Xu, J.; Dedon, P. C. Modulation of enediyne-induced DNA damage by chromatin structures in transcriptionally active genes. *Biochemistry* **1999**, *38*, 15641–15746.

Chapter 7

Drug Resistance and Drug Synergism

Chapter Outline

7.1. DRUG RESISTANCE

7.1.1. What is Drug Resistance?

Drug resistance is when a formerly effective drug dose is no longer effective. This can be a natural resistance or an acquired resistance. Resistance arises mainly by *natural selection*. For example, a drug may destroy all the organisms in a colony that are susceptible to the action of that drug; however, on average, 1 in 10 million organisms in a colony has one or more mutations that make it resistant to that drug.[1] Once all of the susceptible organisms have been killed, the few resistant ones replicate and eventually become the predominant species. Bacteria may also resist antibiotics as a result of chromosomal mutations or inductive expression of a latent chromosomal gene or by exchange of genetic material (gene transfer) through transformation, transduction, or conjugation by *plasmids*, small pieces of circular DNA that carry genes for the survival of the organism, such as antibiotic resistance genes; they can be transmitted to other bacteria.[2] Because mutagenic drugs generally are not used, resistance by drug-induced mutation seldom occurs. Because of the remarkable ability for microorganisms to evolve and adapt, there is a need for new drugs with new mechanisms of action that are not susceptible to mechanisms of resistance. Hopefully, the elucidation of the sequences of genomes of various organisms will present new targets and new mechanisms.

Similar considerations apply to tumor cells. Thus, a drug may destroy most of the cells in a tumor, but those with mutations that are resistant to the drug remain and eventually predominate. Like the cells of microorganisms, tumor cells have a remarkable ability to evolve and adapt to ensure the survival of the cell lineage.

In some cases, a patient may fail to respond to initial treatment with a drug, which is referred to as *primary* or *intrinsic* resistance. In other cases, patients may respond initially to a drug, but later fail to respond, in which case the resistance is *secondary* or *acquired*. Resistance should not be confused with another term, tolerance, which is not related to microorganisms or cancer cell growth. *Tolerance* is when the body adapts to a particular drug and requires more of that drug to attain the same initial effect, typically leading to a decrease in the therapeutic index. This is what occurs with addiction to morphine. When morphine stimulates receptors in the cell membrane that respond to a transmitter, the cells adapt by increasing the number of receptors they produce so that some receptors are available to combine with the transmitter, even in the continued presence of morphine. Because morphine now activates only a fraction of the available receptors, its effects will be less than before, and a higher dose will be needed to achieve the same effect. It also is possible for tolerance to develop to the undesired effects of a drug, which leads to an *increase* in the therapeutic index (such as tolerance to sedation by phenobarbital[3]).

The Organic Chemistry of Drug Design and Drug Action. http://dx.doi.org/10.1016/B978-0-12-382030-3.00007-6

7.1.2. Mechanisms of Drug Resistance

Antibiotic resistance has reached an alarming stage worldwide[4] and resistance to antitumor drugs continues to confound the search for cures to cancer. Many organisms and cells have acquired multiple systems to reduce or avoid the action of drugs designed to treat them.[5] The most threatening mechanisms of resistance involve changes in the target site for drug interaction because that is likely to confer resistance to all compounds that interact similarly with this target. *Exogenous resistance* is when new proteins are developed by the organism or cell to protect it from drugs. *Endogenous resistance* occurs by mutation, even single-point mutations (one amino acid change). Eight significant mechanisms of drug resistance are discussed in the following sections, many of which arise from natural selection.[6]

7.1.2.1. Altered Target Enzyme or Receptor

Mutation of the amino acid residues (replacement of the original amino acid with a different amino acid) in the active site of the target protein can result in poorer binding of the drug to the active site.[7] Similarly, mutations elsewhere in the target protein can cause a conformational change so that the interaction with the drug is altered.

For example, resistance to the antibiotic trimethoprim (**7.1**; Bactrim) derives from a mutation in a single amino acid of its target enzyme, dihydrofolate reductase.[8] The properties of the singly mutated enzyme differ somewhat from those of the normal enzyme, but it still binds its substrate, dihydrofolate.

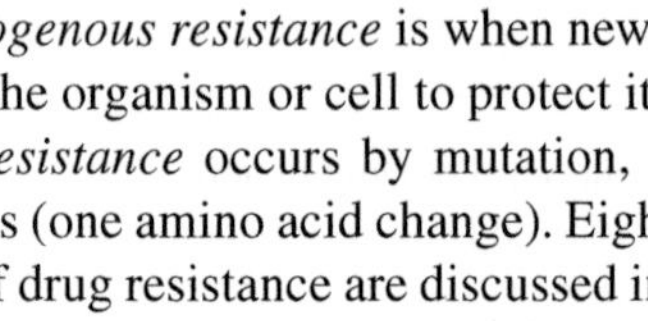

Trimethoprim
7.1

One of several mechanisms for resistance to sulfa drugs (Chapter 5, Section 5.2.2.3) is plasmid-mediated synthesis of a mutated form of the target enzyme, dihydropteroate synthase. The mutated enzyme binds the substrate, *p*-aminobenzoic acid, normally, but binds sulfonamides several thousand times less tightly than the normal enzyme can.[9]

The triphosphate nucleotides of 3′-azidothymidine (AZT, zidovudine; **7.2**; Retrovir) and (−)-2′,3′-dideoxy-3′-thiacytidine (3TC, lamivudine; **7.3**, Epivir) inhibit the reverse transcriptase of human immunodeficiency virus-type 1 (HIV-1) and are used in the treatment of AIDS. However, resistance can develop relatively rapidly.[10] Resistance to AZT arises from mutation of several residues in reverse transcriptase;[11] resistance to 3TC is conferred by a single mutation (different from the ones arising from AZT administration) in which a valine (preferentially) or isoleucine is substituted for Met-184.[12] However, a curious phenomenon occurs when both AZT and 3TC are administered in combination; a much longer delay in resistance occurs, even though Val-184 mutants rapidly emerge.[13] This results because the Val-184 mutant that develops from 3TC administration is much more susceptible to AZT, which therefore suppresses AZT resistance!

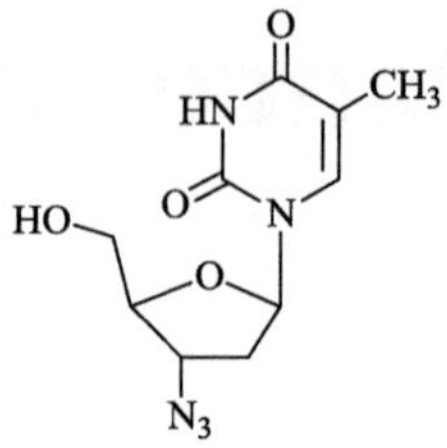

Zidovudine
7.2

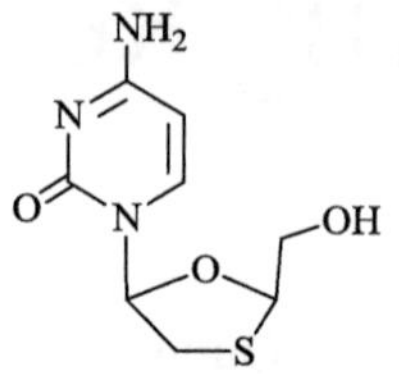

Lamivudine
7.3

An interesting (terrifying!) variant of this mechanism for resistance was elucidated by Walsh and coworkers[14] for the antibiotic vancomycin (**7.4**; Vancocin), which has long been a drug of last defense against resistant streptococcal or staphylococcal organisms. As discussed in Chapter 4 (Section 4.2.7), the bacterial cell wall is constructed by a series of enzyme-catalyzed reactions. One example is shown in Scheme 4.13, depicting the mechanism for cross-linkage of the peptidoglycan. Peptidoglycan is a branched polymer of alternating β-D-*N*-acetylglucosamine (NAG or GlcNAc) and β-D-*N*-acetylmuramic acid (NAM or MurNAc) residues. Attached to what was the carboxylic acid group of the lactyl group of MurNAc is a polypeptide chain that varies in structure according to the strain of bacteria (Figure 7.1).

Vancomycin acts by forming a complex with multiple points of contact to the terminal D-alanyl-D-alanine of the peptidoglycan (Figure 7.2), thereby blocking not only the transpeptidation sequence of Scheme 4.13, but also the transglycosylation (the reaction that builds up the peptidoglycan by formation of the glycosyl linkage between GlcNAc and MurNAc). When bacterial cell wall biosynthesis is blocked, the high internal osmotic pressure (4–20 atm) can no longer be sustained, and the bacteria burst, as shown in Figure 7.3 (for fosfomycin, another inhibitor of cell wall biosynthesis).

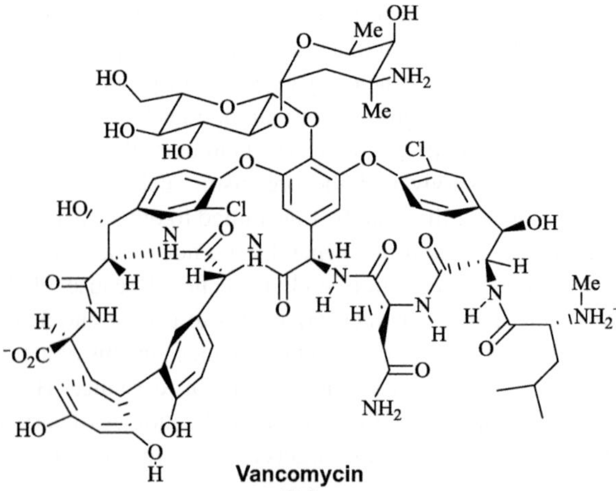

Vancomycin
7.4

A surprising resistance to vancomycin arises in bacteria by induction of five new genes (termed *VanS*, *VanR*, *VanH*, *VanA*, and *VanX*), which are used to construct an altered peptidoglycan in which the terminal D-alanyl-D-alanine is replaced by D-alanyl-D-lactate (**7.5**; Figure 7.4). The *VanS* gene product is a transmembrane histidine kinase that initiates the signal transduction pathway, and the *VanR* gene product is a two-domain response regulator that accepts the phosphate group from phospho-*VanS* and activates *VanH*, *VanA*, and *VanX* transcription. The *VanX* gene product catalyzes the hydrolysis of D-Ala-D-Ala dipeptides so that only D-Ala-D-lactate is available for MurF-catalyzed condensation with UDP-muramyl-L-Ala-D-γ-Glu-L-Lys in the construction of the peptidoglycan strands (Figure 7.4). During transpeptidation (cross-linking of the peptidoglycan strands) in the antibiotic-sensitive organism, the terminal D-alanine residue is released (Scheme 4.13); in the antibiotic-resistant organisms, D-lactate is released instead, but the same cross-linked product as in the wild-type organism is produced. However, substitution of D-alanine by D-lactate in the peptidoglycan leads to the deletion of one hydrogen bond to vancomycin and also produces nonbonded electron repulsion between the lactate ester oxygen and the amide carbonyl of vancomycin (Figure 7.5). This produces a 1000-fold reduction in binding of the drug to the peptidoglycan substrate.[15] Therefore, in this case, the resistant organism has not mutated an essential enzyme, but has mutated the substrate for the enzyme, which is what forms the complex with vancomycin!

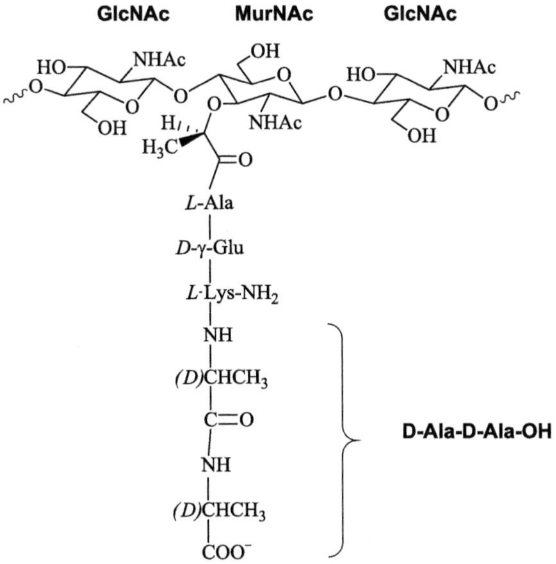

FIGURE 7.1 Structure of a peptidoglycan segment prior to cross-linking with another peptidoglycan fragment catalyzed by peptidoglycan transpeptidase. This structure is an alternative depiction of the transpeptidase substrate shown on the left in Scheme 4.13, graphic **A**.

The discovery of antiviral drug ritonavir (compound **5.90**), an inhibitor of HIV-1 protease, was discussed in Chapter 5 (Section 5.2.5). The development of resistance to ritonavir results from a single-point mutation in HIV-1 protease. Specifically, Val-82 was mutated to Thr, Ala, or Phe, which interferes with the interaction of the isopropyl group attached to the thiazolyl group at P_3 of ritonavir. To avoid this repulsive interaction with the mutant, the (3-isopropylthiazolyl) methyl group of ritonavir was excised to give **7.6** (boldface H shows position of the excised group). This inhibited the mutant enzyme, but only poorly. The pharmacophore was increased by a ring-chain transformation (**7.7**), and to increase potency further, the other thiazolyl group was replaced by a 2,6-dimethylphenyl group, giving lopinavir (**7.8**; an active component of Kaletra).

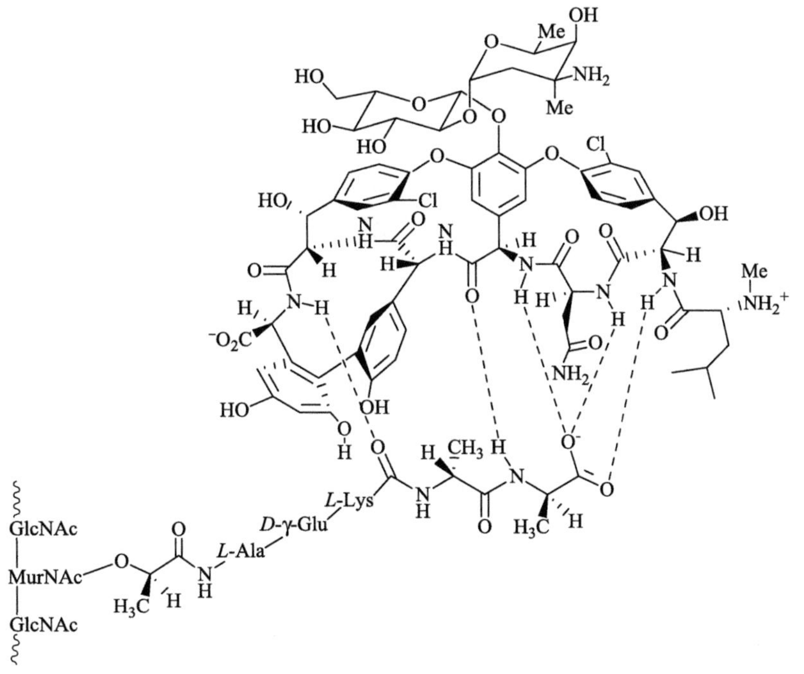

FIGURE 7.2 Complex between vancomycin and the terminal D-alanyl-D-alanine of the peptidoglycan

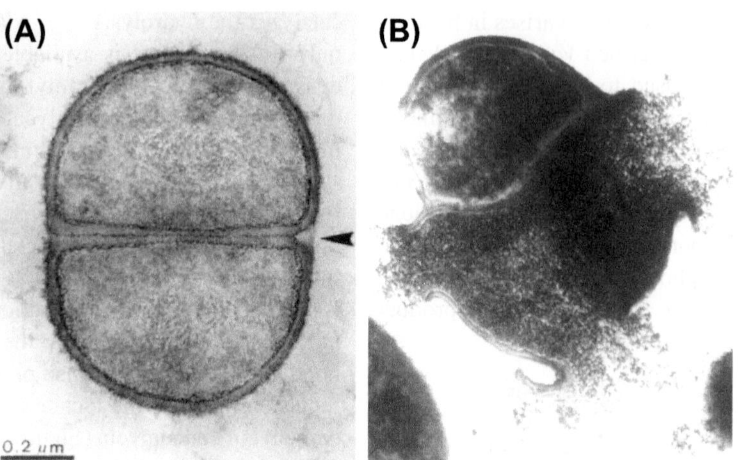

FIGURE 7.3 (A) Intact *Staphylococcus aureus* cell and (B) *Staphylococcus aureus* cell that has undergone lysis due to treatment with fosfomycin. *With permission of Dr David Pompliano, Bioleap, Inc., Pennington, NJ.*

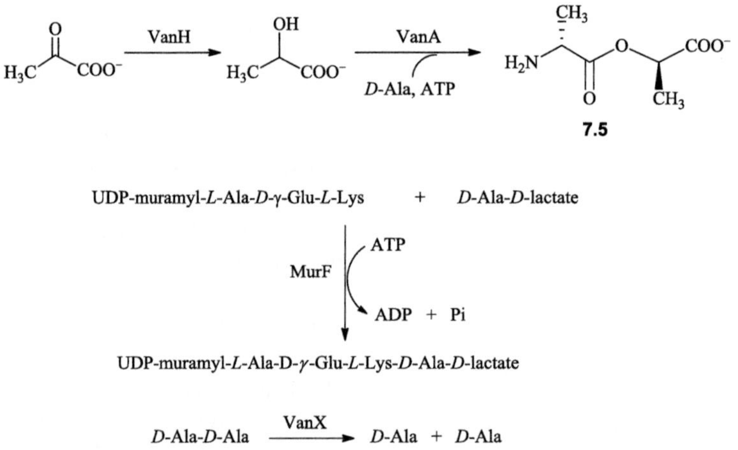

$H_3C-CO-COO^-$ —VanH→ $H_3C-CH(OH)-COO^-$ —VanA (*D*-Ala, ATP)→ **7.5**

UDP-muramyl-*L*-Ala-*D*-γ-Glu-*L*-Lys + *D*-Ala-*D*-lactate

MurF (ATP → ADP + Pi)

UDP-muramyl-*L*-Ala-D-γ-Glu-*L*-Lys-*D*-Ala-*D*-lactate

D-Ala-*D*-Ala —VanX→ *D*-Ala + *D*-Ala

FIGURE 7.4 Biosynthesis of D-alanyl-D-lactate and incorporation into the peptidoglycan of vancomycin-resistant bacteria

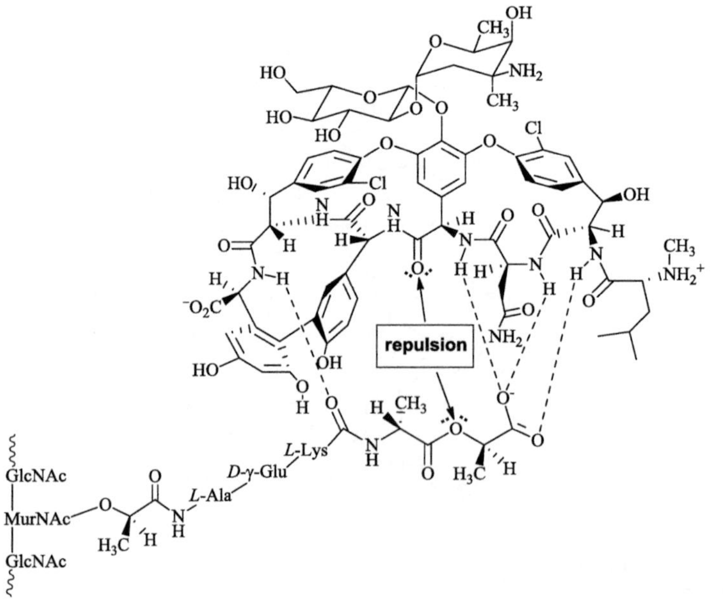

FIGURE 7.5 Complex between vancomycin and the peptidoglycan with terminal D-alanyl-D-lactate instead of D-alanyl-D-alanine in vancomycin-resistant bacteria

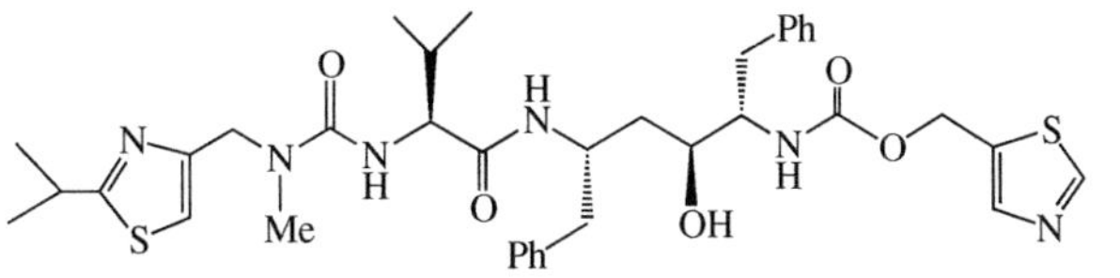

Ritonavir
5.90

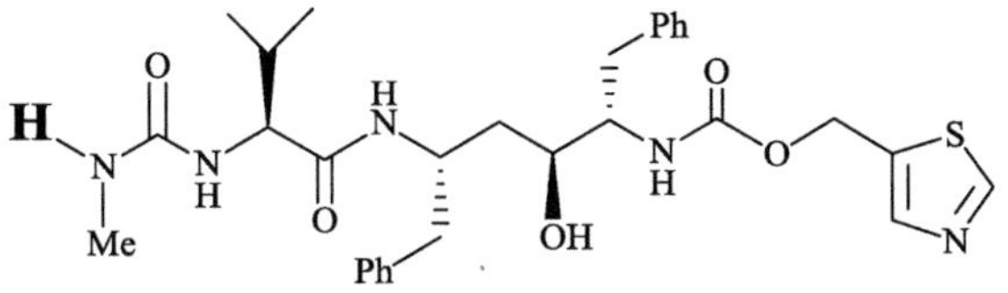

7.6

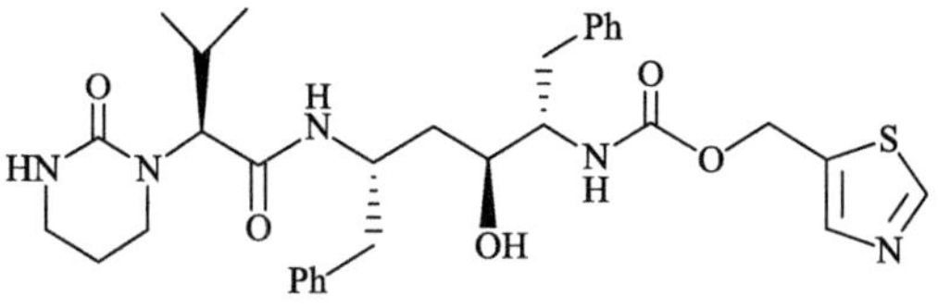

7.7

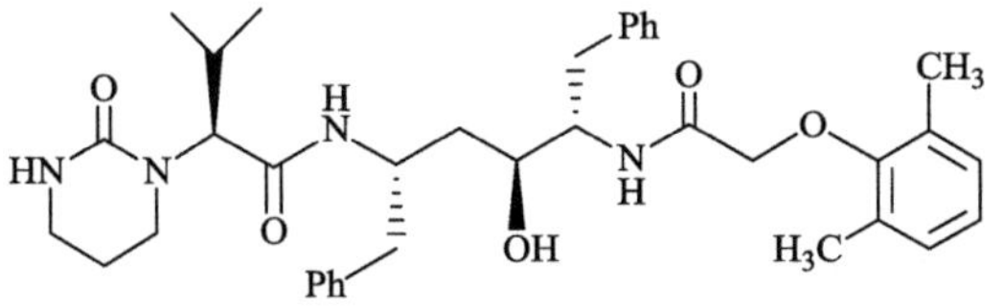

Lopinavir
7.8

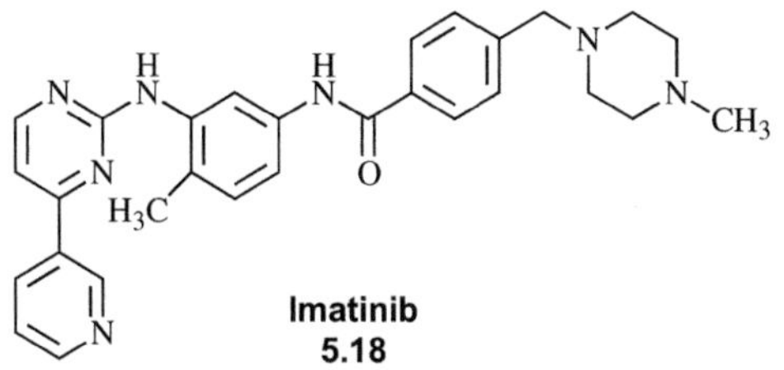

Imatinib
5.18

However, as was noted in Chapter 5 (Section 5.2.5.2), the thiazolyl group in ritonavir was shown to inhibit cytochrome P450 and protect it from metabolism. Removal of the P_2' 5-thiazolyl group to give lopinavir destroyed the cytochrome P450 inhibition activity, and, in fact, the plasma half life for lopinavir is low. To get around that problem, it was found that ritonavir could be added as an inhibitor of cytochrome P450, which protects lopinavir from metabolic degradation. Kaletra is a combination of lopinavir (to inhibit HIV-1 protease in the resistant strains) and ritonavir (to inhibit HIV-1 protease in susceptible strains plus to inhibit cytochrome P450). In clinical trials with this combination, no HIV was detectable in the blood of the patients for three years.

The discovery of imatinib (**5.18**), a Bcr-Abl inhibitor and treatment for leukemia, was discussed in Chapter 5 (Section 5.2.2.2). At least 50 mutations, representing single amino acid changes in Bcr-Abl, have been identified that render Bcr-Abl resistant to imatinib. These mutations are found in various regions of the kinase, for example at H396 (a residue in the activation loop; see Figure 5.4), E255 (a residue in the P-loop, which is important for binding to the triphosphate region of the natural co-substrate ATP), or T315 (a residue directly involved in imatinib binding through formation of a hydrogen bond with the enzyme's T315 hydroxyl group).[16] Drug discovery efforts at Bristol-Myers Squibb and Novartis to find inhibitors of the mutant enzymes resulted in marketed drugs dasatinib (**7.9**, Sprycel) and nilotinib (**7.10**, Tasigna). Inspection of the structure of **7.10** reveals several important differences from **5.18**: (1) reversal of the amide bond linking the two phenyl rings; (2) evolution of the piperizinylmethyl moiety to a substituted imidazole; and (3) addition of a trifluoromethyl group to the distal phenyl ring. Both **7.9** and **7.10** are active against most of the mutant forms responsible for resistance. The design of an inhibitor of multiple mutated forms of an enzyme is pretty impressive because

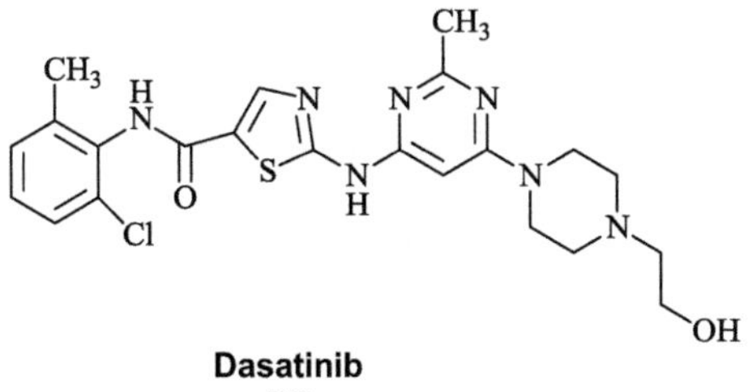

Dasatinib
7.9

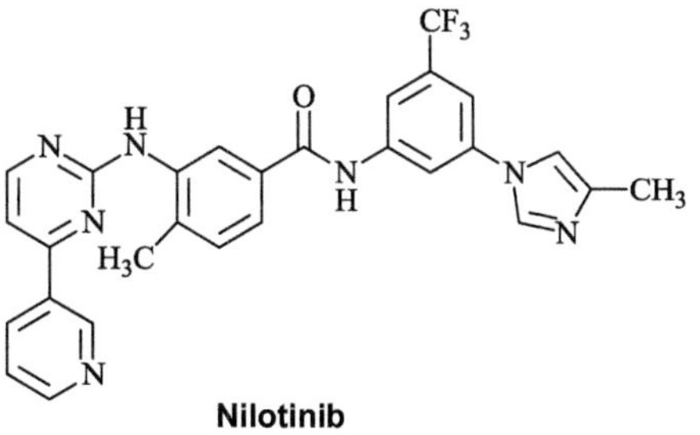

Nilotinib
7.10

the inhibitor of each mutant form can have its own pattern of structure–activity relationships. This is the opposite of the more common situation of trying to achieve a high level of selectivity for one target vs one or more other targets that may be responsible for undesired side effects.

While **7.9** and **7.10** offered hope to many patients whose disease became unresponsive to imatinib, neither compound is active against the T315I mutation, which represents approximately 15–20% of observed mutants. This may result in part from the importance of a hydrogen bond to the hydroxyl group of Thr-315, which is not possible in the mutant containing isoleucine in place of threonine. Thus, a compound that potently inhibits all mutant forms (a pan-Bcr-Abl inhibitor), including Bcr-Abl (T315I), remains an avidly pursued objective.

In the course of investigating inhibitors of unmutated Bcr-Abl, researchers at Ariad Pharmaceuticals screened representative compounds in assays for inhibition of Bcr-Abl (T315I), and identified structure **7.11** (Figure 7.6), which, unlike the structurally similar analog **7.10**, possessed inhibitory potency against the T315I mutant in the triple-digit nanomolar range, together with double-digit potency against other prominent Bcr-Abl mutants.[17] When **7.10** and **7.11** were computationally docked into a Bcr-Abl (T315I) structure, it was shown that there was a steric clash between I315 and the pyrimidine amino group of **7.10**, but this clash was avoided because of the olefinic moiety of **7.11**. Thus, a campaign ensued to optimize the potency of **7.11** against Bcr-Abl (T315I).

Figure 7.6 shows the structural evolution at key stages of lead optimization research starting with **7.11**.[18] Replacement of the vinyl linkage with a less sterically demanding acetylenic linkage and replacement of the purine ring with imidazo[1,2-*a*]pyridine resulted in an approximately fivefold improvement in potency against Bcr-Abl (T315I). For the latter modification, the position of a key nitrogen atom (see arrow to **7.12** in Figure 7.6) was maintained on the basis of the binding model for **7.11**, which indicates that the nitrogen at the analogous position participates in a hydrogen bond to the enzyme. In **7.13**, the substituted imidazole is replaced by a piperazinylmethyl substituent (note that this is a reversion in the evolution from imatinib to nilotinib!), resulting in further potency improvements against both Bcr-Abl and the T315I mutant. The rationale for incorporating the additional nitrogen in **7.14** (imidazo[1,2-b]pyridazine in place of imidazo[1,2-*a*]pyridine) was to reduce lipophilicity. Recall that high lipophilicity can reduce aqueous solubility, decrease membrane permeability, and promote susceptibility to metabolism, all resulting in potentially detrimental effects on pharmacokinetic properties (Chapter 2, Section 2.2.5.2); the magnitude of the effect of this change on CLog *P* is shown in Figure 7.6. X-ray crystallography of **7.14** complexed to Bcr-Abl (T315I) was consistent with the existence of a hydrogen bond between the enzyme and the indicated nitrogen atom (see arrow). This hydrogen bond is to an analogous region of the enzyme as those shown in Figures 5.2(A) and 5.3. Ponatinib (**7.14**, Iclusig) was approved by the FDA in 2012.

Another compound that was designed to inhibit Bcr-Abl and Bcr-Abl (T315I) is DCC-2036 (**7.15**).[19] The design was conceived as one that might be broadly applicable to

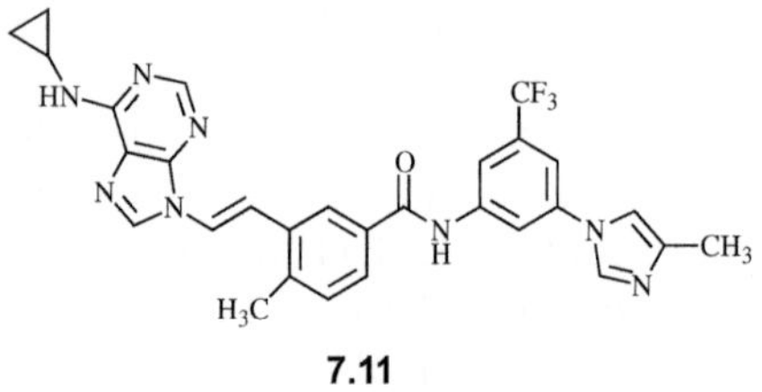

7.11

Bcr-Abl	25 nM
Bcr-Abl (T315I)	478 nM

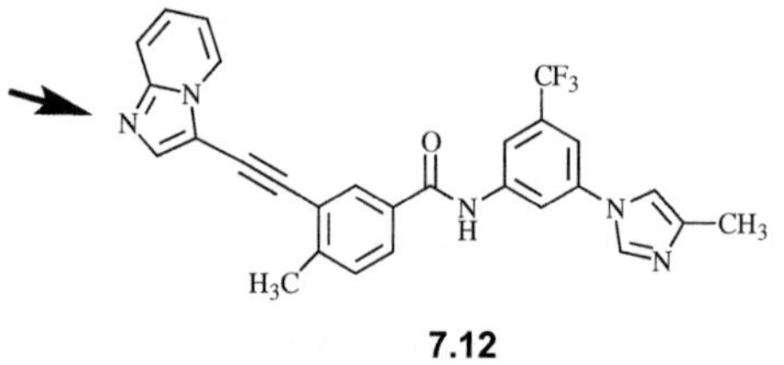

7.12

Bcr-Abl	26 nM
Bcr-Abl (T315I)	102 nM

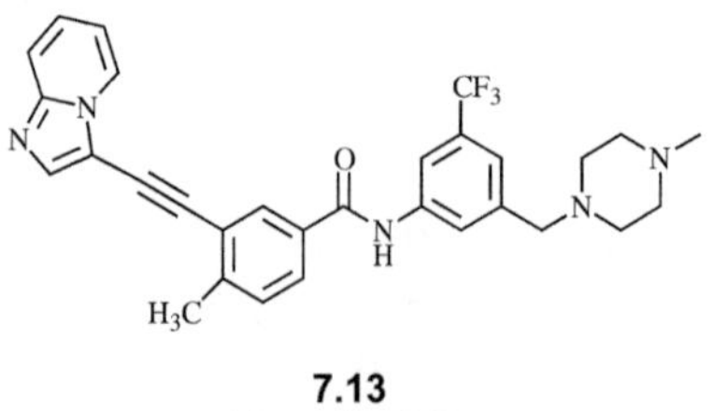

7.13
CLog *P* = 6.7

Bcr-Abl	9 nM
Bcr-Abl (T315I)	56 nM

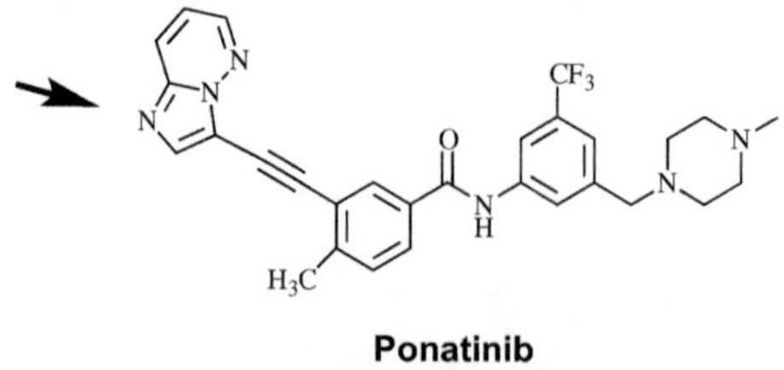

Ponatinib
7.14
CLog *P* = 5.8

Bcr-Abl	9 nM
Bcr-Abl (T315I)	40 nM

FIGURE 7.6 Evolution of **7.11** optimization for inhibition of Bcr-Abl (T315I)

kinase inhibition, since it takes advantage of structural components found across many different kinases. Specifically, a mechanism for controlling the conformational change from inactive to active forms of many kinases involves the interaction of a basic amino acid (e.g., arginine) and an acidic amino acid (e.g., glutamic acid), which are in close proximity in the inactive conformation and therefore stabilize it. However, when the enzyme is phosphorylated (by another kinase) at a specific tyrosine residue near the active site, the basic amino acid moves to form an interaction with the newly installed phosphate group, resulting in stabilization of the active conformation. As shown in Figure 7.7, the quinoline nitrogen atom of DCC-2036 was designed to interact with glutamic acid 282, the amino acid important for stabilization of the inactive form of Bcr-Abl and Bcr-Abl (T315I).

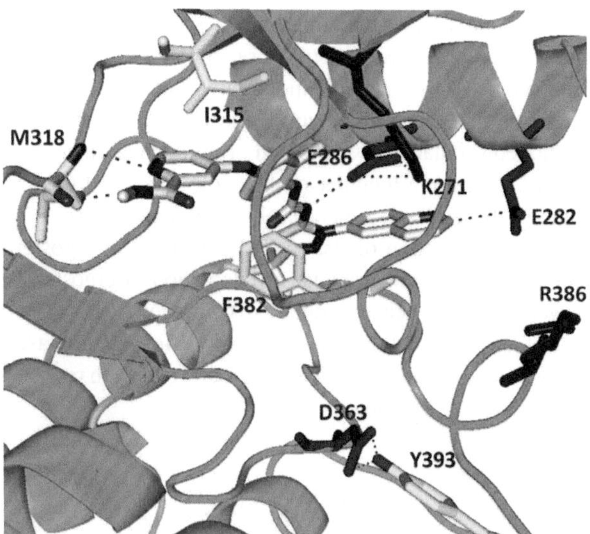

FIGURE 7.7 Image based on X-ray crystal structure of DC-2036 complexed to Bcr-Abl (T315I). Note the position of the I315 residue and the hydrogen bonds to Met318; Met318 is analogous to Met793 in Figures 5.2 and 5.3.

7.15

Gefitinib
5.12

The T315 residue of Bcr-Abl is known as the "gatekeeper residue" because of its position at the "gate" of a binding pocket that is available in the inactive conformation of the kinase. This pocket is occupied by the t-butylpyrazole moiety of **7.15** and by the trifluoromethylphenyl moiety of **7.14**. Mutation of the gatekeeper residue of a target kinase is a frequent cause for resistance. For example, the gatekeeper T790M mutation of epidermal growth factor receptor (EGFR) kinase confers resistance to gefitinib (**5.12**) and erlotinib (**5.13**),[16] and the gatekeeper L1196M mutation of anaplastic lymphoma kinase (ALK) confers resistance to lung cancer drug crizotinib (**7.16**, Xalkori).[20]

Erlotinib
5.13

In Chapter 6, the action of topoisomerase enzymes was described (Section 6.2.3 and associated discussion), in which the enzyme forms a covalent complex ("cleavable complex") with DNA. Several important antitumor drugs, for example amsacrine (**6.18**), act by stabilizing this cleavable complex, thus interrupting progression to the next step in DNA processing and leading to DNA cleavage. The mechanism of resistance to amsacrine and similar drugs has been a question of much interest. Experiments have shown that two specific point mutations in human topoisomerase IIα can occur that diminish the binding of amsacrine to the cleavable complex, leading to a reduction in cell-destroying DNA cleavage.[21]

Crizotinib
7.16

Amsacrine (R=OMe, R'=$NHSO_2Me$)
6.18

Systemic fungal infections are a common problem in immunocompromised patients, for example, those who have been subjected to cancer chemotherapy. Like bacteria, infectious fungi, such as members of the *Candida* family, have developed resistance to antifungal drugs such as fluconazole (**7.17,** Diflucan), a member of the azole class of drugs. The target of the azoles is a lanosterol C14α-demethylase, which catalyzes conversion of the steroid lanosterol to ergosterol. Inhibition of this enzyme leads to depletion of ergosterol in the yeast cell membrane, resulting in restrained fungal growth. More than 80 mutations in lanosterol C14α-demethylase in *Candida krusei* have been detected, many of which reduce its affinity for fluconazole.[22]

Fluconazole
7.17

One way to minimize the effect of target enzyme mutation in drug resistance would be to design a drug that is very close in structure to that of the substrate for the target enzyme. If the organism mutates a residue in the active site to lower the binding affinity of the inhibitor, then it will likely also lower the binding affinity of its substrate, so that the mutant enzyme will be poorly active (inhibited). For example, the structure of the anti-influenza drug zanamivir (Relenza) (Chapter 2, structure **2.127**), which targets a very highly conserved region of the virus's neuraminidase, is very close to that of substrate sialic acid (**2.128**), and it interacts with the same residues at the active site. If the virus loses the neuraminidase function, it cannot replicate, so there is a strong pressure not to mutate that region. This is, therefore, an important strategy to protect the drug from resistance to an organism that tries to mutate its active site.

Another cause for mutations occurs because you decided not to follow your doctor's antibiotic prescription, which warned you to take the prescribed amount, and finish the complete dose regimen. Sometimes you skipped taking a pill or partway through the regimen you felt better, so you did not think it was worth finishing the rest of the prescribed amount of drug. Mistake. It was found that sublethal antibiotic treatment can lead to multidrug resistance. Antibacterial drugs, including β-lactams, quinolones, and aminoglycosides, stimulate bacteria to produce reactive oxygen species (ROS).[23] As we saw in Chapter 6 (Section 6.3.3), ROS, such as hydroxyl radicals, cause DNA damage, leading to gene and, therefore, protein mutations. Protein mutagenesis was found to correlate with the increased ROS produced by sublethal amounts of antibiotics, and this radical-induced mutagenesis resulted in resistance of the microorganism to multiple types of drugs (i.e., multidrug resistance).[24] Maybe next time you'll take the entire antibiotic prescription.

7.1.2.2. Overproduction of the Target Enzyme or Receptor

Increased target enzyme production by induction of extra copies of the gene encoding the enzyme is another mechanism for drug resistance.[25] For example, resistance to inhibitors of the enzyme dihydrofolate reductase by malarial parasites[26] and by malignant white blood cells[27] has been shown to be the result of overproduction of that enzyme in an unaltered form. Similarly, overproduction of a kinase target is a mechanism that cancer cells may employ to overcome the effect of kinase inhibitors. However, because cancer cells can simultaneously employ multiple mechanisms to circumvent the effect of a drug, it is not always straightforward to establish which mechanism is the primary contributor to resistance. For example, numerous studies have been carried out to examine the relationship of the level of EGFR expression to the effectiveness of EGFR inhibitor gefinitib (**5.12**),[28] but a definitive relationship was difficult to establish. In fact, some evidence pointed to *greater* sensitivity to gefitinib in cases of EGFR overexpression, presumably because the growth and survival of these tumor cells were more heavily dependent than others on the cell processes stimulated by EGFR. Similarly, when leukemic cells from a chronic myeloid leukemia patient were cultured with increasing concentrations of imatinib (**5.18**), resistance to imatinib developed, which was accompanied by increased levels of its target enzyme Bcr-Abl.[29] However, because the resistant cells continued to survive even when sufficient additional imatinib was added to completely inhibit the enzyme, it was concluded that the additional factors contributing to resistance were likely at play.

Overproduction of certain protein subunits of the proteasome contributes to resistance to the proteasome inhibitor bortezomib (**7.18**, Velcade), a drug used to treat multiple myeloma.[30] The proteasome is a complex of protease enzymes that are responsible for degrading cellular proteins that have been "tagged" by the cell (through a process known as *ubiquitination*) for degradation. Inhibition of this cellular waste removal process causes cell stress leading to cell death. Since malignant cells are particularly susceptible to this disruption of their waste-processing system, proteasome inhibition has led to successful cancer treatment; however, target overproduction is one mechanism that the cancer cells have implemented to effect resistance to bortezomib.

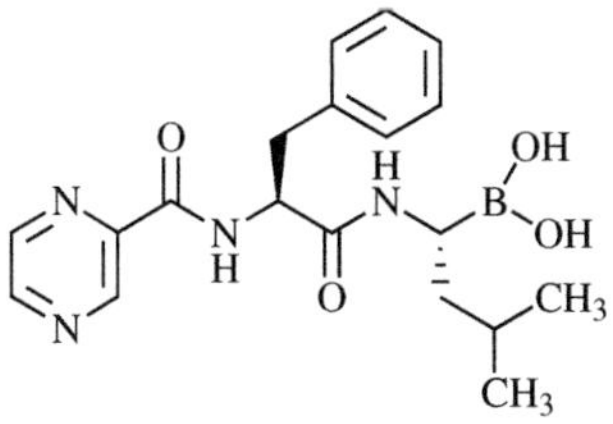

7.18
Bortezomib

The boronic acid moiety of **7.18** was shown by X-ray crystallographic studies to form a boronic ester with the hydroxyl group of the catalytic threonine residue (Figure 7.8).[31] Predecessors of **7.18** had an aldehyde or other carbon-based moiety in place of the boronic acid moiety, and although these were often potent proteasome inhibitors, they lacked selectivity by virtue of also inhibiting cysteine proteases. The boron atom (hard, according to hard-soft acid–base theory[32]) ensures affinity for oxygen (a hard atom) rather than sulfur (a soft atom).

To the extent that resistance to a drug is caused by overproduction of the target, higher doses of the drug may compensate, but in practice this is often limited by toxic effects that become more prominent as drug dose is increased. Alternatively, more potent or irreversible inhibitors could be identified.

7.1.2.3. Overproduction of the Substrate or Ligand for the Target Protein

In Chapter 3 (Section 3.2.3), a competitive antagonist was described as a molecule that is in competition with the natural ligand for binding to a receptor. The same situation applies to inhibitors and substrates of enzymes, such that overproduction of the substrate for a target enzyme would competitively block the ability of the drug to bind at the active site (Section 5.2.1). An example is the overproduction of *p*-aminobenzoic acid as a mechanism of resistance to antibacterial sulfa drugs.[33] As discussed in Chapter 5 (Section 5.2.2.3), sulfa drugs compete with *p*-aminobenzoic acid in the active site of dihydropteroate synthase, which mediates the synthesis of dihydrofolate, a key cofactor for the action of other critical enzymes. Overproduction of *p*-aminobenzoic acid would allow it to compete more effectively in the active site of dihydropteroate synthase, thus diminishing the effect of the sulfa drug.

7.1.2.4. Increased Drug-Destroying Mechanisms

Mechanisms by which drugs are degraded to aid their removal from circulation are the central topic of Chapter 8. It is perhaps not surprising that rapidly dividing tumor or microbial cells, in their quest for survival, would be able to upregulate these drug-destroying mechanisms as a way of fending off drugs that are trying to destroy them. Thus, resistance can occur by induction of genes to produce new enzymes, increase the quantities of existing enzymes, or produce other substances to degrade or sequester the drug. Several known examples are given in this section.

Although penicillins are "wonder drugs" in their activity against a variety of bacteria, many strains of bacteria have become resistant. For example, within 10 years of the introduction of penicillin, half of the strains of *Staphylococcus aureus* had become resistant, and by the 1990s, 90% of these strains were resistant.[34] The principal cause for resistance, as a result of gene transfer and recombination, is destruction of the drug by β-lactamase enzymes.[35] β-Lactamases are classified into types A, B, C, or D. These enzymes catalyze the hydrolysis of the β-lactam ring of

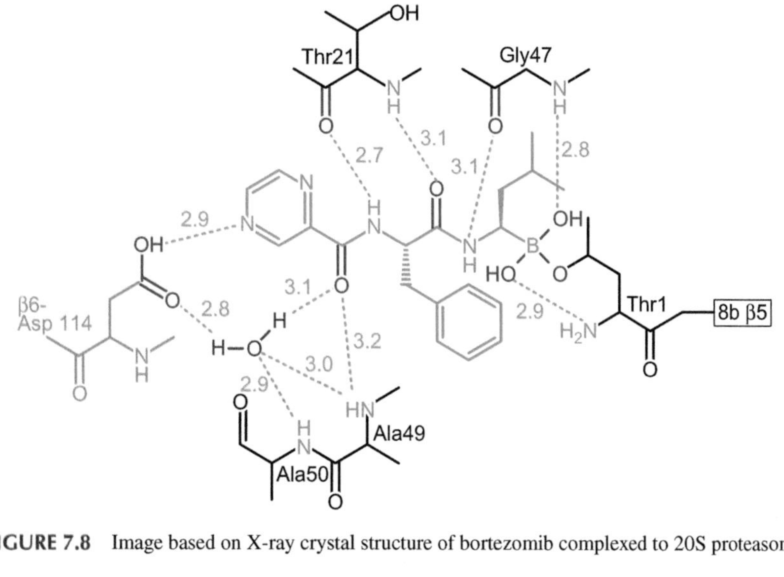

FIGURE 7.8 Image based on X-ray crystal structure of bortezomib complexed to 20S proteasome

penicillin. In a mechanism analogous to that of serine proteases, β-lactamase types A, C, and D possess an active site serine residue that performs nucleophilic attack on the lactam carbonyl group to form a covalent O-acyl serine intermediate, which is subsequently hydrolyzed to regenerate the active enzyme with release of the hydrolyzed lactam. The difference among classes A, C, and D lies in the differences in the residues used as general acids and bases to promote serine acylation and deacylation. Class B β-lactamases are zinc metalloenzymes that act through mechanisms reminiscent of that for angiotensin converting enzyme (Chapter 5, Sections 5.2.3.2.1 and 5.2.4.2). As a consequence of decades of use of penicillins and the structurally related carbapenems, β-lactamases, which are ancient enzymes that were once rare, have evolved into hundreds of different variants, some of which catalyze the hydrolysis of a broad spectrum of substrates. To counteract resistance caused by β-lactamases, β-lactamase inhibitors have been designed that can be coadministered with penicillin or other β-lactamase-sensitive drugs (see Drug Synergism, Section 7.2.2.1).

Aminoglycoside antibiotics, such as kanamycins (**7.19**) and neomycins (**7.20**; Neosporin), function by binding to the organism's ribosomal RNA, thereby inhibiting protein synthesis. Resistance develops because the bacteria acquire enzymes that catalyze the modification of hydroxyl and amino groups of these antibiotics by acetylation, adenylylation, and phosphorylation, thereby blocking their binding to the ribosomal RNA.[36] To avoid this type of resistance, analogs of the antibiotics can be designed that (1) bind poorly to the drug-modifying enzyme (aminoglycoside analogs with one fewer amino group had much diminished binding to and activity for the aminoglycoside 3′-phosphotransferase of resistant organisms, but had the same antibacterial activity as the unmodified compound)[37]; (2) are less susceptible to modification, such as tobramycin (**7.21** TOBI), which lacks the 3′-hydroxyl group that is phosphorylated by resistant organisms; or (3) inhibit the enzymes responsible for destroying the drug. A clever approach to the design of an analog that is less susceptible to modification was reported by Mobashery and coworkers.[38] The 3′-hydroxyl group of kanamycin A (**7.19**) was converted to a carbonyl (**7.22**, Scheme 7.1), which is in equilibrium with the corresponding hydrate (**7.23**). The resistant organism can phosphorylate the hydrate to **7.24** via an ATP-dependent aminoglycoside 3′-phosphotransferase, but the phosphorylated product spontaneously decomposes back to **7.22**. Wong and coworkers[39] designed bifunctional dimeric neamine analogs (**7.25**) that inhibit several of the aminoglycoside-modifying enzymes yet still bind to the organism's ribosomal RNA.

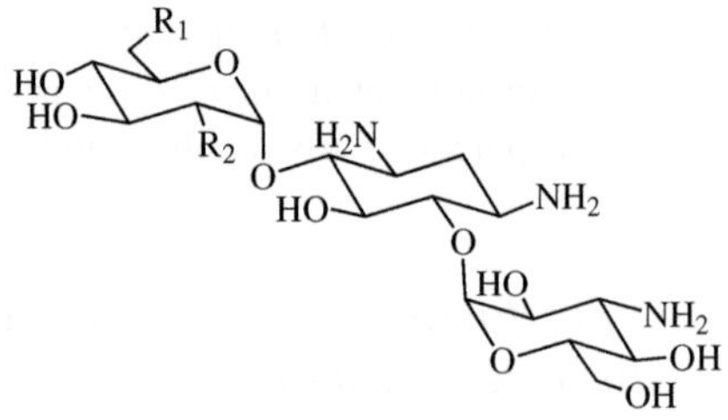

Kanamycin A (R_1=NH_2, R_2=OH)
Kanamycin B (R_1=NH_2, R_2=NH_2)
Kanamycin C (R_1=OH, R_2=NH_2)
7.19

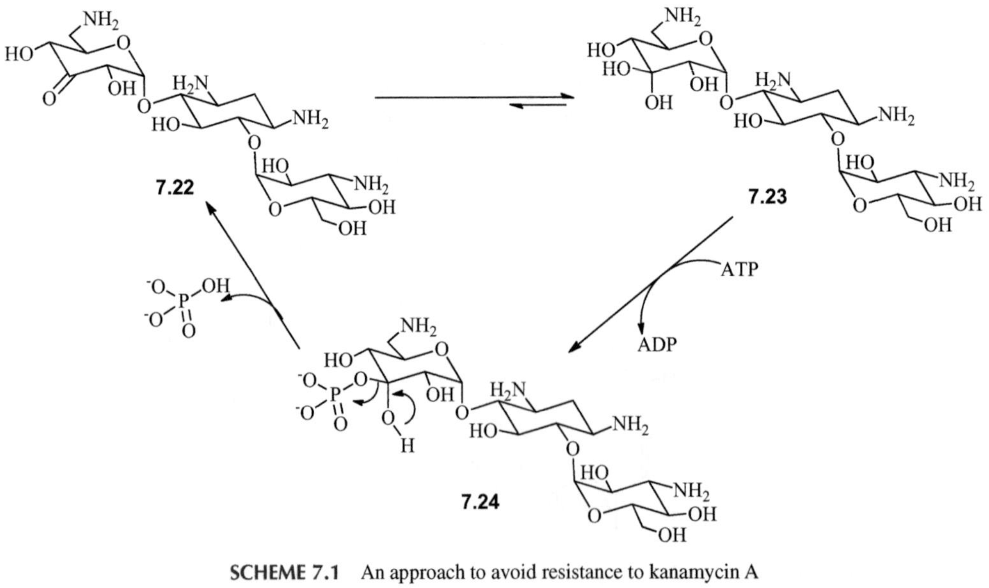

SCHEME 7.1 An approach to avoid resistance to kanamycin A

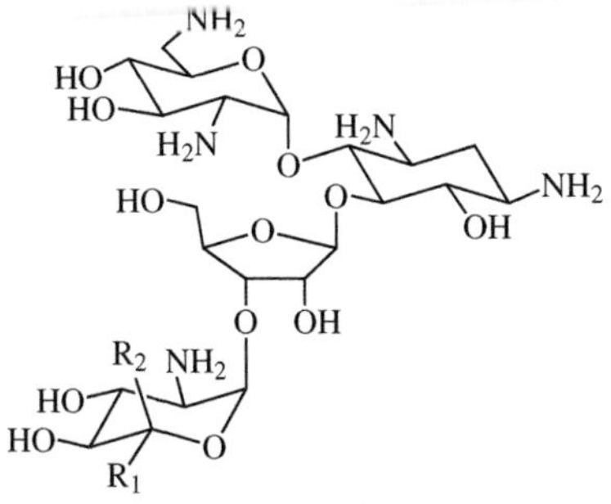

Neomycin B (R_1=H, R_2=CH_2NH_2)
Neomycin C (R_1=CH_2NH_2, R_2=H)
7.20

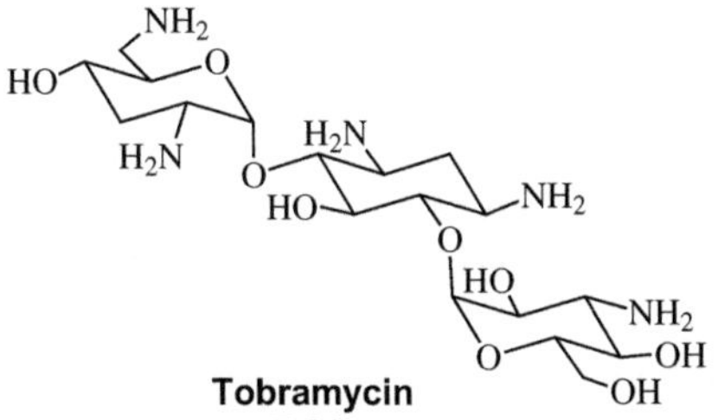

Tobramycin
7.21

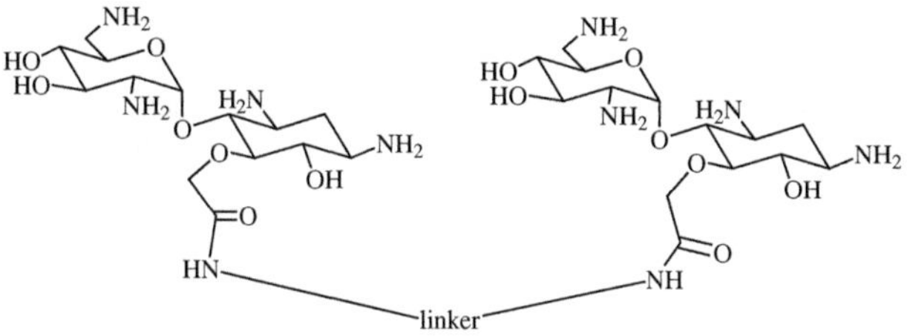

7.25

An analogous mechanism accounts for resistance to the DNA-cleaving antitumor drug bleomycin, discussed in Chapter 6 (Section 6.3.3.2). In this case, the specific resistance mechanism is expression of bleomycin hydrolase, an aminopeptidase that hydrolyzes the carboxamide group of the L-aminoalaninecarboxamide substituent in the metal-free antibiotic (Scheme 7.2).[40] The complex of the bleomycin hydrolysis product with Fe(II) is less effective in the activation of oxygen in the DNA cleavage mechanism shown in Scheme 6.24.[41]

A more general drug-destroying biomolecule is glutathione, which is not an enzyme, but rather a small tripeptide containing cysteine (γ-glutamylcysteinylglycine). Glutathione is often abbreviated as "GSH," where the SH is the sulfhydryl group of the cysteine (the "business end" of the molecule), and G is the rest of the tripeptide. As will be discussed further in Chapter 8 (Sections 8.4.2.1.5 and 8.4.3.5), glutathione can utilize the sulfhydryl group of its cysteine residue to attack an electrophilic center on a drug that is capable of undergoing S_N2, S_NAr, or acylation reactions. Increased secretion of glutathione can thus be a mechanism for resistance to drugs bearing electrophilic centers. On the basis of in vitro studies, increased glutathione has been demonstrated as a mechanism of resistance to alkylating agents such as nitrogen mustards.[42] In this case, the glutathione reacts with the electrophilic center to produce a glutathione-drug adduct (Scheme 7.3).

In many cases, the reaction of an electrophile with glutathione is catalyzed by the enzyme glutathione S-transferase,[43] but these reactions may also occur at a slower rate nonenzymatically. Increased levels of

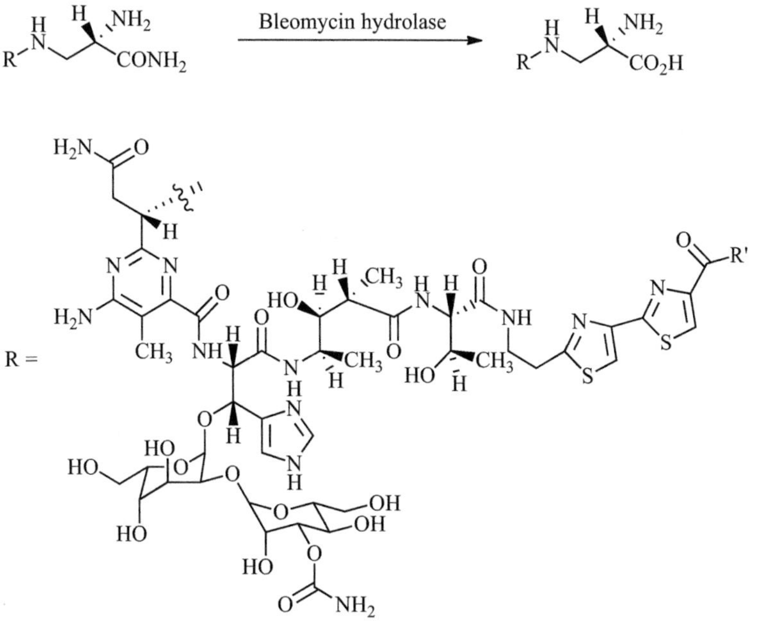

Bleomycin hydrolase

SCHEME 7.2 Action of bleomycin hydrolase to promote tumor resistance to bleomycin

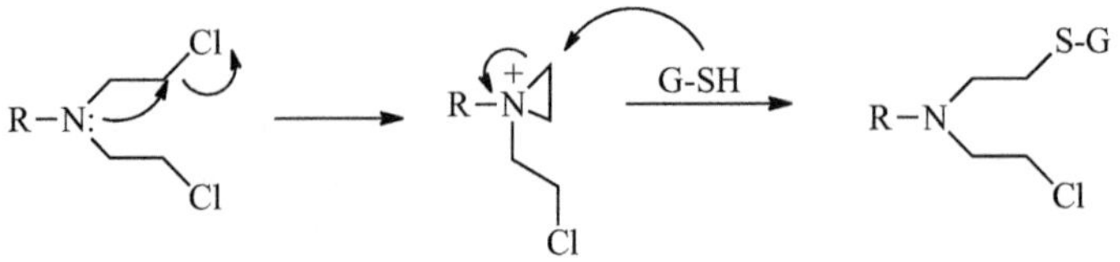

GSH = glutathione = 8.158

SCHEME 7.3 Inactivation of a nitrogen mustard by reaction with glutathione (GSH).

glutathione-S-transferase may thus also constitute a mechanism for resistance to alkylating agents. Glutathione also scavenges free radicals and peroxides, and thus can counteract the effects of anthracycline drugs, for which the mechanism of action involves generation of these reactive species (Chapter 6, Section 6.3.3.1).

7.1.2.5. Decreased Prodrug-Activating Mechanism

Another form of resistance derives from decreased activity of an enzyme needed to convert a prodrug (a compound that is enzymatically converted into the active substance after its administration) into its active form (Chapter 9). Tumor resistance to the antileukemia drug 6-mercaptopurine (**7.26**, R=H, Purinethol) is caused by deletion of hypoxanthine-guanine phosphoribosyltransferase,[44] the enzyme required to convert **7.26** (R=H) into thioinosine monophosphate (**7.26**, R=ribosyl 5′-monophosphate), the active form of the drug. Similarly, resistance to fludarabine and cladrabine, treatments for chronic lymphocytic leukemia, has been associated with low levels of cytidine kinase in cells.[45] Cytidine kinase is the enzyme that converts these prodrugs to their active forms (Scheme 7.4).

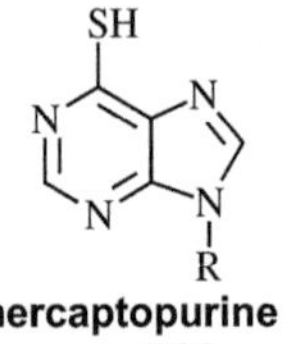

6-mercaptopurine (R=H)
7.26

7.1.2.6. Activation of New Pathways Circumventing the Drug Effect

If the effect of a drug is to block production of a metabolite by enzyme inhibition, the organism could bypass the effect of the drug by inducing a new metabolic pathway that produces the same metabolite. In Chapter 5, we discussed the failure of N-phosphonyl-L-aspartate (PALA, **5.40**) in clinical trials because of tumor resistance. PALA inhibits the enzyme aspartate transcarbamylase early in the pathway for de novo pyrimidine synthesis. When tumor cells acquired the ability to utilize preformed circulating pyrimidine nucleosides, they no longer needed to have a de novo metabolic pathway for pyrimidines, thus rendering PALA ineffective. A similar mechanism contributes to the resistance to EGFR kinase inhibitors erlotinib and gefitinib for treatment of various cancers (discussed in Chapter 5, Section 5.2.2.1).[46] Thus, alternative mechanisms, such as those described in Figure 7.9, for achieving the effects promoted by EGFR (cell survival, proliferation, apoptosis resistance, etc.) might be upregulated by cancer cells, offsetting the effect of the EGFR inhibitors.

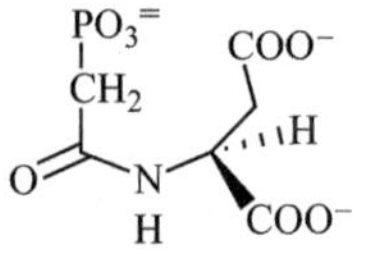

5.40

7.1.2.7. Reversal of Drug Action

As discussed in Chapter 6 (Section 6.3.2.5.1), an important antitumor mechanism of action of some alkylating agents is

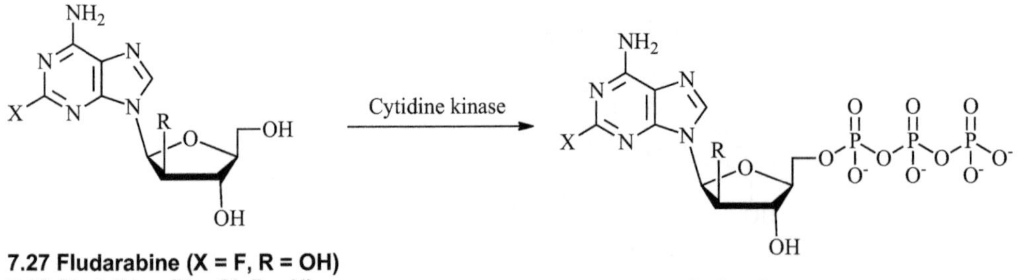

7.27 Fludarabine (X = F, R = OH)
7.28 Cladribine (X = Cl, R = H)

Cytidine kinase

Active form

SCHEME 7.4 Conversion of prodrugs fludarabine and cladrabine to their active form in cells catalyzed by cytidine kinase.

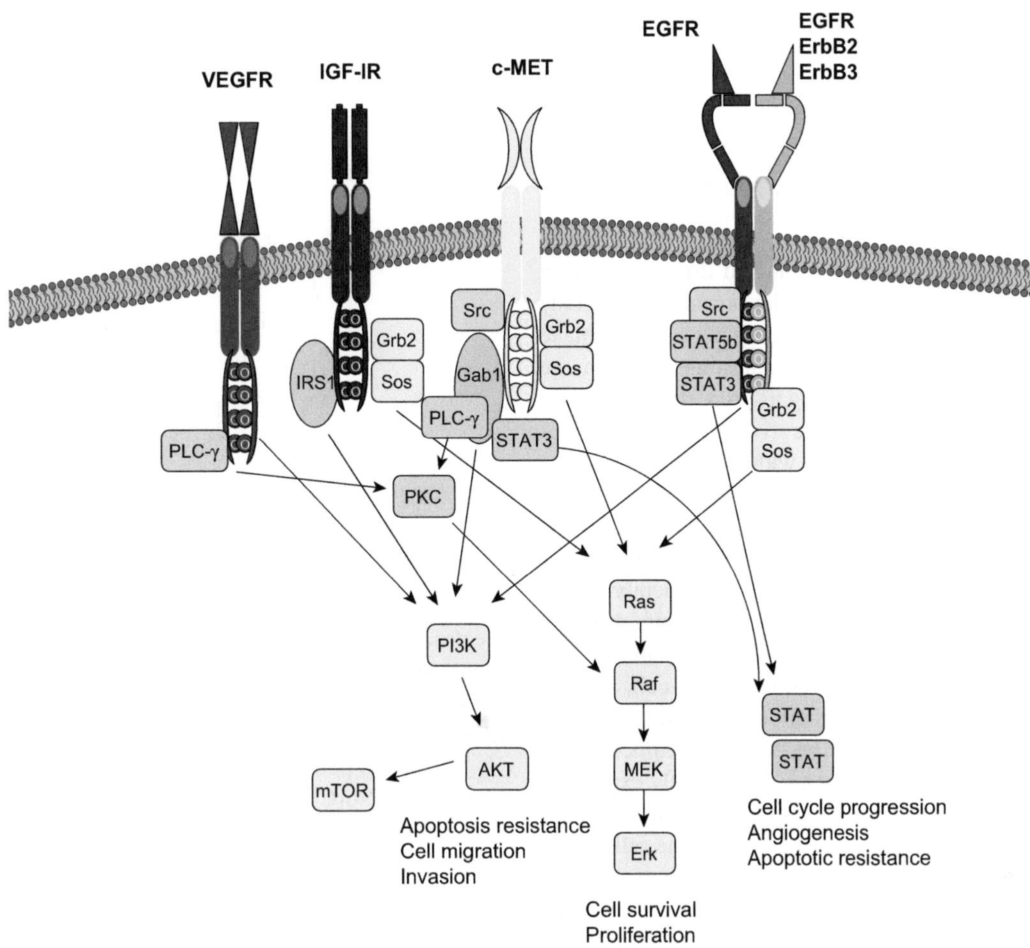

FIGURE 7.9 Example of resistance resulting from activation of alternative pathways. Overactivity of EGFR, which occurs in many cancers, leads to activation of at least two major intracellular processes (RAS–RAF–MEK–ERK pathway and PI3K–AKT–mTOR pathway, yellow boxes) that result in protumor effects. Blockade of EGFR signalling, for example, by inhibition of EGFR kinase (Chapter 5), should inhibit these protumor effects. However, other receptors (c-MET, IGF-1R, and VEGFR) also participate in activation of the same two protumor processes. Some tumor cells have developed resistance to EGFR inhibition by upregulating the activity of one or more of these other receptors to help maintain the protumor activities. Ratushny, V.; Astsaturov, I.; Burness, B. A.; Golemis, E. A.; Silverman, J. S. Targeting EGFR resistance networks in head and neck cancer. *Cell. Signal.* **2009**, *21,* 1255–1268.

formation of guanine-O^6-alkyl adducts on DNA.[47] Dealkylation of these adducts is promoted by O^6-alkylguanine-DNA alkyltransferase (Scheme 7.5), and high levels of the transferase are correlated with resistance to drugs that act by this mechanism.[48] The mechanism for repair is an S_N2 displacement of the O^6-alkyl group by an active-site cysteine residue. Surprisingly, the alkyl group transferred to the active-site cysteine is not subsequently removed; therefore, the "enzyme" can only turnover once and is inactivated by its substrate! By definition, then, O^6-alkylguanine-DNA

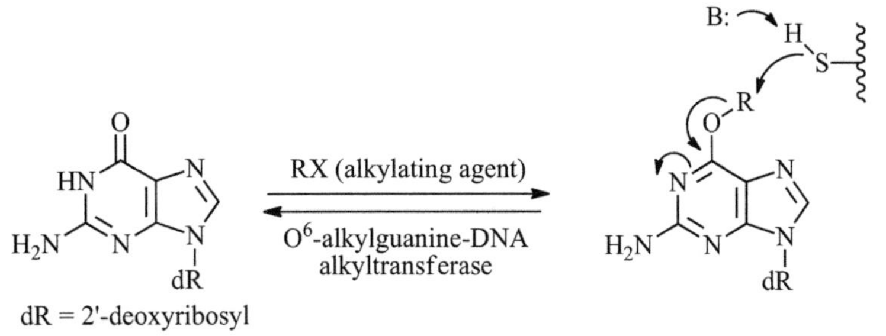

RX (alkylating agent)

O^6-alkylguanine-DNA alkyltransferase

dR = 2'-deoxyribosyl

SCHEME 7.5 O^6-Alkylation of guanine by an alkylating agent and its reversal by O^6-alkylguanine-DNA alkyltransferase

alkyltransferase is not an enzyme (an enzyme is a catalyst, but in this case, the original form of the catalyst is not regenerated). The alkylated form of O^6-alkylguanine-DNA alkyltransferase is rapidly degraded.[49]

7.1.2.8. Altered Drug Distribution to the Site of Action

One type of resistance related to altered drug distribution involves the ability of a cell or cellular organism to exclude the drug from the site of action by preventing cellular uptake of the drug. One mechanism involves adjusting the net charge of the plasma membrane by varying its proportion of anionic (phosphatidylglycerol) to cationic (lysylphosphatidylglycerol) groups.[5] In this way, a drug with the same charge can be repelled from the membrane. Aminoglycoside antibiotic resistance can arise from lack of quinones that mediate the drug transport or from lack of an electrical potential gradient required to drive the drug across the bacterial membrane.[50] One of several mechanisms for resistance to sulfa drugs[51] and penicillins involves altered membrane permeability.

When the site of a drug's action is inside a cell, another mechanism for altering the distribution of a drug to its site of action is increased activity of efflux pumps.[52] *Efflux pumps* are transmembrane proteins that are often highly expressed by microorganisms and tumor cells that act as transporters by binding to drugs and carrying them out of the cell before they can exert their therapeutic effect.[53] Some transporters, known as *multidrug resistance pumps*, are quite broad in their specificity and can efflux a variety of natural and synthetic drugs. Both specific and broad specificity efflux transporter systems have been identified that contribute to bacterial resistance to a broad range of antibiotic drugs.[54] Two categories of transporters, classified by mechanism, are known. The first, associated with multidrug-resistance in cancer, are members of the ATP binding cassette (ABC) family, which require ATP hydrolysis, and include ABCB1 (also called multidrug resistance-1, MDR-1, p-glycoprotein, or P-gp) and ABCG2 (also called breast cancer resistance protein or BCRP).[55] Resistance to dactinomycin (**6.14**), adriamycin, and daunomycin—antitumor drugs discussed in Chapter 6—is associated with overly active efflux mediated by overexpression of the P170 membrane glycoprotein (P-gp), resulting in impaired drug uptake; a correlation has been noted between the ability of the cell to retain dactinomycin and the effectiveness of the drug.[56] Overexpression of P-gp has also been shown to contribute to resistance to the Bcr-Abl kinase inhibitors imatinib (**5.18**), dasatinib (**7.9**), and nilotinib (**7.10**) in a chronic myeloid leukemia cell line.[57] P-gp and related efflux transporters have also been implicated in resistance to antifungal drugs, such as fluconazole (**7.17**).[58] It is believed that these transporters have a role in normal cells to protect the cells from exogenous toxins, as well as endogenous toxins, such as metabolic waste products.[59] One important example is the presence of P-gp in the epithelial cells of the blood–brain barrier, which often contributes to the poor accessibility of drugs to targets in the central nervous system; overexpression of P-gp at the blood–brain barrier has also been implicated in a subset of patients who are resistant to antiepileptic drug therapy.[60]

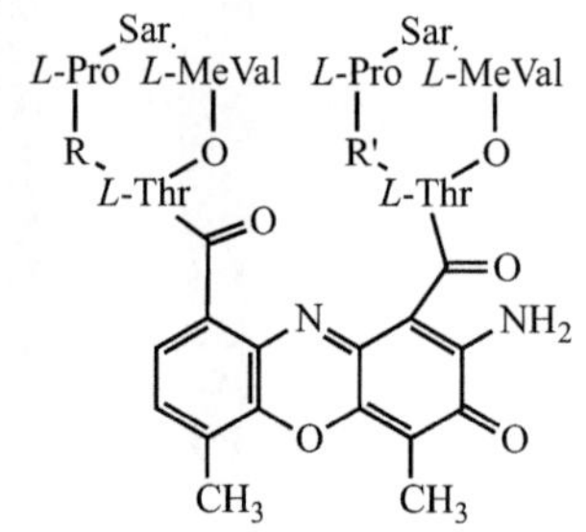

Sar; *L*-Pro; *L*-MeVal; R; O; *L*-Thr; O; Sar; *L*-Pro; *L*-MeVal; R'; O; *L*-Thr; O; N; NH_2; O; O; CH_3; CH_3

Dactinomycin
6.14

The second category of efflux pump proteins operate by electrochemical transmembrane proton or sodium ion gradients and include the major facilitator superfamily, small-multidrug resistance, resistance-modulation-division, and multiple drug and toxin extrusion pumps.[61]

The Gram-positive bacterium *S. aureus*[62] uses NorA, NorB, and NorC efflux pumps to transport structurally diverse drugs, such as fluoroquinolones, quaternary ammonium compounds, reserpine, verapamil, ethidium bromide, and acridines out of the microorganism.[63]

7.2. DRUG SYNERGISM (DRUG COMBINATION)

7.2.1. What is Drug Synergism?

When drugs are given in combination, their effects can be antagonistic, subadditive, additive, or synergistic. *Drug synergism* arises when the therapeutic effect of two or more drugs used in combination is greater than the sum of the effects of the drugs administered individually. From the examples below, it will be apparent that one form of synergism can occur when there is resistance to a drug that is administered alone, whereas coadministration of the drug with a second compound that inhibits the resistance mechanism produces a synergistic response.

7.2.2. Mechanisms of Drug Synergism

7.2.2.1. Inhibition of a Drug-Destroying Enzyme

If the activity of a drug is reduced or shortened because of the action of a drug-destroying enzyme, then an important approach for synergism would be to design a compound that inhibits the drug-destroying enzyme. Such an inhibitor has no real therapeutic effect on its own; it only protects the drug from being destroyed. So, if that compound is administered alone, it obstructs the drug-destroying enzyme, but it has no therapeutic benefit. If the primary drug is administered alone, it is destroyed, and therefore has reduced or no therapeutic effect. But if the primary drug and the inhibitor of the drug-destroying enzyme are administered together, then the drug survives destruction and is thus effective again, a synergistic effect.

Because the major cause for resistance to penicillins is the excretion of β-lactamases, an obvious approach to drug synergism would be the combination of a penicillin and a β-lactamase inhibitor.[64] In the 1970s, certain naturally occurring β-lactams that did not have the general penicillin or cephalosporin structure were isolated from various organisms,[65] and were found to be potent mechanism-based inactivators (Chapter 5, Section 5.3.3) of β-lactamases.[66] These compounds are used in combination with penicillins to destroy penicillin-resistant strains of bacteria, for example, the combination of amoxicillin (**5.95d**, R′=OH) with clavulanate (**7.29**) is sold as Augmentin[67] and ampicillin (**5.95d**, R′=H) plus sulbactam (**7.30**) are in Unasyn (in unison, get it?). The β-lactamase inhibitors have no antibiotic activity, but they protect the penicillin from destruction so that it can interfere with cell wall biosynthesis. The proposed mechanism of action of clavulanate as a β-lactamase inhibitor is shown in Scheme 7.6.[68] Following nucleophilic attack by an enzyme serine hydroxyl group on the carbonyl of the β-lactam, the covalent adduct can partition into at least two pathways, one of which leads to a very slowly hydrolyzable product (**7.31**), and the other leads to a stably inactivated cross-linked product (**7.32**), an example of a mechanism-based inactivator.

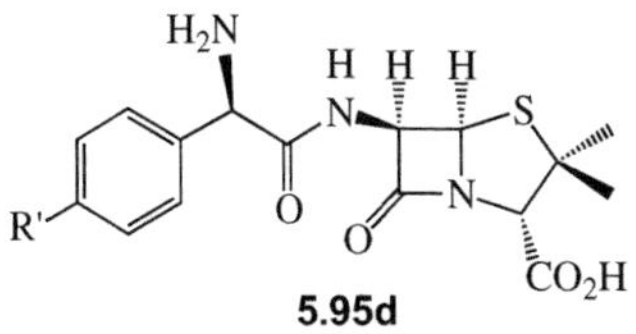

5.95d

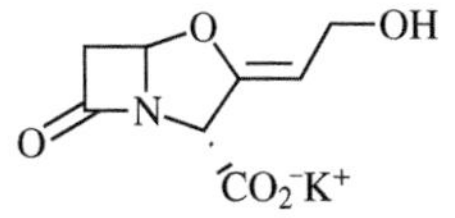

Clavulanate potassium
7.29

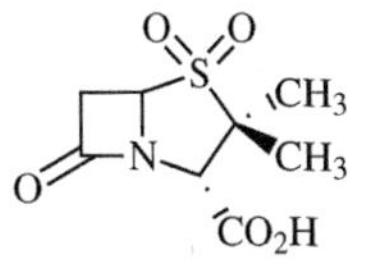

Sulbactam
7.30

The drug-destroying enzymes expressed by bacteria come from plasmids. If these plasmids were eliminated from the bacterium, it would become susceptible to antibiotics. *Plasmid incompatibility* occurs when two plasmids containing the same origin of replication are in the same bacterium.[69] They are forced to compete for proteins and RNA needed for replication, so the most efficient of the two can drive out the other. Hergenrother and coworkers found that a small molecule, the aminoglycoside apramycin (**7.33**, Apralan), could mimic the small piece of RNA that dictates incompatibility of plasmids, thereby evicting the ampicillin-resistant plasmid from the bacterium and restoring its susceptibility to ampicillin.[70]

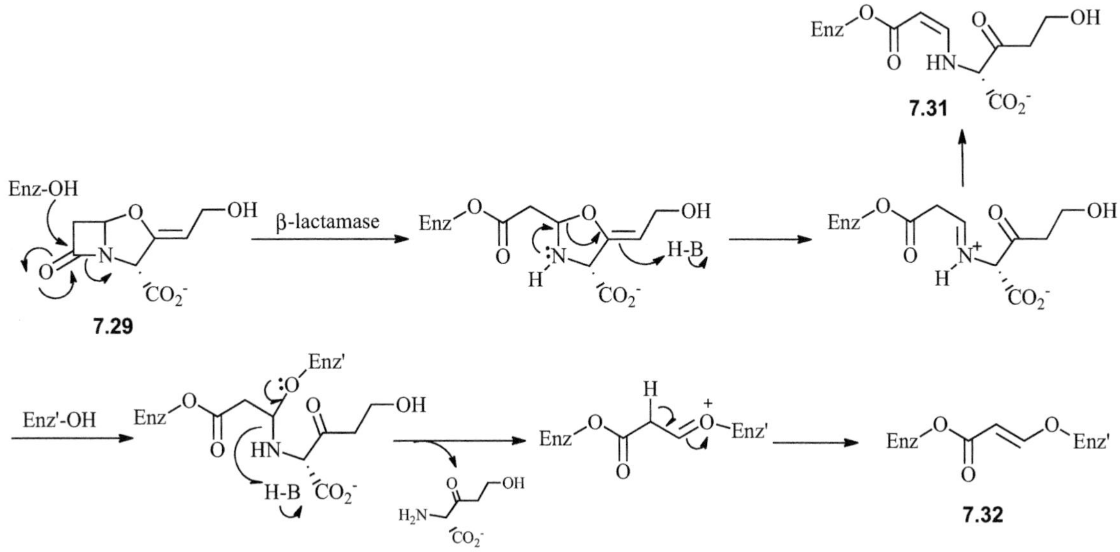

SCHEME 7.6 Proposed mechanism of inactivation of β-lactamase by clavulanate

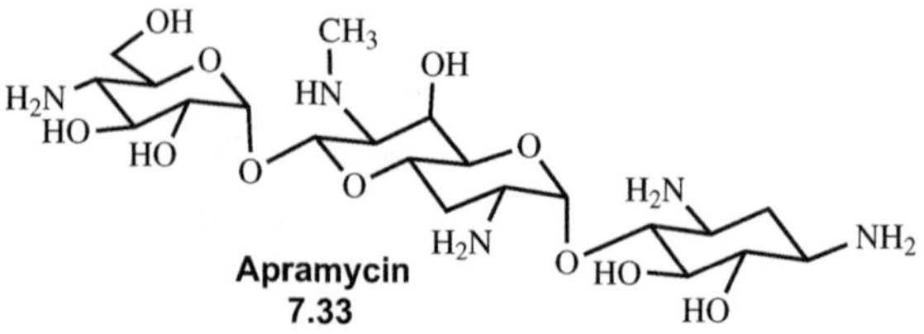

Apramycin
7.33

As another example of synergism, because of the potency of inhibition of cytochrome P450 by ritonavir (Chapter 5, Section 5.2.5.2), it was found that combination of other HIV-1 protease inhibitors with ritonavir was synergistic. For example, the mutant HIV protease inhibitor lopinavir (**7.8**) does not share the cytochrome P450 inhibition activity possessed by ritonavir, and, in fact, the microsomal half life for lopinavir is low. To get around that problem, it was found that ritonavir could be added as an inhibitor of cytochrome P450, thereby protecting lopinavir from metabolic degradation. A combination of lopinavir (to inhibit HIV-1 protease in the resistant strains) and ritonavir (to inhibit HIV-1 protease in susceptible strains plus to inhibit cytochrome P450), known as Kaletra, is the commercial product. In clinical trials with this combination, no HIV was detectable in the blood of the patients for 3 years.

Acquired resistance is a major obstacle to the success of *antipurines*, compounds that mimic purines and either inhibit enzymes on the biosynthetic pathway to DNA or are themselves incorporated into this pathway, to produce dysfunctional variants of DNA. Vidarabine (ara-A; **7.34**) is an antipurine that is degraded by adenosine deaminase. Synergistic effects are observed when **7.34** is used in combination with pentostatin (**5.30**; Chapter 5, Section 5.2.3.2.2), a potent adenosine deaminase inhibitor. Their combination induced remissions of nodular lymphomas and lymphocytic leukemia; however, some unexplained deaths occurred in clinical trials, and consequently further study was restricted.[71]

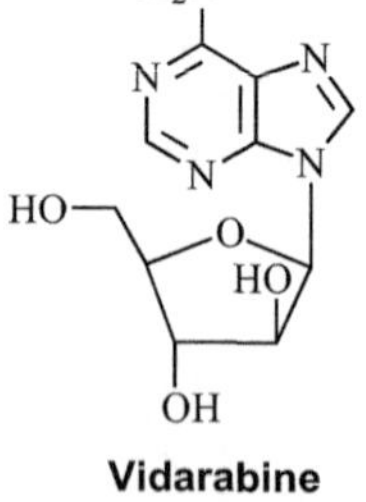

Vidarabine
7.34

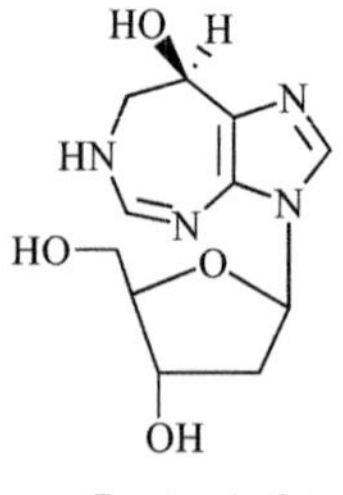

Pentostatin
5.30

7.2.2.2. Sequential Blocking

A third mechanism for synergism is *sequential blocking*, the inhibition of two or more consecutive steps in a metabolic pathway. The reason this is effective is because it is difficult (particularly with a reversible inhibitor) to inhibit an enzyme 100%. If less than 100% of the enzyme activity is blocked, the metabolic pathway has not been shut down. With the combined use of inhibitors of two consecutive enzymes in the pathway, it is possible to block the metabolic pathway virtually completely. Because reversible enzyme inhibition (or receptor antagonism) is hyperbolic in nature, complete inhibition of an enzyme would require a large excess of the drug, which may be toxic. This approach becomes somewhat less important with an irreversible inhibitor that can inhibit an enzyme completely (see Chapter 5, Section 5.3.3.3.2 for an example of an irreversible inhibitor that does not shut down the target enzyme totally).

Examples are combination therapy of sulfadoxine (**7.35**; in combination with pyrimethamine is called Fansidar) with pyrimethamine (**7.36**; Daraprim) or sulfamethoxazole (**7.37**; Gantanol, and in combination with trimethoprim is called Bactrim) with trimethoprim (**7.1**, Proloprim),[72] which have been shown to be quite effective for the treatment of malaria and bacterial infections, respectively.[73]

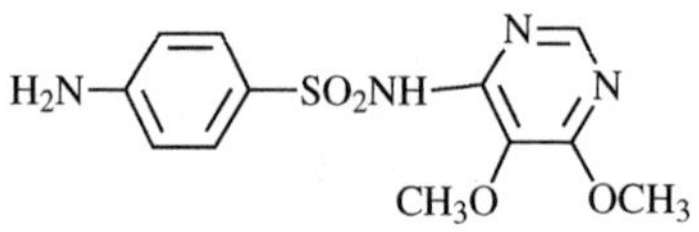

Sulfadoxine
7.35

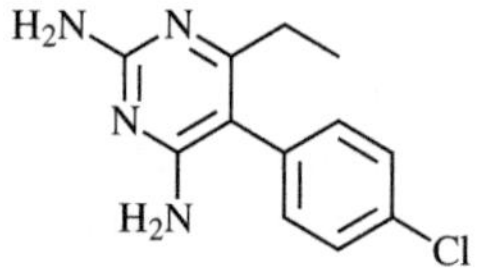

Pyrimethamine
7.36

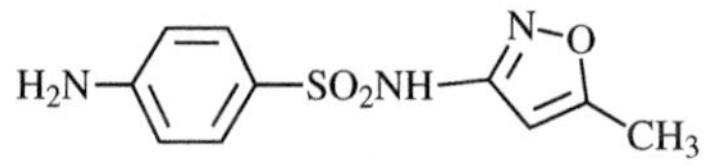

Sulfamethoxazole
7.37

Sulfa drugs **7.35** and **7.37** inhibit dihydropteroate synthase, which catalyzes the synthesis of dihydrofolate, whereas **7.36** and **7.1** inhibit dihydrofolate reductase, which catalyzes the conversion of dihydrofolate (the product of dihydropteroate synthase catalysis) to tetrahydrofolate (Scheme 4.21). Trimethoprim has the desirable property of being a very tight-binding inhibitor of *bacterial* dihydrofolate reductase, but a poor inhibitor of *mammalian* dihydrofolate reductase; the IC_{50} is 5 nM for the former and 2.6×10^5 nM for the latter.[74] As discussed in Chapter 5 (Section 5.3.3.3.5), dihydrofolate reductase is essential for the regeneration of tetrahydrofolate from the dihydrofolate produced in the thymidylate synthase reaction (Scheme 5.30). Drug synergism also occurs when an inhibitor of dihydrofolate reductase (such as methotrexate, **7.38**) is combined with an inhibitor of thymidylate synthase. This is another example of a sequential blocking mechanism.

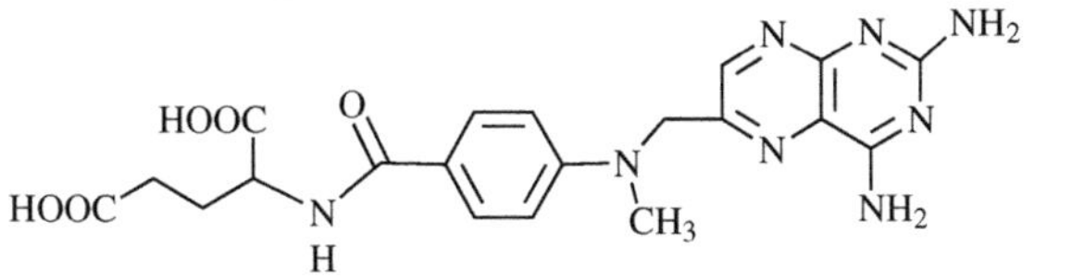

Methotrexate
7.38

Another recent example of synergism from sequential blocking is in the use of kinase inhibitors to treat melanoma. As shown in Figure 7.9, one of the pathways leading to cell proliferation and survival is the Ras–RAF–MEK–ERK pathway, in which RAF, MEK, and ERK are kinase enzymes (RAF catalyzes the phosphorylation of MEK, whereupon MEK is then suitably activated to catalyze phosphorylation of ERK). In some cancers, including prominent types of melanoma and colorectal cancer, the V600E mutation in the RAF enzyme causes it to become hyperactive, leading to overproliferation of the cancer cells. Whereas RAF inhibitor dabrafenib (**7.39**) is effective in treating RAF V600E-positive melanoma, the combination of **7.39** with MEK inhibitor trametinib (**7.40**) led to significant improvements in progression-free survival in a Phase II clinical trial.[75]

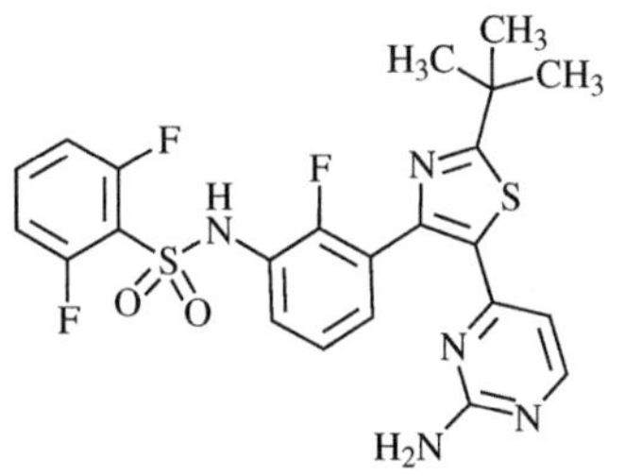

Dabrafenib
7.39

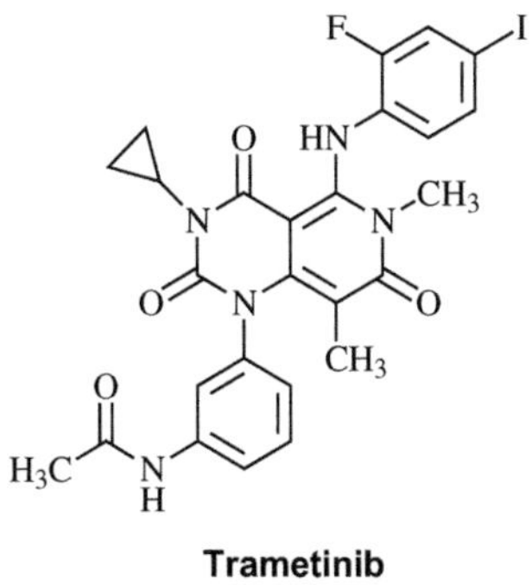

Trametinib
7.40

7.2.2.3. *Inhibition of Targets in Different Pathways*

In Section 7.1.2.6 and Figure 7.9, it was illustrated how cancer cells can activate or upregulate alternative pathways to circumvent the therapeutic effect of EGFR kinase inhibitors. Therefore, clinical trials have been conducted to investigate the effect of combining an EGFR kinase inhibitor with an inhibitor of one or more of the other pathways to attempt to overcome this resistance mechanism.[76] For example, combination of erlotinib (**5.13**) with MET inhibitor tivantinib (**7.41**) showed benefit over erlotinib alone in a Phase II trial; a Phase III study was initiated. Cabozantinib (**7.42**) inhibits both MET and VEGFR, both of which are kinases involved in pathways that can circumvent the action of erlotinib (see Figure 7.9); therefore, combination therapy with erlotinib and cabozantinib is also under investigation.

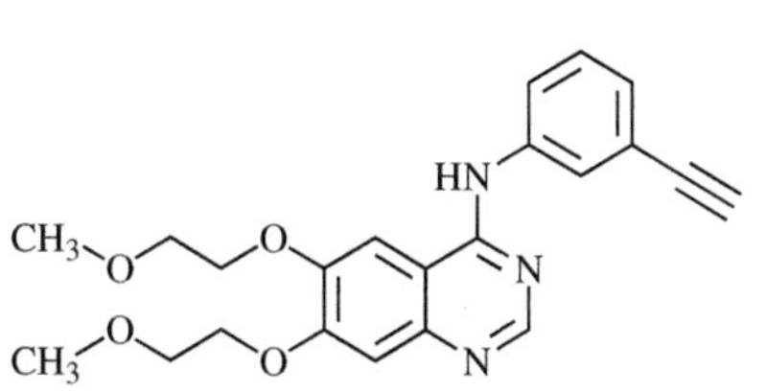

Erlotinib
5.13

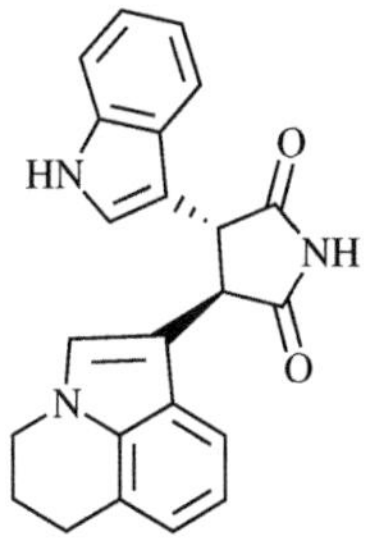

Tivantinib
7.41

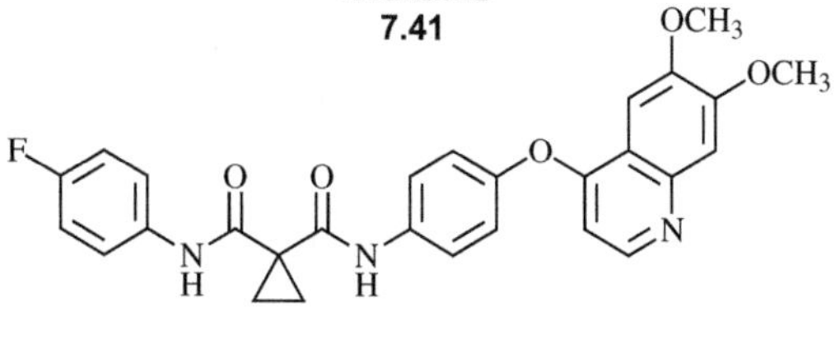

Cabozantinib
7.42

Another example of synergism relates to the mechanism of resistance to PALA (**5.40**) discussed in Section 7.1.2.6. PALA inhibits aspartate transcarbamylase, and thereby blocks the synthesis of pyrimidine building blocks needed for DNA synthesis; resistance developed when cells were able to use a new source of preformed pyrimidine building blocks. To overcome this mechanism of resistance to PALA, nitrobenzylthioinosine, which inhibits the diffusion of nucleoside transport and should block the uptake of the preformed pyrimidines, was tested for its synergistic effect with PALA. These two compounds were indeed synergistic in vitro.[77] However, the combination was too toxic at effective dosages to be used in vivo.

7.2.2.4. Efflux Pump Inhibitors

If a drug is being effluxed from its target cell or organism by a transporter protein, a compound can be designed that inhibits this efflux pump.[78] The efflux pump inhibitor has no therapeutic activity, and without it, the drug intended to treat the disease is ineffective; however, the combination of the two should be synergistic and should exhibit therapeutic activity.[79] Since P-gp is well established as an efflux transporter that is implicated in resistance to cancer treatments, significant effort has been expended toward identifying P-gp inhibitors with the aim of increasing the effectiveness of chemotherapeutic agents that are subject to efflux by this transporter.[80] The first and second generation compounds had only modest potency as P-gp inhibitors, as well as the existence of their primary pharmacological activities, which served to limit how much drug could be administered to achieve a desired level of P-gp inhibition; others were inhibitors of cytochrome P450 3A4 (CYP3A4), which has a principal role in the metabolism of many drugs, and, therefore, have the potential to complicate the pharmacokinetics of any coadministered drug. Third-generation P-gp inhibitors have increased inhibitory potency against P-gp, are selective for P-gp over other transporters, lack pharmacological action on their own, and have minimal interaction with CYP3A4. An example is tariquidar (**7.43**), which was identified following optimization of an earlier, less potent, lead compound XR9051 (**7.44**).[81] In studies in vitro, the potency of doxorubicin to inhibit the proliferation of various breast, ovarian, and lung cancer cell lines was 5- to 150-fold greater in the presence of tariquidar, that is, the compounds acted synergistically. Tariquidar was shown to increase the uptake of P-gp substrates into cells that express P-gp and was well tolerated in Phase I trials as a single agent. However, two Phase III studies in combination with chemotherapeutic agents were terminated early because of toxicity. It is hypothesized that the toxicity was the result of inhibition of P-gp in bone marrow cells, where P-gp plays a protective role. As mentioned previously, P-gp also plays a role in protecting the brain from foreign substances (part of the blood–brain barrier). Therefore, the use of transporter inhibitors to overcome drug resistance can be confounded by their simultaneous effects on the function of normal cells, which rely on the same transporters for protection from foreign substances.

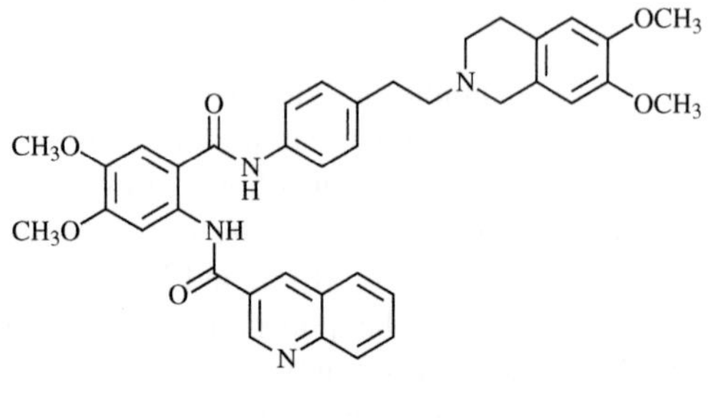

Tariquidar
7.43

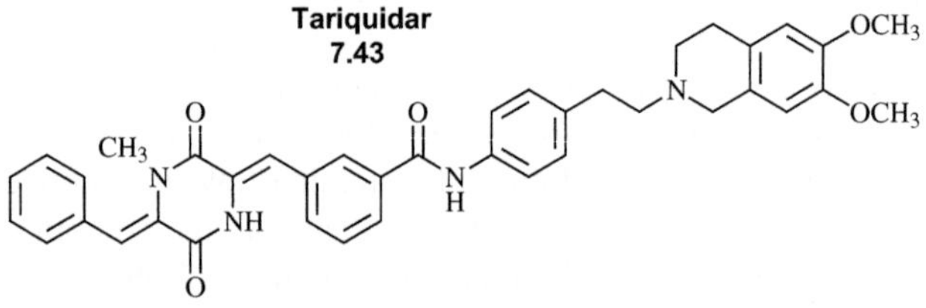

XR9051
7.44

7.2.2.5. Use of Multiple Drugs for the Same Target

De novo guanosine nucleotide biosynthesis depends on inosine 5′-monophosphate dehydrogenase (IMPDH), which catalyzes the NAD^+-dependent oxidation of inosine-5′-monophosphate (IMP) to xanthosine-5′-monophosphate (XMP) (Scheme 7.7).[82] This reaction is important for DNA synthesis in rapidly proliferating cells, such as microbes, cancer cells, and cells of the immune system. Therefore, inhibition of IMPDH has been the target of extensive drug discovery efforts.[83]

Mycophenolic acid (**7.45**) is a natural product inhibitor of IMPDH, and its ester prodrug mycophenolate mofetil (**7.46**, CellCept) has been approved for prevention of organ transplant rejection.[84] The basis of its action is believed to be disruption of the important role of IMPDH (see Scheme 7.7)

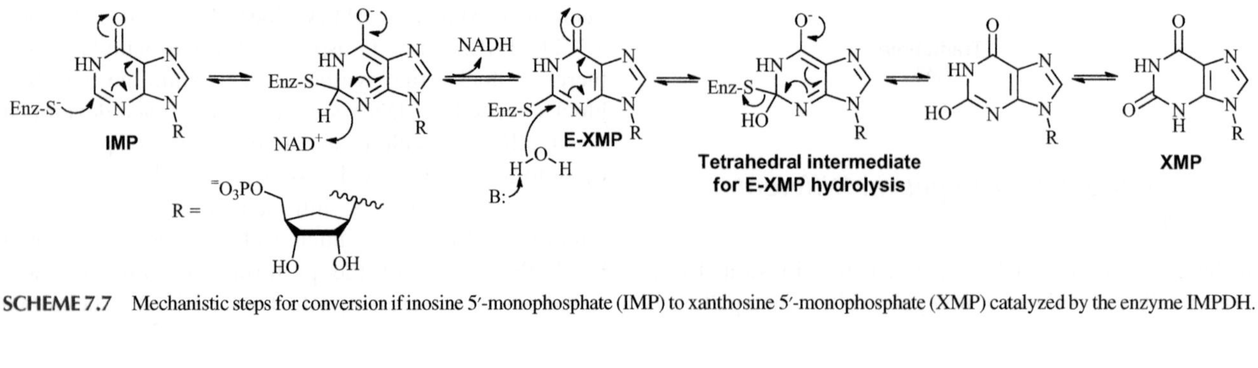

SCHEME 7.7 Mechanistic steps for conversion if inosine 5′-monophosphate (IMP) to xanthosine 5′-monophosphate (XMP) catalyzed by the enzyme IMPDH.

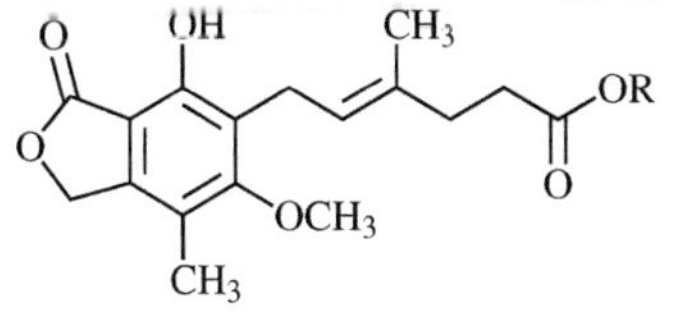

7.45 Mycophenolic acid, R=H
7.46 Mycophenolate mofetil, R=-CH$_2$CH$_2$-N 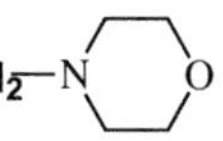

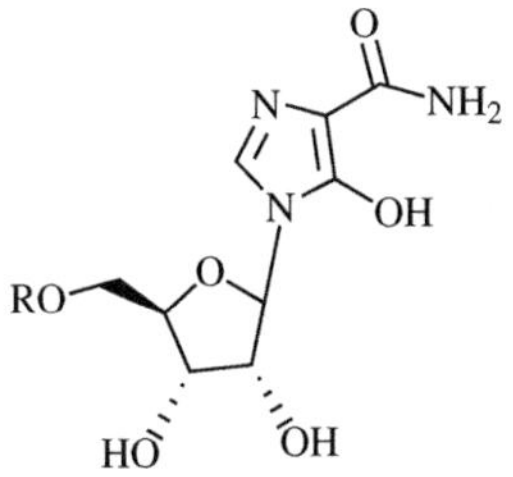

7.47 Mizoribine 5'-monophosphate, R=PO$_3^=$
7.48 Mizoribine, R=H

for DNA synthesis in T- and B-cells (specialized white blood cells of the immune system) that mediate transplant rejection. Mycophenolic acid has been shown to be an uncompetitive inhibitor, and it binds to IMPDH after NADH is released but before XMP is produced. Another IMPDH inhibitor is mizoribine monophosphate (**7.47**); its nonphosphorylated prodrug mizoribine (**7.48**, Bredinin) is also used clinically to prevent transplant rejection and to treat certain autoimmune disorders.[85] On the basis of X-ray crystallographic studies, the complex of mizoribine monophosphate with IMPDH is proposed to mimic the tetrahedral intermediate for E-XMP hydrolysis (shown in Scheme 7.7) or the transition state leading to it (Figure 7.10). In contrast, X-ray crystallographic data indicate that **7.45** binds to IMPDH-XMP (E-XMP in Scheme 7.7) in a site adjacent to the XMP binding site,[84] and, thus, **7.45** and **7.47** must have different mechanisms of IMPDH inhibition. Consequently, in vivo experiments were carried out to determine whether **7.45** and **7.47** would exhibit synergistic actions by administering their respective prodrugs, **7.46** or **7.48**, either alone or in combination.[86] The study measured the mean survival time (MST) of mice that had received heart transplants; the results are summarized in Table 7.1. The combination of the two drugs each dosed at 40 (mg/kg)day^{-1} resulted in an MST of 19.7 days, compared to 12.7 and 8.29 days, respectively, for 80 (mg/kg)day^{-1} of either drug alone. Even more striking, the combination of the two drugs each dosed at 80 (mg/kg)day^{-1} resulted in an MST of 78.4 days, compared to 23.8 or 11.3 days, respectively, for 160 (mg/kg)day^{-1} of either drug alone. These data demonstrate a synergistic effect, but caution is needed when drawing conclusions about the mechanism behind this synergism. Although the data are consistent with a mechanism involving the drugs' inhibition of the same target (IMPDH), alternative mechanisms are that one drug inhibits the metabolism or active cellular efflux of the other. Additional experiments would be needed to distinguish between the possible mechanisms.

TABLE 7.1 Effect on Mean Survival times (MST) from Treatment of Mice with Transplanted Hearts with 7.46 or 7.48 Alone or in Combination

Drug	Dose (mg/kg)day^{-1}	MST (days)
None	–	7.86
7.46	80	12.7
7.46	160	23.8
7.48	80	8.29
7.48	160	11.3
7.46/7.48	40/40	19.7
7.46/7.48	80/80	78.4

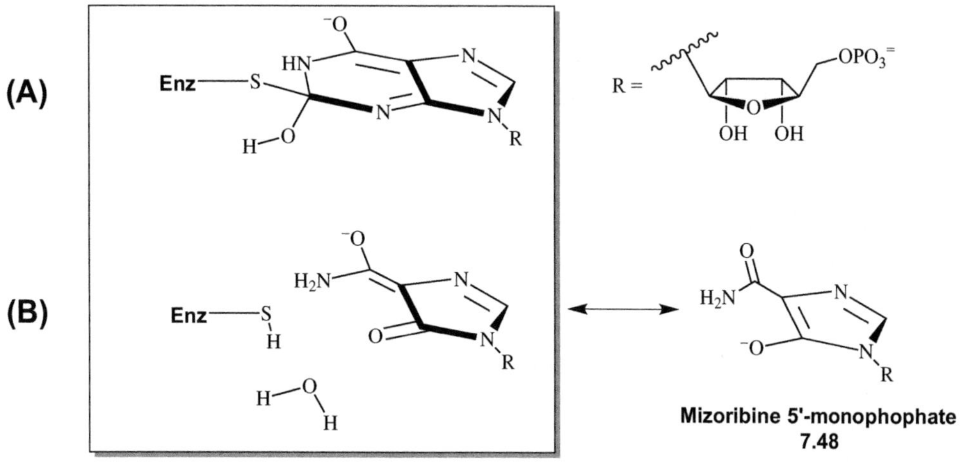

FIGURE 7.10 Schematic drawing showing how the complex (B) of mizoribine monophosphate to IMPDH is believed to mimic the tetrahedral intermediate (A) for E-XMP hydrolysis (compare Scheme 7.7).

Chapters 3–7 have dealt with the structures and functions of various protein and nucleic acid receptors as well as with classes of drugs that interact with these receptors and how the medicinal chemist can begin to design molecules that selectively interact with these receptors while attempting to delay resistance. In Chapter 8, we turn our attention to the heroic efforts our bodies make to destroy and excrete xenobiotics, including drugs, and then in Chapter 9, we discuss the heroic efforts medicinal chemists make to outwit these metabolic processes by protecting the drugs until they reach their desired sites of action.

7.3. GENERAL REFERENCES

Drug resistance

Drlica, K. S.; Perlin, D. S. *Antibiotic Resistance: Understanding and Responding to an Emerging Crisis*, FT Press Science, Saddle River, NJ, 2011.

Goldie, J. H.; Coldman, A. J. *Drug Resistance in Cancer: Mechanisms and Models*, Cambridge University Press, Cambridge, U.K., 2009.

Mayers, D. (Ed.), *Antimicrobial Drug Resistance: Clinical and Epidemiological Aspects*, Vol. 1–2, Humana Press, New York, NY, 2009.

Zhou, J. (Ed.), *Multi-Drug Resistance in Cancer*, Humana Press, New York, NY, 2009.

Walsh, C. *Antibiotics: Actions, Origins, Resistance*, American Society for Microbiology, 2003.

Andersson, B.; Murray, D. (Eds.), *Clinically Relevant Resistance in Cancer Chemotherapy*, Kluwer Academic Publishers, 2002.

***β*-Lactamase Inhibitors**

Biondi, S.; Long, S.; Panunzio, M.; Qin, W. L. Current trends in β-lactam based β-lactamase inhibitors. *Curr. Med. Chem.* **2011**, *18*, 4223–4236.

Bush, K.; Macielag, M. J. New β-lactam antibiotics and β-lactamase inhibitors. *Expert Opin. Therapeut. Patents* **2010**, *20*, 1277–1293.

Knowles, J. R. Penicillin resistance: the chemistry of β-lactamase inhibition. *Acc. Chem. Res.* **1985**, *18*, 97–104.

Cartwright, S. J.; Waley, S. G. Beta-lactamase inhibitors. *Med. Res. Rev.* **1983**, *3*, 341–382.

7.4. PROBLEMS (ANSWERS CAN BE FOUND IN THE APPENDIX AT THE END OF THE BOOK)

1. Your DNA-alkylating agent was found to lead to tumor cell resistance as a result of the induction of a new enzyme that repaired the modified DNA. What do you do next?
2. Resistance to your new potent antibacterial drug (**1**) was shown to be the result of a single-point genetic mutation in the target enzyme such that an important active site lysine residue was mutated to an aspartate residue. Suggest a simple way to proceed toward the design of a new antibacterial drug against the resistant strain.

F COOH H S **1**

3. Augmentin is a medicine that combines amoxicillin, a β-lactam antibiotic, with clavulanate, a β-lactamase inhibitor. Why is it important that the two compounds have similar pharmacokinetics (e.g., oral bioavailability and plasma half-life) in humans?
4. When might it be a disadvantage to inhibit a drug-destroying enzyme?
5. At the conclusion of Section 7.2.2.5, several possible mechanisms were suggested to explain the observed synergism between mycophenolate mofetil and mizoribine. What additional experiments could be proposed to help distinguish which mechanism(s) are at play?
6. If drug resistance to an antibacterial compound arises from expression of a new enzyme in the resistant organism that destroys the drug, what *three* approaches could you take to combat the resistance?
7. One mechanism of drug resistance is that the organism has a genetically altered enzyme. How does this lead to drug resistance?
8. Fosfomycin (see Chapter 5, problem 11 for structure) is a potent antibacterial agent that interferes with cell wall biosynthesis at the enzyme MurA, which contains an active site cysteine residue. Give two reasonable mechanisms for resistance to fosfomycin that do *not* involve the bacterial cell membrane.

REFERENCES

1. Mobashery, S.; Azucena, E. In *Encyclopedia Life Science*, Nature Publishing Group, U.K., 2002, Vol. 2, pp. 472–477.
2. (a) Rivera Gonzales, F. E.; Lopez, M. I. (Eds.), *Plasmids: genetics, applications and health*, Nova Science Publishers, Hauppauge, NY, 2012. (b) Casali, N.; Preston, N. (Eds.), *E. coli* Plasmid Vectors, Methods and Applications, Humana Press, Totowa, NJ, 2010.
3. Michelucci, R.; Pasini, E.; Tassinari, C. S. Phenobarbital, primidone and other barbiturates. In *The Treatment of Epilepsy*, 3rd ed., Shorvon, S.; Perucca, E.; Engel, Jr., J., Wiley-Blackwell, Chichester, U.K., 2009; Chapter 46.
4. (a) Neu, H. C. The crisis in antibiotic resistance. *Science* **1992**, *257*, 1064–1072. (b) Tomasz, A. Multiple-antibiotic-resistant pathogenic bacteria. A report on the Rockefeller University Workshop. *N. Engl. J. Med.* **1994**, *330*, 1247–1251.
5. (a) Gold, H. S.; Moellering, R. C. Antimicrobial-drug resistance. *N. Engl. J. Med.* **1996**, *335*, 1445–1453. (b) Domagala, J. M.; Sanchez, J. P.

New approaches and agents to overcome bacterial resistance. *Annu. Rep. Med. Chem.* **1997**, *32*, 111–120. (c) Levy, S. B. The challenge of antibiotic resistance. *Sci. Am.* **1998**, *278*, 46–53.

6. (a) Albert, A. *Selective Toxicity*, 7th ed, Chapman and Hall, London, 1985, p. 256. (b) Lowe, J. A. III Mechanisms of antibiotic resistance. *Annu. Rep. Med. Chem.* **1982**, *17*, 119–127.
7. Spratt, B. G. Resistance to antibiotics mediated by target alterations. *Science* **1994**, *264*, 388–393.
8. Then, R. L.; Hermann, F. Mechanisms of trimethoprim resistance in Enterobacteria isolated in Finland. *Chemotherapy*, **1981**, *27*, 192–199.
9. (a) Wise, E. M., Jr.; Abou-Donia, M. M. Sulfonamide resistance mechanism in *Escherichia coli*: R plasmids can determine sulfonamide-resistant dihydropteroate synthases. *Proc. Natl. Acad. Sci. U.S.A.* **1975**, *72*, 2621–2625. (b) Skold, O. R-factor-mediated resistance to sulfonamides by a plasmid-borne, drug-resistant dihydropteroate synthase. *Antimicrob. Agents Chemother.* **1976**, *9*, 49–54.
10. (a) Larder, B. A.; Darby, G.; Richman, D. D. HIV with reduced sensitivity to zidovudine (AZT) isolated during prolonged therapy. *Science* **1989**, *243*, 1731–1734. (b) Schuurman, R.; Nijhuis, M.; VanLeeuwen, R.; Schipper, P.; DeJong, D.; Collis, P.; Danner, S. A.; Mulder, J.; Loveday, C.; Christopherson, C.; Kwok, S.; Sninsky, J.; Boucher, C. A. B. Rapid changes in human immunodeficiency virus type 1 RNA load and appearance of drug-resistant virus populations in persons treated with lamivudine (3TC). *J. Infect. Dis.* **1995**, *171*, 1411–1419.
11. Kellam, P.; Boucher, C. A. B.; Larder, B. A. Fifth mutation in human immunodeficiency virus type 1 reverse transcriptase contributes to the development of high-level resistance to zidovudine. *Proc. Natl. Acad. Sci. U.S.A.* **1992**, *89*, 1934–1938.
12. (a) Gao, Q.; Gu, Z.; Parniak, M. A.; Cameron, J.; Cammack, N.; Boucher, C.; Wainberg, M. A. The same mutation that encodes low-level human immunodeficiency virus type 1 resistance to 2',3'-dideoxyinosine and 2',3'-dideoxycytidine confers high-level resistance to the (−)enantiomer of 2',3'-dideoxy-3'-thiacytidine. *Antimicrob. Agents Chemother.* **1993**, *37*, 1390–1392. (b) Tisdale, M.; Kemp, S. D.; Parry, N. R.; Larder, B. A. Rapid in vitro selection of human immunodeficiency virus type 1 resistant to 3'-thiacytidine inhibitors due to a mutation in the YMDD region of reverse transcriptase. *Proc. Natl. Acad. Sci. U.S.A.* **1993**, *90*, 5653–5656.
13. Larder, B. A.; Kemp, S. D.; Harrigan, P. R. Potential mechanism for sustained antiretroviral efficacy of AZT-3TC combination therapy. *Science* **1995**, *269*, 696–699.
14. (a) Bugg, T. D. H.; Walsh, C. T. Intracellular steps of bacterial cell wall peptidoglycan biosynthesis: enzymology, antibiotics, and antibiotic resistance. *Nat. Prod. Rep.* **1992**, *9*, 199–215. (b) Walsh, C. T. Vancomycin resistance: decoding the molecular logic. *Science* **1993**, *261*, 308–309. (c) Walsh, C. T.; Fisher, S. L.; Park, I.-S.; Prahalad, M.; Wu, Z. Bacterial resistance to vancomycin: five genes and one missing hydrogen bond tell the story. *Chem. Biol.* **1996**, *3*, 21–28.
15. Bugg, T. D. H.; Wright, G. D.; Dutka-Malen, S.; Arthur, M.; Courvalin, P.; Walsh, C. T. Molecular basis for vancomycin resistance in Enterococcus faecium BM4147: biosynthesis of a depsipeptide peptidoglycan precursor by vancomycin resistance proteins VanH and VanA. *Biochemistry* **1991**, *30*, 10408–10415.
16. (a) Bikker, J. A.; Brooijmans, N.; Wissner, A.; Mansour, T. S. Kinase Domain Mutations in Caner: Implications for small molecule drug design strategies. *J. Med. Chem.* **2009**, *52*, 1493–1509. (b) Chen, Y.; Fu, L. Mechanisms of acquired resistance to tyrosine kinase inhibitors. *Acta Pharm. Sinica B* **2011**, *1*, 197–207.
17. Huang, W.-S.; Zhu, X.; Wang, Y.; Azam, M. Wen, D.; Sundaramoorthi, R.; Thomas, R. M.; Liu, S.; Banda, G.; Lentini, S. P. 9-(Arenethenyl)purines as dual Src/Abl kinase inhibitors targetin the inactive conformation: design, synthesis, and biological evaluation. *J. Med. Chem.* **2009**, *52*, 4743–4756.
18. Huang, W.-S.; Metcalf, C. A.; Sundaramoorthi, R.; Wang, Y.; Zou, D.; Thomas, R. M.; Zhou, X.; Cai, L.; Wen, D.; Liu, S. Discovery of 3-[2-(imidazo[1,2-b]pyridazin-3-yl)ethynyl]-4-methyl-N-{4-[(4-methylpiperazin-1-yl)-methyl]-3-(trifluoromethyl)phenyl}benzamide (AP24534), a potent, orally active pan-inhibitor of breakpoint cluster region-abelson (Bcr-Abl) kinase including the T315I gatekeeper mutant. *J. Med. Chem.* **2010**, *53*, 4701–4719.
19. Chan, W. W.; Wise, S. C.; Kaufman, M. D.; Ahn, Y. M.; Ensinger, C. L.; Haack, T.; Ahn, Y-M.; Ensinger, C. L.; Haack, T.; Hood, M. M.; Jones, J.; Lord, J. W.; L, W-P. Conformational control inhibition of the Bcr-Abl1 tyrosine kinase, including the gatekeeper T315I mutant, by the switch-control inhibitor DCC-2036. *Cancer Cell* **2011**, *19*, 556–568.
20. Ardini, E.; Galvani, A. ALK inhibitors, a pharmaceutical perspective. *Frontiers Oncol.* **2012**, *2*, 1–8.
21. Patel, S.; Keller, B. A.; Fisher, L. M. Mutations at Arg486 and Glu571 in human topoisomerase IIα confer resistance to amsacrine: relevance for antitumor drug resistance in human cells. *Mol. Pharmacol.* **2000**, *57*, 784–791.
22. Kanafani, Z. A.; Perfect, J. R. Resistance to antifungal agents: mechanisms and clinical impact. *Antimicrob. Resistance* **2008**, *46*, 120–128.
23. (a) Kohanski, M. A.; Dwyer, D. J.; Hayete, B.; Lawrence, C. A., Collins, J. J. A common mechanism of cellular death induced by bactericidal antibiotics. *Cell* **2007**, *130*, 797–810. (b) Dwyer, D. J.; Kohanski, M. A.; Hayete, B.; Collins, J. J. Gyrase inhibitors induce an oxidative damage cellular death pathway in *Escherichia coli*. *Mol. Syst. Biol.* **2007**, *3*, 91.
24. Kohanski, M. A.; DePristo, M. A.; Collins, J. J. Sublethal antibiotic treatment leads to multidrug resistance via radical-induced mutagenesis. *Mol. Cell* **2010**, *37*, 311–320.
25. Alt, F. W.; Kellems, R. E.; Bertino, J. R.; Schimke, R. T. Selective multiplication of dihydrofolate reductase genes in methotrexate-resistant variants of cultured murine cells. *J. Biol. Chem.* **1978**, *253*, 1357–1370.
26. Kan, S.; Siddiqui, W. Comparative studies on dihydrofolate reductases from *Plasmodium falciparum* and *Aotus trivirgatus*. *J. Protozool.* **1979**, *26*, 660–664.
27. Bertino, J.; Cashmore, A.; Fink, N.; Calabresi, P.; Lefkowitz, E. The induction of leukocyte and erythrocyte dihydrofolate reductase by methotrexate. II. Clinical and pharmacologic studies. *Clin. Pharmacol. Ther.* **1965**, *6*, 763–770.
28. Arteaga, C. L. Epidermal growth factor dependence in human tumors: more than just expression? *Oncologist* **2002**, *7* (Suppl. 4), 31–39.
29. Weisberg, E.; Griffin, J. D. Mechanism of resistance to the Abl tyrosine kinase inhibitor STI571 in Bcr-Abl-transformed hematopoietic cell lines. *Blood* **2000**, *95*, 3498–3505.
30. De Wilt, L. H. A. M.; Jansen, G.; Assaraf, Y. G.; van Meerloo, J; Cloos, J.; Schimmer, A. D.; Chan, E. T.; Kirk, C. J.; Peters, G. J.; Kruyt, F. A. E. Proteasome-based mechanisms of intrinsic and acquired bortezomib resistance in non-small cell lung cancer. *Biochem. Pharmacol.* **2012**, *83*, 207–217.
31. Groll, M.; Berkers, C. R.; Ploegh, H. L.; Ovaa, H. Crystal structure of the boronic acid-based proteasome inhibitor bortezomib in complex with the yeast 20S proteasome. *Structure* **2006**, *14*, 451–456.

32. LoPachin, R. M.; Gavin, T.; DeCaprio, A.; Barber, D. S. Application of the hard and soft, acids and bases (HSAB) theory to toxicant-target interactions. *Chem. Res. Toxicol.* **2012**, *25*(2), 239–251.
33. Landy, M.; Gerstung, R. B. p-aminobenzoic acid synthesis by neisseria gonorrhoeae in relation to clinical and cultural sulfonamide resistance. *J. Bacteriol.* **1944**, *47*, 448–480.
34. Jones, R. N. Impact of changing pathogens and antimicrobial susceptibility patterns in the treatment of serious infections in hospitalized patients. *Am J. Med.* **1996**, *100*, 3S–12S.
35. (a) Wilke, M. S.; Lovering, A. L.; Strynadka, N. C. J. β-Lactam antibiotic resistance: a current structural perspective. *Curr. Opin. Microbiol.* **2005**, *8*, 525–533. (b) Jacoby, G. A. AmpC β-lactamases. *Clin. Microbiol. Rev.* **2009**, *22*, 161–182. (c) Queenan, A. M.; Bush, K. Carbapenemases: the versatile β-lactamases. *Clin. Microbiol. Rev.* **2007**, *20*, 440–458.
36. (a) Wright, G. D.; Berghuis, A. M.; Mobashery, S. In *Aminoglycoside Antibiotics: Structures, Functions and Resistance*, Rosen, B. P.; Mobashery, S. (Eds.), Plenum Press, New York, 1998, pp. 27–69. (b) Kondo, S.; Hotta, K. Semisynthetic aminoglycoside antibiotics: development and enzymatic modifications. *J. Infect. Chemother.* **1999**, *5*, 1–9. (c) Mingeot-Leclerco, M.-P.; Glupczynski, Y.; Tulkens, P. M. Aminoglycosides: activity and resistance. *Antimicrob. Agents Chemother.* **1999**, *43*, 727–737. (d) Magnet, S.; Blanchard, J. S. Molecular insights into aminoglycoside action and resistance. *Chem. Rev.* **2005**, *105*, 477–498. (e) De Pascale, G.; Wright, G. D. Antibiotic resistance by enzyme inactivation: from mechanisms to solutions. *Chem. Bio. Chem* **2010**, *11*, 1325–1334.
37. Roestamadji, J.; Grapsas, I.; Mobashery, S. Loss of individual electrostatic interactions between aminoglycoside antibiotics and resistance enzymes as an effective means to overcoming bacterial drug resistance. *J. Am. Chem. Soc.* **1995**, *117*, 11060–11069.
38. Haddad, J.; Vakulenko, S.; Mobashery, S. An antibiotic cloaked by its own resistance enzyme. *J. Am. Chem. Soc.* **1999**, *121*, 11922–11923.
39. Sucheck, S. J.; Wong, A. L.; Koeller, K. M.; Boehr, D. D.; Draker, K.-A.; Sears, P.; Wright, G. D.; Wong, C.-H. Design of bifunctional antibiotics that target bacterial rRNA and inhibit resistance-causing enzymes. *J. Am. Chem. Soc.* **2000**, *122*, 5230–5231.
40. Umezawa, H.; Takeuchi, S.; Hori, T.; Sawa, T.; Ishizuka, T.; Ichikawa, T.; Komai, T. Mechanism of antitumor effect of bleomycin on squamous cell carcinoma. *J. Antibiot.* **1972**, *25*, 409–420.
41. Sugiura, Y.; Muraoka, Y.; Fujii, A.; Takita, T.; Umezawa, H. Chemistry of bleomycin. XXIV. Deamido bleomycin from the viewpoint of mental coordination and oxygen activation. *J. Antibiot.* **1979**, *32*, 756–758.
42. Evans, C. G.; Bodell, W. J.; Tokuda, K.; Doane-Setzer, P.; Smith, M. T. Glutathione and related enzymes in rat brain tumor cell resistance to 1,3-bis(2-chloroethyl)-1-nitrosourea and nitrogen mustard. *Cancer Res.* **1987**, *47*, 2525–2530.
43. (a) Tew, K. D.; Townsend, D. M. Glutathione-S-transferase as determinants of cell survival and death. *Antiox. Redox Signal.* **2012**, *17*(12), 1728–1737. (b) Olvera-Bello, A. E.; Vega, L. Role of glutathione S-transferase enzymes in toxicology, pharmacology and human disease. In *Xenobiotic Metabolizing Enzymes and Xenobiotic Receptors*, Azuela, G. E. (Ed.), 2010, pp. 45–66.
44. Harrap, K. In *Scientific Foundations of Oncology*; Symington, T.; Carter, R. (Eds.), Heinemann: London, 1976, p. 641.
45. Sauter, C.; Lamanna, N.; Weiss, M. A. Pentostatin in chronic lymphocytic leukemia. *Exp. Opin. Drug Metab. Toxicol.* **2008**, *4*, 1217–1222.
46. (a) Ratushny, V.; Astsaturov, I.; Burtness, B. A.; Golemis, E. A.; Silverman, J. S. Targeting EGFR resistance networks in head and neck cancer. *Cell. Signal.* **2009**, *21*, 1255–1268. (b) Giaccone, G.; Wang, Y. Strategies for overcoming resistance to EGFR family tyrosine kinase inhibitors. *Cancer Treat. Rev.* **2011**, *37*, 456–464. (c) Liska, D.; Chen, C.-T.; Bachleitner-Hofmann, T.; Christensen, J. G.; Weiser, M. R. HGF rescues colorectal cancer cells from EGFR inhibition via MET activation. *Clin. Cancer Res.* **2011**, *17*, 472–482.
47. (a) Dolan, M. E.; Pegg, A. E. O^6-Benzylguanine and its role in chemotherapy. *Clin. Cancer Res.* **1997**, *3*, 83847. (b) Karran, P. Mechanisms of tolerance to DNA damaging therapeutic drugs. *Carcinogenesis* **2001**, *22*, 1931–1937.
48. (a) Pegg, A. Repair of O^6-alkylguanine by alkyltransferases. *Mutat. Res.* **2000**, 462, 83–100. (b) Sekiguchi, M.; Nakabeppu, Y.; Sakumi, K.; Tuzuki, T. DNA-repair methyltransferase as a molecular device for preventing mutation and cancer. *J. Cancer Res. Clin. Oncol.* **1996**, *122*, 199–206.
49. (a) Srivenugopal, K. S.; Yuan, X. H.; Bigner, D. D.; Friedman, H. S.; Ali-Osman, F. Ubiquitination-dependent proteolysis of O^6-methylguanine-DNA methyltransferase in human and murine cells following inactivation with O^6-benzylguanine or 1,3-bis(2-chloroethyl)-1-nitrosourea. *Biochemistry* **1996**, *35*, 1328–1934. (b) Xu-Welliver, M.; Pegg, A. E. Degradation of the alkylated form of the DNA repair protein, O^6-alkylguanine-DNA alkyltransferase. *Carcinogenesis* **2002**, *23*, 823–830.
50. (a) Haest, C. W. M.; de Gier, J.; op den Kamp, J. A. F.; Bartels, P.; van Deenen, L. L. M. Changes in permeability of *Staphylococcus aureus* and derived liposomes with varying lipid composition. *Biochim. Biophys. Acta* **1972**, *255*, 720–733. (b) Bryan, C. E.; Kwan, S. Mechanisms of aminoglycoside resistance of anaerobic bacteria and facultative bacteria grown anaerobically. *J. Antimicrob. Chemother.* **1981**, *8* (Suppl. D), 1–8.
51. (a) Swedberg, G.; Skold, O. Characterization of different plasmid-borne dihydropteroate synthases mediating bacterial resistance in sulfonamides. *J. Bacteriol.* **1980**, *142*, 1–7. (b) Nagate, T.; Inoue, M.; Inoue, K.; Mitsuhashi, S. Plasmid-mediated sulfanilamide resistance. *Microbiol. Immunol.* **1978**, *22*, 367–375.
52. Cole, S.; Bagal, S.; El-Kattan, A.; Fenner, K.; Hay, T.; Kempshall, S.; Lunn, G.; Varma, M.; Stupple, P.; Speed, W. Full efficacy with no CNS side-effects: unachievable panacea or reality? DMPK considerations in design of drugs with limited brain penetration. *Xenobiotica* **2012**, *42*, 11–27.
53. (a) Fernandez, L.; Hancock, R. E. W. Adaptive and mutational resistance: role of porins and efflux pumps in drug resistance. *Clin. Microbiol. Rev.* **2012**, *25*(4), 661–681. (b) Kumar, S.; Varela, M. F. Biochemistry of bacterial multi-drug efflux pumps. *Int. J. Mol. Sci.* **2012**, *13*, 4484–4495. (c) Lomovskaya, O.; Watkins, W. J. Efflux pumps: their role in antibacterial drug discovery. *Curr. Med. Chem.* **2001**, *8*(14), 1699–711. (d) Lawrence, L. E.; Barrett, J. F. Efflux pumps in bacteria: overview, clinical relevance, and potential pharmaceutical target. *Exp. Opin. Investig. Drugs* **1998**, *7*, 199–217. (e) Nikaido, H. Multiple antibiotic resistance and efflux. *Curr. Opin. Microbiol.* **1998**, *1*, 516–523.
54. (a) Poole, K. Efflux-mediated antimicrobial resistance. *J. Antimicrob. Chemother.* **2005**, *56*, 20–51. (b) Nikaido, H. Multidrug resistance in bacteria. *Annu. Rev. Biochem.* **2009**, *78*, 119–146.
55. Schinkel, A. H.; Jonker, J. W. Mammalian drug efflux transporters of the APT binding cassette (ABC) family: an overview. *Adv. Drug Deliv. Rev.* **2003**, *55*, 3–29.

56. Schwarz, H. S. Some determinants of the therapeutic efficacy of actinomycin D (NSC 3053), adriamycin (NSC 123127), and daunorubicin (NSC 83142). *Cancer Chemother.* **1974**, *58*, 55–62.
57. Peng, X.-X.; Tiwari, A. K.; Wu, H.-C; Chen, Z.-S. Overexpression of p-glycoprotein induces acquired resistance to imatinib in chronic myelogenous leukemia cells. *Chinese J. Cancer* **2012**, *31*, 110–118.
58. Pfaller, M. A. Antifungal drug resistance: mechanisms, epidemiology, and consequences for treatment. *Am. J. Med.* **2012**, 125, S3–S13
59. Fletcher, J. I.; Haber, M.; Henderson, M. J.; Norris, M. D.; ABC transporters in cancer; more than just drug efflux pumps. *Nat. Rev. Cancer* **2010**, *10*, 147–156.
60. Zhang, C.; Kwan, P.; Zuo, Z.; Baum, L. The transport of antiepileptic drugs by P-glycoprotein. *Adv. Drug Deliv. Rev.* **2012**, *64*, 930–942.
61. Piddock, L. J. V. Clinically relevant chromosomally encoded multidrug resistance efflux pumps in bacteria. *Clin. Microbiol. Rev.* **2006**, *19*, 382–402.
62. Lowy, F. D. *Staphylococcus aureus* infections. *N. Engl. J. Med.* **1998**, *339*, 520–532.
63. (a) Juarez-Verdayes, M. A.; Parra-Ortega, B.; Hernandez-Rodriguez, C.; Betanzos-Cabrera, G.; Rodriguez-Martinez, S.; Cancino-Diaz, M. E.; Cancino-Diaz, J. C. Identification and expression of nor efflux family genes in *Staphylococcus epidermidis* that act against gatifloxacin. *Microb. Pathogen.* **2012**, *52*(6), 318–325. (b) Costa, S. S.; Falcao, C.; Viveiros, M.; Machado, D.; Martins, M.; Melo-Cristino, J.; Amaral, L.; Couto, I. Exploring the contribution of efflux on the resistance to fluoroquinolones in clinical isolates of *Staphylococcus aureus*. *BMC Microbiol.* **2011**, *11*, 241.
64. Drawz, S. M.; Bonomo, R. A. Three decades of β-lactamase inhibitors. *Clin. Microbiol. Rev.* **2010**, *23*, 160–201.
65. (a) Brown, A. G.; Butterworth, D.; Cole, M.; Hanscomb, G.; Hood, J. D.; Reading, C.; Rolinson, G. N. Naturally occurring β-lactamase inhibitors with antibacterial activity. *J. Antibiot.* **1976**, *29*, 668–669. (b) English, A. R.; Retsema, J. A.; Girard, J. A.; Lynch, J. E.; Barth, W. E. CP-45,899, a beta-lactamase inhibitor that extends the antibacterial spectrum of beta-lactams: initial bacteriological characterization. *Antimicrob. Agents Chemother.* **1978**, *14*, 414–419.
66. (a) Silverman, R. B. *Mechanism-Based Enzyme Inactivation: Chemistry and Enzymology*; CRC: Boca Raton, FL, 1988, Vol. 1, p. 135. (b) Cartwright, S. J.; Waley, S. G. β-Lactamase inhibitors. *Med. Res. Rev.* **1983**, *3*, 341–382. (c) Charnas, R. L.; Knowles, J. R. Inactivation of radiolabeled RTEM β-lactamase from *Escherichia coli* by clavulanic acid and 9-deoxyclavulanic acid. *Biochemistry* **1981**, *20*, 3214–3219. (d) Brenner, D. G.; Knowles, J. R. Penicillanic acid sulfone: nature of irreversible inactivation of the RTEM β-lactamase from *Escherichia coli*. *Biochemistry* **1984**, *23*, 5833–5839.
67. Rolinson, G. N. The history and background of Augmentin. *S. Afr. Med. J.* **1982**, *62*, 3A–4A.22
68. Therrien, C.; Levesque, R. C. Molecular basis of antibiotic resistance and β-lactamase inhibition by mechanism-based inactivators: perspectives and future directions. *FEMS Microbiol. Rev.* **2000**, *24*, 251–262.
69. (a) Novick, R. P. Plasmid incompatibility. *Microbiol. Rev.* **1987**, 51, 381–395. (b) Bouet, J.-Y.; Nordstrom, K.; Lane, D. Plasmid partition and incompatibility-the focus shifts. *Mol. Microbiol.* **2007**, *65*(6), 1405–1414.
70. (a) DeNap, J. C. B.; Thomas, J. R.; Musk, D. J.; Hergenrother, P. J. Combating drug-resistant bacteria: small molecule mimics of plasmid incompatibility as antiplasmid compounds. *J. Am. Chem. Soc.* **2004**, *126*, 15402–15404. (b) Thomas, J. R.; DeNap, J. C. B.; Wong, M. L.; Hergenrother, P. J. The relationship between aminoglycosides' RNA binding proclivity and their antiplasmid effect on an IncB plasmid. *Biochemistry* **2005**, *44*(18), 6800–6808.
71. Tritsch, G. L. (Ed.), *Adenosine Deaminase in Disorders of Purine Metabolism and in Immune Deficiency*, New York Academy of Sciences, New York, 1985; Vol. 451.
72. Wormser, G. P.; Keusch, G. T.; Rennie, C. H. Co-trimoxazole (trimethoprim-sulfamethoxazole). An updated review of its antibacterial activity and clinical efficacy. *Drugs* **1982**, *24*, 459–518.
73. (a) Hitchings, G. H.; Burchall, J. J. Inhibition of folate biosynthesis and function as a basis for chemotherapy. *Adv. Enzymol.* **1965**, *27*, 417–468. (b) Anand, N. In Inhibition of Folate Metabolism in Chemotherapy, Hitchings, G. H. (Ed.), Springer-Verlag, Berlin, 1983, p. 25.
74. Ferone, R.; Burchall, J. J.; Hitchings, G. H. Plasmodium berghei dihydrofolate reductase. Isolation, properties, and inhibition by antifolates. *Mol. Pharmacol.* **1969**, *5*, 49–59.
75. Flaherty, K. T.; Infante, J. R.; Daud, A.; Gonazales, R.; Kefford, R. F.; Sosman, J.; Hamid, O.; Schuchter, L.; Cebon, J.; Ibrahim, N.; Kudchadkar, R.; Burris, H. A.; Falchook, G.; Algazi, A.; Lewis, K.; Long, G. V.; Puzanov, I.; Lebowitz, P.; Singh, A.; Little S.; Sun, P.; Allred, A.; Ouellet, D.; Kim, K. B.; Patel, K.; Weber, J. Combined BRAF and MEK inhibition in melanoma with BRAF V600 mutations. *N. Engl. J. Med.* **2012**, *367*, 1694–1703.
76. (a) Dienstmann, R.; De Dosso, S.; Felip, E.; Tabernero, J. Drug development to overcome resistance to EGFR inhibitors in lung and colorectal cancer. *Mol. Oncol.* **2012**, *6*, 15–26. (b) Giaccone, G.; Wang, Y. Strategies of overcoming resistance to EGFR family tyrosine kinase inhibitors. *Cancer Treat. Rev.* **2011**, *37*, 456–464.
77. Erlichman, C.; Vidgen, D. Antitumor activity of N-phosphonacetyl-L-aspartic acid in combination with nitrobenzylthioinosine. *Biochem. Pharmacol.* **1984**, *33*, 3177–3181.
78. Zhang, L.; Ma. S. Efflux pump inhibitors: a strategy to combat p-glycoprotein and the NorA multidrug resistance pump. *Chem. Med. Chem.* **2010**, *5*, 811–822.
79. (a) Renau, T. E.; Leger, R.; Flamme, E. M.; Sangalang, J.; She, M. W.; Yen, R.; Gannon, C. L.; Griffith, D.; Chamberland, S.; Lomovskya, O.; Hecker, S. J.; Lee, V. J.; Ohta, T.; Nakayama, K. Inhibitors of efflux pumps in pseudomonas aeruginosa potentiate the activity of the fluoroquinolone antibacterial levofloxacin. *J. Med. Chem.* **1999**, *42*, 4928–4931. (b) Stermitz, F. R.; Lorenz, P.; Tawara, J. N.; Zenewicz, L. A.; Lewis, K. Synergy in a medicinal plant: antimicrobial action of berberine potentiated by 5'-methoxyhydnocarpin, a multidrug pump inhibitor. *Proc. Natl. Acad. Sci. U.S.A.* **2000**, *97*, 1433–1437.
80. (a) Coley, H. M. Overcoming multidrug resistance in cancer: clinical studies of p-glycoprotein inhibitors. *Meth. Mol. Biol.* **2010**, *596*, 341–358. (b) Schinkel, A. H.; Jonker, J. W. Mammalian drug efflux transporters of the APT binding cassette (ABC) family: an overview. *Adv. Drug Deliv. Rev.* **2003**, *55*, 3–29.
81. Fox, E.; Bates, S. E. Tariquidar (XR9576): a P-glycoprotein drug efflux inhibitor. *Exp. Rev. Anticancer Ther.* **2007**, *7*, 447–459.
82. Hedstrom, L. IMP dehydrogenase: mechanism of action and inhibition. *Curr. Med. Chem.* **1999**, *6*, 545–560.
83. Shu, Q.; Nair, V. Inosine monophosphate dehydrogenase (IMPDH) as a target in drug discovery. *Med. Res. Rev.* **2008**, 28, 219–232.

84. Sintchak, M. D.; Fleming, M. A.; Futer, O.; Raybuck, S. A.; Chambers, S. P.; Caron, P. R.; Murcko, M. A.; Wilson, K. P. Structure and mechanism of inosine monophosphate dehydrogenase in complex with the immunosuppressant mycophenolic acid. *Cell* **1996**, 85, 921–930.
85. (a) Gan, L.; Seyedsayamdost, M. R.; Shuto, S.; Matsuda, A.; Petsko, G. A.; Hedstrom, L. The immunosuppressive agent mizoribine monophosphate forms a transition state analogue complex with inosine monophosphate dehydrogenase. *Biochemistry* 2003, *42*, 857–863. (b) Kawasaki, Y. Mizoribine: a new approach in the treatment of renal disease. *Clin. Dev. Immunol.* **2009**, doi: 10.1155/2009/681482 (epub).
86. Shimmura, H.; Tanabe, K.; Habiro, K.; Abe, R.; Toma, H. Combination effect of mycophenolate mofetil with mizoribine on cell proliferation assays and in a mouse heart transplantation model. *Transplantation* **2006**, *82*, 175–179.

Chapter 8

Drug Metabolism

Chapter Outline

8.1. INTRODUCTION

When a foreign organism enters the body, the immune system produces antibodies to interact with and destroy it. Small molecules, however, do not stimulate antibody production. So how has the human body evolved to protect itself against low-molecular-weight environmental pollutants? The principal mechanism is the use of nonspecific enzymes that transform the foreign compounds (often highly nonpolar molecules) into polar molecules that are excreted by the normal bodily processes. Although this mechanism to rid the body of *xenobiotics* (molecules foreign to the organism) is highly desirable, especially when one considers all the foreign materials to which we are exposed every day, it can cause problems when the foreign agent is a drug that needs to enter and be retained in the body sufficiently long to be effective. The enzymatic biotransformation of drugs is known as *drug metabolism*. Because many drugs have structures similar to those of endogenous compounds, drugs may get metabolized by specific enzymes for the related natural substrates as well as by nonspecific enzymes.

Although the principal site of drug metabolism is the liver, the kidneys, the lungs, and the gastrointestinal (GI) tract are also important metabolic sites. When a drug is taken orally (the most common route of administration), it is usually absorbed through the mucous membrane of the small intestine or from the stomach. Once out of the GI tract it is carried by the bloodstream (via the portal vein) to the liver where it is usually first metabolized. Metabolism by liver enzymes prior to the drug reaching the systemic circulation is called the presystemic or *first-pass effect*, which may result in substantial or even complete deactivation of the drug. If a large fraction of the drug is metabolized, then larger or multiple doses of the drug will be required to get the desired effect. Another undesirable effect of drug metabolism is that occasionally the metabolites of a drug may be toxic, even though the drug is not.

The first-pass effect sometimes can be avoided by changing the route of drug administration. The *sublingual*

The Organic Chemistry of Drug Design and Drug Action. http://dx.doi.org/10.1016/B978-0-12-382030-3.00008-8

route (the drug is placed under the tongue) bypasses the liver. After absorption through the buccal cavity, the drug enters the systemic circulation. This is the route employed for nitroglycerin (**8.1**, Nitrostat), a drug used for the treatment of angina pectoris that is converted by mitochondrial

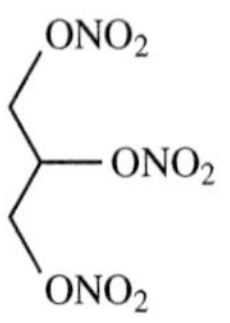

Nitroglycerin
8.1

aldehyde dehydrogenase to nitrite ion, which is then reduced to nitric oxide,[1] a second messenger molecule that dilates blood vessels in the heart. The *rectal route*, in the form of a solid suppository or in solution as an enema, leads to absorption through the colon mucosa. Ergotamine (**8.2**, Ergomar), a drug for migraine headaches, is administered

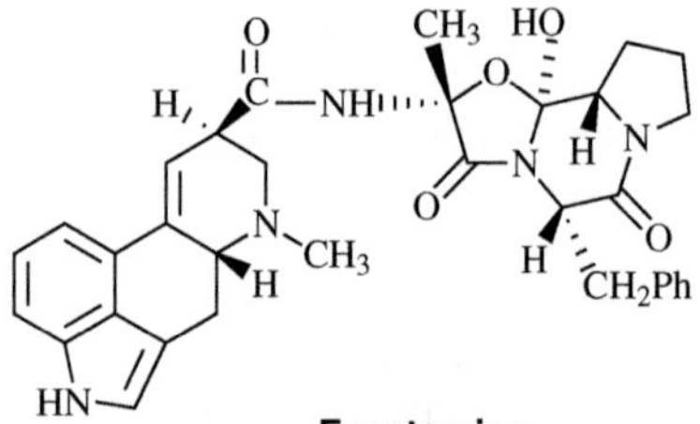

Ergotamine
8.2

this way (who would have guessed?). *Intravenous* (IV) *injection* introduces the drug directly into the systemic circulation to be used when a rapid therapeutic response is desired. The effects are almost immediate when drugs are administered by this route, because the total blood circulation time in man is 15–20 s. *Intramuscular injection* is used when large volumes of drugs need to be administered, if slow absorption is desirable, or if the drug is unstable in the gastric acid of the stomach. A *subcutaneous injection* delivers the drug through the loose connective tissue of the subcutaneous layer of the skin. Another method of administration, particularly for gaseous or highly volatile drugs such as general anesthetics, is by *pulmonary absorption* through the respiratory tract. The asthma drug, isoproterenol (**8.3**, Isuprel), is metabolized in the intestines and liver, but administration by aerosol inhalation is

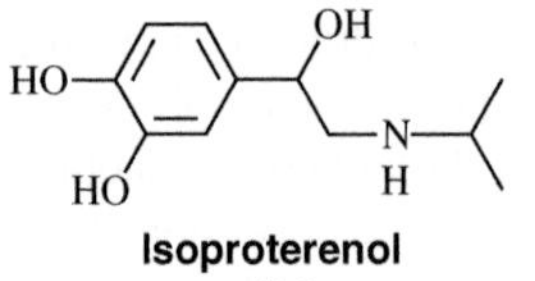

Isoproterenol
8.3

effective in getting the drug directly to the bronchi. *Topical application* of the drug to the skin or a mucous membrane is used for local effects; few drugs readily penetrate the intact skin. Not all drugs can be administered by these alternate routes, so their structures may have to be altered to minimize the first-pass effect or to permit them to be administered by one of these alternate routes. These structural modification approaches in drug design to avoid the first-pass effect are discussed in Chapter 9. Even if the first-pass effect is avoided, a drug in the systemic circulation eventually reaches the liver, and there are many enzymes in tissues other than the liver that are capable of catalyzing drug metabolism reactions. Once a drug has reached its site of action and elicited the desired response, it usually is desirable for the drug to be metabolized and eliminated at a reasonable rate. Otherwise, it may remain in the body and produce the effect longer than desired or it could accumulate and become toxic to the cells.

Drug metabolism studies are essential to evaluate the potential safety and efficacy of drugs. Consequently, prior to approval of a drug for human use, an understanding of the metabolic pathways and disposition of the drug in humans and in preclinical animal species is required. The animal species used for metabolism studies are generally those in which the pharmacokinetic and toxicological evaluations are conducted. Additional toxicological studies have to be carried out on metabolites found in humans that were not observed in the animal metabolism studies. Metabolism studies can also be a useful lead modification approach. For example, after many years on the drug market, terfenadine (**8.4**, R=CH_3; Seldane) was removed because it was found to cause life-threatening cardiac arrhythmias when coadministered with inhibitors of hepatic cytochrome P450, such as erythromycin and ketoconazole.[2] The arrhythmias resulted from binding of terfenadine to the hERG (*h*uman *E*ther à go–go *R*elated *G*ene) channel, a cardiac potassium ion channel that is responsible for the electrical activity that coordinates the heart's beating.[3] The active metabolite of terfenadine, fexofenadine (**8.4**, R=COOH; Allegra), however, produces no arrhythmias (does not bind to the hERG channel),[4] and it has replaced terfenadine in the market.[5] It is known that increasing lipophilicity favors binding to the hERG channel.[6]

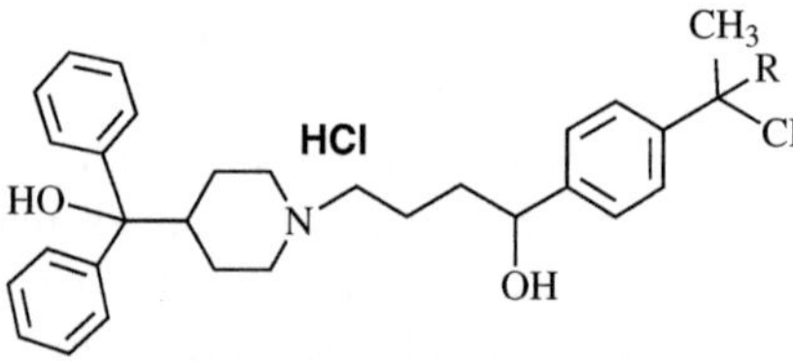

Terfenadine HCl, R = CH_3
Fexofenadine HCl, R = COOH
8.4

With a general understanding of important metabolic pathways, it is often possible to design a compound that is inactive when administered, but which utilizes the metabolic enzymes to convert it into the active form. These compounds are known as *prodrugs* and are discussed in Chapter 9. In the present chapter, we consider the various reactions that are involved in the biotransformations of drugs. Because some metabolites are produced in only very small amounts, it may be difficult to detect all the metabolic products. Moreover, because the biological fluids in which the metabolites are found (e.g., blood or urine) contain many other substances, a way to clearly detect the metabolites vs other substances in the fluids is needed. This can often be accomplished using tandem mass spectrometry/mass spectrometry techniques, whereby one can search for metabolites based on the predicted masses.[7] For example, as we will see shortly, metabolic conversion of C–H to C–OH is a common metabolic transformation; to detect this metabolite, one could use mass spectrometry to search for the presence of a substance with a molecular mass 16 amu higher than the parent drug. To increase the sensitivity of the detection process further, and especially to help with mass balance studies (studies in which one attempts to account for the ultimate fate of all of an administered drug), drug candidates are typically radioactively labeled. Radioactive compounds are useful for studying all aspects of absorption, distribution, metabolism, and excretion.[8] In the next two sections, we will look briefly at how radiolabeling is carried out, and then discuss in more detail the analytical methods commonly used in drug metabolism studies.

8.2. SYNTHESIS OF RADIOACTIVE COMPOUNDS

Because of the sensitivity of detection of particles of radioactive decay, a common approach used for detection, quantification, and profiling of metabolites in whole animal studies is the incorporation of a radioactive label, typically a weak β-emitter such as ^{14}C[9] or ^{3}H,[10] into the drug molecule. When this approach is used, it does not matter how few metabolites or how small the quantities of metabolites are produced, even in the presence of a large number of endogenous compounds. Only the radioactively labeled compounds are isolated from the urine and the feces of the animals, and the structures of these metabolites are elucidated (see Section 8.3). If one of the carbon atoms of a drug is metabolized to carbon dioxide, as is the case with erythromycin (**8.5**, Erythrocin), ^{14}C labeling of the carbon atom that becomes CO_2 (the NMe_2 methyl groups

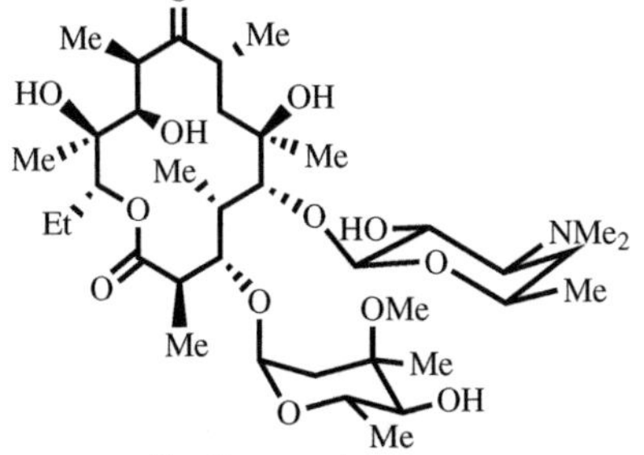

Erythromycin
8.5

of erythromycin are oxidized to CO_2) makes it possible to measure the rate of metabolism of the compound by measuring the rate of exhaled $^{14}CO_2$.[11] To incorporate a radioactive label into a compound, a synthesis must be designed so that a commercially available radioactively labeled compound or reagent can be used in one of the steps. It is highly preferable to incorporate the radioactive moiety in a step at or near the end of the synthesis because once the radioactivity is introduced, the scale of the reaction is generally diminished and special precautions and procedures regarding radiation safety and disposal of radioactive waste must be followed. Often the radioactive synthesis is quite different from or longer than the synthesis of the unlabeled compound in order to use a commercially available radioactive material. It is preferable to prepare a [^{14}C]-labeled analog; when tritium is incorporated into the drug, the site of incorporation must be such that loss of the tritium by exchange with the medium does not occur, even after an early metabolic step. Generally, only one radioactive label is incorporated into a drug because drug metabolism typically leads to a modified structure with little fragmentation of the molecule. If you go back far enough in a synthesis, however, it is possible to synthesize a drug with several carbon atoms radioactively labeled. Radioactive labeling at multiple sites of a molecule would permit the identification of more fragments of the drug and consequently the elucidation of metabolite structures and the fate of the molecule in vivo.

Industrially, the radioactive drug is synthesized with high *specific radioactivity* (a measure of the amount of radioactivity per mole of compound), often >57 mCi/mmol of ^{14}C (the theoretical maximum is 64 mCi/mmol). When needed, the specific radioactivity is diluted with non-[^{14}C]-labeled drug for use in metabolism studies. Typically, commercially available radioactive compounds have relatively low specific radioactivities. This means that possibly only one in 10^6 or fewer molecules actually contain the radioactive tag; the remaining molecules are unlabeled and are carriers of the relatively few radioactive molecules. In the case of ^{14}C there will be no difference in the reactivity of the labeled and unlabeled molecules, so there is the statistical amount of radioactivity in the products formed as was in the starting materials. The specific radioactivity of the metabolites formed during metabolism, then, should be identical to the specific radioactivity of the drug. In the case of tritiated drugs, however, if a carbon–hydrogen bond is broken, the radiolabel will be lost as tritiated water, and satisfactory recovery of total radioactivity in animal studies cannot occur. Also, there will be a *kinetic isotope effect* on those molecules that are tritiated. A kinetic isotope effect occurs because the greater atomic mass of deuterium (one neutron) or tritium (two neutrons) relative to hydrogen decreases the vibrational frequency of the C–^{x}H bond relative to that of the C–H bond, lowering the zero-point energy and increasing its bond strength.[12] This will lead to metabolite formation with a lower specific radioactivity than that of the drug (preferable C–H bond cleavage over C–^{x}H bond cleavage). As a result, quantitation of the various metabolic pathways, where some involve C–H bond cleavage and others do not, may require knowledge of the tritium isotope effect. This, then, is another reason why it is preferable to use [^{14}C] labeling of a drug for metabolism studies rather than tritium labeling.

If the drug is a natural product or derivative of a natural product, the easiest procedure for incorporation of a radioactive label could be a biosynthetic approach, namely, to grow the organism that produces the natural product in the presence of a radioactive precursor, and let nature incorporate the radioactivity into the molecule. Because of the volume of media generally involved, and, therefore, the large amount of radioactive precursor required, this could be a very expensive approach; however, generally the expense is compensated by the ease of the method and an attractive yield of product obtained.

An example of a drug class that could use this approach is the penicillins, which are biosynthesized by *Penicillium* fungi from valine, cysteine, and various carboxylic acids (Scheme 8.1). Valine is commercially available with a ^{14}C label at the carboxylate carbon or it may be obtained uniformly labeled, that is, all the carbon atoms are labeled to some small extent with ^{14}C (albeit very few molecules would contain all the carbon atoms labeled in the same molecule). It also can be purchased with a tritium label at the 2- and 3-positions or at the 3- and 4-positions. Cysteine is available uniformly labeled in ^{14}C or with a ^{35}S label (another weak β-emitter). Penicillin G could be produced if phenylacetic acid (available with a ^{14}C label at either the 1- or 2-position) were inoculated into the *Penicillium* growth medium.

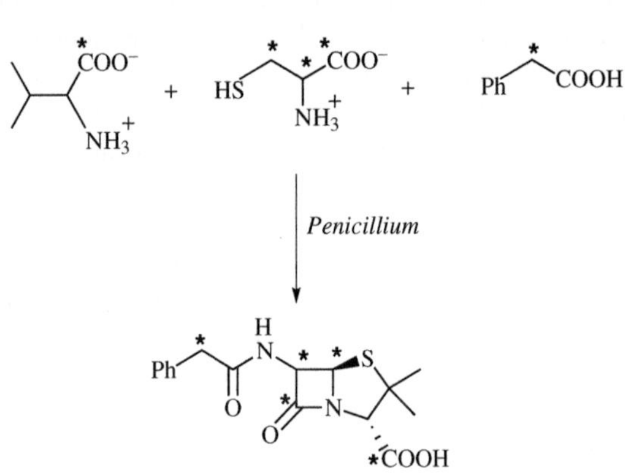

SCHEME 8.1 Biosynthesis of penicillins

If the drug is not a natural product (the more common case) a chemical synthesis must be carried out. For example, the synthesis of the antibacterial drug linezolid (**8.7**, Zyvox), is shown in Scheme 8.2.[13] The last step in the synthesis, acetylation of the primary amine, can be carried out with [^{14}C]acetic anhydride to incorporate a radioactive label in the acetyl group of **8.7**. The oxazolidinone carbonyl carbon also could have been labeled using [*carbonyl*-^{14}C]Cbz-Cl to make **8.6**, but radioactive Cbz-Cl is not commercially available, so it would have had to be synthesized from [^{14}C] phosgene and benzyl alcohol.

Once the radioactive drug has been synthesized, it is used in metabolism studies in preclinical species usually first in rats, mice, or guinea pigs, and then in dogs or monkeys. Typically, the urine and feces are collected from the animals, and the major radioactive compounds are isolated and their structures determined (see Section 8.3.3). After demonstration of drug safety in animals following chronic dosing at elevated doses and satisfactory recovery of the radioactive dose (>95% in the urine and feces; some of the radioactivity may be lost as CO_2, detected in the breath), then the drug can be tested for safety and tolerability in Phase I clinical trials with healthy human subjects. Once the safety is assured, the radioactive drug can be administered to humans during late Phase I or early Phase II clinical trials to obtain the

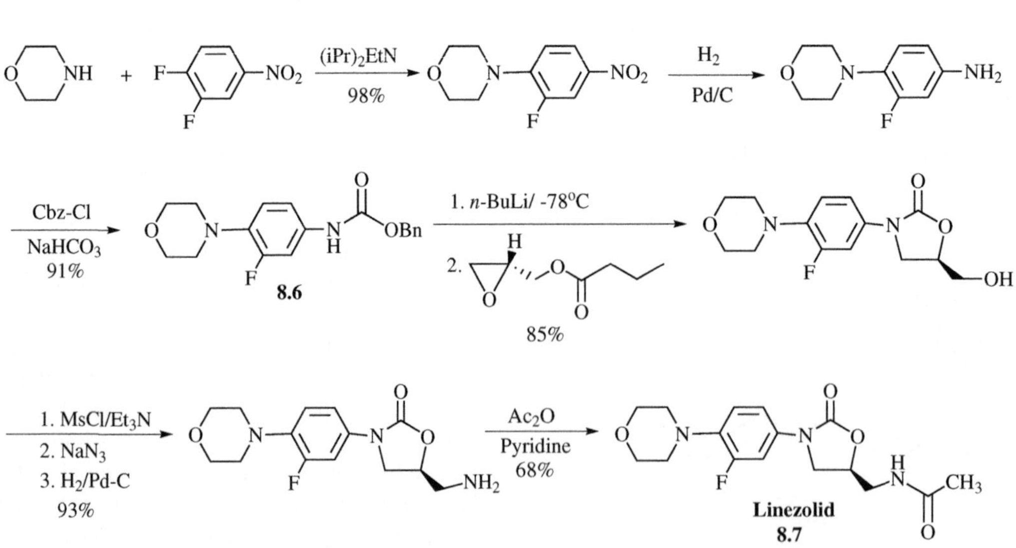

SCHEME 8.2 Chemical synthesis of linezolid

human metabolic profile. In fact, [*acetyl*-^{14}C]-**8.7** was used in a Phase I human metabolism study.[14] The Food and Drug Administration approves a maximum absorbed dose of 3 rem of radioactivity to a specific organ in a healthy adult volunteer for drug metabolism studies.[15] These radioactive levels are estimated from a determination of the absorbed dose in animal models; then, 10–100 times lower amounts are used in the human studies. On rare occasions, other fluids such as saliva, cerebrospinal fluid, eye fluids, perspiration, or breath may be examined as well as various organs and tissues. Generally, the toxicological animal model species used are considered adequate if all the major metabolites observed in humans are also observed in the animal models, even if more metabolites are observed in the animals. If a human metabolite is not formed in the animal toxicological model, a more relevant toxicological animal model has to be identified or additional toxicological studies need to be carried out with the metabolites unique to humans.[16]

If most of the radioactivity administered is not excreted from the animal, then it is dissected to determine the location of the radioactive compounds.[17] A methodology to determine tissue distribution of radiolabeled compounds in whole animals without dissection is quantitative whole-body autoradiography.[18]

From the above discussion it appears that drug metabolism studies are straightforward; however, until relatively recently these studies were difficult, at best, to carry out. The ready commercial availability of radioactively labeled precursors made the synthetic work much less tedious. The advent of *high-performance liquid chromatography* (HPLC) and the advancements in column packing materials permitted the separation of many metabolites very similar in structure. Metabolites that were previously overlooked can now be detected and identified. Structure elucidation by various types of mass spectrometry (see the next section) and by various techniques of nuclear magnetic resonance (NMR) spectrometry has become relatively routine. As a result of these advances in instrumentation, more information can be gleaned from drug metabolism studies than ever before, and this can result in the discovery of new leads or in a basis for prodrug design (see Chapter 9). This also means that the Food and Drug Administration can demand that many more metabolites be identified and their pharmacological and toxicological properties be determined prior to drug approval (which is good news for the consumer, but bad news for the drug companies). The final step in the process to prove the identity of a metabolite is to synthesize it and demonstrate that its spectral and pharmacological properties are identical to those of the metabolite.

8.3. ANALYTICAL METHODS IN DRUG METABOLISM

The four principal steps in drug metabolism studies are sample preparation, separation (chromatography), identification (spectrometry), and quantification of the metabolites. Detection systems are sensitive enough to allow the identification of submicrogram quantities of metabolites. Often, the sample preparation step can be minimized or omitted, and the urine sample or other biological sample injected directly into the HPLC or gas chromatography (GC) for separation. For cleaner results, though, sample pretreatment is recommended.[19] Most pharmaceutical groups now rely most heavily on direct liquid chromatography/tandem mass spectrometry (LC/MS/MS) to identify drug metabolites, as is described below (Section 8.3.3).

8.3.1. Sample Preparation

As discussed in Section 8.1, animals, including man, usually convert drugs into more polar metabolites for excretion. Enzymatic hydrolysis (β-glucuronidase and arylsulfatase) of the Phase II conjugation metabolites releases the less polar drug metabolites for easier extraction and structure identification. A clean sample for analysis is preferred, especially with in vivo drug metabolism studies. Extensive older sample preparation methodologies, such as ion-pair extraction,[20] used to remove hydrophilic ionizable compounds from aqueous solution; salt–solvent pair extraction,[21] to separate metabolites into an ethyl acetate-soluble neutral and basic fraction, ethyl acetate-soluble acidic fraction, and a water-soluble fraction; and various ion exchange resins such as the anion exchange resin diethylaminoethyl (DEAE)-Sephadex,[22] the cation exchange resin Dowex 50,[23] and the nonionic resin Amberlite XAD-2,[24] used to separate acidic, basic, and neutral metabolites, respectively, from body fluids, have been replaced by high-throughput methodologies. With the advent of HPLC/mass spectrometric analyses of metabolites described below (Section 8.3.3), often the sample preparation step can be incorporated as a fast-flow on-line extraction method.[25] Biological samples are injected directly into the liquid chromatography/mass spectrometry (LC-MS) instrument. A narrow-bore HPLC column packed with large-particle-size material extracts small molecule analytes but allows large molecules (such as proteins) to flow to the waste. The adsorbed analytes are then eluted through a column-switching valve onto an analytical column for LC/MS/MS analysis. For many assays, simple protein precipitation or liquid extraction is sufficient.[26] Solid-phase extraction[27] and liquid/liquid extraction[28] have been automated to speed up the process. On-line solid-phase extraction[29] or direct plasma injection into the HPLC/MS[30] are other high-throughput methods of sample preparation.

8.3.2. Separation

The three most important techniques for resolving mixtures of metabolites are HPLC, *capillary* GC,[31] and *capillary electrophoresis* (CE).[32] HPLC is more versatile than GC because the metabolites can be charged or uncharged, they can be thermally unstable, and derivatization is unnecessary. Normal phase columns (silica gel) can be used for uncharged metabolites, and reversed phase columns (silica gel to which C4 to C18 alkyl chains or any of a variety of more exotic lipophilic moieties are attached to give a hydrophobic environment) can be used for neutral or charged metabolites. For GC separation, the metabolites must be volatilized. This often requires prior derivatization[33] in order for the metabolites to volatilize at lower temperatures. Carboxylic acids can be converted into the corresponding methyl esters with diazomethane; hydroxyl groups can be trimethylsilylated with bis-trimethylsilylacetamide or trimethylsilylimidazole in pyridine. Ketone carbonyls can be converted into *O*-substituted oximes. With radiolabeled compounds, the radioactivity can be monitored and separated directly from the HPLC column using an in-line radioactivity detector.

8.3.3. Identification

The two principal methods of metabolite structure identification are mass spectrometry and NMR spectrometry. It is preferable to link the separation and identification steps by running tandem LC-MS, tandem GC-MS, or tandem CE-MS. These methods are sufficiently sensitive to identify subnanogram amounts of material. The most popular methodology is *tandem LC-electrospray ionization mass spectrometry* by which a metabolite extract (or urine directly) can be injected into the HPLC, and each peak run directly into the mass spectrometer.[34] Similarly, *tandem CE-electrospray ionization mass spectrometry* has become a very valuable tool for separation and identification of biomolecules and drug metabolites.[35] In LC-tandem mass spectrometry/mass spectrometry (LC/MS/MS), the HPLC is connected to a mass spectrometer that is capable of not only obtaining parent ion data but also conducting fragmentation of the parent ion and obtaining product ion data. This technique can rapidly provide both a full mass spectrum (MS) and a product ion mass spectrum (MS/MS) for each metabolite.[36] Ultrafast gradient HPLC-MS/MS can produce run times of less than 5 min.[37] In this way there is less chance for metabolite degradation or loss, and workup procedures for mass spectrometry sample preparation are eliminated. Mass spectrometric properties are determined using different ionization techniques. Common vacuum ionization sources include electron impact (EI), chemical ionization (CI), matrix-assisted laser desorption/ionization (MALDI), fast atom bombardment (FAB), and secondary-ion mass spectrometry (SIMS). The development of HPLC coupled to atmospheric pressure ionization sources, namely, electrospray ionization (ESI) and atmospheric pressure chemical ionization (APCI) mass spectrometry, have transformed the status of drug metabolism studies from their former minor role in drug discovery to their current important function during the drug discovery process. These latter LC/MS/MS methods are used not only for drug metabolism studies but also to investigate drug pharmacokinetics (absorption, bioavailability, and clearance), where the focus is less on identification of metabolites and more on quantification of parent drug in the systemic circulation as a function of time (see section on quantification below). The trend in the pharmaceutical industry now is to initiate pharmacokinetic and metabolism studies as early as possible in the drug discovery process to aid in the selection of compounds that have the most druglikeness and best chance for survival through the many drug discovery and drug development hurdles to avoid late attrition of drug candidates.[38] With these HPLC/atmospheric pressure ionization mass spectrometric techniques, assessment of in vivo plasma half-lives and metabolic degradation can be made rapidly on a large number of drug candidates.

A brief description of different mass spectrometric techniques follows. *Electron impact mass spectrometry* (EI-MS) involves the bombardment of the vaporized metabolite by high-energy electrons (0–100 eV), producing a molecular radical cation ($M^{+\cdot}$) having a mass equivalent to the molecular weight of the compound. The electron bombardment causes bond fission and the positively charged fragments produced are detected. The mass spectrum is a plot of the percentage of relative abundance of each ion produced vs the mass-to-charge ratio (m/z).

CI mass spectrometry is important when compounds do not give spectra containing a molecular ion, generally because the molecular ion decomposes to give fragment ions. With CI-MS, a reagent gas such as ammonia, isobutane, or methane is ionized in the mass spectrometer and then ion–molecule reactions, such as protonation, occur instead of electron–molecule reactions. This *soft ionization* process results in little fragmentation. Fragment ions in this case are almost always formed by loss of neutral molecules, and as a result, much less structural information can be gleaned relative to EI-MS.

A variety of mass spectral techniques for nonvolatile or higher mass compounds, including peptides and proteins, also is now available.[39] MALDI is a soft ionization technique that is important for analyzing biopolymers.[40] It has the ability to produce gas-phase ions with little or no molecular fragmentation. FAB ionization involves the bombardment of a liquid film containing the nonvolatile sample with a beam of energized atoms of xenon or argon. This method is also useful for thermally unstable compounds. SIMS is similar to FAB except that energetic ions (Xe^+ and Ar^+) instead of atoms are used in SIMS.[41]

Two important atmospheric pressure ionization techniques arose out of the need for an ionization source that provided even softer ionization (less fragmentation of the molecular ion) and as a convenient interface with a liquid chromatograph. With ESI, ions are generated in solution phase, then the carrier solvent is evaporated, and a gas-phase ion is produced.[42] In contrast to ESI, APCI is a gas-phase ionization process in which gas-phase molecules are isolated from the carrier solvent before ionization.[43] In general, ESI is more applicable to high molecular weight, more polar compounds because it requires less heat and can produce multiply charged ions, whereas APCI is more useful for less polar molecules. Nonetheless, for most compounds with some acidic or basic characteristics and with relatively low molecular weight, either technique is applicable.

In conjunction with HPLC profiling of metabolites, radioactively labeled drugs are useful to pinpoint retention times of metabolites for more focused mass spectrometric characterization. By incorporating a splitter into the HPLC sample stream that directs part of the effluent to a radioactivity detector and the rest to the mass spectrometer, simultaneous radioactivity and mass spectrometry monitoring can be carried out.[44]

In addition to tremendous advances in mass spectrometry, newer technologies in two-dimensional and three-dimensional NMR spectrometry, particularly tandem LC-NMR (which became practical because of advances in solvent suppression techniques), have enhanced this analytical tool for studies in drug metabolism.[45] This continuous-flow method is particularly valuable, allowing ^{1}H and ^{19}F spectra to be obtained with only 5 ng or less of metabolite.[46] A mass spectrometer can be connected in tandem with the LC-NMR to give LC-NMR-MS, which enables high-quality NMR and mass spectra to be obtained simultaneously from a single HPLC injection of biological fluid.[47]

8.3.4. Quantification

Quantification of drug metabolites is carried out by subjecting a solution of separated metabolite to analysis by a detection method that generates a response proportional to the concentration of the metabolite in the solution. In current practice, the detector is usually directly in-line with the separation method, for example an ultraviolet (UV) detector or mass spectrometer that directly receives the effluent from an HPLC or GC column. Most detection methods require the construction of a standard calibration curve using known quantities of the analyte (parent drug or metabolite). The availability of required quantities (typically a few milligrams) of parent drug is rarely an issue once metabolism studies have been initiated. Reference quantities of metabolites, on the other hand, are often available only later in the course of metabolism studies. Therefore until a standard sample of a metabolite is available, quantitation is approximate and often must be roughly estimated using the calibration curve of a structurally similar molecule, for example the parent drug. However, if the change in structure resulting from metabolic transformation has caused a significant change in signal to the detector (for example, a change in chromophore for UV detection or change in ionization potential for mass spectrometric detection), then such estimates may provide inaccurate results and should be interpreted with caution.

Early use of UV absorption as a detection method was hampered by low sensitivity (weak signal for small quantity of metabolite), and thus there was often a need to first derivatize the metabolite with a chromophoric moiety to generate a more robust UV signal. Of course, such derivatization should be close to quantitative to yield accurate results. The much higher sensitivity for detection of radiolabeled metabolites provides a suitable alternative to UV, once the trouble has been taken to prepare the radiolabeled parent drug.[48] To quantitate radioactive metabolites, they are first separated by chromatography, and the rate of radioactive disintegration is determined by liquid scintillation counting methods. The amount of the metabolite isolated can be calculated from the specific radioactivity of the drug (see Section 8.2).

The sensitivity and low volumes necessary for mass spectrometry makes this a detection method of choice for metabolism studies. The use of mass spectrometry reduces the assay development, sample preparation, and analysis times and is also well suited for the 96-well plate format of high-throughput metabolic screens. *Selected ion monitoring (SIM)*[49] is a highly selective method for detection and quantification of small quantities of metabolites, which uses a mass spectrometer as a selective detector of specific components in the effluent from an HPLC or gas chromatograph. By setting the spectrometer to detect characteristic fragment ions at a single *m/z* value, other compounds with the same retention times that do not produce those fragment ions will go undetected. When a full mass spectrum is recorded repetitively throughout a chromatogram, and a SIM profile is reconstructed by computer, it is called *mass fragmentography*. Subpicogram quantities of metabolites in a mixture can be detected by the SIM method. Another important MS/MS mass spectral technique related to SIM, is *selected reaction monitoring*, which monitors the fragmentation reaction of the parent ion mass observed in the first spectrometer and one of its product ions (often the prominent one) in the second mass spectrometer over time for precise quantification.[50] You can think of this method as the SIM of a fragment ion, which generally gives just a single peak for easy and sensitive quantification. When more than a single fragment ion is monitored, it is called *multiple reaction monitoring*.[51]

8.4. PATHWAYS FOR DRUG DEACTIVATION AND ELIMINATION

8.4.1. Introduction

The first mammalian drug metabolite that was isolated and characterized was hippuric acid (**8.8**) from benzoic acid in the early nineteenth century.[52] However, not until the later 1940s, when Mueller and Miller[53] demonstrated

8.8

that the in vivo metabolism of 4-dimethylaminoazobenzene could be studied in vitro (see Section 8.4.2.1), was the discipline of drug metabolism established. As a result of the ready commercial availability of radioisotopes and sophisticated separation, detection, and identification techniques that were developed in the latter half of the twentieth century (see Section 8.3), drug metabolism studies have burgeoned.

The function of drug metabolism is to convert a molecule that can cross biological membranes into one that is cleared, generally in the urine; each progressive metabolic step usually reduces the lipophilicity of the compound. The lipophilicity of the drug molecule will determine whether it undergoes direct renal clearance or is metabolically cleared. As the log $D_{7.4}$ (see Chapter 2, Section 2.2.5.2.2) of the compound increases above zero, there is a marked decrease in direct renal clearance and a sharp increase in metabolic clearance (Figure 8.1 shows the results for a series of chromone-2-carboxylic acid derivatives),[54] indicating the contribution of lipophilicity to drug metabolism. Drug metabolism reactions have been divided into two general categories,[55] termed Phase I and Phase II reactions. *Phase I transformations* involve reactions that introduce or unmask a functional group, such as oxygenation or hydrolysis. *Phase II transformations* mostly generate highly polar derivatives (known as *conjugates*), such as glucuronides and sulfate esters, for excretion in the urine.

The rate and pathway of drug metabolism are affected by species, strain, sex, age, hormones, pregnancy, and liver diseases such as cirrhosis, hepatitis, porphyria, and hepatoma.[56] Drug metabolism can have a variety of profound effects on drugs. It principally causes pharmacological deactivation of a drug by altering its structure so that it no longer interacts appropriately with the target receptor and becomes more susceptible to excretion. Drug metabolism, however, can also convert a pharmacologically inactive prodrug into an active drug (see Chapter 9). The pharmacological response of a drug may be altered if a metabolite has a new activity; in some cases, the metabolite has the same activity and a similar or different potency as the drug. A change in drug absorption and drug distribution (that is, the tissues or organs in which it is concentrated) can also result when it is converted into a much more polar species.

The majority of drug-metabolizing enzymes also catalyze reactions on endogenous compounds. Consequently, the function of these enzymes may be metabolic disposition of endogenous cellular modulators, and it may be fortuitous that they also catalyze the metabolism of drugs and other xenobiotics. The greater affinity of the endogenous substrate over drugs in many cases seems to support this notion; however, many of these enzymes are very broad in specificity or are induced only on the addition of a particular chemical, so it seems likely that at least some of these enzymes have evolved to protect the organism from undesirable exogenous substances.

As was discussed in Chapter 3 (Section 3.2.5.2) and Chapter 4 (Section 4.1.2.1.1), the interaction of a chiral molecule with a receptor or enzyme produces a diastereomeric complex. Therefore, it is not surprising that the processes in drug metabolism are also stereoselective, if not

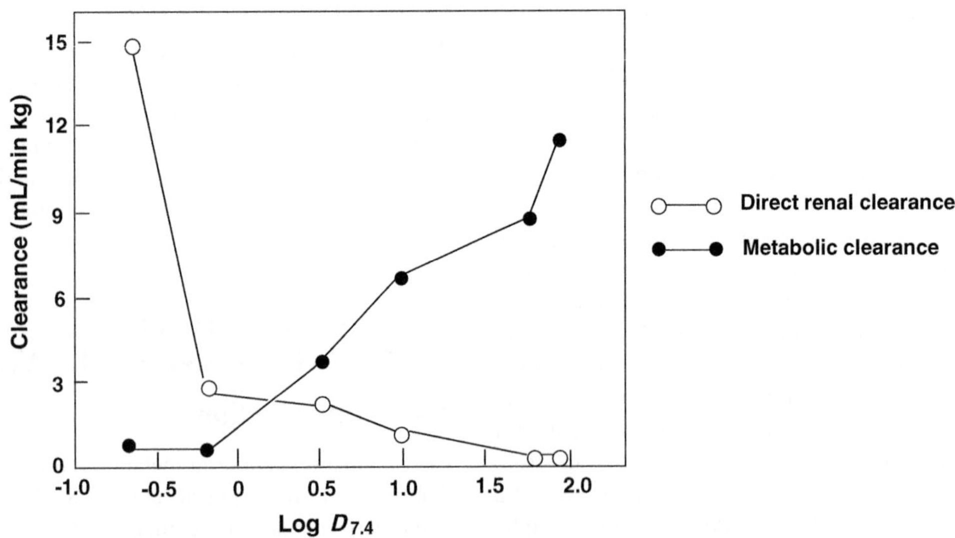

FIGURE 8.1 Effects of lipophilicity on direct renal clearance and on metabolism

stereospecific.[57] *Stereoselectivity* can occur with enantiomers of drugs, in which case one enantiomer may be metabolized to a greater extent by one pathway and the other enantiomer predominantly by another pathway to give two different metabolites. Another type of stereoselectivity is the conversion of an achiral drug into a chiral metabolite. In both of these cases, a difference in rates leads to unequal amounts of metabolites rather than exclusively to one metabolite. In many cases, however, a racemic drug is metabolized as if it were two different xenobiotics, each enantiomer displaying its own pharmacokinetic and pharmacodynamic profile. In fact, it was concluded by Hignite et al.[58] that "warfarin enantiomers should be treated as two drugs"; Silber et al.[59] concluded that "*S*-(−)- and *R*-(+)-propranolol are essentially two distinct entities pharmacologically". These are just two of the many examples of drugs whose enantiomers are metabolized by different routes. In some cases, the inactive enantiomer can produce toxic metabolites, which could be avoided by the administration of only the active enantiomer.[60] In other cases, the inactive isomer may inhibit the metabolism of the active isomer.

The metabolism of enantiomers may depend on the route of administration. For example, the racemic antiarrhythmic drug verapamil (**8.9**, Calan, Verelan, Isoptin) is 16 times more potent when administered intravenously

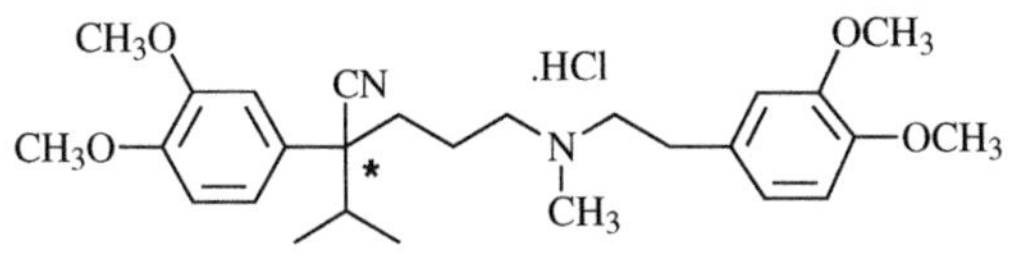

Verapamil hydrochloride
8.9

than when taken orally because of extensive hepatic presystemic elimination that occurs with oral administration (the first-pass effect; see Section 8.1).[61] The (−)-isomer, which is 10 times more potent than the (+)-isomer, is preferentially metabolized during hepatic metabolism and, therefore, there is much more of the less potent (+)-isomer available by the oral route than by the IV route.

In some cases, one enantiomer of a drug can be metabolized to the other enantiomer. The therapeutically inactive *R*-isomer of the analgesic ibuprofen (**8.10**, Advil) is converted

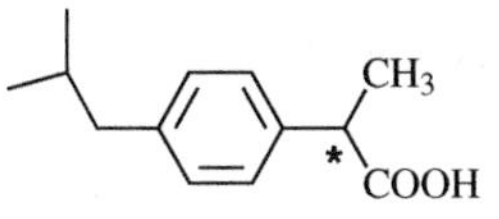

Ibuprofen
8.10

enzymatically[62] in the body to the active *S*-isomer.[63] The racemization occurs by initial conversion of the carboxylic acid group of ibuprofen to the corresponding *S*-CoA thioester, followed by racemization, and then hydrolysis to the two enantiomers of ibuprofen.[64] If a racemic mixture is administered, a 70:30 mixture of *S*:*R* is excreted; if a 6:94 mixture of *S*:*R* is administered, an 80:20 mixture of *S*:*R* is excreted.

In the next two sections, the various types of metabolic biotransformations are described. It is not intended that the reactions discussed are the only ones that occur with those particular drugs, but, rather, they are just examples of the particular metabolic biotransformation under discussion. Some examples are included to show the effect of stereochemistry on drug metabolism. It should be kept in mind, however, that when a racemic mixture is administered, the metabolites observed might be different from those detected when a pure isomer is used. Many of the metabolites that are described are derived from in vitro studies (metabolites produced using either purified enzymes, subcellular fractions, whole cells, or tissue preparations). However, after analysis of a large amount of data on in vitro metabolism and a comparison with in vivo metabolism, it appears that it is valid to look at in vitro data as a reasonable guide to the prediction of in vivo metabolism.[65]

Although metabolic biotransformations are very complex, sufficient data have been accumulated that computer software is now available that makes predictions of the most likely sites on a molecule for metabolism. Some common programs for predicting cytochrome P450 Phase I transformations include MetaSite[66] (Molecular Discoveries, Ltd., Pinner, Middlesex, UK), KnowitAll (Bio-rad; www.biorad.com), Admensa (Inpharmatica; www.inpharmatica.com), and Meteor (Lhasa; www.lhasalimited.org). With MetaSite, the primary site of metabolism was found in the top three predictions in 85% of the cases. A procedure to identify sites of metabolism, combining the in silico prediction capability of MetaSite with mass spectrometric confirmation, increased the predictive power of the primary metabolic site from 50% to 80%.[67] A noncommercial program called SMARTCyp predicts sites of P450 metabolism with a preference for the P450 3A4 isozyme; at least one metabolic site in the two top-ranked positions was predicted 76% of the time.[68]

8.4.2. Phase I Transformations

8.4.2.1. Oxidative Reactions

Mueller and Miller[69] showed that the in vivo metabolism of 4-dimethylaminoazobenzene could be studied in vitro using rat liver homogenates. It was demonstrated that the in vitro system was functional only if nicotinamide adenine dinucleotide phosphate ($NADP^+$), molecular oxygen, and both the microsomal and soluble fractions from the liver homogenates were included. Later, Brodie and coworkers[70] found that the oxidative activity was in the microsomal fraction and that the soluble fraction could be replaced by either

NADPH or an NADPH-generating system. This system was active toward a broad spectrum of structurally diverse compounds. Because it required both O_2 and a reducing system, it was classified as a *mixed function oxidase*,[71] that is, one atom of the O_2 is transferred to the substrate and the other undergoes a two-electron reduction and is converted to water. This classification was confirmed when it was shown[72] that aromatic hydroxylation of acetanilide by liver microsomes in the presence of $^{18}O_2$ resulted in incorporation of one atom of ^{18}O into the product and that a heme protein was an essential component for this reaction.[73] When this heme protein was reduced and exposed to carbon monoxide, a strong absorption in the visible spectrum at 450 nm resulted. Because of this observation, these microsomal oxidases were named *cytochrome P450.*

Cytochrome P450 now represents a superfamily of enzymes containing a heme cofactor but with structurally variable active sites that catalyze the same reaction on different substrates, namely, the oxidation of steroids, fatty acids, and xenobiotics.[74] These related cytochrome P450 enzymes are referred to as *isozymes.* Although about 500 genes encode different cytochrome P450 isozymes, only about 15 of these are very important in drug metabolism. The primary localization of these enzymes is the liver, but they are also present in the lung, kidney, gut, adrenal cortex, skin, brain, aorta, and other epithelial tissues. The heme is noncovalently bound to the apoprotein. Cytochrome P450 is associated with another enzyme, NADPH-cytochrome P450 reductase, a flavoenzyme that contains one molecule each of flavin adenine dinucleotide (FAD) and flavin mononucleotide (FMN).[75] Heme-dependent oxidation reactions were discussed in Chapter 4 (Section 4.3.4). As shown in Scheme 4.35, the NADPH-cytochrome P450 reductase reduces the flavin which, in turn, transfers an electron to the heme–oxygen complex of cytochrome P450. Actually, the FAD accepts electrons from the NADPH, the $FADH^-$ then transfers electrons to the FMN, and the $FMNH^-$ donates the electron to the heme or heme–oxygen complex of cytochrome P450.[76]

In general, cytochrome P450 catalyzes either hydroxylation or epoxidation of various substrates (Table 8.1) and is believed to operate via radical intermediates (see Scheme 4.35). When the concentrations of cytochrome P450 and other drug-metabolizing enzymes are modified, drug metabolism becomes altered. Many drugs and environmental chemicals induce either their own metabolism or the metabolism of other drugs in man as a result of their induction of cytochrome P450 and NADPH-cytochrome P450 reductase. These changes in pharmacokinetics and metabolism of the drugs when multiple drugs are taken together are known as *drug–drug interactions.*[77] One common mechanism leading to drug–drug interactions is the induction of cytochrome P450 isozymes by the administered drugs. For example, St. John's wort (*Hypericum perforatum*) is a herbal remedy for the treatment of depression,[78] which is sometimes referred to as nature's Prozac. The constituent of this herb contributing to the antidepressant activity, hyperforin (**8.11**),[79] was found to activate the pregnane X receptor, which

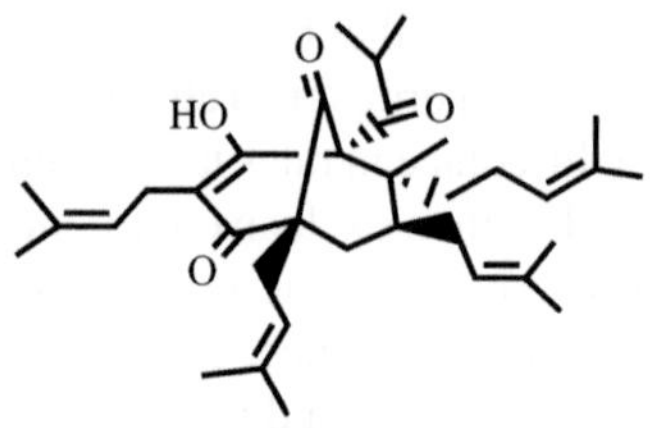

Hyperforin
8.11

is the key regulator of cytochrome P450 3A4 isozyme transcription.[80] Because of the increased expression of this P450 enzyme induced by hyperforin, drugs that are metabolized by this isozyme, which may be more than half of all drugs, are rapidly degraded when taken with St. John's wort.

The other most common mechanism leading to drug–drug interactions arises when multiple drugs are administered and one of the drugs inhibits drug metabolism of the others as a result of its inhibition of cytochrome P450 isozymes or other enzymes. Adverse effects and toxicity of drugs from drug–drug interactions have been two of the major concerns in clinical practices and in new drug approval. A better characterization of the metabolic pathways and the enzymes involved in its metabolism is useful in understanding the underlying mechanisms of drug–drug interactions. On the bright side, as discussed in Chapter 7 (Section 7.2.2.1), a drug known to inhibit cytochrome P450 (such as ritonavir) could be used in combination with another drug (such as lopinavir) to intentionally block metabolism of the other drug.

Some properties of molecules lead to preferred metabolism by cytochrome P450. Lipophilicity is important for the binding of a molecule to cytochrome P450, which explains why increasing lipophilicity leads to an increase in metabolic clearance (see Figure 8.1 above). The next most important physicochemical property of a molecule for cytochrome P450 metabolism is the presence of an ionizable group. This group may have a key role in binding to the active site and determine regioselectivity of the metabolic reaction.[81] If the ionizable group is a secondary or tertiary amine, it may direct the reaction to occur at that part of the molecule (see below). The reaction and site of reaction catalyzed by cytochrome P450s are determined by (1) the topography of the active site of the particular isozyme, (2) the degree of steric hindrance of the heme iron-oxo species to the site of reaction, and (3) the ease of hydrogen atom abstraction or electron transfer from the compound that is metabolized by the isozyme.

Another important family of enzymes involved in drug oxidation is the microsomal *flavin monooxygenases*[82] (see Chapter 4, Section 4.3.3); the mechanism of oxidation was described in Scheme 4.34. According to this scheme, the

TABLE 8.1 Classes of Substrates for Cytochrome P450

Functional Group	Product
R–Ph (aromatic ring)	R–C_6H_4–OH
RR'C=CR' (alkene)	epoxide (O bridging the two carbons; R, R', R')
$ArCH_2R$	ArCHR(OH)
R(R)C=CH–CH_2R'	R(R)C=CH–CHR'(OH)
RC(=O)CH_2R'	RC(=O)CHR'(OH)
RCH_2R'	RCHR'(OH)
RCH_2-X-R' (X = N, O, S, halogen)	(RCH(OH)-XR') → RCHO + R'XH
R-X-R' (X = NR, S)	R-X^+(–O^-)-R'

flavin peroxide intermediate is an electrophilic species, indicating that the substrates for this enzyme are nucleophiles, such as amines, thiols, and related compounds (Table 8.2). The enzyme contains one FAD molecule per subunit and, as in the case of cytochrome P450, requires NADPH to reduce the flavin. It has been found that nucleophilic compounds containing an anionic group are excluded from the active site of this enzyme. Because most endogenous nucleophiles contain negatively charged groups, this may be a way that Nature prevents normal cellular components from being oxidized by this enzyme. Oxygenation of some of these compounds, for example, mercaptopyrimidines and thiocarbamides, leads to reactive intermediates.[83] Metabolism by flavin monooxygenases often can be differentiated from other oxygenases because of its unique stereoselectivity. For example, cytochrome P450s oxidize (*S*)-nicotine (**8.12**) to a mixture of *cis*- (**8.13**) and *trans*-*N*-1′-oxides (**8.14**), but flavin monooxygenase oxidizes it exclusively to the *trans*-*N*-1′-oxide.

Other enzymes involved in oxidative drug metabolism include prostaglandin H synthase (see Section 8.4.2.1.5), alcohol dehydrogenase, aldehyde dehydrogenase (see Section 8.4.2.1.9), xanthine oxidase, monoamine oxidase, and aromatase. These enzymes, however, are involved, for the most part, in the metabolism of endogenous compounds.

(*S*)-Nicotine 8.12 **8.13** **8.14**

TABLE 8.2 Classes of Substrates for Flavin Monooxygenase

Functional Group	Product
R-NR'$_2$	R-N$^+$(O$^-$)R'$_2$
R-NHR'	R-N(OH)R'
R-N(OH)R'	R=N$^+$(O$^-$)R'
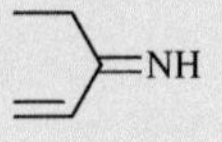 C=NH	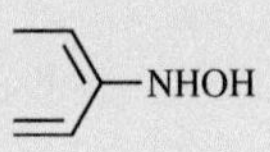 C-NHOH
R-N(R")-NHR'	R-N$^+$(O$^-$)(R")-NHR'
R-C(=S)NH$_2$	R-C(=S$^+$-O$^-$)NH$_2$
RNH-C(SH)=N$^+$HR'	RNH-C(SO$_2^-$)=N$^+$HR'
2RSH	RSSR
RSSR	2 RSO$_2^-$

8.4.2.1.1. Aromatic Hydroxylation

In 1950, Boyland[84] hypothesized that aromatic compounds were metabolized initially to the corresponding epoxides (*arene oxides*). This postulate was confirmed in 1968 by a group at the National Institutes of Health (NIH) that isolated naphthalene 1,2-oxide from the microsomal oxidation of naphthalene[85] (Scheme 8.3).

Subsequent kinetic isotope effect studies,[86] however, indicate that a direct arene epoxidation is a highly unlikely process. Instead, an activated heme iron-oxo species may undergo electrophilic addition to the aromatic ring to give either a tetrahedral intermediate radical (**8.15**) or cation (**8.16**) (Scheme 8.4), similar to the mechanism described for the addition of the corresponding heme iron-oxo species to alkenes, as was discussed in Chapter 4 (Section 4.3.4, Scheme 4.37). Because [1,2]-shifts of hydrogen and alkyl radicals are energetically unfavorable, an electron could be transferred to the Fe^{IV} of the heme, giving a carbocation species bound to Fe^{III} (**8.16**). A [1,2]-shift of hydride to a cation is energetically favorable, leading to the cyclohexadienone (**8.17**, pathway c), which would tautomerize to the phenol (**8.18**). Arene oxide (**8.19**) formation may occur from either **8.15** or **8.16**. Usually, arene oxides can undergo rearrangements to arenols, hydration (catalyzed by epoxide hydrolase) to the corresponding *trans*-diol, reaction with glutathione (catalyzed by glutathione *S*-transferase) to the β-hydroxy sulfide, and reactions with various macromolecular nucleophiles (Scheme 8.5).

The rearrangement of an arene oxide to an arenol is known as the *NIH shift* because a research group at the NIH proposed this mechanism based on studies with specifically deuterated substrates (Scheme 8.6).[87] Ring opening occurs in the direction that gives the more stable carbocation (*ortho*- or *para*-hydroxylation when R is electron donating). Also, because of an isotope effect on the cleavage of the C–D bond (see Section 8.2 about isotope effects), the proton is preferentially removed, leaving the migrated deuterium. Although there is an isotope effect on the cleavage of the C–D vs the C–H bond, this step is not the rate-determining step in the overall reaction; consequently, there is no overall isotope effect on this oxidation pathway when deuterium is incorporated into the substrate.

In competition with the [1,2]-hydride (deuteride) shift is deprotonation (dedeuteronation; Scheme 8.7). The percentage of each pathway depends on the degree of stabilization of the intermediate carbocation by R; the more the stabilization (i.e., the greater electron donation of R), the less is the need for the higher energy hydride shift and the more deprotonation (dedeuteronation) occurs.[88] For example, when R = NH_2, OH, $NHCOCF_3$, and $NHCOCH_3$, only 0–30% of the product phenols retain deuterium (NIH shift), but when R = Br, $CONH_2$, F, CN, and Cl, 40–54% deuterium retention is observed.

Whenever a deuterium is incorporated into a drug for metabolism studies, however, the problem of *metabolic switching* has to be considered. This is when deuteration at one site in the molecule changes the partition between two metabolic pathways because the deuterium isotope effect on C–D bond cleavage at the deuterated site can slow down C–D bond cleavage and favor an increase in metabolism at C–H bonds at a different site in the molecule.[89] However,

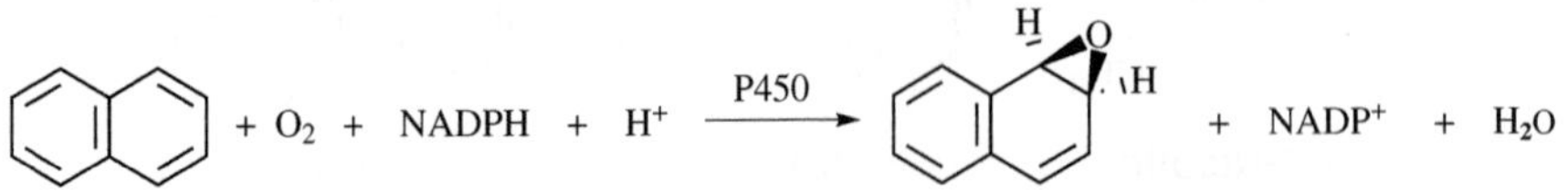

naphthalene + O_2 + NADPH + H^+ —P450→ naphthalene 1,2-oxide + $NADP^+$ + H_2O

SCHEME 8.3 Cytochrome P450 oxidation of naphthalene

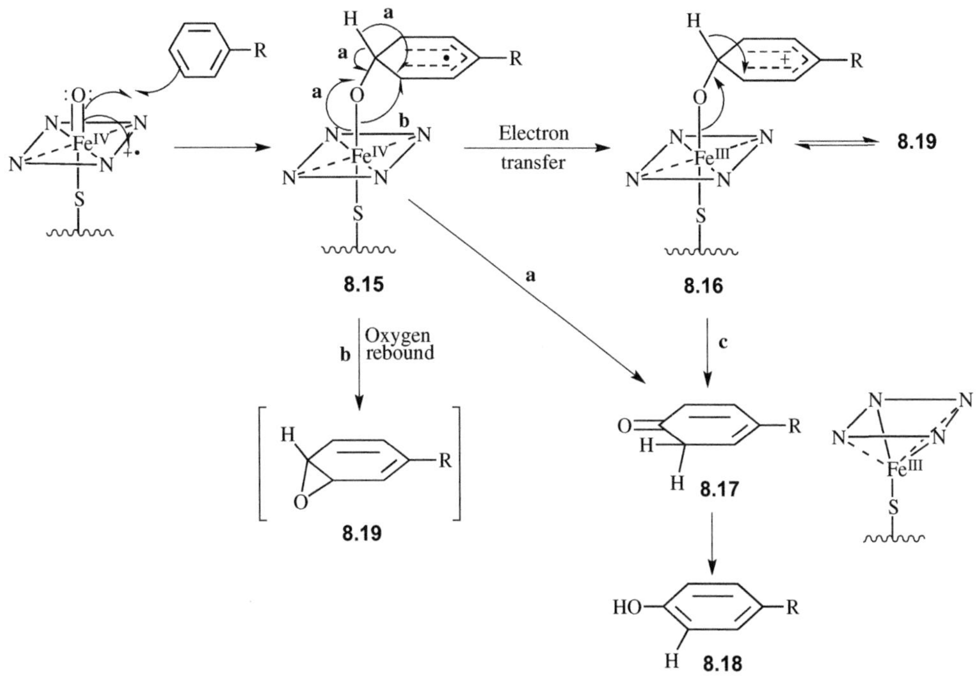

SCHEME 8.4 Addition–rearrangement mechanism for arene oxide formation

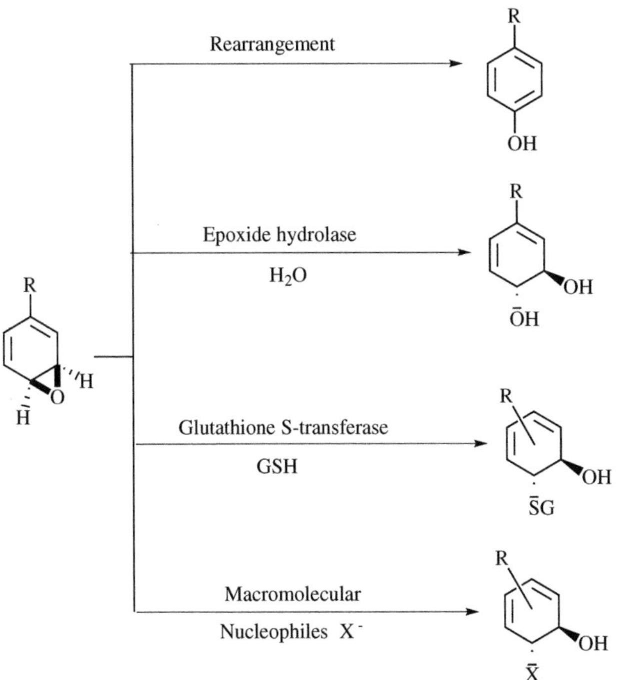

SCHEME 8.5 Possible fates of arene oxides

judicious incorporation of deuterium at a known site of metabolism can also be a useful approach to slow down the rate of metabolism of a drug to allow enhanced absorption, distribution, and half-life.[90] For example, a deuterated analog of the antibacterial drug linezolid (**8.20**, [D_{10}]-Zyvox) had a $t_{1/2}$ in primates of 6.3 h compared with 4.5 h for the parent compound with three times higher concentration in blood after 24 h than the parent molecule. This approach sometimes also allows for novel intellectual property positions to be gained for drugs that were previously disclosed only in proteo form.[91]

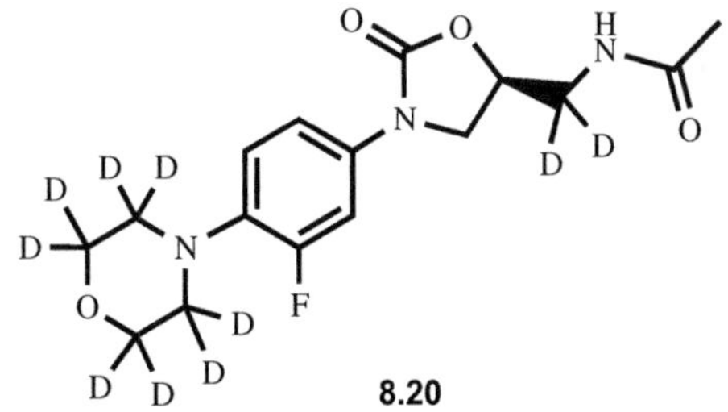

8.20

The NIH shift also occurs with substituents and with hydrogen. For example, rat liver metabolizes *p*-chloroamphetamine (**8.21**) to 3-chloro-4-hydroxyamphetamine[92] (Scheme 8.8). This migration of chlorine is not typical because a common approach to slow down metabolism of aromatic compounds is to substitute the ring with electron-withdrawing *para*-fluoro,[93] *para*-trifluoromethyl, or *para*-chloro groups, which deactivate the ring and decrease the rate of oxidative metabolism.[94] Next to hydrogen, fluorine is the smallest atom, although its van der Waals radius (1.47 Å) and electronegativity (3.98) are closer to those of oxygen (1.52 Å, 3.44) than hydrogen (1.20 Å, 2.20). Therefore, fluorine can act as a bioisostere (see Chapter 2, Section 2.2.4.3) for either hydrogen or a hydroxyl group. The trifluoromethyl group is much larger than a methyl group, closer to the size of an isopropyl[95] or isobutyl[96] group.

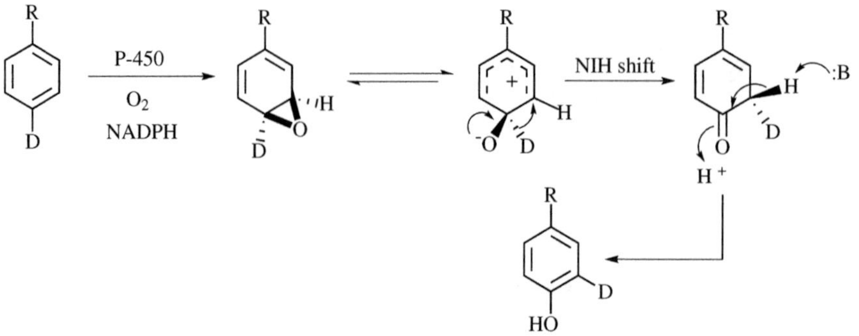

SCHEME 8.6 Rearrangement of arene oxides to arenols (NIH shift)

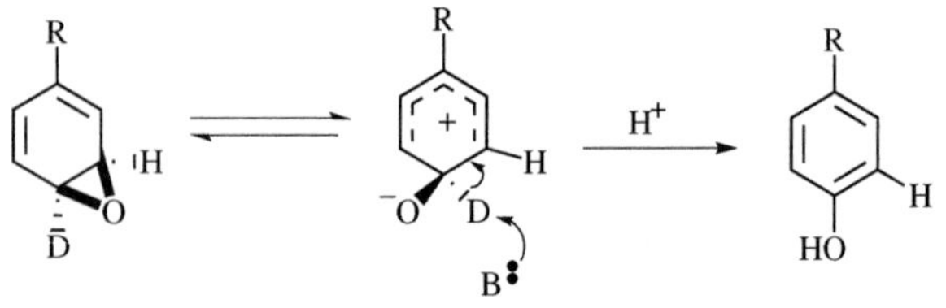

SCHEME 8.7 Competing pathway for NIH shift

As an example of the metabolic protection of halogens, the antiinflammatory drug diclofenac (**8.22**, Voltaren) is metabolized to 4-hydroxydiclofenac with a half-life of 1 h. The related analog with a *para*-chloro substituent, fenclofenac (**8.23**, Flenac), has a half-life of over 20 h.[97] A related substituent NIH shift (migration of the nitro group) was observed in the metabolic oxidation of the antiprotozoal agent tinidazole[98] (**8.24**, Scheme 8.9, Fasigyn).

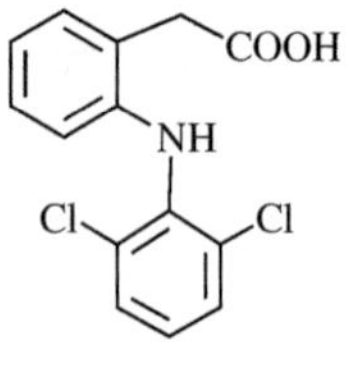

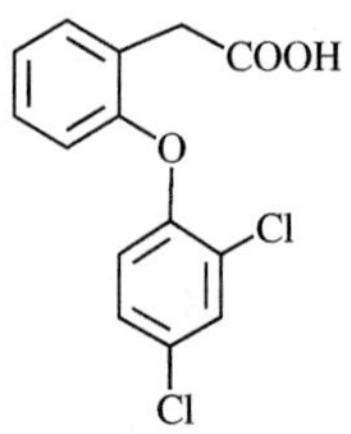

Diclofenac
8.22

Fenclofenac
8.23

As in the case of electrophilic aromatic substitution reactions, it appears that the more electron rich the aromatic ring (R is electron donating), the faster the microsomal hydroxylation will be.[99] Aniline (electron rich), for example, undergoes extensive *ortho*- and *para*-hydroxylation,[100] whereas the strongly electron-poor uricosuric drug probenecid (**8.25**, Benemid) undergoes no detectable aromatic hydroxylation.[101] In the case of drugs with two aromatic rings, the more electron-rich one (or less electron poor), generally, is hydroxylated. The antipsychotic drug chlorpromazine (**8.26**, R=H, Thorazine), for example, undergoes 7-hydroxylation (**8.26**, R=OH).[102] However, the binding orientation in the active site of the hydroxylases, as well as the activation of the aromatic

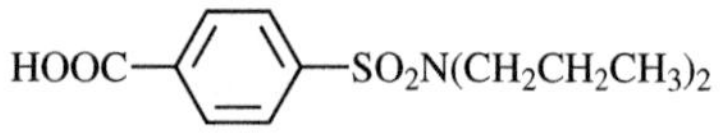

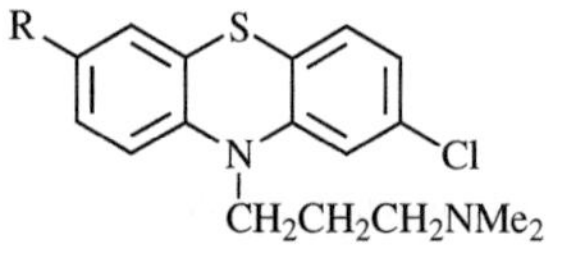

$HOOC-C_6H_4-SO_2N(CH_2CH_2CH_3)_2$

Probenecid
8.25

Chlorpromazine
8.26

8.21

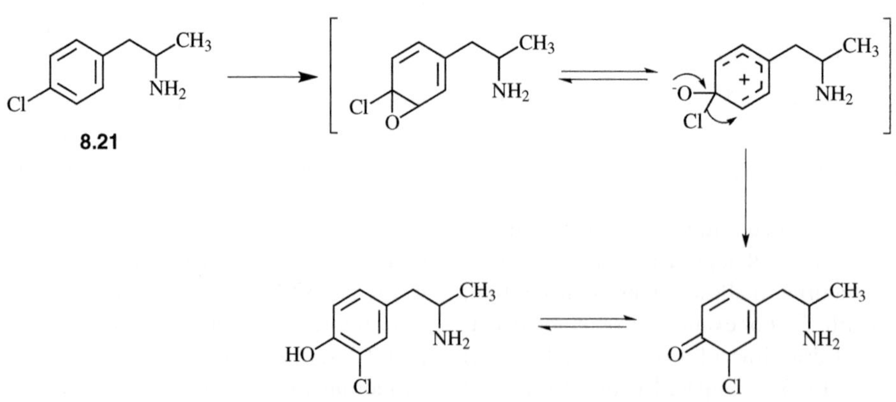

SCHEME 8.8 NIH shift of chloride ion

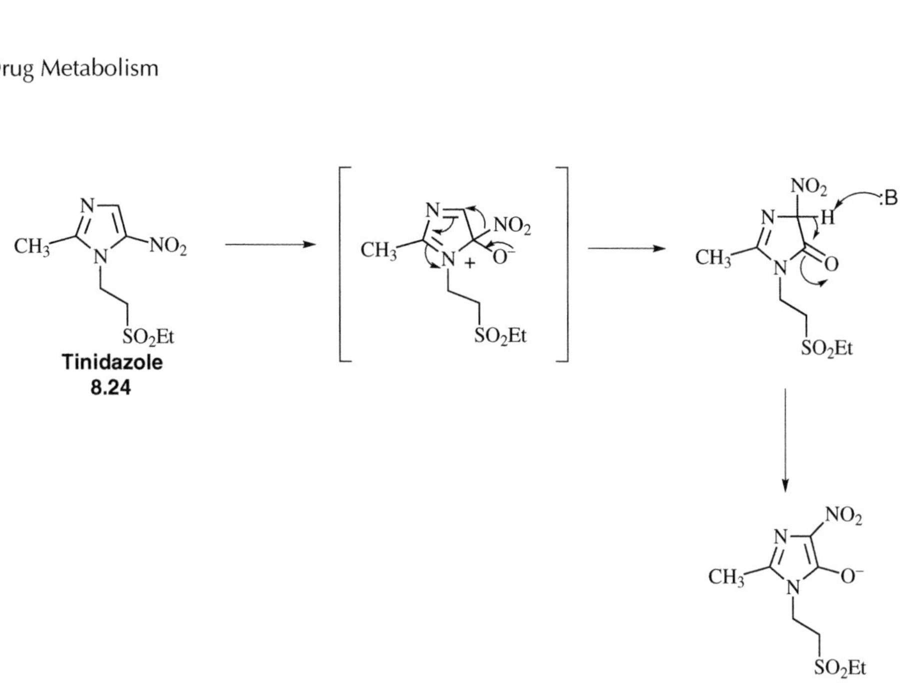

Tinidazole
8.24

SCHEME 8.9 NIH shift of nitro group

activation of the aromatic rings, plays an important role in both the rate and position of hydroxylation on the aromatic ring.

Aromatic hydroxylation, as is the case for all metabolic reactions, is species specific. In man, *para*-hydroxylation is a major route of metabolism for many phenyl-containing drugs. The site and stereoselectivity of hydroxylation depends on the animal studied. In man, the antiepilepsy drug phenytoin (**8.27**; $R_1=R_2=R_3=H$, Dilantin) is *para*-hydroxylated at the *pro-S* phenyl ring (**8.27**; $R_1=OH$, $R_2=R_3=H$) 10 times more often than the *pro-R* ring (**8.27**; $R_3=OH$, $R_1=R_2=H$). In dogs, however, *meta*-hydroxylation of the *pro-R* phenyl ring (**8.27**; $R_2=OH$, $R_1=R_3=H$) is the major pathway. The overall ratio of R_2:R_1:R_3 hydroxylation in dogs is 18:2:1.[103] *Meta*-hydroxylation may be catalyzed by an isozyme of cytochrome P450 that operates by a mechanism different from arene oxide formation.[104] For example, an important metabolite of chlorobenzene is 3-chlorophenol, but it was shown that neither 3- nor 4-chlorophenol oxide gave 3-chlorophenol in the presence of rat liver microsomes, suggesting that a direct oxygen insertion mechanism may be operative in this case. This mechanism is also consistent with the observation of a kinetic isotope effect ($k_H/k_D=1.30–1.75$) on the in vivo 3-hydroxylation.[105] An alternative explanation for *meta*-hydroxylation is that the *meta*-position is the most available site for electrophilic addition by the heme iron-oxo species.

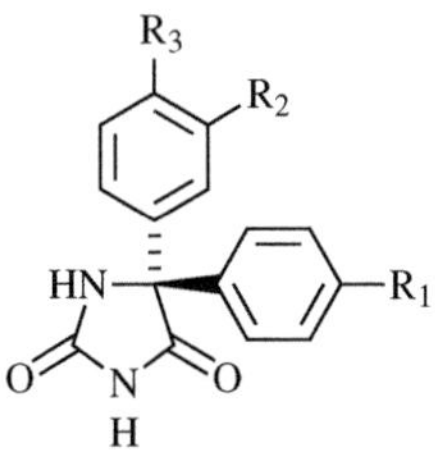

Phenytoin
8.27

Because of the reactivity of arene oxides, they can undergo rapid reactions with nucleophiles. If cellular nucleophiles react with these compounds, toxicity can result. Consequently, there are enzymes that catalyze deactivation reactions on these reactive species (see Scheme 8.5). Epoxide hydrolase (also referred to as epoxide hydratase or epoxide hydrase) catalyzes the hydration of highly electrophilic arene oxides to give *trans*-dihydrodiols.[106] The mechanism of this reaction is initial attack by an active site aspartate to give a covalent catalytic intermediate, which hydrolyzes to the *trans*-glycol (Scheme 8.10)[107]; attack occurs predominantly at the less sterically hindered side.[108] The *trans*-dihydrodiol product can be oxidized further to give catechols; catechols are also generated by hydroxylation of arenols. Because of the instability of catechols to oxidation, they may be converted either to *ortho*-quinones or to semiquinones.

Glutathione *S*-transferase is an important enzyme that protects the cell from the electrophilic arene oxide metabolites.[109] It catalyzes a nucleophilic substitution reaction of glutathione on various electrophiles.[110] Glutathione metabolites of naphthalene are shown in Scheme 8.11; only one of them was not detected.[111] Other reactions catalyzed by glutathione *S*-transferases are discussed in Section 8.4.3.5.

When the arene oxide escapes destruction by these enzymes, toxicity may result. An important example of this is the metabolism of benzo[a]pyrene (**8.28**; Scheme 8.12), a potent carcinogen found in soot and charcoal. The

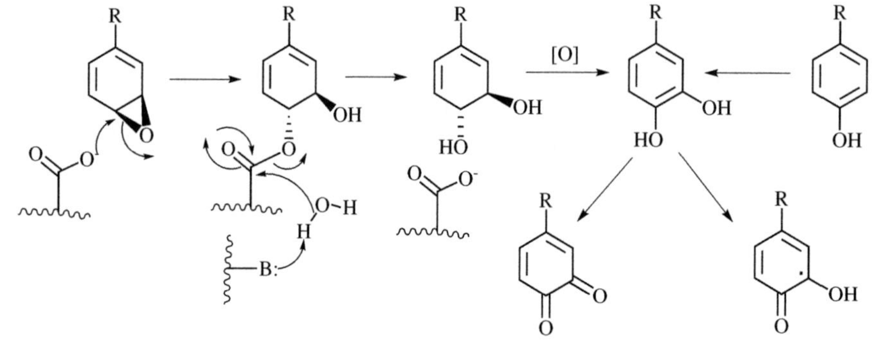

SCHEME 8.10 Metabolic formation and oxidation of catechols

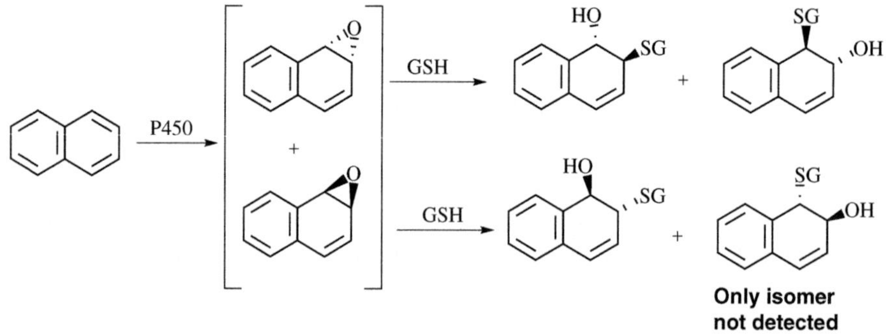

SCHEME 8.11 Formation of glutathione adducts from naphthalene oxides

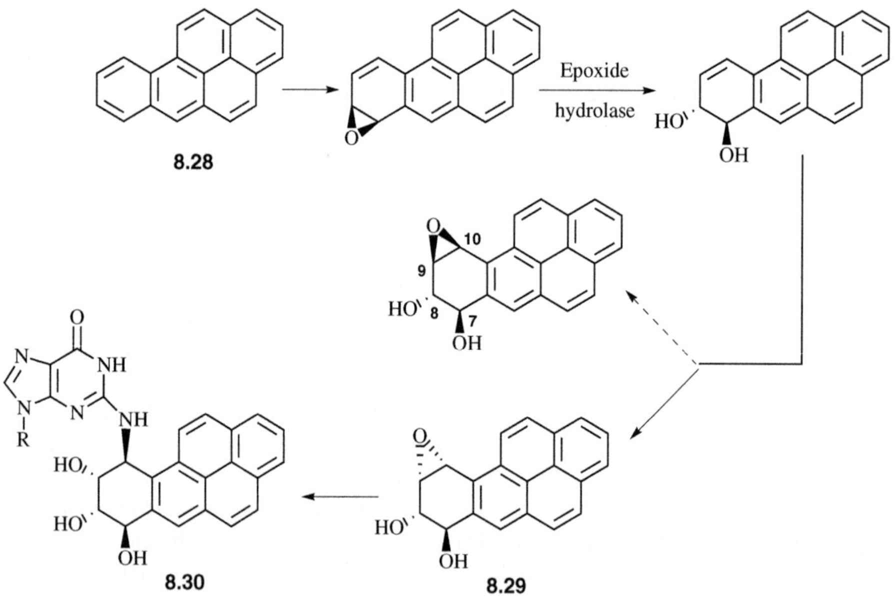

SCHEME 8.12 Deoxyribonucleic acid adduct with benzo[a]pyrene metabolite

relationship between soot and cancer was noted by Sir Percival Scott, a British surgeon, in 1775 when he observed that chimney sweeps (people who cleaned out chimneys) frequently developed skin cancer. Metabolic activation of polyaromatic hydrocarbons can lead to the formation of covalent adducts with ribonucleic acid (RNA), deoxyribonucleic acid (DNA), and proteins.[112] Covalent binding of these metabolites to DNA is the initial event that is responsible for malignant cellular transformation.[113] The key reactive intermediate responsible for alkylation of

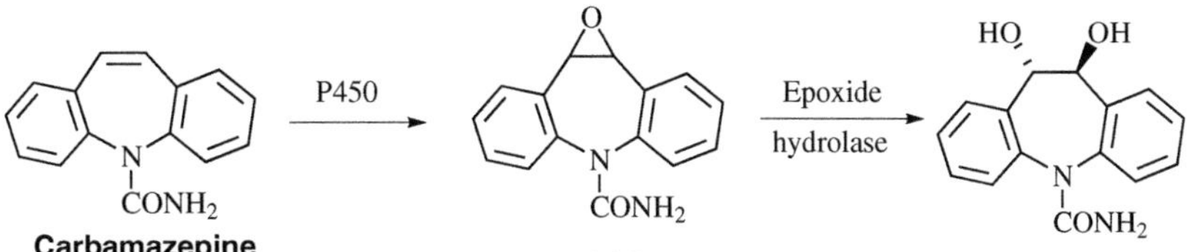

Carbamazepine
8.31

8.32

8.33

SCHEME 8.13 Metabolism of carbamazepine

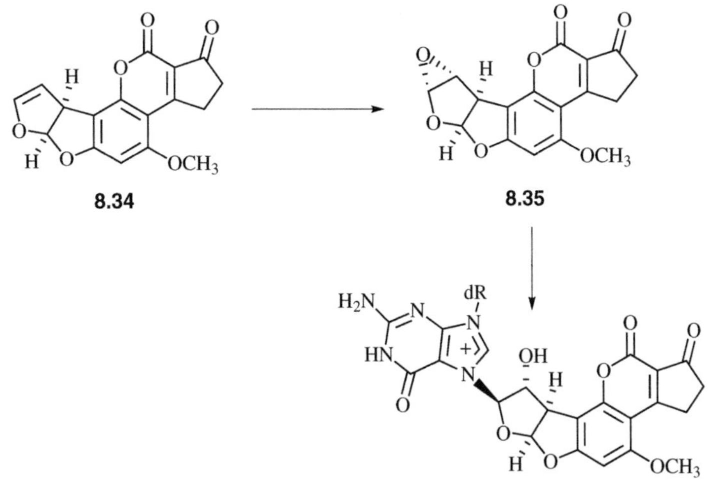

8.34

8.35

8.36

SCHEME 8.14 Metabolic reactions of aflatoxin B_1

nucleic acids is (+)-7*R*,8*S*-dihydroxy-9*R*,10*R*-oxy-7,8,9,10-tetrahydrobenzo[a]pyrene (**8.29**; Scheme 8.12). This metabolite reacts with RNA to form a covalent adduct between the C-2 amino group of a guanosine and C-10 of the hydrocarbon[114] (**8.30**). The reactive metabolite (RM) (**8.29**) also causes nicks in superhelical DNA of *Escherichia coli*; an adduct between the C-2 amino group of deoxyguanosine and the C-10 position of the diol epoxide was isolated.[115]

8.4.2.1.2. Alkene Epoxidation

Because alkenes are more reactive than aromatic π-bonds, it is not surprising that alkenes are also metabolically epoxidized. An example of a drug that is metabolized by alkene epoxidation is the anticonvulsant agent, carbamazepine (**8.31**, Scheme 8.13, Tegretol).[116]

Carbamazepine-10,11-epoxide (**8.32**) has been found to be an anticonvulsant agent as well, so the metabolite may be responsible for the anticonvulsant activity of carbamazepine.[117] The epoxide is converted stereoselectively into the corresponding (10*S*,11*S*)-diol (**8.33**) by epoxide hydrolase.

Metabolic epoxidation of an alkene can also lead to the formation of toxic products. The enol ether mycotoxin aflatoxin B_1 (**8.34**) is a mutagen in several strains of bacteria,[118] a hepatocarcinogen in animals,[119] and causes cancer in humans.[120] Aflatoxin B_1 is metabolized by a cytochrome P450 to the corresponding *exo*-8,9-epoxide (**8.35**),[121] which becomes covalently bound to cellular DNA (Scheme 8.14). With the use of radiolabeled **8.34**,[122] and model studies,[123] it was shown that a covalent bond is formed between C8 of aflatoxin B_1 and the N7 of a guanine residue in DNA (**8.36**). The precursor, epoxide **8.35**, has been synthesized and shown to undergo the expected reaction with a guanine-containing oligonucleotide. Mechanistic studies show that only the exo epoxide reacts with DNA, giving the trans adduct.[124]

8.4.2.1.3. Oxidations of Carbons Adjacent to sp^2 Centers

Carbons adjacent to aromatic, olefinic, and carbonyl or imine groups undergo metabolic oxidations. The oxidation mechanism is not clearly understood, but because the ease of oxidation parallels the C–H bond dissociation energies, it is likely that a typical cytochrome P450 oxidation is responsible. Examples of benzylic oxidation are the metabolism of the antidepressant drug amitriptyline (**8.37**, R=R′=H, Elavil), which is oxidized to **8.37** (R=H, R′=OH and R=OH, R′=H),[125] and the β_1-adrenoreceptor antagonist (β-blocker) antihypertensive drug metoprolol (**8.38**, Toprol-XL).[126] In the case of metoprolol, both enantiomers (1′*R*- and 1′*S*) of the

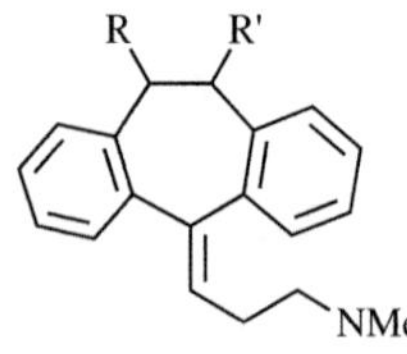

Amitriptyline (R = R' = H)
8.37

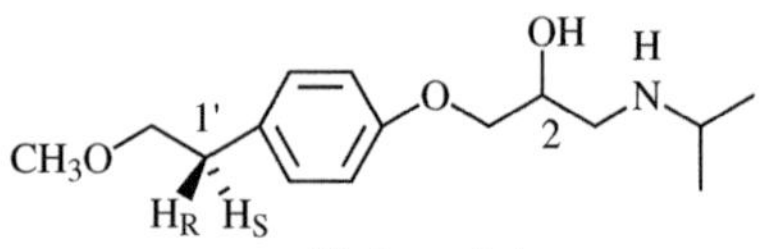

Metaprolol
8.38

hydroxylated drug are formed, but (1′*R*)-hydroxylation occurs to a greater extent. Furthermore, the ratio of the (1'*R*)- to (1'*S*)-isomers depends on the stereochemistry at the 2-position. (2*R*)-Metoprolol gives a ratio of (1'*R*,2*R*)/(1'*S*,2*R*) metabolites of 9.4, whereas (2*S*)-metoprolol gives a ratio of (1'*R*,2*S*)/(1'*S*,2*S*) of 26. Therefore, the stereochemistry in the methoxyethyl side chain is influenced by the stereochemistry in the *para* side chain. Hydroxylations generally are highly stereoselective, if not stereospecific. The stereochemistry at one part of the molecule may influence how the molecule binds in the active site of cytochrome P450, which will influence what part of the molecule is closest to the heme iron-oxo species for reaction, in this case, hydrogen atom abstraction.

The cytochrome P450-catalyzed metabolism of the antiarrhythmic drug quinidine (**8.39**, R=H, Quinidex) leads to allylic oxidation (**8.39**, R=OH).[127] The psychoactive constituent of marijuana (Δ^9-tetrahydrocannabinol) (**8.40**, R=R′=H), is extensively metabolized to **8.40** (R=H, R′=OH) and all stereoisomers of **8.40** (R=OH, R′=H; and R= R′=OH).[128]

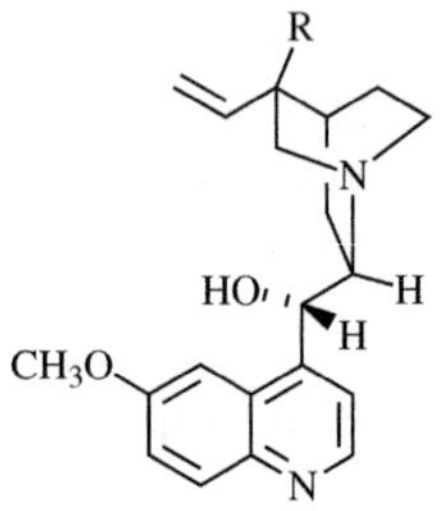

Quinidine (R = H)
8.39

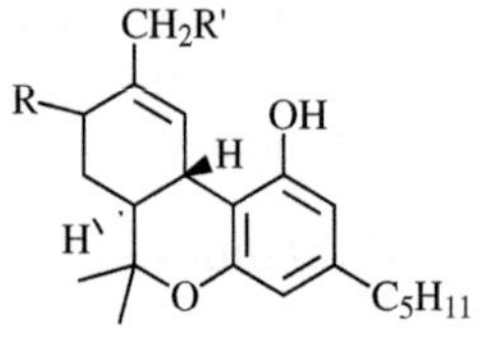

Δ^9-Tetrahydrocannabinol (R = R' = H)
8.40

The sedative–hypnotic, (+)-glutethimide (**8.41**, R=R′=H, Doriden) is converted to 5-hydroxyglutethimide

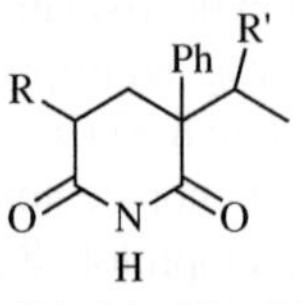

Glutethimide (R = R' = H)
8.41

(**8.41**, R=OH, R′=H).[129] This metabolite is pharmacologically active and may contribute to the comatose state of individuals who have taken toxic overdoses of the parent drug. The (−)-isomer of glutethimide is enantioselectively hydroxylated at the ethyl group to give **8.41** (R=H, R′=OH). The difference in hydroxylation sites for the two enantiomers may reflect the different orientations of binding of each enantiomer in one or more isozymes of cytochrome P450.

8.4.2.1.4. Oxidation at Aliphatic and Alicyclic Carbon Atoms

Metabolic oxidation at the terminal methyl group of an aliphatic side chain is referred to as *ω-oxidation*, and oxidation at the penultimate carbon is *ω*–1 *oxidation*. Because *ω*-oxidation is a chemically unfavorable process (a primary radical/cation would be formed), the active site of the enzyme must be favorably disposed for this particular regiochemistry to proceed. In the case described above of (−)-glutethimide (**8.41**, R=R′=H), *ω*–1 oxidation occurred. The anticonvulsant drug sodium valproate (**8.42**, Depakene) undergoes both *ω*- and *ω*–1 oxidations.[130]

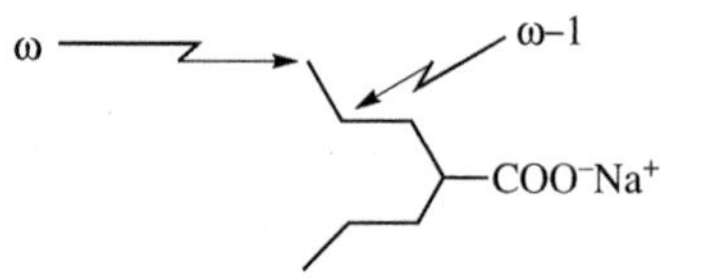

Sodium valproate
8.42

The coronary vasodilator perhexiline (**8.43**, R=H, Pexid) is metabolized to the alicyclic alcohol **8.43** (R=OH).[131]

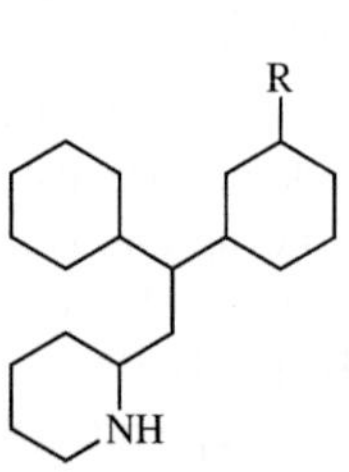

Perhexiline (R = H)
8.43

If a methyl group is beta to a carbonyl, hydroxylation may result in *C*-demethylation by a retro-aldol reaction.[132] For example, the *para*-amino metabolite of flutamide (**8.44**, Eulexin) undergoes a double *C*-demethylation to **8.47** presumably via the *C*-hydroxylated intermediate (**8.45**) to give **8.46** (Scheme 8.15), which may occur with loss of

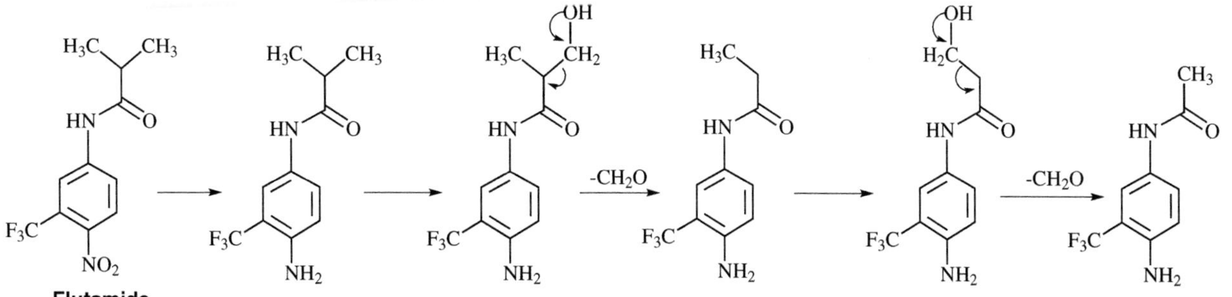

Flutamide **8.44** → → **8.45** → ($-CH_2O$) **8.46** → → ($-CH_2O$) **8.47**

SCHEME 8.15 *C*-demethylation of a flutamide metabolite

TABLE 8.3 Oxidative Reactions of Primary Amines and Amides

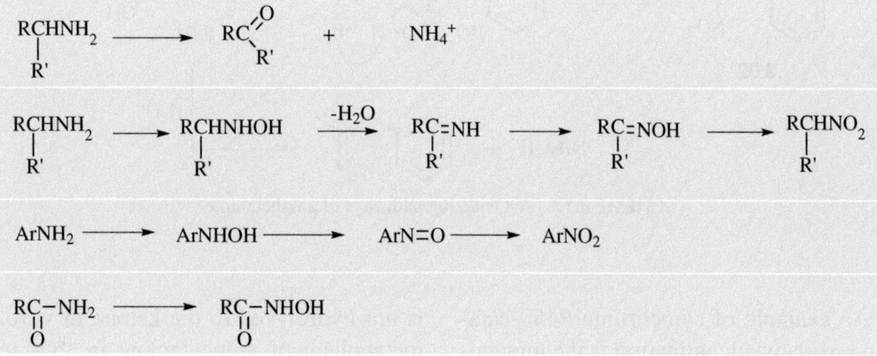

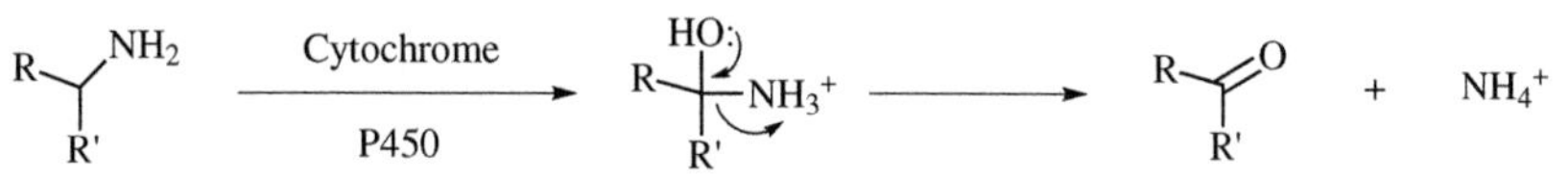

Reaction
$RCH(R')NH_2 \rightarrow RC(=O)R' + NH_4^+$
$RCH(R')NH_2 \rightarrow RCH(R')NHOH \xrightarrow{-H_2O} RC(R')=NH \rightarrow RC(R')=NOH \rightarrow RCH(R')NO_2$
$ArNH_2 \rightarrow ArNHOH \rightarrow ArN=O \rightarrow ArNO_2$
$RC(=O)-NH_2 \rightarrow RC(=O)-NHOH$

$RCH(R')NH_2 \xrightarrow{\text{Cytochrome P450}} RC(R')(OH)NH_3^+$ (**8.48**) $\rightarrow RC(=O)R' + NH_4^+$

SCHEME 8.16 Oxidative deamination of primary amines

formaldehyde. Further *C*-demethylation gives the isolated metabolite (**8.47**).

Often a *tert*-butyl group is added to protect a site, such as an amide, ester, or ether, from metabolism, but oxidation of one of the *tert*-butyl methyl groups to a hydroxymethyl group is commonly observed.[133] Trifluoromethylcyclopropyl (CF_3-C_3H_4-) was found to act as a bioisostere with significantly enhanced metabolic stability.[134]

8.4.2.1.5. Oxidations of Carbon–Nitrogen Systems

The metabolism of organic nitrogen compounds is very complex. Two general classes of oxidation reactions will be considered for primary, secondary, and tertiary amines, respectively, namely, carbon- and nitrogen-oxidation reactions that lead to C–N bond cleavage and *N*-oxidation reactions that do not lead to C–N bond cleavage.

Primary amines and/or amides are metabolized by the oxidation reactions shown in Table 8.3. Primary aliphatic and arylalkyl amines having at least one α-carbon-hydrogen bond undergo cytochrome P450-catalyzed hydroxylation at the α-carbon (see Scheme 4.36 in Chapter 4 for the general mechanism of carbon hydroxylation) to give the carbinolamine (**8.48**), which generally breaks down to the aldehyde or ketone and ammonia (Scheme 8.16). This process of oxidative cleavage of ammonia from the primary amine is known as *oxidative deamination*. As is predicted by this mechanism, primary aromatic amines and α,α-disubstituted aliphatic amines do not undergo oxidative deamination. A variety of endogenous arylalkyl amines, such as the neurotransmitters dopamine, norepinephrine, and serotonin, are oxidized by monoamine oxidase by an electron transfer mechanism (see Chapter 4, Scheme 4.32 for the mechanism). This enzyme also may be involved in drug metabolism of arylalkyl amines with

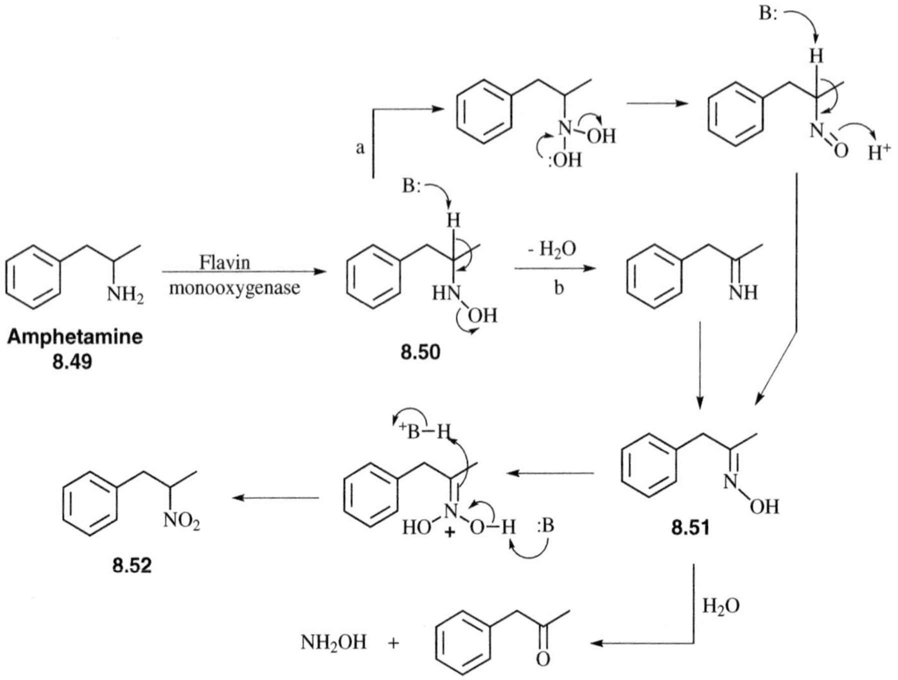

SCHEME 8.17 *N*-Oxidation pathways of amphetamine

no α-substituents. An example of cytochrome P450-catalyzed primary amine oxidative deamination is the metabolism of amphetamine (**8.49**) to give 1-phenyl-2-propanone and ammonia.[135]

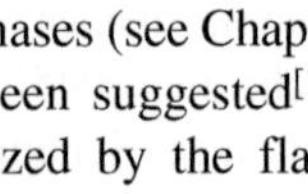

Another important oxidative pathway for primary amines is *N-oxidation* (hydroxylation of the nitrogen atom) to the corresponding hydroxylamine,[136] usually catalyzed by flavin monooxygenases (see Chapter 4, Scheme 4.34 for mechanism). It has been suggested[137] that basic amines (pK_a 8–11) are oxidized by the flavoenzymes; nonbasic nitrogen-containing compounds, such as amides, are oxidized by cytochrome P450 enzymes; and compounds with intermediate basicity, such as aromatic amines, are oxidized by both enzymes. Cytochrome P450 enzymes, however, tend not to catalyze *N*-oxidation reactions when there are α-protons available. Amphetamine undergoes *N*-oxidation to the corresponding hydroxylamine (**8.50**, Scheme 8.17), the oxime (**8.51**), and the nitro compound (**8.52**). It could be argued that the 1-phenyl-2-propanone stated above as the product of oxidative deamination could be derived from hydrolysis of the oxime **8.51** (see Scheme 8.17). However, *N*-hydroxyamphetamine (**8.50**) metabolism to the oxime occurs only to a small extent, and the oxime is not hydrolyzed to the ketone in vitro.[138] Also, in vitro metabolism of amphetamine in $^{18}O_2$ leads to substantial enrichment of 1-phenyl-2-propanone with ^{18}O, indicating the relevance of a cytochrome P450-catalyzed oxidative deamination pathway to the ketone rather than a pathway involving hydrolysis of the oxime.

The amphetamine oxime is derived both from hydroxylation of **8.50** (pathway a) and from dehydration of **8.50** (pathway b) followed by hydroxylation of the imine (see Scheme 8.17). The imine could also be derived from dehydration of the carbinolamine (Scheme 8.18). Hydrolysis of the imine could lead to 1-phenyl-2-propanone.

Like primary amines, secondary amines are metabolized through one of two mechanisms, starting either with hydroxylation of a carbon atom adjacent to the nitrogen or with *N*-oxidation (see examples in Table 8.4). When a small alkyl substituent of a secondary amine is cleaved from the parent amine to give a primary amine, the process is known as *oxidative N-dealkylation* (Scheme 8.19). Although the difference is somewhat a matter of semantics, when a small amino group is cleaved from the parent molecule, leaving an aldehyde or ketone, the process is known as *oxidative deamination*. It is not clear if the precise mechanisms of these two processes are the same, but from a reaction standpoint, they are the same reaction except viewed from whether the nitrogen atom remains with the bulk of the molecule oxidative (*N*-dealkylation) or is cleaved from the bulk of the molecule (oxidative deamination).

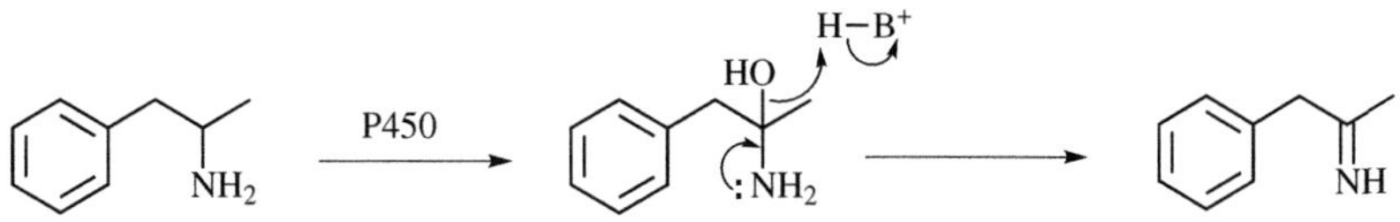

SCHEME 8.18 Amphetamine imine formation via the carbinolamine

TABLE 8.4 Oxidative Reactions of Secondary Amines and Amides

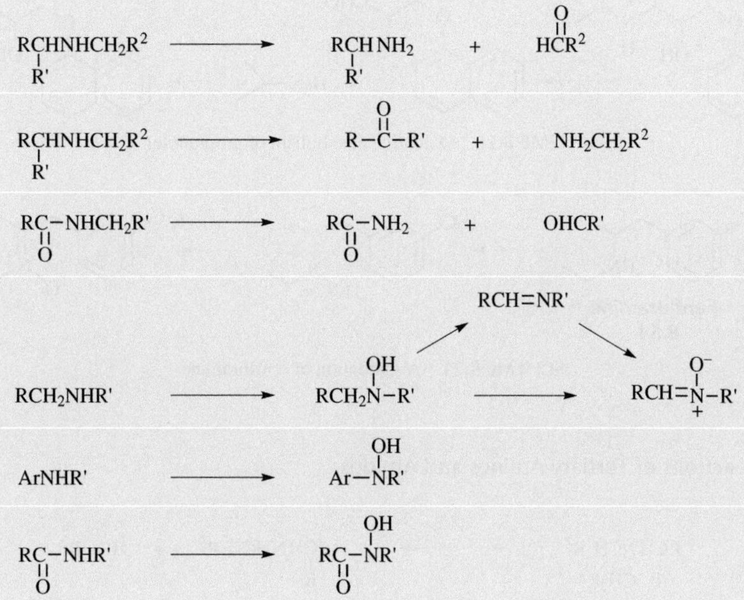

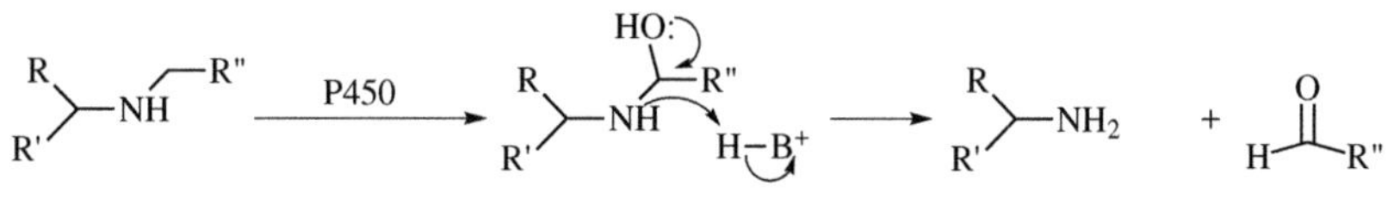

Substrate	Product(s)
$RCH(R')NHCH_2R^2$ →	$RCH(R')NH_2$ + $HC(=O)R^2$
$RCH(R')NHCH_2R^2$ →	$R-C(=O)-R'$ + $NH_2CH_2R^2$
$RC(=O)-NHCH_2R'$ →	$RC(=O)-NH_2$ + $OHCR'$
RCH_2NHR' →	$RCH_2N(OH)-R'$ → $RCH=NR'$ → $RCH=N^+(O^-)-R'$
$ArNHR'$ →	$Ar-N(OH)R'$
$RC(=O)-NHR'$ →	$RC(=O)-N(OH)R'$

SCHEME 8.19 Oxidative *N*-dealkylation of secondary amines

The secondary amine β-blocker propranolol (**8.53**, Inderal) is metabolized both by oxidative *N*-dealkylation (pathway a, Scheme 8.20) and by oxidative deamination (pathway b, Scheme 8.20).[139] Aldehydic metabolites often are oxidized by soluble aldehyde dehydrogenases to the corresponding carboxylic acids (see pathway b).

N-Oxidation of secondary amines leads to a variety of *N*-oxygenated products.[140] Secondary hydroxylamine formation is common, but these metabolites are susceptible to further oxidation to give nitrones. An example of this is in the metabolism of the anorectic drug fenfluramine (**8.54**, Scheme 8.21, Pondimin; withdrawn from the market because of heart valve disease and pulmonary hypertension).[141]

Tertiary aliphatic amines generally undergo two metabolic reactions: oxidative *N*-dealkylation and *N*-oxidation reactions (Table 8.5). Oxidative *N*-dealkylation of tertiary amines (see Scheme 8.19 for secondary amines) leads to the formation of secondary amines. Because of the basicity of tertiary amines relative to secondary amines, oxidative *N*-dealkylation of tertiary amines generally occurs more rapidly than that of secondary amines.[142] For example, the primary amine antihypertensive drug (calcium ion channel blocker) amlodipine (**8.55**, $R=NH_2$, Norvasc) is much more stable metabolically (clearance is 11 mL/min kg) than the corresponding secondary (12.5 mL min^{-1} kg^{-1}) or tertiary (25 mL min^{-1} kg^{-1}) analogs (**8.55**, $R=NHCH_3$ or $N(CH_3)_2$, respectively).[143] The low rate of metabolism may not only be a result of the lower basicity of the primary amino functionality but also because of the increased polarity.[144]

The tricyclic antidepressant drug imipramine (**8.56**, $R=CH_3$, Tofranil), for example, is metabolized to the

Propranolol
8.53

SCHEME 8.20 Oxidative metabolism of propranolol

Fenfluramine
8.54

SCHEME 8.21 *N*-Oxidation of fenfluramine

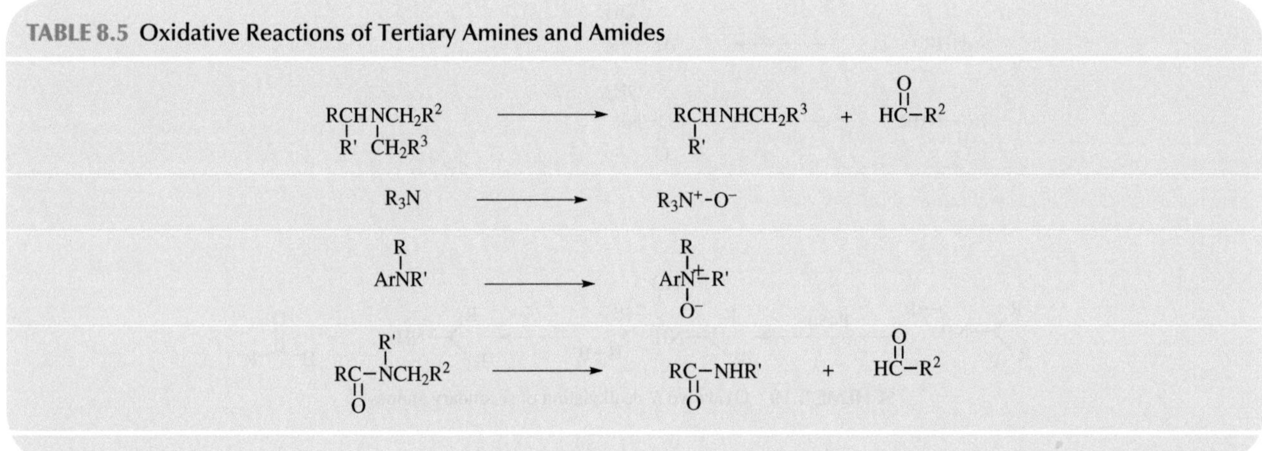

TABLE 8.5 Oxidative Reactions of Tertiary Amines and Amides

Substrate		Products
$RCH(R')N(CH_2R^3)CH_2R^2$	→	$RCH(R')NHCH_2R^3$ + $HC(=O)-R^2$
R_3N	→	$R_3N^+-O^-$
$ArN(R)R'$	→	$ArN^+(R)(O^-)-R'$
$RC(=O)-N(R')CH_2R^2$	→	$RC(=O)-NHR'$ + $HC(=O)-R^2$

corresponding secondary amine, desmethylimipramine (desipramine; **8.56**, R=H, Norpramin), which is also an active antidepressant agent.[145] Very little oxidative

Amlodipine
8.55 (R = NH_2)

Imipramine (R = CH_3)
8.56

N-demethylation of **8.56** (R=H) occurs. Enantioselective oxidative *N*-dealkylation can occur with chiral tertiary amines. The (2*S*,3*R*)-(+)-enantiomer of propoxyphene (**8.57**, Darvon), an analgesic drug, is *N*-demethylated more slowly than the nonanalgesic (−)-enantiomer.[146]

Propoxyphene
8.57

The antiparkinsonian drug selegiline (*R*-(−)-deprenyl, **8.58**, Scheme 8.22; Eldepryl) is a potent inactivator of monoamine oxidase B (MAO B) (see Chapter 5, Section 5.3.3.3.4). It is metabolized by various isozymes

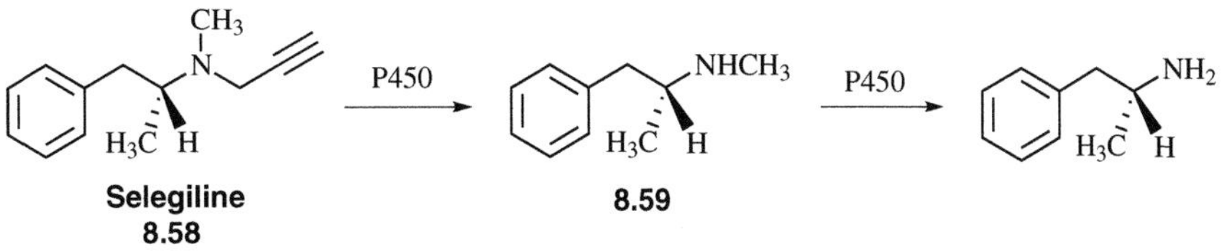

SCHEME 8.22 Metabolism of selegiline (deprenyl)

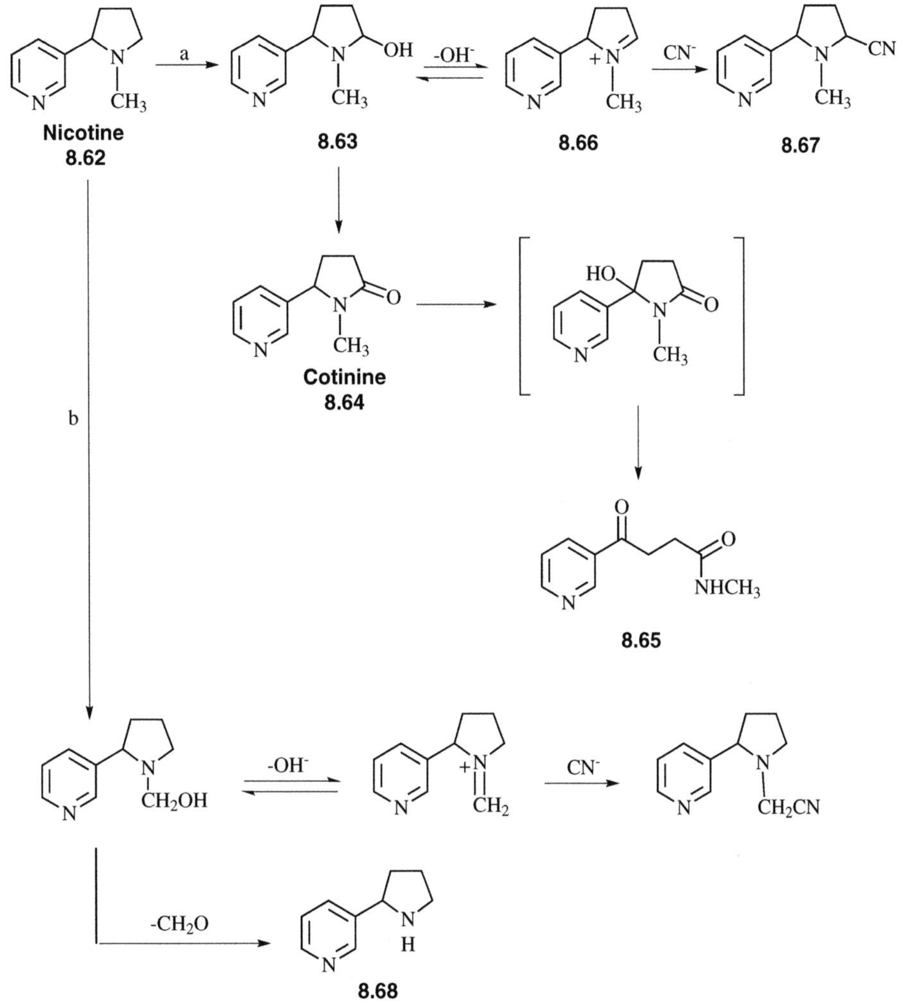

SCHEME 8.23 Oxidative metabolism of nicotine leading to C–N bond cleavage.

of cytochrome P450 to *R*-(−)-methamphetamine (**8.59**) and to *R*-(−)-amphetamine, which have only a weak central nervous system (CNS) stimulant side effect.[147] The *S*-(+)-isomer is only a weak inhibitor of MAO B, but is metabolized by cytochrome P450 to *S*-(+)-methamphetamine, which is converted to *S*-(+)-amphetamine, both of which would produce an undesirable CNS stimulant side effect.[148] Because of this, only the (−)-enantiomer is used in the treatment of Parkinson's disease. Nonetheless, it has been found that *R*-(−)-amphetamine is sufficiently active to cause sympathomimetic effects; the newer drug, rasagiline (**8.60**, Azilect) is metabolized to (*R*)-(+)-1-aminoindan (**8.61**), which is devoid of sympathomimetic effects.[149]

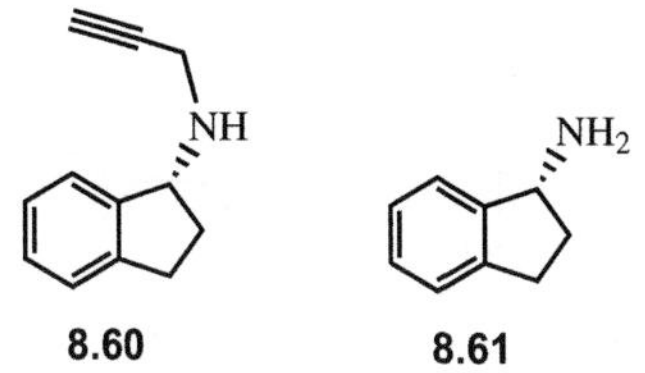

When an alicyclic tertiary amine undergoes oxidation of the alicyclic carbon atoms attached to the nitrogen, a variety of oxidation products can result. Nicotine (**8.62**, Scheme 8.23) is α-hydroxylated both in the pyrrolidine ring (pathway a) and at the methyl carbon (pathway b). Carbinolamine **8.63** is oxidized further to the major metabolite,

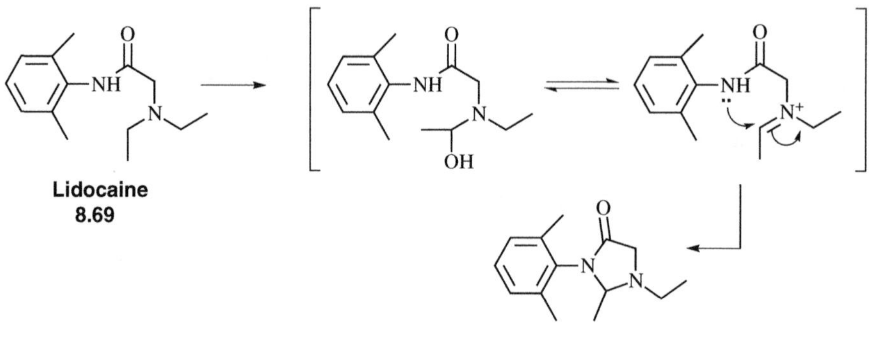

Lidocaine
8.69

8.70

SCHEME 8.24 Metabolism of lidocaine

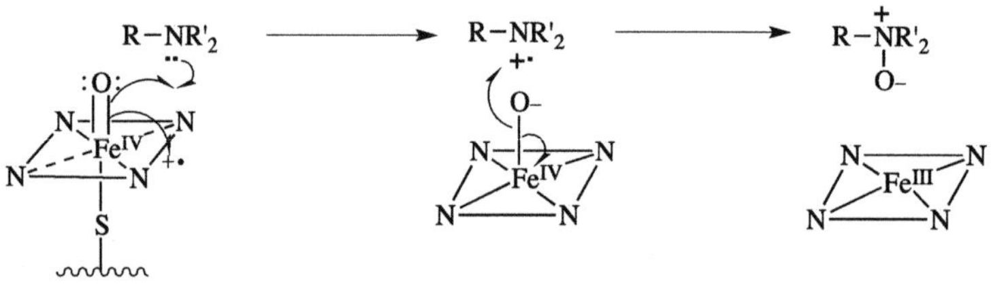

SCHEME 8.25 Mechanism of cytochrome P450-catalyzed *N*-oxidation of tertiary aromatic amines

cotinine (**8.64**, Scotine). Cotinine also can undergo hydroxylation, leading to a minor metabolite, γ-(3-pyridyl)-γ-oxo-*N*-methylbutyramide (**8.65**).[150] Carbinolamine **8.63** is in equilibrium with the iminium ion **8.66**, which has been trapped with cyanide ion in vitro (to confirm its formation), giving **8.67**.[151] Oxidative *N*-demethylation of **8.62** to **8.68** is also observed as a metabolic pathway for nicotine.

Further evidence for metabolic iminium ion intermediates generated during tertiary amine oxidation is the isolation of imidazolidinone **8.70** from the metabolism of the local anesthetic lidocaine[152] (**8.69**, Scheme 8.24, Xylocaine).

N-Oxidation of tertiary amines gives chemically stable tertiary amine *N*-oxides, which do not undergo further oxidation unlike *N*-oxidation of primary and secondary amines. The antihypertensive drug guanethidine (**8.71**, Ismelin) is oxidized at the tertiary cyclic amine nitrogen atom to give the *N*-oxide **8.72**.[153] The pyrrolidine nitrogen of nicotine (**8.62**, Scheme 8.23) is also *N*-oxidized to (1*R*,2*S*)- and (1*S*,2*S*)-nicotine 1-*N*-oxide. One of the major metabolites of the antihistamine cyproheptadine (**8.73**, Periactin) in dogs is the α-*N*-oxide (**8.74**); no β-*N*-oxide was detected.[154]

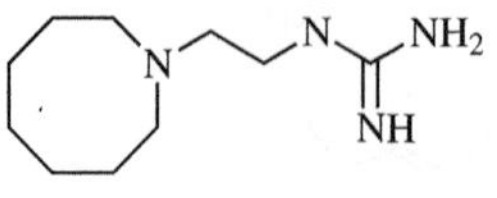

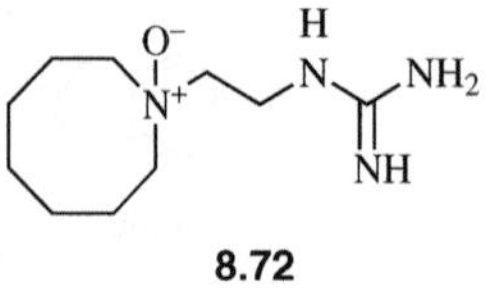

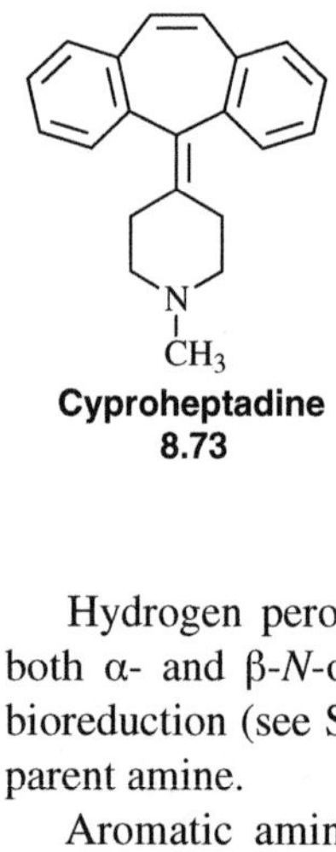

Guanethidine
8.71

8.72

Cyproheptadine
8.73

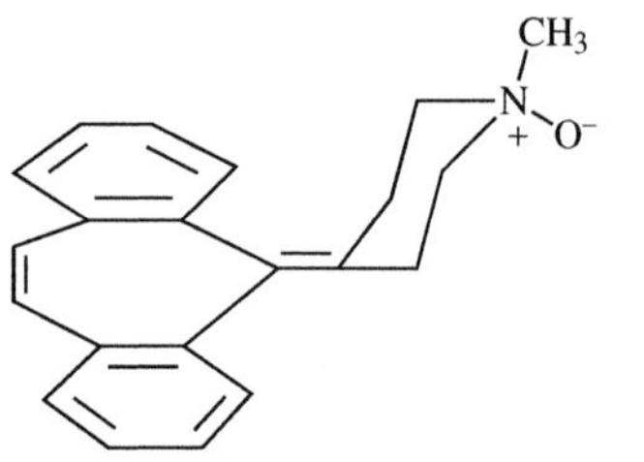

8.74

Hydrogen peroxide oxidation of **8.73**, however, gives both α- and β-*N*-oxides.[155] *N*-Oxides are susceptible to bioreduction (see Section 8.4.2.2.5), which regenerates the parent amine.

Aromatic amine oxidation is similar to that for aliphatic amines. *N*-Oxidation of aromatic amines appears to be an important process related to the carcinogenic and cytotoxic properties of aromatic amines.[156] There are two enzyme systems responsible for *N*-oxidation of tertiary aromatic amines, flavoprotein monooxygenase and cytochrome P450. The mechanism for the flavin-dependent reaction was discussed in Chapter 4 (Section 4.3.3; see Scheme 4.34). Cytochrome P450-catalyzed *N*-oxidation occurs by an electron transfer reaction followed by oxygen rebound (Scheme 8.25). *N*-Oxidation by cytochrome P450 appears to occur only if there are no α-hydrogens available for abstraction, if the iminium radical is stabilized by electron donation, or if Bredt's rule prevents α-hydrogen abstraction.[157]

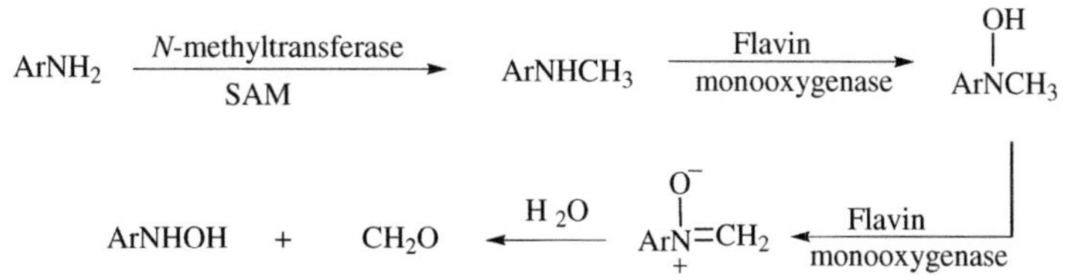

SCHEME 8.26 Possible mechanism for *N*-oxidation of primary arylamines

The mechanism for oxidation of primary arylamines may be different from that of secondary and tertiary arylamines. Most primary arylamines are not substrates of flavin monooxygenases; *N*-oxygenation of secondary and tertiary arylamines occurs readily. Ziegler et al.[158] showed that primary arylamines could be *N*-methylated by *S*-adenosylmethionine (SAM)-dependent *amine N-methyltransferases* and that the secondary arylamines produced were substrates for flavin monooxygenases (Scheme 8.26). The secondary hydroxylamine products of this *N*-oxygenation reaction are then further oxidized to nitrones, which, on hydrolysis, give the primary hydroxylamines and formaldehyde.

N-Demethylation of tertiary aromatic amines is believed to proceed by two mechanisms[159] (Scheme 8.27), one that involves intermediate carbinolamine formation by cytochrome P450 oxidation (pathway a) and one from flavin monooxygenation that produces the *N*-oxide, which rearranges to the carbinolamine (pathway b). Evidence to support carbinolamine formation is the isolation of the ^{18}O-labeled carbinolamine (**8.75**, R=OH) during cytochrome P450-catalyzed oxidation of *N*-methylcarbazole (**8.75**, R=H) under $^{18}O_2$.[160] Intrinsic isotope effects on the cytochrome P450-catalyzed oxidation of tertiary aromatic amines were found to be small.[161] Intrinsic isotope effects associated with deprotonation of aminium radicals are of low magnitude ($k_H/k_D<3.6$), but those associated with direct hydrogen atom abstraction are large; this suggests that the mechanism for carbinolamine formation involves electron transfer and proton transfer (Scheme 8.28) rather than direct hydrogen atom abstraction.

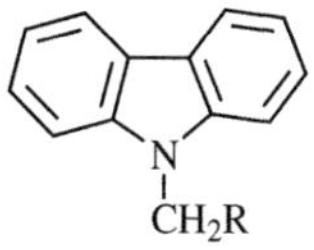

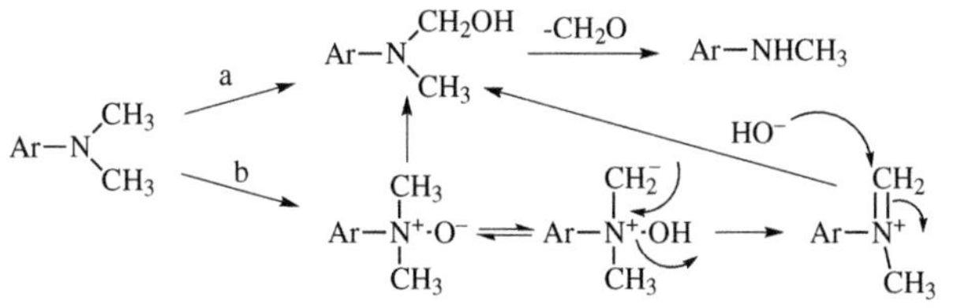

SCHEME 8.27 Two pathways to *N*-demethylation of tertiary aromatic amines

8.75

N-Oxidation of primary and secondary aromatic amines can lead to the generation of reactive electrophilic species that form covalent bonds to cellular macromolecules (Scheme 8.29).[162] The X^+ in Scheme 8.29 represents acetylation or sulfation of the hydroxylamine to give a good leaving group (XO^-) (see Sections 8.4.3.3 and 8.4.3.7). Attachment of aromatic amines to proteins, DNA, and RNA is known.[163]

Although C–N bond cleavage of nitrogen-containing heteroaromatic compounds does not occur, *N*-oxidation to the aromatic *N*-oxide is prevalent. As noted in Chapter 5 (Section 5.2.5.2) for the discovery of the anti-human immunodeficiency virus (HIV) agent ritonavir, compounds with a pyridine substituent underwent extensive metabolic *N*-oxide formation.[164]

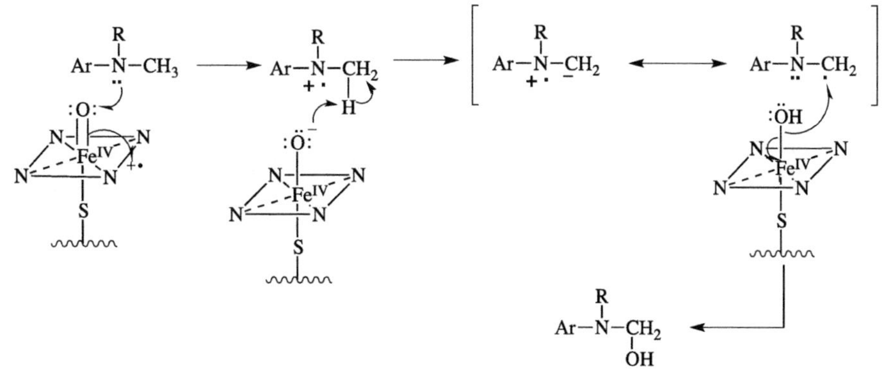

SCHEME 8.28 Mechanism of carbinolamine formation during oxidation of tertiary aromatic amines

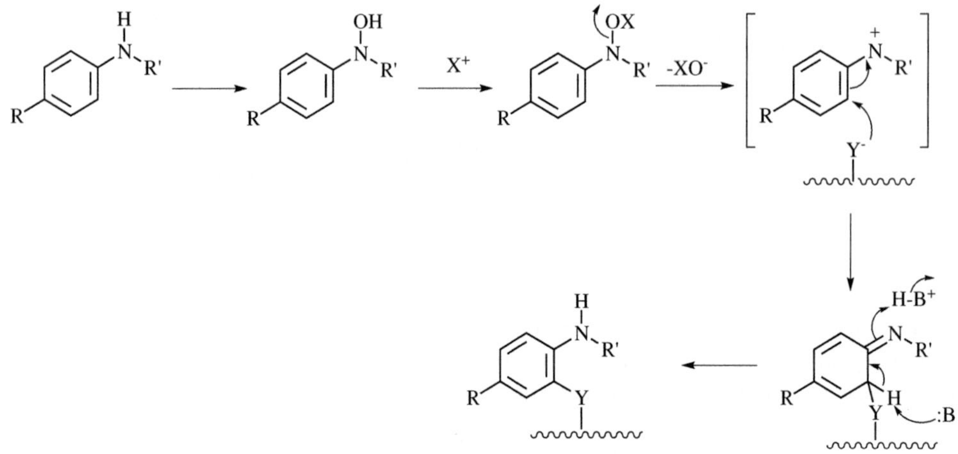

SCHEME 8.29 Metabolic activation of primary and secondary aromatic amines

8.78

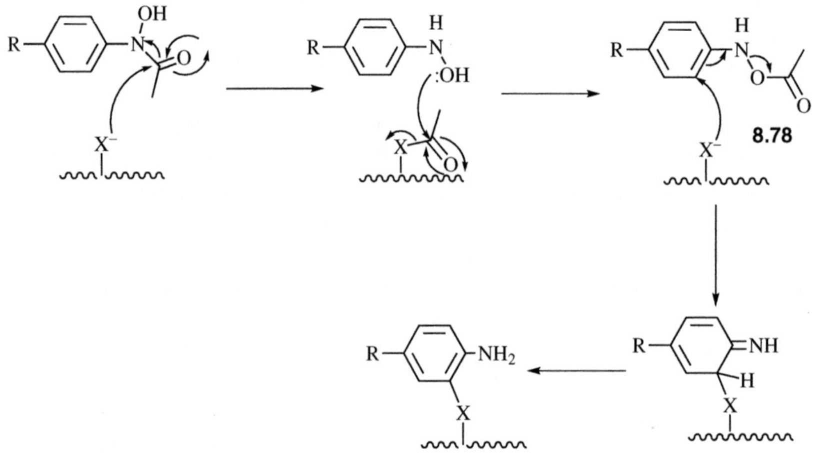

SCHEME 8.30 Arylhydroxamic acid *N,O*-acyltransferase-catalyzed activation of *N*-hydroxy-2-acetylaminoarenes

Amides also are metabolized both by oxidative *N*-dealkylation and *N*-oxidation (see Tables 8.3–8.5). The sedative diazepam (**8.76**, R=CH_3, Valium) undergoes extensive oxidative *N*-demethylation to **8.76** (R=H).

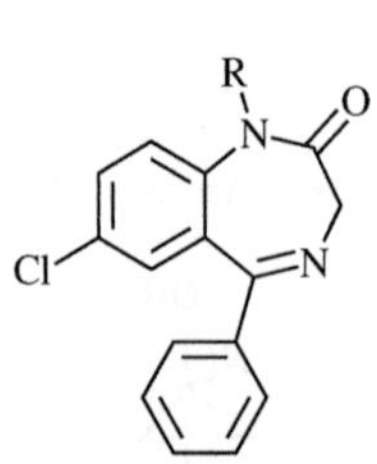

Diazepam (R = CH_3)
8.76

As in the case of primary and secondary aromatic amines, which are activated to cytotoxic and carcinogenic metabolites, *N*-oxidation of primary and secondary aromatic amides also leads to electrophilic intermediates. For example, the carcinogenic agent 2-acetylaminofluorene (**8.77**, R=H) undergoes cytochrome P450-catalyzed oxidation to the

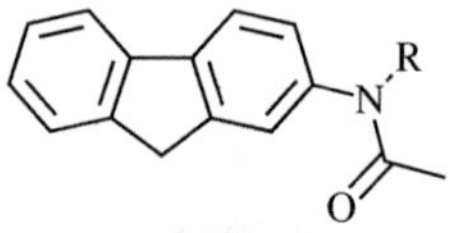

8.77

N-hydroxy analog (**8.77**, R=OH). Activation of the hydroxyl group as in Scheme 8.29 leads to an electrophilic species capable of undergoing attack by cellular nucleophiles. Also, liver arylhydroxamic acid *N,O*-acyltransferase has been shown to catalyze the rearrangement of **8.77** (R=OH) to the corresponding *O*-acetyl hydroxylamine (**8.78**), which is activated for nucleophilic attack by the acyltransferase or by other cellular macromolecules (Scheme 8.30).[165]

Covalent modification of arylhydroxamic acid *N,O*-acyltransferase prior to release of the activated species (**8.78**) is another example of mechanism-based enzyme inactivation (see Chapter 5, Section 5.3.3). Release of **8.78** into the cell can lead to covalent bond formation with other cellular macromolecules, resulting in cytotoxicity and carcinogenicity.

The analgesic agent acetaminophen (**8.79**, Scheme 8.31, Tylenol) is relatively nontoxic at therapeutic doses, but in large doses, it causes severe liver necrosis.[166] The hepatotoxicity arises from the depletion of liver glutathione levels as a result of the reaction of glutathione with an electrophilic metabolite of acetaminophen. When the glutathione levels drop by 80% or greater, then hepatic macromolecules react with the electrophilic metabolite leading to the observed liver necrosis. The original proposal for the mechanism of formation of the electrophilic metabolite (**8.80**) involves *N*-oxidation of the amide (Scheme 8.31, pathway a).[167] The X^- in Scheme 8.31 represents either glutathione or a cellular macromolecular nucleophile (such as a protein amino acid residue). No evidence, however, for the formation of *N*-hydroxyacetaminophen could be found.[168] Another mechanism for the formation of the electrophilic metabolite (**8.80**) involves acetaminophen epoxidation (Scheme 8.31, pathway b). If this were correct, incubation of acetaminophen in the presence of $^{18}O_2$ should result in the incorporation of ^{18}O into metabolites; however, none was incorporated.[169] A third mechanistic possibility for the formation of **8.80** is a hydrogen atom abstraction mechanism via an acetaminophen radical (**8.81a** or **8.81b**, Scheme 8.32).[170] Isozymes of cytochrome P450 were shown to be responsible for the conversion of acetaminophen to electrophile **8.80**.[171] Because ethanol induces an isozyme of cytochrome P450 that could be responsible for the formation of **8.81**,[172] the hepatotoxicity of acetaminophen in ethanol-fed animals was compared with that in normal animals.[173] It was found that the alcoholic animals had increased hepatotoxicity and that radical scavengers protected the animals. Furthermore, there is an increase in acetaminophen hepatotoxicity in human alcoholics.[174] These results support a mechanism that involves cytochrome P450-dependent acetaminophen radical formation followed by second electron transfer (Scheme 8.32) as a mechanism for the generation of the electrophilic metabolite responsible for acetaminophen hepatotoxicity. Because of its reactivity, it was suggested that **8.81** might be responsible for the hepatotoxicity[175]; however, the benzoquinone-imine (**8.80**) has been detected as a metabolite of the oxidation of acetaminophen by purified cytochrome P450[176] and by microsomes, NADPH, and O_2.[177] Furthermore, **8.80** reacts rapidly with glutathione in vitro to give the same conjugate as that found in vivo.

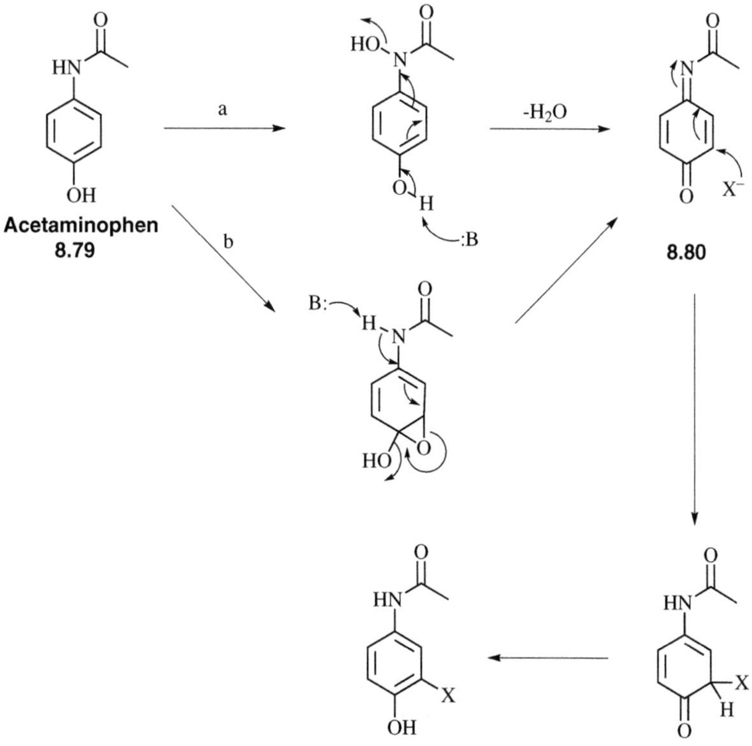

SCHEME 8.31 Initial proposals for bioactivation of acetaminophen

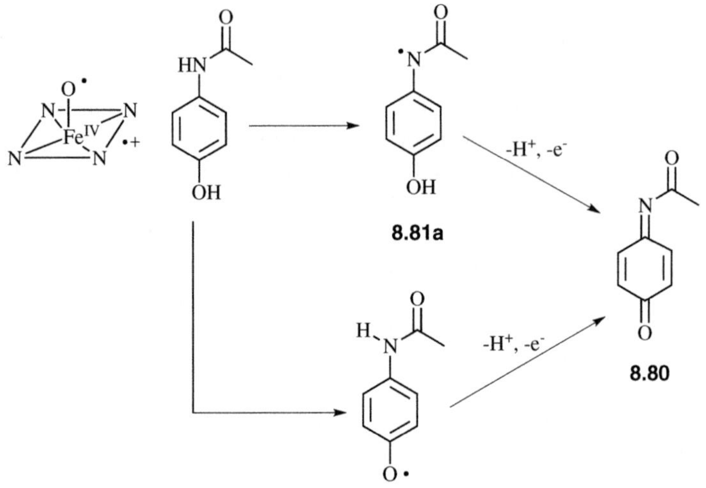

SCHEME 8.32 Bioactivation of acetaminophen via a radical intermediate

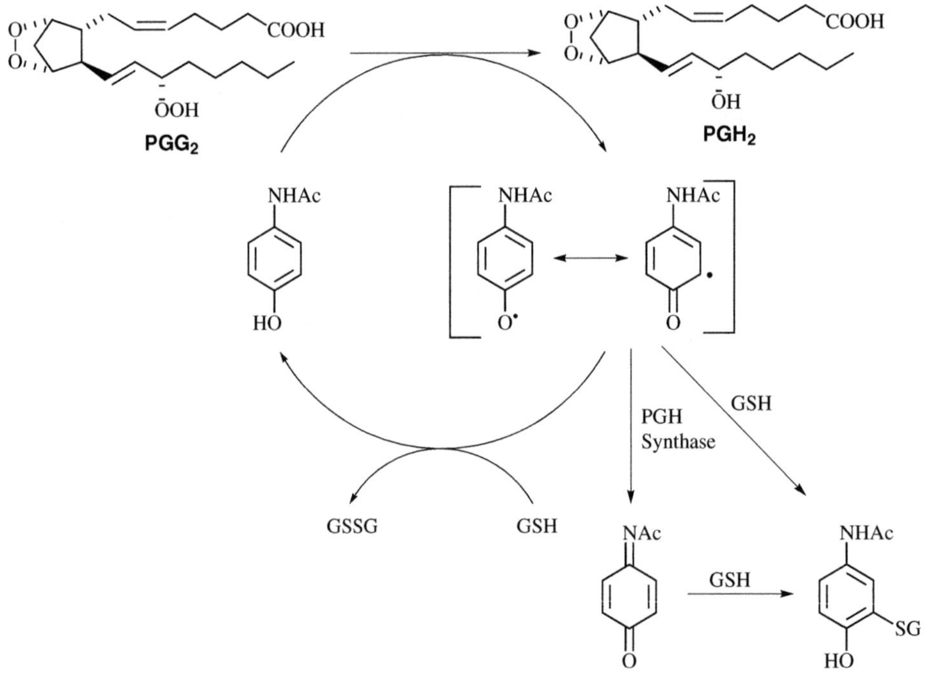

SCHEME 8.33 Proposed bioactivation of acetaminophen by prostaglandin H synthase

High doses of acetaminophen also cause renal damage in humans[178]; however, cytochrome P450 activity is low in the kidneys, and therefore, it may not be the major enzyme responsible for acetaminophen toxicity there. Prostaglandin synthase (also called prostaglandin H synthase, prostaglandin-endoperoxide synthase, or cyclooxygenase), the enzyme that catalyzes the cyclooxygenation of arachidonic acid to prostaglandin G_2 (PGG_2) followed by the reduction of PGG_2 to PGH_2 (see Scheme 5.17 in Chapter 5, Section 5.3.2.2.2), is present in high concentrations in the kidneys and may be important in promoting acetaminophen toxicity in that organ. During the reduction of PGG_2 to PGH_2, prostaglandin H synthase can simultaneously co-oxidize a variety of substrates including certain drugs. Acetaminophen is metabolized to an intermediate that reacts with glutathione to form the glutathione conjugate.[179] By comparison of one- and two-electron reactions in the presence and absence of glutathione, it was found that prostaglandin H synthase can also catalyze the bioactivation of acetaminophen by one- and two-electron mechanisms[180] (Scheme 8.33). Prostaglandin H synthase may be another important drug-metabolizing enzyme in tissues that are rich in this enzyme and have low concentration of cytochrome P450.

8.4.2.1.6. Oxidations of Carbon–Oxygen Systems

Oxidative O-dealkylation is a common biotransformation, which, as in the case of oxidative *N*-dealkylation (see Section 8.4.2.1.5), is catalyzed by microsomal mixed function oxidases. The mechanism appears to be the same as that for oxidative *N*-dealkylation (see Scheme 8.19) and involves hydroxylation on the carbon attached to the oxygen followed by C–O bond cleavage to give the alcohol and the aldehyde or ketone. Although *O*-demethylation is rapid, dealkylation of longer chain *n*-alkyl substituents is generally slow; however, a study of alkoxyethers of 7-hydroxycoumarin showed an increased rate of dealkylation up to propoxyl, then a decrease in rate.[181] Often ω–1 hydroxylation (see Section 8.4.2.1.4) competes with *O*-dealkylation. Also, because of the greater steric bulk of a cyclopropyl group and its greater stability toward hydrogen atom abstraction, addition of this substituent provides compounds with longer plasma half-lives.[182] Lipophilic *N*-cyclopropyl amines, however, can be inactivators of heme and flavin oxidases (for example, see Chapter 5, Section 5.3.3.3.3),[183] which may not be desirable.

A major metabolite of the antiinflammatory drug indomethacin (**8.82**, R=CH_3, Indocin) is the *O*-demethylated

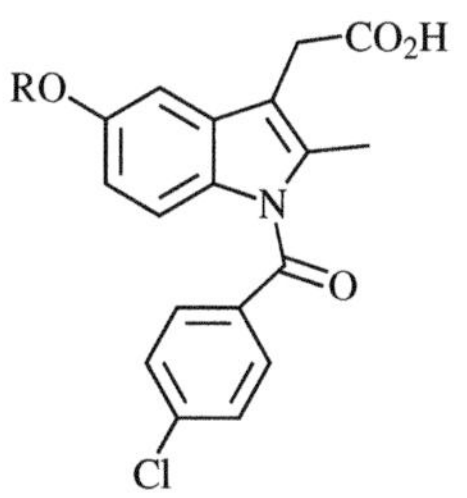

Indomethacin (R = CH_3)
8.82

compound (**8.82**, R=H).[184] Sometimes *O*-dealkylated metabolites are also pharmacologically active, as in the case of the narcotic analgesic codeine (**8.83**, R=CH_3), which is *O*-demethylated to morphine (**8.83**, R=H, Avinza).[185] Nonequivalent methyl groups in drugs can be regioselectively

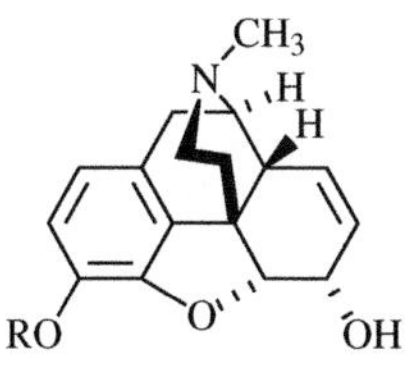

Codeine (R = CH_3)
Morphine (R = H)
8.83

O-demethylated; the blood pressure maintenance drug methoxamine (**8.84**, R=CH_3, Vasoxyl) gives exclusive 2-*O*-demethylation (**8.84**, R=H) in dogs.[186] *O*-Dealkylation can occur with a high degree of stereospecificity in the metabolism of chiral ethers.[187]

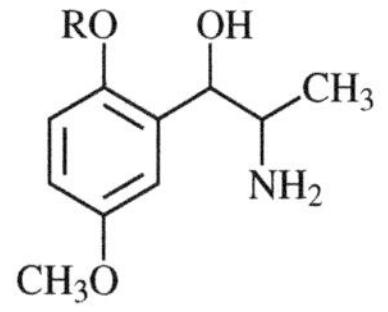

Methoxamine (R = CH_3)
8.84

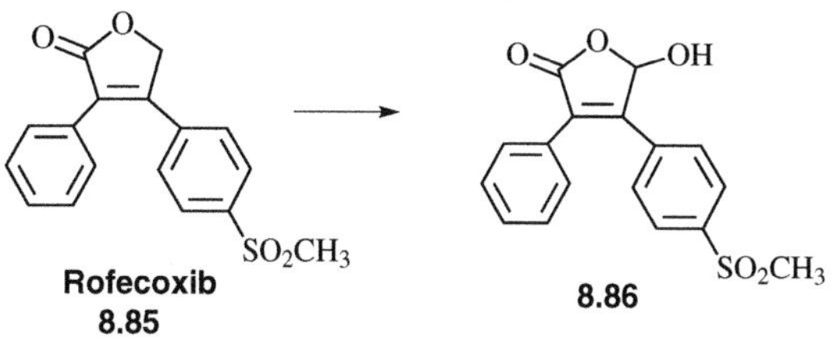

Rofecoxib
8.85

8.86

SCHEME 8.34 Metabolic hydroxylation of rofecoxib

Oxidation on a carbon next to an oxygen leads to a stable hydroxylactone (**8.86**) in the case of the antiarthritis (COX-2 selective inhibitor) drug rofecoxib (**8.85**, Vioxx; withdrawn from market in 2004) (Scheme 8.34).[188]

8.4.2.1.7. Oxidations of Carbon–Sulfur Systems

Fewer drugs contain sulfur than oxygen or nitrogen. The three principal types of biotransformations of carbon–sulfur systems are oxidative *S*-dealkylation, desulfuration, and *S*-oxidation. *Oxidative S-dealkylation* is not nearly as prevalent as the corresponding oxidative *N*- or *O*-dealkylations discussed above. The sedative methitural (**8.87**, R=CH_3) is metabolized by *S*-demethylation to **8.87** (R=H).[189]

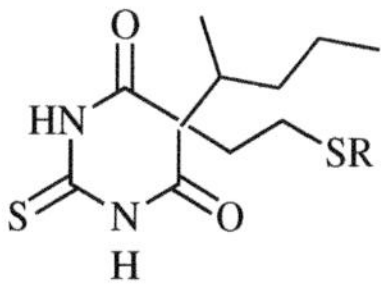

Methitural (R = CH_3)
8.87

Desulfuration is the conversion of a carbon–sulfur double bond (C=S) to a carbon–oxygen double bond (C=O). The anesthetic thiopental (**8.88**, X=S, Pentothal) undergoes desulfuration to pentobarbital (**8.88**, X=O, Nembutal).[190]

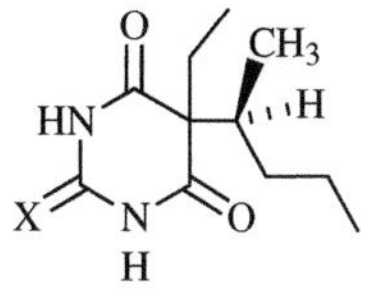

Thiopental (X = S)
Pentobarbital (X = O)
8.88

S-Oxidation of sulfur-containing drugs to the corresponding sulfoxide is a common metabolic transformation,

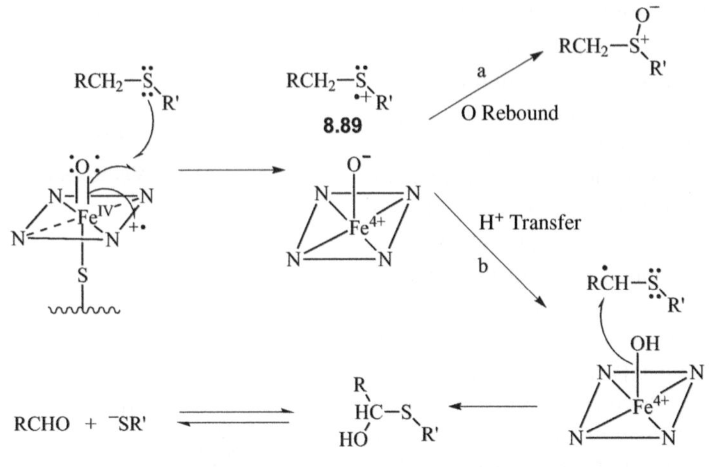

SCHEME 8.35 Cytochrome P450-catalyzed oxidation of sulfides

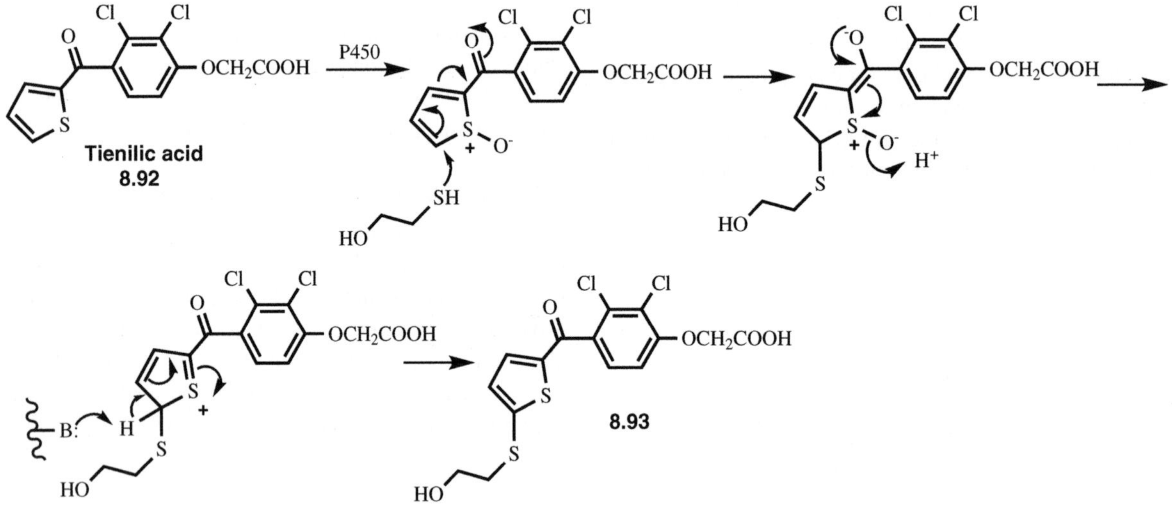

SCHEME 8.36 *S*-Oxidation of tienilic acid

which is catalyzed both by flavin monooxygenase and by cytochrome P450.[191] Flavin monooxygenase produces exclusively the sulfoxide by a mechanism discussed in Chapter 4 (see Scheme 4.34), but cytochrome P450 metabolizes sulfides to *S*-dealkylation products and sulfoxides.[192] The common intermediate for these two metabolites is probably the sulfenium cation radical (**8.89**, Scheme 8.35); the mechanism for *S*-dealkylation may be related to that for cytochrome P450-catalyzed amine oxidations.[193] Both processes occur in the metabolism of the anthelmintic agent albendazole (**8.90**, Albenza).[194] Another example

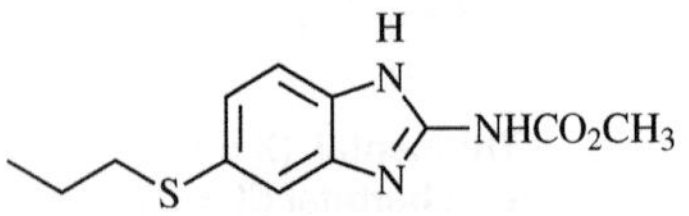

Albendazole
8.90

of *S*-oxidation is the oxidation of the antipsychotic drug thioridazine (**8.91**, R = SCH_3, Mellaril), in which case both sulfur atoms in the drug are metabolized to the corresponding sulfoxides.[195] The metabolite, in which only the methylthio substituent has been converted to the sulfoxide (**8.91**, R = $S(O)CH_3$), is twice as potent as thioridazine, and it also is used as an antipsychotic drug (mesoridazine, Serentil).

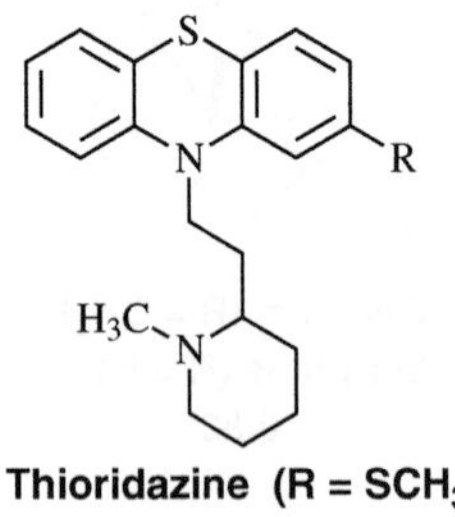

Thioridazine (R = SCH_3)
8.91

Some thiophene[196] and isothiazole[197] derivatives are known to form reactive metabolite (RMs); replacement with an isoxazole or pyrazole can reduce this bioactivation. The diuretic drug tienilic acid (**8.92**, Scheme 8.36, Ticrynafen) is

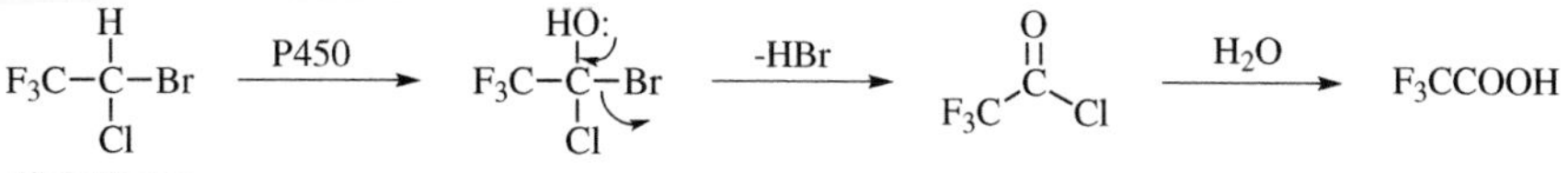

$F_3C-CHClBr$ (Halothane, 8.96) —P450→ $F_3C-C(OH)ClBr$ —-HBr→ $F_3C-C(=O)Cl$ —H_2O→ F_3CCOOH

SCHEME 8.37 Oxidative dehalogenation of halothane

oxidized by cytochrome P450 to give electrophilic metabolites that covalently bind to liver proteins.[198] Incubation of **8.92** with liver microsomes in the presence of 2-mercaptoethanol to trap any electrophilic species generated (and to mimic a cysteine residue in a hepatic protein that might undergo reaction), gave **8.93** as an isolatable product; the mechanism proposed for formation of the RM (Scheme 8.36) involves initial formation of thiophene *S*-oxide.[199]

Further oxidation of sulfoxides to sulfones also occurs; for example, the immunosuppressive drug oxisuran (**8.94**) is metabolized to the corresponding sulfone.[200] The two enantiomeric sulfoxide metabolites of the antiparasitic drug toltrazuril (**8.95**, Toltrazuril) are oxidized further to the corresponding sulfones, but the rate of oxidation of one enantiomer is seven times greater than that of the other.[201] It is not known if the same or different isozymes of cytochrome P450 are responsible for oxidation of the two enantiomers.

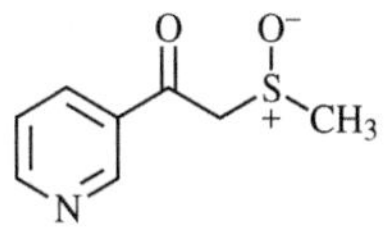

Oxisuran
8.94

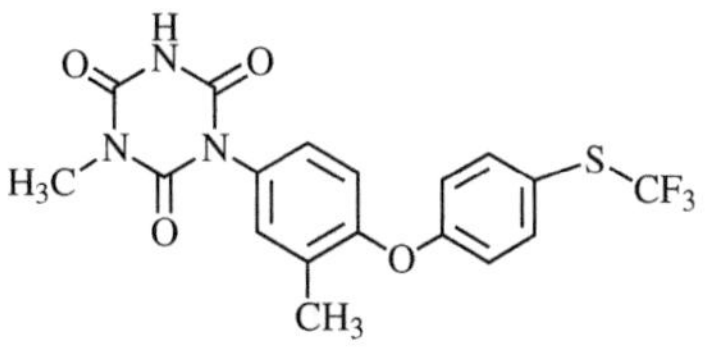

Toltrazuril
8.95

8.4.2.1.8. Other Oxidative Reactions

Oxidative dehalogenation, oxidative aromatization, and oxidation of arenols to quinones are other important drug metabolism pathways that are catalyzed by cytochrome P450 enzymes. *Oxidative dehalogenation* occurs with the volatile anesthetic halothane (**8.96**, Scheme 8.37, Halothane), which is metabolized to trifluoroacetic acid.[202] The acid chloride is responsible for the covalent binding of halothane to liver microsomes.[203]

Oxidative aromatization of the A ring in norgestrel (**8.97**, R = Et, Ovrette), a component of oral contraceptives, by a cytochrome P450 isozyme gives the corresponding phenol (**8.98**, R = Et).[204]

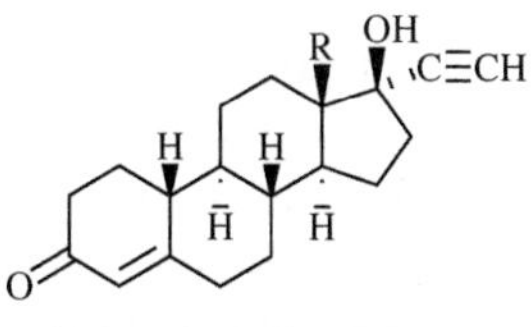

Norgestrel (R = Et)
Norethindrone (R = Me)
8.97

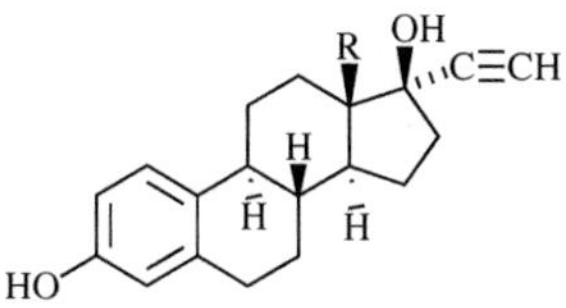

8.98

Catechols may be converted enzymatically to electrophilic *ortho*-quinones. Morphine, for example, is metabolized to a minor extent to its 2,3-catechol (**8.99**), which is oxidized to the *ortho*-quinone (**8.100**).[205]

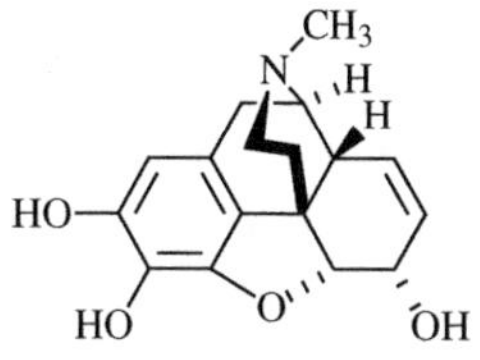

8.99

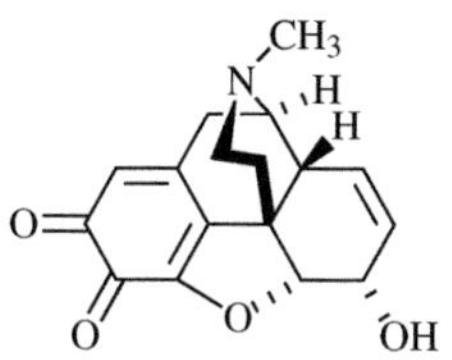

8.100

8.4.2.1.9. Alcohol and Aldehyde Oxidations

The oxidation of alcohols to aldehydes and of aldehydes to carboxylic acids, in general, is catalyzed by alcohol dehydrogenase and aldehyde dehydrogenase, respectively, soluble enzymes that require NAD^+ or $NADP^+$ as the cofactor (see Chapter 4, Section 4.3.2).[206] They are found in highest concentration not only in the liver but also in virtually every other organ. Cytochrome P450 isozymes also catalyze the oxidation of alcohols to aldehydes and aldehydes to carboxylic acids[207]; molybdenum-dependent aldehyde oxidases play a minor role in aldehyde oxidation as well.[208] The reaction catalyzed by alcohol dehydrogenase is the oxidation of primary and secondary alcohols to aldehydes and ketones, respectively, as well as the reverse reaction (Scheme 8.38). As is apparent from this reaction, the equilibrium in solution is pH dependent. The oxidation of alcohols is favored at higher pH (ca. pH10), and the reduction of aldehydes is favored at lower pH (ca. pH 7). On the basis of this equilibrium, at physiological pH, reduction should be favored. However, this is not observed for aldehydes because further oxidation of the aldehyde to the carboxylic acid (catalyzed by aldehyde dehydrogenase) is generally a more rapid process. Aldehydes are usually metabolized to the acid; relatively few examples of aldehyde reduction in vivo are known.[209] Actually, very few drugs contain an aldehyde group. The main exposure to aldehydes is through ingestion of the metabolic precursors such as primary alcohols or various amines (see Section 8.4.2.1.5). The primary alcohol of the anti-HIV (reverse transcriptase inhibitor) drug abacavir sulfate (**8.101**, R = CH_2OH; Ziagen) is oxidized to the corresponding carboxylic acid (**8.101**, R = COOH),[210] presumably via the

$$RR'CHOH + NAD^+ \rightleftharpoons RR'C{=}O + NADH + H^+$$

SCHEME 8.38 Reactions catalyzed by alcohol dehydrogenase

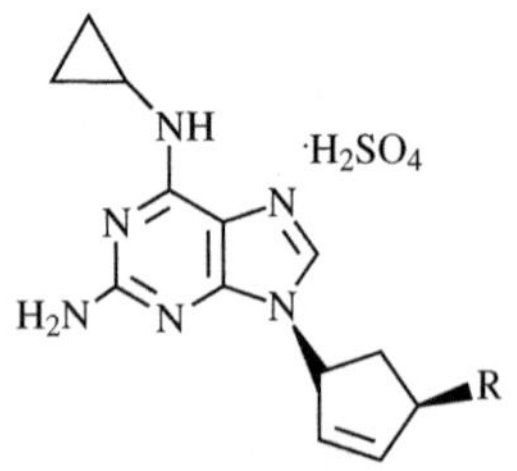

Abacavir sulfate (R = CH_2OH)
8.101

aldehyde, by alcohol and aldehyde dehydrogenases. The antihypertensive drug losartan (**8.102**, R=CH_2OH, Cozaar) is metabolized by an isozyme of cytochrome P450 to the corresponding carboxylic acid (**8.102**, R=COOH).[211] The carboxylic acid metabolite is 10 times more potent as an antagonist of the angiotensin II receptor than losartan.[212]

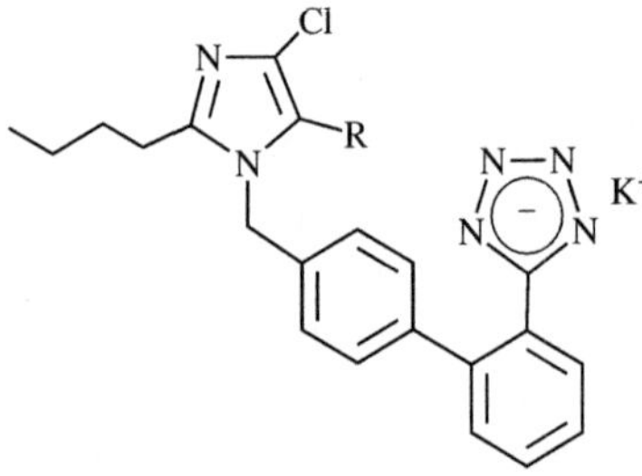

Losartan potassium (R = CH_2OH)
8.102

TABLE 8.6 Classes of Substrates for Reductive Reactions

Functional group	Product
$RC(=O)R'$	$RCH(OH)R'$
RNO_2	RNHOH
RNO	RNHOH
RNHOH	RNH_2
RN=NR'	RNH_2 + $R'NH_2$
$R_3\overset{+}{N}-O^-$	R_3N
quinone (R, R', R, R')	semiquinone radical anion (O^-)
R−X	$R^{\bullet}$ + X^-

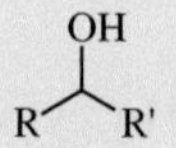

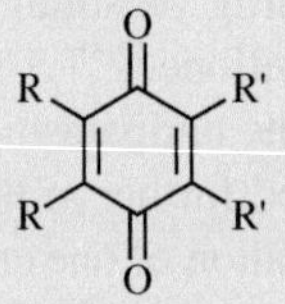

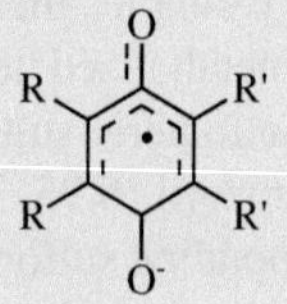

8.4.2.2. Reductive Reactions

Oxidative processes are, by far, the major pathways of drug metabolism, but reductive reactions are important for biotransformations of the functional groups listed in Table 8.6. Reductive reactions are important for the formation of hydroxyl and amino groups that render the drug more hydrophilic and set it up for Phase II conjugation (see Section 8.4.3).

8.4.2.2.1. Carbonyl Reduction

Carbonyl reduction typically is catalyzed by aldo-keto reductases that require NADPH or NADH as the coenzyme. As described in the previous section, alcohol dehydrogenase catalyzes the reduction of aldehydes as well as the oxidation of alcohols. It is not common to observe reduction of aldehydes to alcohols because of the more rapid oxidation of aldehydes by aldehyde dehydrogenase. A large variety of aliphatic and aromatic ketones, however, are reduced to alcohols by NADPH-dependent ketone reductases.[213] As discussed in Chapter 4 (Section 4.3.2), the NADPH carbonyl reductases are stereospecific with regard to hydride transfer from the pyridine nucleotide cofactor. In general, the aldehyde reductases exhibit specificity for the (*pro*-4*R*)-hydrogen (sometimes referred to as the A-hydrogen) and ketone reductases exhibit specificity for the (*pro*-4*S*)-hydrogen (sometimes referred to as B-hydrogen).[214] Stereoselectivity for enantiomer substrates as well as stereospecific reduction of the ketone carbonyl also is typical. The reduction of the anticoagulant drug warfarin (**8.103**, R=H, Coumadin) is selective for the

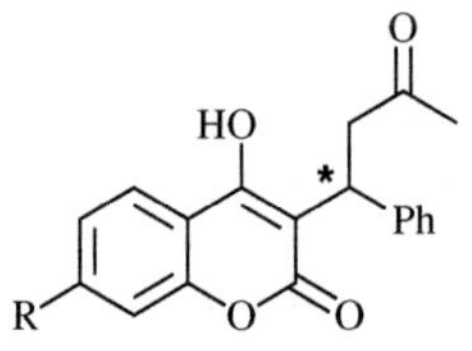

Warfarin (R = H)
8.103

R-(+)-enantiomer; reduction of the *S*-(−)-isomer occurs only at high substrate concentrations.[215] *R*-Warfarin is reduced in humans principally to the *R*,*S*-warfarin alcohol, whereas *S*-warfarin is not only metabolized mainly to 7-hydroxywarfarin (**8.103**, R=OH) but also reduced to a 4:1 mixture of the *S*,*S*-alcohol and the *S*,*R*-alcohol. These studies were carried out by administration of the enantiomers separately to human volunteers. When racemic warfarin was administered to human volunteers, a different picture emerged.[216] Of the 50% of the drug recovered as metabolites, 19% arose from the (*R*)-isomer and 31% from the (*S*)-isomer. The difference in the enantiomeric selectivity is a result of the difference in the rate of clearance from the body of the (*S*)-isomer relative to that of the (*R*)-isomer.[217] The major metabolite was (*S*)-7-hydroxywarfarin (22%); (*S*)-6-hydroxywarfarin (6%) also was obtained. In this study, the main metabolite from (*R*)-warfarin was not the *R,S*-warfarin alcohol (6%), but rather (*R*)-6-hydroxywarfarin (9%); (*R*)-7-hydroxywarfarin (3%) also was obtained. Regioisomeric aromatic hydroxylation results from the selectivity of different isozymes of cytochrome P450.[218] By comparison of the results from the two warfarin studies, it is apparent that one enantiomer can have an effect on the metabolism of the other. This is yet another reason to administer pure enantiomers rather than racemic mixtures as drugs. Enzymatic reduction of ketones, in general, produces the *S*-alcohol as the major metabolite,[219] even in the presence of chiral centers.

Species variation in the stereochemistry of the reduction of ketones is not uncommon. Naltrexone (**8.104**, ReVia), an opioid antagonist used to treat opiate addicts who have been

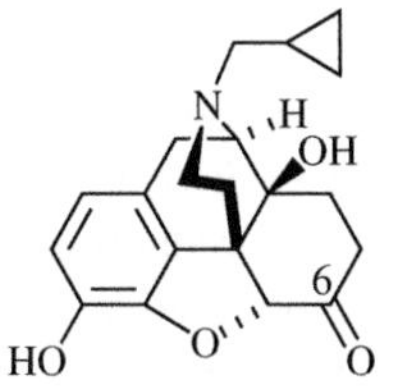

Naltrexone
8.104

withdrawn from opiates in rehabilitation programs, is reduced to the 6α-alcohol (**8.105**, R^1=OH, R^2=H) in chickens[220] and to the 6β-alcohol (**8.105**, R^1=H, R^2=OH) in rabbits and man.[221]

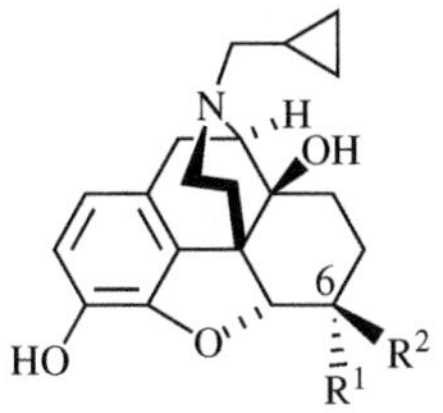

8.105

α,β-Unsaturated ketones can be metabolized to saturated alcohols (reduction of both the carbon–carbon double bond and the carbonyl group). Norgestrel (**8.97**, R=Et, Ovrette) and norethindrone (**8.97**, R=Me, Norlutin), used in oral contraceptives, are reduced in women to the 5β-*H*-3α,17β-diol (**8.106**; R_1=H, R_2=OH, R_3=Et) and 5β-*H*-3β,17β-diol (**8.106**, R_1=OH, R_2=H, R_3=Me) derivatives,

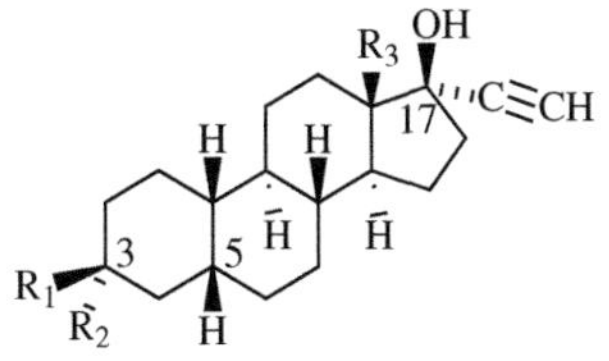

8.106

respectively.[222] The Δ^4-double bond of both drugs is reduced to give the 5β-product, but the 3-keto group is reduced to the 3α-epimer in the case of norgestrel and to the 3β-epimer with norethindrone, even though the only difference in the two molecules is a 13β-ethyl group vs a 13β-methyl group, respectively.

8.4.2.2.2. Nitro Reduction

Aromatic nitro reduction, catalyzed by cytochrome P450 in the presence of NADPH, but under anaerobic conditions (O_2 inhibits the reaction), and by the flavin-dependent NADPH-cytochrome P450 reductase (see Section 8.4.2.2.1) is a multistep process (Scheme 8.39); the reduction of the nitro group to the nitroso group (**8.108**) is the rate-determining step.[223] On the basis of electron paramagnetic resonance spectra and the correlation of rates of radical formation with product formation, it has been proposed[224] that the nitro anion radical (**8.107**) is the first intermediate in the reduction of the nitro group. The reoxidation of this radical by oxygen to give the nitro compound back and superoxide radical anion may explain the inhibition of this metabolic pathway by oxygen.[225] Other enzymes that catalyze nitro group reduction are the bacterial nitro reductase in the GI tract,[226] xanthine oxidase,[227] aldehyde oxidase,[228] and quinone reductase (DT-diaphorase).[229] Metabolic oxidation/reduction cycling of drugs may occur; the balance between the oxidative and reductive pathways is important in determining the pharmacological and toxicological profile of a drug.

An example of nitro reduction is the metabolism of the anticonvulsant drug clonazepam (**8.109**, R=NO_2, Klonopin) to its corresponding amine (**8.109**, R=NH_2).[230] In some in vitro experiments, the reduced metabolite

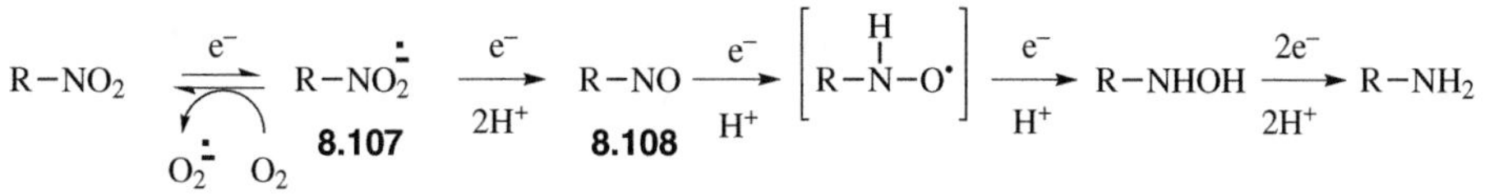

$$R{-}NO_2 \underset{O_2^{\cdot-}\ \ O_2}{\overset{e^-}{\rightleftharpoons}} R{-}NO_2^{\cdot-}\ (\mathbf{8.107}) \xrightarrow[2H^+]{e^-} R{-}NO\ (\mathbf{8.108}) \xrightarrow[H^+]{e^-} \left[R{-}\overset{H}{N}{-}O^{\cdot}\right] \xrightarrow[H^+]{e^-} R{-}NHOH \xrightarrow[2H^+]{2e^-} R{-}NH_2$$

SCHEME 8.39 Nitro group reduction

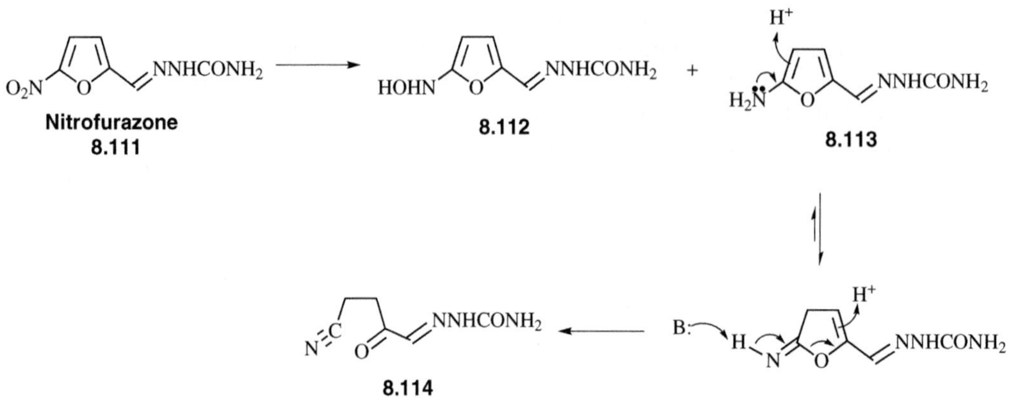

Nitrofurazone
8.111

8.112

8.113

8.114

SCHEME 8.40 Reductive metabolism of nitrofurazone

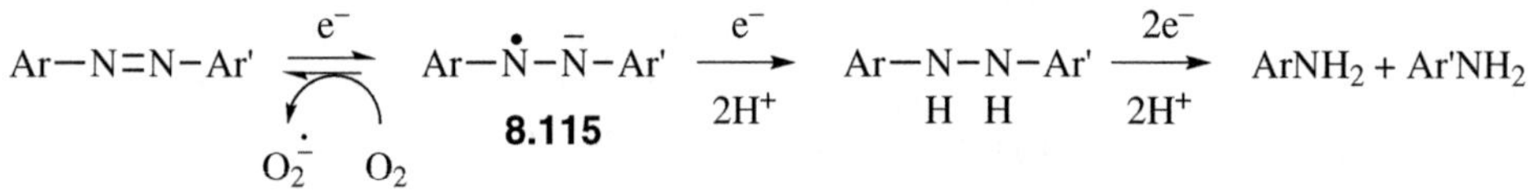

8.115

SCHEME 8.41 Azo group reduction

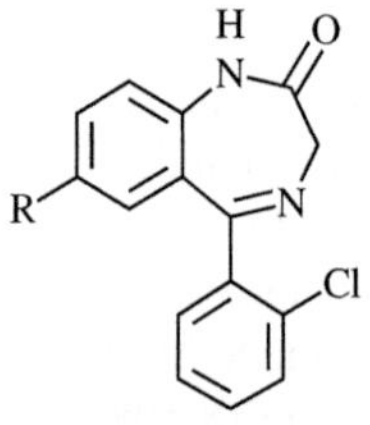

Clonazepam (R = NO_2)
8.109

is not observed because it is easily air-oxidized back to the parent compound. For example, the antiparasitic agent niridazole (**8.110**, R=NO_2, Ambilgar) is reduced to the

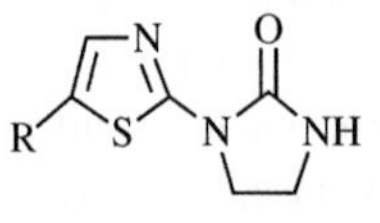

Niridazole (R = NO_2)
8.110

hydroxylamine metabolite (**8.110**, R=NHOH), which is reoxidized in air to **8.110** (R=NO_2).[231] The antibacterial drug nitrofurazone (**8.111**, Scheme 8.40, Furacin) is reduced both to the corresponding 5-hydroxylamino derivative (**8.112**) and the 5-amino derivative (**8.113**). The latter is unstable and tautomerizes to **8.114**.[232]

8.4.2.2.3. Azo Reduction

Azo group (RN=NR) reduction is similar to nitro reduction in many ways. It, too, is mediated both by cytochrome P450 and by NADPH-cytochrome P450 reductase (see Section 8.4.2.2.1), and oxygen often inhibits the reaction. The initial reduction in the oxygen-sensitive metabolism appears to proceed via the azo anion radical (**8.115**, Scheme 8.41)[233]; the oxygen apparently reverses this process with concomitant conversion of the oxygen to the superoxide anion radical.[234] Oxygen-insensitive azoreductases presumably proceed by a two-electron reduction of the azo compound directly to the hydrazo intermediate.

Bacteria in the GI tract are also important in azo reduction.[235] Reduction of sulfasalazine (**8.116**, Scheme 8.42, Azulfidine), used in the treatment of ulcerative colitis, to sulfapyridine (**8.117**, Dagenan) and 5-aminosalicylic acid (**8.118**) occurs primarily in the colon mediated by intestinal bacteria.[236]

8.4.2.2.4. Azido Reduction

The 3-azido group of the anti-acquired immunodeficiency syndrome (AIDS) drug zidovudine (**8.119**, X=N3, Retrovir) is reduced by cytochrome P450 isozymes and NADPH-cytochrome P450 reductase to the corresponding amine (**8.119**, X=NH_2).[237]

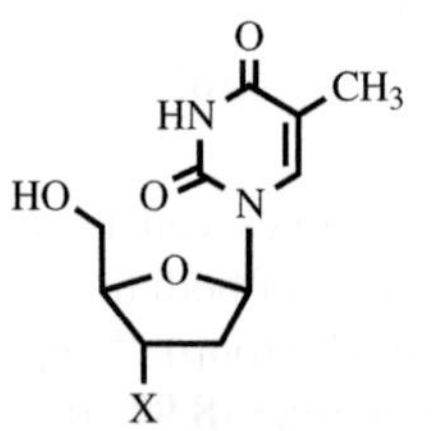

Zidovudine (X = N_3)
8.119

8.4.2.2.5. Tertiary Amine Oxide Reduction

A wide variety of lipophilic aliphatic and aromatic tertiary amine oxides, such as imipramine *N*-oxide (**8.120**), are reduced to the corresponding amine by cytochrome P450 in the absence of oxygen.[238]

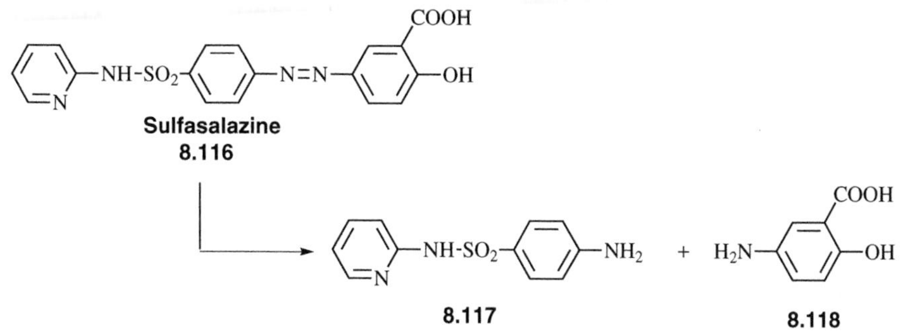

Sulfasalazine
8.116

8.117 8.118

SCHEME 8.42 Reductive metabolism of sulfasalazine

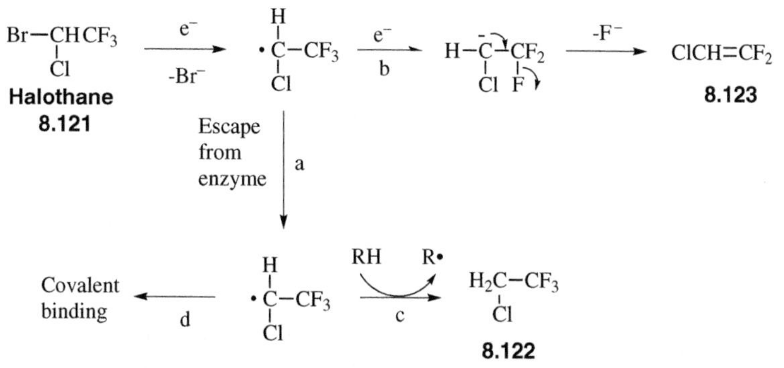

Halothane
8.121

8.123

8.122

SCHEME 8.43 Reductive dehalogenation of halothane

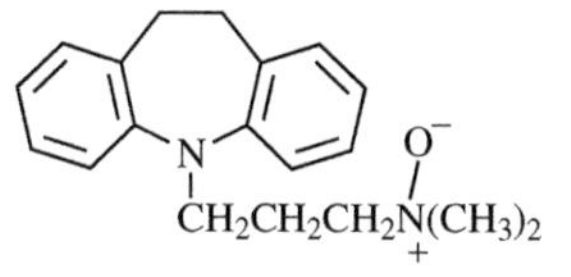

8.120

8.4.2.2.6. Reductive Dehalogenation

Under hypoxic or anaerobic conditions the volatile anesthetic halothane (**8.121**, Halothane) is metabolized by a reductive dehalogenation mechanism by cytochrome P450[239] (Scheme 8.43), which differs from the normal oxidative P450 mechanism discussed in Section 8.4.2.1.8. Thefirst electron is transferred to halothane from cytochrome P450, which is reduced by NADPH-cytochrome P450 reductase. This electron transfer ejects the bromide ion and produces the cytochrome P450-bound 1-chloro-2,2,2-trifluoroethyl radical. If this radical escapes from the active site (pathway a), it either can be reduced by hydrogen atom transfer (pathway c) to give 2-chloro-1,1,1-trifluoroethane (**8.122**) or form a covalent bond to cellular proteins (pathway d). A second electron reduction from **8.121** (pathway b) produces the carbanion; β-elimination of fluoride ion gives chlorodifluoroethylene (**8.123**). Pathway d, resulting in covalent attachment to proteins, has been proposed to mediate halothane hepatitis, a toxic reaction to halothane exposure in the liver. The second electron transfer (pathway b) is thought to be derived from cytochrome b_5; this leads to nonreactive products, and it competes with pathway a, thereby leading to fewer RMs.

8.4.2.3. Carboxylation Reaction

Amino compounds may be susceptible to carboxylation by dissolved carbon dioxide, converting them to the corresponding carbamic acid derivatives, which may be further metabolized. For example, the serotonin reuptake inhibitor/antidepressant drug sertraline (**8.124**, R=H, Zoloft) is metabolized to the carbamic acid (**8.124**, R=COOH), which was detected as the corresponding glucuronide (R=COO-glucuronide).[240]

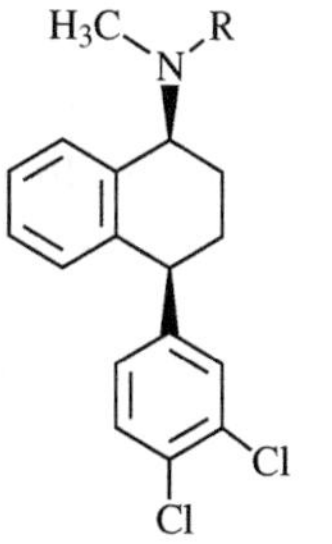

Sertraline (R = H)
8.124

8.4.2.4. Hydrolytic Reactions

The hydrolytic metabolism of esters and amides leads to the formation of carboxylic acids, alcohols, and amines, all of which are susceptible to Phase II conjugation reactions and excretion (see Section 8.4.3). As described in Section 4.2.3 of Chapter 4, enzyme-catalyzed hydrolysis can be acid and/or base catalyzed. Base-catalyzed hydrolysis is accelerated nonenzymatically when electron-withdrawing groups are substituted on either side of the ester or amide bond. When the carbonyl is in conjugation with a π-system, nonenzymatic base hydrolysis is decelerated relative to the aliphatic case. The effects on nonenzymatic hydrolysis rates can also be important in enzyme catalyzed hydrolysis reactions.

A wide variety of nonspecific esterases and amidases involved in drug metabolism are found in the plasma, liver, kidney, and intestine.[241] All mammalian tissues may contribute to the hydrolysis of a drug; however, the liver, the GI tract, and the blood are sites of greatest hydrolytic capacity. Aspirin (**8.125**) is an example of a drug that is hydrolyzed in all human tissues.[242] The hydrolysis of xenobiotics is generally very similar in all mammals, but there are some exceptions, and large species differences can be observed. Some esterases catalyze hydrolysis of aliphatic esters and others aromatic esters. For example, only the benzoyl ester in cocaine (**8.126**, R=CH_3) is hydrolyzed by human liver in vitro, not the alicyclic ester[243]; however, in vivo, the major metabolite of cocaine is the alicyclic ester hydrolysis product, benzoylecgonine (**8.126**, R=H),[244] suggesting that the liver is not the primary site of metabolism in vivo.

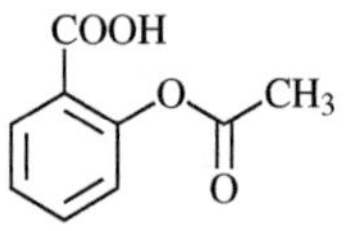

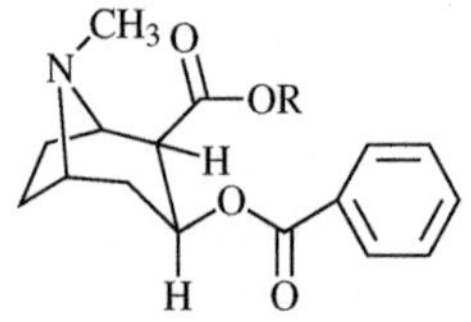

Aspirin
8.125

Cocaine (R = CH_3)
8.126

Generally, amides are more slowly hydrolyzed than esters. For example, the enzymatic hydrolysis of the antiarrhythmic drug procainamide (**8.127**, X=NH, Procanbid) is slow relative to that of the local anesthetic procaine (**8.127**, X=O, Novocain).[245] The ester group, not the amide, is hydrolyzed in the anesthetic propanidid (**8.128**, Epontol).[246] However, the amide bond of the local anesthetic butanilicaine (**8.129**, Hostacain) is hydrolyzed by human liver at rates comparable to those of good ester substrates.

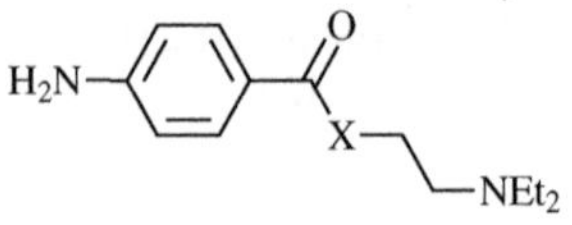

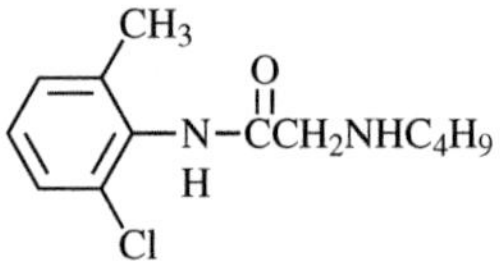

Procainamide (X = NH)
Procaine (X = O)
8.127

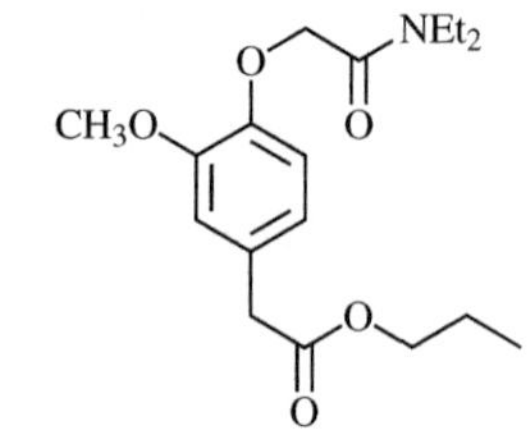

Propanidid
8.128

Butanilicaine
8.129

In some cases, the hydrolysis of an ester or amide bond produces a toxic compound. Aromatic amines generated upon hydrolysis of *N*-acylanilides become methemoglobin-forming agents after *N*-oxidation. Phenacetin (**8.130**, Acetophenetidin) causes methemoglobinemia in rats, which is reduced drastically when the carboxylesterase inhibitor bis(4-nitrophenyl)phosphate is coadministered.[247] The danger of using racemic mixtures as drugs is further exemplified by the observation that although both isomers of prilocaine (**8.131**, Citanest) have local anesthetic action, only the *R*-(−)-isomer is hydrolyzed to toluidine, which causes methemoglobinemia. The *S*-(+)-isomer, which is not hydrolyzed, does not cause this side effect.[248]

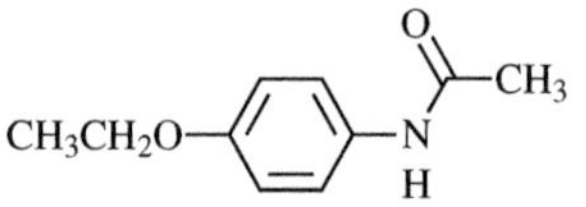

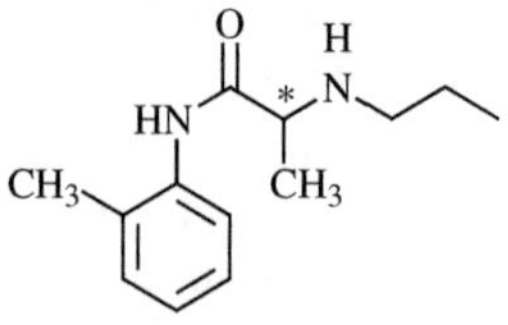

Phenacetin
8.130

Prilocaine
8.131

As shown in the above example, enzymatic hydrolysis often exhibits enantiomeric specificity. A racemic mixture of the anticonvulsant drug phensuximide (**8.132**, Milontin) is enzymatically *N*-demethylated and hydrolyzed stereospecifically to *R*-(−)-2-phenylsuccinamic acid (**8.133**).[249]

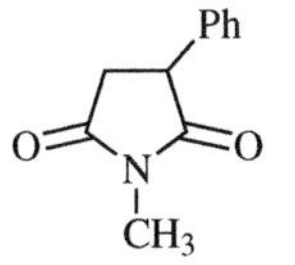

Phensuximide
8.132

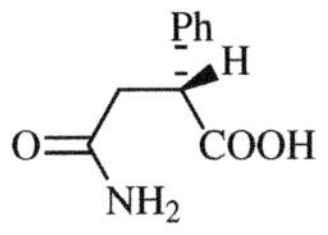

8.133

Enantiomeric selectivity in the hydrolysis of the tranquilizer prodrug oxazepam acetate (**8.134**, Serax) may be organ selective. Preferential hydrolysis of the *R*-(−)-ester occurs in the liver, but the opposite is found in the brain.[250]

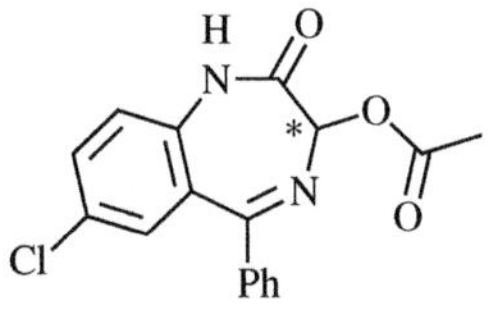

Oxazepam acetate
8.134

Because of the stereoselectivity of various enzymes, one enantiomer may be a preferential substrate for one enzyme and the other enantiomer is a substrate for a different enzyme. Both enantiomers of the hypnotic drug etomidate (**8.135**, Scheme 8.44, Amidate) are metabolized, but by different routes.[251] The active *R*-(+)-isomer is more rapidly hydrolyzed (pathway a) than is the *S*-(−)-isomer, but the *S*-(−)-isomer is more rapidly hydroxylated (pathway b) than is the *R*-(+)-isomer. In vitro, only the *S*-(−)-isomer produces acetophenone (**8.136**).

In addition to carboxylesterases and amidases, hydrolytic reactions are also carried out by various other mammalian enzymes, such as phosphatases, β-glucuronidases, sulfatases, and deacetylases. We now turn our attention to the next phase of drug metabolism, that of transforming the drug or Phase I metabolites into conjugates for excretion.

8.4.3. Phase II Transformations: Conjugation Reactions

8.4.3.1. Introduction

Phase II or *conjugating enzymes*, in general, catalyze the attachment of small polar endogenous molecules such as glucuronic acid, sulfate, or amino acids to drugs or, more often, to metabolites arising from Phase I metabolic processes. This Phase II modification further deactivates the drug, changes its physicochemical properties, and produces water-soluble metabolites that are readily excreted in the urine or bile. Other Phase II processes, such as methylation and acetylation, do not yield more polar metabolites, but serve primarily to terminate or attenuate biological activity. Metabolic reactions with the potent nucleophile glutathione serve to trap highly electrophilic metabolites before they covalently modify biologically important macromolecules such as proteins, RNA, and DNA. Many drugs, such as pregabalin (**8.137**, Lyrica),[252] used to treat fibromyalgia and neuropathic pain, are excreted without any modification at all.

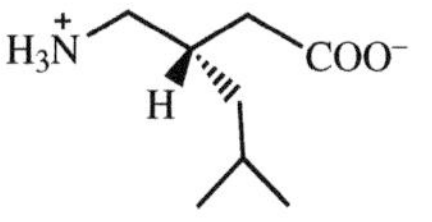

Pregabalin
8.137

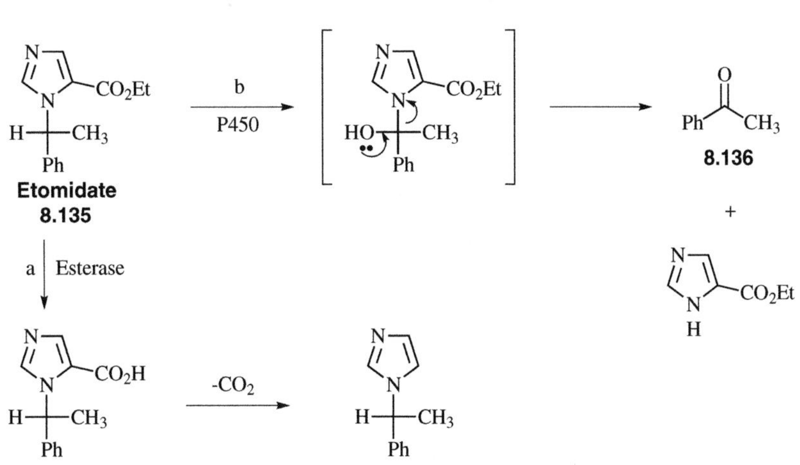

Etomidate
8.135

8.136

SCHEME 8.44 Competitive metabolism of *R*- and *S*-etomidate

TABLE 8.7 Mammalian Phase II Conjugating Agents

Conjugate	Coenzyme Form	Groups Conjugated	Transferase Enzyme
Glucuronide	Uridine-5'-diphospho-α-D-glucuronic acid (UDPGA) 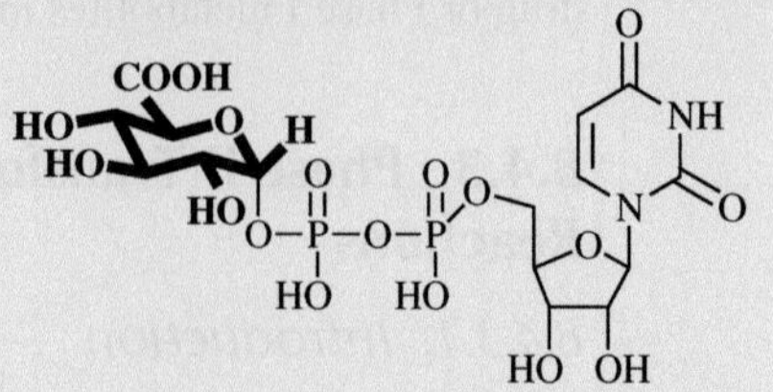	-OH, -COOH, $-NH_2$, $-NR_2$, -SH, C-H	UDP-Glucuronosyl-transferase
Sulfate	3'-Phosphoadenosine-5'-phosphosulfate (PAPS) 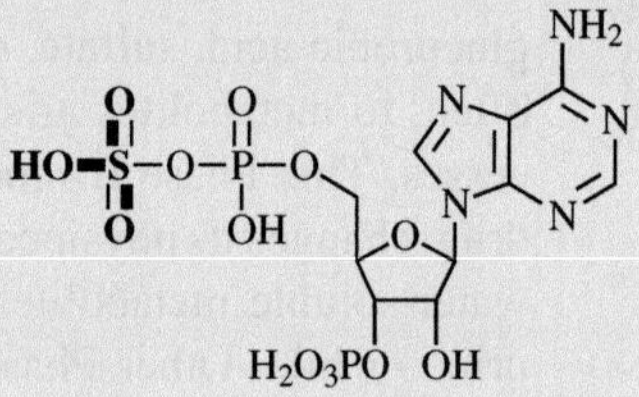	-OH, $-NH_2$	Sulfotransferase
Glycine and glutamine	Activated acyl or aroyl coenzyme A cosubstrate: 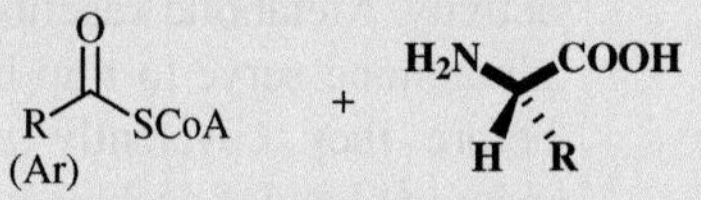 RC(O)SCoA (R = Ar) + H_2N–CH(R)–COOH	-COOH	Glycine *N*-acyltransferase; Glutamine *N*-acyltransferase
Glutathione	Glutathione (GSH) 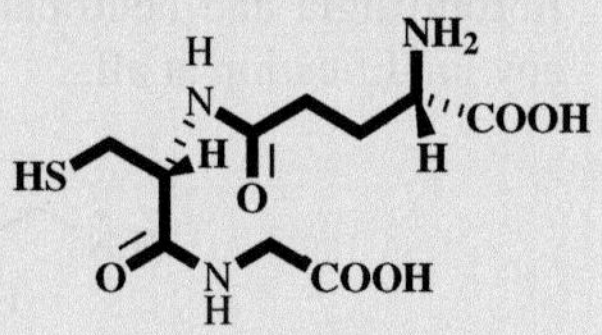	Ar-X, arene oxide, epoxide, carbocation or related	Glutathione *S*-transferase
Acetyl	Acetyl coenzyme A:  H_3CC(O)SCoA	-OH, $-NH_2$	Acetyl-transferase
Methyl	*S*-Adenosyl methionine (SAM) 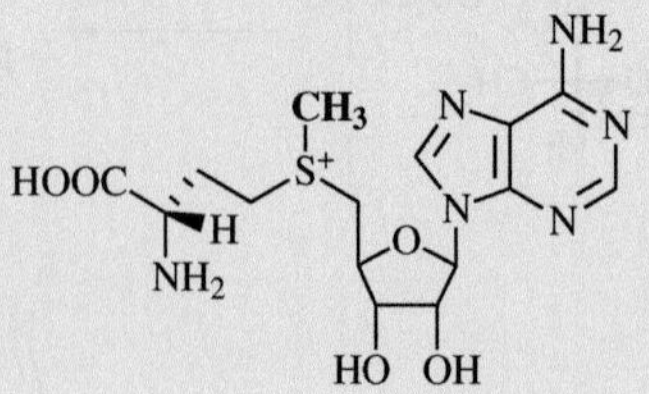	-OH, $-NH_2$, -SH, heterocyclic N	Methyl-transferase

Conjugation reactions take place primarily with hydroxyl, carboxyl, amino, heterocyclic nitrogen, and thiol groups. If these groups are not present in the drug, they are introduced or unmasked by the Phase I reactions. For the most part, the conjugating moiety is an endogenous molecule that is first activated in a coenzyme form prior to its transfer to the acceptor group. The enzymes that catalyze these reactions are known as transferases (Table 8.7).

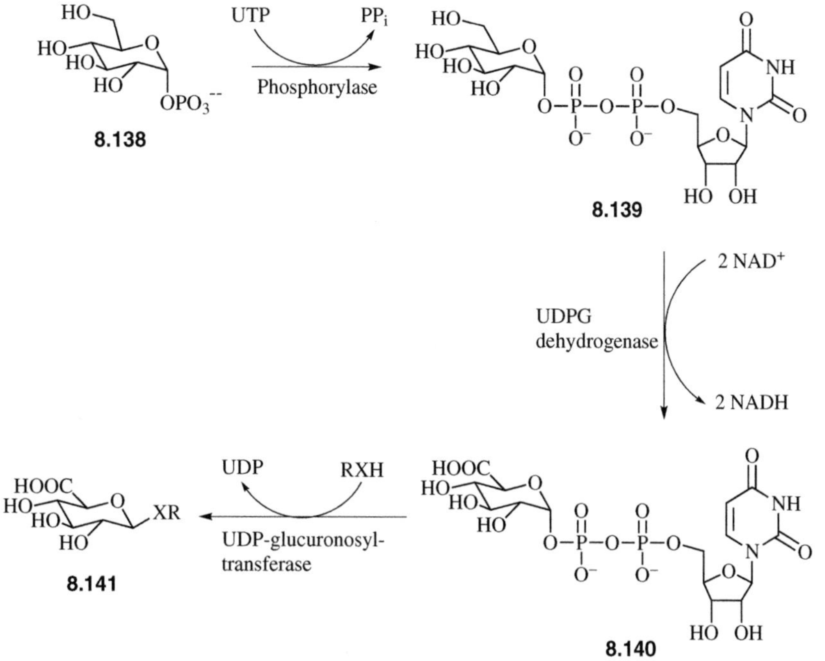

SCHEME 8.45 Biosynthesis and reactions of UDP glucuronic acid

8.4.3.2. Glucuronic Acid Conjugation

Glucuronidation is the most common mammalian conjugation pathway; it occurs in all mammals except cats. The coenzyme form of glucuronic acid, namely, uridine 5′-diphospho-α-D-glucuronic acid (UDP glucuronic acid, **8.140**, Scheme 8.45; see also Table 8.7) is biosynthesized from α-D-glucose-1-phosphate (**8.138**) by phosphorylase-catalyzed conversion to the nucleotide sugar **8.139**, followed by UDP dehydrogenase-catalyzed oxidation. UDP glucuronic acid (**8.140**) contains D-glucuronic acid in the α-configuration at the anomeric carbon, but glucuronic acid conjugates (**8.141**) are β-glycosides. Therefore, the glucuronidation reaction involves an inversion of stereochemistry at the anomeric carbon. Because of the presence of the carboxylate and hydroxyl groups of the glucuronyl moiety, glucuronides are hydrophilic and, therefore, are set up for excretion. Glucuronides generally are excreted in the urine, but when the molecular weight of the conjugate exceeds 300, excretion in the bile becomes significant. There is some evidence that UDP-glucuronosyltransferase is closely associated with cytochrome P450 so that as drugs become oxidized by Phase I cytochrome P450 reactions, the metabolites are efficiently conjugated.[253]

Four general classes of glucuronides have been established, the *O*-, *N*-, *S*-, and *C*-glucuronides; a small sampling of examples is given in Table 8.8; arrows point to the sites of glucuronidation.

There are certain hereditary disease states (inborn errors of metabolism) that are associated with defective glucuronide formation or attachment of the glucuronide to bilirubin, for example, Crigler-Najjar syndrome and Gilbert disease; both are characterized by a deficiency of UDP-glucuronosyltransferase activity.[263] Neonates, who have undeveloped liver UDP-glucuronosyltransferase activity, may exhibit similar metabolic problems. In these cases, there is a greater susceptibility to adverse effects caused by the accumulation of drugs that normally are glucuronidated. An example of this is the inability of neonates to conjugate the antibacterial drug chloramphenicol (**8.142**, Chloroptic), thereby leading to "gray baby syndrome" from the accumulation of toxic levels of the drug.[264]

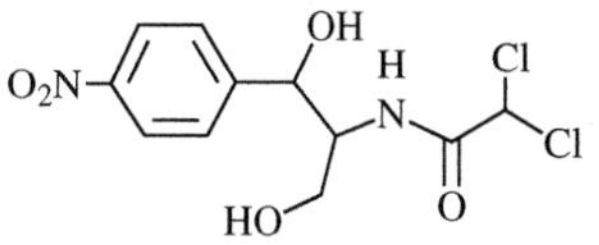

Chloramphenicol
8.142

As in the case of Phase I metabolism, Phase II reactions can also be species specific, regioselective, and stereoselective. The antibacterial drug sulfadimethoxine (see Table 8.8, Albon) is glucuronidated in man but not in rat, guinea pig, or rabbit. The bronchodilator fenoterol (**8.143**, Berotec) is conjugated as two different glucuronides, a *para*-glucuronide and a

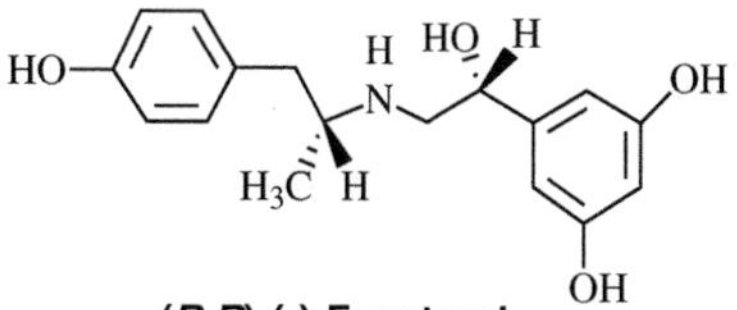

(*R*,*R*)-(-)-Fenoterol
8.143

TABLE 8.8 Classes of Compounds Forming Glucuronides With Examples[254–262]

Type	Example	Structure (Arrow Indicates Site of Glucuronidation)	References
O-Glucuronide			
Hydroxyl			
Phenol	Acetaminophen	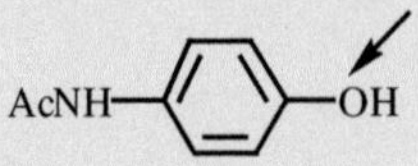 AcNH–C₆H₄–OH	[254]
Alcohol	Chloramphenicol	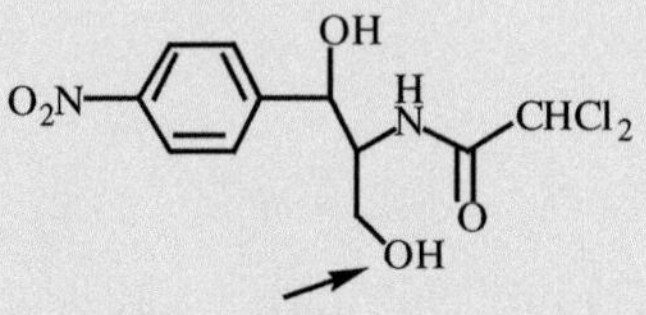	[255]
Carboxyl	Fenoprofen	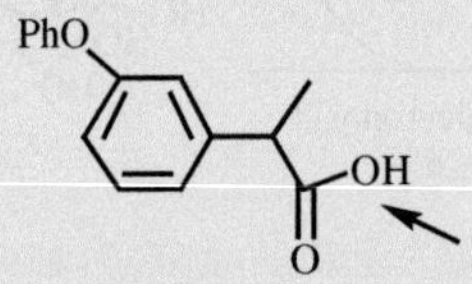	[256]
N-Glucuronide			
Amine	Desipramine	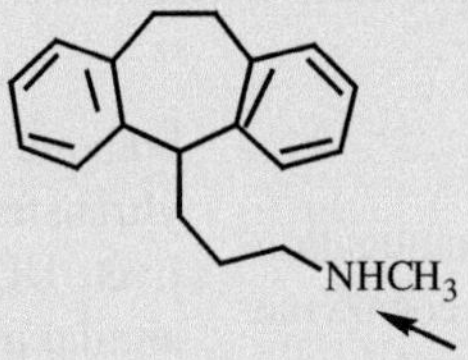	[257]
Amide			
Carbamate	Meprobamate	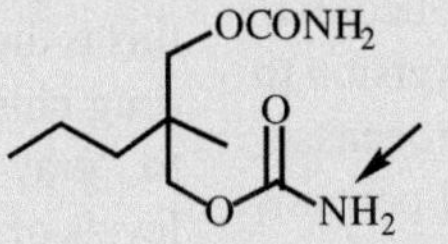	[258]
Sulfonamide	Sulfadimethoxine	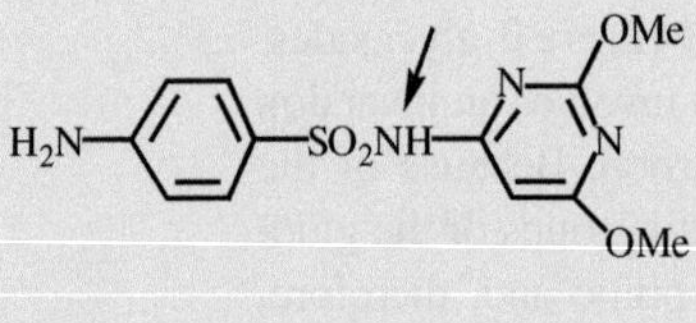	[259]
S-Glucuronide			
Sulfhydryl	Methimazole	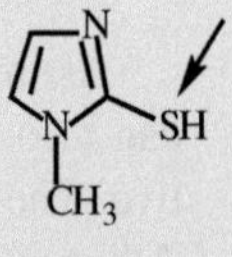	[260]
Carbodithioic acid	Disulfiram	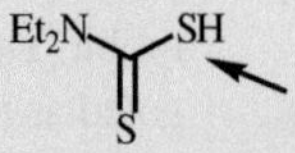 (Reduced metabolite)	[261]
C-Glucuronide			
	Phenylbutazone	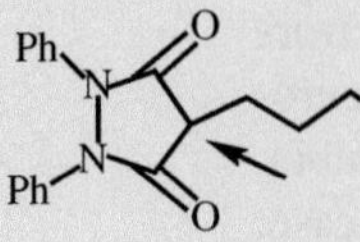	[262]

meta-glucuronide because there are both *para*- and *meta*-hydroxyl groups.[265] The *R*,*R*-(−)-isomer is conjugated with higher affinity but with lower velocity than is the *S*,*S*-(+)-isomer.

The antidepressant drug nortriptyline (**8.144**, R=H, Aventyl) is metabolized (cytochrome P450) predominantly to the *E*-(−)-hydroxy analog (**8.144**, R=OH; the absolute configurations of the (+)- and (−)-enantiomers are not known, so stereochemistry is not specified). This metabolite is converted stereospecifically into the corresponding *O*-glucuronide, but the stereospecificity is organ dependent.[266] The liver and kidney glucuronosyltransferases catalyze the conversion of only the *E*-(+)-isomer of **8.144** (R=OH) into the *O*-glucuronide, whereas the intestinal enzyme metabolizes only the *E*-(−)-isomer. The enantiomer that is not glucuronidated inhibits the glucuronidation of its antipode.

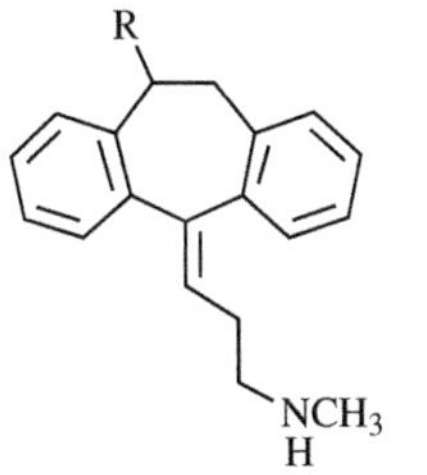

R
NCH_3
H
Nortriptyline (R = H)
8.144

8.4.3.3. Sulfate Conjugation

Sulfate conjugation occurs less frequently than does glucuronidation presumably because of the limited availability of inorganic sulfate in mammals and the fewer number of functional groups (phenols, alcohols, arylamines, and *N*-hydroxy compounds) that undergo sulfate conjugation.[267]

There are three enzyme-catalyzed reactions involved in sulfate conjugation (Scheme 8.46). Inorganic sulfate is activated by the adenosine triphosphate (ATP) sulfurylase-catalyzed reaction with ATP to give adenosine 5′-phosphosulfate (APS, **8.145**), which is phosphorylated in an APS phosphokinase-catalyzed reaction to 3′-phosphoadenosine 5′-phosphosulfate (**8.146**), the coenzyme form used for sulfation.[268] The acceptor molecule (RXH) undergoes sulfotransferase-catalyzed sulfation to **8.147** with release of 3′-phosphoadenosine 5′-phosphate (PAP). There are a variety of sulfotransferases in the liver and other tissues. The main substrates for these enzymes are phenols, but aliphatic alcohols, amines, and to a much lesser extent, thiols are also sulfated. Often both glucuronidation and sulfation occur on the same substrates, but in these cases the K_m for sulfation is usually lower, so it predominates.[269] In addition to substrate-binding differences, sulfotransferases are cytoplasmic (soluble) enzymes and glucuronosyltransferases are microsomal (membrane) enzymes. Sulfate conjugation tends to predominate at low doses, when there is less to diffuse into membranes,[270] and with smaller, less lipid-soluble molecules.[271] However, lipophilicity does not necessarily mean that a compound will be glucuronidated rather than sulfated, because subcellular distances are small, diffusion out of membranes generally is rapid, and the K_m values for sulfotransferases are generally lower than those for glucuronosyltransferases. The bronchodilator albuterol (**8.148**, R=H, Albuterol) is metabolized to the corresponding sulfate ester (**8.148**, R = SO_3^-).[272] Note that although there are three hydroxyl groups in albuterol, phenolic sulfation predominates.

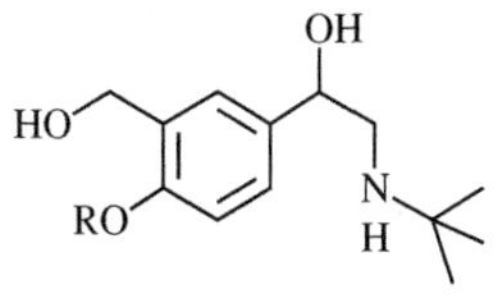

OH
HO
RO
N
H
Albuterol (R = H)
8.148

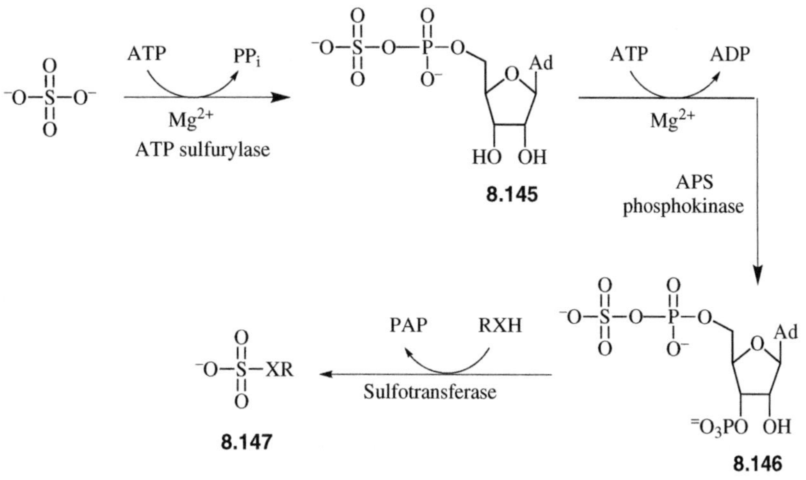

$^-O-SO_2-O^-$ → (ATP, PP_i; Mg^{2+}; ATP sulfurylase) → **8.145** (HO OH; Ad) → (ATP, ADP; Mg^{2+}; APS phosphokinase) → **8.146** ($^=O_3PO$ OH; Ad) → (PAP, RXH; Sulfotransferase) → $^-O-SO_2-XR$ **8.147**

SCHEME 8.46 Sulfate conjugation

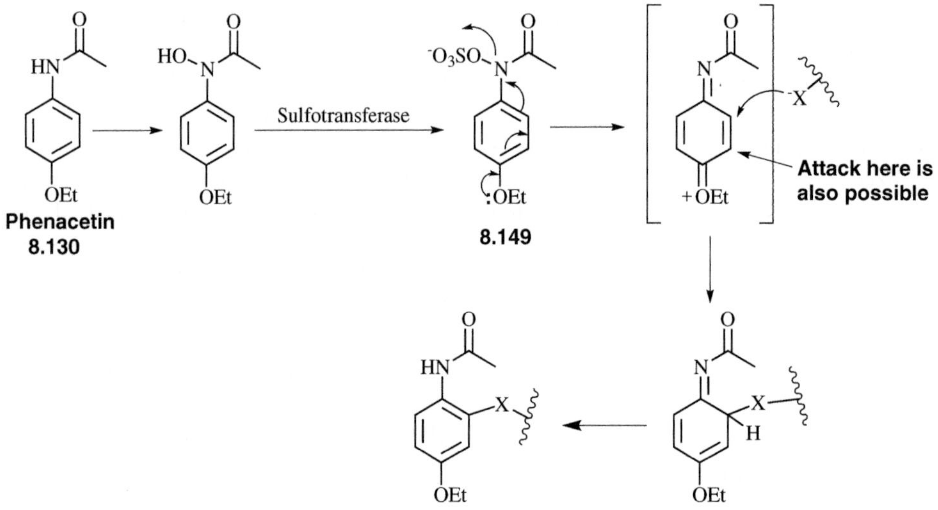

SCHEME 8.47 Bioactivation of phenacetin

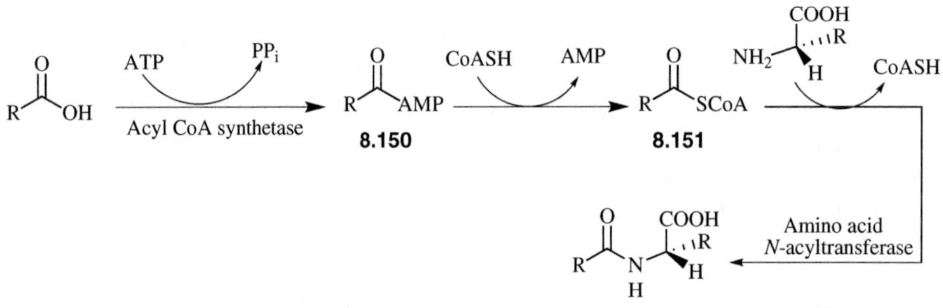

SCHEME 8.48 Amino acid conjugation

Phenolic *O*-glucuronidation often competes favorably with sulfation because of the limited sulfate availability. In some cases the reverse situation occurs. Acetaminophen (Tylenol) is metabolized in adults mainly to the *O*-glucuronide, although some sulfate ester can be detected.[273] However, neonates and children 3–9 years old excrete primarily the acetaminophen sulfate conjugate because they have a limited capacity to conjugate with glucuronic acid.[274] Sulfation of aliphatic alcohols and arylamines occurs, but these are minor metabolic pathways. Sulfate conjugates can be hydrolyzed back to the parent compound by various sulfatases.

Sulfoconjugation plays an important role in the hepatotoxicity and carcinogenicity of *N*-hydroxyarylamides.[275] As described in Section 8.4.2.1.5 (see Schemes 8.29 and 8.30), activated *N*-hydroxyarylamines are quite electrophilic (reactive metabolites; see Section 8.4.4) and can react with protein and DNA nucleophiles. *N*-Hydroxy sulfation also activates these compounds as highly electrophilic nitrenium-like species. For example, sulfoconjugation of the *N*-hydroxylation metabolite of the analgesic phenacetin (**8.130**, Scheme 8.47) produces a reactive species (**8.149**) that may be responsible for hepatotoxicity and nephrotoxicity of that compound.[276]

8.4.3.4. Amino Acid Conjugation

The first mammalian drug metabolite isolated, hippuric acid (see **8.8**), was the product of glycine conjugation of benzoic acid. *Amino acid conjugation* of a variety of carboxylic acids, particularly aromatic, arylacetic, and heterocyclic carboxylic acids, leads to amide bond formation. The specific amino acid involved in conjugation within a class of animals usually depends on the bioavailability of that amino acid from endogenous and dietary sources. Glycine conjugates are the most common amino acid conjugates in animals; glycine conjugation in mammals follows the order herbivores > omnivores > carnivores. Conjugation with L-glutamine is most common in primate drug metabolism[277]; it does not occur to any significant extent in nonprimates. Taurine, arginine, asparagine, histidine, lysine, glutamate, aspartate, alanine, and serine conjugates also have been found in mammals.[278]

The mechanism of amino acid conjugation involves three steps (Scheme 8.48). The carboxylic acid is first activated by ATP to the adenosine monophosphate (AMP) anhydride (**8.150**), which is converted to the corresponding coenzyme A thioester (**8.151**) with CoASH; these first two steps are

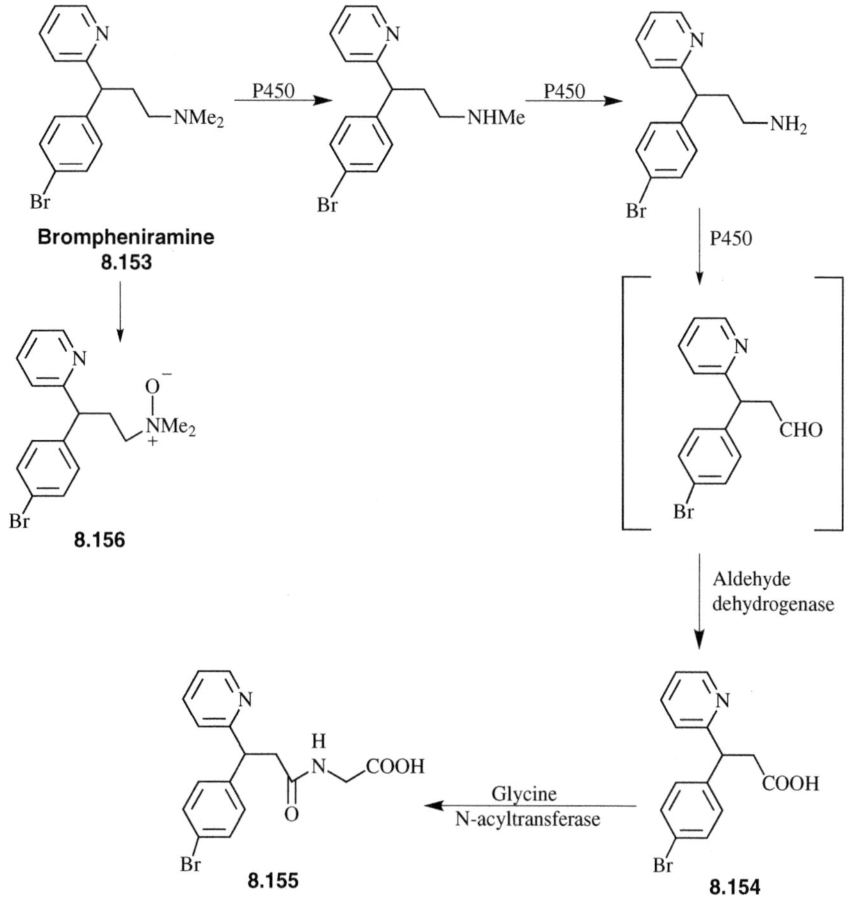

SCHEME 8.49 Metabolism of brompheniramine

catalyzed by acyl CoA synthetases (long-chain fatty acid-CoA ligases). The appropriate amino acid *N*-acyltransferase then catalyzes the condensation of the amino acid and coenzyme A thioester to give the amino acid conjugate (**8.152**). Conjugation does not take place with the AMP anhydride (**8.150**) directly because the AMP anhydride hydrolyzes readily; conversion to the CoA thioester produces a more hydrolytically stable product (**8.151**) that can be transported in the cell readily but is still quite reactive toward the appropriate amine nucleophiles.

As an example of how a drug can undergo both Phase I and Phase II transformations, Scheme 8.49 shows the metabolic pathway for the antihistamine brompheniramine (**8.153**, in cold remedies such as Dimetapp).[279] This drug is converted by Phase I metabolism in both dog and man (oxidative *N*-dealkylation, oxidative deamination, and aldehyde oxidation; see Sections 8.4.2.1.5 and 8.4.2.1.9) to the carboxylic acid (**8.154**), which then is conjugated with glycine (Phase II) to **8.155**. All of the metabolites shown in Scheme 8.49 (including **8.156**), except the aldehyde, were isolated from dog urine, and all except the aldehyde and the *N*-oxide (**8.156**) were isolated from human urine. The related antihistamine diphenhydramine (**8.157**, Benadryl) undergoes similar Phase I oxidation, but is glutamine conjugated.[280]

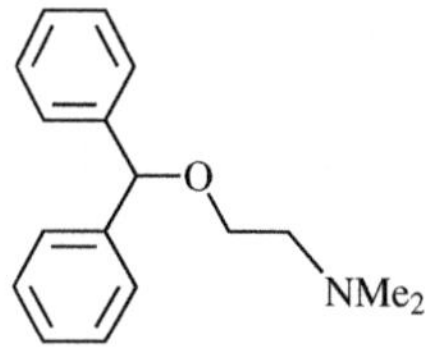

8.4.3.5. Glutathione Conjugation

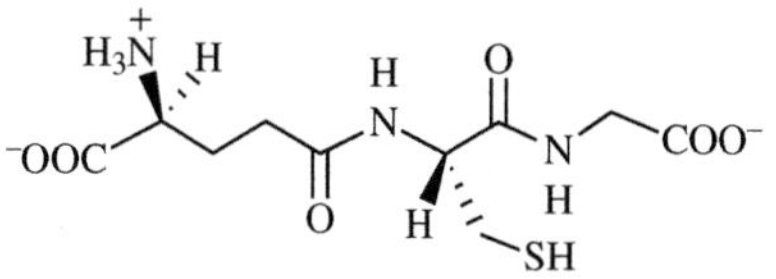

The tripeptide glutathione (**8.158**) is found in virtually all mammalian tissues. It contains a reactive nucleophilic thiol group, and one of its functions appears to be as a scavenger of harmful electrophilic compounds ingested or produced by metabolism. Xenobiotics that are conjugated with glutathione are either highly electrophilic as such or are first metabolized to an electrophilic product prior to conjugation. Toxicity can result from the reaction of cellular nucleophiles with electrophilic metabolites (see Schemes 8.30–8.33), if glutathione does not first intercept these reactive compounds. Electrophilic species include any group capable of undergoing S_N2- or S_NAr-like reactions (e.g., alkyl halides, epoxides, sulfonate esters, and aryl halides), acylation reactions (e.g., anhydrides), conjugate additions (addition to a double or triple bond in conjugation with a carbonyl or related group), and reductions (e.g., disulfides and radicals). All of the reactions catalyzed by glutathione *S*-transferase also occur nonenzymatically, but at a slower rate.

A few examples of glutathione conjugation are given in Scheme 8.50. Examples of S_N2 reactions are the glutathione conjugation of the leukemia drug busulfan (**8.159**, Myleran)[281] and of the coronary vasodilator nitroglycerin (**8.160**, Nitrostat).[282] The reaction of glutathione with the immunosuppressive drug azathioprine (**8.161**, Azathioprine)[283] is an example of an S_NAr reaction. These reactions cause direct deactivations of the drugs. Morphine (**8.162**, Avinza) has been reported to undergo oxidation by two different pathways, both of which lead to potent Michael acceptors that undergo subsequent glutathione conjugation. Pathway a, catalyzed by morphine 6-dehydrogenase, gives morphinone (**8.163**), which undergoes Michael addition with glutathione to **8.164**.[284] Pathway b is a cytochrome P450-catalyzed route that produces an electrophilic quinone methide (**8.165**). Glutathione addition occurs at the sterically less-hindered 10α-position to give **8.166**.[285]

Glutathione conjugates are rarely excreted in the urine; because of their high molecular weight and amphiphilic character, when they are eliminated, it is in the bile. Most typically, however, glutathione conjugates are not excreted; instead they are metabolized further (sometimes referred to as Phase III metabolism)[286] and are excreted ultimately as *N*-acetyl-L-cysteine (also known as mercapturic acid) conjugates (**8.170**, Scheme 8.51).[287] Formation of the mercapturic acid begins from the glutathione conjugate (**8.167**). The γ-glutamyl residue is hydrolyzed to glutamate and the cysteinylglycine conjugate (**8.168**) in a process catalyzed by γ-glutamyltranspeptidase. Cysteinylglycine dipeptidase-catalyzed hydrolysis of **8.168** leads to the release of glycine and the formation of the cysteine conjugate **8.169**, which is *N*-acetylated by acetyl CoA in a reaction catalyzed by cysteine conjugate *N*-acetyltransferase. Conjugation with glutathione occurs in the cytoplasm of most cells, especially in the liver and kidney where the glutathione concentration is 5–10 mM.

8.4.3.6. Water Conjugation

Epoxide hydrolase (also called epoxide hydratase or epoxide hydrase),[288] the enzyme principally involved in *water conjugation* (i.e., *hydration*), was discussed already in the context of hydration of arene oxides (see Section 8.4.2.1.1). Because this enzyme catalyzes the hydration of endogenous epoxides, such as androstene oxide, at much faster rates than exogenous epoxides, it probably plays an important role in endogenous metabolism.[289]

8.4.3.7. Acetyl Conjugation

Acetylation is an important route of metabolism for xenobiotics containing a primary amino group, including aliphatic and aromatic amines, amino acids, sulfonamides, hydrazines, and hydrazides. In all of the previously discussed conjugation reactions, a *more* hydrophilic metabolite is formed. Acetylation, however, converts the ionized primary ammonium group (the aliphatic amine or hydrazine is protonated at physiological pH) to an uncharged amide, which is *less* water soluble. The physicochemical consequences of *N*-acetylation, therefore, are different from those of the other conjugation reactions. The function of acetylation may be to deactivate the drug, although *N*-acetylprocainamide is as potent as the parent antiarrhythmic drug procainamide (Procanbid).[290]

Acetylation occurs widely in the animal kingdom; however, dogs and foxes are unable to acetylate *N*-arylamines or the N^4-amino group of sulfonamides.[291] The extent of *N*-acetylation (and of other metabolic reactions) of a number of drugs in humans is a genetically determined individual characteristic. The genetic variability among individuals in the therapeutic response to the same dose of a drug may be the result of differences in absorption, distribution, metabolism, or elimination. A genetic alteration in an enzyme, known as *polymorphism*, can have serious consequences with regard to the safety of a drug as well as its therapeutic effectiveness.[292] A decrease in the rate of metabolism because of polymorphism of a drug metabolism enzyme would lead to an increase in blood levels of the drug, possibly leading to drug interactions or toxicity; an increase in drug metabolism as a result of an overly active drug metabolism enzyme may cause a reduced therapeutic response. Polymorphisms have been detected for many drug-metabolizing enzymes.[293] These polymorphisms give rise to subgroups in the population that differ in their ability to perform certain enzymatic biotransformations. The three phenotypes in acetylation polymorphism are homozygous fast, homozygous slow, and heterozygous (intermediate) acetylators.[294] The distribution of fast and slow acetylator phenotypes depends on the population studied, varying from mostly slow acetylators (Egyptians) to 50% fast/50% slow in the United States to 90% fast/10% slow in East Asians[295] to 100% fast acetylators in the Canadian Eskimo population.[296] Because of the differences in the rates of *N*-acetylation of certain drugs, there are significant individual variations in the therapeutic responses

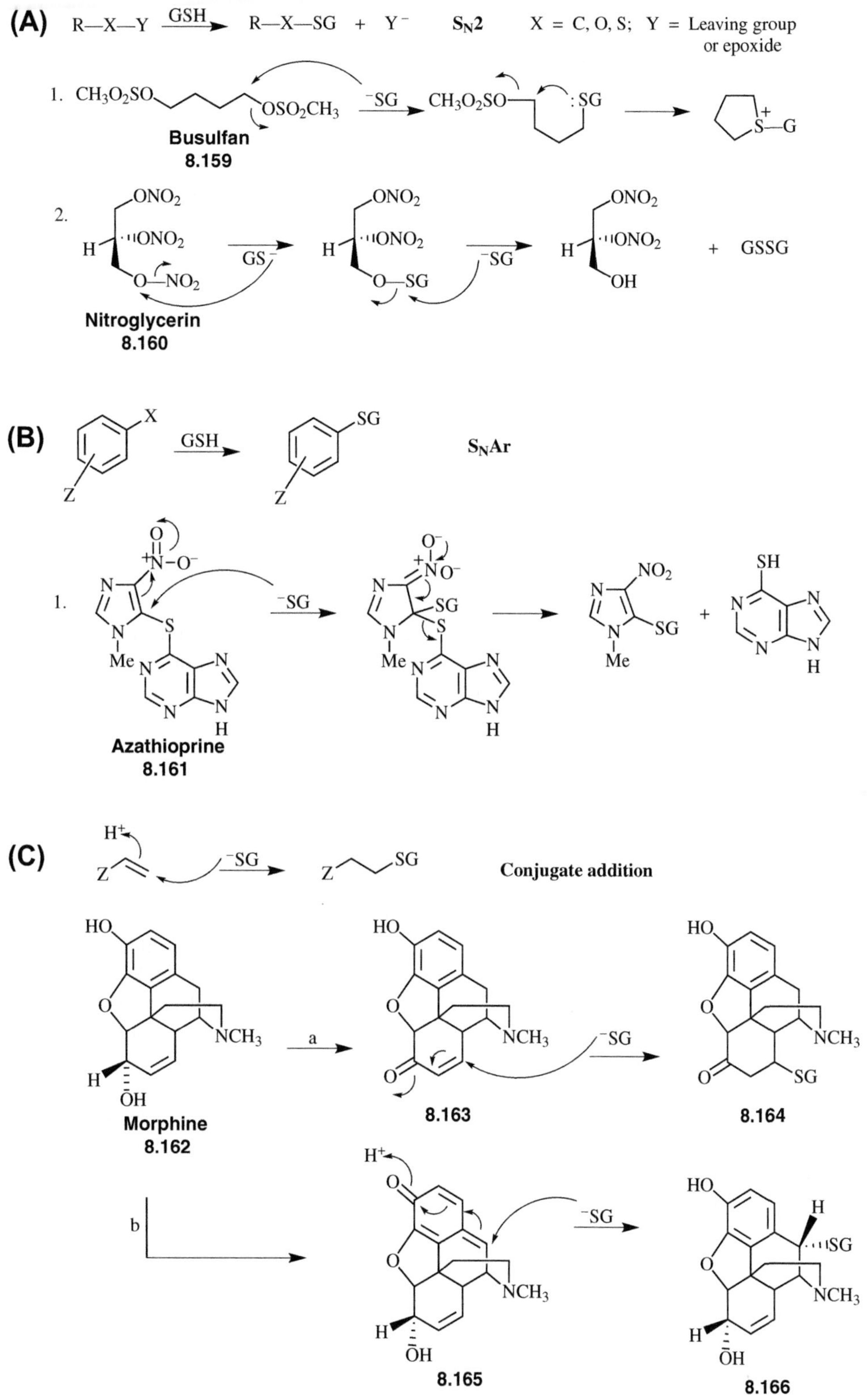

(A) R—X—Y $\xrightarrow{GSH}$ R—X—SG + Y^- S_N2 X = C, O, S; Y = Leaving group or epoxide

1. CH_3O_2SO~~~~OSO_2CH_3 $\xrightarrow{^-SG}$ CH_3O_2SO—⌐—:SG ⟶ S^+—G

Busulfan
8.159

2. Nitroglycerin 8.160 $\xrightarrow{GS^-}$ ... $\xrightarrow{^-SG}$ ... + GSSG

(B) $\xrightarrow{GSH}$ S_NAr

1. Azathioprine 8.161 $\xrightarrow{^-SG}$ ⟶

(C) $\xrightarrow{^-SG}$ Conjugate addition

Morphine 8.162 $\xrightarrow{a}$ 8.163 $\xrightarrow{^-SG}$ 8.164

b ⟶ 8.165 $\xrightarrow{^-SG}$ 8.166

SCHEME 8.50 Examples of glutathione conjugation

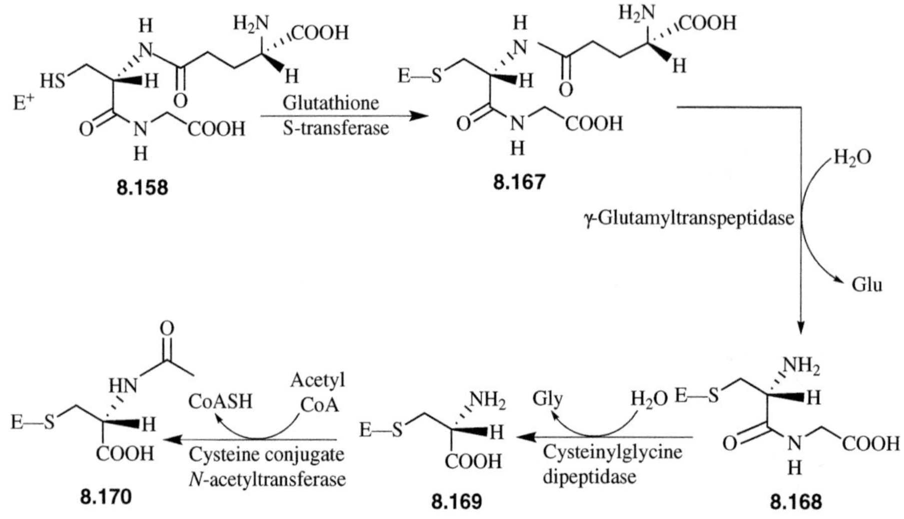

SCHEME 8.51 Metabolism of glutathione conjugates to mercapturic acid conjugates

and toxicity to drugs exhibiting *acetylation polymorphism.* Slow acetylators in general often develop adverse reactions as a result of tissue exposure to higher concentrations of the drugs; however, this also may result in longer drug effectiveness. Fast acetylators are more likely to show an inadequate therapeutic response to standard doses of the drug. Examples of drugs exhibiting acetylation polymorphism are the antibacterial drug sulfamethazine (**8.171**, Sulmet), the antituberculosis drug isoniazid (**8.172**, Rifamate),[297] and dapsone (**8.173**, Dapsone),[298] used in the treatment of leprosy.

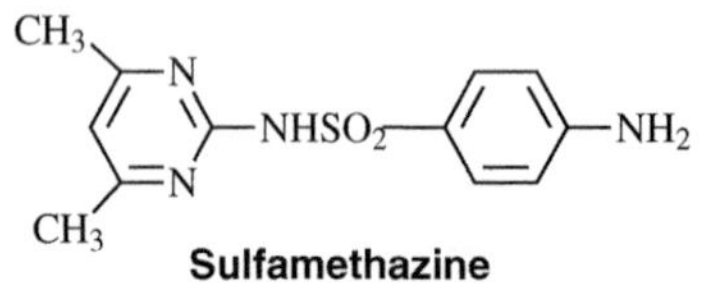

Sulfamethazine
8.171

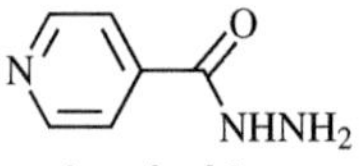

Isoniazid
8.172

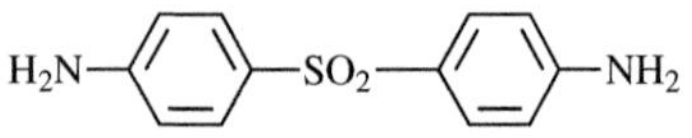

Dapsone
8.173

Acetylation is a two-step covalent catalytic process (Scheme 8.52). First, acetyl CoA acetylates an active site amino acid residue of the soluble hepatic *N*-acetyltransferase (**8.174**), and then the acetyl group is transferred to the substrate amino group.[299] Presumably, this two-step process allows the enzyme to have better control over the catalytic process.

One of the few examples of an aliphatic amine drug that is acetylated is cilastatin (**8.175**, R=H), which is metabolized to **8.175** (R=Ac).[300] Actually, cilastatin is not a drug, but is used in combination with the antibacterial drug imipenem (**8.176**) (the combination is called Primaxin). When imipenem is administered alone, it is rapidly hydrolyzed in the kidneys by dehydropeptidase I.[301] Cilastatin is a potent inhibitor of this enzyme, and it effectively prevents renal metabolism of imipenem when the two compounds are administered in combination (another example of drug synergism; see Chapter 7, Section 7.2.2.1).

Aromatic amines resulting from Phase I reduction of aromatic nitro compounds, such as the amine produced from the anticonvulsant drug clonazepam (see **8.109**, R=NO_2, Klonopin), also may be *N*-acetylated (**8.109**, R=NHAc).[302]

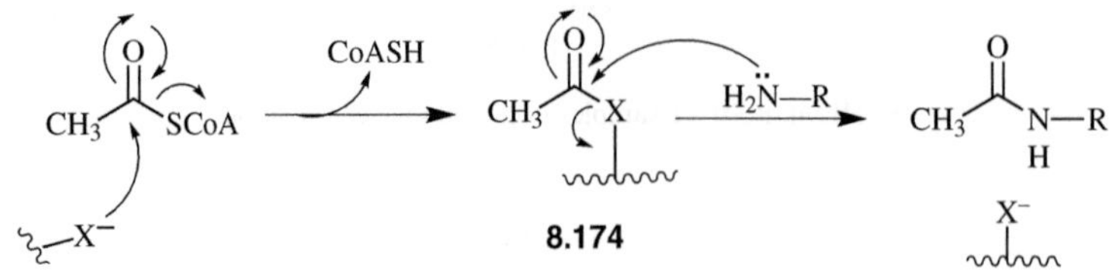

SCHEME 8.52 *N*-Acetylation of amines

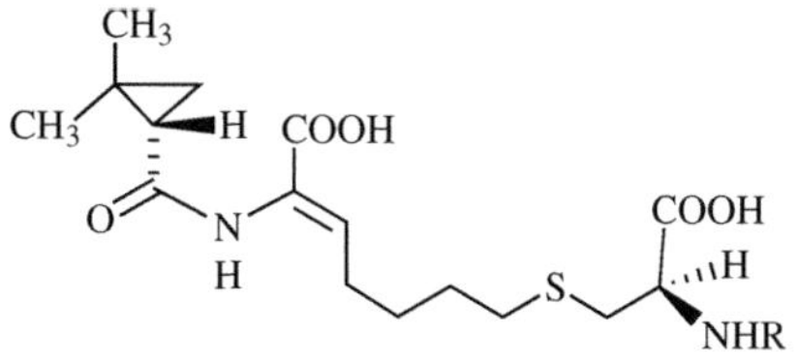

Cilastatin (R = H)
8.175

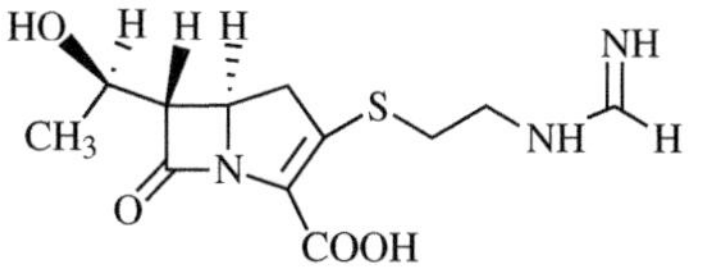

Imipenem
8.176

8.4.3.8. Fatty Acid and Cholesterol Conjugation

Compounds that contain a hydroxyl group can undergo conjugation reactions with a wide range of endogenous fatty acids, such as saturated acids from C_{10} to C_{16} and unsaturated acids such as oleic and linoleic acids.[303] The antiinflammatory agent etofenamate (**8.177**, Rheumon gel) is esterified by oleic, palmitic, linoleic, stearic, palmitoleic, myristic, and lauric acids,[304] presumably via their coenzyme A thioesters.[305] These metabolites account for only 0.1% of the administered dose in rats and humans. Often these lipophilic fatty acid conjugates accumulate in the tissues rather than being excreted. For example, fatty acid ester metabolites of Δ^1-7-hydroxytetrahydrocannabinol (**8.178**) are deposited in the liver, spleen, adipose tissue, and bone marrow and account for the majority of tissue residues of this compound determined 15 days after administration.[306]

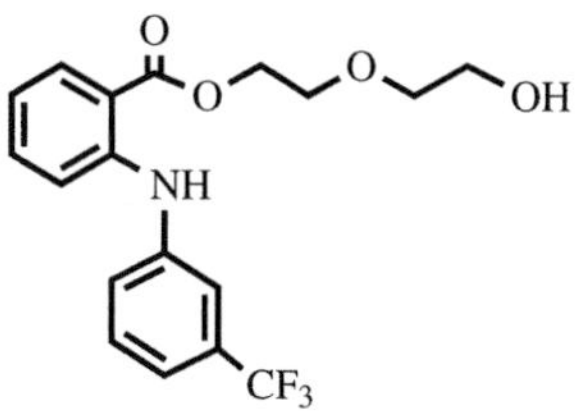

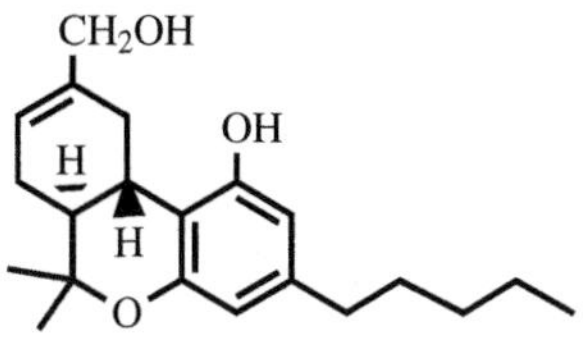

Etofenamate
8.177

Δ^1-7-hydroxytetrahydrocannabinol
8.178

"Flashbacks" by habitual cannabinoid users long after they stop using the drug are believed to be the result of the lipophilic conjugates retained in the tissues for extended periods, which are released and converted back to the cannabinoid.[307]

Cholesterol ester metabolites have been detected for drugs containing either an ester or a carboxylic acid. Transesterification of prednimustine (**8.179**, Sterecyt), an ester of chlorambucil (Leukeran) and prednisolone (Prelone), to the corresponding cholesterol (see **2.104**) ester of chlorambucil was catalyzed by the enzyme lecithin cholesterol acyltransferase.[308] The experimental hypolipidemic drug **8.180** was metabolized to the cholesterol ester, which deposited in the lipids in the liver; as a consequence, the development of this drug had to be discontinued.[309]

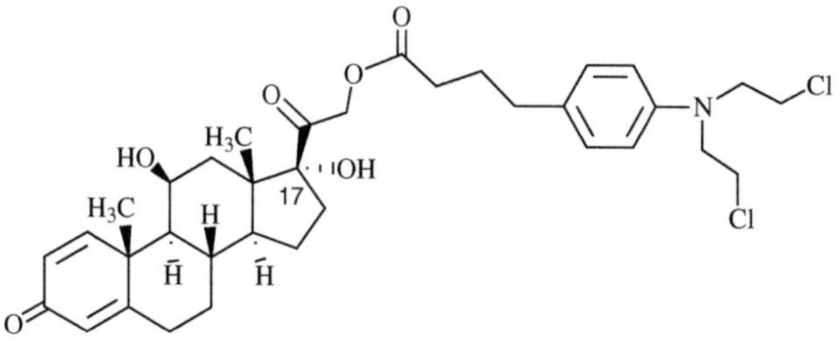

Prednimustine
8.179

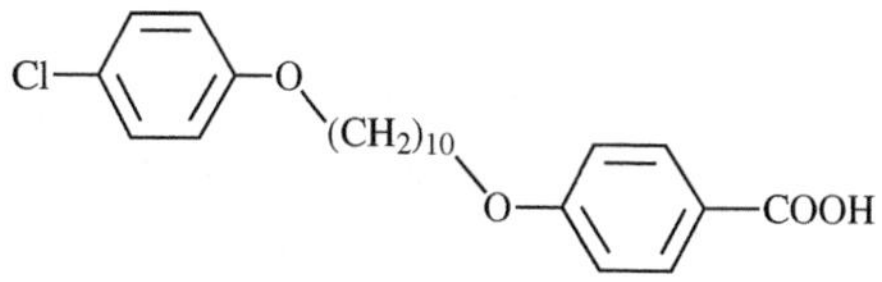

8.180

8.4.3.9. Methyl Conjugation

Methylation is a relatively minor conjugation pathway in drug metabolism, but it is very important in the biosynthesis of endogenous compounds such as epinephrine and melatonin; in the catabolism of biogenic amines such as norepinephrine, dopamine, serotonin, and histamine; and in modulating the activities of macromolecules such as proteins and nucleic acids.[310] Except when tertiary amines are converted to quaternary ammonium salts, methylation differs from almost all other conjugation reactions (excluding acetylation) in that it *reduces* the polarity and hydrophilicity of the substrates. In many cases, this conjugation results in compounds with decreased biological activity. In general, xenobiotics that undergo methylation do share marked structural similarities to endogenous substrates that are methylated.[311]

Methylation is a two-step process (Scheme 8.53). First the methyl-transferring coenzyme, S-adenosylmethionine (SAM, **8.182**), is biosynthesized mostly from methionine (**8.181**) in a reaction catalyzed by methionine adenosyltransferase. Some SAM is also produced by the donation of the methyl group of N^5-methyltetrahydrofolate

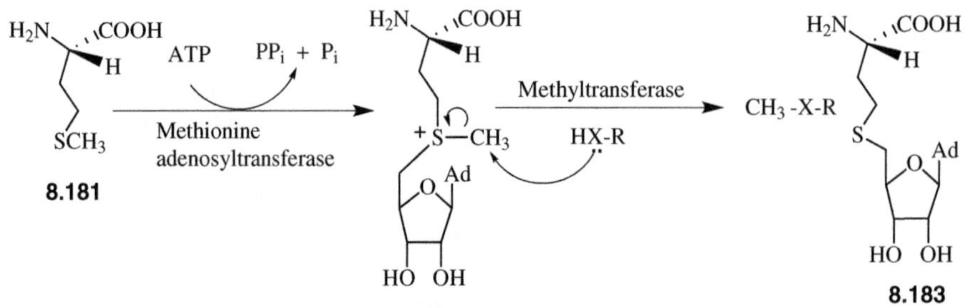

SCHEME 8.53 Methylation of xenobiotics

(see Chapter 4, Section 4.3.2) to *S*-adenosyl-L-homocysteine. Then the SAM is utilized in the transfer of the activated methyl group (Nature's methyl iodide) to the acceptor molecules (RXH), which include catechols and phenols, amines, and thiols. A variety of methyltransferases, such as catechol-*O*-methyltransferase (COMT), phenol-*O*-methyltransferase, phenylethanolamine-*N*-methyltransferase, and nonspecific amine *N*-methyltransferases and thiol *S*-methyltransferases, are responsible for catalyzing the transfer of the methyl group from SAM to RXH, which produces $RXCH_3$ and *S*-adenosylhomocysteine (**8.183**).[312]

COMT-catalyzed *O*-methylation of xenobiotic catechols leads to *O*-monomethylated catechol metabolites. Unlike the free catechols, these metabolites are not oxidized to reactive *ortho*-quinonoid species that can produce toxic effects (see Section 8.4.2.1.8). An example of this reaction is the metabolism of the β_1- and β_2-adrenergic bronchodilator isoproterenol (**8.184**, Isuprel), which is regioselectively methylated at the C-3 catechol OH group.[313] Compounds that are methylated by COMT must contain an aromatic 1,2-dihydroxy (catechol) group.

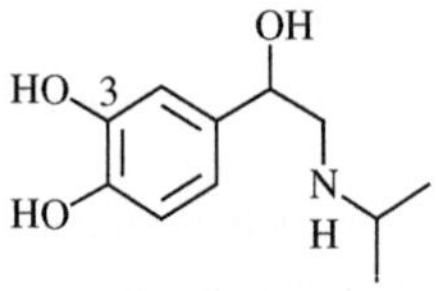

Isoproterenol
8.184

Terbutaline (**8.185**, Brethine), another β_2-adrenergic bronchodilator related in structure to isoproterenol except

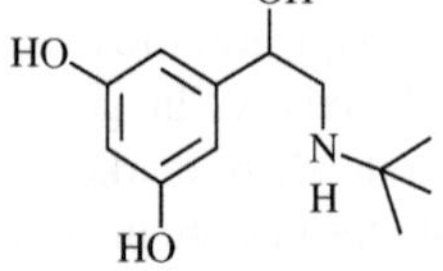

Terbutaline
8.185

that it contains a *meta*-dihydroxy arrangement of hydroxyl groups, does not undergo methylation.[314]

Phenol hydroxyl groups also undergo methylation. Morphine (**8.182**, R=H), for example, is metabolized in man to minor amounts of codeine (**8.186**, $R=CH_3$).[315]

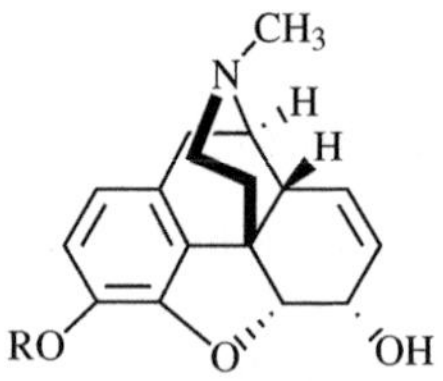

Morphine (R = H)
Codeine (R = CH_3)
8.186

N-Methylation of xenobiotics is less common, but occurs occasionally. The *N*-dealkylated primary amine metabolite (**8.187**, R=H) of the coronary vasodilator (β-adrenergic blocker) oxprenolol (**8.187**, $R=CH(CH_3)_2$, Trasicor) is *N*-methylated to **8.187** ($R=CH_3$).[316]

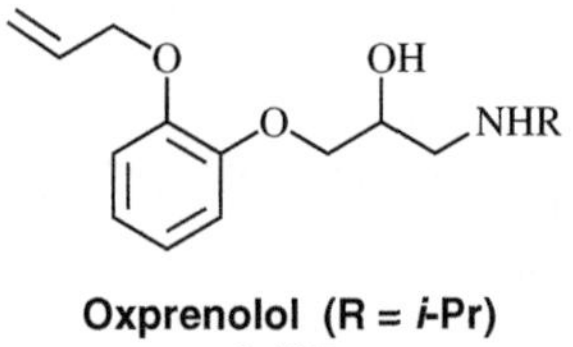

Oxprenolol (R = *i*-Pr)
8.187

Heterocyclic nitrogen atoms are also susceptible to *N*-methylation, such as in the case of the sedative clomethiazole (**8.188**, Heminevrin).[317]

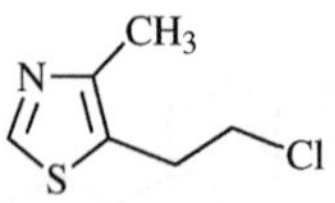

Clomethiazole
8.188

Methylation of sulfhydryl groups in xenobiotics is also known. The thiol group of the antihypertensive drug captopril (**8.189**, Capoten)[318] and the antithyroid drug propylthiouracil (**8.190**, Propylthiouracil)[319] undergo metabolic *S*-methylation.

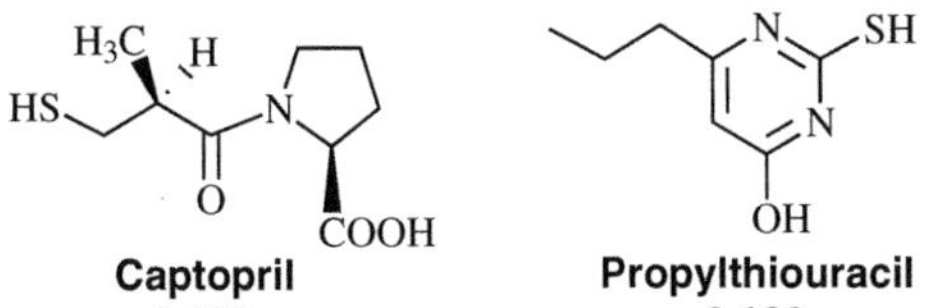

Captopril
8.189

Propylthiouracil
8.190

8.4.4. Toxicophores and Reactive Metabolites (RMs)

As was mentioned in Chapter 2 (Section 2.1.2.3.2), drug metabolism sometimes does not only degrade the drug for excretion but also converts it into a toxic species, often a RM.[320] This RM can be responsible for a variety of toxicities, including mutagenicity,[321] hepatotoxicity, renal toxicity, and cardiotoxicity.[322] If the RM forms a covalent bond with a macromolecule, this can lead to an immune response that causes *idiosyncratic* (specific to certain individuals and are unpredictable) *adverse drug reactions* (IADR) that are rare, do not show up in animal studies or in clinical trials, but, therefore, can be responsible for the withdrawal of a drug after it has been on the broader market.[323] We have already seen some of these IADR in this chapter, for example, mutagenicity of oxidized benzo[a]pyrene (Scheme 8.12) and aflatoxin B_1 (Scheme 8.14), metabolic activation of aromatic amines (Scheme 8.29), the bioactivation of acetaminophen (Schemes 8.32 and 8.33), oxidation of tienilic acid (Scheme 8.36), bioactivation of phenacetin (Scheme 8.47), and glutathione conjugation (Scheme 8.50). There are many drugs that were withdrawn from clinical trials and even the marketplace.[324] However, as pointed out in Chapter 2 (Section 2.1.2.3.2), toxicity appears to be dose related. For example, the largest selling drug to date is the cholesterol-lowering drug atorvastatin (**8.191**, Scheme 8.54, Lipitor). It undergoes metabolic activation to the quinone-imine (**8.192**),[325] which is highly electrophilic and can react with macromolecules; in fact, it is known that atorvastatin binds covalently to human liver microsomes,[326] but, apparently, because of the low dose (about 10 mg/day), this is a safe drug. However, lumiracoxib (**8.193**, Scheme 8.55, Prexige), which undergoes a metabolic activation mechanism similar to that of atorvastatin, is hepatotoxic and has been withdrawn from the market in several countries (it never was approved in the United States); its dose level is about 400 mg/day.

8.4.5. Hard and Soft (Antedrugs) Drugs

Rapid metabolism of a drug candidate can lead to the demise of the compound for further development because of poor bioavailability or short duration. However, sometimes the reverse problem also can be an issue, namely, a plasma half-life that is too long (the compound is not cleared at an adequate rate). Unless the drug can be eliminated, minor toxicities can accumulate and become serious problems. This phenomenon was observed during the discovery of the antiarthritis drug celecoxib (**8.195**; see Chapter 5, Section 5.3.2.2.2, Celebrex). One of the predecessors of **8.195** was **8.196** (in which a *para*-chloro group replaces the methyl group of celecoxib).[327] As we saw earlier, halogens are sometimes incorporated into the *para*-position of

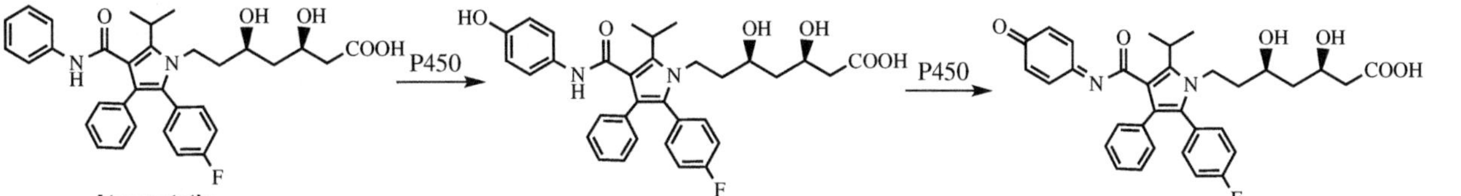

P450 P450

Atorvastatin
8.191

8.192

SCHEME 8.54 Metabolism of atorvastatin to a reactive quinone-imine

P450 P450

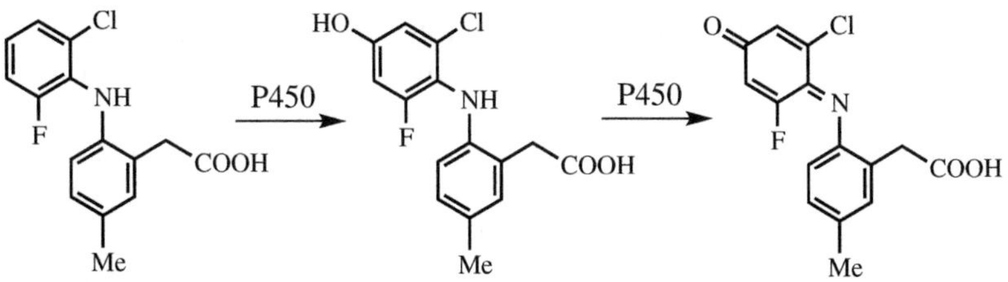

Lumiracoxib
8.193

8.194

SCHEME 8.55 Metabolism of lumiracoxib to a reactive quinone-imine

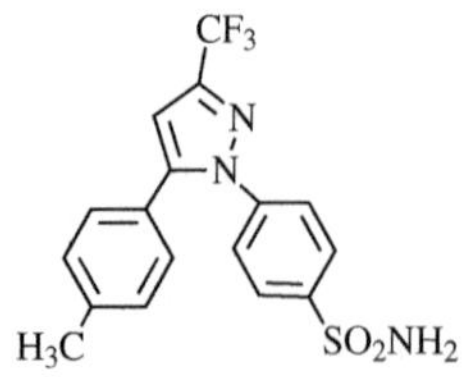

Celecoxib
8.195

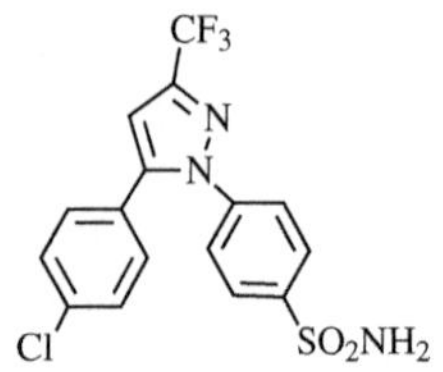

8.196

aromatic compounds to block aromatic oxidation (see Section 8.4.2.1.1). This compound was more COX-2 selective than celecoxib, but had a plasma half-life in dogs of 680 h (about a month)! This may sound like a favorable property, especially for a disease like arthritis in which the medication is taken every day anyway, but the compound accumulated in various tissues, and there was some liver toxicity as a result. To shorten the plasma half-life, the structure was modified to one that would more likely be metabolized and excreted, namely, by a bioisosteric replacement of the *para*-chloro group with a *para*-methyl group (which we saw is readily oxidized via the alcohol, then the aldehyde, to the carboxylic acid; see Sections 8.4.2.1.4 and 8.4.2.1.9). The resulting compound (celecoxib) has a plasma half-life in dogs of 9 h and in man of 10–12 h (peak plasma levels at about 2 h).

Bodor[328] has termed compounds like **8.196** *hard drugs* and those like celecoxib (**8.195**) as *soft drugs*. *Hard drugs* are nonmetabolizable compounds, characterized either by high lipid solubility and accumulation in adipose tissues and organelles or high water solubility. They are poor substrates for the metabolizing enzymes; the potentially metabolically sensitive parts of these drugs are either sterically hindered or the hydrogen atoms are substituted with halogens to block oxidation.

Soft drugs (also referred to as *antedrugs*[329]) are biologically active drugs designed to have a predictable and controllable metabolism to nontoxic products after they have achieved their desired pharmacological effect. By building into the molecule the most desirable way in which the molecule could be deactivated and detoxified shortly after it has exerted its biological effect, the therapeutic index could be increased, providing a safer drug. The advantages of soft drugs can be significant: (1) elimination of toxic metabolites, thereby increasing the therapeutic index of the drug; (2) avoidance of pharmacologically active metabolites that can lead to long-term effects; (3) elimination of drug interactions resulting from metabolite inhibition of enzymes; and (4) simplification of pharmacokinetic problems caused by multiple active species. The general characteristics of a soft drug are[330] (1) that it has a close structural similarity to the lead; (2) that is has a metabolically sensitive moiety built into the lead structure; (3) that the incorporated metabolically sensitive spot does not affect the overall physicochemical or steric properties of the lead compound; (4) that the built-in metabolism is the major, or preferably the only, metabolic route for drug deactivation; (5) that the rate of the predictable metabolism can be controlled; (6) that the products resulting from the metabolism are nontoxic and have no other biological activities; and (7) that the predicted metabolism does not lead to highly reactive intermediates.

Another example of this soft drug concept is the isosteric soft analog (**8.197**) of the hard antifungal cetylpyridinium chloride (**8.198**, Cepacol).[331] The soft analog has a metabolically soft spot (the ester group) built into the

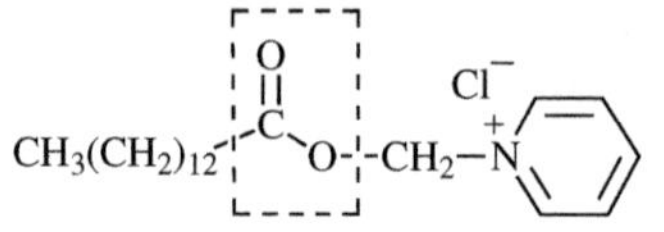

8.197

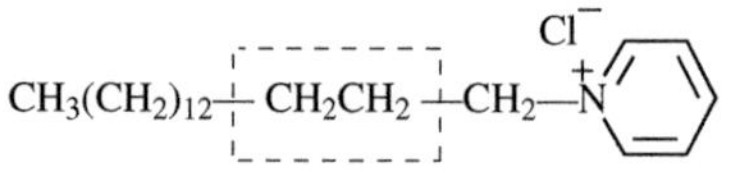

Cetylpyridinium chloride
8.198

structure for detoxification by an esterase-catalyzed hydrolysis. Of course, incorporation of the ester group cannot interfere with binding of the compound to the appropriate receptor if this approach is to be effective.

Another related approach to soft drug design is to utilize, as the basis for the design of the drug, a biologically inactive metabolite, which is modified so that it becomes active, but which can be metabolically degraded to the inactive metabolite after its therapeutic action is completed. An example of this approach is the soft ophthalmic antiinflammatory drug loteprednol etabonate (**8.199**, Alrex).[332] This glucocorticoid drug was designed based on the structure of an inactive metabolite of prednisolone (**8.200**), namely, **8.201**. The unfavorable responses, such as allergies and cataracts, that are observed in patients who undergo long-term glucocorticoid therapy have been proposed to occur because of the α-hydroxycarbonyl group at the 17β-position.[333] Bodor's group changed this to a carboxylic acid (**8.201**), but the compound was inactive. The approach taken, then, involved making compounds that are active, but are metabolically degraded to a biologically inactive metabolite, such as **8.201**, which does not bind to the glucocorticoid receptor. The active compound was found to be **8.199** (a potent binder of the glucocorticoid receptor). Compound **8.199** is metabolically degraded by nonspecific esterases to **8.201**, which is readily cleared. This approach is like doing a retrosynthesis in organic chemistry; you have to think backward from the desired inactive metabolite to an active compound that can be metabolically degraded to the inactive metabolite.

A soft drug, then, is an active drug that is degraded by metabolism *after* it carries out its therapeutic role. Another way of utilizing metabolism in drug design is to modify a

Loteprednol etabonate
8.199

Prednisolone
8.200

8.201

drug so that it is not active, but before it reaches the site of action, is metabolically transformed into the active drug. These compounds are known as *prodrugs*, and they are the topic of the last chapter.

8.5. GENERAL REFERENCES

Synthesis of Radioactive Compounds

Halldin, C.; Högberg, T.; Krogsgaard-Larsen, P.; Liljefors, T.; Madsen, U. (Eds.), Radiotracers: synthesis and use in imaging. *Textbook of Drug Design and Discovery*, 3rd ed., Taylor & Francis, London, U. K., 2002, 229–258.

Muccino, R. R. (Ed.), *Synthesis and Applications of Radioactive Compounds*, Elsevier, Amsterdam, 1986.

Analytical Methods in Drug Metabolism

NMR Spectroscopy

Claridge, T. D. W. *High-Resolution NMR Techniques in Organic Chemistry*, 2nd ed., Elsevier, Amsterdam, The Netherlands, 2009.

Friebolin, H. *Basic One- and Two-Dimensional NMR Spectroscopy*, 5th ed., Wiley-VCH, Weinheim, Germany, 2011.

Jacobsen, N. E. *NMR Spectroscopy Explained*, Wiley, Hoboken, NJ, 2007.

Keeler, J. *Understanding NMR Spectroscopy*, 2nd ed., Wiley, 2010.

Lambert, J. B.; Mazzola, E. P. *Nuclear Magnetic Resonance Spectroscopy: An Introduction to Principles, Applications, and Experimental Methods*, Prentice-Hall, Upper Saddle River, NJ, 2003.

Mass Spectrometry

Cole, R. B. (Ed.), *Electrospray and MALDI Mass Spectrometry*, 2nd ed., John Wiley & Sons, Hoboken, NJ, 2010.

de Hoffmann, E.; Stroobant, V. *Mass Spectrometry: Principles and Applications*, 2nd ed., Wiley, Chichester, U. K., 2007.

Gross, J. H. *Mass Spectrometry: A Textbook*, 2nd ed.,Springer-Verlag, Berlin, Germany, 2011.

Korfmacher, W. A. (Ed.), *Mass Spectrometry for Drug Metabolism Studies*, 2nd ed., CRC Press, Boca Raton, FL, 2010.

Lee, M. S.; Zhu, M. (Eds.), *Mass Spectrometry in Drug Metabolism and Disposition: Basic Principles and Applications*, Wiley, Hoboken, N. J., 2011.

McMaster, M. C. *LC/MS, A Practical User's Guide*, Wiley, Hoboken, NJ, 2005.

Ramanathan, R. (Ed.), *Mass Spectrometry in Drug Metabolism and Pharmacokinetics*; John Wiley & Sons: Hoboken, NJ, 2009.

Rossi, D. T.; Sinz, M. W. (Eds.), *Mass Spectrometry in Drug Discovery*, Marcel Dekker, New York, 2002.

Testa, B.; Krämer, S. D. *The Biochemistry of Drug Metabolism: Two Volume Set*, Wiley-VCH, Weinheim, Germany, 2010.

Watson, J. T.; Sparkman, O. D. *Introduction to Mass Spectrometry*, 4th ed., Wiley, Chichester, U. K., 2007.

HPLC

Ahuja, S.; Rasmussen, H. (Eds.), *HPLC Method Development for Pharmaceuticals*, Elsevier: Amsterdam, The Netherlands, 2007.

Dong, M. W. *Modern HPLC for Practicing Scientists*, Wiley & Sons, Hoboken, NJ, 2006.

Kazakevich, Y. V.; LoBrutto, R. (Ed.), *HPLC for Pharmaceutical Scientists*, Wiley & Sons, Hoboken, NJ, 2007.

Mascher, H. *HPLC Methods for Clinical Pharmaceutical Analysis: A User's Guide*, Wiley-VCH, Weinheim, Germany, 2012.

Pathways for Drug Deactivation and Elimination

Cheng, X.; Blumenthal, R. M. (Ed.), *S-Adenosylmethionine-Dependent Methyltransferases: Structures and Functions*, World Scientific Pub, 2000.

Danielson, P. B. The Cytochrome P450 Superfamily: Biochemistry, Evolution and Drug Metabolism in Humans. *Curr. Drug Metab.* 2002, *3*, 561–597.

Dolphin, D.; Poulson, R.; Avramovic, O. *Glutathione*, Wiley, New York, 1989, Vols. 1–2.

Guengerich, F. P. (Ed.), *Mammalian Cytochromes P450*, CRC Press, Boca Raton, FL, 1987; Vols. 1–2.
Hawkins, D. R. (Ed.), *Biotransformations*, Royal Society of Chemistry, Cambridge, U.K., annual series, beginning in 1989.
Ioannides, C. (Ed.), *Enzyme Systems That Metabolise Drugs and Other Xenobiotics (Current Toxicology Series)*, John Wiley & Sons, New York, 2002.
Ionescu, C.; Caira, M. R. *Drug Metabolism. Current Concepts*, Springer, Dordrecht, The Netherlands, 2010.
Khojasteh, S. C.; Wong, H.; Hop, C. E. C. *A. Drug Metabolism and Pharmacokinetics Quick Guide*, Springer, New York, 2011.
Mulder, G. J. (Ed.), *Conjugation Reactions in Drug Metabolism*, Taylor & Francis, London, 1990.
Nassar, A. F. (Ed.), *Biotransformation and Metabolite Elucidation of Xenobiotics*, Wiley & Sons, Hoboken, NJ, 2010.
Nassar, A. F.; Hollenberg, P. F.; Scatina, J. (Eds.), *Drug Metabolism Handbook: Concepts and Applications*, Wiley, Hoboken, N. J., 2009.
Ortiz de Montellano, P. R. (Ed.), *Cytochrome P450: Structure, Mechanism, and Biochemistry*, 3rd ed., Kluwer Academic, New York, NY, 2005.
Pearson, P. G.; Wienkers, L. C. (Eds.), Handbook of Drug Metabolism, 2nd ed., Informa Healthcare, New York, NY, 2009.
St. Jean, Jr., D. J.; Fotsch, C. Mitigating heterocycle metabolism in drug discovery. *J. Med. Chem.* **2012**, *55*, 6002–6020.
Steventon, G.; Mitchell, S.; Hutt, A. *Drug Metabolism and Pharmacokinetics: The Journey of a Chemical Through the Body*, Wiley, Hoboken, NJ, 2012.
Testa, B.; Krämer, S. D. *The Biochemistry of Drug Metabolism (Two-volume set)*, Wiley-VCH, Weinheim, Germany, 2010.
Uetrecht, J. P.; Trager, W. *Drug Metabolism. Chemical and Enzymatic Aspects*, Informa Healthcare, New York, 2007.
Weber, W. W. *The Acetylator Genes and Drug Response*, Oxford, New York, 1997.
Zhang, D.; Zhu, M.; Humphreys, W. G. *Drug Metabolism in Drug Design and Development*; Wiley, Hoboken, N. J., 2008.

Journals

Chemico-Biological Interactions
Computer Methods and Programs in Biomedicine
Current Drug Metabolism
Drug Metabolism & Disposition
Drug Metabolism & Drug Interactions
Drug Metabolism Letters
Drug Metabolism and Pharmacokinetics
Drug Metabolism Reviews
European Journal of Drug Metabolism and Pharmacokinetics
Frontiers in Drug Metabolism and Transport
Journal of Drug Metabolism & Toxicology
Open Drug Metabolism Journal
The Pharmacogenomics Journal

Software

PredictIt™ Metabolism (Bio-Rad)
PredictIt™ Toxicity (Bio-Rad)
MetaPred (http://crdd.osdd.net/raghava/metapred/)
MetaSite (Molecular Discovery; http://www.moldiscovery.com/soft_metasite.php)
Meteor (Lhasa Limited; https://www.lhasalimited.org/meteor/general_information/)
Optdes (Graham, G.; Gueorguieva, I.; Dickens, K. *Comp. Meth. Prog. Biomed.* **2005**, *78*, 237–249.
StarDrop (Optibrium; http://www.optibrium.com/stardrop/stardrop-p450-models.php)

8.6. PROBLEMS (ANSWERS CAN BE FOUND IN THE APPENDIX AT THE END OF THE BOOK)

1. How do you explain the observation that two enantiomers can be metabolized to completely different products?
2. Why do the two enantiomers of glutethimide (**8.41**) undergo P450-catalyzed hydroxylation at two different sites?

Ph O N H O **8.41** (+)- Isomer HO Ph O N H O (-)- Isomer Ph OH O N H O

3. Draw a mechanism to account for the following metabolic reaction:

Cl → Cl OH

4. Compound **1** is metabolized to **2**; however, the metabolic enzyme that acts on **1** is slowly inactivated. Draw a mechanism for the conversion of **1** to **2** that also rationalizes the loss of enzyme activity (indicate at what point inactivation could occur).

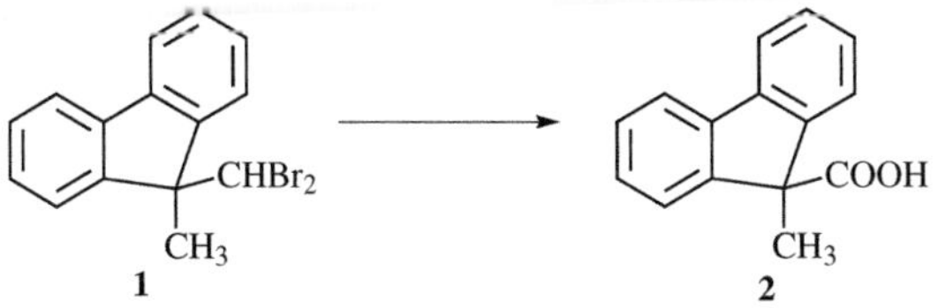

CHBr$_2$ CH$_3$ 1 → COOH CH$_3$ 2

5. Draw two mechanisms for oxidative aromatization (**8.97** to **8.98** in the text).
6. Show all of the steps (not the mechanisms) of this metabolic pathway, and note over each arrow used in your scheme the general types of enzymes involved.

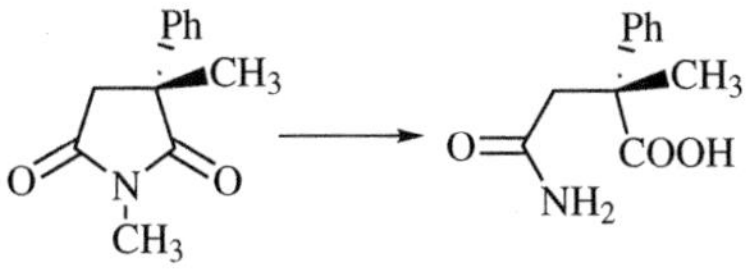

Ph CH$_3$ O N O CH$_3$ → O Ph CH$_3$ COOH NH$_2$

7. The antiestrogen drug toremifene (**3**) is metabolized into six major metabolites. Draw six reasonable metabolites for this drug, and note over each arrow what kind of reaction and enzyme are involved.

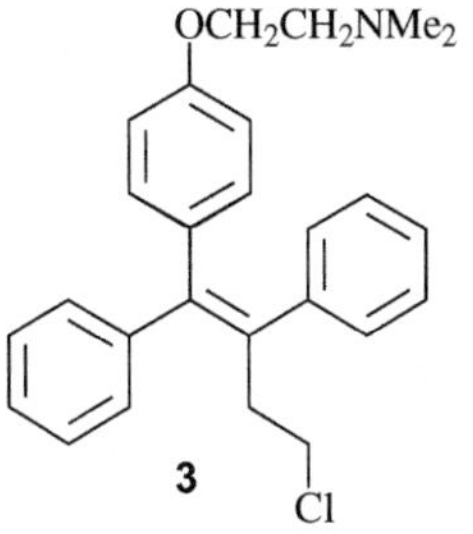

OCH$_2$CH$_2$NMe$_2$ 3 Cl

8. Compound **4** undergoes P450-catalyzed conversion to **5** and **6**. Draw a mechanism to account for this metabolic reaction.

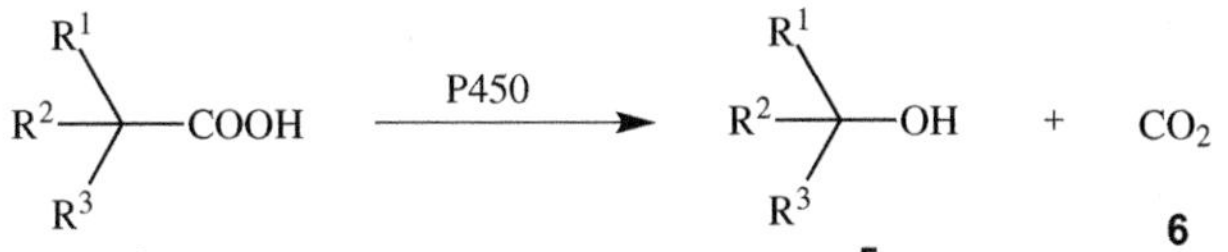

R^1 R^2 COOH R^3 4 —P450→ R^1 R^2 OH R^3 5 + CO$_2$ 6

9. Compound **7** is metabolized by *O*-dealkylation to a dicarboxylic acid. Draw a mechanism.

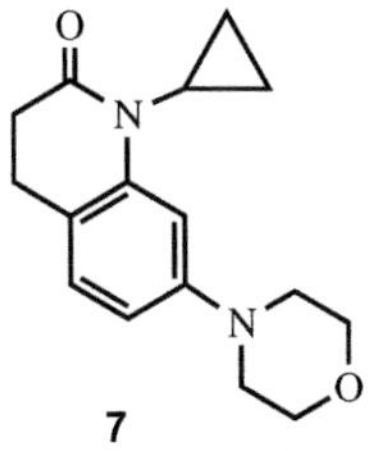

O N N O 7

10. Why is glutathione in high concentrations (5–10 mM) in many tissues?
11. Which of the hypothetical metabolic reactions shown in Scheme I on the next page are Phase I and which are Phase II? Name the reactions, and note the cofactors (no mechanisms needed).
12. Compound **8** was metabolized to **9** (GSH is glutathione). Draw a reasonable mechanism to account for this metabolic reaction.

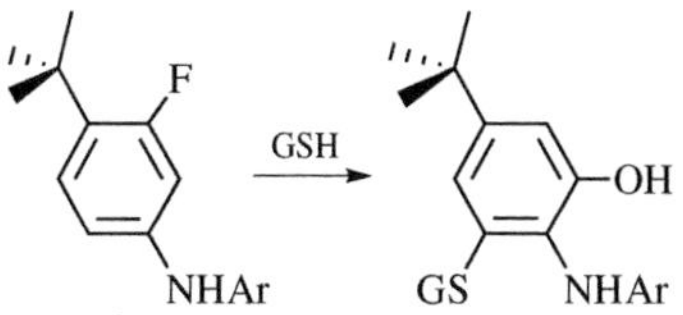

F NHAr 8 —GSH→ OH GS NHAr 9

13. Predict the structures of the compounds that produce the metabolites shown in (A), (B), and (C) below (you need to work backward from the metabolite to the compound). Show the steps (not mechanisms), and name the enzymes.

(A) O O COOH + NH$_2$ **(2 steps)**

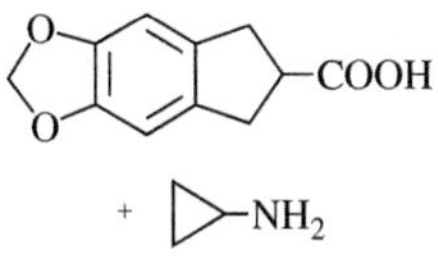

(B) CO$_2$^{t}Bu O O OH OH OH COOH **(2 steps)**

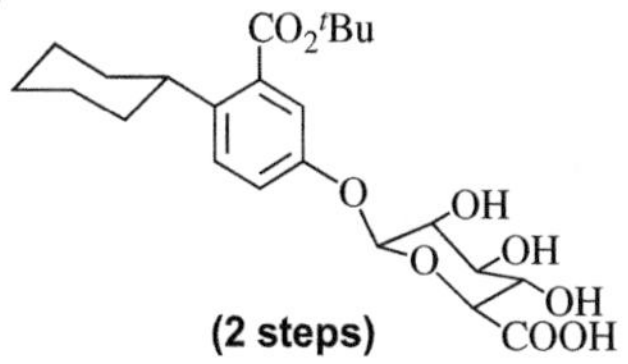

(C) OH CH$_3$ OH HO$_2$C SG + EtOH **(4 steps)**

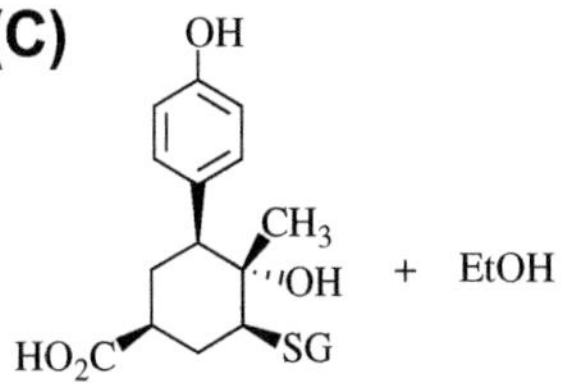

14. Statistics show that about 50% of AIDS patients in the United States experience serious diarrhea (AIDS-related diarrhea). N^1, N^{14}-Diethylhomospermine (**10**) prevents diarrhea in rodents, but the metabolite, homospermine (**11**), accumulates and persists in tissues for a protracted period, which accounts for the chronic toxicity of **10**.

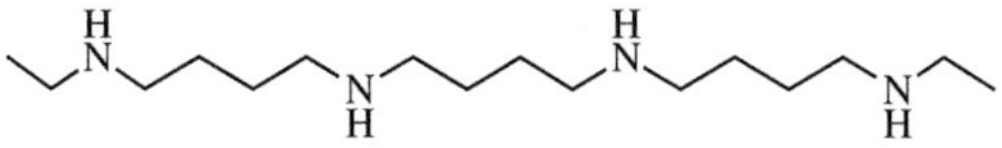

H N N H H N N H 10

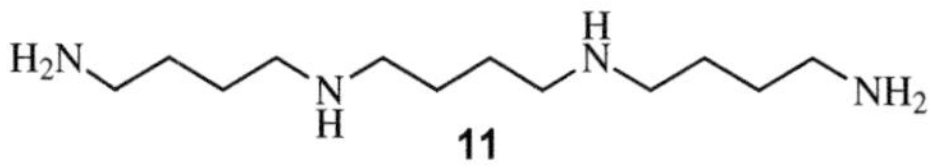

H$_2$N N H H N NH$_2$ 11

Scheme I

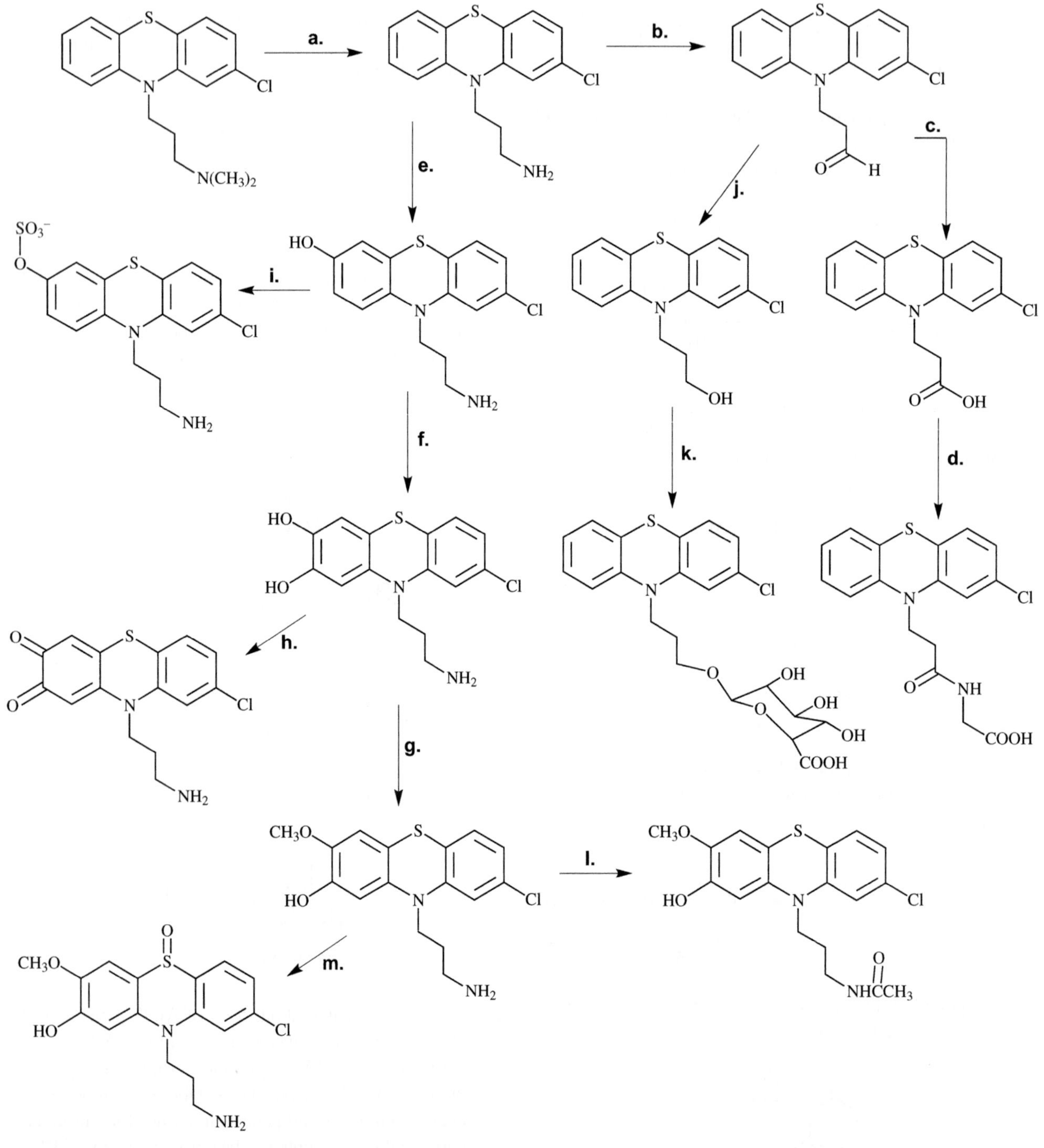

a. Suggest an approach you would take to get around the toxicity problem.
b. Draw a compound you might try to overcome the long plasma half-life, and note how you think it might work.

15. The thiazole ring of **12** is metabolized to a toxic metabolite. Suggest two approaches you would take to prevent this toxicity.

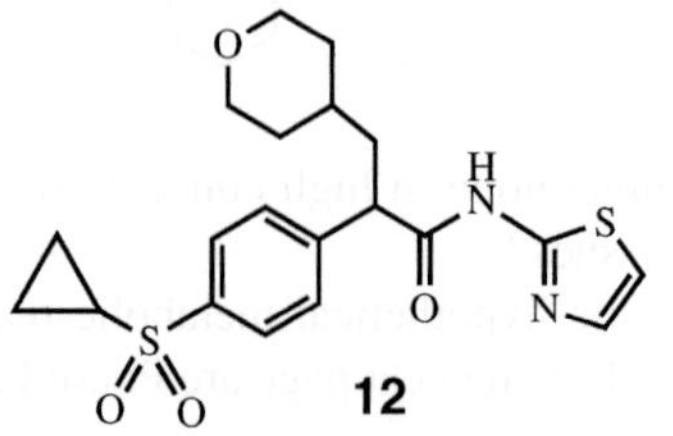

12

REFERENCES

1. Chen, Z.; Zhang, J.; Stamler, J. S. Identification of the enzymatic mechanism of nitroglycerin bioactivation. *Proc. Natl. Acad. Sci. U.S.A.* **2002**, *99*, 8306–8311.
2. (a) Honig, P. K.; Woosley, R. L.; Zamani, K.; Conner, D.P.; Cantilena, L. R., Jr. Changes in the pharmacokinetics and electrocardiograph pharmacodynamics of terfenadine with concomitant administration of erythromycin. *Clin. Pharmacol. Ther.* **1992**, *52*, 231–238. (b) Honig, P. K.; Wortham, D. C.; Zamani, K.; Conner, D. P.; Mullin, J. C.; Cantilena, L.R. Terfenadine-ketoconazole interaction. Pharmacokinetic and electrocardiographic consequences. *J. Am. Med. Assoc.* **1993**, *269*, 1513–1518.
3. Roy, M.-L.; Dumaine, R.; Brown, A. M. HERG, a primary human ventricular target of the nonsedating antihistamine terfenadine. *Circulation* **1996**, *94*, 817–823.
4. Scherer, C. R.; Lerche, C.; Decher, N.; Dennis, A. T.; Maier, P.; Ficker, E.; Busch, A. E.; Wollnik, B.; Steinmeyer, K. The antihistamine fexofenadine does not affect IKr currents in a case report of drug-induced cardiac arrhythmia. *Br. J. Pharmacol.* **2002**, *137*(6), 892–900.
5. Markham, A.; Wagstaff, A. J. Fexofenadine. *Drugs* **1998**, *55*, 269–274.
6. Waring, M. J.; Johnstone, C. A. A quantitative assessment of hERG liability as a function of lipophilicity. *Bioorg. Med. Chem. Lett.* **2007**, *17*, 1759–1764.
7. (a) Lim, C.-K.; Lord, G. Current developments in LC-MS for pharmaceutical analysis. *Biol. Pharm. Bull.* **2002**, *25*, 547–557. (b) Jackson, P. J.; Brownsill, R. D.; Taylor, A. R.; Walther, B. Use of electrospray ionization and neutral loss liquid chromatography/tandem mass spectrometry in drug metabolism studies. *J. Mass Spectrom.* **1995**, *30*, 446–451.
8. (a) Dalvie, D. Recent advances in the applications of radioisotopes in drug metabolism, toxicology and pharmacokinetics. *Curr. Pharm. Des.* **2000**, *6*, 1009–1028. (b) Dain, J. G.; Collins, J. M.; Robinson, W. T. A regulatory and industrial perspective of the use of carbon-14 and tritium isotopes in human ADME studies. *Pharm. Res.* **1994**, *11*, 925–928.
9. McCarthy, K. E. Recent advances in the design and synthesis of carbon-14 labelled pharmaceuticals from small molecule precursors. *Curr. Pharm. Des.* **2000**, *6*, 1057–1083.
10. Saljoughian, M.; Williams, P. G. Recent developments in tritium incorporation for radiotracer studies. *Curr. Pharm. Des.* **2000**, *6*, 1029–1056.
11. Hirth, J.; Watkins, P. B.; Strawderman, M.; Schott, A.; Bruno, R.; Baker, L. H. The effect of an individual's cytochrome CYP3A4 activity on docetaxel clearance. *Clin. Cancer Res.* **2000**, *6*, 1255–1258.
12. Wiberg, K. B. The deuterium isotope effect. *Chem. Rev.* **1955**, *55*, 713–743.
13. Brickner, S. J.; Hutchinson, D. K.; Barbachyn, M. R.; Manninen, P. R.; Ulanowicz, D. A.; Garmon, S. A.; Grega, K. C.; Hendges, S. K.; Toops, D. S.; Ford, C. W.; Zurenko, G. E. Synthesis and antibacterial activity of U-100592 and U-100766, two oxazolidinone antibacterial agents for the potential treatment of multidrug-resistant gram-positive bacterial infections. *J. Med. Chem.* **1996**, *39*, 673–679.
14. Slatter, J. G.; Stalker, D. J.; Feenstra, K. L.; Welshman, I. R.; Bruss, J. B.; Sams, J. P.; Johnson, M. G.; Sanders, P. E.; Hauer, M. J.; Fagerness, P. E.; Stryd, R. P.; Peng, G. W.; Shobe, E. M. Pharmacokinetics, metabolism, and excretion of linezolid following an oral dose of [^{14}C]linezolid to healthy human subjects. *Drug Metab. Dispos.* **2001**, *29*, 1136–1145.
15. U. S. Code of Federal Regulations, Title 21, Drugs Used in Research; Part 361.1, U. S. Government Printing Office, Washington, DC, 1986, p. 160.
16. deSousa, G.; Florence, N.; Valles, B.; Coassolo, P.; Rahmani, R. Relationships between in vitro and in vivo biotransformation of drugs in humans and animals: pharmacotoxicological consequences. *Cell Biol. Toxicol.* **1995**, *11*, 147–153.
17. Coe, R. A. J. Quantitative whole-body autoradiography. *Regul. Toxicol. Pharmacol.* **2000**, *31*, S1–S3.
18. Solon, E. G.; Kraus, L. Quantitative whole-body autoradiography in the pharmaceutical industry: Survey results on study design, methods, and regulatory compliance. *J. Pharmacol. Toxicol. Meth.* **2002**, *46*, 73–81.
19. Henion, J.; Brewer, E.; Rule, G. Sample preparation for LC/MS/MS: analyzing biological and environmental samples. *Anal. Chem.* **1998**, *70*, 650A–656A.
20. (a) Schill, G.; Borg, K. O.; Modin, R.; Persson, B. A. In Progress in Drug Metabolism; Bridges, J. W.; Chasseaud, L. F. (Eds.), Wiley, New York, 1977; Vol. 2, p. 219. (b) Schill, G.; Modin, R.; Borg, K. O.; Persson, B. A. In *Drug Fate and Metabolism: Methods and Techniques*, Garrett, E. R.; Hirtz, J. L. (Eds.), Marcel Dekker, New York, 1977; Vol. 1, p. 135.
21. Horning, M. G.; Gregory, P.; Nowlin, J.; Stafford, M.; Letratanangkoon, K.; Butler, C.; Stillwell, W. G.; Hill, R. M. Isolation of drugs and drug metabolites from biological fluids by use of salt-solvent pairs. *Clin. Chem.* **1974**, *20*, 282–287.
22. Thompson, J. A.; Markey, S. P. Quantitative metabolic profiling of urinary organic acids by gas chromatography-mass spectrometry. Comparison of isolation methods. *Anal. Chem.* **1975**, *47*, 1313–1321.
23. Brodie, B. B.; Cho, A. K.; Gessa, G. L. In *Amphetamines and Related Compounds*; Costa, E.; Garattini, S. (Eds.), Raven, New York, 1970, p. 217.
24. Stolman, A.; Pranitis, P. A. F. XAD-2 resin drug extraction methods for biologic samples. *Clin. Toxicol.* **1977**, *10*, 49–60.
25. (a) Xia, Y. Q.; Whigan, D. B.; Powell, M. L.; Jemal, M. Ternary-column system for high-throughput direct-injection bioanalysis by liquid chromatography/tandem mass spectrometry. *Rapid Commun. Mass Spectrom.* **2000**, *14*, 105–111. (b) Wu, J. T.; Zeng, H.; Qian, M.; Brogdon, B. L.; Unger, S. E. Direct plasma sample injection in multiple-component LC-MS-MS assays for high-throughput pharmacokinetic screening. *Anal. Chem.* **2000**, *72*, 61–67.
26. (a) Bennett, P.; Li, Y. T.; Edom, R.; Henion, J. Quantitative determination of orlistat (tetrahydrolipostatin, Ro 18-0647) in human plasma by high-performance liquid chromatography coupled with ion spray tandem mass spectrometry. *J. Mass Spectrom.* **1997**, *32*, 739–749. (b) Xia, Y.; Whigan, D.; Jemal, M. A simple liquid-liquid extraction with hexane for low-picogram determination of drugs and their metabolites in plasma by high-performance liquid chromatography with positive ion electrospray tandem mass spectrometry. *Rapid Commun. Mass Spectrom.* **1999**, *13*, 1611–1621.
27. Janiszewski, J.; Swyden, M.; Fouda, H. High-throughput method development approaches for bioanalytical mass spectrometry. *J. Chromatogr. Sci.* **2000**, *38*, 255–258.
28. (a) Zhang, N.; Hoffman, K. L.; Li, W.; Rossi, D. T. Semi-automated 96-well liquid-liquid extraction for quantitation of drugs in biological fluids. *J. Pharm. Biomed. Anal.* **2000**, *22*, 131–138. (b) Steinborner, S.; Henion, J. Liquid-liquid extraction in the 96-well plate format with SRM LC/MS quantitative determination of methotrexate and its major metabolite in human plasma. *Anal. Chem.* **1999**, *71*, 2340–2345.

29. Marchese, A.; McHugh, C.; Kehler, J.; Bi, H. Determination of pranlukast and its metabolites in human plasma by LC/MS/MS with PROSPEKTTM online solid-phase extraction. *J. Mass Spectrom.* **1998**, *33*, 1071–1079.
30. (a) Needham, S. R.; Cole, M. J.; Fouda, H. G. Direct plasma injection for high-performance liquid chromatographic-mass spectrometric quantitation of the anxiolytic agent CP-93,393. *J. Chromatogr. B* **1998**, *718*, 87–94. (b) Jemal, M.; Huang, M.; Jiang, X.; Mao, Y.; Powell, M. Direct injection vs liquid-liquid extraction for plasma sample analysis by high performance liquid chromatography with tandem mass spectrometry. *Rapid Commun. Mass Spectrom.* **1999**, *13*, 2125–2132.
31. Maurer, H. H. Systematic toxicological analysis procedures for acidic drugs and/or metabolites relevant to clinical and forensic toxicology and/or doping control. *J. Chromatogr. B* **1999**, *733*(1–2), 3–25.
32. (a) Hadley, M. R.; Camilleri, P.; Hutt, A. J. Enantiospecific analysis by capillary electrophoresis: applications in drug metabolism and pharmacokinetics. *Electrophoresis* **2000**, *21*, 1953–1976. (b) Naylor, S.; Benson, L. M.; Tomlinson, A. J. Application of capillary electrophoresis and related techniques to drug metabolism studies. *J. Chromatogr. A* **1996**, *735*(1–2), 415–438.
33. (a) Wells, R. J. Recent advances in non-silylation derivatization techniques for gas chromatography. *J. Chromatogr. A* **1999**, *843*, 1–18. (b) Little, J. L. Artifacts in trimethylsilyl derivatization reactions and ways to avoid them. *J. Chromatogr. A* **1999**, *844*, 1–22.
34. (a) Lim, C.-K.; Lord, G. Current developments in LC-MS for pharmaceutical analysis. *Biol. Pharm. Bull.* **2002**, *25*, 547–557. (b) Cole, M. J.; Janiszewski, J. S.; Fouda, H. G. Electrospray mass spectrometry in contemporary drug metabolism and pharmacokinetics. *Pract. Spectrosc.* **2002**, *32*, 211–249. (c) Gibbs, B.; Masse, R. MS analytical tools in drug discovery and development - mass spectrometry (MS) has underpinned many of the recent advances in drug discovery and development technologies. *Innov. Pharm. Technol.* **2001**, *1, 76, 78, 80, 82, 84*.
35. (a) Von Brocke, A.; Nicholson, G.; Bayer, E. Recent advances in capillary electrophoresis/electrospray-mass spectrometry. *Electrophoresis* **2001**, *22*, 1251–1266. (b) Cherkaoui, S.; Rudaz, S.; Varesio, E.; Veuthey, J.-L. Onlne capillary electrophoresis-electrospray mass spectrometry for the analysis of pharmaceuticals. *Chimia* **1999**, *53*, 501–505.
36. (a) Deng, Y.; Zeng, H.; Wu, J.-T. New analytical approaches for high-throughput quantitation of drug candidates in biological fluids. *Recent Res. Develop. Anal. Chem.* **2001**, *1*, 45–60. (b) Jemal, M. High-throughput quantitative bioanalysis by LC/MS/MS. *Biomed. Chromatogr.* **2000**, *14*, 422–429.
37. Miller, J. D.; Chang, S. Y. ADME studies using fast gradient HPLC-tandem MS. *PharmaGenomics* **2001**, (August 1), *46*, 48–50, 52–53.
38. Cole, M. J.; Janiszewski, J. S.; Fouda, H. G. In *Practical Spectroscopy*; Pramanik, B. N.; Ganguly, A. K.; Gross, M. L. (Eds.), Marcel Dekker, New York, 2002, Vol. 32, pp 211–249.
39. Pramanik, B. N.; Bartner, P. L.; Chen, G. The role of mass spectrometry in the drug discovery process. *Curr. Opin. Drug Dis. Develop.* **1999**, *2*, 401–417.
40. (a) Beavis, R. C.; Chait, B. T. Matrix-assisted laser desorption ionization mass-spectrometry of proteins. *Meth. Enzymol.* **1996**, *270*, 519–551. (b) Hillenkamp, F.; Karas, M.; Beavis, R. C.; Chait, B. T. Matrix-assisted laser desorption/ionization mass spectrometry of biopolymers. *Anal. Chem.* **1991**, *63*, 1193A–1203A.
41. Clerc, J.; Fourre, C.; Fragu, P. SIMS microscopy: methodology, problems and perspectives in mapping drugs and nuclear medicine compounds. *Cell Biol. Internat.* **1997**, *21*, 619–633.
42. Cole, R. B. *Electrospray Ionization Mass Spectrometry*, Wiley, New York, 1997.
43. Willoughby, R.; Sheehan, E.; Mitrovich, S. *A Global View of LC/MS*, Global View, Pittsburgh, PA, 1998.
44. (a) Schneider, R. P.; Fouda, H. G.; Inskeep, P. B. Tissue distribution and biotransformation of zopolrestat, an aldose reductase inhibitor, in rats. *Drug Metab. Dispos.* **1998**, *26*, 1149–1159. (b) Prakash, C.; Kamel, A.; Anderson, W.; Howard, H. Metabolism and excretion of the novel antipsychotic drug ziprasidone in rats after oral administration of a mixture of ^{14}C- and ^{3}H-labeled ziprasidone. *Drug Metab. Dispos.* **1997**, *25*, 206–218.
45. (a) Pochapsky, S. S.; Pochapsky, T. C. Nuclear magnetic resonance as a tool in drug discovery, metabolism and disposition. *Curr. Top. Med. Chem.* **2001**, *1*, 427–441. (b) Lindon, J. C.; Nicholson, J. K.; Wilson, I. D. Directly coupled HPLC-NMR and HPLC-NMR-MS in pharmaceutical research and development. *J. Chromatogr.* B **2000**, *748*, 233–258. (c) Peng, S. X. Hyphenated HPLC-NMR and its applications in drug discovery. *Biomed. Chromatogr.* **2000**, *14*, 430. (d) Wilson, I. D.; Nicholson, J. K.; Lindon, J. C. In *Handbook of Drug Metabolism*; Woolf, T. F. (Ed.), Marcel Dekker, New York, 1999, p. 523.
46. (a) Wu, N.; Webb, L.; Peck, T. L.; Sweedler, J. V. Online NMR detection of amino acids and peptides in microbore LC. *Anal. Chem.* **1995**, *67*, 3101–3107. (b) Behnke, B.; Schlotterbeck, G.; Tallarck, U.; Strohschein, S.; Tseng, L.-H.; Keller, T.; Albert, K.; Bayer, E. Capillary HPLC-NMR coupling: high-resolution ^{1}H NMR spectroscopy in the nanoliter scale. *Anal. Chem.* **1996**, *68*, 1110–1115.
47. (a) Sandvoss, M.; Weltring, A.; Preiss, A.; Levsen, K.; Wuensch G. Combination of matrix solid-phase dispersion extraction and direct on-line liquid chromatography-nuclear magnetic resonance spectroscopy-tandem mass spectrometry as a new efficient approach for the rapid screening of natural products: Application to the total asterosaponin fraction of the starfish Asterias rubens. *J. Chromatog. A* **2001**, *917*, 75–86. (b) Dear, G. J.; Plumb, R. S.; Sweatman, B. C.; Ayrton, J.; Lindon, J. C.; Nicholson, J. K.; Ismail, I. M. Mass directed peak selection, an efficient method of drug metabolite identification using directly coupled liquid chromatography-mass spectrometry-nuclear magnetic resonance spectroscopy. *J. Chromatog. B* **2000**, *748*, 281–293.
48. Hiller, D. L.; Zuzel, T. J.; Williams, J. A.; Cole, R. O. Rapid scanning technique for the determination of optimal tandem mass spectrometric conditions for quantitative analysis. *Rapid Commun. Mass Spectrom.* **1997**, *11*, 593–597.
49. Watson, J. T. Selected-ion measurements. *Meth. Enzymol.* **1990**, *193*(Mass Spectrom.), 86–106.
50. Sherman, J.; McKay, M. J.; Ashman, K.; Molloy, M. P. How specific is my SRM? The issue of precursor and product ion redundancy. *Proteomics* **2009**, *9*(5), 1120–1123.
51. (a)Ackermann, B. L.; Berna, M. J.; Eckstein, J. A.; Ott, L. W.; Chaudhary, A. K. Current applications of liquid chromatography/mass spectrometry in pharmaceutical discovery after a decade of innovation. *Annu. Rev. Anal. Chem.* **2008**, *1*, 357–396. (b) Gao, H.; Materne, O. L.; Howe, D. L.; Brummel, C. L. Method for rapid metabolite profiling of drug candidates in fresh hepatocytes using liquid chromatography coupled with a hybrid quadrupole linear ion trap. *Rap. Commun. Mass Spectrom.* **2007**, *21*, 3683–3693. (c) Prakash, C.; Shaffer, C. L.; Nedderman, A. Analytical strategies for identifying drug metabolites. *Mass Spectrom. Rev.* **2007**, *26*, 340–369.

52. (a) Liberg, J. Ueber die säure welche in dem harn der grasfressenden vierfufsigen thiere enthalten ist. *Poggendorff's Ann. Phys. Chem.* **1829**, *17*, 389. (b) Lehmann, C. G. Vorkommen von harnbenzoesäure im diabetischen urine. *J. Prakt. Chem.* **1835**, *6*, 113–120. (c) Ure, A. On gouty concretions: with a new method of treatment. *Pharm. J. Trans.* **1841**, *1*, 24.
53. (a) Mueller, G. C.; Miller, J. A. The metabolism of 4-dimethylaminoazobenzene by rat liver homogenates. *J. Biol. Chem.* **1948**, *176*, 535–544. (b) Mueller, G. C.; Miller, J. A. The reductive cleavage of 4-dimethylaminoazobenzene by rat liver: the intracellular distribution of the enzyme system and its requirement for triphosphopyridine nucleotide. *J. Biol. Chem.* **1949**, *180*, 1125–1136. (c) Mueller, G. C.; Miller, J. A. The metabolism of methylated aminoazo dyes. *J. Biol. Chem.* **1953**, *202*, 579–587.
54. Smith, D. A.; Brown, K.; Neale, M.G. Chromone-2-carboxylic acids: roles of acidity and lipophilicity in drug disposition. *Drug Metab. Rev.* **1985**, *16*, 365–388.
55. (a) Woolf, T. F. (Ed.), Handbook of Drug Metabolism; Marcel Dekker, New York, 1999. (b) Williams, R. T. *Detoxification Mechanisms*, 2nd ed., Chapman & Hall, London, 1959.
56. Rendic, S.; Guengerich, F. P. Update information on drug metabolism systems-2009, part II. Summary of information on the effects of diseases and environmental factors on human cytochrome P450 (CYP) enzymes and transporter *Curr. Drug Metab.* **2010**, *11*(1), 4–84.
57. (a) Skarydova, L.; Skarka, A.; Solich, P.; Wsol, V. Enzyme stereospecificity as a powerful tool in searching for new enzymes. *Curr. Drug Metab.* **2010**, *11*(6), 547–559. (b) Jamali, F.; Mehvar, R.; Pasutto, F. M. Enantioselective aspects of drug action and disposition: therapeutic pitfalls. *J. Pharm. Sci.* **1989**, *78*, 695–715.
58. Hignite, C.; Utrecht, J.; Tschang, C.; Azarnoff, D. Kinetics of R and S warfarin enantiomers. *Clin. Pharmacol. Ther.* **1980**, *28*, 99–105.
59. Silber, B.; Holford, N. H. G.; Riegelman, S. Stereoselective disposition and glucuronidation of propranolol in humans. *J. Pharm. Sci.* **1982**, *71*, 699–704.
60. (a) Ariëns, E. J. Stereochemistry: a source of problems in medicinal chemistry. *Med. Res. Rev.* **1986**, *6*, 451–466. (b) Simonyi, M. On chiral drug action. *Med. Res. Rev.* **1984**, *4*, 359–413.
61. Eichelbaum, M. Pharmacokinetic and pharmacodynamic consequences of stereoselective drug metabolism in man. *Biochem. Pharmacol.* **1988**, *37*, 93–96.
62. Adams, S. S.; Bresloff, P.; Mason, C. G. Pharmacological differences between the optical isomers of ibuprofen: evidence for metabolic inversion of the (-)-isomer. *J. Pharm. Pharmacol.* **1976**, *28*, 256–257.
63. (a) Kaiser, D. G.; Vangeissen, G. J.; Reischer, R. J.; Wechter, W. J. Isomeric inversion of ibuprofen (R)-enantiomer in humans. *J. Pharm. Sci.* **1976**, *65*, 269–273. (b) Lee, E.; Williams, K.; Day, R.; Graham, G.; Champion, D. Stereoselective disposition of ibuprofen enantiomers in man. *Br. J. Clin. Pharmacol.* **1985**, *19*, 669–674.
64. Caldwell, J.; Hutt, A. J.; Fournel-Gigleux, S. The metabolic chiral inversion and dispositional enantioselectivity of the 2-arylpropionic acids and their biological consequences. *Biochem. Pharmacol.* **1988**, *37*, 105–114.
65. Houston, J. B. Utility of in vitro drug metabolism data in predicting in vivo metabolic clearance. *Biochem. Pharmacol.* **1994**, *47*, 1469–1479.
66. Cruciani, G.; Carosati, E.; De Boeck, B.; Ethirajulu, K.; Mackie, C.; Howe, T.; Vianello, R. MetaSite: Understanding metabolism in human cytochromes from the perspective of the chemist. *J. Med. Chem.*, **2005**, *48*(22), 6970–6979.
67. Trunzer, M.; Faller, B.; Zimmerlin, A. Metabolic soft spot identification and compound optimization in early discovery phases using MetaSite and LC-MS/MS validation. *J. Med. Chem.* **2009**, *52*, 329–335.
68. Rydberg, P.; Gloriam, D. E.; Zaretzki, J.; Breneman, C.; Olsen, L. SMARTCyp: a 2D method for prediction of cytochrome P450-mediated drug metabolism. *ACS Med. Chem. Lett.* **2010**, *1*, 96–100.
69. (a) Mueller, G. C.; Miller, J. A. The metabolism of 4-dimethylaminoazobenzene by rat liver homogenates. *J. Biol. Chem.* **1948**, *176*, 535–544. (b) Mueller, G. C.; Miller, J. A. Reductive cleavage of 4-dimethylaminoazobenzene by rat liver tissue. Intracellular distribution of the enzyme system and its requirement for triphosphopyridine nucleotide. *J. Biol. Chem.* **1949**, 180, 1125–1136. (c) Mueller, G. C.; Miller, J. A. The metabolism of methylated aminoazo dyes. *J. Biol. Chem.* **1953**, *202*, 579–587.
70. Brodie, B. B.; Axelrod, J.; Cooper, J. R.; Gaudette, L.; LaDu, B. N.; Mitoma, C.; Udenfriend, S. Detoxicfiation of drugs and other foreign compounds by liver microsomes. *Science* **1955**, *121*, 603–604.
71. Mason, H. S. Mechanisms of oxygen metabolism. *Science* **1957**, *125*, 1185–1188.
72. Posner, H. S.; Mitoma, C.; Rothberg, S.; Udenfriend, S. Enzymatic hydroxylation of aromatic compounds. II. Studies on the mechanism of microsomal hydroxylation. *Arch. Biochem. Biophys.* **1961**, *94*, 280–290.
73. (a) Klingenberg, M. Pigments of rat-liver microsomes. *Arch. Biochem. Biophys.* **1958**, 75, 376–386. (b) Garfinkel, D. Studies on pig liver microsomes. I. Enzymic and pigment composition of different microsomal fractions. *Arch. Biochem. Biophys.* **1958**, *77*, 493–509.
74. (a) Lamb, D. C.; Waterman, M. R.; Kelly, S. L.; Guengerich, F. P. Cytochromes P450 and drug discovery. *Curr. Opin. Biotechnol.* **2007**, *18*(6), 504–512. (b) Guengerich, F. P. Common and uncommon cytochrome P450 reactions related to metabolism and chemical toxicity. *Chem. Res. Toxicol.* **2001**, *14*, 611–650. (c) Danielson, P. B. The cytochrome P450 superfamily: biochemistry, evolution and drug metabolism in humans. *Curr. Drug Metab.* **2002**, *3*, 561–597. (d) Guengerich, F. P. Mechanisms of drug toxicity and relevance to pharmaceutical development. *Drug Metab. Pharmacokin.* **2011**, *26*(1), 3–14.
75. (a) Masters B. S.; Marohnic C. C Cytochromes P450–a family of proteins and scientists-understanding their relationships *Drug Metab. Rev.* **2006**, *38*(1–2), 209–25. (b) Kim, J.-J. P.; Roberts, D. L.; Djordjevic, S.; Wang, M.; Shea, T. M.; Masters, B. S. S. Crystallization studies of NADPH-cytochrome P450 reductase. *Meth. Enzymol.* **1996**, *272*, 368–377.
76. (a) Vermilion, J. L.; Ballou, D. P.; Massey, V.; Coon, M. J. Separate roles for FMN and FAD in catalysis by liver microsomal NADPH-cytochrome P-450 reductase. *J. Biol. Chem.* **1981**, *256*, 266–267. (b) Oprian, D. D.; Coon, M. J. Separate roles for FMN and FAD in catalysis by liver microsomal NADPH-cytochrome P-450 reductase. *J. Biol. Chem.* **1982**, *257*, 8935–8944. (c) Strobel, H. W.; Dignam, J. D.; Gum, J. R. NADPH cytochrome P-450 reductase and its role in the mixed function oxidase reaction. *Int. Encycl. Pharmacol. Ther.* **1982**, *108*, 361–373.
77. Levy, R. H.; Thummel, K. E.; Trager, W. F. (Eds.), *Metabolic Drug Interactions*; Lippincott Williams & Wilkins, Philadelphia, 2000.
78. (a) Volz, H. P. Controlled clinical trials of hypericum extracts in depressed patients–an overview. *Pharmacopsychiatry* **1997**, *30*(Suppl. 2), 72–76. (b) Wheatley, D. LI 160, an extract of St. John's wort, vs amitriptyline in mildly to moderately depressed outpatients - a controlled 6-week clinical trial. *Pharmacopsychiatry* **1997**, *30*(Suppl. 2), 77–80.

79. Laakmann, G.; Schule, C.; Baghai, T.; Kieser, M. St. John's wort in mild to moderate depression: the relevance of hyperforin for the clinical efficacy. *Pharmacopsychiatry* **1998**, *31*(Suppl.), 54–59.
80. Moore, L. B.; Goodwin, B.; Jones, S. A.; Wisely, G. B.; Serabjit-Singh, C. J.; Willson, T. M.; Collins, J. L.; Kliewer, S. A. St. John's wort induces hepatic drug metabolism through activation of the pregnane X receptor. *Proc. Natl. Acad. Sci. U.S.A.* **2000**, *97*, 7500–7502.
81. Smith, D. A.; Jones, B. C. Speculations on the substrate structure-activity relationship (SSAR) of cytochrome P450 enzymes. *Biochem. Pharmacol.* **1992**, *44*, 2089–2098.
82. (a) Ziegler, D. M. An overview of the mechanism, substrate specificities, and structure of FMOs. *Drug Metab. Rev.* **2002**, *34*, 503–511. (b) Ziegler, D. M. Recent studies on the structure and function of multisubstrate flavin-containing monooxygenases. *Annu. Rev. Pharmacol. Toxicol.* **1993**, *33*, 179–199.
83. Hines, R. N.; Cashman, J. R.; Philpot, R. M.; Williams, D. E.; Ziegler, D. M. The mammalian flavin-containing monooxygenases: molecular characterization and regulation of expression. *Toxicol. Appl. Pharmacol.* **1994**, *125*, 1–6.
84. Boyland, E. The biological significance of metabolism of polycyclic compounds. *Biochem. Soc. Symp.* **1950**, *5*, 40–54.
85. Jerina, D. M.; Daly, J. W.; Witkop, B.; Zaltzman-Nirenberg, P.; Udenfriend, S. 1,2-Naphthalene oxide as an intermediate in the microsomal hydroxylation of naphthalene. *Biochemistry* **1970**, *9*, 147–156.
86. Korzekwa, K. R.; Swinney, D. C.; Trager, W. F. Isotopically labeled chlorobenzenes as probes for the mechanism of cytochrome P-450 catalyzed aromatic hydroxylation. *Biochemistry* **1989**, *28*, 9019–9027.
87. (a) Guroff, G.; Daly, J. W.; Jerina, D. M.; Renson, J.; Witkop, B.; Udenfriend, S. Hydroxylation-induced migration: the NIH shift. Recent experiments reveal an unexpected and general result of enzymatic hydroxylation of aromatic compounds. *Science* **1967**, *157*, 1524–1530. (b) Jerina, D. Hydroxylation of aromatics. Chemical models for the biological processes. *Chem. Technol.* **1973**, *4*, 120–127.
88. (a) Daly, J.; Jerina, D.; Witkop, B. Migration of deuterium during hydroxylation of aromatic substrates by liver microsomes. I. Influence of ring substituents. *Arch. Biochem. Biophys.* **1968**, *128*, 517–527. (b) Daly, J. W.; Jerina, D. M.; Witkop, B. Arene oxides and NIH shift. Metabolism, toxicity, and carcinogenicity of aromatic compounds. *Experientia* **1972**, *28*, 1129–1149.
89. Atkinson, J. K.; Hollenberg, P. F.; Ingold, K. U.; Johnson, C. C.; Le Tadic, M.-H.; Newcomb, M.; Putt, D. A. Cytochrome P450-catalyzed hydroxylation of hydrocarbons: kinetic deuterium isotope effects for the hydroxylation of an ultrafast radical clock. *Biochemistry* **1994**, *33*, 10630–10637.
90. (a) Harbeson, S. L.; Tung, R. D. Deuterium in drug discovery and development. *Ann. Rep. Med. Chem.* **2011**, *46*, 403–417. (b) Shao, L.; Hewitt, M. C. The kinetic isotope effect in the search for deuterated drugs. *Drug News Perspect.* **2010**, *23*(6), 398–404.
91. Graham, P.; Liu, J. F.; Turnquist, D.; Deuterium modification as a new branch of medicinal chemistry to develop novel, highly differentiated drugs. *Drug Devel. Deliv.* **2012** (July/August), http://www.drug-dev.com
92. Parli, C. J.; Schmidt, B. Metabolism of 4-chloroamphetamine to 3-chloro-4-hydroxyamphetamine in rat. Evidence for an in vivo NIH shift of chlorine. *Res. Commun. Chem. Pathol. Pharmacol.* **1975**, *10*, 601–604.
93. (a) Park, B. K.; Kitteringham, N. R.; O'Neill, P. M. Metabolism of fluorine-containing drugs. *Annu. Rev. Pharmacol. Toxicol.* **2001**, *41*, 443–470. (b) Hagmann, W. K. The many roles for fluorine in medicinal chemistry. *J. Med. Chem.* **2008**, *51*, 4359–4369.
94. Marchetti, P.; Navalesci, R. Pharmacokinetic-pharmacodynamic relationships of oral hypoglycaemic agents. An update. *Clin. Pharmacokinet.* **1989**, *16*, 100–128.
95. Wolf, C.; Konig, W. A.; Roussel, C. Influence of substituents on the rotational energy barrier of atropisomeric biphenyls. Studies by polarimetry and dynamic gas chromatography. *Leibigs Ann.* **1995**, 781–786. (b) Leroux, F. Atropisomerism, biphenyls, and fluorine: a comparison of rotational barriers and twist angles. *ChemBioChem* **2004**, *5*, 644–649.
96. Molteni, M.; Pesenti, C.; Sani, M.; Volonterio, A.; Zanda, M. Fluorinated peptidomimetics: synthesis, conformational and biological features. *J. Fluorine Chem.* **2004**, *125*, 1735–1743.
97. Verbeek, R. K.; Blackburn, J. L.; Loewen, G. R. Clinical pharmacokinetics of non-steroidal antiinflammatory drugs. *Clin. Pharmacokinet.* **1983**, *8*, 297–331.
98. Wood, S. G.; Scott, P. W.; Chasseaud, L. F.; Faulkner, J. K.; Matthews, R. W.; Henrick, K. A novel metabolite of tinidazole involving nitro-group migration. *Xenobiotica* **1985**, *15*, 107–113.
99. Daly, J. In *Concepts in Biochemical Pharmacology*; Brodie, B. B.; Gillette, J. R. (Eds.), Springer-Verlag: Berlin, 1971, Part 2, p. 285.
100. Parke, D. V. Detoxication. LXXXIV. The metabolism of aniline-C14 in the rabbit and other animals. *Biochem. J.* **1960**, *77*, 493–503.
101. Dayton, P. G.; Perel, J. M.; Cummingham, R. F.; Israeli, Z. H.; Weiner, I. M. Studies of the fate of metabolites and analogs of probenecid. The significance of metabolic sites, especially lack of ring hydroxylation. *Drug Metab. Dispos.* **1973**, *1*, 742–751.
102. Perry. T. L.; Culling, C. F. A.; Berry, K.; Hansen, S. 7-Hydroxychlorpromazine; potential toxic drug metabolite in psychiatric patients. *Science* **1964**, *146*, 81–83.
103. Butler, T. C.; Dudley, K. H.; Johnson, D.; Roberts, S. B. Studies of the metabolism of 5,5-diphenylhydantoin relating principally to the stereoselectivity of the hydroxylation reactions in man and the dog. *J. Pharmacol. Exp. Ther.* **1976**, *199*, 82–92.
104. (a) Selander, H. G.; Jerina, D. M.; Piccolo, D. E.; Berchtold, G. A. Synthesis of 3- and 4-chlorobenzene oxides. Unexpected trapping results during metabolism of carbon-14-labeled chlorobenzene by hepatic microsomes. *J. Am. Chem. Soc.* **1975**, *97*, 4428–4430. (b) Billings, R. E.; McMahon, R. E. Microsomal biphenyl hydroxylation: the formation of 3-hydroxybiphenyl and biphenyl catechol. *Mol. Pharmacol.* **1978**, *14*, 145–154.
105. Tomaszewski, J. E.; Jerina, D. M.; Daly, J. W. Deuterium isotope effects during formation of phenols by hepatic monoxygenases. Evidence for an alternative to the arene oxide pathway. *Biochemistry* **1975**, *14*, 2024–2031.
106. Fretland, A. J.; Omiecinski, C. J. Epoxide hydrolases: biochemistry and molecular biology. *Chem.-Biol. Interact.* **2000**, *129*, 41.
107. (a) Armstrong, R. N.; Cassidy, C. S. New structural and chemical insight into the catalytic mechanism of epoxide hydrolases. *Drug Metab. Rev.* **2000**, *32*, 327. (b) Lacourciere, G. M.; Armstrong, R. N. The catalytic mechanism of microsomal epoxide hydrolase involves an ester intermediate. *J. Am. Chem. Soc.* **1993**, *115*, 10466–10467. (c) Borhan, B.; Jones, A. D.; Pinot, F.; Grant, D. F.; Kurth, M. J.; Hammock, B. D. Mechanism of soluble epoxide hydrolase. Formation of an α -hydroxy ester-enzyme intermediate through Asp-333. *J. Biol. Chem.* **1995**, *270*, 26923–26930. (d) Hammock, B. D.; Pinot, F.;

Beetham, J. K.; Grant, D. F.; Arand, M. E.; Oesch, F. Isolation of a putative hydroxyacyl enzyme intermediate of an epoxide hydrolase. *Biochem. Biophys. Res. Commun.* **1994**, 198, 850–856.

108. (a) Hanzlik, R. P.; Edelman, M.; Michaely, W. J.; Scott, G. Enzymatic hydration of [^{18}O]epoxides. Role of nucleophilic mechanisms. *J. Am. Chem. Soc.* **1976**, *98*, 1952–1955. (b) Hanzlik, R. P.; Hiedeman, S.; Smith, D. Regioselectivity in enzymatic hydration of cis-1,2-disubstituted [^{18}O]-epoxides. *Biochem. Biophys. Res. Commun.* **1978**, *82*, 310–315.
109. (a) Rushmore, T. H.; Pickett, C. B. Glutathione S-transferases, structure, regulation, and therapeutic implications. *J. Biol. Chem.* **1993**, *268*, 11475–11478. (b) Eaton, D. L.; Bammler, T. K. Concise review of the glutathione *S*-transferases and their significance to toxicology. *Toxicol. Sci.* **1999**, *49*, 156–164. (c) Schipper, D. L.; Wagenmans, M. J. H.; Wagener, D. J. T.; Peters, W. H. M. Glutathione *S*-transferases and cancer (review). *Int. J. Oncol.* **1997**, *10*, 1261–1264.
110. (a) Liu, S.; Zhang, P.; Ji, X.; Johnson, W. W.; Gilliland, G. L.; Armstrong, R. N. Contribution of tyrosine 6 to the catalytic mechanism of isoenzyme 3-3 of glutathione *S*-transferase. *J. Biol. Chem.* **1992**, *267*, 4296–4299. (b) Thorson, J. S.; Shin, I.; Chapman, E.; Stenberg, G.; Mannervik, B.; Schultz, P. G. Analysis of the role of the active site tyrosine in human glutathione transferase A1-1 by unnatural amino acid mutagenesis. *J. Am. Chem. Soc.* **1998**, *120*, 451–452.
111. Buckpitt, A. R.; Castagnoli, N., Jr.; Nelson, S. D.; Jones, A. D.; Bahnson, L. S. Stereoselectivity of naphthalene epoxidation by mouse, rat, and hamster pulmonary, hepatic, and renal microsomal enzymes. *Drug Metab. Dispos.* **1987**, *15*, 491–498.
112. (a) Nebert, D. W.; Boobis, A. R.; Yagi, H.; Jerina, D. M; Khouri, R. E. In Biological Reactive Intermediates; Jollow, D. J.; Kocsis, J. J.; Snyder, R.; Vainio, H., (Eds.), Plenum, New York, 1977, p. 125. (b) Yamamoto, J.; Subramaniam, R.; Wolfe, A. R.; Meehan, T. The formation of covalent adducts between benzo[a]pyrenediol epoxide and RNA: structural analysis by mass spectrometry. *Biochemistry* **1990**, *29*, 3966–3972.
113. Heidelberger, C. Chemical carcinogenesis. *Annu. Rev. Biochem.* **1975**, *44*, 79–121.
114. (a) Weinstein, I. B.; Jeffrey, A. M.; Jennette, K. W.; Blobstein, S. H.; Harvey, R. G.; Harris, C.; Autrup, H.; Kasai, H.; Nakanishi, K. Benzo[a]pyrene diol epoxides as intermediates in nucleic acid binding in vitro and in vivo. *Science* **1976**, *193*, 592–595. (b) Koreeda, M.; Moore, P. D.; Yagi, H.; Yeh, H. J. C.; Jerina, D. M. Alkylation of polyguanylic acid at the 2-amino group and phosphate by the potent mutagen (±)-7β ,8α -dihydroxy-9β ,10β -epoxy-7,8,9,10-tetrahydrobenzo[a]pyrene. *J. Am. Chem. Soc.* **1976**, *98*, 6720–6722.
115. Straub, K. M.; Meehan, T.; Burlingame, A. L.; Calvin, M. Identification of the major adducts formed by reaction of benzo[a] pyrene diol epoxide with DNA in vitro. *Proc. Natl. Acad. Sci. U.S.A.* **1977**, *74*, 5285–5289.
116. Bellucci, G.; Berti, G.; Chiappe, C.; Lippi, A.; Marioni, F. The metabolism of carbamazepine in humans: steric course of the enzymatic hydrolysis of the 10,11-epoxide. *J. Med. Chem.* **1987**, *30*, 768–773.
117. Johannessen, S. I.; Gerna, N. M.; Bakke, J.; Strandjord, R. E.; Morselli, P. L. CSF concentrations and serum protein binding of carbamazepine and carbamazepine-10,11-epoxide in epileptic patients. *Br. J. Clin. Pharmacol.* **1976**, *3*, 575–582.
118. McCann, J.; Spingtain, N. E.; Ikobori, J.; Ames, B. N. Detection of carcinogens as mutagens. Bacterial tester strains with R factor plasmids. *Proc. Natl. Acad. Sci. U.S.A.* **1975**, *72*, 979–983.
119. McMahon, G.; Davis, E. F.; Huber, L. J.; Kim, Y.; Wogan, G. N. Characterization of c-Ki-ras and N-ras oncogenes in aflatoxin B1-induced rat liver tumors. *Proc. Natl. Acad. Sci. U.S.A.* **1990**, *87*, 1104.
120. Wogan, G. N. Aflatoxin carcinogenesis: interspecies potency differences and relevance for human risk assessment. *Prog. Clin. Biol Res.* **1992**, *374*, 123–137.
121. Iyer, R.; Harris, T. M. Preparation of aflatoxin B1 8,9-epoxide using m-chloroperbenzoic acid. *Chem. Res. Toxicol.* **1993**, *6*, 313–316.
122. (a) Essigmann, J. M.; Croy, R. G.; Nadzan, A. M.; Busby, W. F., Jr.; Reinhold, V. N.; Büchi, G.; Wogan, G. N. Structural identification of the major DNA adduct formed by aflatoxin B1 in vitro. *Proc. Natl. Acad. Sci. U.S.A.* **1977**, *74*, 1870–1874. (b) Croy, R. G.; Essigmann, J. M.; Reinhold, V. N.; Wogan, G. N. Identification of the principal aflatoxin B1-DNA adduct formed in vivo in rat liver. *Proc. Natl. Acad. Sci. U.S.A.* **1978**, *75*, 1745–1749.
123. Gopalakrishnan, S.; Stone, M. P.; Harris, T. M. Preparation and characterization of an aflatoxin B1 adduct with the oligodeoxynucleotide d(ATCGAT)2. *J. Am. Chem. Soc.* **1989**, *111*, 7232–7239.
124. Iyer, R. S.; Coles, B. F.; Raney, K. D.; Thier, R.; Guengerich, F. P.; Harris, T. M. DNA adduction by the potent carcinogen aflatoxin B1: mechanistic studies. *J. Am. Chem.* Soc. **1994**, *116*, 1603–1609.
125. Hucker, H. B. Metabolism of amitriptyline. *Pharmacologist* **1962**, *4*, 171.
126. Shetty, H. U.; Nelson, W. L. Chemical aspects of metoprolol metabolism. Asymmetric synthesis and absolute configuration of the 3-[4-(1-hydroxy-2-methoxyethyl)phenoxy]-1-(isopropylamino)-2-propanols, the diastereomeric benzylic hydroxylation metabolites. *J. Med. Chem.* **1988**, *31*, 55–59.
127. Guengerich, F. P.; Müller-Enoch, D.; Blair, I. A. Oxidation of quinidine by human liver cytochrome P-450. *Mol. Pharmacol.* **1986**, *30*, 287–295.
128. Nakahara, Y.; Cook. C. E. Confirmation of cannabis use. III. Simultaneous quantitation of six metabolites of Δ 9-tetrahydrocannabinol in plasma by high-performance liquid chromatography with electrochemical detection. *J. Chromatogr.* **1988**, *434*, 247–252.
129. Keberle, H.; Reiss, W.; Hoffman, K. The stereospecific metabolism of the optical antipodes of α phenyl-α-ethylglutarimide (Doriden). *Arch. Int. Pharmacodyn.* **1963**, *142*, 117–124.
130. Ponchaut, S.; van Hoof, F.; Veitch, K. In vitro effects of valproate and valproate metabolites on mitochondrial oxidations. Relevance of CoA sequestration to the observed inhibitions. *Biochem. Pharmacol.* **1992**, *43*, 2435–2442.
131. Cooper, R. G.; Evans, D. A. P.; Whibley, E. J. Polymorphic hydroxylation of perhexiline maleate in man. *J. Med. Genet.* **1984**, *21*, 27–33.
132. Katchen, B.; Buxbaum, S. Disposition of a new, nonsteroid, antiandrogen, α,α,α-trifluoro-2-methyl-4'-nitro-*m*-propionotoluidide (flutamide), in men following a single oral 200 mg dose. *J. Clin. Endocrinol. Metabol.* **1975**, *41*, 373–379.
133. Ram, N.; Kalasz, H.; Adeghate, E.; Darvas, F.; Hashemi, F.; Tekes, K. Medicinal chemistry of drugs with active metabolites (N-, O-, and S-desalkylation and some specific oxidative alterations). *Curr. Med. Chem.* **2012**, *19*, 5683–5704.
134. Barnes-Seeman, D.; Jain, M.; Bell, L.; Ferreira, S.; Cohen, S.; Chen, X.-H.; Amin, J.; Snodgrass, B.; Hatsis, P. Metabolically stable tert-butyl replacement. *ACS Med. Chem. Lett.* **2013**, *4*, 514–516.
135. Wright, J.; Cho, A. K.; Gal, J. The role of N-hydroxyamphetamine in the metabolic deamination of amphetamine. *Life Sci.* **1977**, *20*, 467–473.

136. Coutts, R. T.; Beckett, A. H. Metabolic N-oxidation of primary and secondary aliphatic medicinal amines. *Drug Metab. Rev.* **1977**, *6*, 51–104.
137. Gorrod, J. W. Differentiation of various types of biological oxidation of nitrogen in organic compounds. *Chem.Biol. Interact.* **1973**, *7*, 289–303.
138. Parli, C. H.; McMahon, R. E. Mechanism of microsomal deamination. Heavy isotope studies. *Drug Metab. Dispos.* **1973**, *1*, 337–341.
139. Bakke, O. M.; Davies, D. S.; Davies, L.; Dollery, C. T. Metabolism of propranolol in rat. Fate of the N-isopropyl group. *Life Sci.* **1973**, *13*, 1665–1675.
140. Beckett, A. H.; Al-Sarraj, S. N-oxidation of primary and secondary amines to give hydroxylamines. General metabolic route. *J. Pharm. Pharmacol.* **1972**, *24*, 916–917.
141. Beckett, A. H.; Coutts, R. T.; Ogunbona, F. A. Structure of nitrones derived from amphetamines. *J. Pharm. Pharmacol.* **1974**, *26*, 312–316.
142. Floyd, D. M.; Kimball, S. D.; Krapcho, J.; Das, J.; Turk, C. F.; Moquin, R. V.; Lago, M. W.; Duff, K. J.; Lee, V. G. et al. Benzazepinone calcium channel blockers. 2. Structure activity and drug metabolism studies leading to potent antihypertensive agents. Comparison with benzothiazepinones. *J. Med. Chem.* **1992**, *35*, 756–772.
143. Smith, D. A. Species differences in metabolism and pharmacokinetics: are we close to an understanding? *Drug Metab. Rev.* **1991**, *23*, 355–373.
144. Atwal, K. S.; Swanson, B. N.; Unger, S. E.; Floyd, D. M.; Moreland, S.; Hedberg, A.; O'Reilly, B. C. Dihydropyrimidine calcium channel blockers. 3. 3-Carbamoyl-4-aryl-1,2,3,4-tetrahydro-6-methyl-5-pyrimidinecarboxylic acid esters as orally effective antihypertensive agents. *J. Med. Chem.* **1991**, *34*, 806–811.
145. Gram, T. E.; Wilson, J. T.; Fouts, J. R. Some characteristics of hepatic microsomal systems which metabolize aminopyrine in the rat and rabbit. *J. Pharmacol. Exp. Ther.* **1968**, *159*, 172–181.
146. Anders, M. W.; Cooper, M. J.; Takemori, A. E. Kinetics of microsomal metabolism and binding of enantiomerically related substrates. *Drug Metab. Dispos.* **1973**, *1*, 642–644.
147. (a) Yasar, S.; Justinova, Z.; Lee, S.-H.; Stefanski, R.; Goldberg, S. R.; Tanda, G. Metabolic transformation plays a primary role in the psychostimulant-like discriminative-stimulus effects of selegiline [(R)-(-)-deprenyl]. *J. Pharmacol. Exp. Ther.* **2006**, *317*(1), 387–394. (b) Dragoni, S.; Bellik, L.; Frosini, M.; Sgaragli, G.; Marini, S.; Gervasi, P. G.; Valoti, M. l-Deprenyl metabolism by the cytochrome P450 system in monkey (Cercopithecus aethiops) liver microsomes. *Xenobiotica* **2003**, *33*, 181–195. (c) Haberle, D.; Szoko, E.; Magyar, K. The influence of metabolism on the MAO-B inhibitory potency of selegiline. *Curr. Med. Chem.* **2002**, *9*, 47–51. (d) Shin, H.-S. Metabolism of selegiline in humans: identification, excretion, and stereochemistry of urine metabolites. *Drug Metab. Dispos.* **1997**, *25*, 657–662. (e) Bach, M. V.; Coutts, R. T.; Baker, G. B. Metabolism of N, N-dialkylated amphetamines, including deprenyl, by CYP2D6 expressed in a human cell line. *Xenobiotica* **2000**, *30*, 297–306.
148. (a) Tarjanyi, Zs.; Kalasz, H.; Szebeni, G.; Hollosi, I.; Bathori, M.; Furst, S. Gas-chromatographic study on the stereoselectivity of deprenyl metabolism. *J. Pharm. Biomed. Anal.* **1998**, *17*, 725–731. (b) Baker, G. B.; Urichuk, L. J.; McKenna, K. F.; Kennedy, S. H.; Scheinin, H. Anttila, M.; Dahl, M. L.; Karnani, H.; Nyman, L. Metabolism of monoamine oxidase inhibitors. *Cell. Mol. Neurobiol.* **1999**, *19*, 411–426. (c) Bach, M. V.; Coutts, R. T.; Baker, G. B. Metabolism of N, N-dialkylated amphetamines, including deprenyl, by CYP2D6 expressed in a human cell line. *Xenobiotica* **2000**, *30*, 297–306.
149. Glezer, S.; Finberg, J. P. M. Pharmacological comparison between the actions of methamphetamine and 1-aminoindan stereoisomers on sympathetic nervous function in rat vas deferens. *Eur. J. Pharmacol.* **2003**, *472*, 173–177.
150. (a) Langone, J. J.; Van Vunakis, H. Radioimmunoassay of nicotine, cotinine, and γ-(3-pyridyl)-γ-oxo-N-methylbutyramide. *Meth. Enzymol.* **1982**, *84*, 628–640. (b) Schievelbein, H. Nicotine, resorption and fate. *Int. Encyl. Pharmacol. Ther.* **1984**, *114*, 1–15.
151. Nguyen, T.-L.; Gruenke, L. D.; Castagnoli, N., Jr. Metabolic N-demethylation of nicotine. Trapping of a reactive iminium species with cyanide ion. *J. Med. Chem.* **1976**, *19*, 1168–1169.
152. Nelson, S. D.; Garland, W. A.; Breck, G. D.; Trager, W. F. Quantification of lidocaine and several metabolites utilizing chemical-ionization mass spectrometry and stable isotope labeling. *J. Pharm. Sci.* **1977**, *66*, 1180–1190.
153. McMartin, C.; Simpson, P. The absorption and metabolism of guanethidine in hypertensive patients requiring different doses of the drug. *Clin. Pharmacol. Ther.* **1971**, *12*, 73–77.
154. Hucker, H. B.; Balletto, A. J.; Stauffer, S. C.; Zacchei, A. G.; Arison, B. H. Physiological disposition and urinary metabolites of cyproheptadine in the dog, rat, and cat. *Drug Metab. Dispos.* **1974**, *2*, 406–415.
155. Christy, M. E.; Anderson, P. S.; Arison, B. H.; Cochran, D. W.; Engelhardt, E. L. Stereoisomerism of cyproheptadine N-oxide. *J. Org. Chem.* **1977**, *42*, 378–379.
156. (a) Weisburger, J. H.; Weisburger, E. K. Biochemical formation and pharmacological, toxicological, and pathological properties of hydroxylamines and hydroxamic acids. *Pharmacol. Rev.* **1973**, *25*, 1–66. (b) Miller, J. A. Carcinogenesis by chemicals: an overview--G. H. A. Clowes memorial lecture. *Cancer Res.* **1970**, *30*, 559–576.
157. Bondon, A.; Macdonald, T. L.; Harris, T. M.; Guengerich, F. P. Oxidation of cycloalkylamines by cytochrome P-450. Mechanism-based inactivation, adduct formation, ring expansion, and nitrone formation. *J. Biol. Chem.* **1989**, *264*, 1988–1997.
158. Ziegler, D. M.; Ansher, S. S.; Nagata, T.; Kadlubar, F. F.; Jakoby, W. B. N-Methylation: potential mechanism of metabolic activation of carcinogenic primary arylamines. *Proc. Natl. Acad. Sci. U.S.A.* **1988**, *85*, 2514–2517.
159. (a) Barker, E. A.; Smuckler, E. A. Altered microsome function during acute thioacetamide poisoning. *Mol. Pharmacol.* **1972**, *8*, 318–326. (b) Willi, P.; Bickel, M. H. Liver metabolic reactions. Tertiary amine N-dealkylation, tertiary amine N-oxidation, N-oxide reduction, and N-oxide N-dealkylation. II. N, N-Dimethylaniline. *Arch. Biochem. Biophys.* **1973**, *156*, 772–779.
160. Gorrod, J. W.; Temple, D. J. The formation of an N-hydroxymethyl intermediate in the N-demethylation of N-methylcarbazole in vivo and in vitro. *Xenobiotica* **1976**, *6*, 265–274.
161. Miwa, G. T.; Walsh, J. S.; Kedderis, G. L.; Hollenberg, P. F. The use of intramolecular isotope effects to distinguish between deprotonation and hydrogen atom abstraction mechanisms in cytochrome P-450- and peroxidase-catalyzed N-demethylation reactions. *J. Biol. Chem.* **1983**, *258*, 14445–14449.
162. Weisburger, E. K. Mechanisms of chemical carcinogenesis. *Ann. Rev. Pharmacol. Toxicol.* **1978**, *18*, 395–415.
163. (a) Lin, J.-K.; Miller, J. A.; Miller, E. C. Structures of polar dyes derived from the liver proteins of rats fed N-methyl-4-aminoazobenzene. III. Tyrosine and homocysteine sulfoxide polar dyes. *Biochemistry* **1969**, *8*, 1573–1582. (b) Lin, J.-K.; Miller, J. A.; Miller, E. C. Structure of hepatic nucleic acid-bound dyes in rats given the carcinogen N-methyl-4-aminoazobenzene. *Cancer Res.* **1975**, *35*, 844–850.

164. Kempf, D. J.; Marsh, K. C.; Fino, L. C.; Bryant, P.; Craig-Kennard, A.; Sham, H. L.; Zhao, C.; Vasavanonda, S.; Kohlbrenner, W. E. Design of orally bioavailable, symmetry-based inhibitors of HIV protease. *Bioorg. Med. Chem.* **1994**, *2*, 847–858.

165. Wick, M. J.; Jantan, I.; Hanna, P. E. Irreversible inhibition of rat hepatic transacetylase activity by N-arylhydroxamic acids. *Biochem. Pharmacol.* **1988**, *37*, 1225–1231.

166. (a) Mitchell, J. R.; Jollow, D. J.; Potter, W. Z.; Gillette, J. R.; Brodie, B. B. Acetaminophen-induced hepatic necrosis. IV. Protective role of glutathione. *J. Pharmacol. Exp. Ther.* **1973**, *187*, 211–217. (b) Potter, W. Z.; Davis, D. C.; Mitchell, J. R.; Jollow, D. J.; Gillette, J. R.; Brodie, B. B. Acetaminophen-induced hepatic necrosis. III. Cytochrome P 450-mediated covalent binding in vitro. *J. Pharmacol. Exp. Ther.* **1973**, *187*, 203–210.

167. Jollow, D. J.; Thorgeirsson, S. S.; Potter, W. Z.; Hashimoto, M.; Mitchell, J. R. Acetaminophen-induced hepatic necrosis. VI. Metabolic disposition of toxic and nontoxic doses of acetaminophen. *Pharmacology* **1974**, *12*, 251–271.

168. (a) Hinson, J. A.; Pohl, L. R.; Gillette, J. R. N-Hydroxyacetaminophen: a microsomal metabolite of N-hydroxyphenacetin but apparently not of acetaminophen. *Life Sci.* **1979**, *24*, 2133–2138. (b) Nelson, S. D.; Forte, A. J.; Dahlin, D. C. Lack of evidence for N-hydroxyacetaminophen as a reactive metabolite of acetaminophen in vitro. *Biochem. Pharmacol.* **1980**, *29*, 1617–1620.

169. (a) Nelson, S. D.; McMurty, R. J.; Mitchell, J. R. In *Biological Oxidation of Nitrogen*; Gorrod, J. W. (Ed.), Elsevier/North-Holland, Amsterdam, 1978, p. 319. (b) Hinson, J. A.; Pohl, L. R.; Monks, T. J.; Gillette, J. R.; Guengerich, F. P. 3-Hydroxyacetaminophen: a microsomal metabolite of acetaminophen. Evidence against an epoxide as the reactive metabolite of acetaminophen. *Drug Metab. Dispos.* **1980**, *8*, 289–294.

170. Nelson, S. D. Mechanisms of the formation and disposition of reactive metabolites that can cause acute liver injury. *Drug Metab. Rev.* **1995**, *27*, 147–177.

171. (a) Thummel, K. E.; Lee, C. A.; Kunze, K. L.; Nelson, S. D.; Slattery, J. T. Oxidation of acetaminophen to N-acetyl-p-aminobenzoquinone imine by human CYP3A4. *Biochem. Pharmacol.* **1993**, *45*, 1563–1569. (b) Patten, C. J.; Thomas, P. E.; Guy, R. L.; Lee, M.; Gonzalez, F. J.; Guengerich, F. P.; Yang, C. S. Cytochrome P450 enzymes involved in acetaminophen activation by rat and human liver microsomes and their kinetics. *Chem. Res. Toxicol.* **1993**, *6*, 511–518. (c) Raucy, J. L.; Lasker, J. M.; Lieber, C. S.; Black, M. Acetaminophen activation by human liver cytochromes P450IIE1 and P450IA2. *Arch. Biochem. Biophys.* **1989**, *271*, 270–283.

172. Ryan, D. E.; Koop, D. R.; Thomas, P. E.; Coon, M. J.; Levin, W. Evidence that isoniazid and ethanol induce the same microsomal cytochrome P-450 in rat liver, an isozyme homologous to rabbit liver cytochrome P-450 isozyme 3a. *Arch. Biochem. Biophys.* **1986**, *246*, 633–644.

173. Rosen, G. M.; Singletary, W. V. Jr; Rauckman, E. J.; Killenberg, P. G. Acetaminophen hepatotoxicity. An alternative mechanism. *Biochem. Pharmacol.* **1983**, *32*, 2053–2059. (b) Vendemiale, G.; Altomare, E.; Lieber, C. S. Altered biliary excretion of acetaminophen in rats fed ethanol chronically. *Drug Metab. Dispos.* **1984**, *12*, 20–24.

174. (a) McClain, C. J.; Kromhout, J. P.; Peterson, F. J.; Holtzman, J. L. Potentiation of acetaminophen hepatotoxicity by alcohol. *J. Am. Med. Assoc.* **1980**, *244*, 251–253. (b) Hall, A. H.; Kulig, K. W.; Rumack, B. H. Acetaminophen hepatotoxicity in alcoholics. *Ann. Intern. Med.* **1986**, *105*, 624–625.

175. West, P. R.; Harman, L. S.; Josephy, P. D.; Mason, R. P. Acetaminophen: enzymatic formation of a transient phenoxyl free radical. *Biochem. Pharmacol.* **1984**, *33*, 2933–2936.

176. Dahlin, D. C.; Miwa, G. T.; Lu, A. Y. H.; Nelson, S. D. N-Acetyl-p-benzoquinone imine: a cytochrome P-450-mediated oxidation product of acetaminophen. *Proc. Natl. Acad. Sci. U.S.A.* **1984**, *81*, 1327–1331.

177. Potter, D. W.; Hinson, J. A. Mechanisms of acetaminophen oxidation to N-acetyl-p-benzoquinone imine by horseradish peroxidase and cytochrome P 450. *J. Biol. Chem.* **1987**, *262*, 966–973.

178. (a) Mitchell, J. R.; Jollow, D. J.; Potter, W. Z.; Gillette, J. R.; Brodie, B. B. Acetaminophen-induced hepatic necrosis. IV. Protective role of glutathione. *J. Pharmacol. Exp. Ther.* **1973**, *187*, 211–217. (b) Potter, W. Z.; Davis, D. C.; Mitchell, J. R.; Jollow, D. J.; Gillette, J. R.; Brodie, B. B. Acetaminophen-induced hepatic necrosis. III. Cytochrome P 450-mediated covalent binding in vitro. *J. Pharmacol. Exp. Ther.* **1973**, *187*, 203–210.

179. Moldéus, P.; Andersson, B.; Rahimtula, A.; Berggren, M. Prostaglandin synthetase-catalyzed activation of paracetamol. *Biochem. Pharmacol.* **1982**, *31*, 1363–1368.

180. Potter, D. W.; Hinson, J. A. The 1- and 2-electron oxidation of acetaminophen catalyzed by prostaglandin H synthase. *J. Biol. Chem.* **1987**, *262*, 974–980.

181. Reen, R. K.; Ramakanth, S.; Wiebel, F. J.; Jain, M. P.; Singh, J. Dealkylation of 7-methoxycoumarin as assay for measuring constitutive and phenobarbital-inducible cytochrome P450s. *Anal. Biochem.* **1991**, *194*, 243–249.

182. Manoury, P. M.; Binet, J. L.; Rousseau, J.; Leferre-Borg, F. M.; Cavero, I. G. Synthesis of a series of compounds related to betaxolol, a new β 1-adrenoceptor antagonist with a pharmacological and pharmacokinetic profile optimized for the treatment of chronic cardiovascular diseases. *J. Med. Chem.* **1987**, *30*, 1003–1011.

183. (a) Tullman, R. H.; Hanzlik, R. P. Inactivation of cytochrome P 450 and monoamine oxidase by cyclopropylamines. *Drug Metab. Rev.* **1984**, *15*, 1163–1182. (b) Silverman, R. B. Radical ideas about monoamine oxidase. *Acc. Chem. Res.* **1995**, *28*, 335–342.

184. Duggan, D. E.; Hogans, A. F.; Kwan, K. C.; McMahon, F. G. Metabolism of indomethacin in man. *J. Pharmacol. Exp.* Ther. **1972**, *181*, 563–575.

185. Adler, T. K.; Fujimoto, J. M.; Way, E. L.; Baker, E. M. Metabolic fate of codeine in man. *J. Pharmacol. Exp. Ther.* **1955**, *114*, 251–262.

186. Klutch, A.; Bordun, M. Metabolic fate of methoxamine and N-isopropylmethoxamine. *J. Med. Chem.* **1967**, *10*, 860–863.

187. Davis, P. J.; Abdel-Maksoud, Hamdy; Trainor, T. M.; Vouros, Paul; Neumeyer, J. L. Stereospecific microbiological 10-O-demethylation of R-(-)-10,11-dimethoxyaporphines. *Biochem. Biophys. Res. Commun.* **1985**, 127, 407–412.

188. (a) Baillie, T. A.; Halpin, R. A.; Matuszewski, B. K.; Geer, L. A.; Chavez-Eng, C. M.; Dean, D.; Braun, M.; Doss, G.; Jones, A.; Marks, T.; Melillo, D.; Vyas, K. P. Mechanistic studies on the reversible metabolism of rofecoxib to 5-hydroxyrofecoxib in the rat: evidence for transient ring opening of a substituted 2-furanone derivative using stable isotope-labeling techniques. *Drug Metab. Dispos.* **2001**, *29*, 1614–1628. (b) Nicoll-Griffith, D. A.; Yergey, J. A.; Trimble, L. A.; Silva, J. M.; Li, C.; Chauret, N.; Gauthier, J. Y.; Grimm, E.; Leger, S.; Roy, P.; Therien, M.; Wang, Z.; Prasit, P.; Zamboni, R.; Young, R. N.; Brideau, C.; Chan, C.-C.; Mancini, J.; Riendeau, D. Synthesis, characterization, and activity of metabolites derived from the cyclooxygenase-2 inhibitor rofecoxib (MK-0966, Vioxx). *Bioorg. Med. Chem. Lett.* **2000**, *10*, 2683–2686.

189. Mazel, P.; Henderson, J. F.; Axelrod, J. S-Demethylation by microsomal enzymes. *J. Pharmacol. Exp. Ther.* **1964**, *143*, 1–6.
190. Spector, E.; Shideman, F. E. Metabolism of thiopyrimidine derivatives: thiamylal, thiopental, and thiouracil. *Biochem. Pharmacol.* **1959**, *2*, 182–196.
191. Mitchell, S. C.; Waring. R. H. The early history of xenobiotic sulfoxidation. *Drug Metab. Rev.* **1986**, *16*, 255–284.
192. Oae, S.; Mikami, A.; Matsuura, T.; Ogawa-Asada, K.; Watanabe, Y.;Fujimori, K.; Iyanagi, T. Comparison of sulfide oxygenation mechanism for liver microsomal FAD-containing monooxygenase with that for cytochrome P-450. *Biochem. Biophys. Res. Commun.* **1985**, *131*, 567–573.
193. Goto, Y.; Matsui, T.; Ozaki, S.-i.; Watanabe, Y.; Fukuzumi, S. Mechanisms of sulfoxidation catalyzed by high-valent intermediates of heme enzymes: electron-transfer vs oxygen-transfer mechanism. *J. Am. Chem. Soc.* **1999**, *121*, 9497–9502.
194. Souhaili el Amri, H.; Fargetton, X.; Delatour, P.; Batt, A. M. Sulfoxidation of albendazole by the FAD-containing and cytochrome P-450 dependent mono-oxygenases from pig liver microsomes. *Xenobiotica* **1987**, *17*, 1159–1168.
195. Gruenke, L. D.; Craig, J. C.; Dinovo, E. C.; Gottschalk, L. A.; Noble, E. P.; Biener, R. Identification of a metabolite of tioridazine and mesoridazine from human plasma. *Res. Commun. Chem. Pathol. Pharmacol.* **1975**, *10*, 221–225.
196. (a) Beaune, P.; Dansette, P. M.; Mansuy, D.; Kiffel, L.; Finck, M.; Amar, C.; Leroux, J. P.; Homberg, J. C. Human anti-endoplasmic reticulum autoantibodies appearing in a drug-induced hepatitis are directed against a human liver cytochrome P-450 that hydroxylates the drug. *Proc. Natl. Acad. Sci. U.S.A.* **1987**, *84*, 551–555. (b) Vaughan, D. P.; Tucker, G. F. *Pharm. J.* **1987**, *577.*
197. Teffera, Y.; Choquette, D.; Liu, J.; Bolletti, A. E.; Hollis, L. S.; Lin, M. H. J.; Zhao, Z. Bioactivation of isothiazoles: minimizing the risk of potential toxicity in drug discovery. *Chem. Res. Toxicol.* **2010**, *23*, 1743–1752.
198. Dansette, P. M.; Amar, C.; Valadon, P.; Pons, C.; Beaune, P. H.; Mansuy, D. Hydroxylation and formation of electrophilic metabolites of tienilic acid and its isomer by human liver microsomes: catalysis by a cytochrome P450 IIC different from that responsible for mephenytoin hydroxylation. *Biochem. Pharmacol.* **1991**, *40*, 553–560.
199. Mansuy, D.; Valadon, P.; Erdelmeier, I.; Lopez-Garcia, P.; Amar, C.; Girault, J.-P.; Dansette, P. M. Thiophene S-oxides as new reactive metabolites: formation by cytochrome P-450 dependent oxidation and reaction with nucleophiles. *J. Am. Chem. Soc.* **1991**, *113*, 7825–7826.
200. Crew, M. C.; Melgar, M. D.; Haynes, L. J.; Gala, R. L.; DiCarlo, F. J. Disposition and biotransformation of ^{14}C-oxisuran in the dog. *Xenobiotica* **1972**, *2*, 431–440.
201. Benoit, E.; Buronfosse, T.; Moroni, P.; Delatour, P.; Riviere-J.-L. Stereoselective S-oxygenation of an aryl-trifluoromethyl sulfoxide to the corresponding sulfone by rat liver cytochromes P450. *Biochem. Pharmacol.* **1993**, *46*, 2337–2341.
202. (a) Pohl, L. R.; Pumford, N. R.; Martin, J. L. Mechanisms, chemical structures and drug metabolism. *Eur. J. Haematol.* **1996**, Suppl. 60, 98–104. (b) Bourdi, M.; Amouzadeh, H. R.; Rushmore, T. H.; Martin, J. L.; Pohl, L. R. Halothane-induced liver injury in outbred guinea pigs: role of trifluoroacetylated protein adducts in animal susceptibility. *Chem. Res. Toxicol.* **2001**, *14*, 362–370.
203. Martin, J. L.; Meinwald, J.; Radford, P.; Liu, Z.; Graf, M. L. M.; Pohl, L. R. Stereoselective metabolism of halothane enantiomers to trifluoroacetylated liver proteins. *Drug Metab. Rev.* **1995**, *27*, 179–189.
204. Sisenwine, S. F.; Kimmel, H. B.; Lin, A. L.; Ruelius, H. W. Excretion and stereoselective biotransformations of dl-, d- and l-norgestrel in women. *Drug Metab. Dispos.* **1975**, *3*, 180–188.
205. Misra, A. L.; Vadlamani, N. L.; Pontani, R. B.; Mulé, S. J. New metabolite of morphine-N-methyl-^{14}C in the rat. *Biochem. Pharmacol.* **1973**, *22*, 2129–2139.
206. Dolphin, D.; Poulson, R.; Avramovic, O. (Eds.), *Pyridine Nucleotide Coenzymes, part A and B*; Wiley, New York, 1987.
207. Yun, C.-H.; Lee, H. S.; Lee, H.; Rho, J. K.; Jeong, H. G.; Guengerich, F. P. Oxidation of the angiotensin II receptor antagonist losartan (DuP 753) in human liver microsomes: role of cytochrome P4503A(4) in formation of the active metabolite EXP3174. *Drug Metab. Dispos.* **1995**, *23*, 285–289.
208. Orbach, R. S.; Huynh, P.; Allen, M. C.; Beedham, C. Human liver aldehyde oxidase: inhibition by 239 drugs. *J. Clin. Pharmacol.* **2004**, *44*, 7–19.
209. (a) Kanamori, T.; Inoue, H.; Iwata, Y.; Ohmae, Y.; Kishi, T. In vivo metabolism of 4-bromo-2,5-dimethoxyphenethylamine (2C-B) in the rat: identification of urinary metabolites. *J. Anal. Toxicol.* **2002**, *26*, 61–66. (b) Lien E. A.; Solheim E.; Lea O. A.; Lundgren S.; Kvinnsland S.; Ueland P. M. Distribution of 4-hydroxy-N-desmethyltamoxifen and other tamoxifen metabolites in human biological fluids during tamoxifen treatment. *Cancer Res.* **1989**, *49*, 2175–2183.
210. Ravitch, J. R.; Moseley, C. G. High-performance liquid chromatographic assay for abacavir and its two major metabolites in human urine and cerebrospinal fluid. *J. Chromatogr. B* **2001**, *762*, 165–173.
211. (a) Stearns, R. A.; Chakravarty, P. K.; Chen, R.; Chiu, S. H. L. Biotransformation of losartan to its active carboxylic acid metabolite in human liver microsomes: role of cytochrome P4502C and 3A subfamily members. *Drug Metab. Dispos.* **1995**, *23*, 207–215. (b) Tamaki, T.; Nishiyama, A.; Kimura, S.; Aki, Y.; Yoshizumi, M.; Houchi, H.; Morita, K.; Abe, Y. EXP3174: the major active metabolite of losartan. Cardiovasc. *Drug Rev.* **1997**, *15*, 122–136. (c) Yun, C.-H.; Lee, H. S.; Lee, H.; Rho, J. K.; Jeong, H. G.; Guengerich, F. P. Oxidation of the angiotensin II receptor antagonist losartan (DuP 753) in human liver microsomes: role of cytochrome P4503A(4) in formation of the active metabolite EXP3174. *Drug Metab. Dispos.* **1995**, *23*, 285–289.
212. Sachinidis, A.; Ko, Y.; Weisser, P.; Meyer zu Brickwedde, M. K.; Dusing, R.; Christian, R.; Wieczorek, A. J.; Vetter, H. EXP3174, a metabolite of losartan (MK 954, DuP 753) is more potent than losartan in blocking the angiotensin II-induced responses in vascular smooth muscle cells. *J. Hypertens.* **1993**, *11*, 155–162.
213. (a) Bachur, N. R. Cytoplasmic aldo-keto reductases: a class of drug metabolizing enzymes. *Science* **1976**, *193*, 595–597. (b) Wermuth, B. Aldo-keto reductases. *Prog. Clin. Biol. Res.* **1985**, *174*, 209–230.
214. Felsted, R. L.; Richter, D. R.; Jones, D. M.; Bachur, N. R. Isolation and characterization of rabbit liver xenobiotic carbonyl reductases. *Biochem. Pharmacol.* **1980**, *29*, 1503–1516.
215. (a) Chan, K. K.; Lewis, R. J.; Trager, W. F. Absolute configurations of the four warfarin alcohols. *J. Med. Chem.* **1972**, *15*, 1265–1270. (b) Sutcliffe, F. A.; MacNicoll, A. D.; Gibson, G. G. Aspects of anticoagulant action: a review of the pharmacology, metabolism and toxicology of warfarin and congeners. *Rev. Drug Metab. Drug Interact.* **1987**, *5*, 225–272. (c) Park, B. K. Warfarin: metabolism and mode of action. *Biochem. Pharmacol.* **1988**, *37*, 19–27. (d) Hermans, J. J. R.; Thijssen, H. H. W. Properties and stereoselectivity of carbonyl reductases involved in the ketone reduction of warfarin and analogs. *Adv. Exp. Med. Biol.* **1993**, *328*, 351–360.

216. Toon, S.; Low, L. K.; Gibaldi, M.; Trager, W. F.; O'Reilly, R. A.; Mottey, C. H.; Goulart, D. A. The warfarin-sulfinpyrazone interaction: stereochemical considerations. *Clin. Pharmacol. Ther.* **1986**, *39*, 15–24.
217. Holford, N. H. G. Clinical pharmacokinetics and pharmacodynamics of warfarin. Understanding the dose-effect relationship. *Clin. Pharmacokinet.* **1986**, *11*, 483–504.
218. (a) Kaminsky, L. S.; Zhang, Z.-Y. Human P450 metabolism of warfarin. *Pharmacol. Ther.* **1997**, *73*, 67–74. (b) Yamazaki, H.; Shimada, T. Human liver cytochrome P450 enzymes involved in the 7-hydroxylation of R- and S-warfarin enantiomers. *Biochem. Pharmacol.* **1997**, *54*, 1195–1203.
219. (a) Prelog, V. Specification of the stereospecificity of some oxidoreductases by diamond lattice sections. *Pure Appl. Chem.* **1964**, *9*, 119–130. (b) Horjales, E.; Brändén, C.-I. Docking of cyclohexanol-derivatives into the active site of liver alcohol dehydrogenase. Using computer graphics and energy minimization. *J. Biol. Chem.* **1985**, *260*, 15445–15451.
220. Roerig, S.; Fujimoto, J. M.; Wang, R. I. H.; Lange, D. Preliminary characterization of enzymes for reduction of naloxone and naltrexone in rabbit and chicken liver. *Drug Metab. Dispos.* **1976**, *4*, 53–58.
221. Dayton, H. E.; Inturrisi, C. E. The urinary excretion profiles of naltrexone in man, monkey, rabbit, and rat. *Drug Metab. Dispos.* **1976**, *4*, 474–478.
222. Gerhards, E.; Hecker, W.; Hitze, H.; Nieuweboer, B.; Bellmann, O. Metabolism of norethisterone(17-hydroxy-19-nor-17α -pregn-4-en-20-yn-3-one) and of DL- and D-norgestrel(13-ethyl-17-hydroxy-18,19-dinor-17α -pregn-4-en-20-yn-3-one) man. *Acta Endocrinol.* **1971**, *68*, 219–248.
223. (a) Uehleke, H. Nitrosobenzene and phenylhydroxylamine as intermediates in the biological reduction of nitrobenzene. *Naturwissenschaften* **1963**, *50*, 335. (b) Gillette, J. R. Metabolism of drugs and other foreign compounds by enzymatic mechanisms. *Forschr. Arzneim-Forsch.* **1963**, *6*, 11–73.
224. (a) Mason, R. P.; Josephy, P. D. In *Toxicity of Nitroaromatic Compounds*; Rickert, D. (Ed.), Hemisphere, New York, 1985, p. 121. (b) Mason, R. P.; Holtzman, J. L. Mechanism of microsomal and mitochondrial nitroreductase. Electron spin resonance evidence for nitroaromatic free radical intermediates. *Biochemistry* **1975**, *14*, 1626–1632. (c) Moreno, S. N. J. The reductive metabolism of nifurtimox and benznidazole in *Crithidia fasciculata* is similar to that in *Trypanosoma cruzi*. *Comp. Biochem. Physiol.* **1988**, *91C*, 321–325.
225. Mason, R. P.; Holtzman, J. L. Role of catalytic superoxide formation in the oxygen inhibition of nitroreductase. *Biochem. Biophys. Res. Commun.* **1975**, *67*, 1267–1274.
226. (a) Scheline, R. R. Metabolism of foreign compounds by gastrointestinal microorganisms. *Pharmacol. Rev.* **1973**, *25*, 451–523. (b) Wheeler, L. A.; Soderberg, F. B.; Goldman, P. Relation between nitro group reduction and the intestinal microflora. *J. Pharmacol. Exp. Ther.* **1975**, *194*, 135–144.
227. Morita, M.; Feller, D. R.; Gillette, J. R. Reduction of niridazole by rat liver xanthine oxidase. *Biochem. Pharmacol.* **1971**, *20*, 217–226.
228. Wolpert, M. K.; Althaus, J. R.; Johns, D. G. Nitroreductase activity of mammalian liver aldehyde oxidase. *J. Pharmacol. Exp. Ther.* **1973**, *185*, 202–213.
229. Poirier, L. A.; Weisburger, J. H. Enzymatic reduction of carcinogenic aromatic nitro compounds by rat and mouse liver fractions. *Biochem. Pharmacol.* **1974**, *23*, 661–669.
230. Garattini, S.; Marcucci, F.; Mussini, E. In *Psychotherapeutic Drugs*, Part 2; Usdin, E.; Forrest, I. S. (Eds.), Marcel Dekker, New York, 1977; p. 1039.
231. Feller, D. R.; Morita, M.; Gillette, J. R. Enzymatic reduction of niridazole by rat liver microsomes. *Biochem. Pharmacol.* **1971**, *20*, 203–215.
232. Tatsumi, K.; Kitamura, S.; Yoshimura, H. Reduction of nitrofuran derivatives by xanthine oxidase and microsomes. Isolation and identification of reduction products. *Arch. Biochem. Biophys.* **1976**, *175*, 131–137.
233. (a) Mason, R. P.; Peterson, F. J.; Holtzman, J. L. The formation of an azo anion free radical metabolite during the microsomal azo reduction of sulfonazo III. *Biochem. Biophys. Res. Commun.* **1977**, *75*, 532–540. (b) Peterson, F. J.; Holtzman, J. L.; Crankshaw, D.; Mason, R. P. The formation of an azo anion free radical metabolite during the microsomal azo reduction of sulfonazo III. *Mol. Pharmacol.* **1988**, *34*, 597–603.
234. Mason, R. P.; Peterson, F. J.; Holtzman, J. L. Inhibition of azoreductase by oxygen. The role of the azo anion free radical metabolite in the reduction of oxygen to superoxide. *Mol. Pharmacol.* **1978**, *14*, 665–671.
235. (a) Scheline, R. R. Metabolism of foreign compounds by gastrointestinal microorganisms. *Pharmacol. Rev.* **1973**, *25*, 451–523. (b) Wheeler, L. A.; Soderberg, F. B.; Goldman, P. Relation between nitro group reduction and the intestinal microflora. *J. Pharmacol. Exp. Ther.* **1975**, *194*, 135–144.
236. (a) Peppercorn, M. A.; Goldman, P. Role of intestinal bacteria in the metabolism of salicylazosulfapyridine. *J. Pharmacol. Exp. Ther.* **1972**, *181*, 555–562. (b) Schröder, H.; Gustafsson, B. E. Azo reduction of salicyl-azo-sulphapyridine in germ-free and conventional rats. *Xenobiotica* **1973**, *3*, 225–231.
237. (a) Eagling, V. A.; Howe, J. L.; Barry, M. J.; Back, D. J. The metabolism of zidovudine by human liver microsomes in vitro: formation of 3'-amino-3'-deoxythymidine. *Biochem. Pharmacol.* **1994**, *48*, 267–276. (b) Stagg, M. P.; Cretten, E. M.; Kidd, L. Clinical pharmacokinetics of 3'-azido-3'-deoxythymidine (zidovudine) and catabolites with formation of a toxic catabolite, 3'-amino-3'-deoxythymidine. *Clin. Pharmacol. Ther.* **1992**, *51*, 668–676.
238. Kato, R.; Iwasaki, K.; Noguchi, H. Reduction of tertiary amine N-oxides by cytochrome P-450. Mechanism of the stimulatory effect of flavins and methyl viologen. *Mol. Pharmacol.* **1978**, *14*, 654–664.
239. Tamura, S.; Kawata, S.; Sugiyama, T.; Tarui, S. Modulation of the reductive metabolism of halothane by microsomal cytochrome b5 in rat liver. *Biochim. Biophys. Acta* **1987**, *926*, 231–238.
240. Tremaine, L. M.; Welch, W. M.; Ronfeld, R. A. Metabolism and disposition of the 5-hydroxytryptamine uptake blocker sertraline in the rat and dog. *Drug Metab. Dispos.* **1989**, *17*, 542–550.
241. (a) Williams, F. M. Serum enzymes of drug metabolism. *Pharmacol. Ther.* **1987**, *34*, 99–109. (b) Heymann, E. In *Enzymatic Basis of Detoxification*; Jakoby, W. B. (Ed.), Academic Press, New York, 1980, Vol. 2, p. 291.
242. Puetter, J. Extramicrosomal drug metabolism. *Eur. J. Drug Metab. Pharmacokinet.* **1979**, *4*, 1–7.
243. Steward, D. J.; Inaba, T.; Lucassen, M.; Kalow, W. Cocaine metabolism: cocaine and norcocaine hydrolysis by liver and serum esterases. *Clin. Pharmacol. Ther.* **1979**, *25*, 464–468.
244. Kogan, M. J.; Verebey, K. G.; DePace, A. C.; Resnick, R. B.; Mulé, S. J. Quantitative determination of benzoylecgonine and cocaine in human biofluids by gas-liquid chromatography. *Anal. Chem.* **1977**, *49*, 1965–1969.
245. Mark, L. C.; Kayden, H. J.; Steele, J. M.; Cooper, J. R.; Berlin, I.; Rovenstein, E. A.; Brodie, B. B. The physiological disposition and cardiac effects of procaine amide. *J. Pharmacol. Exp. Ther.* **1951**, *102*, 5–15.

246. Junge, W.; Krisch, K. The carboxylesterases/amidases of mammalian liver and their possible significance. *CRC Crit. Rev. Toxicol.* **1975**, *3*, 371–435.
247. Heymann, E.; Krisch, K.; Buch, H.; Buzello, W. Inhibition of phenacetin- and acetanilide-induced methemoglobinemia in the rat by the carboxylesterase inhibitor bis(p-nitrophenyl) phosphate. *Biochem. Pharmacol.* **1969**, *18*, 801–811.
248. Akerman, B.; Ross, S. Stereospecificity of the enzymatic biotransformation of the enantiomers of prilocaine (Citanest). *Acta Pharmacol. Toxicol.* **1970**, *28*, 445–453.
249. Dudley, K. H.; Roberts, S. B. Dihydropyrimidinase. Stereochemistry of the metabolism of some 5-alkylhydantoins. *Drug Metab. Dispos.* **1978**, *6*, 133–139.
250. Maksay, G.; Tegyey, Z.; Ötvös, L. Stereospecificity of esterases hydrolyzing oxazepam acetate. *J. Pharm. Sci.* **1978**, *67*, 1208–1210.
251. (a) Heykants, J. J. P.; Meuldermans, W. E. G.; Michiels, L. J. M.; Lewi, P. J.; Janssen, P. A. J. Distribution, metabolism, and excretion of etomidate, a short-acting hypnotic drug, in the rat. Comparative study of (R)-(+) and (S)-(-)-etomidate. *Arch. Int. Pharmacodyn. Ther.* **1975**, *216*, 113–129. (b) Meuldermans, W. E. G.; Lauwers, W. F. J.; Heykants, J. J. P. In-vitro metabolism of etomidate by rat liver fractions. *Arch. Int. Pharmacodyn. Ther.* **1976**, *221*, 140–149.
252. Ben-Menachem, E. Pregabalin pharmacology and its relevance to clinical practice. *Epilepsia* **2004**, *45*(Suppl. 6), 13–18.
253. Vainio, H. In *Mechanisms of Toxicity and Metabolism*, Proceedings of the 6th International Congress of Pharmacology; Karki, N. T. (Ed.), Pergamon, Oxford, 1976, Vol. 6, p. 53.
254. Cummings, A. J.; King, M. L.; Martin, B. K. A kinetic study of drug elimination: the excretion of paracetamol and its metabolites in man. *Br. J. Pharmacol. Chemother.* **1967**, *29*, 150–157.
255. Nakagawa, T.; Masada, M.; Uno, T. Gas chromatographic determination and gas chromatographic-mass spectrometric analysis of chloramphenicol, thiamphenicol, and their metabolites. *J. Chromatogr.* **1975**, *111*, 355–364.
256. Rubin, A.; Warrick, P.; Wolen, R. L.; Chernish, S. M.; Ridolfo, A. S.; Gruber, C. M., Jr. Physiological disposition of fenoprofen in man. III. Metabolism and protein binding of fenoprofen. *J. Pharmacol. Exp. Ther.* **1972**, *183*, 449–457.
257. Bickel, M. H.; Minder, R.; diFrancesco, C. Formation of N-glucuronide of demethylimipramine in the dog. *Experientia* **1973**, *29*, 960–961.
258. Tsukamoto, H.; Yoshimura, H.; Tatsumi, K. Metabolism of drugs. XXXV. Metabolic fate of meprobamate. 3. A new metabolic pathway of carbamate group. The formation of meprobamate N-glucuronide in the animal body. *Chem. Pharm. Bull.* **1963**, *11*, 421–426.
259. Adamson, R. H.; Bridges, J. W.; Kibby, M. R.; Walker, S. R.; Williams, R. T. Metabolism of drugs. XXXV. Metabolic fate of meprobamate. 3. A new metabolic pathway of carbamate group. The formation of meprobamate N-glucuronide in the animal body. *Biochem. J.* **1970**, *118*, 41–45.
260. Sitar, D. S.; Thornhill, D. P. Methimazole. Absorption, metabolism, and excretion in the albino rat. *J. Pharmacol. Exp. Ther.* **1973**, *184*, 432–439.
261. Dutton, G. J.; Illing, H. P. A. Mechanism of biosynthesis of thio-β-D-glucuronides and thio-β-D-glucosides. *Biochem. J.* **1972**, *129*, 539–550.
262. Dieterle, W.; Faigle, J. W.; Frueh, F.; Mory, H.; Theobald, W.; Alt, K. O.; Richter, W. J. Metabolism of phenylbutazone in man. *Arzneim-Forsch.* **1976**, *26*, 572–577.
263. (a) Kadakol, A.; Ghosh, S. S.; Sappal, B. S.; Sharma, G.; Chowdhury, J. R.; Chowdhury, N. R. Genetic lesions of bilirubin uridine-diphosphoglucuronate glucuronosyltransferase (UGT1A1) causing Crigler-Najjar and Gilbert syndromes: Correlation of genotype to phenotype. *Hum. Mutat.* **2000**, *16*, 297–306. (b) Tukey, R. H.; Strassburg, C. P. Human UDP-glucuronosyltransferases. Metabolism, expression, and disease. *Annu. Rev. Pharmacol. Toxicol.* **2000**, *40*, 581–616.
264. (a) Kasten, M. J. Clindamycin, metronidazole, and chloramphenicol. *Mayo Clin. Proc.* **1999**, *74*, 825–833. (b) Knight, M. Adverse drug reactions in neonates. *J. Clin. Pharmacol.* **1994**, *34*, 128–135.
265. Koster, A. Sj.; Frankhuijzen-Sierevogel, A. C.; Mentrup, A. Stereoselective formation of fenoterol-para-glucuronide and fenoterol-meta-glucuronide in rat hepatocytes and enterocytes. *Biochem. Pharmacol.* **1986**, *35*, 1981–1985.
266. Dahl-Puustinen, M.-L.; Dumont, E.; Bertilsson, L. Glucuronidation of E-10-hydroxynortriptyline in human liver, kidney, and intestine. Organ-specific differences in enantioselectivity. *Drug Metab. Dispos.* **1989**, *17*, 433–436.
267. (a) Mulder, G. J.; Jakoby, W. B. In *Conjugation Reactions in Drug Metabolism*; Mulder, G. J. (Ed.), Taylor & Francis: London, 1990, p. 107. (b) Levy, G. Sulfate conjugation in drug metabolism: role of inorganic sulfate. *Fed. Proc.* **1986**, *45*, 2235–2240. (c) Mulder, G. J. (Ed.), *Sulfation of Drugs and Related Compounds*, CRC Press, Boca Raton, FL, 1981.
268. (a) Jakoby, W. B.; Ziegler, D. M. The enzymes of detoxication. *J. Biol. Chem.* **1990**, *265*, 20715–20718. (b) Falany, C. N. Enzymology of human cytosolic sulfotransferases. *FASEB J.* **1997**, *11*, 206–216.
269. Pang, K. S. In Conjugation Reactions in Drug Metabolism; Mulder, G. J. (Ed.), Taylor & Francis, London, 1990, p. 5.
270. Capel, I. D.; French, M. R.; Milburn, P.; Smith, R. I.; Williams, R. T. Fate of [^{14}C]phenol in various species. *Xenobiotica* **1972**, *2*, 25–34.
271. (a) Mulder, G. J. In *Conjugation Reactions in Drug Metabolism*; Mulder, G. J. (Ed.), Taylor & Francis, London, 1990, p. 41. (b) Whitmer, D. I.; Ziurys, J. C.; Gollan, J. L. Hepatic microsomal glucuronidation of bilirubin in unilamellar liposomal membranes. Implications for intracellular transport of lipophilic substrates. *J. Biol. Chem.* **1984**, *259*, 11969–11975.
272. (a) Lin, C.; Li, Y.; McGlotten, J.; Morton, J. B.; Symchowicz, S. Isolation and identification of the major metabolite of albuterol in human urine. *Drug Metab. Dispos.* **1977**, *5*, 234–238. (b) Walle, T.; Walle, U. K.; Thornburg, K. R.; Schey, K. L. Stereoselective sulfation of albuterol in humans. Biosynthesis of the sulfate conjugate by Hep G2 cells. *Drug Metab. Dispos.* **1993**, *21*, 76–80.
273. Albert, K. S.; Sedman, A. J.; Wagner, J. G. Pharmacokinetics of orally administered acetaminophen in man. *J. Pharmacokinet. Biopharm.* **1974**, *2*, 381–393.
274. (a) Miller, R. P.; Roberts, R. J.; Fischer, L. J. Acetaminophen elimination kinetics in neonates, children, and adults. *Clin. Pharmacol. Ther.* **1976**, *19*, 284–294. (b) Levy, G.; Khana, N. N.; Soda, D. M.; Tsuzuki, O.; Stern, L. Pharmacokinetics of acetaminophen in the human neonate. Formation of acetaminophen glucuronide and sulfate in relation to plasma bilirubin concentration and D-glucaric acid excretion. *Pediatrics* **1975**, *55*, 818–825.
275. Mulder, G. J.; Meerman, J. H. H.; van den Goorbergh, A. M. In Xenobiotic Conjugation Chemistry; Paulson, G. D.; Caldwell, J.; Hutson, D. H.; Menn, J. J. (Eds.), American Chemical Society, Washington, D.C., 1986, p. 282.

276. Mulder, G. J.; Hinson, J. A.; Gillette, J. R. Generation of reactive metabolites of N-hydroxyphenacetin by glucuronidation and sulfation. *Biochem. Pharmacol.* **1977**, *26*, 189–196.
277. Smith, R. L.; Caldwell, J. In *Drug Metabolism: From Microbe to Man*; Parke, D. V.; Smith, R. L. (Eds.), Taylor & Francis, London, 1977, p. 331.
278. Killenberg, P. G.; Webster, L. T., Jr. In *Enzymatic Basis of Detoxification*, Vol. 2, Jakoby, W. B. (Ed.), Academic, New York, 1980; p. 141.
279. Bruce, R. B.; Turnbull, L. B.; Newman, J. H.; Pitts, J. E. Metabolism of brompheniramine. *J. Med. Chem.* **1968**, *11*, 1031–1034.
280. Drach, J. C.; Howell, J. P.; Borondy, P. E.; Glazko, A. J. Species differences in the metabolism of diphenhydramine (Benadryl). *Proc. Soc. Exp. Biol. Med.* **1970**, *135*, 849–853.
281. Marchand, D. H.; Remmel, R. P.; Abdel-Monem, M. M. Biliary excretion of a glutathione conjugate of busulfan and 1,4-diiodobutane in the rat. *Drug Metab. Dispos.* **1988**, *16*, 85–92.
282. Needleman, P. In *Organic Nitrates*; Needleman, P. (Ed.), Springer-Verlag, Berlin, 1975, p. 57.
283. de Miranda, P.; Beacham, L. M., III; Creagh, T. H.; Elion, G. B. Metabolic fate of the methylnitroimidazole moiety of azathioprine in the rat. *J. Pharmacol. Exp. Ther.* **1973**, *187*, 588–601.
284. Ishida, T.; Kumagai, Y.; Ikeda, Y.; Ito, K.; Yano, M.; Toki, S.; Mihashi, K.; Fujioka, T.; Iwase, Y.; Hachiyama, S. (8S)-(Glutathion-S-yl) dihydromorphinone, a novel metabolite of morphine from guinea pig bile. *Drug Metab. Dispos.* **1989**, *17*, 77–81.
285. Correia, M. A.; Krowech, G.; Caldera-Munoz, P.; Yee, S. L.; Straub, K.; Castagnoli, N., Jr. Morphine metabolism revisited. II. Isolation and chemical characterization of a glutathionylmorphine adduct from rat liver microsomal preparations. *Chem. Biol. Interact.* **1984**, *51*, 13–24.
286. Suzuki, T.; Nishio, K.; Tanabe, S. The MRP family and anticancer drug metabolism. *Curr. Drug Metab.* **2001**, *2*, 367–377.
287. Stevens, J. L.; Jones, D. P. In *Glutathione*; Dolphin, D.; Poulson, R.; Avramovib, O. (Eds.), Wiley, New York, 1989, Part B, p. 45.
288. (a) Guenthner, T. M. In *Conjugation Reactions in Drug Metabolism*; Mulder, G. J. (Ed.), Taylor & Francis, London, 1990, p. 365. (b) Fretland, A. J.; Omiecinski, C. J. Epoxide hydrolases: biochemistry and molecular biology. *Chem. Biol. Interact.* **2000**, *129*, 41–59.
289. (a) Vogel-Bindel, U.; Bentley, P.; Oesch, F. Endogenous role of microsomal epoxide hydrolase. Ontogenesis, induction, inhibition, tissue distribution, immunological behavior and purification of microsomal epoxide hydrolase with 16α,17α-epoxyandrostene-3-one as substrate. *Eur. J. Biochem.* **1982**, *126*, 425–431. (b) Faendrich, F.; Degiuli, B.; Vogel-Bindel, U.; Arand, M.; Oesch, F. Induction of rat liver microsomal epoxide hydrolase by its endogenous substrate 16α ,17α -epoxyestra-1,3,5-trien-3-ol. *Xenobiotica* **1995**, *25*, 239–244.
290. Elson, J.; Strong, J. M.; Atkinson, A. J., Jr. Antiarrhythmic potency of N-acetylprocainamide. *Clin. Pharmacol. Ther.* **1975**, *17*, 134–140.
291. Williams, R. T. In *Biogenesis of Natural Compounds*, 2nd ed., Bernfeld, P. (Ed.), Pergamon: Oxford, 1967, p. 589.
292. Meyer, U. A. Pharmacogenetics and adverse drug reactions. *Lancet* **2000**, *356*, 1667–1671.
293. (a) Daly, A. K.; Cholerton, S.; Gregory, W.; Idle, J. R. Metabolic polymorphisms. *Pharmacol. Ther.* **1993**, *57*, 129–160. (b) Meyer, U. A; Zanger, U. M. Molecular mechanisms of genetic polymorphisms of drug metabolism. *Annu. Rev. Pharmacol. Toxicol.* **1997**, *37*, 269–296. (c) Lee, C. R.; Goldstein, J. A.; Pieper, J. A. Cytochrome P450 2C9 polymorphisms: a comprehensive review of the in-vitro and human data. *Pharmacogenetics* **2002**, *12*, 251–263. (d) Nagata, K.; Yamazoe, Y. Genetic polymorphism of human cytochrome P450 involved in drug metabolism. *Drug Metab. Pharmacokinet.* **2002**, *17*, 167–189.
294. Drayer, D. E.; Reidenberg, M. M. Clinical consequences of polymorphic acetylation of basic drugs. *Clin. Pharmacol. Ther.* **1977**, *22*, 251–258.
295. (a) Kalow, W. *Pharmacogenetics, Heredity and the Response to Drugs*; Saunders: Philadelphia, 1962. (b) Weber, W. W. In Metabolic Conjugation and Metabolic Hydrolysis; Fishman, W. H. (Ed.), Academic, New York, 1973, Vol. 3, p. 250.
296. (a) Kalow, W. *Pharmacogenetics, Heredity and the Response to Drugs*; Saunders: Philadelphia, 1962. (b) Weber, W. W. In *Metabolic Conjugation and Metabolic Hydrolysis*; Fishman, W. H. (Ed.), Academic Press, New York, 1973, Vol. 3, p. 250. (c) Lunde, P. K. M.; Frislid, K.; Hansteen, V. Disease and acetylation polymorphism. *Clin. Pharmacokinet.* **1977**, *2*, 182–197.
297. (a) Lunde, P. K. M.; Frislid, K.; Hansteen, V. Disease and acetylation polymorphism. *Clin. Pharmacokinet.* **1977**, *2*, 182–197. (b) Weber, W. W. In Therapeutic Drugs; Dollery, C. T. (Ed.), Churchill Livingstone, New York, 1986.
298. Patterson, E.; Radtke, H. E.; Weber, W. W. Immunochemical studies on rabbit N-acetyltransferases. *Mol. Pharmacol.* **1980**, *17*, 367–373.
299. Dyda, F.; Klein, D. C.; Hickman, A. B. GCN5-related N-acetyltransferases: a structural overview. *Annu. Rev. Biophys. Biomol. Struct.* **2000**, *29*, 81–103.
300. Lin, J. H.; Chen, I.-W.; Ulm, E. H. Dose-dependent kinetics of cilastatin in laboratory animals. *Drug Metab. Dispos.* **1989**, *17*, 426–432.
301. Kropp, H.; Sundelof, J. G.; Hajdu, R.; Kahan, F. M. Metabolism of thienamycin and related carbapenem antibiotics by the renal dipeptidase, dehydropeptidase-I. *Antimicrobial. Agents Chemother.* **1982**, *22*, 62–70.
302. Eschenhof, E. Fate of the anticonvulsant clonazepam in the organism of rat, dog, and men. *Arzneim-Forsch.* **1973**, *23*, 390–400.
303. Caldwell, J.; Marsh, M. V. Interrelationships between xenobiotic metabolism and lipid biosynthesis. *Biochem. Pharmacol.* **1983**, *32*, 1667–1672.
304. Dell, H. D.; Fiedler, J.; Kamp, R.; Gau, W.; Kurz, J.; Weber, B.; Wuensche, C. Etofenamate fatty acid esters. An example of a new route of drug metabolism. *Drug Metab. Dispos.* **1982**, *10*, 55–60.
305. Caldwell, J. Xenobiotic acyl-coenzymes A: critical intermediates in the biochemical pharmacology and toxicology of carboxylic acids. *Biochem. Soc. Trans.* **1984**, *12*, 9–11.
306. Leighty, E. G.; Fentiman, A. F., Jr.; Foltz, R. L. Long-retained metabolites of delta9- and delta8-tetrahydrocannabinols identified as novel fatty acid conjugates. *Res. Commun. Chem. Pathol. Pharmacol.* **1976**, *14*, 13–28.
307. (a) Leighty, E. G. Metabolism and distribution of cannabinoids in rats after different methods of administration. *Biochem. Pharmacol.* **1973**, *22*, 1613–1621. (b) Caldwell, J.; Parkash, M. K. In *Perspectives in Medicinal Chemistry*; Testa, B.; Kyburz, E.; Fuhrer, W.; Giger, R. (Eds.), VCH: Weinheim, 1993, p. 595.
308. Gunnarsson, P. O.; Johansson, S.-A.; Svensson, L. Cholesterol ester formation by transesterification of chlorambucil: a novel pathway in drug metabolism. *Xenobiotica* **1984**, *14*, 569–574.
309. Fears, R.; Baggaley, K. H.; Walker, P.; Hindley, R. M. Xenobiotic cholesteryl ester formation. *Xenobiotica* **1982**, *12*, 427–433.
310. (a) Thakker, D. R.; Creveling, C. R. In *Conjugation Reactions in Drug Metabolism*; Taylor & Francis, London, 1990, p. 193. (b) Ansher, S. S.; Jakoby, W. B. In *Conjugation Reactions in Drug Metabolism*; Taylor & Francis, London, 1990; p. 233. (c) Stevens, J. L.; Bakke, J. E. In *Conjugation Reactions in Drug Metabolism*; Taylor & Francis, London, 1990, p. 251.

311. Bonifacio, M. J.; Archer, M.; Rodrigues, M. L.; Matias, P. M.; Learmonth, D. A.; Carrondo, M. A.; Soares-Da-Silva, P. Kinetics and crystal structure of catechol-O-methyltransferase complex with co-substrate and a novel inhibitor with potential therapeutic application. *Mol. Pharmacol.* **2002**, *62*, 795–805.
312. (a) Clarke, S.; Banfield, K. Homocysteine in Health and Disease; Carmel, R.; Jacobsen, D. W. (Eds.), Cambridge University Press, Cambridge, UK, 2001, pp 63–78. (b) Martin, J. L.; McMillan, F. M. SAM (dependent) I AM: the S-adenosylmethionine-dependent methyltransferase fold. *Curr. Opin. Struct. Biol.* **2002**, *12*, 783–793. (c) Usdin, E.; Borchardt, R. T.; Creveling, C. R. (Eds.), *The Biochemistry of S-Adenosylmethionine and Related Compounds*; Macmillan Press, London, 1982.
313. Morgan, C. D.; Sandler, M.; Davies, D. S.; Connolly, M.; Paterson, J. W.; Dollery, C. T. The metabolic fate of DL-[7-^{3}H] isoprenaline in man and dog. *Biochem. J.* **1969**, *114*, 8 p.
314. Persson, K.; Persson, K. Metabolism of terbutaline in vitro by rat and human liver o-methyltransferases and monoamine oxidases. *Xenobiotica* **1972**, *2*, 375–382.
315. Börner, U.; Abbott, S. New observations in the metabolism of morphine. The formation of codeine from morphine in man. *Experientia* **1973**, *29*, 180–181.
316. Leeson, G. A.; Garteiz, D. A.; Knapp, W. C.; Wright, G. J. N-methylation, a newly identified pathway in the dog for the metabolism of oxprenolol, a β-receptor blocking agent. *Drug Metab. Dispos.* **1973**, *1*, 565–568.
317. Herbertz, G.; Metz, T.; Reinauer, H.; Staib, W. Metabolism of chlormethiazole in the perfused liver of the rat. *Biochem. Pharmacol.* **1973**, *22*, 1541–1546.
318. Drummer, O. H.; Miach, P.; Jarrott, B. S-Methylation of captopril. Demonstration of captopril thiol methyltransferase activity in human erythrocytes and enzyme distribution in rat tissues. *Biochem. Pharmacol.* **1983**, *32*, 1557–1562.
319. Lindsay, R. H.; Hulsey, B. S.; Aboul-Enein, H. Y. Enzymatic S-methylation of 6-propyl-2-thiouracil and other antithyroid drugs. *Biochem. Pharmacol.* **1975**, *24*, 463–468.
320. (a) Guengerich, F. P. Metabolism-based toxicology prediction. *RSC Drug Discovery Series* **2012**, *12*, 542–562. (b) Williams, D. P. Toxicophores: investigations in drug safety. *Toxicology* **2006**, *226*, 1–11.
321. Kazius, J.; McGuire, R.; Bursi, R. Derivation and validation of toxicophores for mutagenicity prediction. *J. Med. Chem.* **2005**, *48*, 312–320.
322. Hakimelahi, G. H.; Khodarahmi, G. A. The identification of toxicophores for the prediction of mutagenicity, hepatotoxicity and cardiotoxicity. *J. Iran. Chem. Soc.* **2005**, *2*, 244–267.
323. Kalgutkar, A. S.; Fate, G.; Didiuk, M. T.; Bauman, J. Toxicophores, reactive metabolites and drug safety: when is it cause for concern? *Exp. Rev. Clin. Pharmacol.* **2008**, *1*, 515–531.
324. Stepan, A. F.; Walker, D. P.; Bauman, J.; Price, D. A.; Baillie, T. A.; Kalgutkar, A. S.; Aleo, M. D. Structural alert/reactive metabolite concept as applied in medicinal chemistry to mitigate the risk of idiosyncratic drug toxicity: A perspective based on the critical examination of trends in the top 200 drugs marketed in the United States. *Chem. Res. Toxicol.* **2011**, *24*, 1345–1410.
325. Lennernas, H. Clinical pharmacokinetics of atorvastatin. *Clin. Pharmacokinet.* **2003**, *42*, 1141–1160.
326. Nakayama, S., Atsumi, R., Takakusa, H., Kobayashi, Y., Kurihara, A., Nagai, Y., Nakai, D., and Okazaki, O. A zone classification system for risk assessment of idiosyncratic drug toxicity using daily dose and covalent binding. *Drug Metab. Dispos.* **2009**, *37*, 1970–1977.
327. Penning, T.D.; Talley, J. J.; Bertenshaw, S. R.; Carter, J. S.; Collins, P. W.; Docter, S.; Graneto, M. J.; Lee, L. F.; Malecha, J. W.; Miyashiro, J. M.; Rogers, R. S.; Rogier, D. J.; Yu, S. S.; Anderson, G. D.; Burton, E. G.; Cogburn, J. N.; Gregory, S. A.; Koboldt, C. M.; Perkins, W. E.; Seibert, K.; Veenhuizen, A. W.; Zhang, Y. Y.; Isakson, P. C. Synthesis and biological evaluation of the 1,5-diarylpyrazole class of cyclooxygenase-2 inhibitors: identification of 4-[5-(4-methylphenyl)-3-(trifluoromethyl)-1H-pyrazol-1-yl]benzenesulfonamide (SC-58635, celecoxib). *J. Med. Chem.* **1997**, *40*, 1347–1365.
328. (a) Bodor, N. In *Design of Biopharmaceutical Properties through Prodrugs and Analogs*; Roche, E. B. (Ed.), American Pharmaceutical Association, Washington, D.C, 1977, Chap. 7. (b) Bodor, N. Novel approaches to the design of safer drugs: soft drugs and site-specific chemical delivery systems. *Adv. Drug Res.* **1984**, *13*, 255–331. (c) Bodor, N. Soft drugs: principles and methods for the design of soft drugs. *Med. Res. Rev.* **1984**, *4*, 449.
329. Lee, H. J.; Cooperwood, J. S.; You, Z.; Ko, D.-H. Prodrug and antedrug: two diametrical approaches in designing safer drugs. *Arch. Pharm. Res.* **2002**, *25*, 111–136.
330. Bodor, N.; Buchwald, P. Soft drug design: general principles and recent applications. *Med. Res. Rev.* **2000**, *20*, 58–101.
331. Bodor, N.; Kaminski, J. J.; Selk, S. Soft drugs. 1. Labile quaternary ammonium salts as soft antimicrobials. *J. Med. Chem.* **1980**, *23*, 469–474.
332. Druzgala, P.; Hochhaus, G.; Bodor, N. Soft drugs. 10. Blanching activity and receptor binding affinity of a new type of glucocorticoid: loteprednol etabonate. *J. Steroid Biochem. Mol. Biol.* **1991**, *38*, 149–154.
333. Bucala, R.; Fishman, J.; Cerami, A. Formation of covalent adducts between cortisol and 16α-hydroxyestrone and protein: possible role in the pathogenesis of cortisol toxicity and systemic lupus erythematosus. *Proc. Natl. Acad. Sci. U.S.A.* **1982**, *79*, 3320–3324.

Chapter 9

Prodrugs and Drug Delivery Systems

Chapter Outline

9.1. ENZYME ACTIVATION OF DRUGS

The term *prodrug*, which was used initially by Albert,[1] is a pharmacologically inactive compound (or at least 1000 times less potent than the parent drug[2]) that is converted into an active drug by a metabolic biotransformation. A prodrug can also be activated by a nonenzymatic process such as hydrolysis, but in this case, the compounds generally are inherently unstable and may cause stability problems, for example, during storage. The prodrug to drug conversion can occur before, during, or after absorption or at a specific site in the body, especially if the activating enzyme is more abundant at the target site than anywhere else.[3] In the ideal case, a prodrug is converted to the drug as soon as the desired goal for designing the prodrug has been achieved. In 2006, about 16% of small-molecule drugs were prodrugs.[4]

As noted in Chapter 8 (Section 8.4.5), the concepts of prodrugs and soft drugs (antedrugs) are opposite. Prodrugs are inactive compounds that require a metabolic conversion to the active form, whereas a soft drug is pharmacologically active and uses metabolism as a means of promoting deactivation and excretion. However, it is possible to design a *pro-soft drug*, a modified soft drug that requires metabolic activation for conversion to the active soft drug.

The Organic Chemistry of Drug Design and Drug Action. http://dx.doi.org/10.1016/B978-0-12-382030-3.00009-X

9.1.1. Utility of Prodrugs

Prodrug design is a lead modification approach that is used to correct a flaw in a viable drug candidate. Typically, it is considered as an option after lead optimization has been carried out and when the desired drug candidate has failed in pharmacokinetic/preclinical studies. However, it has been argued that a prodrug approach should be considered in early stages of lead optimization as well,[5] particularly when there are pharmacokinetic defects in the molecule, such as poor absorption, very short or very long half-life, high first-pass effect, or off-target inhibition.[6] Below are briefly discussed numerous reasons why you may want to utilize a prodrug strategy in drug design.

9.1.1.1. Aqueous Solubility

Consider an injectable drug that is so insoluble in water that it would need to be taken up in more than a liter of saline to administer the appropriate dose! Or what if each dose of your ophthalmic drug required a liter of saline for dissolution, but it was to be administered as eye drops? These drugs could be safe, effective, and potent, but they would not be viable for their applications. In these cases, a water-solubilizing group, which is metabolically released after drug administration, could be attached to the drugs.

9.1.1.2. Absorption and Distribution

If the desired drug is not absorbed and transported to the target site in sufficient concentration, it can be made more water soluble or lipid soluble depending on the desired site of action. Once absorption has occurred or when the drug is at the appropriate site of action, the water- or lipid-soluble group is removed enzymatically.

9.1.1.3. Site Specificity

Specificity for a particular organ or tissue can be made if there are high concentrations of or uniqueness of enzymes present at that site that can cleave the appropriate appendages from the prodrug and unmask the drug. Alternatively, something that directs the drug to a particular type of tissue, which is released after the drug reaches the target tissue, could be attached to the drug.

9.1.1.4. Instability

A drug may be rapidly metabolized and rendered inactive before it reaches the site of action. The structure may be modified to block that metabolism until the drug is at the desired site.

9.1.1.5. Prolonged Release

It may be desirable to have a steady low concentration of a drug released over a long period of time. The drug may be altered so that it is metabolically converted to the active form slowly.

9.1.1.6. Toxicity

A drug may be toxic in its active form and would have a greater therapeutic index if it were administered in a non-toxic inactive form that was converted into the active form only at the site of action.

9.1.1.7. Poor Patient Acceptability

An active drug may have an unpleasant taste or odor, produce gastric irritation, or cause pain when administered (for example, when injected). The structure of the drug could be modified to alleviate these problems, but once administered, the prodrug would be metabolized to the active drug.

9.1.1.8. Formulation Problems

If the drug is a volatile liquid, it would be more desirable to have it in a solid form so that it could be formulated as a tablet. An inactive solid derivative could be prepared, which would be converted in the body to the active drug.

9.1.2. Types of Prodrugs

There are several classifications of prodrugs. Some prodrugs are not designed as such; the biotransformations are fortuitous, and it is discovered only after isolation and testing of the metabolites that activation of the drug had occurred. In most cases, a specific modification in a drug has been made on the basis of known metabolic transformations. It is expected that, after administration, the prodrug will be appropriately metabolized to the active form. This has been termed *drug latentiation* to signify the rational design approach rather than *serendipity*.[7] The term drug latentiation has been refined even further by Wermuth[8] into two classes, which he called carrier-linked prodrugs and bioprecursors.

A *carrier-linked prodrug* is a compound that contains an active drug linked to a carrier group that can be removed enzymatically, such as an ester, which is hydrolyzed to an active carboxylic acid-containing drug. The bond to the carrier group must be labile enough to allow the active drug to be released efficiently in vivo, and the carrier group must be nontoxic and biologically inactive when detached from the drug. Carrier-linked prodrugs can be subdivided even further into bipartite, tripartite, and mutual prodrugs. A *bipartite prodrug* is a prodrug composed of one carrier attached to the drug. When a carrier is connected to a linker that is connected to the drug, it is called a *tripartite prodrug*. A *mutual prodrug* (also known as a *codrug*) consists of two, usually synergistic, drugs attached to each other (one drug is the carrier for the other, and vice versa).

A *bioprecursor prodrug* is a compound that is metabolized by molecular modification into a new compound, which is the active principle or which can be metabolized further to the active drug. For example, if the drug contains a carboxylic acid group, the bioprecursor may be a primary amine, which is metabolized by oxidation to the aldehyde and then further

A. $RCO_2H \xrightarrow[\text{HCl}, \Delta]{\text{EtOH}} RCO_2Et \xrightarrow[\text{on R}]{\text{Reaction}} R'CO_2Et \xrightarrow[\Delta]{H_3O^+} R'CO_2H$

B. $RCH{=}CH_2 \xrightarrow[\text{on R}]{\text{Reaction}} R'CH{=}CH_2 \xrightarrow[2.\ H_2O_2]{1.\ O_3} R'CO_2H$

SCHEME 9.1 Protecting group analogy for a prodrug

metabolized to the carboxylic acid drug (see Chapter 8, Section 8.4.2.1.5). Unlike the carrier-linked prodrug, which is the active drug linked to a carrier, a bioprecursor contains a different structure that cannot be converted into the active drug by simple cleavage of a group from the prodrug.

The concept of prodrugs can be analogized to the use of protecting groups in organic synthesis.[9] If, for example, you wanted to carry out a reaction on a compound that contained a carboxylic acid group, it may be necessary first to protect the carboxylic acid as, say, an ester, so that the acidic proton of the carboxylic acid did not interfere with the desired reaction. After the desired synthetic transformation was completed, the carboxylic acid analog could be unmasked by deprotection, i.e., hydrolysis of the ester (Scheme 9.1). This is analogous to a carrier-linked prodrug; an ester functionality can be used to make the properties of the drug more desirable until it reaches the appropriate biological site where it is "deprotected". Another type of protecting group in organic synthesis is one that has no resemblance to the desired functional group. For example, a terminal alkene can be oxidized with ozone to an aldehyde,[10] and the aldehyde can be oxidized to a carboxylic acid with hydrogen peroxide (Scheme 9.1). As in the case of a bioprecursor prodrug a drastic structural change is required to unmask the desired group. Oxidation is a common metabolic biotransformation for bioprecursor prodrugs.

Approximately 49% of marketed prodrugs are activated by hydrolysis, and 23% are bioprecursor prodrugs.[11] When designing a prodrug, you should keep in mind that a particular metabolic transformation may be species specific (see Chapter 8). Therefore, a prodrug whose design was based on rat metabolism studies may not necessarily be effective in humans. Typically, a combination of animal studies in vivo and human and animal cellular studies in vitro might be used to study the conversion of a candidate prodrug to its active form prior to the ability to test the prodrug directly in humans.

9.2. MECHANISMS OF DRUG INACTIVATION

9.2.1. Carrier-Linked Prodrugs

An ideal drug carrier must (1) protect the drug until it is at the site of action; (2) localize the drug at the site of action; (3) allow for release of the drug chemically or enzymatically; (4) minimize host toxicity; (5) be biodegradable, biochemically inert, and nonimmunogenic; (6) be easily prepared inexpensively; and (7) be chemically and biochemically stable in its dosage form.

The most common reaction for activation of carrier-linked prodrugs is hydrolysis. First, we will consider the general functional groups involved and then look at specific examples for different types of prodrugs.

9.2.1.1. Carrier Linkages for Various Functional Groups

9.2.1.1.1. Alcohols, Carboxylic Acids, and Related

There are several reasons why the most common prodrug form for drugs containing alcohol or carboxylic acid functional groups is an ester.[12] First, esterases are ubiquitous, so metabolic regeneration of the drug is a facile process. Also, it is possible to prepare ester derivatives with virtually any degree of hydrophilicity or lipophilicity. Finally, a variety of stabilities of esters can be obtained by appropriate manipulation of electronic and steric factors; the use of carbonate and carbamate linkages increases the range of reactivities.[13] Therefore, a multitude of ester prodrugs can be prepared to accommodate a wide variety of problems that require the prodrug approach.

Alcohol-containing drugs can be acylated with aliphatic or aromatic carboxylic acids to decrease water solubility (increase lipophilicity) or with carboxylic acids containing amino or additional carboxylate groups to increase water solubility (Table 9.1).[14] Conversion to phosphate or sulfate esters also increases water solubility. By using these approaches, a wide range of solubilities can be achieved that will affect the absorption and distribution properties of the drug. These derivatives can also have an important effect on the dosage form, that is, whether used in tablet form or in aqueous solution. One problem with the use of this prodrug approach is that in some cases certain esters are not very good substrates for the endogenous esterases, sulfatases, or phosphatases and may not be hydrolyzed at a rapid enough rate. For example, a series of amino acid ester prodrugs of the dual-acting (see Chapter 5, Section 5.2.4.2.4) vascular endothelial growth factor receptor-2/fibroblast growth factor receptor-1 antitumor agent brivanib (**9.1** is the general prodrug form), which failed in Phase III clinical trials (when RCO is L-Ala), was prepared; depending on the amino acid used, there was a great disparity in hydrolysis rates (Table 9.2).[15] When that occurs, however, a different ester can be tried. Another approach to accelerate

TABLE 9.1 Ester Analogs of Alcohols as Prodrugs

X	Effect on Water Solubility
$C(=O)R$	(R=aliphatic or aromatic); decreases
$C(=O)CH_2\overset{+}{N}HMe_2$	Increases (pK_a~8)
$C(=O)CH_2CH_2COO^-$	Increases (pK_a~5)
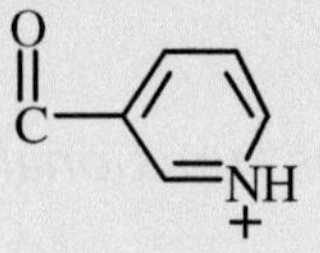 $C(=O)$–(3-pyridinium, $\overset{+}{N}H$)	Increases (pK_a~4)
$PO_3^=$	Increases (pK_a~2 and ~6)
$C(=O)CH_2SO_3^-$	Increases (pK_a~1)

TABLE 9.2 Percentage of Different Amino Acid Prodrugs of 9.1 Metabolized after 30 min Incubation With Mouse or Human Microsomes

Amino Acid Ester Prodrug (RCO)	% Metabolized after 30 min Incubation	
	Mouse	Human
L-Alanine	100	100
L-Valine	28	64
L-Leucine	100	100
L-Isoleucine	26	12

the hydrolysis rate could be to attach electron-withdrawing groups (if a base hydrolysis mechanism is relevant) or electron-donating groups (if an acid hydrolysis mechanism is important)[16] to the carboxylate side of the ester. Succinate esters can be used to accelerate the rate of hydrolysis by intramolecular catalysis (Scheme 9.2). If the ester is too reactive, substituents can be appended that cause steric hindrance to hydrolysis or esters of long-chain fatty acids can be employed. Alcohol-containing drugs can also be converted into the corresponding acetals or ketals for rapid hydrolysis in the acidic medium of the gastrointestinal tract.

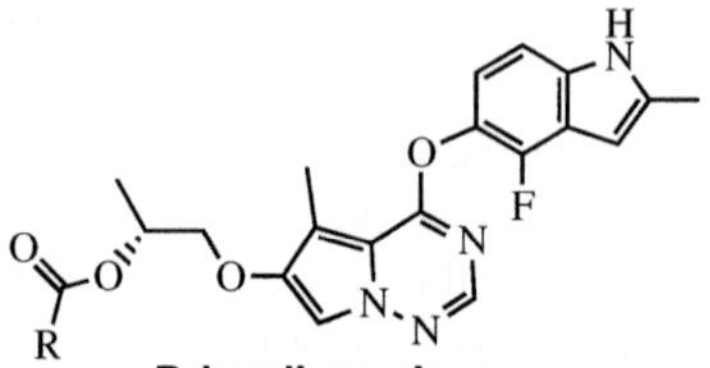

Brivanib prodrugs
9.1

Enolic hydroxyl groups can be esterified to prodrug forms as well. A series of enol esters of the antirheumatic oxindole **9.2** was prepared, and it was found that the hemifumarate derivative (**9.3**) was more stable than the corresponding nonionizable esters at neutral pH.[17]

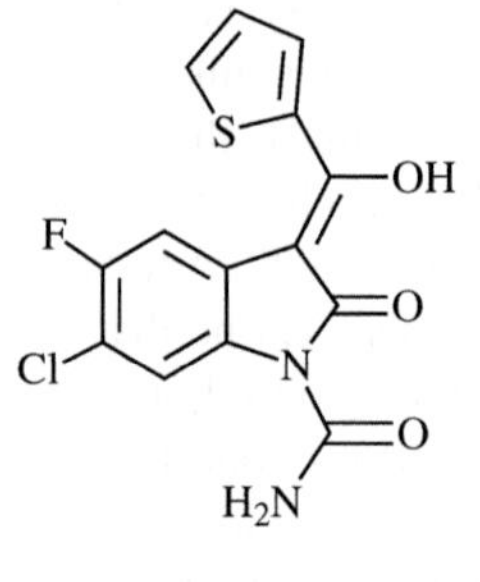

9.2 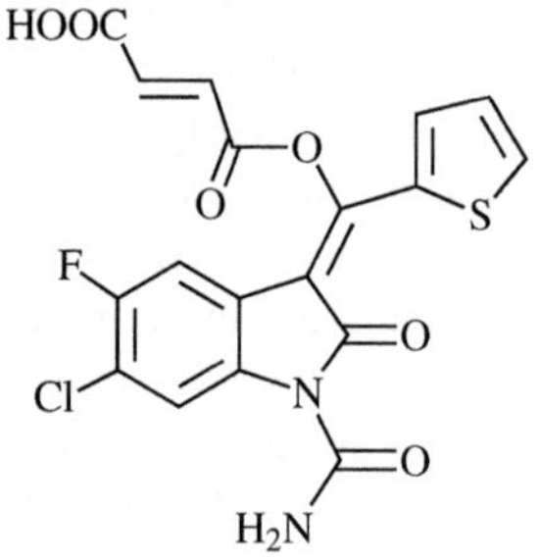 **9.3**

Carboxylic acid-containing drugs can be esterified with various alcohols. The reactivity of the derivatized drug can be adjusted by appropriate structural manipulations as discussed above for ester prodrugs of alcohol-containing drugs. The anionic character of a carboxylic acid at physiological pH can be changed to cationic character by conversion to a choline ester (**9.4**, R = R′ = Me) or an amino ester (**9.4**, R = H, R′ = H or Me; the pK_a of the protonated ammonium salt is ~9). Likewise, phosphate- or phosphonate-containing drugs can be converted into ester prodrug forms,[18] such as bis(pivaloyloxymethyl) adefovir dipivoxil (**9.5**, Hepsera), a drug for the treatment of hepatitis B,[19] selected on the basis of its favorable oral bioavailability in rats.

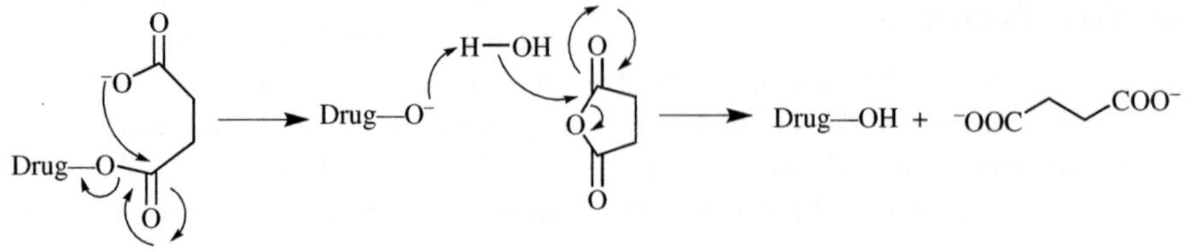

Drug—O⁻ ... Drug—OH + ⁻OOC–CH₂CH₂–COO⁻

SCHEME 9.2 Intramolecular catalysis of succinate esters

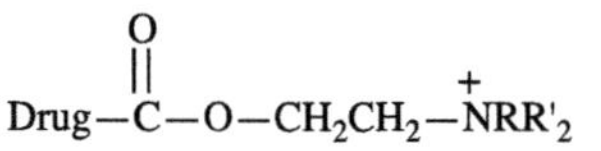

$Drug{-}C(=O){-}O{-}CH_2CH_2{-}\overset{+}{N}RR'_2$

9.4

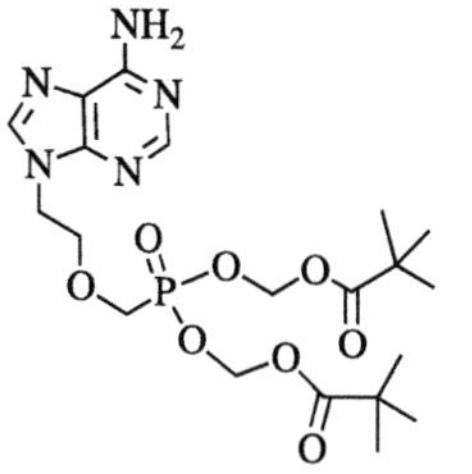

Adefovir dipivoxil
9.5

Phosphate prodrugs can be used to enhance either intravenous (IV) or oral delivery of drugs.[20] In cases when IV is the preferred route of administration, low solubility of a drug is disadvantageous because there are limits to the volume of a drug solution that can be administered over a given time period, and if the solution is too dilute owing to low solubility, then insufficient drug can be administered. Phosphate prodrugs administered intravenously can be converted to the parent drug by alkaline phosphatases in the liver. Somewhat less obvious is the use of a phosphate prodrug to enhance oral delivery, since the highly polar phosphate group is expected to impede permeability through the intestinal membranes. In this case, the role of the phosphate group is to improve initial solubilization of the drug (as the prodrug), which is followed by phosphate cleavage by phosphatases at the intestinal brush border. This places the parent drug in intimate contact with the intestine, resulting in enhanced intestinal membrane permeation of the parent drug. An example of this approach is the phosphoryloxymethyl prodrug fostamatinib (**9.6**),[21] an inhibitor of spleen tyrosine kinase that is in clinical trials for treatment of autoimmune disease. This example, in which a phosphoryloxymethyl moiety is a substituent on a heterocyclic nitrogen atom, also illustrates that phosphate prodrugs need not be limited to phosphorylated alcohols.

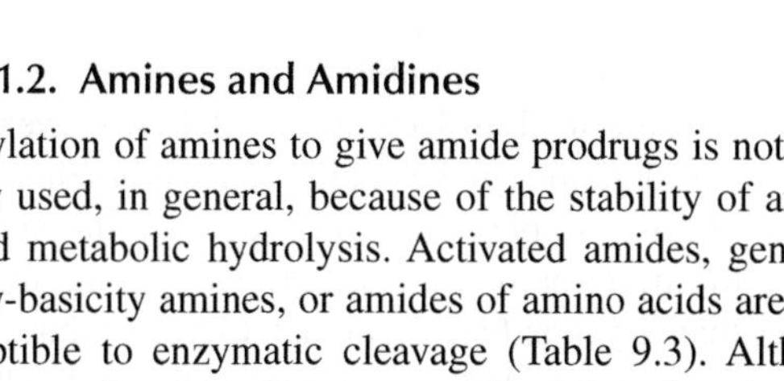

Fostamatinib
9.6

9.2.1.1.2. Amines and Amidines

N-Acylation of amines to give amide prodrugs is not commonly used, in general, because of the stability of amides toward metabolic hydrolysis. Activated amides, generally of low-basicity amines, or amides of amino acids are more susceptible to enzymatic cleavage (Table 9.3). Although carbamates in general are too stable, phenyl carbamates ($RNHCO_2Ph$) are rapidly cleaved by plasma enzymes,[22] and, therefore, they can be used as prodrugs.

The pK_a of amines can be lowered by approximately three units by conversion to their *N-Mannich bases* (Table 9.3, $X{=}CH_2NHCOAr$). This lowers the basicity of the amine so that at physiological pH few of the prodrug molecules are protonated, thereby increasing its lipophilicity. For example, the partition coefficient (see Chapter 2, Section 2.2.5.2.2) between octanol and phosphate buffer (pH 7.4) for the *N*-Mannich base **9.7** ($R{=}CH_2NHCOPh$), derived from benzamide and the decongestant phenylpropanolamine hydrochloride (**9.7**, R=H•HCl; in several cold remedies, such as Entex), is almost 100 times greater than for the parent amine.[23] However, the rate of hydrolysis of *N*-Mannich bases depends on the amide carrier group; salicylamide and succinimide are more susceptible to hydrolysis than is benzamide.[24]

TABLE 9.3 Prodrug Analogs of Amines

$Drug{-}NH_2 \longrightarrow Drug{-}NHX$

X					
$-C(=O)R$	$-C(=O)CH(R)\overset{+}{N}H_3$	$-C(=O)-OPh$	$-CH_2NHC(=O)Ar$	$=CHAr$	$=NAr$

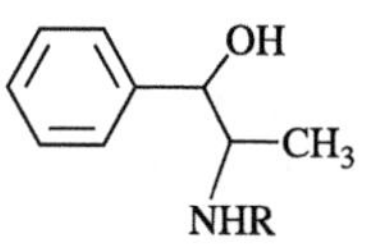

Phenylpropanolamine hydrochloride (R = H·HCl)
9.7

Another approach for lowering the pK_a of amines and, thereby, making them more lipophilic, is to convert them to *imines (Schiff bases)*; however, imines often are too labile in aqueous solution. The anticonvulsant agent progabide (**9.8**; Gabrene)[25] is a prodrug form of γ-aminobutyric acid (GABA), an important inhibitory neurotransmitter (see Chapter 5, Section 5.3.3.3.1). The lipophilicity of **9.8** allows the compound to cross the blood–brain barrier; once inside the brain it is hydrolyzed to GABA.[26]

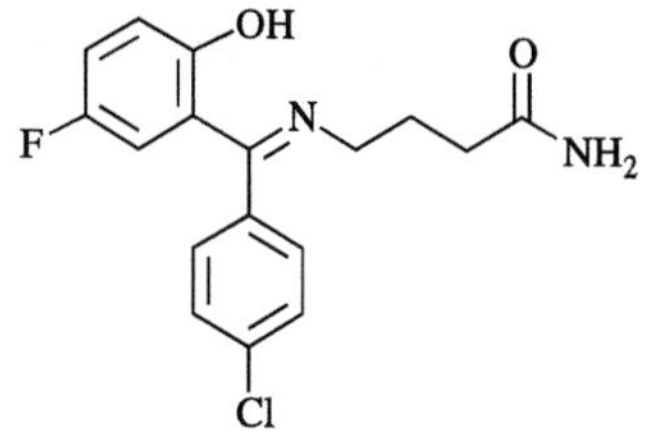

Progabide
9.8

Amidines also can be acylated to give orally active prodrugs. For example, dabigatran etexilate (**9.9**, Pradaxa) is

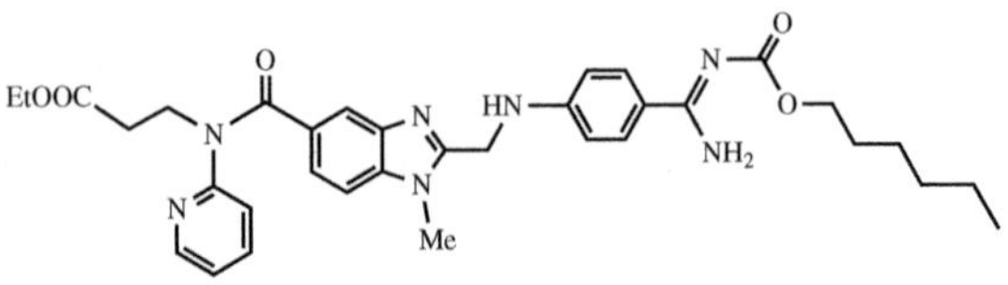

Dabigatran etexilate
9.9

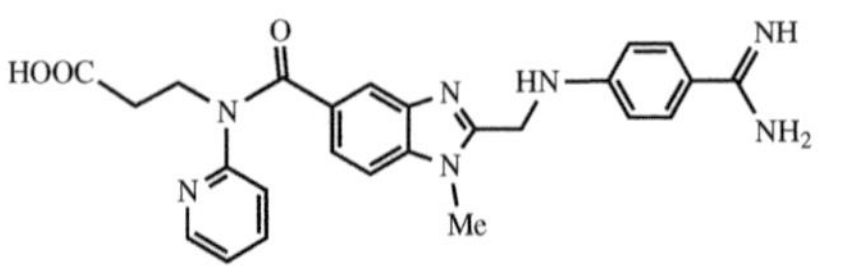

Dabigatran
9.10

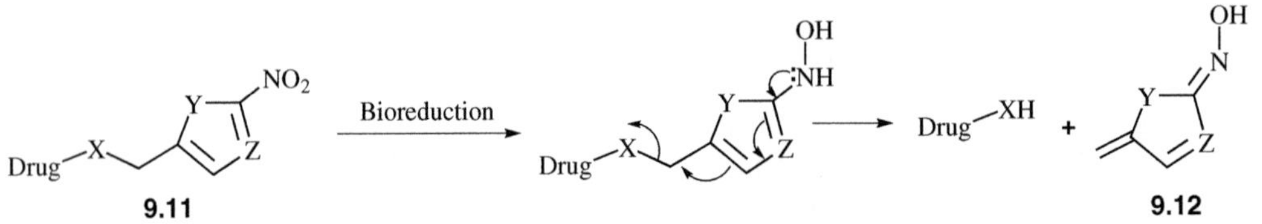

9.11 **9.12**

SCHEME 9.3 Bioreductively activated carrier-linked prodrug

an orally active antithrombotic and anticoagulant prodrug of the thrombin inhibitor dabigatran (**9.10**),[27] which must be administered intravenously.

Because most solid tumors are *hypoxic* (have a low oxygen concentration), these cells are resistant to radiation therapy and to many chemotherapeutic approaches.[28] To take advantage of the reductive milieu, prodrugs that require reductive mechanisms are beneficial. Although most carrier-linked prodrugs require hydrolytic activation mechanisms, a reductive activation mechanism is also feasible. A general approach is the reduction of a nitroaromatic prodrug (**9.11**, X=NH or O; Y=O, S, NCH_3; Z=CH, N), converting the electron-withdrawing nitro group to an electron-donating hydroxylamino group (or amino group, depending on the rate of elimination versus the in vivo rate of reduction of the hydroxylamino to an amino group), which initiates release of the drug from the carrier. This can be used for release of amines (X=NH) or alcohols (X=O) from the carrier (Scheme 9.3).[29] The drawback of this approach is that **9.12** may be electrophilic and will react with whatever nucleophiles may be present, including other proteins, unless it is trapped possibly by the reducing agent.

9.2.1.1.3. Sulfonamides

Just like amines, sulfonamides can be acylated, but this generates an acidic proton, which makes these compounds amenable to conversion to water-soluble sodium salts. For example, the second-generation COX-2 inhibitor valdecoxib (**9.13**; Bextra) (see Chapter 5, Section 5.3.2.2.2) has been converted to parecoxib sodium (**9.14**; Dynastat), an injectable analgesic drug.[30] The plasma half-life in man for conversion back to valdecoxib is about 5 minutes.

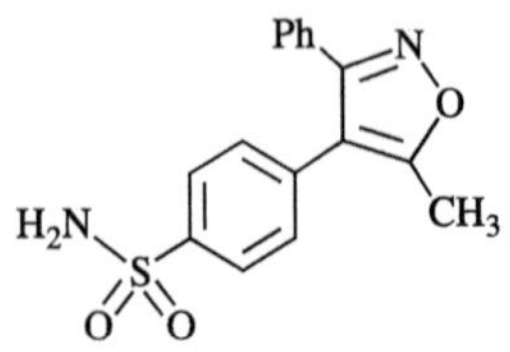

Valdecoxib
9.13

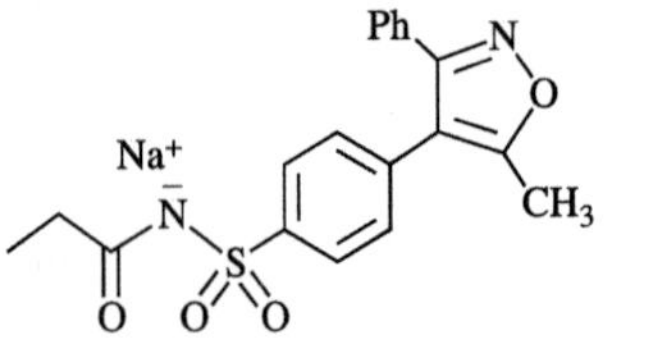

Parecoxib sodium
9.14

9.2.1.1.4. Carbonyl Compounds

The most important prodrug forms of aldehydes and ketones are Schiff bases: oximes, acetals (ketals), enol esters, oxazolidines, and thiazolidines (Table 9.4).

TABLE 9.4 Prodrug Analogs of Carbonyl Compounds

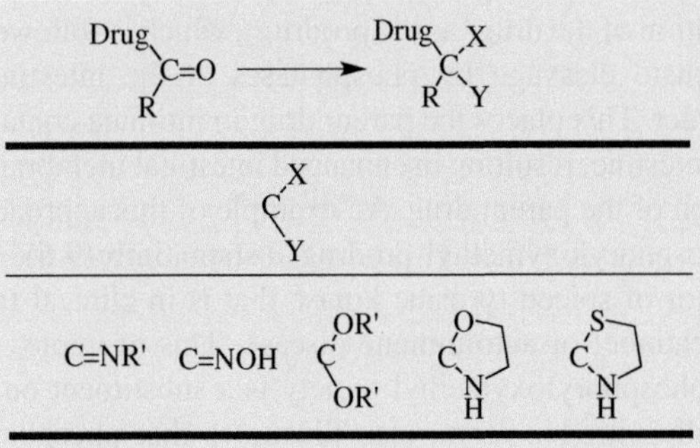

C=NR'	C=NOH	$C(OR')_2$	oxazolidine	thiazolidine

9.2.1.2. *Examples of Carrier-Linked Bipartite Prodrugs*

9.2.1.2.1. Prodrugs for Increased Water Solubility

Prednisolone (**9.15**; R=R′=H; Prelone) and methylprednisolone (**9.15**; R=CH_3, R′=H; Depo-Medrol) are poorly water-soluble corticosteroid drugs. To permit aqueous injection or ophthalmic delivery of these drugs, they must be converted into water-soluble forms, such as one of the ionic esters described in Section 9.2.1.1.1. However, there are two considerations in the choice of a solubilizing group: the ester must be stable enough in aqueous solution so that a ready-to-inject solution has a reasonably long shelf life (greater than 2 years; half-life about 13 years), but it must be hydrolyzed in vivo with a reasonably short half-life after administration (less than 10 min). For this optimal situation to occur, the in vivo/in vitro lability ratio would have to be on the order of 10^6. This is possible when the biotransformation is enzyme catalyzed.

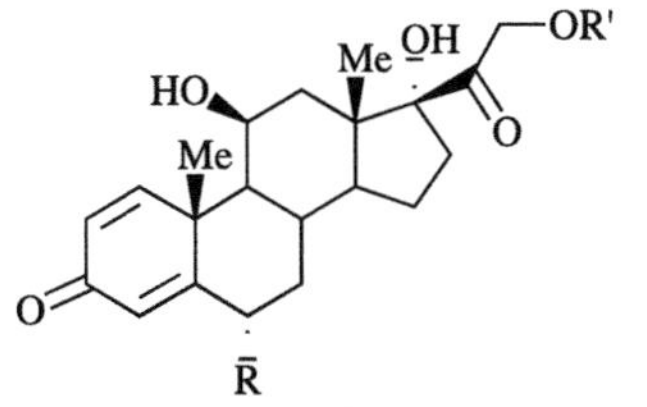

Prednisolone (R=R'=H)
Methylprednisolone (R=CH_3, R'=H)
Methylprednisolone sodium succinate (R=CH_3, R'=$COCH_2CH_2CO_2Na$)
Prednisolone phosphate (R=H, R'=PO_3Na_2)
9.15

The water-soluble prodrug form of methylprednisolone that is in medical use is methylprednisolone sodium succinate (**9.15**; R=CH_3, R′ $COCH_2CH_2CO_2Na$; Solu-Medrol). However, the in vitro stability is low, probably because of intramolecular catalysis; consequently, it is distributed as a lyophilized (freeze-dried) powder that must be reconstituted with water and then used within 48 h. The lyophilization process adds to the cost of the drug and makes its use less convenient. On the basis of physical-organic chemical rationalizations, a series of more stable water-soluble methylprednisolone esters was synthesized, and several of the analogs were shown to have shelf lives in solution of greater than 2 years at room temperature.[31] Ester hydrolysis studies of these compounds in human and monkey serum indicated that derivatives having an anionic solubilizing moiety, such as carboxylate or sulfonate, are poorly or not hydrolyzed, but compounds with a cationic (tertiary amino) solubilizing moiety are hydrolyzed rapidly by serum esterases.[32] Prednisolone phosphate (**9.15**; R=H, R′=PO_3Na_2; Pediapred) is prescribed as a water-soluble prodrug for prednisolone that is activated in vivo by phosphatases.

Because of the poor water solubility of the antitumor drug etoposide (**9.16**, R=H, VePesid), it has to be formulated with the detergent Tween 80, polyethylene glycol (PEG), and ethanol, all of which have been shown to be toxic.[33] Conversion to the corresponding phosphate ester, etoposide phosphate (**9.16**, R=PO_3H_2; Etopophos), allows the drug to be delivered in a more concentrated form over a much shorter period of time without the detrimental vehicle.[34]

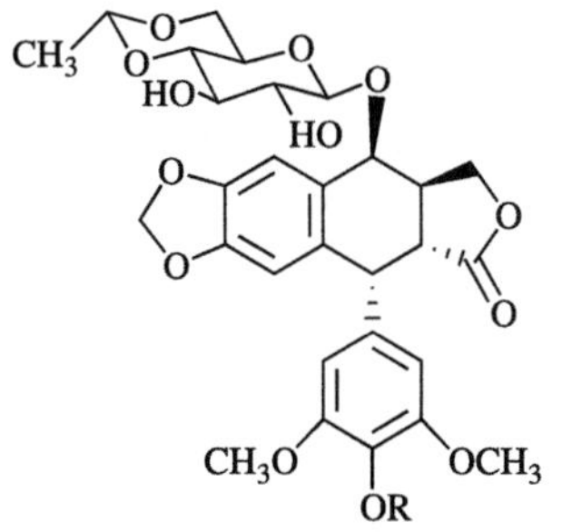

Etoposide (R = H)
Etoposide phosphate (R = PO_3H_2)
9.16

The local anesthetic benzocaine (**9.17**, R=H; one trade name is Americaine) has been converted into water-soluble amide prodrug forms with various amino acids (**9.17**, R=$^+NH_3CHR'CO$); amidase-catalyzed hydrolysis in human serum occurs rapidly.[35]

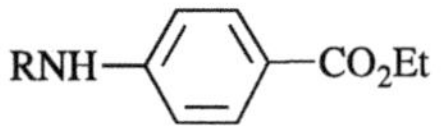

Benzocaine (R = H)
9.17

9.2.1.2.2. Prodrugs for Improved Absorption and Distribution

The skin is designed to maintain the body fluids and prevent absorption of xenobiotics into the general circulation. Consequently, drugs applied to the skin are poorly absorbed.[36] Even steroids have low dermal permeability, particularly if they contain hydroxyl groups that can interact with the skin or binding sites in the keratin. Corticosteroids for the topical treatment of inflammatory, allergic, and pruritic skin conditions can be made more suitable for topical absorption by esterification. For example, fluocinonide (**9.18**, R=$COCH_3$; Lidex) is a prodrug of fluocinolone acetonid (**9.18**, R=H, one trade name is Synalar) used for inflammatory and pruritic manifestations. Once absorbed through the skin an esterase releases the drug.

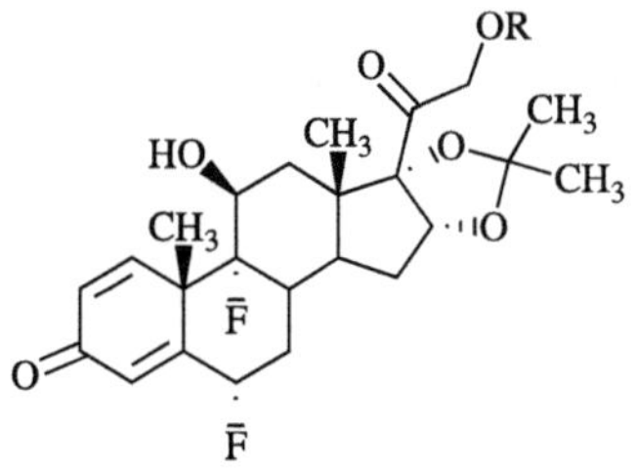

Fluocinolone acetonide (R = H)
Fluocinonide (R = $COCH_3$)
9.18

Dipivaloylepinephrine (dipivefrin; **9.19**, R=Me_3CCO; Propine), a prodrug for the antiglaucoma drug epinephrine (**9.19**, R=H; one trade name is Epifrin), is able to penetrate the cornea better than epinephrine. The cornea and aqueous humor have significant esterase activity.[37]

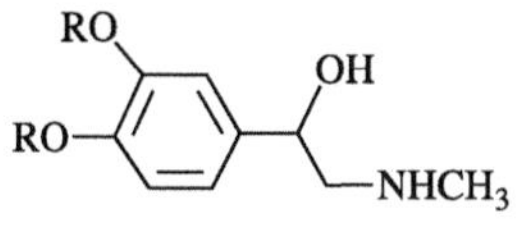

Dipivefrin (R = Me_3CCO)
Epinephrine (R = H)
9.19

9.2.1.2.3. Prodrugs for Site Specificity

The targeting of drugs for a specific site in the body by conversion to a prodrug is plausible when the physicochemical properties of the parent drug and prodrug are optimal for the target site. It should be kept in mind, however, that when the lipophilicity of a drug is increased, it will improve passive transport of the drug nonspecifically to all tissues.

Oxyphenisatin (**9.20**, R=H) is a bowel sterilant that is active only when administered rectally. However, when the hydroxyl groups are acetylated (oxyphenisatin acetate, **9.20**, R=Ac, one trade name is Lavema), the prodrug can be administered orally, and it is hydrolyzed at the site of action in the intestines to oxyphenisatin.

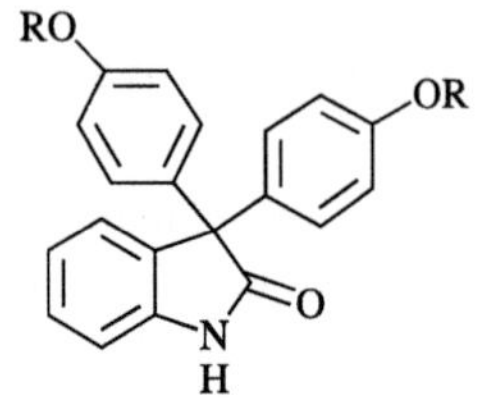

Oxyphenisatin (R = H)
9.20

One important membrane that must be traversed for drug delivery into the brain is the *blood–brain barrier*,[38] a unique lipid-like protective barrier that prevents hydrophilic compounds from entering the brain unless they are actively transported. The blood–brain barrier also contains active enzyme systems and efflux transporters to protect the central nervous system even further. Consequently, molecular size and lipophilicity are often necessary, but not sufficient, criteria for gaining entry into the brain.[39]

As was discussed in Chapter 5 (Section 5.3.3.3.1), increasing the brain concentration of the inhibitory neurotransmitter GABA results in anticonvulsant activity. However, GABA is too polar to cross the blood–brain barrier, so it is not an effective anticonvulsant drug. As mentioned above, progabide (**9.8**) is an effective lipophilic analog of GABA that crosses the blood–brain barrier, releases GABA inside the brain, and exhibits anticonvulsant activity.[40] Another related example of anticonvulsant drug design is a glyceryl lipid (**9.21**, R=linolenoyl) containing one GABA molecule and one vigabatrin (Sabril) molecule, a mechanism-based inactivator of GABA aminotransferase and anticonvulsant drug (see Chapter 5, Section 5.3.3.3.1).[41] This compound inactivates GABA aminotransferase in vitro only if brain esterases are added to cleave the vigabatrin from the glyceryl lipid. It is also 300 times more potent than vigabatrin in vivo presumably because of its increased ability to enter the brain.

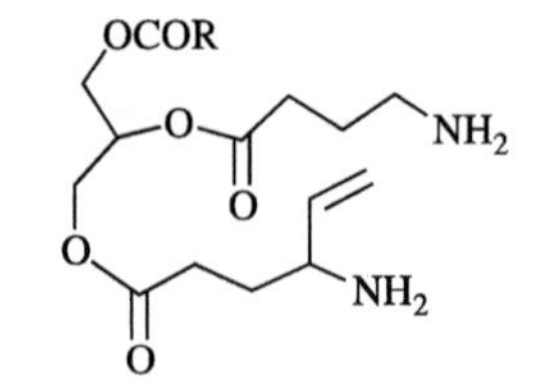

9.21

In the above examples, the lipophilicity of the drugs was increased so that they could diffuse through various membranes. Another approach for site-specific drug delivery is to design a prodrug that requires activation by an enzyme found predominantly at the desired site of action. Given the high concentration of metabolic enzymes in the liver and the role of the liver as the first organ exposed to any drug absorbed from the gastrointestinal tract, this approach has been applied to potential treatments for liver diseases such as hepatitis and hepatocellular carcinoma. In addition to the examples immediately below, further examples for releasing the parent drug selectively in the kidney or in viral vs normal cells are presented later in this chapter.

As an example of liver-directed delivery, a general approach to targeting drugs that contain a phosphate or phosphonate moiety to the liver has been developed (Scheme 9.4).[42] In this approach, oxidative hydroxylation at the benzylic position of an aryl-substituted cyclic phosphodiester **9.22** is mediated by a 3A isoform of cytochrome P450, which is expressed predominantly in the liver and to a lesser extent in the small intestine. Breakdown of tetrahedral intermediate **9.23** leads to the transient formation of **9.24**, which is largely retained within the hepatocyte (liver cell) owing to its anionic nature. β-Elimination releases the phosphate or phosphonate parent drug (**9.25**), which may itself be a prodrug if further phosphorylation is needed to produce the active substance; for example, see Section 9.2.2.8. The β-elimination reaction produces α,β-unsaturated ketone **9.26**, which is trapped by the glutathione present in millimolar concentration in liver cells. This system was applied to targeting reverse transcriptase inhibitor adefovir (**9.27**) selectively to the liver for treatment of hepatitis B. The approved form of adefovir is the bispivaloyloxymethylene prodrug adefovir dipivoxil (**9.5**, Hepsera) discussed above, which can only be administered at suboptimal doses because of renal toxicity. Toward circumventing this renal toxicity, the liver-targeted variant pradefovir (**9.28**) was developed.[43] Levels of radioactivity in liver versus kidney were measured after administration of radiolabeled adefovir either

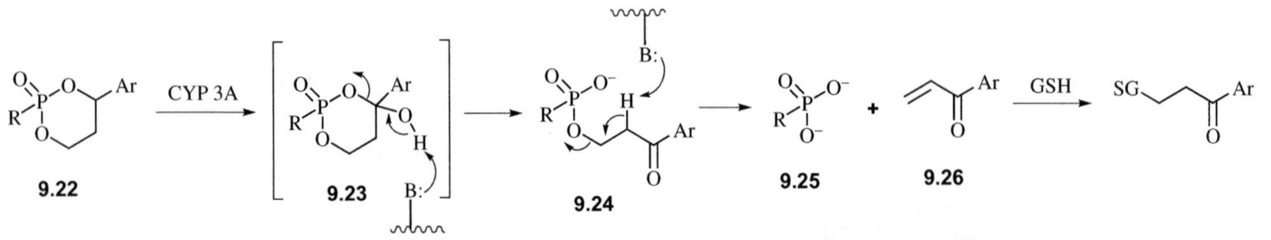

SCHEME 9.4 Liver metabolism of drugs containing phosphate and phosphonate moieties

as **9.5** or **9.28**. In monkeys, 24 hours after administration of identical oral doses, 60-fold higher radioactivity was found in the liver after administration of **9.28** compared to **9.5**, whereas in kidney, radioactivity was 33% lower for **9.28** vs **9.5**.

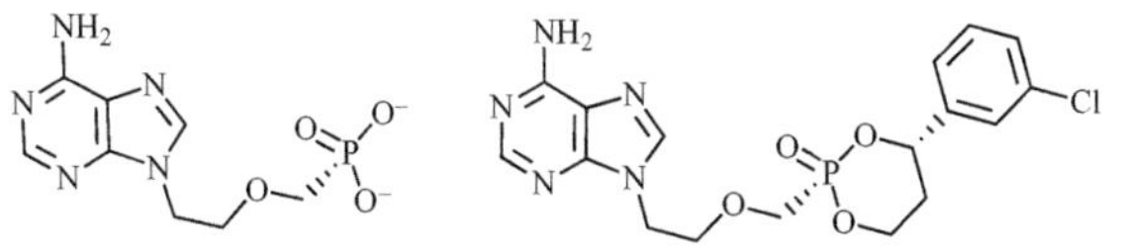

Adefovir 9.27 **Pradefovir 9.28**

Another example, which proved to be unsuccessful, was targeted to tumor cells, which contain a higher concentration of phosphatases and amidases than do normal cells. Consequently, a prodrug of a cytotoxic agent might be directed to tumor cells if either of these enzymes were important to the prodrug activation process. Diethylstilbestrol diphosphate (**9.29**, R = $PO_3^=$) was designed for site-specific delivery of diethylstilbestrol (**9.29**, R = H; one trade name is Stilbestrol) to prostatic carcinoma tissue.[44] However, this tumor-selective approach has not been very successful for several reasons: the appropriate prodrugs are too polar to reach the enzyme site, the relative enzymatic selectivity is insufficient (that is, the phosphatase and amidase levels in normal cells are still sufficient to cleave most of the prodrug), and the tumor cell perfusion rate is too poor.

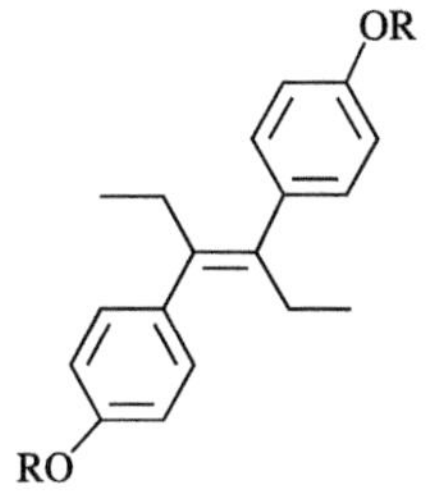

Diethylstilbestrol diphosphate (R = $PO_3^=$)
Diethylstilbestrol (R = H)
9.29

Several strategies, under the rubric of *enzyme prodrug therapies*, have been developed to achieve selective activation of prodrugs at a desired site, typically in tumor cells.[45] All of these approaches involve two steps: in the first step a prodrug-activating enzyme is incorporated into the target tumor cells, and in the second step, a nontoxic prodrug, which is a substrate of the exogenous enzyme that was incorporated into the tumor cells, is administered systemically. The prodrug is selectively converted into the active anticancer drug in a high local concentration inside the tumor cell. Certain criteria are important for this general approach to be effective:[46] (1) the prodrug-activating enzyme either should be of nonhuman origin or should be a human protein that is absent or expressed only at low concentrations in normal tissues, (2) the enzyme must achieve adequate expression in the targeted tumor cells and have high catalytic activity, (3) the prodrug should be a good substrate for the enzyme incorporated in the tumors but not be activated by endogenous enzymes outside of the tumors, (4) the prodrug must be able to cross the tumor cell membrane for intracellular activation, and (5) the cytotoxicity difference between the prodrug and its corresponding active drug should be high. Preferably, the activated drug should be highly diffusible or be actively taken up by adjacent nonexpressing cancer cells for what is known as a *bystander killing effect*, the ability of the drug to kill neighboring nonexpressing cells.[47] Furthermore, the half-life of the active drug should be long enough to induce a bystander effect but short enough to avoid the drug leaking out of the tumor cells and causing damage elsewhere.

One strategy of this type is called *a*ntibody-*d*irected *e*nzyme *p*rodrug *t*herapy (abbreviated *ADEPT*). In the first step, an antibody raised against a particular tumor cell line is conjugated with (attached to) the enzyme that is needed to activate an antitumor prodrug. After the antibody–enzyme conjugate is administered and has accumulated on the tumor cell, the excess conjugate not bound to the tumor cell is given enough time to clear from the blood and normal tissues. The prodrug is then administered. The enzyme conjugated with the antibody at the tumor cell surface catalyzes the conversion of the prodrug to the drug when it reaches the tumor cell. The advantage of this method, relative to direct administration of the prodrug, is the increased selectivity for release of high concentrations of the drug at the targeted cells. This advantage is only evident if enough time is allowed for clearance of the antibody–enzyme conjugate that is not bound to the tumor cells. An increase in the clearance rate of unbound antibody–enzyme conjugate would permit the administration of a higher concentration of the prodrug for a longer period. Galactosylation of the antibody leads to more rapid and efficient clearance of the unbound antibody–enzyme conjugate by galactose receptors in the liver.[48]

The drawbacks to ADEPT are the potential for immunogenicity and rejection of the antibody–enzyme conjugate, the potential for leakback of the active drug formed at the tumor, and the requirement of IV administration as well as the complexity of the treatment. Nonetheless, this approach is in the clinical trial stage.

An example of ADEPT is the delivery of a nitrogen mustard as a glutamic acid conjugate (**9.30**) after administration of a humanized monoclonal antibody[49] conjugated to the bacterial enzyme carboxypeptidase G2 (Scheme 9.5).[50] Humanization of monoclonal antibodies that are raised from nonhuman species occurs by modifying protein sequences to increase the similarity of the nonhuman antibody to human antibody variants, which minimizes immunogenicity. Note that the enzyme selected for prodrug activation is a bacterial enzyme so that there may be selectivity for prodrug activation by this enzyme in preference to that by human carboxypeptidase at sites other than at the tumor cells. The drawback of this approach is the potential for rejection as a result of using a bacterial enzyme. This problem could be alleviated with the use of a humanized catalytic antibody[51] in place of the bacterial enzyme for activation of the prodrug. The reaction shown in Scheme 9.5 was also effected using a humanized

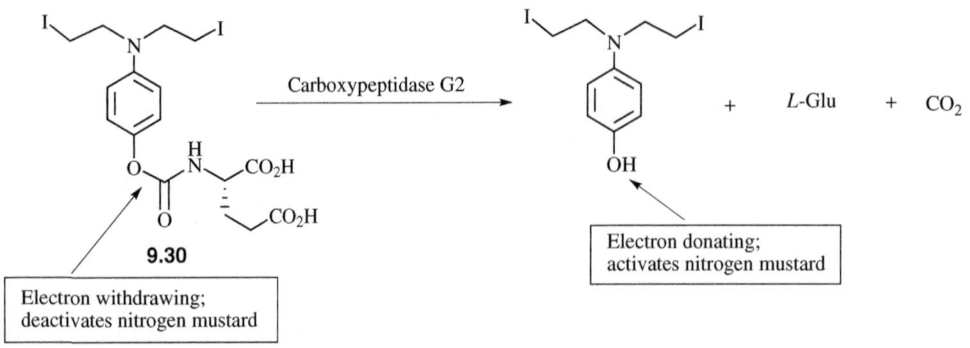

SCHEME 9.5 Nitrogen mustard activation by carboxypeptidase G2 for use with ADEPT

catalytic antibody in a process termed *antibody-directed abzyme*[52] *prodrug therapy* (*ADAPT*).[53]

An even more effective approach to attain selectivity for prodrug activation at the tumor cell would be to use a humanized catalytic antibody that not only does not exist in humans but also catalyzes a reaction not known to occur in humans. That way, the *only* site where the prodrug could be activated would be where the catalytic antibody resides, presumably directed with a monoclonal antibody to the targeted tumor cells. Antibody 38C2[54] is a broad specificity catalytic antibody that catalyzes sequential retro-aldol and retro-Michael reactions, a combination of reactions not catalyzed by any known human enzyme. The abzyme was found to be long lived in vivo, was shown to activate prodrugs selectively, and potentiated the killing of colon and prostate cancer cell lines.[55] An example of the activation reactions catalyzed by this abzyme is shown in Scheme 9.6 for a doxorubicin prodrug (**9.31**).

Another bipartite strategy for directing the prodrug-activating enzyme to a specific tumor cell line is by attaching a sugar to the surface of the enzyme that is recognized by a tumor cell, known as *l*ectin-directed *e*nzyme-*a*ctivated *p*rodrug *t*herapy (*LEAPT*).[56] Carbohydrates have a high specificity of interaction with a wide variety of cellular receptors, including tumor cell surfaces, so glycoproteins have been developed for drug delivery.[57] LEAPT is initiated by targeting a glycosylated prodrug-activating enzyme to a specific cell type or tissue. Then prodrugs capped with the same sugar are administered, and the glycosylated enzyme activates the prodrug at the target site. Selective enzyme targeting can be achieved using carbohydrates that bind to specific lectins on the surface of the targeted cells. Receptor-mediated endocytosis is triggered at the surface of cells to transfer bound ligands inside the cell, which leads to more effective targeting.

A related strategy for improving the selectivity of cancer chemotherapy is called *g*ene-*d*irected *e*nzyme *p*rodrug *t*herapy (*GDEPT*; also called *suicide gene therapy*).[58] In this approach a gene encoding the prodrug-activating enzyme is integrated into the genome of the target cancer cells under the control of tumor-selective promoters or by viral transfection.

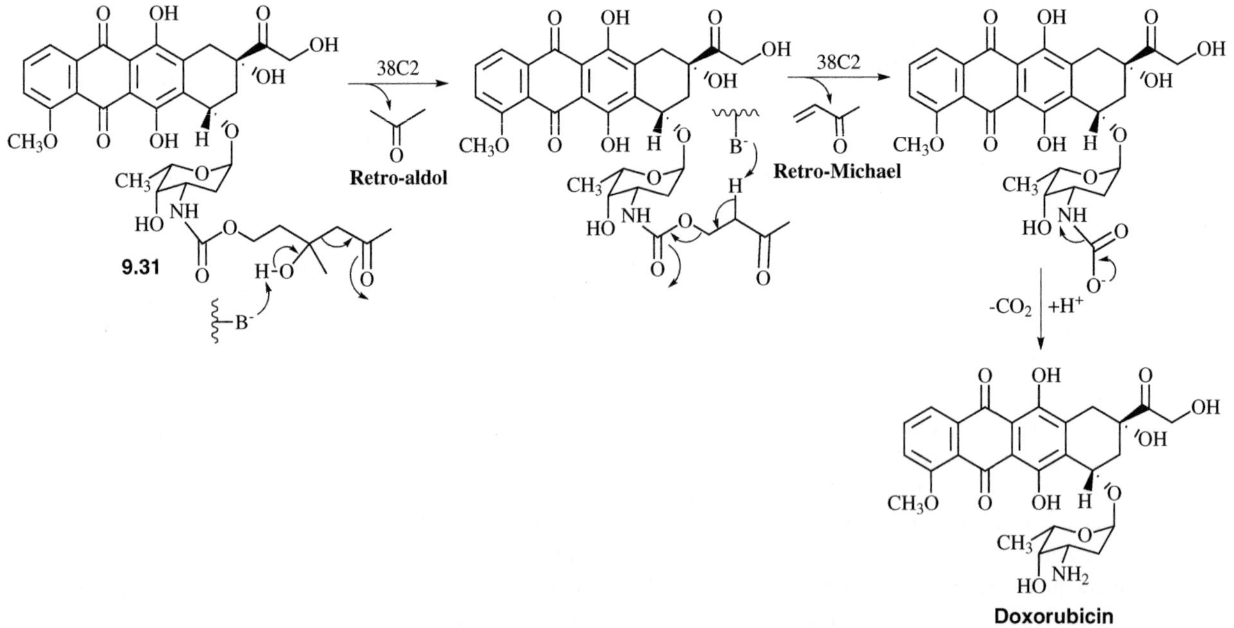

SCHEME 9.6 Catalytic antibody 38C2 activation of a doxorubicin prodrug by tandem retro-aldol/retro-Michael reactions for use with ADAPT

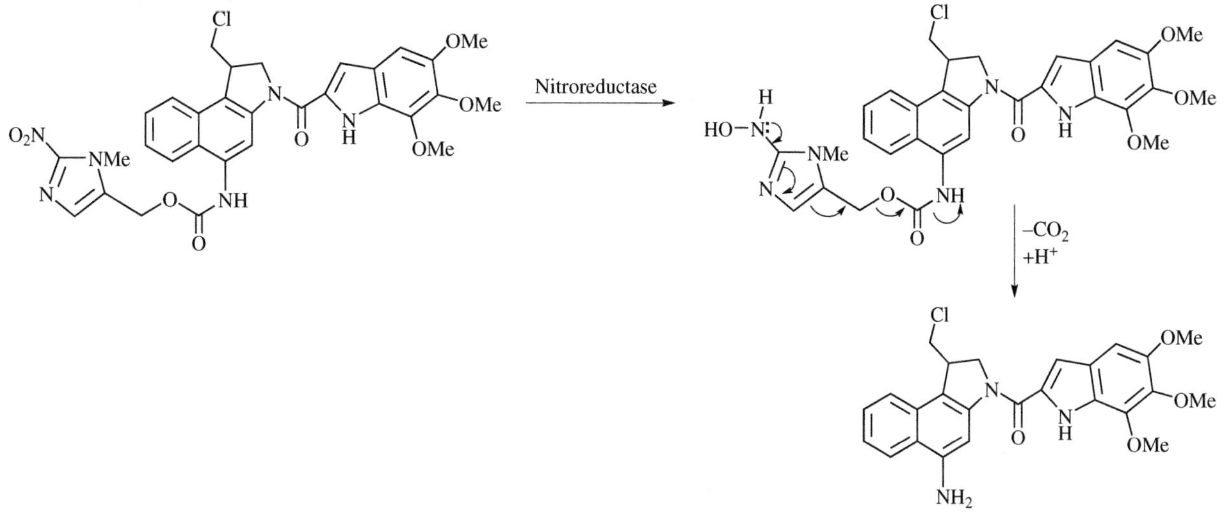

amino-*seco*-CBI-TMI
9.32

SCHEME 9.7 Nitroreductase activation of a prodrug for use with GDEPT

These cells, then, express the enzyme that activates the prodrug added in the second step, as described above for ADEPT.

A common enzyme used for activation of a prodrug by GDEPT is an aerobic flavin-dependent nitroreductase from *Escherichia coli* B, which catalyzes the reduction of aromatic nitro groups to the corresponding hydroxylamino group.[59] As an illustration of the general concept introduced in Scheme 9.3, the (2-nitroimidazol-5-yl)methyl carbamate prodrug of the minor groove alkylating agent amino-*seco*-CBI-TMI (**9.32**, Scheme 9.7) is stable until nitroreductase reduces the aromatic nitro group to the corresponding hydroxylamino group. This initiates the elimination of the carbamate to give CO_2 and the free alkylating agent.[60]

Virus-directed enzyme prodrug therapy (*VDEPT*) is another gene therapy strategy that uses viral vectors to deliver a gene that encodes a prodrug-activating enzyme.[61] Despite extensive use of retroviral and adenoviral vectors to deliver prodrug-activating enzyme genes, both vectors have some disadvantages. The principal disadvantage of a retroviral vector is that recombinant retroviruses only target dividing cells,[62] but most human tumor cells divide slowly. Even in a rapidly growing tumor nodule, only 6–20% of the cells are in a proliferating state. Therefore, the majority of the tumor would not be sensitive to killing by retroviral VDEPT. However, this drawback could be beneficial in some cases, such as for brain tumors, because in the brain only the tumor cells would be proliferating. This would allow for a high tumor:normal transfection differential for retroviral delivery. Although most viral vectors are engineered to be replication deficient, there is a slight risk of reversion to the wild-type virus (yikes!). Furthermore, retrovirus vectors are inserted into the host cell deoxyribonucleic acid (DNA), which may cause mutagenesis of the host's genome (double yikes!).[63]

GDEPT and VDEPT have an advantage over ADEPT in that many enzymes need cofactors that are present only inside the cells. Therefore, enzymes delivered by ADEPT may need to gain access to the inside of tumor cells before they can optimally activate prodrugs. This is a problem because of the poor cell penetration of antibody–enzyme conjugates. In GDEPT, gene-encoding enzymes can be specifically delivered to target tissues, which allows for the expression of the enzyme within the target cells.[64] Problems associated with GDEPT include insertional mutagenesis, anti-DNA antibody formation, local infection, and tumor nodule ulceration,[65] as well as difficulties with the selective delivery and expression of the genes.[66]

9.2.1.2.4. Prodrugs for Stability

Some prodrugs protect the drug from the *first-pass effect* (see Chapter 8, Section 8.1). Propranolol (**9.33**, R=R′=H; Inderal) is a widely used antihypertensive drug, but because of first-pass elimination, an oral dose has a much lower bioavailability than does an IV injection. The major metabolites (see Chapter 8) are propranolol *O*-glucuronide (**9.33**, R=H, OR'=glucuronide) and *p*-hydroxypropranolol (**9.33**, R=OH, R′=H) and its *O*-glucuronide (**9.33**, R=OH, OR'=glucuronide). The hemisuccinate ester of propranolol (**9.33**, R=H, R′=$COCH_2CH_2COOH$) was prepared to block glucuronide formation; following oral administration of propranolol hemisuccinate, the plasma levels of propranolol were eight times greater than when propranolol was used.[67]

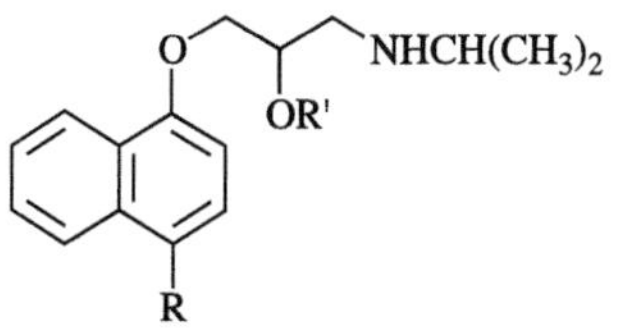

Propanolol (R = R' = H)
9.33

Naltrexone (**9.34**, R=H; Trexan), used in the treatment of opioid addiction, is nonaddicting and is well absorbed from the gastrointestinal tract. However, it undergoes extensive first-pass metabolism when given orally. Ester prodrugs, the anthranilate (**9.34**, R=CO-*o*-NH_2Ph) and the acetylsalicylate (**9.34**, R=CO-*o*-AcO-Ph), enhanced the bioavailability 45- and 28-fold, respectively, relative to **9.34** (R=H).[68]

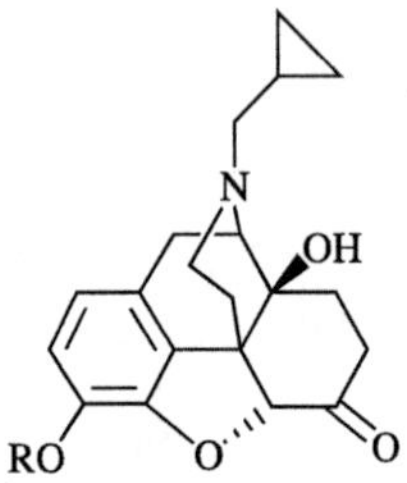

Naltrexone (R=H)
9.34

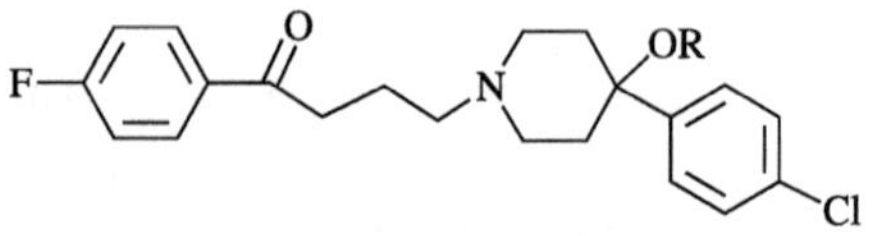

9.2.1.2.5. Prodrugs for Slow and Prolonged Release

The utility of slow and prolonged release of drugs is severalfold. (1) It reduces the number and frequency of doses required. (2) It eliminates nighttime administration of drugs. (3) Because the drug is taken less frequently, it minimizes patient noncompliance. (4) When a fast-release drug is taken, there is a rapid surge of the drug throughout the body. As metabolism of the drug proceeds, the concentration of the drug diminishes. A slow-release drug would eliminate these peaks and valleys of fast-release drugs, which are a strain on cells. (5) Because a constant lower concentration of the drug is being released, it reduces the possibility of toxic levels of drugs. (6) It reduces gastrointestinal side effects. A common strategy in the design of slow-release prodrugs is to make a long-chain aliphatic ester or polyethylene glycolated esters,[69] because these esters hydrolyze slowly, and to inject them intramuscularly.

Prolonged-release drugs are quite important in the treatment of psychoses because these patients require medication for extended periods of time and often show high patient noncompliance rates. Haloperidol (**9.35**, R=H; Haldol) is a potent, orally active central nervous system depressant, sedative, and tranquilizer. However, peak plasma levels are observed between 2 and 6 h after administration. The ester prodrug haloperidol decanoate (**9.35**, R=$CO(CH_2)_8CH_3$; Haldol decanoate) is injected intramuscularly as a solution in sesame oil, and its antipsychotic activity lasts for about 1 month.[70] The antipsychotic fluphenazine (**9.36**, R=H; one trade name is Prolixin) also has a short duration of activity (6–8 h). Fluphenazine enanthate (**9.36**, R=$CO(CH_2)_5CH_3$; Prolixin Enan) and fluphenazine decanoate (**9.36**, R=$CO(CH_2)_8CH_3$; Prolixin Dec), however, have durations of activity of about a month.[71]

Haloperidol (R = H)
Haloperidol decanoate (R = $CO(CH_2)_8CH_3$)
9.35

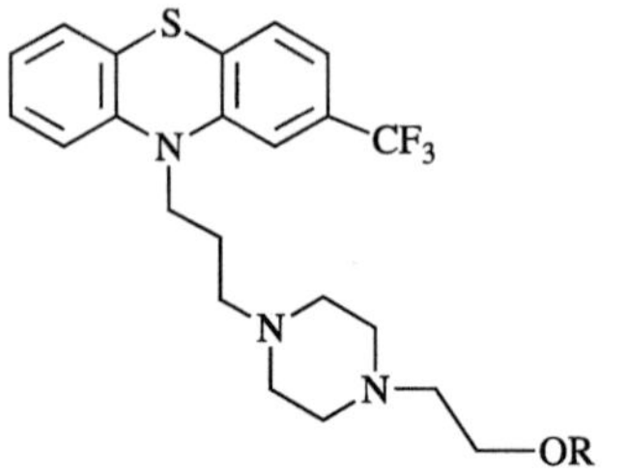

Fluphenazine (R=H)
Fluphenazine ethanate (R=$CO(CH_2)_5CH_3$)
Fluphenazine decanoate (R=$CO(CH_2)_8CH_3$)
9.36

Conversion of the nonsteroidal antiinflammatory (antiarthritis) drug tolmetin sodium (**9.37**, R=O^-Na^+; Tolectin) to the corresponding glycine conjugate (**9.37**, R=$NHCH_2$-COOH) increases the potency and extends the peak concentration of tolmetin from 1 h to about 9 h because of the slow hydrolysis of the prodrug amide linkage.[72]

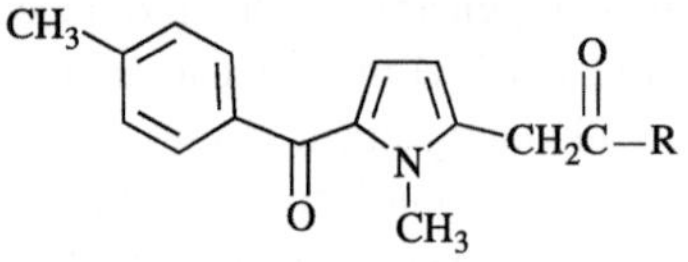

Tolmetin sodium (R = O^-Na^+)
9.37

9.2.1.2.6. Prodrugs to Minimize Toxicity

The prodrugs that were designed for improved absorption (Section 9.2.1.2.2), for site specificity (Section 9.2.1.2.3), for stability (Section 9.2.1.2.4), and for slow release (Section 9.2.1.2.5) also lowered the toxicity of the drug. For example, epinephrine (**9.19**, R=H) (see Section 9.2.1.2.2), used in the treatment of glaucoma, has a number of ocular and systemic side effects associated with its use. The prodrug dipivaloylepinephrine (**9.19**, R=Me_3CCO) has been shown to be more potent than epinephrine in dogs and rabbits and nearly as effective in humans[73] with a significantly improved toxicological profile compared with epinephrine.

Another example of the utility of the prodrug approach to lower toxicity of a drug can be found in the design of aspirin (**9.38**, R=H) analogs.[74] Side effects associated with the use of aspirin are gastric irritation and bleeding. The gastric irritation and ulcerogenicity associated with aspirin use may result from an accumulation of the acid in the gastric mucosal cells. Esterification of aspirin (**9.38**, R=alkyl) and other nonsteroidal antiinflammatory agents greatly suppresses gastric ulcerogenic activity. However, esterification also renders the acetyl ester of aspirin extremely susceptible to enzymatic hydrolysis (the $t_{1/2}$ for deacetylation of aspirin in human plasma is about 2h, but for deacylation of aspirin esters is 1–3min). Esters of certain *N,N*-disubstituted 2-hydroxyacetamides (**9.38**, $R=CH_2CONR_1R_2$) were found to be chemically highly stable, but were hydrolyzed very rapidly by pseudocholinesterase in plasma and therefore are well suited as aspirin prodrugs to lower the gastric irritation effects of aspirin.

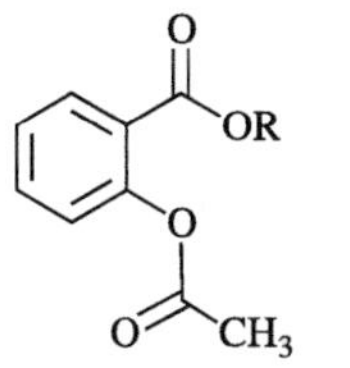

Aspirin (R = H)
9.38

9.2.1.2.7. Prodrugs to Encourage Patient Acceptance

A fundamental tenet in medicine is that in order for a drug to be effective, the patient has to take it! Painful injections and unpleasant taste or odor are the most common reasons for the lack of patient acceptance of a drug. An excellent example of how a prodrug can increase the potential for patient acceptance is related to the antibacterial drug clindamycin (**9.39**, R=H; one trade name is Cleocin). Clindamycin causes pain on injection, whereas the prodrug clindamycin phosphate (**9.39**, $R=PO_3H_2$; one trade name is Dalacin) is well tolerated; hydrolysis of the prodrug in vivo occurs with a $t_{1/2}$ of approximately 10min.[75] Also, clindamycin has a bitter taste, so it is not well accepted orally by children. However, it was found that by increasing the chain length of 2-acyl esters at the sugar moiety of clindamycin, the taste improved from bitter (acetate ester) to no bitter taste (palmitate ester).[76] Of course, when dealing with young children, it is not sufficient for a drug to be just tasteless; consequently, clindamycin palmitate (**9.39**, $R=CO(CH_2)_{14}CH_3$; Cleocin Pediatric) is sold for pediatric use in a cherry-flavored syrup. Bitter taste results from a compound dissolving in the saliva and interacting with a bitter taste receptor in the mouth. Esterification with long-chain fatty acids makes the drug less water soluble and unable to dissolve in the saliva. It also may alter the interaction of the compound with the taste receptor.

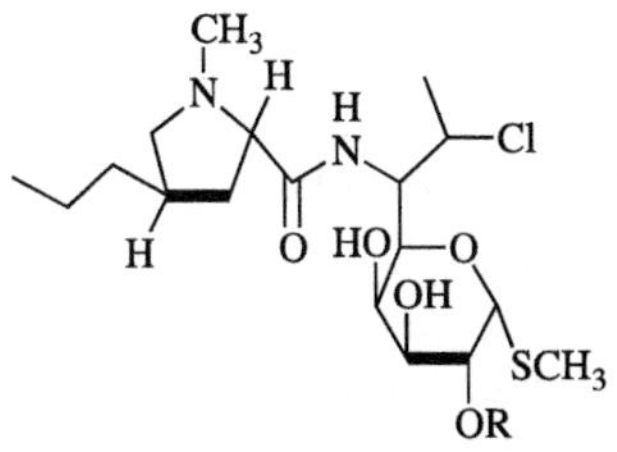

Clindomycin (R = H)
Clindomycin phosphate (R = PO_3H_2)
Clindomycin palmitate (R = $O(CH_2)_{14}CH_3$)
9.39

The antibacterial sulfa drug sulfisoxazole (**9.40**, R=H; Gantrisin) is also bitter tasting, but sulfisoxazole acetyl (**9.40**, $R=COCH_3$) is tasteless. For pediatric use this prodrug is combined with the tasteless prodrug form of erythromycin, i.e., erythromycin ethylsuccinate, in a strawberry-banana-flavored suspension (Pediazole).

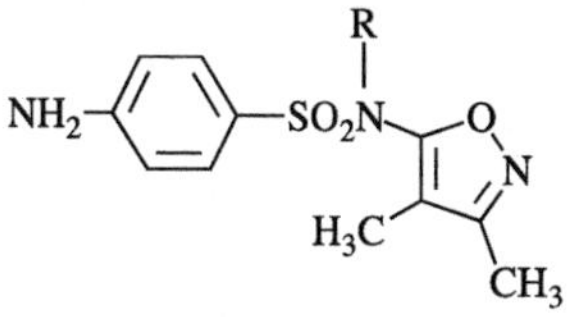

Sulfisoxazole (R = H)
Sulfisoxazole acetyl (R = $COCH_3$)
9.40

9.2.1.2.8. Prodrugs to Eliminate Formulation Problems

Formaldehyde (CH_2O) is a flammable, colorless gas with a pungent odor that is used as a disinfectant. Solutions of high concentrations of formaldehyde are toxic. Consequently, it cannot be used directly in medicine. However, the reaction of formaldehyde with ammonia produces a stable adamantane-like solid compound, methenamine (**9.41**; one trade name is Hiprex). In acidic pH media, methenamine hydrolyzes to formaldehyde and ammonium ions. Because the pH of urine in the bladder is mildly acidic, methenamine is used as a urinary tract antiseptic.[77] To prevent hydrolysis of this prodrug in the acidic environment of the stomach, the tablets are enteric coated.

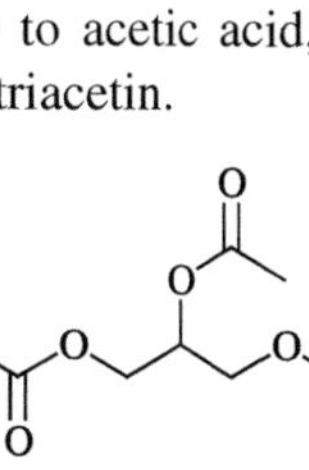

Methenamine
9.41

The topical fungistatic prodrug triacetin (**9.42**; Captex 500) owes its activity to acetic acid, the product of skin esterase hydrolysis of triacetin.

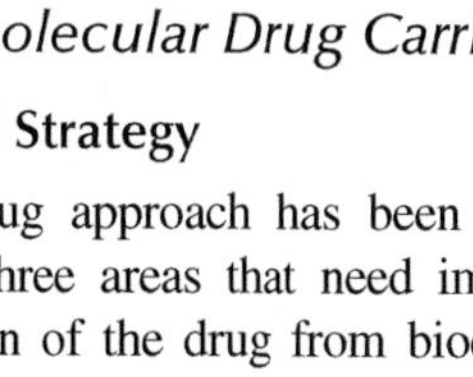

Triacetin
9.42

9.2.1.3. *Macromolecular Drug Carrier Systems*

9.2.1.3.1. General Strategy

Although the prodrug approach has been very fruitful in general, there are three areas that need improvement: site specificity, protection of the drug from biodegradation, and minimization of side effects. Another carrier-linked bipartite prodrug approach that has been utilized to address these shortcomings is *macromolecular drug delivery*.[78] This is a drug carrier system in which the drug is covalently attached to a macromolecule, such as a synthetic polymer, a glycoprotein, a lipoprotein, a lectin, a hormone, albumin, a liposome, DNA, dextran, an antibody, or a cell. The pharmacokinetic characteristics of these drugs change dramatically because the absorption and distribution of the drug depends on the physicochemical properties of the macromolecular carrier, not the drug. These parameters can be altered by manipulation of the properties of the carrier. This approach has the potential advantage of targeting drugs for a specific site and improving the therapeutic index by minimizing interactions with nontarget tissues (i.e., lowering the toxicity) as well as reducing premature drug metabolism and excretion. However, it has the disadvantages that the macromolecules may not be well absorbed after oral administration, that alternative means of administration are required, and that they may be immunogenic. Although polymer conjugates generally cannot pass through membranes, they can gain access to the interior of a cell by *pinocytosis*, a type of endocytosis by which the cell membrane invaginates the particle and then pinches itself off to form an intracellular vesicle, which moves into the cell and eventually fuses with lysosomes. Because the breakdown of proteins and other macromolecules is believed to occur in the lysosomes,[79] and because this breakdown then liberates the drug, the design of a macromolecular drug carrier system should be a fruitful approach to deliver the drug inside a cell. This approach has already been taken in Nature, although in this case, the drug is not covalently bound to the macromolecular carrier. The antitumor antibiotic C-1027 consists of an enediyne (see Chapter 6, Section 6.3.3.4) bound to a carrier protein; the protein protects the labile enediyne from destroying the host.[80]

Some of the macromolecular drug carrier systems exert their effects while the drug is still attached to the carrier, but these are not prodrugs. Several examples of macromolecular drug carrier systems follow.

9.2.1.3.2. Synthetic Polymers

Aspirin linked to poly(vinyl alcohol) (**9.43**) was shown to have the same potency as aspirin, but was less toxic. Another antiinflammatory agent, ibuprofen (the carboxylic acid from hydrolysis of **9.44**; Advil), was attached as a poly(oxyethylene) diester (**9.44**).[81] This macromolecular carrier system resulted in a sustained release of ibuprofen, giving prolonged antiinflammatory activity and a higher plasma half-life relative to the free drug.

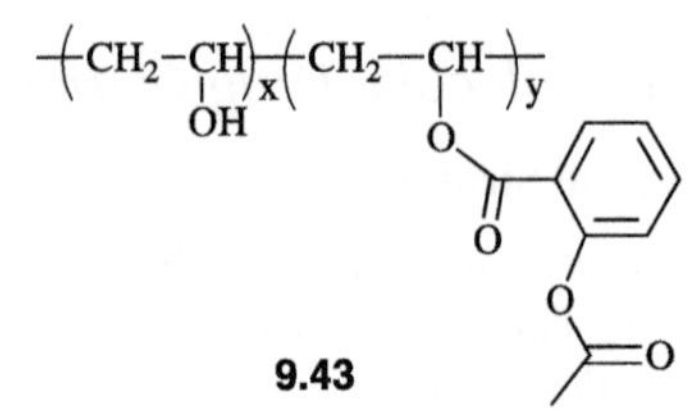

9.43

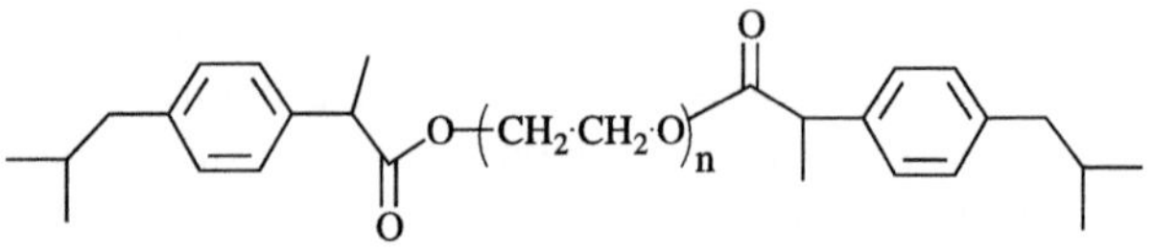

9.44

Because it is necessary for the drug to be released from the polymer backbone, steric hindrance by the polymer to chemical or enzymatic hydrolysis may cause problems. To avoid steric hindrance by the polymer in the first synthetic step, a spacer was incorporated between the polymer and the first building block. When the steroid hormone testosterone was linked to poly(methacrylate) (**9.45**), no androgenic effect was observed, but when a spacer arm was inserted between the polymer and the testosterone (**9.46**), this macromolecular drug carrier was as effective as testosterone. The 3-thiabutyl oxide chain was attached to the polymer to enhance water solubility.

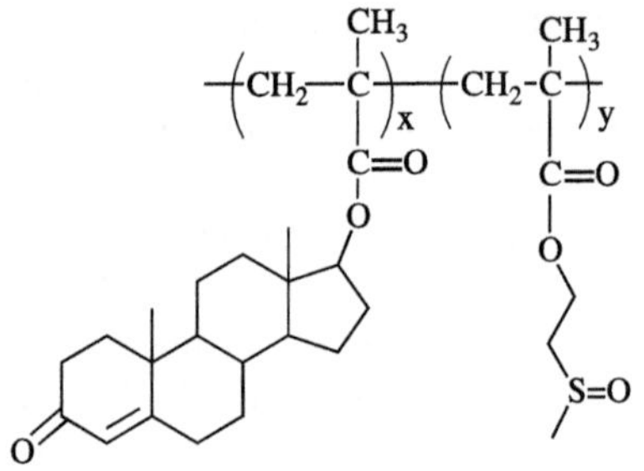

9.45

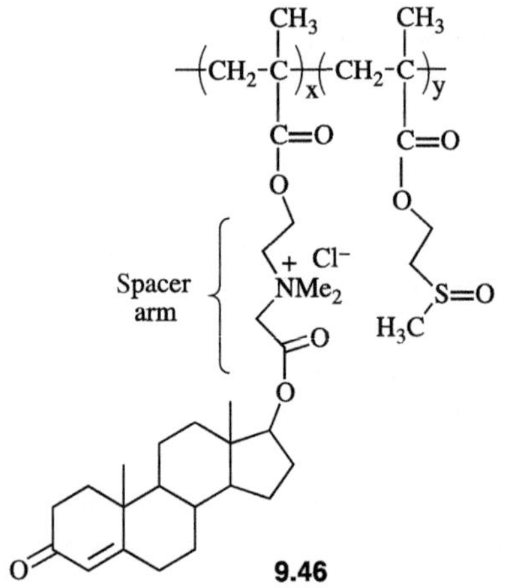

Spacer arm

9.46

9.2.1.3.3. Poly(α-Amino Acids)

The disadvantage of using synthetic polymers is that they are generally not biodegradable, and they can take 5–12 months to be eliminated from the body. Poly(α-amino acids) are biodegradable (at least the L-isomers are); the rate of biodegradability depends on the choice of amino acid.[82]

Conjugation of the antitumor drug methotrexate (Rheumatrex) to poly(L-lysine) (**9.47**; attachment of the polymer also may be to the α-carboxyl group of the terminal lysine residue) markedly increased the cellular uptake of the drug and provided a new way to overcome drug resistance related to deficient drug transport.[83] Because the activity of methotrexate is a function of its inhibitory properties of dihydrofolate reductase (see Chapter 4, Section 4.3.2), and **9.47** is a poor inhibitor of this enzyme in vitro, the methotrexate must become detached from the polymer backbone inside the cell. Furthermore, attachment of methotrexate to poly(D-lysine), which, unlike poly(L-lysine) does not undergo proteolytic digestion inside the cell, gave a conjugate devoid of activity with resistant or normal cell lines. Methotrexate attached to poly(L-lysine) is also more inhibitory to the growth of human solid tumor cell lines than to the growth of human lymphocytes; free methotrexate is equally toxic to both kinds of cells.[84]

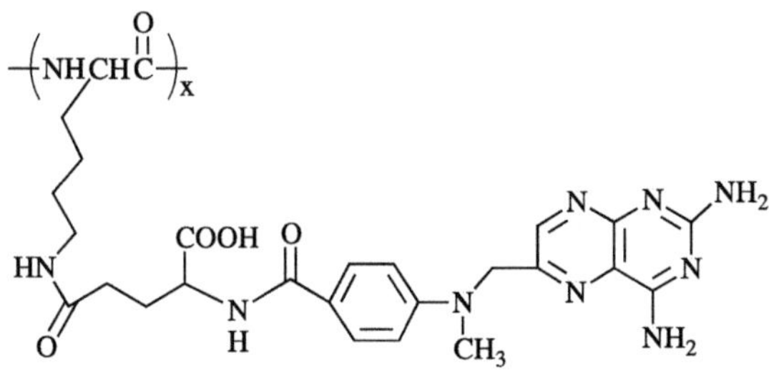

9.47

Research directed at a sustained-release contraceptive resulted in the macromolecular drug delivery system **9.48**.[85] The contraceptive norethindrone (Nor-QD) was attached via a 17-carbonate linkage to poly-N^5-(3-hydroxypropyl)-L-glutamine. In rats the contraceptive agent was slowly released over a nine-month period.

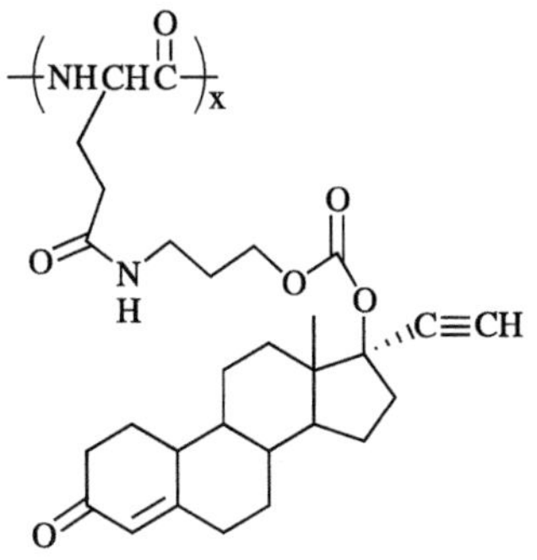

9.48

A general scheme (**9.49**) for the design of a site-specific macromolecular drug delivery system was described by Ringsdorf.[86] A drug is attached to the polymer backbone usually through a spacer, so that it can be cleaved hydrolytically or enzymatically without steric hindrance. The desired solubility of the drug–polymer conjugate can be adjusted by attachment of an appropriate hydrophilic or hydrophobic "solubilizer". Finally, site specificity, for example, to a particular cancer cell line, can be manipulated by attachment of a "homing device" such as an antibody raised against that cell line.

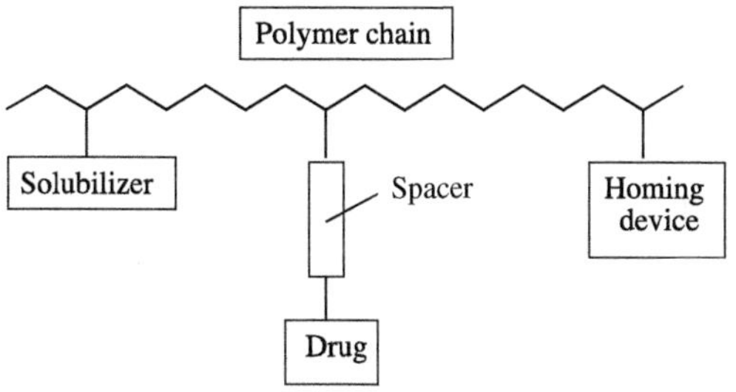

9.49

An elegant example of this approach in which a nitrogen mustard was delivered to tumor cells is shown in **9.50**.[87] Poly(L-glutamate) was used as the polymeric backbone so that the side-chain carboxylic acid groups could be functionalized appropriately. The water-solubilizing groups are the unsubstituted glutamate side-chain carboxylate groups; the antitumor alkylating agent (the *p*-phenylenediamine mustard) is attached to the built-in spacer arm, i.e., the glutamate side chain; and the homing device is an immunoglobulin derived from a rabbit antiserum against mouse lymphoma cells. This macromolecular drug delivery system was much more effective than the individual components or a mixture of the components. None of the five control mice was alive and tumor free after 60 days, whereas all five of the polymer prodrug-treated mice were. Also, the therapeutic index of *p*-phenylenediamine mustard was greatly enhanced (40-fold) when it was attached to the polymer system, because it is less toxic to normal proliferating cells. Similar results were obtained when the neutral and water-soluble polymer dextran was used.[88]

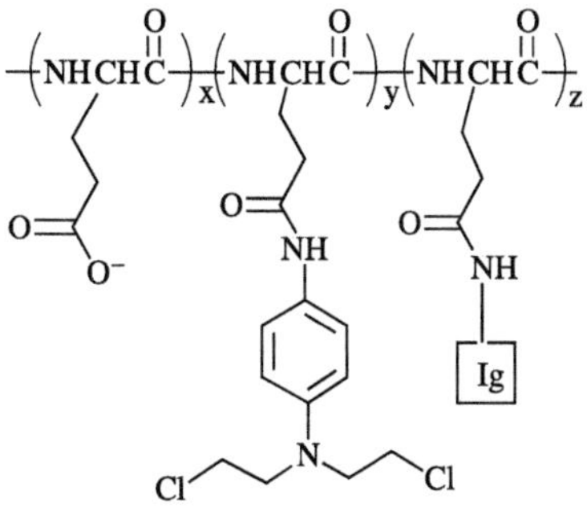

9.50

9.2.1.3.4. Other Macromolecular Supports

Because inhibitors of DNA synthesis generally are toxic to normal rapidly proliferating cells as well as to cancer cells, a targeted macromolecular approach to the delivery of the antitumor agents floxuridine (**9.51**, R=H; sterile FUDR) and cytosine arabinoside (cytarabine; **9.52**; R=H; Cytosar-U) was taken to decrease their toxicity.[89] These drugs were conjugated to albumin because once proteins enter cells, they are rapidly broken down by lysosomal enzymes, and this would release the drugs from the albumin inside the cells. As certain neoplastic proliferating cells are highly endocytic (high protein uptake) and normal cells with high protein uptake do not proliferate, selective toxicity to neoplastic or to DNA viruses that replicate in cells with high protein uptake could be accomplished. Both conjugates (**9.51** and **9.52**, R=albumin-CO) were shown to inhibit the growth of *Ectromelia* virus in mouse liver, whereas the free inhibitors were ineffective. The conjugates exert their antiviral activity in liver macrophages (cells with high protein uptake), suggesting that the drugs are concentrated in these cells.

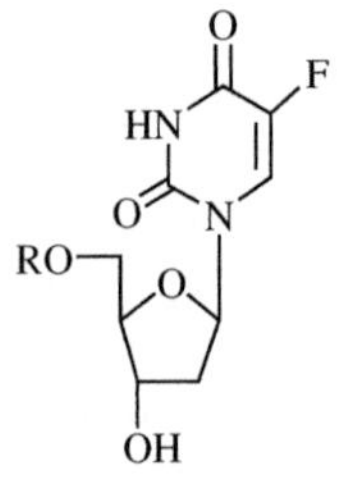

Floxuridine (R=H)
9.51

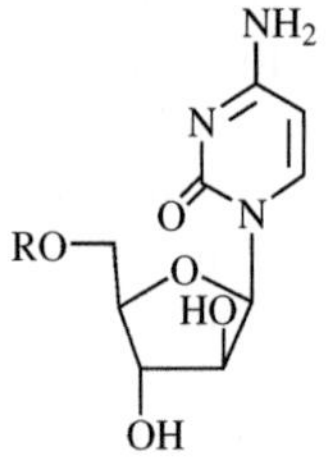

Cytosine arabinoside (R=H)
9.52

The enediyne antitumor antibiotic calicheamicin (Chapter 6, Section 6.3.3.4.1) is too toxic for use in cancer chemotherapy. To minimize its toxicity, *antibody-targeted chemotherapy* was undertaken using a slightly modified version of calicheamicin (a disulfide instead of a trisulfide), which was attached to a humanized antibody through a spacer to give the drug called gemtuzumab ozogamicin (Scheme 9.8, **9.53**, Mylotarg; withdrawn from the market in 2010). This *antibody–drug conjugate (ADC)*, used for the treatment of CD33-positive acute myeloid leukemia (the antibody is specific for the CD33 antigen, a protein commonly expressed by myeloid leukemic cells),[90] passes through healthy tissue and does not exhibit an immune response. Reduction of the disulfide bond at the tumor cell releases the calicheamicin intermediate (**9.54**), which leads to DNA strand breakage as shown in Chapter 6 (Section 6.3.3.4.1, Scheme 6.28). The principal reason gemtuzumab ozogamicin was withdrawn from the market was because the linker between the drug and the antibody was unstable and the drug was released into the patient's bloodstream, leading to toxicity. New linkers have been developed that are sufficiently stable so that the drug is released only at the tumor cell.[91] One of the first new ADCs, brentuximab vedotin (Figure 9.1A, Adcetris), is directed to CD30, which is expressed in Hodgkin lymphoma and systemic anaplastic large cell lymphoma. The drug attached to the monoclonal antibody brentuximab is monomethyl auristatin E (also known as vedotin), an antimitotic agent (inhibits cell division by blocking polymerization of tubulin), which is too toxic to be used by itself. The linker, containing valine–citrulline, is cleaved by the proteolytic enzyme cathepsin inside the tumor cell, which releases vedotin. Trastuzumab emtansine[92] (Figure 9.1B, Kadcyla) was the second ADC approved, which links the monoclonal antibody trastuzumab (Herceptin) to mertansine. Trastuzumab can stop tumor cell growth by binding to the HER2/neu receptor, and mertansine destroys the cells by binding to tubulin. Because HER2 is only overexpressed in tumor cells (e.g., HER2-postive metastatic breast cancer), this ADC delivers mertansine only to cancerous cells. Other ADCs are in the drug pipeline.[93]

9.2.1.4. Tripartite Prodrugs

Bipartite prodrugs may be ineffective because the prodrug linkage is too labile (e.g., certain esters) or too stable (because of steric hindrance to hydrolysis). Katzenellenbogen and coworkers[94] designed a *tripartite* (also known as a *self-immolative*) *prodrug* to remedy this problem. In a tripartite prodrug, the carrier is not connected directly

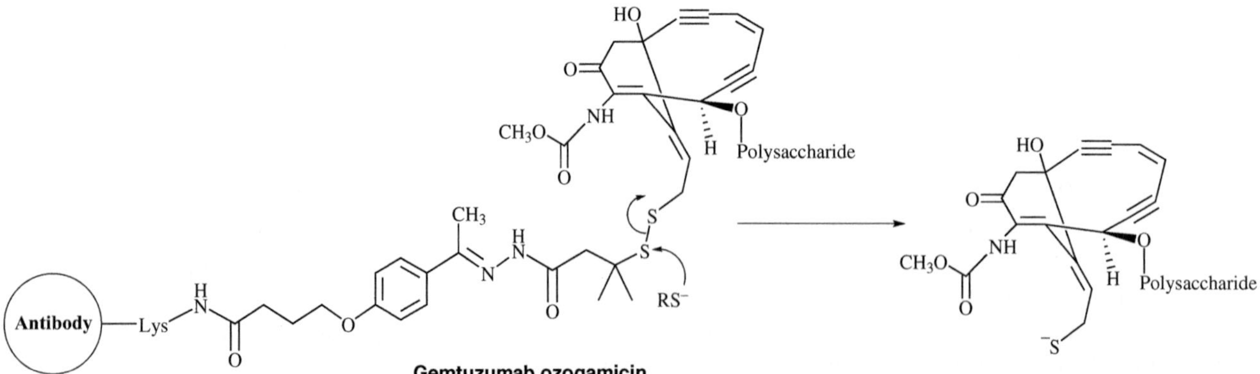

Gemtuzumab ozogamicin
9.53

9.54

SCHEME 9.8 Antibody-targeted chemotherapy, a prodrug for calicheamicin

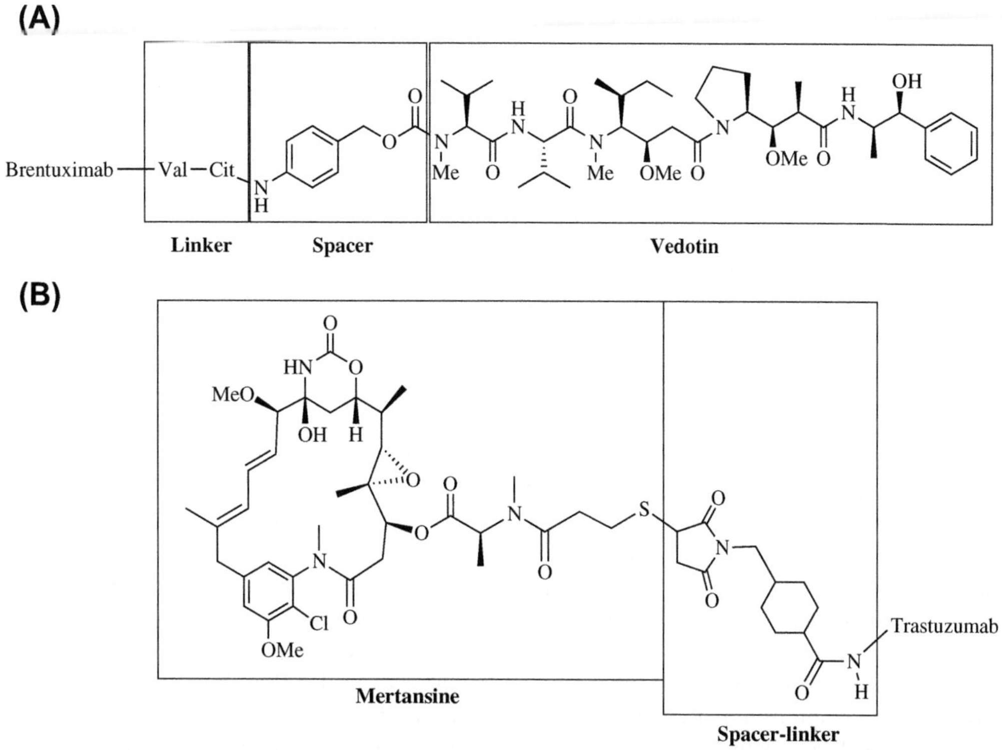

FIGURE 9.1 (**A**) Brentuximab vedotin, the first antibody-drug conjugate with a more stable linker; (**B**) Trastuzumab emtansine, the second antobody-drug conjugate approved

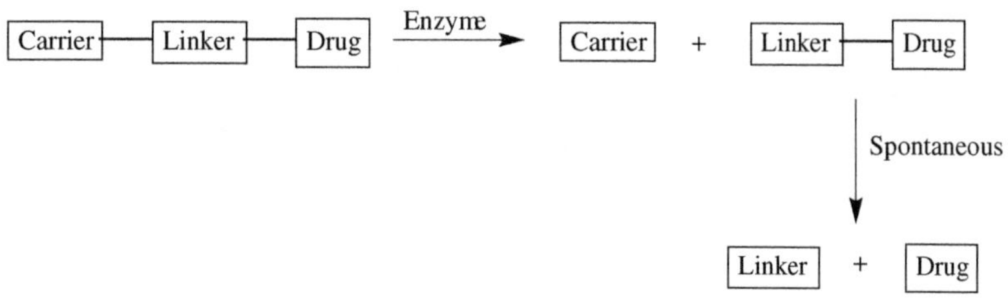

SCHEME 9.9 Tripartite prodrugs

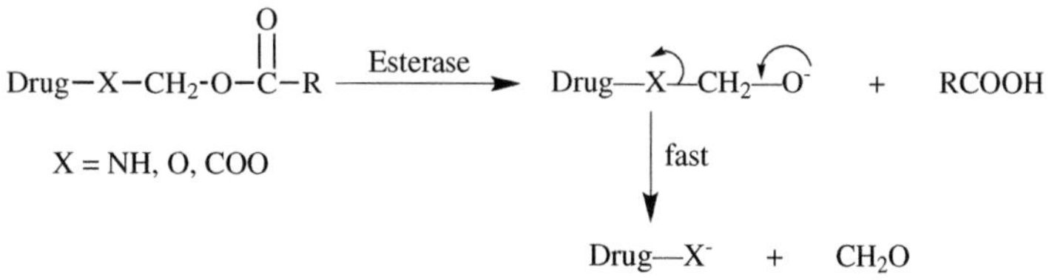

$$\text{Drug-X-CH}_2\text{-O-C(=O)-R} \xrightarrow{\text{Esterase}} \text{Drug-X-CH}_2\text{-O}^- + \text{RCOOH}$$

X = NH, O, COO

$$\xrightarrow{\text{fast}} \text{Drug-X}^- + \text{CH}_2\text{O}$$

SCHEME 9.10 Double prodrug concept

to the drug, but rather to a linker, which is attached to the drug (Scheme 9.9). This allows for different kinds of functional groups to be incorporated for varying stabilities, and it also displaces the drug farther from the hydrolysis site, which decreases the steric interference by the carrier. The drug–linker connection, however, must be designed so that it cleaves spontaneously (i.e., is *self-immolative*) *after* the carrier has been detached. One approach to accomplish this generalized in Scheme 9.10 [95] has been termed the *double prodrug* or, in the case where X=COO, the *double ester* concept.

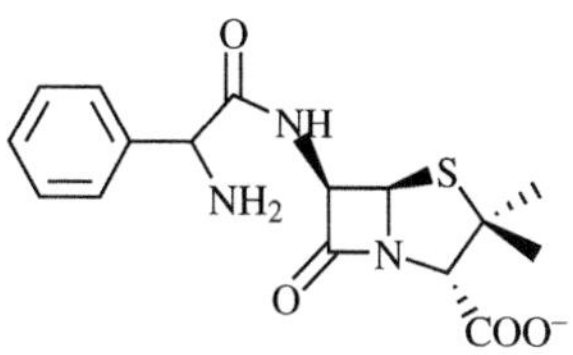

This strategy was employed in the design of prodrugs of ampicillin (**9.55**), a β-lactam antibiotic that is poorly absorbed when administered orally. Because only 40% of the drug is absorbed, 2.5 times more drug must be administered orally than by injection. Having to take extra antibiotic can lead to a more rapid onset of resistance, and the nonabsorbed antibiotic may destroy important intestinal bacteria used in digestion and for the biosynthesis of cofactors.[96] A lipid-soluble prodrug of ampicillin would be a useful approach to increase absorption of this drug. However, although various simple alkyl and aryl esters of the thiazolidine carboxyl group are hydrolyzed rapidly to ampicillin in rodents, they are too stable in man to be therapeutically useful. This suggests that the esterases in rodents and man are different and that, most likely, steric hindrance of the ester carbonyl by the thiazolidine ring is important in the human esterase. A solution to the problem was the construction of a "double ester", an acyloxymethyl ester[97] such as **9.56** (R=CH_3, R′=OEt; bacampicillin; Penglobe)[98] or **9.56** (R=H, R′=*t*-Bu; pivampicillin; Pondocillin)[99] (Scheme 9.11), which would extend the terminal ester carbonyl away from the thiazolidine ring and eliminate the inherent steric hindrance with the enzyme. Hydrolysis of the terminal ester (or carbonate, in the case of bacampicillin) gives an unstable hydroxymethyl ester (**9.57**) that spontaneously decomposes to ampicillin and either acetaldehyde (bacampicillin) or formaldehyde (pivampicillin). Bacampicillin is a nontoxic prodrug because it decomposes to ampicillin and all natural metabolites in the body: CO_2, acetaldehyde, and ethanol (the usual recommended dose of bacampicillin is 400 mg twice a day, therefore only about 50 μl of ethanol would be released with each dose, so do not expect to get drunk). Unlike ampicillin, bacampicillin is absorbed to the extent of 98–99%, and ampicillin is liberated into the bloodstream in less than 15 min. Because of the excellent absorption properties of bacampicillin, only one-half to one-third of the ampicillin dose is required orally.

As mentioned in Section 9.2.1.2.3, the blood–brain barrier is an important membrane for protection of the brain from polar, hydrophilic molecules that do not belong there. Bodor and coworkers have devised a reversible redox drug delivery system for getting drugs into the central nervous system, and then once in, from preventing their passive efflux.[100] The approach is based on the attachment of a hydrophilic drug to a lipophilic carrier, a dihydropyridine (**9.58**), thereby making the prodrug overall sufficiently lipophilic to cross the blood–brain barrier passively (Scheme 9.12). Furthermore, the nitrogen atom in the dihydropyridine ring is conjugated

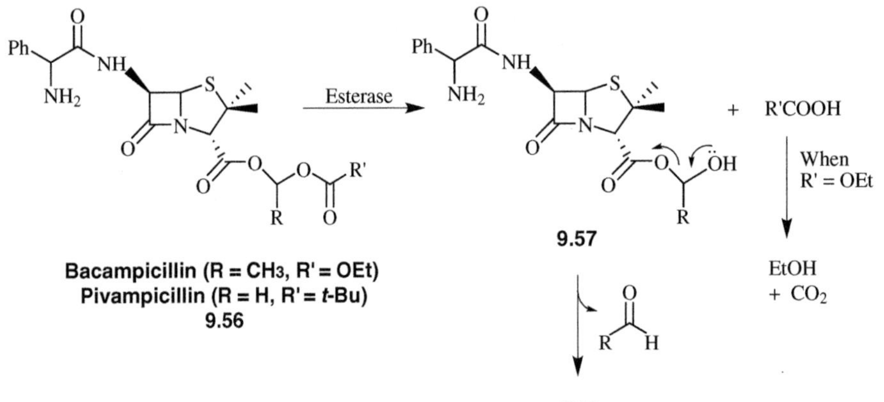

Ph NH₂ NH S N O O O O R' R O Esterase + R'COOH When R' = OEt EtOH + CO_2 OH R 9.57 R H O 9.55

Bacampicillin (R = CH_3, R' = OEt)
Pivampicillin (R = H, R' = *t*-Bu)
9.56

SCHEME 9.11 Tripartite prodrugs of ampicillin

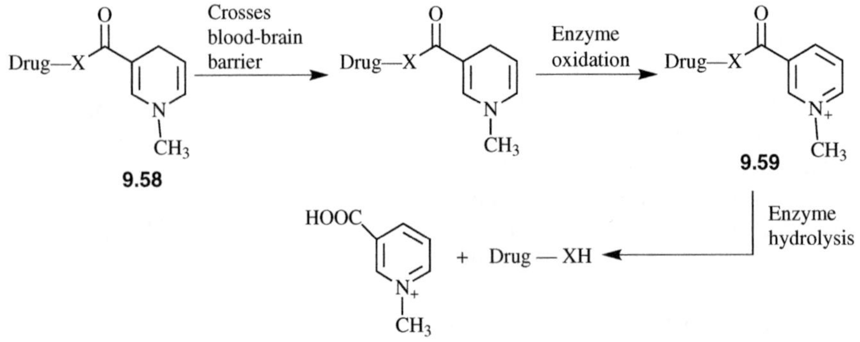

Drug—X O N CH_3 **9.58** Crosses blood-brain barrier Drug—X O N CH_3 Enzyme oxidation Drug—X O N+ CH_3 **9.59** Enzyme hydrolysis HOOC N+ CH_3 **9.60** + Drug — XH

SCHEME 9.12 Redox drug delivery system to cross the blood–brain barrier through the double bond into the carbonyl (a vinylogous amide), thereby making the carbonyl less reactive toward nucleophiles, such as water, and therefore more stable to hydrolysis. Once inside the brain, the lipophilic carrier is converted enzymatically into a highly hydrophilic species (**9.59**) in which the pyridinium nitrogen atom is no longer conjugated into the carbonyl; in fact, the pyridinium group is electron withdrawing, so it now activates the carbonyl for nucleophilic attack. The drug is readily released by enzymatic hydrolysis, and the *N*-methylnicotinic acid (**9.60**) is relatively nontoxic and actively transported out of the brain. The XH group on the drug is an amino, hydroxyl, or carboxyl group. When it is a carboxylic acid, the linkage is an acyloxymethyl ester (**9.61**), which decomposes by the self-immolative reaction shown in Scheme 9.13. The oxidation of the dihydropyridine (**9.58**) to the pyridinium ion (**9.59**) (half-life generally 20–50 min) prevents the drug from escaping out of the brain because it becomes charged. This drives the equilibrium of the lipophilic precursor (**9.58**) throughout all the tissues of the body to favor the brain. Any oxidation occurring outside the brain produces a hydrophilic species that can be rapidly eliminated from the body (see Chapter 8). Although this is a carrier-linked prodrug, it requires enzymatic oxidation to target the drug to the brain. The oxidation reaction is a bioprecursor reaction (see Section 9.2.2).

A tripartite example of this approach is the brain delivery of β-lactam antibiotics for the possible treatment of bacterial meningitis. The difficulty in purging the central nervous system of infections is that the cerebrospinal fluid contains less than 0.1% of the number of immunocompetent leukocytes found in the blood and almost no immunoglobulins; consequently, antibody generation to these foreign organisms is not significant. Because β-lactam antibiotics are hydrophilic, they enter the brain very slowly, and they are actively transported out of the brain back into the bloodstream. Therefore, they are not as effective in the treatment of brain infections as elsewhere. Bodor and coworkers[101] prepared a variety of penicillin prodrugs **9.62** (Scheme 9.14) in which the drug is attached to the dihydropyridine carrier through various linkers, and showed that β-lactam antibiotics could be delivered in high concentrations into the brain, presumably aided by the mechanism in Scheme 9.14.

The antitumor agent 5-fluorouracil (**9.63**, R=H; one trade name is Adrucil) also has been used in the treatment of certain skin diseases. However, because of its low lipophilicity, it does not produce optimal topical bioavailability. *N*-1-Acyloxymethyl derivatives (**9.63**, R=CH_2O-COR) were prepared for increased lipophilicity. These prodrugs were shown to penetrate the skin about five times faster than **9.63** (R=H) and to be metabolized to **9.63** (R=H) rapidly.[102] The mechanism for conversion of **9.63**

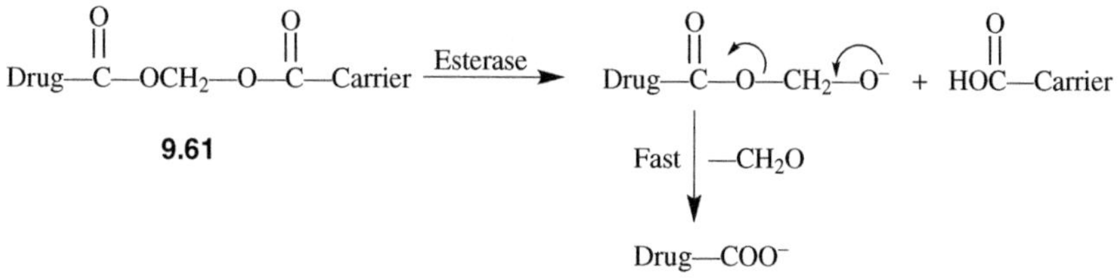

SCHEME 9.13 Redox tripartite drug delivery system

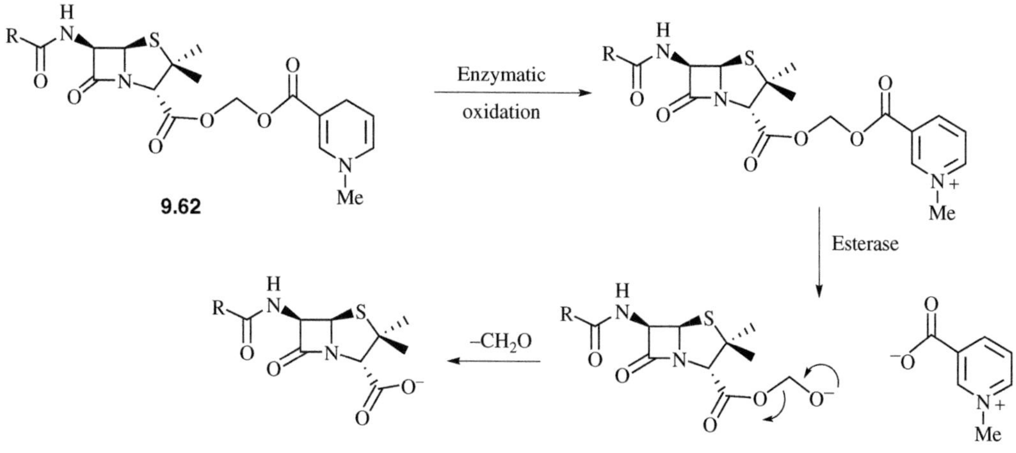

SCHEME 9.14 Redox tripartite drug delivery of β-lactam antibiotics

(R = CH_2OCOR) to **9.63** (R = H) is the same as that shown in Scheme 9.11 for ampicillin derivatives.

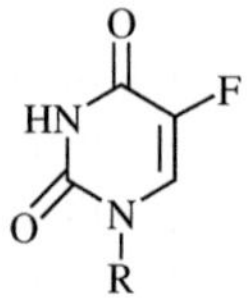

5-fluorouracil (R = H)
9.63

Microorganisms have specialized transport systems for the uptake of peptides (permeases), and these transport systems generally have little side-chain specificity. Consequently, peptidyl derivatives of 5-fluorouracil (**9.64**) were designed as potential antifungal and antibacterial agents that would be substrates for both microbial permeases and peptidases.[103] In accord with the known stereochemical selectivity of peptide permeases, only the peptidyl prodrug with the *L,L* configuration was active. The mechanism for release of 5-fluorouracil after peptidase action is shown in Scheme 9.15.

Amine-containing drugs can be solubilized as acid salts, but their rate of renal excretion is often high. If these drugs are converted into small amide prodrugs, they are no longer able to make salts, so the aqueous solubility decreases. A tripartite macromolecular drug delivery system was designed to retain water solubility of amine-containing antitumor agents without high renal excretion (**9.65**, Scheme 9.16).[104] In this system, 40-kDa PEG is incorporated in the carrier to retain water solubility, and the linker is an *o*-hydroxyphenyl-3,3-dimethylpropionic acid, which, after carrier hydrolysis, undergoes a rapid intramolecular lactonization[105] to release the amine-containing drug. The rate of hydrolysis of the carrier–linker bond can be controlled by varying

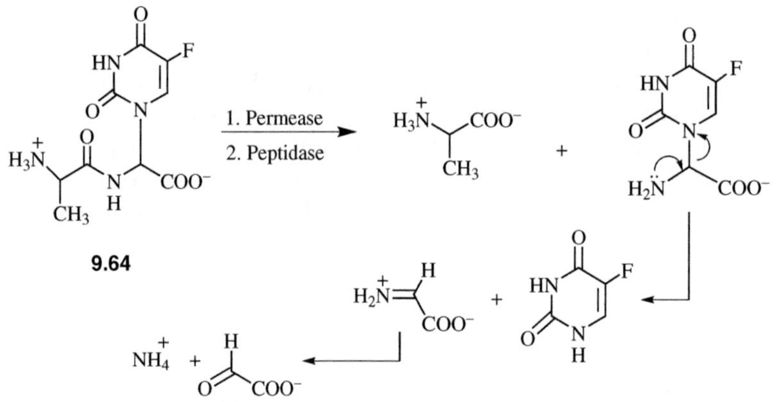

9.64

1. Permease
2. Peptidase

SCHEME 9.15 Activation of peptidyl derivatives of 5-fluorouracil

9.65

Esterase

Spontaneous

+ Drug—NH_2

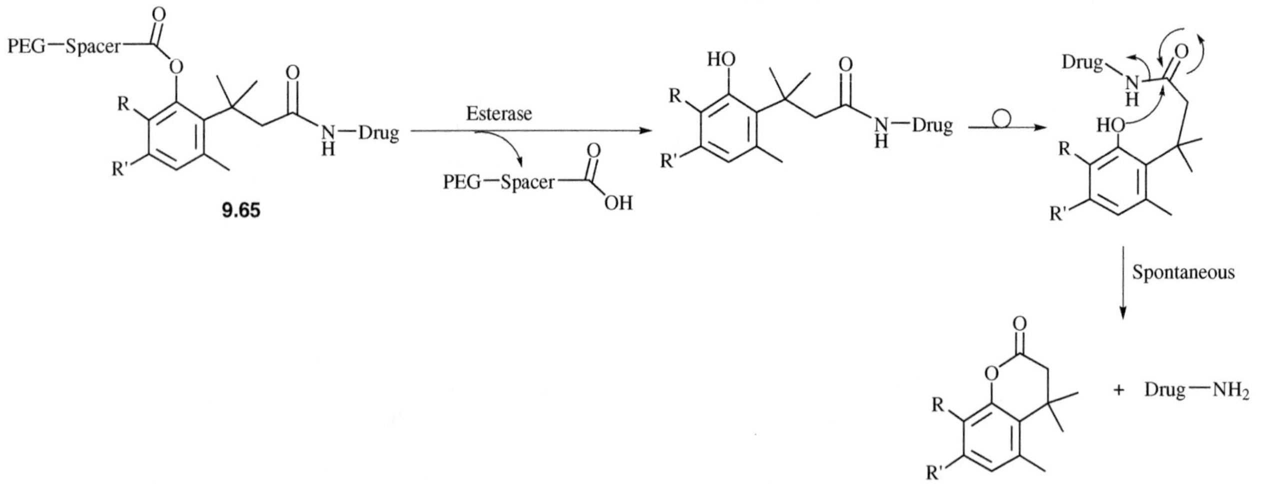

SCHEME 9.16 Tripartite macromolecular drug delivery system

the spacer and the substituents around the linker aromatic ring. This approach was taken for the drugs daunorubicin (**9.66**; Cerubidine) and for cytarabine (**9.52**, R = H). In an in vivo solid tumor panel, one of the PEG-daunorubicin prodrugs was more efficacious against ovarian tumors than daunorubicin.

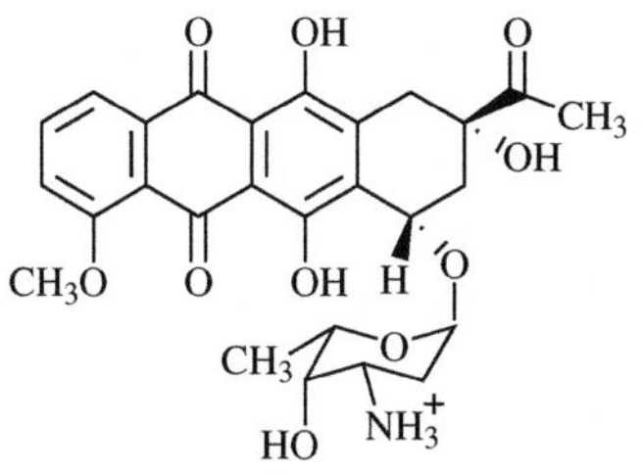

Daunorubicin
9.66

9.2.1.5. *Mutual Prodrugs (also Called Codrugs)*

When it is necessary for two synergistic drugs to be at the same site at the same time, a mutual prodrug approach should be considered. A *mutual prodrug* (or *codrug*) is a bipartite or tripartite prodrug in which the carrier is a synergistic drug with the drug to which it is linked.[106] The problems that arise from this type of prodrug relate to the much larger molecular weight of the combined drugs and the associated side effects from each. In Chapter 7 (Section 7.1.2.4), a form of resistance to β-lactam antibacterial drugs was discussed in which these bacteria excrete a high concentration of the enzyme β-lactamase, which deactivates these antibiotics by hydrolysis of the β-lactam ring. For resistant bacteria, compounds that inhibit β-lactamase are given in combination with a β-lactam antibacterial drug. For example, the combination of the penicillin derivative amoxicillin (**9.67**; Amoxil) and the β-lactamase inactivator potassium clavulanate (**9.68**) (the combination is called Augmentin) is used for oral treatment of infections caused by β-lactamase-producing bacteria. Another combination used is the ampicillin prodrug, pivampicillin (**9.56**, R = H; R′ = *t*-Bu) plus the double ester (**9.69**, R = $CH_2OCOCMe_3$) of the β-lactamase inactivator penicillanic acid sulfone

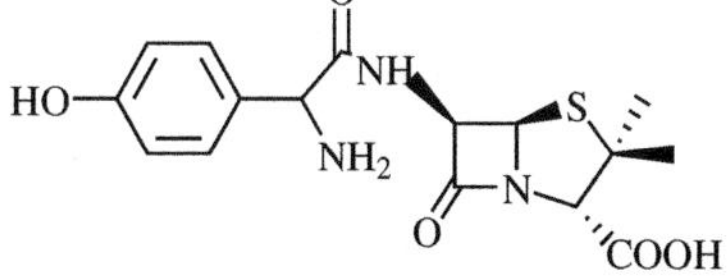

Amoxicillin
9.67

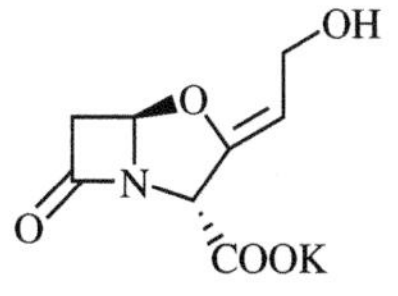

Potassium clavulanate
9.68

(**9.69**, R = H; Zosyn). However, if the two prodrugs are given separately, it is not clear that they are absorbed and transported to the site of action at the same time and in equivalent amounts. An example of a tripartite mutual prodrug is sultamicillin (**9.70**; Unasyn Oral), which upon hydrolysis by an esterase produces ampicillin, penicillanic acid sulfone, and formaldehyde in a reaction like that shown in Scheme 9.11.[107] A mutual prodrug would have a high probability of success provided it is well absorbed, both components are released concomitantly and quantitatively after absorption, the maximal effect of the combination of the two drugs occurs at a 1:1 ratio, and the distribution and elimination of the two components are similar.

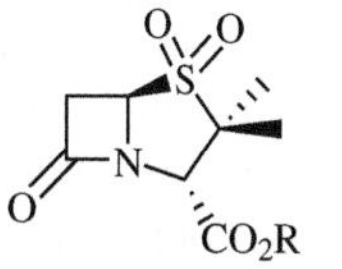

Penicillanic acid sulfone (R = H)
9.69

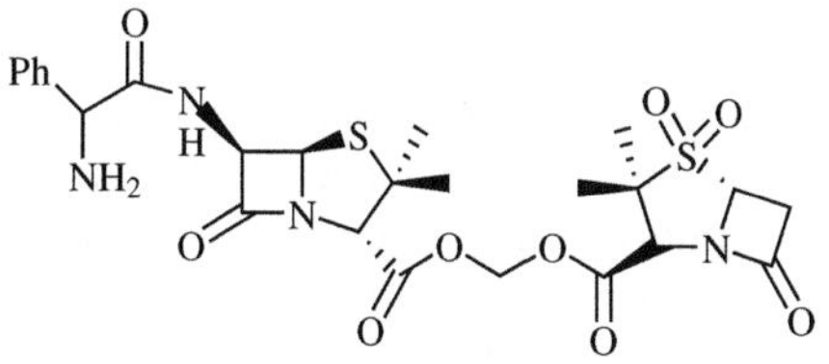

Sultamicillin
9.70

9.2.2. Bioprecursor Prodrugs

9.2.2.1. *Origins*

The birth of bioprecursor prodrugs occurred when it was demonstrated that the antibacterial agent Prontosil was active only in vivo because it was metabolized to the actual drug sulfanilamide (see Chapter 5, Section 5.2.2.3.1). In this case, the azo prodrug Prontosil was reduced to the amine sulfa drug. This exemplifies the bioprecursor strategy. The compounds discussed in Chapter 6, Section 6.3.2.5, metabolically activated alkylating agents, are also examples of bioprecursor prodrugs, but because of their eventual alkylation of DNA, they were placed in that part of the book instead of here. Some of the examples here lead to DNA modification and could have equally been placed in Chapter 6 as well.

Carrier-linked prodrugs rely largely on hydrolysis reactions for their effectiveness, whereas bioprecursor prodrugs mostly utilize either oxidative or reductive activation reactions. The examples given below are arranged according to the type of metabolic activation reaction involved. The first example is the simplest of metabolic transformations, namely, protonation as a mechanism for prodrug activation.

9.2.2.2. Proton Activation: An Abbreviated Case History of the Discovery of Omeprazole

In Chapter 3, we discussed the development of the antiulcer drugs cimetidine and ranitidine (Section 3.2.6). These compounds lowered gastric acid secretion by antagonizing the H_2 histamine receptor. Another approach for lowering gastric acid secretion is by inhibition of the enzyme H^+,K^+-ATPase (also known as the *proton pump*), which is responsible for acid secretion by the *parietal cell*, the cell in the gastric mucosa responsible for acidification of the stomach. This enzyme catalyzes a one-to-one exchange of proton and potassium ions, increasing the acidity in the stomach[108]; inhibition of this enzyme could lead to a new class of antiulcer drugs.[109]

In 1972, the Swedish pharmaceutical company Hässle was searching for a compound that could block gastric acid secretion and discovered a lead compound (**9.71**) in a random screen.[110] The liver toxicity caused by this compound was attributed to the thioamide group, so other sulfur-containing analogs were made, and **9.72** emerged with good antisecretory activity. A series of analogs of **9.72** with different heterocycles led to **9.73** with high activity. A metabolism study in dogs demonstrated that the corresponding sulfoxide (**9.74**, timoprazole) was more potent, but it also blocked the uptake of iodine into the thyroid gland, so it could not be used in humans. A variety of analogs of timoprazole were synthesized, and **9.75** (picoprazole) was found to have antisecretory activity without the iodine blockage activity. In 1977, it was found that picoprazole inhibited the enzyme H^+,K^+-ATPase. The SAR of analogs of picoprazole showed that electron-donating groups on the pyridine ring, which increased the pK_a of the pyridine ring, also increased the potency as an inhibitor of H^+,K^+-ATPase. The best analog was omeprazole (**9.76**, Prilosec).[111]

Studies with ^3H-labeled omeprazole showed that the compound concentrated in the gastric mucosa.[112] Later it was found to be bound to the enzyme H^+,K^+-ATPase in parietal cells.[113] Omeprazole is a relatively weak base, having a pK_a of only about 4. Therefore, the pyridine ring is not protonated at physiological pH, so it is lipid permeable and able to diffuse into the secretory canaliculus of the parietal cell. However, the pH in the parietal cell is below 1, so omeprazole becomes protonated *inside* the canaliculus of the cell, where it becomes trapped and then undergoes a proton-initiated transformation to **9.77**, which reacts covalently with a cysteine residue of H^+,K^+-ATPase (Scheme 9.17).[114] Omeprazole also inhibits human carbonic anhydrase isozymes I and II in erythrocytes and isozyme IV selectively in gastric mucosa.[115] Inhibition of carbonic anhydrase has been shown to be another mechanism for lowering gastric acid secretion.[116] This indicates that omeprazole may have a twofold mechanism of action, which may explain the greater effectiveness of the substituted benzimidazole class of antiulcer drugs compared to other classes of antiulcer drugs. Related analogs that are comparable to omeprazole include lansoprazole (**9.78**, Prevacid),[117] rabeprazole (**9.79**, Aciphex),[118] and pantoprazole sodium (**9.80**, Protonix).[119] Racemic switch analogs (see Chapter 3, Section 3.2.5.2) include esomeprazole (the *S*-isomer of omeprazole, Nexium) and dexlansoprazole (the *R*-isomer of lansoprazole, Kapidex).

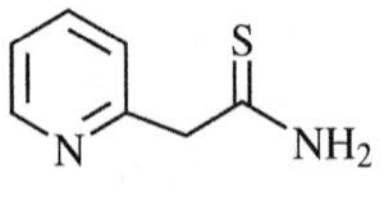

9.71 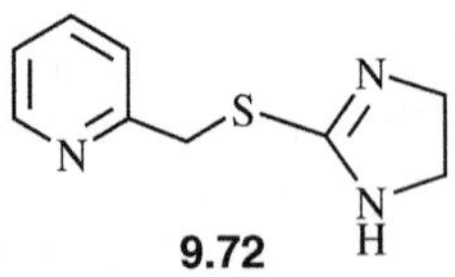 **9.72** 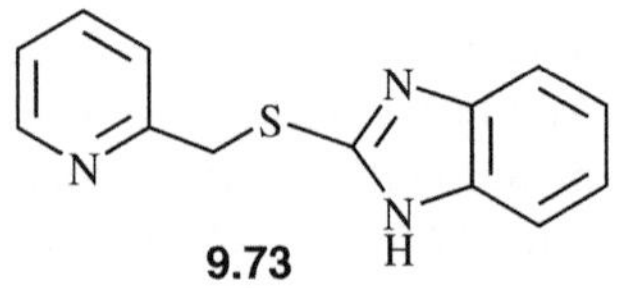 **9.73**

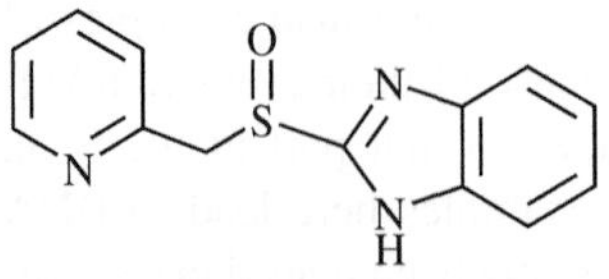

Timoprazole 9.74 **Picoprazole 9.75** **Omeprazole 9.76**

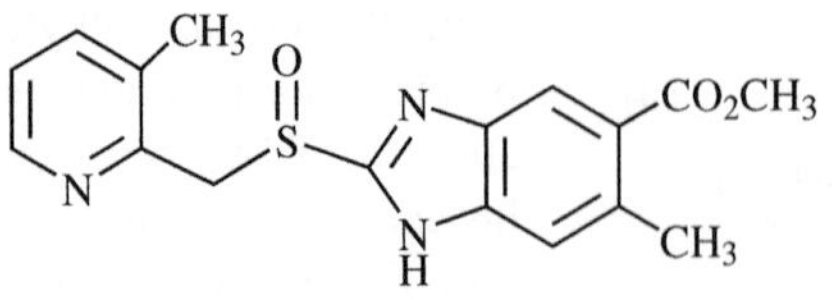

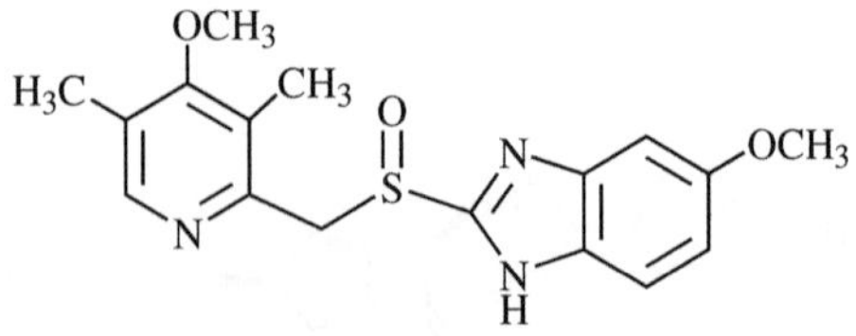

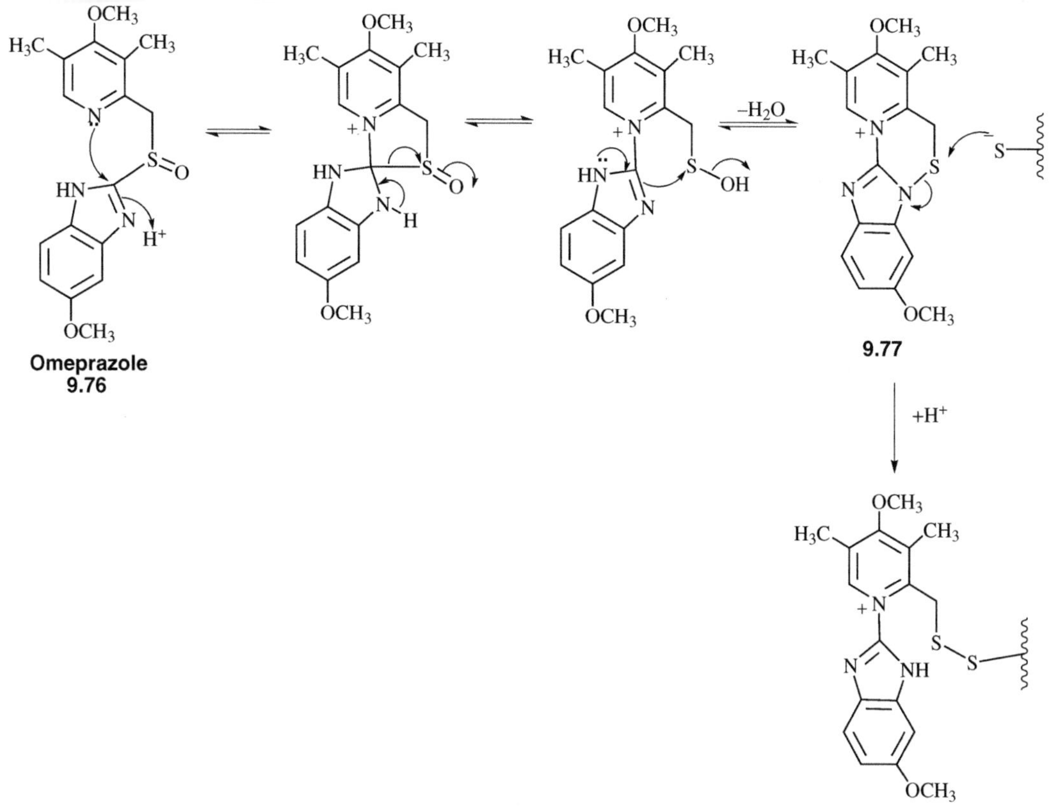

Omeprazole
9.76

9.77

SCHEME 9.17 Mechanism of inactivation of H^+,K^+-ATPase by omeprazole

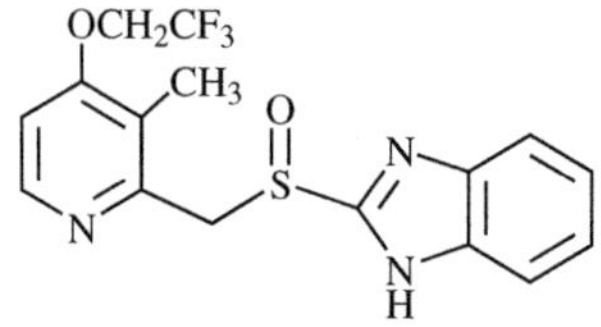

Lansoprazole
9.78

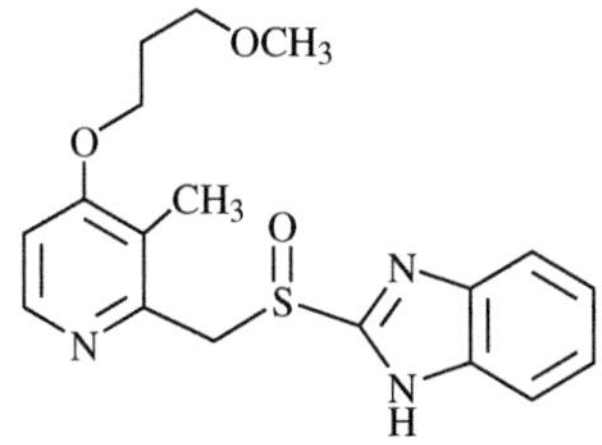

Rabeprazole
9.79

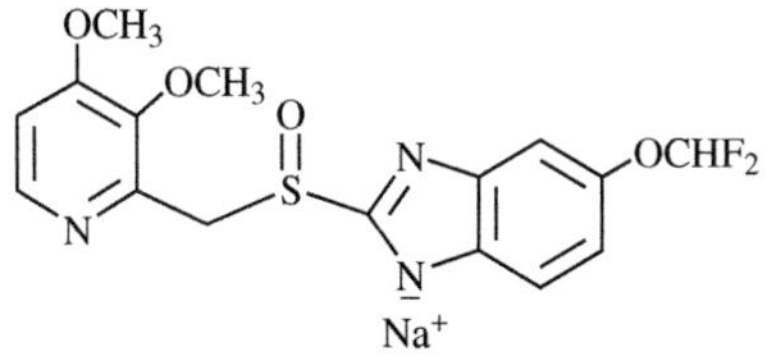

Pantoprazole sodium
9.80

9.2.2.3. Hydrolytic Activation

Hydrolysis can be a mechanism for bioprecursor prodrug activation, if the product of hydrolysis requires additional activation to become the active drug. Leinamycin (**9.81**, Scheme 9.18) is a potent antitumor agent described in Chapter 6 (Section 6.3.2.5.4), which is unstable and toxic. The half-life of aqueous stability was increased by a factor of up to fivefold by conversion into a series of prodrugs (**9.82**).[120] Incubation of analog **9.83** (Scheme 9.19; the most potent analog, in which the macrocyclic ring hydroxyl group also is protected as an (*R*)-2-tetrahydropyranyl prodrug) with fetal calf serum gave **9.85** as the major metabolite. The same metabolite was produced from the (*R*)-2-tetrahydropyranyl ether of leinamycin (**9.84**), supporting the proposal that **9.84** is an intermediate in the hydrolytic activation of **9.83**. The prodrug showed increased antitumor activity relative to leinamycin, presumably as a result of its increased metabolic stability.

9.2.2.4. Elimination Activation

Another relatively simple prodrug activation mechanism is elimination. The rheumatoid arthritis drug leflunomide (**9.86**, Arava) is an immunomodulatory agent shown to

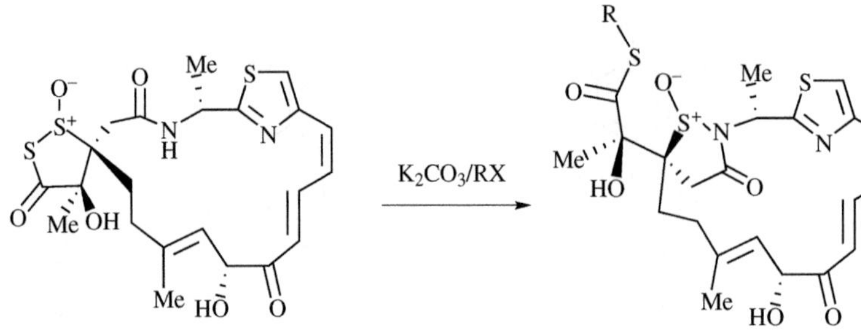

Leinamycin
9.81

K_2CO_3/RX

9.82

SCHEME 9.18 Conversion of leinamycin into a series of prodrugs

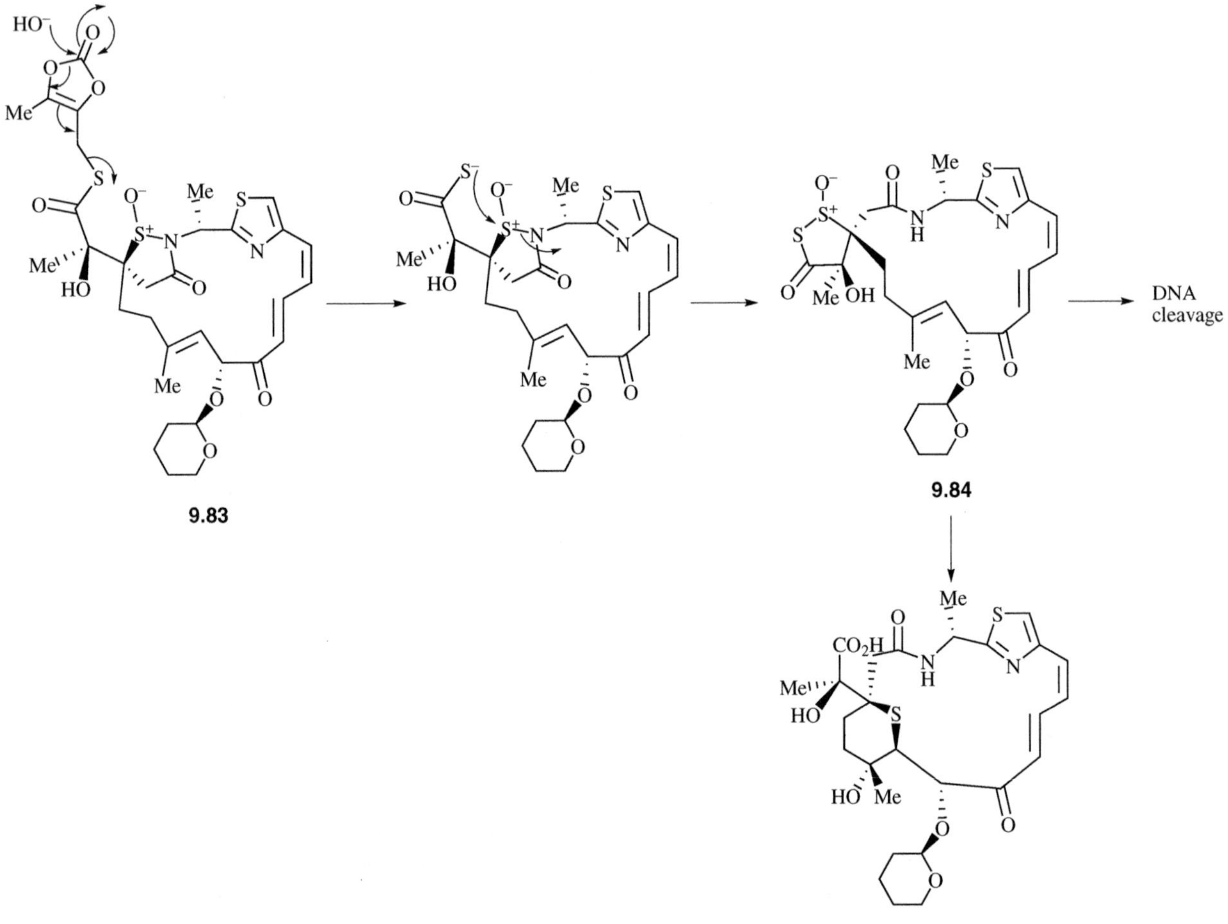

9.83

9.84

DNA cleavage

9.85

SCHEME 9.19 Hydrolytic activation of leinamycin prodrugs

inhibit pyrimidine biosynthesis in human T-lymphocytes by blocking the enzyme dihydroorotate dehydrogenase.[121] Leflunomide shows no inhibitory effect on dihydroorotate dehydrogenase at 1 μM concentration, whereas its metabolite, **9.87**, is a potent inhibitor (K_i 179 nM).[122] Isoxazoles that are unsubstituted at the 3-position are known to undergo facile elimination to nitriles (Scheme 9.20).[123]

9.2.2.5. *Oxidative Activation*

9.2.2.5.1. *N*- and *O*-Dealkylations

Open ring analogs of benzodiazepines, such as the anxiolytic drug alprazolam (**9.88**, X=H; Xanax) and the sedative triazolam (**9.88**, X=Cl; Halcion), undergo metabolic *N*-dealkylation and spontaneous cyclization (Scheme 9.21).[124]

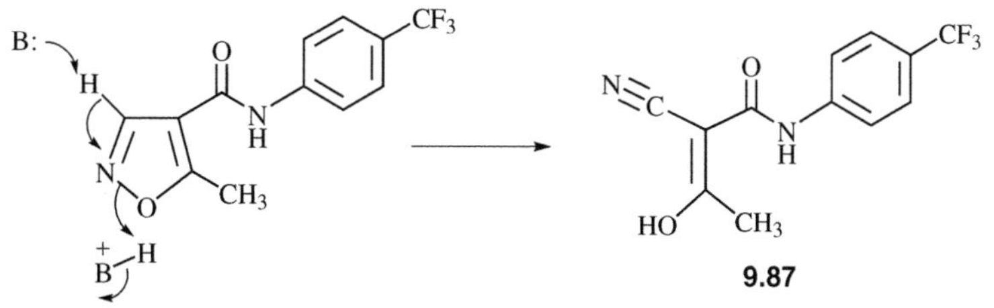

Leflunomide
9.86

9.87

SCHEME 9.20 Activation of leflunomide to the active drug

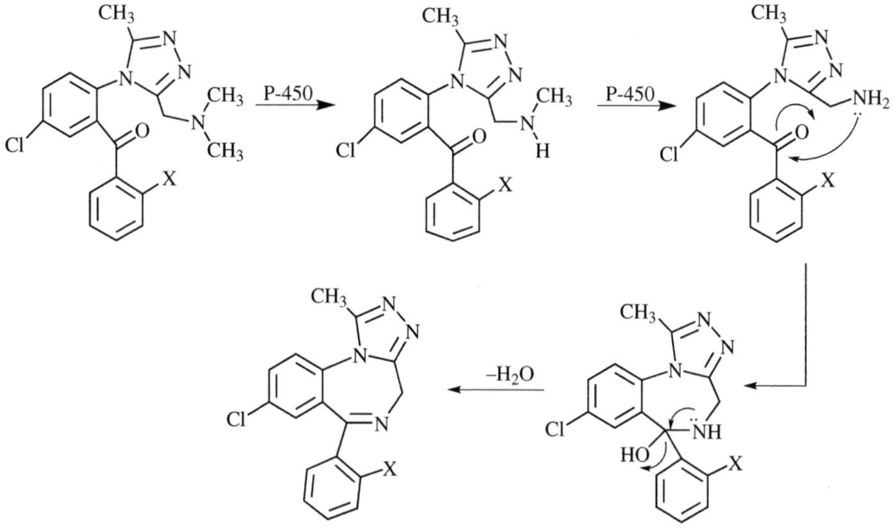

Alprazalam (X = H)
Triazolam (X = Cl)
9.88

SCHEME 9.21 Bioprecursor prodrugs for alprazolam and triazolam

An example of a bioprecursor prodrug that is activated by *O*-dealkylation is the analgesic and antipyretic agent phenacetin (**9.89**, R=CH_2CH_3), which owes its activity to its conversion by *O*-dealkylative metabolism to acetaminophen (**9.89**, R=H; Tylenol).[125]

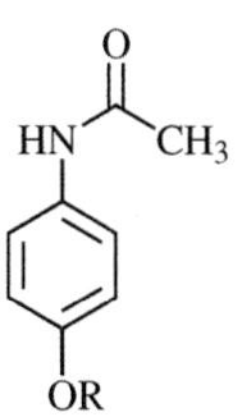

Phenacetin (R = CH_2CH_3)
Acetaminophen (R = H)
9.89

9.2.2.5.2. Oxidative Deamination

Because of the high concentration of phosphoramidases in neoplastic cells, hundreds of phosphamide analogs of nitrogen mustards were synthesized and tested as carrier-linked antitumor prodrugs. Cyclophosphamide (**9.90**, Scheme 9.22; Cytoxan) emerged as an important drug for the treatment of a wide variety of malignant diseases; however, it was later found that it was inactive in tissue culture, suggesting that simple hydrolysis was not involved. Preincubation of the compound with liver homogenates, however, activated it, suggesting that cyclophosphamide is a prodrug requiring an oxidative mechanism (see Chapter 8, Section 8.4.2.1.5).[126] The activation mechanism is believed to be that shown in Scheme 9.22 (there are other metabolites that are not shown in this scheme derived from each of the intermediates). It is not clear which of the toxic metabolites, the phosphoramide mustard (**9.93**) or the parent nitrogen mustard (**9.95**), is responsible for the therapeutic action; the major adduct isolated by high-performance liquid chromatography from in vitro and in vivo studies in rat is *N*-(2-hydroxyethyl)-*N*-[2-(7-guaninyl)ethyl]amine (**9.97**),[127] the hydrolysis product of **9.96**. The reaction of nitrogen mustards with DNA was discussed in Chapter 6 (Section 6.3.2.1). Acrolein (**9.94**) is a potent Michael

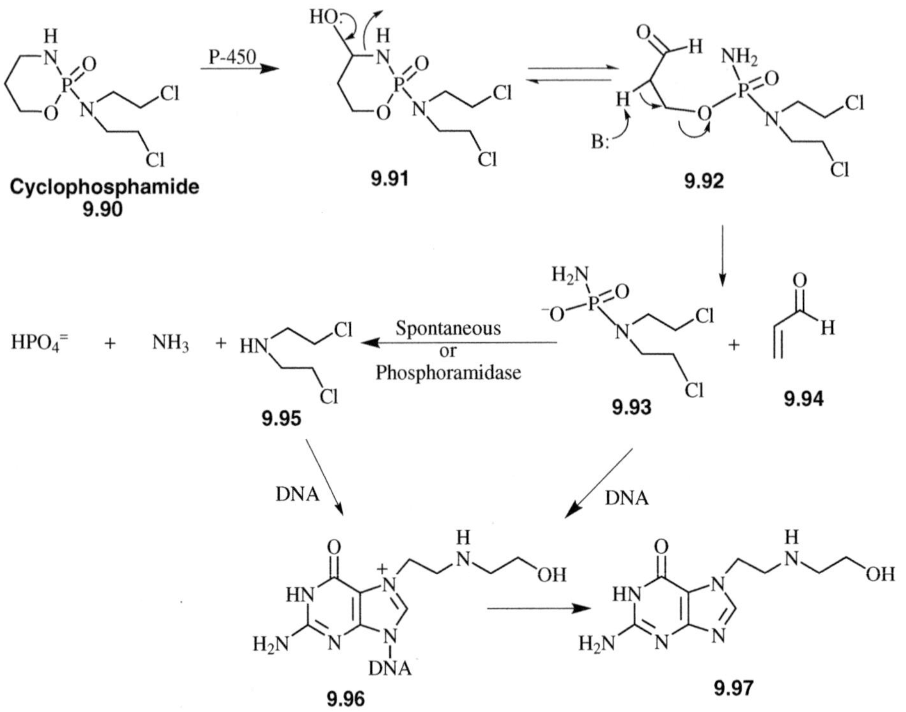

SCHEME 9.22 Cytochrome P450-catalyzed activation of cyclophosphamide

acceptor that may be responsible for the hemorrhagic cystitis side effect; administration of sulfhydryl compounds, which react readily with acrolein, can prevent this side effect. Aldehyde dehydrogenase catalyzes the oxidation of **9.91** to the corresponding cyclic amide and of **9.92** to the corresponding carboxylic acid; however, both of these metabolites are inactive. It has been suggested that these detoxification reactions occur to a greater extent in normal cells than with cancer cells and may account for the selective toxicity of cyclophosphamide.[128]

9.2.2.5.3. *N*-Oxidation

The antitumor drug used against advanced Hodgkin disease, procarbazine (**9.98**, Matulane), is believed to be activated by *N*-oxidation (Scheme 9.23); it is inert unless treated with liver homogenates or oxidized in neutral solution.[129] Those of you who have studied Chapter 8 are probably wondering why this circuitous mechanism to **9.100** and methylhydrazine, starting with an *N*-oxidation reaction, was written instead of a direct conversion of **9.98** to these same metabolites by an oxidative deamination mechanism, starting with C-oxidation of the methylene adjacent to the phenyl ring. The reason is that azoprocarbazine (**9.99**) was identified as the initial metabolic product.[130] 7-Methylguanine (**9.101**) was identified in the urine of mice given procarbazine,[131] which suggests that an activated methylating agent such as methyl diazonium or methyl radical[132] is the reactive intermediate.

Another *N*-oxidation prodrug activation reaction is based on the reversible redox drug delivery strategy of Bodor and coworkers for getting drugs into the brain (see Section 9.2.1.4). In the case of pralidoxime chloride (**9.102**; Protopam chloride), an antidote for poisoning by organophosphorus pesticides and nerve toxins, the oxidation reaction converts prodrug **9.103** into the active drug as well as prevents efflux of the drug from the brain. The neurotoxic organophosphorus compounds exert their effects by reacting with acetylcholinesterase, the enzyme found in nervous tissue of all species of animals, which catalyzes the hydrolysis of the excitatory neurotransmitter acetylcholine after this neurotransmitter has served its neurohumoral transmission function (Chapter 1, Section 1.3.1, Scheme 1.2). Inhibition of the enzyme results in accumulation of acetylcholine, which continues to act on the receptors in various muscles. Muscle cells in the airways contract and secrete mucus, both of which cause choking and difficulty in breathing, and eventually the muscles become paralyzed. Excess acetylcholine at nerve terminals causes over secretion of epinephrine, which increases the blood pressure and heart rate. Excess acetylcholine at nerve cells in the brain causes severe confusion, dizziness, and convulsions.

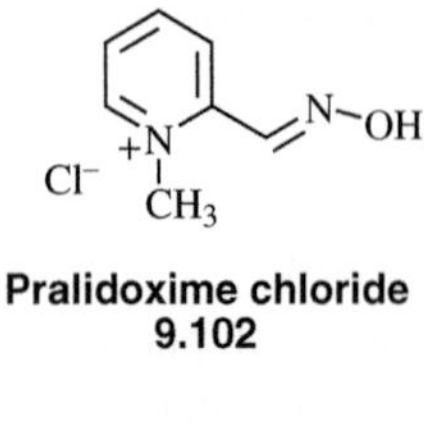

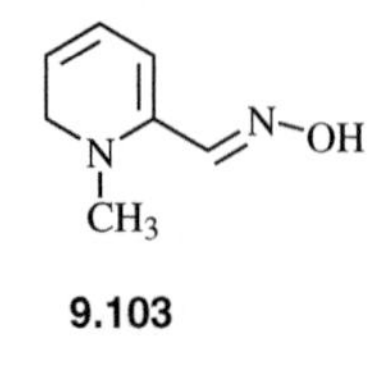

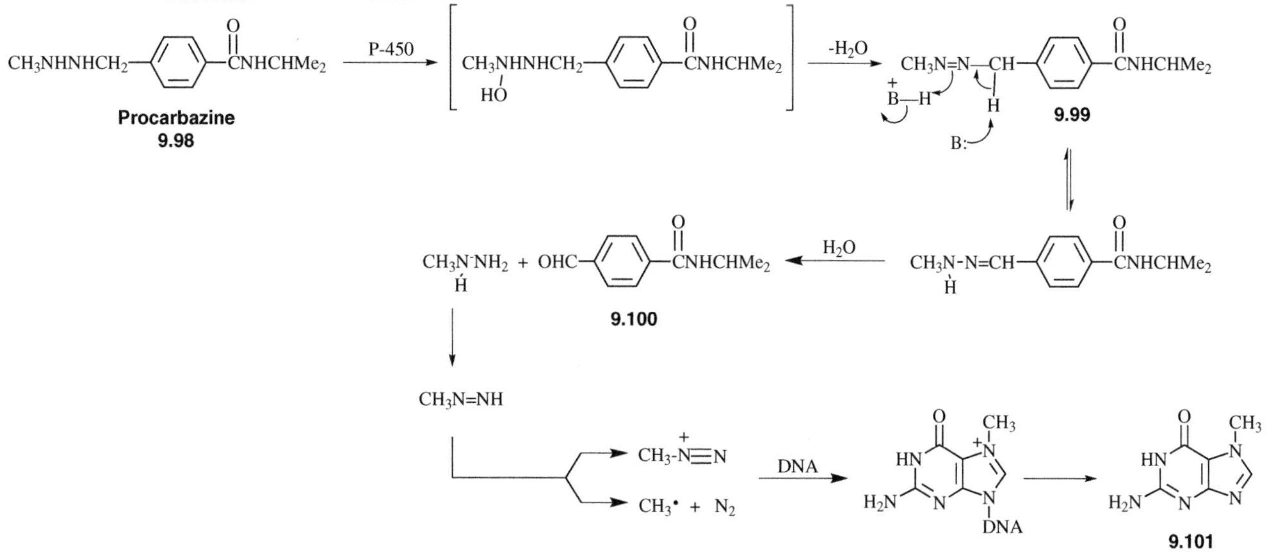

Procarbazine
9.98

9.99

9.100

9.101

SCHEME 9.23 Cytochrome P450-catalyzed activation of procarbazine

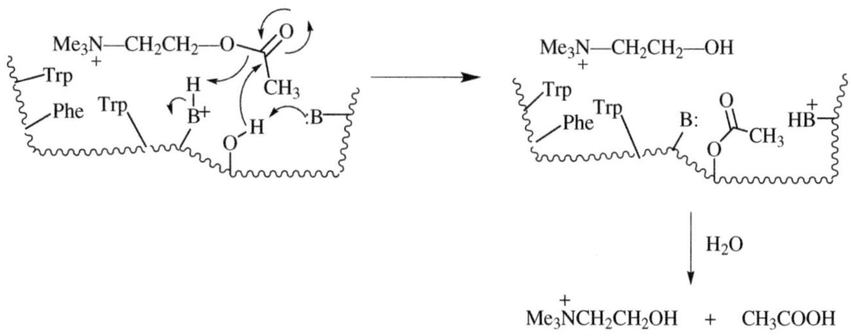

SCHEME 9.24 Acetylcholinesterase-catalyzed hydrolysis of acetylcholine

The active site of the enzyme contains two important binding sites: the site that binds the quaternary ammonium cation of acetylcholine and the ester site where the catalytic hydrolysis of the acetyl group occurs (Scheme 9.24).[133] An X-ray crystal structure of acetylcholinesterase revealed that the ammonium cation does not interact with an anionic residue, but with the π-electrons of a group of aromatic residues.[134] Stabilization of cations by aromatic π-electrons was discussed in Chapter 3 (Section 3.2.2.7).[135]

Organophosphorus compounds, such as the nerve poison diisopropyl phosphorofluoridate (**9.104**), phosphorylate acetylcholinesterase at the ester site[136] (Scheme 9.25). It was thought that a nucleophilic agent may be capable of dephosphorylating the ester site, and this would reactivate the enzyme. Hydroxylamine not only appeared to be effective but also was quite toxic. Because acetylcholinesterase has a cation binding site, quaternary amine analogs were designed, and 2-formyl-1-methylpyridinium chloride oxime (pralidoxime chloride, **9.102**) was found to be an effective reactivator of the enzyme (Scheme 9.26). However, **9.102** is very poorly soluble in lipids, so its generation is most likely restricted to the peripheral nervous system; little reactivation of brain acetylcholinesterase was observed in vivo.[137] Apparently, the effectiveness of **9.102** as an antidote for organophosphorus nerve poisons results from the fact that the primary damage done by these poisons is to the peripheral nervous system. To improve the permeability of **9.102** into the central nervous system, Bodor and coworkers[138] prepared the 5,6-dihydropyridine analog **9.103**. Because it is uncharged, its permeability through the blood–brain barrier is quite good. Once inside the brain it is oxidized to **9.102**.

It is interesting to note that whereas irreversible inactivators of acetylcholinesterase, such as organophosphorus nerve gases, are highly toxic, compounds that form weakly stable covalent bonds to the serine residue in the ester-binding site are useful therapeutic agents. This inhibition of acetylcholinesterase results in the enhancement of cholinergic action by facilitating the transmission of impulses across neuromuscular junctions, which has a cholinomimetic

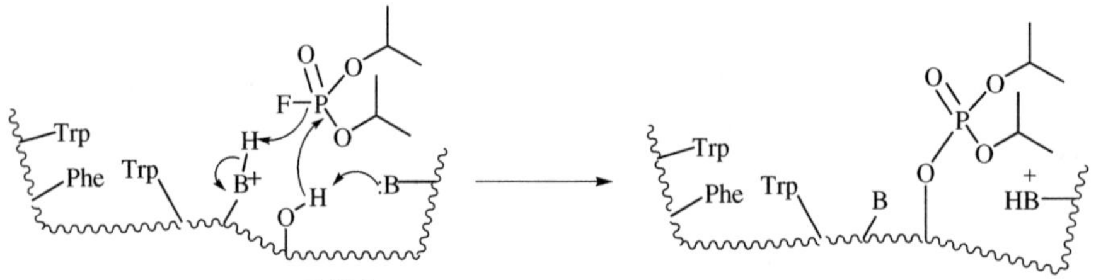

9.104

SCHEME 9.25 Phosphorylation of acetylcholinesterase by diisopropyl phosphorofluoridate

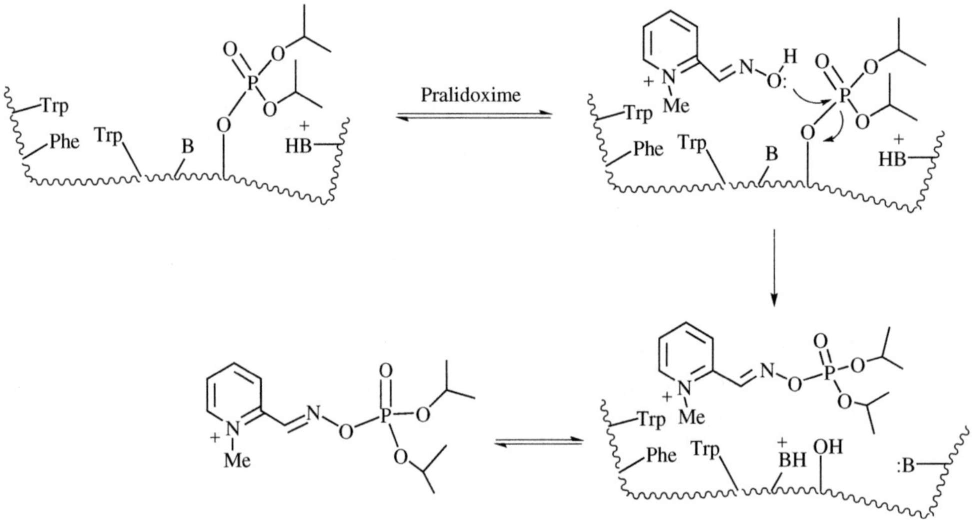

SCHEME 9.26 Reactivation of phosphorylated acetylcholinesterase by pralidoxime chloride

effect on skeletal muscle. An example of this is neostigmine (**9.105**; Prostigmin),[139] a drug used in the treatment of the neuromuscular disease myasthenia gravis. This drug carbamylates the active site serine residue of acetylcholinesterase; the carbamate, however, hydrolyzes slowly so that, in effect, **9.105** acts as a reversible inhibitor of the enzyme (Scheme 9.27). Therefore, the difference in effects of the acetylcholinesterase substrates and inhibitors is derived from the stabilities of the covalent adducts. The acetylated serine formed from acetylcholine (a substrate) hydrolyzes readily, the carbamylated serine produced from neostigmine (an inhibitor) hydrolyzes slowly, and the phosphorylated serine from organophosphorus compounds (inactivators) is stable to hydrolysis.

Reversible inhibitors of acetylcholinesterase, such as donepezil hydrochloride (**9.106**; Aricept)[140] and tacrine hydrochloride (**9.107**; Cognex),[141] which interact with the enzyme noncovalently, and rivastigmine tartrate (**9.108;** Exelon), which transiently carbamylates the enzyme, are used to treat dementia from Alzheimer disease (rivastigmine is also approved for dementia from Parkinson's disease). The increase in acetylcholine as a result of acetylcholinesterase inhibition enhances cholinergic neurotransmission involved in the memory circuit.

9.2.2.5.4. *S*-Oxidation

An important prodrug that uses *S*-oxidation as the means of activation is the antithrombotic drug clopidogrel

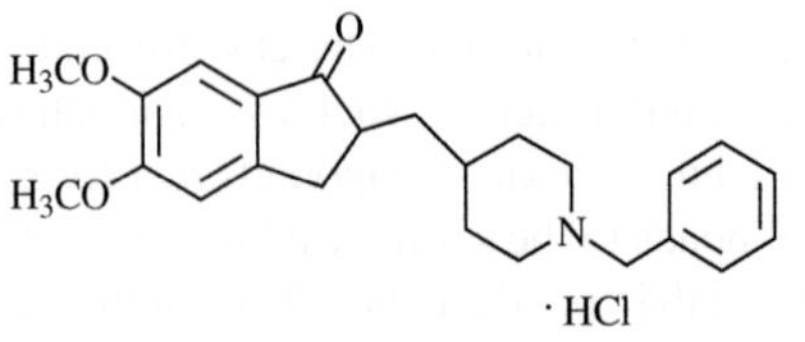

Donepezil hydrochloride
9.106

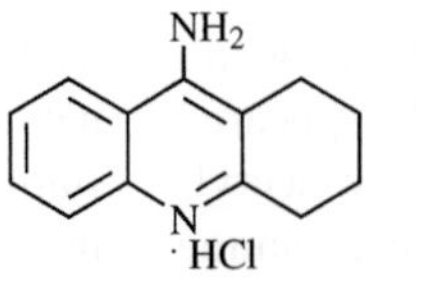

Tacrine hydrochloride
9.107

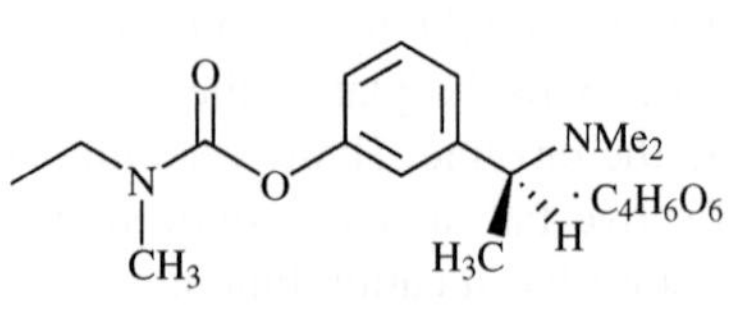

Rivastigmine tartrate
9.108

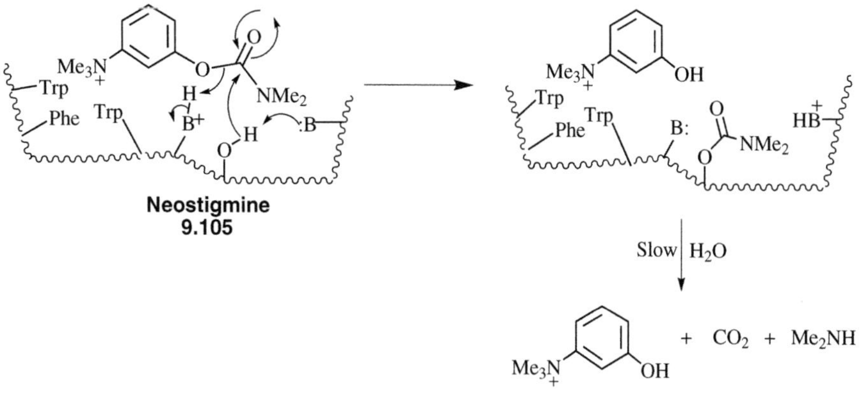

SCHEME 9.27 Carbamylation of acetylcholinesterase by neostigmine

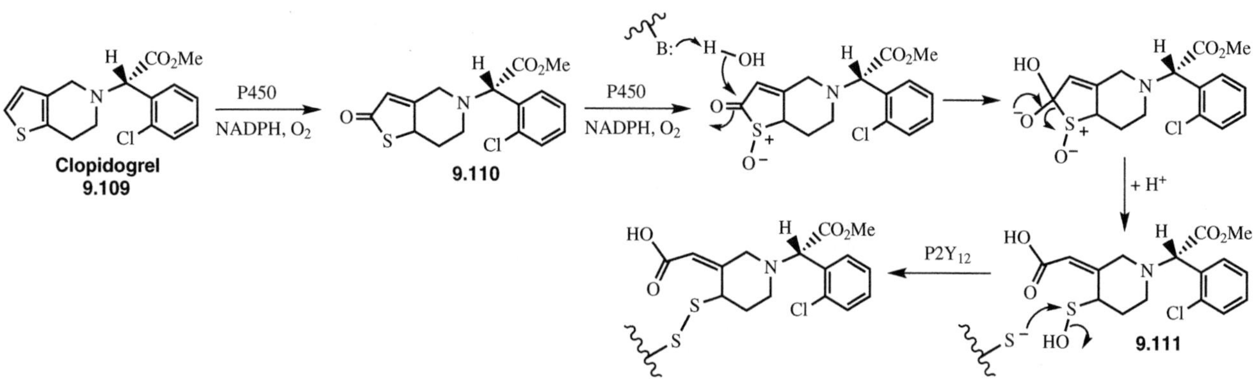

SCHEME 9.28 *S*-Oxidation of clopidogrel, which initiates its inactivation of $P2Y_{12}$

(**9.109**; Scheme 9.28, Plavix). This drug is metabolized by microsomes (P450)[142] and plasma (esterases), but it is activated by P450-catalyzed thiophene oxidation of known metabolite **9.110** to the sulfenic acid (**9.111**),[143] which undergoes a covalent reaction with cysteine residues of $P2Y_{12}$,[144] an adenosine diphosphate-dependent G protein-coupled receptor in platelets important to platelet aggregation.[145]

9.2.2.5.5. Aromatic Hydroxylation

Cyclohexenone **9.112** (Scheme 9.29) was activated in vivo by hydroxylation to the potent dopaminergic catecholamine **9.113**. There is no evidence for the reactions shown in Scheme 9.29, but it depicts one possible pathway. The conversion of **9.112** to **9.113** suggests that cyclohexenones can be prodrugs for catechols with increased bioavailability.[146]

9.2.2.5.6. Other Oxidations

Carbamazepine (**9.114**; one trade name is Cabatrol) is an anticonvulsant drug that is the metabolic precursor of the active agent, carbamazepine-10,11-oxide (**9.115**).[147]

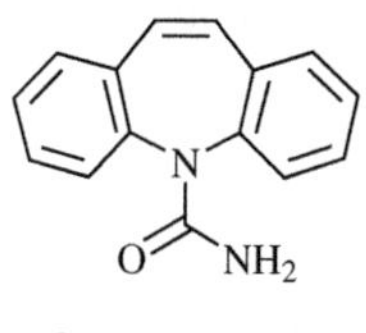

Stimulation of pyruvate dehydrogenase results in a change of myocardial metabolism from fatty acid to glucose utilization. Because the latter requires less oxygen consumption, glucose utilization is beneficial to patients with ischemic heart disease in which arterial blood flow is blocked and therefore less oxygen is available.[148] Arylglyoxylic acids (**9.116**) are important stimulators of pyruvate dehydrogenase, but they have short durations of action. *L*-(+)-2-(4-Hydroxyphenyl)glycine (oxfenicine; **9.117**, R=OH) is a stable amino acid that is actively transported across lipid membranes and is rapidly transaminated (see Chapter 4, Section 4.3.1.3) to 4-hydroxyphenylglyoxylic acid (**9.116**, R=OH).[149] This active transport system and rapid conversion of the prodrug to the drug allow a higher

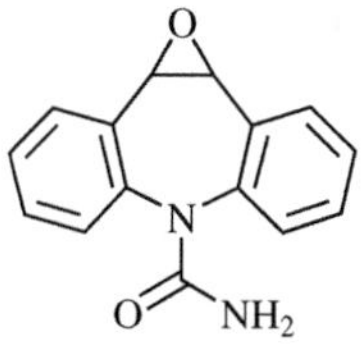

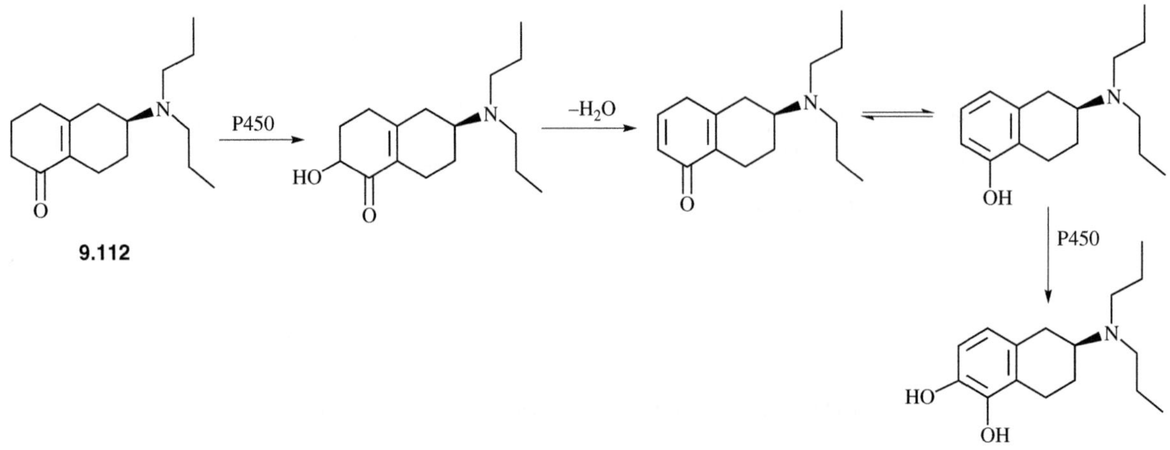

P450 $-H_2O$ P450

9.112 9.113

SCHEME 9.29 Cytochrome P450-catalyzed conversion of a cyclohexenone to a catechol

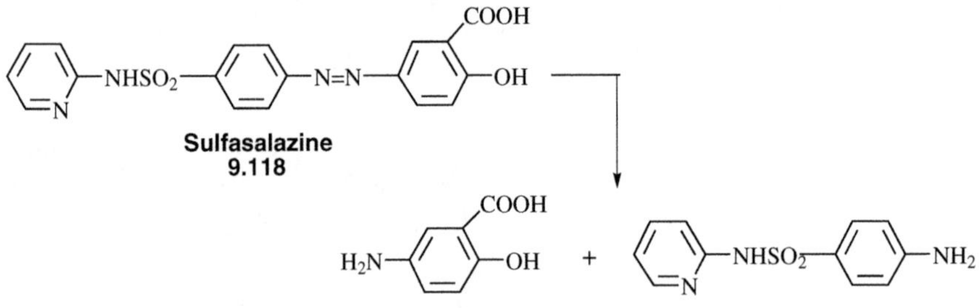

Sulfasalazine 9.118

9.119 9.120

SCHEME 9.30 Reductive activation of sulfasalazine

concentration of the active drug to remain at the desired site of action longer.

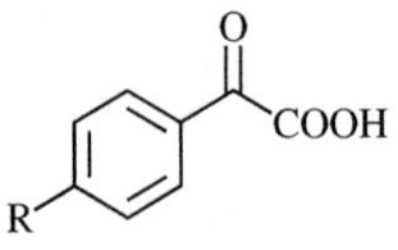

9.116 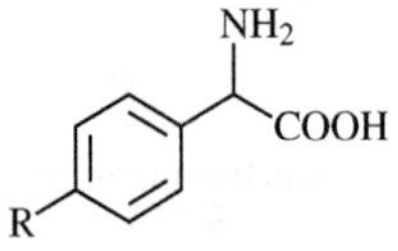 Oxfenicine (R = OH) 9.117

9.2.2.6. Reductive Activation

9.2.2.6.1. Azo Reduction

As described in Section 9.2.2.1 the paradigm for bioprecursor prodrugs, Prontosil, is activated by reduction of its azo linkage to the true bacteriostatic agent, sulfanilamide. Sulfasalazine (**9.118**; Azulfidine), which is used in the treatment of inflammatory bowel disease (ulcerative colitis) and rheumatoid arthritis,[150] is reductively cleaved by anaerobic bacteria in the lower bowel to 5-aminosalicylic acid (**9.119**) and sulfapyridine (**9.120**) (Scheme 9.30); **9.119** is the therapeutic agent, and **9.120** produces adverse side effects.[151] A macromolecular drug delivery system was developed to improve the therapeutic index of this drug. The drug (**9.119**) was azo-linked at the 5-position through a spacer to poly(vinyl amine) to give **9.121**.[152] The advantages of this polymeric drug delivery system are that it is not absorbed or metabolized in the small intestine, **9.119** can be released by reduction at the disease site, and the carrier polymer is not absorbed or metabolized. The water-soluble polymer-linked drug (**9.121**) was more potent than **9.118** or **9.119** in the guinea pig ulcerative colitis model.

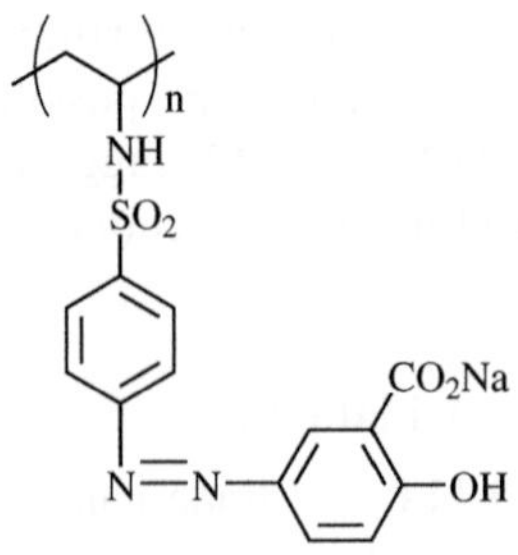

9.121

9.2.2.6.2. Azido Reduction

Vidarabine (**9.122**, R=NH_2; Vira-A) was originally discovered as an antitumor agent,[153] then later it was shown to be active against herpes simplex virus types 1 and 2.[154] However, the clinical use of vidarabine is limited because of its rapid deamination by adenosine deaminase[155] and its poor aqueous solubility. 9-(β-D-Arabinofuranosyl)-6-azidopurine, the 6-azido analog of vidarabine (**9.122**, R=N_3), however, is not a substrate for adenosine deaminase, and it is considerably more stable in vivo.[156] The 6-azido prodrug is activated by cytochrome P450-catalyzed reduction to vidarabine; in vivo, the half-life for vidarabine produced from the corresponding azide is 7–14 times higher than for vidarabine administered directly. Furthermore, whereas vidarabine was not found in the brain after its direct IV administration, significant levels of vidarabine were found in the brain after either oral or IV administration of the azido prodrug, which is useful for the treatment of brain infections.

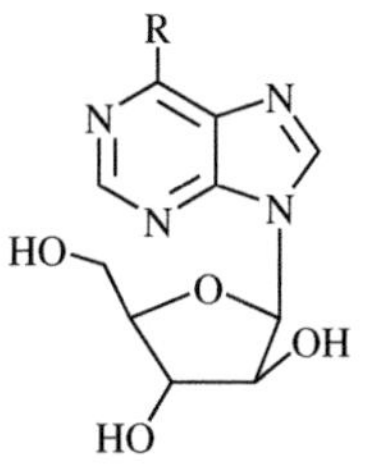

Vidarabine (R = NH_2)
9.122

9.2.2.6.3. Sulfoxide Reduction

The antiarthritis drug sulindac (**9.123**; Clinoril) is an indene isostere (see Chapter 2, Section 2.2.4.3) of the nonsteroidal antiinflammatory (antiarthritis) drug indomethacin (**9.124**; Indocin), which originally was designed as a serotonin analog. Sulindac is less irritating to the gastrointestinal tract and produces many fewer and more mild central nervous system effects than does indomethacin.[157] The 5-fluoro group was substituted for the methoxyl group to improve the analgesic properties, and the *p*-methylsulfinyl group was substituted for the chlorine atom to increase the solubility. Sulindac is inactive in vitro, but is highly active in vivo. The corresponding sulfide, however, is active in vitro and in vivo. Therefore, sulindac is a prodrug for the sulfide, the metabolic reduction product.

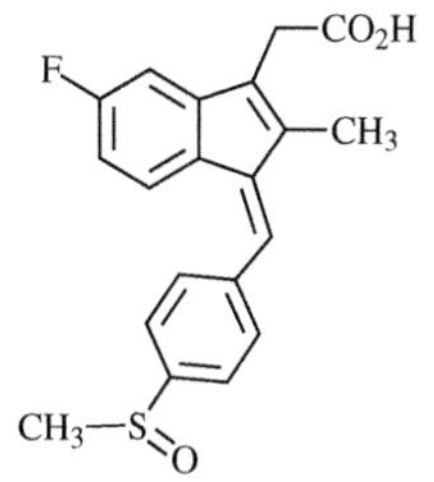

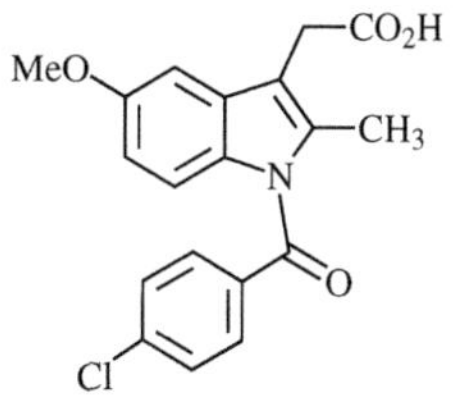

Sulindac
9.123

Indomethacin
9.124

9.2.2.6.4. Disulfide Reduction

Because thiamin (vitamin B_1; **9.126**) is a quaternary ammonium salt, it is poorly absorbed into the central nervous system and from the gastrointestinal tract. To increase its lipophilicity, thiamin tetrahydrofurfuryl disulfide (**9.125**; Scheme 9.31) was designed as a lipid-soluble prodrug of thiamin.[158] The prodrug permeates rapidly through red blood cell membranes (as a model for other membranes), where it reacts with glutathione to produce thiamin.[159]

To diminish the toxicity of the antimalarial drug primaquine (**9.127**; Primaquine) and target it for cells that contain the malarial parasite, a macromolecular drug delivery

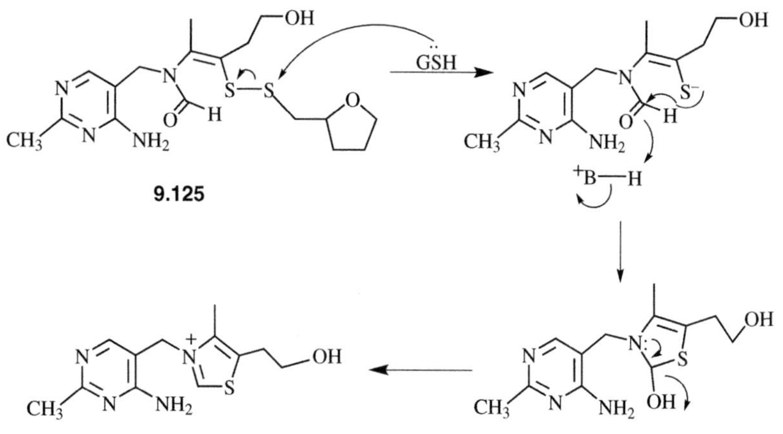

9.125

Thiamin
9.126

SCHEME 9.31 Conversion of thiamin tetrahydrofurfuryl disulfide to thiamin system was designed[160] (**9.128**). The lactose-linked albumin was used for improved uptake in the liver via the asialoglycoprotein receptor system. Because the concentration of free thiol in the blood is relatively low, but is high intracellularly, it was expected that thiol reduction of the disulfide linkage would occur mostly inside the cell. It is not known if after disulfide reduction the cysteinyl residue is detached by hydrolysis or remains attached to primaquine. The therapeutic index of **9.128**, however, is 12 times higher than that of the free drug in *Plasmodium*-infected mice.

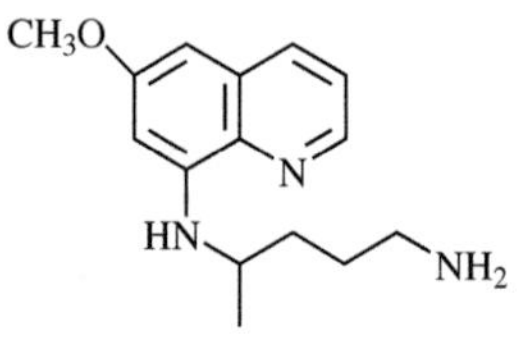

Primaquine
9.127

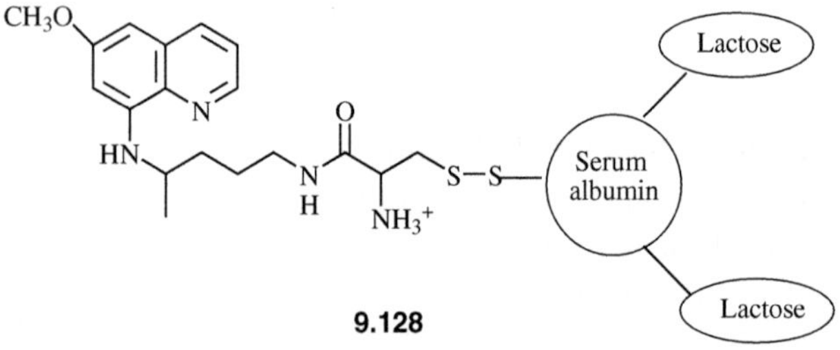

9.128

9.2.2.6.5. Nitro Reduction

The mechanism of action of the antiprotozoal agent ronidazole (**9.129**; Dugro) is not known, but on the basis of metabolism studies using several radioactively-labeled analogs, it was suggested that **9.129** is activated by initial four-electron reduction of the 5-nitro group to the corresponding hydroxylamine, which can react with protein thiols by one of two mechanisms (Scheme 9.32).[161]

A phosphoramidate-based prodrug system that acts through initial nitro group reduction was designed for intracellular delivery of nucleotides.[162] A prodrug for the delivery of the anticancer nucleotide 5-fluoro-2′-deoxyuridine 5′-monophosphate (**9.133**) is **9.130** (Scheme 9.33). Bioreduction of **9.130** gives **9.131**, which undergoes elimination to **9.132**, followed by spontaneous cyclization, elimination, and hydrolysis to **9.133**. Compound **9.130** gave excellent inhibition of cell proliferation and thymidylate synthase inhibition (see Chapter 5, Section 5.3.3.3.5) via intracellular conversion to **9.133**.

9.2.2.7. Nucleotide Activation

The antineoplastic agent 6-mercaptopurine (**9.134**; Purinethol) produces a 50% remission rate for acute childhood leukemias. Although **9.134** inhibits several enzyme systems, these inhibitions are irrelevant to its anticancer activity. Only tumors that convert the drug to its nucleotide are affected. 6-Mercaptopurine is activated

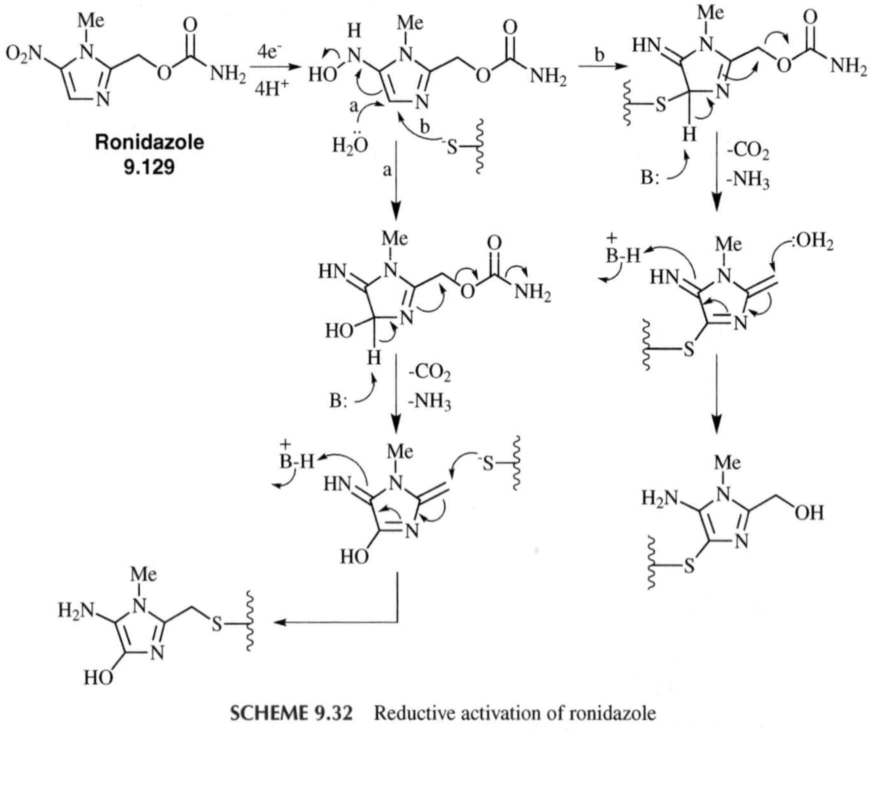

Ronidazole
9.129

SCHEME 9.32 Reductive activation of ronidazole

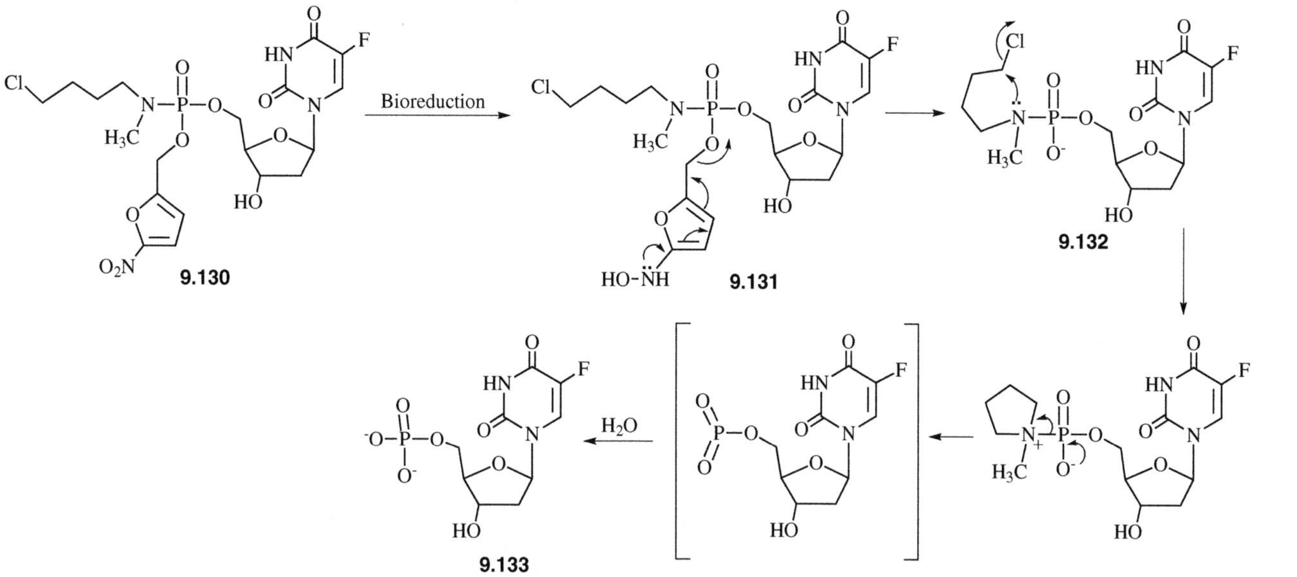

SCHEME 9.33 Phosphoramidite-based prodrug for intracellular delivery of nucleotides

by a hypoxanthine-guanine phosphoribosyltransferase-catalyzed reaction with 5-phosphoribosylpyrophosphate (Scheme 9.34). The nucleotide (**9.135**) inhibits several enzymes in the purine nucleotide biosynthetic pathway, but the most prominent site is one of the early enzymes in the de novo pathway, namely, phosphoribosylpyrophosphate amidotransferase, which catalyzes the conversion of phosphoribosylpyrophosphate to phosphoribosylamine.[163] 5-Fluorouracil (see Chapter 5, Section 5.3.3.3.5) is similar to 6-mercaptopurine in the sense that it must first be converted to the corresponding deoxyribonucleotide for it to be active.

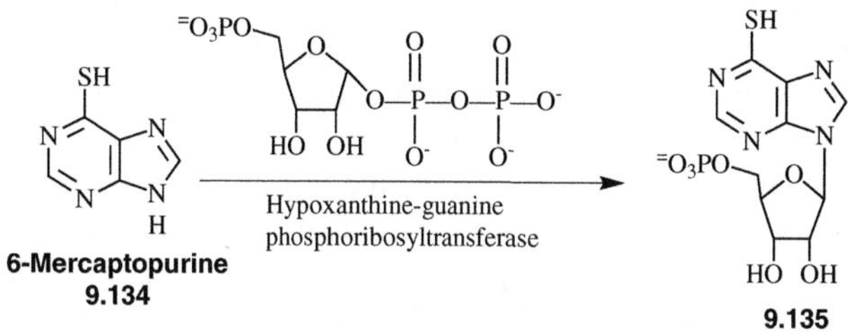

SCHEME 9.34 Nucleotide activation of 6-mercaptopurine

9.2.2.8. Phosphorylation Activation

The antiviral drug acyclovir (**9.136**, R=H, Zovirax), one of the drugs for which Gertrude Elion and George Hitchings received the Nobel Prize in 1988, is highly effective against genital herpes simplex virus and varicella zoster virus infections.[164] Its structure can be drawn so that it closely resembles the structure of 2′-deoxyguanosine (**9.137**), the nucleoside that is metabolically converted into 2′-deoxyguanosine triphosphate and is incorporated into the viral DNA. Acyclovir itself is inactive, but it is selectively phosphorylated by a viral thymidine kinase to the corresponding monophosphate (**9.136**, $R=PO_3^{=}$).[165] Uninfected cells do not phosphorylate acyclovir, and this accounts for the selective toxicity of acyclovir toward viral cells. The second step in the activation of acyclovir is the conversion of the monophosphate (**9.136**, $R=PO_3^{=}$) to the diphosphate (**9.136**, $R=P_2O_6^{3-}$), catalyzed by guanylate kinase.[166] The final activation step is the conversion of the diphosphate to the triphosphate (**9.136**, $R=P_3O_9^{4-}$), which could be accomplished by a variety of enzymes, particularly phosphoglycerate kinase.[167] Further selective toxicity is derived from the fact that acyclovir triphosphate is selectively taken up by viral α-DNA polymerases because its structure resembles that of the essential DNA precursor, 2′-deoxyguanosine triphosphate. The K_i for viral α-DNA polymerase is up to 40 times lower than that for normal cellular α-DNA polymerase.[168] Acyclovir triphosphate is a substrate for the viral α-DNA polymerase but not for the normal cellular α-DNA polymerase; however, incorporation of acyclovir triphosphate into the viral DNA leads to the formation of a *dead-end complex* (an enzyme–substrate complex that is no longer active) after the next deoxynucleotide triphosphate unit is incorporated.[169] This disrupts the replication cycle of the virus and destroys it. Even if the phosphorylated acyclovir were released from the virus cell, it would be too polar to be taken up by normal cells, and, as indicated above, the triphosphate is a poor substrate for normal human α-DNA polymerase anyway. Therefore, this drug exhibits a high degree of selective toxicity against viral cells.

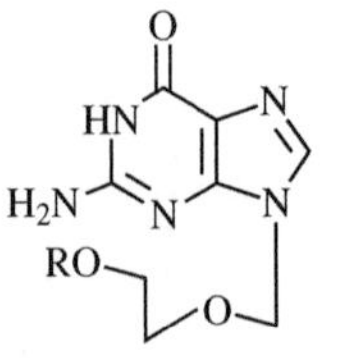

Acyclovir (R = H)
9.136

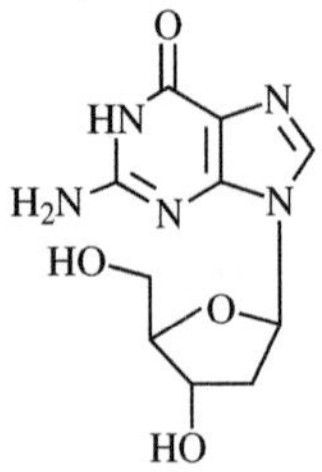

9.137

Valaciclovir
9.140

As might be predicted from knowledge of the mechanism of acyclovir, acquired resistance to the drug can occur by three different mechanisms. Because of the importance of the thymidine kinase to the activation of acyclovir, resistance arises from a deletion of this enzyme or a change in its substrate specificity.[170] The third mechanism is an altered viral α-DNA polymerase.[171] The degree of inhibition by acyclovir triphosphate of several different α-DNA polymerase mutants encoded by drug-resistant viruses correlated with the degree of resistance conferred by the mutation in vivo.[172] In some cases, the enzyme mutation resulted in a decrease in binding of acyclovir triphosphate (higher K_m), and in others, the mutation caused a reduction in catalytic activity for incorporation of acyclovir triphosphate into DNA (lower k_{cat}).

The largest shortcoming to the use of acyclovir is the fact that only 15–20% of acyclovir is absorbed after oral administration. Consequently, prodrugs for the prodrug acyclovir have been designed to improve gastrointestinal absorption and to protect acyclovir against biotransformations to inactive metabolites. 2,6-Diamino-9-(2-hydroxyethoxymethyl)purine (**9.138**) is converted to acyclovir by the enzyme adenosine deaminase[173] (catalyzes the hydrolysis of adenosine to inosine) and 6-amino-9-(2-hydroxyethoxymethyl)purine (**9.139**, 6-deoxyacyclovir) is oxidized to acyclovir by xanthine oxidase.[174] The latter compound is 18 times more water soluble than acyclovir. In humans urinary excretion of acyclovir is 5–6 times greater when **9.139** is given than an equivalent dose of acyclovir. Valacyclovir (**9.140**; Valtrex), the L-valyl ester of acyclovir, is a bipartite carrier-linked prodrug of acyclovir that has a three- to fivefold higher oral bioavailability while retaining the excellent safety profile.[175] The enzyme that catalyzes the hydrolysis of valacyclovir has been isolated and characterized.[176]

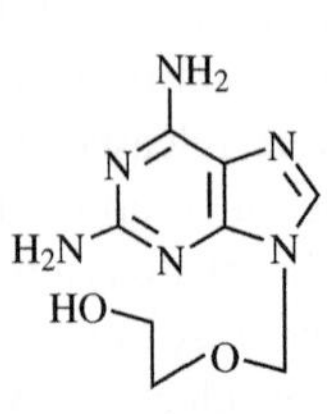

9.138

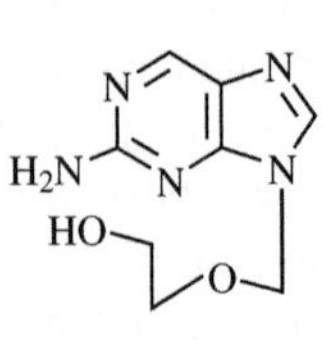

9.139

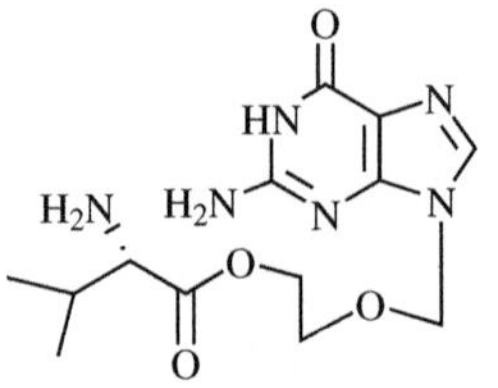

Ganciclovir (**9.141**; one trade name is Cytovene)[177] is an analog of acyclovir that has a structure and conformation resembling those of 2′-deoxyguanosine even closer than does acyclovir. This compound is about as potent as acyclovir against herpes simplex viruses and varicella zoster virus, but is much more inhibitory than acyclovir against human cytomegalovirus,[178] an important pathogen in immunocompromised and acquired immune deficiency syndrome patients.

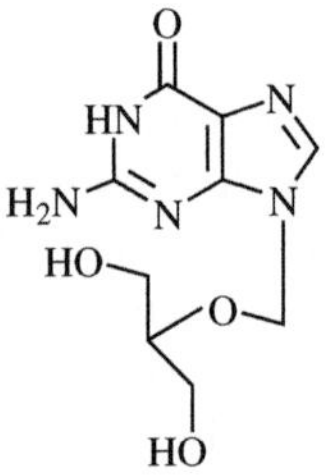

Ganciclovir
9.141

The carbon isostere of ganciclovir, penciclovir (**9.142**; Denavir), is more potent and longer acting than acyclovir.[179] However, it is still poorly absorbed when given orally.[180] To increase absorption, a prodrug of the prodrug was designed called famciclovir (**9.143**; Famvir),[181] the diacetyl and 6-deoxy analog of penciclovir, which is used orally.[182] Greater than 75% absorption is attained with rapid conversion to penciclovir.[183] Metabolism studies showed that the diacetyl groups are hydrolytically cleaved prior to oxidation of the purine.[184] The *pro-S* acetoxyl group is hydrolyzed before the *pro-R* acetoxyl group.[185]

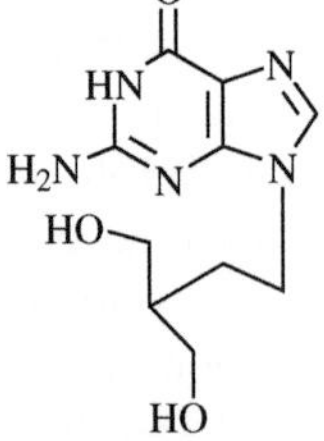

Penciclovir
9.142

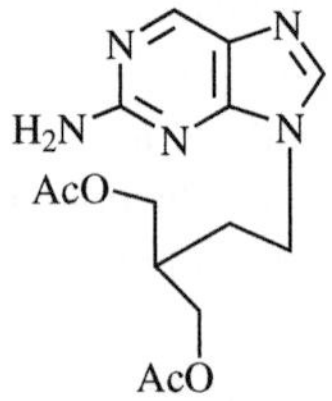

Famciclovir
9.143

9.2.2.9. Sulfation Activation

The hypotensive[186] and hair growth[187] activities of minoxidil (**9.144**, R=$^-$, Rogaine) require a sulfotransferase-catalyzed sulfation to minoxidil sulfate (**9.144**, R=SO_3^-).[188] Because minoxidil sulfate is more potent than minoxidil, and inhibitors of the sulfotransferase inhibit minoxidil activity but not minoxidil sulfate activity, it is apparent that minoxidil is a prodrug for minoxidil sulfate. As described in Chapter 8 (Section 8.4.3.3), sulfation is a common mechanism in drug metabolism for the deactivation and excretion of drugs, so it is quite unusual for sulfation to be involved in prodrug activation.

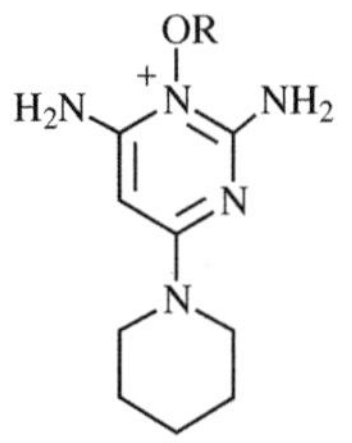

Minoxidil (R = $^-$)
Minoxidil sulfate (R = SO_3^-)
9.144

9.2.2.10. Decarboxylation Activation

The striatal tracts in the brain, which are important for the control of voluntary movements, contain a balance of the inhibitory neurotransmitter dopamine and the excitatory neurotransmitter acetylcholine. An imbalance in the dopaminergic and cholinergic components produces disorders of movement. In Parkinson's disease there is a marked deficiency in the dopaminergic component, which is attributed to the loss of dopaminergic neurons and a low concentration of dopamine in the *substantia nigra*. The obvious treatment for Parkinson's disease would be to give high doses of dopamine (**9.145**, R=H), but this does not work because dopamine does not cross the blood–brain barrier. However, there is an active transport system for L-amino acids; consequently, L-dopa (levodopa) (**9.145**, R=COOH) is transported into the brain where it is decarboxylated by the pyridoxal 5′-phosphate-dependent enzyme (see Chapter 4, Section 4.3.1.2) aromatic L-amino acid decarboxylase (also called dopa decarboxylase) to dopamine. Because the D,L-mixture produces unwanted side effects, L-dopa (levodopa; Sinemet is the combination of levodopa and carbidopa; see below) is used as a prodrug for dopamine. Unfortunately, because dopaminergic neurons cannot be rejuvenated, levodopa does not reverse the course of the disease; it merely treats the symptoms of the disease.[189]

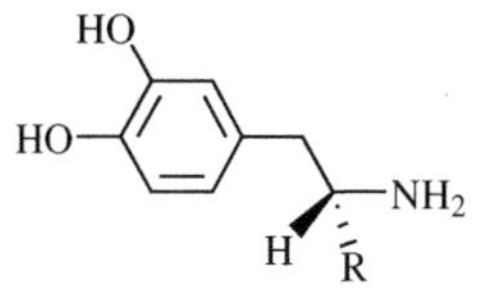

Levodopa (R = COOH)
9.145

As discussed in Chapter 5 (Section 5.3.3.3.4), dopamine is a substrate for monoamine oxidase B; consequently, as levodopa is being converted to dopamine in the brain, monoamine oxidase B is degrading the dopamine. Inactivators of monoamine oxidase B, selegiline (Eldepryl) and rasagiline (Azilect) are now used in combination with levodopa to minimize the degradation of the dopamine generated from levodopa.[190]

One major complication with the use of levodopa therapy arises from the fact that aromatic L-amino acid decarboxylase also exists in the periphery (outside of the central nervous system), and greater than 95% of the orally administered levodopa is decarboxylated in its first pass through the liver and kidneys. Possibly only 1% of the levodopa taken actually penetrates into the central nervous system. If the peripheral aromatic L-amino acid decarboxylase could be inhibited without inhibition of the same enzyme in the brain, the levodopa would be protected from the undesired metabolism. This, in fact, is possible because inhibitors of aromatic L-amino acid decarboxylase are charged molecules, and unless they are actively transported, they will not cross the blood–brain barrier. Carbidopa (**9.146**; Lodosyn) is used in the United States and benserazide (**9.147**; one trade name is Prolopa) is used in Europe and Canada in combination with levodopa for the treatment of Parkinson's disease.[191] With the combined use of a peripheral aromatic L-amino acid decarboxylase inhibitor, the optimal effective dose of levodopa can be reduced by greater than 75%.

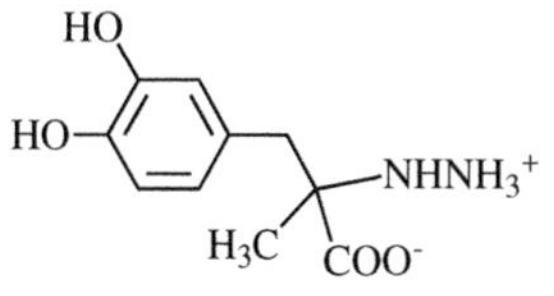

Carbidopa
9.146

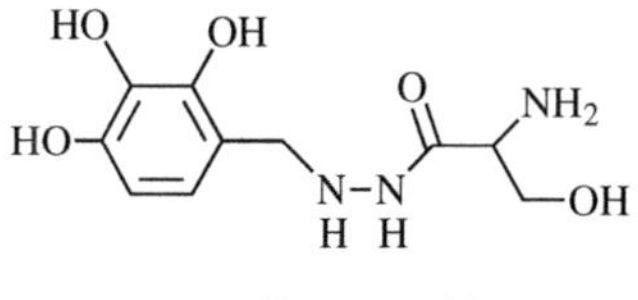

Benserazide
9.147

In Chapter 5 (Section 5.3.3.3.3), the application of inactivators of monoamine oxidase A (MAO A) as antidepressant agents was discussed. Although monoamine oxidase (MAO) inactivators are used in the treatment of depression, a severe cardiovascular side effect can result unless the diet is controlled to minimize the intake of tyramine-containing foods. This side effect results from the concurrent inactivation of the peripheral MAO A along with brain MAO A. A brain-specific MAO A inactivator would give the desired antidepressant effect without the undesirable cardiovascular effect. A prodrug approach for the brain-selective delivery of an MAO A-selective inactivator was developed at the former company Marion Merrell Dow (now Sanofi-Aventis).[192] This particular type of prodrug was termed a *dual enzyme-activated inhibitor* because the activating enzyme is, by design, part of the same metabolic pathway as the enzyme that is targeted for inhibition.[193] In this case, the activating enzyme is aromatic L-amino acid decarboxylase, and the target enzyme is MAO A. (*E*)-β-Fluoromethylene-*m*-tyramine (**9.148**, R=H) is a mechanism-based inactivator (see Chapter 5, Section 5.3.3) of MAO with selectivity for MAO A.[194] The corresponding amino acid, (*E*)-β-fluoromethylene-*m*-tyrosine (**9.148**, R=COOH) is not an inhibitor of MAO, but it is a good substrate for aromatic L-amino acid decarboxylase, which converts **9.148** (R=COOH) to **9.148** (R=H). The amino acid (**9.148**, R=COOH) is actively transported into the central nervous system and is concentrated in the synaptosomes. Because brain aromatic L-amino acid decarboxylase is located predominantly in monoamine nerve endings, **9.148** (R=COOH) is decarboxylated to **9.148** (R=H) at the desired site of action. To prevent inactivation of peripheral MAO A, **9.148** (R=COOH) is administered with carbidopa, which blocks peripheral L-aromatic amino acid decarboxylase-catalyzed decarboxylation of **9.148** (R=COOH). This results in brain-selective MAO A inactivation with little or no peripheral MAO A inhibition and only a minimal tyramine effect.

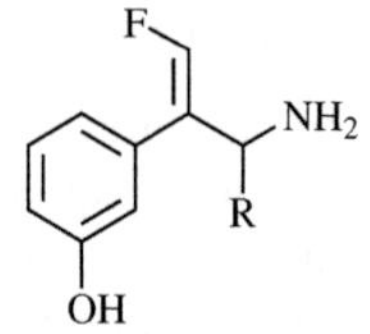

F NH_2 R OH

9.148

Dopamine is not only a major inhibitory neurotransmitter but also plays an important role in the kidneys. Dopamine increases systolic and pulse blood pressure and renal blood flow. If it is desired to have selective delivery of dopamine to the kidneys to attain renal vasodilation without a blood pressure effect, a prodrug for dopamine can be used. There is a high concentration of L-γ-glutamyltranspeptidase, the enzyme that catalyzes the transfer of the L-glutamyl group from the *N*-terminus of one peptide to another, in kidney cells. Consequently, an L-γ-glutamyl derivative of an amino acid or amine drug could be cleaved selectively in the kidneys.[195] L-γ-Glutamyl-L-dopa (**9.149**) is selectively accumulated in the kidneys, and the L-dopa released by L-glutamyltranspeptidase is decarboxylated to dopamine by aromatic L-amino acid decarboxylase, which also is abundant in kidneys (Scheme 9.35).[196] Even at high.concentrations of this compound little central nervous system effect is apparent. This, then, is an example of a site-selective carrier-linked prodrug of a bioprecursor prodrug for dopamine.

Drug design is typically initiated with approaches to maximize the pharmacodynamic properties of molecules (increased binding to a receptor). A compound may be found that has the desired in vitro properties, but has unfavorable in vivo properties. It should be apparent, then, from the discussion in this chapter, that it may be possible to alter the structure of the compound to improve its pharmacokinetic properties and, thereby, transform it into a promising drug candidate.

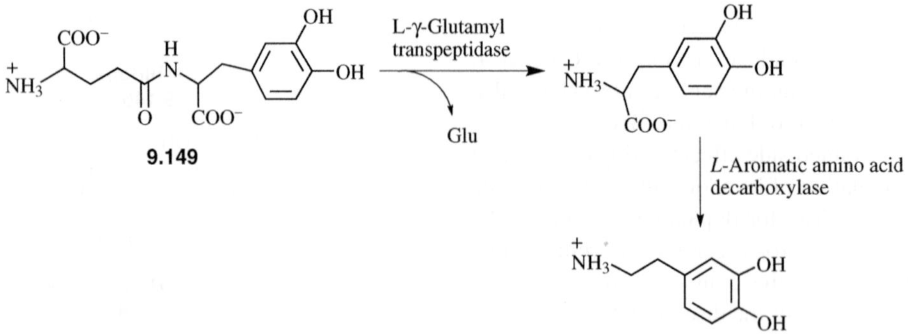

COO^- $\overset{+}{N}H_3$ H N O COO^- OH OH

9.149

L-γ-Glutamyl transpeptidase

Glu

$\overset{+}{N}H_3$ OH OH COO^-

L-Aromatic amino acid decarboxylase

$\overset{+}{N}H_3$ OH OH

SCHEME 9.35 Metabolic activation of L-γ-glutamyl-L-dopa to dopamine

9.3. GENERAL REFERENCES

Prodrugs

Hsieh-P.-W.; Hung, C.-F.; Fang, J.-Y. Current prodrug design for drug discovery. *Curr. Pharm. Des.* **2009**, *15*, 2236–2250.

Huttunen, K. M.; Rautio, J. Prodrugs-an efficient way to breach delivery and targeting barriers. *Curr. Top. Med. Chem.* **2011**, *11*(18), 2265–2287.

Huttunen, K. M.; Raunio, H.; Rautio, J. Prodrugs-from serendipity to rational design. *Pharmacol. Rev.* **2011**, *63*(3), 750–771.

Pavan, B.; Dalpiaz, A.; Ciliberti, N.; Biondi, C.; Manfredini, S.; Vertuani, S. Progress in drug delivery to the central nervous system by the prodrug approach. *Molecules* **2008**, *13*, 1035–1065.

Rautio, J.; Kumpulainen, H.; Heimbach, T.; Oliyai, R.; Oh, D.; Järvinen, T.; Savolainen, J. Prodrugs: design and clinical applications. *Nature Rev.* **2008**, *7*, 255–270.

Rautio, J.; Laine, K.; Gynther, M.; Savolainen, J. Prodrug approaches for CNS delivery. *AAPS J.* **2008**, *10*(1), 92–102.

Stella, V. J.; Borchardt, R.; Hagerman, M.; Oliyai, R.; Maag, H.; Tilley, J., Eds. *Prodrugs: Challenges and Rewards*; Springer/AAPS Press: New York, NY, 2007.

Stella, V. J.; Nti-Addae, K. W. Prodrug strategies to overcome poor water solubility. *Adv. Drug Deliv. Rev.* **2007**, *59*(7), 677–694.

Stella, V. J. Prodrugs as therapeutics. *Exp. Opin. Ther. Pat.* **2004**, *14*(3), 277–280.

Testa, B.; Mayer, J. M. *Hydrolysis in Drug and Prodrug Metabolism*; Wiley-VCH: Weinheim, Germany, 2003.

Macromolecular Drug Carrier Systems

Luo, Y.; Prestwich, G. D. Cancer-targeted polymeric drugs. *Curr. Cancer Drug Targets* **2002**, *2*, 209–226.

Narang, A. S.; Mahato, R. I., Eds. *Targeted Delivery of Small and Macromolecular Drugs*; CRC Press: Boca Raton, FL, 2010.

Pathak, Y.; Benita, S. *Antibody-Mediated Drug Delivery Systems: Concepts, Technology, and Applications*; Wiley: Hoboken, NJ, 2012.

Ranade, V. V.; Cannon, J. B. *Drug Delivery Systems*, 3rd edition; CRC Press: Boca Raton, FL, 2011.

Takakura, Y.; Hashida, M. Macromolecular carrier systems for targeted drug delivery: pharmacokinetic considerations on biodistribution. *Pharm. Res.* **1996**, *13*, 820–831.

Vyas, S. P.; Khar, R. K. *Targeted & Controlled Drug Delivery. Novel Carrier Systems*; CBS Publishers, 2006.

9.4. PROBLEMS (ANSWERS CAN BE FOUND IN THE APPENDIX AT THE END OF THE BOOK)

1. The antiinflammatory agent fluocinolone (**1**) is too hydrophilic for topical application. Suggest a prodrug (other than one used for a steroid in the text).

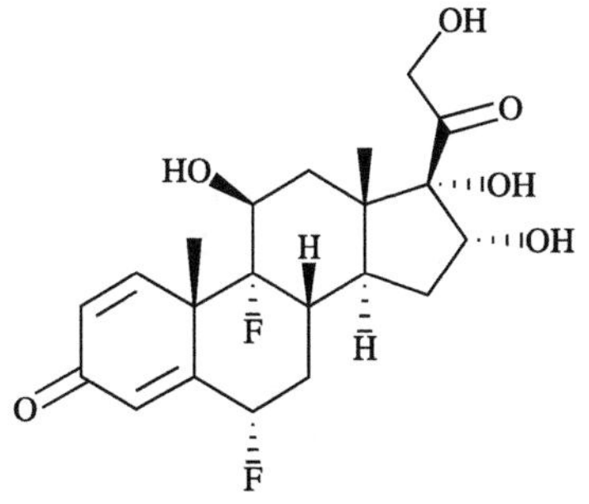

1

2. The cornea has significant esterase activity. Epinephrine (**2**) is an antiglaucoma agent that does not penetrate the cornea well. Suggest a prodrug (other than the one used for epinephrine in the text).

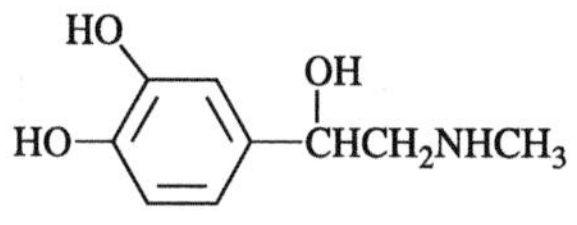

2

3. Draw a mechanism for the activation of the prodrug fostamatinib (**9.6**).
4. You have discovered a highly potent new drug (**3**) to treat West Nile virus, but it is soluble only to the extent of 50 mg/liter, and it must be administered by injection. The prescribed dose is 400 mg. Design two prodrugs (one at each functional group) to get around this problem.

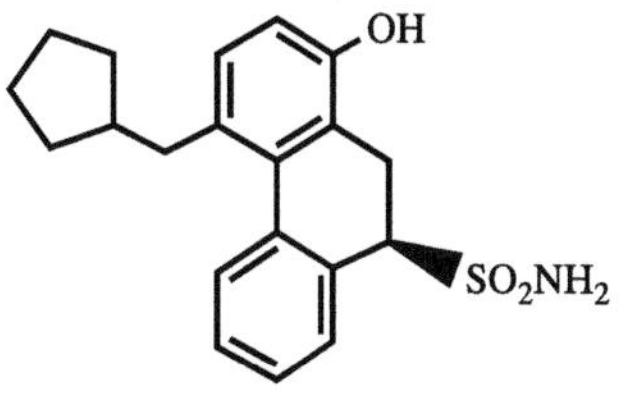

3

5. Resistance develops to your company's new antiviral drug (**4**) as a result of the encoding of a new viral urease that catalyzes the hydrolysis of the urea in your drug. As senior group leader you need to devise a plan to rectify this catastrophe. Briefly describe two strategies that involve different prodrug approaches.

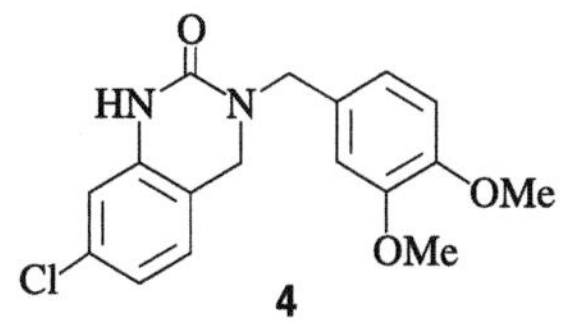

4

6. Design a prodrug to get **5** into the brain using a strategy similar to (but not the same as) the one by Bodor (see Scheme 9.12 in the text). In your prodrug strategy the prodrug activation should require monoamine oxidase for activation (see Chapter 5, Section 5.3.3.3.4 regarding MPTP activation).

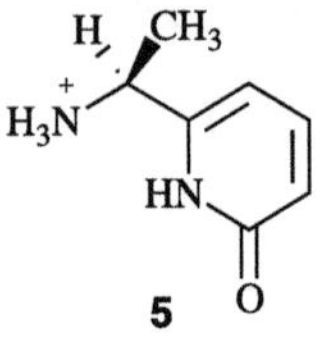

5

7. Design a copolymer-linked prodrug of ampicillin using a copolymer of poly(vinyl alcohol) and poly(vinyl amine).

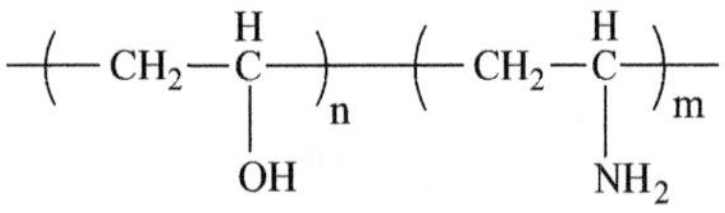

8. Compound **6** is a type of a tripartite antitumor prodrug that is activated by the enzyme β-glucuronidase. β-Glucuronidase catalyzes the hydrolysis of β-glucuronic acid acetals (**7**) to glucuronic acid (**8**). Give a reasonable mechanism for the activation of **6.**
9. Draw a mutual prodrug for antitumor agents having different mechanisms of action.
10. Design bioprecursor prodrugs for (A) acetaminophen, (B) cimetidine, and (C) captopril based on whatever hypothetical pathway you want. Show the enzymes involved in the conversions of your prodrugs to the drugs.
11. An antibody was raised against a tumor cell line and was conjugated to a β-lactamase. A nitrogen mustard was conjugated to a cephalosporin (**9**) for use in ADEPT. Draw a mechanism for the activation of the prodrug by the ADEPT conjugate.

(A) Acetaminophen

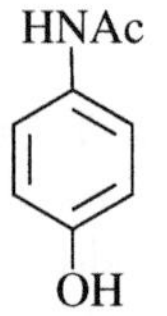

(B) Cimetidine

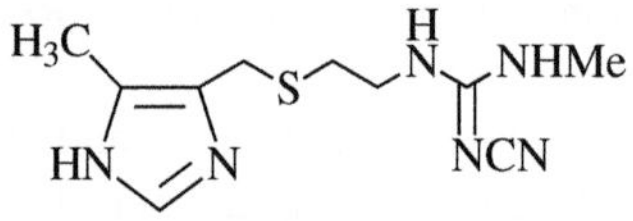

(C) Captopril

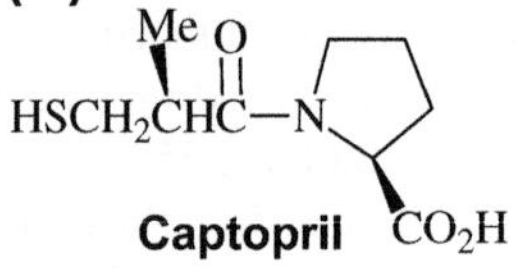

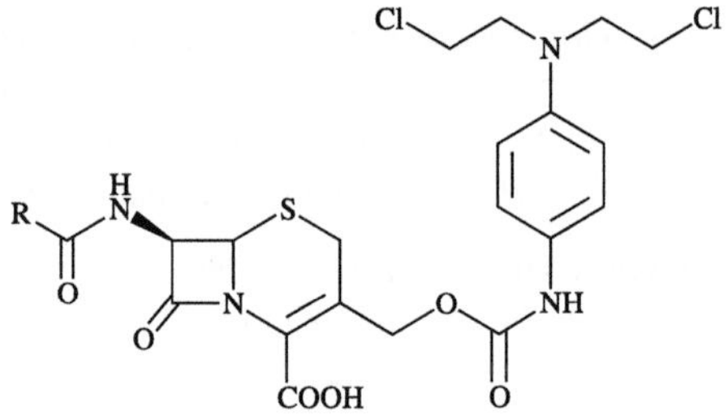

9

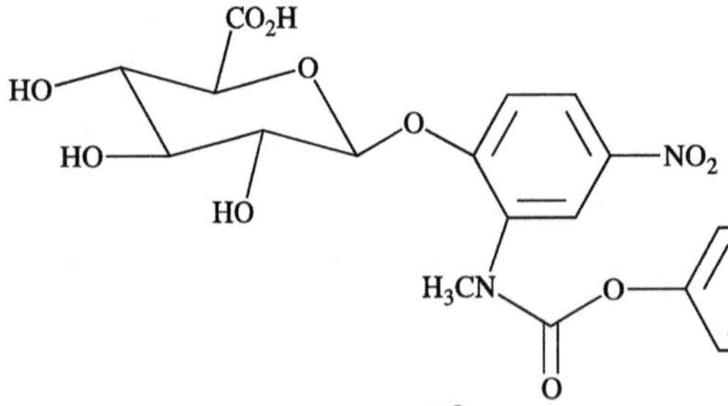

6

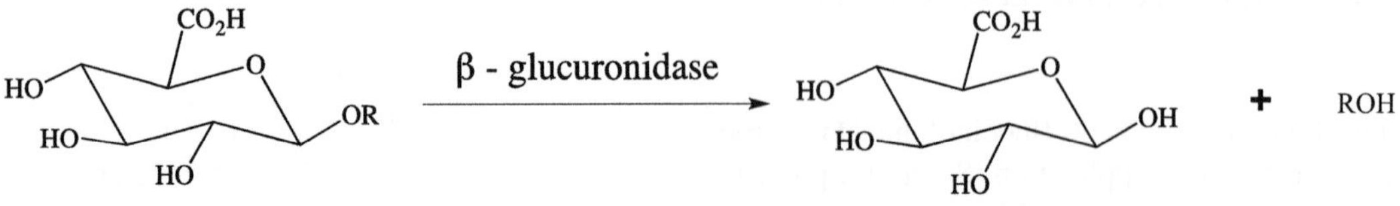

β - glucuronidase

7 **8**

12. Draw an activation mechanism for release of two drug molecules from tripartite prodrug **10** to produce the metabolites shown.

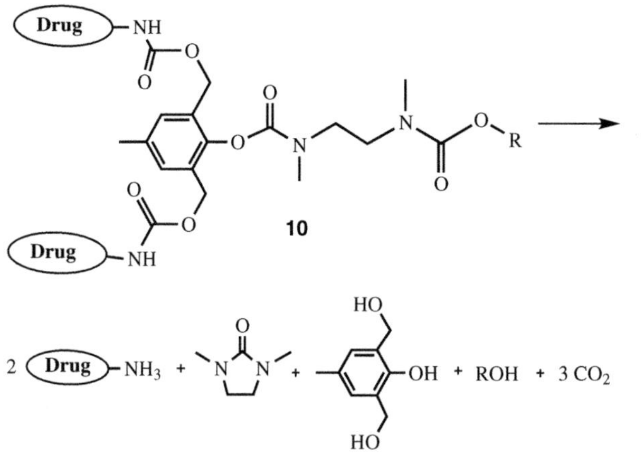

13. Draw a mechanism for the release of all-*trans*-retinoic acid (**11**) and antitumor histone deacetylase inhibitor **12** from tripartite mutual prodrug **13.**

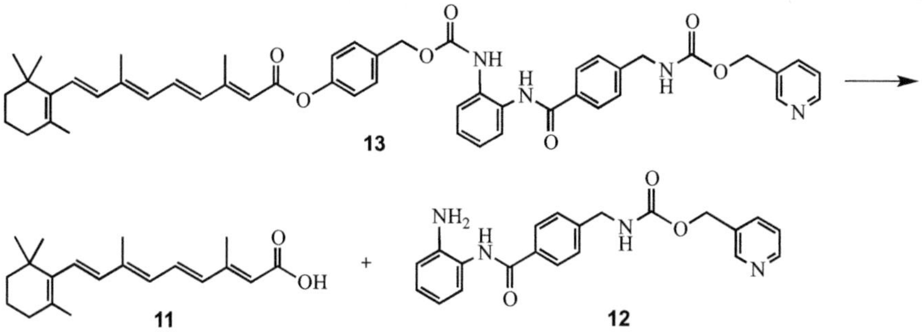

14. In Scheme 9.28, clopidogrel (**9.109**) is oxidized to **9.110.** Draw a reasonable mechanism for that metabolic conversion.

15. Prodrug **14** is converted into a duocarmycin-type antitumor agent (**15**) only under the hypoxic reducing environment of a tumor cell. Draw a mechanism for the prodrug conversion.

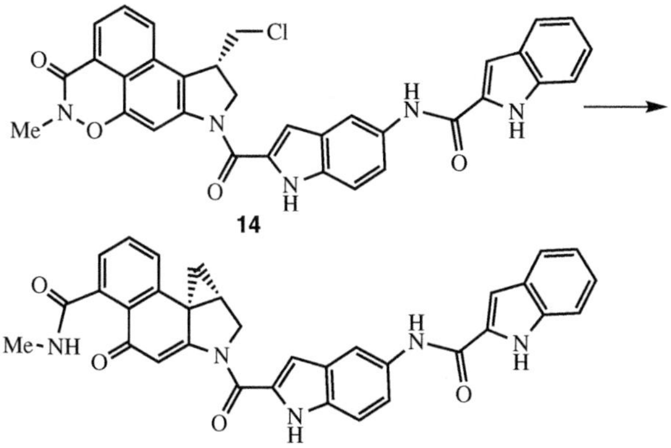

REFERENCES

1. (a) Albert, A. *Selective Toxicity*, Chapman and Hall, London, 1951. (b)Albert, A. Chemical aspects of selective toxicity. *Nature (London)* **1958**, *182*, 421–423.
2. Sloan, K. B.; Wasdo, S. Designing for topical delivery: prodrugs can make the difference. *Med. Res. Rev.* **2003**, *23*, 763–793.
3. Pavan, B.; Dalpiaz, A.; Ciliberti, N.; Biondi, C.; Manfredini, S.; Vertuani, S. Progress in drug delivery to the central nervous system by the prodrug approach. *Molecules* **2008**, *13*, 1035–1065.
4. Overington, J. P.; Al-Lazikani, B.; Hopkins, A. L. How many drug targets are there? *Nat. Rev. Drug Discov.* **2006**, *5*, 993–996.
5. Testa, B.; Caldwell, J. Prodrugs revisited: the "ad hoc" approach as a complement to ligand design. *Med. Res. Rev.* **1996**, *16*, 233–241.
6. Caldwell, J. The role of drug metabolism in drug discovery and development: opportunities to enhance time- and cost-efficiency. *Pharm. Sci.* **1996**, *2*, 117–119.
7. Harper, N. J. Drug latentiation. *J. Med. Pharm. Chem.* **1959**, *1*, 467–500.
8. Wermuth, C. G. In *Drug Metabolism and Drug Design: Quo Vadis?* Briot, M.; Cautreels, W.; Roncucci, R. (Eds.), Sanofi-Clin-Midy, Montpellier, 1983, p. 253.
9. Wuts, G. M.; Greene, T. W. *Greene's Protective Groups in Organic Synthesis*. 4th ed., Wiley, Hoboken, NJ, 2007.
10. Long, L., Jr. The ozonization reaction. *Chem. Rev.* **1940**, *27*, 437–493.
11. Ettmayer, P.; Amidon, G. L.; Clement, B.; Testa, B. Lessons learned from marketed and investigational prodrugs. *J. Med. Chem.* **2004**, *47*, 2393–2404.
12. Beaumont, K.; Webster, R.; Gardner, I.; Dack, K. Design of ester prodrugs to enhance oral absorption of poorly permeable compounds: challenges to the discovery scientist. *Curr. Drug Metab.* **2003**, *4*, 461–485.
13. (a) Potter, P. M.; Wadkins, R. M. Carboxylesterases-detoxifying enzymes and targets for drug therapy. *Curr. Med. Chem.* **2006**, *13*, 1045–1054. (b) Ferriz, J.M.; Vinsova, J. Prodrug design of phenolic drugs. *Curr. Pharm. Des.* **2010**, *16*, 2033–2052. (c) Guarino, V. R.; Stella, V. J. Prodrugs of amides, imides and other NH-acidic compounds. *Biotechnol. Pharm. Aspects* **2007**, *5* (Pt. 2, Prodrugs: Challenges and Rewards, Part 2), 133–187.
14. Bundgaard, H. In *Design of Prodrugs*, Bundgaard, H. (Ed.), Elsevier: Amsterdam, 1985, p. 1.
15. Cai, Z.-w.; Zhang, Y.; Borzilleri, R. M.; Quian, L.; Barbosa, S.; Wei, D.; Zheng, X.; Wu, L.; Fan, J.; Shi, Z.; et al. Discovery of brivanib alaninate ((S)-R-1-1(4-(4-fluoro-2-methyl-1H-indol-5-yloxy)-5-methylpyrrolo[2,1-f][1,2,4]triazin-6-yloxy)propan-2-yl)2-aminopropanoate), a novel prodrug of dual vascular endothelial growth factor receptor-2 and fibroblast growth factor receptor-1 kinase inhibitor (BMS-540215). *J. Med. Chem.* **2008**, *51*, 1976–1980.
16. Reynolds, W. F. Polar substituent effects. *Prog. Phys. Org. Chem.* **1983**, *14*, 165–203.

17. Robinson, R. P.; Reiter, L. A.; Barth, W. E., Campeta, A. M.; Cooper, K.; Cronin, B. J.; Destito, R.; Donahue, K. M.; Falkner, F. C.; Fiese, E. F.; Johnson, D. L.; Kuperman, A. V.; Liston, T. E.; Malloy, D.; Martin, J. J.; Mitchell, D. Y.; Rusek, F. W.; Shamblin, S. L.; Wright, C. F. Discovery of the hemifumarate and (alpha-L-alanyloxy)methyl ether as prodrugs of an antirheumatic oxindole: prodrugs for the enolic OH group. *J. Med. Chem.* **1996**, *39*, 10–18.
18. Hecker, S. J.; Erion, M. D. Prodrugs of phosphates and phosphonates. *J. Med. Chem.* **2008**, *51*, 2328–2345.
19. Starrett, J. E., Jr.; Synthesis, oral bioavailability determination, and in vitro evaluation of prodrugs of the antiviral agent 9-[2-(phosphonomethoxy)ethyl]adenine (PMEA). *J. Med. Chem.* **1994**, *37*, 1857–1864.
20. Rautio, J.; Kumpilainen, H.; Heimbach, T.; Oliyzi, R.; Oh, D.; Järvinen, T.; Savolainen, J. Prodrugs: design and clinical applications. *Nat. Rev. Drug Discov.* **2008**, *7*, 255–270.
21. Sweeny, D. J.; Li, W.; Clough, J.; Bhamidipati, S.; Singh, R.; Park, G.; Baluom, M.; Grossbard, E.; Lau, D. T.-W. Metabolism of fostamatinib, the oral methylene phosphate prodrug of the spleen tyrosine kinase inhibitor R406 in humans: contribution of hepatic and gut bacterial processes to the overall biotransformation. *Drug Metab. Dispos.* **2010**, *38*, 1166–1176.
22. Bundgaard, H. In *Bioreversible Carriers in Drug Design*, Roche, E. B. (Ed.), Pergamon Press, New York, 1987, p. 13.
23. Johansen, M.; Bundgaard, H. Pro-drugs as drug delivery systems. XXIV. N-Mannich bases as bioreversible lipophilic transport forms for ephedrine, phenethylamine and other amines. *Arch. Pharm. Chem. Sci. Ed.* **1982**, *10*, 111–121.
24. (a) Johansen, M.; Bundgaard, H. Pro-drugs as drug delivery systems. XIII. Kinetics of decomposition of N-Mannich bases of salicylamide and assessment of their suitability as possible pro-drugs for amines. *Int. J. Pharm.* **1980**, *7*, 119–127. (b) Bundgaard, H.; Johansen, M. Prodrugs as drug delivery systems. XIX. Bioreversible derivatization of aromatic amines by formation of N-Mannich bases with succinimide. *Int. J. Pharm.* **1981**, *8*, 183–192.
25. Bergmann, K. Progabide: a new GABA-mimetic agent in clinical use. *J. Clin. Neuropharmacol.* **1985**, *8*, 13–26.
26. Kaplan, J.-P.; Raizon, B. M.; Desarmenien, M.; Feltz, P.; Headley, P. M.; Worms, P.; Lloyd, K. G.; Bartholini, G. New anticonvulsants: Schiff bases of γ-aminobutyric acid and γ-aminobutyramide. *J. Med. Chem.* **1980**, *23*, 702–704.
27. (a) Ma, T. K.; Yan, B. P.; Lam, Y.-Y. Dabigatran etexilate versus warfarin as the oral anticoagulant of choice? A review of clinical data. *Pharmacol. Ther.* **2011**, *129*, 185–194. (b) Mehta, H. R.; Patel, P. B.; Galani, V. J.; Patel, K. B. Dabigatran etexilate: new direct thrombin inhibitors anticoagulants. *Int. Res. J. Pharm.* **2011**, *2*, 50–55.
28. Vaupel, P.; Kallinowski, F.; Okunieff, P. Blood flow, oxygen and nutrient supply, and metabolic microenvironment of human tumors: a review. *Cancer Res.* **1989**, *49*, 6449–6465.
29. Parveen, E.; Naughton, D. P.; Whish, W. J. D.; Threadgill, M. D. 2-Nitroimidazol-5-ylmethyl as a potential bioreductively activated prodrug system: reductively triggered release of the PARP inhibitor 5-bromoisoquinolinone. *Bioorg. Med. Chem. Lett.* **1999**, *9*, 2031–2036.
30. Talley, J. J.; Bertenshaw, S. R.; Brown, D. L.; Carter, J. S.; Graneto, M. J.; Kellogg, M. S.; Koboldt, C. M.; Yuan, J.; Zhang, Y. Y.; Seibert, K. N-[[(5-Methyl-3-phenylisoxazol-4-yl)-phenyl]sulfonyl]propanamide, sodium salt, parecoxib sodium: a potent and selective inhibitor of COX-2 for parenteral administration. *J. Med. Chem.* **2000**, *43*, 1661–1663.
31. (a) Anderson, B. D.; Conradi, R. A.; Knuth, K. E. Strategies in the design of solution-stable, water-soluble prodrugs. I: A physical-organic approach to pro-moiety selection for 21-esters of corticosteroids. *J. Pharm. Sci.* **1985**, *74*, 365–374. (b) Anderson, B. D.; Conradi, R. A.; Knuth, K. E.; Nail, S. L. Strategies in the design of solution-stable, water-soluble prodrugs. II: Properties of micellar prodrugs of methylprednisolone. *J. Pharm. Sci.* **1985**, *74*, 375–381.
32. Anderson, B. D.; Conradi, R. A.; Spilman, C. H.; Forbes, A. D. Strategies in the design of solution-stable, water-soluble prodrugs. III: Influence of the pro-moiety on the bioconversion of 21-esters of corticosteroids. *J. Pharm. Sci.* **1985**, *74*, 382–387.
33. Masini, E.; Planchenault, J.; Pezziardi, F.; Gautier, P.; Gangnol, J. P. Histamine-releasing properties of Polysorbate 80 in vitro and in vivo: correlation with its hypotensive action in the dog. *Agents Actions* **1985**, *16*, 470–477.
34. Saulnier, M. G.; Langley, D. R.; Kadow, J. F.; Senter, S. D.; Knipe, J. O.; Tun, M. M.; Vyas, D. M.; Doyle, T. W. Synthesis of etoposide phosphate, BMY-40481: a water-soluble clinically active prodrug of etoposide. *Bioorg. Med. Chem. Lett.* **1994**, *4*, 2567–2572.
35. Slojkowska, Z.; Krakuska, H. J.; Pachecka, J. Enzymic hydrolysis of amino acid derivatives of benzocaine. *Xenobiotica* **1982**, *12*, 359–364.
36. Hadgraft, J. In *Design of Prodrugs*, Bundgaard, H. (Ed.), Elsevier, Amsterdam, 1985, p. 271.
37. Mandell, A. I.; Stentz, F.; Kitabuchi, A. E. Dipivalyl epinephrine: a new pro-drug in the treatment of glaucoma. *Ophthalmology* **1978**, *85*, 268–275.
38. (a) Pardridge, W. M. *Brain Drug Targeting: The Future of Brain Drug Development*. Cambridge University Press, UK, 2001. (b) Begley, D. J.; Bradbury, M. W.; Kreuter, J. (Eds.), *The Blood-Brain Barrier and Drug Delivery to the CNS*, Mercel Dekker, NY, 2000.
39. Bodor, N.; Brewster, M. Problems of delivery of drugs to the brain. *Pharmacol. Ther.* **1983**, *19*, 337–386.
40. Worms, P.; Depoortere, H.; Durand, A.; Morselli, P. L.; Lloyd, K. G.; Bartholini, G. γ-Aminobutyric acid (GABA) receptor stimulation. I. Neuropharmacological profiles of progabide (SL 76002) and SL 75102, with emphasis on their anticonvulsant spectra. *J. Pharmacol. Exp. Ther.* **1982**, *220*, 660–671.
41. Jacob, J. N.; Hesse, G. W.; Shashoua, V. E. Synthesis, brain uptake, and pharmacological properties of a glyceryl lipid containing GABA and the GABA-T inhibitor γ-vinyl-GABA. *J. Med. Chem.* **1990**, *33*, 733–736.
42. Erion, M. D.; van Poelje, P. D.; MacKenna, D. A.; Colby, T. J.; Montag, A. C.; Fujitaki, J. M.; Linemeyer, D. L.; Bullough, D. A.; Liver-targeted drug delivery using HepDirect prodrugs. *J. Pharmacol. Exp. Ther.* **2005**, *312*, 554–560.
43. (a) Reddy, K. R.; Matelich, M. C.; Ugarkar, B. G.; Gomez-Galeno, J. E.; DaRe, J.; Ollis, K.; Sun, Z.; Craigo, W.; Colby, T. J.; Fujitaki, J. M.; Boyer, S. H.; van Poelje, P. D.; Erion, M. D.; Pradefovir: a prodrug that targets adefovir to the liver for the treatment of hepatitis B. *J. Med. Chem.* **2008**, *51*, 666–676. (b) Tillman, H. L. Drug evaluation: Pradefovir, a liver-targeted prodrug of adefovir against HBV infection. *Curr. Opin. Investig. Drugs* **2007**, *8*, 682–690.
44. (a) Harper, N. J. Drug latentiation. *J. Med. Pharm. Chem.* **1959**, *1*, 467–500. (b) Brandes, D.; Bourne, G. H. Stilbestrol phosphate and prostatic carcinoma. *Lancet* **1955**, *1*, 481–482.
45. (a) Niculescu-Duvaz, D.; Negoita-Giras, G.; Niculescu-Duvaz, I.; Hedley, D.; Springer, C. J. Directed enzyme prodrug therapies. *Meth. Prin. Med. Chem.* **2011**, *47* (Prodrugs and Targeted Delivery), 271–344. (b) Tietze, L. F.; Schmuck, K. Prodrugs for targeted tumor therapies: recent developments in ADEPT, GDEPT, and PMT. *Curr. Pharm. Des.* **2011**, *17*(32), 3527–3547. (c) Schellmann, N.; Deckert, P. M.; Bachran, D.; Fuchs, H.; Bachran, C. Targeted enzyme prodrug therapies. *Mini Rev. Med. Chem.* **2010**, *10*(10), 887–904. (d) Both, G. W. Recent progress in gene-directed enzyme prodrug therapy: an emerging cancer treatment. *Curr. Opin. Mol. Therap.* **2009**, *11*(4), 421–432.

46. Xu, G.; McLeod, H. L. Strategies for enzyme/prodrug cancer therAPY. *CLIN. CANCER RES.* **2001**, *7*, 3314–3324.
47. Dachs, G. U.; Hunt, M. A.; Syddall, S.; Singleton, D. C.; Patterson, A. V. Bystander or no bystander for gene directed enzyme prodrug therapy. *Molecules* **2009**, *14*(11), 4517–4545.
48. (a) Sharma, S. K.; Bagshawe, K. D.; Burke, P. J.; Boden, J. A.; Rogers, G. T.; Springer, C. J.; Melton, R. G.; Sherwood, R. F. Galactosylated antibodies and antibody-enzyme conjugates in antibody-directed enzyme prodrug therapy. *Cancer* **1994**, *73*, 1114–1120. (b) Rogers, G. T.; Burke, P. J.; Sharma, S. K.; Koodie, R.; Boden, J. A. Plasma clearance of an antibody-enzyme conjugate in ADEPT by monoclonal anti-enzyme: its effect on prodrug activation in vivo. *Br. J. Cancer* **1995**, *72*, 1357–1363.
49. (a) Vaswani, S. K.; Hamilton, R. G. Humanized antibodies as potential therapeutic drugs. *Ann. Allergy Asthma Immunol.* **1998**, *81*, 105–115. (b) Bernard-Marty, C.; Lebrun, F.; Awada, A.; Piccart, M. J. Monoclonal antibody-based targeted therapy in breast cancer: current status and future directions. *Drugs* **2006**, *66*(12), 1577–1591.
50. Springer, C. J.; Dowell, R.; Burke, P. J.; Hadley, E.; Davies, D. H.; Blakey, D. C.; Melton, R. G.; Niculescu-Duvaz, I. Optimization of alkylating agent prodrugs derived from phenol and aniline mustards: a new clinical candidate prodrug (ZD2767) for antibody-directed enzyme prodrug therapy. *J. Med. Chem.* **1995**, *38*, 5051–5065.
51. (a) Wentworth, P., Jr.; Janda, K. D. Catalytic antibodies: structure and function. *Cell Biochem. Biophys.* **2001**, *35*, 63–87. (b) Hilvert, D. Critical analysis of antibody catalysis. *Annu. Rev. Biochem.* **2000**, *69*, 751–793.
52. Abzyme is the generic term for a catalytic antibody.
53. (a) Wentworth, P.; Datta, A.; Blakey, D.; Boyle, T.; Partridge, L. J.; Blackburn, G. M. Toward antibody-directed "abzyme" prodrug therapy, ADAPT: carbamate prodrug activation by a catalytic antibody and its in vitro application to human tumor cell killing. *Proc. Natl. Acad. Sci. U.S.A.* **1996**, *93*, 799–803. (b) Bagshawe, K. D. Towards generating cytotoxic agents at cancer sites. *Br. J. Cancer* **1989**, *60*, 275–281.
54. (a) Barbas, C. F., III; Heine, A.; Zhong, G.; Hoffmann, T.; Gramatikova, S.; Björnstedt, R.; List, B.; Anderson, J.; Stura, E. A.; Wilson, I. A.; Lerner, R. A. Immune versus natural selection: antibody aldolases with enzymic rates but broader scope. *Science* **1997**, *278*, 2085–2092. (b) List, B.; Barbas, C. F. III; Lerner, R. A. Aldol sensors for the rapid generation of tunable fluorescence by antibody catalysis. *Proc. Natl. Acad. Sci. U.S.A.* **1998**, *95*, 15351–15355.
55. Shabat, D.; Rader, C.; List, B.; Lerner, R. A.; Barbas, C. F. III Multiple event activation of a generic prodrug trigger by antibody catalysis. *Proc. Natl. Acad. Sci. U.S.A.* **1999**, *96*, 6925–6930.
56. (a) Robinson, M. A.; Charlton, S. T.; Garnier, P.; Wang, X. T.; Davis, S. S.; Perkins, A. C.; Frier, M.; Duncan, R.; Savage, T. J.; Wyatt, D. A.; Watson, S. A.; Davis, B. G. LEAPT: lectin-directed enzyme-activated prodrug therapy. *Proc. Natl. Acad. Sci. U.S.A.* **2004**, *101*, 14527–14532. (b) Garnier, P.; Wang, X.-T.; Robinson, M. A.; van Kasteren, S.; Perkins, A. C.; Frier, M.; Fairbanks, A. J.; Davis, B. G. Lectin-directed enzyme activated prodrug therapy (LEAPT): synthesis and evaluation of rhamnose-capped prodrugs. *J. Drug Target* **2010**, *18*, 794–802.
57. (a) Singh, Y.; Palombo, M.; Sinko, P. J. Recent trends in targeted anticancer prodrug and conjugate design. *Curr. Med. Chem.* **2008**, *15*, 1802–1826. (b) Davis, B. G.; Robinson, M. A. Drug delivery systems based on sugar-macromolecule conjugates. *Curr. Opin. Drug Discov. Devel.* **2002**, *5*, 279–288.
58. (a) Both, G. W. Recent progress in gene-directed enzyme prodrug therapy: an emerging cancer treatment. *Curr. Opin. Mol. Therap.* **2009**, *11*(4), 421–432. (b) Niculescu-Duvaz, I.; Spooner, R. A.; Marais, R.; Springer, C. J. Gene-directed enzyme prodrug therapy. *Bioconjug. Chem.* **1998**, *9*, 4–22.
59. Denny, W. A. Nitroreductase-based GDEPT. *Curr. Pharm. Des.* **2002**, *8*, 1349–1361.
60. Hay, M. P.; Sykes, B. M.; Denny, W. A.; Wilson, W. R. A 2-nitroimidazole carbamate prodrug of 5-amino-1-(chloromethyl)-3-[(5,6,7-trimethoxyindol-2-yl)carbonyl]-1,2-dihydro-3H-benz[e]indole (amino-seco-CBI-TMI) for use with ADEPT and GDEPT. *Bioorg. Med. Chem. Lett.* **1999**, *9*, 2237–2242.
61. (a) Grove, J. I.; Searle, P. F.; Weedon, S. J.; Green, N. K.; McNeish, I. A.; Kerr, D. J. Virus-directed enzyme prodrug therapy using CB1954. *Anticancer Drug Des.* **1999**, *14*, 461–472. (b) Huber, B. E.; Richards, C. A.; Austin, E. A. VDEPT: an enzyme/prodrug gene therapy approach for the treatment of metastatic colorectal cancer. *Adv. Drug Deliv. Rev.* **1995**, *17*, 279–292. (c) Houston T. A. Painting the target around the arrow: two-step prodrug therapies from a carbohydrate chemist's perspective. *Curr. Drug Deliv.* **2007**, *4*(4), 264–8.
62. Weedon, S. J.; Green, N. K.; McNeish, I. A.; Gilligan, M. G.; Mautner, V.; Wrighton, C. J.; Mountain, A.; Young L. S.; Kerr, D. J.; Searle, P. F. Sensitization of human carcinoma cells to the prodrug CB1954 by adenovirus vector-mediated expression of *E. coli* nitroreductase. *Int. J. Cancer* **2000**, *86*, 848–854.
63. Rigg, A.; Sikora, K. Genetic prodrug activation therapy. *Mol. Med. Today* **1997**, *3*, 359–366.
64. Muller, P.; Jesnowski, R.; Karle, P.; Renz, R.; Saller, R.; Stein, H.; Puschel, K.; Rombs, K.; Nizze, H.; Liebe, S.; Wagner, T.; Gunzburg, W. H.; Salmons, B.; Lohr, M. Injection of encapsulated cells producing an ifosfamide-activating cytochrome P450 for targeted chemotherapy to pancreatic tumors. *Ann. N. Y. Acad. Sci.* **1999**, *880*, 337–351.
65. Panhda, H.; Martin, L. A.; Rigg, A.; Hurst, H. C.; Stamp, G. W. H.; Sikora, K.; Lemoine, N. R. Genetic prodrug activation therapy for breast cancer: a phase I clinical trial of erbB-2-directed suicide gene expression. *J. Clin. Oncol.* **1999**, *17*, 2180–2189.
66. Denny, W. A.; Wilson, W. R. The design of selectively-activated anti-cancer prodrugs for use in antibody-directed and gene-directed enzyme-prodrug therapies. *J. Pharm. Pharmacol.* **1998**, *50*, 387–394.
67. Garceau, Y.; Davis, I.; Hasegawa, J. Plasma propranolol levels in beagle dogs after administration of propranolol hemisuccinate ester. *J. Pharm. Sci.* **1978**, *67*, 1360–1363.
68. Hussain, M. A.; Koval, C. A.; Myers, M. J.; Shami, E. G.; Shefter, E. Improvement of the oral bioavailability of naltrexone in dogs: a prodrug approach. *J. Pharm. Sci.* **1987**, *76*, 356–358.
69. Filpula, D.; Zhao, H. Releasable PEGylation of proteins with customized linkers. *Adv. Drug Deliv. Rev.* **2008**, *60*(1), 29–49.
70. Deberdt, R.; Elens, P.; Berghmans, W.; Heykants, J.; Woestenborghs, R.; Driesens, F.; Reyntjens, A.; Van Wijngaarden, I. Intramuscular haloperidol decanoate for neuroleptic maintenance therapy. Efficacy, dosage schedule and plasma levels. An open multicenter study. *Acta Psychiat. Scand.* **1980**, *62*, 356–363.
71. Chouinard, G.; Annable, L.; Ross-Chouinard, A. Fluphenazine enanthate and fluphenazine decanoate in the treatment of schizophrenic outpatients: extrapyramidal symptoms and therapeutic effect. *Am. J. Psychiat.* **1982**, *139*, 312–318.
72. Persico, F. J.; Pritchard, J. F.; Fischer, M. C.; Yorgey, K.; Wong, S.; Carson, J. Effect of tolmetin glycinamide (McN-4366), a prodrug of tolmetin sodium, on adjuvant arthritis in the rat. *J. Pharmacol. Exp. Ther.* **1988**, *247*, 889–896.
73. Mandell, A. I.; Stentz, F.; Kitabuchi, A. E. Dipivalyl epinephrine: a new pro-drug in the treatment of glaucoma. *Ophthalmology* **1978**, *85*, 268–275.

74. Nielsen, N. M.; Bundgaard, H. Evaluation of glycolamide esters and various other esters of aspirin as true aspirin prodrugs. *J. Med. Chem.* **1989**, *32*, 727–734.
75. De Haan, R. M.; Metzler, C. M.; Schellenberg, D.; Vanderbosch, W. D. Pharmacokinetic studies of clindamycin phosphate. *J. Clin. Pharmacol.* **1973**, *13*, 190–209.
76. Sinkula, A. A.; Morozowich, W.; Rowe, E. L. Chemical modification of clindamycin. Synthesis and evaluation of selected esters. *J. Pharm. Sci.* **1973**, *62*, 1106–1111.
77. Notari, R. E. Pharmacokinetics and molecular modification. Implications in drug design and evaluation. *J. Pharm. Sci.* **1973**, *62*, 865–881.
78. (a) Onishi, H.; Machida, Y. In vitro and in vivo evaluation of microparticulate drug delivery systems composed of macromolecular prodrugs. *Molecules* **2008**, *13*(9), 2136–2155. (b) Tavora de Albuquerque S., Antonio; C., Man C.; Castro, L. F.; Guido, R. V. C.; Ferreira, E. I. Advances in prodrug design. *Mini Rev. Med. Chem.* **2005**, *5*(10), 893–914. (c) Kratz, F.; Mueller, I. A.; Ryppa, C.; Warnecke, A. Prodrug strategies in anticancer chemotherapy. *ChemMedChem* **2008**, *3*(1), 20–53.
79. de Duve, C.; de Barsy, T.; Poole, B.; Trouet, A.; Tulkens, P.; Van Hoof, F. Lysosomotropic agents. *Biochem. Pharmacol.* **1974**, *23*, 2495–2531.
80. (a) Wang, X.-W.; Xie, H. C-1027: Antineoplastic antibiotic. *Drugs Future* **1999**, *24*, 847–852. (b) Okuno, Y.; Iwashita, T.; Sugiura, Y. Structural basis for reaction mechanism and drug delivery system of chromoprotein antitumor antibiotic C-1027. *J. Am. Chem. Soc.* **2000**, *122*, 6848–6854.
81. Cecchi, R.; Rusconi, L.; Tanzi, M. C.; Danusso, F.; Ferruti, P. Synthesis and pharmacological evaluation of poly(oxyethylene) derivatives of 4-isobutylphenyl-2-propionic acid (ibuprofen). *J. Med. Chem.* **1981**, *24*, 622–625.
82. Cavallaro, G.; Pitarresi, G.; Giammona, G. Macromolecular prodrugs based on synthetic polyaminoacids: drug delivery and drug targeting in antitumor therapy. *Curr. Top. Med. Chem.* **2011**, *11*(18), 2382–2389.
83. Shen, W.-C.; Ryser, H. J.-P. Poly(L-lysine) and poly(D-lysine) conjugates of methotrexate: different inhibitory effect on drug resistant cells. *Mol. Pharmacol.* **1979**, *16*, 614–622.
84. Chu, B. C. F.; Howell, S. B. Differential toxicity of carrier-bound methotrexate toward human lymphocytes, marrow, and tumor cells. *Biochem. Pharmacol.* **1981**, *30*, 2545–2552.
85. Zupon, M. A.; Fang, S. M.; Christensen, J. M.; Petersen, R. V. In vivo release of norethindrone coupled to a biodegradable poly(α-amino acid) drug delivery system. *J. Pharm. Sci.* **1983**, *72*, 1323–1326.
86. Ringsdorf, H. Structure and properties of pharmacologically active polymers. *J. Polym. Sci., Polym. Symp.* **1975**, *51*, 135–153.
87. Rowland, G. F.; O'Neill, G. J.; Davies, D. A. L. Suppression of tumor growth in mice by a drug-antibody conjugate using a novel approach to linkage. *Nature* **1975**, *255*, 487–488.
88. Rowland, G. F. Effective antitumor conjugates of alkylating drug and antibody using dextran as the intermediate carrier. *Eur. J. Cancer* **1977**, *13*, 593–596.
89. Balboni, P. G.; Minia, A.; Grossi, M. P.; Barbarti-Brodano, G.; Mattioli, A.; Fiume, L. Activity of albumin conjugates of 5-fluorodeoxyuridine and cytosine arabinoside on poxviruses as a lysosomotropic approach to antiviral chemotherapy. *Nature* **1976**, *264*, 181–183.
90. Voutsadakis, I. A. Gemtuzumab ozogamicin (CMA-676, Mylotarg) for the treatment of CD33+ acute myeloid leukemia. *Anticancer Drugs* **2002**, *13*, 685–692.
91. Stable linker (technologies). *J. ADC Rev.* **2013**. http://adcreview.com/page/stable-linker-technologies
92. (a) Peddi, P. F.; Hurvitz, S. A.; Trastuzumab emtansine: the first targeted chemotherapy for treatment of breast cancer. *Future Oncol.* **2013**, *9*, 319–326. (b) Niculescu-Duvaz, I. Trastuzumab emtansine, an antibody-drug conjugate for the treatment of HER2+ metastatic brest cancer. *Curr. Opin. Mol. Ther.* **2010**, *12(3)*, 350–360.
93. (a) Zolot, R. S.; Basu, S.; Million, R. P. Antibody-drug conjugates. *Nature. Rev. Drug Disc.* **2013**, *12*, 259–260. (b) Mullard, A. Maturing antibody-drug conjugate pipeline hits 30. *Nature. Rev. Drug Disc.* **2013**, *12*, 329–332.
94. Carl, P. L.; Chakravarty, P. K.; Katzenellenbogen, J. A. A novel connector linkage applicable in prodrug design. *J. Med. Chem.* **1981**, *24*, 479–480.
95. Bundgaard, H. In *Bioreversible Carriers in Drug Design*, Roche, E. B. (Ed.), Pergamon Press, New York, 1987, p. 13.
96. (a) Metges, C. C. Contribution of microbial amino acids to amino acid homeostasis of the host. *J. Nutrition* **2000**, *130*, 1857S–1864S. (b) White, A.; Bardocz, S. In *Polyamines in Health and Nutrition*, Kluwer Academic Publ., Hingham, Mass., 1999, pp. 117–122. (c) Roth, J. R.; Lawrence, J. G.; Bobik, T. A. Cobalamin (coenzyme B12): synthesis and biological significance. *Annu. Rev. Microbiol.* **1996**, *50*, 137–181. (d) Conly, J. M.; Stein, K. The production of menaquinones (vitamin K2) by intestinal bacteria and their role in maintaining coagulation homeostasis. *Prog. Food Nutr. Sci.* **1992**, *16*, 307–343.
97. Jansen, A. B. A.; Russell, T. J. Some novel penicillin derivatives. *J. Chem. Soc.* **1965**, *65*, 2127–2132.
98. Bodin, N. D.; Ekström, B.; Forsgren, U.; Jalar, L. P.; Magni, L.; Ramsey, C. H.; Sjöberg, B. Bacampicillin: a new orally well-absorbed derivative of ampicillin. *Antimicrob. Agents Chemother.* **1975**, *8*, 518–525.
99. Daehne, W. V.; Frederiksen, E.; Gundersen, E.; Lund, F.; Mørch, P.; Petersen, H. J.; Roholt, K.; Tybring, L.; Godtfredsen, W. O. Acyloxymethyl esters of ampicillin. *J. Med. Chem.* **1970**, *13*, 607–612.
100. Bodor, N. Redox drug delivery systems for targeting drugs to the brain. *Ann. N. Y. Acad. Sci.* **1987**, *507*, 289–306.
101. (a) Pop, E.; Wu, W.-M.; Shek, E.; Bodor, N. Improved delivery through biological membranes. 38. Brain-specific chemical delivery systems for β-lactam antibiotics. Synthesis and properties of some dihydropyridine and dihydroisoquinoline derivatives of benzylpenicillin. *J. Med. Chem.* **1989**, *32*, 1774–1781. (b) Wu, W.-M.; Pop, E.; Shek, E.; Bodor, N. Brain-specific chemical delivery systems for beta-lactam antibiotics. In vitro and in vivo studies of some dihydropyridine and dihydroisoquinoline derivatives of benzylpenicillin in rats. *J. Med. Chem.* **1989**, *32*, 1782–1788.
102. Møllgaard, B.; Hoelgaard, A.; Bundgaard, H. Prodrugs as drug delivery systems. XXIII. Improved dermal delivery of 5-fluorouracil through human skin via N-acyloxymethyl prodrug derivatives. *Int. J. Pharm.* **1982**, *12*, 153–162.
103. Kingsbury, W. D.; Boehm, J. C.; Mehta, R. J.; Grappel, S. F.; Gilvarg, C. A novel peptide delivery system involving peptidase activated prodrugs as antimicrobial agents. Synthesis and biological activity of peptidyl derivatives of 5-fluorouracil. *J. Med. Chem.* **1984**, *27*, 1447–1451.
104. Greenwald, R. B.; Choe, Y. H.; Conover, C. D.; Shum, K.; Wu, D.; Royzen, M. Drug delivery systems based on trimethyl lock lactonization: poly(ethylene glycol) prodrugs of amino-containing compounds. *J. Med. Chem.* **2000**, *43*, 475–487.

105. (a) Shan, D.; Nicholaou, M. G.; Borchardt, R. T.; Wang, B. Prodrug strategies based on intramolecular cyclization reactions. *J. Pharm. Sci.* **1997**, *86*, 765–767. (b) Testa, B.; Mayer, J. M. Design of intramolecularly activated prodrugs. *Drug Metab. Rev.* **1998**, *30*, 787–807. (c) Wang, W.; Jiang, J.; Ballard, C. E.; Wang, B. Prodrug approaches to the improved delivery of peptide drugs. *Curr. Pharm. Des.* **1999**, *5*, 265–287.
106. (a) Lau, W. M.; White, A. W.; Gallagher, S. J.; Donaldson, M.; McNaughton, G.; Heard, C. M. Scope and limitations of the co-drug approach to topical drug delivery. *Curr. Pharm. Des.* **2008**, *14*(8), 794–802. (b) Das, N.; Dhanawat, M.; Dash, B.; Nagarwal, R. C.; Shrivastava, S. K. Codrug: an efficient approach for drug optimization. *Eur. J. Pharm. Sci.* **2010**, *41*(5), 571–588.
107. (a) Hartley, S.; Wise, R. A three-way crossover study to compare the pharmacokinetics and acceptability of sultamicillin at two dose levels with that of ampicillin. *J. Antimicrob. Chemother.* **1982**, *10*, 49–55. (b) Baltzer, B.; Binderup, E.; Von Daehne, W.; Godtfredsen, W. O.; Hansen, K.; Nielsen, B.; Sørensen, H.; Vangedal, S. Mutual prodrugs of β-lactam antibiotics and β-lactamase inhibitors. *J. Antibiot.* **1980**, *33*, 1183–1192.
108. (a) Van Uem, T. J. F.; De Pont, J. J. H. H. M. Structure and function of gastric hydrogen ion-potassium-ATPase. *New Compr. Biochem.* **1992**, *21* (Molecular Aspects of Transport Proteins), 27–55. (b) Rabon, E. C.; Reuben, M. A. The mechanism and structure of the gastric H, K-ATPase. *Annu. Rev. Physiol.* **1990**, *52*, 321–344.
109. (a) Bamford, M. H^+/K^+ ATPase inhibitors in the treatment of acid-related disorders. *Prog. Med. Chem.* **2009**, *47*, 75–162. (b) Shin, J. M.; Sachs, G. Long-lasting inhibitors of gastric H, K-ATPase. *Exp. Rev. Clin. Pharmacol.* **2009**, *2*(5), 461–468. (c) Herling, A. W.; Weidmann, K. Gastric H^+/K^+-ATPase inhibitors. *Prog. Med. Chem.* **1994**, *31*, 233–64.
110. (a) Lee, Y.-H.; Phillips, E.; Sause, S. W. Antigastrin activities of 2-phenyl-2-(2-pyridyl)thioacetamide (SC-15396) and structurally-related compounds in experimental animals. *Arch. Int. Pharmacodyn. Ther.* **1972**, *195*, 402–410. (b) Brändström, A.; Lindberg, P.; Junggren, U. Structure activity relationships of substituted benzimidazoles. *Scand. J. Gastroenterol.* **1985**, *108*(Suppl.), 15–22.
111. Lindberg, P.; Brändström, A.; Wallmark, B.; Mattsson, H.; Rikner, L.; Hoffmann, K.-J. Omeprazole: the first proton pump inhibitor. *Med. Res. Rev.* **1990**, *10*, 1–54.
112. Helander, H. F.; Ramsay, C.-H.; Regärdh, C.-G. Localization of omeprazole and metabolites in the mouse. *Scand. J. Gastroenterol.* **1985**, *108*(Suppl.), 95–104.
113. Fryklund, J.; Gedda, K.; Wallmark, B. Specific labeling of gastric proton-potassium ATPase by omeprazole. *Biochem. Pharmacol.* **1988**, *37*, 2543–2549.
114. (a) Brändström, A.; Lindberg, P.; Bergman, N.-Å.; Alminger, T.; Ankner, K.; Junggren, U.; Lamm, B.; Nordberg, P.; Erickson, M.; Grundevik, I.; Hagin, I.; Hoffmann, K. J.; Johansson, S.; Larsson, S.; Löftberg, I.; Ohlson, K.; Persson, B.; Skånberg, I.; Tekenbergs-Hjelte, L. Chemical reactions of omeprazole and omeprazole analogs. I. A survey of the chemical transformations of omeprazole and its analogs. *Acta Chem. Scand.* **1989**, *43*, 536–548. (b) Brändström, A.; Bergman, N.-Å.; Lindberg, P.; Grundevik, I.; Johansson, S.; Tekenbergs-Hjelte, L.; Ohlson, K. Chemical reactions of omeprazole and omeprazole analogs. II. Kinetics of the reaction of omeprazole in the presence of 2-mercaptoethanol. *Acta Chem. Scand.* **1989**, *43*, 549–568. (c) Brändström, A.; Bergman, N. Å.; Grundevik, I.; Johansson, S.; Tekenbergs-Hjelte, L.; Ohlson, K. Chemical reactions of omeprazole and omeprazole analogs. III. Protolytic behavior of compounds in the omeprazole system. *Acta Chem. Scand.* **1989**, *43*, 569–576. (d) Brändström, A.; Lindberg, P.; Bergman, N. Å.; Tekenbergs-Hjelte, L.; Ohlson, K. Chemical reactions of omeprazole and omeprazole analogs. IV. Reactions of compounds of the omeprazole system with 2-mercaptoethanol. *Acta Chem. Scand.* **1989**, *43*, 577–586. (e) Brändström, A.; Lindberg, P.; Bergman, N. Å.; Tekenbergs-Hjelte, L.; Ohlson, K.; Grundevik, I.; Nordberg, P.; Alminger, T. Chemical reactions of omeprazole and omeprazole analogs. V. The reaction of N-alkylated derivatives of omeprazole analogs with 2-mercaptoethanol. *Acta Chem. Scand.* **1989**, *43*, 587–594. (f) Brändström, A.; Lindberg, P.; Bergman, N. Å.; Grundevik, I.; Tekenbergs-Hjelte, L.; Ohlson, K. Chemical reactions of omeprazole and omeprazole analogs. VI. The reactions of omeprazole in the absence of 2-mercaptoethanol. *Acta Chem. Scand.* **1989**, *43*, 595–611.
115. Puscas, I.; Coltau, M.; Baican, M.; Domuta, G. Omeprazole has a dual mechanism of action: it inhibits both H(+)K(+)ATPase and gastric mucosa carbonic anhydrase enzyme in humans (in vitro and in vivo experiments). *J. Pharmacol. Exp. Ther.* **1999**, *290*, 530–534.
116. (a) Puscas, I. In *Carbonic Anhydrase and Modulation of Physiologic and Pathologic Processes in the Organism*, Puscas, I. (Ed.), Helicon Publishing House, Timisoara, Romania, 1994, pp 373–530. (b) Puscas, I. In *New Pharmacology of Ulcer Disease*, Szabo, S.; Mozsik, G. (Eds.),; Elsevier, New York, 1987, pp 164–179.
117. (a) Gremse, D. A. Lansoprazole: Pharmacokinetics, pharmacodynamics and clinical use. *Exp. Opin. Pharmacother.* **2001**, *2*, 1663–1670. (b) Matheson, A. J.; Jarvis, B. Lansoprazole: An update of its place in the management of acid-related disorders. *Drugs* **2001**, *61*, 1801–1833.
118. Carswell, C. I.; Goa, K. L. Rabeprazole: An update of its use in acid-related disorders. *Drugs* **2001**, *61*, 2327–2356.
119. Jungnickel, P. W. Pantoprazole: A new proton pump inhibitor. *Clin. Ther.* **2000**, *22*, 1268–1293.
120. Kanda, Y.; Ashizawa, T.; Kakita, S.; Takahashi, Y.; Kono, M.; Yoshida, M.; Saitoh, Y. Okabe, M. Synthesis and antitumor activity of novel thioester derivatives of leinamycin. *J. Med. Chem.* **1999**, *42*, 1330–1332.
121. (a) Rückemann, K.; Fairbanks, L. D.; Carrey, E. A.; Hawrylowicz, C. M; Richards, D. F.; Kirschbaum, B.; Simmonds, H. A. Leflunomide inhibits pyrimidine de novo synthesis in mitogen-stimulated T-lymphocytes from healthy humans. *J. Biol. Chem.* **1998**, *273*, 21682–21691. (b) Bruneau, J.-M.; Yea, C. M.; Spinella-Jaegle, S.; Fudali, C.; Woodward, K.; Robson, P. A.; Sautes, C.; Westwood, R.; Kuo, E. A.; Williamson, R. A.; Ruuth, E. Purification of human dihydro-orotate dehydrogenase and its inhibition by A77 1726, the active metabolite of leflunomide. *Biochem. J.* **1998**, 336, 299–303.
122. Davis, J. P.; Cain, G. A.; Pitts, W. J.; Magolda, R. L.; Copeland, R. A. The immunosuppressive metabolite of leflunomide is a potent inhibitor of human dihydroorotate dehydrogenase. *Biochemistry* **1996**, *35*, 1270–1273.
123. Sutharchanadevi, M.; Murugan, R. In *Comprehensive Heterocyclic Chemistry II*, Shinkai, I. (Ed.), Elsevier, Oxford, UK, 1996, Vol. 3, pp. 221–260.
124. (a) Lahti, R. A.; Gall, M. Conversion of N-alkylaminobenzophenones to benzodiazepines in vivo. *J. Med. Chem.* **1976**, *19*, 1064–1067. (b) Gall, M.; Hester, J. B., Jr.; Rudzik, A. D.; Lahti, R. A. Synthesis and pharmacology of novel anxiolytic agents derived from 2-[(dialkylamino)methyl-4H-triazol-4-yl]benzophenones and related heterocyclic benzophenones. *J. Med. Chem.* **1976**, *19*, 1057–1064.

125. Brodie, B. B.; Axelrod, J. Fate of acetophenetidin (phenacetin) in man and methods for the estimation of acetophenetidin and its metabolites in biological material. *J. Pharmacol. Exp. Ther.* **1949**, *97*, 58–67.
126. (a) Colvin, M.; Chabner, B. A. In *Cancer Chemotherapy: Principles and Practice*, Chabner, B. A.; Collins, J. M. (Eds.), J. B. Lippincott, Philadelphia, 1990, p. 276. (b) Connors, T. A.; Cox, P. J.; Foster, A. B.; Jarman, M. Some studies of the active intermediates formed in the microsomal metabolism of cyclophosphamide and isophosphamide. *Biochem. Pharmacol.* **1974**, *23*, 115–129. (c) Hill, D. L. *A Review of Cyclophosphamide*, Thomas: Springfield, IL, 1975.
127. Benson, A. J.; Martin, C. N.; Garner, R. C. N-(2-hydroxyethyl)-N-[2-(7-guaninyl)ethyl]amine, the putative major DNA adduct of cyclophosphamide in vitro and in vivo in the rat. *Biochem. Pharmacol.* **1988**, *37*, 2979–2985.
128. Connors, T. A.; Cox, P. J.; Farmer, P. B.; Foster, A. B.; Jarman, M. Active intermediates formed in the microsomal metabolism of cyclophosphamide and isophosphamide. *Biochem. Pharmacol.* **1974**, *23*, 115–129.
129. (a) Oliverio, V. T. In *Cancer Medicine*, 2nd ed., Holland, J. F.; Frei, E., III. (Eds.), Lea & Febiger, Philadelphia, 1982, p. 850. (b) Weinkam, R. J.; Shiba, D. A.; Chabner, B. A. In *Pharmacologic Principles of Cancer Treatment*, Chabner, B. E., (Ed.), W. B. Saunders, Philadelphia, 1982, p. 340.
130. Raaflaub, J.; Schwartz, D. E. Metabolism of a cytostatically active methylhydrazine derivative (Natulan). *Experientia* **1965**, *21*, 44–45.
131. Kreis, W.; Piepho, S. B.; Bernhard, H. V. The metabolic fate of the 14C-labeled methyl group of a methylhydrazine derivative in P815 mouse leukemia. *Experientia* **1966**, *22*, 431–433.
132. Tsuji, T.; Kosower, E. M. Diazenes. VI. Alkyldiazenes. Diazenes. VI. Alkyldiazenes. *J. Am. Chem. Soc.* **1971**, *93*, 1992–1999.
133. Froede, H. C.; Wilson, I. B. In *The Enzymes*, 3rd ed.; Boyer, P. (Ed.), Academic Press, New York, 1971, Vol. 5, p. 87.
134. Sussman, J. L.; Harel, M.; Frolow, F.; Oefner, C.; Goldman, A.; Toker, L.; Silman, I. Atomic structure of acetylcholinesterase from *Torpedo californica*: a prototypic acetylcholine-binding protein. *Science* **1991**, *253*, 872–879.
135. Dougherty, D. A. The cation-π interaction. *Chem. Rev.* **1997**, *97*, 1303–1324.
136. Jansen, E. F.; Nutting, M.-D. F.; Balls, A. K. Mode of inhibition of chymotrypsin by diisopropyl fluorophosphate. I. Introduction of phosphorus. *J. Biol. Chem.* **1949**, *179*, 201–204.
137. Wilson, I. B. Designing of a new drug with antidotal properties against the nerve-gas sarin (isopropyl methylphosphonofluoridate). *Biochim. Biophys. Acta* **1958**, *27*, 196–199.
138. Shek, E.; Higuchi, T.; Bodor, N. Improved delivery through biological membranes. 3. Delivery of N-methylpyridinium-2-carbaldoxime chloride through the blood-brain barrier in its dihydropyridine prodrug form. *J. Med. Chem.* **1976**, *19*, 113–117.
139. (a) Lee, C.; Katz, R. L. Clinical implications of new neuromuscular concepts and agents: so long, neostigmine! So long, sux! *J. Crit. Care* **2009**, *24*(1), 43–49. (b) Habib, A. S.; Gan, T. J. Use of neostigmine in the management of acute postoperative pain and labour pain: a review. *CNS Drugs* **2006**, *20*(10), 821–839.
140. Shigeta, M.; Homma, A. Donepezil for Alzheimer's disease: Pharmacodynamic, pharmacokinetic, and clinical profiles. *CNS Drug Rev.* **2001**, *7*, 353–368.
141. (a) Summers, W. K. Tacrine (THA, Cognex). *J. Alzheimer's Dis.* **2000**, *2*, 85–93. (b) Kurz, A. The therapeutic potential of tacrine. *J. Neural Trans.* (Suppl.) **1998**, *54*, 295–299.
142. Savi, P.; Herbert J. M.; Pflieger, A. M.; Dol, F.; Delebassee, D.; Combalbert, J.; Defreyn, G.; Maffrand, J. P. Importance of hepatic metabolism in the antiaggregating activity of the thienopyridine clopidogrel. *Biochem. Pharmacol.* **1992**, *44*, 527–532.
143. (a) Savi, P.; Pereillo, J. M.; Uzabiaga, M. F. ; Combalbert, J.; Picard, C.; Maffrand, J. P.; Pascal, M.; Herbert, J. M. Identification and biological activity of the active metabolite of clopidogrel. *Thromb. Haemost.* **2000**, *84*, 891–896. (b) Pereillo, J. M.; Maftouh, M.; Andrieu, A.; et al. Structure and stereochemistry of the active metabolite of clopidogrel. *Drug Metab. Dispos.* **2002**, *30*, 1288–1295.
144. Ding, Z.; Kim, S.; Dorsam, R. T.; Jin, J.; Kunapuli, S. P. Inactivation of the human $P2Y_{12}$ receptor by thiol reagents requires interaction with both extracellular cysteine residues, Cys17 and Cys270. *Blood* **2003**, *101*, 3908–3914.
145. Savi, P.; Labouret, C.; Delesque, N.; Guette, F.; Lupker, J.; Herbert, J. M. $P2Y_{12}$, a new platelet ADP receptor, target of clopidogrel. *Biochem. Biophys. Res. Commun.* **2001**, *283*, 379–383.
146. Venhuis, BG. J.;Wikström, H. V.; Rodenhuis, N.; Sundell, S.; Dijkstra, D. A new type of prodrug of catecholamines: an opportunity to improve the treatment of Parkinson's disease. *J. Med. Chem.* **2002**, *45*, 2349–2351.
147. (a) Frigerio, A.; Fanelli, R.; Biandrate, P.; Passerini, G.; Morselli, P. L.; Garattini, S. Mass spectrometric characterization of carbamazepine-10,11-epoxide, a carbamazepine metabolite isolated from human urine. *J. Pharm. Sci.* **1972**, *61*, 1144–1147. (b) Johannessen, S. I.; Gerna, N. M.; Bakke, J.; Strandjord, R. E.; Morselli, P. L. CSF concentrations and serum protein binding of carbamazepine and carbamazepine-10,11-epoxide in epileptic patients. *Br. J. Clin. Pharmacol.* **1976**, *3*, 575–582.
148. Neely, J. R.; Morgan, H. E. Relation between carbohydrate and lipid metabolism and the energy balance of heart muscle. *Annu. Rev. Physiol.* **1974**, *36*, 413–459.
149. Barnish, I. T.; Cross, P. E.; Danilewicz, J. C.; Dickinson, R. P.; Stopher, D. A. Promotion of carbohydrate oxidation in the heart by some phenylglyoxylic acids. *J. Med. Chem.* **1981**, *24*, 399–404.
150. Bhosle, D.; Bharambe, S.; Gairola, N.; Dhaneshwar, S. S. Mutual prodrug concept: fundamentals and applications. *Ind. J. Pharm. Sci.* **2006**, *68*, 286–294.
151. (a) Kirsner, J. B. Observations on the medical treatment of inflammatory bowel disease. *J. Am. Med. Assoc.* **1980**, *243*, 557–564. (b) Eastwood, M. A. Pharmacokinetic patterns of sulfasalazine. *Ther. Drug Monit.* **1980**, *2*, 149–152.
152. Brown, J. P.; McGarraugh, G. V.; Parkinson, T. M.; Wingard, R. E., Jr.; Onderdonk, A. B. A polymeric drug for treatment of inflammatory bowel disease. *J. Med. Chem.* **1983**, *26*, 1300–1307.
153. Reist, E. J.; Benitez, A.; Goodman, L.; Baker, B. L.; Lee, W. W. Potential anticancer agents. LXXVI. Synthesis of purine nucleosides of β-D-arabinofuranose. *J. Org. Chem.* **1962**, *27*, 3274–3279.
154. Andrei, G.; Snoeck, R.; Goubou, P.; Desmyter, J.; DeClercq, E. Comparative activity of various compounds against clinical strains of herpes simplex virus. *Eur. J. Clin. Microbiol. Infect. Diseases* **1992**, *11*, 143–151.
155. Whitley, R.; Alford, C.; Hess, F.; Buchanan, R. Vidarabine: a preliminary review of its pharmacological properties and therapeutic use. *Drugs* **1980**, *20*, 267–282.
156. Kotra, L. P.; Manouilof, K. K.; Cretton-Scott, E.; Sommadossi, J.-P.; Boridinot, F. D.; Schinazi, R. F.; Chu, C. K. Synthesis, biotransformation, and pharmacokinetic studies of 9-(β-D-arabinofuranosyl)-6-azidopurine: a prodrug for ara-a designed to utilize the azide reduction pathway. *J. Med. Chem.* **1996**, *39*, 5202–5207.

157. Shen, T. Y.; Winter, C. A. Chemical and biological studies on indomethacin, sulindac and their analogs. *Adv. Drug Res.* **1977**, *12*, 90–245.
158. Matsukawa, T.; Yurugi, S.; Oka, Y. The synthesis of S-acrylthiamine derivatives and their stability. *Ann. N. Y. Acad. Sci.* **1962**, *98*, 430–444.
159. Stella, V. J.; Himmelstein, K. J. In *Design of Prodrugs*. Bundgaard, H. (Ed.), Elsevier: Amsterdam, 1985, p. 177.
160. Hofsteenge, J.; Capuano, A.; Altszuler, R.; Moore, S. Carrier-linked primaquine in the chemotherapy of malaria. *J. Med. Chem.* **1986**, *29*, 1765–1769.
161. Miwa, G. T.; Wang, R.; Alvaro, R.; Walsh, J. S.; Lu, A. Y. H. The metabolic activation of ronidazole [(1-methyl-5-nitroimidazole-2-yl)-methyl carbamate] to reactive metabolites by mammalian, cecal bacterial and *T. foetus* enzymes. *Biochem. Pharmacol.* **1986**, *35*, 33–36.
162. Tobias, S. C.; Borch, R. F. Synthesis and biological studies of novel nucleoside phosphoramidate prodrugs. *J. Med. Chem.* **2001**, *44*, 4475–4480.
163. (a) McCollister, R. J.; Gilbert, W. R., Jr.; Ashton, D. M.; Wyngaarden, J. B. Pseudofeedback inhibition of purine synthesis by 6-mercaptopurine ribonucleotide and other purine analogs. *J. Biol. Chem.* **1964**, *239*, 1560–1563. (b) Caskey, C. T.; Ashton, D. M.; Wyngaarden, J. B. The enzymology of feedback inhibition of glutamine phosphoribosylpyrophosphate amidotransferase by purine ribonucleotides. *J. Biol. Chem.* **1964**, *239*, 2570–2579. (c) Henderson, J. F.; Khoo, M. K. Y. The mechanism of feedback inhibition of purine biosynthesis de novo in Ehrlich ascites tumor cells in vitro. *J. Biol. Chem.* **1965**, *240*, 3104–3109.
164. (a) Elion, G. B. Acyclovir: discovery, mechanism of action, and selectivity. *J. Med. Virol.* **1993**, (Suppl. 1), 2–6. (b) Richards, D. M.; Carmine, A. A.; Brogden, R. N.; Heel, R. C.; Speight, T. M.; Avery, G. S. Acyclovir. A review of its pharmacodynamic properties and therapeutic efficacy. *Drugs* **1983**, *26*, 378–438. (c) Elion, G. B. The biochemistry and mechanism of action of acyclovir. *J. Antimicrob. Chemother.* **1983**, *12* (Suppl. B), 9–17. (d) Vanpouille, C.; Lisco, A.; Margolis, L. Acyclovir: a new use for an old drug. *Curr. Opin. Infect. Dis.* **2009**, *22*(6), 583–587.
165. Furman, P. A.; McGuirt, P. V.; Keller, P. M.; Fyfe, J. A.; Elion, G. B. Inhibition by acyclovir of cell growth and DNA synthesis of cells biochemically transformed with herpes virus genetic information. *Virology* **1980**, *102*, 420–430.
166. Miller, W. H.; Miller, R. L. Phosphorylation of acyclovir (acycloguanosine) monophosphate by GMP kinase. *J. Biol. Chem.* **1980**, *255*, 7204–7207.
167. Miller, W. H.; Miller, R. L. Phosphorylation of acyclovir diphosphate by cellular enzymes. *Biochem. Pharmacol.* **1982**, *31*, 3879–3884.
168. Furman, P. A.; St. Clair, M. H.; Fyfe, J. A.; Rideout, J. L.; Keller, P. M.; Elion, G. B. Inhibition of herpes simplex virus-induced DNA polymerase activity and viral DNA replication by 9-(2-hydroxyethoxymethyl)guanine and its triphosphate. *J. Virol.* **1979**, *32*, 72–77.
169. Reardon, J. E.; Spector, T. Herpes simplex virus type 1 DNA polymerase. Mechanism of inhibition by acyclovir triphosphate. *J. Biol. Chem.* **1989**, *264*, 7405–7411.
170. Larder, B. A.; Cheng, Y.-C.; Darby, G. Characterization of abnormal thymidine kinases induced by drug-resistant strains of herpes simplex virus type 1. *J. Gen. Virol.* **1983**, *64*, 523–532.
171. (a) Coen, D. M.; Schaffer, P. A.; Furman, P. A.; Keller, P. M.; St. Clair, M. H. Biochemical and genetic analysis of acyclovir-resistant mutants of herpes simplex virus type 1. *Am. J. Med.* **1982**, *73*(1A), 351–560. (b) Schnipper, L. E.; Crumpacker, C. S. Resistance of herpes simplex virus to acycloguanosine: Role of viral thymidine kinase and DNA polymerase loci. *Proc. Natl. Acad. Sci. U.S.A.* **1980**, *77*, 2270–2273.
172. Huang, L.; Ishii, K. K.; Zuccola, H.; Gehring, A. M.; Hwang, C. B. C.; Hogle, J.; Coen, D. M. The enzymological basis for resistance of herpesvirus DNA polymerase mutants to acyclovir: relationship to the structure of α-like DNA polymerases. *Proc. Natl. Acad. Sci. U.S.A.* **1999**, *96*, 447–452.
173. Good, S. S.; Krasny, H. C.; Elion, G. B.; de Miranda, P. Disposition in the dog and the rat of 2,6-diamino-9-(2-hydroxyethoxymethyl) purine (A134U), a potential prodrug of acyclovir. *J. Pharmacol. Exp. Ther.* **1983**, *227*, 644–651.
174. Krenitsky, T. A.; Hall, W. W.; de Miranda, P.; Beauchamp, L. M.; Schaeffer, H. J.; Whiteman, P. D. 6-Deoxyacyclovir: a xanthine oxidase-activated prodrug of acyclovir. *Proc. Natl. Acad. Sci. U.S.A.* **1984**, *81*, 3209–3213.
175. (a) Weller, S.; Blum, M. R.; Doucette, M.; Burnette, T.; Cederberg, D. M.; de Miranda, P.; Smiley, M. L. Pharmacokinetics of the acyclovir pro-drug valaciclovir after escalating single-and multiple-dose administration to normal volunteers. *Clin. Pharmacol. Ther.* **1993**, *54*, 595–605. (b) Antman, M. D.; Gudmundsson, O. S. Valacyclovir: a prodrug of acyclovir Biotechnol.: *Pharm. Asp.* **2007**, *5* (Pt. 2, Prodrugs: Challenges and Rewards, Part 2), 669–676.
176. Burnette, T. C.; Harrington, J. A.; Reardon, J. E.; Merrill, B. M.; de Miranda, P. Purification and characterization of a rat liver enzyme that hydrolyzes valaciclovir, the L-valyl ester prodrug of acyclovir. *J. Biol. Chem.* **1995**, *270*, 15827–15831.
177. Bailey, S. M.; Hart, I.; Lohmeyer, M. Ganciclovir and the HSV-tk enzyme-prodrug system in cancer therapy. *Drugs Future* **1998**, *23*, 401–413.
178. (a) McGavin, J. K.; Goa, K. L. Ganciclovir: An update of its use in the prevention of cytomegalovirus infection and disease in transplant recipients. *Drugs* **2001**, *61*, 1153–1183. (b) Spector, S. A. Oral ganciclovir. *Adv. Exp. Med. Biol.* **1999**, *458*, 121–127. (c) Elion, G. B. In *Antiviral Chemotherapy: New Directions for Clinical Application and Research*. Mills, J.; Corey, L. (Eds.), Elsevier, New York, 1986, p. 118.
179. (a) Sutton, D.; Kern, E.R. Activity of famciclovir and penciclovir in HSV-infected animals: A review. *Antiviral Chem. Chemother.* **1993**, *4*, 37–46. (b) Vere Hodge, R. A.; Cheng, Y. C. The mode of action of penciclovir. *Antiviral Chem. Chemother.* **1993**, *4*, 13–24. (c) Earnshaw, D. L.; Bacon, T. H.; Darlison, S. J.; Edmonds, K.; Perkins, R. M.; Vere Hodge, R. A. Mode of antiviral action of penciclovir in MRC-5 cells infected with herpes simplex virus type 1 (HSV-1), HSV-2, and varicella-zoster virus. *Antimicrob. Agents Chemother.* **1992**, *36*, 2747–2757.
180. Boyd, M. R.; Bacon, T. H.; Sutton, D. Antiherpesvirus activity of 9-(4-hydroxy-3-hydroxymethylbut-1-yl)guanine (BRL 39123) in animals. *Antimicrob. Agents Chemother.* **1988**, *32*, 358–363.
181. (a) Bacon, T. H. Famciclovir, from the bench to the patient - a comprehensive review of preclinical data. *Int. J. Antimicrob. Agents* **1996**, *7*, 119–134. (b) Jarvest, R. L. Discovery and characterization of famciclovir (Famvir), a novel anti-herpesvirus agent. *Drugs Today* **1994**, *30*, 575–588.
182. Vere Hodge, R. A. Antiviral portraits series. Number 3. Famciclovir and penciclovir. The mode of action of famciclovir including its conversion to penciclovir. *Antiviral Chem. Chemother.* **1993**, *4*, 67–84.
183. (a) Cirelli, R.; Herne, K.; McCrary, M.; Lee, P.; Tyring, S. K. Famciclovir: review of clinical efficacy and safety. *Antiviral Res.* **1996**, *29*, 141–151. (b) Vere Hodge, R. A.; Sutton, D.; Boyd, M. R.; Harnden, M. R.; Jarvest, R. L. Selection of an oral prodrug (BRL 42810; famciclovir) for the antiherpes virus agent BRL 39123 [9-(4-hydroxy-3-hydroxymethylbut-1-yl)guanine; penciclovir]. *Antimicrob. Agents Chemother.* **1989**, *33*, 1765–1773.

184. (a) Winton, C. F.; Fowles, S. E.; Pierce, D. M.; Vere Hodge, R. A. Gradient high-performance liquid chromatographic method for the analysis of the prodrug famciclovir and its metabolites, including the active antiviral agent penciclovir, in plasma and urine. *Anal. Proc.* **1990**, *27*, 181–182. (b) Winton, C. F.; Fowles, S. E.; Vere Hodge, R. A.; Pierce, D. M. In *Analysis of Drugs and Metabolites Including Anti-Infective Agents*. Reid, E.; Wilson, I. D. (Eds.), Royal Society of Chemistry, Cambridge, UK., 1990, pp. 163–171.
185. Vere Hodge, R. A.; Earnshaw, D. L.; Jarvest, R. L.; Readshaw, S. A. Use of isotopically chiral [4′-13C] penciclovir (BRL 39123) and its oral prodrug [4′-13C] Famciclovir (BRL 42810) to determine the absolute configuration of their metabolites. *Antiviral Res.* **1990** (Suppl.), *1*, 87.
186. McCall, J. M.; Aiken, J. W.; Chidester, C. G.; DuCharme, D. W.; Wending, M. G. Pyrimidine and triazine 3-oxide sulfates: a new family of vasodilators. *J. Med. Chem.* **1983**, *26*, 1791–1793.
187. Buhl, A. E.; Waldon, D. J.; Baker, C. A.; Johnson, G. A. Minoxidil sulfate is the active metabolite that stimulates hair follicles. *J. Invest. Dermatol.* **1990**, *95*, 553.
188. Meisheri, K. D.; Johnson, G. A.; Puddington, L. Enzymic and nonenzymic sulfation mechanisms in the biological actions of minoxidil. *Biochem. Pharmacol.* **1993**, *45*, 271–279.
189. (a) Navarro, A.; Boveris, A. Brain mitochondrial dysfunction and oxidative damage in Parkinson's disease. *J. Bioenerg. Biomembr.* **2009**, *41*(6), 517–521. (b) Langston, J. W. Progress in understanding the mechanisms of neuronal dysfunction and degeneration in Parkinson's disease. *Protein Rev.* **2007**, *6* (Protein Misfolding, Aggregation, and Conformational Diseases, Part B), 49–59. (c) Gupta, A.; Dawson, V. L.; Dawson, T. M. What causes cell death in Parkinson's disease? *Ann. Neurol.* **2008**, *64*(6, Suppl.), S3–S15.
190. (a) Deleu, D.; Northway, M. G.; Hanssens, Y. Clinical pharmacokinetic and pharmacodynamic properties of drugs used in the treatment of Parkinson's disease. *Clin. Pharmokin.* **2002**, *41*, 261–309. (b) Myllyla, V. V.; Sotaniemi, K. A.; Hakulinen, P.; Mki-Ikola, O.; Heinonen, E. H. Selegiline as the primary treatment of Parkinson's disease - a long-term double-blind study. *Acta Neurol. Scand.* **1997**, *95*, 211–218.
191. (a) Galler, R. M.; Hallas, B. H.; Fazzini, E. Current trends in the pharmacologic and surgical treatment of Parkinson's disease. *J. Am. Osteopath. Assoc.* **1996**, *96*, 228–232. (b) Lieberman, A. Treatment of Parkinson's disease. *Curr. Opin. Neurol. Neurosurg.* **1993**, *6*, 339–343.
192. (a) Palfreyman, M. G.; McDonald, I. A.; Fozard, J. R.; Mely, Y.; Sleight, A. J.; Zreika, M.; Wagner, J.; Bey, P.; Lewis, P. J. Inhibition of monoamine oxidase selectively in brain monoamine nerves using the bioprecursor (E)-β-fluoromethylene-m-tyrosine (MDL 72394), a substrate for aromatic L-amino acid decarboxylase. *J. Neurochem.* **1985**, *45*, 1850–1860. (b) McDonald, I. A.; Lacoste, J. M.; Bey, P.; Wagner, J.; Zreika, M.; Palfreyman, M. G. Dual enzyme-activated irreversible inhibition of monoamine oxidase. *Bioorg. Chem.* **1986**, *14*, 103–118.
193. McDonald, I. A.; Lacoste, J. M.; Bey, P.; Wagner, J.; Zreika, M.; Palfreyman, M. G. Dual enzyme-activated irreversible inhibition of monoamine oxidase. *Bioorg. Chem.* **1986**, *14*(2), 103–18.
194. McDonald, I. A.; Lacoste, J. M.; Bey, P.; Palfreyman, M. G.; Zreika, M. Enzyme-activated irreversible inhibitors of monoamine oxidase: phenylallylamine structure-activity relationships. *J. Med. Chem.* **1985**, *28*, 186–193.
195. Magnan, S. D. J.; Shirota, F. N.; Nagasawa, H. T. Drug latentiation by γ-glutamyl transpeptidase. *J. Med. Chem.* **1982**, *25*, 1018–1021.
196. (a) Wilk, S.; Mizoguchi, H.; Orlowski, M. γ-Glutamyl dopa: a kidney-specific dopamine precursor. *J. Pharmacol. Exp. Ther.* **1978**, *206*, 227–232. (b) Kyncl, J. J.; Minard, F. N.; Jones, P. H. L-γ-Glutamyl dopamine, an oral dopamine prodrug with renal selectivity. *Adv. Biosci.* **1979**, *20*, 369–380.

Appendix

Answers to Chapter Problems

CHAPTER 1

1. a. A prototype compound that has a number of attractive characteristics, including the desired biological or pharmacological activity, but may have other undesirable characteristics.
   b. Assay that rapidly measures activity (often binding affinity) at a target of interest.
   c. Several screening methods using, for example, X-ray crystallography or NMR spectrometry developed to identify simple molecules (fragments) possessing typically modest affinity for a target, with the intent of connecting two or more of these fragments to create a useful lead compound.
   d. Drug degradation products produced in vivo by enzymes.
   e. Proteins that span cell membranes, where their role is to carry or transport molecules or ions from one side of the cell to the other.
   f. Enzymes are biological catalysts that facilitate the conversion of one or more reactants ("substrates") to one or more new products.
   g. The process by which a drug reaches the bloodstream from its site of administration. Frequently, the term refers to absorption from the gastrointestinal tract after oral administration.
2. Natural ligand for a receptor or substrate for an enzyme; another substance known to interact with the target of interest; random or targeted screening; fragment-based screening; computational methods.
3. Nature; serendipity; through drug metabolism studies; through clinical observations.
4. Ionic interactions, ion–dipole interactions, dipole–dipole interactions, hydrogen bonding, charge–transfer complexes, hydrophobic interactions, cation-π interactions, halogen bonding, and van der Waals forces.
5. The less potent a drug, the higher the dose that will need to be administered to achieve the desired effect. Administering more drug increases the cost per dose of the drug, minimizes the convenience of administration, increases the probability of off-target effects.
6. The less selective a drug, the more the off-target interactions, including closely related targets.
7. A patent allows you to block others from selling your drug for a period of time. If a drug is not patented, companies will not be interested in investing hundreds of millions, if not billions, of dollars to advance it through the drug discovery process.

CHAPTER 2

1. Advantages
   - You do not have to be creative.
   - You do not have to understand what causes the disease state.
   - You can test whatever becomes available.
   - You may get hits with compounds that would not have been tested in a rational approach because structures are unrelated to what you think should be important.

   Disadvantages
   - There are an almost infinite number of possibilities.
   - You may be testing compounds completely unrelated to the biological system that is important.
   - Time and expense may be excessive.
2. The endogenous ligand for a new biological target may not be well characterized.
   The only known ligand may not have desirable properties of a lead compound.

The endogenous ligand may be a complex molecule that is not readily amenable to synthetic modification or has some other undesirable properties that cannot be removed.

3. **a.** As a peptide, it would likely have poor pharmacokinetic properties: low metabolic stability (peptidases), low permeability through the gut and especially into the brain (high charge and polarity; gut peptidase metabolism), rapid excretion. Also, peptides tend to bind to multiple targets (in the case of CCK-4, there is also activity at the gastrin receptor).
   b. Identify receptor to which CCK-4 binds and screen a library of compounds for an antagonist of that receptor. Identify the peptidase that clips off the tetrapeptide from its polypeptide precursor and screen compounds for inhibitors of that enzyme. Identify the protein that causes the release of CCK in the brain, and screen for antagonists/inhibitors of that protein. Make a peptidomimetic of CCK-4 as a lead.
4. **a.** More compound was needed for animal studies, so that could be the driving force to switch to screening methods.
   b. The expense of animals is considerable and would encourage methods (like high-throughput screening) that were less expensive.
   c. The early information about pharmacokinetics is a significant reason not to change to modern methods. Improving PK is generally the difficult part of optimizing a lead compound, so animal studies would be beneficial for getting druglike molecules earlier on.
   d. In vivo studies do not give direct pharmacodynamic information because there are too many pharmacokinetic interferences. Screening assays that use isolated proteins give the best direct indication of pharmacodynamic properties.
5. The optimized structures in Table 2.2 are closely related to the screening hits, which supports the assumption.
6. Pros:
   A large percentage of drugs are derived or inspired by natural product structures.
   Natural products often have complex structures that are difficult to synthesize; modification of the natural product lessens the synthetic challenge.
   Natural products often have desirable pharmacokinetic properties, so they are good starting points.
   Cons:
   You can only make a limited number of analogues because of the interference of many other functional groups in the molecule with many different synthetic reactions; selective protection may be a large challenge.
   You can usually only isolate small amounts of compound from nature, so not much starting material to modify.
   The amount of each product obtained may be sufficient to do in vitro tests, but in vivo tests require larger amounts of compound.
7. Start with analogues of the substrate, arginine, and make small changes.

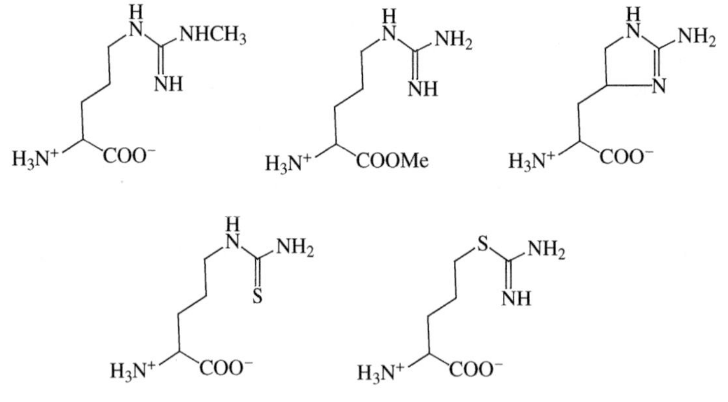

H N NHCH$_3$ NH H_3N^+ COO^-
H N NH$_2$ NH H_3N^+ COOMe
H N NH$_2$ N H_3N^+ COO^-
H N NH$_2$ S H_3N^+ COO^-
S NH$_2$ NH H_3N^+ COO^-

8. **a.** Cut pieces of **1** away and determine the effect on binding. If binding is decreased, then the group may have been in the pharmacophore. If binding is increased, then the group may have been preventing binding. If no binding effect, then the group may not be involved.
   b. **2** - one or both of the CH_3 groups interferes with binding.
   3 - only top CH_3 interferes.
   4 - phenyl is in the pharmacophore.
   5 - carbonyl is important; maybe because without it the amine is protonated, and a cation interferes with binding or maybe F interferes with binding.
   6 - F not interfering.
   7 - less conformational flexibility is better.

9. A random screen would give you the ability to find compounds without having to know the details of the target.
10. $TI = LD_{50}/ED_{50} = 5$
    Because this is not a potentially lethal disease, the therapeutic index would not be acceptable; this is not a safe compound.
11. There are many good answers; below is a possibility.

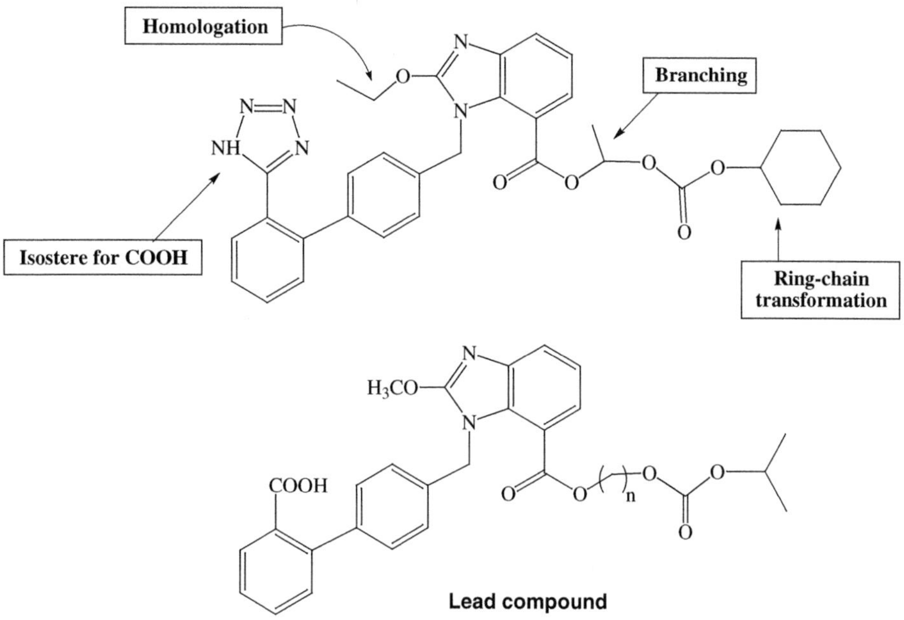

Lead compound

12. Most bioisosteric changes result in a change in size, shape, electronic distribution, lipid solubility, water solubility, pK_a, chemical reactivity, and/or hydrogen bonding capacity. These modifications can affect both the pharmacokinetics and pharmacodynamics, so a simple bioisosteric replacement may destroy activity.
13. **a.** There are still many side reactions that occur on solid phase, which are retained as impurities on the polymer support throughout the synthesis. Monitoring reactions on solid phase also is more difficult than in solution phase.
    b. Reactivities of various reagents can differ, so different rates of reaction occur depending on the temperature and time of reaction. The products could have different solubilities, and the same workup may not have the same efficiency for all products formed.
    c. A huge effort is required to isolate and analyze large numbers of compounds, which may not be an efficient use of resources.
    Diversity in commercially available building blocks does not always translate into a high level of diversity in the final products because the building blocks that are successfully incorporated into final products may be biased toward those containing simpler, less reactive, functionality (like substituted phenyl instead of a heterocycle).
    Large numbers of compounds generated may preclude individual purification and weighing of final products; therefore, the screening samples may be of only approximate purity and concentration.
    Although the incorporation of three or more diversity elements in a library contributes greatly to combinatorial power and the number of compounds in the library, this also tends to yield compounds of molecular weight higher than that of most orally active drugs (violation of Rule of 5).
14. Reactions in solution generally occur more smoothly and efficiently than on a solid support, a heterogeneous mixture. Monitoring reactions in solution is easier than on a solid support; in many cases, if the product is covalently attached to a solid support, an aliquot of the product needs to be cleaved from the solid support and analyzed in solution.
15. **a.** (1) These properties are generally directed at suitable pharmacokinetics and lack of toxicity.
    (2) These structures occur within many different drugs and are good starting points, especially if nothing is known about the target structure.

(3) Toxicophores can cause side effects, so by not including them in your screening library, you can avoid obvious undesirable effects.

b. The disadvantage is that you may overlook some important substructures, which themselves may not be desirable, but they may give you ideas for other structures that have more desirable properties.

16. Ligand-based molecules are designed from known active ligands or computational methods and SAR information; the 3D structure of the target protein is not known. Structure-based molecules are designed from a known crystal or NMR structure or homology model of the target protein, and, generally, designed ligands are docked into the known binding cavity of the target protein to determine calculated binding energies.

17. You need to discuss this with regard to the principles of SAR by NMR.

a. The carboxylic acid group may be essential for binding to the receptor. Therefore, compounds **8** and **9** are not oriented correctly on the receptor or the two compounds were connected in the wrong place.

b. There is more than one correct answer. The structures below are possibilities.

These structures allow the carboxylate to remain free to bind to the receptor.

18. a. For Compound 1:

$IC_{50} = 8\ \mu M = 8 \times 10^{-6}$ M
$pIC_{50} = -\log IC_{50} = -\log (8 \times 10^{-6}) = 5.1$
MW = 168.2 Da = 0.1682 kDa
BEI = pIC_{50}/MW = 5.1/0.1682 = **30.3**

For Compound 2:

$IC_{50} = 0.3\ \mu M = 3 \times 10^{-7}$ M
$pIC_{50} = -\log IC_{50} = -\log (3 \times 10^{-7}) = 6.5$
MW = 374.4 Da = 0.3744 kDa
BEI = pIC_{50}/MW = 6.5/0.3744 = **17.4**

b. LELP = log P/LE; therefore BEI-LP = log P/BEI

For Compound 1:
BEI-LP = 0.93/30.3 = **0.03**
For Compound 2:
BEI-LP = 2.08/17.4 = **0.12**

c. Compound 2

d. Compound 1 (higher BEI is better)

e. Compound 1 (analogous to LELP, lower BEI-LP is better)

19. There are many correct answers; below are some of them.

(a)

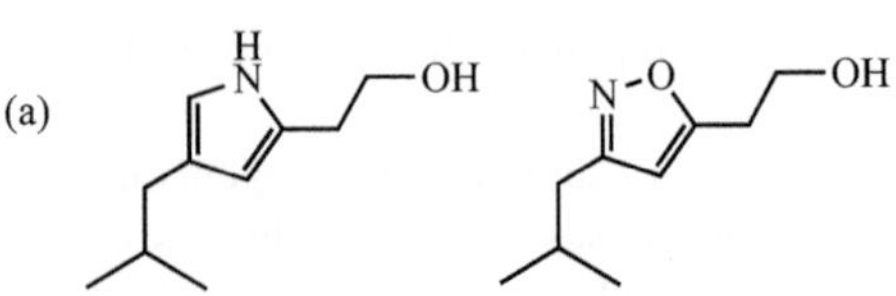

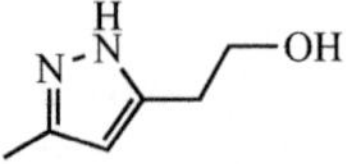

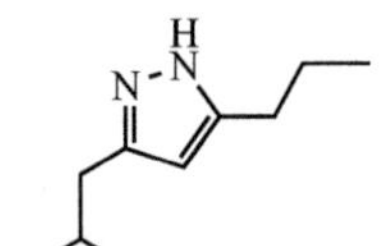

To see the importance of the pyrazole N

To see the importance of the pyrazole NH and if H-bond donor necessary

To see the importance of the pyrazole NH and if H-bond donor or acceptor is necessary

To see the importance of the isobutyl group

To see the importance of the hydroxyl group

(b)

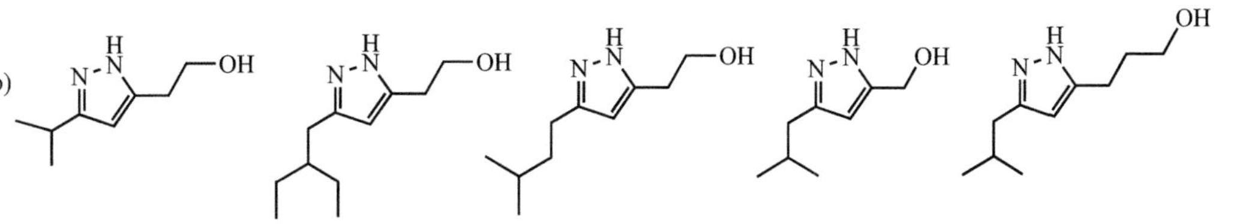

(c)

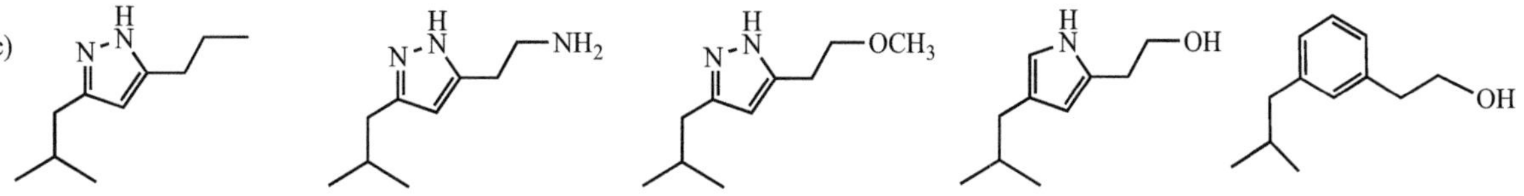

20. **a.** There are many correct answers. Although all the compounds below are peptidomimetics, none may work because the bioactive conformation is not known, and none of these may be in the correct conformation for binding to the receptor.

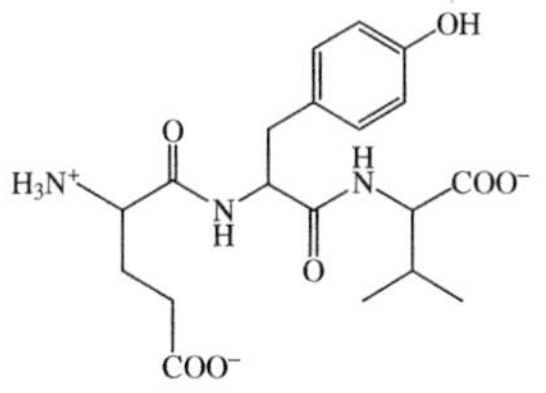

Glu-Tyr-Val (EYV)

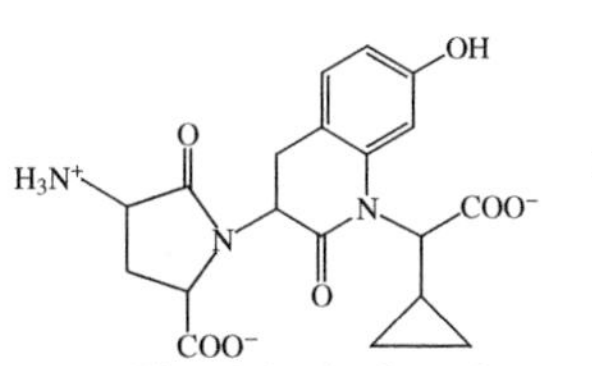

Ring-chain transformations

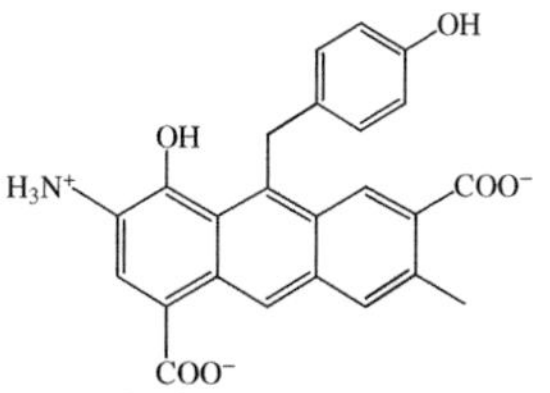

Scaffold peptidomimetic

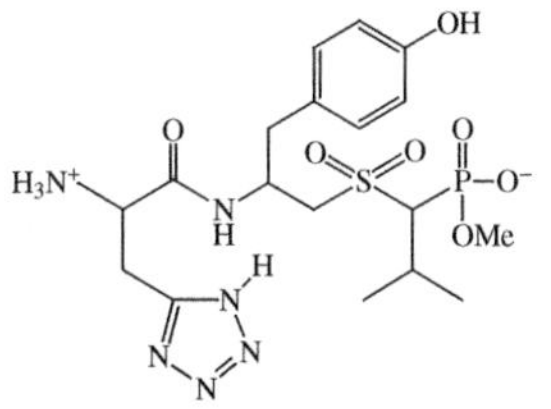

Bioisosteric replacements

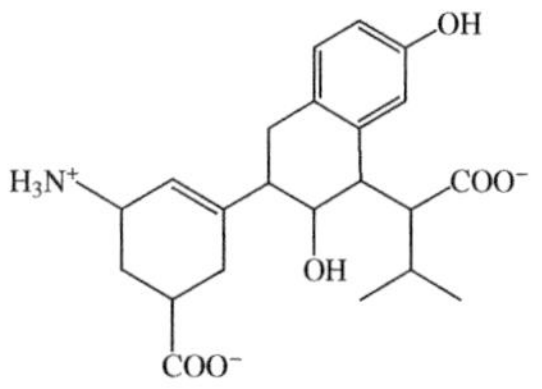

All three

b. You might expect an improvement, but there is generally no assurance that a bioisosteric replacement will improve any of those properties; nonetheless, it might, so it is certainly worth trying. However, it is possible to calculate the extent to which a particularly bioisosteric replacement will improve calculable properties, such as ClogP.

21. The intermediates in brackets are highly electron deficient. Therefore, X = electron donating would increase rate, but X = electron withdrawing would decrease rate.

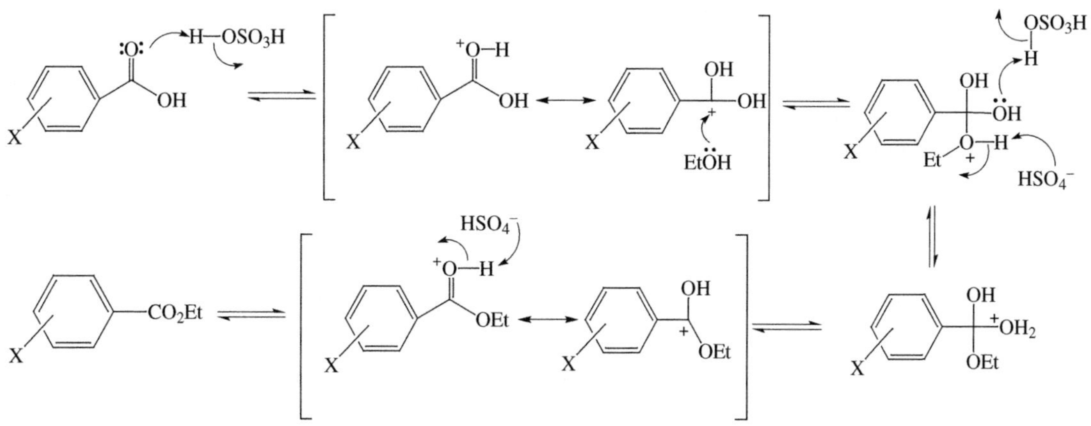

22. a.

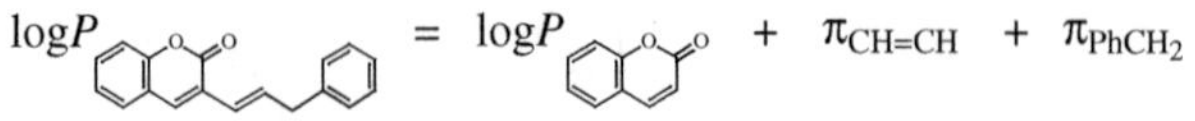

$$\log P_{\text{(3-styrylcoumarin)}} = \log P_{\text{(coumarin)}} + \pi_{CH=CH} + \pi_{PhCH_2}$$

$$\pi_{CH=CH} = 1/3 \log P_{benzene} = 1/3\ (2.13) = 0.71$$
$$\pi_{PhCH_2} = \pi_{Ph} + \pi_{CH_2} = 2.13 + 0.50 = 2.63$$

$$\text{(Note: } \pi_{PhCH_2} = \log P_{PhCH_2NH_2} - \pi_{NH_2} = \log P_{PhCH_2NH_2} - (\log P_{MeNH_2} - \pi_{Me})$$
$$= 1.09 - (-0.57 - 0.50) = 2.16)$$

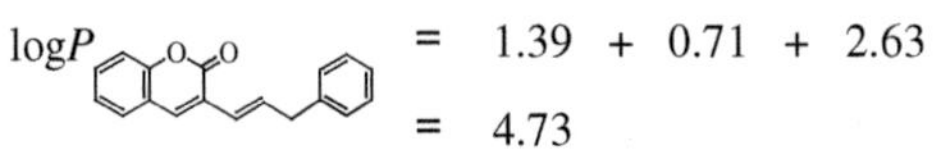

$$\log P_{\text{(3-styrylcoumarin)}} = 1.39 + 0.71 + 2.63$$
$$= 4.73$$

b.

$$\log P_{\text{(iodo vinyl ether)}} = \pi_{CH_3CH_2CH_2} + \pi_{CH_3I} + \pi_{OCH=CH_2} - 0.2 \text{ (for branching)}$$

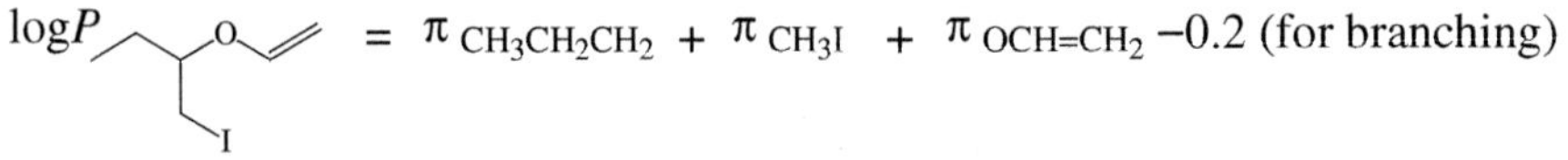

$$\pi_{OCH=CH_2} = \log P_{CH_2=CHOCH_2CH_3} - \pi_{CH_2CH_3} = 1.04 - 1.00 = 0.04$$

$$\pi_{CH_3CH_2CH_2} = 3\pi_{CH_3} = 1.50$$

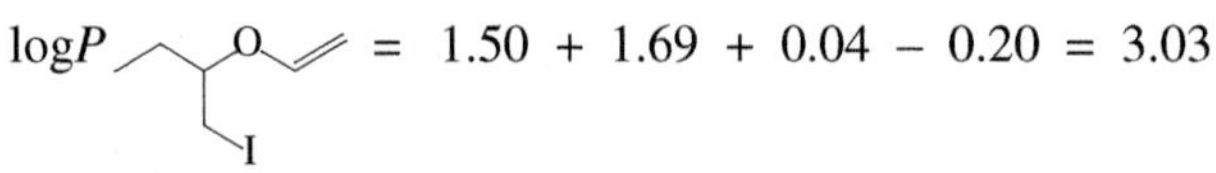

$$\log P_{\text{(iodo vinyl ether)}} = 1.50 + 1.69 + 0.04 - 0.20 = 3.03$$

c. The log$P$ and π values are constitutive, so the log$P$ values obtained will depend on which log$P$ or π values of pieces of the molecule were used to do the calculation for the whole molecule. To demonstrate the constitutive properties of lipophilicity, consider the following examples:

$\log P_{ICH_2COOH} - \log P_{CH_3I}$ should give $\pi_{COOH} = 0.87 - 1.69 = -0.82$

But $\log P_{NCCH_2COOH} - \log P_{CH_3CN}$ also should give $\pi_{COOH} = -0.33 - (-0.34) = 0.01$

Therefore, use log$P$ values of molecules as close in structure as possible to the one for which you need a log$P$ or π value.

Also, $\pi_{CH_3O} \neq \log P_{CH_3OH}$. However, π_{CH_2OH} does equal $\log P_{CH_3OH}$. Because $\pi_{Me_2N} = \log P_{Me_2NH}$ ($\pi_{Me_2N} = \log P_{Me_3N} - \pi_{Me} = 0.27 - 0.50 = -0.23 = \log P_{Me_2NH}$) and the p$K_a$ of amine is much greater than alcohol, the ionization factor may be important.

Therefore, there will be different correct answers depending on which molecules you choose to determine your π values.

23. a. Log D measures the log$P$ at a given pH. If the molecule does not ionize, then pH will not have an effect, and the log D will not change with pH.

b. At low pH (<pH 5), the pyridine ring is protonated, producing an equilibrium between the charged form and the neutral form; the more in the charged form, the larger negative the log D value. As the pH increases, the equilibrium changes more and more to favor the neutral form, thereby making the log D less negative.

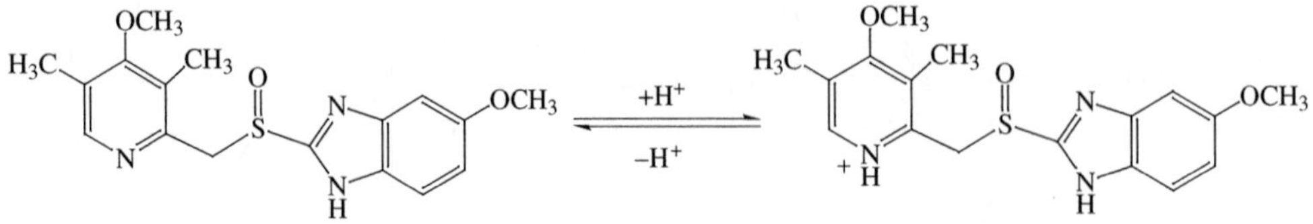

(Structures: neutral pyridine form $\underset{-H^+}{\overset{+H^+}{\rightleftharpoons}}$ protonated pyridinium form; labels: OCH_3, H_3C, CH_3, O, S, N, N, H, OCH_3)

Between pH 5 and 9 the molecule is completely in the neutral form.
At higher pH values (>pH 9) deprotonation of the imidazole ring occurs, giving an equilibrium between the neutral form and the stabilized anion, which again lowers the log *D*.

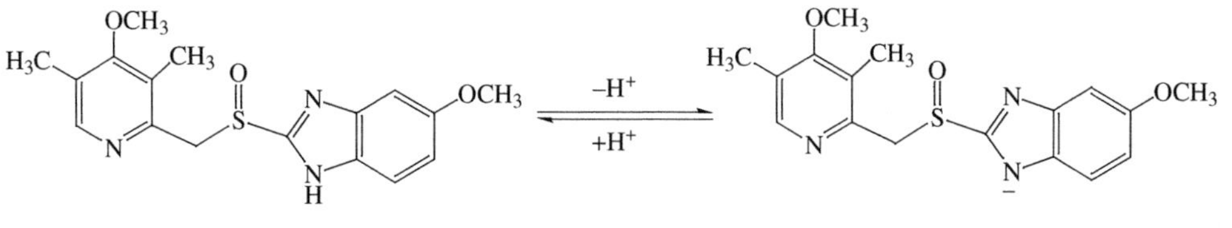

$-H^+$ / $+H^+$

24. No. The molecular weight is too high (663), there are too many hydrogen bond donors (7) and acceptors (13), and too many rotatable bonds (21).

25. a. A possible reason is that the molecule is too basic, so the equilibrium strongly favors the protonated form, which cannot cross membranes and therefore exhibits low activity.

b. Change the pK_a of the pyridine by adding electron-withdrawing groups to it or using less electron-donating groups that are isosteric:

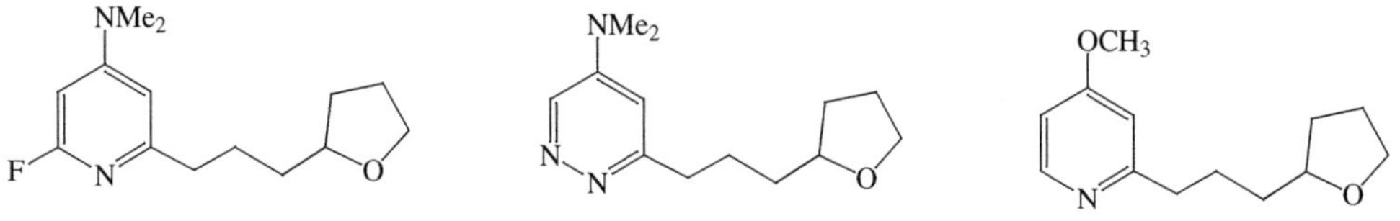

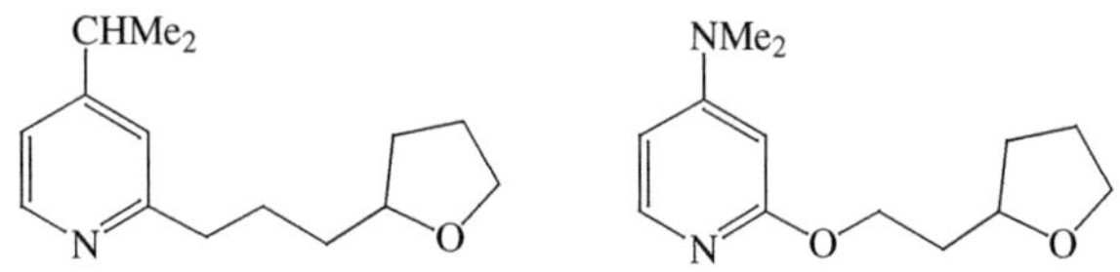

26. a. Either there is a steric problem at position 4 or the optimum π and/or σ value has been exceeded.
Try the 3-Br compound to check for steric hindrance.
Try the 4-CN compound ($-\pi$)
Try the 4-CH_3 compound ($-\sigma$)
Try the 4-NH_2 compound ($-\pi$, $-\sigma$)

b. Two variables have been introduced: a change in position and a change in π. Therefore, you cannot determine if it is a steric effect or a π effect.
Try the 3-Br compound to see if steric.

27. DOCK—An algorithm for novel drug discovery that determines the best fit of a large number of random compounds into a known receptor structure. Each molecule is docked into the receptor in a number of geometrically allowable orientations to determine shape complementarity. First, the Connolly molecular surface (Connolly, M. L. *Science* **1983**, *221*, 709) for the receptor is developed, then a space-filling negative image of the receptor site is created, then a database of compounds (potential ligands for the receptor) is used to match the molecular structure to the negative image of the receptor. See Kuntz *et al.*, *Acc. Chem. Res.* **1994**, *27*, 117 for applications.
CoMFA—A 3-D QSAR methodology that involves the use of partial least-squares data analysis to compute separately the contributions of steric (shape) and electrostatic (electronic) molecular mechanic force fields of a set of molecules. These results provide parameters that can be correlated with noncovalent receptor binding and the biological properties of the molecules.

28. When a ligand binds to a receptor, it can change the receptor conformation. If you use the unliganded crystal structure, you could be docking compounds into the wrong binding site conformation. The liganded structure also gives you an idea of which interactions are important.

29.

- Steric effects affect the surface area and volume.
- Electronic effects affect the pK_a, the charge, and the electrostatic potential.

- Lipophilic effects affect the log*P*.
- H-bonding affects the number of H-bond acceptors, donors, log*P*, and conformation.

CHAPTER 3

1. **a** electrostatic interaction/hydrogen bonding
   b halogen bonding
   c hydrophobic effect
   d dipole–dipole interaction
   e ion–dipole interaction (NH_3^+ with O–C bond) and cation-π interaction (NH_3^+ with phenyl ring)
2. There may be another lysine close to the lysine, lowering its pK_a.

   $LysH^+$ Lys^+H disfavored so one lysine becomes more acidic (lower pK_a).

   There may be a compensating aspartate near the histidine, raising its pK_a.

   $HisH^+$ Asp^- this is favorable, so the pK_a of histidine increases (to keep it protonated).
3.

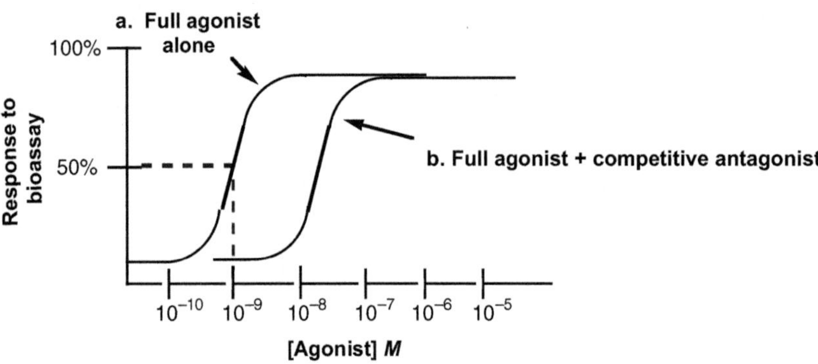

4.

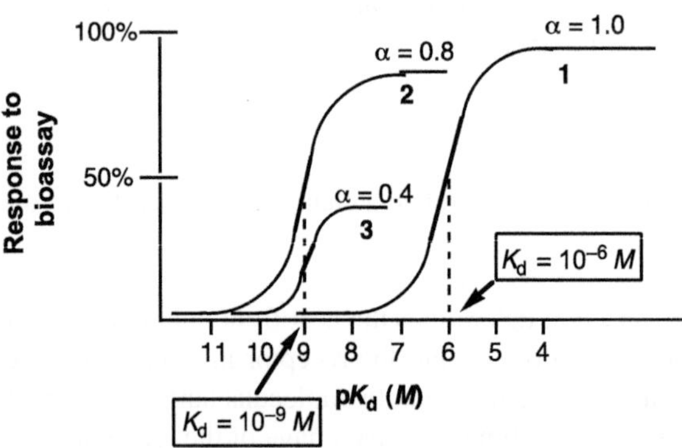

5. **a.** Affinities are measured by the K_d. The order of increasing affinity is **1** < **2** < **4** < dopamine < **3**
   b. **2** < **4** < **3** = dopamine. Inverse agonist **1** has efficacy roughly comparable to that of **4** but in the opposite direction; put another way, **1** and **4** are equally efficacious for stimulating their respective responses.
   c. **1** inverse agent
   2 antagonist
   3 full agonist
   4 partial agonist
6. **a.** Generally, only one enantiomer is active or, at least, much more potent (the eutomer) than the other enantiomer (the distomer). The distomer may be responsible for side effects or toxicities. The distomer may be the eutomer for a different activity. The distomer may block the eutomer from binding to the receptor. One enantiomer may be an agonist, and the other an antagonist for the same receptor.

b. The best way to increase the eudismic ratio is to administer only the eutomer. Otherwise, incorporate the chiral center in the pharmacophore.

7. a.

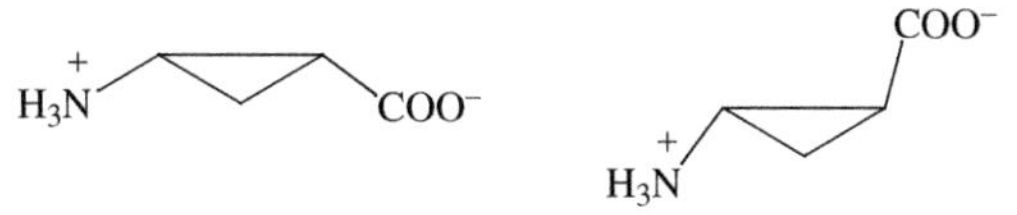

(Many other ring systems)

b.

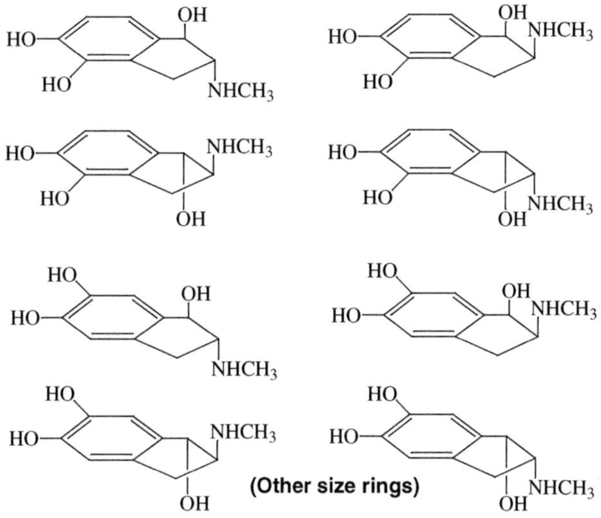

(Other size rings)

c.

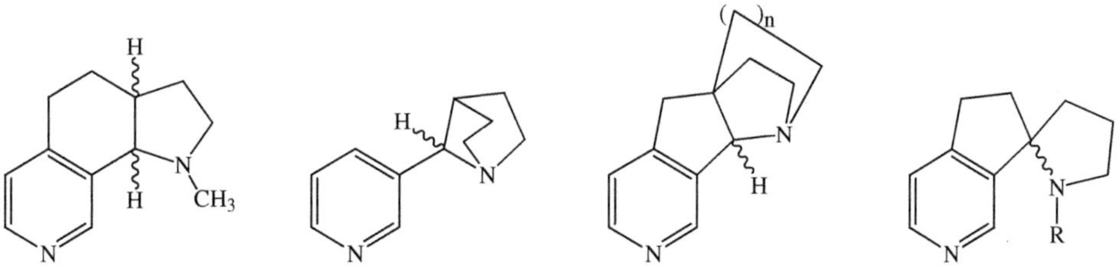

Ulrich, T.; Krich, S.; Binder, D.; Mereiter, K.; Anderson, D. J.; Meyer, M. D.; Pyerin, M. *J. Med. Chem.* **2002**, *45*, 4047.

8. It should be a pure antidepressant. It has an α angle because all of the central atoms have sp^3 hybridization. It has a β angle because the center ring is 7-membered and the other two are 6-membered. It has a γ angle because the top two center atoms are sp^3.

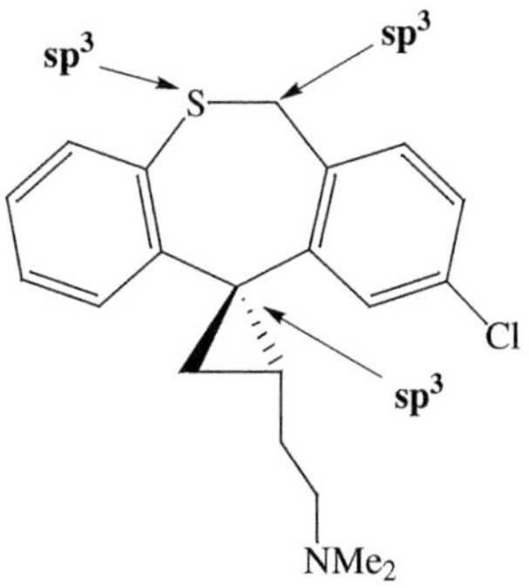

9. B is more stable than A when X = CH_2 because of smaller 1,3-diaxial interactions. When X = S, the ring size is larger (atomic size of S > CH_2), so the equilibrium does not favor B as strongly.

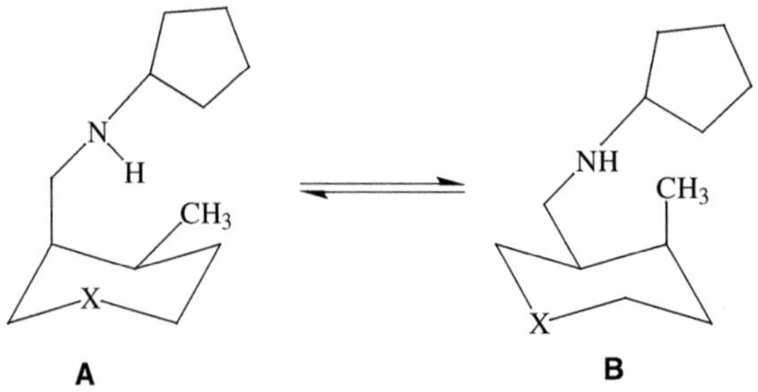

When X = O, H-bonding to NH changes equilibrium to favor A.

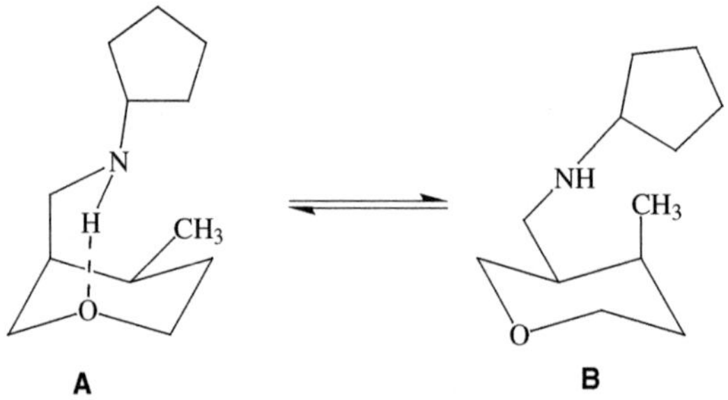

When X = NH, the H bond is even stronger, favoring A more. The bioactive conformation is A.

10. This is supposed to be related to the discussion of the discovery of cimetidine. Without knowledge of the structure of the receptor or the structure of any lead compounds, a lead must be discovered. A random lead discovery approach could be tried. A more rational approach would be to prepare analogues of tyramine. The phenolic OH and the protonated amino groups may be important to agonist activity. Therefore, you may want to modify these groups to avoid agonism, for example, by replacing the OH with other groups, such as OCH_3, CH_3, SH, NH_2. CH_3 and NH2 are isosteres; SH would be more ionized; OCH_3 cannot ionize. The phenyl can be replaced by $–(CH_2)_3–$ or by another ring system or be further substituted. The $–CH_2–CH_2–$ can be lengthened or shortened or substituted. The amino group can be replaced by amidino, guanidino, ureido, etc. At each stage, antagonist and agonist activities need to be evaluated by bioassay.

11. The most stable atropisomer has the wrong geometry for good binding. Below are four atropisomers.

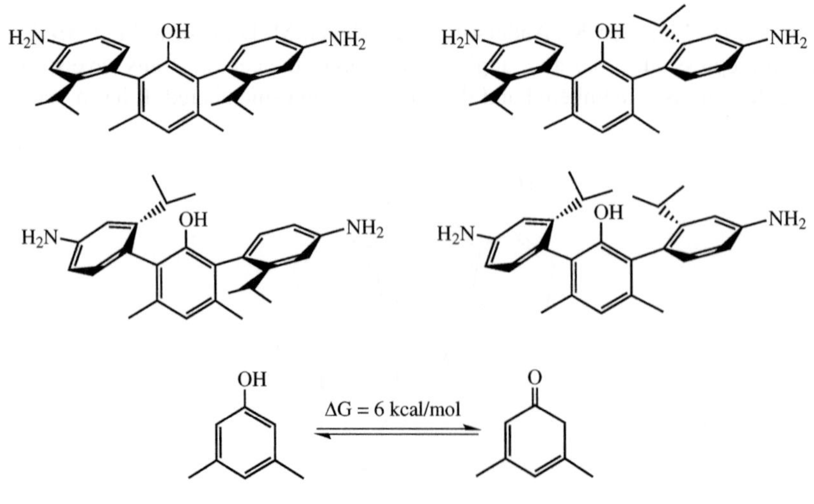

$\Delta G = 6$ kcal/mol

CHAPTER 4

1. Substrate recognition and catalysis of a reaction or specificity and rate acceleration.
2.

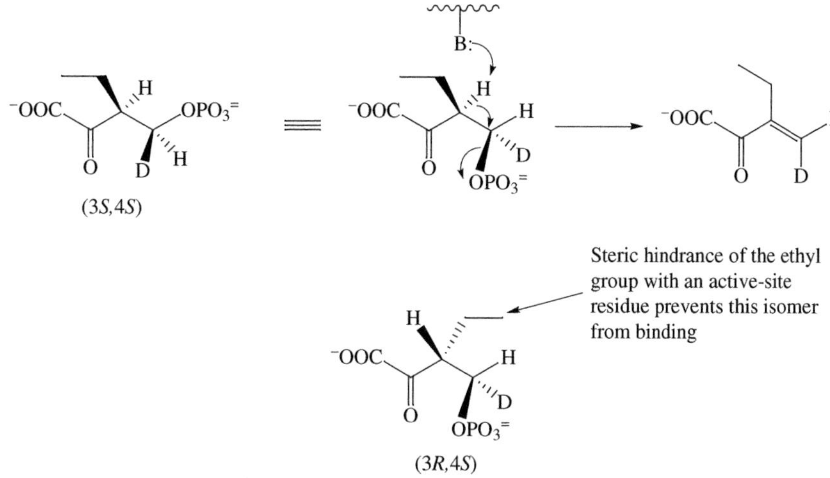

(3S,4S)

Steric hindrance of the ethyl group with an active-site residue prevents this isomer from binding

(3R,4S)

3.

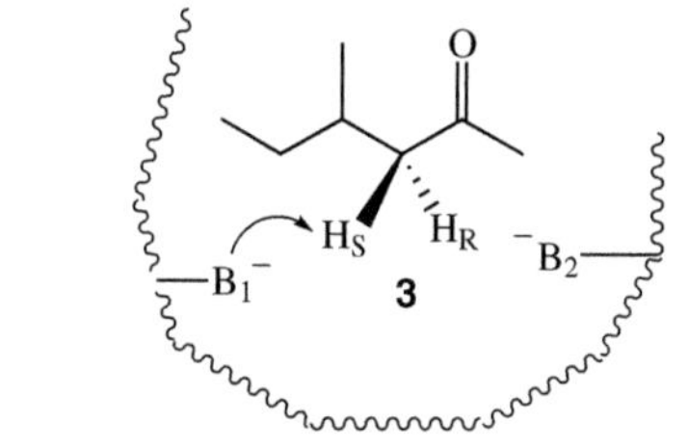

B_1 is the active site base responsible for deprotonation of H_S. B_2 is not a strong enough base and/or not close enough to H_R of **3** to remove it.

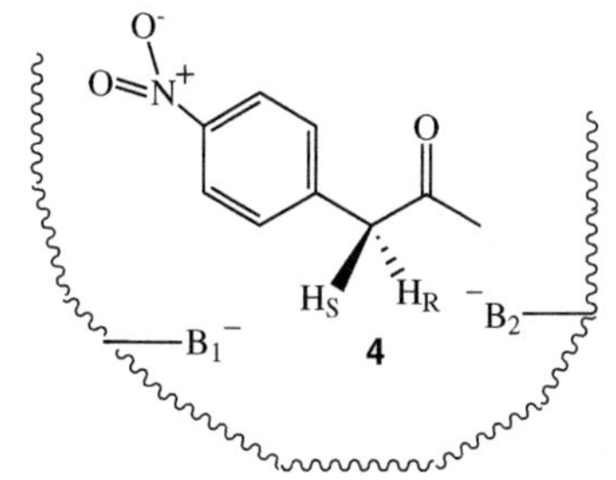

The *p*-nitrophenyl group of **4** lowers the pK_a of both H_S and H_R. Now, maybe B_2 is a strong enough base to remove H_R some of the time. Or maybe there is a small conformational change when it binds, which moves B_2 closer to H_R.

4. Simultaneous protonation of the ketone carbonyl and deprotonation of the α-proton gives the enol, which can be coupled with protonation of the β-hydroxyl group to effect β-elimination. Because the pK_a values of residues inside

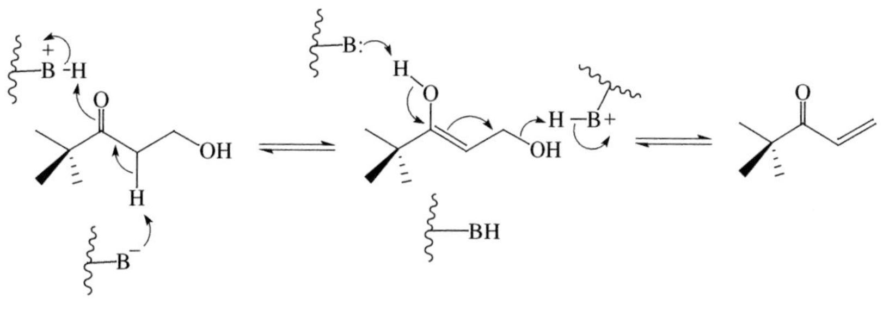

an enzyme can be quite different from those in solution, these protonations and deprotonations may be more facile than expected.

5.

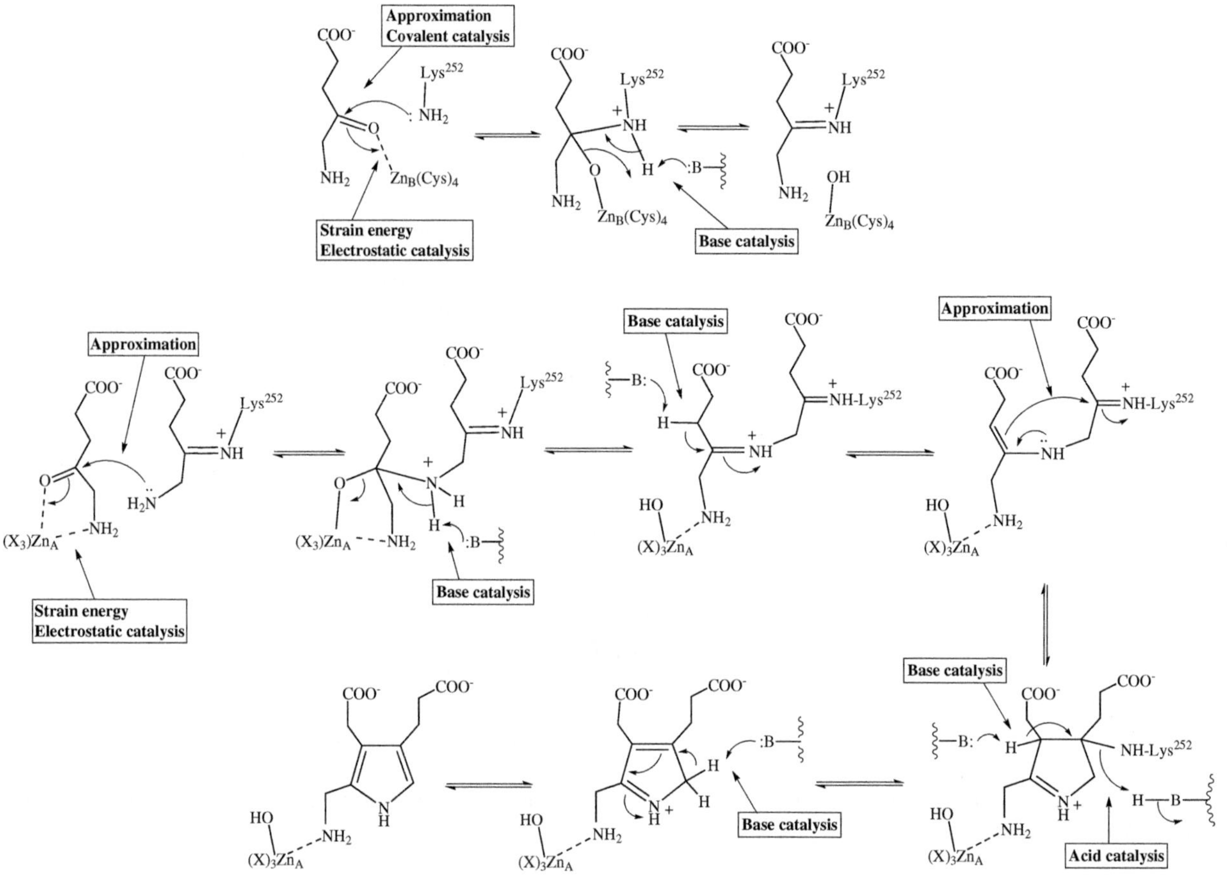

Approximation
Covalent catalysis
Strain energy
Electrostatic catalysis
Base catalysis
Approximation
Strain energy
Electrostatic catalysis
Base catalysis
Base catalysis
Approximation
Base catalysis
Base catalysis
Acid catalysis

6.

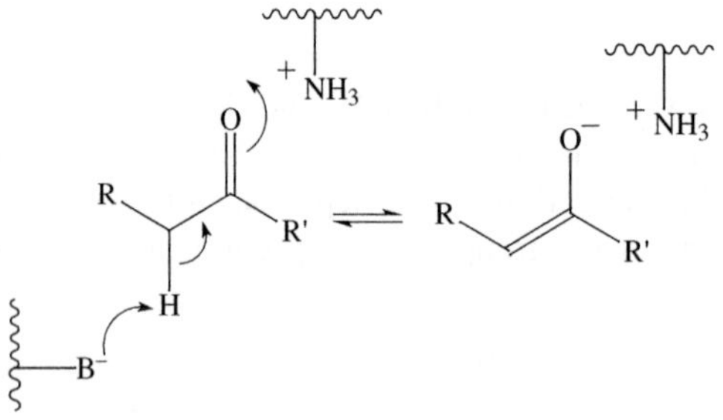

7.

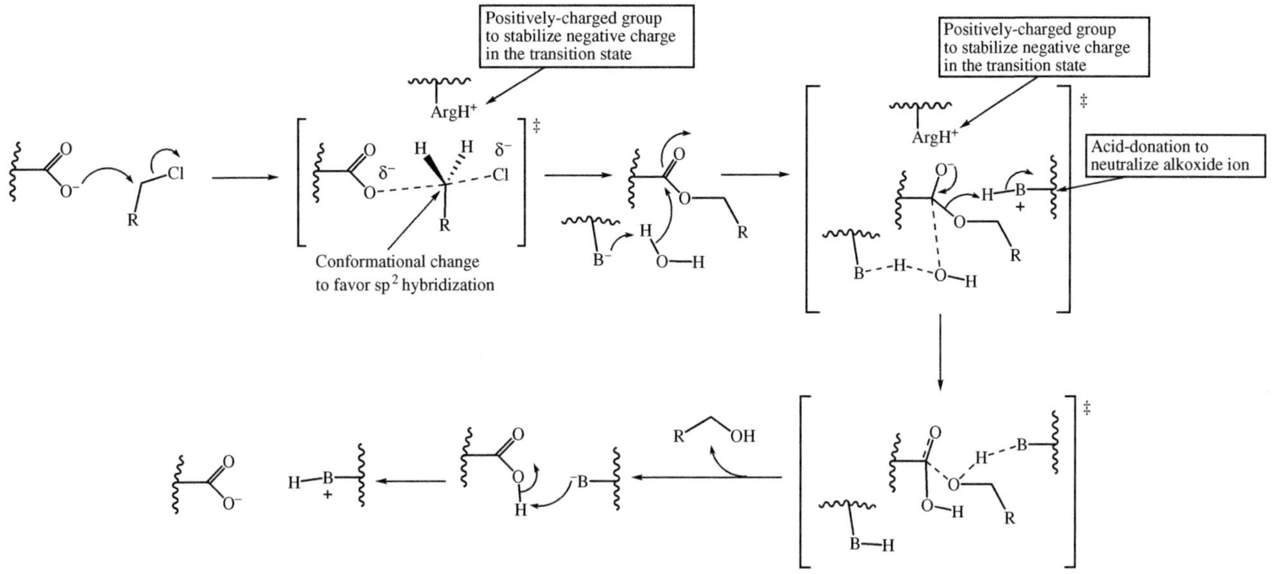
Positively-charged group to stabilize negative charge in the transition state
Conformational change to favor sp2 hybridization
Positively-charged group to stabilize negative charge in the transition state
Acid-donation to neutralize alkoxide ion

8. Both the substrate (aspartate) concentration and the α-keto acid (pyruvate) concentration. One of them may have been completely consumed.

9.

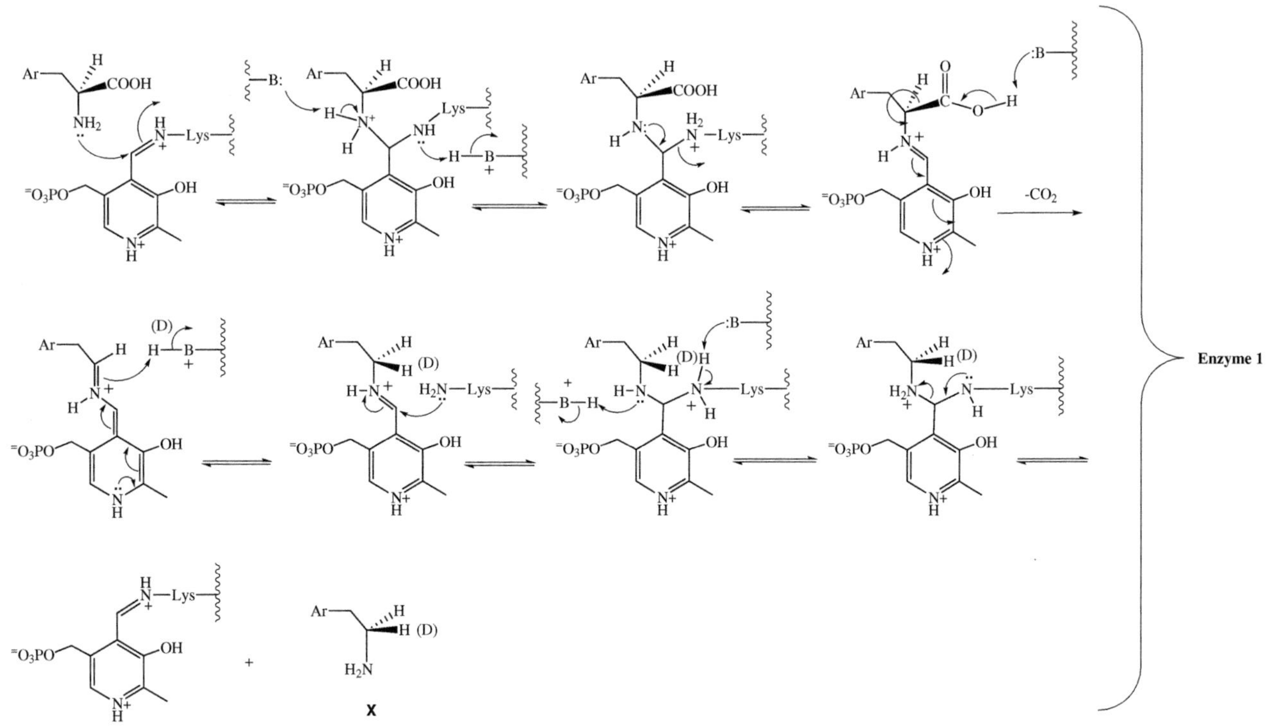
-CO2
Enzyme 1
X

NADPH
NADP+ Fl_ox
FlH⁻ FlH•
O₂
FlH• Fl_ox
+H⁺
-H₂O
Enzyme 2

10. a.

PMP
SSA

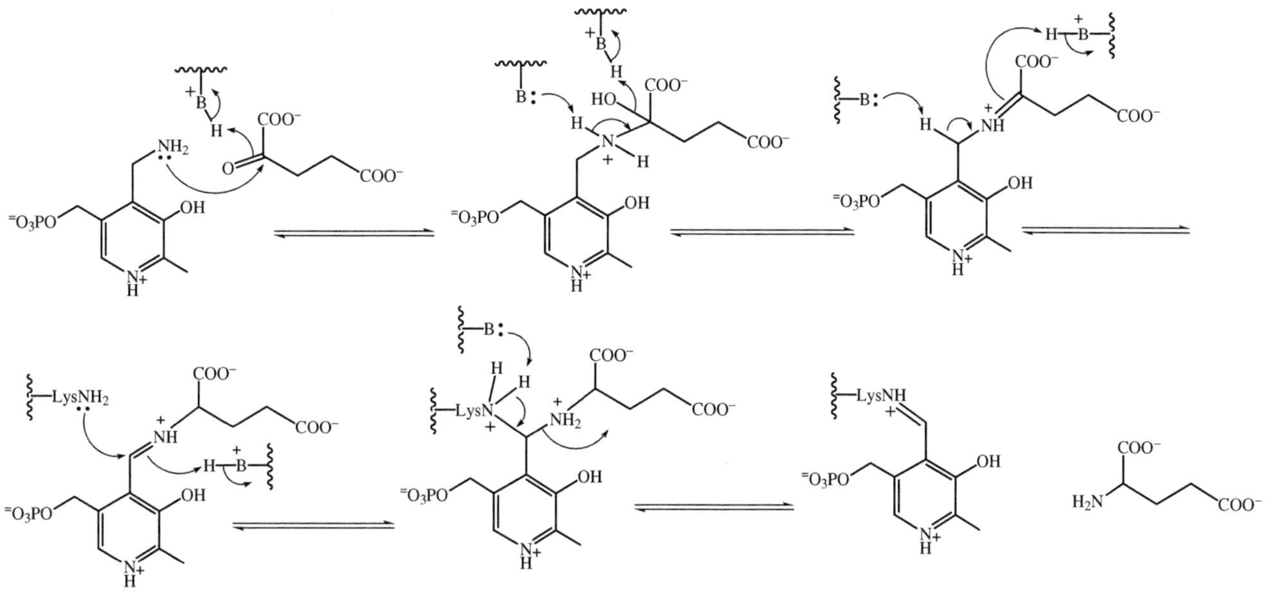

b. Without α-ketoglutarate, the coenzyme cannot get back into the active PLP form, so one turnover is all you get.

c. See mechanism in part a. The ^{18}O is incorporated into the carbonyl oxygen of succinic semialdehyde. Because of the acidity of protons adjacent to imines and carbonyls, there could be exchange with deuterium, but the answer based on the mechanism shown is that the succinic semialdehyde would have ^{18}O in the aldehyde carbonyl, and the glutamic acid would have a deuterium at the α-carbon.

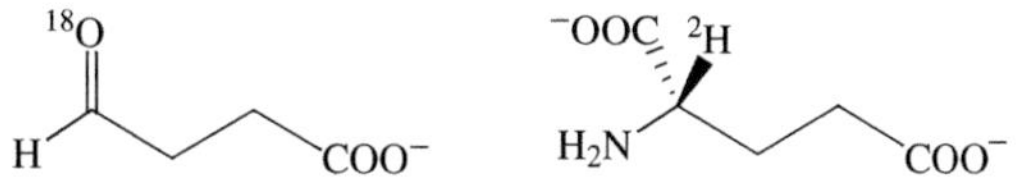

11.

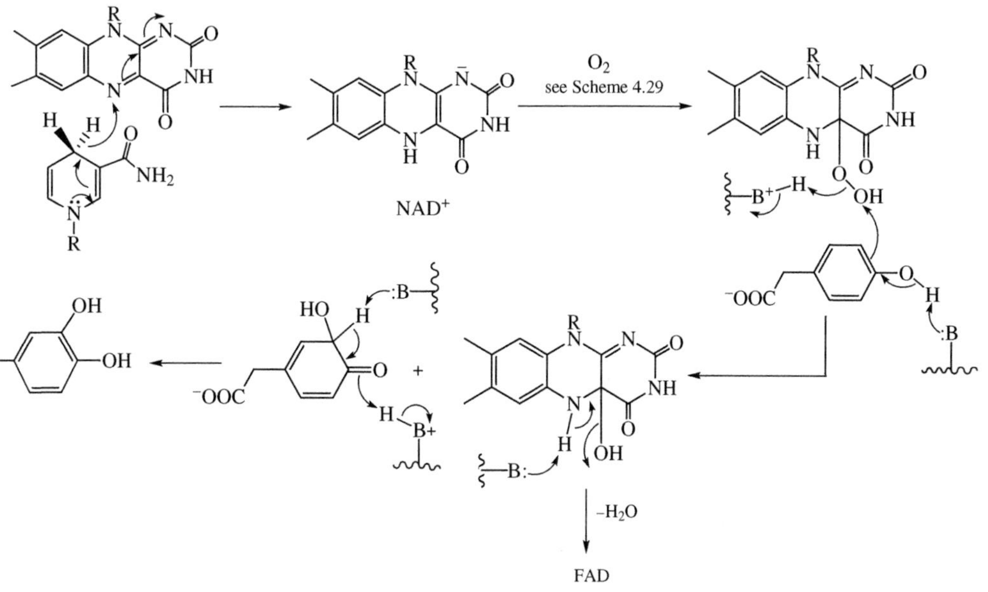

12.

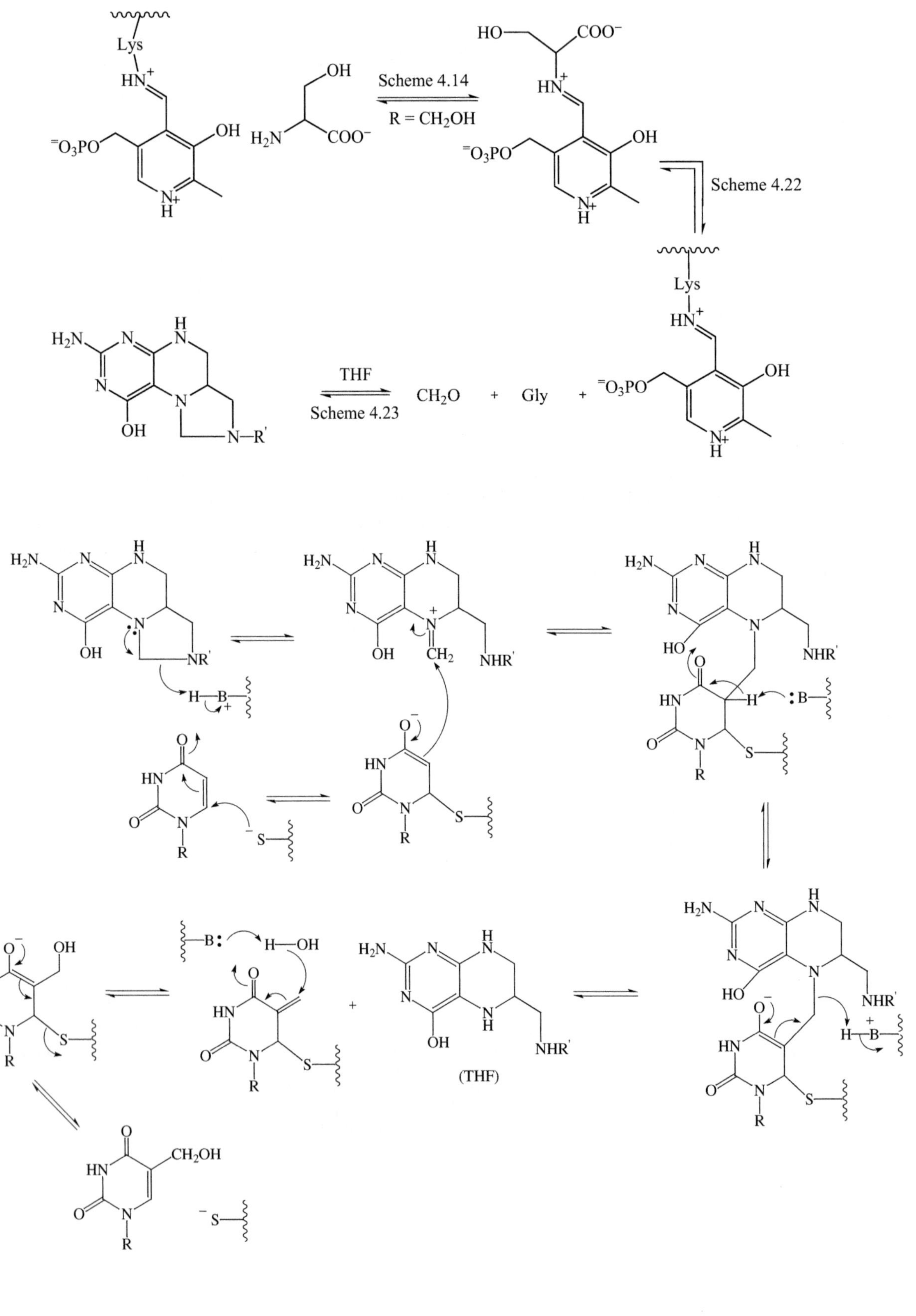

Scheme 4.14

R = CH_2OH

Scheme 4.22

THF

Scheme 4.23

CH_2O + Gly +

(THF)

13.

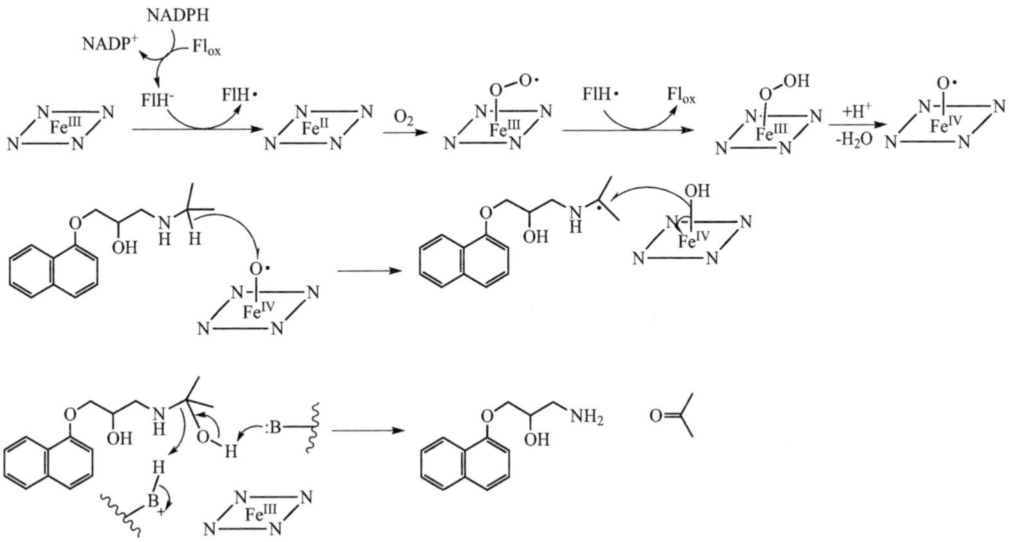

NADPH
NADP$^+$
Fl_{ox}
FlH$^-$
FlH•
Fe^{III}
Fe^{II}
O_2
Fe^{III}
FlH•
Fl_{ox}
Fe^{III}
+H$^+$
-H$_2$O
Fe^{IV}

14.

ATP
–ADP
NAD$^+$

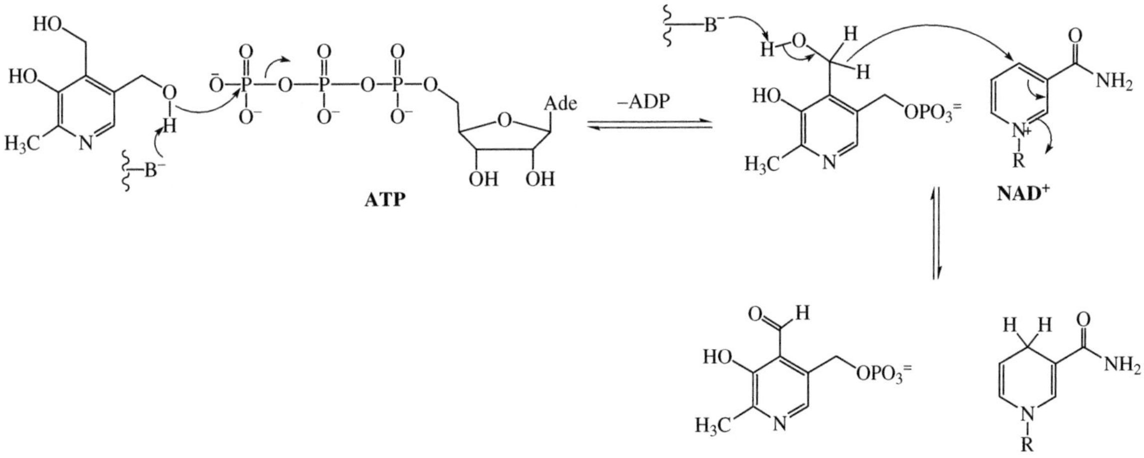

15. Enzyme reaction (1):
Requires L-serine and tetrahydrofolate (or folate and NADPH)

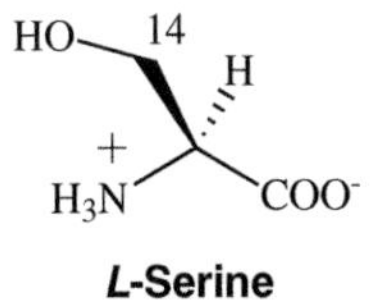

L-Serine
The starting amino acid

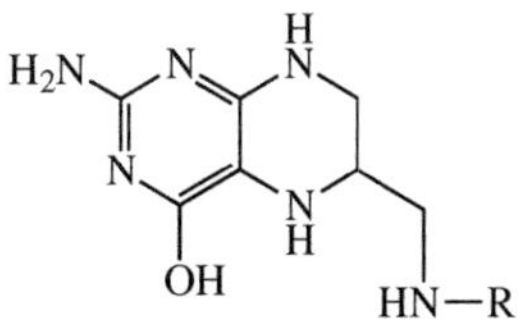

Tetrahydrofolate

These give [^{14}C]methylenetetrahydrofolate

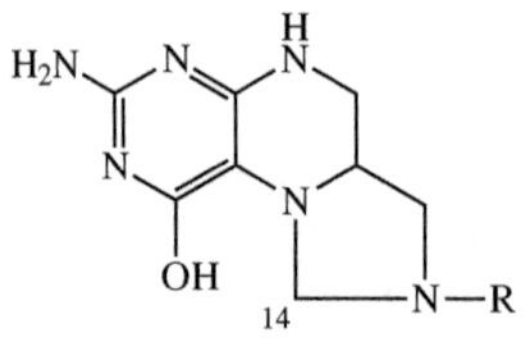

Methylenetetrahydrofolate

Enzyme reaction (2):

Requires NAD$^+$ to oxidize [^{14}C]methylenetetrahydrofolate to [^{14}C]formyltetrahydrofolate

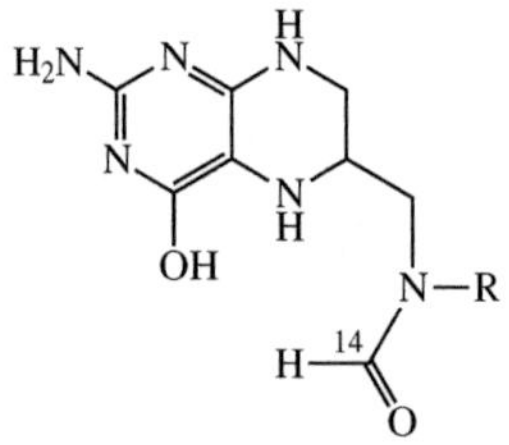

N^{10}**-[^{14}C]Formyltetrahydrofolate**

Enzyme reaction (3):

Transfers the [^{14}C]-formyl group from [^{14}C]formyltetrahydrofolate to methionine.

16. There are two half reactions in a PLP-dependent aminotransferase reaction.

Step 1:

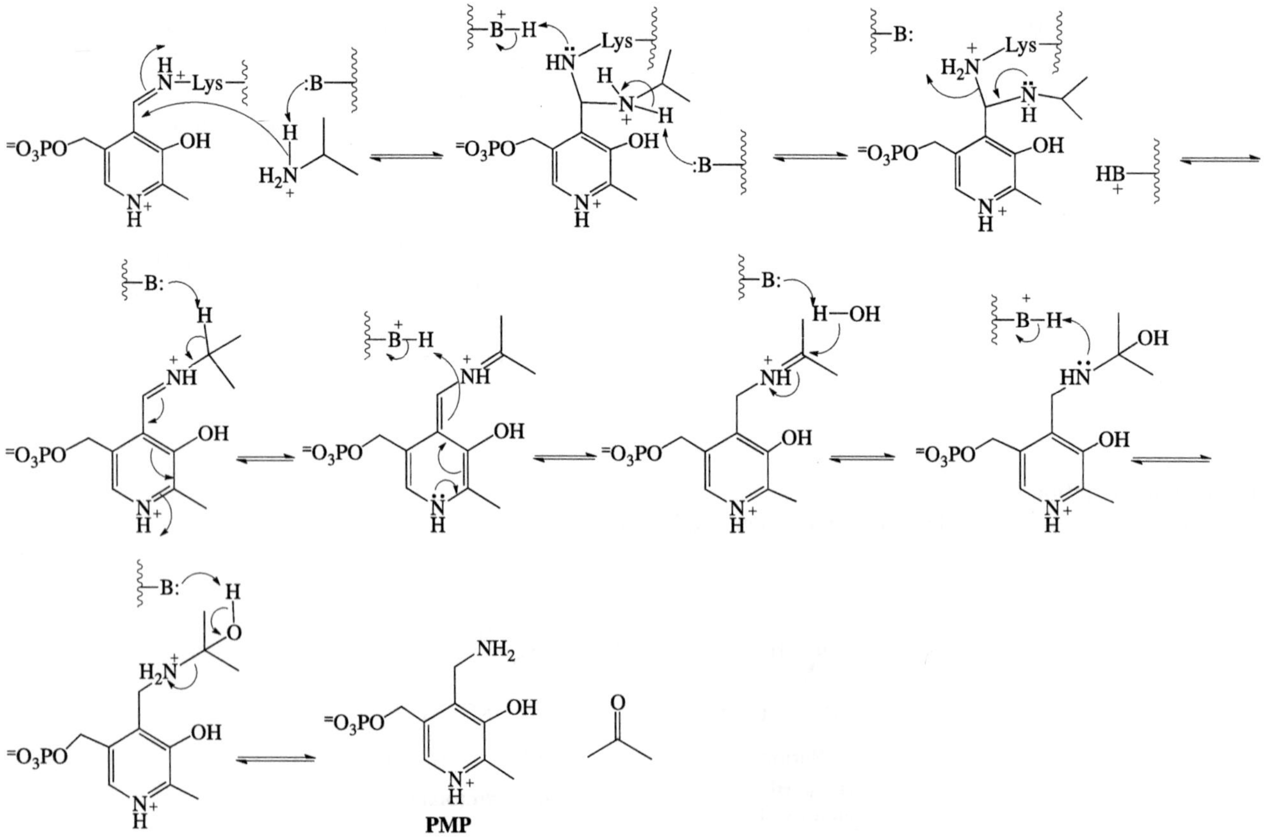

PMP

Step 2:

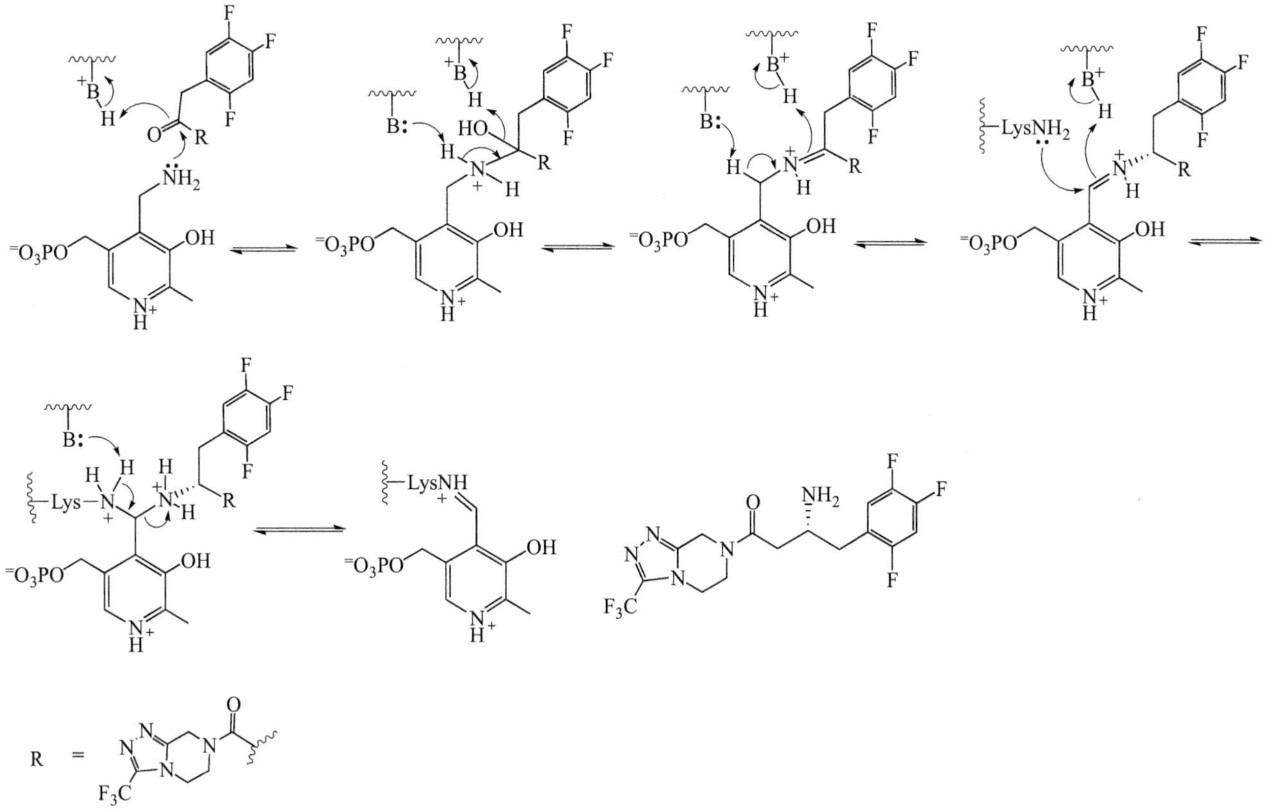

CHAPTER 5

1. To block the metabolism of a substrate, if the substrate produces the desired therapeutic effect; to block the production of a product, if the product causes the problem; to block a metabolic pathway in a foreign organism or tumor cell.
2. Try to determine thermodynamic or kinetic differences in the enzymes from the two sources so that selective inhibition might be possible. Synthesize substrate analogues and measure binding constants for the two enzymes to see if there is a difference in binding affinities.
3. There are many correct answers for both parts.

   a.

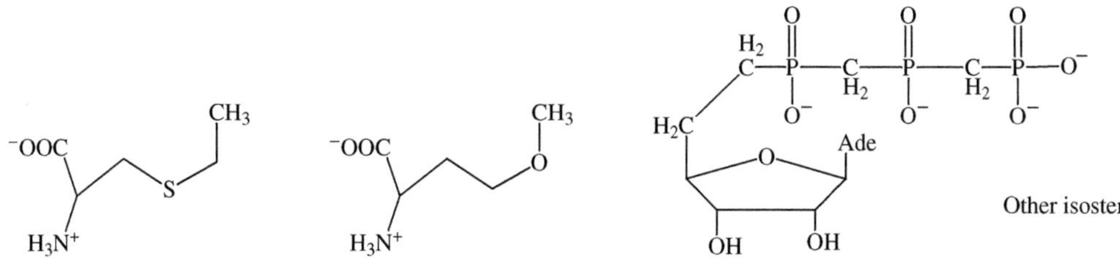

Other isosteres

b.

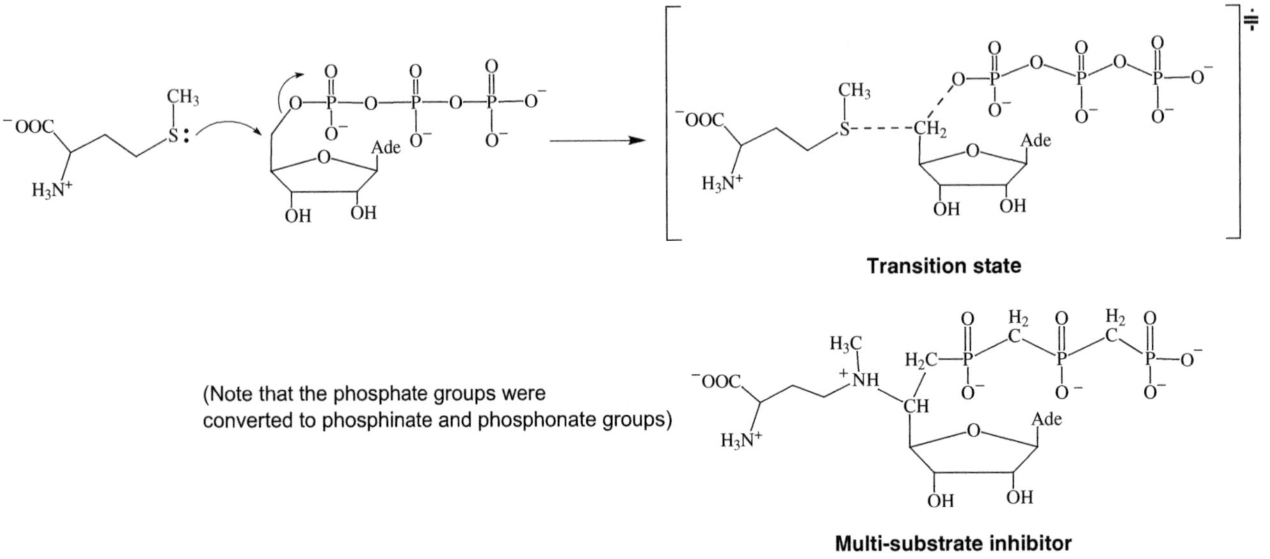

Transition state

(Note that the phosphate groups were converted to phosphinate and phosphonate groups)

Multi-substrate inhibitor

4. In a slow, tight-binding inhibitor the E·I complex is more stable. The enzyme has changed conformation so it has to change back to release the inhibitor.
5. a. Design an inhibitor of TXA_2 synthase, an antagonist of the TXA_2 receptor, or a dual-acting enzyme inhibitor/receptor antagonist.
   b. Design a dual-acting drug.
   c. The idea is to combine parts of each molecule into a single compound that could bind to both receptors. Examples would be:

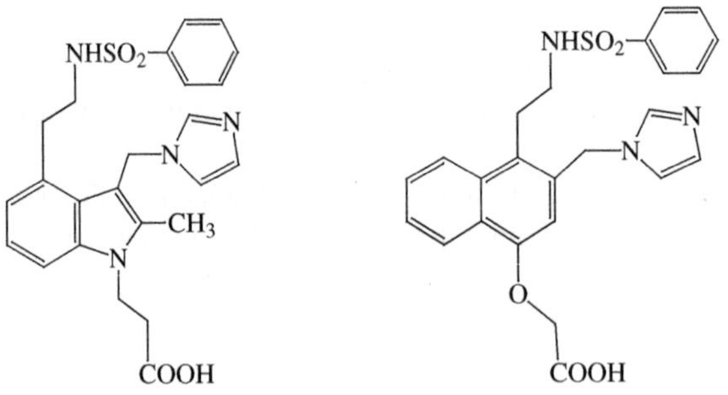

The following compound was shown to be active:

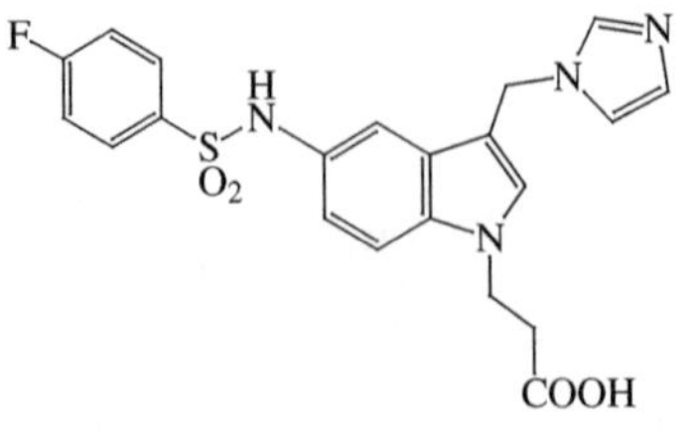

Dickinson, R. P.; Dack, K. N.; Steele, J. Tute, M. S. *Bioorg. Med. Chem. Lett.* **1996**, *6*, 1691. See also, Dickinson, R. P.; Dack, K. N.; Long, C. J.; Steele, J. *J. Med. Chem.* **1997**, *40*, 3442.

6.

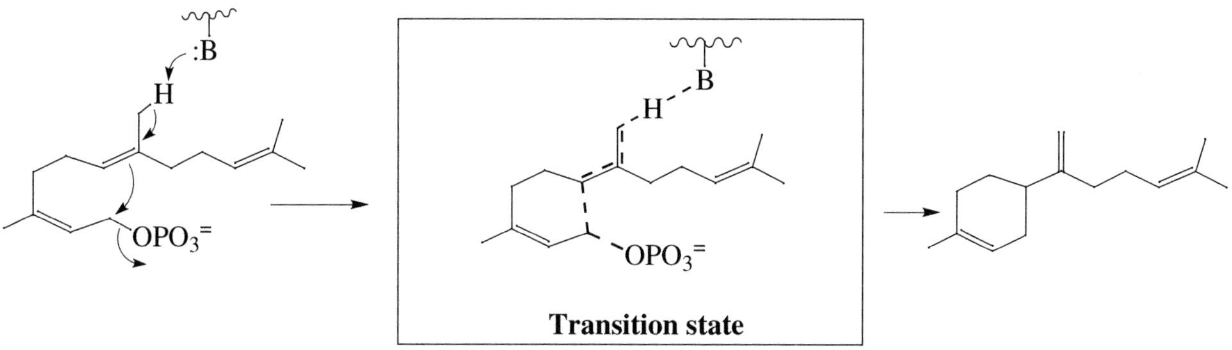

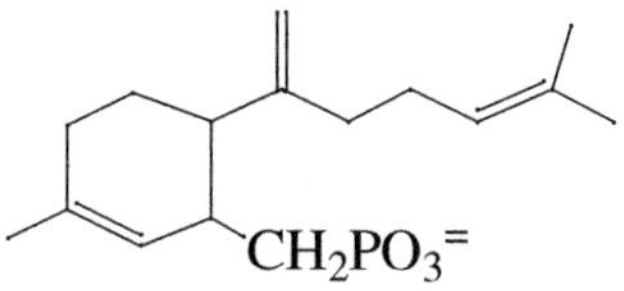

Transition state

$CH_2PO_3^=$

Use a phosphonate so it is not very reactive

7. A dihydrofolate reductase inhibitor, because dihydrofolate reductase is needed to reduce the dihydrofolate to tetrahydrofolate produced from methylene tetrahydrofolate in the thymidylate synthase reaction.

8. It coordinates to the ferric ion.

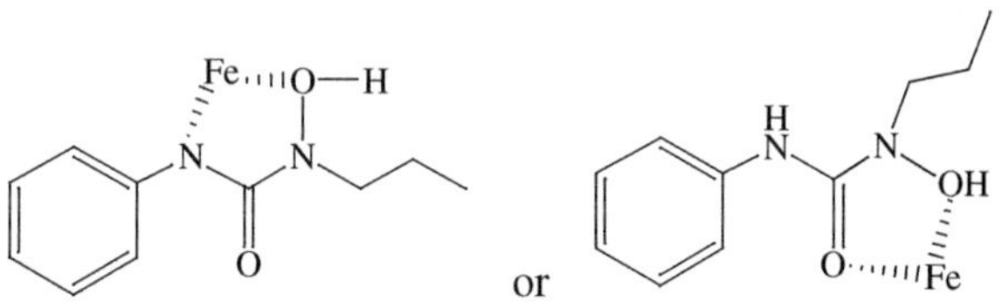

or

9.

- A—water soluble isostere of phenyl; the methyl group may impart a hydrophobic effect and expand the pharmacophore
- B—ketomethylene peptidomimetic for an amide
- C—hydrophobic group; the fluorine prevents metabolism (see Chapter 8)
- D—ring-chain glutamine mimic; more lipophilic
- E—Michael acceptor for affinity labeling with an active-site nucleophile. This is an example of a quiescent affinity labeling agent. The Michael acceptor does not react with thiols, such as glutathione under physiological conditions, but the active site cysteine residue in rhinovirus protease is exceptionally nucleophilic and undergoes Michael addition with irreversible inactivation of the enzyme.

10. **a.** Design a selective inhibitor of isoform-1.

b. Try to make a selective reversible inhibitor that has a group large enough to bind at the cysteine bonding site, but not in the phenylalanine binding site.

Design an affinity labeling agent that reacts with the cysteine residue.

11.

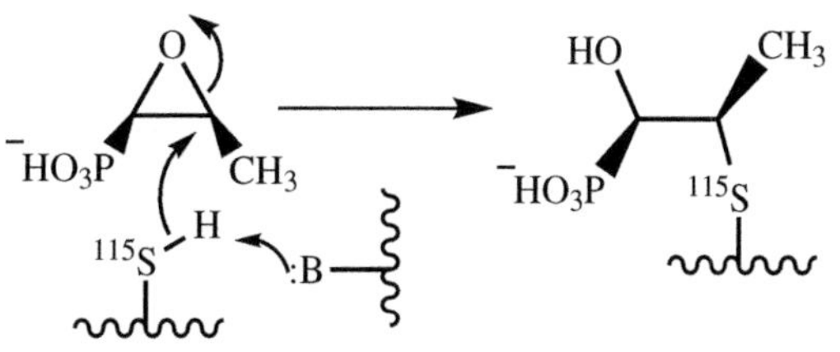

12.

5a

Base catalyzed to make a better nucleophile

Acid catalyzed to activate the carbonyl

Acid catalyzed to allow cyclization at tertiary carbon more readily

5b

Attack at tertiary carbon, suggesting S_N1-like, requiring acid catalysis. This gives 6-membered ring; attack at secondary carbon gives less favored 7-membered ring

13. Only bortezomib, orlistat, and neratinib are covalent inhibitors. Threonine-1 of the 20S proteasome reacts with the boron atom of bortezomib,[1] serine-2308 of fatty acid synthase is acylated by the β-lactone ring,[2] and cysteine-797 of epidermal growth factor receptor undergoes Michael addition to neratinib.[3]

Bortezomib

Orlistat

Neratinib

1. Groll, M.; Berkers, C. R.; Ploegh, H, L.; Ovaa, H. Crystal structure of the boronic acid-based proteasome inhibitor bortezomib in complex with the yeast 20S proteasome. *Structure* **2006**, *14*, 451–456.
2. Pemble IV, C. W.; Johnson, L. C.; Kridel, S. J.; Lowther, W. T. Crystal structure of the thioesterase domain of human fatty acid synthase inhibited by orlistat. *Nature Struct. Mol. Biol.* **2007**, *14*, 704–709.
3. Sogabe, S. et al Structure-based approach for the discovery of pyrrolo[3.2-*d*]pyrimidine-based EGFR T790M/L858R mutant inhibitors. *ACS Med. Chem. Lett.* **2013**, *4*, 201–205.

14. a. This is a mechanism-based inactivator.

Substituents on the PLP omitted for clarity

This mechanism shown to be incorrect

This mechanism shown to be correct

Silverman, R.B.; Invergo, B.J. *Biochemistry* **1986**, *25*, 6817–6820.

b. Give glutamine, which can cross the blood–brain barrier and get hydrolyzed to Glu and may shift the Glu-GABA imbalance. Give a Glu decarboxylase inhibitor to prevent degradation of the Glu and block GABA formation.

15.

Bull, H. G.; et al. *J. Am. Chem. Soc.* **1996**, *118*, 2359–2365.

CHAPTER 6

1. See Scheme 6.1 in the text.
2.

Sugar N N O H N H H N H O N N N Sugar

3. Incorporate a flat, aromatic, or heteroaromatic portion to stack between DNA base pairs.
4. Cancer cells replicate much faster than most normal cells. Therefore, they have a greater need for DNA precursors. Their uptake systems for these metabolites, and also antimetabolites, are very efficient. Therefore, there is selective toxicity for the cancer cells. Because of rapid cell division cancer cell mitosis can be stopped preferentially to normal cells where there is sufficient time for repair mechanisms to be effective.
5.

MeO_2SO OSO_2Me HN H_2N N N O=P–O O^- ^-O–P=O → HN H_2N N N R OSO_2Me NH NH_2 R' → HN H_2N R NH NH_2 R'

6.

N=N OH O N N NH_2 O=P–O O^- ^-O–P=O **6.51** → N≡N O N dR NH_2 → N dR NH_2 :NH_2 N O dR

6.44 ← H_2O / $-NH_4^+$ — N N dR NH_2 $^+NH_2$ N O N dR

7.

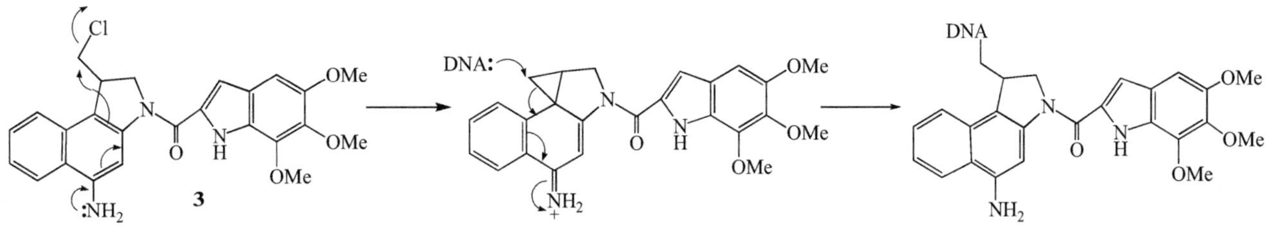

8.

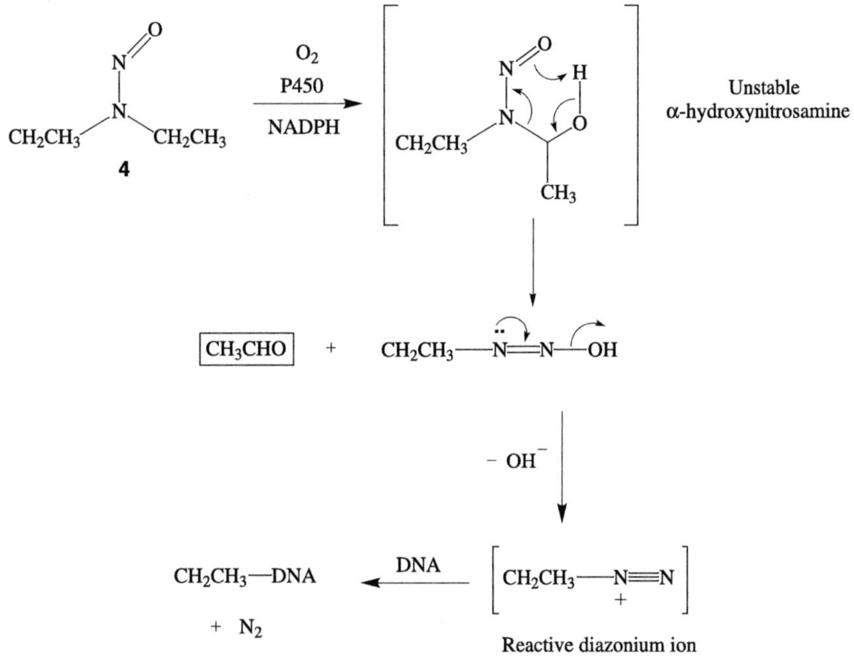

Loeppky, R. N. *Drug. Metab. Rev.* **1999**, *31*, 175–193.

9. The quinone lowers the pK_a of the aziridine nitrogens by resonance, so it is not protonated. Reduction of the quinone to the hydroquinone prevents that delocalization of the nonbonded electrons of the aziridine nitrogen, so it becomes more basic, is protonated, and activated for DNA attack.

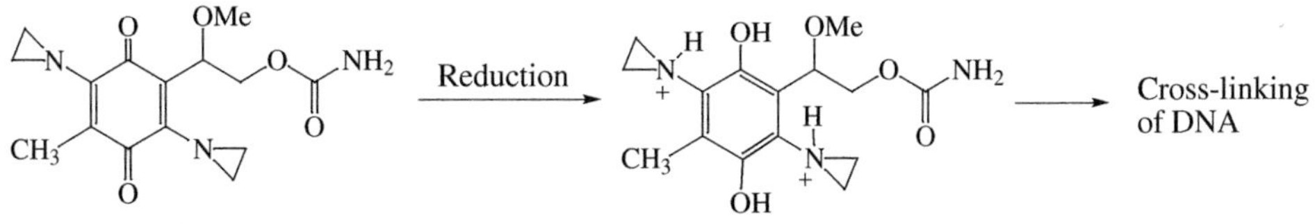

10. The geometry of the molecule is such that the enediyne orbitals do not interact until reduction of an adjacent moiety takes place, which leads to the enediyne orbitals coming close enough to react, giving the 1,4-dehydrobenzene diradical (Bergman rearrangement).

11.

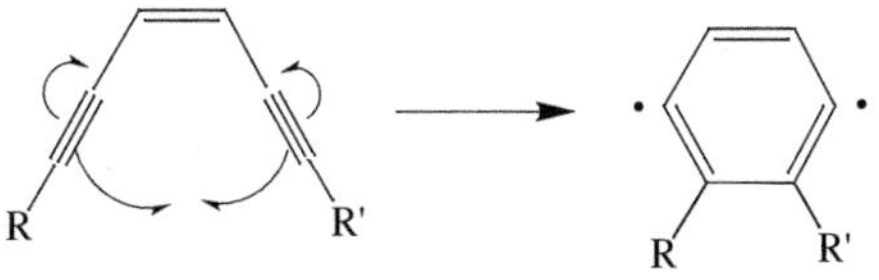

12. Any structure with a flat, aromatic part and an enediyne.

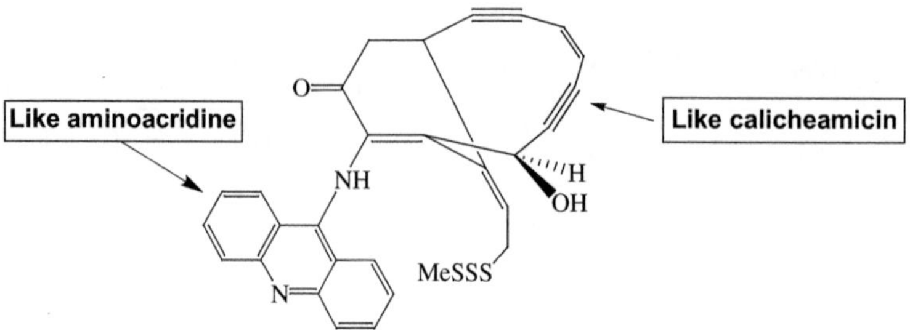

13.

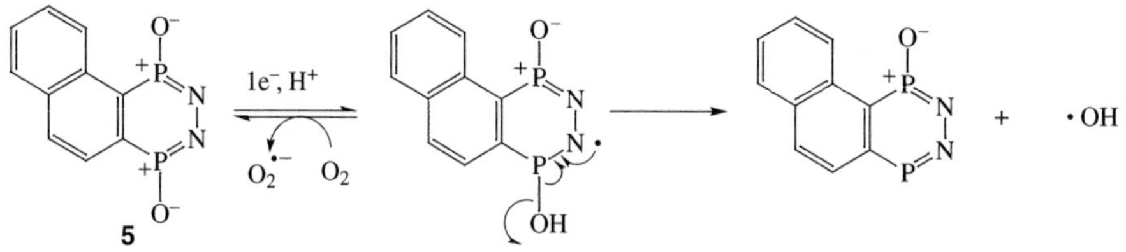

14.

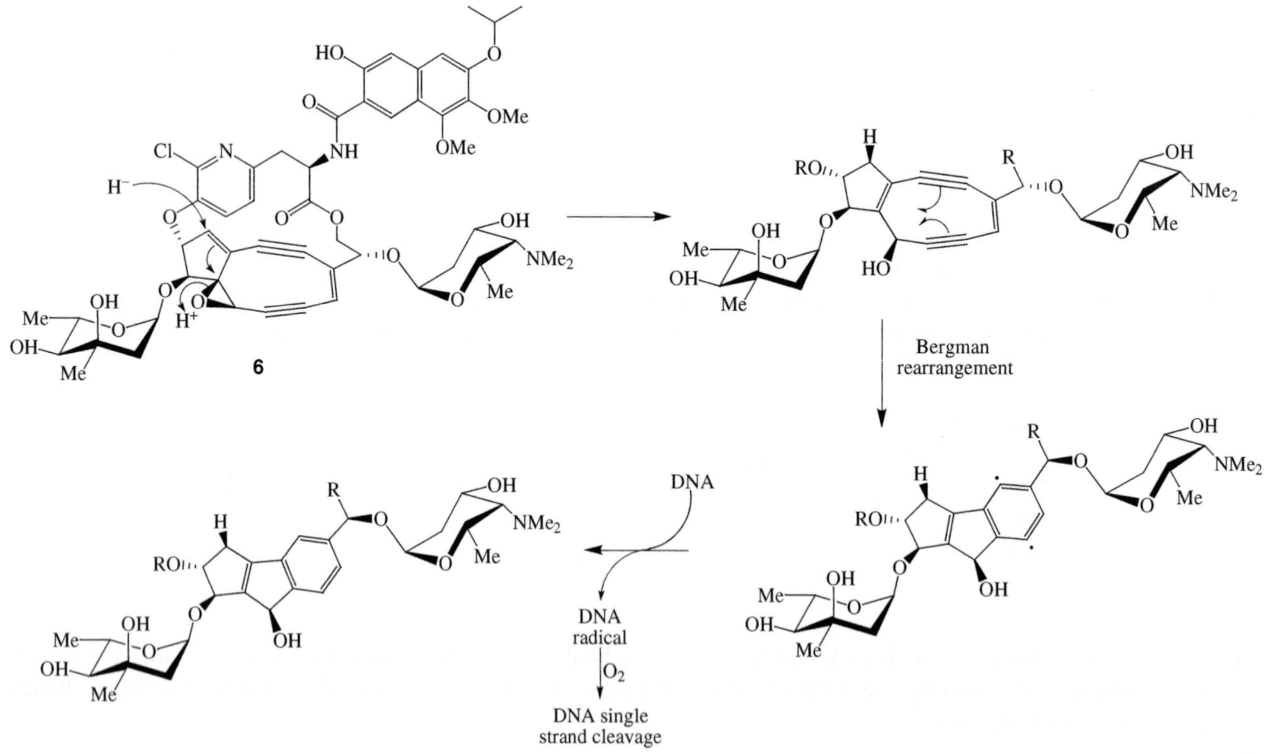

15.

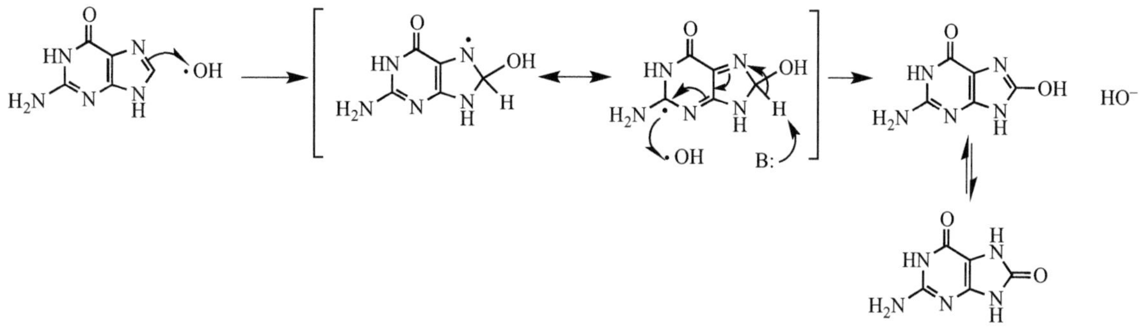

7

CHAPTER 7

1. Design a compound to inhibit the new enzyme, and give both the DNA alkylating agent and the new inhibitor in combination.
2. Convert the carboxylic acid, which presumably is deprotonated and forms an electrostatic interaction with the active site lysine residue, into an amino group, which will be protonated and could form an electrostatic interaction with the aspartate of the resistant enzyme.
3. So that they achieve maximal benefit of the synergistic effect, the two compounds should be exerting their activities to the desired extent during similar time periods.
4. If the drug has a very narrow therapeutic window, then inhibition of its breakdown could lead to toxic levels of the drug.
5. Check for a change in blood levels of each drug when coadministered with the other or new metabolites. See if either inhibits CYP enzymes or P-gp or other major known transporters/efflux pumps.
6. Determine what modification the new enzyme is making to your drug, then make a compound that does not have that group that is modified, and see if it is still active.
   Make an inhibitor of the new enzyme, and give it in combination with the drug.
   Make a drug that binds poorly to the new enzyme but is still active.
7. The drug can no longer bind to the altered enzyme, but presumably the substrate can. However, if the mutation is at an important residue in the active site, the enzyme may not bind to its substrate either, resulting in the equivalent of an inhibited enzyme.
8. Mutation of the Cys.
   Production of a new enzyme (FosA) that catalyzes the glutathione-dependent opening of the epoxide (see Armstrong and coworkers, *J. Am. Chem. Soc.* **2002**, *124*, 11001).

CHAPTER 8

1. Metabolism is enzyme catalyzed, and therefore will be dependent on chirality. The enantiomers can be substrates for two different enzymes.
2. By changing the stereochemistry at one site in the molecule there will be an effect on how the molecules bind at the active site of P450. The oxygenation occurs at the heme-oxo species, so whatever part of the substrate molecule that is closest to this reactive species will be oxygenated.

3.

Cl
O
N
Fe^{IV}
N
N
N
O
Cl
N
Fe^{IV}
N
N
N
H
O
Cl
+
H
$^{-}$O
Cl
NIH shift
Cl
H
O
B:
H^+
Cl
OH

4.

Br
Br
H_3C
H
O•
N
Fe^{IV}
N
N
N
Br
Br
H_3C
OH
N
Fe^{IV}
N
N
N
Br
Br
H_3C
O—H
:B
N
Fe^{III}
N
N
N
Br
a
H_3C
O
$^{-}$X
H_2O
hydrolysis
COOH
H_3C
inactivation a
O
H_3C
X

5.

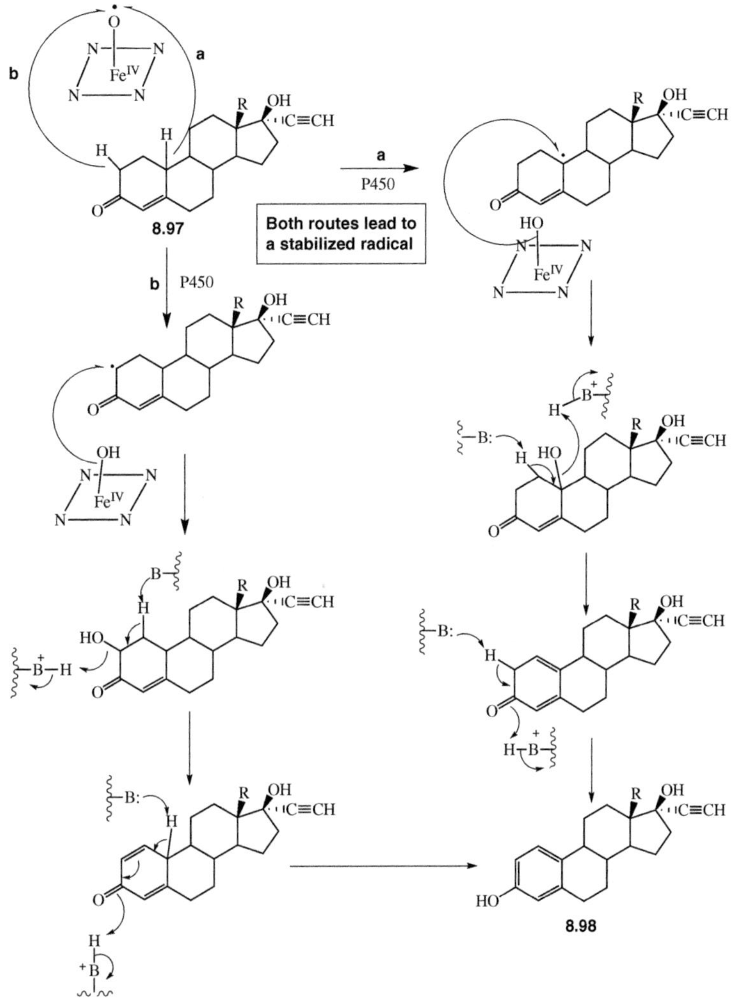

b
a
O
N
N
Fe^{IV}
N
N
H
H
H
R
OH
C≡CH
O
8.97
a
P450
Both routes lead to a stabilized radical
b
P450
HO
OH
B
HO
B:
8.98

6.

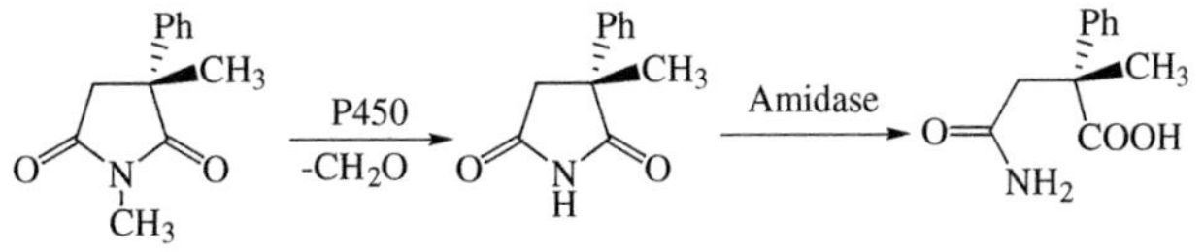

Ph
CH$_3$
N
CH$_3$
P450
-CH$_2$O
H
Amidase
COOH
NH$_2$

7.

OCH$_2$CH$_2$NMe$_2$

3

Cl

N-dealkylation

P450

OCH$_2$CH$_2$NHMe

Cl

N-dealkylation

P450

OCH$_2$CH$_2$NH$_2$

Cl

Aromatic hydroxylation P450

Aromatic hydroxylation P450

1. Oxidative deamination
2. Reduction

P450

NADH

OCH$_2$CH$_2$NMe$_2$

HO

Cl

OCH$_2$CH$_2$NHMe

HO

Cl

OCH$_2$CH$_2$OH

Cl

Aromatic hydroxylation

P450

OCH$_2$CH$_2$OH

HO

Cl

Bethou, F.; et al. *Biochem. Pharmacol.* **1994**, *47*, 1883–1895.

8.

R^1, R^2, R^3 C–C(=O)–O–H ; O• N–Fe^{IV}–N

$-CO_2$

OH N–Fe^{IV}–N

R^2–C(R^1)(R^3)–OH

Komuro, M.; Higuchi,T.; Hirobe, M. *Bioorg. Med. Chem.* **1995**, *3*, 55–65.

9.

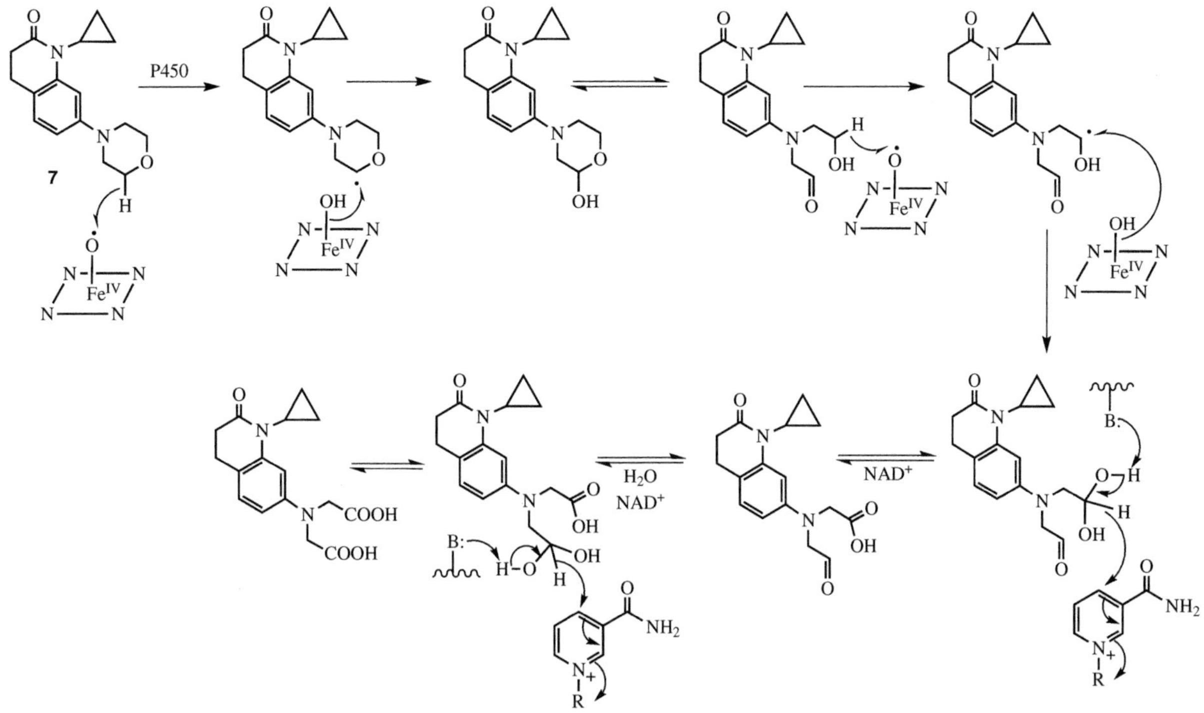

The aldehydes also could be oxidized by P450.

10. Glutathione protects cells from potent electrophiles by a variety of nucleophilic reactions.

11.

Reaction	Phase	Type of Reaction	Cofactor
a	I	*N*-Dealkylation	Heme
b	I	Oxidative deamination	Heme
c	I	Oxidation	NAD^+
d	II	Amino acid conjugation	ATP/CoASH
e	I	Aromatic hydroxylation	Heme
f	I	Aromatic hydroxylation	Heme
g	II	Methylation	*S*-Adenosylmethionine
h	I	Oxidation	Heme
i	II	Sulfate conjugation	PAPS
j	I	Reduction	NADPH
k	II	Glucuronidation	UDP-glucuronic acid
l	II	Acetylation	Acetyl CoA
m	I	*S*-Oxidation	Heme

12.

P450

$+H^+$

$+H^+$

Hydride shift

$-F^-$

$-H^+$

Tautomerize

B:

GS

GS^-

NH·Ar

8

9

13. a.

Aldehyde dehydrogenase

Cytochrome P450

COOH

$+$ ▷$-NH_2$

b.

UDP-glucuronosyl transferase UDPGA

Cytochrome P450

c.

Esterase

Glutathione *S*-transferase GSH

Cytochrome P450

Cytochrome P450

(Note: anti attack at less hindered side)

(The first two steps can be switched)

14. a. Soft drug approach; or try to circumvent the dealkylation to **11**, for example, by changing Et groups to CH_2CF_3 or other fluorinated or branched analogs; or design an analog where the same metabolic pathway (bis-dealkylation) generates a more rapidly cleared metabolite.

b. A compound found effective is shown below, but there are many correct answers where the compound is modified to facilitate clearance. In this case, the hydroxyl groups could be conjugated, but the bis-deethylated metabolite (assuming that it forms in analogy to **11**) might simply be eliminated more rapidly because of its higher polarity.

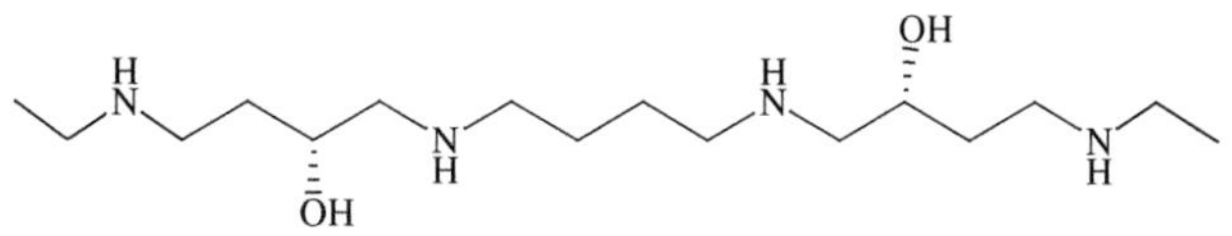

Bergeron, R. J.; et al. *J. Med. Chem.* **1996**, *39*, 2451.

15. One approach would be to protect the thiazole ring from metabolism by addition of a fluorine atom.*

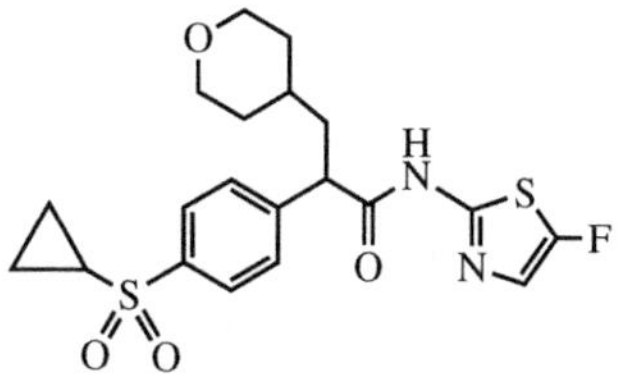

*Bertram, L. S. et al., *J. Med. Chem.* **2008**, *51*, 4340-4345.

Another approach would be to use a bioisostere of the thiazole, such as an oxazole or benzene, which do not have a sulfur to oxidize. In the case of the latter, a *para*-fluorine (R = F) would be reasonable to protect the ring from oxidation.

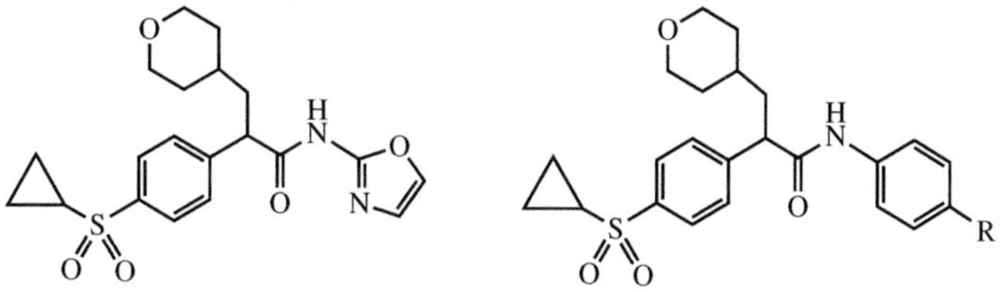

CHAPTER 9

1.

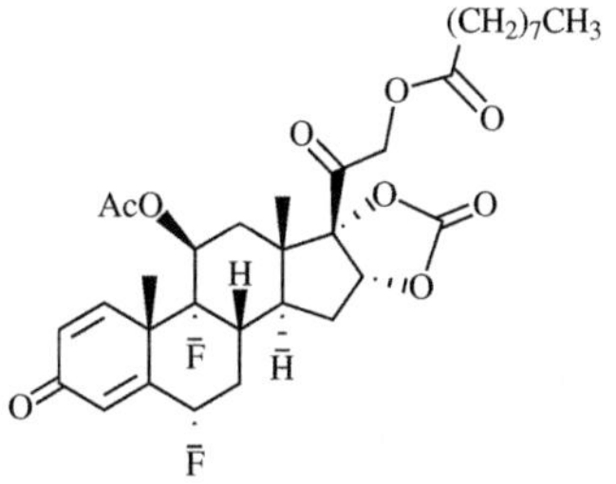

Any one of these changes might suffice, but possibly more than one would be needed.

2.

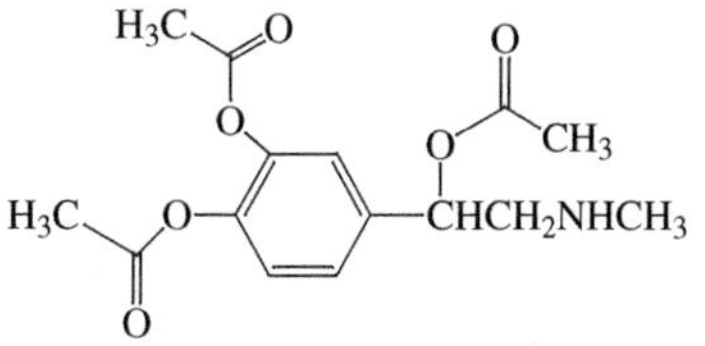

Or any ester at these positions.

3.

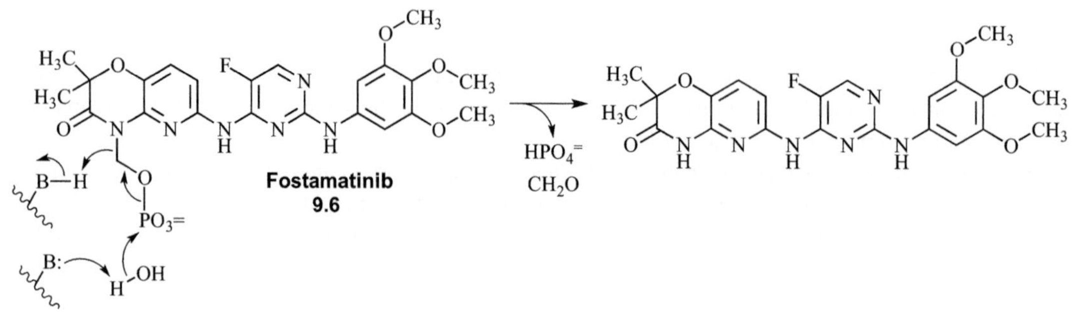

Fostamatinib
9.6

$HPO_4^{=}$
CH_2O

4.

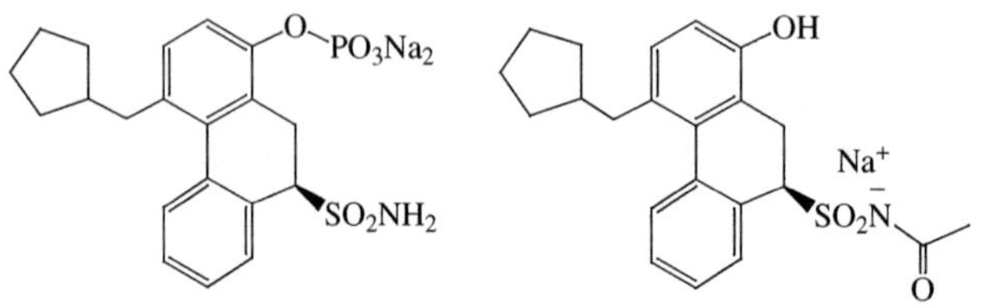

5. Make a prodrug that prevents hydrolysis by the new enzyme, but can be converted to the drug after activation. In this case, the additional RCO group may not be a substrate for the resistant enzyme or may inhibit the resistant enzyme. This may give the prodrug the ability to get to a different site for activation (hydrolysis of the RCO group).

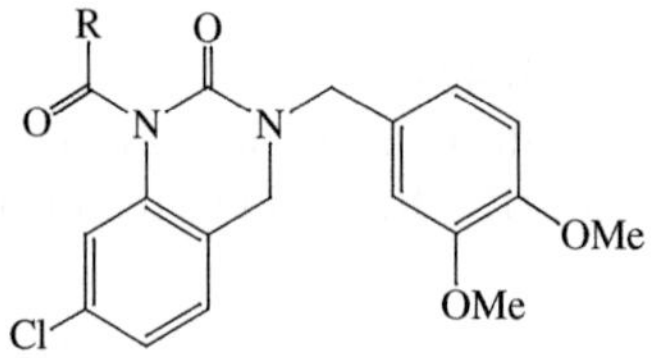

Design a molecule (**XH**) to inhibit the new enzyme and then make a mutual prodrug that can undergo hydrolysis to the two drugs. A possible example is shown below.

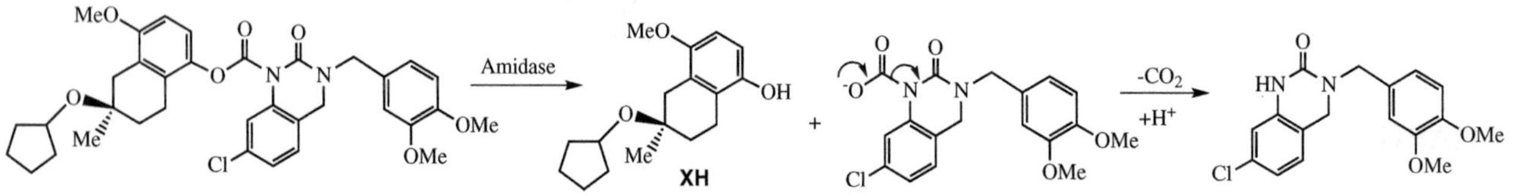

Amidase

XH

$-CO_2$
$+H^+$

6.

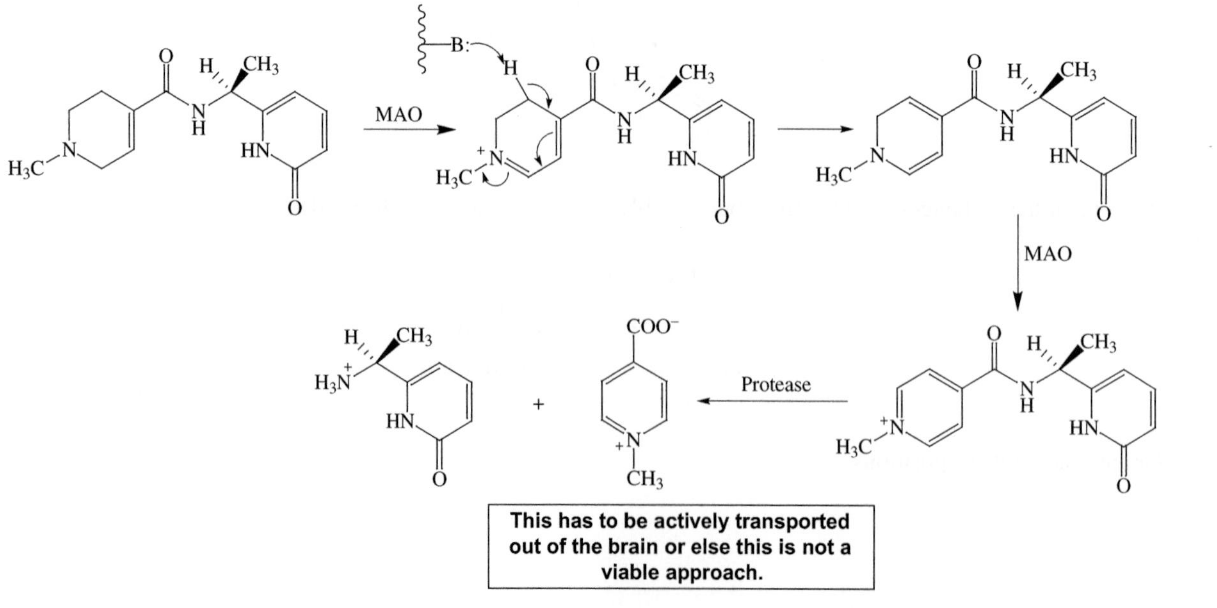

MAO

MAO

Protease

This has to be actively transported out of the brain or else this is not a viable approach.

7. Ampicillin will be released by an amidase.

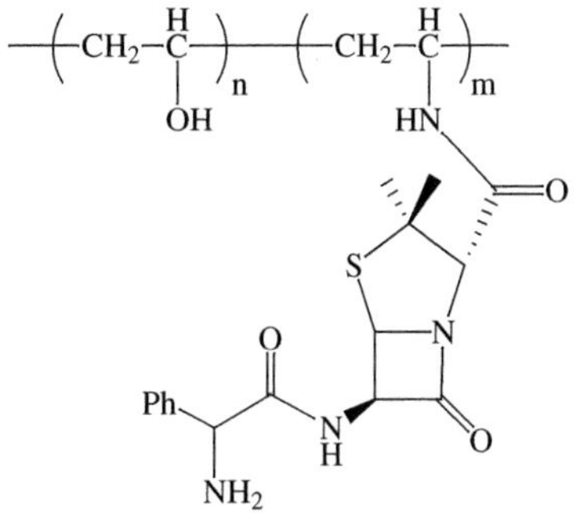

8.

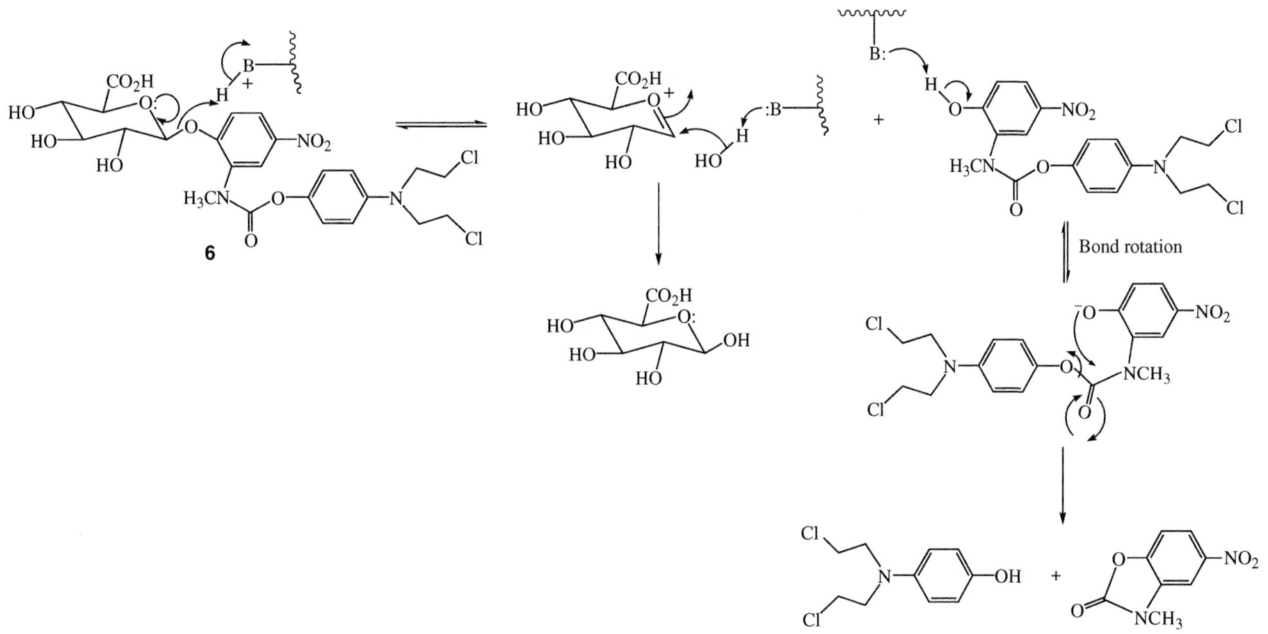

Schmidt, F.; et al. *Bioorg. Med. Chem. Lett.* **1997**, *7*, 1071–1076.

9. There are many correct answers. Here is shown 5-fluoro-2′-deoxyuridine, an inactivator of thymidylate synthase, linked to a nitrogen mustard through a phosphoramidate bond (tumor cells have high phosphoamidase activity).

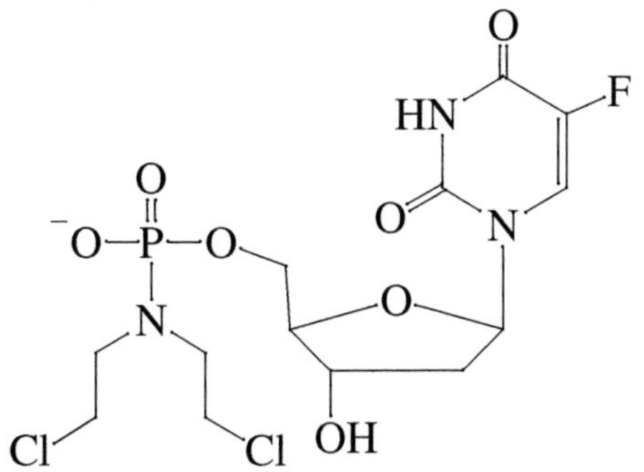

10. a.

NHAc, O-Dealkylation, Oxidase, NHAc, OH, +, H, O

b.

H, N, NHMe, S, O, HN, N, NCN, Reductase, H, N, NHMe, S, HN, N, NCN

c.

Me O, H_3CSCH_2CHC—N, CO_2H, S-Dealkylation, P450, Me O, $HSCH_2CHC$—N, CO_2H

$NADP^+$ dehydrogenase

Me O, $HSCH_2CHC$—N, NH_2, Oxidative deamination P450, Me O, $HSCH_2CHC$—N, H, O

11.

Cl, N, Cl, R, H, N, S, O, O, N, O, NH, O^-, COOH, O, β-lactamase, Cl, N, Cl, NH_2

Svensson, H.P.; Frank, I. S.; Berry, K. K.; Senter, P. D. *J. Med. Chem.* **1998**, *41*, 1507–1512.

12.

Drug NH O O O N N O O R ROH H B B: HO H HO H 2 CO_2 + Drug NH_3

H OH Drug NH_3 + CO_2 HO OH

Shamis, M. et al. *J. Med. Chem. Soc* **2004**, *126*, 1726–1731.

13.

H OH NH HO H B: + CO_2 OH + NH_2 N N

11 12

Gediya, L. K. et al. *J. Med. Chem.* **2008**, *51*, 3895–3904.

14.

H CO_2Me N S Cl **Clopidogrel** **9.109** P450 NADPH, O_2 $+H^+$ HO Fe^{IV} OH B: HO H 9.110

15.

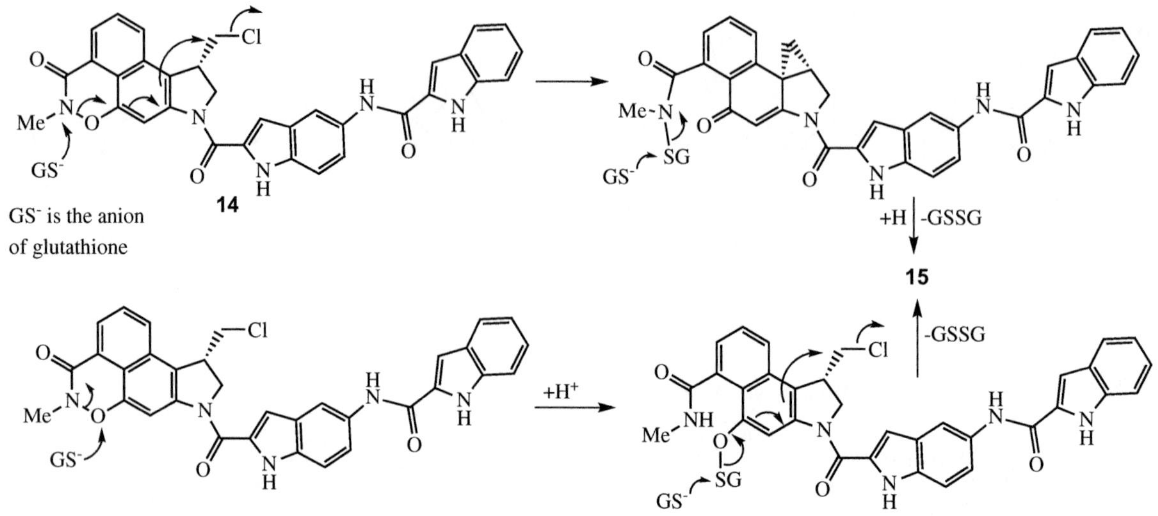

GS⁻ is the anion of glutathione

14

+H | -GSSG

15

-GSSG

+H⁺

Boger, D. L. and coworkers, *J. Med. Chem.* **2013**, *56*, 4104-4115.

Alternatively, NADPH could be used.

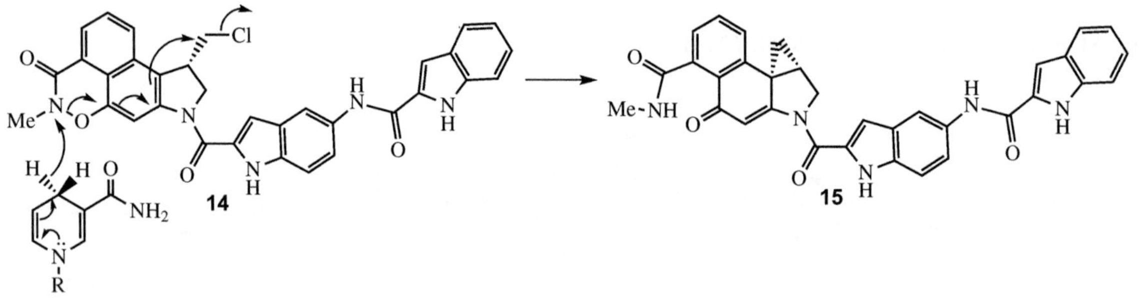

14

15

Index

Note: Page Numbers followed by *f* indicate figures; *t*, tables; *b*, boxes.

E

F

N

O

P

Printed and bound by CPI Group (UK) Ltd, Croydon, CR0 4YY
03/10/2024
01040316-0018